SMALL WATER SYSTEM OPERATION AND MAINTENANCE

Fourth Edition

A Field Study Training Program

prepared by

California State University, Sacramento
College of Engineering and Computer Science
Office of Water Programs

in cooperation with the
National Environmental Training Association

★★★★★★★★★★★★★★★★★★★★★★★★★★★★★★★

Kenneth D. Kerri, Project Director

★★★★★★★★★★★★★★★★★★★★★★★★★★★★★★★

for the

California Department of Health Services
Sanitary Engineering Branch
Standard Agreement #80-64652

and

U.S. Environmental Protection Agency
Office of Drinking Water
Grant No. T-901361-01-0

2002

NOTICE

This manual is revised and updated before each printing based on comments from persons using the manual.

FIRST EDITION — **WATER SUPPLY SYSTEM OPERATION**

First Printing, 1983	7,000

SECOND EDITION — **SMALL WATER SYSTEM O & M**

First Printing, 1987	7,000
Second Printing, 1990	8,000

THIRD EDITION

First Printing, 1993	8,000
Second Printing, 1995	10,000

FOURTH EDITION

First Printing, 1999	9,000
Second Printing, 2001	5,000
Third Printing, 2002	17,000

Copyright © 2002 by
California State University, Sacramento Foundation

ISBN 1-884701-29-9

OPERATOR TRAINING MANUALS

OPERATOR TRAINING MANUALS AND VIDEOS IN THIS SERIES are available from the Office of Water Programs, California State University, Sacramento, 6000 J Street, Sacramento, CA 95819-6025, phone: (916) 278-6142, e-mail: wateroffice@csus.edu or FAX: (916) 278-5959.

1. *SMALL WATER SYSTEM OPERATION AND MAINTENANCE,**
2. *WATER DISTRIBUTION SYSTEM OPERATION AND MAINTENANCE,*
3. *WATER TREATMENT PLANT OPERATION,* 2 Volumes,
4. *UTILITY MANAGEMENT,*
5. *SMALL WASTEWATER SYSTEM OPERATION AND MAINTENANCE,* 2 Volumes,
6. *OPERATION OF WASTEWATER TREATMENT PLANTS,* 2 Volumes,
7. *OPERATION AND MAINTENANCE OF WASTEWATER COLLECTION SYSTEMS,* 2 Volumes,**
8. *ADVANCED WASTE TREATMENT,*
9. *INDUSTRIAL WASTE TREATMENT,* 2 Volumes,
10. *TREATMENT OF METAL WASTESTREAMS,* and
11. *PRETREATMENT FACILITY INSPECTION.****

The Office of Water Programs at California State University, Sacramento, has been designated by the U.S. Environmental Protection Agency as a *SMALL PUBLIC WATER SYSTEMS TECHNOLOGY ASSISTANCE CENTER.* This recognition will provide funding for the development of training videos for the operators and managers of small public water systems. Additional training materials will be produced to assist the operators and managers of small systems.

* *SMALL WATER SYSTEM TRAINING VIDEOS.* This series of ten training videos was prepared for the operators, managers, owners, and board members of small drinking water systems. Topics covered include operators' roles and responsibilities; safety, operation, and maintenance of surface and groundwater treatment, distribution, and storage facilities; monitoring; administration; financial management; and also emergency preparedness and response.

** Other materials and training aids developed by the Office of Water Programs to assist operators in improving collection system operation and maintenance and overall performance of their systems include:

1. *COLLECTION SYSTEMS: METHODS FOR EVALUATING AND IMPROVING PERFORMANCE.* This handbook presents detailed benchmarking procedures and worksheets for using performance indicators to evaluate the adequacy and effectiveness of existing O & M programs. It also describes how to identify problems and suggests many methods for improving the performance of a collection system.
2. *OPERATION AND MAINTENANCE TRAINING VIDEOS.* This series of six 30-minute videos demonstrates the equipment and procedures collection system crews use to safely and effectively operate and maintain their collection systems. These videos complement and reinforce the information presented in Volumes I and II of *OPERATION AND MAINTENANCE OF WASTEWATER COLLECTION SYSTEMS.*

*** *PRETREATMENT FACILITY INSPECTION TRAINING VIDEOS.* This series of five 30-minute videos demonstrates the procedures to effectively inspect an industry, measure flows, and collect samples. These videos complement and reinforce the information presented in *PRETREATMENT FACILITY INSPECTION.*

PREFACE TO THE FIRST EDITION

The purposes of this water supply system field study training program are to:

1. Develop new qualified water supply system operators,
2. Expand the abilities of existing operators, permitting better service to both their employers and the public, and
3. Prepare operators for civil service and *CERTIFICATION EXAMINATIONS.*[1]

To provide you with the knowledge and skills needed to operate and maintain water supply systems as efficiently and effectively as possible, experienced water supply system operators prepared the material in each chapter of this manual.

Water supply systems vary from city to city and from region to region. The material contained in this program is presented to provide you with an understanding of the basic operation and maintenance aspects of your system and with information to help you analyze and solve operation and maintenance problems. This information will help you operate and maintain your system in a safe and efficient manner.

Water supply operation and maintenance is a rapidly advancing field. To keep pace with scientific and technological advances, the material in this manual must be periodically revised and updated. *THIS MEANS THAT YOU, THE OPERATOR, MUST RECOGNIZE THE NEED TO BE AWARE OF NEW ADVANCES AND THE NEED FOR CONTINUOUS TRAINING BEYOND THIS PROGRAM.*

The Project Director is indebted to the many operators and other persons who contributed to this manual. Every effort was made to acknowledge material from the many excellent references in the water supply field. Reviewers Leonard Ainsworth, Jack Rossum, and Joe Monscvitz deserve special recognition for their extremely thorough review and helpful suggestions. John Trax, Chet Pauls, and Ken Hay, Office of Drinking Water, U.S. Environmental Protection Agency, and John Gaston, Bill MacPherson, Bert Ellsworth, Clarence Young, Ted Bakker, and Beverlie Vandre, Sanitary Engineering Branch, California Department of Health Services, all performed outstanding jobs as resource persons, consultants, and advisers. Larry Hannah served as Educational Consultant. Illustrations were drawn by Martin Garrity. Charlene Arora helped type the field test and final manuscript for printing. Special thanks are well deserved by the Program Administrator, Gay Kornweibel, who typed, administered the field test, managed the office, administered the budget, and did everything else that had to be done to complete this project successfully.

KENNETH D. KERRI
PROJECT DIRECTOR

1983

[1] *Certification Examination. An examination administered by a state agency that small water supply system operators take to indicate a level of professional competence. In the United States, certification of small water system operators is mandatory.*

PREFACE TO THE SECOND EDITION

SMALL WATER SYSTEM OPERATION AND MAINTENANCE is the title of this portion of the Second Edition. When the First Edition, **WATER SUPPLY SYSTEM OPERATION**, was developed, the objective was to produce one comprehensive training manual for water supply system operators. The resulting manual covered all aspects of water supply systems from the source of the water supply to the consumer's tap. Operators using the First Edition, **WATER SUPPLY SYSTEM OPERATION**, expressed a valid concern that too much material was covered in one training manual. In response to this concern the manual was split approximately in half to produce the Second Edition as two manuals:

- **SMALL WATER SYSTEM OPERATION AND MAINTENANCE**, and
- **WATER DISTRIBUTION SYSTEM OPERATION AND MAINTENANCE**.

SMALL WATER SYSTEM OPERATION AND MAINTENANCE contains information on the responsibilities of being an operator, water sources, wells, small water treatment plants, disinfection, safety, lab procedures, and a new chapter on rates. **WATER DISTRIBUTION SYSTEM OPERATION AND MAINTENANCE** describes the responsibilities of being an operator, storage facilities, distribution systems, water quality considerations, distribution system operation and maintenance, disinfection, and safety. The chapters on disinfection and safety are in both manuals because of their extreme importance. Both of these operator training manuals are equally important if you are responsible for a water supply system. If your agency is large enough to employ two or more crews of operators, the crews may specialize in a particular aspect of operation and maintenance. For example, one crew may deal primarily with wells and disinfection, while the other crew is assigned the responsibility of the distribution system. The splitting of the original manual, **WATER SUPPLY SYSTEM OPERATION**, into two manuals will allow operators to concentrate their studies on the subject areas related directly to their jobs. At the same time, it is hoped that operators will realize the importance of understanding all aspects of water supply systems and attempt to successfully complete both of the manuals.

1987

KENNETH D. KERRI
PROJECT DIRECTOR

TECHNICAL CONSULTANTS

John Brady
Gerald Davidson
Larry Hannah
Jim Sequeira
R. Rhodes Trussell
Mike Young

NATIONAL ENVIRONMENTAL TRAINING ASSOCIATION REVIEWERS

George Kinias, Project Coordinator

E.E. "Skeet" Arasmith
Terry Engelhardt
Dempsey Hall
Jerry Higgins
Andrew Holtan
Deborah Horton
Kirk Laflin
Rich Metcalf
William Redman
Kenneth Walimaa
Anthony Zigment

PROJECT REVIEWERS

Leonard Ainsworth
Ted Bakker
Jo Boyd
Dean Chausee
Walter Cockrell
Fred Fahlen
David Fitch
Richard Haberman
Lee Harry
Jerry Hayes
Ed Henley
Charles Jeffs
Chet Latif
Frank Lewis
Perry Libby
D. Mackay
William Maguire
Nancy McTigue
Joe Monscvitz
Angela Moore
Harold Mowry
Theron Palmer
Eugene Parham
Catherine Perman
David Rexing
Jack Rossum
William Ruff
Gerald Samuel
Carl Schwing
David Sorenson
Russell Sutphen
Robert Wentzel
James Wright
Mike Yee
Clarence Young

OBJECTIVES OF THIS MANUAL

Proper installation, inspection, operation, maintenance, repair, and management of small water systems have a significant impact on the operation and maintenance costs and effectiveness of the systems. The objective of this manual is to provide small water system operators with the knowledge and skills required to operate and maintain these systems effectively, thus eliminating or reducing the following problems:

1. Health hazards created by the delivery of unsafe water to the consumer's tap;
2. System failures that result from the lack of proper installation, inspection, preventive maintenance, surveillance, and repair programs designed to protect the public investment in these facilities;
3. Tastes and odors caused by water system problems;
4. Corrosion damages to pipes, equipment, tanks, and structures in the water system;
5. Complaints from the public or local officials due to the unreliability or failure of the water system to perform as designed; and
6. Fire damage caused by insufficient water and/or inadequate pressures at a time of need.

SCOPE OF THIS MANUAL

Operators with the responsibility for wells, pumps, disinfection, and small water treatment plants will find this manual very helpful. This manual contains information on:

1. What small water system operators do,
2. Sources and uses of water,
3. How to operate and maintain wells and pumps,
4. Operation and maintenance of small water treatment plants,
5. Disinfection of new and repaired facilities as well as water delivered to consumers,
6. Techniques for recognizing hazards and developing safe procedures and safety programs,
7. Laboratory procedures for analyzing samples of water, and
8. Procedures to develop a reasonable rate structure.

Material in this manual furnishes you with information concerning situations encountered by most small water system operators in most areas. These materials provide you with an understanding of the basic operational and maintenance concepts for small water systems and with an ability to analyze and solve problems when they occur. Operation and maintenance programs for small water systems will vary with the age of the system, the extent and effectiveness of previous programs, and local conditions. You will have to adapt the information and procedures in this manual to your particular situation.

Technology is advancing very rapidly in the field of operation and maintenance of small water systems. To keep pace with scientific advances, the material in this program must be periodically revised and updated. This means that you, the small water system operator, must be aware of new advances and recognize the need for continuous personal training reaching beyond this program. ***MANY TRAINING OPPORTUNITIES EXIST IN YOUR DAILY WORK EXPERIENCE, FROM YOUR ASSOCIATES, AND FROM ATTENDING MEETINGS, WORKSHOPS, CONFERENCES, AND CLASSES.***

USES OF THIS MANUAL

This manual was developed to serve the needs of operators in several different situations. The format used was developed to serve as a home-study or self-paced instruction course for operators in remote areas or persons unable to attend formal classes either due to shift work, personal reasons, or the unavailability of suitable classes. This home-study training program uses the concepts of self-paced instruction where you are your own instructor and work at your own speed. In order to certify that a person has successfully completed this program, an objective test is included at the end of each chapter.

Also, this manual can serve effectively as a textbook in the classroom. Many colleges and universities have used this manual as a text in formal classes (often taught by operators). In areas where colleges are not available or are unable to offer classes in the operation of small water systems, operators and utility agencies can join together to offer their own courses using the manual.

Cities or utility agencies can use the manual in several types of on-the-job training programs. In one type of program, a manual is purchased for each operator. A senior operator or a group of operators are designated as instructors. These operators help answer questions when the persons in the training program have questions or need assistance. The instructors grade the objective tests at the end of each chapter, record scores, and notify California State University, Sacramento, of the scores when a person successfully completes this program. This approach eliminates any waiting while papers are being graded and returned by CSUS.

This manual was prepared to help operators operate and maintain their small water systems. Please feel free to use the manual in the manner which best fits your training needs and the needs of your operators. We will be happy to work with you to assist you in developing your training program. Please feel free to contact:

Project Director
Office of Water Programs
California State University, Sacramento
6000 J Street
Sacramento, California 95819-6025

Phone (916) 278-6142

INSTRUCTIONS TO PARTICIPANTS IN HOME-STUDY COURSE

Procedures for reading the lessons and answering the questions are contained in this section.

To progress steadily through this program, you should establish a regular study schedule. For example, many operators in the past have set aside two hours during two evenings a week for study.

The study material is contained in eight chapters. Some chapters are longer and more difficult than others. For this reason, many of the chapters are divided into two or more lessons. The time required to complete a lesson will depend on your background and experience. Some people might require an hour to complete a lesson and some might require three hours; but that is perfectly all right. *THE IMPORTANT THING IS THAT YOU UNDERSTAND THE MATERIAL IN THE LESSON!*

Each lesson is arranged for you to read a short section, write the answers to the questions at the end of the section, check your answers against suggested answers; and then *YOU* decide if you understand the material sufficiently to continue or whether you should read the section again. You will find that this procedure is slower than reading a normal textbook, but you will remember much more when you have finished the lesson.

Some discussion and review questions are provided following each lesson in some of the chapters. These questions review the important points you have covered in the lesson. Write the answers to these questions in your notebook.

At the end of each chapter, you will find an Objective Test. Mark your answers on the special answer sheet provided for each chapter. The Objective Test contains True-False, Best Answer, and Multiple Choice types of questions. The purposes of this exam are to review the chapter and to give experience in taking different types of exams. *MAIL TO THE PROGRAM DIRECTOR ONLY YOUR ANSWERS TO OBJECTIVE TESTS ON THE PROVIDED ANSWER SHEETS.*

The answer sheets you mail to the Project Director will be computer graded in the following manner. True-False questions have only one possible correct answer and they have a value of one point. Multiple Choice questions, however, are worth five points because there may be more than one correct answer and all correct answers must be filled in for full credit. The computer deducts a point for each correct answer not filled in and each incorrect response filled in. For example, if 1, 3, and 5 are the correct responses and you fill in 1 and 5, a point is deducted for failing to fill in 3. If you fill in 1, 3, 4, and 5, a point is deducted because answer number 4 is filled in and should not be. Best Answer questions are worth four points. Only one of the four possible answers is correct. For example, if answer number 2 is correct and marked, you receive four points. If answer number 2 is correct and answer number 3 is incorrectly marked, the computer will deduct one point from your score for not marking the correct answer 2 and will also deduct one point for incorrectly marking the wrong answer 3.

After you have completed the last objective test, you will find a final examination. This exam is provided for you to review how well you remember the material. You may wish to review the entire manual before you take the final exam. Some of the questions are essay-type questions which are used by some states for higher-level certification examinations. After you have completed the final examination, grade your own paper and determine the areas in which you might need additional review before your next certification or civil service examination.

You are your own teacher in this program. You could merely look up the suggested answers from the answer sheet or copy them from someone else, but you would not understand the material. Consequently, you would not be able to apply the material to the operation of your facilities nor recall it during an examination for certification or a civil service position.

YOU WILL GET OUT OF THIS PROGRAM WHAT YOU PUT INTO IT.

SUMMARY OF PROCEDURE

A. OPERATOR (YOU)

1. Read what you are expected to learn in each chapter (the chapter objectives).
2. Read sections in the lesson.
3. Write your answers to questions at the end of each section in your notebook. You should write the answers to the questions just as you would if these were questions on a test.
4. Check your answers with the suggested answers.
5. Decide whether to reread the section or to continue with the next section.
6. Write your answers to the discussion and review questions at the end of each lesson in your notebook.
7. Mark your answers to the objective test on the answer sheet provided by the Project Director or by your instructor.
8. Mail material to the Project Director. (Send *ONLY* your completed answer sheet.)

   Project Director
   Office of Water Programs
   California State University, Sacramento
   6000 J Street
   Sacramento, California 95819-6025

B. PROJECT DIRECTOR

1. Mails answer sheet for each chapter to operator.
2. Corrects tests, answers any questions, and returns results to operators, including explanations and solutions for missed questions.

C. ORDER OF WORKING LESSONS

To complete this program you will have to work all of the lessons. You may proceed in numerical order, or you may wish to work some lessons sooner.

SAFETY IS A VERY IMPORTANT TOPIC. Everyone working in a small water system must always be safety conscious. Operators daily encounter situations and equipment that can cause a serious disabling injury or illness if the operator is not aware of the potential danger and does not exercise adequate precautions. For these reasons, you may decide to work on Chapter 6, "Safety," early in your studies. In each chapter, SAFE PROCEDURES ARE ALWAYS STRESSED.

SMALL WATER SYSTEM OPERATION AND MAINTENANCE
COURSE OUTLINE

(Wells, Small Treatment Plants, and Rates)

Other similar operator training programs that may be of interest to you are our manuals and courses on water distribution systems and water treatment plant operation (two volumes).

WATER DISTRIBUTION SYSTEM OPERATION AND MAINTENANCE
COURSE OUTLINE

WATER TREATMENT PLANT OPERATION, VOLUME I
COURSE OUTLINE

WATER TREATMENT PLANT OPERATION, VOLUME II
COURSE OUTLINE

CHAPTER 1

THE SMALL WATER SYSTEM OPERATOR

by

Ken Kerri

TABLE OF CONTENTS

Chapter 1. THE SMALL WATER SYSTEM OPERATOR

OBJECTIVES

Chapter 1. THE SMALL WATER SYSTEM OPERATOR

At the beginning of each chapter in this manual you will find a list of *OBJECTIVES*. The purpose of this list is to stress those topics in the chapter that are most important. Contained in the list will be items you need to know and skills you must develop to operate, maintain, repair, and manage a small water system as efficiently and as safely as possible.

Following completion of Chapter 1, you should be able to:

1. Explain the type of work done by small water system operators,
2. Describe where to look for jobs in this profession, and
3. Find sources of further information on how to do the jobs performed by small water system operators.

CHAPTER 1. THE SMALL WATER SYSTEM OPERATOR

Chapter 1 is prepared especially for new operators or people interested in becoming small water system operators. If you are an experienced small water system operator, you may find some new viewpoints in this chapter.

In these training manuals, the water supply system has been divided into two parts: (1) the water system, and (2) the water distribution system. The water system begins with the source of water such as groundwater or surface water. Wells, pumps, and treatment facilities are part of the water system which delivers water to the distribution system. The water distribution system consists of tanks and pipes which store and deliver treated water to the consumers.

1.0 NEED FOR SMALL WATER SYSTEM OPERATORS

People need safe and pleasant water to drink. Many sources of water are not directly suitable for drinking purposes. Engineers have designed water supply facilities to collect, store, transport, treat, and distribute water to people and industries. Once these facilities are built, the operators make sure the facilities do their intended job. Small water system operators operate, maintain, repair, and manage these facilities. Operators have the responsibility of ensuring that safe and pleasant drinking water is delivered to everyone's tap. Another responsibility is to be sure that adequate amounts of water and pressure are available during times of emergency, such as a fire. Cities and towns need qualified, capable, and dedicated operators to do these jobs.

The need for *RESPONSIBLE* small water system operators cannot be overstressed. You, as a small water system operator, have the responsibility for the health and well-being of the community you serve. Yes, you are responsible for the drinking water of your community and any time you fail to do your job, you could be responsible for an outbreak of a waterborne disease which could even result in death. As an operator, you do not want the knowledge that you were negligent in your duty and, as a result, were responsible for the death of a fellow human being.

QUESTIONS

Below are some questions for you to answer. You should have a notebook in which you can write the answers to the questions. By writing down the answers to the questions, you are helping yourself learn and retain the information. After you have answered all the questions, compare your answers with those given in the Suggested Answers section on page 10. Reread any sections you do not understand and then proceed to the next section. You are your own teacher in this training program, and *YOU* should decide when you understand the material and are ready to continue with new material.

1.0A What do small water system operators do with small water facilities?

1.0B What is the responsibility of small water system operators?

1.1 WHAT IS A WATER SUPPLY SYSTEM?

1.10 Sources of Water

A water supply system delivers water from its source to the ultimate consumer — homes, businesses, or industrial water users (Figure 1.1). The source of a water supply may be either groundwater or surface waters. Surface waters may be natural lakes, streams, rivers, or reservoirs behind dams. Groundwater is a water source below the ground surface.

Water is usually transported from its source through a transmission system consisting of either open channels (canals) or large-diameter pipes (called raw water conduits) to the water treatment plant. After treatment the water is usually pumped into transmission lines which are connected to a distribution grid system. From the grid system the water is delivered to individual service lines which serve private homes, businesses, factories, and industries. Groundwaters may not have transmission lines, but are often disinfected with chlorine and pumped directly to service storage and distribution systems.

1.11 Storage Facilities

Storage facilities may consist of large reservoirs behind dams (impoundments) or service storage reservoirs located at water treatment plants and/or at various places in distribution systems. Storage facilities for treated water at water treatment plants are called clear wells. Operational service storage tanks in distribution systems may be pressure tanks, elevated tanks, ground-level tanks or reservoirs, or underground facilities.

1.12 Treatment Facilities

Surface waters require treatment to remove suspended and dissolved materials and also to kill disease-causing organisms. Groundwater may require treatment for the removal of excessive hardness, taste- and odor-causing substances, dissolved gases, and impurities such as iron and manganese. Water treatment plants provide the necessary treatment to make the waters safe and suitable for drinking purposes. All

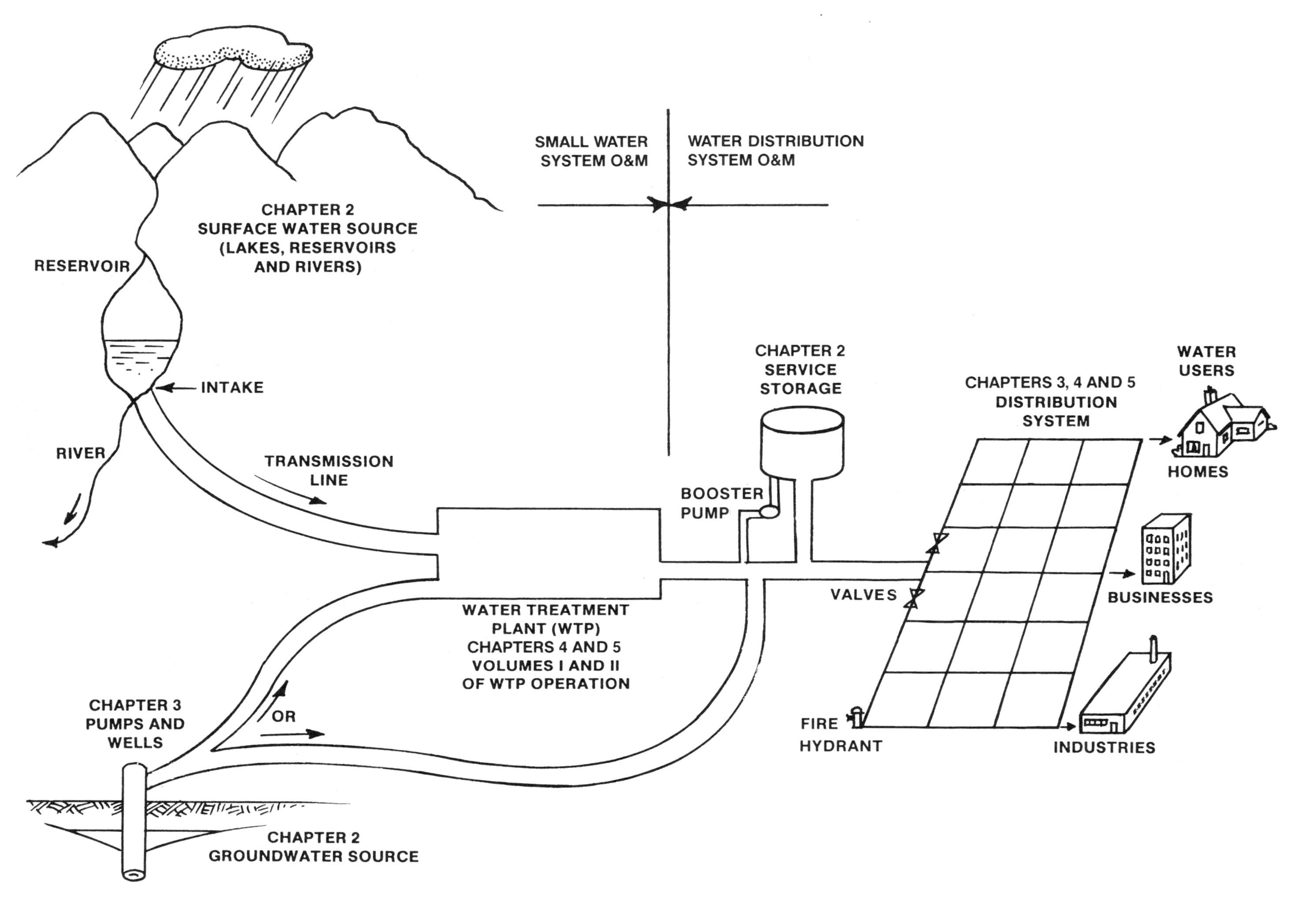

Fig. 1.1 Typical water supply system

waters, regardless of source or treatment received, should be disinfected to prevent the spread of disease-causing organisms. Chlorination is the most common means of disinfection used today to protect the public's health.

1.13 Distribution Systems

A distribution system consists of a network of pipes, valves, fire hydrants, service lines, meters, and pumping stations. The system delivers water to homes, businesses, and industries for drinking and other uses. This water also is used for fire protection. The network of pipes, pumping stations, and service storage reservoirs must have sufficient capacity to meet maximum water demands plus firefighting requirements and still maintain adequate water pressures throughout the water distribution system. Valves are necessary to isolate portions of the distribution system for cleaning, maintenance, repairs, and making additions to the system. The distribution system should be free of cross connections with unapproved water supplies which could allow contamination to be introduced into the system.

QUESTIONS

Write your answers in a notebook and then compare your answers with those on page 10.

1.1A List the two major sources of water supplies.

1.1B Why should surface waters be treated?

1.2 WHAT DOES A SMALL WATER SYSTEM OPERATOR DO?

1.20 Operation and Maintenance

Simply described, small water system operators keep the system operating efficiently to deliver safe and pleasant water. They inspect the system to keep the water flowing today and in the future. Physically, they have manual and power-operated equipment to help them do the job, including repairs and minor additions to the system. Other jobs include lubricating and maintaining equipment, collecting water samples, and recordkeeping. Typical duties performed by small water system operators are summarized in Table 1.1.

To describe your duties in detail, let's start at the beginning. Let us say that the need for new wells and pumps, and also for new or improved storage tanks and distribution pipe network, has long been recognized by your community. The community has voted to issue the necessary bonds to finance the project and the consulting engineers have been asked to submit plans and specifications. In the best interests of the community and the consulting engineer, you should be in on the ground floor planning and design.

You and the engineer should discuss the proposed locations and layout of wells, service reservoirs, and pipe networks. Your job is to discuss with the engineer how these new facilities could be operated, maintained, and repaired and also to make suggestions on how these jobs could be done more easily. Be sure there is adequate room for maintenance and repair equipment even during adverse weather conditions. Together with the consulting engineers, you can be a member of an expert team able to advise your water utility.

Ultimately you want to operate your pumps and reservoirs in such a fashion that your consumers will always have sufficient, safe, and pleasant water at adequate pressures.

TABLE 1.1 TYPICAL DUTIES OF A SMALL WATER SYSTEM OPERATOR

1. Place barricades, signs, and traffic cones around work sites to protect operators and public.
2. Excavate trenches and install shoring.
3. Lay, connect, test, and disinfect water mains.
4. Tap into water mains.
5. Flush and clean water mains.
6. Read and update water distribution system maps and record drawings (also called "as built" plans).
7. Operate and maintain well pumps and hydropneumatic pressure tanks.
8. Collect and transport water samples.
9. Clean and disinfect storage tanks and reservoirs.
10. Protect water mains and storage facilities from corrosion effects.
11. Observe pump motors to detect unusual noises, vibrations, or excessive heat.
12. Adjust and clean pump seals and packing glands and also clean mechanical seals.
13. Repair and overhaul pumps, motors, chlorinators, and control valves.
14. Safely load and unload chlorine cylinders and other dry and liquid chemicals.
15. Keep records and prepare reports.
16. Estimate and justify budget requests for supplies and equipment.

17. Start up or shut down pumps as necessary to regulate system flows and pressures.
18. Perform efficiency tests on pumps and related equipment.
19. Troubleshoot minor electrical and mechanical equipment problems and correct.
20. Detect hazardous atmospheres and correct before entry.
21. Conduct safety inspections, follow safety rules for waterworks facilities, and also develop and conduct tailgate safety meetings.
22. Troubleshoot to locate the causes of water quality complaints.
23. Discuss with the public their concerns regarding the quality of the water they receive.
24. Communicate effectively with other operators and supervisors on the technical level expected for your position.

1.21 Supervision and Administration

In addition to operation and maintenance duties for your small water system, you may also be responsible for supervision of personnel. Chief operators or supervisors frequently have the responsibility of training new operators and should encourage all operators to strive for higher levels of certification.

As a small water system administrator, you may be in charge of recordkeeping. In this case, you will be responsible for operating and maintaining the system as efficiently as possible, keeping in mind that the primary objective is to deliver safe and pleasant drinking water to your consumers. Without adequate, reliable records of important phases of operation and maintenance, the effectiveness of your operation has not been documented (recorded). Also, accurate records are required by regulatory agencies for compliance with the Drinking Water Regulations of the Safe Drinking Water Act.

You may also be the budget administrator. Here you will be in the best position to give advice on budget requirements, management problems, and future planning. You should be aware of the necessity for additional expenditures, including funds for new facilities and enlargement of existing facilities, equipment replacement, and additional employees. You should recognize these and define them clearly for the proper officials. Early planning and budgeting of this type will contribute greatly to the continued smooth operation of your facility.

1.22 Public Relations

As an operator you are in the field of public relations and must be able to explain the purpose and operation of your facilities to civic organizations, school classes, representatives of news media, and even to city council members or directors of your district. Lots of people want to know about the large elevated storage tank or those "funny things" sticking out the top of an underground storage reservoir. One of the best results from a well-guided tour is gaining support from your city council and the public to obtain the funds necessary to run a good operation.

The overall appearance of your pump stations and elevated tanks indicates to the public the type of operation you maintain. If the facilities are dirty and rundown, in need of painting, and overgrown with weeds, you will be unable to convince the public that you are doing a good job. *YOUR RECORDS SHOWING THAT YOU ARE DELIVERING SAFE DRINKING WATER TO YOUR CONSUMERS WILL MEAN NOTHING TO VISITORS AND NEIGHBORS OF YOUR FACILITIES UNLESS YOUR FACILITIES APPEAR CLEAN AND WELL MAINTAINED.*

Another aspect of your job may be handling complaints. When someone contacts you complaining that their drinking water looks muddy, tastes bad, or smells bad, you may have a serious problem. Whenever someone complains, record all of the necessary information (name, date, location, and phone number) and have the complaint thoroughly investigated. Be sure to notify the person who complained of the results of your investigation and what corrective action was or will be taken.

1.23 Safety

Safety is a very important operator responsibility. Unfortunately, too many small water system operators take safety for granted. *YOU* have the responsibility to be sure that your facilities are a safe place to work and that everyone follows safe procedures. Everyone must follow safe procedures and understand why safe procedures must be followed at all times. All operators must be aware of safety hazards in and around water system facilities. Work in traffic and excavating for the installation or repair of pipes can be extremely hazardous. Explosive conditions can develop when painting the inside of an elevated tank if adequate ventilation is not provided. Most accidents result from carelessness or negligence. You should plan or be a part of an active safety program. Also, you may have the responsibility of training new operators and safe procedures must be stressed.

Clearly, today's small water system operator must be capable of doing many jobs — *AND DOING THEM ALL SAFELY.*

QUESTIONS

Write your answers in a notebook and then compare your answers with those on page 10.

1.2A Why should small water system operators discuss proposed facilities with engineers?

1.2B Why are adequate and reliable records very important?

1.2C To whom might you have to explain the purpose and operation of your facilities?

1.2D Why is the appearance of pump stations and elevated tanks and the grounds around them important?

1.2E Why is safety important?

1.3 JOB OPPORTUNITIES

1.30 Staffing Needs

The small water system field is changing rapidly. New facilities and systems are being constructed, and old facilities and systems are being modified and enlarged to meet the water system demands of our growing population and industries. Towns, municipalities, special districts, and industries all employ water system operators. Operators, maintenance personnel, foremen, managers, instrumentation experts, and laboratory technicians are sorely needed now and will be in the future.

1.31 Who Hires Small Water System Operators?

Operators' paychecks usually come from a city, water agency or district, or a private utility company. The operator also may be employed by one of the many large industries which operate their own water system. As an operator, you are always responsible to your employer for operating and maintaining an economical and efficient water system. An even greater obligation rests with the operator because of the great number of people who drink the water from the distribution system. In the final analysis, the operator is really working for the people who depend on the operator to provide them with safe and pleasant drinking water.

1.32 Where Do Small Water System Operators Work?

Jobs are available for small water system operators wherever people live and need someone to deliver water to their homes, offices, or industrial processes. The different types and locations of small water systems offer a wide range of working conditions. From the mountains to the sea, wherever people gather together into communities, small water systems will be found. From a small system foreman or a computer control center operator at a complex municipal storage and distribution system to a one-person manager of a small town water supply system, you can select your own special place in water system operation.

1.33 What Pay Can a Small Water System Operator Expect?

In dollars? Prestige? Job satisfaction? Community service? In opportunities for advancement? By whatever scale you use, returns are mainly what you make them. If you choose a large municipality, the pay is good and advancement prospects are tops. Choose a small town and the pay may not be as good, but job satisfaction, freedom from time-clock hours, community service, and prestige may well add up to a more desirable outstanding personal achievement. If you have the ability and take advantage of the opportunities, you can make this field your career and advance to an enviable position. Many of these positions are or will be represented by an employee organization that will try to obtain higher pay and other benefits for you. Total reward depends on you and how *YOU APPLY YOURSELF.*

1.34 What Does It Take To Be a Small Water System Operator?

DESIRE. First you must make the serious decision to enter this fine profession. You can do this with a high school or a college education. While some jobs will always exist for manual labor, the real and expanding need is for *QUALIFIED OPERATORS.* You must be willing to study and take an active role in upgrading your capabilities. New techniques, advanced equipment, and increasing use of complex instrumentation and computers require a new breed of small water system operator: one who is willing to learn today, and gain tomorrow, for surely your small water system will move toward newer and more effective operation and maintenance procedures. Indeed, the truly service-minded operator assists in adding to and improving the performance of the small water system on a continuing basis.

You can be a small water system operator tomorrow by beginning your learning today; or you can be a better operator, ready for advancement, by accelerating your learning today.

This training course, then, is your start toward a better tomorrow, both for you and for the public who will receive better water from your efforts.

QUESTIONS

Write your answers in a notebook and then compare your answers with those on page 10.

1.3A Who hires small water system operators?

1.3B What does it take to be a good small water system operator?

1.4 PREPARING YOURSELF FOR THE FUTURE

1.40 Your Qualifications

What do you know about your job or the job you'd like to obtain? Perhaps a little, and perhaps a lot. You must evaluate the knowledge, skills, and experience you already have and what you will need to achieve future jobs and advancement.

The knowledge and skills required for your job depend to a large degree on the size and type of water system where you work. You may work on a large, complex system employing several hundred or more operators. In this case, you are probably a specialist in one or more phases of the distribution system or storage reservoirs (such as being responsible for the valves).

On the other hand, you may operate and maintain a small water system serving only a thousand people or even fewer. You may be the only operator for the entire system or, at best, have only one or two helpers. If this is the case, you must be a "jack-of-all-trades" because of the diversity of your tasks.

1.41 Your Personal Training Program

Beginning on this page you are starting a training course which has been carefully prepared to help you improve your knowledge and skills to operate and maintain small water systems.

You will be able to proceed at your own pace; you will have the opportunity to learn a little or a lot about each topic. This training manual has been prepared this way to meet the various needs of small water system operators, depending on the size and type of system for which you are responsible. To study for certification and civil service exams, you may have to cover most of the material in this manual. You will never know everything about small water systems and the equipment and procedures available for operation and maintenance. However, you will be able to answer some very important questions about how, why, and when certain things happen in small water systems. You can also learn how to manage your small water system to provide a reliable supply of safe and pleasant drinking water to your customers while minimizing costs in the long run.

This training course is not the only one available to help you improve your abilities. Some state water utility associations, vocational schools, community colleges, and universities offer training courses on both a short- and long-term basis. Many state, local, and private agencies have conducted training programs and informative seminars. Most state health departments can be very helpful in providing training programs or directing you to good programs.

Some libraries can provide you with useful journals and books on small water systems. Listed below are several very good references in the field of small water systems. Prices listed were those available when this manual was published; they will probably increase in the future.

1. *MANUAL OF INSTRUCTION FOR WATER TREATMENT PLANT OPERATORS (NEW YORK MANUAL).* Obtain from Health Education Services, Inc., PO Box 7126, Albany, NY 12224. Price, $19.50, includes cost of shipping and handling.

2. *MANUAL OF WATER UTILITIES OPERATIONS (TEXAS MANUAL).* Obtain from Texas Water Utilities Association, 1106 Clayton Lane, Suite 101 East, Austin, TX 78723-1093. Price to members, $22.85; nonmembers, $34.85; price includes cost of shipping and handling.

3. *WATER DISTRIBUTION OPERATOR TRAINING HANDBOOK.* Obtain from American Water Works Association (AWWA), Bookstore, 6666 West Quincy Avenue, Denver, CO 80235. Order No. 20428. ISBN 1-58321-014-8. Price to members, $35.50; nonmembers, $50.50; price includes cost of shipping and handling.

4. *WATER SOURCES.* Obtain from American Water Works Association (AWWA), Bookstore, 6666 West Quincy Avenue, Denver, CO 80235. Order No. 1955. ISBN 0-89876-778-5. Price to members, $53.50; nonmembers, $81.50; price includes cost of shipping and handling.

5. *WATER TRANSMISSION AND DISTRIBUTION.* Obtain from American Water Works Association (AWWA), Bookstore, 6666 West Quincy Avenue, Denver, CO 80235. Order No. 1957. ISBN 0-89867-821-8. Price to members, $72.50; nonmembers, $106.50; price includes cost of shipping and handling.

6. *OPERATOR CERTIFICATION STUDY GUIDE—A WORKBOOK FOR TREATMENT PLANT OPERATORS AND DISTRIBUTION SYSTEM PERSONNEL.* Obtain from American Water Works Association (AWWA), Bookstore, 6666 West Quincy Avenue, Denver, CO 80235. Order No. 20206. ISBN 0-89867-303-8. Price to members, $29.50; nonmembers, $43.50; price includes cost of shipping and handling.

Throughout this manual we will be recommending American Water Works (AWWA) publications. Members of AWWA can buy some publications at reduced prices. You can join AWWA by writing to the headquarters office in Denver or by contacting a member of AWWA. Headquarters can help you contact your own state or regional AWWA Section. This professional organization can offer you many helpful training opportunities and educational materials when you join and actively participate with your associates in the field.

1.42 Certification

Certification examinations are usually administered by state regulatory agencies or professional associations. Operators take these exams in order to obtain certificates which indicate a level of professional competence. You should continually strive to achieve higher levels of certification. Successful completion of this operator training program will help you achieve your certification goals.

1.5 ACKNOWLEDGMENTS

Many of the topics and ideas discussed in this chapter were based on similar work written by Larry Trumbull and Walt Driggs.

SUGGESTED ANSWERS

Chapter 1. THE SMALL WATER SYSTEM OPERATOR

You are not expected to have the exact answers suggested for questions requiring written answers, but you should have the correct idea. The numbering of the questions refers to the section in the chapter where you can find the information to answer the questions. Answers to questions numbered 1.0A and 1.0B can be found in Section 1.0, "Need for Small Water System Operators."

Answers to questions on page 4.

1.0A Small water system operators make sure small water system facilities do their intended job. Operators operate, maintain, repair, and manage these facilities.

1.0B Small water system operators have the responsibility of ensuring that safe and pleasant drinking water is delivered to everyone's tap. Also adequate amounts of water and pressure must be available during times of emergency, such as a fire.

Answers to questions on page 6.

1.1A The two major sources of water supplies are (1) groundwater, and (2) surface water.

1.1B Surface water is usually treated to remove suspended and dissolved materials and to kill disease-causing organisms.

Answers to questions on page 7.

1.2A Small water system operators should discuss proposed facilities with engineers to find out how the engineer intends for these facilities to be operated, maintained, and repaired and also to make suggestions on how these jobs could be done more easily. Be sure there is adequate room for maintenance and repair equipment even during adverse weather conditions.

1.2B Adequate and reliable records are very important to document the effectiveness of your operation.

1.2C Small water system operators must be able to explain the purpose and operation of their facilities to civic organizations, school classes, representatives of news media, and even to city council members or directors of their district.

1.2D The appearance of pump stations and elevated tanks and the grounds around them indicates to the public the type of operation you maintain.

1.2E Safety is a very important operator responsibility. Most accidents result from carelessness or negligence. Safe procedures must be stressed at all times.

Answers to questions on page 8.

1.3A Small water system operators may be hired by cities, water agencies or districts, private utility companies, or industries.

1.3B *DESIRE*, if you want to be a qualified small water system operator, you can do it.

SMALL WATER SYSTEM OPERATION AND MAINTENANCE
4th Edition

IMPORTANT

PLEASE READ INSTRUCTIONS ON REVERSE SIDE BEFORE COMPLETING THIS FORM.

Name: OPERATOR, (Last) JOE (First) B. (MI)

Address: 711 MAIN ST.

CLEARWATER, (City) CA (State) 98765 (Zip Code)

Home Phone: (707) 123-4567 Work: (707) 123-7654

Mail to: Office of Water Programs
California State University, Sacramento
6000 J Street
Sacramento, California 95819-6025

○ If your address has changed, please fill in this circle with a Number 2 pencil.

For help in preparing this box, see example on reverse side.

SOCIAL SECURITY NUMBER	CHAPTER NUMBER
123-45-6789	01

FOLD

IMPORTANT DIRECTIONS FOR MARKING ANSWERS

Use black lead pencil only (#2 or softer).
Make heavy black marks that fill the circle completely.
Erase clearly any answer you wish to change.
Make no stray marks on this answer sheet.

1. **MULTIPLE CHOICE QUESTIONS: Fill in the correct answers. If 2 and 3 are correct for question 1, mark:**

   1 2 3 4 5
   1 Ⓣ ● ● ○ ○

2. **TRUE — FALSE QUESTIONS: If true, fill in circle in column 1; if false, fill in column 2. If question 3 is true, mark:**

   1 2 3 4 5
   3 ● Ⓕ ○ ○ ○

URGENT!

ANSWER SHEET WILL BE GRADED INCORRECTLY IF:

- THIS FORM IS USED FOR ANY COURSE OTHER THAN ***Small Water System Operation and Maintenance, 4th Edition***
- A NUMBER 2 PENCIL IS NOT USED *(Do not use ink)*
- CIRCLES FOR CHAPTER NUMBER & SOCIAL SECURITY NUMBER ARE NOT FILLED IN ON THE BOX ABOVE.

BEFORE completing this form, please read ALL instructions, including those printed on the REVERSE SIDE.

SAMPLE DO NOT USE

FOLD

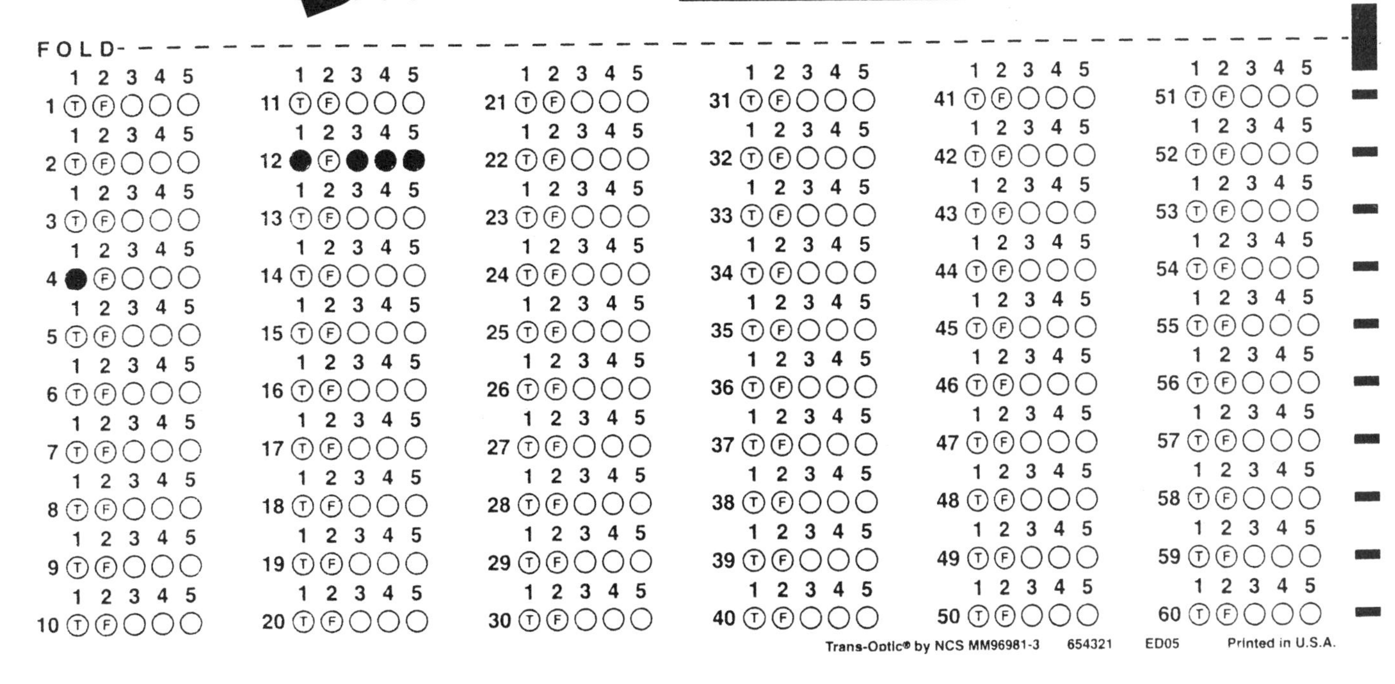

1 2 3 4 5	1 2 3 4 5	1 2 3 4 5	1 2 3 4 5	1 2 3 4 5	1 2 3 4 5
1 Ⓣ Ⓕ ○ ○ ○	11 Ⓣ Ⓕ ○ ○ ○	21 Ⓣ Ⓕ ○ ○ ○	31 Ⓣ Ⓕ ○ ○ ○	41 Ⓣ Ⓕ ○ ○ ○	51 Ⓣ Ⓕ ○ ○ ○
2 Ⓣ Ⓕ ○ ○ ○	12 ● Ⓕ ● ● ●	22 Ⓣ Ⓕ ○ ○ ○	32 Ⓣ Ⓕ ○ ○ ○	42 Ⓣ Ⓕ ○ ○ ○	52 Ⓣ Ⓕ ○ ○ ○
3 Ⓣ Ⓕ ○ ○ ○	13 Ⓣ Ⓕ ○ ○ ○	23 Ⓣ Ⓕ ○ ○ ○	33 Ⓣ Ⓕ ○ ○ ○	43 Ⓣ Ⓕ ○ ○ ○	53 Ⓣ Ⓕ ○ ○ ○
4 ● Ⓕ ○ ○ ○	14 Ⓣ Ⓕ ○ ○ ○	24 Ⓣ Ⓕ ○ ○ ○	34 Ⓣ Ⓕ ○ ○ ○	44 Ⓣ Ⓕ ○ ○ ○	54 Ⓣ Ⓕ ○ ○ ○
5 Ⓣ Ⓕ ○ ○ ○	15 Ⓣ Ⓕ ○ ○ ○	25 Ⓣ Ⓕ ○ ○ ○	35 Ⓣ Ⓕ ○ ○ ○	45 Ⓣ Ⓕ ○ ○ ○	55 Ⓣ Ⓕ ○ ○ ○
6 Ⓣ Ⓕ ○ ○ ○	16 Ⓣ Ⓕ ○ ○ ○	26 Ⓣ Ⓕ ○ ○ ○	36 Ⓣ Ⓕ ○ ○ ○	46 Ⓣ Ⓕ ○ ○ ○	56 Ⓣ Ⓕ ○ ○ ○
7 Ⓣ Ⓕ ○ ○ ○	17 Ⓣ Ⓕ ○ ○ ○	27 Ⓣ Ⓕ ○ ○ ○	37 Ⓣ Ⓕ ○ ○ ○	47 Ⓣ Ⓕ ○ ○ ○	57 Ⓣ Ⓕ ○ ○ ○
8 Ⓣ Ⓕ ○ ○ ○	18 Ⓣ Ⓕ ○ ○ ○	28 Ⓣ Ⓕ ○ ○ ○	38 Ⓣ Ⓕ ○ ○ ○	48 Ⓣ Ⓕ ○ ○ ○	58 Ⓣ Ⓕ ○ ○ ○
9 Ⓣ Ⓕ ○ ○ ○	19 Ⓣ Ⓕ ○ ○ ○	29 Ⓣ Ⓕ ○ ○ ○	39 Ⓣ Ⓕ ○ ○ ○	49 Ⓣ Ⓕ ○ ○ ○	59 Ⓣ Ⓕ ○ ○ ○
10 Ⓣ Ⓕ ○ ○ ○	20 Ⓣ Ⓕ ○ ○ ○	30 Ⓣ Ⓕ ○ ○ ○	40 Ⓣ Ⓕ ○ ○ ○	50 Ⓣ Ⓕ ○ ○ ○	60 Ⓣ Ⓕ ○ ○ ○

Trans-Optic® by NCS MM96981-3 654321 ED05 Printed in U.S.A.

DIRECTIONS FOR WORKING OBJECTIVE TEST

Chapter 1. THE SMALL WATER SYSTEM OPERATOR

1. You have been provided with a special answer sheet (see page 11) for each chapter. Be sure you follow the special directions provided with the answer sheets. If you lose an answer sheet or have any problems, please notify the Project Director.

2. Mark your answers on the answer sheet with a dark lead pencil. Do not use ink. Be certain to mark *ALL* correct answers.

   For example, Question 12 has four correct answers (1, 3, 4, and 5); you should place a mark under Columns 1, 3, 4, and 5 on the answer sheet.

   Questions 1 through 5 are true or false questions. If a question is true, then mark Column 1, and if false mark Column 2. The correct answer to Question 4 is true; therefore, place a mark in Column 1.

   Please mark your answers in your workbook for your record because answer sheets will not be returned to you.

3. Mail your answer sheet to the Project Director immediately after you have completed the test.

4. Answer sheets may be folded (but not into more than 3 equal parts) and mailed in a 4 x 9 1/2-inch standard white envelope.

OBJECTIVE TEST

Chapter 1. THE SMALL WATER SYSTEM OPERATOR

Please write your name and mark the correct answers on the answer sheet. There may be more than one correct answer to each multiple-choice question.

True-False

1. Operators are responsible for the drinking water of their community.
   1. True
   2. False

2. All waters, regardless of source or treatment received, should be disinfected to prevent the spread of disease-causing organisms.
   1. True
   2. False

3. Water distribution systems should be free of cross connections with unapproved water supplies which could allow contamination to be introduced into the system.
   1. True
   2. False

4. Handling complaints is an important part of an operator's job.
   1. True
   2. False

5. All small water system operators are very safety conscious.
   1. True
   2. False

Best Answer (Select only the closest or best answer.)

6. The water supply system in this training manual has been divided into what two parts?
   1. Storage system and the distribution system
   2. Water system and the collection system
   3. Water system and the distribution system
   4. Well system and the treatment system
7. What do water system operators do?
   1. Design water supply facilities
   2. Develop demands for system water
   3. Make the water facilities do their intended job
   4. Plan for future expansion of the system
8. What is the name of storage facilities for treated water at water treatment plants?
   1. Basins
   2. Clear wells
   3. Reservoirs
   4. Tanks
9. What does the overall appearance of pump stations and elevated tanks indicate to the public?
   1. Adequate funding is not necessary to produce safe drinking water
   2. The neighbors accept the presence of the facility
   3. The quality of your painting and landscaping program
   4. The type of operation you maintain
10. Why do operators take certification examinations?
    1. To certify that their water is safe to drink
    2. To certify that they have an operator's license
    3. To gain experience taking certification examinations
    4. To indicate a level of professional competence

Multiple Choice (Select all correct answers.)

11. Cities and towns need operators with what kinds of skills or talents?
    1. Capable
    2. Dedicated
    3. Disinterested
    4. Qualified
    5. Uninformed
12. What do consumers expect from the operators of water systems?
    1. Adequate water pressure
    2. High-cost water
    3. Pleasant water
    4. Safe water
    5. Sufficient water
13. Why are records important for small water system operators?
    1. Document compliance with Safe Drinking Water Act
    2. Document effectiveness of operation
    3. Justify a record management program
    4. Record important phases of operation and maintenance
    5. Show operator's ability to maintain records
14. Why is public relations an important task for operators?
    1. To eliminate the need for operators to perform public speaking
    2. To explain purpose of large elevated tanks
    3. To explain purpose of operation of facilities
    4. To gain support to obtain funds necessary for good operation
    5. To help operators avoid public meetings
15. What safety hazards are encountered on the job by small water system operators?
    1. Drinking treated water
    2. Explosive conditions in elevated tanks
    3. Lack of ventilation in elevated tanks
    4. Work in excavations
    5. Work in traffic

End of Objective Test

CHAPTER 2

WATER SOURCES AND TREATMENT

by

Bert Ellsworth

TABLE OF CONTENTS

Chapter 2. WATER SOURCES AND TREATMENT

OBJECTIVES

Chapter 2. WATER SOURCES AND TREATMENT

Following completion of Chapter 2, you should be able to:

1. Describe the importance of water,
2. Identify various sources of water,
3. Outline the procedures of a sanitary survey,
4. Evaluate the suitability of a water source for drinking purposes and as a general water supply, and
5. Identify water quality problems and treatment processes to solve the problems.

PROJECT PRONUNCIATION KEY

by Warren L. Prentice

The Project Pronunciation Key is designed to aid you in the pronunciation of new words. While this key is based primarily on familiar sounds, it does not attempt to follow any particular pronunciation guide. This key is designed solely to aid operators in this program.

You may find it helpful to refer to other available sources for pronunciation help. Each current standard dictionary contains a guide to its own pronunciation key. Each key will be different from each other and from this key. Examples of the difference between the key used in this program and the *WEBSTER'S NEW WORLD COLLEGE DICTIONARY** "Key" are shown below.

In using this key, you should accent (say louder) the syllable that appears in capital letters. The following chart is presented to give examples of how to pronounce words using the Project Key.

	SYLLABLE				
WORD	1st	2nd	3rd	4th	5th
acid	AS	id			
coliform	COAL	i	form		
biological	BUY	o	LODGE	ik	cull

The first word, *ACID*, has its first syllable accented. The second word, *COLIFORM*, has its first syllable accented. The third word, *BIOLOGICAL*, has its first and third syllables accented.

We hope you will find the key useful in unlocking the pronunciation of any new word.

Term
Project Key
Webster Key
acid
AS-id
aś id
coliform
COAL-i-form
kō′ lə fôrm
biological
BUY-o-LODGE-ik-cull
bī ə läj′ i kəl

* *The WEBSTER'S NEW WORLD COLLEGE DICTIONARY, Fourth Edition, 1999, was chosen rather than an unabridged dictionary because of its availability to the operator. Other editions may be slightly different.*

WORDS

Chapter 2. WATER SOURCES AND TREATMENT

ACID RAIN — ACID RAIN

Precipitation which has been rendered (made) acidic by airborne pollutants.

AESTHETIC (es-THET-ick) — AESTHETIC

Attractive or appealing.

APPROPRIATIVE — APPROPRIATIVE

Water rights to or ownership of a water supply which is acquired for the beneficial use of water by following a specific legal procedure.

AQUIFER (ACK-wi-fer) — AQUIFER

A natural underground layer of porous, water-bearing materials (sand, gravel) usually capable of yielding a large amount or supply of water.

ARTESIAN (are-TEE-zhun) — ARTESIAN

Pertaining to groundwater, a well, or underground basin where the water is under a pressure greater than atmospheric and will rise above the level of its upper confining surface if given an opportunity to do so.

CAPILLARY FRINGE — CAPILLARY FRINGE

The porous material just above the water table which may hold water by capillarity (a property of surface tension that draws water upward) in the smaller void spaces.

CISTERN (SIS-turn) — CISTERN

A small tank (usually covered) or a storage facility used to store water for a home or farm. Often used to store rainwater.

CONTAMINATION — CONTAMINATION

The introduction into water of microorganisms, chemicals, toxic substances, wastes, or wastewater in a concentration that makes the water unfit for its next intended use.

CROSS CONNECTION — CROSS CONNECTION

A connection between a drinking (potable) water system and an unapproved water supply. For example, if you have a pump moving nonpotable water and hook into the drinking water system to supply water for the pump seal, a cross connection or mixing between the two water systems can occur. This mixing may lead to contamination of the drinking water.

DETENTION TIME — DETENTION TIME

(1) The theoretical (calculated) time required for a small amount of water to pass through a tank at a given rate of flow.

(2) The actual time in hours, minutes or seconds that a small amount of water is in a settling basin, flocculating basin or rapid-mix chamber. In storage reservoirs, detention time is the length of time entering water will be held before being drafted for use (several weeks to years, several months being typical).

$$\text{Detention Time, hr} = \frac{(\text{Basin Volume, gal})(24\ \text{hr/day})}{\text{Flow, gal/day}}$$

DIRECT RUNOFF — DIRECT RUNOFF

Water that flows over the ground surface or through the ground directly into streams, rivers, or lakes.

DRAWDOWN — DRAWDOWN

(1) The drop in the water table or level of water in the ground when water is being pumped from a well.

(2) The amount of water used from a tank or reservoir.

(3) The drop in the water level of a tank or reservoir.

EPIDEMIOLOGY (EP-uh-DE-me-ALL-o-gee) — EPIDEMIOLOGY

A branch of medicine which studies epidemics (diseases which affect significant numbers of people during the same time period in the same locality). The objective of epidemiology is to determine the factors that cause epidemic diseases and how to prevent them.

EVAPORATION — EVAPORATION

The process by which water or other liquid becomes a gas (water vapor or ammonia vapor).

EVAPOTRANSPIRATION (ee-VAP-o-TRANS-purr-A-shun) — EVAPOTRANSPIRATION

(1) The process by which water vapor passes into the atmosphere from living plants. Also called TRANSPIRATION.

(2) The total water removed from an area by transpiration (plants) and by evaporation from soil, snow and water surfaces.

GEOLOGICAL LOG — GEOLOGICAL LOG

A detailed description of all underground features discovered during the drilling of a well (depth, thickness and type of formations).

HYDROLOGIC (HI-dro-LOJ-ick) CYCLE — HYDROLOGIC CYCLE

The process of evaporation of water into the air and its return to earth by precipitation (rain or snow). This process also includes transpiration from plants, groundwater movement, and runoff into rivers, streams and the ocean. Also called the WATER CYCLE.

IMPERMEABLE (im-PURR-me-uh-BULL) — IMPERMEABLE

Not easily penetrated. The property of a material or soil that does not allow, or allows only with great difficulty, the movement or passage of water.

INFILTRATION (IN-fill-TRAY-shun) — INFILTRATION

The seepage of groundwater into a sewer system, including service connections. Seepage frequently occurs through defective or cracked pipes, pipe joints, connections or manhole walls.

MICROORGANISMS (MY-crow-OR-gan-IS-zums) — MICROORGANISMS

Living organisms that can be seen individually only with the aid of a microscope.

NONPOTABLE (non-POE-tuh-bull) — NONPOTABLE

Water that may contain objectionable pollution, contamination, minerals, or infective agents and is considered unsafe and/or unpalatable for drinking.

PALATABLE (PAL-uh-tuh-bull) — PALATABLE

Water at a desired temperature that is free from objectionable tastes, odors, colors, and turbidity. Pleasing to the senses.

PATHOGENIC (PATH-o-JEN-ick) ORGANISMS — PATHOGENIC ORGANISMS

Organisms, including bacteria, viruses or cysts, capable of causing diseases (giardiasis, cryptosporidiosis, typhoid, cholera, dysentery) in a host (such as a person). There are many types of organisms which do *NOT* cause disease. These organisms are called non-pathogenic.

PERCOLATION (PURR-co-LAY-shun) PERCOLATION

The slow passage of water through a filter medium; or, the gradual penetration of soil and rocks by water.

POLLUTION POLLUTION

The impairment (reduction) of water quality by agricultural, domestic, or industrial wastes (including thermal and radioactive wastes) to a degree that has an adverse effect on any beneficial use of water.

POTABLE (POE-tuh-bull) WATER POTABLE WATER

Water that does not contain objectionable pollution, contamination, minerals, or infective agents and is considered satisfactory for drinking.

PRECIPITATION (pre-SIP-uh-TAY-shun) PRECIPITATION

(1) The process by which atmospheric moisture falls onto a land or water surface as rain, snow, hail, or other forms of moisture.

(2) The chemical transformation of a substance in solution into an insoluble form (precipitate).

PRESCRIPTIVE (pre-SKRIP-tive) PRESCRIPTIVE

Water rights which are acquired by diverting water and putting it to use in accordance with specified procedures. These procedures include filing a request (with a state agency) to use unused water in a stream, river or lake.

RAW WATER RAW WATER

(1) Water in its natural state, prior to any treatment.

(2) Usually the water entering the first treatment process of a water treatment plant.

RIPARIAN (ri-PAIR-ee-an) RIPARIAN

Water rights which are acquired together with title to the land bordering a source of surface water. The right to put to beneficial use surface water adjacent to your land.

SAFE DRINKING WATER ACT SAFE DRINKING WATER ACT

Commonly referred to as SDWA. An Act passed by the U.S. Congress in 1974. The Act establishes a cooperative program among local, state and federal agencies to ensure safe drinking water for consumers. The Act has been amended several times, including the 1980, 1986, and 1996 Amendments.

SAFE WATER SAFE WATER

Water that does not contain harmful bacteria, or toxic materials or chemicals. Water may have taste and odor problems, color and certain mineral problems and still be considered safe for drinking.

SAFE YIELD SAFE YIELD

The annual quantity of water that can be taken from a source of supply over a period of years without depleting the source permanently (beyond its ability to be replenished naturally in "wet years").

SANITARY SURVEY SANITARY SURVEY

A detailed evaluation and/or inspection of a source of water supply and all conveyances, storage, treatment and distribution facilities to ensure protection of the water supply from all pollution sources.

SEWAGE SEWAGE

The used household water and water-carried solids that flow in sewers to a wastewater treatment plant. The preferred term is WASTEWATER.

SHORT-CIRCUITING SHORT-CIRCUITING

A condition that occurs in tanks or basins when some of the water travels faster than the rest of the flowing water. This is usually undesirable since it may result in shorter contact, reaction, or settling times in comparison with the theoretical (calculated) or presumed detention times.

STRATIFICATION (STRAT-uh-fuh-KAY-shun) STRATIFICATION

The formation of separate layers (of temperature, plant, or animal life) in a lake or reservoir. Each layer has similar characteristics such as all water in the layer has the same temperature.

SUBSIDENCE (sub-SIDE-ence) SUBSIDENCE

The dropping or lowering of the ground surface as a result of removing excess water (overdraft or overpumping) from an aquifer. After excess water has been removed, the soil will settle, become compacted and the ground surface will drop and can cause the settling of underground utilities.

TOPOGRAPHY (toe-PAH-gruh-fee) TOPOGRAPHY

The arrangement of hills and valleys in a geographic area.

TRANSPIRATION (TRAN-spur-RAY-shun) TRANSPIRATION

The process by which water vapor is released to the atmosphere by living plants. This process is similar to people sweating. Also see EVAPOTRANSPIRATION.

TRIHALOMETHANES (THMs) (tri-HAL-o-METH-hanes) TRIHALOMETHANES (THMs)

Derivatives of methane, CH_4, in which three halogen atoms (chlorine or bromine) are substituted for three of the hydrogen atoms. Often formed during chlorination by reactions with natural organic materials in the water. The resulting compounds (THMs) are suspected of causing cancer.

TURBIDITY (ter-BID-it-tee) TURBIDITY

The cloudy appearance of water caused by the presence of suspended and colloidal matter. In the waterworks field, a turbidity measurement is used to indicate the clarity of water. Technically, turbidity is an optical property of the water based on the amount of light reflected by suspended particles. Turbidity cannot be directly equated to suspended solids because white particles reflect more light than dark-colored particles and many small particles will reflect more light than an equivalent large particle.

TURBIDITY UNITS (TU) TURBIDITY UNITS (TU)

Turbidity units are a measure of the cloudiness of water. If measured by a nephelometric (deflected light) instrumental procedure, turbidity units are expressed in nephelometric turbidity units (NTU) or simply TU. Those turbidity units obtained by visual methods are expressed in Jackson Turbidity Units (JTU) which are a measure of the cloudiness of water; they are used to indicate the clarity of water. There is no real connection between NTUs and JTUs. The Jackson turbidimeter is a visual method and the nephelometer is an instrumental method based on deflected light.

WASTEWATER WASTEWATER

A community's used water and water-carried solids (including used water from industrial processes) that flow to a treatment plant. Storm water, surface water, and groundwater infiltration also may be included in the wastewater that enters a wastewater treatment plant. The term "sewage" usually refers to household wastes, but this word is being replaced by the term "wastewater."

WATER CYCLE WATER CYCLE

The process of evaporation of water into the air and its return to earth by precipitation (rain or snow). This process also includes transpiration from plants, groundwater movement, and runoff into rivers, streams and the ocean. Also called the HYDROLOGIC CYCLE.

WATER TABLE WATER TABLE

The upper surface of the zone of saturation of groundwater in an unconfined aquifer.

WATERSHED WATERSHED

The region or land area that contributes to the drainage or catchment area above a specific point on a stream or river.

YIELD YIELD

The quantity of water (expressed as a rate of flow — GPM, GPH, GPD, or total quantity per year) that can be collected for a given use from surface or groundwater sources. The yield may vary with the use proposed, with the plan of development, and also with economic considerations. Also see SAFE YIELD.

ZONE OF AERATION ZONE OF AERATION

The comparatively dry soil or rock located between the ground surface and the top of the water table.

ZONE OF SATURATION ZONE OF SATURATION

The soil or rock located below the top of the groundwater table. By definition, the zone of saturation is saturated with water. Also see WATER TABLE.

CHAPTER 2. WATER SOURCES AND TREATMENT

2.0 IMPORTANCE OF WATER

For decades Americans have used water as though their supply would never fail. In recent years, drought conditions have forcibly brought the need to conserve and properly budget our water resources to the minds of water supply managers. Even in the driest years, though, rain across the country enormously exceeds water use. The trouble is that the nation's water resources are poorly distributed. The Pacific Northwest has a big surplus. The agricultural states of the Southwest fight for the last salty drop of water from the Lower Colorado River. The Federal Government has spent billions of dollars building and operating facilities to divert water for use in arid and water-short areas. Contamination is a problem, too. Mineral residues from irrigation have damaged once fertile soil. *ACID RAIN*[1] is killing the fish in mountain lakes. America's drinking water has been tainted with substances as exotic as trichloroethylene (TCE) and as commonplace as highway salt. Vast underground basins of water, deposited over many years, have been seriously depleted in a matter of decades.

All water comes as rain or precipitation from the sky, but 92 percent of the water either evaporates immediately or runs off eventually into the oceans. One-quarter of the water that irrigates, powers, and bathes America is taken from an ancient network of underground aquifers. In 1950, the United States took some 12 trillion gallons (45 billion cubic meters) of water out of the ground; by 1980 the figure had more than doubled.

Water shortages directly influence energy consumption. As groundwater levels fall, more energy is required to pump water from deeper levels in the basin. In several areas, vast water projects use large amounts of electricity to pump water many miles along the project. As energy becomes more expensive, the users of the water will see the increased cost reflected in their water rates. This link between water demands and pumping costs is a driving force behind water conservation programs.

Water is regarded as commonplace because it is the most plentiful liquid on earth and because of our familiarity with it. All of the tissues of our bodies are bathed in it. Whatever may be the thing which we call life on earth, it requires a water environment. Our foods must be suspended or dissolved in water solutions to be carried to the different parts of the body. Also, most waste products are eliminated from the body as water-soluble substances.

Both plant life and animal life depend upon water for survival. A plant receives the greater part of its food from the soil in water solutions and manufactures the rest of its food in the presence of water.

Water is present in almost all natural objects and in almost every part of the earth that people can reach. There is water vapor in the air, and liquid water in rocks and soil. In addition to the water that wets them, clay and certain kinds of rocks contain water in chemical combination with other substances.

Water may be commonplace, but useful water is not always readily available. Even before the discovery of America by Europeans, one of the common causes of war between Native American tribes was water rights. Among the first considerations of any new land development is water. Useful water is only rarely free, and it is not very abundant in many parts of the United States. There are not many places left where a person can feel safe in drinking water from a spring, stream, or pond. Even the groundwater produced by wells must be tested regularly. In some areas, human activities have made it difficult to locate a safe water supply of any sort. In certain coastal areas, for example, overpumping from the ground has depleted the groundwater basins. As a result, the intrusion of sea water is ruining the basin for most useful purposes. Other sources of groundwater contamination include seepage from septic tank leaching systems, agricultural drainage systems, and the improper disposal of hazardous wastes in sanitary landfills and dumps. Some human activities clearly pose a serious threat to life on this planet.

[1] *Acid Rain. Precipitation which has been rendered (made) acidic by airborne pollutants.*

QUESTIONS

Write your answers in a notebook and then compare your answers with those on page 34.

2.0A Why has it become necessary to conserve and properly budget our water resources?

2.0B How are water shortages and energy consumption linked together?

2.0C Name three ways groundwater may become contaminated.

2.1 SOURCES OF WATER

2.10 The Water (Hydrologic) Cycle (Figure 2.1)

All water comes in the form of precipitation. Water evaporates from the ocean by the energy of the sun at an overall rate of about six feet (1.8 m) of water annually. The water which is evaporated is salt-free water, since the heavier mineral salts are left behind. This water vapor rises, is carried along by winds, and eventually condenses into clouds. When clouds become chilled, the small particles of water collect into larger droplets which may precipitate over land or water. As the water falls in the form of rain, snow, sleet, or hail, it clings to and carries with it all the dust and dirt in the air. Needless to say, the first water that falls during a storm picks up the greatest concentration of contamination. After a short period of fall, the precipitation is relatively free of pollutants. A large part of the evaporated water is carried over land masses by the winds and the droplets that fall there make up our supply of fresh water. These droplets may soak into the ground, fall as snow on the mountain tops, or collect in lakes, but in one way or another, all of the droplets seek to return to the ocean from where they came. This, in brief, is the framework of the hydrologic cycle (also called the water cycle).

2.11 Rights to the Use of Water

The rights of an individual to use water for domestic, irrigation, or other purposes varies in different states. Some water rights stem from ownership of the land bordering or overlying the source, while others are acquired by a performance of certain acts required by law.

There are three basic types of water rights:

1. *RIPARIAN* — rights which are acquired with title to the land bordering a source of surface water.
2. *APPROPRIATIVE* — rights which are acquired for the beneficial use of water by following a specific legal procedure.
3. *PRESCRIPTIVE* — rights which are acquired by diverting water and putting to use, for a period of time specified by statute, water to which other parties may or may not have prior claims. The procedure necessary to obtain prescriptive rights must conform with the conditions established by the water-rights laws of individual states.

When there is any question regarding the right to the use of water, a property owner should consult with the appropriate authority and clearly establish rights to its use.

2.12 Ocean

At some time in its history, virtually all water resided in the oceans. By evaporation, moisture is transferred from the ocean surface to the atmosphere, where winds carry the moisture-laden air over land masses. Under certain conditions, this water vapor condenses to form clouds, which release their moisture as precipitation in the form of rain, hail, sleet, or snow.

When rain falls toward the earth, part of it may reevaporate and return immediately to the atmosphere. Precipitation in excess of the amount that wets a surface or evaporates immediately is available as a potential source of water supply.

2.13 Surface Water

2.130 Direct Runoff

Surface water accumulates mainly as a result of direct runoff from precipitation (rain or snow). Precipitation that does not enter the ground through infiltration or is not returned to the atmosphere by evaporation flows over the ground surface and is classified as direct runoff. Direct runoff is water that drains off of saturated or *IMPERMEABLE*[2] surfaces, into stream channels, and then into natural or artificial storage sites (or into the ocean in coastal areas).

The amount of available surface water depends largely upon rainfall. When rainfall is limited, the supply of surface water will vary considerably between wet and dry years. In areas of scant rainfall, people build individual *CISTERNS*[3] for the storage of rain which drains from the catchment areas of roofs. This type of water supply is used extensively in areas such as the Bermuda Islands, where groundwater is virtually nonexistent and there are no streams.

Surface water supplies may be further divided into river, lake, and reservoir supplies. In general, they are characterized by turbidity, suspended solids, some color, and microbiological contamination. Groundwaters, on the other hand, are characterized by higher concentrations of dissolved solids, lower levels of color, and freedom from microbiological contamination.

2.131 Rivers and Streams

Many of the largest cities in the world depend entirely upon large rivers for their water supplies. In using a river or stream supply, one should always be concerned with upstream conditions. Some cities drink water from a stream or river into which the treated wastewater (sewage) from upstream cities has been discharged. This can present very serious problems in

---

[2] *Impermeable (im-PURR-me-uh-BULL). Not easily penetrated. The property of a material or soil that does not allow, or allows only with great difficulty, the movement or passage of water.*

[3] *Cistern (SIS-turn). A small tank (usually covered) or a storage facility used to store water for a home or farm. Often used to store rainwater.*

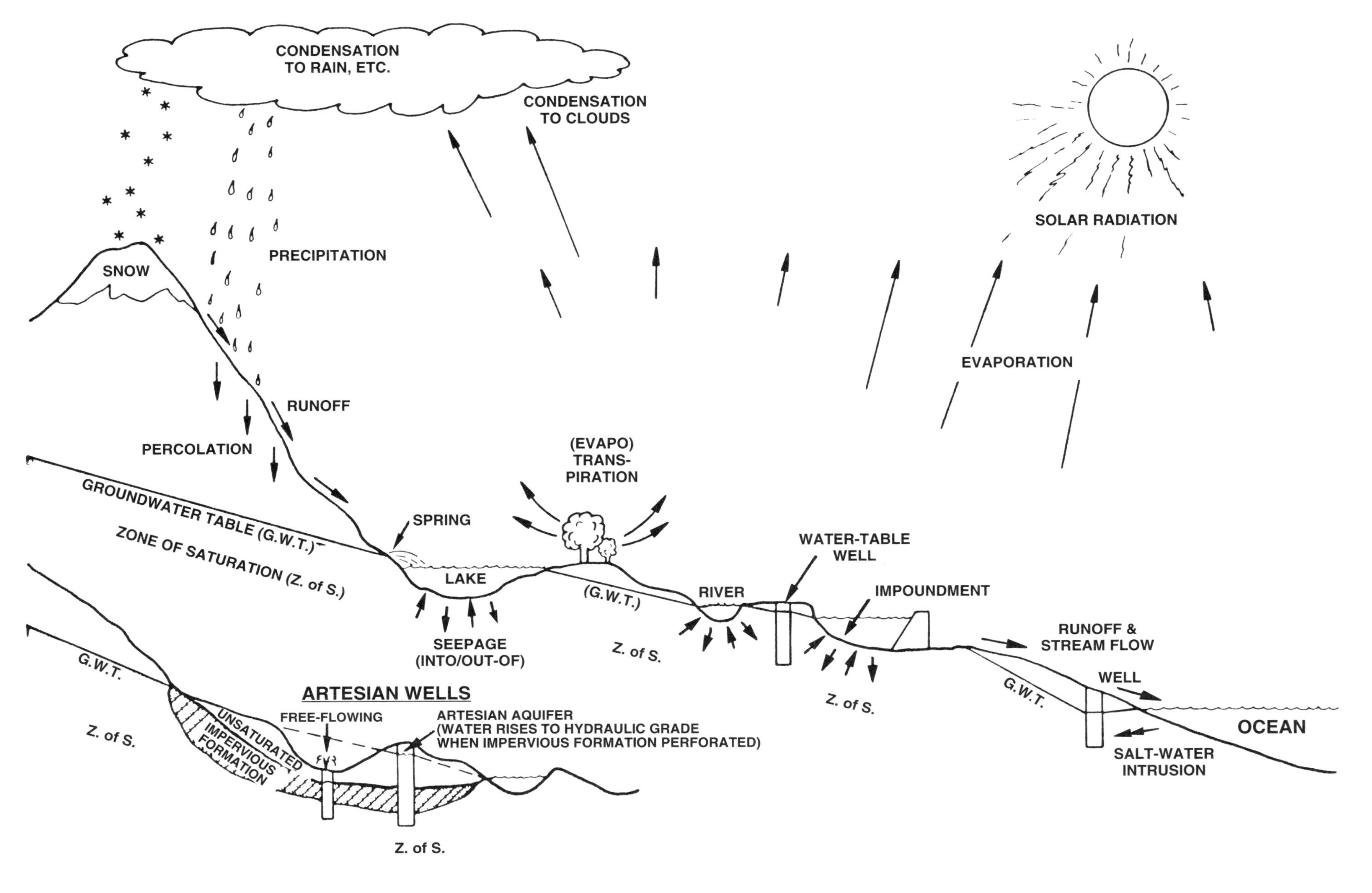

Fig. 2.1 Water (hydrologic) cycle as related to water supply
(Source: *BASIC WATER TREATMENT OPERATOR TRAINING COURSE I*, by Leonard Ainsworth, by permission of California-Nevada Section, AWWA)

water treatment. Because of upstream pollution (wastewater, agricultural drainage, or industrial waste), the proper treatment of river and stream supplies is extremely important. Rivers and streams are also susceptible to scouring of the bottom, changing channels, and silting. Before the intake for a water supply is located in a river or stream, a careful study must be made of the stream bottom, its degree of scour, and the settling out of silt. Provisions must be made in the design of the intake to make sure that it can withstand the force which will act upon it during times of flood, heavy silting, ice conditions, and adverse runoff conditions. Because of variations in the quality of water supplied by a river or stream, purification effectiveness must be continually checked. This is especially true if there are industries upstream from the intake which may dump undesirable wastes into the supply. Sudden pollutant loads might not be discovered unless constant monitoring of the raw water is maintained by the treatment plant operator.

2.132 Lakes and Reservoirs

The selection and use of water from any surface storage source requires considerable study and thought. When ponds, lakes, or open reservoirs are used as sources of water supply, the danger of contamination and the resulting spread of diseases such as typhoid, hepatitis, dysentery, giardiasis, and cryptosporidiosis exists. Clear water is not always safe water and the old saying that running water "purifies itself" to drinking water quality within a stated distance is false.

The potential for contamination of surface water makes it necessary to regard such sources of supply as unsafe for domestic use unless properly treated, including filtration and disinfection. Ensuring the delivery of a constant, safe drinking water to consumers also requires diligent attention to the operation and maintenance of the distribution system.

Lakes and reservoirs are subject to seasonal changes in water quality such as those brought about by *STRATIFICATION*[4] and the possible increase of organic and mineral contamination that occurs when a lake "turns over." In any body of water, the surface water will be warmed by the sun in spring and summer causing higher temperatures on the surface. Then in the fall, the cooler air temperatures cool the surface water until it reaches the same temperature as the subsurface waters. At this point, the water temperature is fairly uniform (the same) throughout the entire depth of the lake or reservoir. A breeze will start the surface water circulating and cause the lake to "turn over," thus bringing poor quality deeper water to the surface.

Lakes and reservoirs are susceptible to algal blooms, especially after fall or spring turnovers. The rapid growth of algae (blooms) will occur when the temperature is right and the water contains enough nutrients for the rapid growth of algae. In any given body of water, blooms of various types of algae can occur several times during a season depending on what algae is present and whether the conditions are right for algae growth.

Water supplies drawn from large lakes and reservoirs through multiport intake facilities (openings at several depths) are generally of good quality since the water can be drawn from a depth where algal growths are not prevalent. A large lake or reservoir also dilutes any contamination that may have been discharged into it or one of its tributaries.

Large bodies of water are generally attractive recreation areas. If the water is also used for domestic supplies, however, it must be protected from contamination. This will require proper construction and location of recreation facilities such as boat launching ramps, boat harbors, picnic and camping areas, fishing and open beach areas away from the intake area. The location and construction of wastewater collection, treatment, and disposal facilities must also be carefully studied to protect domestic water supplies from contamination.

Write your answers in a notebook and then compare your answers with those on pages 34 and 35.

2.1A What is the water (hydrologic) cycle?

2.1B List the three basic types of water rights.

2.1C What are the general water quality characteristics of surface water supplies?

2.1D What are the general water quality characteristics of groundwater supplies?

2.1E What items should be considered before selecting a location and constructing a water supply intake located in a river or stream?

2.1F What water treatment processes are considered essential to reliably treat physically and bacteriologically contaminated surface waters for domestic use?

2.1G How can provisions be made to allow recreation on water supply lakes and reservoirs without endangering water quality?

---

[4] *Stratification (STRAT-uh-fuh-KAY-shun). The formation of separate layers (of temperature, plant, or animal life) in a lake or reservoir. Each layer has similar characteristics such as all water in the layer has the same temperature.*

2.14 Groundwater

2.140 Sources (refer to Figure 2.1, page 25)

Part of the precipitation that falls infiltrates the soil. This water replenishes the soil moisture, or is used by growing plants and returned to the atmosphere by *TRANSPIRATION*.[5] Water that drains downward (percolates) below the root zone finally reaches a level at which all of the openings or voids in the earth's materials are filled with water. This zone is known as the zone of saturation. Water in the zone of saturation is referred to as groundwater. The upper surface of the zone of saturation, if not confined by impermeable material, is called the water table. When an overlying, impermeable formation confines the water in the zone of saturation under pressure, the groundwater is said to be under artesian pressure. The name "artesian" comes from the ancient province of Artesium in France where, in the days of the Romans, water flowed to the surface of the ground from a well. However, not all water from wells that penetrate artesian formations flows to ground level. For a well to be artesian, the water in the well must rise above the top of the aquifer. (An aquifer, or water-bearing formation, is an underground layer of rock or soil which permits the passage of water.)

The porous material just above the water table may contain water by capillarity in the smaller void spaces. This zone is referred to as the capillary fringe. Since the water held in the capillary fringe will not drain freely by gravity, this zone is not considered a true source of supply.

Because of the irregularities in underground deposits or layers and in surface *TOPOGRAPHY*,[6] the water table occasionally intersects (meets) the surface of the ground at a spring or in the bed of a stream, lake, or the ocean. As a result, groundwater moves to these locations as seepage out of the aquifer (groundwater reservoir). Thus, groundwater is continually moving within aquifers even though the movement may be very slow (see Figure 2.1). The water table (artesian pressure surface) thus may slope from areas of recharge to lower areas of discharge. The total head difference represented by these slopes causes the flow of groundwater within the aquifer. Seasonal variations in the supply of water to the underground reservoir cause considerable changes in the elevation and slope of the water table and the artesian pressure level.

2.141 Wells

A well that penetrates the water table can be used to extract water from the groundwater basin (see Figure 2.1). The removal of water by pumping will naturally cause a lowering of the water table near the well. If pumping continues at a rate that exceeds the rate of replacement by the water-bearing formations, the "sustained yield" of the well or group of wells has been exceeded. The "safe yield" will be exceeded if wells extract water from an aquifer over a period of time at a rate that will deplete the aquifer and bring about other undesired results (such as sea water intrusion and land *SUBSIDENCE*[7]). This situation is a poor practice, but occurs quite frequently in many areas of the United States.

2.142 Springs

Groundwater that flows naturally from the ground is called a spring. Depending upon whether the discharge is from a water table or an artesian aquifer, springs may flow by gravity or by artesian pressure. The flow from a spring may vary considerably; when the water table or artesian pressure fluctuates, so does the flow from the spring.

2.15 Reclaimed Water

The use of treated wastewater as a source of water for nonfood crop irrigation is an established practice in many regions of the world. The type of crop which can be safely irrigated depends somewhat on the quality of wastewater and method of irrigation. At the present time, more than 20,000 acres (8,000 hectares) of agricultural lands in California are irrigated, all or in part, with reclaimed water. Two of the largest operations are at Bakersfield and at Fresno, California. The City of Bakersfield has used wastewater effluent for irrigation since 1912. At the present time, approximately 2,400 acres (1,000 hectares) of alfalfa, cotton, barley, sugar beets, and pasture are irrigated. Fresno irrigates 3,500 acres (1,400 hectares) of the same types of crops. These two operations use almost 30,000 acre-feet (37 million cubic meters) of reclaimed water per year. Almost 90 percent of the 2,000,000 acre-feet (2.5 billion cubic meters) of reclaimed wastewater in California is used for crop irrigation.

Other uses for reclaimed wastewater include:

1. Greenbelt (parks) areas,
2. Golf course irrigation,
3. Landscape irrigation,
4. Industrial reuse,
5. Groundwater recharge,
6. Landscape impoundments, and
7. Wetlands — marsh enhancement.

Reclaimed water can be used safely for any of these purposes, with the possible exception of groundwater recharge.

---

[5] *Transpiration (TRAN-spur-RAY-shun). The process by which water vapor is released to the atmosphere by living plants. This process is similar to people sweating. Also called evapotranspiration.*

[6] *Topography (toe-PAH-gruh-fee). The arrangement of hills and valleys in a geographic area.*

[7] *Subsidence (sub-SIDE-ence). The dropping or lowering of the ground surface as a result of removing excess water (overdraft or overpumping) from an aquifer. After excess water has been removed, the soil will settle, become compacted and the ground surface will drop and can cause the settling of underground utilities.*

Health experts have serious questions regarding organic compounds that are present in wastewater and about our ability to reduce them to safe levels. These doubts increase further when we realize that a large number of new and potentially toxic chemicals are developed each year. Laboratories are unable to adequately detect all of these chemicals without expensive monitoring programs. To protect groundwater resources, regulations require that "reclaimed water used for groundwater recharge of domestic water supply aquifers by surface spreading shall be at all times of a quality that fully protects public health." Proposed groundwater recharge projects must be investigated on an individual basis where the use of reclaimed water involves a potential risk to public health.

Treatment of reclaimed water should be appropriate for the intended use. The greater the potential exposure to the public, the more extensive the treatment needs to be. Regulations often specify not only the degree of treatment for the usage of water, but also the reliability features that must be incorporated into the treatment processes to ensure a continuous high degree of finished water quality. Studies indicate that the average reclamation plant does not achieve the quality of treatment expected on a continuous basis. This is of concern to the water supplier and also the regulatory agencies that are charged with the responsibility of ensuring that the health of the public is protected.

QUESTIONS

Write your answers in a notebook and then compare your answers with those on page 35.

2.1H What causes the flow of groundwater within an aquifer?

2.1I How can the "safe yield" of an aquifer be exceeded?

2.1J List some of the possible uses of reclaimed wastewater.

2.1K How much treatment should reclaimed water receive before use?

2.2 SELECTION OF A WATER SOURCE

2.20 Sanitary Survey[8]

The importance of detailed sanitary surveys of water supply sources cannot be overemphasized. With a new supply, the sanitary survey should be made during the collection of initial engineering data covering the development of a given source and its capacity to meet existing and future needs. The sanitary survey should include the location of all potential and existing health hazards and the determination of their present and future importance. Persons trained in public health engineering and the *EPIDEMIOLOGY*[9] of waterborne diseases should conduct the sanitary survey. In the case of an existing supply, sanitary surveys should be made frequently enough to control health hazards and to maintain high water quality.

The information furnished by a sanitary survey is essential to evaluating the bacteriological and chemical water quality data. The following outline lists the essential factors which should be investigated or considered in a sanitary survey. These items are essential to (1) identify potential hazards, (2) determine factors which affect water quality, and (3) select treatment requirements. Not all of the items are important to any one supply and, in some cases, items not in the list could be found to be significant during the field investigation.

GROUNDWATER SUPPLIES:

a. Character of local geology; slope (topography) of ground surface.

b. Nature of soil and underlying porous material; whether clay, sand, gravel, rock (especially porous limestone); coarseness of sand or gravel; thickness of water-bearing stratum; depth of water table; location and *GEOLOGICAL LOG*[10] of nearby wells.

c. Slope of water table, preferably as determined from observation wells or as indicated by slope of the ground surface.

d. Extent of the drainage area likely to contribute water to the supply.

e. Nature, distance, and direction of local sources of pollution.

f. Possibility of surface-drainage water entering the supply and of wells becoming flooded.

g. Methods used for protecting the supply against contamination from wastewater collection and treatment facilities and industrial waste disposal sites.

---

[8] *Sanitary Survey. A detailed evaluation and/or inspection of a source of water supply and all conveyances, storage, treatment and distribution facilities to ensure protection of the water supply from all pollution sources.*

[9] *Epidemiology (EP-uh-DE-me-ALL-o-gee). A branch of medicine which studies epidemics (diseases which affect significant numbers of people during the same time period in the same locality). The objective of epidemiology is to determine the factors that cause epidemic diseases and how to prevent them.*

[10] *Geological Log. A detailed description of all underground features discovered during the drilling of a well (depth, thickness and type of formations).*

h. Well construction: materials, diameter, depth of casing and concrete collar; depth to well screens or perforations; length of well screens or perforations.

i. Protection of well head at the top and on the sides.

j. Pumping station construction (floors, drains); capacity of pumps; storage or direct to distribution system.

k. Drawdown when pumps are in operation; recovery rate when pumps are off.

l. Presence of an unsafe supply nearby, and the possibility of *CROSS CONNECTIONS*[11] causing a danger to the public health.

m. Disinfection: equipment, supervision, test kits, or other types of laboratory control.

SURFACE WATER SUPPLIES:

a. Nature of surface geology; character of soils and rocks.

b. Character of vegetation; forests; cultivated and irrigated land.

c. Population and wastewater collection, treatment, and disposal on the *WATERSHED.*[12]

d. Methods of wastewater disposal, whether by diversion from watershed or by reclamation treatment.

e. Closeness of sources of fecal pollution (especially birds) to intake of water supply.

f. Proximity and character of sources of industrial wastes, oil field brines, acid waters from mines, and agricultural drain waters.

g. Adequacy of supply as to quantity (safe yield).

h. For lake or reservoir supplies: wind direction and velocity data; drift of pollution; and algae growth potential.

i. Character and quality of raw water: typical coliform counts (MPN or membrane filter), algae, turbidity, color, and objectionable mineral constituents.

j. Normal period of *DETENTION TIME.*[13]

k. Probable minimum time required for water to flow from sources of pollution to reservoir and through the reservoir to the intake tower.

l. The possible currents of water within the reservoir (induced by wind or reservoir discharge) which could cause *SHORT-CIRCUITING*[14] to occur.

m. Protective measures in connection with the use of the watershed to control fishing, boating, landing of airplanes, swimming, wading, ice cutting, and permitting animals on shoreline areas.

n. Efficiency and constancy of policing activities on the watershed and around the lake.

o. Treatment of water: kind and adequacy of equipment; duplication of parts for reliable treatment; effectiveness of treatment; numbers and competency of supervising and operating personnel; contact period after disinfection; free chlorine residuals and monitoring of the water supply both during treatment and following treatment.

p. Pumping facilities: pump station design, pump capacity and standby unit(s).

q. Presence of an unsafe supply nearby, and the possibility of cross connections causing a danger to the public health.

2.21 Precipitation

Precipitation in the form of rain, snow, hail, or sleet contains very few impurities. (However, there are exceptions such as acid rain and dust from dust bowl areas.) Trace amounts of mineral matter, gases, and other substances may be picked up by precipitation as it forms and falls through the earth's atmosphere. Precipitation, however, has virtually no microbiological content.

Once precipitation reaches the earth's surface, many opportunities are presented for the introduction of mineral and organic substances, microorganisms, and other forms of contamination. When water runs over or through the ground surface it may pick up particles of soil. This is noticeable in the water as cloudiness or turbidity. This water also picks up particles of organic matter and microorganisms. As surface water seeps downward into the soil and through the underlying material to the water table, most of the suspended particles are filtered out. This natural filtration may be partially effective in removing microorganisms and other particulate materials; however, the chemical characteristics of the water usually change considerably when it comes in contact with underground mineral deposits.

The widespread use of synthetically produced chemical compounds, especially pesticides, has raised concern for their potential to contaminate water. Many of these materials are known to be toxic (poisonous), some cause cancer, and others have certain undesirable characteristics even when present in a relatively small concentration.

Agents which alter the quality of water as it moves over or below the surface of the earth may be classified under four major headings:

---

[11] *Cross Connection. A connection between a drinking (potable) water system and an unapproved water supply. For example, if you have a pump moving nonpotable water and hook into the drinking water system to supply water for the pump seal, a cross connection or mixing between the two water systems can occur. This mixing may lead to contamination of the drinking water.*

[12] *Watershed. The region or land area that contributes to the drainage or catchment area above a specific point on a stream or river.*

[13] *Detention Time. In storage reservoirs, detention time is the length of time entering water will be held before being drafted for use (several weeks to years, several months being typical).*

[14] *Short-Circuiting. A condition that occurs in tanks or basins when some of the water travels faster than the rest of the flowing water. This is usually undesirable since it may result in shorter contact, reaction, or settling times in comparison with the theoretical (calculated) or presumed detention times.*

A. *PHYSICAL* — Physical characteristics relate to the sensory qualities of water for domestic use; for example, the water's observed color, turbidity, temperature, taste, and odor.

B. *CHEMICAL* — Chemical differences between waters include mineral content and the presence or absence of constituents such as fluoride, sulfide, and acids. The comparative performance of hard and soft waters in laundering is one visible effect of chemical differences.

C. *BIOLOGICAL* — The presence of organisms (viruses, bacteria, algae, mosquito larvae), alive or dead, and their metabolic products determine the biological character of water. These may also be significant in modifying the physical and chemical characteristics of water.

D. *RADIOLOGICAL* — Radiological factors must be considered because there is a possibility that the water may have come in contact with radioactive substances.

Consequently, in the development of water supply systems, it is necessary to examine carefully all the factors which might adversely affect the water supply.

QUESTIONS

Write your answers in a notebook and then compare your answers with those on page 35.

2.2A What is the purpose of a sanitary survey?

2.2B How frequently should a sanitary survey be conducted for an existing water supply?

2.2C When conducting a sanitary survey, what protective measures should be investigated regarding use of the watershed?

2.2D List the common physical characteristics of water.

2.22 Physical Characteristics

To be suitable for human use, water should be free from all impurities which are offensive to the senses of sight, taste, and smell. The physical characteristics which might be offensive include turbidity, color, taste, odor, and temperature.

TURBIDITY: The presence of suspended material in water causes cloudiness which is known as turbidity. Clay, silt, finely divided organic material, plankton, and other inorganic materials give water this appearance. Turbidities in excess of 5 turbidity units are easily visible in a glass of water, and this level is usually objectionable for *AESTHETIC*[15] reasons. Turbidity's major danger in drinking water is that it can harbor bacteria as well as exert a high demand on chlorine. Water that has been filtered to remove the turbidity should have considerably less than one turbidity unit. Good treatment plants consistently obtain finished water turbidity levels of 0.05 to 0.3 units.

COLOR: Dissolved organic material from decaying vegetation and certain inorganic matter cause color in water. Occasionally, excessive blooms of algae or the growth of other aquatic microorganisms may also impart color. Iron and manganese may be the cause of consumer complaints (red or black water). While the color itself is not objectionable from a strict standpoint of health, its presence is aesthetically objectionable and suggests that the water needs better treatment. In some instances, however, a color in the water indicates more than an aesthetic problem. For example, an amber color in the water could indicate the presence of humic substances which could later be formed into *TRIHALOMETHANES*[16] or it could indicate acid waters from mine drainage.

TEMPERATURE: The most desirable drinking waters are consistently cool and do not have temperature fluctuations of more than a few degrees. Groundwater and surface water from mountainous areas generally meet these requirements. Most individuals find that water having a temperature between 50° and 60°F (10° and 15°C) is most pleasing while water over 86°F (30°C) is not acceptable. The temperature of groundwaters varies with the depth of the aquifer. Water from very deep wells (more than 1,000 ft or 300 m) may be quite warm. Temperature also affects sensory perception of tastes and odors.

TASTES: Each area's natural waters have a distinctive taste related to the dissolved mineral characteristics of local geology. Occasionally, algal growths also impart a distinctive taste. However, taste is rarely measured since most water treatment plants cannot alter a water's mineral characteristics.

ODORS: Growths of algae in a water supply can give the water an unpleasant odor. Some groundwaters may contain hydrogen sulfide which will produce a disagreeable rotten egg odor.

2.23 Chemical Characteristics

The nature of the materials that form the earth's crust affects not only the quantity of water that may be recovered, but also its chemical makeup. As surface water infiltrates and percolates downward to the water table, it dissolves some of the minerals contained in soils and rocks. Groundwater, therefore, sometimes contains more dissolved minerals than surface water. The use and disposal of chemicals by society can also affect water quality.

---

[15] *Aesthetic (es-THET-ick). Attractive or appealings.*

[16] *Trihalomethanes (THMs) (tri-HAL-o-METH-hanes). Derivatives of methane, CH_4, in which three halogen atoms (chlorine or bromine) are substituted for three of the hydrogen atoms. Often formed during chlorination by reactions with natural organic materials in the water. The resulting compounds (THMs) are suspected of causing cancer.*

Chemical analysis of a domestic water supply is broken down into three areas:

1. Inorganic chemicals which include the toxic (poisonous) metals — arsenic, barium, cadmium, chromium, lead, mercury, selenium, and silver; and the non-metals — fluoride and nitrate;
2. Organic chemicals which include the pesticides (chlorinated hydrocarbons) — Endrin, Lindane, Methoxychlor and Toxaphene, and the Chlorophenoxys; and
3. The general mineral constituents which include alkalinity, calcium, chloride, copper, foaming agents (MBAS), iron, magnesium, manganese, pH, sodium, sulfate, zinc, specific conductance, total dissolved solids, and hardness (calcium and magnesium).

Upper limits for the concentrations of the chemicals listed in this section have been established by the Safe Drinking Water Act. For a summary of the Drinking Water Regulations established by the Act, see the poster included with this manual. Chapter 22, "Drinking Water Regulations," in *WATER TREATMENT PLANT OPERATION*, Volume II, contains a detailed discussion of the Safe Drinking Water Act.

2.24 Biological Factors

Water for domestic purposes must be made free from disease-producing (pathogenic) organisms. These organisms include bacteria, protozoa, spores, viruses, cysts, and helminths (parasitic worms).

Many organisms that cause disease in humans originate with the fecal discharges of infected individuals. Monitoring and controlling the activities of human disease-carriers is seldom practical. For this reason, it is necessary to take precautions to prevent contamination of a normally safe water source or to institute treatment methods which will produce a safe water.

Unfortunately, the specific disease-producing organisms present in water are not easily isolated and identified. The techniques for comprehensive bacteriological examination are complex and time-consuming. Therefore, it has been necessary to develop tests which indicate the relative degree of contamination in terms of an easily defined quality. The most widely used test involves estimation of the number of bacteria of the coliform group, which are always present in fecal wastes and vastly outnumber disease-producing organisms. Coliform bacteria normally inhabit the human intestinal tract, but are also found in most animals and birds, as well as in the soil. The Drinking Water Standards in the Safe Drinking Water Act have established upper limits for the concentration of coliform bacteria in a series of water samples with a goal of zero coliforms in all samples. The Maximum Contaminant Level (MCL) for coliforms for systems analyzing fewer than 40 samples per month is no more than one sample per month may be total coliform positive.

To further ensure protection against the spread of waterborne diseases, the Surface Water Treatment Rule (SWTR), which took effect December 31, 1990, requires that all public water systems disinfect their water. If the source of supply (the raw water) is surface water or groundwater under the influence of surface water, the SWTR also requires water suppliers to install filtration equipment unless the source water meets very high standards for purity.

2.25 Radiological Factors

The development and use of nuclear energy as a power source and the mining of radioactive materials have made it necessary to examine the safe limits of exposure for humans. Such limits include concentrations of radioactive material taken into the body in drinking water.

QUESTIONS

Write your answers in a notebook and then compare your answers with those on page 35.

2.2E What causes turbidity in water?

2.2F Chemical analysis of a domestic water supply measures what three general types of chemical concentrations?

2.2G Why are coliform bacteria used to measure the bacteriological quality of water?

2.2H Why must upper limit concentrations of radioactive materials in drinking water be established?

2.3 THE SAFE DRINKING WATER ACT

Water has many important uses and each requires a certain specific level of water quality. The major concern of the operators of water treatment plants and water supply systems is to produce and deliver to consumers water that is safe and pleasant to drink. The water should be acceptable to domestic and commercial water users and many industries. Some industries, such as food and drug processors and the electronics industry, require higher quality water. Many industries will locate where the local water supply meets their specific needs while other industries may have their own water treatment facilities to produce water suitable for their needs.

On December 16, 1974, the Safe Drinking Water Act (SDWA) was signed into law. The SDWA gave the federal government, through the U.S. Environmental Protection Agency (EPA), the authority to:

- Set national standards regulating the levels of contaminants in drinking water;
- Require public water systems to monitor and report their levels of identified contaminants; and
- Establish uniform guidelines specifying the acceptable treatment technologies for cleansing drinking water of unsafe levels of pollutants.

While the SDWA gave EPA responsibility for developing drinking water regulations, it gave state regulatory agencies the opportunity to assume primary responsibility for enforcing those regulations.

Implementation of the SDWA has greatly improved basic drinking water purity across the nation. However, recent EPA surveys of surface water and groundwater indicate the presence of synthetic organic chemicals in 20 percent of the nation's water sources, with a small percentage at levels of concern. In addition, research studies suggest that some naturally occurring contaminants may pose even greater risks to human health than the synthetic contaminants. Further, there is growing concern about microbial and radon contamination.

In the years following passage of the SDWA, Congress felt that EPA was slow to regulate contaminants and states were lax in enforcing the law. Consequently, in 1986 and again in

1996 Congress enacted amendments designed to strengthen the 1974 SDWA. These amendments set deadlines for the establishment of maximum contaminant levels, placed greater emphasis on enforcement, authorized penalties for tampering with drinking water supplies, mandated the complete elimination of lead from drinking water, and placed considerable emphasis on the protection of underground drinking water sources.

All public water systems must comply with the SDWA regulations. A PUBLIC WATER SYSTEM is any publicly or privately owned water supply system that:

1. Has at least 15 service connections, or
2. Regularly serves an average of at least 25 individuals daily at least 60 days out of the year.

Drinking water regulations also take into account the type of population served by the system and classify water systems as community or noncommunity systems. Therefore, in order to understand what requirements apply to any specific system, it is first necessary to determine whether the system is considered a community system or a noncommunity system. A COMMUNITY WATER SYSTEM is defined as one which:

1. Has at least 15 service connections used by all-year residents, or
2. Regularly serves at least 25 all-year residents.

Any public water system that is not a community water system is classified as a NONCOMMUNITY WATER SYSTEM. Restaurants, campgrounds, and hotels could be considered noncommunity systems for purposes of drinking water regulations.

In addition to distinguishing between community and noncommunity water systems, EPA identifies some small systems as NONTRANSIENT NONCOMMUNITY systems if they regularly serve at least 25 of the same persons over 6 months per year. This classification applies to water systems for facilities such as schools or factories where the consumers served are nearly the same every day but do not actually live at the facility. In general, nontransient noncommunity systems must meet the same requirements as community systems.

A TRANSIENT NONCOMMUNITY water system is a system that does not regularly serve drinking water to at least 25 of the same persons over 6 months per year. This classification is used by EPA only in regulating nitrate levels and total coliform. Examples of a transient noncommunity system might be campgrounds or service stations if those facilities do not meet the definition of a community, noncommunity, or nontransient noncommunity system.

The maximum contaminant levels (MCLs) and also the sampling and testing requirements established by the Safe Drinking Water Act are summarized on the poster included with this manual. For additional information on the Safe Drinking Water Act, see *WATER TREATMENT PLANT OPERATION*, Volume II, Chapter 22, "Drinking Water Regulations."

QUESTIONS

Write your answers in a notebook and then compare your answers with those on page 35.

2.3A Which industries require extremely high-quality water?

2.3B Who must comply with Safe Drinking Water Act (SDWA) regulations?

2.3C Define "community water system."

2.4 WATER TREATMENT

In the operation of water treatment plants, three basic objectives are controlling:

1. Production of a safe drinking water,
2. Production of an aesthetically pleasing drinking water, and
3. Production of drinking water at a reasonable cost with respect to capital and also operation and maintenance costs.

From a public health perspective, production of a safe drinking water, one that is free of harmful bacteria and toxic materials, is the first priority. But, it is also important to produce a high-quality water which appeals to the consumer. Generally, this means that the water must be clear (free of turbidity), colorless, and free of objectionable tastes and odors. Consumers also show a preference for water supplies that are nonstaining (plumbing fixtures and washing clothes), noncorrosive to plumbing fixtures and piping, and ones that do not leave scale deposits or spot glassware.

Consumer sensitivity to the environment (air quality, water quality, noise) has significantly increased in recent years. With regard to water quality, consumer demands have never been greater. In some instances, consumers have substituted bottled water to meet specific needs, namely, for drinking water and cooking purposes.

Design engineers select water treatment processes on the basis of the type of water source, source water quality, and desired finished water quality established by drinking water regulations and consumer desires. Table 2.1 is a summary of typical water treatment processes and plants depending on the source and quality of the raw water.

Operators of water treatment facilities must be very conscientious in order to produce a high-quality finished water. Also they must realize that water can degrade in the distribution or delivery system. The remainder of this manual contains chapters written by operators on how to produce and deliver high-quality drinking water to consumers.

QUESTIONS

Write your answers in a notebook and then compare your answers with those on page 35.

2.4A What is the first priority for operating a water treatment plant?

2.4B What type of water is appealing to consumers?

TABLE 2.1 SOURCES AND TREATMENT OF WATER

WATER QUALITY PROBLEM	TREATMENT[a]
GROUNDWATER	
1. Coliforms or Microbiological Contamination	1. Disinfection (Chlorination)
2. Sulfide Odors (Rotten Egg)	2a. Aeration 2b. Oxidation (Chlorination) 2c. Desulfuration (Sulfur Dioxide)
3. Excessive Hardness (Calcium and Magnesium)	3a. Ion Exchange Softening 3b. Lime (& Soda) Softening
4. Iron and/or Manganese	4a. Sequestration (Polyphosphates) 4b. Removal by Special Ion Exchange 4c. Permanganate and Greensand 4d. Oxidation by Aeration* 4e. Oxidation with Chlorine* 4f. Oxidation with Permanganate* * Filtration Must Follow Oxidation
5. Dissolved Minerals (High Total Dissolved Solids)	5a. Ion Exchange 5b. Reverse Osmosis
6. Corrosivity (Low pH)	6a. pH Adjustment with Chemicals 6b. Carbon Dioxide Stripping by Aeration 6c. Corrosion Inhibitor Addition (Zinc Phosphate, Silicate)
7. Preventive Treatment (Fluoridation)	7. Add Fluoride Chemicals
8. Sand	8. Sand Separators
9. Nitrate	9. Anion Exchange
SURFACE WATER	
WATER QUALITY PROBLEM	*TREATMENT*[a]
1. Coliforms or Microbiological Contamination	1a. Disinfection (Chlorination) 1b. Disinfection (Other Oxidants — Ozone, Chlorine Dioxide, Chloramination) 1c. Coagulation, Flocculation, Sedimentation, Filtration, and Disinfection
2. Turbidity, Color	2. Coagulation, Flocculation, Sedimentation, and Filtration
3. Odors (Organic Materials)	3a. Clarification (Coagulation, Flocculation, Sedimentation, and Filtration) 3b. Oxidation (Chlorination or Permanganate) 3c. Special Oxidation (Chlorine Dioxide) 3d. Adsorption (Granular Activated Carbon)
4. Iron and/or Manganese	4a. Sequestration (Polyphosphates) 4b. Removal by Special Ion Exchange 4c. Permanganate and Greensand 4d. Oxidation by Aeration* 4e. Oxidation with Chlorine* 4f. Oxidation with Permanganate* * Filtration Must Follow Oxidation
5. Excessive Hardness (Calcium and Magnesium)	5a. Ion Exchange Softening 5b. Lime (& Soda) Softening
6. Dissolved Minerals (High Total Dissolved Solids)	6a. Ion Exchange 6b. Reverse Osmosis
7. Corrosivity (Low pH)	7a. pH Adjustment with Chemicals 7b. Corrosion Inhibitor Addition (Zinc Phosphate, Silicate)
8. Preventive Treatment a. Fluoridation b. Trihalomethanes (THMs)	8a. Add Fluoride Chemicals 8b. (1) Do Not Prechlorinate Disinfect with Ozone, Chlorine Dioxide, or Chloramination (2) Remove THM Precursors (3) Remove THMs After They Are Formed

[a] For details on the treatment processes, refer to the appropriate chapters in this series of operator training manuals.

2.5 ARITHMETIC ASSIGNMENT

A good way to learn how to solve arithmetic problems is to work on them a little bit at a time. In this operator training manual we are going to make a short arithmetic assignment at the end of every chapter. If you will work this assignment at the end of every chapter, you can easily learn how to solve waterworks arithmetic problems.

Turn to the Appendix, "How to Solve Small Water System Arithmetic Problems," at the back of this manual and read the following sections:

1. *OBJECTIVES*,
2. A.0, *HOW TO STUDY THIS APPENDIX*, and
3. A.1, *BASIC ARITHMETIC*.

Solve all of the problems in Sections A.10, Addition; A.11, Subtraction; A.12, Multiplication; A.13, Division; A.14, Rules for Solving Equations; and A.15, Actual Problems; on an electronic pocket calculator.

2.6 ADDITIONAL READING

1. AWWA. *WATER SOURCES*, Chapter 1,* "Water Supply Hydrology," Chapter 2,* "Groundwater Sources," and Chapter 3,* "Surface Water Sources."
2. *NEW YORK MANUAL*, Chapter 2,* "Water Sources and Water Uses."
3. *TEXAS MANUAL*, Chapter 2,* "Groundwater Supplies"; Chapter 3,* "Surface Water Supplies"; and Chapter 4,* "Raw Water Quality Management."

* Depends on edition.

Please answer the discussion and review questions before continuing with the Objective Test.

DISCUSSION AND REVIEW QUESTIONS

Chapter 2. WATER SOURCES AND TREATMENT

Write the answers to these questions in your notebook before continuing with the Objective Test on page 36. The purpose of these questions is to indicate to you how well you understand the material in the chapter.

1. What has been the impact of drought conditions on the managers of water supplies?
2. Why has the quality of many water supplies deteriorated?
3. How does the water (hydrologic) cycle work?
4. What are water rights?
5. What are the differences between the water quality characteristics of groundwater and surface water supplies?
6. Lakes turn over under what conditions?
7. Algal blooms in lakes and reservoirs occur under what general conditions?
8. How can a lake used for a water supply also be used for recreation without endangering water quality?
9. The elevation and slope of water tables and artesian pressure levels can change due to what factors?
10. Who should conduct a sanitary survey? Why?

SUGGESTED ANSWERS

Chapter 2. WATER SOURCES AND TREATMENT

Answers to questions on page 24.

2.0A Drought conditions, pollution, and increasing energy costs have forced water supply managers to recognize the need to conserve and properly budget our water resources.

2.0B Energy consumption and water shortages are linked together because (1) with falling groundwater levels energy is required to pump from deeper levels, and (2) water projects that use large amounts of electricity to pump water over long distances are requiring water users to pay increased water rates to cover higher energy costs.

2.0C Groundwater becomes contaminated from (1) sea water intrusion, (2) seepage from septic tank leaching systems, (3) agricultural drainage systems, and (4) seepage from improper disposal of hazardous wastes.

Answers to questions on page 26.

2.1A The water (hydrologic) cycle is the cycle or path water follows from evaporation from oceans, to formation of clouds, to precipitation, to runoff, to evaporation and transpiration back to the atmosphere, and eventually back to the ocean.

2.1B The three basic types of water rights are:

1. Riparian,
2. Appropriative, and
3. Prescriptive.

2.1C In general, surface water supplies are characterized by suspended solids, turbidity, some color, and microbiological contamination.

2.1D In general, groundwater supplies are characterized by higher concentrations of dissolved solids and low color, but freedom from microbiological contamination.

2.1E Before selecting a location and constructing a water supply intake in a river or stream, careful consideration must be given to (1) the stream bottom, its degree of scour, and the settling out of silt, and (2) design of the intake to make sure that it can withstand the forces which will act upon it during times of flood, heavy silting, ice conditions, and adverse runoff conditions.

2.1F Filtration and disinfection are considered essential water treatment processes to reliably treat physically and bacteriologically contaminated surface waters for domestic use.

2.1G Lakes and reservoirs used for domestic water supplies can be protected from contamination by (1) proper construction and location of recreation facilities; and (2) careful evaluation of adverse effects of the construction of wastewater collection, treatment, and disposal facilities near the reservoir or its tributaries.

Answers to questions on page 28.

2.1H Groundwater flows within an aquifer because of the total head differences between the areas of recharge and discharge.

2.1I The "safe yield" of an aquifer can be exceeded if wells extract water from an aquifer over a period of time at a rate such that the aquifer will become depleted or bring about other undesired results, such as sea water intrusion and land subsidence.

2.1J Possible uses of reclaimed wastewater include: (1) crop irrigation, (2) greenbelt irrigation, (3) golf course irrigation, (4) landscape irrigation, (5) industrial reuse, (6) groundwater recharge, (7) landscape impoundments, and (8) wetlands — marsh enhancement.

2.1K Reclaimed water should receive treatment that is appropriate for the intended use of the water.

Answers to questions on page 30.

2.2A The purpose of a sanitary survey is to detect all health hazards and to evaluate their present and future importance.

2.2B For an existing water supply, a sanitary survey should be made frequently enough to control all health hazards and to maintain good sanitary water quality.

2.2C When conducting a sanitary survey, protective measures that should be investigated include control of sources of waste discharges (municipal wastewater treatment plants, sources of fecal pollution, industrial wastes, oil field brines, acid waters from mines, and agricultural drain waters), fishing, boating, landing of airplanes, swimming, wading, ice cutting, and permitting animals on shoreline areas.

2.2D The common physical characteristics of water are observable color, turbidity, temperature, taste, and odor.

Answers to questions on page 31.

2.2E Turbidity in water is caused by the presence of suspended material such as clay, silt, finely divided organic material, plankton, and other inorganic material.

2.2F The three general types of chemicals measured by a chemical analysis of domestic water supplies are inorganic, organic, and general mineral concentrations.

2.2G Coliforms are used to measure the bacteriological quality of drinking water because the test indicates the fecal contamination where specific disease-producing organisms are not easily isolated and identified.

2.2H The development and use of nuclear energy as a power source and the mining of radioactive materials have made it necessary to establish upper limit concentrations for drinking water.

Answers to questions on page 32.

2.3A Food and drug processors and also electronics industries require extremely high-quality water.

2.3B All public water systems must comply with the Safe Drinking Water Act (SDWA) regulations.

2.3C A community water system is one which:

1. Has at least 15 service connections used by all year residents, or
2. Regularly serves at least 25 all-year residents.

Answers to questions on page 32.

2.4A The first priority for operating a water treatment plant is the production of a safe drinking water, one that is free of harmful bacteria and toxic materials.

2.4B Water that appeals to consumers must be clear (free of turbidity), colorless, and free of objectionable tastes and odors. Consumers also show a preference for water supplies that are nonstaining (plumbing fixtures and washing clothes), noncorrosive to plumbing fixtures and piping, and ones that do not leave scale deposits or spot glassware.

OBJECTIVE TEST

Chapter 2. WATER SOURCES AND TREATMENT

Please write your name and mark the correct answers on the answer sheet, as directed at the end of Chapter 1. There may be more than one correct answer to each multiple-choice question.

True-False

1. As groundwater levels fall, less energy is required to pump water from deeper levels in the basin.
   1. True
   2. False
2. Groundwater produced by wells must be tested regularly to ensure safe drinking water.
   1. True
   2. False
3. Clear water is always safe to drink.
   1. True
   2. False
4. Water flows to ground level from all artesian wells.
   1. True
   2. False
5. The chemical characteristics of water usually change considerably when it comes in contact with underground mineral deposits.
   1. True
   2. False
6. The specific disease-producing organisms present in water are easily isolated and identified.
   1. True
   2. False

Best Answer (Select only the closest or best answer.)

7. How can sudden pollutant loads be discovered in a raw water supply?
   1. Compliance with the Safe Drinking Water Act
   2. Constant monitoring of raw water
   3. Monitoring of treated water
   4. Networking with waste dischargers
8. How can the delivery of a constant, safe drinking water to consumers be ensured?
   1. Attention to the operation and maintenance of the distribution system
   2. Recommend consumers boil water
   3. Suggest bottled water be used for drinking purposes
   4. Use of clean delivery facilities
9. What is a sanitary survey?
   1. A detailed evaluation of a source of water supply and all water facilities
   2. A detailed survey of the boundaries of a watershed
   3. A survey of consumers' sanitary disposal practices
   4. A survey of sanitary landfills located over an aquifer
10. What is a cross connection?
    1. A connection between a drinking water system and a raw water source
    2. A connection between a drinking water system and an unapproved supply
    3. A connection between a pump and a distribution system
    4. A connection between two drinking (potable) water systems
11. What is the most widely used test to indicate the bacteriological quality of water?
    1. Coliform
    2. *Cryptosporidium*
    3. *Giardia*
    4. Hepatitis

Multiple Choice (Select all correct answers.)

12. What are potential sources of groundwater contamination?
    1. Agricultural drainage systems
    2. Evapotranspiration from ponds
    3. Improper disposal of hazardous wastes
    4. Sea water intrusion
    5. Seepage from septic tank leaching systems
13. How do surface water supplies compare with groundwater supplies?
    1. Higher concentrations of dissolved solids
    2. Higher microbiological contamination
    3. Lower levels of color
    4. Lower suspended solids
    5. Lower turbidity
14. What undesirable results can occur if wells pump water from an aquifer at a rate that will deplete the aquifer?
    1. Artesian flows
    2. Land subsidence
    3. Replenishment of aquifers
    4. Sea water intrusion
    5. Sustained yields

15. Which physical characteristics of water might be offensive to humans?
    1. Color
    2. Odors
    3. Tastes
    4. Temperature
    5. Turbidity

16. What objectives control the operation of water treatment plants?
    1. Production of a safe drinking water
    2. Production of an esthetically pleasing drinking water
    3. Production of drinking water at a reasonable cost
    4. Production of water with sufficient pressure
    5. Production of water with sufficient quantity

CHAPTER 3

WELLS

by

Al Ward

and

John German

TABLE OF CONTENTS

Chapter 3. WELLS

OBJECTIVES

Chapter 3. WELLS

Chapter 3 is divided into two sections. In the first section we have assumed that you are responsible for the operation and maintenance of a well. In the second section (the Appendix) we have assumed that your responsibilities have been expanded and require the knowledge and skills necessary to select a well site, work with a well driller, test and evaluate pumps and wells, and also abandon and plug wells.

Following completion of Chapter 3, you should be able to:

1. Protect a well from contamination,
2. Identify the parts of a well and pump system,
3. Maintain and rehabilitate a well,
4. Operate and maintain a well pump and hydropneumatic pressure tank,
5. Inspect a well and pumping system,
6. Disinfect wells and pumps,
7. Keep accurate records of a well and pumping system,
8. Remove sand from water and mains, and
9. Troubleshoot a well and pumping system.

Following completion of the Appendix of Chapter 3 (optional), you should be able to:

1. Select a well site,
2. Describe the types of wells and drilling methods,
3. Test and evaluate a well and pump, and
4. Abandon and plug a well no longer productive or needed.

WORDS

Chapter 3. WELLS

AIR GAP — AIR GAP

An open vertical drop, or vertical empty space, that separates a drinking (potable) water supply to be protected from another water system in a water treatment plant or other location. This open gap prevents the contamination of drinking water by backsiphonage or backflow because there is no way raw water or any other water can reach the drinking water supply.

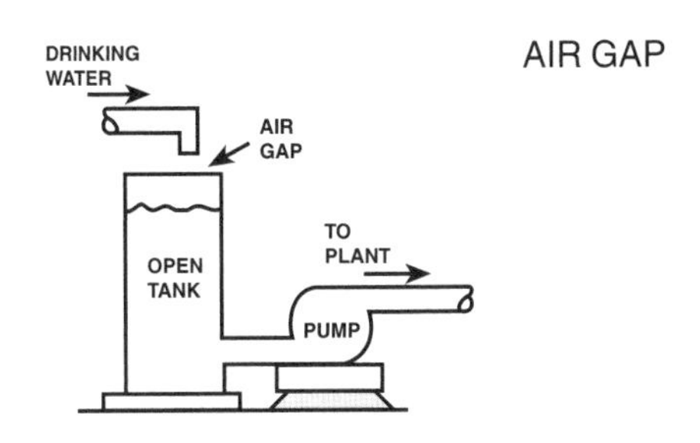

ALARM CONTACT — ALARM CONTACT

A switch that operates when some preset low, high or abnormal condition exists.

ALLUVIAL (uh-LOU-vee-ul) — ALLUVIAL

Relating to mud and/or sand deposited by flowing water. Alluvial deposits may occur after a heavy rainstorm.

ANALYZER — ANALYZER

A device which conducts periodic or continuous measurement of some factor such as chlorine, fluoride or turbidity. Analyzers operate by any of several methods including photocells, conductivity or complex instrumentation.

ANNULAR (AN-you-ler) SPACE — ANNULAR SPACE

A ring-shaped space located between two circular objects, such as two pipes.

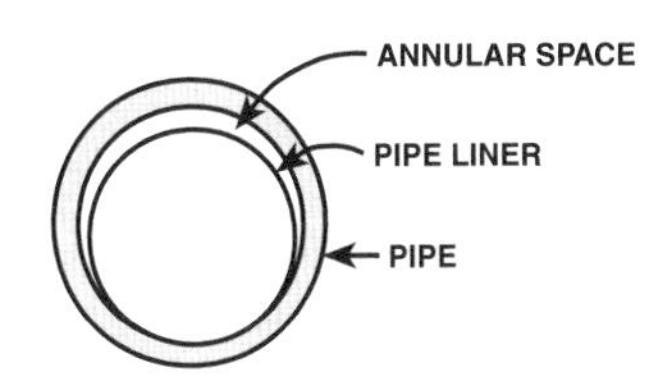

APPURTENANCE (uh-PURR-ten-nans) — APPURTENANCE

Machinery, appliances, structures and other parts of the main structure necessary to allow it to operate as intended, but not considered part of the main structure.

AQUIFER (ACK-wi-fer) — AQUIFER

A natural underground layer of porous, water-bearing materials (sand, gravel) usually capable of yielding a large amount or supply of water.

ARTESIAN (are-TEE-zhun) — ARTESIAN

Pertaining to groundwater, a well, or underground basin where the water is under a pressure greater than atmospheric and will rise above the level of its upper confining surface if given an opportunity to do so.

AVAILABLE CHLORINE — AVAILABLE CHLORINE

A measure of the amount of chlorine available in chlorinated lime, hypochlorite compounds, and other materials that are used as a source of chlorine when compared with that of elemental (liquid or gaseous) chlorine.

BAILER (BAY-ler) — BAILER

A 10- to 20-foot-long pipe equipped with a valve at the lower end. A bailer is used to remove slurry from the bottom or the side of a well as it is being drilled.

BOWL, PUMP — BOWL, PUMP

The submerged pumping unit in a well, including the shaft, impellers and housing.

BRAKE HORSEPOWER — BRAKE HORSEPOWER

(1) The horsepower required at the top or end of a pump shaft (input to a pump).

(2) The energy provided by a motor or other power source.

CAISSON (KAY-sawn) — CAISSON

A structure or chamber which is usually sunk or lowered by digging from the inside. Used to gain access to the bottom of a stream or other body of water.

CAPILLARY ACTION — CAPILLARY ACTION

The movement of water through very small spaces due to molecular forces.

CAPILLARY FORCES — CAPILLARY FORCES

The molecular forces which cause the movement of water through very small spaces.

CENTRIFUGAL (sen-TRIF-uh-gull) PUMP — CENTRIFUGAL PUMP

A pump consisting of an impeller fixed on a rotating shaft that is enclosed in a casing, and having an inlet and discharge connection. As the rotating impeller whirls the liquid around, centrifugal force builds up enough pressure to force the water through the discharge outlet.

CHECK VALVE — CHECK VALVE

A special valve with a hinged disc or flap that opens in the direction of normal flow and is forced shut when flows attempt to go in the reverse or opposite direction of normal flows.

CIRCLE OF INFLUENCE — CIRCLE OF INFLUENCE

The circular outer edge of a depression produced in the water table by the pumping of water from a well. Also see CONE OF INFLUENCE and CONE OF DEPRESSION.

CONDUCTOR CASING — CONDUCTOR CASING

The outer casing of a well. The purpose of this casing is to prevent contaminants from surface waters or shallow groundwaters from entering a well.

CONE OF DEPRESSION — CONE OF DEPRESSION

The depression, roughly conical in shape, produced in the water table by the pumping of water from a well. Also called the CONE OF INFLUENCE. Also see CIRCLE OF INFLUENCE.

CONE OF INFLUENCE — CONE OF INFLUENCE

The depression, roughly conical in shape, produced in the water table by the pumping of water from a well. Also called the CONE OF DEPRESSION. Also see CIRCLE OF INFLUENCE.

CONFINING UNIT — CONFINING UNIT

A layer of rock or soil of very low hydraulic conductivity that hampers the movement of groundwater in and out of an aquifer.

CONSOLIDATED FORMATION — CONSOLIDATED FORMATION

A geologic material whose particles are stratified (layered), cemented or firmly packed together (hard rock); usually occurring at a depth below the ground surface. Also see UNCONSOLIDATED FORMATION.

CONTACTOR — CONTACTOR

An electric switch, usually magnetically operated.

CONTAMINATION — CONTAMINATION

The introduction into water of microorganisms, chemicals, toxic substances, wastes, or wastewater in a concentration that makes the water unfit for its next intended use.

CONTROLLER — CONTROLLER

A device which controls the starting, stopping, or operation of a device or piece of equipment.

CROSS CONNECTION — CROSS CONNECTION

A connection between a drinking (potable) water system and an unapproved water supply. For example, if you have a pump moving nonpotable water and hook into the drinking water system to supply water for the pump seal, a cross connection or mixing between the two water systems can occur. This mixing may lead to contamination of the drinking water.

DATUM LINE DATUM LINE

A line from which heights and depths are calculated or measured. Also called a datum plane or a datum level.

DISCHARGE HEAD DISCHARGE HEAD

The pressure (in pounds per square inch or psi) measured at the centerline of a pump discharge and very close to the discharge flange, converted into feet. The pressure is measured from the centerline of the pump to the hydraulic grade line of the water in the discharge pipe.

Discharge Head, ft = (Discharge Pressure, psi)(2.31 ft/psi)

DRAWDOWN DRAWDOWN

(1) The drop in the water table or level of water in the ground when water is being pumped from a well.

(2) The amount of water used from a tank or reservoir.

(3) The drop in the water level of a tank or reservoir.

EFFECTIVE SIZE (E.S.) EFFECTIVE SIZE (E.S.)

The diameter of the particles in a granular sample (filter media) for which 10 percent of the total grains are smaller and 90 percent larger on a weight basis. Effective size is obtained by passing granular material through sieves with varying dimensions of mesh and weighing the material retained by each sieve. The effective size is also approximately the average size of the grains.

ENERGY GRADE LINE (EGL) ENERGY GRADE LINE (EGL)

A line that represents the elevation of energy head (in feet) of water flowing in a pipe, conduit or channel. The line is drawn above the hydraulic grade line (gradient) a distance equal to the velocity head ($V^2/2g$) of the water flowing at each section or point along the pipe or channel. Also see HYDRAULIC GRADE LINE.

[SEE DRAWING ON NEXT PAGE]

EVAPORATION EVAPORATION

The process by which water or other liquid becomes a gas (water vapor or ammonia vapor).

EVAPOTRANSPIRATION (ee-VAP-o-TRANS-purr-A-shun) EVAPOTRANSPIRATION

(1) The process by which water vapor passes into the atmosphere from living plants. Also called TRANSPIRATION.

(2) The total water removed from an area by transpiration (plants) and by evaporation from soil, snow and water surfaces.

FOOT VALVE FOOT VALVE

A special type of check valve located at the bottom end of the suction pipe on a pump. This valve opens when the pump operates to allow water to enter the suction pipe but closes when the pump shuts off to prevent water from flowing out of the suction pipe.

GEOPHYSICAL LOG GEOPHYSICAL LOG

A record of the structure and composition of the earth encountered when drilling a well or similar type of test hole or boring.

HEAD HEAD

The vertical distance (in feet) equal to the pressure (in psi) at a specific point. The pressure head is equal to the pressure in psi times 2.31 ft/psi.

HEADER HEADER

A large pipe to which the ends of a series of smaller pipes are connected. Also called a MANIFOLD.

HEAT SENSOR HEAT SENSOR

A device that opens and closes a switch in response to changes in the temperature. This device might be a metal contact, or a thermocouple which generates a minute electric current proportional to the difference in heat, or a variable resistor whose value changes in response to changes in temperature. Also called a temperature sensor.

HYDRAULIC CONDUCTIVITY (K) HYDRAULIC CONDUCTIVITY (K)

A coefficient describing the relative ease with which groundwater can move through a permeable layer of rock or soil. Typical units of hydraulic conductivity are feet per day, gallons per day per square foot, or meters per day (depending on the unit chosen for the total discharge and the cross-sectional area).

HYDRAULIC GRADE LINE (HGL) HYDRAULIC GRADE LINE (HGL)

The surface or profile of water flowing in an open channel or a pipe flowing partially full. If a pipe is under pressure, the hydraulic grade line is at the level water would rise to in a small vertical tube connected to the pipe. Also see ENERGY GRADE LINE.

[SEE DRAWING ON NEXT PAGE]

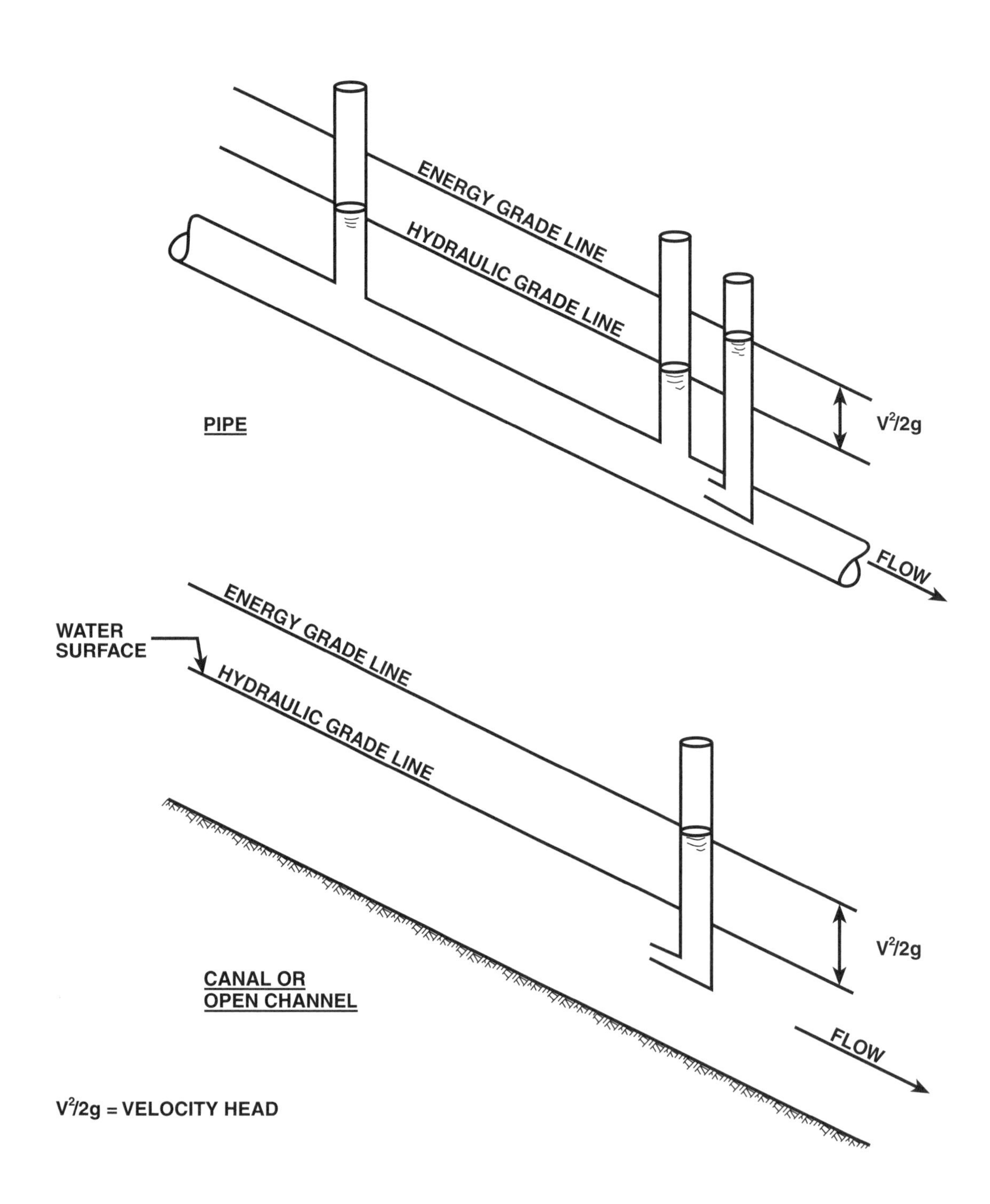

ENERGY GRADE LINE and HYDRAULIC GRADE LINE

HYDRAULIC GRADIENT — HYDRAULIC GRADIENT

The slope of the hydraulic grade line. This is the slope of the water surface in an open channel, the slope of the water surface of the groundwater table, or the slope of the water pressure for pipes under pressure.

HYDROGEOLOGIST (HI-dro-gee-ALL-uh-gist) — HYDROGEOLOGIST

A person who studies and works with groundwater.

HYDROLOGIC (HI-dro-LOJ-ick) CYCLE — HYDROLOGIC CYCLE

The process of evaporation of water into the air and its return to earth by precipitation (rain or snow). This process also includes transpiration from plants, groundwater movement, and runoff into rivers, streams and the ocean. Also called the WATER CYCLE.

HYDROPNEUMATIC (HI-dro-new-MAT-ick) — HYDROPNEUMATIC

A water system, usually small, in which a water pump is automatically controlled (started and stopped) by the air pressure in a compressed-air tank.

HYDROSTATIC (HI-dro-STAT-ick) PRESSURE — HYDROSTATIC PRESSURE

(1) The pressure at a specific elevation exerted by a body of water at rest, or

(2) In the case of groundwater, the pressure at a specific elevation due to the weight of water at higher levels in the same zone of saturation.

IMPELLER — IMPELLER

A rotating set of vanes in a pump or compressor designed to pump or move water or air.

INDICATOR (INSTRUMENT) — INDICATOR (INSTRUMENT)

A device which indicates the result of a measurement. Most indicators in the water utility field use either a fixed scale and movable indicator (pointer) such as a pressure gage or a movable scale and movable indicator like those used on a circular flow-recording chart. Also called a RECEIVER.

INPUT HORSEPOWER — INPUT HORSEPOWER

The total power used in operating a pump and motor.

$$\text{Input Horsepower, HP} = \frac{(\text{Brake Horsepower, HP})(100\%)}{\text{Motor Efficiency, \%}}$$

INTERLOCK — INTERLOCK

An electric switch, usually magnetically operated. Used to interrupt all (local) power to a panel or device when the door is opened or the circuit is exposed to service.

INTERSTICE (in-TUR-stuhz) — INTERSTICE

A very small open space in a rock or granular material. Also called a PORE, void, or void space.

KELLY — KELLY

The square section of a rod which causes the rotation of the drill bit. Torque from a drive table is applied to the square rod to cause the rotary motion. The drive table is chain or gear driven by an engine.

KINETIC ENERGY — KINETIC ENERGY

Energy possessed by a moving body of matter, such as water, as a result of its motion.

LEATHERS — LEATHERS

O-rings or gaskets used with piston pumps to provide a seal between the piston and the side wall.

LEVEL CONTROL — LEVEL CONTROL

A float device (or pressure switch) which senses changes in a measured variable and opens or closes a switch in response to that change. In its simplest form, this control might be a floating ball connected mechanically to a switch or valve such as is used to stop water flow into a toilet when the tank is full.

LOGGING, ELECTRICAL — LOGGING, ELECTRICAL

A procedure used to determine the porosity (spaces or voids) of formations in search of water-bearing formations (aquifers). Electrical probes are lowered into wells, an electric current is induced at various depths, and the resistance measured of various formations indicates the porosity of the material.

MANIFOLD MANIFOLD

A large pipe to which the ends of a series of smaller pipes are connected. Also called a HEADER.

MANOMETER (man-NAH-mut-ter) MANOMETER

An instrument for measuring pressure. Usually, a manometer is a glass tube filled with a liquid that is used to measure the difference in pressure across a flow measuring device such as an orifice or a Venturi meter. The instrument used to measure blood pressure is a type of manometer.

MEASURED VARIABLE MEASURED VARIABLE

A characteristic or component part that is sensed and quantified (reduced to a reading of some kind) by a primary element or sensor.

MECHANICAL JOINT MECHANICAL JOINT

A flexible device that joins pipes or fittings together by the use of lugs and bolts.

MEG MEG

A procedure used for checking the insulation resistance on motors, feeders, bus bar systems, grounds, and branch circuit wiring. Also see MEGGER.

MEGGER (from megohm) MEGGER

An instrument used for checking the insulation resistance on motors, feeders, bus bar systems, grounds, and branch circuit wiring. Also see MEG.

MESH MESH

One of the openings or spaces in a screen or woven fabric. The value of the mesh is usually given as the number of openings per inch. This value does not consider the diameter of the wire or fabric; therefore, the mesh number does not always have a definite relationship to the size of the hole.

MOTOR EFFICIENCY MOTOR EFFICIENCY

The ratio of energy delivered by a motor to the energy supplied to it during a fixed period or cycle. Motor efficiency ratings will vary depending upon motor manufacturer and usually will be near 90.0 percent.

NIOSH (NYE-osh) NIOSH

The **N**ational **I**nstitute of **O**ccupational **S**afety and **H**ealth is an organization that tests and approves safety equipment for particular applications. NIOSH is the primary federal agency engaged in research in the national effort to eliminate on-the-job hazards to the health and safety of working people. The NIOSH Publications Catalog contains a listing of NIOSH publications concerning industrial hygiene and occupational health. To obtain a copy of the catalog, write to National Technical Information Service (NTIS), 5285 Port Royal Road, Springfield, VA 22161. NTIS Stock No. PB-86-116-787. Price $103.50.

NAMEPLATE NAMEPLATE

A durable metal plate found on equipment which lists critical operating conditions for the equipment.

NOMINAL DIAMETER NOMINAL DIAMETER

An approximate measurement of the diameter of a pipe. Although the nominal diameter is used to describe the size or diameter of a pipe, it is usually not the exact inside diameter of the pipe.

OSHA (O-shuh) OSHA

The Williams-Steiger **O**ccupational **S**afety and **H**ealth **A**ct of 1970 (OSHA) is a federal law designed to protect the health and safety of industrial workers and also the operators of water supply systems and treatment plants. The Act regulates the design, construction, operation and maintenance of water supply systems and water treatment plants. OSHA also refers to the federal and state agencies which administer the OSHA regulations.

OPERATING PRESSURE DIFFERENTIAL OPERATING PRESSURE DIFFERENTIAL

The operating pressure range for a hydropneumatic system. For example, when the pressure drops below 40 psi the pump will come on and stay on until the pressure builds up to 60 psi. When the pressure reaches 60 psi the pump will shut off.

ORIFICE (OR-uh-fiss) ORIFICE

An opening (hole) in a plate, wall, or partition. An orifice flange or plate placed in a pipe consists of a slot or a calibrated circular hole smaller than the pipe diameter. The difference in pressure in the pipe above and at the orifice may be used to determine the flow in the pipe.

OVERALL EFFICIENCY, PUMP — OVERALL EFFICIENCY, PUMP

The combined efficiency of a pump and motor together. Also called the WIRE-TO-WATER EFFICIENCY.

OVERDRAFT — OVERDRAFT

The pumping of water from a groundwater basin or aquifer in excess of the supply flowing into the basin. This pumping results in a depletion or "mining" of the groundwater in the basin.

PACKER ASSEMBLY — PACKER ASSEMBLY

An inflatable device used to seal the tremie pipe inside the well casing to prevent the grout from entering the inside of the conductor casing.

PERCOLATING (PURR-co-LAY-ting) WATER — PERCOLATING WATER

Water that passes through soil or rocks under the force of gravity.

PERMEABILITY (PURR-me-uh-BILL-uh-tee) — PERMEABILITY

The property of a material or soil that permits considerable movement of water through it when it is saturated.

PET COCK — PET COCK

A small valve or faucet used to drain a cylinder or fitting.

PITLESS ADAPTER — PITLESS ADAPTER

A fitting which allows the well casing to be extended above ground while having a discharge connection located below the frost line. Advantages of using a pitless adapter include the elimination of the need for a pit or pump house and it is a watertight design, which helps maintain a sanitary water supply.

POLLUTION — POLLUTION

The impairment (reduction) of water quality by agricultural, domestic, or industrial wastes (including thermal and radioactive wastes) to a degree that has an adverse effect on any beneficial use of water.

PORE — PORE

A very small open space in a rock or granular material. Also called an INTERSTICE, void, or void space.

POROSITY — POROSITY

(1) A measure of the spaces or voids in a material or aquifer.

(2) The ratio of the volume of spaces in a rock or soil to the total volume. This ratio is usually expressed as a percentage.

$$\text{Porosity, \%} = \frac{(\text{Volume of Spaces})(100\%)}{\text{Total Volume}}$$

POTABLE (POE-tuh-bull) WATER — POTABLE WATER

Water that does not contain objectionable pollution, contamination, minerals, or infective agents and is considered satisfactory for drinking.

PRESSURE CONTROL — PRESSURE CONTROL

A switch which operates on changes in pressure. Usually this is a diaphragm pressing against a spring. When the force on the diaphragm overcomes the spring pressure, the switch is actuated (activated).

PRIMARY ELEMENT — PRIMARY ELEMENT

(1) A device that measures (senses) a physical condition or variable of interest. Floats and thermocouples are examples of primary elements. Also called a SENSOR.

(2) The hydraulic structure used to measure flows. In open channels, weirs and flumes are primary elements or devices. Venturi meters and orifice plates are the primary elements in pipes or pressure conduits.

PRIME — PRIME

The action of filling a pump casing with water to remove the air. Most pumps must be primed before start-up or they will not pump any water.

PUMP BOWL — PUMP BOWL

The submerged pumping unit in a well, including the shaft, impellers and housing.

PUMPING WATER LEVEL PUMPING WATER LEVEL

The vertical distance in feet from the centerline of the pump discharge to the level of the free pool while water is being drawn from the pool.

PURVEYOR (purr-VAY-or), WATER PURVEYOR, WATER

An agency or person that supplies water (usually potable water).

RANNEY COLLECTOR RANNEY COLLECTOR

This water collector is constructed as a dug well from 12 to 16 feet (3.5 to 5 m) in diameter that has been sunk as a caisson near the bank of a river or lake. Screens are driven radially and approximately horizontally from this well into the sand and the gravel deposits underlying the river.

[SEE DRAWING ON NEXT PAGE]

RECEIVER RECEIVER

A device which indicates the result of a measurement. Most receivers in the water utility field use either a fixed scale and movable indicator (pointer) such as a pressure gage or a movable scale and movable indicator like those used on a circular flow-recording chart. Also called an INDICATOR.

RECORDER RECORDER

A device that creates a permanent record, on a paper chart or magnetic tape, of the changes in a measured variable.

SAFE YIELD SAFE YIELD

The annual quantity of water that can be taken from a source of supply over a period of years without depleting the source permanently (beyond its ability to be replenished naturally in "wet years").

SENSOR SENSOR

A device that measures (senses) a physical condition or variable of interest. Floats and thermocouples are examples of sensors. Also called a PRIMARY ELEMENT.

SET POINT SET POINT

The position at which the control or controller is set. This is the same as the desired value of the process variable.

SLURRY (SLUR-e) SLURRY

A watery mixture or suspension of insoluble (not dissolved) matter; a thin, watery mud or any substance resembling it (such as a grit slurry or a lime slurry).

SOLENOID (SO-luh-noid) SOLENOID

A magnetically (electric coil) operated mechanical device. Solenoids can operate small valves or electric switches.

SOUNDING TUBE SOUNDING TUBE

A pipe or tube used for measuring the depths of water.

SPECIFIC CAPACITY SPECIFIC CAPACITY

A measurement of well yield per unit depth of drawdown after a specific time has passed, usually 24 hours. Typically expressed as gallons per minute per foot (GPM/ft or cu m/day/m).

SPECIFIC CAPACITY TEST SPECIFIC CAPACITY TEST

A testing method used to determine the adequacy of an aquifer or well by measuring the specific capacity.

SPECIFIC GRAVITY SPECIFIC GRAVITY

(1) Weight of a particle, substance, or chemical solution in relation to the weight of an equal volume of water. Water has a specific gravity of 1.000 at 4°C (39°F). Particulates in raw water may have a specific gravity of 1.005 to 2.5.

(2) Weight of a particular gas in relation to the weight of an equal volume of air at the same temperature and pressure (air has a specific gravity of 1.0). Chlorine has a specific gravity of 2.5 as a gas.

SPECIFIC YIELD SPECIFIC YIELD

The quantity of water that a unit volume of saturated permeable rock or soil will yield when drained by gravity. Specific yield may be expressed as a ratio or as a percentage by volume.

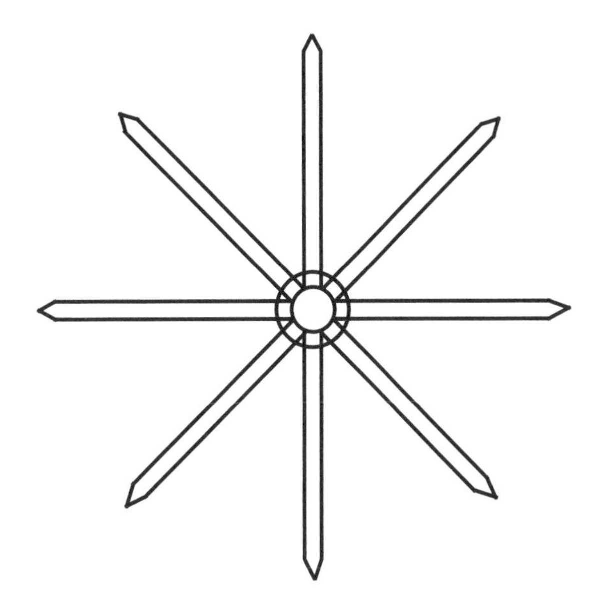

PLAN VIEW OF COLLECTOR PIPES

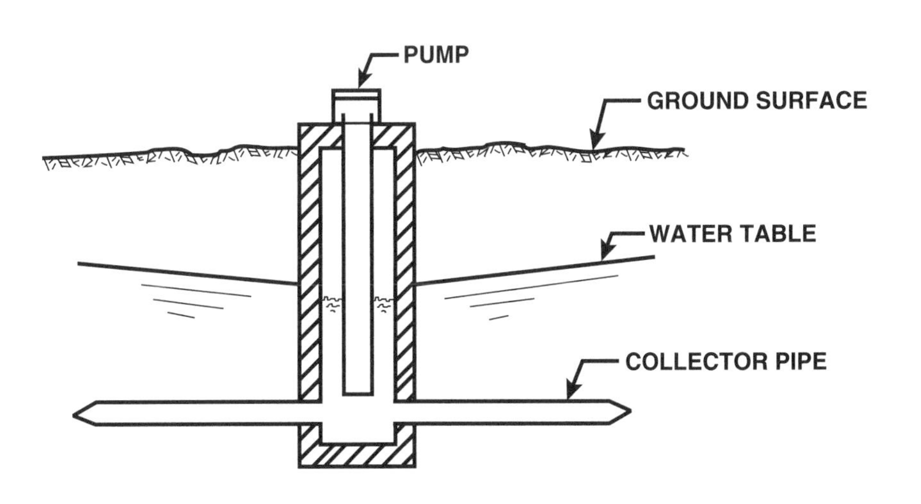

ELEVATION VIEW

RANNEY COLLECTOR

STARTERS (MOTOR) STARTERS (MOTOR)

Devices used to start up large motors gradually to avoid severe mechanical shock to a driven machine and to prevent disturbance to the electrical lines (causing dimming and flickering of lights).

STATIC WATER DEPTH STATIC WATER DEPTH

The vertical distance in feet from the centerline of the pump discharge down to the surface level of the free pool while no water is being drawn from the pool or water table.

STORATIVITY (S) STORATIVITY (S)

The volume of groundwater an aquifer releases from or takes into storage per unit surface area of the aquifer per unit change in head. Also called the storage coefficient.

STATIC WATER LEVEL STATIC WATER LEVEL

(1) The elevation or level of the water table in a well when the pump is not operating.

(2) The level or elevation to which water would rise in a tube connected to an artesian aquifer, basin, or conduit under pressure.

SUBSIDENCE (sub-SIDE-ence) SUBSIDENCE

The dropping or lowering of the ground surface as a result of removing excess water (overdraft or overpumping) from an aquifer. After excess water has been removed, the soil will settle, become compacted and the ground surface will drop and can cause the settling of underground utilities.

SUCTION LIFT SUCTION LIFT

The *NEGATIVE* pressure [in feet (meters) of water or inches (centimeters) of mercury vacuum] on the suction side of a pump. The pressure can be measured from the centerline of the pump *DOWN TO* (lift) the elevation of the hydraulic grade line on the suction side of the pump.

TIME LAG TIME LAG

The time required for processes and control systems to respond to a signal or to reach a desired level.

TIMER TIMER

A device for automatically starting or stopping a machine or other device at a given time.

TRANSDUCER (trans-DUE-sir) TRANSDUCER

A device that senses some varying condition measured by a primary sensor and converts it to an electrical or other signal for transmission to some other device (a receiver) for processing or decision making.

TRANSMISSIVITY (TRANS-miss-SIV-it-tee) TRANSMISSIVITY

A measure of the ability to transmit (as in the ability of an aquifer to transmit water).

TRANSPIRATION (TRAN-spur-RAY-shun) TRANSPIRATION

The process by which water vapor is released to the atmosphere by living plants. This process is similar to people sweating. Also see EVAPOTRANSPIRATION.

TREMIE (TREH-me) TREMIE

A device used to place concrete or grout under water.

UNCONSOLIDATED FORMATION UNCONSOLIDATED FORMATION

A sediment that is loosely arranged or unstratified (not in layers) or whose particles are not cemented together (soft rock); occurring either at the ground surface or at a depth below the surface. Also see CONSOLIDATED FORMATION.

UNIFORMITY COEFFICIENT (U.C.) UNIFORMITY COEFFICIENT (U.C.)

The ratio of (1) the diameter of a grain (particle) of a size that is barely too large to pass through a sieve that allows 60 percent of the material (by weight) to pass through, to (2) the diameter of a grain (particle) of a size that is barely too large to pass through a sieve that allows 10 percent of the material (by weight) to pass through. The resulting ratio is a measure of the degree of uniformity in a granular material such as filter media.

$$\text{Uniformity Coefficient} = \frac{\text{Particle Diameter}_{60\%}}{\text{Particle Diameter}_{10\%}}$$

VARIABLE FREQUENCY DRIVE — VARIABLE FREQUENCY DRIVE

A control system that allows the frequency of the current applied to a motor to be varied. The motor is connected to a low-frequency source while standing still; the frequency is then increased gradually until the motor and pump (or other driven machine) are operating at the desired speed.

VARIABLE, MEASURED — VARIABLE, MEASURED

A factor (flow, temperature) that is sensed and quantified (reduced to a reading of some kind) by a primary element or sensor.

VELOCITY HEAD — VELOCITY HEAD

The energy in flowing water as determined by a vertical height (in feet or meters) equal to the square of the velocity of flowing water divided by twice the acceleration due to gravity ($V^2/2g$).

VOLATILE (VOL-uh-tull) — VOLATILE

(1) A volatile substance is one that is capable of being evaporated or changed to a vapor at relatively low temperatures. Volatile substances also can be partially removed by air stripping.

(2) In terms of solids analysis, volatile refers to materials lost (including most organic matter) upon ignition in a muffle furnace for 60 minutes at 550°C. Natural volatile materials are chemical substances usually of animal or plant origin. Manufactured or synthetic volatile materials such as ether, acetone, and carbon tetrachloride are highly volatile and not of plant or animal origin.

WATER CYCLE — WATER CYCLE

The process of evaporation of water into the air and its return to earth by precipitation (rain or snow). This process also includes transpiration from plants, groundwater movement, and runoff into rivers, streams and the ocean. Also called the HYDROLOGIC CYCLE.

WATER HAMMER — WATER HAMMER

The sound like someone hammering on a pipe that occurs when a valve is opened or closed very rapidly. When a valve position is changed quickly, the water pressure in a pipe will increase and decrease back and forth very quickly. This rise and fall in pressures can cause serious damage to the system.

WATER TABLE — WATER TABLE

The upper surface of the zone of saturation of groundwater in an unconfined aquifer.

WELL ISOLATION ZONE — WELL ISOLATION ZONE

The surface or zone surrounding a water well or well field, supplying a public water system, with restricted land uses to prevent contaminants from a not permitted land use to move toward and reach such water well or well field. Also see WELLHEAD PROTECTION AREA (WHPA).

WELL LOG — WELL LOG

A record of the thickness and characteristics of the soil, rock and water-bearing formations encountered during the drilling (sinking) of a well.

WELLHEAD PROTECTION AREA (WHPA) — WELLHEAD PROTECTION AREA (WHPA)

The surface and subsurface area surrounding a water well or well field, supplying a public water system, through which contaminants are reasonably likely to move toward and reach such water well or well field. Also see WELL ISOLATION ZONE.

WIRE-TO-WATER EFFICIENCY — WIRE-TO-WATER EFFICIENCY

The combined efficiency of a pump and motor together. Also called the OVERALL EFFICIENCY.

YIELD — YIELD

The quantity of water (expressed as a rate of flow — GPM, GPH, GPD, or total quantity per year) that can be collected for a given use from surface or groundwater sources. The yield may vary with the use proposed, with the plan of development, and also with economic considerations. Also see SAFE YIELD.

ZONE OF SATURATION — ZONE OF SATURATION

The soil or rock located below the top of the groundwater table. By definition, the zone of saturation is saturated with water. Also see WATER TABLE.

CHAPTER 3. WELLS

(Lesson 1 of 4 Lessons)

3.0 GROUNDWATER — CRITICAL LINK TO WELLS

3.00 Importance of Groundwater

The function of a well is to intercept groundwater moving through *AQUIFERS*[1] and bring water to the surface for use by people. Although we are concerned with wells and their construction and maintenance, we must also be concerned with the lifeline to these wells. The aquifers and the quality of water in the aquifers must be maintained if we are going to preserve this precious source of water. Pollution and misuse of the aquifers can seriously affect a good water source. Pollution can render a supply useless while excessive use of our wells can affect both volume and rate of output causing permanent damage to the water-bearing formations of the earth's surface.

Approximately 45 percent of the water used in the United States comes from underground sources. In many locations, water from wells or springs is the only water available to a community. Estimates indicate that there are between 10,000,000 and 20,000,000 water wells scattered throughout the United States. Most are situated in valleys or river-bottom land, although many are located in hilly and mountainous regions. They range from shallow hand-dug wells to carefully designed, large production wells.

The principal reasons for using groundwater are:

1. Groundwater is generally available in most localities although quantities may be very limited in certain areas;
2. Well and pumping facilities cost less than surface treatment facilities;
3. Groundwater is usually clear and, with few exceptions, can meet turbidity requirements;
4. Conditions for growth and survival of bacteria and viruses in groundwater are generally unfavorable when compared with surface waters;
5. The mineral content for a given well is usually uniform;
6. Well water usually has a more constant and lower temperature during the summer; and
7. *WELL SUPPLIES ARE PARTICULARLY SUITED TO THE NEEDS OF SMALLER COMMUNITIES IN MANY AREAS.*

3.01 Water (Hydrologic) Cycle[2] (Figure 3.1)

The earth's water cycle, or hydrologic cycle, is the continuous circulation of water (including moisture) on our planet. The cycle has neither a beginning nor an end, but the concept of the hydrologic cycle commonly begins with the waters of the ocean, since they cover about three-fourths of the earth's surface (see Section 2.10, "The Water (Hydrologic) Cycle," for details).

Water that infiltrates the soil is called subsurface water, but not all of it becomes groundwater. Basically, three things may happen to that water. First, it may be pulled back to the surface by capillary forces and be *EVAPORATED*[3] into the atmosphere, thus skipping much of the journey through the water cycle we have described. Second, it may be absorbed by plant roots growing in the soil and then re-enter the atmosphere by a process known as *TRANSPIRATION*.[4] Third, water that has infiltrated the soil deeply enough may be pulled on downward by gravity until it reaches the level of the *ZONE OF SATURATION*[5] — *THE GROUNDWATER RESERVOIR THAT SUPPLIES WATER TO WELLS.* Groundwater conditions directly affect the design, operation, and performance of a water well.

---

[1] *Aquifer (ACK-wi-fer). A natural underground layer of porous, water-bearing materials (sand, gravel) usually capable of yielding a large amount or supply of water.*

[2] *Hydrologic (HI-dro-LOJ-ick) Cycle. The process of evaporation of water into the air and its return to earth by precipitation (rain or snow). This process also includes transpiration from plants, groundwater movement, and runoff into rivers, streams and the ocean. Also called the water cycle.*

[3] *Evaporation. The process by which water or other liquid becomes a gas (water vapor or ammonia vapor).*

[4] *Transpiration (TRAN-spur-RAY-shun). The process by which water vapor is released to the atmosphere by living plants. This process is similar to people sweating.*

[5] *Zone of Saturation. The soil or rock located below the top of the groundwater table. By definition, the zone of saturation is saturated with water.*

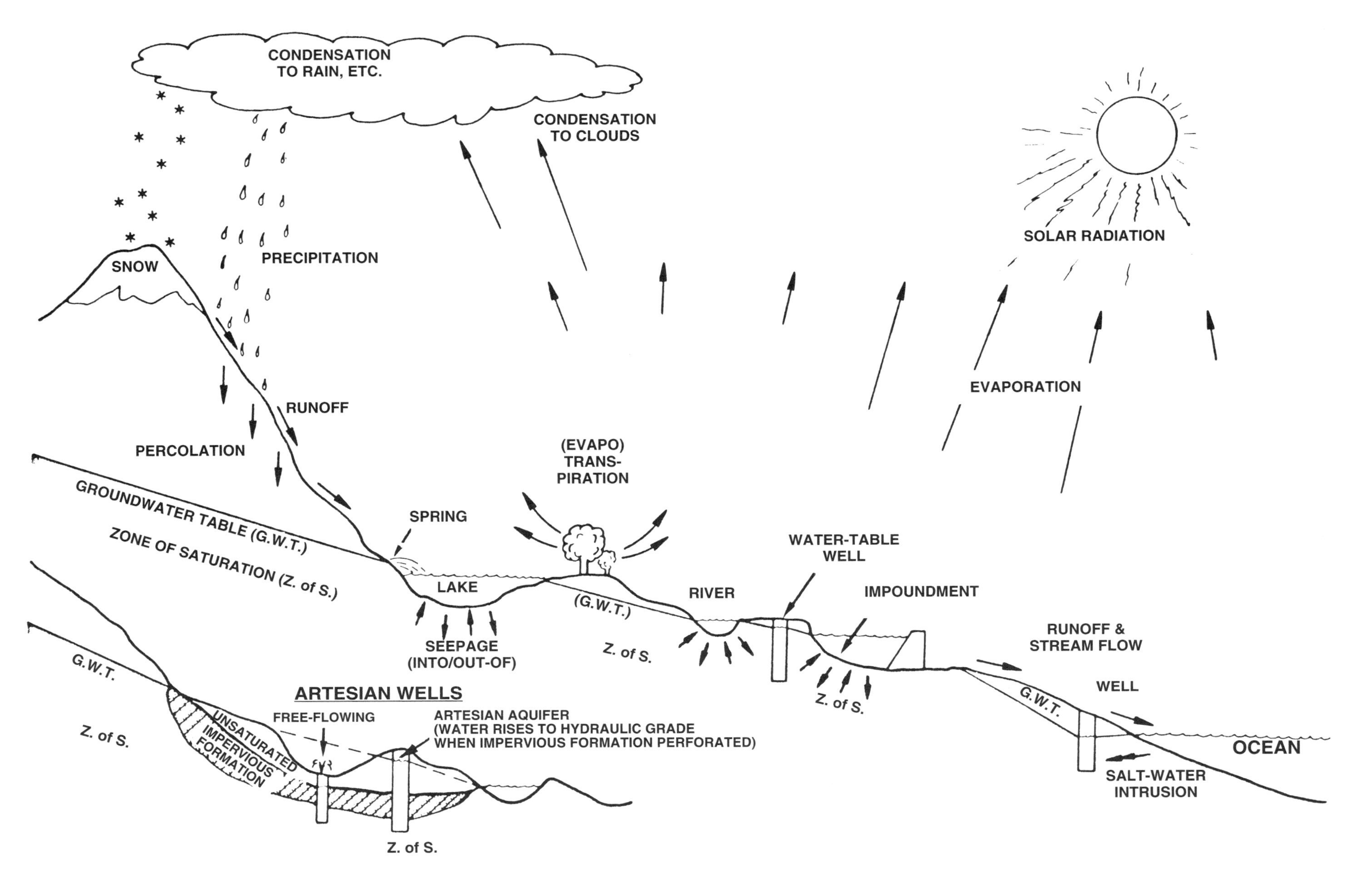

Fig. 3.1 The water (hydrologic) cycle as related to water supply

(Source: *BASIC WATER TREATMENT OPERATOR TRAINING COURSE I*, by Leonard Ainsworth, by permission of California-Nevada Section AWWA)

QUESTIONS

Write your answers in a notebook and then compare your answers with those on page 125.

3.0A What is the purpose of a well?

3.0B List two potential problems that can affect wells.

3.0C List as many of the principal reasons as you can recall for using groundwater.

3.0D What is the hydrologic cycle?

3.02 Aquifers

An aquifer, shown in Figure 3.2, is an underground layer of pervious material, such as sand, gravel, cracked rock, limestone, and/or other porous soil material capable of transporting and storing water. At the bottom of the aquifer will be an impervious layer of clay or rock which holds the aquifer in place.

If the water table is unconfined, the aquifer is known as a water table aquifer and a well in this aquifer is called a table well. If the aquifer is confined between two impervious layers, it is known as an artesian aquifer and from this type of aquifer we get what is referred to as an artesian well. Both types of aquifers and wells are shown in Figure 3.2. The artesian aquifer is under pressure and quite often this natural pressure reduces pumping requirements and thereby reduces pumping costs.

For a saturated material to qualify as an aquifer, it must have: (1) *POROSITY,*[6] area, and thickness sufficient to store an adequate water supply; (2) sufficient *SPECIFIC YIELD*[7] to allow the stored water to drain into a well; and (3) hydraulic *TRANSMISSIVITY*[8] to permit a well to drain water from the aquifer fast enough to meet flow requirements.

3.020 Porosity and Specific Yield

Porosity (usually expressed as a percentage) is a measure of the openings or voids (*PORES*[9]) in a particular soil. Porosity measurements represent (quantify) the amount of water that a particular soil type or rock can store.

$$\text{Porosity, \%} = \frac{\text{(Volume of Voids)(100\%)}}{\text{Total Volume of Soil Sample}}$$

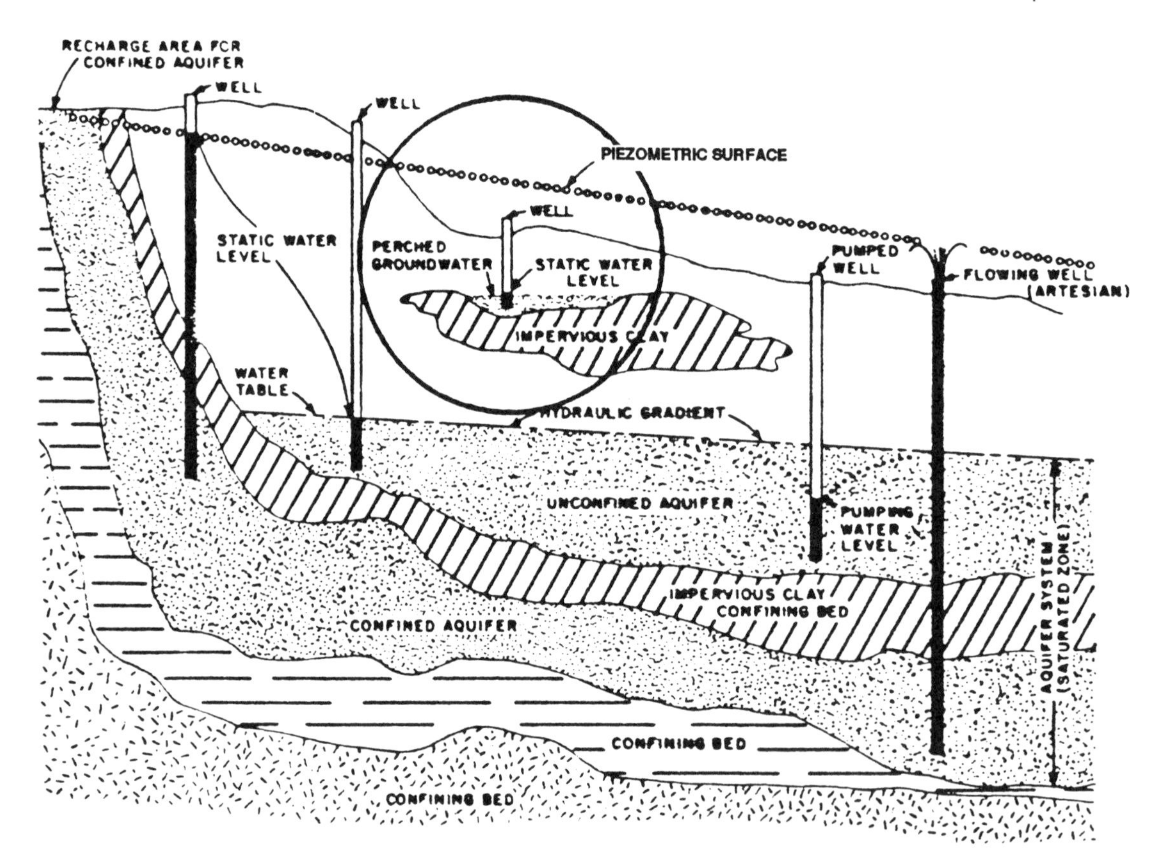

Note perched aquifer inside circle

Fig. 3.2 Water (hydrologic) cycle as related to groundwater

(Source: *WATER WELLS AND PUMPS: THEIR DESIGN, CONSTRUCTION, OPERATION AND MAINTENANCE,* Division of Agricultural Sciences, University of California, Davis)

---

[6] *Porosity. (1) A measure of the spaces or voids in a material or aquifer. (2) The ratio of the volume of spaces in a rock or soil to the total volume. This ratio is usually expressed as a percentage.*

$$\text{Porosity, \%} = \frac{\text{(Volume of Spaces)(100\%)}}{\text{Total Volume}}$$

[7] *Specific Yield. The quantity of water that a unit volume of saturated permeable rock or soil will yield when drained by gravity. Specific yield may be expressed as a ratio or as a percentage by volume.*

[8] *Transmissivity (TRANS-miss-SIV-it-tee). A measure of the ability to transmit (as in the ability of an aquifer to transmit water).*

[9] *Pore. A very small open space in a rock or granular material. Also called an interstice, void, or void space.*

EXAMPLE:

V_V = Volume of Voids = 0.2 cu ft

V_S = Total Volume of Soil Sample = 1.0 cu ft

$$\text{Porosity, \%} = \frac{(V_V)}{(V_S)}(100\%)$$

$$= \frac{(0.2 \text{ cu ft})}{(1.0 \text{ cu ft})}(100\%)$$

$$= 20\%$$

Large porosities are usually associated with fine-grained, highly sorted materials while small porosities are representative of dense rock and soils.

Only a portion of the stored water can be used to supply water to wells or aquifers. A certain amount of water will be retained and not affected by forces of gravity. The water volume that can move through the pores in the rock is affected by gravitational forces and is termed the specific yield (Y_{sp}). Table 3.1 shows the relationship in percentage between porosity and specific yield of various types of soil. Notice that although high porosity may be a good indicator of water storage capacity, it does not necessarily guarantee a high specific yield. Clay, for example, has a high porosity but a low specific yield.

TABLE 3.1 SELECTED VALUES OF POROSITY AND SPECIFIC YIELD[a]

Materials	Porosity[b]	Specific Yield[b]
Soil	55	40
Clay	50	2
Sand	25	22
Gravel	20	19
Limestone	20	18
Sandstone (semiconsolidated)	11	6
Granite	0.1	0.09
Basalt (young)	11	8

[a] *BASIC GROUNDWATER HYDROLOGY*, U.S. Geological Survey Water Supply Paper 2220. Prepared by Ralph C. Heath in cooperation with the North Carolina Department of Natural Resources and Community Development.

[b] Values in percent by volume.

3.021 Overdraft[10]

Aquifers have a certain *YIELD*[11] that can normally be replaced each year through recharge due mainly to precipitation. This yield, commonly called safe yield, is determined during well development through analysis by a qualified *HYDROGEOLOGIST.*[12] Overdraft (overpumping) of the aquifer can cause permanent damage to the water storage and transmitting properties of the aquifers. If an excessive amount of water is removed (overdraft) from an aquifer, the soil may settle and cause compaction of the aquifers, which results in closing the pores through which water moves and in which the water is stored. This compaction and closing of pores also produces what we call subsidence (sub-SIDE-ence) of the land. Over the past three decades, ground subsidence has been measured from 0.91 foot (0.27 meter) in Savannah, Georgia, to over 27 feet (8 meters) in the San Joaquin Valley in California. Sink holes in Florida can also be attributed to overdraft of the groundwater supplies.

QUESTIONS

Write your answers in a notebook and then compare your answers with those on page 125.

3.0E What does porosity measure?

3.0F What causes water to move through pores in the soil or rocks?

3.0G What problems can be caused by the overdraft of a groundwater supply?

3.022 Salt Water Intrusion

A special phenomenon occurs where salt water comes in contact with fresh water. A natural boundary (Figure 3.3) exists because of the differences in *SPECIFIC GRAVITIES*[13] of the two waters. The specific gravity of salt water is greater than the specific gravity of fresh water. This boundary prevents mixing of fresh and salty waters. As long as the natural movement of fresh water within the aquifer replaces water drawn by a well, the boundary between fresh and salty water will remain stable and little mixing will occur. When more fresh water is removed (by overpumping or overdraft) than can be replaced naturally, the salt water will intrude into the aquifer and may be drawn into the well.

Care must be exercised when installing or operating wells near any salt water source to prevent intrusion and pollution of those and neighboring wells. Any well operated in the fresh water area of Figure 3.3(a) will create a *CONE OF DEPRESSION*[14] from the salty groundwater upward toward the well as shown in Figure 3.3(b); a second cone of depression will form from the surface of the water table downward toward the aquifer. Excessive drawdown can bring the salt water into the aquifer and close enough to the well that salt water will be drawn into the cone of depression and into the well. As a general operating guideline, the ratio of "d" to "h" (shown in Figure 3.3(a)) should be at least 40 to 1; for example, if "h" equals 1 foot, "d" would equal 40 feet.

[10] *Overdraft. The pumping of water from a groundwater basin or aquifer in excess of the supply flowing into the basin. This pumping results in a depletion or "mining" of the groundwater in the basin.*

[11] *Yield. The quantity of water (expressed as a rate of flow — GPM, GPH, GPD, or total quantity per year) that can be collected for a given use from surface or groundwater sources. The yield may vary with the use proposed, with the plan of development, and also with economic considerations.*

[12] *Hydrogeologist (HI-dro-gee-ALL-uh-gist). A person who studies and works with groundwater.*

[13] *Specific Gravity. (1) Weight of a particle, substance, or chemical solution in relation to the weight of an equal volume of water. Water has a specific gravity of 1.000 at 4°C (39°F). Particulates in raw water may have a specific gravity of 1.005 to 2.5. (2) Weight of a particular gas in relation to the weight of an equal volume of air at the same temperature and pressure (air has a specific gravity of 1.0). Chlorine has a specific gravity of 2.5 as a gas.*

[14] *Cone of Depression. The depression, roughly conical in shape, produced in the water table by the pumping of water from a well. Also called the cone of influence.*

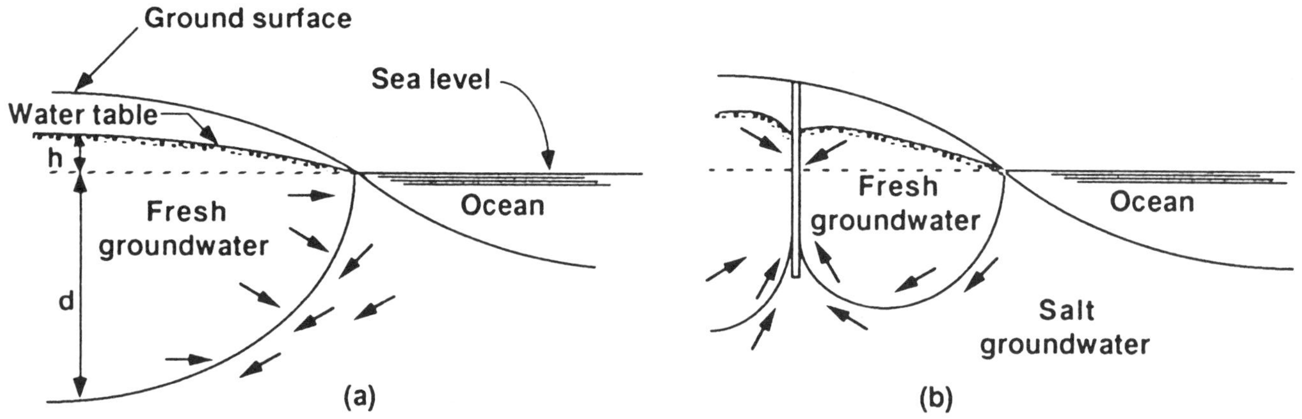

Fig. 3.3 (a) Salt water intrusion along coastal areas;
(b) Effect of excessive drawdown near a salt water source

Salt water intrusion is a common problem along coastlines, but also occurs inland where groundwater supplies contain more than 1,000 mg/*L* of total dissolved solids. Figure 3.4 shows areas within the United States where salt water intrusion could be a problem. Local geologic conditions need to be known prior to placing wells in salt water areas.

QUESTIONS

Write your answers in a notebook and then compare your answers with those on page 125.

3.0H What usually keeps salt water from mixing with fresh water in an aquifer or underground basin?

3.0I How can excessive drawdown of a well lead to salt water intrusion?

3.0J Where can salt water intrusion develop?

3.03 Pollution

3.030 Pollution Control

Groundwater moves very slowly through the soil purifying itself of suspended particles as it travels. If water should be contaminated on its journey to a well and the distance from the source of contamination is far, there is a possibility that we will not see adverse effects at the well. The soil mantle acts as a natural filter for suspended material to bring us a good, clean product. Unfortunately, the soil mantle does not always filter out or remove dissolved pollutants and contaminants (such as organic chemicals) found in water. Chemicals used for farming, waste disposal pits, wastewater disposal, mining, leaks from gas storage tanks, oil spills, hazardous waste disposal, and urban drainage are a few of the factors contributing to groundwater pollution. Human wastes containing pathogenic organisms are also a source of pollution.

If sources of pollution are allowed to increase without limit and without regard to proximity to our wells, the earth's surface will become saturated with pollutants which no known treatment process can remove. Since water moves so slowly through the soil, we may end up with a water supply that cannot be used for years, even after the sources of pollution are eliminated.

Cleaning a polluted aquifer is difficult and expensive at best. Aquifers contaminated with oil spills and other chemical problems are almost impossible to clean up. The most reasonable course against groundwater pollution is prevention. Common sense and strict adherence to health agencies' requirements will go a long way in protecting this vital resource. The Environmental Protection Agency, and others, are making positive strides in keeping our groundwater supplies clean and safe, but a vigorous effort and awareness from all the public (and especially the public water supplier) are necessary.

The well operator has a responsibility to preserve the quality of wells through preventive maintenance of the aquifers. Be alert for any construction that might result in wastes entering the groundwater stream. Septic tanks, subsurface leaching systems, mining operations, agricultural practices, solid waste disposal sites, and wastewater collection facilities are all potential sources of groundwater pollution. Any such activity should be reported to the proper authorities so that adequate safeguards can be taken. Also you must make sure that none of your wells allow direct contamination of an aquifer due to inadequate grouting or seals. Although unsanitary conditions are not as common in the United States as in some countries, they still pose potential problems to our groundwater supplies.

Operators must always be alert for potential sources of groundwater pollution to the aquifers providing drinking water to their communities. Some of the possible sources include[15]:

1. Leaks from oil and gas pipelines,
2. Leaks from storage tanks,
3. Fertilizers, pesticides, and irrigation water,
4. Improper management of animal wastes,
5. Liquid wastes and solid-waste tailing piles from surface and underground mining operations,

[15] *Source: The Johnson Drillers Journal, Second Quarter, 1982.*

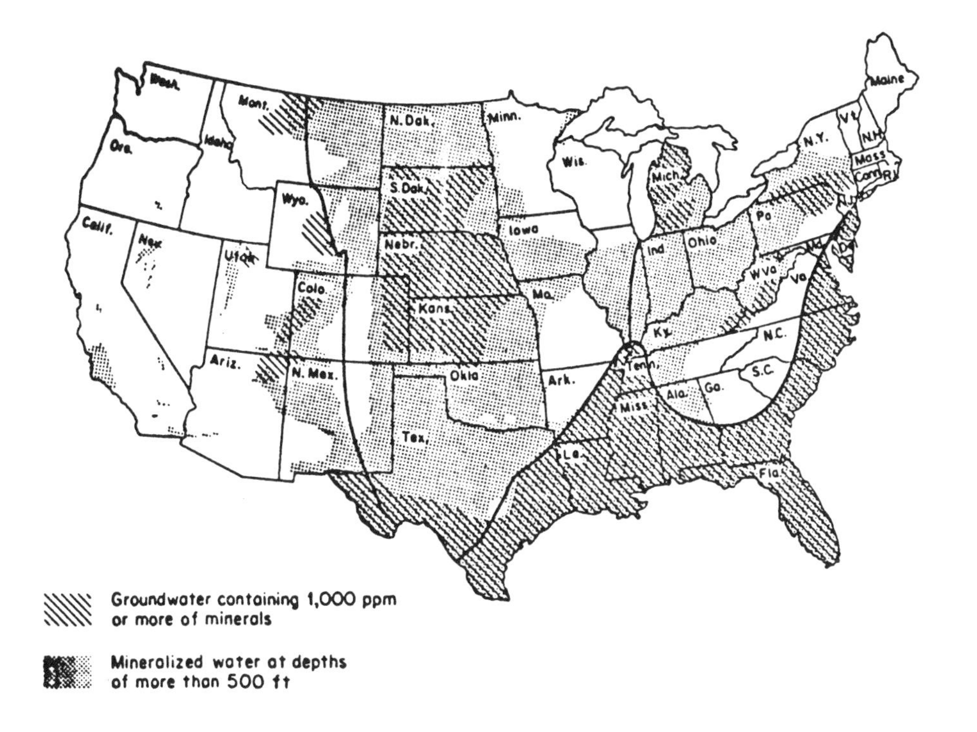

Fig. 3.4 Map of the conterminous United States showing depth to groundwater containing more than 1,000 mg/L total dissolved solids

(From paper by Bill Katz, *TREATING BRACKISH WATER FOR COMMUNITY SUPPLIES*, published in Proceedings in "Role of Desalting Technology," a series of Technology Transfer Workshops presented by the US EPA Office of Water Research and Technology)

6. Drainage from abandoned mines,
7. Salt water encroachment from saline aquifers and the ocean,
8. Improper construction and use of injection wells for the disposal of industrial, nuclear, and hazardous wastes,
9. Accidental spills or liquid wastes, toxic fuels, gasoline and oil, and
10. Abandoned wells which were not properly plugged.

3.031 Geologic and Hydrologic Data

If operators are going to protect aquifers, something must be known about them. Recordkeeping can be an invaluable tool for understanding the aquifer and its performance. Geologic data from the well driller's logs (records of depths of various layers of soil and type of soil in each layer), mining operations in the area, and data from the U.S. Geological Surveyor, U.S. Natural Resources Conservation Service, and from other sources will help operators to understand how water is getting into a well. Consultation with a geologist is recommended for a more complete understanding of the underground formation.

Records of pumping amounts and dates, depths of water at start and end of pumping, time of pumping, and water quality results remain the best sources of information about a well's performance. Using past performance data, an engineer, hydrologist, geologist, and well operator can predict the well's performance with a high degree of accuracy. Recordkeeping is very important and it is not possible for operators to accumulate too much information. So, if records are not now being kept, a program should be started immediately.

QUESTIONS

Write your answers in a notebook and then compare your answers with those on page 125.

3.0K As groundwater moves through the soil, what type of pollutant is removed by the soil mantle?

3.0L What types of groundwater pollutants are not usually removed by the soil mantle?

3.0M What actions should an operator take to preserve the quality of wells through preventive maintenance of the aquifers?

3.0N What types of records should be kept by the operator to document a well's performance?

3.04 Wellhead Protection

3.040 Statement of the Problem

Groundwater contamination can originate on the surface of the ground, in the ground above the water table, or in the ground below the water table. Table 3.2 shows the types of activities that can cause groundwater contamination at each level. The contaminant may be microorganisms or toxic substances.

TABLE 3.2 ACTIVITIES THAT CAN CAUSE GROUNDWATER CONTAMINATION [a]

Origin	Activity	
GROUND SURFACE	Infiltration of polluted surface water Land disposal of wastes Stockpiles Dumps Sewage sludge disposal	De-icing salt use and storage Animal feedlots Fertilizers and pesticides Accidental spills Airborne source particulates
ABOVE WATER TABLE	Septic tanks, cesspools, and privies Holding ponds and lagoons Sanitary landfills Underground storage tank leaks	Underground pipeline leaks Artificial recharge Sumps and dry wells Graveyards Waste disposal in excavations
BELOW WATER TABLE	Waste disposal in wells Drainage wells and canals Underground storage	Exploratory wells Abandoned wells Water-supply wells Groundwater withdrawal Mines

[a] *CITIZEN'S GUIDE TO GROUNDWATER PROTECTION*, Office of Groundwater Protection, U.S. Environmental Protection Agency, Washington, DC 20460. EPA 440/6-90-004, April 1990.

Where a contaminant originates is a factor that can affect its actual impact on groundwater quality. For example, if a contaminant is spilled on the surface of the ground or injected into the ground above the water table, it may have to move through numerous layers of soil or other materials before it reaches the groundwater. As the contaminant moves through these layers of soil, a number of treatment processes are in operation in the soil (for example, filtration, ion exchange, adsorption, dilution, oxidation, biological decay) that can lessen the eventual impact of the contaminant once it finally reaches the groundwater. The effectiveness of these processes also is affected by both the distance between the groundwater and where the contaminant is introduced and the amount of time it takes the substance or microorganism to reach the groundwater. If the contaminant is introduced directly into the area below the water table, the primary process that can affect the impact of the contaminant is dilution by the surrounding groundwater.

Cleaning up a contaminated groundwater supply is a complicated, costly, and sometimes impossible process. In general, a community whose groundwater supply has been contaminated has five options:

1. Contain the contaminants to prevent their migration from their source,
2. Withdraw the pollutants from the aquifer,
3. Treat the groundwater where it is withdrawn or at its point of use,
4. Rehabilitate the aquifer by either immobilizing or detoxifying the contaminants while they are still in the aquifer, and
5. Abandon the use of the aquifer and find alternative sources of water.

Given the importance of groundwater as a source of drinking water for so many communities and individuals and the cost and difficulty of cleaning up groundwater, the best way to guarantee continued supplies of clean groundwater is to prevent contamination. The National Program for Wellhead Protection is designed to protect groundwater supplies of drinking water.

3.041 National Program

The 1986 Amendments to the Safe Drinking Water Act (SDWA) formalized the concept of State Wellhead Protection (WHP) Programs to protect wellhead areas of public water supply wells from contamination. The Act contains requirements for the minimum program elements necessary to develop and implement a WHP program. Each state program may be slightly different depending on the sources and types of groundwater supplies a state is attempting to protect. This section provides an overview of important aspects of all wellhead protection programs, and emphasizes the program components that an operator can control or influence.

Operators have the responsibility of protecting the drinking water supply of their community and delivering safe drinking water to all of their consumers. A wellhead protection program is designed to protect the source of a groundwater supply. Important components of a successful wellhead protection program that an operator can be involved in include:

1. Community involvement and support,
2. Surveys to identify potential and actual sources of contaminants,
3. Determination of characteristics of basin and groundwater supplies, and
4. Development, implementation, and enforcement of land use regulations to protect groundwater supplies.

3.042 Community Involvement and Support

Effective wellhead protection programs require control of sources of contaminants; this requires land use planning and control of certain types of development activities on private and public lands. Control of land use is difficult to achieve and requires support of the public and also public officials. Success in this effort must start with a program to educate the public and public officials on the nature of the problem and ways the problem can be controlled. Once everyone understands the problem, support to control or prevent the problem should be easier to achieve.

"What are the threats to our groundwater drinking supply?" is a question of concern to all responsible citizens in a community. Community involvement can start with this question and an explanation of the existing situation. Citizens should be informed that the quality of a groundwater is at risk (1) if a land use presents the capacity to leach contaminants into a groundwater supply which is used for drinking water, and (2) if the contaminants are of sufficient quantity and type to endanger human health. Contamination of groundwater which is not used for drinking does not present an immediate threat because there would not be the potential for health consequences unless the water could be used for drinking in the future. Similarly, just because a land use *MAY* contaminate groundwater does not necessarily mean that there is a threat to public health; the contamination may be in quantities small enough that no adverse health consequences would result.

The importance of broad public support and the fullest possible participation cannot be overstressed. Time and effort spent at this stage educating the community about the goals and process of establishing a wellhead protection program will greatly improve the chance of successfully implementing the program.

3.043 Step One — Community Participation

If a community does not have the financial resources to hire a hydrologist, engineer, or other qualified water professional, the community must rely on the leadership of its water supply system operation and on volunteer community efforts to conduct the risk assessment. This is central to the success of the project. Representation from *ALL* community boards is essential, as well as the highway superintendent, the fire chief, and other concerned citizens.

While broad representation will help the process along, the most important members of the risk assessment committee will be volunteers with backgrounds in geology, planning, and engineering. Without some in-house expertise in geology and groundwater dynamics, the committee's efforts will be significantly more difficult. Often, local colleges and universities can provide students who may assist the community effort in return for academic credit. In this way, the necessary technical support can be supplied without significant financial commitment.

The committee should plan to meet as a group at least once a month, with substantive work being done between meetings. While this is a volunteer effort, it will not be cost free. The community should budget enough money for the committee to mail surveys, meeting reminders, and press releases; purchase office supplies and mapping supplies; photocopy documents and surveys; and print their report. Ideally, there will also be funding for water quality testing.

3.044 Step Two — Collecting Existing Data

The community already has access to a surprising amount of information concerning the community's geology, hydrology, water supplies, and risks to water supplies. Once the committee is formed their first task will be to start collecting this information. Their goals will be to:

1. Understand the geology of the community,
2. Gather information on the local aquifer(s),
3. Identify groundwater information already available in the community,
4. Locate underground storage tanks,
5. Locate septic tanks and identify their ultimate disposal methods,
6. Locate the wells in the community,
7. Locate businesses which may present threats to groundwater quality,
8. Identify what community regulations and by-laws are already in place, and
9. Identify town and state public works practices.

3.045 Step Three — Filling in the Data Gaps

Despite the reams of valuable information which can be accumulated from the above procedure, it is unlikely that this process will have provided all of the information needed about threats to a particular community's groundwater. A community-wide survey can be very helpful. The goal of the survey is to learn as much about specific sites as possible in order to determine what risks to groundwater quality may exist. For example, the survey will try to identify patterns of land use (relative to siting of wells), underground storage tanks, septic systems, proximity of roads to wells, location of abandoned wells, and other important information.

Conducting surveys can present several problems. There can be considerable community resistance to providing information about private property to anyone — especially to community officials. Residents frequently fear that the information will somehow "be used against them." Therefore, it is important that there be adequate publicity about the survey, the reasons for it, and assurances that the information is not intended to be used as a basis for action against any individual.

Surveys in search of contaminated well water should be limited to areas affecting public water supplies unless there is evidence to suggest that private wells are also contaminated. It could be extremely costly to test home wells and analyze all the data. Public support could erode quickly and bog down the process, and many people simply won't believe *ANY* assurances that the results won't be used against them.

Mapping all the information accumulated through the processes described above is very difficult and not necessarily helpful. Key pieces of information plotted on map overlays can be very productive. Important overlays include:

1. Sand and gravel deposits,
2. Locations and depths of wells,
3. Locations and identification of land uses which present threats to groundwater quality,
4. Locations of underground storage tanks,
5. Septic system locations,
6. Wells within 100 feet of a septic system, and
7. Wells within 20 feet of a roadway.

These overlays are a critical first step in determining what threats to groundwater quality may exist in the community. Through the use of the overlays it is possible to begin to see the relationship between activities and wells, and the overlays can highlight areas of critical concern in a clear and simple way.

QUESTIONS

Write your answers in a notebook and then compare your answers with those on page 125.

3.0O When water passes through a soil, what kinds of wastewater treatment processes might occur which could lessen the eventual impact of the contaminant once it reaches groundwater?

3.0P How can a wellhead protection program achieve control of sources of contaminants and certain types of development activities on private and public lands?

3.0Q List the various types of data that need to be gathered to develop a wellhead protection program.

3.046 Step Four — Which Land Uses Constitute a Threat to Groundwater Quality?

Not all threats to groundwater quality are of equal significance. Some land uses, while appearing on a list (Table 3.3) of businesses which present risks for groundwater contamination, present a low risk. This determination is based upon the types of contaminants, how they interact with the environment, how they degrade in soil and water, how fast they move, and how dangerous they are to human health. It is important in the analysis of threats to groundwater that the specific characteristics of the community are taken into consideration. Many types of activities identified in Table 3.3 may not occur in a specific community. Therefore, it is an important step to identify which land uses with potential risk to groundwater do occur in a given community and to determine the level of risk which they present. Railroad tracks, yards, and maintenance stations are listed as a *SLIGHT* risk in Table 3.3. However, in some communities in certain locations a railroad yard could be considered a *SEVERE* risk. Also, any other types of activities that there is a reason to be concerned about but which have not appeared in the Table should be identified and included.

3.047 Step Five — Learning From What Has Been Collected

There should now be enough information gathered to critically assess the potential for groundwater contamination in the community. It will be important to study the geological maps, location of sand and gravel deposits, clay layers, and depths of wells to determine general patterns of groundwater use in the community. It is in this step that the assistance of someone with geologic and hydrogeologic expertise is critical.

TABLE 3.3 RELATIVE LEVELS OF RISK FROM LAND USES TO GROUNDWATER

Taking Into Consideration Volume, Likelihood of Release,
Toxicity of Contaminants, and Mobility

(Compiled and Analyzed by Vermont Department of Health, 1988)

SEVERE
- Dry Cleaners
- Gas Stations
- Car Wash With Gas Station
- Service Station — full or minor repairs
- Painting and Rust Proofing
- Junk Yards
- Highway De-Icing — application and storage
- Right-of-Way Maintenance
- Dust Inhibitors
- Parking Lot Runoff
- Commercial Size Fuel Tanks
- Underground Storage Tanks
- Injection Wells: automobile service station disposal wells; industrial process water, and waste disposal wells
- Hazardous Waste Disposal
- Landfills
- Salt Stockpiles

SEVERE TO MODERATE
- Machine Shops: metal working; electroplating, machining
- Chemical and Allied Products
- Industrial Lagoons and Pits
- Septic Tanks, Cesspools, and Privies
- Septic Cleaners
- Septage
- Household Cleaning Supplies
- Commercial Size Septic Systems
- Chemical Stockpiles
- Clandestine Dumping

MODERATE
- Carpet and Upholstery Cleaners
- Printing and Publishing
- Photography and X-Ray Labs
- Funeral Homes
- Pest Control
- Oil Distributors
- Paving and Roofing
- Electrical Component Industry
- Fertilizers and Pesticides
- Paint Products
- Automotive Products
- Home Heating Oil Tanks, Greenhouses, and Nurseries
- Golf Courses
- Landscaping
- Above Ground Fuel Tanks
- Agricultural Drainage Wells
- Raw Wastewater Disposal Wells, Abandoned Drinking Wells

MODERATE TO SLIGHT
- Water Softeners
- Research Laboratories
- Above Ground Manure Tanks
- Storm Water and Industrial Drainage Wells
- Stump Dumps
- Construction

SLIGHT
- Beauty Salons
- Car Wash
- Taxidermists
- Dyeing/Finishing of Textiles
- Paper and Allied Products
- Tanneries
- Rubber and Miscellaneous Plastic Products
- Stone, Glass, Clay, and Concrete Products
- Soft Drink Bottlers
- Animal Feedlots, Stables, and Kennels
- Animal Burial
- Dairy Wastes
- Poultry and Egg Processing
- Railroad Tracks, Yards, and Maintenance Stations
- Electrical Power Generation Plants and Powerline Corridors
- Mining of Domestic Stones
- Meat Packing, Rendering, and Poultry Plants
- Open Burning and Detonation Sites
- Aquifer Recharge Wells
- Electric Power and Direct Heat Reinjection Wells
- Domestic Wastewater Treatment Plant and Effluent Disposal Wells
- Radioactive Waste

That person will be able to study the well logs, geologic maps, survey results, and USGS (United States Geological Survey) studies to describe the hydrogeologic setting of the community.

Once the hydrogeologic setting is understood, the information that has been collected about types of land use and risks that they may present can be analyzed. Some land uses present great risks to shallow wells, but may present little to no risk to bedrock/deep wells. For example, a problem could develop if a town stores its winter sand and salt piles outdoors. This may present a groundwater contamination threat to shallow wells in the vicinity. By knowing that there are shallow wells nearby, rational judgments can be made about water testing programs and appropriate town response to protect against contamination.

There may also be locations where septic system failures are more likely due to hydrogeologic settings, and this can be analyzed by the expert in geology. Small-capacity home wells are particularly susceptible to organic chemical contamination from nearby septic systems. Domestic wastewater has frequently been found to contain trace amounts of petroleum distillates as well as benzene, toluene, chlorinated hydrocarbons such as trichloroethane, trichloroethylene, tetrachloroethylene, dichlorobenzene, and alkyl phenols. Municipal wells can be less susceptible because they derive some protection from their large volumes of mixing and withdrawal, whereas private wells are not so protected.

Unlike nitrogen and biological pollutants, organic chemicals do not enter household wastewater continuously. Gasoline, paint thinners, solvents, and even pesticides are typically discharged only at irregular intervals. Furthermore, once discharged into wastewater, some proportion of these chemicals is absorbed into soil particles or broken down by bacterial action. However, some organic chemicals are very persistent. If a neighbor's septic system lies within the recipient well's zone of contribution, a quantity of a discharged chemical will probably enter that well and the volume may be sufficient to exceed recommended maximum concentration levels of the contaminant in drinking water.

3.048 Step Six — Now What?

By studying the proximity of various land uses with wells, the depth of those wells, and the degree of risk presented by the land use, decisions can be made about what to do next. At this point it should be apparent what activities in the community may have already contaminated groundwater, or which may in the future. If there is reason to believe that groundwater may already be contaminated, the community should immediately implement a water quality testing program.

The water quality testing program should be designed to test for the contaminants which are believed to be present. For example, if the study indicated that septic tank system failure may have occurred in proximity to shallow wells, the tests would be looking for nitrate and coliform contamination. If the threat is presented by road salt, saline concentration would be highlighted. On the other hand, perhaps the threat is seen to be from agriculture. It will be necessary to identify the types of pesticides believed to have been used in order to test for them in the groundwater. An annual community water testing program will help monitor long-term and changing groundwater conditions.

Without water quality testing it will not be possible to know if there is an existing problem with groundwater quality. If there is a problem, it is important to take remedial action to protect the health of the water users. If there is no contamination, then it is important to take steps to ensure against future contamination. Town regulations and by-laws, as well as public education, can be key in safeguarding future water quality. Examples of town regulations and by-laws which can work to protect groundwater quality include: aquifer protection zoning, private well regulations, hazardous materials and underground storage tank regulations, required septic tank pumping regulations, unregistered motor vehicle by-laws, and wetlands protection by-laws. Other community actions can include: covering salt piles, decreasing use of road salt, land acquisition and protection activities, community supported water quality testing, public education efforts, hiring professional health agents and planning staff, and cooperation in regional efforts to protect groundwater.

3.049 Step Seven — Establishing Wellhead Protection Areas (WHPA)[16]

Contact your state regulatory agency for recommended procedures to define *WELL ISOLATION ZONES*,[17] which are the areas that require land use controls to protect a wellhead or well field. Some states have issued minimum distance guidelines or requirements. EPA's *GUIDELINES FOR DELINEATION OF WELLHEAD PROTECTION AREAS*[18] lists the following six primary methods in order of increasing technical sophistication:

1. Arbitrary fixed radii,
2. Calculated fixed radii,
3. Simplified variable shapes,
4. Analytical methods,
5. Hydrogeologic mapping, and
6. Numerical flow/transport methods.

These methods range from simple, inexpensive methods to highly complex and costly ones.

Factors that need to be considered when determining the wellhead protection area include:

1. Circle or area of influence around a well or well field,
2. Depth of drawdown of the water table by such well or well field at any given point,
3. Time or rate of travel of various contaminants in various hydrologic conditions,

[16] *Wellhead Protection Area (WHPA). The surface and subsurface area surrounding a water well or well field, supplying a public water system, through which contaminants are reasonably likely to move toward and reach such water well or well field. Also see WELL ISOLATION ZONE.*

[17] *Well Isolation Zone. The surface or zone surrounding a water well or well field, supplying a public water system, with restricted land uses to prevent contaminants from a not permitted land use to move toward and reach such water well or well field. Also see WELLHEAD PROTECTION AREA (WHPA).*

[18] *Available from National Technical Information Service (NTIS), 5285 Port Royal Road, Springfield, VA 22161. Order No. PB93-215861. Price, $54.50, plus $5.00 shipping and handling per order.*

4. Distance of source of contaminants from well or well field, and
5. Other factors affecting the likelihood of contaminants reaching the well or well field.

Threats to wells or well fields include direct introduction of contaminants into well casings, microbial contamination, and chemical contamination. The fate, degradation, and/or assimilation of both microbial and chemical contaminants as they move from their sources toward the well or well field are important considerations. Also it is important to realize that water flowing underground may follow unusual paths. Recharge water carrying contaminants that enters an aquifer through an area of influence of a well will not necessarily travel to the well, and recharge water that enters the aquifer outside of the area of influence may travel to the well.

The procedures used by a community to delineate wellhead protection areas (WHPA) will depend on available state regulatory requirements, guidelines, and technical assistance. Also the community will need to assess the seriousness of contaminant threats to its wells and/or well field and make optimum use of the resources available. Also see Section 3.12, "Well Site Selection," in the Appendix to this chapter for additional information on minimum recommended distances between wells and potential sources of contamination.

3.0410 Step Eight — FINALLY

The results of the volunteer risk assessment must be well publicized and presented to the community. Public education and support for the effort are essential if behavior is to be changed and town regulations and by-laws passed. Copies of any reports should be distributed around the community and made generally available.

Unless the community takes affirmative steps as a result of the study about threats to groundwater quality, it is inviting contamination to occur. Perhaps the most important result of this study will be to educate the community officials about the threats to public safety which exist in the community and the actions they can take to protect the public health.

Major portions of this section were adapted from a paper titled "Model for Rural Communities to Assess Threats to Groundwater Quality," by Lynn Rubinstein, Land Use Planner, Franklin County Planning Department, Greenfield, Massachusetts. The information is used with her permission and is sincerely appreciated.

3.05 Groundwater Protection Tips

The preventive practices listed below are effective means for operators to work with various groups in their community to prevent contamination of groundwater.

3.050 Agriculture

THE POTENTIAL PROBLEM

Pesticides, such as insecticides, and herbicides may contaminate groundwater if improperly applied or spilled in a wellhead protection area. Fertilizers applied in greater amounts than can be taken up by plants may release nitrates into groundwater. Storage and improper application of manure in wellhead protection areas may result in nitrates and bacteria leaching into groundwater.

PREVENTIVE PRACTICES

- Do not mix, rinse, or store pesticides in a wellhead protection area.
- Minimize or avoid pesticide use in wellhead protection areas by using integrated pest management or alternate methods for pest control.
- Move manure storage out of wellhead protection areas.
- Apply nitrogen according to a crop nutrient budget which includes analysis of soil suitability and cropping practices.

3.051 Gas Stations

THE POTENTIAL PROBLEM

Gas stations pose a threat to groundwater because of the many possibilities of incidental, small spills or leaks of hazardous materials. While major leaks from tanks or transfers are now under much improved legal and technical control, small spills can occur as customers fill their gas tanks. Waste oil, solvents, and hydrocarbons from repair facilities are also a hazard to the groundwater.

PREVENTIVE PRACTICES

- Use available technology to prevent overfilling of tanks.
- Install a secondary containment structure around tanks to prevent leaks and install a monitoring system to detect water inflow.
- Test tanks and pipes regularly.
- Connect floor drains in repair stalls to a holding tank.
- Dispose of waste fluids in licensed facilities, *NEVER* on the ground or in a septic system.
- Use absorbents to clean up minor fluid leaks and spills.

3.052 Fuel Storage

THE POTENTIAL PROBLEM

Corrosion may cause fuel oil to leak, go through the soil and contaminate groundwater. Gasoline may float on top of groundwater and move along with it. Heating oils are less likely to be soluble and mobile; however, they are both persistent in groundwater.

Underground tanks, because they are out of sight and closer to the groundwater, pose more of a threat than aboveground tanks. They are tightly regulated and are not likely to be a threat if replaced since 1986. Small, domestic, aboveground tanks are not now regulated, but owners can monitor their condition and prevent extensive loss of fuel and contamination. Buried, abandoned tanks may continue to cause problems far removed in time and place from the original leak.

It has been demonstrated that the cost of prevention is dramatically lower than the cost of remediation and cleanup.

PREVENTION TIPS

- Be sure you have a complete inventory of all tanks in your wellhead protection area.
- Inform the owners that the tank is in a sensitive area. Ask if they are aware of the programs available for replacement, if such programs are available.
- Surround above-ground tanks with a dike or berm with a concrete base. Protect the area from rainwater by installing a permanent roof or install a manually operated valve to drain the diked area.
- Monitor the condition of above-ground tanks and replace them at the first sign of corrosion.

3.053 Photo Labs/Printers, Dry Cleaners, Furniture Strippers/Painting, Medical Labs

THE POTENTIAL PROBLEM

Several organic compounds used in developing color film and printers' inks contain resins, solvents, and heavy metals. If flushed into septic systems, these chemicals may destroy the bacteria essential to proper functioning of the system. Common sources of contamination include improper outdoor storage, accidental spills, disposal of spent fluids, or wash water in the ground or in the septic system.

Solvents such as trichloroethylene, methylene, chloride, naphthalene, and benzene used in dry cleaning and furniture staining are among the most hazardous for wells. Even the smallest amount, once in the groundwater, may sink to deep locations in the aquifer, then migrate in unpredictable ways. Many solvents are very persistent.

PREVENTIVE PRACTICES

- Install permanent covers over outdoor storage facilities for chemicals and construct berms to hold spills or leaks.
- Separate chemical wash water from domestic wastewater.
- Connect floor drains and sinks used for washing chemicals to holding tanks, not sewer or septic systems.
- Dispose of waste chemicals at licensed hazardous waste disposal facilities.
- Equip exhaust fans to catch dripping liquid.
- Store spent acids and caustic bottles containing methylene chloride, ink, and sludges in secure, clearly labeled containers and dispose of them at licensed facilities.
- Contact a rag rental/cleaning service to recycle cleaning rags contaminated with solvents.
- Recycle wastes (for example, spent fixer to recover silver).

3.054 Septic Systems, Subsurface Waste Disposal

THE POTENTIAL PROBLEM

Properly designed, installed, and maintained septic systems will remove pathogenic organisms, such as bacteria and viruses, before effluent reaches the groundwater. Systems that don't allow for percolation of wastewater wastes through a sufficient amount of unsaturated soil may not effectively remove viruses and bacteria.

Nitrate and many other chemicals and solvents are not removed by septic systems. Moving easily through the soils, they may enter groundwater. More than 10 mg/*L* nitrate is particularly toxic to fetuses and babies.

PREVENTIVE PRACTICES

- Do not flush *ANY* chemicals or cleaners down the drain, not even septic tank cleaners. They are not only ineffective, but kill beneficial bacteria. If you don't know what it is, don't flush it, *PLEASE!*
- Avoid garbage disposals; they can cause costly drain field clogging.
- If there are many small, old systems within the wellhead protection area, consider piping the wastewater to a community treatment system.
- Since the least costly way to treat nitrate is by dilution, ensure adequate setbacks between septic systems, wells, and surface water.
- Have septic tanks pumped regularly, every two to three years, to prevent solids from clogging the disposal field.

3.055 What to Tell Homeowners

DEAR NEIGHBOR

It may come as a surprise to you that your home is on top of a public water supply. We have been asked by Congress and the State Department of Health Engineering to protect our water supply by preventing contaminants from reaching the well. Your house/lot/yard happens to be on the land that recharges our well.

We certainly respect your privacy and the right to go about your life unencumbered by any regulations. We only ask your cooperation in conducting your activities in a way that will least likely cause our well to become contaminated. These tips will also protect your water supply.

If you garden, bear in mind that fertilizers, pesticides, fungicides, and herbicides might be quite soluble in water and "leach" into the groundwater.

PREVENTIVE PRACTICES

- Try to use organic compost material. It is preferable to nitrate-releasing commercial fertilizer. If you use the latter, a soil analysis will help you use just the right amount.
- Try to use "natural" pesticides, such as rotenone and BT. Used according to directions, they are unlikely to cause problems.
- If you spill pesticides that are poison, please try to clean up spills as completely as possible. Sweep up granular material and soak up liquids, disposing of them in a ziplock bag.

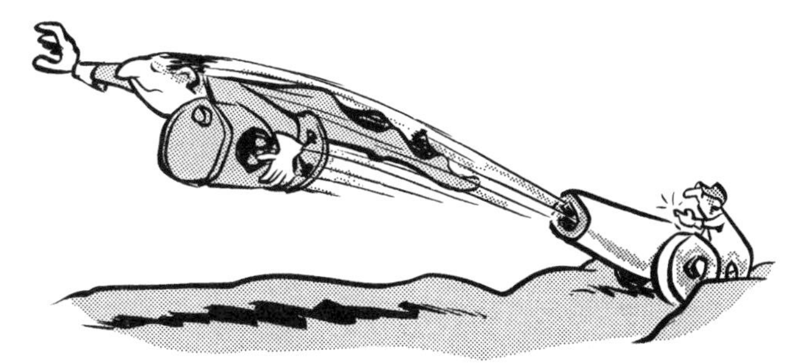

DISPOSING OF WASTE OIL, PAINT THINNERS, AND SOLVENTS

Remember, these substances pose a high risk to groundwater and drinking water pumped out of a well. They may float or sink to the bottom, but they don't disappear and they make the water unsafe to drink for many years to come. Cleaning it up is expensive and finding an alternate water supply may be impossible.

PREVENTIVE PRACTICES

- Do not dispose of solvents, degreasers, waste oil, cleaning fluids, gasoline, paint thinners, or radiator fluid by pouring into sinks, toilet, or by burying them in the ground.
- If you work on your car, please pick up any spills with absorbent material, place in a ziplock bag, and dispose of it properly (ask your landfill or waste pickup operator how to dispose of it).

Your SEPTIC SYSTEM, if properly installed, maintained, and located at least 300 feet from a well, should not pose a problem.

PREVENTIVE PRACTICES

- If you have an old system, please have it inspected and consider replacing it, if warranted.
- Please do not flush chemicals, grease, or septic tank cleaners into your system.
- Have your system pumped every two or three years.

STORING CHEMICALS

Please be aware that chemicals or fuel oil improperly stored and/or exposed to the elements — rain, snow, freezing, and thawing — can pose a threat to our drinking water. When containers give way, hazardous substances or fuel oil may leak into the soil and our groundwater.

PREVENTIVE PRACTICES

- Store chemicals in sound containers, off the ground.
- If they must be outdoors, try to fashion a secure cover and/or
- Place a solid container under a tank, for instance, from which you can empty the rain or snow periodically.
- Periodically check containers for corrosion and replace if necessary.
- Ask your landfill or garbage pickup operator how you can safely dispose of household chemicals.

3.056 Acknowledgment

The information in this section was provided by Maine's Wellhead Protection Program. The contribution is greatly appreciated.

3.06 Summary

YOU, as an operator, have as much of a responsibility to preserve the quality and capacities of your wells as you do to oil the bearings on your pumps. The damage from neglect and excessive use of aquifers is sometimes irreversible. You can and should:

1. Be on the lookout for any possible sources of pollution of groundwater in the area around the well,
2. Be careful not to overtax the capability of aquifers to supply water by installing oversized pumps,
3. Ensure that all wells are properly sealed to prevent direct contamination of the aquifer, and
4. Develop a recordkeeping program of well production and quality. These records should include geologic as well as hydrologic and hydraulic data gathered during the construction and testing of the well.

3.07 Additional Reading

1. *EPA's MANUAL OF INDIVIDUAL WATER SUPPLY SYSTEMS*, Office of Water Programs, Water Supply Division, 1975. Obtain from National Technical Information Service (NTIS), 5285 Port Royal Road, Springfield, VA 22161. Order No. PB-258402/7. EPA No. 430-9-74-007. Price, $45.00, plus $5.00 shipping and handling per order.
2. *GROUNDWATER AND WELLS*, Second Edition, 1986. Published by U.S. Filter/Johnson Screens. Obtain from Mower House Color Graphics, Inc., 508 10th Street, NE, Austin, MN 55912. ISBN 0-9616456-0-1. Price, $62.00, plus $4.25 shipping and handling.
3. *WELLHEAD PROTECTION: A GUIDE FOR SMALL COMMUNITIES*, EPA Seminar Publication, February 1993. Published by Office of Research and Development, Office of Water, U.S. Environmental Protection Agency, Washington, DC 20460. Obtain from National Technical Information Service (NTIS), 5285 Port Royal Road, Springfield, VA 22161. Order No. PB93-215580. EPA No. 625-R-93-002. Price, $45.00, plus $5.00 shipping and handling per order.

QUESTIONS

Write your answers in a notebook and then compare your answers with those on page 126.

3.0R List at least five land use activities that are considered to have a severe risk level to groundwater.

3.0S What kinds of organic chemical contaminants may enter well water from domestic wastewater?

3.0T If a septic tank system may have failed in the vicinity of shallow wells, what water quality tests should be performed?

3.0U What could be the most important result of a community groundwater study?

3.1 SURFACE FEATURES OF A WELL

3.10 Purpose of Surface Features

A number of openings are found in the top of a well. These openings are designed to provide access to the well for taking water level measurements, permitting entrance or escape of air or gas, adding gravel, or for applying disinfection or well cleaning agents. Table 3.4 lists the major surface features of a well and the purpose of each component. The openings in the top of a well must be protected against the entrance of surface waters or foreign matter. Next to the sanitary seal, this is the most important sanitary feature of well construction.

Figure 3.5 is a cross-sectional drawing showing the typical surface features of a domestic water well. Many of the component parts we will be discussing in the following pages are illustrated in this drawing. Study this drawing and become familiar with the basic parts and their relationship to the entire facility.

TABLE 3.4 PURPOSE OF SURFACE FEATURES OF A WELL

Component	Purpose
Well-Casing Vent	Allows air to enter well during drawdown to prevent vacuum conditions; vents excess air during well recovery period.
Gravel Tube	Permits operator to see level of gravel and add gravel as necessary.
Sounding Tube	Permits insertion of water level measuring device; also used to add chlorine or well cleaning agents.
Air Line Water Level Measuring Device	Permits measurement of water level by means of air pressure measurements.
Pump Pedestal	Supports the weight of the pumping unit.
Pump Motor Base Seal	Provides watertight seal between the motor base and the concrete support pedestal.
Sampling Taps	Permit sampling of pumped water.
Air-Release Vacuum Breaker Valve	Permits discharge of air in column pipe during start-up and admits air during shutdown.
Pump Blowoff (or Drain Line)	Used to remove first water (usually sandy) pumped at start-up.

3.11 Well-Casing Vent

A well-casing vent is provided to prevent vacuum conditions inside a well by admitting air during the *DRAWDOWN*[19] period when the well pump is first started. If vacuum conditions are allowed to develop, contamination may be sucked into the well through some hidden defect in the well or *CONDUCTOR CASING*,[20] or through the top of the well at the pump base.

The well-casing vent also prevents pressure buildup inside the well casing by allowing excess air to escape during the well recovery (refilling) period after the well pump shuts off. If pressures are allowed to build up, they will loosen and blow out the sealing compound around the pump base.

A properly sized and constructed vent should allow the unrestricted flow of air into and out of the well interior. A vent of at least three inches (75 mm) in diameter should be provided. Dual venting is desirable on wells over 14 inches (350 mm) in diameter.

All well vents should be constructed so that openings are in a vertical downward position. Openings should be a minimum of 36 inches (900 mm) in height above the finished surface of the well lot (yard) and should be covered with a fine mesh screen or similar device to keep insects from the well interior.

3.12 Gravel Tube

On gravel-envelope wells, a gravel tube must be provided to monitor the level of gravel and to add gravel as necessary. A typical gravel tube is four inches (100 mm) in diameter and is capped at the top end. Such openings must be elevated above ground level to prevent well contamination during flooding, and must be kept tightly sealed.

3.13 Sounding Tube[21]

A sounding tube is necessary so that the water level in the well can be periodically determined. This tube can also be used for the addition of chlorine and well cleaning agents. The sounding tube is generally a minimum of two inches (50 mm) in diameter and must be kept tightly sealed. The water level is determined by inserting a disinfected rope or measuring tape into the tube, lowering it down to the water level, and recording the distance. Often, the well-casing vent is also used as a sounding tube by installing a tee in the well-casing vent.

All vent pipes, sounding lines, and gravel fill pipes must be one continuous conduit through the concrete pedestal. Also, all conduits which penetrate the casings must be provided with a continuous watertight weld at the point of entry into the well interior.

If a water level sounding line is to be incorporated, it may be installed either through the upper portion of the concrete pump pedestal or through the pump base. If installed within the concrete pump pedestal, a conduit of sufficient diameter to allow the passage of the sounding line is placed in the pedestal form prior to the placement of concrete. One-quarter inch (6 mm) galvanized pipe or schedule 80 PVC plastic pipe is frequently used as a sounding line. The space between the sounding line and the conduit or the openings around the line passing through the pump base must be provided with a watertight seal.

A quick-disconnect fitting on the sounding line is preferable to a permanent gage installation. To determine the distance to the water level, air pressure is applied to the sounding line to force out all water. The pressure is measured and the measurement is converted to the depth of water in the line.

$$\text{Water Depth in Line, ft} = (\text{Air Pressure, psi})(2.31 \text{ ft water/psi})$$

The length of the line must be known to determine the distance down to the surface of the water.

$$\text{Distance Down to Water Surface, ft} = \text{Length of Line, ft} - \text{Water Depth in Line, ft}$$

3.14 Pump Pedestal

The well-casing vent pipe and the gravel tube are generally encased within a concrete pump pedestal designed to support the full weight of the pumping unit.

The pump pedestal should be constructed of continuously poured concrete to a minimum height of 18 inches (450 mm) above the finished elevation of the well lot. A lower pedestal height may be used due to lack of space (entrance to tank less than 18 inches above floor), but in no case should the pedestal height be less than 12 inches (300 mm).

---

[19] *Drawdown.* (1) The drop in the water table or level of water in the ground when water is being pumped from a well. (2) The amount of water used from a tank or reservoir. (3) The drop in the water level of a tank or reservoir.

[20] *Conductor Casing.* The outer casing of a well. The purpose of this casing is to prevent contaminants from surface waters or shallow groundwaters from entering a well.

[21] *Sounding Tube.* A pipe or tube used for measuring the depths of water.

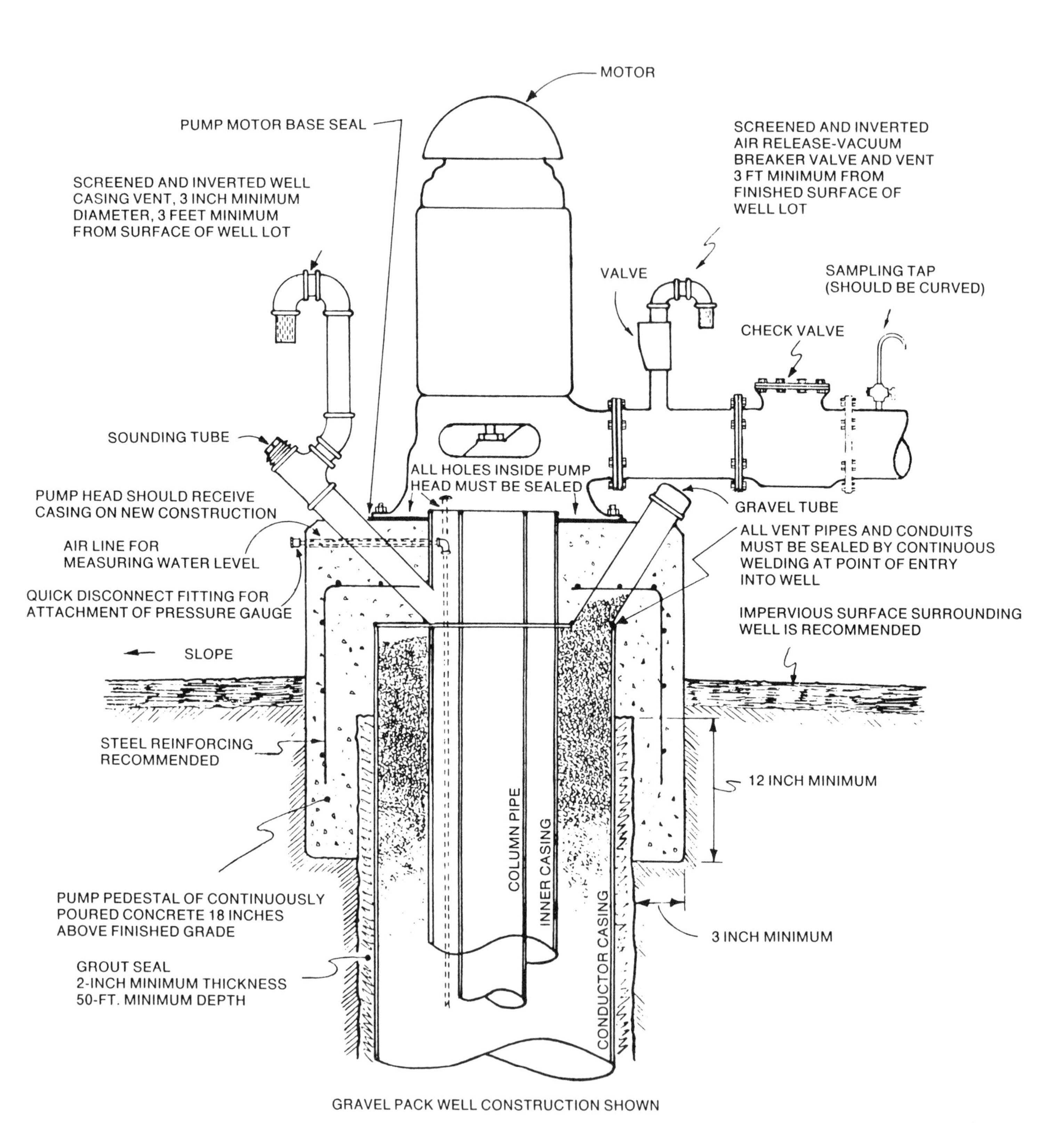
MOTOR
PUMP MOTOR BASE SEAL
SCREENED AND INVERTED AIR RELEASE-VACUUM BREAKER VALVE AND VENT 3 FT MINIMUM FROM FINISHED SURFACE OF WELL LOT
SCREENED AND INVERTED WELL CASING VENT, 3 INCH MINIMUM DIAMETER, 3 FEET MINIMUM FROM SURFACE OF WELL LOT
VALVE
SAMPLING TAP (SHOULD BE CURVED)
CHECK VALVE
SOUNDING TUBE
ALL HOLES INSIDE PUMP HEAD MUST BE SEALED
PUMP HEAD SHOULD RECEIVE CASING ON NEW CONSTRUCTION
GRAVEL TUBE
AIR LINE FOR MEASURING WATER LEVEL
ALL VENT PIPES AND CONDUITS MUST BE SEALED BY CONTINUOUS WELDING AT POINT OF ENTRY INTO WELL
QUICK DISCONNECT FITTING FOR ATTACHMENT OF PRESSURE GAUGE
IMPERVIOUS SURFACE SURROUNDING WELL IS RECOMMENDED
SLOPE
STEEL REINFORCING RECOMMENDED
12 INCH MINIMUM
COLUMN PIPE
INNER CASING
CONDUCTOR CASING
PUMP PEDESTAL OF CONTINUOUSLY POURED CONCRETE 18 INCHES ABOVE FINISHED GRADE
3 INCH MINIMUM
GROUT SEAL
2-INCH MINIMUM THICKNESS
50-FT. MINIMUM DEPTH
GRAVEL PACK WELL CONSTRUCTION SHOWN
DRAWING NOT TO SCALE

Fig. 3.5 Surface features of a domestic water well

The size of the pump pedestal may vary depending upon the given installation, but in all cases must provide a minimum of three inches (75 mm) of concrete around the outside of the conductor casing grout seal, if one is used. The pedestal should extend down to enclose the top of the grout seal a minimum depth of 12 inches (300 mm). Steel reinforcement of the pedestal is recommended.

3.15 Pump Motor Base Seal

A watertight seal should be provided between the pump motor base and the concrete pedestal. If the two surfaces are parallel, a neoprene rubber seal cut from flat stock can be used. In many cases, however, these two surfaces are not parallel because wedges may have been installed to correctly align the pump head shaft in the center of the motor hollow-shaft unit. When this occurs, an oil-resistant, non-hardening material such as latex rubber can be used to create a seal. Cement grout is not recommended because it will eventually crack and create openings through the seal. A neoprene rubber or regular gasket material seal is very satisfactory for a submersible pump installation.

3.16 Sampling Taps

A sampling tap is generally provided on the downstream side of the pump check valve. A *PET COCK*[22] valve fitted with a three-eighths inch (9 mm) copper line with the outlet turned down toward the ground is recommended. No hose bib, faucet, or any other threaded valve should be installed between the well pump and the check valve. This is to prevent any contaminated water from being drawn into the well when the pump shuts down and the water column drains back into the well.

3.17 Air Release-Vacuum Breaker Valves

If a pump is not equipped with a *FOOT VALVE*[23] and you want to discharge the air in the column pipe, install an air release-vacuum breaker valve in the piping between the pump head and well discharge check valve. The valve functions as follows:

1. When the well pump is initially started, air inside the pump column is expelled (forced out) through the valve to the atmosphere instead of into the system;
2. When the well pump is shut down, air is admitted into the pump column, thus freely allowing the column to dewater into the well;
3. The valve should be mounted as close as possible to the check valve in a vertical position on top of the discharge piping; and
4. The opening in the top of the valve must be equipped with a downturned, screened assembly, and the opening should be protected from flooding.

3.18 Pump Blowoff

The pump blowoff or drain line is used to remove initially pumped water containing sand away from the system. When there is a blowoff or drain line from the pump discharge or from a pump control valve, the waste line must be located above any known flood levels and protected against the possibility of backsiphonage or back pressure. The blowoff or drain line should not be directly connected to any sewer or storm drain.

QUESTIONS

Write your answers in a notebook and then compare your answers with those on page 126.

3.1A Why are there openings into the top of a well?

3.1B What is the purpose of the well-casing vent?

3.1C How would you determine the distance down to the water level in a well?

End of Lesson 1 of 4 Lessons on WELLS

Please answer the discussion and review questions next.

[22] *Pet Cock. A small valve or faucet used to drain a cylinder or fitting.*

[23] *Foot Valve. A special type of check valve located at the bottom end of the suction pipe on a pump. This valve opens when the pump operates to allow water to enter the suction pipe but closes when the pump shuts off to prevent water from flowing out of the suction pipe.*

DISCUSSION AND REVIEW QUESTIONS

Chapter 3. WELLS

(Lesson 1 of 4 Lessons)

At the end of each lesson in this chapter you will find some discussion and review questions. The purpose of these questions is to indicate to you how well you understand the material in the lesson. Write the answers to these questions in your notebook before continuing.

1. What is the term used to describe the water volume that can move through the pores in a rock and is affected by gravitational forces?
2. Identify the sources of pollutants that can contaminate a groundwater supply.
3. What kind of preventive maintenance program would you develop for an aquifer serving your wells?
4. How would you protect an aquifer from overdraft damage?
5. What options are available for a community whose groundwater supply has been contaminated?
6. Why is community involvement and support important for a wellhead protection program?
7. What factors need to be considered when determining the wellhead protection area?
8. What are the major responsibilities of an operator with regard to an aquifer and well pump system?
9. Why should hose bibs, faucets, or threaded valves not be installed between the well pump and the check valve?
10. What is the purpose of a pump blowoff?

CHAPTER 3. WELLS

(Lesson 2 of 4 Lessons)

3.2 WELL APPURTENANCES[24]

3.20 Typical Well Pump and Pressure Tank

The well appurtenances described and shown in this section are illustrated in Figure 3.6 which is an elevation or profile drawing of a well pump and tank installation. This is a *TYPICAL* well pump-tank installation showing the general features and location of the appurtenances under discussion. Many variations of this layout are possible depending upon geographic location, state and local building and plumbing codes, hydraulic factors, and individual preferences.

3.21 Valves

3.210 Check Valves[25]

Most well pump stations are equipped with some type of check valve installed on the discharge side of the pump. The purpose of the check valve is to act as the automatic shutoff valve when the pump stops. This valve prevents draining of the system or tank being pumped to, and keeps pressurized water from flowing back down the pump column into the well. This reversal of pressurized flow into a well can cause serious problems that may include removal of the pump from service and redevelopment of the well. (This flow reversal problem does not occur in well pump installations equipped with a foot valve.)

Several different types of check valves are available such as swing check, lift check, foot check, slant disc check, flap check, double disc check, and automatic control check. Four different types of check valves are illustrated in Figure 3.7.

The important features of any check valve are that the valve must (1) close watertight; (2) be able to safely withstand hydraulic shock loads; (3) have a low head loss; and, most importantly, (4) be equipped to close the valve disc in advance of flow reversal.

There are numerous ways to ensure rapid shutoff of the valve disc within the recommended time of $\frac{1}{20}$ to $\frac{1}{10}$ of a second. This can be accomplished with an outside lever and weight (most commonly used), spring loaded (either internal or external) disc (spring closes the disc), and compressed air assisted device (compressed air moves arm and closes disc).

3.211 Pump Control Valves

A pump control valve is a diaphragm-type valve designed to eliminate pipeline surges caused by the starting and stopping of deep well pumps. This valve is available in a "normally closed" type, or a "normally open" type.

The "normally closed" type valve is designed for installation in the main discharge line of the pump. The pump starts against a closed valve. When it is started, the *SOLENOID*[26] control valve is energized and the valve begins to open slowly, gradually increasing line pressure to full pumping head. When the pump is signalled to shut off, the solenoid control valve is de-energized and the valve begins to close slowly, gradually reducing flow while the pump continues to run. When the valve is closed, a limit switch assembly, which serves as an electrical interlock between the valve and the pump, releases the pump starter and the pump stops. Should a power failure occur, a built-in lift-type check valve closes the moment flow stops, preventing reverse flow regardless of solenoid or diaphragm assembly position.

The "normally open" type valve is designed for installation in a bypass line on the discharge side of the pump. Operation is completely automatic, fully hydraulic, and electrically controlled by a solenoid control valve. With the pump off, the valve is wide open. When the pump is started, the solenoid is energized and the valve begins to close slowly, discharging air and the initial rush of water and any sand from the pump column to the atmosphere. As the check valve closes, the pump output is gradually diverted into the main line as the main line check valve opens, thus preventing the development of starting surges.

When it is time to shut off the pump, the solenoid is de-energized. A limit-switch on the valve keeps the pump motor circuit closed (power stays on). The pump continues to run while the pump control valve opens slowly and diverts pump output to

---

[24] *Appurtenance (uh-PURR-ten-nans). Machinery, appliances, structures and other parts of the main structure necessary to allow it to operate as intended, but not considered part of the main structure.*

[25] *Check Valve. A special valve with a hinged disc or flap that opens in the direction of normal flow and is forced shut when flows attempt to go in the reverse or opposite direction of normal flows.*

[26] *Solenoid (SO-luh-noid). A magnetically (electric coil) operated mechanical device. Solenoids can operate small valves or electric switches.*

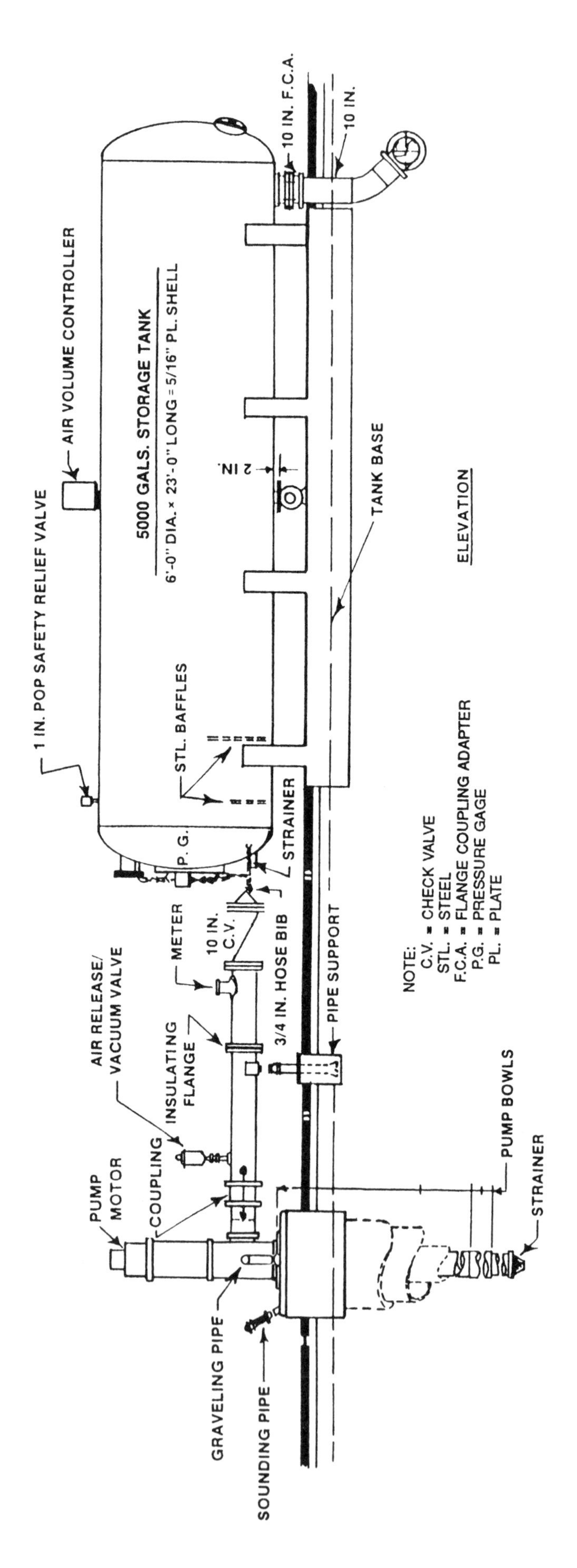

Fig. 3.6 Typical well pump and appurtenance installation

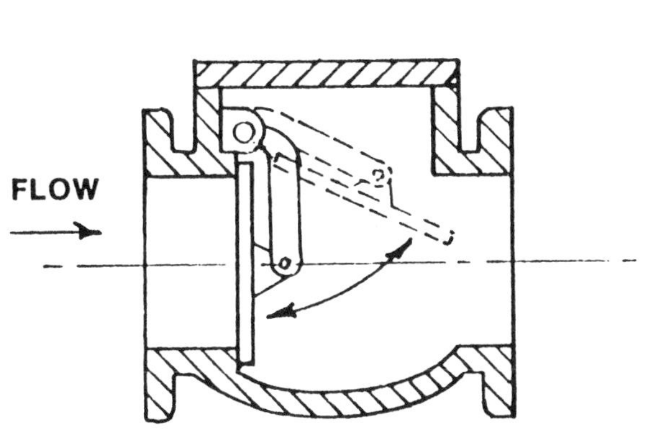

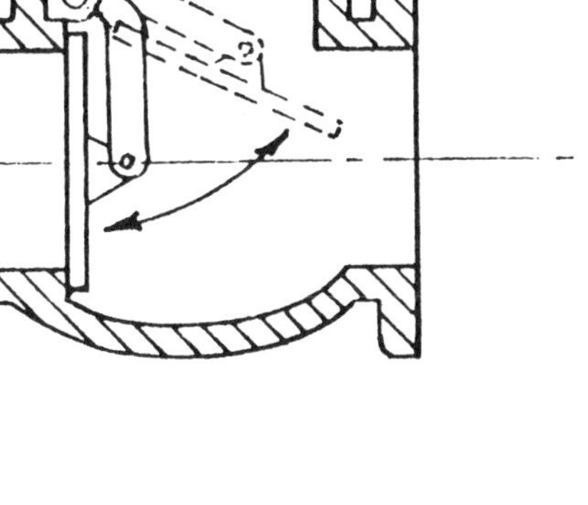

1. SWING CHECK

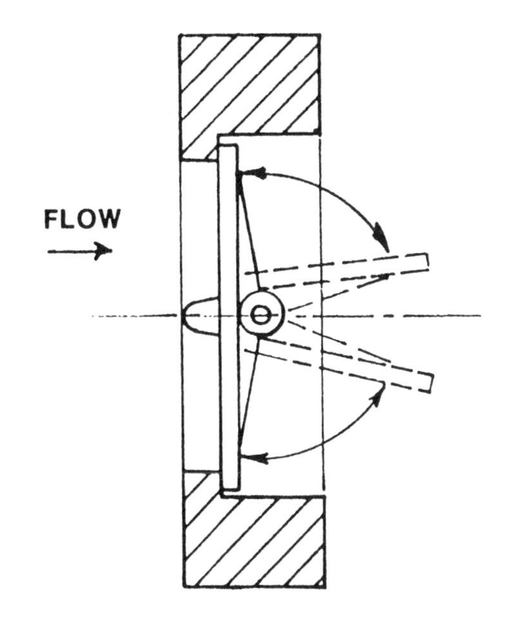

2. DOUBLE DISC CHECK

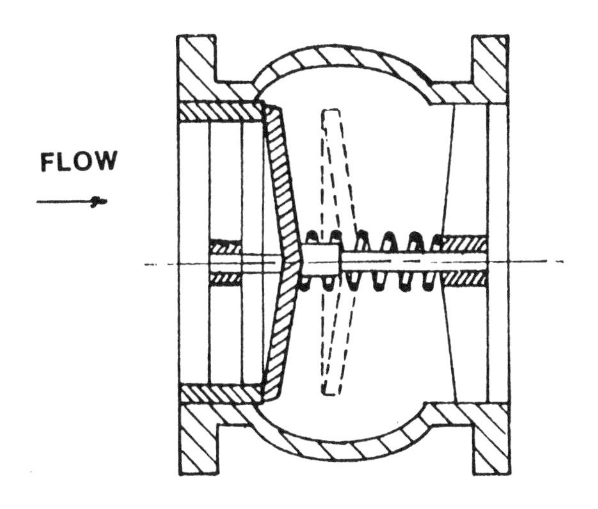

3. GLOBE CHECK

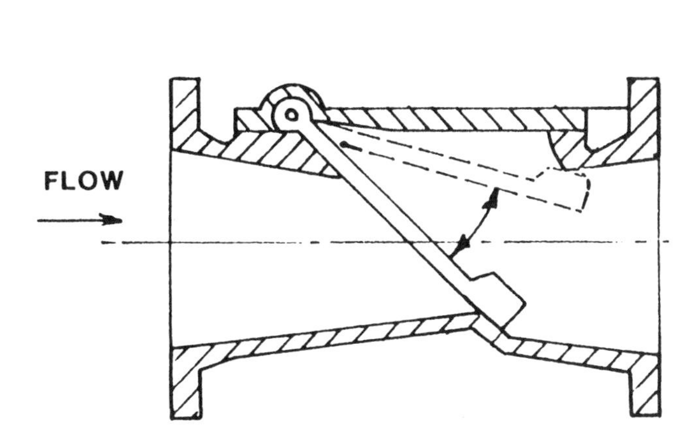

4. SLANTING DISC CHECK

**NOTE: ALL VALVES ARE IN THE CLOSED POSITION.
THE DASHED LINES SHOW THE OPEN POSITIONS.**

Fig. 3.7 Types of check valves

the atmosphere. As pump pressure gradually decreases, the main line check valve closes slowly, effectively preventing shock or slam during the pump stopping cycle. When the pump control valve is wide open, the limit-switch assembly breaks the pump motor circuit and the pump stops.

Both the normally closed and normally open pump control valves are hydraulically operated, diaphragm-type globe or angle valves powered by line pressure. While somewhat different in installation and operation, the purpose of both is much the same. They eliminate line surge when the pump is started and stopped. Both valves are electrically interlocked to the pump motor so that the valve is gradually opened or closed in coordination with the pump, thus protecting the line from uncontrolled surge.

3.212 Foot Valves

A foot valve is often placed at the inlet to the pump suction line in a well. The foot valve can serve several important purposes:

1. Maintains the *PRIME*[27] on the pump,
2. Prevents reversal of flow into the well (which could stir up sand) when the pump shuts off, and
3. Eliminates problems created when there is no means of discharging air in the pump column and the air enters the system.

3.22 Flowmeters

Good waterworks practice requires you to know the amount of water being delivered to the system. If chemicals are being applied at the well pumping station, then it is necessary to know the rate at which water is being pumped to the system in order to calculate the correct feed rate. A flowmeter is used to measure this flow.

Well pumps are usually designed for a certain rated capacity (flow rate) pumping against a specified head (system pressure). Since the system pressure against which a well pump must operate changes with system demand (usage), the amount of water delivered to the system usually deviates from the rated capacity given for the pump. Thus, a 1,000 gallon per minute (5,540 cu m/day) pump is supposed to deliver, under normal conditions of specified head and pipe friction, 1,000 gallons of water per minute to the discharge end of the pump. Since this ideal condition is seldom experienced in actual operation, the rated capacity of the pump is not an accurate measurement of the water delivered. Flowmeters are used to determine the amount of water actually being delivered by the pump.

Flowmeters in pipes will produce accurate flowmeter readings when the meter is located at least five pipe diameters distance downstream from any pipe bends, elbows, or valves and also at least two pipe diameters distance upstream from any pipe bends, elbows, or valves. Flowmeters also should be calibrated in place to ensure accurate flow measurements.

Many different types of flowmeters are available today such as positive displacement, propeller, turbine, orifice plate, and electronic sensor types. The most commonly used flowmeter for well pump applications is the propeller or turbine type, with magnetic drive, equipped with a register for total flow and a rate of flow indicator. This type of flowmeter has an accuracy of plus or minus 2 percent of actual flow within a specified range. Accessory features include remote indicator/totalizer plus an onsite strip recorder. In variable-flow well pump installations, this type of meter can be equipped to pace (regulate) auxiliary chemical feeders or chlorinators.

QUESTIONS

Write your answers in a notebook and then compare your answers with those on page 126.

3.2A What is the purpose of a check valve?

3.2B List the important features of a check valve.

3.2C What is the purpose of pump control valves?

3.2D Why must flows be known when chemicals are being applied at a well pumping station?

3.23 Sand Traps and Sand Separators

Sand particles should be prevented from entering the distribution system. Nearly all wells produce some sand and tests have shown it is the sand particles larger than 200 *MESH*[28] (74 microns) that cause the most trouble in a distribution system. Almost all sand found in well water is this size or coarser. These particles can cause reduced pump efficiencies due to worn impellers, sanded water mains, excessive meter wear and plugging, wearing of household plumbing fixtures and appliances, and customer complaints.

3.230 Sand Traps

Conventional sand traps usually consist of a large tank equipped with a series of internal baffles or chambers. The tank is mounted on the discharge side of the well pump. Sand particles settle to the bottom of the tank by the force of gravity and are removed by means of blowoff valves or underdrain systems. *HYDROPNEUMATIC*[29] pressure tanks are sometimes modified with internal baffles and can function, to a limited degree, as a sand trap. Generally speaking, sand traps are costly and inefficient.

3.231 Sand Separators

Most sand separators are designed around the principle of centrifugal force and are relatively effective in removing fine sand, scale, and similar materials from the water.

The centrifugal sand separator is a specifically designed hydraulic centrifuge operating on energy supplied by the well pump. Take a close look at Figure 3.8 as you read about how the separator works. Sand-laden water enters the cylindrical chamber through the feed entry. Inside the chamber the water

[27] *Prime. The action of filling a pump casing with water to remove the air. Most pumps must be primed before start-up or they will not pump any water.*

[28] *Mesh. One of the openings or spaces in a screen or woven fabric. The value of the mesh is usually given as the number of openings per inch. This value does not consider the diameter of the wire or fabric; therefore, the mesh number does not always have a definite relationship to the size of the hole.*

[29] *Hydropneumatic (HI-dro-new-MAT-ick). A water system, usually small, in which a water pump is automatically controlled (started and stopped) by the air pressure in a compressed-air tank.*

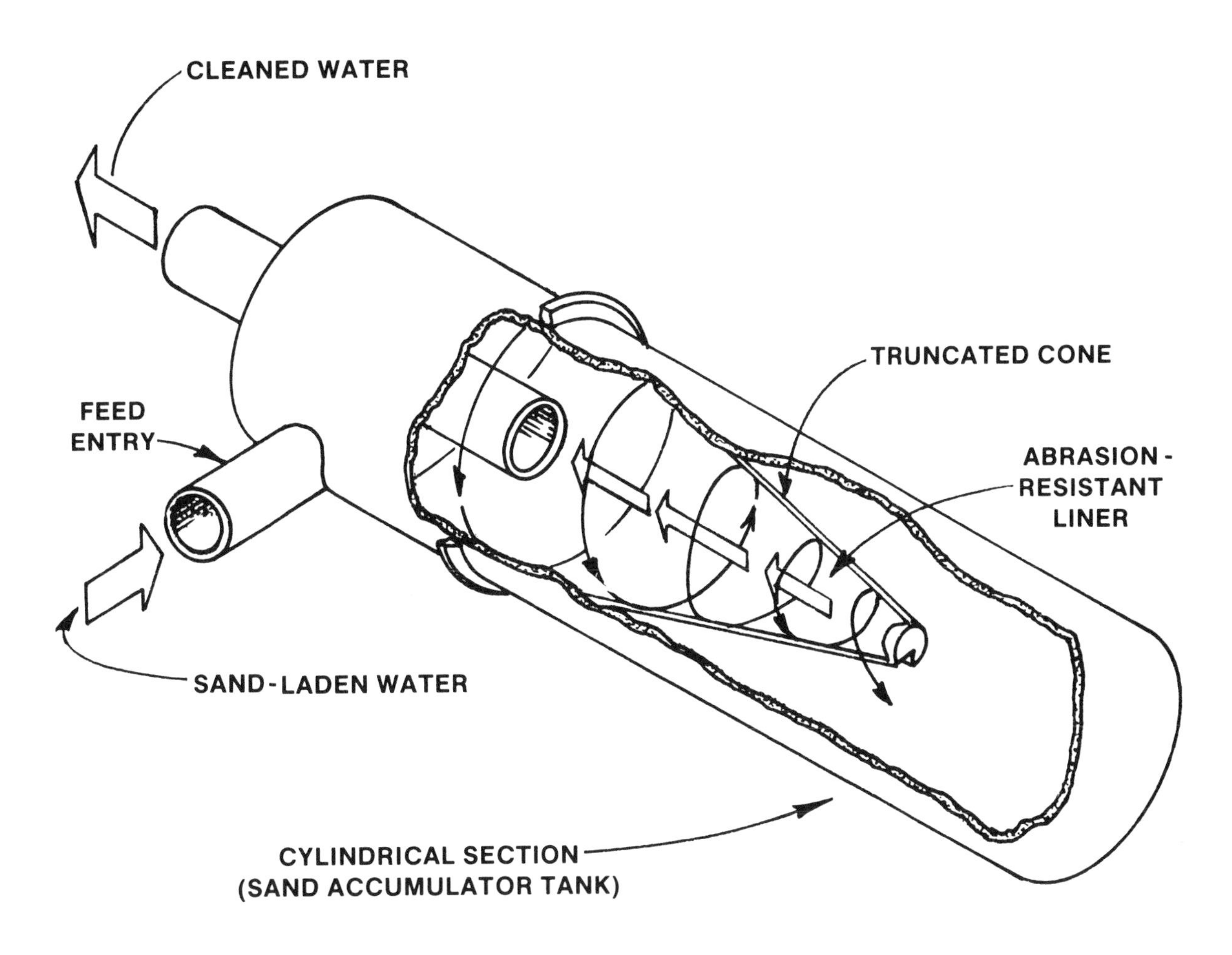

Fig. 3.8 Typical sand separator

flows in a circular motion around the outside of the cylinder. As the water spirals inward toward the entrance to the cleaned water exit pipe, extremely high forces act on the sand particles to rapidly force them toward the wall of the cylinder. Once the sand particles reach the cylinder wall, they progress by a downward spiraling action toward the tip of the truncated cone and are finally discharged through an *ORIFICE*[30] (slit in cone) into the accumulator tank, or underflow chamber. Sand is periodically drained from the accumulator tank or chamber by either manual or automatic means. Centrifugal sand separators are usually capable of removing 95 percent of the plus 200 mesh sand particles entering the unit.

Sanitary hazards associated with sand separators or sand traps are generally limited to the discharge system from the accumulator tank. Typically, the sand/water mixture from the accumulator tank is discharged to a sand recovery basin where the sand particles settle out and the sand-free water is disposed of to the community storm sewer or sanitary sewer main. To provide the necessary backflow protection, an *AIR GAP*[31] is required on the discharge line between the accumulator tank and the sand disposal basin or structure. Backflow protection is usually not required if the discharge line from the accumulator has a free discharge to the atmosphere and ends at least 12 inches (300 mm) above the surrounding ground elevation.

For additional information on how to solve problems caused by sand, see Section 3.7, "Sand in Well Water Systems."

3.24 Tank Coatings

Steel water storage tanks and hydropneumatic pressure tanks are generally painted on the interior surface with some type of protective coating to extend the useful life of the tank, prevent discolored water from entering the distribution system, and to facilitate cleaning of the tank.

Numerous interior tank coating materials are currently available. Many of these coatings give excellent results if the metal surfaces are properly prepared, the proper material for the job is selected, the material is applied according to the manufacturer's specifications, and it is allowed to dry thoroughly between coats and after the final coat. Adequate curing time is essential with all types of coatings and may require special

[30] *Orifice (OR-uh-fiss). An opening (hole) in a plate, wall, or partition. An orifice flange or plate placed in a pipe consists of a slot or a calibrated circular hole smaller than the pipe diameter. The difference in pressure in the pipe above and at the orifice may be used to determine the flow in the pipe.*

[31] *Air Gap. An open vertical drop, or vertical empty space, that separates a drinking (potable) water supply to be protected from another water system in a water treatment plant or other location. This open gap prevents the contamination of drinking water by backsiphonage or backflow because there is no way raw water or any other water can reach the drinking water supply.*

techniques (such as forced ventilation and extended drying time) to ensure that *VOLATILES*[32] have been dissipated from the coating.

Before applying any coating, the water system operator must verify that the selected paint, lining, coating, thinners, and conditioners contain no toxic ingredients which might contaminate the water in the tank. The American Water Works Association has developed an excellent standard for outside and inside paint systems for tanks titled *AWWA D102.*

QUESTIONS

Write your answers in a notebook and then compare your answers with those on page 126.

3.2E Why should sand particles be prevented from entering the distribution system?

3.2F Why are sand traps not recommended for the removal of sand from well water?

3.2G What types of materials may be removed from water by a sand separator?

3.2H What sanitary hazards are associated with sand traps or sand separators?

3.2I Why are steel water storage tanks and hydropneumatic pressure tanks painted on the interior surface with some type of protective coating?

3.25 Surge Suppressors

In certain situations a booster pump is used in conjunction with the main well pump. This type of operation usually involves pumping a small amount of water to a specific pressure zone in the distribution system that operates at a much higher pressure or higher elevation than the main well pump system.

Frequent start/stop operation of the booster pump, combined with the higher system pressure, could create problems such as pulsation, vibration, and possible damage to the piping network. To correct this problem, surge suppressors are sometimes installed on the discharge side of the booster pump. Their function is to absorb shock waves in the fluid (water) system and prevent their transmittal through the line.

One relatively simple surge suppressor design consists of a flexible, fully enclosed rubber bladder or diaphragm mounted within a steel shell and connected to the piping system with flanged fittings. The upper chamber is filled with air or nitrogen gas. When a surge occurs, flow from the main system line enters the surge suppressor through a specially designed metering port where a portion of the *KINETIC ENERGY*[33] is dispelled. The dampened pressure surge then continues into the lower fluid chamber where the remaining kinetic energy is dissipated by compressing the air or gas in the upper pressurized chamber. When the compressed air or gas expands forcing the fluid back into the main system, more residual kinetic energy is lost in the controlled flow back through the metering port.

Hydropneumatic tanks also serve as surge suppressors and can be used in a direct line of flow mode or connected to the booster pump discharge line by means of a tee fitting.

3.26 Air and Vacuum Valves

An air and vacuum valve, commonly referred to as an air release/vacuum breaker valve, has a large venting orifice and is used to exhaust large quantities of air very rapidly from a deep well pump column when the pump is started. After the air has been exhausted from the pump column, the valve closes and remains closed. When the pump stops, the valve will immediately open to allow air to re-enter the pump column and prevent a vacuum from developing.

The orifice of the valve must be equipped with an approved vent assembly consisting of a downturned, screened opening mounted at least 12 inches (300 mm) above the finished surface of the well lot or above any anticipated high-water (flood) level. The open end of the vent must be equipped with holes or fine mesh screen to keep insects out of the well.

The valve should be properly sized (usually a minimum of two inches (50 mm)) to allow the unrestricted flow of air into and out of the pump column. Improperly sized or poorly constructed vents could cause the well pump base gasket or sealing material to be blown out on pump start-up. This in turn could allow insects and foreign debris to be drawn into the well when the pump stops.

3.27 Pressure Relief Valves

Pressure relief valves should be installed on all hydropneumatic tanks. Excessive internal pressures (surge pressures), often referred to as *WATER HAMMER*,[34] could cause the tank to explode or rupture. A typical pressure relief valve is illustrated in Figure 3.9.

---

[32] *Volatile (VOL-uh-tull). (1) A volatile substance is one that is capable of being evaporated or changed to a vapor at relatively low temperatures. Volatile substances also can be partially removed by air stripping. (2) In terms of solids analysis, volatile refers to materials lost (including most organic matter) upon ignition in a muffle furnace for 60 minutes at 550°C. Natural volatile materials are chemical substances usually of animal or plant origin. Manufactured or synthetic volatile materials such as ether, acetone, and carbon tetrachloride are highly volatile and not of plant or animal origin.*

[33] *Kinetic Energy. Energy possessed by a moving body of matter, such as water, as a result of its motion.*

[34] *Water Hammer. The sound like someone hammering on a pipe that occurs when a valve is opened or closed very rapidly. When a valve position is changed quickly, the water pressure in a pipe will increase and decrease back and forth very quickly. This rise and fall in pressures can cause serious damage to the system.*

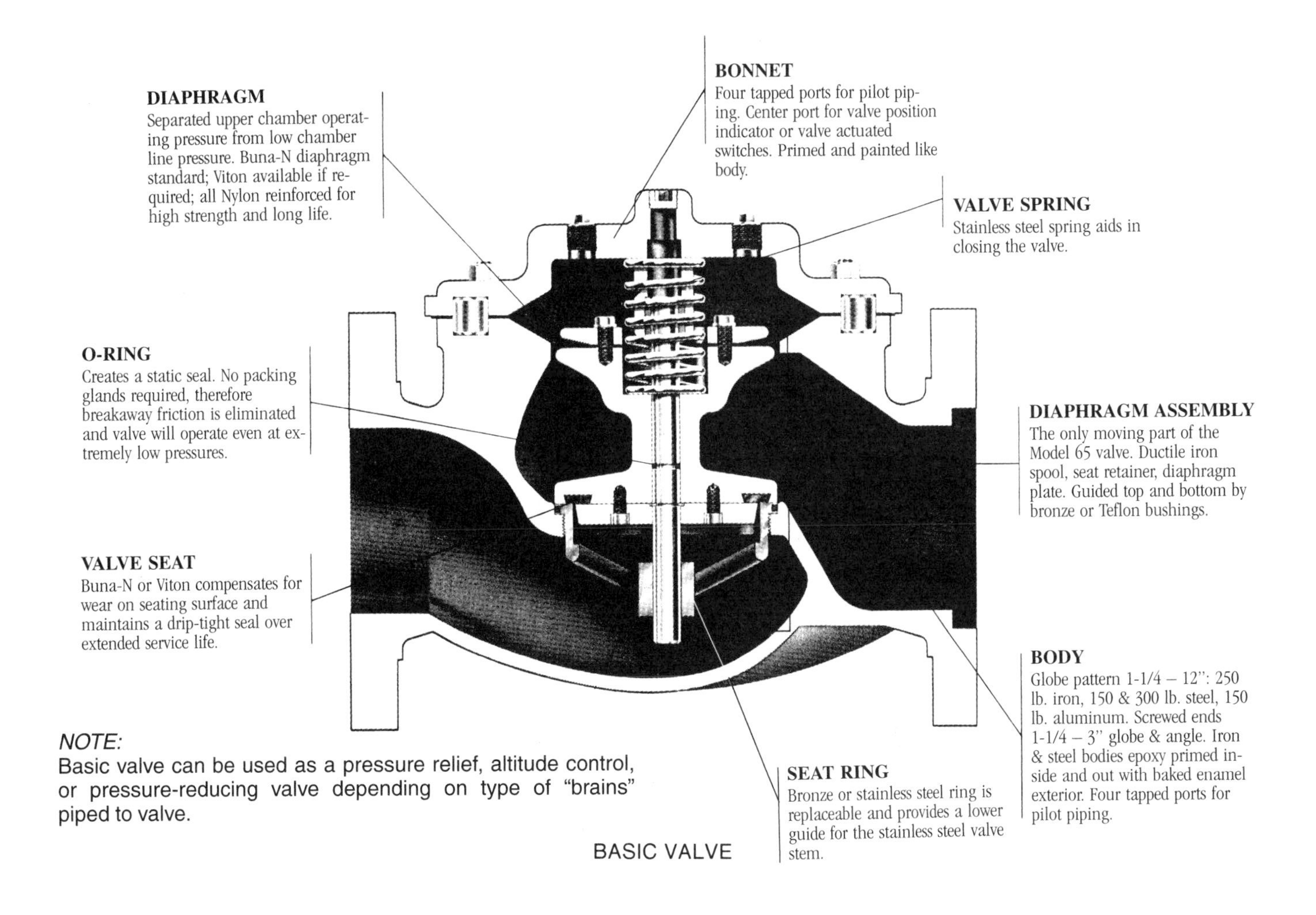

DIAPHRAGM
Separated upper chamber operating pressure from low chamber line pressure. Buna-N diaphragm standard; Viton available if required; all Nylon reinforced for high strength and long life.
BONNET
Four tapped ports for pilot piping. Center port for valve position indicator or valve actuated switches. Primed and painted like body.
VALVE SPRING
Stainless steel spring aids in closing the valve.
O-RING
Creates a static seal. No packing glands required, therefore breakaway friction is eliminated and valve will operate even at extremely low pressures.
DIAPHRAGM ASSEMBLY
The only moving part of the Model 65 valve. Ductile iron spool, seat retainer, diaphragm plate. Guided top and bottom by bronze or Teflon bushings.
VALVE SEAT
Buna-N or Viton compensates for wear on seating surface and maintains a drip-tight seal over extended service life.
BODY
Globe pattern 1-1/4 – 12": 250 lb. iron, 150 & 300 lb. steel, 150 lb. aluminum. Screwed ends 1-1/4 – 3" globe & angle. Iron & steel bodies epoxy primed inside and out with baked enamel exterior. Four tapped ports for pilot piping.
NOTE:
Basic valve can be used as a pressure relief, altitude control, or pressure-reducing valve depending on type of "brains" piped to valve.
SEAT RING
Bronze or stainless steel ring is replaceable and provides a lower guide for the stainless steel valve stem.
BASIC VALVE

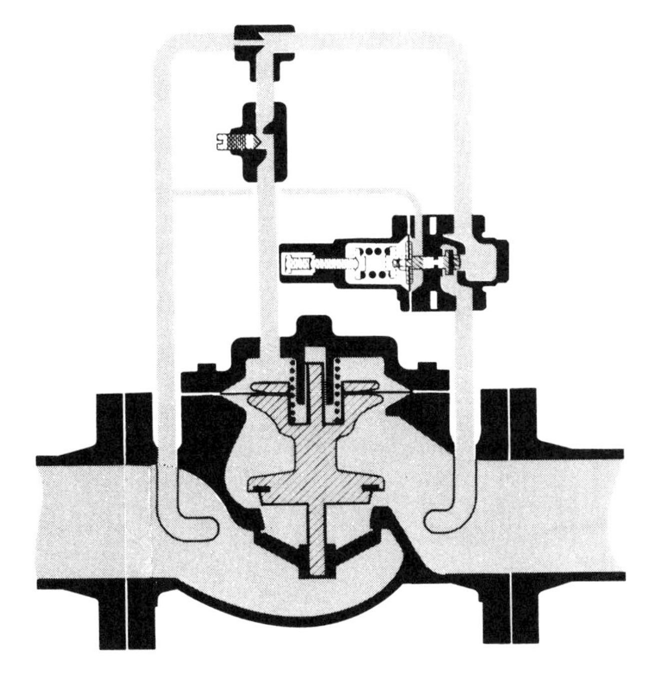

Fig. 3.9 Pressure relief valve
(Permission of OCV Control Valves)

Water hammer is frequently encountered in water supply and treatment facilities. Water hammer may be caused by the quick opening or closing of valves such as fire hydrant valves, the start-up or shutdown of pumps, and other conditions that could cause a pulsation of pressures above and below the operating pressure, and a rapid acceleration or deceleration of the velocity of flow. The increased pressure may be several times the normal operating pressure, which can seriously damage hydropneumatic tanks, valves, and the piping network.

Water hammer can be controlled or eliminated by using pressure relief valves. These automatic valves have a spring tension preset to a certain operating pressure. In the case of well pump installations, the valve is usually set at 125 pounds per square inch (862 kPa or 8.8 kg/sq cm). Pressure in excess of this setting will open the valve, allowing air or water to escape and preventing an excess pressure buildup.

3.28 Air Chargers

3.280 Methods of Air Charging

Several methods are available for adding air to a hydropneumatic tank. *ONE METHOD* is to use the air in the well pump column. Instead of discharging the air in the column to atmosphere when the pump starts, the column of air is forced into the tank. With this method, the tank must be equipped with an air release valve and a riser pipe extending down to the midpoint of the tank. The riser pipe is designed to maintain an approximate 50:50 air:water ratio in the tank. This method should be used with caution due to the possibility that during periods when the well pump is cycling on/off at a frequent rate, excess air may not be discharged fast enough through the air release valve and the tank may become oversaturated with air. If this occurs, air could be pumped out into the distribution system with resultant problems and consumer complaints.

A *SECOND METHOD* of air charging a tank is to manually add air to the tank as needed using a portable air compressor. This technique can be very time consuming and costly.

The *MOST SATISFACTORY METHOD* is to automatically add air by means of an on-site, permanent, air-charging device. Two general types are currently available: the hydraulic principle type and the motor-driven air compressor type.

3.281 Hydraulic Principle Air Charger (Figure 3.10)

The hydraulic principle air charger uses the water pressure of the tank to force air into the tank. The unit normally mounts vertically on one end of a horizontal pressure tank. This unit consists of a closed cylindrical tube section, an upper and lower float, a pilot valve, a one-way air valve, plus a water supply line with sand strainer connected near the bottom of the tank, and an air discharge line connected to the top of the tank.

The unit adds air to the tank on the upward compression stroke and releases water on the downward exhaust stroke. The unit continues to cycle until the preset air-to-water ratio is reached. Approximately 2 $^{2}/_{3}$ gallons (10 liters) of water is discharged from the unit during each complete cycle. The drain line from this unit should not be directly connected to a sewer line. An air-gap separation is recommended to protect the tank from backsiphonage.

3.282 Air Compressor Air Charger (Figure 3.11)

This type of air charger is composed of three primary elements: the air compressor, the liquid level switch, and the pressure switch. The liquid level switch senses the water level in the tank by an electrode suspended into the tank. When the water level in the tank exceeds the preset level, the liquid level switch starts the compressor and air is pumped into the tank. The compressor runs until either the pressure rises sufficiently to open the unit's pressure switch or the water level descends below the preset level.

The unit mounts directly on top of the hydropneumatic tank and is protected by a weatherproof cover. No drain lines are required and the air entering the compressor is filtered.

The two types of air chargers require routine maintenance and have not been trouble-free. However, the desirability and advantages of automatically maintaining the proper air-to-water ratio in a hydropneumatic tank far offset the operational problems of these two types of air charging devices.

Routine maintenance on the hydraulic principle air charger consists of draining the "Y" pattern sand strainer on the inlet line as needed and maintaining the freeze-protection equipment if mounted outdoors.

Routine maintenance on the air compressor air charger consists of lubrication, cleaning the filter element on the compressor air inlet as needed, and maintaining the correct tension on the compressor belts. Set up a maintenance program based on the specific type of facilities that you are working with. Use the manufacturers' bulletins and maintenance procedures as guidelines.

3.29 Hydropneumatic Pressure Tank Systems

3.290 Applications

Hydropneumatic pressure tank systems can be used very successfully in well pump operations to maintain adequate

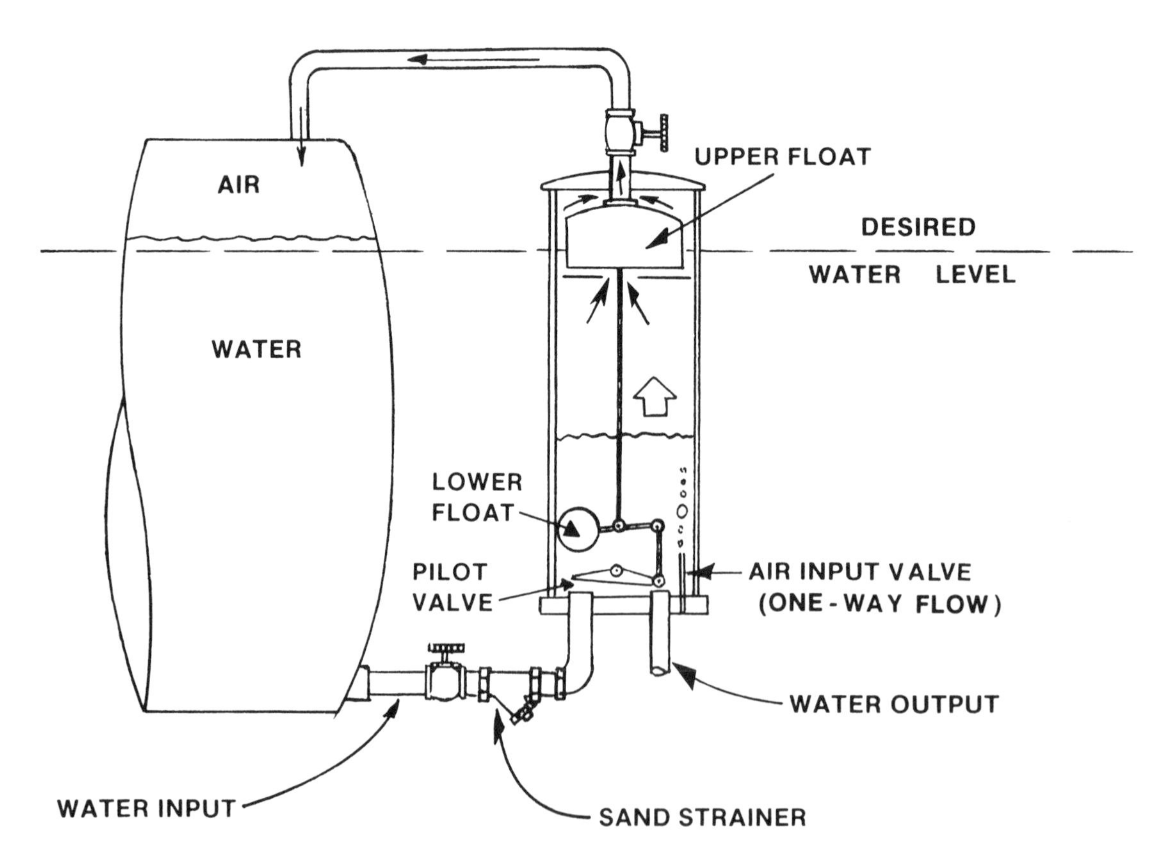

Fig. 3.10 Typical hydraulic air charger

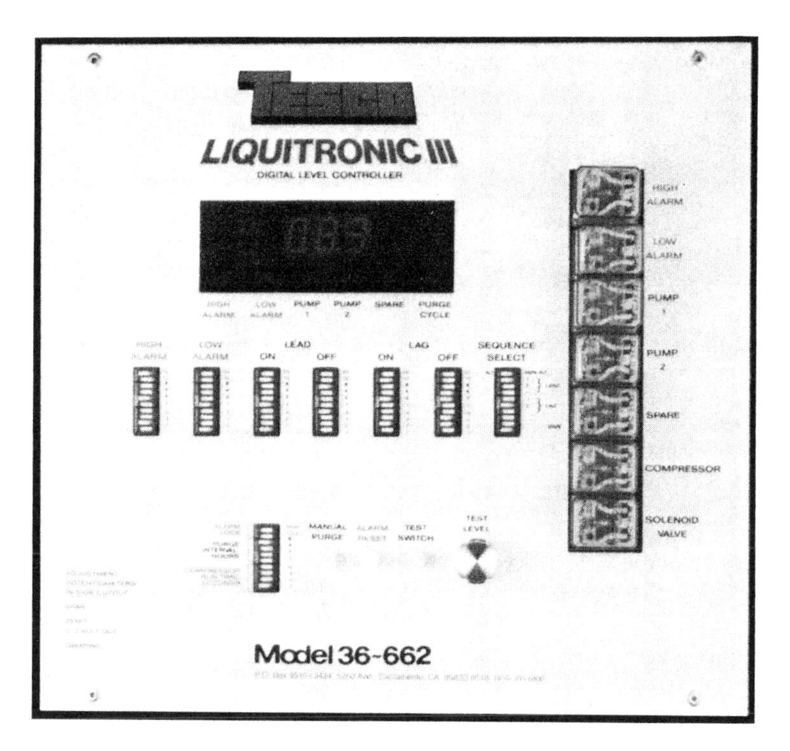

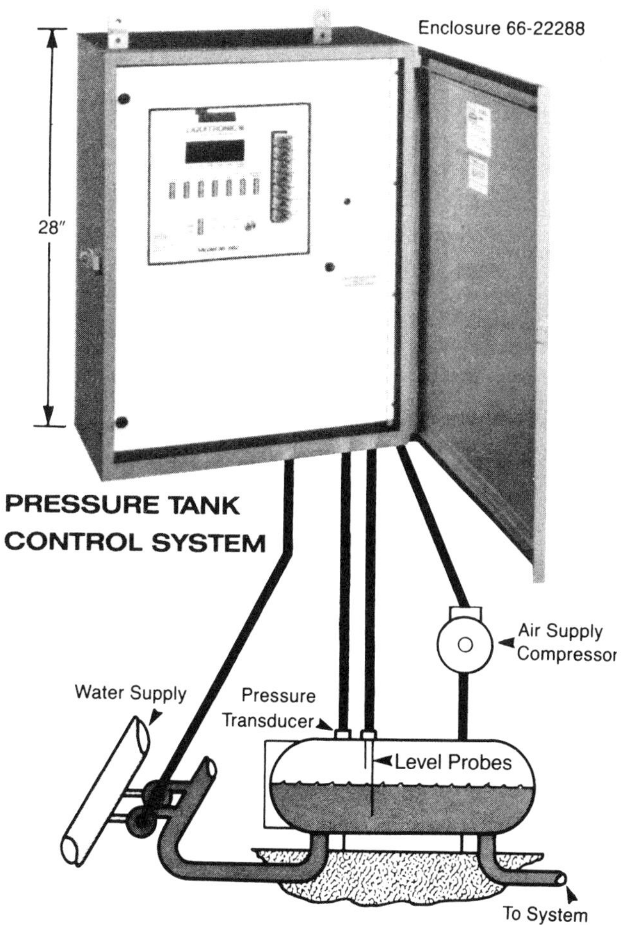

Fig. 3.11 Air compressor air charger
(Permission of TESCO, Sacramento, CA)

pressure throughout distribution systems. In areas where the topography is generally flat, hydropneumatic systems are quite common. They can be used for small water systems with one or two wells, for larger water systems with many wells strategically located in a grid system, or to supply service to isolated areas. Hydropneumatic pressure tank systems can also be integrated with gravity storage systems, although this is not a widespread practice.

3.291 Tank Size

Hydropneumatic tanks range in capacity from 40-gallon (150-liter) tanks serving single families to 21,000-gallon (80,000-liter) or larger size tanks for public water systems. The tank size will depend on many factors such as geographic location, pump capacity, peak demand, pump cycling rate, economics, past operational experience, and the amount of property available to the water supplier for plant facilities.

3.292 Operation

When hydropneumatic pressure tank systems are designed, the best operating pressure differential, the control levels in the tank, the pumping differential and the tank efficiency, and the tank size are determined by the designer. This information and procedures for any necessary adjustments should be provided in the operation and maintenance instructions for the system.

QUESTIONS

Write your answers in a notebook and then compare your answers with those on page 126.

3.2J Under what circumstances might a booster pump be installed?

3.2K What is the purpose of surge suppressors that are sometimes installed on the discharge side of a booster pump?

3.2L What is the purpose of an air release/vacuum breaker valve?

3.2M Why are pressure relief valves installed on hydropneumatic tanks?

3.2N What is the purpose of an air charger?

3.2O The size of a hydropneumatic tank is influenced by what factors?

3.3 WELL MAINTENANCE AND REHABILITATION

3.30 Importance of Well Maintenance

If the well casing, well screen, filtering material, and other appurtenances of a well are properly designed and correctly installed by a qualified well driller, maintenance of a well will be minimized. However, wells, like all other waterworks facilities, need periodic routine maintenance in the interest of a continuous high level of performance and a maximum useful life. The usual attitude toward maintenance of wells is "out of sight — out of mind." Consequently, very little or no attention is paid to wells until problems reach crisis levels, often resulting in the complete loss of the well. The importance of a routine maintenance program for the prevention, early detection, and correction of problems that reduce well performance and useful life cannot be overemphasized. A maintenance program can pay handsome dividends to a well owner and will certainly result in long-term benefits that exceed the cost of implementation and continuation of the program.

3.31 Factors Affecting the Maintenance of Well Performance

3.310 Adverse Conditions

The factors affecting the maintenance of well performance or yield are numerous. Care should be taken to distinguish between factors associated with the normal wearing of pump parts and those directly associated with changing conditions in and around the well. A perfectly functioning well, for example, can show a reduced yield because of a reduction in the capacity of the pump due to excessively worn parts. On the other hand, the excessive wearing of pump parts may be due to the pumping of sand that entered the well through a corroded well screen. Corrosion may reduce the pump capacity but, at the same time have little or no effect on a properly designed well. There will be some overlap in what is called well maintenance and what is called pump maintenance. Conditions (which we'll consider well maintenance problems) that should be guarded against are overpumping and lowering of the water table, clogging or collapse of a screen or perforated section, and corrosion or incrustation.

3.311 Overpumping

Overpumping was discussed in Section 3.021, "Overdraft," so we'll only summarize the problems here. Overpumping an aquifer can damage the aquifer by reducing the storage and production capacity of groundwater systems. The net result is consolidation of the water-bearing formations which provides less storage space, a lowering of the water table, and a reduced yield from the well. Other problems that may develop from overpumping or installing the pump suction too high include pumping air and also water cascading into the well. The well production rate is normally determined by the well driller and hydrogeologist at the time the well is drilled and developed.

Overpumping may cause sand pumping. This will subject the pump to excessive wear, which over time can reduce its operating efficiency. Under severe conditions, the pump may become sand locked, either during pumping or after shut-off. Should sand locking occur, the pump must be pulled, disassembled, cleaned, and repaired, if necessary, before being placed back into service.

3.312 Clogging or Collapse of Screen

A decrease in the capacity of a well most commonly results from clogging of the well screen openings and the water-bearing formation immediately around the well screen by incrusting deposits. These incrusting deposits may be of the hard, cement-like form typical of the carbonate and sulfate compounds of calcium and magnesium, the soft, sludge-like forms of the iron and manganese hydroxides, or the gelatinous slimes of iron bacteria. Iron may also be deposited in the form of ferric oxide with a reddish-brown, scale-like appearance. Less common is the deposit of soil materials such as silt and clay. More will be said about incrustation later.

Well screens are designed to fit individual well formations and are composed of numerous slits as shown on Figure 3.12. As can be seen from this figure, if the slits in the casing become incrusted and blocked, the available area for water to move through is severely limited. If corrosion should occur, the slits may enlarge enough for grit and sand to be carried into the well along with water. This, in turn, will damage the pump and appurtenances. If the corrosion is severe enough, total collapse of the screen could occur.

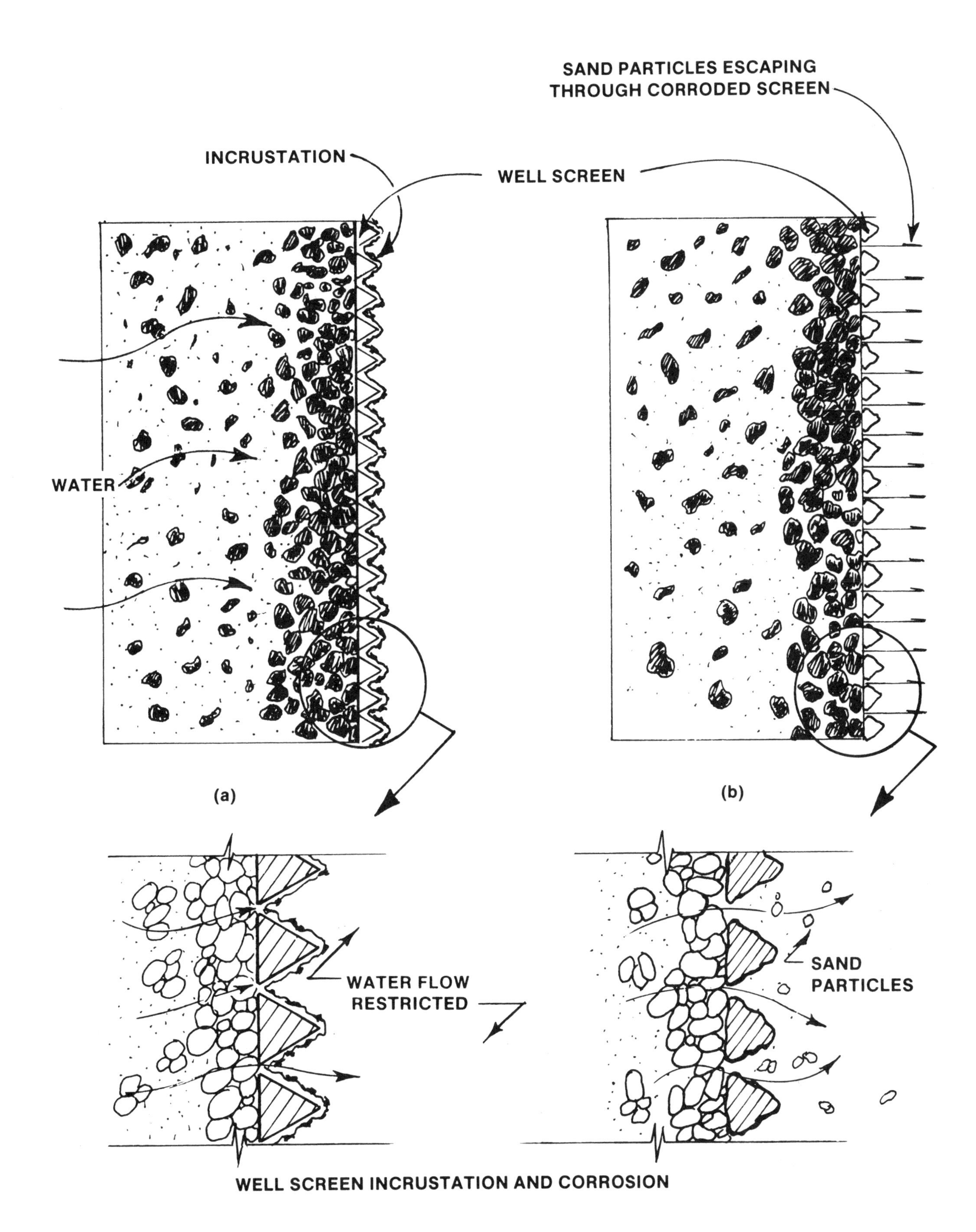

Fig. 3.12 Well screen incrustation and corrosion (also see Figure 3.52, page 149)

3.313 Corrosion or Incrustation

Corrosion is a process which results in the gradual decomposition or destruction of metals. Corrosive waters are usually acidic and may contain relatively high concentrations of dissolved oxygen which is often necessary for and increases the rate of corrosion. High concentrations of carbon dioxide, total dissolved solids, and hydrogen sulfide (with its characteristic odor of rotten eggs) are other indications of a possibly corrosive water.

Besides water quality, other factors such as velocity of flow and dissimilarity of metals also contribute to the corrosion process. The greater the velocity of flow, the greater is the removal of the protective corrosion end products from the surface of the metal and hence the exposure of that surface to further corrosion. This is another important reason for keeping the velocity through screen openings within acceptable limits. The use of two or more different types of metals such as stainless steel and ordinary steel, or steel and brass or bronze should be avoided whenever possible. Corrosion is usually greatest at the points of contact of the different metals or where they come closest to contact.

Casing failure by corrosion ruins a well as fast as failure of the screen. Failure can cause the introduction of clay and polluted or otherwise unsatisfactory water into the well. Corrosive well waters have been known to destroy steel casings in less than six months, thus ruining many wells.

Incrustation, unlike corrosion, results not in the destruction of metal, but in the deposition of minerals on the metal and in the aquifer immediately around a well. Physical and chemical changes in the water in the well and the adjacent formation cause dissolved minerals to change to their insoluble states and settle out as deposits. These deposits block the screen openings and the pore spaces immediately around the screen with a resulting reduction in the yield of the well.

Incrusting waters are usually alkaline or the opposite of corrosive waters, which are acidic. Excessive carbonate hardness is a common source of incrustation in wells. Scale deposits of calcium carbonate (lime scale) occur in pipes carrying hard waters. Iron and manganese, to a lesser extent, are other common sources of incrustation in wells. Iron causes characteristic reddish-brown deposits while those of manganese are black.

Iron bacteria are often associated with groundwater that contains iron. These tiny living organisms aid in the deposition of iron but are not dangerous to health. However, iron bacteria produce accumulations of slimy, jelly-like material which block well screen openings and aquifer pore spaces.

QUESTIONS

Write your answers in a notebook and then compare your answers with those on pages 126 and 127.

3.3A List three major well maintenance problems.

3.3B How can overpumping damage a well?

3.3C What kinds of deposits can develop on well screens?

3.3D Why should the use of two or more different types of metals be avoided in a well?

3.3E What is the difference between incrusting and corrosive waters?

3.32 Preventive Maintenance and Repairs

The best possible procedure for a good maintenance program is adequate recordkeeping. This aspect of the operator's responsibility cannot be overemphasized. Regular water level measurements in the well before and after pumping, flow rates, water quality samples, length of time pumping, and accurate data on pump repairs and causes are a few of the records that should be routine. The records should be kept neatly and in logical order. This collected data will be an invaluable tool in the hands of an engineer, hydrogeologist, chemist, or skilled operator. Operators should monitor this data to look for early warnings of potential problems such as overpumping. In areas where extensive oil and/or gas exploration is being conducted, operators should be alert for changes in water quality which could indicate pollution of an aquifer.

3.33 Casing and Screen Maintenance

The types of materials that go into the well are very important. Wells composed of materials with little or no resistance to corrosion can be destroyed beyond usefulness by a highly corrosive water within a few months of completion. This will be the case no matter how excellent the other aspects of design, construction, and maintenance. A poor selection of materials can also result in collapse of the well due to inadequate strength. All these factors have a considerable influence on the useful life of a well.

3.330 Surging

Surging, which is a form of plunging, is a procedure used for opening pores in the screen and for cleaning the gravel pack around the screen. A typical surge arrangement is shown in Figure 3.13.

Surging is commonly used for new well development to remove sand from the area around the well screen. However, it is also an effective procedure in combating incrustation when used with acid treatment. Care must be exercised when using surge plungers within the screen area itself and in wells in which aquifers contain large amounts of clay. In cleaning screens, the plunger can become sand-locked by the settling of sand above it. In cleaning the gravel pack around the screen, the action of the plunger can cause the clay to plaster over the screen surface. Plungers should be used only under the supervision of someone experienced in their use.

3.331 High-Velocity Jetting

High-velocity jetting is the spraying of water at a high velocity. This is an effective form of backwashing which, when used with acid treatment, is another procedure for removing incrustation from well screens and casings. Jetting can also be used for re-opening the pores of the aquifer and removing sand from the immediate vicinity of the well screen.

A simple form of jetting tool for use in wells is shown in Figure 3.14. An appropriately sized coupling with a steel plate

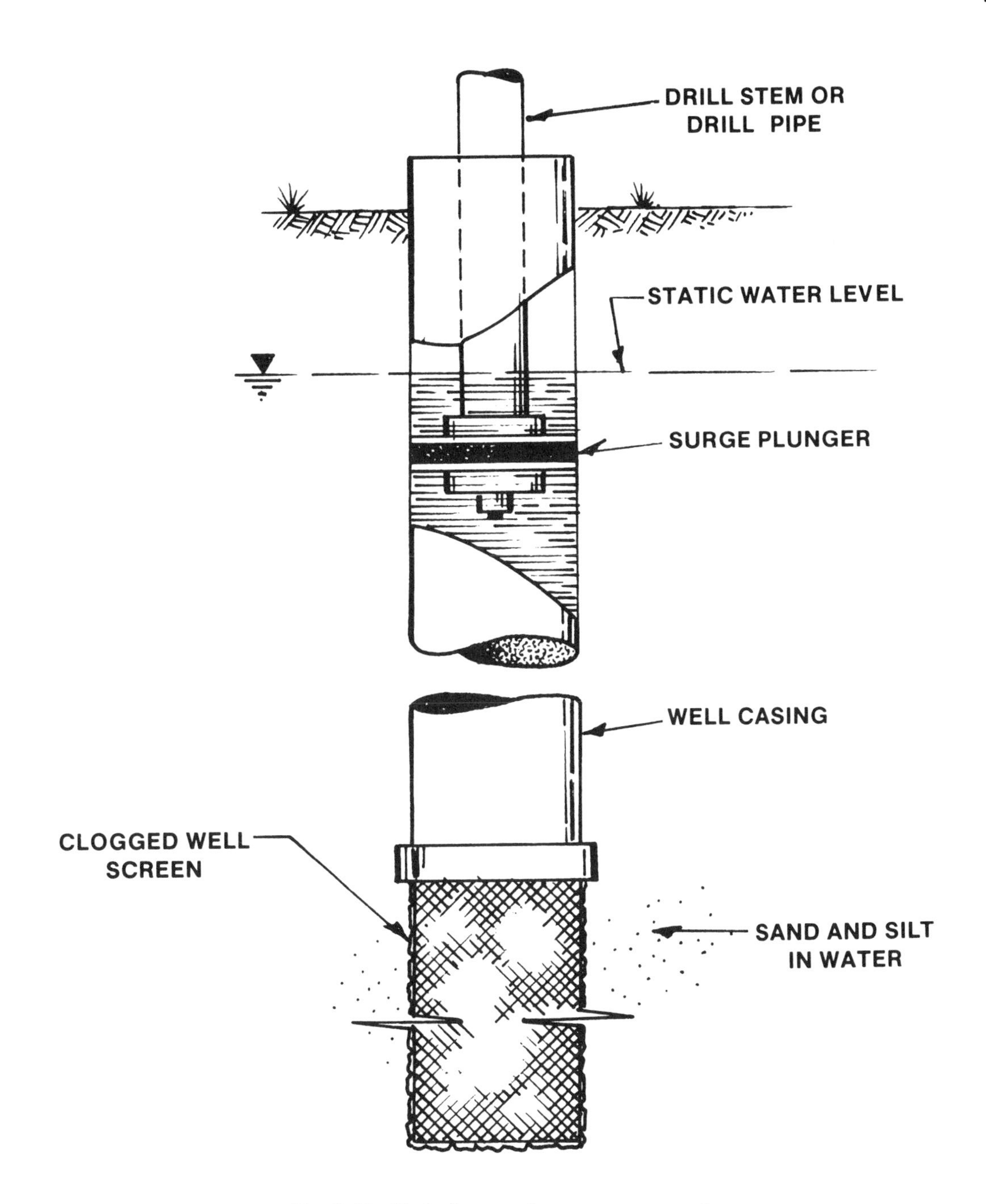

Fig. 3.13 Typical surge plunger arrangement

welded over one end is screwed or welded to a pipe. The jetting tool's outside diameter will vary depending on the inside diameter of the well. The maximum and minimum difference in diameter between the jet tool and well screen are two inches (50 mm) and one inch (25 mm) respectively. That is, if the well screen has a 10-inch (250-mm) inside diameter, then the jet tool should have an outside diameter between eight and nine inches (200 and 225 mm). Two to four 3/16- or 1/4-inch (5- or 6-mm) diameter holes, equally spaced around the circumference, are drilled through the full thicknesses of the coupling and the jetting pipe at a fixed distance along the coupling from the near surface of the steel plate. Better results can be obtained if properly shaped nozzles are used instead of the straight, drilled holes shown in Figure 3.14, but these are also acceptably effective.

The procedure is to lower the tool on the jetting pipe to a point near the bottom of the screen. The upper end of the pipe is connected through a swivel and hose to the discharge end of a high-pressure pump such as the mud pump used for hydraulic rotary drilling. The pump should be capable of operating at a pressure of at least 100 pounds per square inch (psi) (690 kPa or 7 kg/sq cm) and preferably at about 150 psi (1,040 kPa or 10.5 kg/sq cm) while delivering 10 to 12 gallons per minute (GPM) (0.6 to 0.8 *L*/sec) for each 3/16-inch (5-mm) nozzle or 16 to 20 GPM (1.0 to 1.25 *L*/sec) for each 1/4-inch (6-mm) nozzle on the tool. For example, a tool with two 3/16-inch (5-mm) diameter nozzles would require a pumping rate of about 20 to 24 GPM (1.25 to 1.5 *L*/sec), while a tool with three 1/4-inch (6-mm) diameter nozzles would require a pumping rate of 48 to 60 GPM (3.0 to 3.8 *L*/sec). While pumping water through the nozzles and screen into the formation, the jetting tool is slowly rotated, thus washing and developing the formation near the bottom of the well screen. The jetting tool is then raised at intervals of a few inches and the process

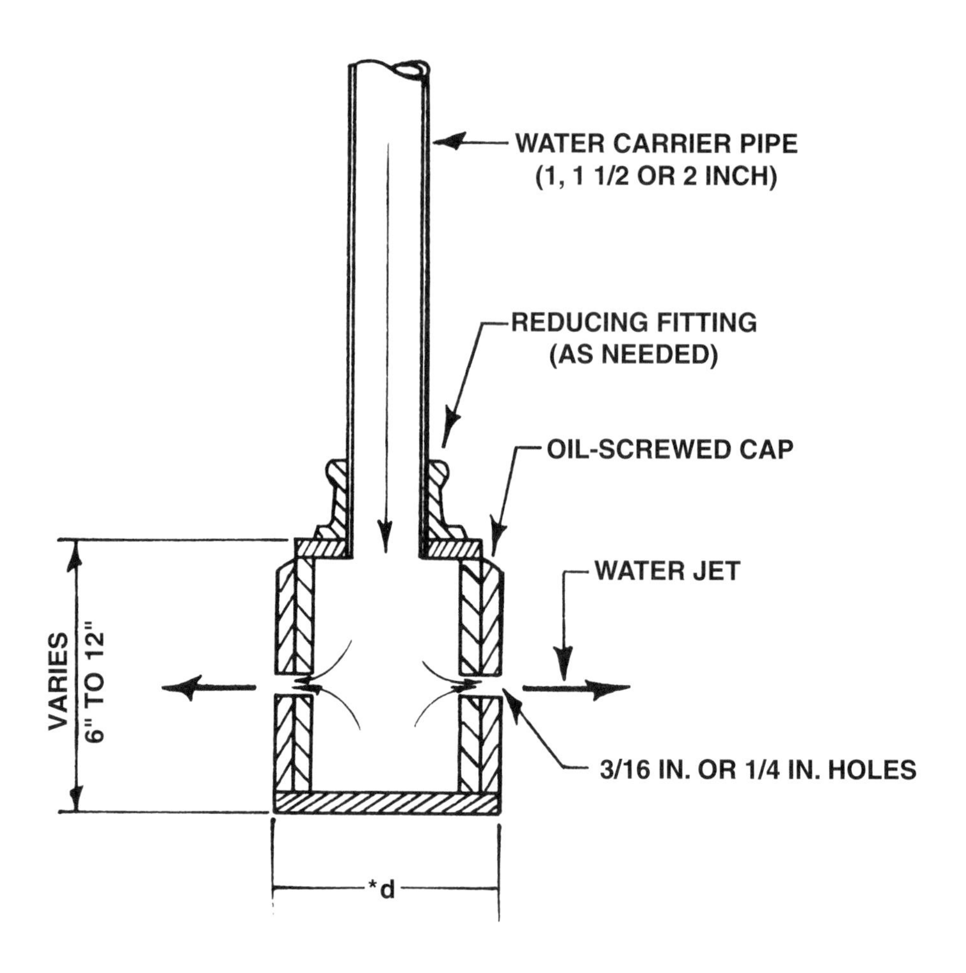

Fig. 3.14 Simple jetting tool

repeated until the entire length of screen has been backwashed and fully developed. Where possible, it is very desirable to pump the well at the same time the jetting operation is in progress.

As shown in Figure 3.15, the area of concentration of the spray is very small. Because of this concentration, the jet spraying procedure becomes one of the most effective procedures for screen cleaning and well development.

3.332 Acid Treatment

Acid treatment may be the only effective procedure available to loosen incrustation so that it may be removed from the well casing and well. Acids normally used are hydrochloric or sulfamic. Both of these acids readily dissolve calcium and magnesium carbonate, though hydrochloric acid does so at a faster rate. Strong hydrochloric acid solutions also dissolve iron and manganese hydroxides. The simultaneous use of an inhibitor serves to slow up the tendency of the acid to attack steel casing. The use of chemicals in a well requires the proper selection of well materials at the time of construction to avoid damage to the materials by the chemicals.

Hydrochloric acid should be used at full strength. Each treatment usually requires 1 1/2 to 2 times the volume of water in the screen or pipe to be cleaned. This provides enough acid to fill the area to be cleaned and additional acid to maintain adequate strength as the chemical reacts with the incrusting materials. Figure 3.16 illustrates a method of placing acid in a well. Acid can be introduced into the well by means of a wide-mouthed funnel and 3/4- or 1-inch (18- or 25-mm) plastic pipe. Acid is heavier than water which it tends to displace but with which it also mixes readily to become diluted.

The acid solution in the well should be agitated by means of a surge plunger or other suitable means for 1 to 2 hours. Following this, the well should be bailed until the water is relatively clear. The driller usually can detect an improvement in the yield of the well while running the bailer. The well may, however, be pumped to determine the extent of improvement. If the results are less than expected, the treatment may be repeated using a longer period of agitation before bailing. Additional treatment may even be undertaken.

The procedure is sometimes varied to alternate acid treatment and chlorine treatment. The chlorine helps to remove the slime deposited by iron bacteria. Sulfamic acid offers a number of advantages over hydrochloric acid as a means of treating incrustation in wells. Sulfamic acid can be added to a well in either its original granular form or as an acid solution

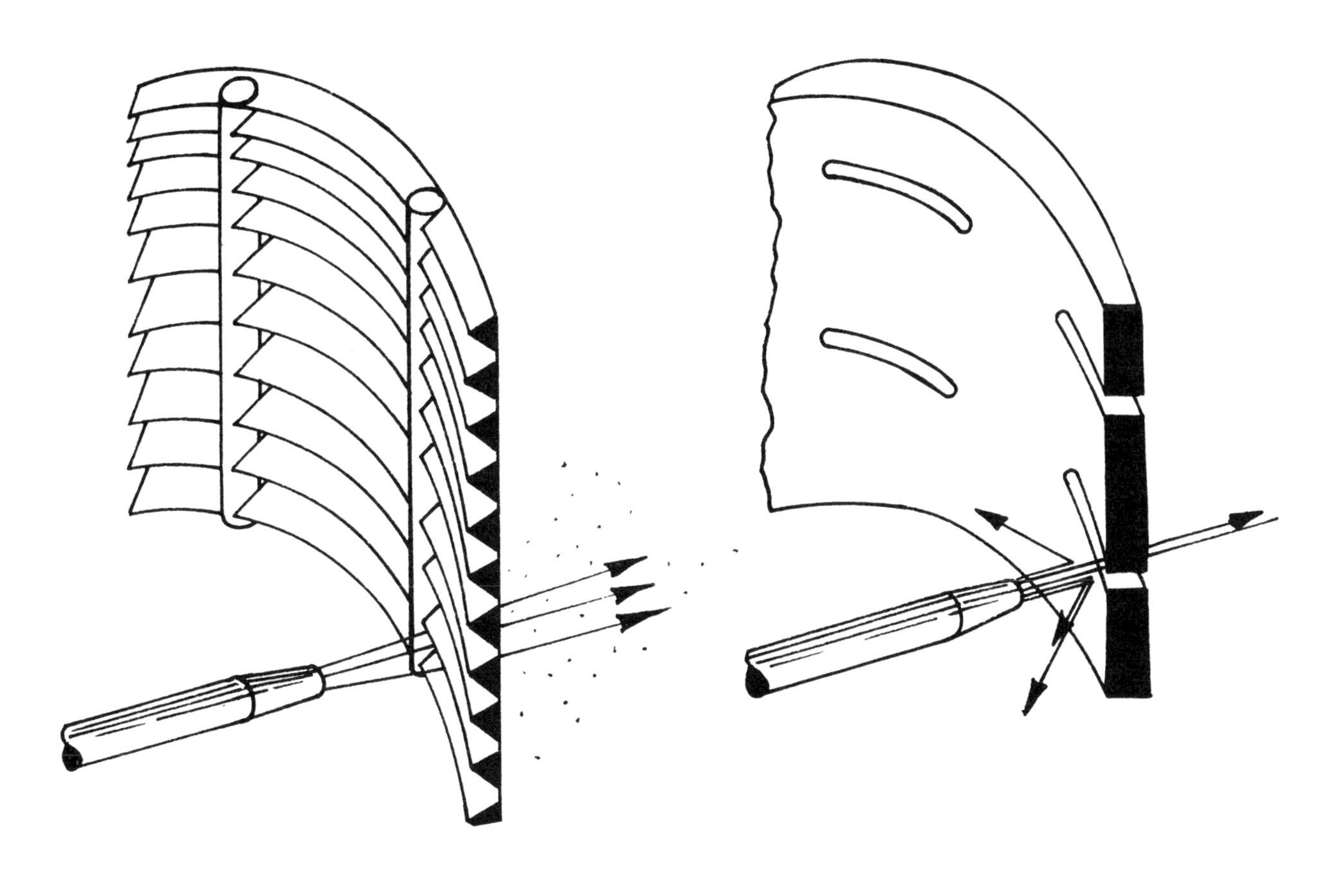

Fig. 3.15 Use of jet spray on well screen

mixed on site. Granular sulfamic acid is nonirritating to dry skin and its solution gives off no fumes except when reacting with incrusting materials. Spillage, therefore, presents no hazards and handling is easier, cheaper, and safer. Sulfamic acid also has a markedly less corrosive effect on well casing and pumping equipment. Sulfamic acid dissolves calcium and magnesium carbonate compounds to produce very soluble products. The reaction is, however slower than that using hydrochloric acid and a somewhat longer contact period in the well is required. Again, agitation must be provided by some sort of plunger. The quantity of acid added in this case should be based on the total volume of water standing in the well — *NOT* on the volume of just the part of the well to be cleaned (as is the case if the acid is applied in solution form). A little extra granular sulfamic acid may be added to keep the solution up to maximum strength while it is being used up through reaction with the incrusting material. The addition of a low-foaming, non-ionic wetting agent improves the cleansing action to some extent.

A number of precautions must be exercised in using any strong acid solution. Goggles and waterproof gloves should be worn by all persons handling the acid. When preparing an acid solution, *ALWAYS POUR THE ACID SLOWLY INTO THE WATER.* In view of the variety of gases, some of them very toxic, produced by the reaction of acid with incrusting materials, adequate ventilation must be provided in pump houses or other confined spaces around treated wells. Do not allow personnel to stand in a pit or depression around the well during treatment because some of the toxic gases (such as hydrogen sulfide) are heavier than air and will tend to settle in the lowest areas. After a well has been treated, it should be pumped to waste to ensure the complete removal of all acid (measure pH) before it is returned to normal service.

3.333 Chlorine Treatment

Chlorine treatment of wells is more effective than acid treatment in loosening bacterial growths and slime deposits which often accompany the deposition of iron oxide. Because of the very high concentrations required, 100 to 200 mg/*L* of *AVAILABLE CHLORINE*,[35] the process is often referred to as shock treatment with chlorine. Calcium or sodium hypochlorite may be used as the source of chlorine.

> *NOTE:* Either calcium or sodium hypochlorite can be used. They are *NEVER* used together. Extreme heat or an explosion could occur.

The chlorine solution in the well must be agitated. This may be done by using the high-velocity jetting technique or by surging with a surge plunger or other suitable technique. The recirculation provided with the use of the jetting technique greatly improves the effectiveness of the treatment.

[35] *Available Chlorine. A measure of the amount of chlorine available in chlorinated lime, hypochlorite compounds, and other materials that are used as a source of chlorine when compared with that of elemental (liquid or gaseous) chlorine.*

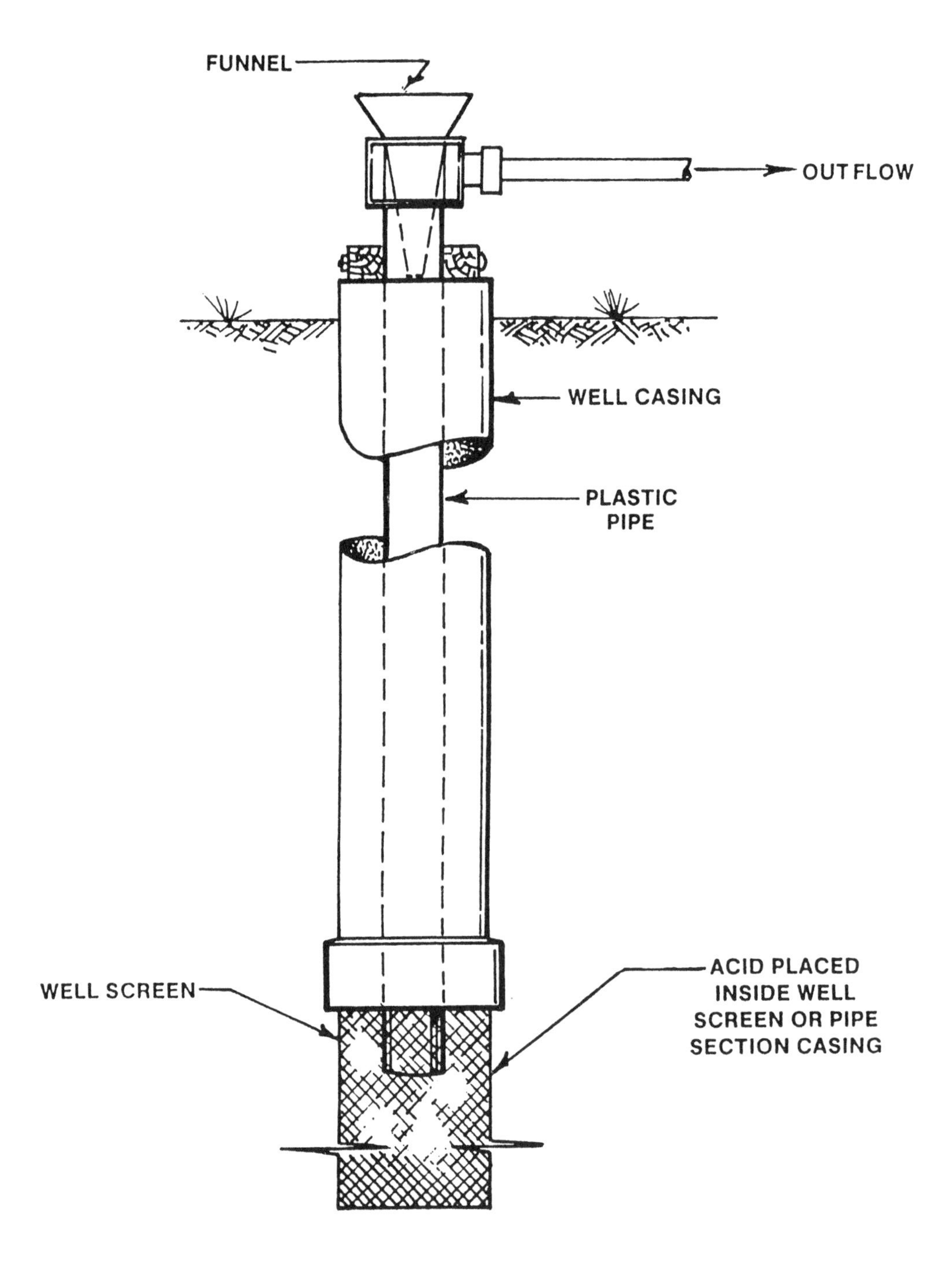

Fig. 3.16 Procedure for introducing acid into well for purpose of cleaning incrustations from screen and casing

The treatment should be repeated 3 or 4 times in order to reach every part of the formation that may be affected. This treatment may also be alternated with acid treatment, but use the acid first.

3.334 Polyphosphates

Polyphosphates, or glassy phosphates as they are commonly called, effectively disperse silts, clays, and the oxides and hydroxides of iron and manganese. The dispersed materials can be easily removed by pumping. In addition, the polyphosphates are safe to handle and are often used in the chemical treatment of wells.

3.335 Explosive Charges

Small explosive charges have been used to clean plugged well screens. Experts in this field should be consulted before attempting this procedure.

3.336 Summary

Although many forms of well cleaning and maintenance have been discussed here, you are cautioned to exercise care in the application of any of these. Always observe manufacturers' recommendations. Obtain expert advice if you are not sure which chemical to use, how to apply the chemical or how to remove it from the well.

QUESTIONS

Write your answers in a notebook and then compare your answers with those on page 127.

3.3F What records should be kept regarding a well?

3.3G What is the purpose of surging?

3.3H How can the pores in a well screen and the gravel pack around the screen be cleaned?

3.3I How can incrustation be removed from the well casing and well?

3.3J How can bacterial growths and slime deposits be removed from well screens?

3.34 Water Quality Monitoring

Well performance can be affected by the quality of the groundwater. Similarly, water quality, both physical and chemical, is a good indicator of existing or potential problem areas and should be monitored regularly for clues to problems. For example, excessive sand production indicates problems with the well completion procedure and will produce excessive pump wear. The chemical quality of water indicates the type of dissolved minerals in the groundwater and will help in the design of a maintenance program if mineral deposition is suspected to be a cause of decreased well performance.

3.35 Downhole Video Inspection

Downhole video inspection can aid in well maintenance. Cameras take both still photographs and motion pictures of the well with the pumping equipment removed. The tape recorded "video log" of the well can then be reviewed as an aid in designing a maintenance or rehabilitation program. Specific problem areas, such as mineral deposition or other incrustation, corrosion of screen perforations or casing, and mechanical collapse or other failure, can be identified, permitting more precise procedures to correct problems.

3.36 Troubleshooting

3.360 Decline in Yield

The yield of any water supply well depends on three factors: the aquifer, the well, and the pump. A decline in yield is due to a change in at least one of these factors, and correction of the problem depends on identification of the factor that is involved. This identification in many cases can be made only if data are available on the depth to the water level in the well and the pumping rate. Inability to identify reasons for a decline in yield frequently results in discontinuing the use of the groundwater and developing more expensive supplies from either groundwater or surface water sources. Table 3.5 is a summary analysis of the causes of declines in well yields and potential corrective actions.

The specific capacity test is a measure of the adequacy of an aquifer or well. Specific capacity of a well is determined measuring the yield of a well in gallons per minute per unit of drawdown during a specific time period, usually 24 hours. For example, if a well yield was 150 GPM and the drawdown was 15 feet, the specific capacity would be 150 GPM/15 ft or 10 GPM/ft of drawdown. Specific capacity generally varies with the duration of pumping; as pumping time increases, specific capacity decreases. Also, specific capacity decreases and discharge increases in the same well.

3.361 Changes in Water Quality

Deterioration in water quality may result from changes in the quality of water in the aquifer or changes in the well. These changes may affect the biological quality, the chemical quality, or the physical quality of the water. Deterioration in biological and chemical quality generally results from conditions in the aquifer whereas changes in physical quality result from changes in the well. Both the biological and chemical quality of water from new public water supply wells must be analyzed before the wells are placed in use to determine if the water meets drinking water standards and, if it does not, what treatment is required. Table 3.6 is a summary analysis of the causes of changes in water quality and possible corrective actions.

3.37 Summary

Maintenance operations should not be put off until problems become serious. When this happens, rehabilitation of a well becomes more difficult and sometimes impossible or impractical. Incrustation not treated early enough can so thoroughly clog the well screen and the formation around it that it becomes extremely difficult (and even impossible) to diffuse or circulate a chemical solution to all affected points in the formation. At this point, any attempts at rehabilitation would most likely prove unsuccessful.

No methods have yet been developed for the complete prevention of incrustation in wells. Various steps can be taken to delay the process and reduce the magnitude of its effects. Among these are the proper design of well screens and the reduction of pumping rates, both aimed at reducing entrance velocities into screens and drawdown in wells. For example, it may be worthwhile to share the pumping load among a larger number of wells in order to reduce the rate of incrustation. However, the ultimate or final solution will be in a regular cleaning program. Incrusting wells are usually treated with chemicals which either dissolve the incrusting deposits or loosen them from the surfaces of the well screen and formation materials so that the deposits may be easily removed by bailing. Corkscrew-shaped brushes have been rotated in wells to remove incrustations.

Recordkeeping is a must. Only with data on the well's performance can problems be identified or predictions be accurately estimated. Start a recordkeeping program when a well is constructed. If such steps were not taken at the time of construction, they should be started as soon as possible.

QUESTIONS

Write your answers in a notebook and then compare your answers with those on page 127.

3.3K How does the chemical quality of groundwater influence a well maintenance program?

3.3L How can the inside of a well be inspected for incrustation and/or corrosion of a well screen?

3.3M When the yield of a water well declines, what three factors should be investigated to determine the cause?

End of Lesson 2 of 4 Lessons on WELLS

Please answer the discussion and review questions next.

TABLE 3.5 ANALYSIS OF DECLINES IN WELL YIELDS[a]

Symptom	Cause	Corrective Action
Decline in available drawdown, no change in specific capacity.	The aquifer, due to a decline in groundwater level resulting from depletion of storage caused by decline in recharge or excessive withdrawals.	Increase spacing of new supply wells. Institute measures of artificial recharge.
No change in available drawdown, decline in specific capacity.	The well, due to increase in well head loss resulting from blockage of screen by rock particles or by deposition of carbonate or iron compounds; or reduction in length of the open hole by movement of sediment into the well.	Redevelop the well through the use of a surge block or other means. Use acid to dissolve incrustations.
No change in available drawdown, no change in specific capacity.	The pump, due to wear of impellers and other moving parts or loss of power from the motor.	Recondition or replace motor, or pull pump and replace worn or damaged parts.

[a] *BASIC GROUNDWATER HYDROLOGY.* United States Geological Survey (USGS) Water Supply Paper 2220.

TABLE 3.6 ANALYSIS OF CHANGES IN WATER QUALITY[a]

Change in Quality	Cause of the Change	Corrective Action
Biological	Movement of polluted water from the surface or near-surface layers through the *ANNULAR SPACE*.[b]	Seal annular space with cement grout or other impermeable material and mound dirt around the well to deflect surface runoff.
Chemical	Movement of polluted water into the well from the land surface or from shallow aquifers.	Seal the annular space. If sealing does not eliminate pollution, extend the casing to a deeper level (by telescoping and grouting a smaller diameter casing inside the original casing).
	Upward movement of water from zones of salty water.	Reduce the pumping rate and (or) seal the lower part of the well.
Physical	Migration of rock particles into the well through the screen or from water-bearing fractures penetrated by open-hole wells.	Remove pump and redevelop the well.
	Collapse of the well screen or rupture of the well casing.	Remove screen, if possible, and install new screen. Install smaller diameter casing inside the original casing.

[a] *BASIC GROUNDWATER HYDROLOGY.* United States Geological Survey (USGS) Water Supply Paper 2220.
[b] Annular (AN-you-ler) Space. A ring-shaped space located between two circular objects, such as two pipes.

DISCUSSION AND REVIEW QUESTIONS

Chapter 3. WELLS

(Lesson 2 of 4 Lessons)

Write the answers to these questions in your notebook before continuing. The question numbering continues from Lesson 1.

11. What could happen when the check valve on a well pump station fails?
12. What is the relationship between system pressure and the flow delivered to the system by a pump?
13. What problems may be caused by sand entering a distribution system?
14. What items would you include in your well maintenance program?
15. What procedures are available for cleaning a well screen?
16. What problem might cause a change in the biological quality of well water and how would you correct the problem?

CHAPTER 3. WELLS

(Lesson 3 of 4 Lessons)

3.4 WELL PUMPS AND SERVICE GUIDELINES

3.40 Purpose of Well Pumps

3.400 Well Pumps

Once a well is completed and water is available from an aquifer, some type of pump must be installed to lift the water from the well and deliver it to the point of use. The intent of this section is to discuss the general characteristics of well pumps operators are likely to encounter.

Well pumps are generally classified into two basic groups:

1. *POSITIVE DISPLACEMENT* pumps which deliver the same volume or flow of water against any *HEAD*[36] within their operating capacity. Typical types are piston (reciprocating) pumps, and screw or squeeze displacement (diaphragm) pumps; and
2. *VARIABLE DISPLACEMENT* pumps which deliver water with the volume or flow varying inversely with the head (the *GREATER* the head, the *LESS* the volume or flow) against which they are operating. The major types are centrifugal, jet, and air-lift pumps.

Either of these types of pumps can be used for pumping water from a well. However, centrifugal pumps are by far the most commonly used pump in the waterworks field because of their capability to deliver water in large quantities, against high as well as low heads, and with high efficiencies.

3.401 Shallow Well Pump

A pump installed above a well is often called a *SHALLOW WELL PUMP*; this pump takes water from the well by *SUCTION LIFT.*[37] Such a pump can be used for either a deep well or a shallow well providing the pumping level is within the suction lift capability of the pump (maximum of 20-feet (6-m) lift).

3.402 Deep Well Pump

A pump installed in the well with the *PUMP BOWL*[38] inlet submerged below the pumping level in the well is generally referred to as a *DEEP WELL PUMP.* This type of pump may be used for any well, regardless of depth, where the pumping level is below the limit of suction lift.

3.41 Types of Pumps

3.410 Centrifugal Pumps

A centrifugal pump raises the water by a centrifugal force which is created by a wheel, referred to as an *IMPELLER*, revolving inside a tight *CASING.* In operation, the water enters the pump at the center of the impeller, called the *EYE.* The impeller throws the water outward toward the inside wall of the casing by the centrifugal force resulting from the revolution of the impeller. The water passes through the channel or diffuser vanes between the rim of the impeller and the casing, and emerges at the discharge under pressure. Centrifugal pumps are used almost exclusively in the waterworks field. Advantages of centrifugal pumps include: (1) relatively small space needed for any given capacity, (2) rotary rather than reciprocating motion, (3) adaptability to high-speed driving mechanisms such as electric motors and gas engines, (4) low initial cost, (5) simple mechanism, (6) simple operation and repair, and (7) safety against damage from high pressure because of limited maximum pressure that can be developed. Centrifugal pumps are generally classed as *VOLUTE* or *TURBINE* pumps.

3.411 Volute-Type Pumps

This type of centrifugal pump has no diffusion vanes (see Figure 3.17). The impeller is housed in a spiral-shaped case in which the velocity of the water is reduced upon leaving the impeller, with a resultant increase in pressure. Ordinarily, the volute-type pump is of single-stage design and used in the water utility field for large-capacity, low-head application, and for low- to mid-range booster pump operations.

3.412 Turbine-Type Pumps

This type of centrifugal pump is the one most commonly used for well pump operations (see Figure 3.18). In the turbine-type pump, the impeller is surrounded by diffuser vanes which provide gradually enlarging passages in which the velocity of the water leaving the impeller is gradually reduced, thus transforming velocity head to pressure head.

Use of multi-stage pumps is standard practice in well pumping operations. The stages are bolted together to form a pump bowl assembly and it is not uncommon to assemble a pump bowl assembly with 10 or more stages. The function of each stage is to add pressure head capacity; the volume capacity

---

[36] *Head. The vertical distance (in feet) equal to the pressure (in psi) at a specific point. The pressure head is equal to the pressure in psi times 2.31 ft/psi.*

[37] *Suction Lift. The NEGATIVE pressure [in feet (meters) of water or inches (centimeters) of mercury vacuum] on the suction side of a pump. The pressure can be measured from the centerline of the pump DOWN TO (lift) the elevation of the hydraulic grade line on the suction side of the pump.*

[38] *Pump Bowl. The submerged pumping unit in a well, including the shaft, impellers and housing.*

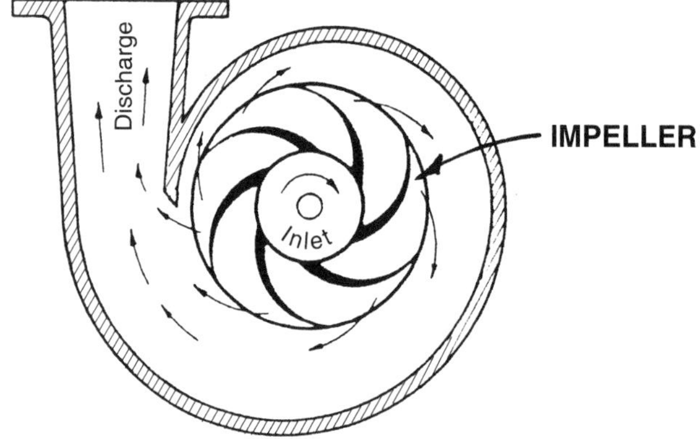

Volute-type centrifugal pump has no diffuser vanes or guides.

Fig. 3.17 Volute-type pump
(Source: *GROUNDWATER AND WELLS*, permission of Johnson Division, UOP, St. Paul, Minn.)

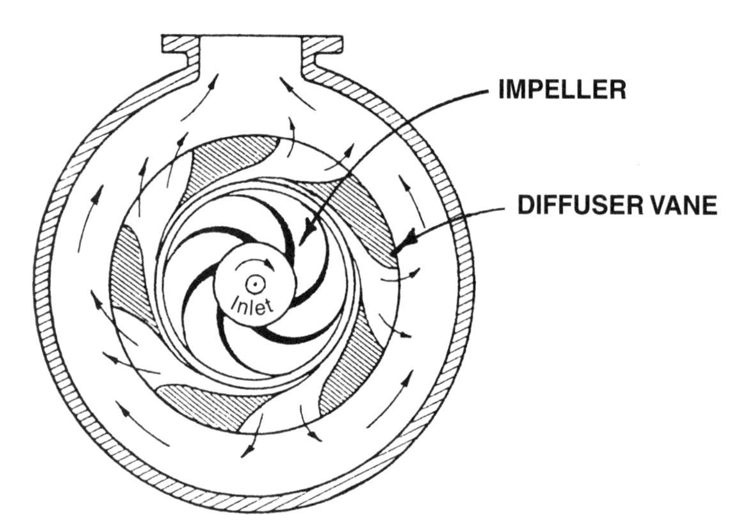

In turbine-type pump, water leaving the impeller moves out through the curved passages between diffuser vanes.

Fig. 3.18 Turbine-type pump
(Source: *GROUNDWATER AND WELLS*, permission of Johnson Division, UOP, St. Paul, Minn.)

and efficiency are almost identical for each stage. As an example, in the case of a 10-stage pump rated at 500 gallons per minute (32 liters/sec) at 250 feet (75 m) of required head, utilizing 40 *BRAKE HORSEPOWER*[39] (30 kW), the first stage would pump 500 gallons per minute (32 liters/sec) at 25 feet (7.5 m) of head, the next stage would not increase the GPM but would add 25 feet (7.5 m) more of head; each of the remaining eight stages would also add 25 feet (7.5 m) of head making the total 250 feet (75 m) of head. The capacity would remain at 500 GPM (32 *L*/sec). However, the brake horsepower for each stage is also additive (as is head). Therefore, if each stage requires 4 BHP (3 kilowatts), then the total for the 10 stages would amount to 40 BHP (30 kilowatts).

Well pumps for water utility operation are generally of the turbine design and often are referred to as variable-displacement deep well centrifugal pumps or more simply, *DEEP WELL TURBINE* pumps.

3.413 Deep Well Turbine Pumps

There are two classifications of deep well turbine pumps, depending upon the location of the prime mover (electric motor or engine).

1. *STANDARD DEEP WELL TURBINE* pumps are driven through a rotating shaft (lineshaft) connected to an electric motor or engine mounted on top of the well (see Figure 3.19). This type of pump requires lubrication of the lineshaft connecting the motor and the pump. Manufacturers have incorporated both *OIL LUBRICATION* and *WATER LUBRICATION* into this design.

   a. In *WATER LUBRICATED* models, the lineshaft is supported in the center of the pump column pipe by means of stainless-steel lineshaft sleeves equipped with neoprene bearings which are lubricated by the water as it flows upward in the column pipe (see Figure 3.20).

      This type of pump is most commonly used for large capacity wells and is designed specifically for each well and for its intended function.

      Pumping capacities generally range from 200 to 2,000 gallons per minute (12.6 to 126 liters per second).

   b. *OIL LUBRICATED* models have a watertight oil tube surrounding the lineshaft and oil is fed from the surface (see Figure 3.21).

Although both oil and water lubricated pumps are used in water utility operation, oil lubricated pumps are most often used.

2. *SUBMERSIBLE DEEP WELL TURBINE* pumps use a pumping bowl assembly similar to the standard deep well turbine except that the motor is mounted directly beneath the bowl assembly. This eliminates the need for the lineshaft and oil tube (see Figure 3.22). Unit efficiency approaches that of a lineshaft turbine pump.

Submersible pumps are available in a wide range of capacities from 5 to 2,000 gallons per minute (0.3 to 126 liters/sec), and are used by individual well owners as well as by small and large water system operators.

[39] *Brake Horsepower. (1) The horsepower required at the top or end of a pump shaft (input to a pump). (2) The energy provided by a motor or other power source.*

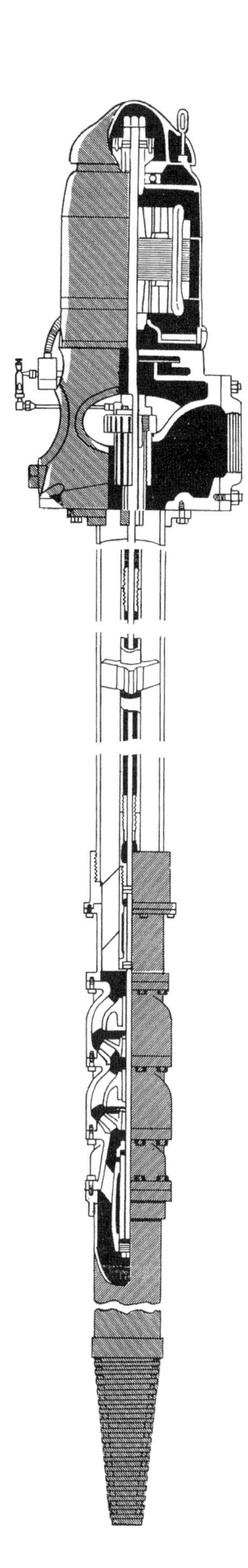

Fig. 3.19 Standard deep well turbine pump

(Permission of Jacuzzi Brothers, Inc.)

Submersible pumps are ideally suited to small water system operation where source capacities range from 25 to 1,000 gallons per minute (1.5 to 63 liters/sec). Maintenance is minimal, the noise level is very low, and they are suited to installations that have limited building areas.

3.414 Other Pumps

There are numerous types of pumps that may be used in the water utility field such as those discussed in the following paragraphs.

1. *JET PUMPS*

   These were used for years on individual wells prior to the development of submersible pumps. They are low in efficiency and are generally restricted to lifts of 100 feet (30 m) or less (see Figure 3.23).

2. *PISTON PUMPS*

   All piston pumps function by means of a piston movement which displaces water in a cylinder. The flow is controlled by valves. They are restricted to low capacity-high pressure applications and are being phased out in favor of turbine-type pumps (see Figure 3.24).

3. *ROTARY PUMPS*

   The rotary pump uses cogs or gears, rigid vanes and flexible vanes. When rotated, a gear-type pump squeezes the water from between the close-fitting gear teeth, moving the water from the inlet side to the outlet side of the pump. A typical rigid-vane rotary pump has a series of dividers or vanes fitted into a slotted rotor. When rotated, these vanes move radially to conform to the contour of the pump housing. The pump housing is eccentric with relation to the rotor, so that the water is pushed from the pump in a continuous flow ahead of the vanes. In flexible-vane rotary pumps, the vanes are elastic blades (usually rubber) which bend to provide the change in displacement volume which forces the water along its path. Rotary pumps are usually used for booster purposes and are generally used in conjunction with another pump (see Figure 3.25).

The above-mentioned pumps have a very limited application and are discussed in this section for the purpose of familiarizing operators with pumps that could be found in water utility operations.

QUESTIONS

Write your answers in a notebook and then compare your answers with those on page 127.

3.4A What are the two basic groups of well pumps and the differences between them?

3.4B List as many advantages of a centrifugal pump as you can recall.

3.4C Why is the installation of jet pumps not too common today?

3.42 Column Pipe

Operators should be aware of the functions of the pump column pipe in a deep well turbine pumping installation. The column pipe is an integral part of the pump assembly and serves three basic purposes: (1) the column pipe connects to the bottom of the surface discharge head, extends down into the well and connects to the top of the well pump (bowl unit) thereby supporting the pump in the well, (2) the column pipe delivers water under pressure from the well pump to the sur-

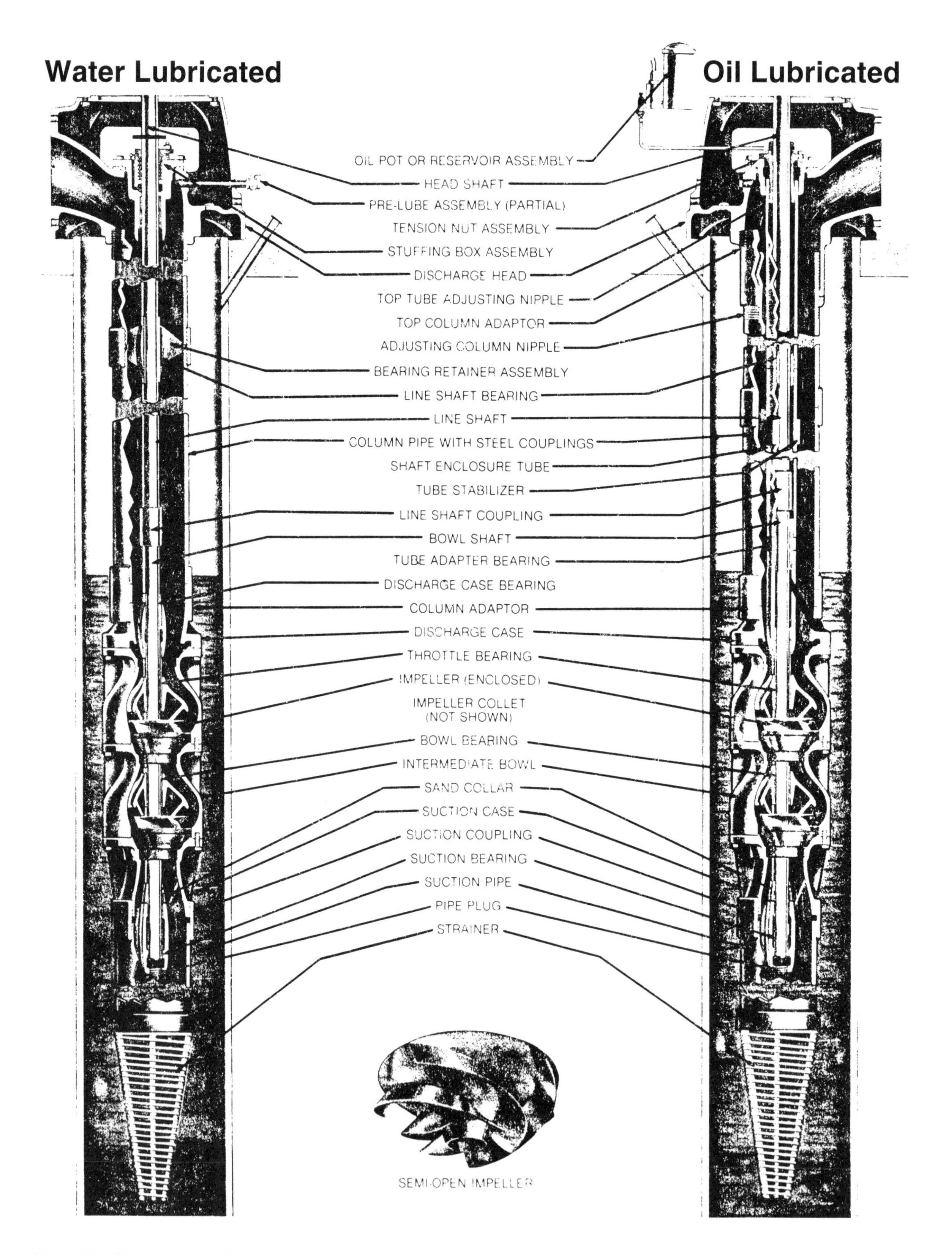

Fig. 3.20 Water lubricated pump

Fig. 3.21 Oil lubricated pump

(Permission of Peabody Floway, Inc., Fresno, CA)

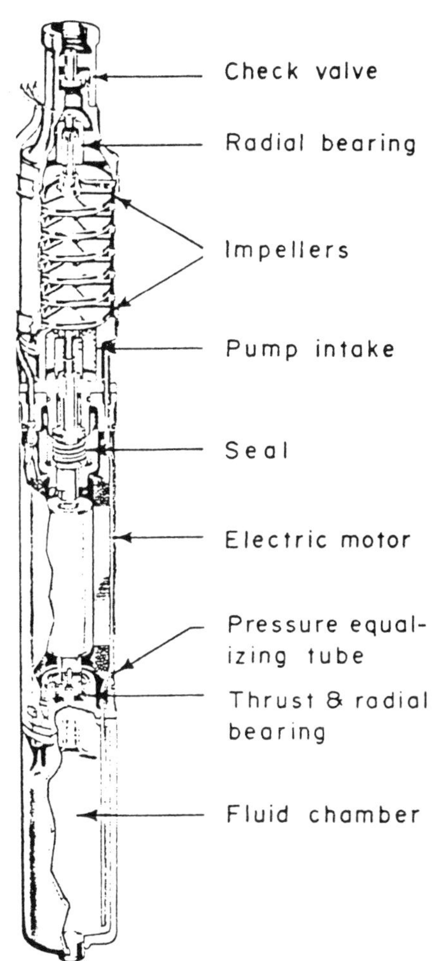

Fig. 3.22 Submersible pump
(Source: *GROUNDWATER AND WELLS*, Johnson Division, UOP, Inc., St. Paul, Minn.)

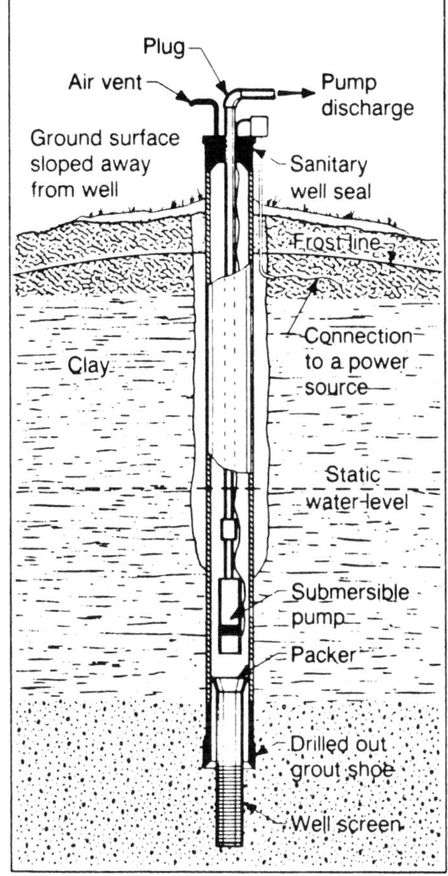

This well has been vented and sealed properly. The groundwater surface around the top of the casing has been graded to slope away in all directions. *(After U.S. Environmental Protection Agency, 1973)*

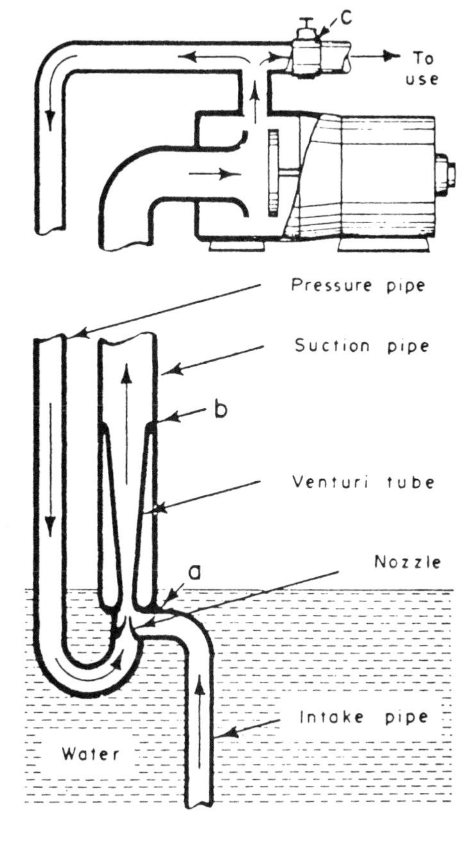

Fig. 3.23 Typical jet pump
(Source: *GROUNDWATER AND WELLS*, Johnson Division, UOP, Inc., St. Paul, Minn.)

face, and (3) keeps the lineshaft and shaft enclosing (oil) tube assembly in straight alignment. Column pipe assemblies for both water and oil lubricated pumps are shown in Figures 3.26 and 3.27.

3.43 Right-Angle Gear Drives (Figures 3.28 and 3.29)

Right-angle gear drives for water utility operations have two distinct applications and provide an economical, efficient, and positive power transmission from a horizontal prime mover (electric motor or engine) to a vertical shaft.

In one application, the right-angle gear drive replaces the electric motor on top of the well and is used on either a full-time or part-time basis.

In a second application, the right-angle gear drive is used with the electric motor. The gear drive is mounted on top of the well discharge head and the electric motor is connected to the right-angle gear drive. An extra long headshaft (an extension of the lineshaft) connects both prime movers to the bowl unit in the well. In most applications, the electric motor is the lead prime mover and the right-angle gear drive unit is used for standby or emergency purposes only. The unit is usually set up for automatic operation.

In either application, the prime mover could be in the form of a gasoline, natural gas, diesel, or propane-powered engine connected to the gear head by means of a flexible drive shaft.

3.44 Selecting a Pump

Before a pump can be intelligently selected for any installation, accurate information about required capacity, location and operating conditions, and total head is needed. With this data available, a selection of the type, class, and size of pump can be made.

After the best type of pump has been determined on the basis of available data, an individual pump must be selected which will best fit each situation. This selection is particularly important if a well is the source of water supply and must take into account differences in pumping head caused by seasonal variations of static water level, temporary lowering of the pumping water level as a result of long periods of continuous pumping, and interference from other wells in the area.

3.45 Service Guidelines

DEEP WELL TURBINE, OIL LUBRICATED pumps are usually equipped with an automatic electric oiler system that is activated by means of an electric solenoid valve when the well pump starts. An adjusting needle and sight glass are part of this assembly. The needle valve is adjusted to feed approxi-

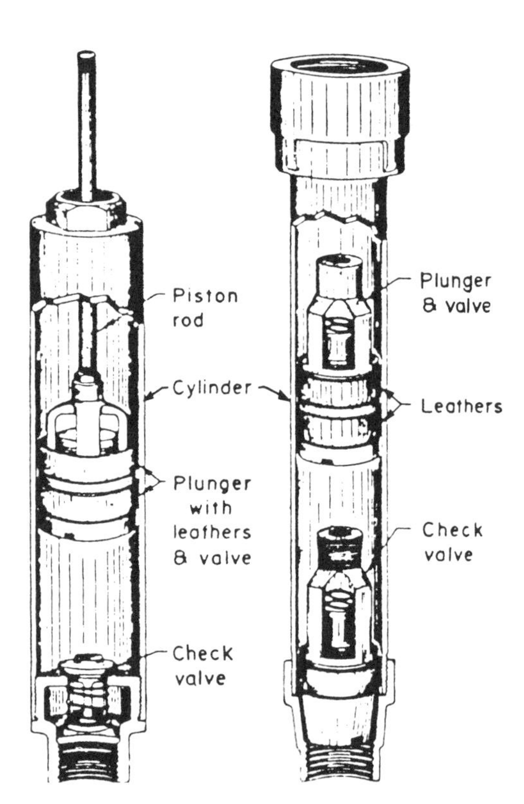

NOTE: Leathers are O-rings or gaskets used to provide a seal between the piston and the side wall.

(Source: *GROUNDWATER AND WELLS*, permission of Johnson Division, UOP, St. Paul, Minn.)

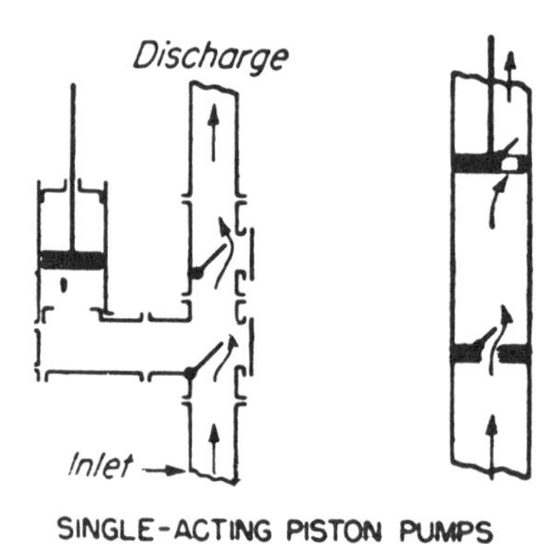

SINGLE-ACTING PISTON PUMPS

Fig. 3.24 Typical piston pumps

(Reprinted from *WATER SUPPLY ENGINEERING* by Babbit, Dolan, and Cleasby, Sixth Edition, by permission. Copyright 1962, McGraw-Hill Book Company)

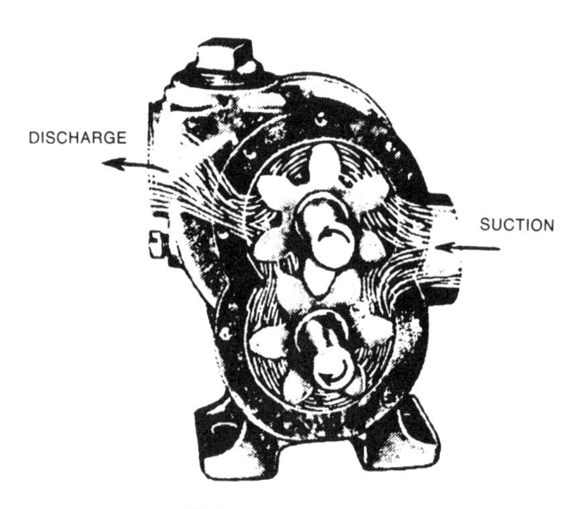

Rotary gear pump

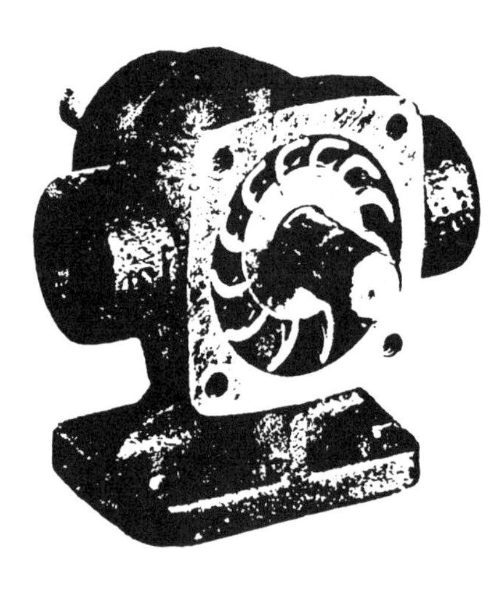

(Source: *GROUNDWATER AND WELLS*, permission of Johnson Division, UOP, St. Paul, Minn.)

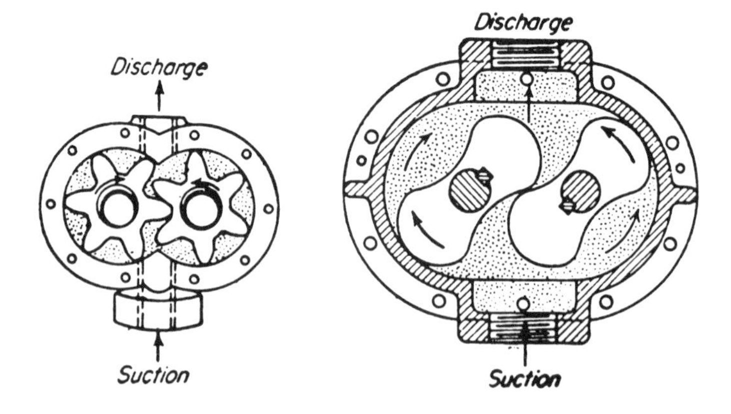

Fig. 3.25 Rotary pumps

(Reprinted from *WATER SUPPLY ENGINEERING* by Babbit, Dolan, and Cleasby, Sixth Edition, by permission. Copyright 1962, McGraw-Hill Book Company)

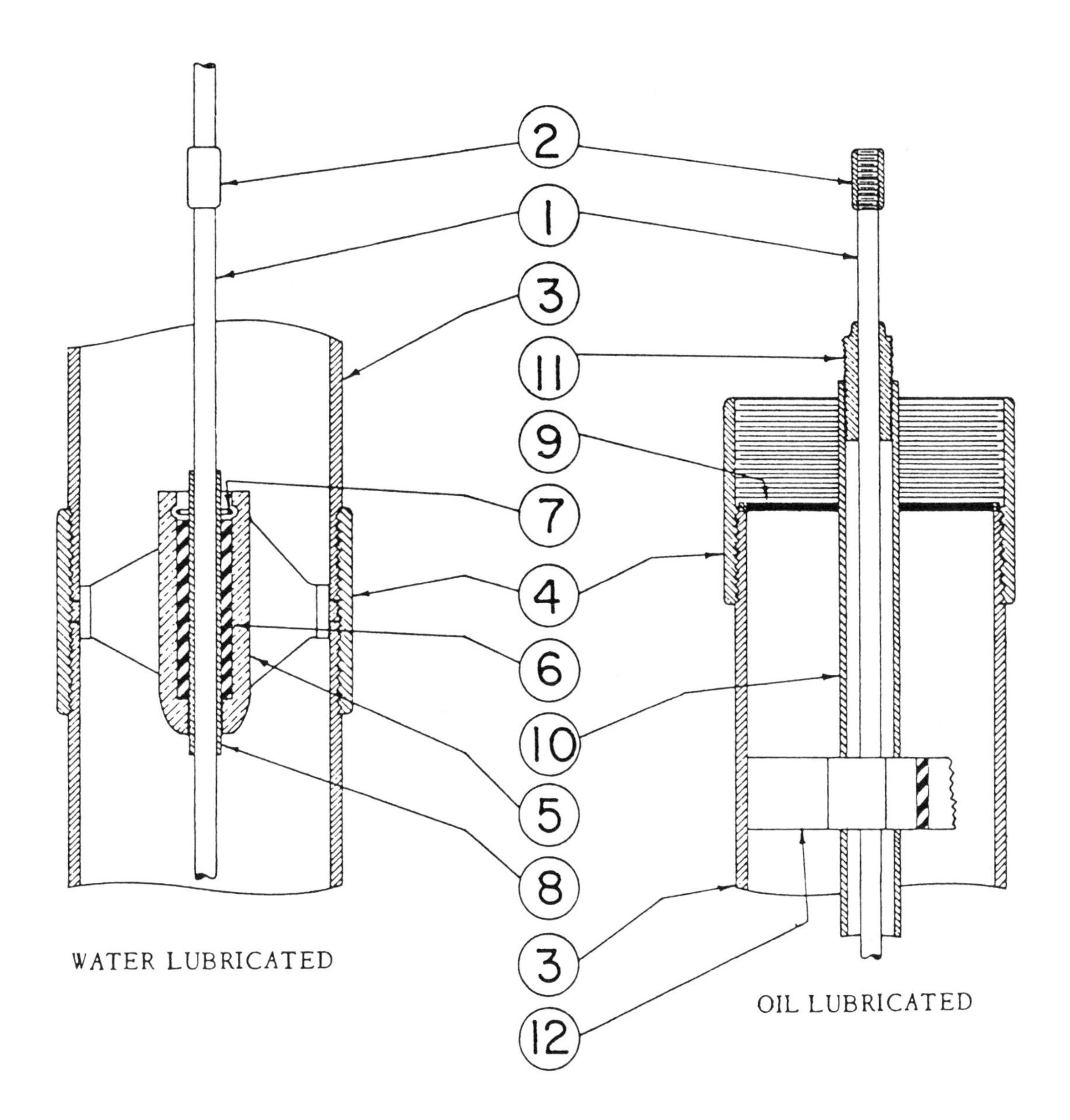

1 LINE SHAFT
2 SHAFT COUPLING
3 COLUMN PIPE
4 COLUMN PIPE COUPLING
5 BEARING CAGE
6 RUBBER SHAFT BEARING
7 SNAP RING
8 SHAFT SLEEVE
9 COLUMN PIPE SPACER RING (OPTIONAL)
10 OIL TUBE
11 LINE SHAFT BEARING
12 TUBE STABILIZER

Fig. 3.26 Column pipe assembly for water lubricated pumps

Fig. 3.27 Column pipe assembly for oil lubricated pumps

(Permission of Peabody Floway, Inc., Fresno, CA)

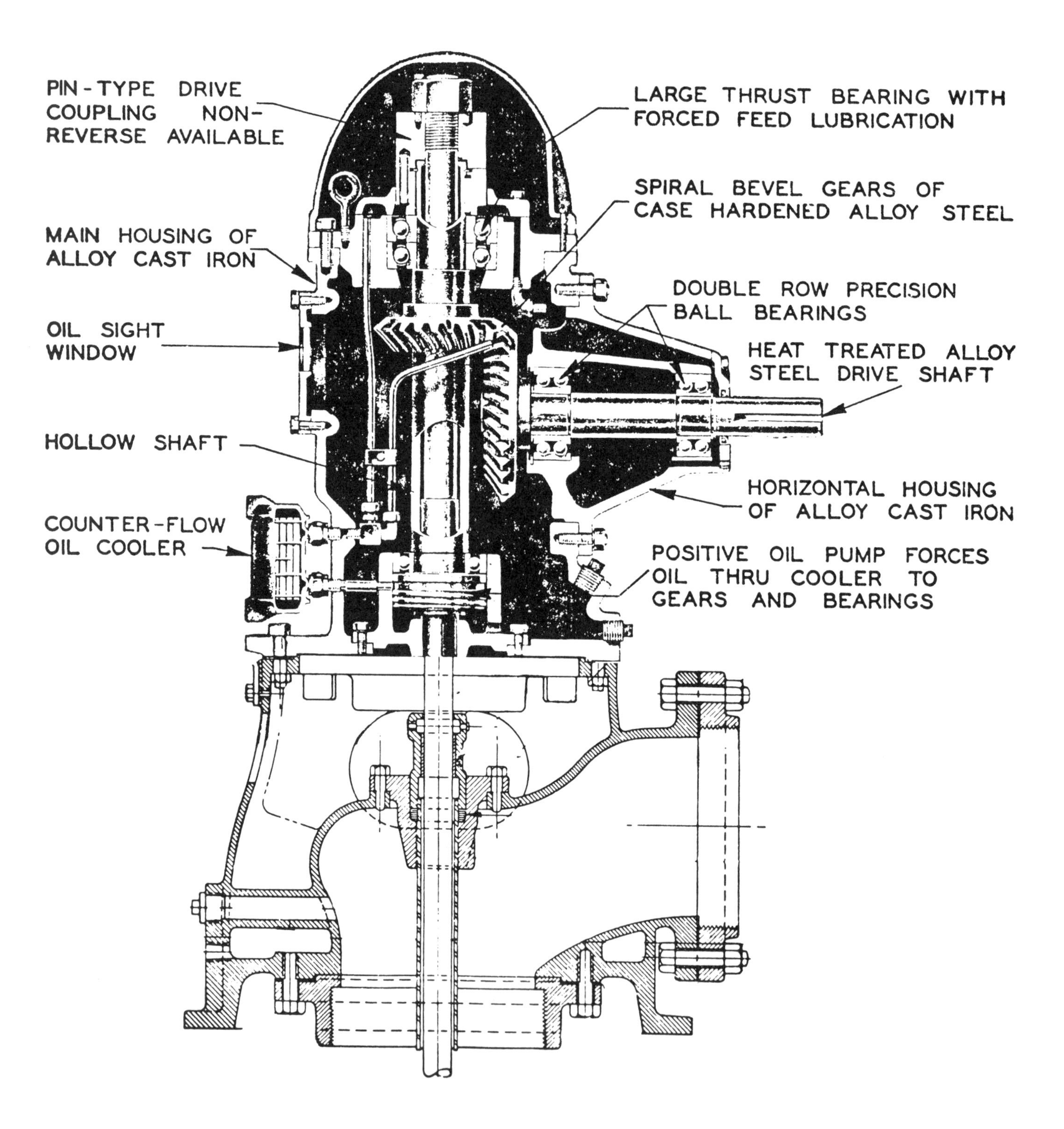
PIN-TYPE DRIVE COUPLING NON-REVERSE AVAILABLE
LARGE THRUST BEARING WITH FORCED FEED LUBRICATION
SPIRAL BEVEL GEARS OF CASE HARDENED ALLOY STEEL
MAIN HOUSING OF ALLOY CAST IRON
DOUBLE ROW PRECISION BALL BEARINGS
OIL SIGHT WINDOW
HEAT TREATED ALLOY STEEL DRIVE SHAFT
HOLLOW SHAFT
HORIZONTAL HOUSING OF ALLOY CAST IRON
COUNTER-FLOW OIL COOLER
POSITIVE OIL PUMP FORCES OIL THRU COOLER TO GEARS AND BEARINGS

Fig. 3.28 Right-angle gear drive

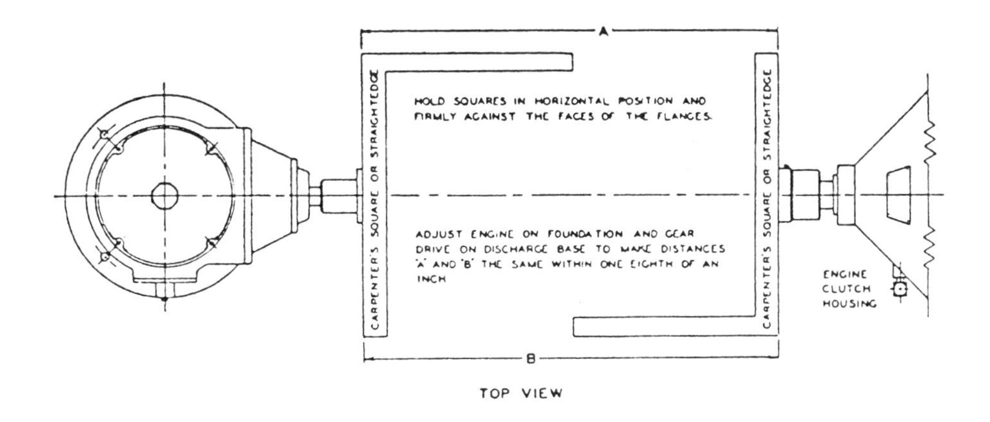

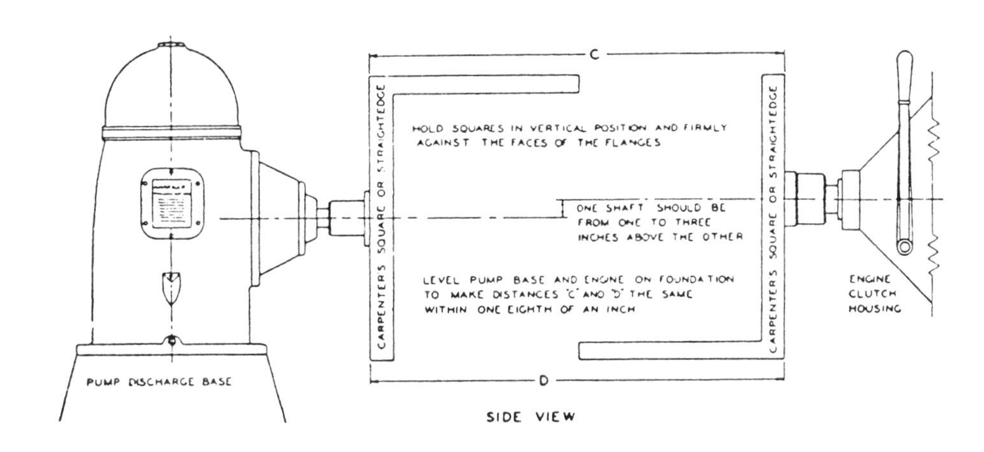

Fig. 3.29 Alignment of right-angle gear drive
(Source: Amarillo Right Angle Pump Drive, Amarillo, Texas)

mately five drops per minute plus one drop of oil per minute for each 20 feet (6 m) of column during the first week of operation. The drip rate may thereafter be reduced to one drop per minute for each 40 feet (12 m) of column. Under no circumstances should the drip rate be less than five drops per minute, regardless of the length of column.

A good grade of turbine oil (mineral base) SAE 10 is used as the lubricant. Automotive or diesel engine lubricating oils cannot be used as a lineshaft lubricant. The oils listed in Table 3.7 are recommended for lubricating the lineshaft bearings. This list does not include all acceptable oils.

TABLE 3.7 OILS RECOMMENDED FOR LUBRICATING THE LINESHAFT BEARINGS [a]

OIL	*SOURCE*
Turbine Oil, Light	Atlantic Refining
Teresso, #43	ESSO Standard
Gulfcrest, #44	Gulf
Turbo, #27	Shell
DTE, #797	Socony Mobil
Chevron OC Turbine, #9	Standard Oil of California
Nonpareil, Medium L5803	Standard Oil of Indiana
Sunvise, #916	Sun Oil
Regal, R & O	The Texas Company
Pacemaker, #1	Cities Service

[a] *JOHNSTON VERTICAL TURBINE PUMPS, INSTALLATION, OPERATION AND MAINTENANCE MANUAL*, Glendora, CA.

DEEP WELL TURBINE, WATER LUBRICATED pumps are self-lubricating and normally require little or no lineshaft maintenance. In a few cases where the static water level is over 100 feet (30 m) and the pump is operated on an intermittent basis, a special small-diameter, pressurized water line may be used to keep the bearings above the water level lubricated.

3.46 Motors

Vertical, hollow-shaft motors for deep well turbine pumps (motor on top of well) require some degree of routine maintenance. The motor bearings at the top and bottom of the motor are enclosed within a weatherproof oil bath container. The oil in this container should be changed annually. Most motors are equipped with a lubrication instruction plate attached to the motor that specifies the proper type and viscosity of oil required for various operating temperatures.

On small motors, the bearings are generally grease lubricated and require weekly attention during the heavy pumping season. Do not use excess grease because the bearings will overheat. The motor manufacturer's instruction manual should specify the type of grease recommended for various applications.

QUESTIONS

Write your answers in a notebook and then compare your answers with those on page 127.

3.4D List the three basic purposes of the pump column pipe in a deep well turbine pumping installation.

3.4E What are the prime movers used with right-angle gear drives?

3.4F How are deep well turbine, oil lubricated pumps lubricated?

3.5 DISINFECTION OF WELLS AND PUMPS

(Also see Chapter 5, "Disinfection.")

3.50 New Wells — During Construction

During the drilling of new wells, contamination could be introduced into the well from the drilling tools and mud, make-up water, topsoil falling in or sticking to tools, and from the gravel itself.

The procedure described below is generally satisfactory for disinfecting a well; however, other methods may be used provided it can be demonstrated that they will produce comparable results. Disinfection should take place following development, testing for yield, and before the test pump is removed from the well. This will ensure that the well is purged of drilling mud, dirt, and other debris that reduces the effectiveness of the disinfecting solution.

1. Add to the well a chlorine solution strong enough to produce a chlorine concentration of 50 mg/*L* *IN* the well casing. Table 3.8 lists quantities of various chlorine compounds required to dose 100 feet (30 meters) of water-filled casing at 50 mg/*L* for diameters ranging from 6 to 24 inches (150 to 600 mm). Organic matter such as oil may need an initial concentration of 1,000 mg/*L* before being injected into the well. This is 20 times the values in Table 3.8.
2. Turn the pump on and off several times so as to thoroughly mix the disinfectant with the water in the well. Pump until the water discharged has the odor of chlorine. Repeat this procedure several times at one-hour intervals.
3. Allow the well to stand without pumping for 24 hours.
4. The water should then be pumped to waste until the odor of chlorine is no longer detectable. Use a chlorine test kit to determine the absence of a chlorine residual.
5. Collect a bacteriological sample in a sterile container and submit it to a laboratory for examination.
6. If the laboratory analysis shows the water is not free of bacterial contamination, repeat the disinfection procedure and retest the water. If repeated attempts to disinfect the well are unsuccessful, a detailed investigation to determine the cause of the contamination should be undertaken.

3.51 New Wells — After Construction

Prior to placing a new well pumping installation in service, the well disinfection procedure previously described should be repeated. This will protect against contamination caused during the construction of the pump base and related appurtenances, and the installation of the permanent pumping unit.

3.52 Existing Wells — After Well or Pump Repairs

Disinfection of existing wells following repairs to the well or replacement of pumping equipment may require special disinfection methods. During the repair work, deposits of slime, bacterial growth, and other debris are dislodged from the inside surfaces of the well casing and from the outside surfaces of the well pump column pipe. These deposits generally settle to the bottom of the well, but some are also smeared on the inside surfaces of the well casing, particularly above the water line. Those deposits above the water line are difficult to disinfect by typical well disinfection procedures. The following special procedures are recommended.

TABLE 3.8 CHLORINE COMPOUNDS REQUIRED TO DOSE 100 FEET[a] OF WATER-FILLED CASING AT 50 MILLIGRAMS PER LITER[b]

	Chlorine Compounds		
Diameter of Casing (inches)	Calcium Hypochlorite (65%), Sold as HTH or Perchloron (Dry Weight)[c,d]	Chloride of Lime (25%) (Dry Weight)[c,d]	Sodium Hypochlorite (5.25%), Sold as Purex or Clorox (Liquid Measure)[e]
6	2 ounces	4 ounces	20 ounces
8	3 ounces	7 ounces	$2^1/_8$ pints
10	4 ounces	11 ounces	$3^1/_2$ pints
12	6 ounces	1 pound	5 pints
16	11 ounces	$1^3/_4$ pounds	1 gallon
20	1 pound	3 pounds	$1^2/_3$ gallons
24	$1^1/_2$ pounds	4 pounds	$2^2/_3$ gallons

[a] *WATER WELL STANDARDS*, Bulletin 74, California Department of Water Resources, Sacramento, CA.
[b] Some authorities recommend a minimum concentration of 100 mg/*L*. To obtain this concentration, double the amounts shown.
[c] Where a dry chemical is used, it should be mixed with water to form a chlorine solution prior to placing it into the well. HTH stands for **H**igh **T**est **H**ypochlorite which is calcium hypochlorite or $Ca(OCl)_2$.
[d] 16 ounces = 1 pound (Dry Weight)
[e] 16 fluid ounces = 1 pint; 2 pints = 1 quart; 4 quarts = 1 gallon

1. Swab inside of well casing with a strong non-foaming detergent such as trisodium phosphate or Calgon.
2. Calculate amount of chlorine based on diameter of well and water depth. The chlorine dosage rate should produce a free chlorine concentration of at least *100* mg/*L* in the water-filled casing.
3. Add chlorine solution to the well, preferably through a hose raised and lowered to reach all areas of the well, including the well casing above the water line.
4. Clean and disinfect pump, pump column pipe, cable, and other equipment before the units are lowered into the well.
5. Follow procedures 2 through 6 for new well disinfection (Section 3.50).

3.53 Contaminated Wells

If the well and pumping unit must be disinfected because of contamination or bacterial problems and *THE WELL PUMP IS LEFT IN PLACE*, then follow these *SPECIAL METHODS*:

1. To the well* add a chlorine solution, strong enough that it produces a chlorine concentration of *200* mg/*L IN THE WELL CASING.* (* See Sections A and B below for procedures on how to add the chlorine solution to the well.)
2. Turn the pump on and off several times to thoroughly mix the disinfectant with the water in the well. Pump until the water discharged has the odor of chlorine. Repeat this procedure several times at one-hour intervals.
3. Allow the well to stand without pumping for 24 hours.
4. Pump the water to waste until the odor of chlorine is no longer detectable. A chlorine test kit should be used to determine the absence of a chlorine residual.
5. Take a bacteriological sample and submit it to a laboratory for examination.
6. If the laboratory analysis shows the water is not free of bacterial contamination, repeat the disinfection procedure and retest the water. If repeated attempts to disinfect the well are unsuccessful, a detailed investigation to determine the cause or source of the contamination should be made.

* A *SUBMERSIBLE PUMPS AND DEEP WELL TURBINE PUMPS — WATER LUBRICATED*

The chlorine solution should be introduced into the well through either the air release valve piping assembly (remove air release valve) or the well-casing vent, and water added to flush the chlorine solution back into the well. If the well is equipped with a foot valve, then the chlorine solution must be introduced into the well by the well-casing vent.

* B *DEEP WELL TURBINE PUMPS — OIL LUBRICATED*

Introduce the chlorine solution into the well through the air release valve piping assembly as described in (A) above. Chlorine solution or chlorine powder should not be added to the well by means of the well-casing vent. Nearly all oil lubricated pumps have a certain amount of oil floating on the surface of the water within the well casing, ranging from a thin film on the surface to slight traces of oil down to 20 feet (6 m) or more below the water surface. Chlorine added through the well-casing vent strikes the oil floating on the surface of the well and carries oil and debris down into the well and makes *CLEANUP DIFFICULT.* If it is absolutely necessary to apply chlorine by means of the well-casing vent, then a chlorine solution should be introduced into the well through a hose raised and lowered to reach all areas of the well below the level of the oil. If this method cannot be used, then resort to chlorine tablets. If tablets are used, allow sufficient dissolving time and follow Section 3.50, Step 2, procedure for new wells.

QUESTIONS

Write your answers in a notebook and then compare your answers with those on page 127.

3.5A How could contamination be introduced during the drilling of a new well?

3.5B When should a well be disinfected?

3.5C When disinfecting a new well, what is the desired chlorine concentration in the well casing?

3.5D Why does the disinfection of existing wells after well or pump repairs require special disinfection methods?

3.5E What would you do if repeated attempts to disinfect a well are unsuccessful?

3.54 Chlorine Requirement Calculation

The amount of a chlorine compound required to disinfect a well may be determined by two different methods. The first method is the easier and uses values from Table 3.8. The second method calculates the amount of chlorine required based on the volume of water in the casing being disinfected, the desired chlorine dose, and the amount of chlorine available in the disinfecting chlorine compound. Both methods will give the same results with the second method being slightly more accurate.

FORMULAS

Method 1

$$\text{Chlorine Required} = \frac{(\text{Table 3.8 Value})(\text{Casing Length, ft})}{100 \text{ ft}}$$

To find the value in Table 3.8, you need to know:

1. Diameter of casing in inches, and
2. Type of chlorine compound.

The length of water-filled casing in feet is used to calculate the chlorine required. Also, Table 3.8 assumes the desired chlorine dose is 50 mg/*L*. If the dose is not 50 mg/*L*, use the following formula:

$$\begin{array}{l}\text{Chlorine Required}\\ \text{(If dose not}\\ \text{50 mg/}L\text{)}\end{array} = \frac{(\text{Chlorine Required})(\text{Desired Dose, mg/}L)}{50 \text{ mg/}L}$$

Calculate the "Chlorine Required" using Table 3.8 and then adjust this answer by multiplying by the "Desired Dose, mg/*L*" and dividing by "50 mg/*L*."

Method 2

$$\text{Casing Volume, gal} = \frac{(0.785)(\text{Casing Diameter, in})^2(\text{Casing Length, ft})(7.48 \text{ gal/cu ft})}{144 \text{ sq in/sq ft}}$$

$$\text{Chlorine Required, gal} = \frac{(\text{Casing Volume, gal})(\text{Desired Dose, mg/}L)}{\text{Chlorine Compound Solution, mg/}L}$$

To determine the chlorine required using Method 2 requires two steps. The first step calculates the casing volume in gallons. The second step calculates the chlorine required by multiplying the casing volume in gallons times the desired chlorine dose in mg/*L* and dividing by the chlorine concentration in the chlorine compound solution in mg/*L*. A one percent (1%) chlorine solution is the same as a 10,000 mg/*L* chlorine solution.

EXAMPLE 1

How much sodium hypochlorite is required to dose a well at 50 mg/*L*? The casing diameter is 6 inches and the length of water-filled casing is 150 feet. Sodium hypochlorite is 5.25 percent or 52,500 mg/*L* chlorine.

Known		Unknown
Casing Diameter, in	= 6 in	Chlorine Required, gal
Casing Length, ft	= 150 ft	
Chlorine Compound, mg/*L* (Sodium Hypochlorite)	= 52,500 mg/*L*	

Method 1

1. Find the chlorine required from Table 3.8 for a 6-inch diameter well casing when using sodium hypochlorite.

   Table 3.8 Value = 20 ounces

2. Calculate the ounces of chlorine required.

$$\text{Chlorine Required, ounces} = \frac{(\text{Table 3.8 Value})(\text{Casing Length, ft})}{100 \text{ ft}}$$

$$= \frac{(20 \text{ ounces})(150 \text{ ft})}{100 \text{ ft}}$$

$$= 30 \text{ ounces}$$

3. Convert the chlorine required from ounces to gallons of sodium hypochlorite.

$$\text{Chlorine Required, gal} = \frac{(\text{Chlorine Required, ounces})}{(16 \text{ ounces/pint})(8 \text{ pints/gallon})}$$

$$= \frac{30 \text{ ounces}}{(16 \text{ ounces/pint})(8 \text{ pints/gallon})}$$

$$= 0.23 \text{ gallons}$$

Method 2

1. Calculate the water-filled casing volume in gallons.

$$\text{Casing Volume, gal} = \frac{(0.785)(\text{Casing Diam, in})^2(\text{Casing Length, ft})(7.48 \text{ gal/cu ft})}{144 \text{ sq in/sq ft}}$$

$$= \frac{(0.785)(6 \text{ in})^2(150 \text{ ft})(7.48 \text{ gal/cu ft})}{144 \text{ sq in/sq ft}}$$

$$= 220 \text{ gal}$$

2. Calculate the required gallons of sodium hypochlorite.

$$\text{Chlorine Required, gal} = \frac{(\text{Casing Volume, gal})(\text{Desired Dose, mg}/L)}{\text{Chlorine Compound Solution, mg}/L}$$

$$= \frac{(220 \text{ gal})(50 \text{ mg}/L)}{52{,}500 \text{ mg}/L}$$

$$= 0.21 \text{ gallons}$$

NOTE: We obtained essentially the same answer by either method. Table values will tend to be slightly higher and on the safe side.

QUESTION

Work this problem in a notebook and then compare your solution with the one on page 128.

3.5F How much sodium hypochlorite is required to dose a well at 50 mg/*L*? The casing diameter is 12 inches and the length of water-filled casing is 200 feet.

End of Lesson 3 of 4 Lessons on WELLS

Please answer the discussion and review questions next.

DISCUSSION AND REVIEW QUESTIONS

Chapter 3. WELLS

(Lesson 3 of 4 Lessons)

Write the answers to these questions in your notebook before continuing. The question numbering continues from Lesson 2.

17. What is the difference between a shallow well pump and a deep well pump?
18. How does a centrifugal pump work?
19. Under what conditions are submersible deep well turbine pumps commonly used?
20. Under what circumstances should a well be disinfected?
21. What chlorine compounds are commonly used to disinfect wells?

CHAPTER 3. WELLS

(Lesson 4 of 4 Lessons)

3.6 OPERATOR RESPONSIBILITY AND RECORDKEEPING

3.60 Health and Safety

Properly designed and constructed well pumping stations should produce water in sufficient quantity and quality to meet the needs of the entire community. If the pumping stations fail to do so, the fault can often be traced to poor operation or neglect. The health and safety of the water users depends on good operation.

3.61 Operator Responsibility

All operators have the responsibility to exercise due care and diligence to protect the water sources under their surveillance; to effectively operate and maintain the water production facilities; and to take corrective action as necessary to ensure that safe and potable water in adequate quantities and pressure is continuously supplied to the water users in the community.

3.62 Knowledge of the Water Production Facilities

All operators should have a working knowledge of all of the component parts of a well pumping station and completely understand the role each individual part performs in the overall operation of the facility.

3.63 Routine Facility Servicing

The frequency of routine service calls to a well pumping facility usually depends on the nature and importance of the pumping facility and the availability of personnel. In small utility operations the manager may also perform the duties of operator, meter reader, and maintenance personnel. In large utility operations a full-time staff of qualified operators is usually available including personnel trained in removing and replacing pumping equipment and electrical components.

Regardless of the size of the water utility, the pumping stations must be checked often enough to ensure that the facility components are receiving proper servicing, that routine maintenance functions are performed, that the plant is operational and capable of producing safe water, and the security of the plant is intact.

The frequency of service calls for small utility operations may be two or three per week, while a large utility may check their facilities daily including weekends and holidays.

3.64 What To Look For

LOOK, LISTEN, and *FEEL*! When the operator enters the pumping facility, a complete visual inspection of the facility should be made, including listening for any unusual noises and feeling for vibrations on pieces of equipment such as motors and pumps.

3.65 Forms

Some type of inspection form should be available for the operator to record pertinent data relative to the operation of the pumping facility. Various types of forms have been developed based on the individual preferences of the utility and the characteristics of the facility. Figure 3.30 is an example of a typical form that could be used by the average water utility.

A monthly operational record should be maintained for each well pumping facility. This record should be filled out each time the operator visits the well pumping facility either for routine service or for other purposes.

The items of information to record will vary depending on the type of installation, but generally speaking the following items are necessary:

1. Date and time of service visit.
2. Water production meter reading.
3. Electric power meter reading. (Optional)
4. Amount of lineshaft oil added.
5. Oil level if motor is equipped with motor bearing oil reservoir.
6. Greasing frequency.
7. Air level in pressure tank.
8. Water level in storage tank if well pumps directly to tank.
9. Status of chlorination equipment, including feed rate and amount of chlorine used. (Scales or graduated solution tank required.)
10. Standing water level in well. (Recommend monthly, but could be performed quarterly.)
11. Water level in well before and after pumping and well yield.
12. General operation and appearance of facility. Note any unusual conditions or observations such as noise, vibration, and signs of vandalism.

DATE	TIME	METER READING IN GALLONS	WATER PRODUCTION IN GALLONS	DISCHARGE PRESSURE	AIR LEVEL IN TANK	LUBE OIL ADDED	STANDING WATER LEVEL	OPERATOR	REMARKS
31									
30									
29									
28									
27									
26									
25									
24									
23									
22									
21									
20									
19									
18									
17									
16									
15									
14									
13									
12									
11									
10									
9									
8									
7									
6									
5									
4									
3									
2									
1									
LAST DAY OF PREVIOUS MONTH									
TOTAL									

Fig. 3.30 Pumping facility inspection form

3.66 Records

In previous sections we have discussed the importance of maintaining good records. We cannot overstress that recordkeeping is an extremely important part of any water purveyor's operation. Often, recordkeeping for the small water system operator is minimal at best. The operator must realize that there is *NO SUBSTITUTE FOR GOOD RECORDS*.

Due to the variety and size of the various types of water utilities, it is difficult to set forth a specific guideline as to what type of records should be maintained. As a minimum, recordkeeping should cover the following areas of operation:

1. *NEW CONSTRUCTION RECORDS*

   New water facilities added to the system (wells, pumps, boosters, storage tanks, water mains, chlorination equipment) should be adequately recorded on an "as built" (record) drawing showing type and location of facilities installed.

   For well pumping stations, this should also include copies of the well drillers' report, well log, pump suppliers' equipment sheet with performance curves, construction details of the pump base and related appurtenances, motor specifications, pump controls, wiring diagrams, parts list, and operating instructions.

2. *EQUIPMENT RECORDS*

   Individual equipment items should have the manufacturers' specifications and literature available for review along with any warranty documents.

3. *REPAIRS AND MODIFICATIONS*

   Records of repairs and modifications to well pumping facilities should be accurately maintained with a drawing showing what work was done.

Many states require that well drillers file a "Water Well Drillers' Report" upon completion of the well. The well owner receives a copy of this report. If the well driller is accurate in filling out this report, then it could be a very important document for the well owner's file. Figure 3.31 is a copy of a report required in California by the State Department of Water Resources.

Figures 3.32 and 3.33 are examples of forms used for recording pump information.

Figure 3.34 is a self-inspection form that could be used by the operator as a starting point toward maintaining adequate records.

The length of time that records must be kept from a legal standpoint is generally considered to be seven years. However, records pertaining to wells, equipment, water production, and other pertinent data should be kept longer than seven years.

QUESTIONS

Write your answers in a notebook and then compare your answers with those on page 128.

3.6A What might cause a properly designed and constructed well pumping station to fail to produce water in sufficient quantity and quality to meet the needs of the entire community?

3.6B What should an operator do when entering a pumping facility for an inspection?

3.6C Recordkeeping should cover what areas of operation?

3.7 SAND IN WELL WATER SYSTEMS

3.70 Sources of Sand

Nearly all wells produce a certain amount of sand. Every reasonable effort should be made to prevent sand particles from entering the distribution system.

Wells drilled in *ALLUVIAL*[40] formations where the water-bearing aquifers consist of numerous layers of sand and gravel deposits are susceptible to sand production. In many localities, formations of sand and gravel are the only water-bearing formations of sufficient yield available to a community. Properly designed and constructed wells can be drilled in these types of formations that produce high yields while at the same time are virtually sand free. A carefully designed gravel envelope well, with selected louvers or well screen, supported by an engineered filter pack, should operate many years without producing any significant quantities of sand. However, the typical perforated casing or open-bottom well that has penetrated water-bearing sand formations is likely to produce sand. Some wells produce sand from the very beginning while others may be in use for some length of time before it is evident that the quantity of sand produced by the well is causing problems.

3.71 Problems Associated With Sand

The abrasive action of sand can damage well pumping facilities, consumers' fixtures and appliances, water meters, and precision equipment. In addition, sand can accumulate in the mains in the distribution system, thereby reducing their carrying capacity and increasing friction loss. Sand can also be carried into the consumers' premises with resultant complaints. Excessive sand production from a well could create cavities in the water-bearing formations and result in the eventual collapse of unstable overlying strata (layer of soil) and damage to the well. Tests have demonstrated that it is the sand particles larger than 200 mesh (74 microns) that cause the most trouble. Almost all sand contained in well water is this size or coarser.

Several methods are available to the well system operator to reduce sand production to an acceptable level. One method is to install a sand separator designed to remove the objectionable sand and other solids from the well water. Another method would be to pump the well at various rates of flow to determine if a lower rate of flow would reduce the amount of sand to an acceptable level. Many times a small change in water production will make a large change in sand production. In addition, if the system hydraulics permit, the well or wells producing objectionable amounts of sand could be operated on a continuous basis by either raising the cutoff pressure, or

[40] *Alluvial (uh-LOU-vee-ul). Relating to mud and/or sand deposited by flowing water. Alluvial deposits may occur after a heavy rainstorm.*

ORIGINAL

File with DWR

Notice of Intent No.______

Local Permit No. or Date______

STATE OF CALIFORNIA
THE RESOURCES AGENCY
DEPARTMENT OF WATER RESOURCES
WATER WELL DRILLERS REPORT

Do not fill in

No. 104475

State Well No.______

Other Well No.______

(1) **OWNER:** Name______
Address______
City______ Zip______

(2) **LOCATION OF WELL** (See instructions):
County______ Owner's Well Number______
Well address if different from above______
Township______ Range______ Section______
Distance from cities, roads, railroads, fences, etc.______

WELL LOCATION SKETCH

(3) **TYPE OF WORK:**
New Well ☐ Deepening ☐
Reconstruction ☐
Reconditioning ☐
Horizontal Well ☐
Destruction ☐ (Describe destruction materials and procedures in Item 12)

(4) **PROPOSED USE:**
Domestic ☐
Irrigation ☐
Industrial ☐
Test Well ☐
Stock ☐
Municipal ☐
Other ☐

(5) **EQUIPMENT:**
Rotary ☐ Reverse ☐
Cable ☐ Air ☐
Other ☐ Bucket ☐

(6) **GRAVEL PACK:**
Yes ☐ No ☐ Size______
Diameter of bore______
Packed from______ to______ ft.

(7) **CASING INSTALLED:**
Steel ☐ Plastic ☐ Concrete ☐

From ft.	To ft.	Dia. in.	Gage or Wall

(8) **PERFORATIONS:**
Type of perforation or size of screen______

From ft.	To ft.	Slot size

(9) **WELL SEAL:**
Was surface sanitary seal provided? Yes ☐ No ☐ If yes, to depth______ ft.
Were strata sealed against pollution? Yes ☐ No ☐ Interval______ ft.
Method of sealing______

(10) **WATER LEVELS:**
Depth of first water, if known______ ft.
Standing level after well completion______ ft.

(11) **WELL TESTS:**
Was well test made? Yes ☐ No ☐ If yes, by whom?______
Type of test Pump ☐ Bailer ☐ Air lift ☐
Depth to water at start of test______ ft. At end of test______ ft
Discharge______ gal/min after______ hours Water temperature______
Chemical analysis made? Yes ☐ No ☐ If yes, by whom?______
Was electric log made? Yes ☐ No ☐ If yes, attach copy to this report

(12) **WELL LOG:** Total depth______ ft. Depth of completed well______ ft.
from ft. to ft. Formation (Describe by color, character, size or material)

Work started______ 19____ Completed______ 19____

WELL DRILLER'S STATEMENT:
This well was drilled under my jurisdiction and this report is true to the best of my knowledge and belief.

SIGNED______
(Well Driller)

NAME______
(Person, firm, or corporation) (Typed or printed)

Address______
City______ Zip______
License No.______ Date of this report______

DWR 188 (REV. 7-76) IF ADDITIONAL SPACE IS NEEDED, USE NEXT CONSECUTIVELY NUMBERED FORM

CONFIDENTIAL
NOT FOR PUBLIC USE
WATER CODE SEC. 13752

Fig. 3.31 Water well drillers' report

(Permission of California Department of Water Resources)

SAMPLE RECORD OF PUMP INFORMATION

Owner ______________________________

Pump purchased from ______________________________

Pump installed by ______________________________

Well Number ____________ Date pump installed ______________________________

Pump: Make ______________ Type ______________

Model ______________ Serial No. ______________

Pump performance curve number ______________

Bowl diameter __________ Capacity __________ Total head __________

Number of bowls __________ Depth of setting __________

Column diameter __________ Shaft diameter __________ Length __________

Length of strainer or tailpipe __________ Length of airline __________

Motor: Make ______________ Type ______________

Model ______________ Serial No. ______________

Horsepower ______________ RPM ______________

Voltage __________ Phase __________ Cycles __________

REPAIR RECORD		
Date	Type of Repairs	Repairs made by

Fig. 3.32 Sample record of pump information

Well No. ______________

Location______________

MOTOR Cat. No__________ Date______________
Serial__________Frame________Thr Brg.________Oil Cap.____
City Inv. No.______Mfg.______HP______Rad.Brg.______Oil Cap_
Voltage__________MPM__________Make__________Type__________
Yes____
Type of Shaft______________Weatherproof
No_____ Drive Nut
Thd.____

HEAD Cat. No. ______________
Mfg.__________City Inv. No.__________Serial No.________
Col. Size__________Max. Col. Size.______Col. Bolt Cir_____
Disch.Size __________Disch. Bolt Cir.__________
Tension Nut Type______________Thd.__________

COLUMN
Diameter______________Type of Thread______________

TUBE
Inside Dia.______________Type of Thread______________

SHAFT
Outside Dia__________Type of Thread______________

BOWLS Cat. No.______________
Mfg.____ ________Size or Type__________No. Stages______
Serial No.______________Impeller O.D.________ ________
Performance Curve No.______________Pump Level________
Draw Down__________ ____Capacity________GPM at_

SUCTION
Diameter______________Type of Thread______________
Adaoter______________X______________

STRAINER
Size______________

AIR LINE
Length______________

MISCELLANEOUS
Well Dia.________Sounding Depth______S.W.L.__________
Spiders At______________Joints

REMARKS ______________________________

Pulled By______________Installed By__________

Fig. 3.33 Sample record of well pump

(Source: City of Phoenix, Arizona)

WELL INSPECTION FORM

System: ______________________ IBM No. ____________

Source of Information: ______________________________

Evaluated By: ____________________ Date: ____________

Item or Component	Well No.	Well No.	Well No.	Well No.
Site fenced				
Lot paved				
Size of lot				
Nearest property line				
Wastewater disposal				
Nearest abandoned well				
Flood hazard, describe				
Year well drilled				
Cable tool				
Gravel envelope				
Hard rock or other				
Casing size and depth				
Conductor size and depth				
Sanitary grout seal, depth				
Perforations, depth, type				
Pedestal height				
Pump base sealed				
Slab around pump base, type				
Pump in pit, special construction				
Well casing vent, approved type				
Air-vac relief valve, approved type				
Well head seal, approved type				
Gravel tube, approved type				

Fig. 3.34 Well inspection form

Item or Component	Well No.	Well No.	Well No.	Well No.
Sampling tap, approved type				
Type of meter				
Type pump, HP				
Pump capacity, GPM				
Pump setting				
Lubrication, oil or water				
Power, electric or gas				
Auxiliary power, gearhead/generator				
Pump control				
Pressure range, psi				
Frequency of use				
Pump discharges to -				
Pressure tank, capacity				
Pressure relief valve				
Pressure tank drains to -				
Air release valve discharge				
Air charger, approved type				
Housing type				
Floor material				
Housekeeping				
Chemicals fed at well				
Chemical application point				
Well records maintained				
Pump efficiency tests				
Well log				
Water level measured				

Fig. 3.34 Well inspection form (continued)

preferably, by controlling the operation of the well pump by means of a time clock. An alternative, and perhaps less desirable, approach would be to keep the sand-producing well in an emergency or standby mode. This can be done by setting the cut-in (start) and cutoff (stop) pressure 10 pounds (psi) below the other pumps in the system.

If sand production from a well cannot be controlled by the methods described, then rehabilitation of the well or eventual abandonment may be necessary.

Many water utilities have adopted new well specifications that define a sand-free well as one having a sand content of less than five milligrams per liter. This is measured by a properly located Rossum sand tester when it is surged and pumped at 1,000 gallons per minute (63 liters/sec). An alternative to this specification would be a sand content of less than five milligrams per liter for a 10-minute period after the well pump starts.

3.72 Flushing Mains[41]

Numerous factors must be considered when a flushing program is established, such as hydraulic gradient from sources of supply, size of mains within the distribution system, location of main line valves, location of flushing outlets, and the ability to dispose of large quantities of water in a short period of time.

Generally speaking, the majority of sand complaints are usually associated with smaller mains and deadend lines. Repeated flushing of small diameter mains may be ineffective and costly. Sand problems associated with small mains may originate from the larger mains supplying them; unless the larger mains are adequately flushed, sand complaints are likely to continue from consumers served by the small mains.

Available data suggest that flushing velocities should be at least five feet per second (1.5 m/sec) or higher, and the total volume of water flushed should be 10 times the volume of the main. Velocities in this range are more readily obtainable in smaller mains, but may be difficult to achieve in larger mains unless special flushing outlets are provided.

The flushing guidelines shown in Table 3.9 are for each 1,000 feet (300 m) of water main and are based on a velocity of five feet per second (1.5 m/sec).

TABLE 3.9 FLUSHING GUIDELINES

Main Size in Inches	Flow in GPM	Flushing Time in Minutes	Volume of Water to Dispose of in Gallons
4	220	30	6,000
6	450	30	13,500
8	775	30	23,250
10	1,225	30	36,750
12	1,750	30	52,500

Flushing techniques and procedures should be tailored to the characteristics of the distribution system. In order to obtain the desired velocity of five ft/sec (1.5 m/sec) or more, it may be necessary to alter the normal water flow to a problem area by closing off selected valves and concentrating the flow in a given section of main. If the sand problem area is extensive, flushing should start on the source side of the problem area and progress toward the extremities of the distribution system.

Several minutes of flushing may be required before there is any noticeable amount of sand, but once sand is observed, the flushing operation should continue as long as sand is evident, even if the time period exceeds the 30 minute criterion shown in column three of the guideline (Table 3.9). In other words, don't shut off a flushing outlet if any significant amount of sand is present unless there is danger of flooding property. Where it is not possible to dispose of the entire quantity of water needed, it is more effective to flush at the required velocity for a shorter period of time than it is to reduce the velocity.

3.73 Test for Sand, Volumetric Method

Numerous devices are available for collecting sand samples, such as settling basins, filtration units, and centrifugal separators. The most frequently used device is the centrifugal sand separator which is designed for continuous sampling.

The centrifugal sand sampler most often used for evaluating new wells and monitoring existing wells is the Rossum sampler developed in the early fifties by John R. Rossum, Engineer of Sanitation for the California Water Service Company. Rossum's work on the quantitative measurement of sand and his theoretical and experimental investigations of hydraulic transportation of sand through water mains have become a standard in the water utility field. The Rossum sand sampler is illustrated in Figure 3.35.

The sampler is usually installed on the discharge piping from the well pump. Samples taken immediately downstream from valves, meters, and other fittings that create turbulence are representative if the water velocity is five feet per second (1.5 m/sec) or more. The water enters the sampler tangentially (along the outside edge), and the entrained sand, thrown to the outside by centrifugal force, falls into the graduated tube. The sand-free water flows out through an orifice (opening) in the top of the unit. Flow through the unit is maintained by a flow control valve rated at 0.5 gallon per minute (0.03 liter/sec). Periodically, the graduated tube is checked and the volume of sand recorded, along with the time period the pump has operated. With this information, it is possible to compute the average sand concentration based on the flow through the tester.

Studies conducted by Rossum have shown that the sand concentration is considerably greater when a well is first started than it is after a period of continuous operation. This effect is illustrated in Figure 3.36.

The sand tester is a valuable tool. The tester is used to determine if new wells meet the contract specifications for sand volume. When used as a daily operational tool, any increase in sand production can be readily detected and remedial steps taken before any significant amount of sand enters the distribution system.

3.74 Acceptable Concentrations

Studies have been made that suggest that permissible concentration of sand production from a well should not exceed 0.3 cubic foot per million gallons of water (2.2 cubic centimeters of sand per cubic meter of water) pumped. Sand concen-

[41] *See Chapter 5, "Distribution System Operation and Maintenance," in WATER DISTRIBUTION SYSTEM OPERATION AND MAINTENANCE for additional information on how to flush water mains.*

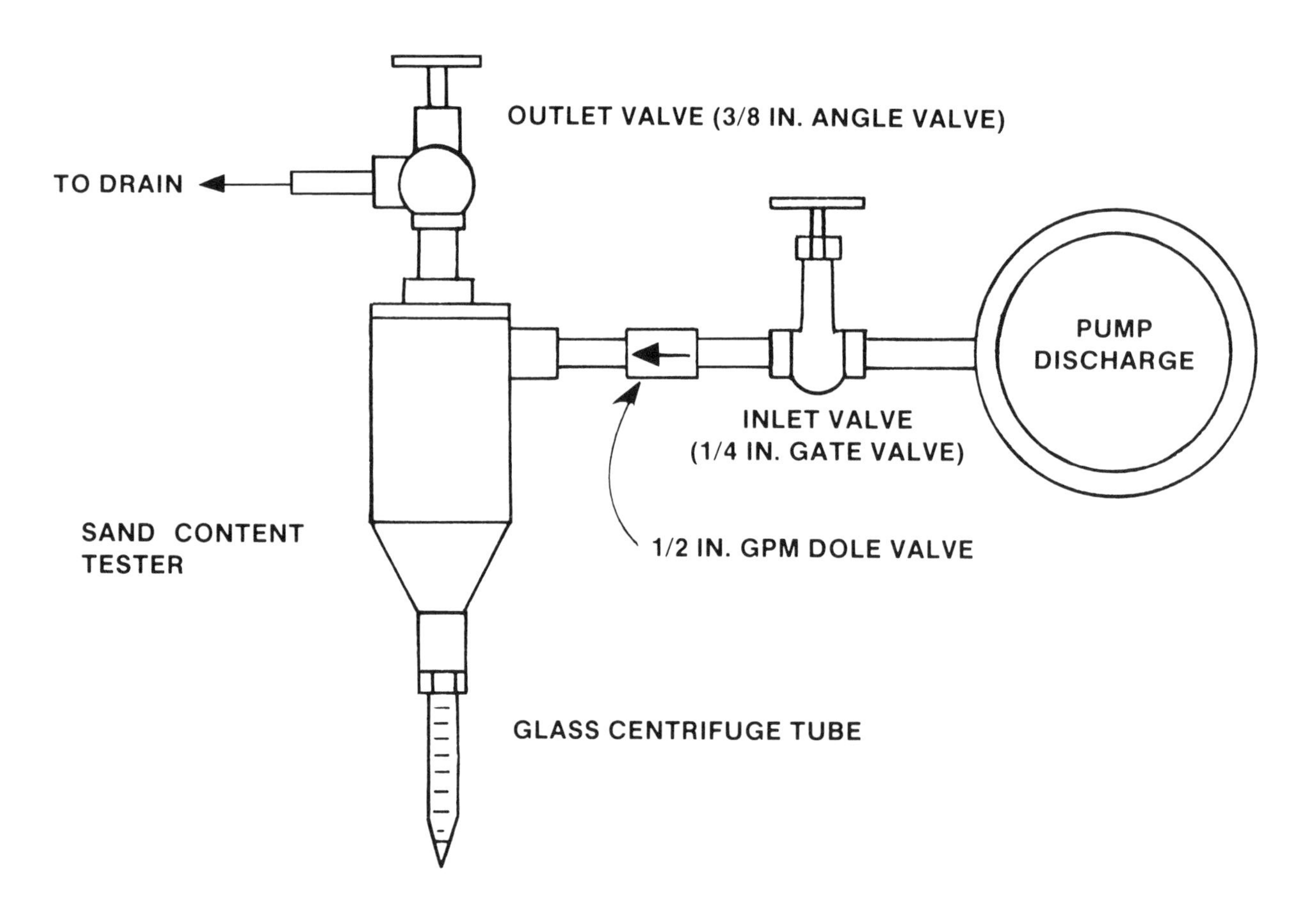

Fig. 3.35 Rossum sand sampler

trations above this value will almost always result in sanded meters and consumer complaints. Ideally, the sand concentration should not exceed 0.1 cubic foot per million gallons (0.7 cubic centimeter of sand per cubic meter of water).

3.75 Responding to Complaints

Respond to a sand complaint by obtaining the name, address, and phone number of the person calling. Try to determine the source of the sand (well, main) and how many other people have similar problems. Institute corrective action such as removing sand at the well, flushing the mains, or relocating the consumer's tap at or near the top of the main.

Detailed records should be maintained showing the date and location of sand complaints, including estimated concentrations. From this record, you can establish a routine flushing program to flush these problem areas before sand builds up to a nuisance level.

When flushing is necessary, notify consumers in the immediate area not to use water during this flushing period. Heavy concentrations of sand in suspension could be drawn into the consumers' premises causing malfunction of appliances, fixture damage, plugging of lawn sprinklers, and numerous other problems leading to additional complaints and callbacks.

Sand complaints may be isolated to only one or two residences in a block. This condition can occur if the consumer's service lateral is tapped into the side of the main near a fitting or valve which creates turbulence. This type of situation can generally be corrected by relocating the tap at or near the top of the main.

QUESTIONS

Write your answers in a notebook and then compare your answers with those on page 128.

3.7A How can a well in a sand formation be constructed so that it will be almost sand free?

3.7B What problems are associated with sand in well water?

3.7C The majority of sand complaints concern what part of the distribution system?

3.7D How long should a flushing operation be continued?

3.7E What is the permissible concentration of sand in water from a well?

3.8 ELECTRICAL SUPPLY AND CONTROLS

3.80 Purpose of Electrical Supply and Controls

Control, reduced to its basic definition, is the influence over the action of a device resulting from the measurement and decision of another device or control.

Many options are available for electrical and electronic control of well production. These range from a simple manual start/stop system to automated facilities complete with recorded drawdown and water quality analysis. In this section, we will limit our discussion to the most common and useful types of controls and automation.

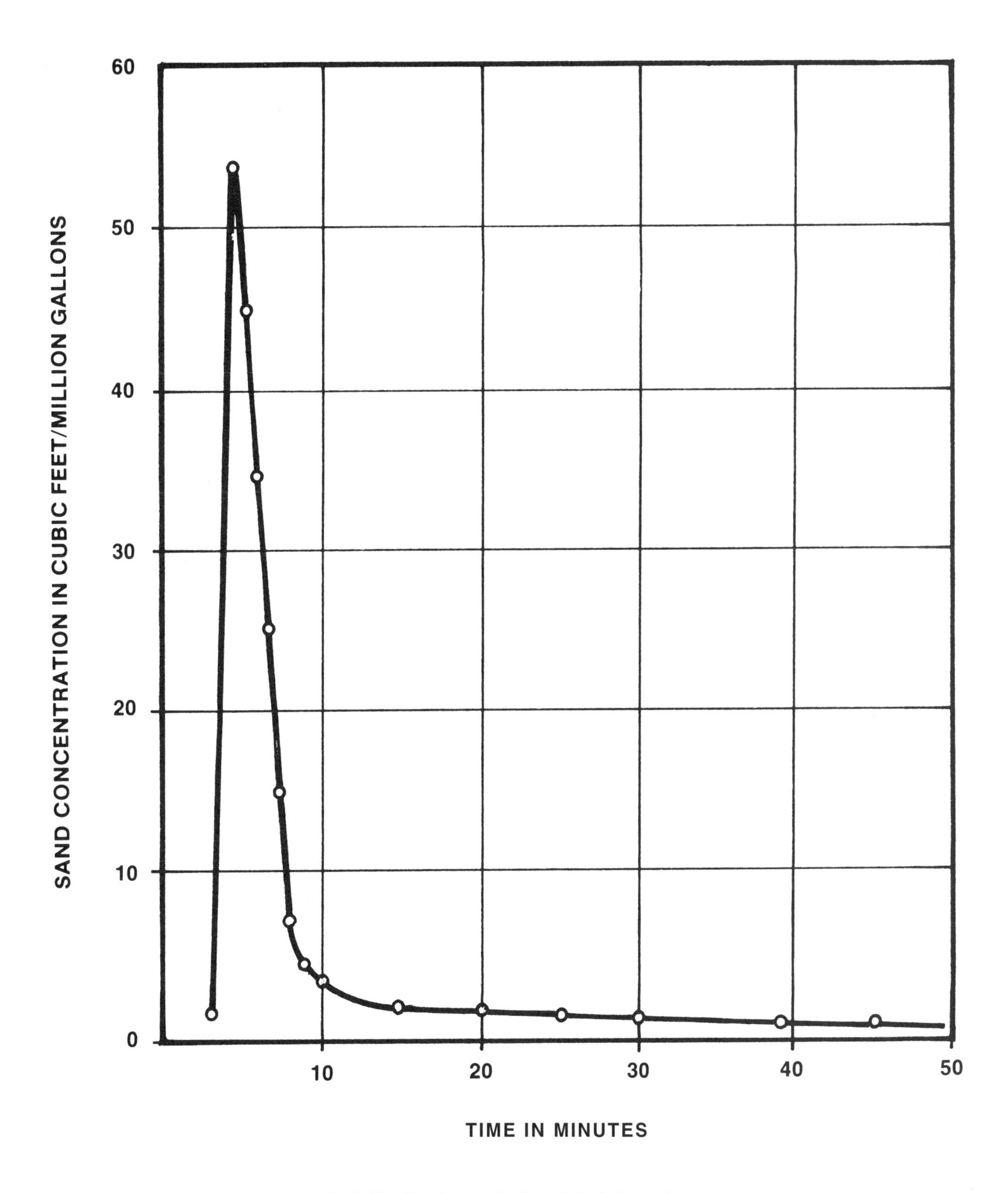

Fig. 3.36 Sand concentration at start of pumping

3.81 Electrical Supply

3.810 Electricity

Electricity is normally supplied as an alternating current (A.C.) at 120, 240, or 480 volts. The voltage is considered the driving force and quite often the amount of voltage available depends on the location of the well installation and the ability of the utility company to provide power. The voltage required is related to the size of the pump; larger motors require higher voltages. High horsepower (HP) motors may require large surges of power at start-up.

3.811 Motor Starters

Motor starters are basically controls for starting and stopping motors used with large pumps. Upon receiving a signal to start, current is fed into the motor causing the pump to run. Starters can be direct, across-the-line types or what is referred to as step starters. In step starters power to the motor is increased slowly (in steps) allowing the pump to come up to speed gradually. This prevents pump damage and disturbances along electrical lines. A typical across-the-line starter arrangement for a three-phase pump is shown in Figure 3.37.

3.812 Auxiliary Power

Auxiliary power is an important consideration to many water districts. If a water supply system depends on wells and hydropneumatic pressure tanks for both supply and storage, a power outage could create very severe water shortages. Auxiliary power is not difficult to provide at well sites, but it is expensive. These systems can be installed to start and stop automatically and are quite dependable. A typical gasoline-driven engine generator is shown in Figure 3.38. There are numerous manufacturers of diesel or gasoline-powered generators and they can be obtained in almost any size.

The manufacturers of auxiliary generators also produce automated control packages, but the most commonly used control device is a time delay starter. If a power failure occurs, a relay will sense the loss of power and drop out. A timer then measures a definite time delay that power is off and, after a predetermined period, the auxiliary engine is activated. Shutdown usually occurs in the same manner.

Although we are not going to describe auxiliary power systems in detail here, they should be an important consideration for any water district. These systems can be a real asset under very adverse conditions.

3.82 Pump Controls

3.820 Types of Controls

There are three major types of pump controls, (1) ON/OFF, (2) proportional, and (3) derivative (sometimes called "reset" or "rate"). Electrical controls fall mostly into the first category. ON/OFF, the simplest form, consists of a measured variable such as the level in a pond, or the pressure in a tank, which upon falling to some preset level, closes a switch contact. This engages a motor which drives a pump to increase the level until it reaches a preset cutoff point. For most applications, the use of the simple ON/OFF set of controls is quite acceptable, and it has the advantages of being low in cost, having few parts, and performing reliably.

As drinking water regulations become more stringent, some form of proportional control will be needed. Proportional control provides more corrective effort as the measured variable gets farther from the *SET POINT.*[42] In the case of the reservoir level, this might mean that as the level gets low, and then lower, several additional pumps may be called upon to pump into it. As the level approaches the desired level, the extra pumps are turned off, and eventually, as it arrives at the set point, all pumps stop. Figure 3.39 shows the typical start/stop arrangement for both ON/OFF and proportional controls.

The derivative or rate controls are used to maintain water levels or pressures within very close tolerances. This type of control is normally coupled with variable-speed motor drives. As the need arises, the controller can cause the pump to increase or decrease its speed to keep water levels or pressures within closely confined limits. Unless the system is highly sophisticated or is restricted by critical operating guidelines, rate controls are seldom used.

3.821 Control Systems

Control systems vary but practically all pumping facilities have some kind of automatic start/stop arrangement. Various pump start/stop sequences are associated with one or more of the following: pressure, water level, time sequences, heat protection, backspin protection, flow, and water quality. The control system is quite often coupled with recordings of flow, bearing temperatures, pressure, water levels, and alarms. The two most common pump control elements are pressure and water level.

3.822 Pressure

Pressure is a necessary element of a water system and can easily be monitored at various locations in the system. The most common such pressure regulation system is that associated with hydropneumatic tanks. As a pressure reduction is sensed, a signal is sent to the pumps to start. The subsequent pump start-up pushes water into the system which results in an increase in pressure. Once the desired pressure is reached and sustained for a predetermined length of time, a signal is sent to the pump ordering a shutdown. High or low pressure cutoffs (signals to shut off the pumps if pressure is too high or too low) are common safety features built into the system. These are often coupled with alarms which alert the operator to unusual conditions at the pump. High pressure might indicate valve failure or blockage in the discharge lines. Low pressure would be an indicator of excessive water use or a broken water line. In either case, the automatic pressure controls provide a mechanism for shutting down the pumps until the problem is located and corrected.

3.823 Water Level

The most frequently used pump control system is the water level controller in reservoirs and storage tanks. A sensor

---

[42] *Set Point. The position at which the control or controller is set. This is the same as the desired value of the process variable.*

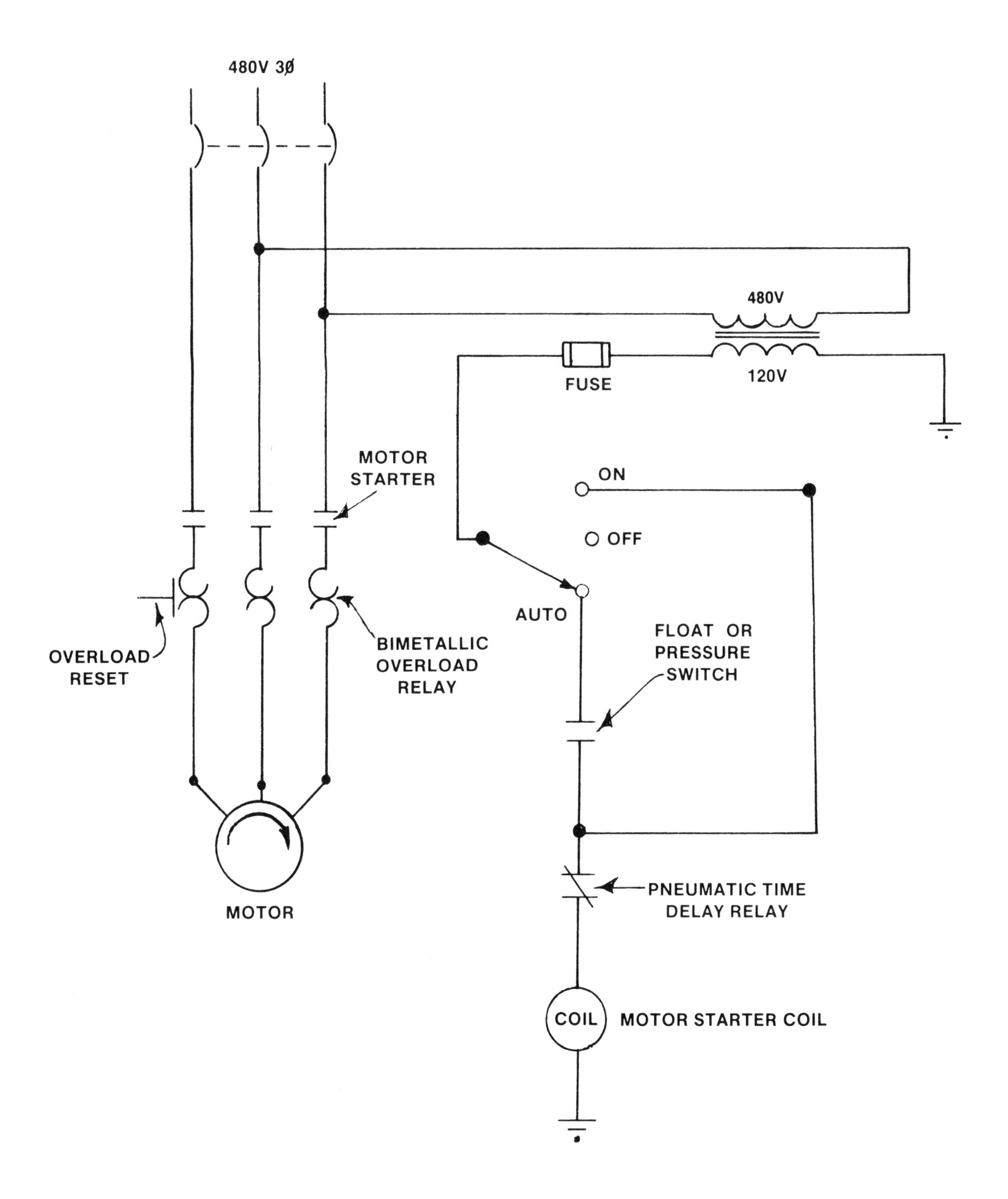

Fig. 3.37 Schematic of typical 3-phase pump control starter circuit

Fig. 3.38 Gasoline-powered auxiliary power generator
(Photo courtesy of Onan Corporation, Minneapolis, Minnesota)

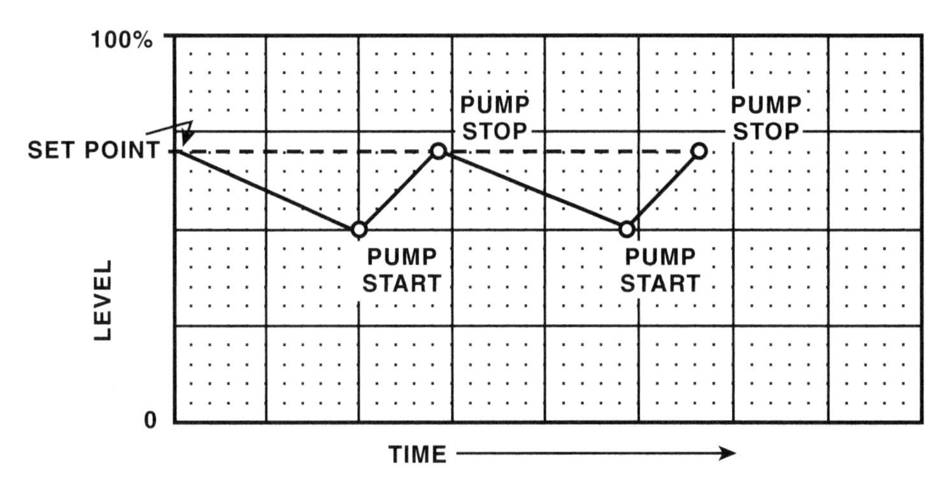

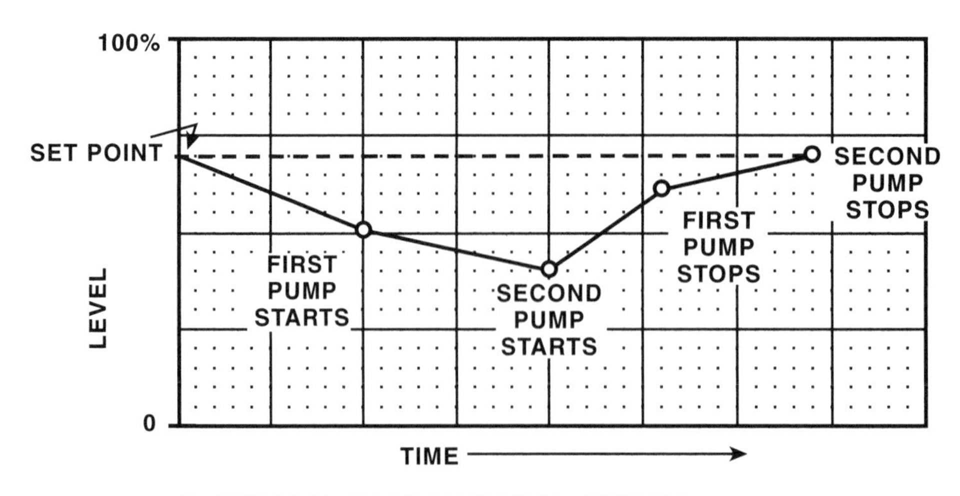

Fig. 3.39 Typical pump start/stop control systems

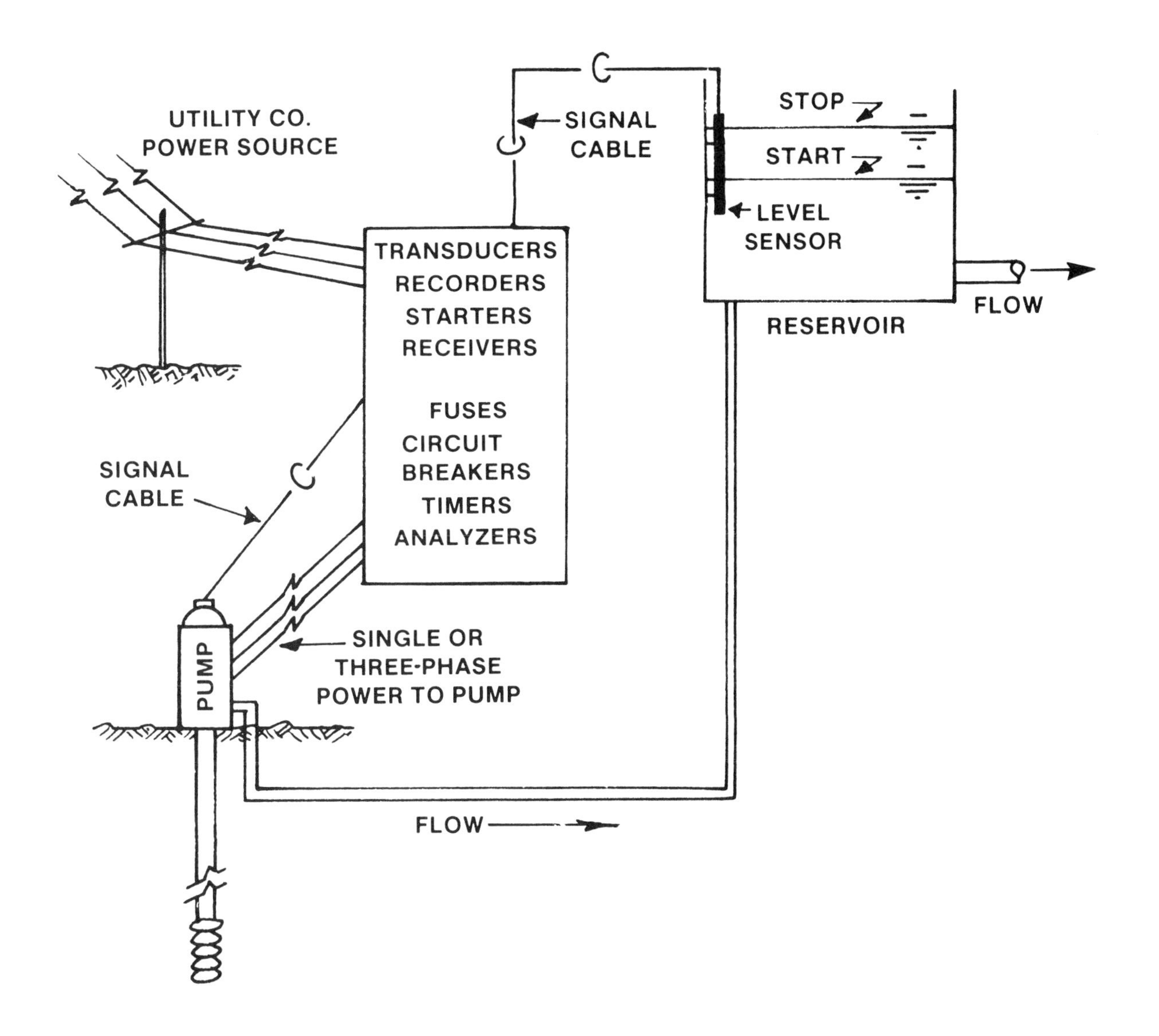

Fig. 3.40 Typical pump control system

measures the water level and signals the pump to start or stop. As the water level is lowered in the tank, the pump is instructed to start. Once the tank has filled, the pump will be ordered to stop. A typical pump control system from a reservoir is shown in Figure 3.40. However, well pump starts/stops can also be controlled by the water level in a well.

In the case of wells, water level operation would only be implemented in the case of low-yield wells that had to be protected from overdraft. Damage to the pumps can occur if the water level is allowed to be drawn below the pump bowls. If low water level appears to be a real problem associated with a particular well, then some provision for low level shutdown must be made.

Other items that are often monitored or controlled automatically at pump installations are excessive power demand by the pump, bearing heat on the motor and/or pump, turbidity measurements, and flow. Backspin protection[43] is provided for with time delays built into the automatic controls. High head or high volume pumps are sometimes shut down through the use of automatic pump control valves. The total number and combination of controls of pumps is extremely large but automation can provide a high measure of reliability along with lower cost for operation and maintenance.

3.83 Equipment

Control equipment ranges from direct, connected units from tanks to pumps joined by signal wires, to units separated by several miles from the pumps which transmit signals through telephone lines or by radio. Definitions of equipment commonly used in control systems are listed in Section 3.84. Figure 3.41 shows a typical pump control operation.

3.84 Common Electrical Control Definitions

1. *ALARM CONTACT.* A switch that operates when some preset low, high or abnormal condition exists.

[43] *If pump and motor are restarted while backspinning (water driving pump in reverse), the shaft may break causing expensive repairs. Provision must be made to prevent this from occurring.*

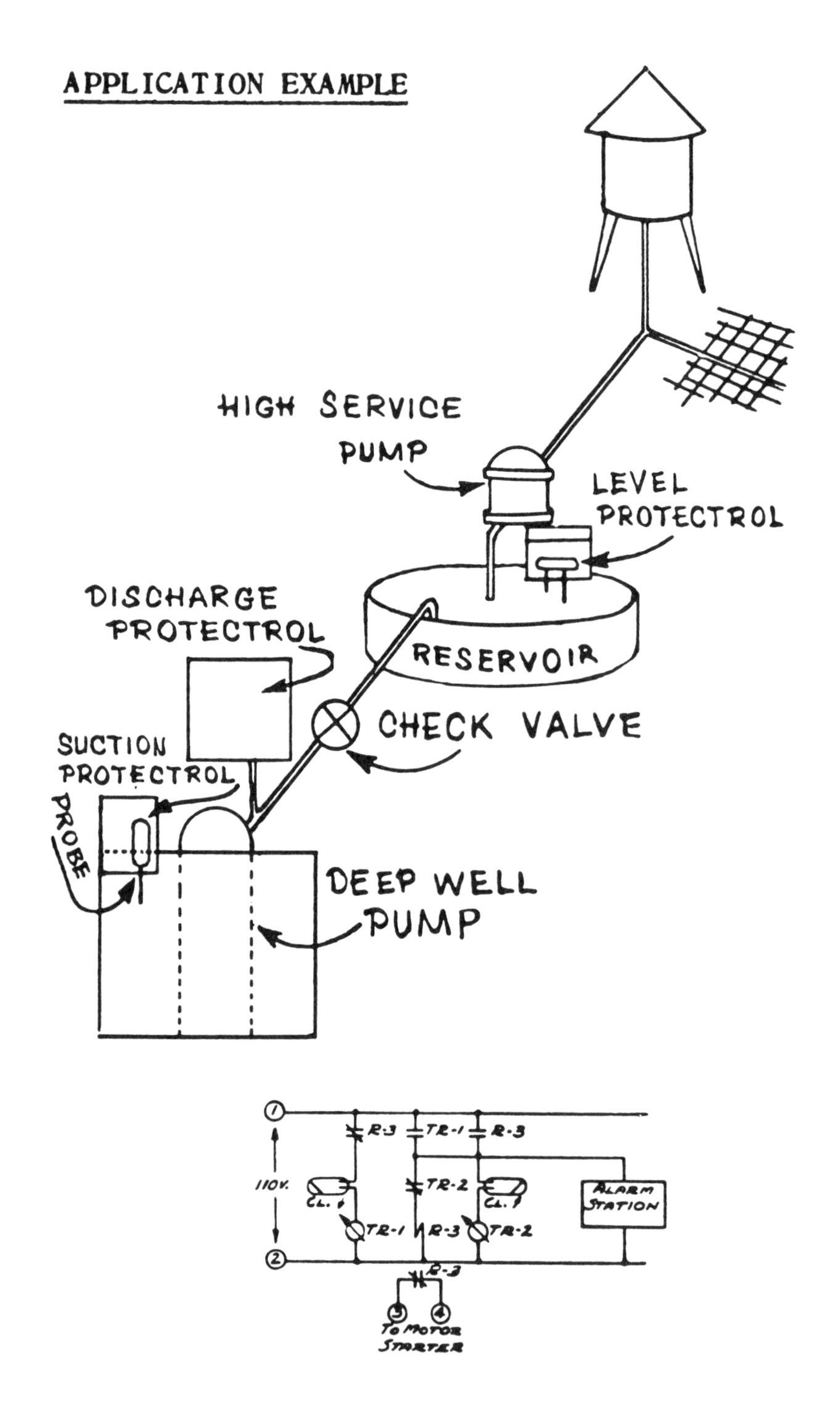

Two pressure elements are manifolded and tapped directly into the suction main. If pressure falls to a danger point, the "cutout" element will close its switch, completing the circuit to the timer (TR-1), which begins to time out. If pressure remains low long enough for timer to time out, its contact will close, energizing the relay (R-3). A normally closed relay contact then opens, stopping the pump. When the pressure rises, the "restore" switch closes, energizing the second timer (TR-2). After it has timed out, its normally closed contact will open, de-energizing the relay, allowing the pump to restart if required by primary control.

Fig. 3.41 Typical pump control operation

(Permission of Automatic Control Company)

2. *ANALYZER.* A device which conducts periodic or continuous measurement of some factor such as chlorine, fluoride or turbidity. Analyzers operate by any of several methods including photocells, conductivity or complex instrumentation. A pH meter is a type of analyzer.

3. *CONTACTOR.* An electric switch, usually magnetically operated.

4. *CONTROLLER.* A device which controls the starting, stopping, or operation of a piece of equipment.

5. *HEAT SENSOR.* A device that opens and closes a switch in response to changes in the temperature. This device might be a metal contact, or a thermocouple which generates a minute electric current proportional to the difference in heat, or a variable resistor whose value changes in response to changes in temperature. Also called a temperature sensor.

6. *INTERLOCK.* An electric switch, usually magnetically operated. Used to interrupt all (local) power to a panel or device when the door is opened or the circuit is exposed to service.

7. *LEVEL CONTROL.* A float device (or pressure switch) which senses changes in a measured variable and opens or closes a switch in response to that change. In its simplest form, this control might be a floating ball connected mechanically to a switch or valve such as is used to stop water flow into a toilet when the tank is full.

8. *MEASURED VARIABLE.* A characteristic or component part that is sensed and quantified (reduced to a reading of some kind) by a primary element or sensor.

9. *PRESSURE CONTROL.* A switch which operates on changes in pressure. Usually this is a diaphragm pressing against a spring. When the force on the diaphragm overcomes the spring pressure, the switch is actuated (activated).

10. *PRIMARY ELEMENT.* A device that measures (senses) a physical condition or variable of interest. Floats and thermocouples are examples of primary elements. Also called a sensor.

11. *RECEIVER.* A device which indicates the result of a measurement. Most receivers in the water utility field use either a fixed scale and movable indicator (pointer) such as a pressure gage or a movable scale and movable indicator like those used on a circular flow-recording chart. Also called an indicator.

12. *RECORDER.* A device that creates a permanent record, on a paper chart or magnetic tape, of the changes in a measured variable.

13. *SENSOR.* A device that measures (senses) a physical condition or variable of interest. Floats and thermocouples are examples of sensors. Also called a primary element.

14. *SET POINT.* The position at which the control or controller is set. This is the same as the desired value of the process variable or level in a tank.

15. *SOLENOID.* A magnetically (electric coil) operated mechanical device. Solenoids can operate small valves or electric switches.

16. *STARTERS.* Most starters for small motors are "across-the-line" which means that they simply connect the motor terminals to the incoming line. Large motors would impose severe mechanical shock to the driven machine if started this way, as well as creating a severe disturbance to the electrical lines, causing dimming and flickering lights. Therefore, motors over 100 HP (75 kW) are usually started by "reduced voltage" starters or two-step starters. The voltage reduction can be accomplished by the use of an auto-transformer with taps that provide 50 percent voltage until the motor and load are moving, then go to full voltage "across-the-line." These can be operated manually, in response to a pair of switches, or automatically in response to motor current or a definite time delay which allows the assembly time to come up to speed.

17. *TIME LAG.* The time required for processes and control systems to respond to a signal or reach a desired level.

18. *TIMER.* A device for automatically starting or stopping a machine or other device at a given time.

19. *TRANSDUCER.* A device which senses some varying condition measured by a primary sensor and converts it to an electrical or other signal for transmission to some other device (a receiver) for processing or decision making. Flowmeters and heat sensors are examples of transducers.

20. *VARIABLE FREQUENCY DRIVE (V.F.D.).* A control system which allows the frequency of the current applied to the motor to be varied. The motor is connected to a low frequency source while standing still, and the frequency is increased gradually until the motor and pump, or other driven machine, is at the desired speed. This system offers the additional advantage of continuous control of speed, in accord with some measurement.

3.85 Instrumentation

For additional information on controls and instrumentation, see Chapter 19, "Instrumentation," in Volume II, *WATER TREATMENT PLANT OPERATION*, of this series of operator training manuals.

QUESTIONS

Write your answers in a notebook and then compare your answers with those on page 128.

3.8A The voltage required for a pump depends on what factors?

3.8B Why do large pumps need motor starters?

3.8C List the three major types of pump controls.

3.8D How does a proportional pump control work?

3.9 TROUBLESHOOTING

3.90 Need for Troubleshooting

Approximately 75 percent of well pump and control problems are associated with electricity. The well pump operator should have a good working knowledge of electric circuits and circuit testing instruments before attempting to service or troubleshoot the electric circuits and components commonly used in well pump operations. The operator should not undertake any electrically related troubleshooting or repair job until instructions have been received on how to do it properly and until the operator has been authorized to perform that job.

Small water utilities that do not have a knowledgeable operator or electrician on their staff should arrange with a local electrical firm or pump service company to perform this service.

3.91 Troubleshooting Guide

The "Troubleshooting Guide" on the following pages is designed to assist the operator or service personnel in diagnosing and correcting the most common problems associated with well pumping facilities.

TO USE THE GUIDE:

1. Find the appropriate condition in the SYMPTOM column or section.
2. Find the cause in the PROBABLE CAUSE column.
3. Perform the CORRECTIVE ACTION listed in the third column. The remedy is listed briefly and the procedures may not be detailed enough to cover every possibility.

The operator should not proceed if there is any doubt as to what is meant or what course of action should be taken to correct a given problem.

QUESTIONS

Write your answers in a notebook and then compare your answers with those on page 128.

3.9A What is the greatest cause of well pump and control problems?

3.9B If you are neither authorized nor qualified to perform electrical repairs, who should do it?

3.9C How would you correct a waterlogged hydropneumatic tank?

3.9D What could cause a pump to draw too much power?

3.10 ARITHMETIC ASSIGNMENT

Turn to the Appendix, "How to Solve Small Water System Arithmetic Problems," at the back of this manual and read the following sections:

1. A.2, *AREAS*,
2. A.3, *VOLUMES*,
3. A.10, *BASIC CONVERSION FACTORS*,
4. A.11, *BASIC FORMULAS*, and
5. A.12, *HOW TO USE THE BASIC FORMULAS*.

Check all of the arithmetic in Sections A.2, *AREAS* (A.20, A.21, A.22, A.23, A.24, A.25, and A.26) and A.3, *VOLUMES* (A.30, A.31, A.32, A.33, and A.34) on an electronic pocket calculator. You should be able to get the same answers.

3.11 ADDITIONAL READING

TEXAS MANUAL, Chapter 13,* "Pumps and Measurement of Pumps."

* Depends on edition.

End of Lesson 4 of 4 Lessons on WELLS

Please answer the discussion and review questions before continuing with the Objective Test.

DISCUSSION AND REVIEW QUESTIONS

Chapter 3. WELLS

(Lesson 4 of 4 Lessons)

Write the answers to these questions in your notebook before continuing with the Objective Test on page 129. The question numbering continues from Lesson 3.

22. What are the general responsibilities of operators of well systems?
23. How frequently should operators make routine service calls to a well pumping facility?
24. What procedures can be followed to reduce sand problems in wells drilled in sand formations?
25. How would you respond to a complaint by a consumer about sand in the water?
26. If a power failure occurs, how are auxiliary power generators started?
27. What are the advantages of ON/OFF pump controls?
28. Why do pumps have low-pressure cutoff controls?
29. What types of problems might an operator encounter when working with pumps?

TROUBLESHOOTING GUIDE

Symptom	Probable Cause	Corrective Action
3.910 Pump Will Not Start	Circuit breaker or overload relay tripped, motor cold.	Reset breaker or reset manual overload relay.
	Fuses burned out.	Check for cause and correct, replace fuses.
	No power to switch box.	Confirm with multimeter by checking incoming power source, notify power company.
	Motor is hot and overload relay has tripped.	Allow motor to cool. Check supply voltage. If low, notify power company. If normal, reset overload relay, start motor, check amperage; if above normal, call electrician.
	Loose or broken wire, or short.	Tighten wiring terminal, replace any broken wires, check for shorts and correct.
	Low line voltage.	Check incoming power, use multimeter; if low, notify power company.
	Defective motor.	*MEG*[44] out motor; if bad, replace.
	Defective pressure switch.	With contact points closed, check for voltage through switch; if no voltage, replace switch; if low voltage, clean contact points; if full voltage, proceed to next item.
	Line to pressure switch is plugged or valve in line has accidentally been shut off.	Open valve if closed. Clean or replace line.
	Pump control valve malfunctioning.	Check limit switch for proper travel and contact. Adjust or replace as required.
	Defective time delay relay or pump start timer.	Check for voltage through relay or timer — replace as necessary — check for loose linkage.
	Float switch or transducer malfunctioning.	If pump is activated by float switch or pressure transducer on storage tank, check for incoming signal; if no signal, check out switch or transducer with multimeter. If OK, look for broken cable between storage tank and pump station.
3.911 Pump Will Not Shut Off	Defective pressure switch.	Points in switch stuck or mechanical linkage broken, replace switch.
	Line to pressure switch is plugged or valve in line has been accidentally shut off.	Open valve if closed. Clean or replace plugged line.
	Cutoff pressure setting too high.	Adjust setting.
	Pump control valve malfunctioning.	Check limit switch for proper travel and contact. Adjust or replace as required.
	Float switch or transducer malfunctioning.	Defective incoming signal, check and replace components as required. Check cable.
	Defective timer in pump stop mode.	Check for voltage through pump stop timer, replace if defective.
3.912 Pump Starts Too Frequently	Pressure switch cut-in and cutoff settings too close.	Adjust settings, maintain minimum 20 psi (138 kPa or 1.4 kg/sq cm) differential.
	Waterlogged tank.	Add air to tank. Check air charging system and air release valve. Also check tank and connections for air leaks.
	Leaking foot valve.	Check for backflow into well; if excessive or if pump shaft is turning backward, correct problem as soon as possible.
	Time delay relay or pump start/stop timers are malfunctioning.	Check relay or timers for proper operation, replace defective components.

[44] *Meg. A procedure used for checking the insulation resistance on motors, feeders, bus bar systems, grounds, and branch circuit wiring.*

TROUBLESHOOTING GUIDE (continued)

Symptom	Probable Cause	Corrective Action
3.913 Fuses Blow, Circuit Breaker or Overload Relays Trip When Pump Is in Operation	Switch box or control not properly vented, or in full sunshine or dead air location, overload relay may be tripping due to external heat.	Provide adequate ventilation (may require small fan). Provide shelter from sun. Paint box or panel with heat reflective paint, preferably white.
	Incorrect voltage.	Check incoming power source. If not within prescribed limits, notify power company.
	Overload relays tripped.	Check motor running amperage, verify that thermal relay components are correctly sized to operating conditions. Repeated tripping will weaken units, replace if necessary.
	Motor overloaded and running very hot.	Modern motors are designed to run hot and if the hand can be held on the motor for 10 seconds without extreme discomfort, the temperature is not damaging. Motor current should not exceed *NAMEPLATE*[45] rating. Fifteen percent overload reduces motor life by 50 percent.
3.914 Pump Will Not Deliver Normal Amount of Water	Pump breaking suction.	Check water level to be certain water is above pump bowls when operating. If not, lower bowls.
	Pump impeller improperly adjusted.	Check adjustment and lower impellers (qualified personnel only).
	Rotation incorrect.	Check rotation.
	Impellers worn.	If well pumps sand, impeller could be excessively worn thus reducing amount of water pump can deliver. Evaluate and recondition pump bowls if required.
	Pump control valve malfunctioning.	Check limit switch for proper travel and contact. Adjust or replace as required.
	Impeller or bowls partially plugged.	Wash down pump by forcing water back through discharge pipe. Evaluate sand production from well.
	DRAWDOWN[46] more than anticipated.	Check pumping water level. Reduce production from pump or lower bowls.
	Pump motor speed too slow.	Check speed and compare with performance curves. Also check lift and discharge pressure for power requirements.
3.915 Pump Takes Too Much Power	Impellers not properly adjusted.	Refer to manufacturer's bulletin for adjustment of open or closed impellers.
	Well is pumping sand.	Check water being pumped for presence of sand. Restrict discharge until water is clean. Care should be taken not to shut down pump if it is pumping very much sand.
	Crooked well, pump shaft binding.	Reshim between pump base and pump head to center shaft in motor quill. Never shim between pump head and motor.
	Worn bearings or bent shaft.	Check and replace as necessary.
3.916 Excessive Operating Noise	Motor bearings worn.	Replace as necessary.
	Bent line shaft or head shaft.	Check and replace.
	Line shaft bearings not receiving oil.	Make sure there is oil in the oil reservoir and the oiler solenoid is opening. Check sight gage drip rate, adjust drip feed oiler for 5 drops per minute plus 1 drop per minute for each 40 feet (12 m) of column.

[45] *Nameplate. A durable metal plate found on equipment which lists critical operating conditions for the equipment.*
[46] *Drawdown. The drop in the water table or level of water in the ground when water is being pumped from a well.*

SUGGESTED ANSWERS

Chapter 3. WELLS

ANSWERS TO QUESTIONS IN LESSON 1

Answers to questions on page 60.

3.0A The purpose of a well is to intercept groundwater moving through aquifers and bring water to the surface for use by people.

3.0B Two potential problems that can affect wells are (1) pollution, and (2) excessive use (overdraft).

3.0C The principal reasons for using groundwater include:

1. Generally available in most localities,
2. Well and pumping facilities cost less than surface treatment facilities,
3. Water is clearer and can usually meet turbidity requirements,
4. Conditions for growth and survival of bacteria and viruses are unfavorable when compared with surface waters,
5. Mineral content for given well is usually uniform,
6. Temperature is usually constant and lower in summer, and
7. Well supplies are particularly suited to the needs of smaller communities in many areas.

3.0D The hydrologic cycle is the continuous circulation of water on our planet. This cycle is the process of evaporation of water into the air and its return to earth by precipitation (rain or snow).

Answers to questions on page 61.

3.0E Porosity is a measure of the openings or voids in a particular soil. Porosity measurements quantify the amount of water that a particular soil type or rock can store.

3.0F Gravitational forces cause water movement through the pores in soil and rocks.

3.0G Problems that can be caused by overdraft include:

1. Permanent damage to the water storage and transmitting properties of aquifers, and
2. Subsidence of land.

Answers to questions on page 62.

3.0H The differences in specific gravities of fresh water and salt water usually keep them from mixing in aquifers and underground basins.

3.0I Excessive drawdown of a well can bring salt water into the aquifer and close enough to the well that salt water will be drawn into the cone of depression and into the well.

3.0J Salt water intrusion can develop along coastlines and inland where underground water supplies contain more than 1,000 mg/*L* of total dissolved solids.

Answers to questions on page 63.

3.0K The soil mantle acts as a natural filter and removes suspended particles from water flowing through it.

3.0L The soil mantle does not always filter out dissolved pollutants and contaminants such as organic chemicals.

3.0M To preserve the quality of aquifers, operators should take the following actions:

1. Be alert to any construction that might result in wastes entering the groundwater stream and report any such activity to the proper authorities,
2. Make sure that wells under the operator's control do not contaminate the aquifer through inadequate grouting or seals, and
3. Watch for and report any other potential sources of groundwater pollution so that appropriate action can be taken.

3.0N Records of well performance that should be kept by the operator include pumping amounts and dates, depths of water at start and end of pumping, time of pumping, and water quality results.

Answers to questions on page 66.

3.0O Kinds of wastewater treatment processes that can occur when water passes through soil include filtration, ion exchange, adsorption, dilution, oxidation, and biological decay.

3.0P A wellhead protection program can achieve control of sources of contaminants and certain types of development activities on private and public lands by gaining community involvement and support.

3.0Q Types of data that must be gathered for a wellhead protection program include:

1. Geology of community,
2. Information on aquifer(s),
3. Information on groundwater,
4. Location of underground storage tanks,
5. Location of septic tanks and identification of ultimate disposal methods,
6. Location of wells in community,
7. Location of businesses which may present threats to groundwater quality,
8. Community regulations and by-laws already in place, and
9. Town and state public works practices.

Answers to questions on page 70.

3.0R Land use activities considered to have a severe risk level to groundwater include dry cleaners, gas stations, junk yards, highway de-icing, underground storage tanks, injection wells, hazardous waste disposal, and landfills. See Table 3.3, page 66.

3.0S Organic chemical contaminants found in domestic wastewater include trace amounts of petroleum distillates as well as benzene, toluene, chlorinated hydrocarbons such as trichloroethane, trichloroethylene, tetrachloroethylene, dichlorobenzene, and alkyl phenols.

3.0T Water quality tests for nitrate and coliform contamination should be performed if a septic tank system may have failed in the vicinity of shallow wells.

3.0U The most important result of a community groundwater study could be to educate the community officials about the threats to public safety which exist in the community and the actions they can take to protect the public health.

Answers to questions on page 73.

3.1A Openings are installed in the top of the well to provide access to the well for taking water level measurements, permitting entrance or escape of air or gas, adding gravel, or for applying disinfection or chemical cleaning agents.

3.1B The purpose of the well-casing vent is to prevent vacuum conditions inside a well by admitting air into the well during the drawdown period when the well pump is first started. (If vacuum conditions are allowed to develop, contamination may be sucked into the well through a possible defective well or conductor casing, or through the top of the well at the pump base.) The well-casing vent also prevents pressure buildup inside the well casing by allowing excess air to escape during the well recovery (refilling) period after the well pump shuts off.

3.1C The water level in a well is determined by inserting a measuring tape into the sounding tube, lowering it down the tube to the water level, and recording the distance. Or, air pressure in a sounding line may be used; the pressure necessary to force water out of the line is recorded and converted to feet. Then,

$$\text{Distance Down to Water Surface, ft} = \text{Length of Line, ft} - \text{Water Depth in Line, ft}$$

ANSWERS TO QUESTIONS IN LESSON 2

Answers to questions on page 78.

3.2A Check valves act as an automatic shutoff valve when the pump stops to prevent draining of the system or the tank being pumped to, and to prevent pressurized water from flowing back down the pump column into the well.

3.2B A check valve must (1) close watertight, (2) be able to safely withstand hydraulic shock loads, (3) have a low head loss, and (4) be equipped to close the valve disc in advance of flow reversal.

3.2C Pump control valves serve to eliminate line surge when the pump is started and stopped.

3.2D Flows must be known in order to calculate the correct chemical feed rate.

Answers to questions on page 80.

3.2E Sand particles should be prevented from entering the distribution system because sand particles can cause (1) reduced pump efficiencies due to worn impellers, (2) sanded water mains, (3) excessive meter wear and plugging, (4) wearing of household plumbing fixtures and appliances, and (5) customer complaints.

3.2F Sand traps are not recommended for the removal of sand from well water because they are costly and inefficient.

3.2G Most sand separators are relatively effective in removing fine sand, scale, and similar materials from water.

3.2H Sanitary hazards associated with sand separators or sand traps are generally limited to the discharge system from the accumulator tank and the sand disposal basin or structure.

3.2I The interior surfaces of tanks are painted to (1) extend the useful life of the tank, (2) prevent discolored water from entering the distribution system, and (3) to facilitate cleaning of the tank.

Answers to questions on page 84.

3.2J Booster pumps are installed to pump a smaller amount of water to a pressure zone that operates at a much higher pressure or elevation than the main well pump system.

3.2K Surge suppressors are installed to absorb shock waves in the fluid (water) system and prevent their transmittal through the line.

3.2L The purpose of an air release/vacuum breaker valve is to (1) exhaust large quantities of air very rapidly from a deep well pump column when the pump is started, and (2) allow air to re-enter the pump column and prevent a vacuum from developing when the pump stops.

3.2M Pressure relief valves should be installed on all hydropneumatic tanks to prevent excessive internal pressures (surge pressures) from causing the tank to explode or rupture.

3.2N Air chargers are used for adding air to a hydropneumatic tank.

3.2O The size of a hydropneumatic tank will depend on the geographic location, pump capacity, peak demand, pump cycling rate, economics, past operational experience, and the amount of property available.

Answers to questions on page 86.

3.3A Three major well maintenance problems are:

1. Overpumping and lowering of the water table,
2. Clogging or collapse of a screen or perforated section, and
3. Corrosion or incrustation.

3.3B Overpumping an aquifer can damage a well by reducing the storage and production capacity of groundwater systems.

3.3C Deposits that develop on well screens include the hard, cement-like form typical of the carbonate and sulfate compounds of calcium and magnesium, the soft, sludge-like forms of the iron and manganese hydroxides, or the gelatinous slimes of iron bacteria. Silt and clay may also be deposited on well screens.

3.3D The use of two or more different types of metals such as stainless steel and ordinary steel, or steel and brass or bronze should be avoided because corrosion is usually greatest at the points of contact of different metals or where they come closest to contact.

3.3E Incrusting waters are usually alkaline while corrosive waters are usually acidic.

Answers to questions on page 91.

3.3F Records that should be kept regarding a well include (1) well depth measurements before and after pumping, (2) flow rates, (3) water quality, (4) length of time pumping, and (5) accurate data on pump repairs and causes.

3.3G Surging is a form of plunging which can be used for opening pores in the screen and for cleaning the gravel pack around the screen.

3.3H High-velocity jetting can be used to remove sand from pores and the immediate vicinity of the well screen.

3.3I Incrustation can be removed from the well casing and well by acid treatment.

3.3J Bacterial growths and slime deposits can be removed from well screens by chlorine treatment.

Answers to questions on page 91.

3.3K The chemical quality of groundwater indicates the type of dissolved minerals and will help in the design of a maintenance program if mineral deposition is suspected to be a cause of decreased well performance.

3.3L Downhole video inspection can be used to look for incrustation and/or corrosion of a well screen.

3.3M When the yield of a well declines, the aquifer, the well, and the pump should be investigated to determine the cause.

ANSWERS TO QUESTIONS IN LESSON 3

Answers to questions on page 95.

3.4A Well pumps are generally classified into two basic groups:

1. *POSITIVE DISPLACEMENT* pumps which deliver the same volume or flow of water against any head within their operating capacity, and
2. *VARIABLE DISPLACEMENT* pumps which deliver water with the volume or flow varying inversely with the head (the *GREATER* the head, the *LESS* the volume or flow).

3.4B Advantages of centrifugal pumps include: (1) relatively small space needed for any given capacity, (2) rotary rather than reciprocating motion, (3) adaptability to high-speed mechanisms such as electric motors and gas engines, (4) low initial cost, (5) simple mechanism, (6) simple operation and repair, and (7) safety against damage from high pressure because of limited maximum pressure that can be developed.

3.4C Jet pumps are not commonly installed today because of their low efficiency and the fact that they are being replaced by submersible pumps.

Answers to questions on page 102.

3.4D The three basic purposes of the pump column pipe in a deep well turbine pumping installation are:

1. Support of the pump in the well,
2. Delivery of water under pressure from the well pump to the surface, and
3. Maintenance of the lineshaft and shaft enclosing (oil) tube assembly in straight alignment.

3.4E Prime movers used with right-angle gear drives include electric motors and gasoline, natural gas, diesel, or propane-powered engines.

3.4F Deep well turbine, oil lubricated pumps are lubricated by an automatic electric oiler system that is activated by means of an electric solenoid valve when the well pump starts. The needle valve is adjusted to feed the necessary drops of oil per minute for the appropriate column length.

Answers to questions on page 103.

3.5A During the drilling of new wells, contamination could be introduced into the well from the drilling tools and mud, makeup water, topsoil falling in or sticking to tools, and from the gravel itself.

3.5B Disinfection of a well should take place following development, testing for yield, and before the test pump is removed from the well, or when there is evidence of contamination.

3.5C When disinfecting a new well, the desired chlorine concentration in the well casing is 50 mg/*L*.

3.5D Special disinfection methods may be required for existing wells because during repair work deposits of slime, bacterial growth, and other debris are dislodged from the inside surfaces of the well casing and from the outside surfaces of the well pump column pipe. These deposits generally settle to the bottom of the well, but some are also smeared on the inside surfaces of the well casing, particularly above the water line which require swabbing of the inside of the well casing.

3.5E If repeated attempts to disinfect a well are unsuccessful, a detailed investigation to determine the cause or source of the contamination should be undertaken.

Answer to question on page 104.

3.5F

Known		Unknown
Casing Diameter, in	= 12 in	Chlorine Required, pints
Casing Length, ft	= 200 ft	
Chlorine Compound, mg/*L* (Sodium Hypochlorite)	= 52,500 mg/*L*	

1. Find the chlorine required from Table 3.8 for a 12-inch diameter well casing when using sodium hypochlorite.

   Table 3.8 Value = 5 pints

2. Calculate the pints of sodium hypochlorite required.

$$\text{Chlorine Required, pints} = \frac{(\text{Table 3.8 Value})(\text{Casing Length, ft})}{100 \text{ ft}}$$

$$= \frac{(5 \text{ pints})(200 \text{ ft})}{100 \text{ ft}}$$

$$= 10 \text{ pints}$$

ANSWERS TO QUESTIONS IN LESSON 4

Answers to questions on page 107.

3.6A Poor operation and neglect could be reasons why properly designed and constructed well pumping stations fail to produce water in sufficient quantity and quality to meet the needs of the entire community.

3.6B When entering a pumping facility for an inspection, the operator should (1) *LISTEN* for any unusual noises, (2) *FEEL* for vibrations on pieces of equipment, and (3) *LOOK* for anything unusual.

3.6C Areas that should be covered by recordkeeping include (1) new construction records, (2) equipment records, and (3) repairs and modifications.

Answers to questions on page 114.

3.7A A well in a sand formation can be constructed so that it is sand free by properly constructing a gravel envelope, with selected louvers or well screen, and supported by an engineered filter pack.

3.7B The abrasive action of sand can damage well pumping facilities, consumers' fixtures and appliances, water meters, and precision equipment. Also, sand can reduce the carrying capacity of water mains and cause consumers to complain.

3.7C The majority of sand complaints are usually associated with smaller mains and deadend lines.

3.7D Flushing operations should be continued as long as sand is evident in the water.

3.7E The permissible concentration of sand in water from a well is 0.3 cubic foot per million gallons of water pumped and 0.1 cubic foot per million gallons is the desirable value.

Answers to questions on page 121.

3.8A The voltage required for a pump depends on the size (motor horsepower) of the pump.

3.8B Large pumps need motor starters so the power is induced gradually (in steps) into the motor allowing the pump to come up to speed gradually. This prevents pump damage and disturbances along electrical lines.

3.8C The three major types of pump controls are (1) ON/OFF, (2) proportional, and (3) derivative (sometimes called "reset" or "rate").

3.8D Proportional pump control provides more corrective effort as the measured variable gets farther from the set point. In the case of reservoir level, this might mean that as the level gets low, and then lower, several additional pumps may be called upon to pump into it. As the level approaches the desired level, the extra pumps are turned off, and eventually, as it arrives at the set point, all pumps stop.

Answers to questions on page 122.

3.9A Approximately 75 percent of well pump and control problems are associated with electricity.

3.9B Electrical repairs should be done by someone who is authorized and qualified to do the job. Small water utilities may have to make arrangements with a local electrical firm or pump service company to perform this service.

3.9C To correct a waterlogged tank, add air to the tank. Also check the air charging system, the air release valve, and the tank and connections for air leaks.

3.9D Causes of a pump drawing too much power include:

1. Impellers not properly adjusted,
2. Well is pumping sand,
3. Crooked well causing pump shaft binding, and
4. Worn bearings or bent shaft.

OBJECTIVE TEST

Chapter 3. WELLS

Please write your name and mark the correct answers on the answer sheet, as directed at the end of Chapter 1. There may be more than one correct answer to each multiple-choice question.

True-False

1. Conditions for growth and survival of bacteria and viruses in groundwater are generally favorable when compared with surface waters.
   1. True
   2. False
2. Wells can allow direct contamination of an aquifer due to inadequate grouting or seals.
   1. True
   2. False
3. Floor drains and sinks used for washing chemicals should be connected to the sewer.
   1. True
   2. False
4. Cement grout is recommended for a pump motor base seal.
   1. True
   2. False
5. Flowmeters should be calibrated in place to ensure accurate flow measurements.
   1. True
   2. False
6. Corrosion is usually lowest at points of contact of different metals.
   1. True
   2. False
7. Never use calcium and sodium hypochlorite together.
   1. True
   2. False
8. Use of excess grease in pump motor bearings will keep the bearings cool.
   1. True
   2. False
9. During the drilling of new wells, contamination could be introduced into the well.
   1. True
   2. False
10. The health and safety of the water users depends on good operation of the well system.
    1. True
    2. False
11. There are few substitutes for good records.
    1. True
    2. False
12. Flushing sand is more difficult in smaller mains than in larger mains.
    1. True
    2. False
13. Auxiliary power is difficult to provide at well sites.
    1. True
    2. False
14. A pump will shut off when the cutoff pressure setting is too high.
    1. True
    2. False
15. A pump will deliver the normal amount of water when the impeller is worn.
    1. True
    2. False

Best Answer (Select only the closest or best answer.)

16. An aquifer is a natural underground layer of what?
    1. Nonporous, nonwater-bearing materials usually capable of yielding very little water
    2. Porous, water-bearing materials usually capable of yielding a large amount of water
    3. Porous, water-bearing materials usually capable of yielding a small amount of water
    4. Water which is directly related to the rate of evapotranspiration.
17. Which pollutant is removed as groundwater travels through the soil mantle?
    1. Dissolved pollutants
    2. Farming chemicals
    3. Organic chemicals
    4. Suspended particles
18. What is the most reasonable course against groundwater pollution?
    1. Filtration
    2. Prevention
    3. Regulation
    4. Treatment
19. What type of drain field problem can occur when garbage disposals are used in septic tank systems?
    1. Adsorption
    2. Clogging
    3. Leaching
    4. Nitrification

20. Where are sampling taps provided on well systems?
    1. At the water table
    2. Between well pump and check valve
    3. Downstream side of pump check valve
    4. On pump suction

21. Why must pumps be primed?
    1. To allow pumping of water
    2. To avoid backsiphonage
    3. To conserve water
    4. To prevent cross connections

22. What is the best possible procedure for a good maintenance program?
    1. Adequate recordkeeping
    2. Controlling corrosion
    3. Periodic screen cleaning
    4. Scheduled lubrication

23. What is the function of each stage in multi-stage turbine type pumps?
    1. Add pressure head capacity
    2. Add volume capacity
    3. Decrease brake horsepower requirements
    4. Increase efficiency

24. When should existing wells be disinfected?
    1. After an intense, long-duration storm
    2. After well or pump repairs
    3. Every year
    4. Never. Existing wells don't need to be disinfected

25. How can an operator determine if a well has been properly disinfected?
    1. No customer complaints are received
    2. Proper disinfection procedures were followed
    3. Water from well appears clear
    4. Well water sample passes bacteriological test

26. What could be the cause of sand complaints from only one or two residences in a block?
    1. Customers are using a sand sampler
    2. Lateral is tapped into side of main near valve, which creates turbulence
    3. Lateral is tapped into top of main near valve, which creates turbulence
    4. Sand is entering pipe near residences

27. What could be the cause of high pressure in a pump discharge line?
    1. Broken water line
    2. Excessive water use
    3. Pump shut down
    4. Valve failure

28. How can an operator determine that there is no power to a switch box?
    1. Check incoming power source with a voltmeter
    2. Clean contact points
    3. Reset circuit breaker
    4. Tighten wiring terminal

29. What could cause a pump to start too frequently?
    1. Cutoff pressure setting too high
    2. Impellers worn
    3. Leaking foot valve
    4. Motor bearings worn

30. What could cause a pump to take too much power?
    1. Circuit breaker tripped
    2. Crooked well, pump shaft binding
    3. Pump breaking suction
    4. Waterlogged tank

Multiple Choice (Select all correct answers.)

31. What problems can result from overdraft of a groundwater basin or aquifer?
    1. Cleansing of pollutants in the aquifer
    2. Closing of pores in which water is stored
    3. Closing of pores through which water moves
    4. Excessive replenishment of the aquifer
    5. Subsidence of the land

32. What are potential sources of groundwater pollution?
    1. Animal wastes
    2. Fertilizers
    3. Septic tanks
    4. Solid waste disposal sites
    5. Wastewater collection systems

33. Which treatment processes are in operation in the soil to treat contaminants?
    1. Adsorption
    2. Biological decay
    3. Filtration
    4. Ion exchange
    5. Oxidation

34. Which land use factors contribute to the level of risk to groundwater?
    1. Likelihood of contaminant release
    2. Mobility of contaminants
    3. Toxicity of contaminants
    4. Type of business
    5. Volume of wastewater

35. What water quality tests should be performed on groundwater samples if a study indicates that septic tank system failures may have occurred near shallow wells?
    1. Coliform
    2. Nitrate
    3. Pesticides
    4. Petroleum
    5. Saline

36. Gas stations pose which of the following threats to groundwater?
    1. Hazardous materials
    2. Hydrocarbons
    3. Nutrients
    4. Solvents
    5. Waste oils

37. Which preventive practices can be used to protect groundwater from chemicals stored in outdoor storage facilities?
    1. Connect drain to sewer
    2. Cover the area
    3. Install berms
    4. Inventory stored chemicals
    5. Label containers
38. What is the purpose of a well-casing vent?
    1. To allow release of air from water hammer
    2. To prevent vacuum conditions inside a well
    3. To replace discharge check valves
    4. To stop contamination from being drawn into the well
    5. To vent air released from cavitation
39. What problems can be caused by sand found in well water?
    1. Customer complaints
    2. Excessive meter wear
    3. Reduced pump efficiencies
    4. Sanded water mains
    5. Worn pump impellers
40. What causes water hammer?
    1. Pressure relief valves
    2. Quick closing of valves
    3. Quick opening of valves
    4. Shutdown of pumps
    5. Start-up of pumps
41. Which of the following are considered well maintenance problems?
    1. Clogging of screen
    2. Collapse of screen
    3. Overpumping
    4. Screen corrosion
    5. Screen incrustation
42. What could be the cause(s) of a decline in well yield and specific capacity, but not change in available drawdown?
    1. Blockage of screen by rock particles
    2. Deposition of iron compounds on screen
    3. Excessive withdrawals from well
    4. Movement of sediment into well
    5. Wear of pump impeller
43. How can contamination be introduced during the drilling of a new well?
    1. Contamination can't be introduced during the drilling of a new well
    2. From drilling tools and mud
    3. From gravel placed in the well
    4. From makeup water
    5. From topsoil falling in or sticking to tools
44. As a minimum, recordkeeping should cover which of the following areas of operation?
    1. Equipment records
    2. New construction records
    3. Newspaper records
    4. Repair and modification records
    5. Visitor records
45. What problems are associated with sand in water?
    1. Consumer complaints
    2. Damage to water meters
    3. Damage to well pumping facilities
    4. Decreasing friction loss in pipes
    5. Reducing pipe carrying capacity
46. What could be the probable causes of a pump not starting?
    1. Circuit breaker tripped
    2. Float switch malfunctioning
    3. No power to switch box
    4. Waterlogged tank
    5. Well is pumping sand
47. What could be the cause of excessive operating noise?
    1. Bent line shaft
    2. Line shaft bearings not receiving oil
    3. Motor bearings worn
    4. Pump breaking suction
    5. Well is pumping sand
48. What is the surface area of a rectangular settling basin 40 feet long and 8 feet wide?
    1. 230 square feet
    2. 320 square feet
    3. 360 square feet
    4. 460 square feet
    5. 540 square feet
49. What is the surface area of a circular clarifier 32 feet in diameter? Select the closest answer.
    1. 25 square feet
    2. 100 square feet
    3. 250 square feet
    4. 400 square feet
    5. 800 square feet
50. Calculate the volume in cubic feet of a rectangular settling basin 30 feet long, 10 feet wide, and 8 feet deep.
    1. 240 cu ft
    2. 2,400 cu ft
    3. 3,000 cu ft
    4. 3,200 cu ft
    5. 4,800 cu ft
51. Calculate the volume of water in cubic feet in a water-filled well casing 12 inches (1 foot) in diameter and 120 feet long. Select the closest answer.
    1. 38 cu ft
    2. 94 cu ft
    3. 118 cu ft
    4. 225 cu ft
    5. 377 cu ft
52. Calculate the volume of water in gallons in a water-filled well casing 10 inches in diameter and 75 feet long. Select the closest answer.
    1. 30 gallons
    2. 40 gallons
    3. 300 gallons
    4. 400 gallons
    5. 440 gallons

53. Calculate the gallons of sodium hypochlorite required to disinfect a well if the well casing contains 1,200 gallons of water. The chlorine dose is 50 mg/*L*. Sodium hypochlorite is 5.25 percent or 52,500 mg/*L* chlorine. Select the closest answer.
    1. 0.60 gallon
    2. 1.14 gallons
    3. 2.28 gallons
    4. 3.36 gallons
    5. 6.0 gallons

54. How much sodium hypochlorite is required to dose a well at 50 mg/*L*? The casing diameter is 16 inches and the length of the water-filled casing is 120 feet. Sodium hypochlorite is 5.25 percent or 52,500 mg/*L* chlorine. Select the closest answer.
    1. 0.8 gallon
    2. 1.0 gallons
    3. 1.1 gallons
    4. 1.2 gallons
    5. 1.5 gallons

Review Questions (Select all correct answers.)

55. Which of the following items cause(s) turbidity in water?
    1. Finely divided organic material
    2. Hardness
    3. pH
    4. Plankton
    5. Suspended material

56. What types of activities should be controlled in watersheds to protect a water supply source?
    1. Boating
    2. Disposal of wastewater
    3. Ice cutting
    4. Lake and stream fishing
    5. Operation of sanitary landfills

End of Objective Test

APPENDIX

Chapter 3. WELLS

3.12 WELL SITE SELECTION

This section is intended to be used as a guideline only. Local county health departments or state health agencies should have sanitary setback requirements and regulations pertaining to the location of wells in relation to sanitary hazards. *PLEASE CONTACT YOUR LOCAL AUTHORITIES FOR THEIR REQUIREMENTS AND REGULATIONS BEFORE YOU SELECT A SITE AND DRILL.*

3.120 Factors to Be Considered

Perhaps the most important consideration in developing a groundwater source is the selection of the well site. In many cases, wells located too close to sources of contamination must be abandoned and destroyed long before their useful life has been utilized.

Safety of a groundwater source depends primarily on three factors: (1) the proper selection of the well site, (2) good well design and construction, and (3) geology. These factors should be the guides used in determining safe distances for different situations. The guidelines in this appendix apply only to properly constructed wells. A poorly constructed well is potentially unsafe regardless of the distance from a source of contamination.

Geologic and hydrologic (hydrogeologic) data need to be considered when siting a well to maximize the chances of getting a firm yielding well with acceptable water quality, keep costs to a minimum (both for construction and operation), for well site protection, and for wellhead protection. If the location and type of aquifer are known, the quality and quantity of the aquifer should be considered when choosing a well site. Nothing is more frustrating than drilling dry holes or finding water of poor quality requiring extensive treatment. Contact and utilize the services of a competent hydrogeologist, or geologist, familiar with the local area. But even in the absence of such a consultant there are some simple guidelines to follow that will increase your chances of obtaining a firm potable yield.

1. Drill where aquifers are known to exist. Numerous agencies have some information regarding existing aquifers, their yield, and water quality:
   - Water Agencies
   - Health Departments
   - U.S. Natural Resources Conservation Service
   - U.S. Geological Survey
   - Local Public Works Department
   - Bureau of Reclamation
   - U.S. Army Corps of Engineers

   This is not a complete list and these agencies are not guaranteed to know all about the local groundwater conditions, but they certainly are a good start.

2. Look for areas where you believe groundwater might exist. Water tends to gravitate downward toward lower points. When rain occurs on a watershed, some water percolates downward through aquifers to water tables. As the water table rises, water is deposited in lakes or streams as shown in Figure 3.1. If groundwater can seep into lakes and streams, then aquifers probably exist close to lakes and streams. When the water table in the aquifer is high, water is fed into lakes and streams. Conversely aquifers can be charged from high water levels in lakes and streams.

   If reliable sources of surface water exist in streams and lakes, there should be groundwater nearby. Hence, it is probable that you will be able to develop a usable well near these sources.

3. Check with area well drillers. Wells are their business and they may be your very best sources to determine a good well location.

Regardless of where you drill your wells, you should do your research. Hydrology and geology are the keys to selecting productive, unpolluted sites.

If you know you have a firm yield of usable water, you should carefully examine several other elements before proceeding to drill. Some of the factors can be evaluated from a cost standpoint, while others must be subjectively weighed.

1. *ACCESS* is important for ease of maintenance and construction. Can you get to the well on a daily basis to make repairs or monitor your supply? Being able to drive to the site has definite advantages. Can the drill equipment get to the site and will the driller be able to set up and easily operate the equipment? Easy access means efficient installation, operation, and repair.
2. *POWER* sources should be readily available. Utility companies must be able to provide the necessary power at affordable prices. In a remote area the cost of getting power is sometimes prohibitive. Alternative sources of power, such as gas generators, are always a possibility, but they are expensive both initially and on a long-term basis.
3. *RELATIONSHIP TO STORAGE AND TREATMENT FACILITIES* has to do with the cost of pumping, transmission lines, and deliverability. This may not be as important as some other considerations, but installation of wells close to your other facilities could simplify maintenance and reduce long-term delivery costs. Longer transmission lines (and initial costs for same) tend to be secondary to some of the other location factors, but distance is definitely a cost factor and a serious look at detailed costs for all materials must be considered.
4. *POLLUTION POTENTIAL* is a serious consideration. The adequate separation of a domestic water supply well away from sources of contamination or pollution is a primary factor in ensuring the continued safety of the water produced by the well. Facilities using or transporting hazardous chemicals near a well are objectionable because leakage of contaminants is possible. Such contaminants may gain entry to the well through unprotected openings, poor well design or construction, failure of the well casing from corrosion or age, or leakage directly into the zone of influence of the well. Well installation upstream (higher portion of the basin) of industrial manufacturing facilities and urban areas with a large number of septic tanks is more desirable than downstream of these facilities. Groundwater tends to gravitate downward and a well located downstream of potential pollutants has a higher probability of contamination. All cases are not the same and you will not necessarily have to avoid downstream installations because pollution may not be a problem. You need to make sure you address pollution potential when considering a new well location.
5. *DRAINAGE* should definitely move away from the proposed well site. Any runoff from urban areas will be carrying pollutants of some kind. Keep away from areas where drainage might affect the groundwater.
6. *AESTHETICS* is sometimes more of a factor than cost. *(NOTHING IS OF GREATER IMPORTANCE THAN PROTECTION AGAINST POLLUTION.)* Wells are unsightly to some people and they can be noisy. Having maintenance people around a well at frequent intervals may be undesirable to some neighbors. Aesthetics may seem like a feeble complaint to you, but to more and more people it is becoming a major factor. And you live in a world where human factors must be dealt with — "Out of sight, out of mind" may be a suitable approach when locating new facilities.
7. *SECURITY* of some kind needs to be considered. Vandalism is a fact of life. You don't need to install razor tape around the facilities, but a good chain link fence or pump house is often warranted. Can the site be inspected easily without actually visiting the site? Is the well near residents who can see people coming and going from the site, for example, only one road into and out of the site?
8. *EASEMENTS AND LAND* acquisitions are expensive items and clearly need to be factored into site selection.

These guidelines are for the location of new wells, evaluation of existing wells, and maintenance of sanitary control of future construction in the vicinity of wells. The design and location of a new well should be reviewed with the local health officer or other regulatory authority for approval before construction begins.

3.121 Safe Horizontal Distance

The safe distance of a well from sources of contamination depends upon numerous local factors. To determine the safe distance, evaluate the following factors: (1) character and location of sources of possible contamination, (2) type of well construction, (3) natural *HYDRAULIC GRADIENT*[47] of the water table, (4) permeability of the soil overlying the water-bearing formation, (5) extent of the *CONE OF DEPRESSION*[48] formed in the water table due to pumping the well, and (6) the nature of the soil or rock structure.

3.122 Minimum Horizontal Distance

The minimum safe horizontal distance between a well and potential sources of contamination should be maintained in accordance with Table 3.10 and as shown in Figure 3.42. No lesser distances should be considered unless approved by the local health officer and unless the special protection requirements described in Section 3.125, "Special Construction of Sanitary or Storm Sewers Under Gravity Flow," are met.

Different states have different minimum acceptable safe horizontal distances between a well and potential sources of contamination. For example, the State of Vermont has the following requirements:

1. The minimum radius, in feet, of the area that must be owned by the water purveyor should be at least 200 feet. Acceptable land uses within this 2.88-acre plot would be skiing, organized sports, forest or conservation land, or similar non-threatening uses.

[47] *Hydraulic Gradient. The slope of the hydraulic grade line. This is the slope of the water surface in an open channel, the slope of the water surface of the groundwater table, or the slope of the water pressure for pipes under pressure.*

[48] *Cone of Depression. The depression, roughly conical in shape, produced in the water table by the pumping of water from a well. Also called the cone of influence.*

TABLE 3.10 MINIMUM HORIZONTAL DISTANCE TO WELL[a]

Wastewater Facilities	Feet	Meters
Septic Tank (Watertight)	50	15
Horizontal Leach Lines	100	30
Seepage Pit or Cesspool	150[b]	45
Pit Privy	50	15
Vault Privy (Pumpout)	50	15
Sanitary Sewers and House Laterals	100[c]	30
Storm Sewers	100[c]	30
Wastewater Pumping Station	100	30
Effluent Discharge Channel	200	60
Drainage Channel	50	15
Wastewater Treatment Plant	*	*
Wastewater Lagoon	*	*
Wastewater Irrigation Area	*	*
Industrial Facilities		
Barnyard, Feedlot, Grazing Area	*	*
Waste Sewers	100[c]	30
Waste Holding (Watertight)	50	15
Petroleum Storage	*	*
Petroleum Transmission	500[d]	150
Solid Waste Disposal Site		
Class 1	*	*
Class 2	2,000	600
Class 3	1,000	300
Other		
Dwellings	50	15
Pond, Lake, Stream	100	30
Abandoned Conduit	50	15
Cathodic Protection Well		
Cased	50	15
No Casing	200	60
Abandoned and Destroyed Wells	*	*
Wells Destroyed in Accordance With State or Local Health Department Guidelines	None required	

NOTES:

* Case-by-case evaluation.

[a] *STANDARDS REGARDING LOCATION OF PUBLIC WATER SUPPLY WELLS WITH RESPECT TO SOURCES OF CONTAMINATION OR POLLUTION*, California Department of Health Services, Sanitary Engineering Branch, Berkeley, California.

[b] Bottom of pit should be more than 10 feet above groundwater, preferably separated by an impervious stratum.

[c] See Section 3.125, "Special Construction of Sanitary or Storm Sewers Under Gravity Flow."

[d] Underground storage and transmission facilities should be pressure tested annually.

2. Any subsurface wastewater treatment systems should be well away from this protective zone. They must be at a distance that provides at least a two-year residence flow time to permit die-off of viruses that may be pathogenic.

3. Occasionally the 200-foot radius (protective zone) can be reduced to 125 feet, but only after an extensive study has been conducted by a hydrogeologist which shows no hydraulic connection between the upper soils and the water-bearing strata for the well.

4. Only water supply grade ductile iron pipe (leak-tested) is permitted for sewers within the well isolation zone up to within 75 feet of the well. Manholes are not permitted within the 200-foot protection zone and must be leak tested.

5. Permitted and non-permitted land uses in the well isolation zone within a radius of 200 feet are as follows:

   a. Land uses permitted within the well isolation zones include:

      (1) playground, ballfields, tennis courts; and

      (2) seasonal light-duty roads.

   b. Land uses *NOT* permitted within the well isolation zone include:

      (1) application of pesticides and herbicides;

      (2) buildings other than those required for the water system;

      (3) parking of motor vehicles;

      (4) chemical storage other than that required by the water system;

      (5) swimming pools;

      (6) salted or paved roads passing through the area;

      (7) leach fields, septic tanks, and sewer lines; and

      (8) any other activity which may contaminate the water supply.

These standards apply only to wells serving more than 10 living units. This is because public wells are apt to be stressed far more than individual home wells, and thus warrant the extra protective actions and zones because of the tendency to pull water toward them.

3.123 Design of Well Fields

Sometimes when developing water supplies, especially when large supplies are needed, it becomes necessary to design a system of wells or a well field. In addition to the siting considerations already discussed, planning for the development of a well field must take into account well spacing and location for economics and best use of the aquifer. Other design considerations include the quantity of water needed, size of well, depths of pumps, drawdown effects, and nearby pollution sources. Important information needed to design well fields includes groundwater hydraulic characteristics, such as aquifer transmissivity, storativity, and confining-unit leakage, all of which can be derived from the "comprehensive aquifer tests" described in Section 3.160, "Well Performance Testing." It will also be necessary to conduct a thorough evaluation of the potential environmental effects of the well field, for example, overdraft and possible subsidence, and the effects of the well field on existing wells nearby.

According to Ralph C. Heath, author of the USGS Paper 2220, some of the basic hydrologic considerations relative to well spacing are:

1. The minimum distance between pumping wells should be at least twice the aquifer thickness if the wells are open to less than about half the aquifer thickness;

2. Wells near recharging boundaries should be located along a line parallel to the boundary and as close to the boundary as possible; and

3. Wells near impermeable boundaries should be located along a line perpendicular to the boundary and as far from the boundary as possible.

Beyond these basic rules, a qualified hydrologist or geologist should be consulted for assistance in the design and spacing of well fields.

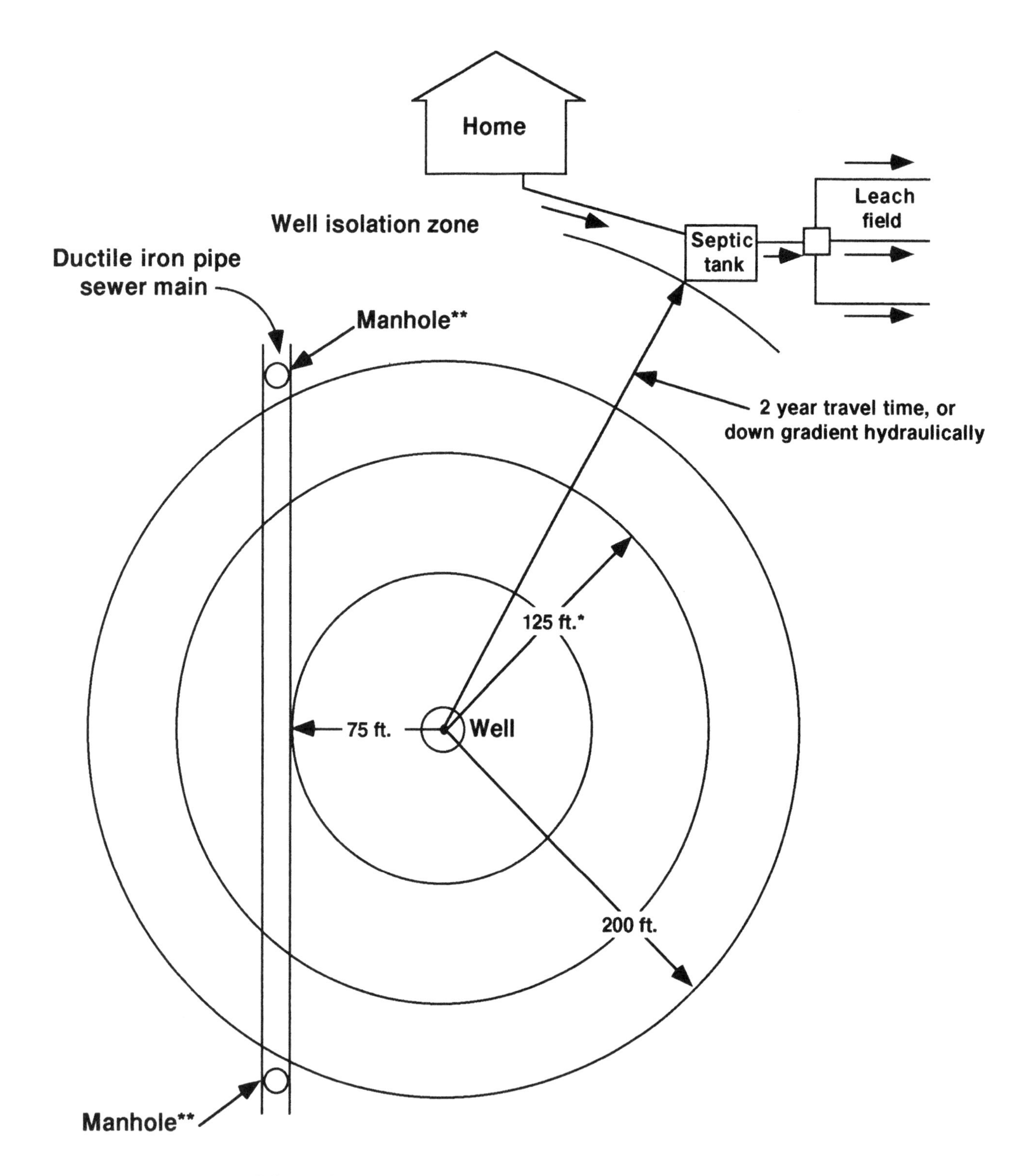

NOTE: Check with your state water supply agency for required distances.

* The 200-foot isolation zone may be reduced to 125 feet on approval by reviewing authority if a hydrogeologist's report shows no hydraulic connection between the upper soil layers and the aquifer.

** Manholes and sewers must be tested to show zero leakage prior to use.

Fig. 3.42 Recommended safe distances from a well

3.124 Adverse Conditions

In areas where adverse conditions exist, such as coarse gravel formation, lava, fractured or limestone rock, high groundwater, or unusual well construction or topography, the local health officer should approve the well site and the design of the well *BEFORE* construction.

3.125 Special Construction of Sanitary or Storm Sewers Under Gravity Flow

Where 100 feet (30 meters) of separation of the well from a sanitary or storm sewer is not feasible, the sewer should be constructed or reconstructed as specified below.

However, in no case should sewers be permitted within 50 feet (15 meters) of any well.

1. The sewer should be constructed of ductile iron pipe — Class 150 — with either rubber ring bell and spigot ends or *MECHANICAL JOINTS*.[49] This type of construction should be adequate under most conditions.

2. Under certain conditions, extra-heavy-wall plastic pipe with rubber ring joints could be used.

3. Special construction methods may be considered case by case. For existing sewers, it may be possible to use a 1/4-inch (6-mm) thick continuous steel casing to enclose the sewer pipe. All voids between the sewer pipe and casing must then be pressure grouted with sand-cement grout.

4. The sewer should be of adequate strength to withstand the weight of both backfill and any live load (load from a moving vehicle) that may be imposed on the pipe. The trench should be excavated only to the depth required. This will provide uniform, continuous support for the pipe on solid and undisturbed ground at every point. Any part of the bottom of the trench excavated below the specific grade (such as at the joints) should be corrected with suitable material and thoroughly compacted before the pipe is laid. Backfilling should be accomplished by hand from the bottom of the trench to the centerline of the pipe, using suitable material, placed in layers of three inches (75 mm), and compacted by tamping. The pipe should be pressure tested after installation.

3.126 Sanitary Control of Future Construction

Unless future construction in the vicinity of a well is controlled, the continued use of a well for domestic water supplies may be threatened. The best way to prevent the encroachment of hazardous facilities is by owning enough land around the well to guarantee the required separation whenever possible. Where the utility or water company has limited ownership, it should request notification from the building department, public works, and local wastewater utility agencies of any proposed construction near the well. The water agency will then be assured of an opportunity to suggest modification of any proposed facilities.

3.13 TYPES OF WELLS

There are numerous well construction methods ranging from low-yield, hand-dug wells to high-yield, gravel-envelope wells. Let's briefly review the distinguishing characteristics of wells you are likely to encounter.

3.130 Dug Wells

Dug wells are commonly excavated with hand tools. The excavator descends into the well as the excavation progresses. To prevent the native material from caving in, one must place a crib or lining in the excavation and move it downward as the pit is deepened. The space between the lining and the undisturbed embankment should be filled with cement grout down to the water-bearing strata to prevent entrance of surface water along the well lining (see Figure 3.43). Dug wells usually have a very limited yield and tend to fail in times of drought. Also dug wells are more subject to contamination hazards due to their shallow construction.

3.131 Bored Wells

Bored wells are commonly constructed with earth augers (similar to an oversized drill bit) turned either by hand or by power equipment. Such wells are usually regarded as practical at depths of less than 100 feet (30 meters) when the water requirement is low and the material overlying the water-bearing formation has non-caving properties and contains few large boulders. In suitable material, holes from 2 to 30 inches (50 to 750 mm) in diameter can be bored to about 100 feet (30 m) without caving.

In general, bored wells have the same characteristics as dug wells, but they may be extended deeper into the water-bearing formation due to the fact they are less apt to cave in.

Bored wells may be cased with vitrified clay tile, concrete pipe, standard wrought iron, steel casing, or other suitable material capable of sustaining imposed loads. The well may be completed by installing well screens or perforated casing in the water-bearing sand and gravel. Proper protection from surface drainage should be provided by sealing the casing with cement grout to the depth necessary to protect the well from contamination (see Figure 3.44).

3.132 Driven Wells

The simplest and least expensive of all well types is the driven well. This type of well is constructed by driving into the ground a well drive point which is fitted to the end of a series of pipe sections (see Figure 3.44). The drive point is of forged or cast steel. Drive points are usually 1 1/4 or 2 inches (30 or 50 mm) in diameter. The well is driven with the aid of a maul, or a special shafting or falling weight (see Figure 3.45). For deeper wells, the well points are sometimes driven into water-bearing strata at the bottom of a bored or dug well.

The yield of driven wells is generally small to moderate. Where they can be driven an appreciable depth below the water table, they are no more likely than bored wells to be seriously affected by water table fluctuations. The most suitable

[49] *Mechanical Joint. A flexible device that joins pipes or fittings together by the use of lugs and bolts.*

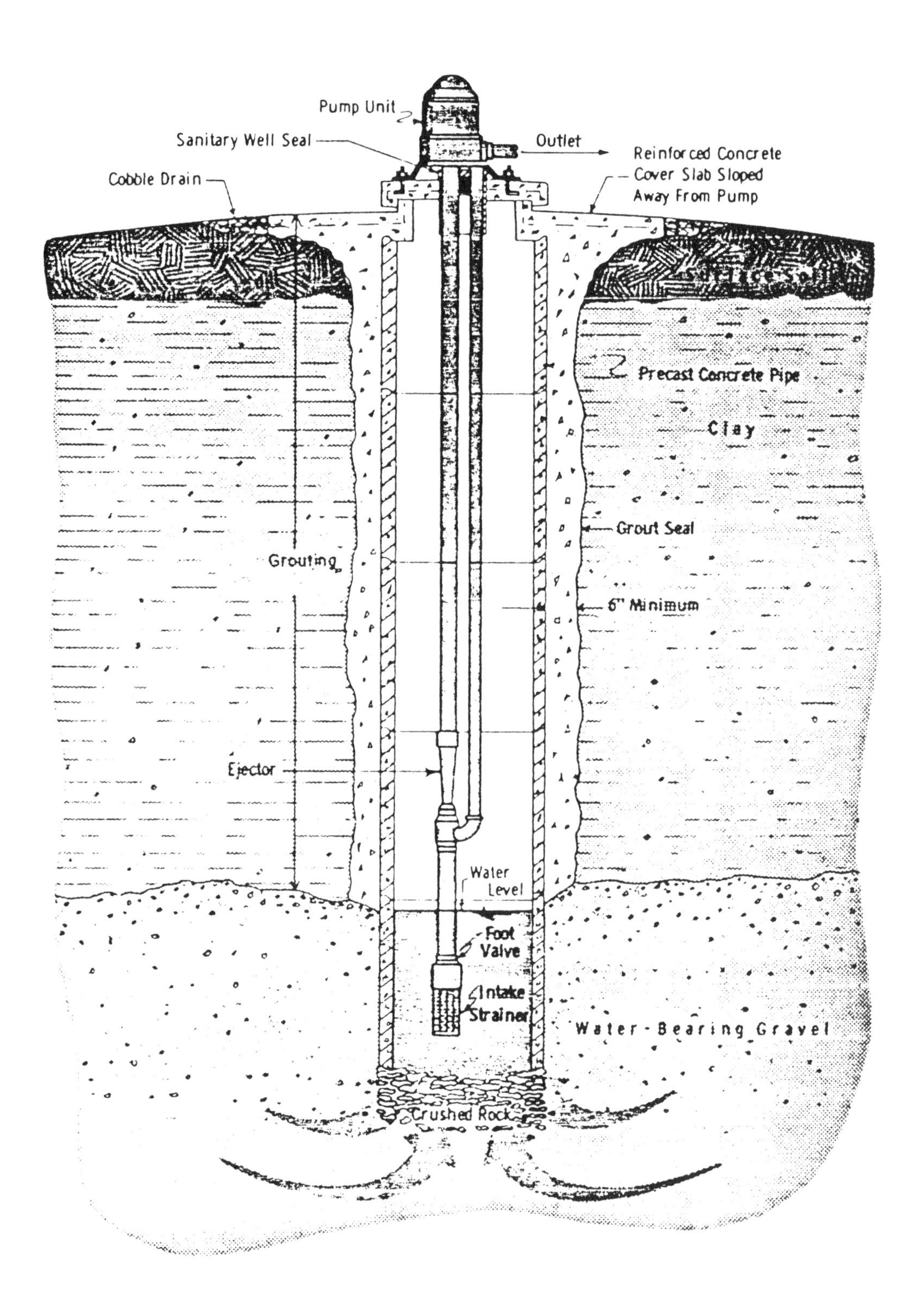

NOTES:
1. Pump screen is placed below point of maximum draw-down.
2. The grouting should be at least six inches thick, form a seal, and extend from the surface cover slab down to the water-bearing gravel.

Fig. 3.43 Dug well with two-pipe jet pump installation

(Source: *MANUAL OF INDIVIDUAL WATER SUPPLY SYSTEMS,*
U.S. Environmental Protection Agency,
Office of Water Programs, Washington, D.C.)

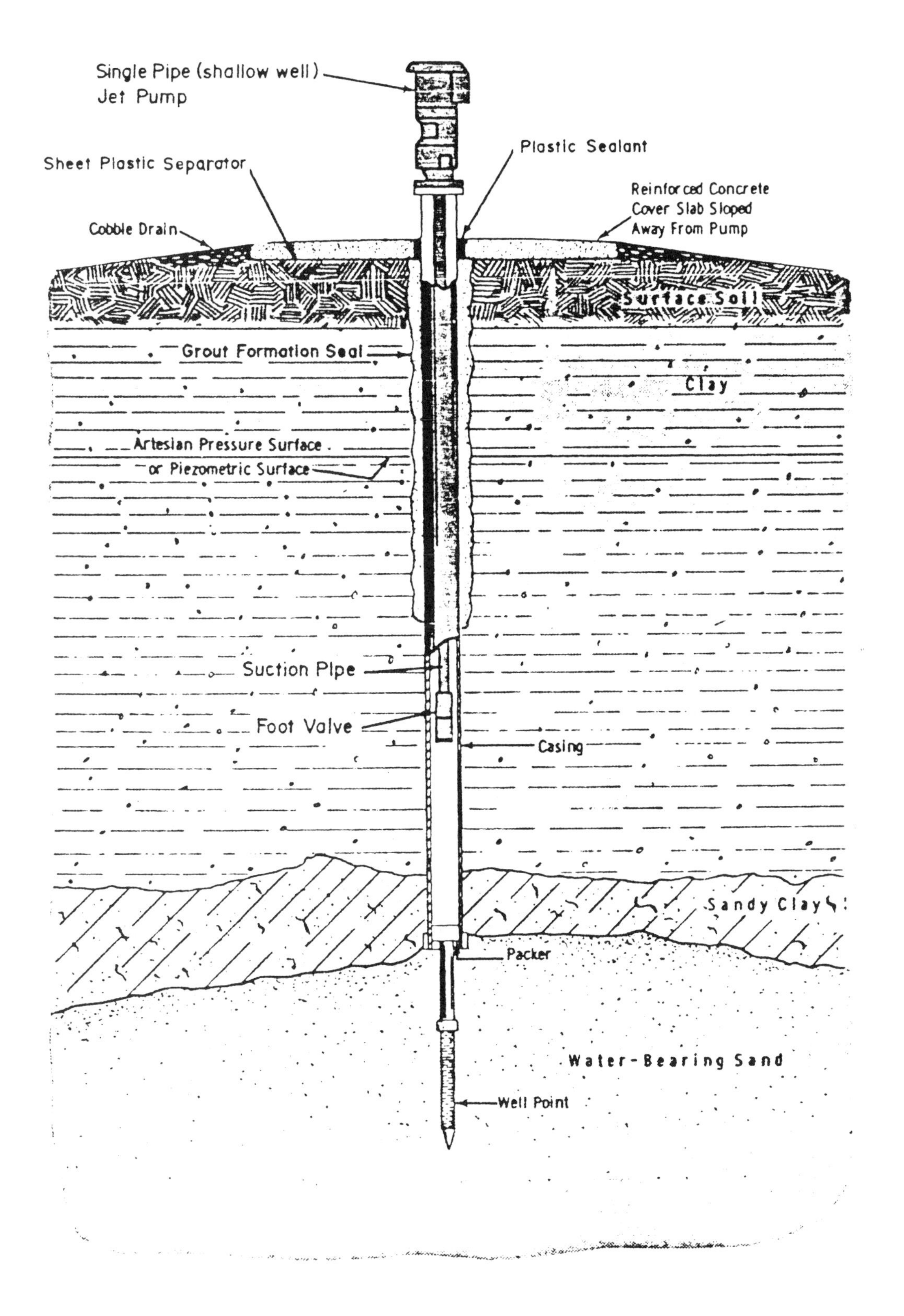

Fig. 3.44 Hand-bored well with driven well point and "Shallow Well" pump

(Source: *MANUAL OF INDIVIDUAL WATER SUPPLY SYSTEMS*, U.S. Environmental Protection Agency, Office of Water Programs, Washington, D.C.)

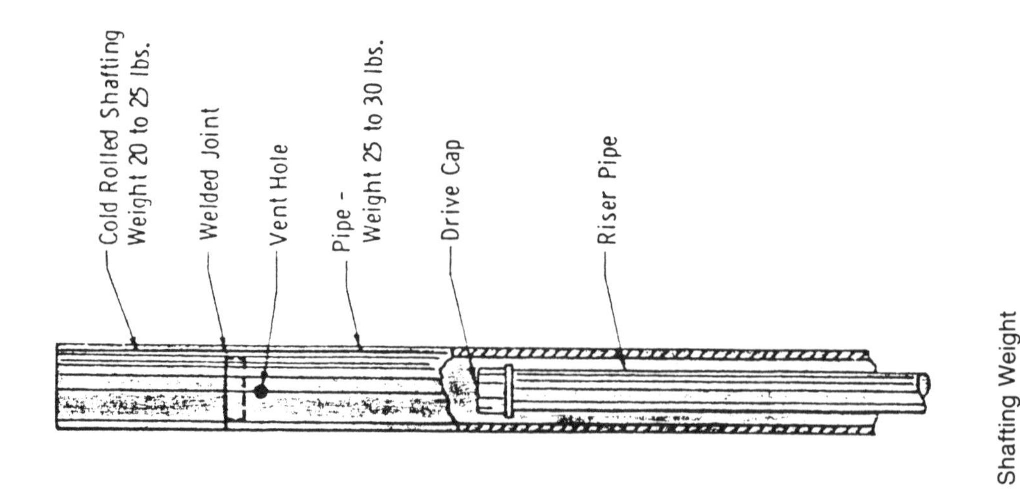

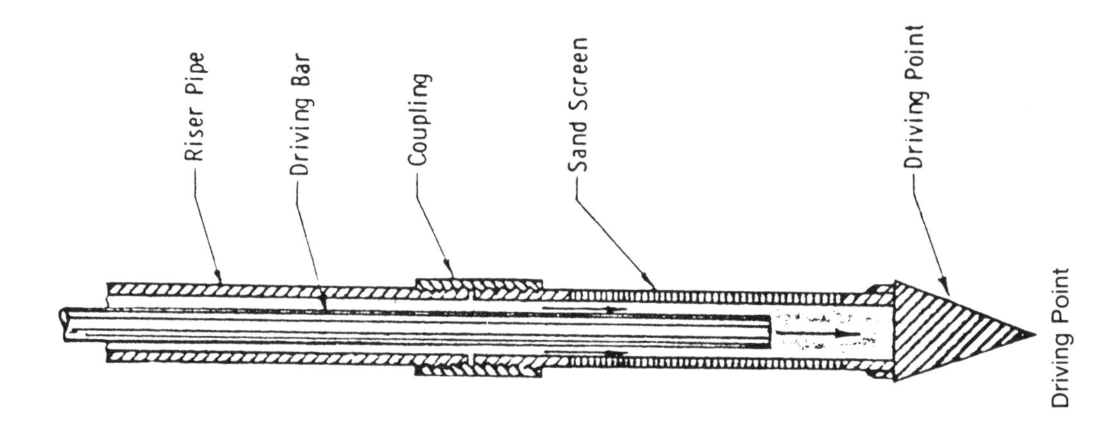

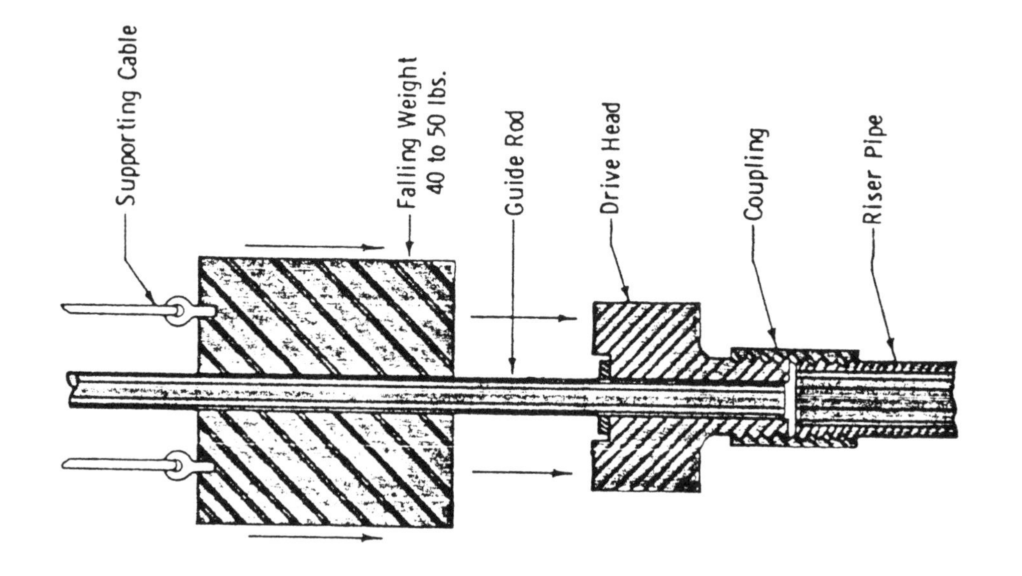

Fig. 3.45 Typical well point for driven wells
(Source: *MANUAL OF INDIVIDUAL WATER SUPPLY SYSTEMS,*
U.S. Environmental Protection Agency, Office of Water Programs, Washington, D.C.)

locations for driven wells are areas containing *ALLUVIAL*[50] deposits of high *PERMEABILITY*.[51] The presence of coarse gravels, cobbles, or boulders interferes with sinking the well point and may damage the sand screen (Figure 3.45) or wire mesh jacket.

3.133 Drilled Wells

Construction of a drilled well is ordinarily accomplished by one of two techniques — percussion or rotary hydraulic drilling. The selection of the method depends primarily on the geology of the site and the preference of the well owner or driller.

3.1330 Percussion (Cable-Tool) Method

Drilling by the cable-tool or percussion method (Figure 3.46 on next page) is accomplished by raising and dropping a heavy drill bit and stem. The impact of the bit crushes and dislodges pieces of the formation. The reciprocating motion of the drill tools mixes the drill cuttings with water into a *SLURRY*[52] at the bottom of the hole. This is periodically brought to the surface with a bailer, a 10- to 20-foot-long (3- to 6-m) pipe equipped with a valve at the lower end.

Caving is prevented as drilling progresses by driving or sinking into the ground a casing slightly larger in diameter than the bit. When wells are drilled in hard rock, casing is usually necessary only through the overburden or upper layer of uncompacted material. A casing may be necessary in hard rock formations to prevent caving of beds of softer material.

When good drilling practices are followed, water-bearing beds are readily detected in cable-tool holes because the slurry does not tend to seal off the water-bearing formation. A rise or fall in the water level in the hole during drilling, or more rapid recovery of the water level during bailing, indicates that a permeable bed has been entered. Crevices or soft streaks in hard formations are often water-bearing. Sand, gravel, limestone, and sandstone formations are generally permeable and yield the most water.

3.1331 Hydraulic Rotary Drilling Method

The direct rotary drilling method may be used in most formations. The essential parts of the drilling assembly include a derrick and hoist, a revolving table through which the drill pipe passes, a series of drill sections, a cutting bit at the lower end of the drill pipe, a pump for circulation of drilling fluid, and a power source to drive the drill.

In the drilling operation, the bit breaks up the materials as it rotates and advances. The drilling fluid (called mud) pumped down the drill pipe picks up the drill cuttings and carries them up the *ANNULAR SPACE*[53] between the rotating pipe and the wall of the hole (Figure 3.47). The mixture of mud and cuttings is discharged to a settling pit where the cuttings drop to the bottom and mud is recirculated to the drill pipe.

Two types of bits are used in direct rotary drilling (Figure 3.48). A roller or rock type is used to cut the borehole in hard *CONSOLIDATED FORMATIONS*.[54] This type of bit uses a crushing and chipping action between rotating gears to produce fine cuttings that can be carried in suspension by the

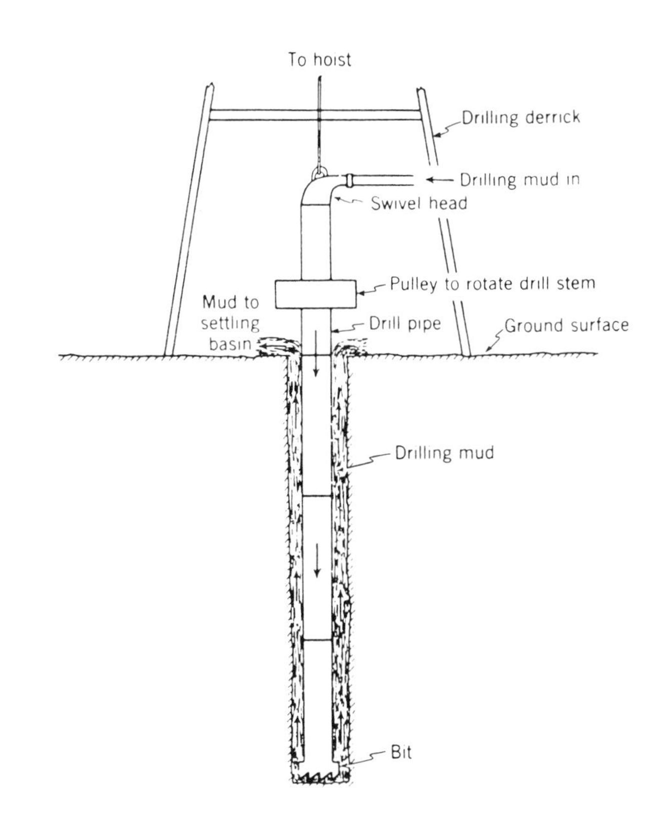

Fig. 3.47 Rotary drilling rig

(Reprinted from *WATER RESOURCES ENGINEERING*, Third Edition, by R.K. Linsley and J.B. Franzini, by permission. Copyright 1979, McGraw-Hill Book Company, New York, New York)

---

[50] *Alluvial (uh-LOU-vee-ul). Relating to mud and/or sand deposited by flowing water. Alluvial deposits may occur after a heavy rainstorm.*

[51] *Permeability (PURR-me-uh-BILL-uh-tee). The property of a material or soil that permits considerable movement of water through it when it is saturated.*

[52] *Slurry (SLUR-e). A watery mixture or suspension of insoluble (not dissolved) matter; a thin, watery mud or any substance resembling it (such as a grit slurry or a lime slurry).*

[53] *Annular (AN-you-ler) Space. A ring-shaped space located between two circular objects, such as two pipes.*

[54] *Consolidated Formation. A geologic material whose particles are stratified (layered), cemented or firmly packed together (hard rock); usually occurring at a depth below the ground surface.*

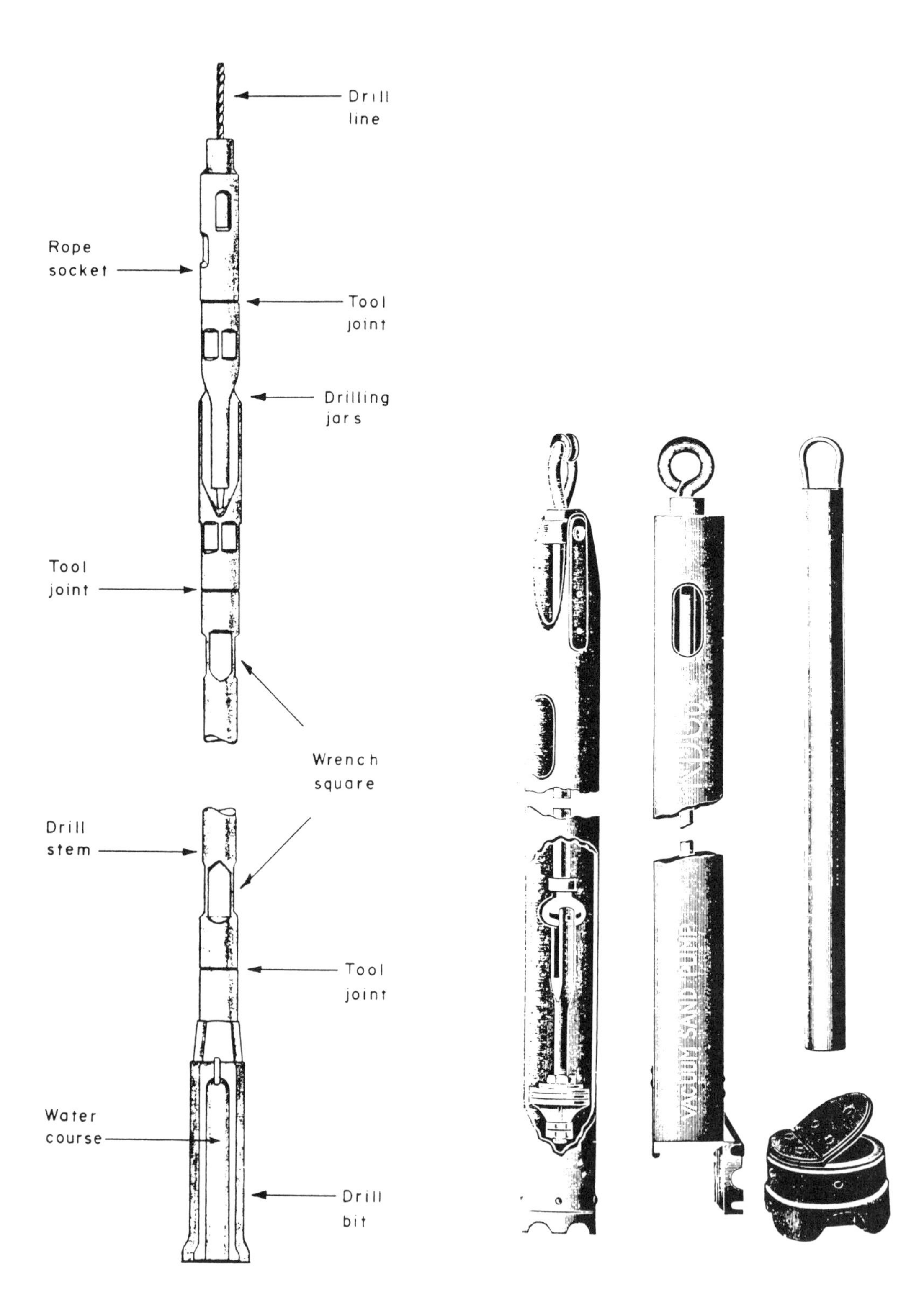

Four components of the string of drill tools for cable-tool percussion drilling.

Sand pumps and regular bailer, with details of flat-valve bottoms.

Fig. 3.46 Percussion (Cable-Tool Method)

(Source: *GROUNDWATER AND WELLS*, permission of Johnson Division, UOP, Inc., St. Paul, Minn. 55164)

Roller or rock type

Drag or fishtail type

Fig. 3.48 Rotary drilling bits
(Permission of Varel Manufacturing Company)

drilling fluid. The other type of bit, used in sand and clay formations, is the drag type, which has blades arranged in a fishtail or three-way design to cut the borehole.

3.1332 Reverse Circulation Rotary Drilling Method

The reverse circulation rotary drilling method is normally used in drilling large-diameter wells in consolidated formations. As the drill bit rotates, a suction pump is used to pull the cuttings through the hollow drill stem to the surface. Water from a supply pit near the rig circulates through a trench and back to the open hole. This water raises the water level in the drill hole to pit level so that *HYDROSTATIC PRESSURE*[55] is applied against the wall of the open hole to prevent caving. Figure 3.49 shows how the direction of circulation is reverse to that employed in the direct rotary drilling method.

3.1333 Air Rotary Drilling Method

Rotary drilling equipment using compressed air as the drilling fluid, rather than drilling mud, is fairly popular. Air is circulated through the center of the drill pipe, out through ports in the drill bit, and upward in the annular space around the drill pipe. The air rotary method requires that air be supplied at pressures from 100 to 250 pounds per square inch (690 to 1,725 kPa or 7 to 17 kg/sq cm). To remove the cuttings, ascending air velocities of at least 3,000 feet per minute (900 m/min) are necessary. *CAUTION:* Never use compressed air to dust off clothing or any part of your body because air can enter the tissues of your body and/or bloodstream and cause serious injury.

Penetration rates of 20 to 30 feet per hour (6 to 9 m/hr) in hard rock are common with air rotary drilling. However, drilling with air as the circulating fluid can be done only in consolidated materials. Air rotary drilling rigs are usually equipped with a conventional mud pump in addition to a high-capacity air compressor. Drilling mud can be used, therefore, in drilling through caving materials above bedrock. Casings may have to be installed through the overburden (looser materials) before continuing with the air rotary method.

3.134 Shallow Collector Wells — Ranney Type

The water collector is a dug well from 12 to 16 feet (3.5 to 5 m) in diameter that has been sunk as a *CAISSON*[56] near the bank of a river or lake. Screen pipes are driven radially and approximately horizontally from this well into the sand and gravel deposits underlying the river (Figure 3.50). The length of these horizontal screens varies from 100 to 300 feet (30 to 90 m). With proper design features, this can be constructed in the flood plain area of certain rivers. Such wells have large capacities, some up to 10 million gallons per day (37,850 cu m/day).

3.14 WELL STRUCTURE AND COMPONENTS

3.140 Rules and Regulations Regarding Well Construction

The size, shape, and structure of a well can vary significantly from one region to another. State or local ordinances may establish the criteria that must be followed when constructing a well. In some cases, local customs or unusual subsurface formations will dictate the construction method, while in other areas drilling methods may be left entirely up to the well driller.

In this section we will discuss the component parts and construction materials used in typical cable-tool, gravel-envelope, and hard-rock wells. Figure 3.51 shows typical cross-sectional views of the three different types of wells mentioned above. This section will discuss the portion of the well located below the ground surface. See Section 3.1, "Surface Features of a Well," for a description of the above-ground components of a well.

3.141 Subsurface Features of a Well

The important *SUBSURFACE* component parts of a typical well include the conductor casing, well casing, well screen or perforated intake section, sanitary grout seal, and the gravel pack or filtering media.

We will review each of these parts in detail, but before we do that, let's discuss the importance of the sanitary grout seal. From a sanitary health and water utility viewpoint, the most important part of any well is the portion from ground level down to the point where the sanitary grout seal ends. This upper zone of the well must be effectively sealed to protect the well against contamination or pollution by surface or shallow subsurface waters, and against improper disposal of liquid wastes and wastewater-related hazards. If this section of the well is not properly designed and installed, then the useful life of the well and also the aquifer is in jeopardy.

The component parts of the sanitary seal include the conductor casing, the upper portion of the well casing, and the grouting material itself.

Now let us review each of the *SUBSURFACE* components in more detail, keeping in mind that construction, design, and materials can vary significantly from one geographic area to another.

3.142 Conductor Casing

The conductor casing is the outer casing of a well (see Figure 3.51, Gravel Packed Wells, B. With Conductor Casing). The purpose of the conductor casing is to prevent contaminants from surface waters or shallow groundwaters from entering the well.

For community water supply wells, the minimum thickness of the steel conductor casing should be 1/4 inch (6 mm) for single casing and 10-gauge for double casing. Steel used for conductor casing should be mild low-carbon steel similar to ASTM-A 139 Grade B. The casing should extend 18 inches (450 mm) above the ground surface.

1. Cable-Tool Wells: The conductor casing should be at least 4 inches (100 mm) larger in diameter than the well casing.
2. Gravel-Envelope Wells: The conductor casing should be at least 12 inches (300 mm) larger in diameter than the well casing.

---

[55] *Hydrostatic (HI-dro-STAT-ick) Pressure. (1) The pressure at a specific elevation exerted by a body of water at rest, or (2) in the case of groundwater, the pressure at a specific elevation due to the weight of water at higher levels in the same zone of saturation.*

[56] *Caisson (KAY-sawn). A structure or chamber which is usually sunk or lowered by digging from the inside. Used to gain access to the bottom of a stream or other body of water.*

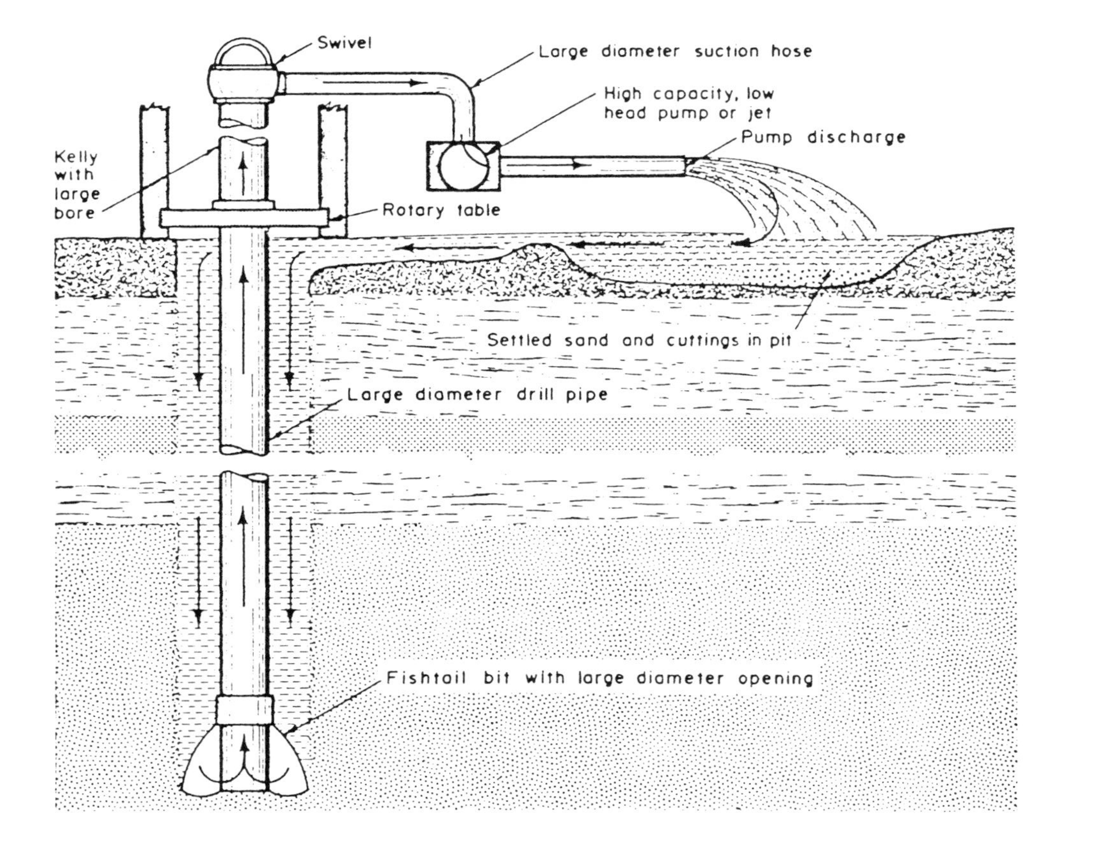

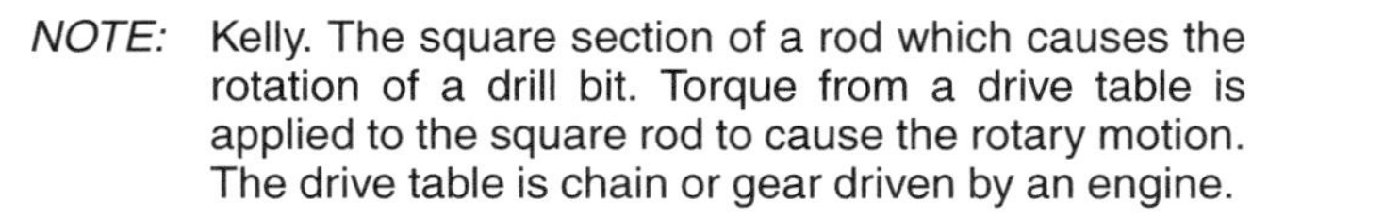
NOTE: Kelly. The square section of a rod which causes the rotation of a drill bit. Torque from a drive table is applied to the square rod to cause the rotary motion. The drive table is chain or gear driven by an engine.

Fig. 3.49 Basic principles of reverse circulation rotary drilling are shown by this schematic diagram. Cuttings are lifted by upflow inside drill pipe.
(Source: *GROUNDWATER AND WELLS*, permission of Johnson Division, UOP, Inc., St. Paul, Minn. 55164)

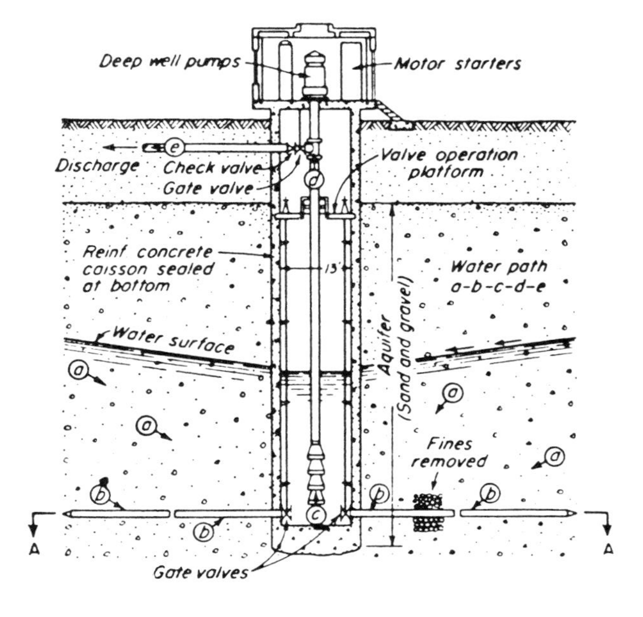

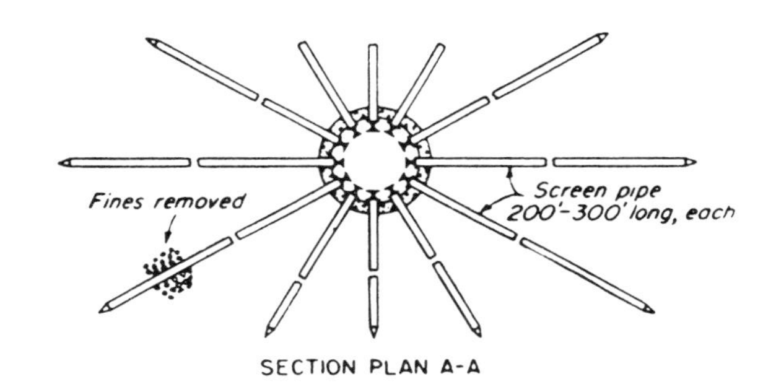

Fig. 3.50 Typical shallow collector-type well (often referred to as a Ranney collector)
(From R. Nebolsine, J. New Engl. Water Works Assoc., Vol. 57, p. 191)

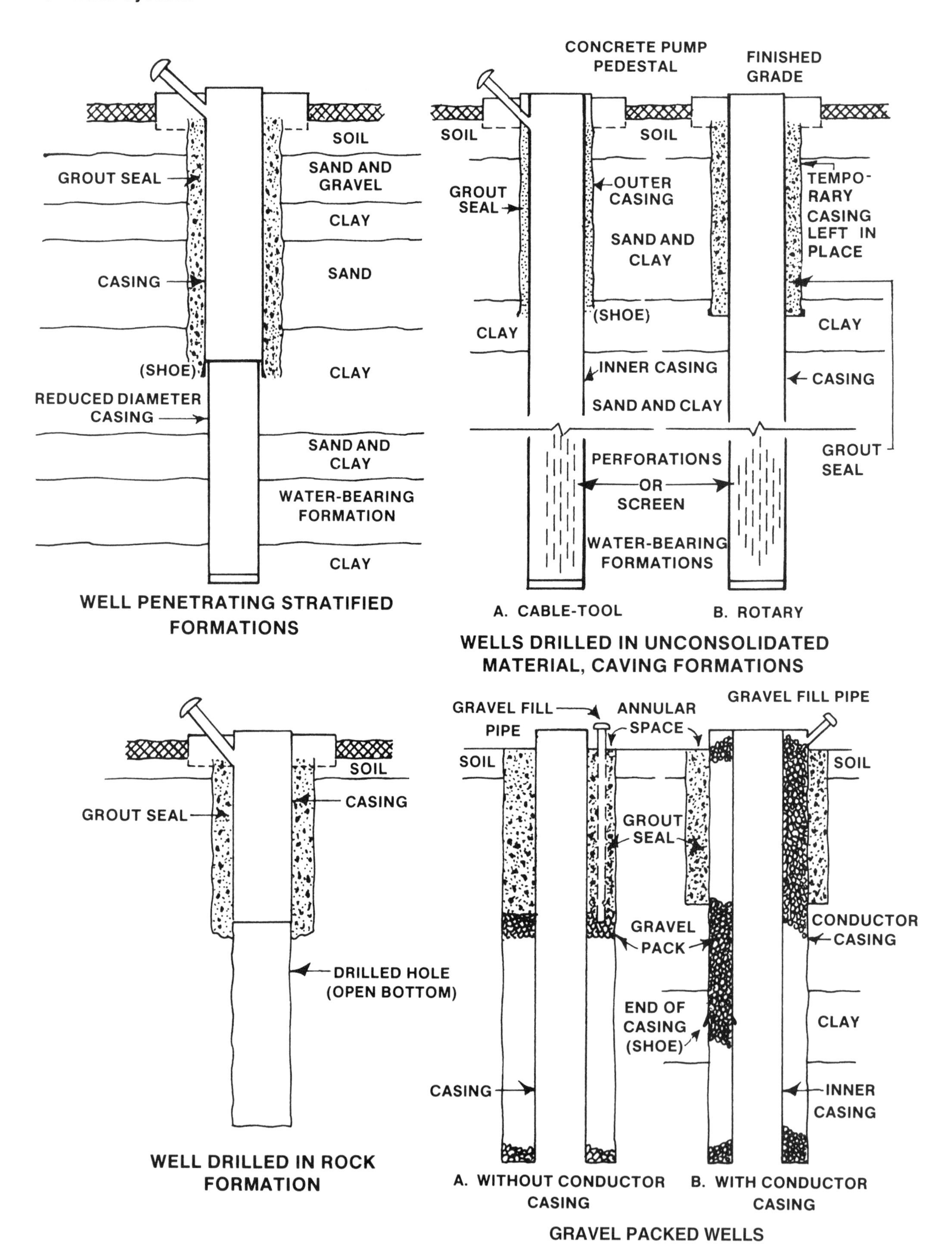

Fig. 3.51 Typical cross-sectional views of different types of wells

3. Hard-Rock Wells: The conductor casing should be at least 4 inches (100 mm) larger in diameter than the well casing. However, the well casing also functions as the conductor casing in many instances.

4. Depth of Conductor Casing: The conductor casing should extend down and into an impervious stratum where no liquid can flow in or out. As a minimum, the conductor casing should be 50 feet (15 m) in depth. In some cases, the casing might extend down to 150 feet (45 m) or more if the nature of the upper stratum and the degree of hazard require the extra protection. State or county health authorities often establish guidelines for the minimum depth of a conductor casing and sanitary seal.

3.143 Well Casing

The functions of the well casing are: (1) to maintain the well hole by preventing its walls from collapsing, (2) to provide a way to get the water to the pumping unit, (3) to form a chamber or housing for the well pump, and (4) to protect the quality of water pumped.

The diameter of the well casing is generally determined by the amount of water that the water utility desires to pump from the well and by the safe yield of the well. Table 3.11 shows the recommended casing diameters for various ranges of well yields or pumping rates.

A careful review of this table shows that, in most cases, the casing should be at least 4 inches (100 mm) larger in diameter than the *PUMP BOWL*.[57] The two-inch (50-mm) clearance on all sides allows the pump bowl and column assembly to move freely in the well during installation and removal. This much clearance is necessary in case a well is slightly crooked or out of plumb. In some cases, however, the size of the well casing may be determined by the drilling capability of drilling rigs in the area and availability of well casing.

3.144 Intake Section of a Well

3.1440 Purpose of Intake Section

Another important subsurface feature of a well is the intake section located at or near the bottom of the well and attached to the end of the well casing. The intake section of the well permits the free flow of water from water-bearing formations into the well itself, while at the same time it supports the water-bearing formations and prevents the drill hole from collapsing. In most cases, the intake section also performs the important function of preventing sand from entering the well.

The intake section may take the form of (1) a well screen, (2) mill-cut slots, (3) formed louvers, (4) torch-cut/chisel-cut slots, and (5) slots made by a mechanical perforator after the well has been completed. These five types of intake sections are described in the following paragraphs.

3.1441 Well Screens

The three basic types of well screens are continuous slot, bar, and wire-wound. They are generally constructed of stainless steel, monel metal, special nickel alloys, silicon red brass, red brass, special alloy steel, and plastic.

The length and diameter of the screen section are typically based on the expected yield of the well. Frequently, screen sections are installed at more than one location in the well to fully utilize all of the desirable water-producing formations. The size of the screen openings may vary from formation to formation, depending on the various aquifer materials that might clog them. The size of the screen openings is determined by

TABLE 3.11 RECOMMENDED WELL DIAMETERS

Anticipated Well Yield, GPM	Nominal Size of Pump Bowl, inches	Optimum Size of Well Casing, inches	Smallest Size of Well Casing, inches
Less than 100	4	6 ID[a]	5 ID
75 to 175	5	8 ID	6 ID
150 to 400	6	10 ID	8 ID
350 to 650	8	12 ID	10 ID
600 to 900	10	14 ID	12 ID
850 to 1,300	12	16 OD[b]	14 OD
1,200 to 1,800	14	20 OD	16 OD
1,600 to 3,000	16	24 OD	20 OD

[a] ID. **I**nside **D**iameter
[b] OD. **O**utside **D**iameter

[57] *Pump Bowl. The submerged pumping unit in a well, including the shaft, impellers and housing.*

the *EFFECTIVE SIZE*[58] and *UNIFORMITY COEFFICIENT*[59] of the sands in the water-bearing strata.

The slots or openings in the screen should have sufficient open area so that the water flowing from the water-bearing formations through the screen openings does not exceed a recommended velocity of 0.1 foot per second (0.03 m/sec). This recommended velocity may vary according to local conditions, health departments, and manufacturers. Figure 3.52 illustrates a few typical sections of well screens.

3.1442 Mill-Cut Slots

Mill-slotted intake sections are typically made at the well casing manufacturer's plant from the same type and diameter of well casing that was selected for the well. The openings are machine milled (cut) into the wall of the casing pipe parallel to the axis of the casing and uniformly spaced around the casing pipe at approximately 2-inch (50-mm) intervals. The slot openings are typically 1/16 to 1/8 inch (1.5 to 3 mm) wide by 2 to 6 inches (50 to 300 mm) long. The slot width is determined by the size of the aquifer material it must keep out. The rows are sometimes staggered and placed several inches apart. Figure 3.53 illustrates several different types of mill-slotted intake sections. Mill-cut slots are still used but may be on the decline due to the increased popularity of the louvered casing.

3.1443 Formed Louvers

Research and development of the formed louver has made this type of intake section a very popular product. The openings or louvers in the well casing are machine made, horizontal to the axis of the casing, with the openings facing downward. The louvers are shaped to create an upward flow as water enters the well. The louvers are spaced close together in vertical rows as shown in Figure 3.54. As with other types of intake sections, the size of the openings is determined by the aquifer material.

Louvered casing is available in a variety of materials ranging from stainless steel to mild steel. The diameter of the louvered casing is usually the same diameter as the well casing and is available in 10-, 15-, and 20-foot (3-, 4.5- and 6-m) lengths.

Use of properly sized openings and properly selected and sized gravel will produce optimum results. The louvered sections are usually installed directly opposite the desired water-bearing strata. However, in situations where the water-bearing strata are separated by thin layers of consolidated material, a continuous length of louvered casing is often used rather than inserting short lengths of blank (non-louvered) well casing. This practice allows for better placing and setting of the gravel pack, quicker well cleanup and shorter development time, and tends to draw the maximum water yield from the selected water-bearing formations.

3.1444 Torch-Cut Slots

Torch-cut and chisel-cut slots should be avoided. Well drillers find it next to impossible to control the size of the openings in the well casing and to maintain the correct open area per foot of casing. Without this type of control, the well is likely to produce sand in excessive quantities. See Figure 3.55 for examples of unacceptable torch-cut and chisel-cut slots.

3.1445 Hydraulic or Mechanical Slots

Cable-tool wells are usually slotted after the well has been drilled. The openings are made opposite the water-bearing formations by means of a casing perforator tool lowered into the well and activated from the drill rig.

This method of producing openings or slots in the well casing has serious limitations because the openings cannot be closely spaced, the percentage of open area is low, the opening size and shape can vary, and the correct size openings to control fine or medium sand are almost impossible to produce.

3.145 Annular Grout Seal (Figure 3.56)

The purpose of the annular grout seal is to seal out water of any unsatisfactory chemical or bacterial quality, to protect the well casing or conductor casing pipe against exterior corrosion, and to stabilize soil formations which are of a caving nature.

Grouting or cementing is simply filling with cement or other suitable material (1) the annular space between the well casing and the conductor casing; or (2) the space between the conductor casing and the bore hole; or (3) the space between the well casing and the bore hole.

However, an *OPTIONAL* sanitary sealing method (Figure 3.63, page 170) also has been used for gravel-envelope wells. The typical outside conductor casing is eliminated and the annular space between the well casing and the oversize bore

[58] *Effective Size (E.S.). The diameter of the particles in a granular sample (filter media) for which 10 percent of the total grains are smaller and 90 percent larger on a weight basis. Effective size is obtained by passing granular material through sieves with varying dimensions of mesh and weighing the material retained by each sieve. The effective size is also approximately the average size of the grains.*

[59] *Uniformity Coefficient (U.C.). The ratio of (1) the diameter of a grain (particle) of a size that is barely too large to pass through a sieve that allows 60 percent of the material (by weight) to pass through, to (2) the diameter of a grain (particle) of a size that is barely too large to pass through a sieve that allows 10 percent of the material (by weight) to pass through. The resulting ratio is a measure of the degree of uniformity in a granular material such as filter media.*

$$\text{Uniformity Coefficient} = \frac{\text{Particle Diameter}_{60\%}}{\text{Particle Diameter}_{10\%}}$$

Fabricating the Johnson Well Screen — an all-welded, continuous-slot well screen for water wells and oil wells. Screen may be made of any metal or alloy that can be resistance welded.

(Source: *GROUNDWATER AND WELLS*, permission of Johnson Division, UOP, St. Paul, Minn. 55164)

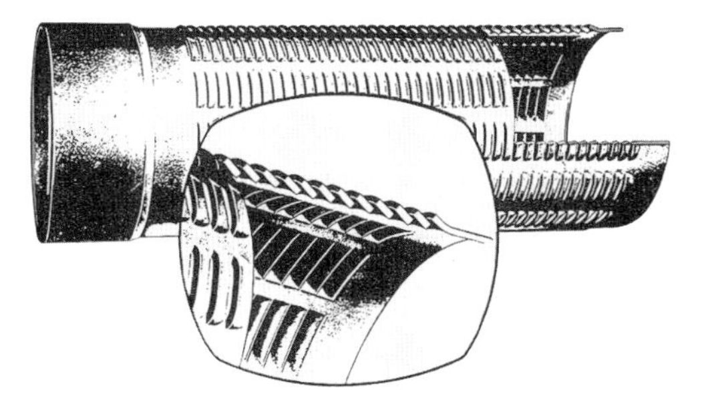

Sections through well screens.

(Permission of Layne and Bowler, Inc. Memphis, Tennessee)

Fig. 3.52 Typical sections of well screens

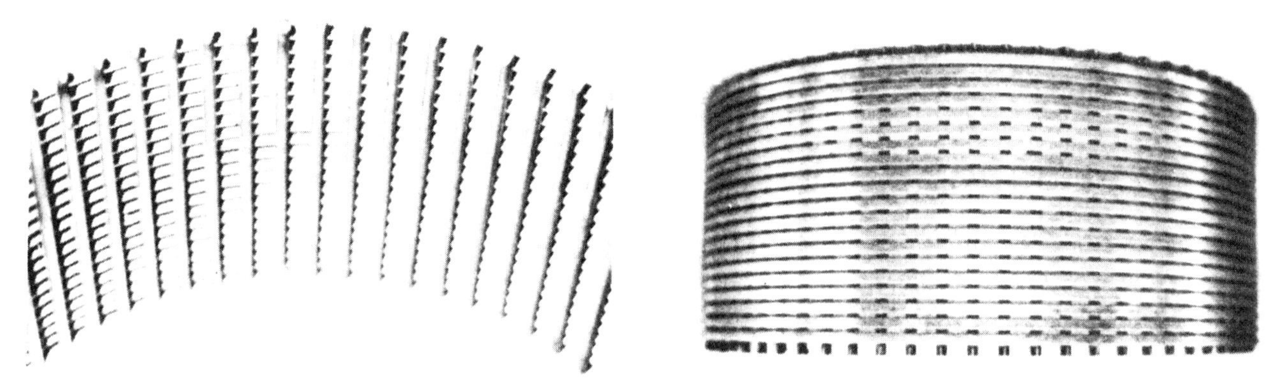

Continuous slot well screen and perforations

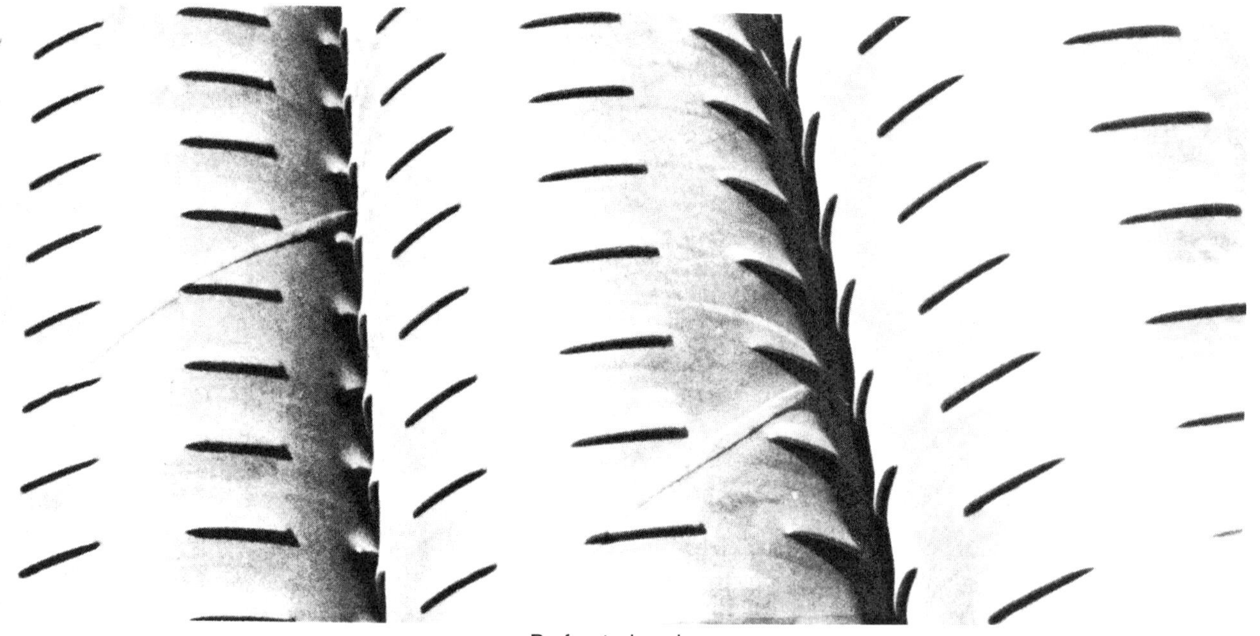

Perforated casing

Fig. 3.53 Different types of intake sections

(Source: *WATER WELLS AND PUMPS: THEIR DESIGN, CONSTRUCTION, OPERATION AND MAINTENANCE*, Division of Agricultural Sciences, University of California, Davis)

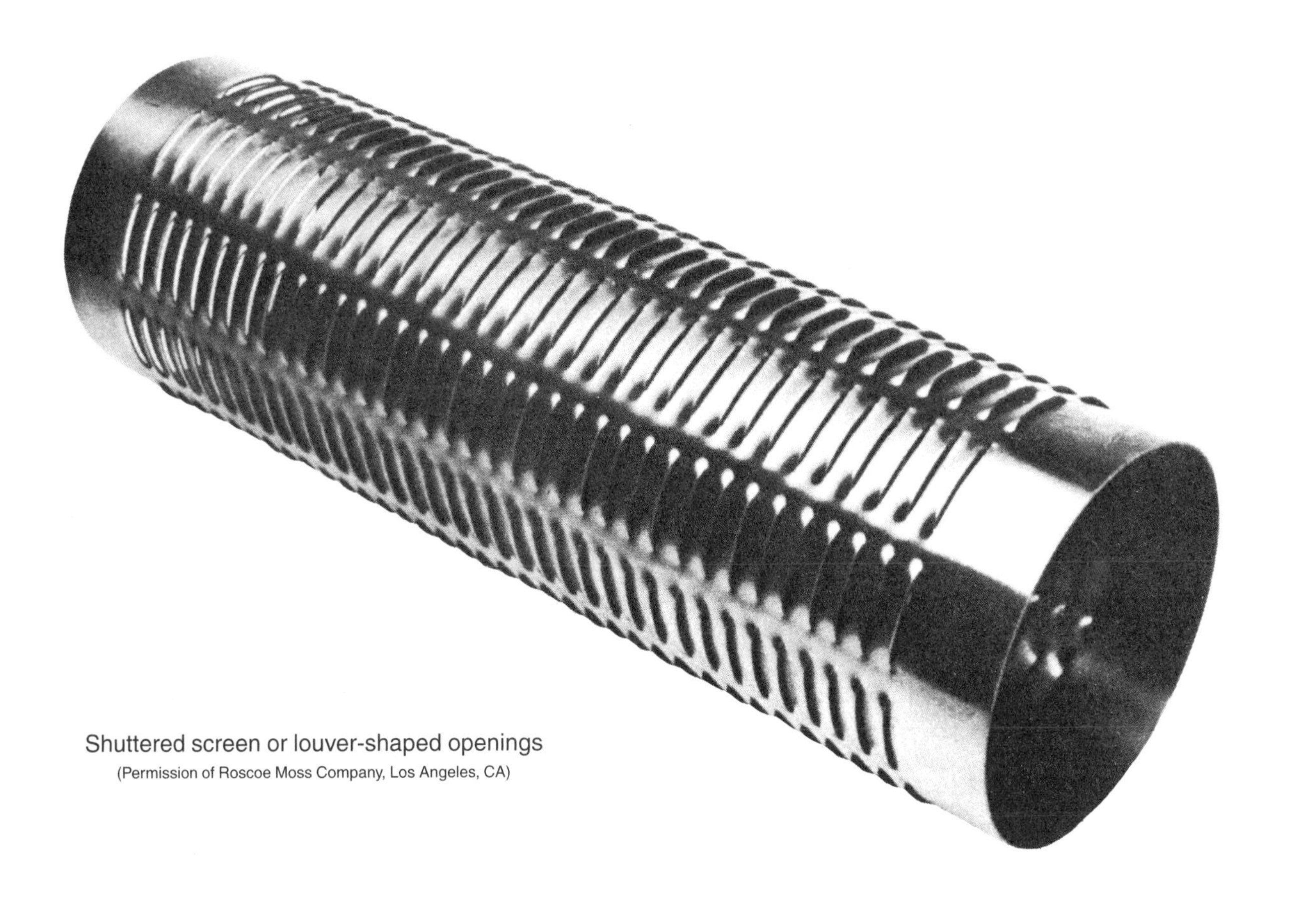

Shuttered screen or louver-shaped openings
(Permission of Roscoe Moss Company, Los Angeles, CA)

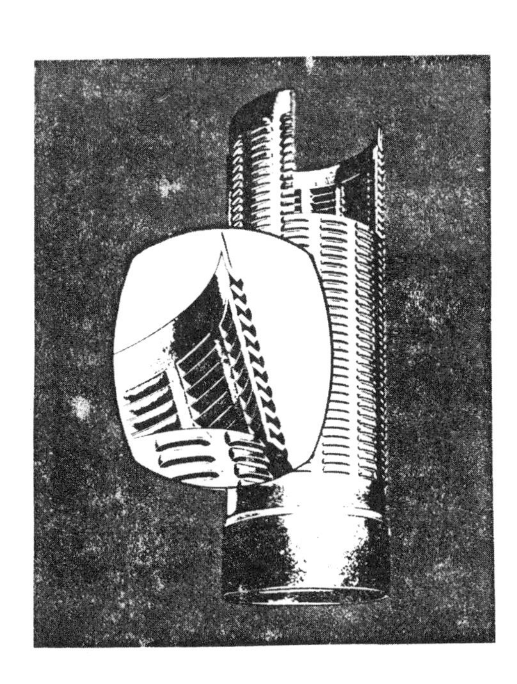

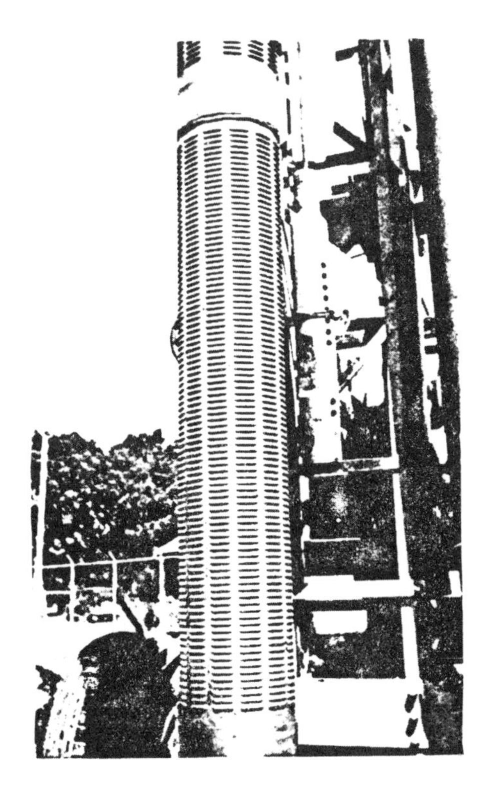

Well casing louvers
(Permission of Layne & Bowler, Inc., Memphis, Tennessee)

Fig. 3.54 Well casing openings or louvers

In torch-cut, slotted pipe, percent open area is low and width of slots is too large to permit developing well to sand-free condition.

Casing perforator produces crude, jagged openings of uncontrolled size.

Fig. 3.55 Examples of unacceptable torch-cut and chisel-cut slots
(Source: *GROUNDWATER AND WELLS*, permission of Johnson Division, UOP, Inc., St. Paul, Minn.)

hole is cemented. In order to provide access for further graveling, a 4-inch (100-mm) standard steel pipe is placed in the annular space extending from above ground down to the area where the gravel envelope is needed. In order to ensure a good seal around the 4-inch (100-mm) graveling pipe, the thickness of the grout is usually increased to 6 or 8 inches (150 to 200 mm).

This new optional method requires an extra thick grout seal and a 4-inch (100-mm) pipe for adding gravel, but does eliminate the expensive outside conductor casing and bore hole. On a typical 14-inch (350-mm) diameter well, the conductor casing is 30 inches (750 mm) in diameter and the bore hole for the conductor casing is 36 inches (900 mm) across.

Hard-Rock Wells: In consolidated formations, the general practice is to extend the well casing down and into the water-bearing rock formation for a minimum distance of 5 feet (1.5 m). Then the annular space between the well casing and the bore hole is cemented. This grout seal should also be a minimum of 2 inches (50 mm) thick. The balance of the bore hole through the water-bearing rock formations is left uncased. Water enters the uncased area of the well by way of the fissures and voids in the rock formations.

3.146 Sealing Material

The material used for sealing grout should consist of neat cement grout, sand-cement grout, or bentonite clay. Cement used for sealing mixtures must meet the requirements of ASTM C-150 "Standard Specifications for Portland Cement" Type I (common construction cement) or Type III (high early strength).

3.147 Placement of Sealing Material

Before placing the sealing material in position, all loose cuttings, drilling mud, or other obstructions should be removed from the annular space by flushing with water.

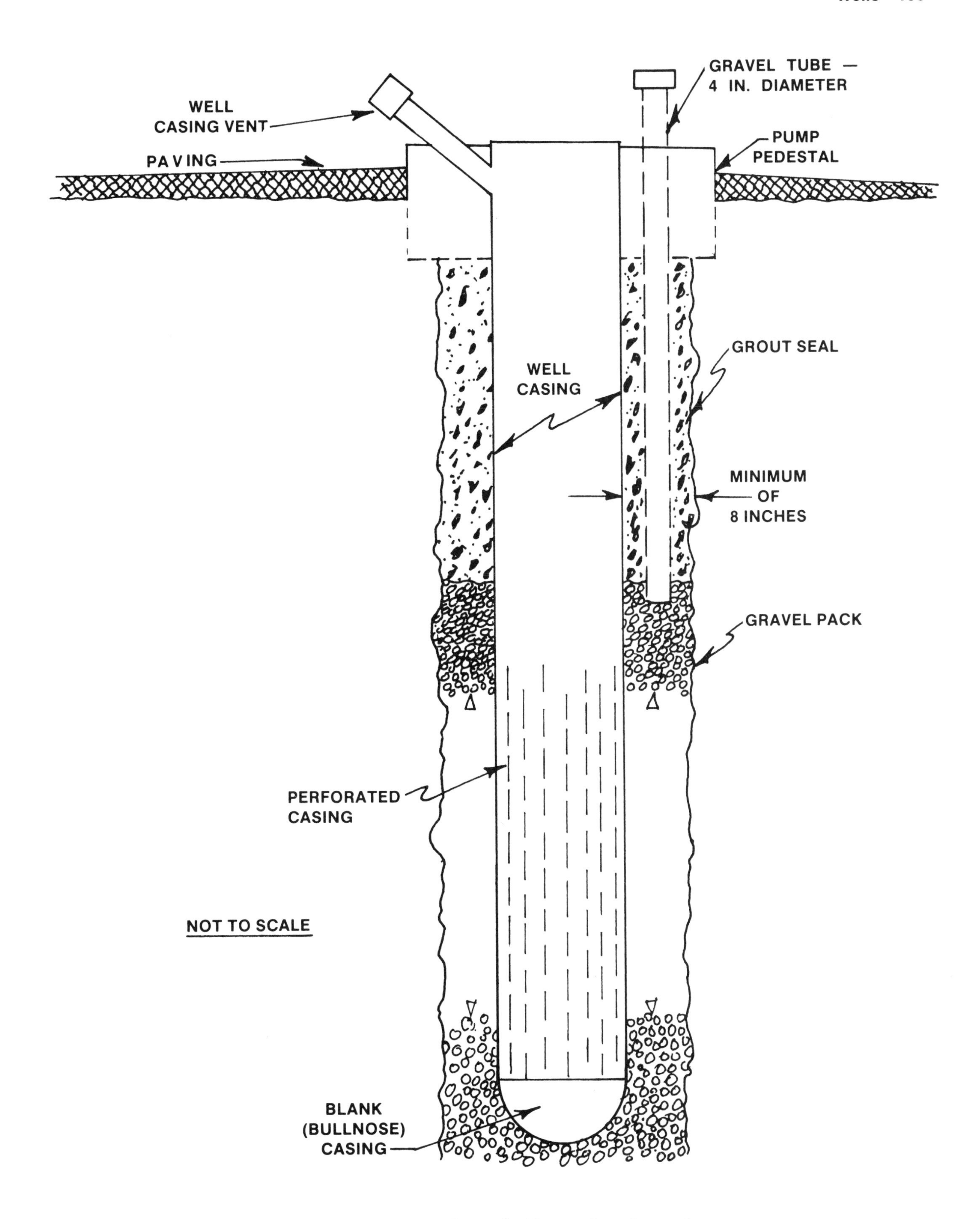

Fig. 3.56 Sanitary sealing method for gravel-envelope wells

There are three basic methods for placing the sealing material in the annular space when a conductor casing is used. The methods are grout pipe outside casing, grout pipe inside casing, and the Halliburton Method.

No drilling or work in the well should be permitted until the cement grouting materials have developed sufficient strength to prevent damage to the seal.

3.148 Gravel Pack

The purpose of the gravel pack is to control the entrance of sand into the well. The size (or combination of sizes) of gravel selected for the gravel pack is determined by an analysis of the grain size of the water-bearing aquifer materials encountered during the drilling operation. The goal is to select a gravel pack which, while still maintaining sand control, will allow the largest possible screen or slot opening and thus increase the flow into the well.

WHENEVER PLACING GRAVEL IN A WELL, THE GRAVEL MUST BE DISINFECTED WITH A STRONG CHLORINE SOLUTION (50 mg/L) BEFORE INSTALLATION TO AVOID CONTAMINATING THE WELL.

Another type of gravel-envelope construction is the "Layne Underreamed Gravel Wall" method developed by Layne and Bowler, Inc., of Memphis, Tennessee. With this method, the water-bearing formation is underreamed (the hole is made larger under the clay layer, Figure 3.57), after the permanent upper casing is set in place. A screen or louvered section of the proper diameter and length is centered in the larger diameter of the underreamed hole. Selected and graded gravel is placed around the screen or louvered section, resulting in a gravel envelope held in place by a long-lasting intake section. This design allows for a maximum flow of water into the well and is a very efficient groundwater-producing unit.

3.15 DETERMINATION OF WORKING PRESSURE

Hydropneumatic pressure tanks are used to provide a consistent range of suitable pressures in the distribution system despite changing demands and pump cycling. Other reasons for using a hydropneumatic pressure tank system are:

1. Gravity storage may not be practical or available due to the lack of elevation differentials within the service area.
2. By using the best operating pressure differential and control levels in the tank (referred to as "tank efficiency"), the time period between start/stop operation of the pump can be extended considerably, thereby reducing the cycling frequency of the pump. Excessive pump cycling increases the wear on the pump and its parts, plus using additional energy.
3. Hydropneumatic tanks maintain distribution system pressures within predetermined limits by maintaining a cushion of compressed air in the top portion of the tank. In effect, this provides stored energy to force the water out as the pressure in the distribution system drops.
4. The compressed air in the tank greatly reduces the hydraulic shock load on the system when the well pump starts and stops.

The effective (usable) water storage capacity in a horizontal hydropneumatic tank is limited to approximately 25 percent of the total capacity of the tank.

You can select the best operating pressure differential, the control levels in the tank, the pumping differential, and the tank efficiency and tank size by studying Figures 3.58 and 3.59 and working through the following examples.

As an example, let's study a typical 5,000-gallon capacity horizontal tank, six feet in diameter by approximately 25 feet in length, with the inlet and outlet connections in the bottom of the tank, and an *OPERATING PRESSURE DIFFERENTIAL*[60] of 20 psi (from 40 to 60 psi). For this example, the most efficient tank operating criteria would require the pump start point or low water level (LWL) to be set at 12 inches (30 cm) above the bottom of the tank; set the pump stop point or high water level (HWL) at 27 inches (67 cm) above the bottom of the tank. If the tank is constructed with an outlet connection that extends up into the tank (generally 6 to 10 inches (15 to 25 cm)) as is a common practice if the well produces sand, then the LWL in the tank must be maintained at a minimum of 10 inches (25 cm) above the top rim of the outlet connection so that the possibility of air loss into the piping system will be minimized. This would require the HWL to be raised to the optimum level (27 inches or 67 cm).

Since we have the guidelines for our tank, let us proceed through an example of how we derived the HWL and check to see if the 20 psi (138 kPa or 1.4 kg/sq cm) differential is the most desirable for our situation.

EXAMPLE 1: PRESSURE DIFFERENTIAL AND EFFICIENCY DETERMINATION

Refer to Figure 3.58. Start at the point indicating a reserve of 10 percent by volume in the tank and follow this line horizontally to where it intersects the vertical 40 psi pressure line. Follow the closest pressure curve (in this case the 35 psi curve) to where it intersects the vertical 60 psi line. Then by interpolation determine the point which indicates that the water will occupy approximately 34 percent of the total tank capacity when the air has been compressed from 40 psi to 60 psi. The water level equivalent to 34 percent of the tank volume established the desired HWL.

The pumping differential is the difference in volume between the HWL and LWL in the tank. This differential expressed in percent also indicates the tank efficiency. Thus, 34 percent minus 10 percent indicates that the pumping differential is 24 percent of the total tank volume. With 24 percent of the total tank volume available for pumping, the tank efficiency also is 24 percent.

The actual HWL and LWL in the tank may now be established. The volume in a cylindrical, vertical tank is proportional to the height. Assume that a vertical tank is 72 inches high and the tank discharge is located in the tank bottom. Then the LWL is (10/100) 72 or 7.2 inches above the bottom of the tank and the HWL is (34/100) 72 or 24.28 inches above the bottom of the tank.

The volume in a cylindrical, horizontal tank is not proportional to the diameter (height) so volume height calculations

[60] *Operating Pressure Differential. The operating pressure range for a hydropneumatic system. For example, when the pressure drops below 40 psi the pump will come on and stay on until the pressure builds up to 60 psi. When the pressure reaches 60 psi the pump will shut off.*

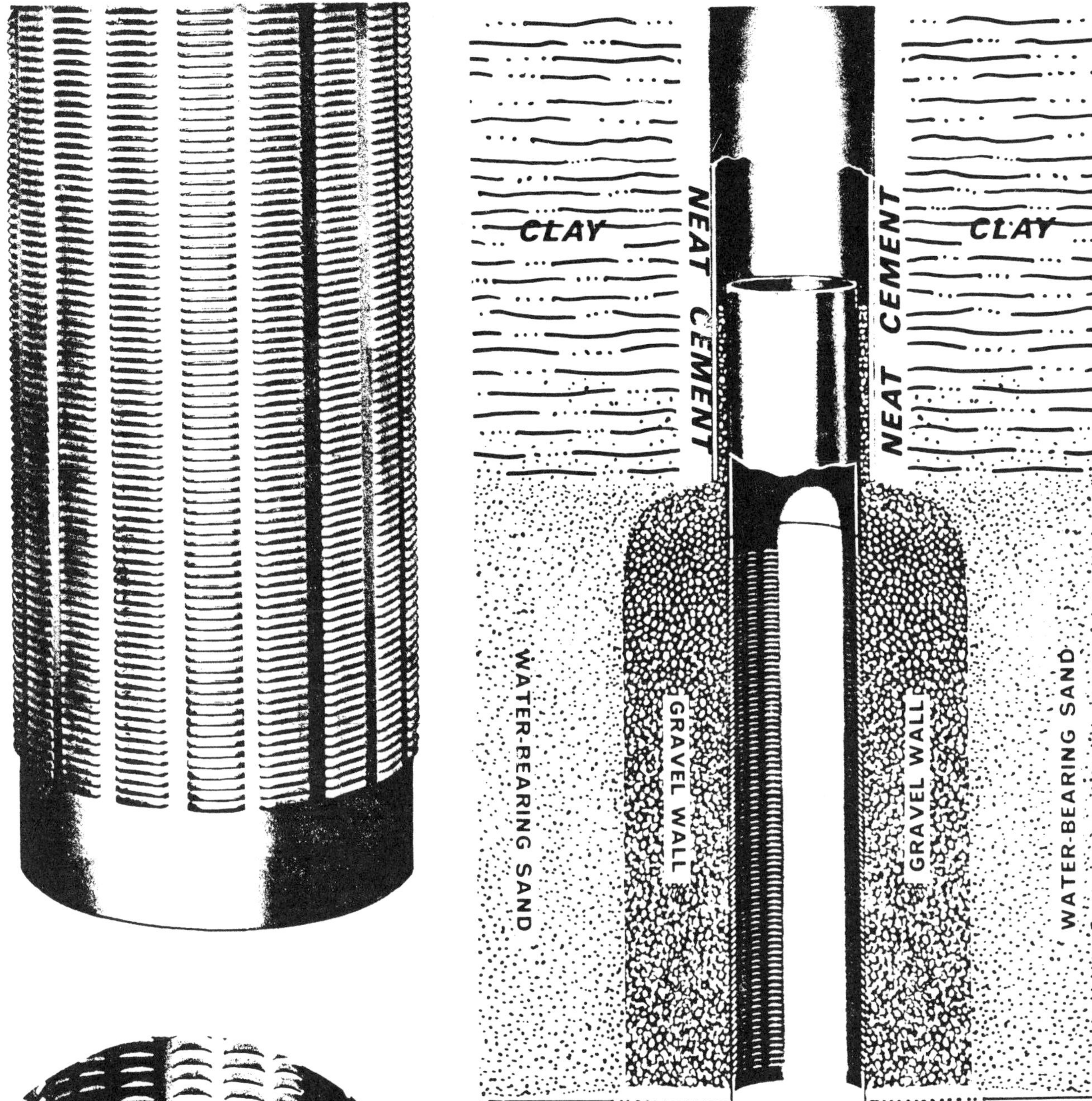

Fig. 3.57 Layne Underreamed Gravel Wall

(Permission of Layne and Bowler, Inc., Memphis, Tenn.)

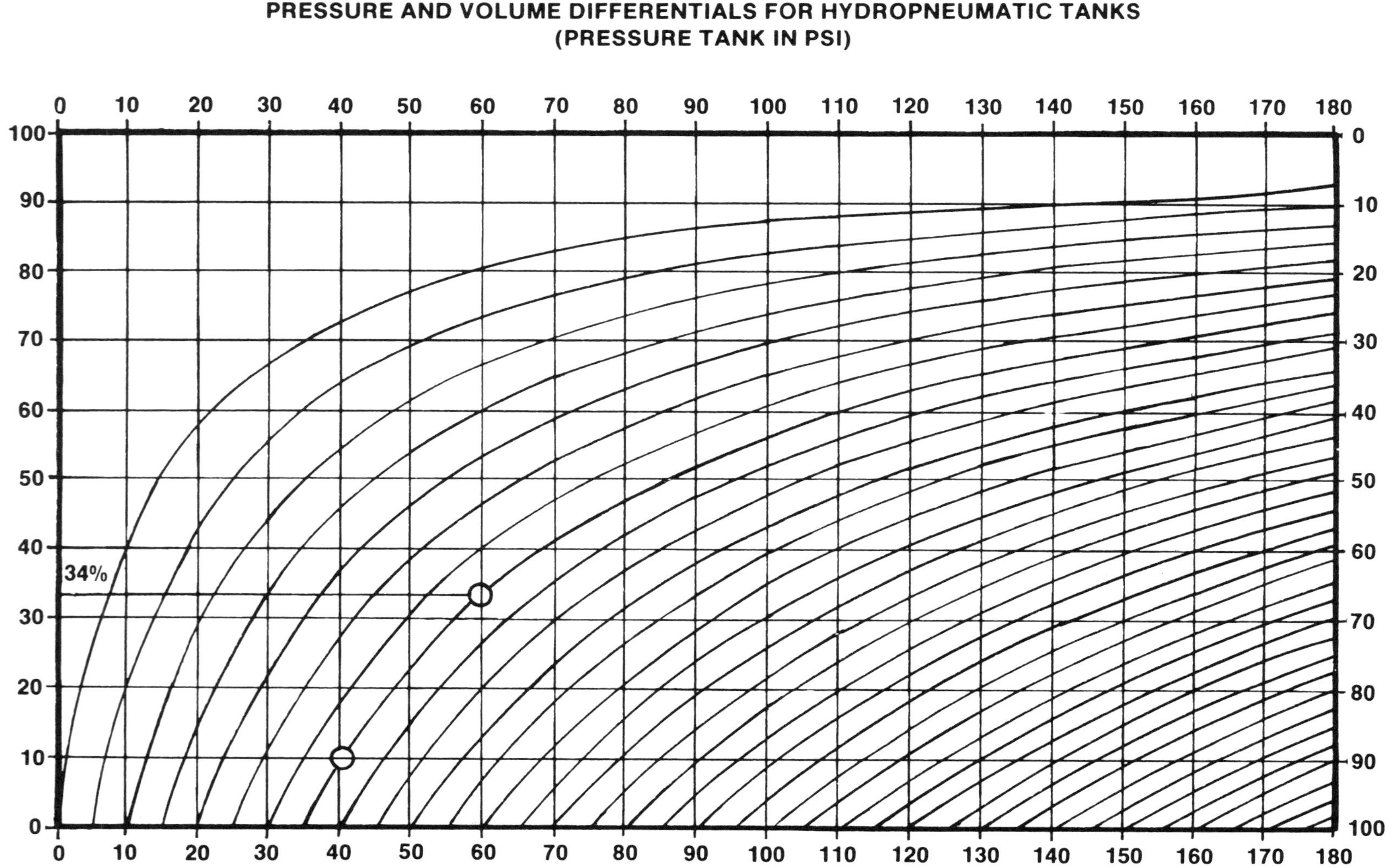

Fig. 3.58 Pressure-volume determination chart

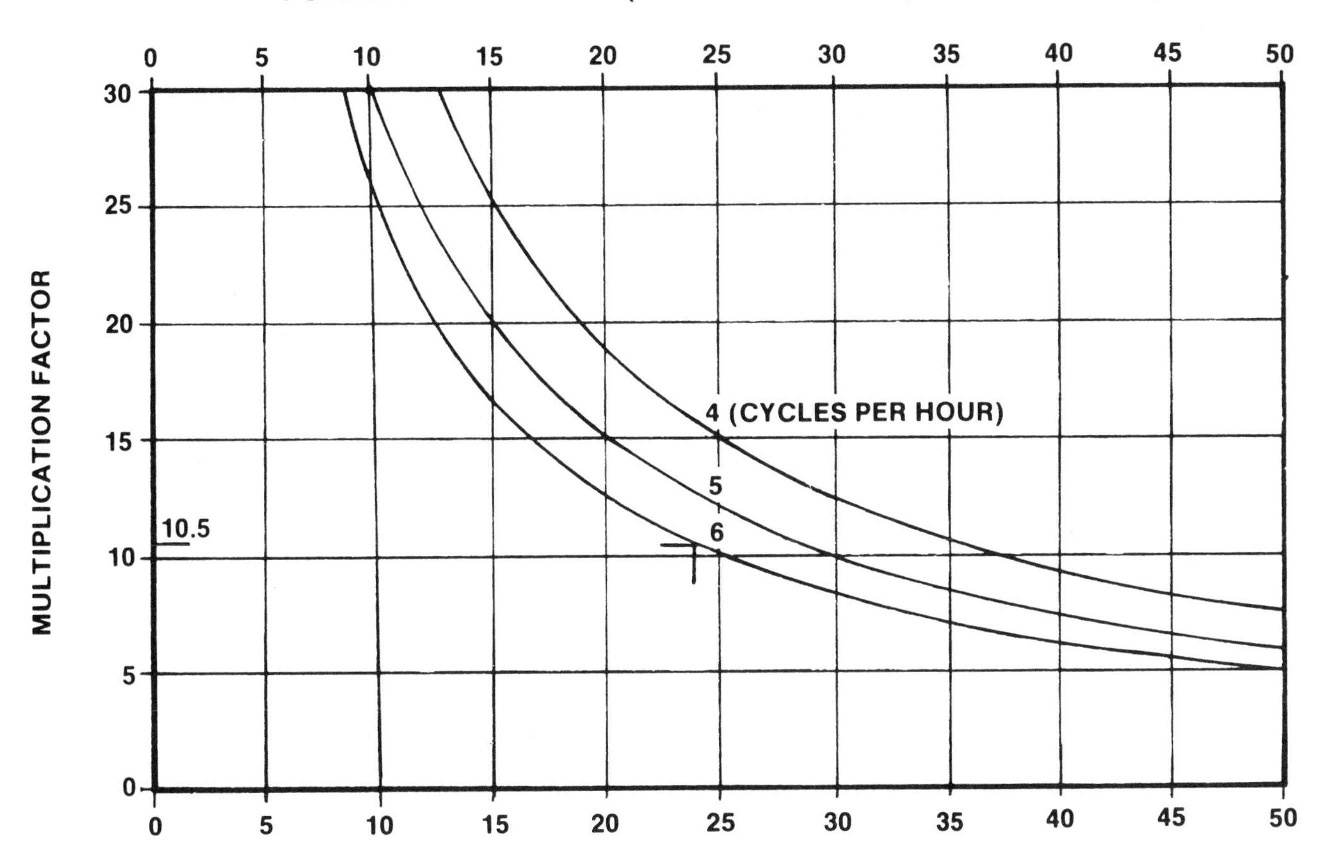

Fig. 3.59 Pressure tank volume requirements

must be made. These calculations will not be done here. To summarize, we have now established how to determine the LWL, the HWL, the pumping differential, and the tank efficiency. The remaining consideration is for determining the most desirable operating pressure differential.

Assume, for example, that the pressure differential is to be 30 psi with 40 psi at LWL and 70 psi at HWL. Proceed as described above and determine that at 70 psi the water will occupy approximately 42 percent of the total tank volume. The pumping differential is 32 percent of the total tank volume and the tank efficiency is also 32 percent. This is a gain of 8 points over the 40 to 60 psi pressure selection. The pumping differential governs the size of the tank which will be required and also may affect the size of the pump because of the range in the pressure differential. Costs of each arrangement must be evaluated to determine the most efficient system.

EXAMPLE 2: TANK VOLUME DETERMINATION

The size of the tank is governed by both the established pumping volume differential and the number of pumping cycles desired. Experience indicates that the average number of pumping cycles need never be greater than six per hour and very seldom is it necessary to provide for fewer than four cycles per hour. However, cases of 30 cycles per hour have been encountered.

The greater the number of pumping cycles, the smaller will be the size of the required tank. This must be given serious consideration when the initial cost of an installation is of prime importance.

Fewer cycles will require the use of a larger tank but sometimes other important considerations besides initial cost assume greater importance. Fewer pumping cycles are recommended for installations in hospitals, sanitariums, and hotels where frequent starting and stopping may be annoying, and when greater reserve is desired or required, for example, when the installation is used for fire protection.

To determine the tank size, let us assume the required pump capacity has been determined at 50 GPM, the pumping differential is 24 percent of the total tank capacity, and that six pumping cycles per hour are desired.

Refer to Figure 3.59. This curve is based on the assumption that the average system demand is equivalent to one-half of the pump capacity. So with a pumping differential of 24 percent, we determine from the curve for six pumping cycles per hour that the multiplication factor is 10.5 x 50 = 525 gallons or the total volume for the required tank.

Some very small well pumping installations may be equipped with one or more small conventional vertical mounted hydropneumatic tanks *MANIFOLDED*[61] to the pump discharge. These tanks could range from 40 to 330 gallons (150 to 1,250 liters) capacity and be equipped with a flexible or bag-type diaphragm.

3.16 WELL TESTING AND EVALUATION

3.160 Well Performance Testing

Well testing for quantity of water is an important phase of well construction. Information obtained during pumping tests provides data necessary to determine well and pump efficiency, pump installation depth settings, aquifer characteristics, capacity of the well, well recharge potential, and other factors which will be of use in the long-term operation and maintenance of the well. The type of test for a specific well depends on intended use, size of well, and costs of the test. Performance tests should be consistent with the dimensions and capacity of the well and the rate at which it will be pumped when placed in service. Care must be taken to avoid excessive pumping rates which would damage the aquifer.

Depending on the size and proposed use of the well, performance testing may range from a simple bailing test of short duration to tests lasting 72 hours or longer. The driller normally has sole responsibility for small facilities. For larger wells, the operation will usually be under the direction of a competent engineer or geologist. In the latter case, the responsibility of the driller will be to operate the equipment and make the necessary measurements of flow and drawdown.

Measurements of water levels must be taken before, during, and after performance testing. A static or non-pumping level must be established for comparison of the measurements made during pumping and recovery. This static level can be determined by making periodic measurements of water level for a period of time equal to, or longer than, the duration of the proposed test prior to its start. Water levels can be measured with a steel tape, by flagging (marking distances on) the bailer line, by an electric sounder, or by reading pressure on an air line. Recovery readings are also an important part of well testing and should be started immediately upon shutdown of the pumping test and continued at some specific interval.

With small-capacity wells that operate intermittently or irregularly, testing should continue until an apparent stability of bailing or pumping level is achieved. Ideally, with large-capacity wells, pumping should be continued at a uniform rate of discharge until the *CONE OF INFLUENCE*[62] reflects any boundary condition which could affect future performance of the well. This probably will not exceed 24 hours for an artesian well, and 72 hours for a water table well. Large well testing should be supervised by someone with adequate qualifications and experience and may require use of observation wells.

Comprehensive aquifer tests require a minimum of one or two observation wells, depending on the purpose of the test results or the well. In typical situations, observation wells may be from 100 to 300 feet (30 to 90 m) from the production well and about the same depth. Observation wells may be smaller in diameter, however. For testing relatively thick artesian aquifers, observation well distances of 300 to 700 feet (90 to 210 m) from the pumped well are occasionally used.

The amount and rate of drawdown and recovery of the water level with time are the most critical items of data needed to evaluate the initial efficiency of the well and the hydraulic characteristics of the aquifer.

3.161 Types of Pumping Tests

Several procedures for testing wells are available. They range from the very simple bailing procedure to the more complex step-continuous method. The bailing test method and the air blow test method are used to make rough estimates of the yield of a well without installing a pump. The variable rate method uses a pump in a well. These methods are not adequate for a well-acceptance test for well yield for public water supply wells in some states.

3.1610 Bailing Test Method

The first step is to determine the *STATIC WATER LEVEL*[63] in the well. First, use a bailer of known volume to bail (remove water from) the well until the water level reaches a static level below which it cannot be lowered by further bailing. Lower the bailer until it just touches the water (if the bailer is dropped, it will make a splash as it hits the water surface). Mark the bailing line clearly at two points: one point at the top of the casing when the bailer is just on top of the static water level, and at a second point one bailer length above the first point. At even time intervals, lower the bailer until the second mark is just level with the top of the casing. The bailer must be full each trip. If the bailer is not full, increase the time per round trip until the bailer comes out full each time. Once this procedure continuously produces the same amount of water in equal time periods, the yield of that particular depth can be determined. The gallons per minute for that well at that depth is equal to the volume of the bailer (in gallons) divided by the time (in minutes) per round trip.

If the bailer cannot be completely submerged, the same procedure is used except that the volume must be measured and the time period adjusted for each trip until a constant volume is retrieved at regular intervals.

Bailing can be used successfully in low-yield wells, but other procedures are commonly used when the well has a high yield.

---

[61] *Manifold. A large pipe to which a series of smaller pipes are connected. Also called a header.*

[62] *Cone of Influence. The depression, roughly conical in shape, produced in the water table by the pumping of water from a well. Also called the cone of depression.*

[63] *Static Water Level. (1) The elevation or level of the water table in a well when the pump is not operating. (2) The level or elevation to which water would rise in a tube connected to an artesian aquifer, basin, or conduit under pressure.*

3.1611 Air Blow Test Method

This procedure requires the injection of air at the bottom of the well in sufficient quantity to blow water out of the well. As pressures must be high, some kind of deflector will be needed at the top of the well to capture and measure the water. Again, as with the bailing procedure, accurate measurements of water levels must be kept. For a measured period of time and volume of water taken out of the well, the rate of flow (GPM) can be calculated.

3.1612 Variable Rate Method

This method requires pumping the well at a series of constant rates and measuring the variation in drawdown during pumping. Install the pump at the lowest production point in the well and pump at a constant rate until the pump breaks suction. If the pump runs for more than 24 hours without breaking suction, further testing is probably not needed. (Essentially, the test would then be a constant rate test.) If the pump breaks suction, decrease the pumping rate until the water level in the well stabilizes approximately two feet (0.6 m) above the pump intake. Then decrease the pump rate by 5 percent and pump until the water level has been stabilized for at least four hours. The drawdown and pumping rate at that level are considered to be representative of the production rate and pumping level for that particular well.

3.1613 Constant Rate Method

When a well is pumped at a constant rate, the water level will continuously drop at a decreasing rate until a certain water level is reached and maintained. As the water level is lowered during the test, the flow rate of the pump may change due to the head/flow characteristics of the pump. Since this is a constant rate test, however, a valve for throttling and a meter for measuring flow must be used to ensure that the pumping rate remains constant. Adjustments during testing should be made as needed to keep the flow constant.

Again, as in other test procedures, the static or non-pumping level should be established before testing starts. Frequent measurements should be made during the drawdown and recovery periods according to the following schedule: 0 to 10 minutes — every minute; 10 to 45 minutes — every 5 minutes; 45 to 105 minutes — every 10 minutes; 105 to 180 minutes — every 15 minutes; and from 3 to 24 hours — every hour. This schedule can be modified somewhat, but the purpose is to develop a good relationship between time and drawdown at a constant pumping rate.

If more than one drawdown test is to be conducted, the pump should be capable of providing flow rates from approximately 125 percent of design capacity to about 60 percent capacity. For a thorough understanding of the well and aquifer characteristics, more than one test is recommended. This procedure should result in some very accurate information about the well and its production capacity.

The constant rate method is used for well-acceptance testing for public water supply wells. This method is commonly used to determine specific capacity. The constant rate test is typically used to determine several aquifer and confining unit hydraulic characteristics (aquifer transmissivity (T), hydraulic conductivity (K), and storativity (S), and confining unit vertical hydraulic conductivity (Kv)) that are required to design a well field (Section 3.123). Where nearly all the groundwater comes from aquifer storage, aquifer T and K can be determined from a single well test (only the pumping well is used). The problem is that you must know this condition exists before the test. Storativity (S) cannot be determined from this test, but an experienced hydrogeologist can make a good estimate so that the required hydraulic characteristics needed for well field design are available. Where nearly all of the groundwater comes from leakage through the overlying confining unit, aquifer T, K, and S, and confining unit vertical hydraulic conductivity (Kv) can be determined, but a multiple well test (observation wells included) is required. From these constant rate test data, drawdown over time and distance at various pumping rates can be determined. A well field design can be made from this information.

3.1614 Step-Continuous Composite Method

There are several variations or modifications of this test method. In this procedure, the well is tested at approximately 1/2, 3/4, 1, and 1 1/2 times the pump design capacity and usually runs for 24 hours or more. The test would start at 1/2 capacity for 6 hours, then 3/4 capacity for 6 hours, then design capacity for 6 hours, and then 1 1/2 capacity for 6 hours. Pumping times must be the same at each step in this test.

Also important to this test are the measurement periods which should be taken at even time intervals and in the same manner as for the constant rate method with one modification. Each time the pumping rate is increased, a new set of measurements is to be started. Recovery measurements should be taken once the drawdown is completed.

This method is often run by the well driller to see if the well has been properly developed and is used as a method of well development. Specific capacity can be calculated from data obtained from this method.

3.17 PUMP TESTING AND EVALUATION

3.170 Mechanical Wear of Pumps

Well pumps, like other pieces of moving machinery, are subject to some degree of mechanical wear. This wear can be accelerated if the pump was not correctly installed, the well is crooked, or the pump received inadequate lubrication. In addition, sand in the water can cause excessive and rapid wear of the bowl unit. If the well water is highly corrosive, a chemical reaction may occur that results in localized pitting and possible penetration of the well casing, bowl unit, or pump column pipe.

3.171 Guidelines for Testing

Well pump performance tests should be made on a routine basis to determine if excessive wear is occurring. These tests also reflect the yield characteristics of the well and are indicators of potential well problems. Well pumps should be tested once every other year. If there are clues that pump performance is rapidly changing, then more frequent testing is strongly advised. Tests should be run at about the same time each year to minimize the impact of seasonal variations in groundwater conditions. The same test point on the pump curve should be used. Therefore, if the discharge pressure at the pump was maintained at 50 psi (345 kPa or 3.5 kg/sq cm)(115.5 feet or 35.2 m of discharge head) during the previous test, then subsequent tests should also be made at the 50 psi point.

If the well pump installation is equipped with a flowmeter, discharge pressure gage, and air line or probe device to measure water levels, then the operator can run one or more simple field tests to check well and pump performance. Tests for flow, pressure, drawdown, well yield, and gallons per minute per foot of drawdown can be readily made, and the results compared with previous tests to determine any changes in performance. Figure 3.60 illustrates the basic configuration and mechanical apparatus used during the field testing of a well pump.

If the operator is unable to perform these tests, then assistance should be requested from the local power supplier. Many power companies offer this testing service free of charge. In larger communities some of the water well drillers or pump suppliers can perform this service for a small fee.

3.172 Operator Responsibility

Most small or medium-sized water companies are not equipped to perform full scale accurate field pump efficiency tests, nor is the average operator expected to have the knowledge and skills to perform these tests. However, a good operator should be capable of understanding the *TERMINOLOGY* (Section 3.173) and how the *CALCULATIONS* (Section 3.174) are made, and be able to evaluate the test results to the degree that the operator is aware when *PERFORMANCE VALUES* (Section 3.175) indicate that repair work to the pump is justified.

3.173 Terminology

The operator should have a working knowledge of the definitions in this section.

1. *STATIC WATER DEPTH.* The vertical distance in feet from the centerline of the pump discharge down to the surface level of the free pool while no water is being drawn from the pool or water table.
2. *DRAWDOWN.* The drop in the water table or level of water in the ground when water is being pumped from a well. The difference in feet between the pumping water level and the static water level.
3. *PUMPING WATER LEVEL.* The vertical distance in feet from the centerline of the pump discharge to the level of the free pool while water is being drawn from the pool.
4. *DISCHARGE HEAD.* The pressure (in pounds per square inch or psi) measured at the centerline of discharge and very close to the discharge flange, converted into feet.

   Discharge Head, ft = (Pressure, psi)(2.31 ft/psi)
5. *TOTAL PUMPING HEAD* (Field Head). The total pumping head equals the lift below the discharge plus the head (or pressure (psi) for water x 2.31 ft/psi) above the discharge. The latter item (head above the discharge) must include friction losses through any discharge pipe and fittings.

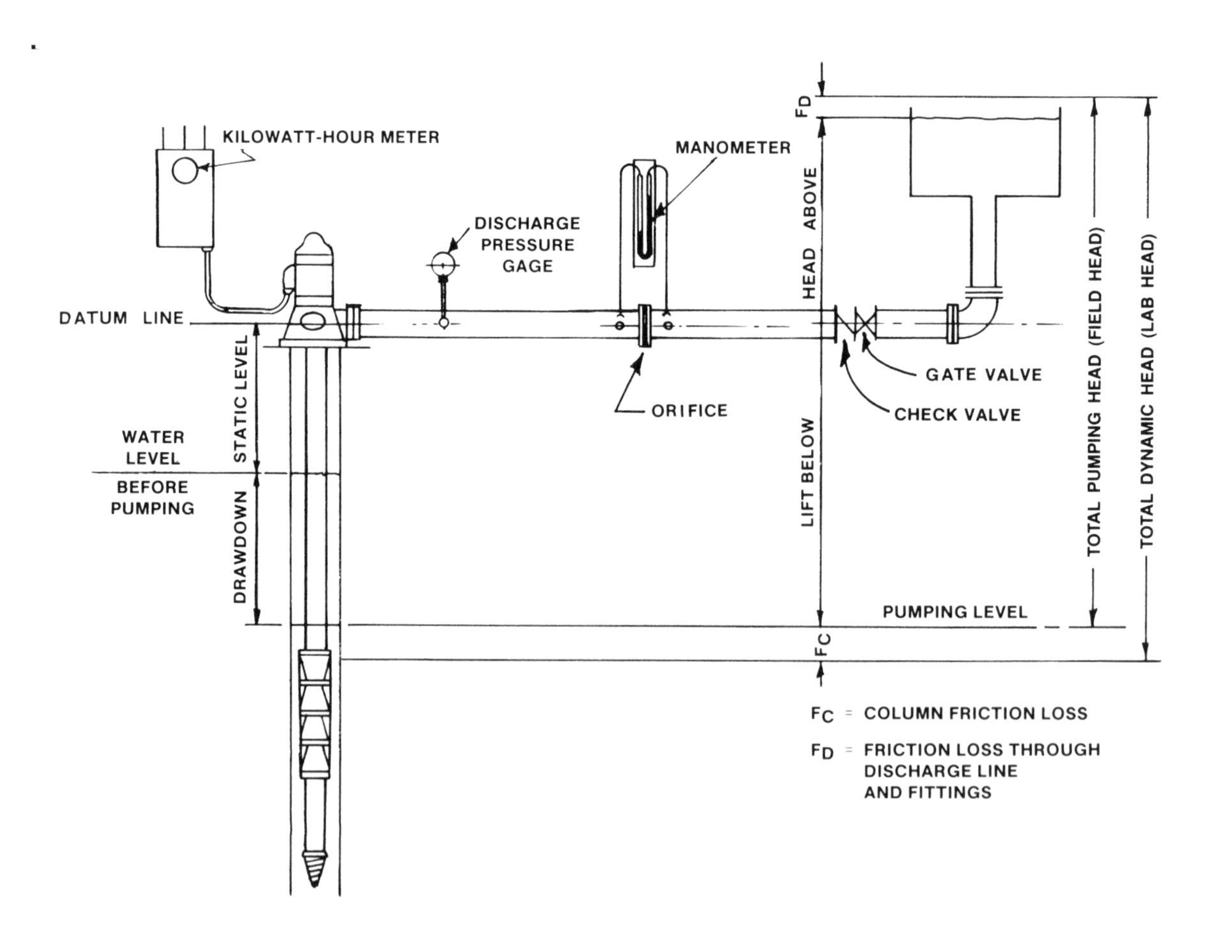

Fig. 3.60 Layout for field testing a well pump

6. *TOTAL DYNAMIC HEAD* (Lab Head). The total head on the pump bowl; equal to total pumping head plus column friction loss.

   Total Dynamic Head = Total Pumping Head + Column Friction Loss

7. *CAPACITY.* The rate of flow in liquid measure per unit of time (gallons per minute).

8. *LABORATORY EFFICIENCY* (Bowl Efficiency). This value is read directly from the performance curve based on total head in feet and pumping capacity in gallons per minute.

9. *LABORATORY HORSEPOWER.* The horsepower required at the impeller shaft to deliver the required capacity against the total dynamic head. Laboratory horsepower is calculated from the formula,

$$\text{Laboratory HP} = \frac{(\text{Total Dynamic Head, ft})(\text{Capacity, GPM})(100\%)}{(3{,}960)(\text{Laboratory Efficiency, }\%)}$$

10. *COLUMN FRICTION LOSS.* The feet of head of friction loss in the column. This is dependent upon the length and size of column and shaft used. Column friction loss is tabulated in feet of head per hundred feet of column and shaft in column friction loss charts.

11. *SHAFT LOSS.* The energy lost in the shaft measured in horsepower and determined by the length and the size of shaft and the speed of the pump. To find the shaft loss, refer to shaft loss charts. The values on these charts are expressed in horsepower per hundred feet of shafting.

12. *BRAKE HORSEPOWER* (Field Horsepower). The horsepower required at the top of a pump shaft (input to a pump). The brake horsepower is the SUM of Laboratory Horsepower and Shaft Loss.

    Brake Horsepower = Laboratory Horsepower + Shaft Loss

13. *PUMP FIELD EFFICIENCY.* The efficiency of the complete pump with all losses between laboratory performance and field performance accounted for. Pump field efficiency is calculated by means of the formula,

$$\text{Pump Field Efficiency, }\% = \frac{(\text{Total Pumping Head, ft})(\text{Capacity, GPM})}{(3{,}960)(\text{Brake Horsepower, HP})}(100\%)$$

14. *TOTAL PUMP THRUST.* The weight of the rotating parts in the pump bowl, the weight of the line shaft, and the hydraulic thrust of the water being pumped.

15. *THRUST BEARING LOSS.* The horsepower lost in the thrust bearings. This is dependent upon the type of thrust bearing and the total thrust load on the bearing. Bearing manufacturers indicate the loss in an angular contact ball bearing to be approximately 0.0075 HP per 100 RPM per 1,000 lb thrust load.

16. *MOTOR EFFICIENCY.* The ratio of energy delivered by a motor to the energy supplied to it during a fixed period or cycle. Motor efficiency ratings will vary depending upon motor manufacturer and usually will be near 90.0 percent.

17. *OVERALL EFFICIENCY.* (Wire-to-Water Efficiency.) The combined efficiency of a pump and motor together.

    Overall Efficiency = (Pump Field Efficiency)(Motor Efficiency)

18. *INPUT HORSEPOWER.* The total power used in operating a pump and motor.

$$\text{Input Horsepower, HP} = \frac{(\text{Brake Horsepower, HP})(100\%)}{\text{Motor Efficiency, }\%}$$

3.174 Calculations

Operators should be aware of the calculations involved in determining well pump efficiencies. At first glance you may feel that the calculations are too difficult, but if taken step by step, you should be able to understand the procedure and, in time, become comfortable with the method.

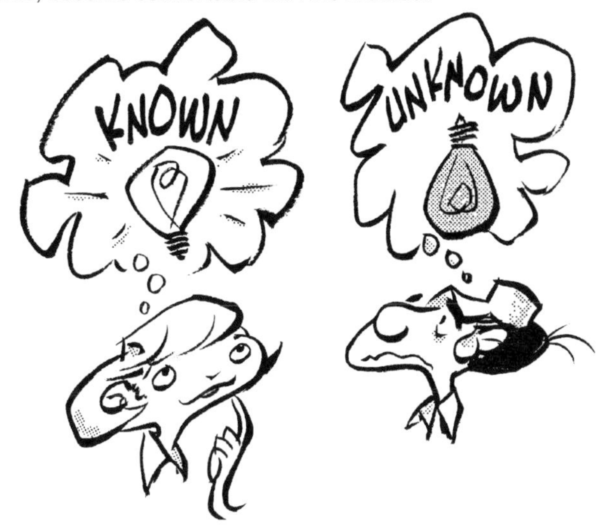

As an example, let's calculate the *OVERALL PUMP EFFICIENCY* based on specifications as follows:

Known

Pump — 12 DKL, 4 stage, 1,760 RPM

Capacity — 990 GPM

Static Water Level — 150 ft

Drawdown — 30 ft

Pumping Level — 180 ft

Total Head — 190 ft

Discharge Pressure — 10 ft

BHP — 59.8 total

Column — 8 inches, tube size is 2 inches

Motor — 60 HP, 4 pole, 440 volt, 3 phase, 60 cycle, vertical hollow shaft

Shaft — size, 1 1/4 inches; length, 200 ft

To calculate the *OVERALL PUMP EFFICIENCY,* try to obtain the above *"KNOWN"* information for the pump using Table 3.12 as a guide. After identifying the *"KNOWN"* information, continue with Section 3.175.

TABLE 3.12 PROCEDURE TO COMPLETELY EVALUATE PUMP PERFORMANCE IN WATER WELL APPLICATIONS

Note	Line	Item	Value	Explanation
	1.	Static water level	150 ft	Measured from centerline of discharge head to water surface *BEFORE* the pump is started.
	2.	Drawdown	30 ft	Measured from the water surface *BEFORE* the pump is started to the water surface after the pump has operated at a specified capacity for a specified time.
	3.	Pumping level	180 ft	Sum of lines 1 and 2.
(a)	4.	Discharge pressure ______ psi x 2.31	10 ft	Gage reading in psi times 2.31 plus correction to centerline of discharge head. Correction is plus distance in feet from gage down to centerline of discharge head or minus distance in feet up to centerline of discharge head.
(b)	5.	Field head	190 ft	Sum of lines 3 and 4.
	6.	Column friction loss	6 ft	From Table 3.13 for a tube size of 2", a flow of 990 GPM and a column size of 8".
	7.	Laboratory head	196 ft	Sum of lines 5 and 6. Use this value for selection from curves in Figure 3.61. *NOTE:* Since we have a four stage pump, the TOTAL HEAD, ft/stage = 196 ft/4 stages = 49 ft/stage.
	8.	Capacity	990 GPM	
(c)	9.	Laboratory efficiency	84.5%	From curves in Figure 3.61 for a total head of 49 ft/stage and a capacity of 990 GPM. From a flow of 990 GPM draw a vertical line and head of 49 ft/stage draw a horizontal line. Where these lines meet, read the efficiency of 84.5%.
	10.	Laboratory BHP	58.0 HP	$\frac{\text{Line 7 x Line 8 x 100\%}}{\text{3,960* x Line 9}}$ or $\text{BHP} = \frac{(196\text{ ft})(990\text{ GPM})(100\%)}{(3{,}960)(84.5\%)}$; $^*3{,}960 = \frac{33{,}000\text{ ft-lb/min-HP}}{8.34\text{ lbs/gal}}$
	11.	Shaft loss	1.58 HP	From Table 3.14, for a shaft size of 1¼ inches and a shaft RPM of 1,760, friction is 0.79 BHP/100 ft of shaft. Shaft loss for 200 feet of shaft is 1.58 HP.
(d)	12.	Field BHP	59.58 HP	Sum of lines 10 and 11.
(e)	13.	Field efficiency	79.7%	$\frac{\text{Line 5 x Line 8 x 100\%}}{\text{3,960 x Line 12}}$ or $\text{Efficiency, \%} = \frac{(190\text{ ft})(990\text{ GPM})(100\%)}{(3{,}960)(59.58\text{ BHP})}$
(f)	14.	Motor efficiency without thrust load	90%	From motor manufacturer.
	15.	HP input to motor without thrust bearing losses	66.20 HP	$\frac{\text{Line 12 x 100\%}}{\text{Line 14}}$ or $\text{Input HP} = \frac{(59.58\text{ HP})(100\%)}{90\%}$
	16.	Thrust bearing losses	0.42 HP	0.0075 HP per 100 RPM per 1,000 lbs thrust - - - - (g)
	17.	Motor efficiency with thrust load	89.4%	$\frac{(\text{Line 12})(100\%)}{(\text{Line 15} + \text{Line 16})}$ or $\text{Efficiency, \%} = \frac{(59.58\text{ HP})(100\%)}{(66.20\text{ HP} + 0.42\text{ HP})}$
(h)	18.	Overall plant efficiency	71.3%	$\frac{\text{Line 13 x Line 17}}{100\%}$ or $\text{Eff, \%} = \frac{(79.7\%)(89.4\%)}{100\%}$

(a) This is zero when pump discharges directly into the atmosphere through no more than 10 feet of horizontal discharge pipe.

(b) Sometimes called "total head." The term "field head" is to be preferred and its use is encouraged.

(c) Caution should be exercised in adjusting values shown on curves for number of stages used.

(d) This value should be used for selection of motors or gears, which should not be overloaded more than 15%.

(e) Frequently called "water-to-water" efficiency.

(f) These values are normally given for full, ¾, and ½ load, but vary so little near full load that the full load value can be used for slight under- or overloads on the motor.

(g) Thrust load value is approximate only. The values obtained should be checked against thrust bearing capacities given for motor (or gear) to avoid overload. Load is weight of water in column.

(h) Frequently called "wire-to-water" efficiency.

$$\text{Wire-to-Water Efficiency, \%} = \frac{(\text{Flow, GPM})(\text{TDH, ft})(100\%)}{(\text{Voltage, volts})(\text{Current, amps})(5.308)}$$

TABLE 3.13 FRICTION LOSS TABLE FOR STANDARD PIPE COLUMN[a]

COLUMN FRICTION LOSS (IN FEET) PER 100 FEET OF COLUMN

Column Size	4"			5"			6"			8"			10"				12"				14" OD			
Tube Size	1¼"	1½"	2"	1¼"	1½"	2"	1½"	2"	2½"	2"	2½"	3"	2"	2½"	3"	3½"	2"	2½"	3"	3½"	2½"	3"	3½"	4"
50[b]	.65	.86	1.6																					
75	1.3	1.7	3.3																					
100	2.2	2.8	5.3	.54	.65	.94																		
125	3.2	4.2	7.8	.81	.96	1.4																		
150	4.4	5.8		1.1	1.3	1.9																		
175	5.8	7.5		1.5	1.7	2.5																		
200	7.3	9.4		1.8	2.2	3.1	.73	.96	1.4															
225				2.3	2.7	3.9	.90	1.2	1.7															
250				2.7	3.3	4.7	1.1	1.4	2.0															
275				3.3	3.9	5.6	1.3	1.7	2.4															
300				3.8	4.5	6.4	1.5	2.0	2.8															
325				4.4	5.2	7.4	1.7	2.3	3.2															
350				5.0	6.0	8.4	2.0	2.6	3.6															
375				5.6	6.7	9.5	2.2	2.9	4.1															
400				6.3	7.5		2.5	3.3	4.6	.61	.74	1.0												
450				7.8	9.3		3.1	4.1	5.7	.77	.91	1.3												
500							3.7	5.0	6.9	.93	1.1	1.5												
550							4.4	5.8		1.1	1.3	1.8												
600							5.2	6.8		1.3	1.5	2.1												
650							6.0			1.5	1.8	2.5												
700										1.7	2.0	2.8												
750										1.9	2.3	3.2												
800										2.2	2.6	3.6	.57	.65	.77	.95								
850										2.4	2.9	4.0	.63	.72	.86	1.1								
900										2.7	3.2	4.5	.70	.80	.96	1.2								
950										2.9	3.5	4.9	.77	.88	1.1	1.3								
1000										3.2	3.9	5.4	.85	.97	1.2	1.4	.34	.38	.44	.50				
1200										4.5	5.4	7.6	1.2	1.4	1.6	2.0	.47	.54	.62	.71				
1400										6.0	7.2	10.	1.6	1.8	2.2	2.7	.62	.71	.82	.94				
1600										7.6	9.1	13.	2.0	2.3	2.8	3.4	.80	.90	1.1	1.2	.47	.53	.59	.67
1800										9.4	11.		2.5	2.8	3.4	4.3	.99	1.1	1.3	1.5	.58	.65	.73	.84
2000										11.	13.		3.0	3.5	4.2	5.2	1.2	1.4	1.6	1.8	.71	.80	.89	1.0
2200													3.6	4.1	5.0	6.1	1.4	1.6	1.9	2.1	.85	.95	1.1	1.2
2400													4.2	4.9	5.8	7.2	1.7	1.9	2.2	2.5	.99	1.1	1.2	1.4
2600													4.9	5.6	6.8	8.2	1.9	2.2	2.5	2.9	1.1	1.3	1.4	1.6
2800													5.6	6.4	7.8	9.6	2.2	2.5	2.8	3.3	1.3	1.5	1.6	1.9
3000													6.4	7.4	8.8	10.	2.5	2.9	3.3	3.8	1.5	1.7	1.9	2.1
3200																	2.8	3.2	3.7	4.3	1.7	1.9	2.1	2.4
3400																	3.2	3.6	4.2	4.8	1.9	2.1	2.4	2.7
3600																	3.5	4.0	4.7	5.3	2.1	2.4	2.6	2.9
3800																	3.9	4.4	5.1	5.9	2.3	2.6	2.9	3.3
4000																	4.3	4.9	5.6	6.4	2.5	2.9	3.2	3.6
4200																	4.7	5.3	6.2	7.1	2.8	3.1	3.5	3.9
4400																	5.1	5.8	6.7	7.7	3.0	3.4	3.8	4.3
4600																	5.6	6.3	7.4	8.4	3.3	3.7	4.1	4.6
4800																	6.0	6.8	7.9	9.0	3.5	4.0	4.4	5.0
5000																					3.8	4.3	4.8	5.4
5200																					4.2	4.7	5.2	5.9
5500																					4.6	5.1	5.7	6.4
5750																					5.0	5.5	6.2	6.9
6000																					5.4	6.0	6.7	

[a] Floway Turbine Pumps, Peabody Floway Inc., Fresno, CA.
[b] This column is flow in GPM

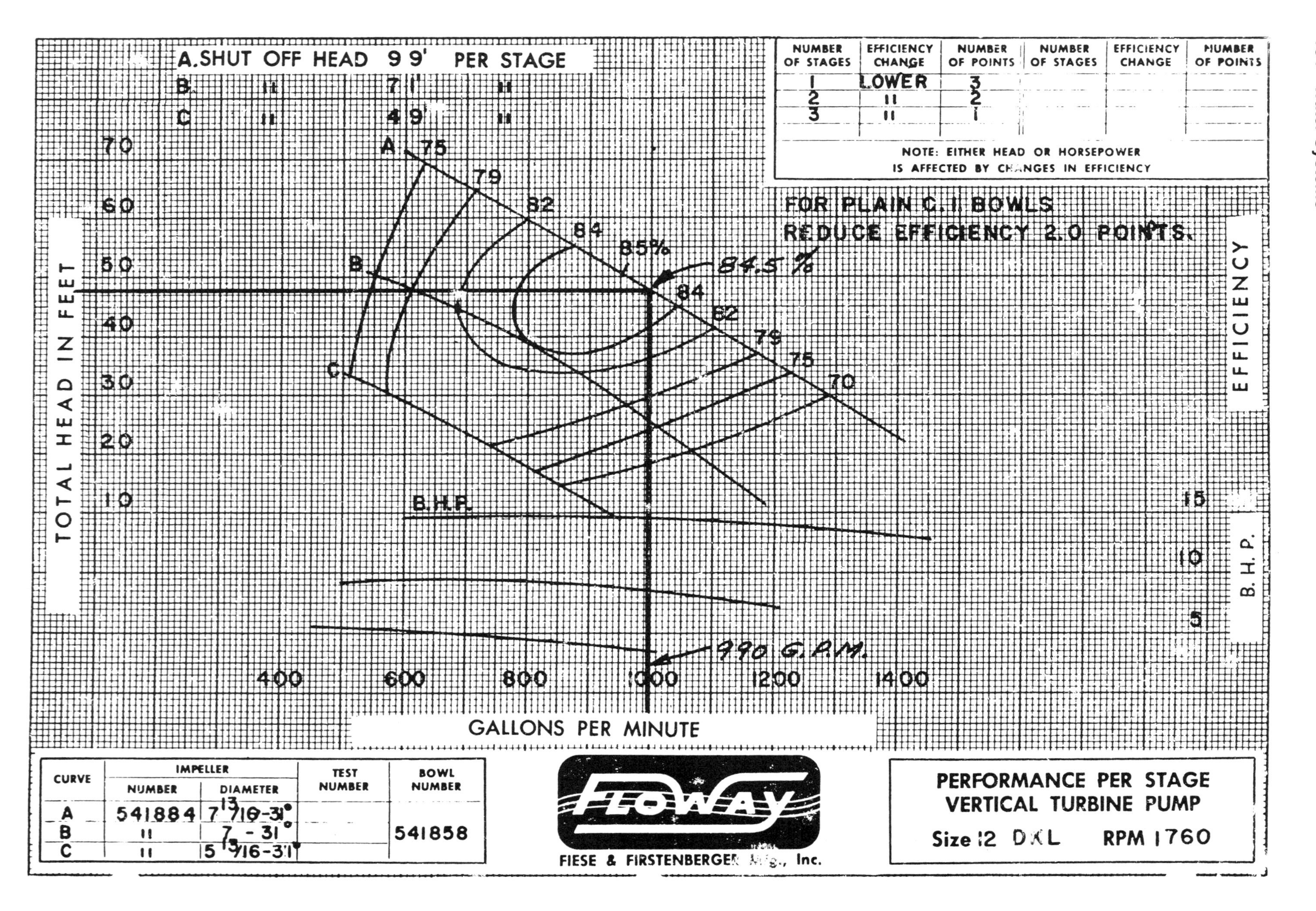

Fig. 3.61 Performance per stage for vertical turbine pump
(Permission of Peabody Floway Inc., Fresno, CA)

TABLE 3.14 MECHANICAL FRICTION IN BHP PER 100 FEET OF SHAFT[a,b]

Inches Shaft Size	RPM of Shaft								
	3,460	2,900	1,760	1,450	1,160	960	860	720	690
3/4	.60	.51	.31	.26	.20	.17			
1	1.05	.87	.53	.44	.35	.29	.26	.25	
1 1/4	1.60	1.33	.79	.67	.52	.44	.39	.34	
1 1/2	2.20	1.90	1.14	.96	.74	.63	.56	.47	.44
1 3/4		2.50	1.50	1.25	.97	.83	.74	.64	.59
2			1.90	1.60	1.25	1.05	.95	.81	.76
2 1/4			2.40	2.00	1.55	1.35	1.20	1.00	.96

[a] The horsepower loss due to mechanical friction in column shaft rotation may be determined from the chart.
[b] Floway Turbine Pumps, Peabody Floway Inc., Fresno, CA.

3.175 Performance Values

The operator should be able to evaluate the pump performance test.

Figure 3.62 is a copy of a typical well pump test report. Using the data and computations shown on this report, we can make a good performance evaluation of this well pumping facility. Under remarks, the test engineer has already summarized the general conditions at this pumping facility. Two important items to consider on this report are: (1) the *OVERALL PLANT EFFICIENCY*, and (2) the *HORSEPOWER INPUT TO MOTOR*. Let us look at these two items in more detail.

1. *OVERALL PLANT EFFICIENCY (O.P.E.)*

   The O.P.E. is 62.8 percent and the test engineer has commented that this is fair and rightly so; a new 100 HP pumping plant would have an O.P.E. in excess of 72 percent. This plant, at 62.8 percent, is fair but should be tested annually for a further reduction in O.P.E. If the O.P.E. was between 50 and percent, then you should consider replacing the pumping unit.

2. *HORSEPOWER INPUT TO MOTOR*

   The test engineer has noted that the motor is underloaded approximately 16 percent at 91.6 HP. If we assume that the correct bowl unit was installed at this plant, then the horsepower input should be from 100 to 110 HP. This report does not indicate if the bowl unit is a *SEMI-OPEN* or *FULLY ENCLOSED* impeller type. If the impeller is *SEMI-OPEN*, then it would have been possible to adjust the impeller clearance and perhaps increase the load on the motor which in turn would also raise the O.P.E. to a higher value. If the impeller was *FULLY ENCLOSED*, then no impeller adjustment could be made, and we would assume that the reason for the fair O.P.E. and the 16 percent motor underload is a result of wear to the pumping unit.

Two other important factors must be considered in the overall performance evaluation of a well pumping station: (1) the frequency of use, and (2) the energy use. These two factors are interrelated and should be carefully considered before determining if pump replacement is economically justified.

1. *FREQUENCY OF USE*

   If the pumping station is used only to meet peak summer demands or emergencies, then an O.P.E. rating between 50 and 55 percent may be acceptable. An O.P.E. rating under 50 percent would qualify almost any deep well turbine pump for replacement.

   If the pumping station is a lead facility operating between 20 and 40 percent of the time, then an O.P.E. rating of 55 to 60 percent could justify replacing the pump.

2. *ENERGY USE*

   The high cost of energy may be the major factor in determining when a pumping unit should be replaced.

   Using the pump values as shown on the pump test report (Figure 3.62) and the energy use data (from Table 3.15), we can estimate power consumption for various O.P.E. ratings.

   Assuming that the well pump in this report is a lead pump operating 30 percent of the time and we use O.P.E. ratings of 72 percent (new facility), 63 percent (pump under evaluation), and 52 percent (a worn pump), then we can estimate what the annual energy use would be based on the above-mentioned O.P.E. ratings. The calculations are as follows:

Overall Efficiency Pump Unit	Kilowatts per 1,000 Gallons at one foot T.H.*		Total Lift in feet		Kilowatts per 1,000 Gallons at Total Lift
72%	.00435	x	173	=	.753
63%	.00498	x	173	=	.862
52%	.00603	x	173	=	1.043

* Values from Table 3.15.

   Therefore, 30 percent pumping time for 1 year (365 days x 24 hr/day x 60 min/hr x 0.30 = 157,680 minutes) x pumping rate (1,319 GPM) = 207,979,920 gallons or 207,980 thousand gallons pumped in 1 year. With the three O.P.E. ratings we are using, we will have yearly power consumptions as follows:

(1) O.P.E.	(2) Annual Water Production in 1,000 Gallons	(3) Kilowatts per 1,000 Gallons at Total Lift	(4) Annual Kilowatts Used*	(5) % Additional Energy Required**
72%	207,980	.753	156,609	0
63%	207,980	.862	179,279	14.48
52%	207,980	1.043	216,923	38.51

* (2)(3) = (207,980)(0.753) = 156,609

** $\frac{((4) - 156{,}609)(100\%)}{156{,}609} = \frac{(179{,}279 - 156{,}609)(100\%)}{156{,}609} = 14.48\%$

   Using the values in column (5) we can see that the well pump tested in Figure 3.62 with an O.P.E. of 63 percent will

PUMP TEST REPORT

TEST NO. M P A -2

NAME Good Water DATE 2-20-98

ADDRESS 777 Moon Lake Road, Nicetown, USA

PLANT LOCATION Metropolitan Airport Pumping Plant #2

EQUIPMENT

METER NO. ____ MOTOR US H.P. 100 VOLTS 400 R.P.M. 1800 SERIAL NO 382838

PUMP US TYPE DWT ____ SERIAL NO. 34052

TEST RESULTS

Standing Water Level Below Surface of Ground		Ft.
Standing Water Level Below Centerline of Discharge	12.05	Ft.
Drawdown from Standing to Pumping Level	43.3	Ft.
Pumping Water Level Below Centerline of Discharge	55.35	Ft.
Discharge Level Above Centerline of Discharge	117.4	Ft.
TOTAL LIFT (Water to Water)	172.75	Ft.
WATER PUMPED	* 1319	G.P.M.
Flow of Well (G.P.M. per Foot Drawdown)	30.46	G.P.M./Ft.
Water Pumped in 24 Hours	1.89	MG
HORSEPOWER INPUT TO MOTOR	91.6	H.P.
Kilowatt Input to Motor	68.3	KW.
KILOWATT HRS PER MILLION GALLONS WATER PUMPED	863	KWH/MG
OVERALL PLANT EFFICIENCY	62.8	%

Line Volts 1-2 480 2-3 480 1-3 480 Amps 1 95 2 92 3 92

REMARKS The overall efficiency of this pumping plant is fair under the existing water level conditions.

*From the line meter we determined a flow of 1351 GPM. The motor is underloaded approximately 16%.

____________________ Test Engineer

Fig. 3.62 Well pump test report

TABLE 3.15 ENERGY USE DATA[a]

Overall Efficiency Pump Unit[b]	Kilowatts per 1,000 Gallons at one foot T.H.	Overall Efficiency Pump Unit[b]	Kilowatts per 1,000 Gallons at one foot T.H.	Overall Efficiency Pump Unit[b]	Kilowatts per 1,000 Gallons at one foot T.H.
32	.00980	52	.00603	72	.00435
33	.00951	53	.00592	73	.00430
34	.00922	54	.00581	74	.00424
35	.00896	55	.00570	75	.00418
36	.00871	56	.00560	76	.00413
37	.00848	57	.00550	77	.00407
38	.00826	58	.00541	78	.00402
39	.00804	59	.00532	79	.00397
40	.00784	60	.00523	80	.00392
41	.00765	61	.00514	81	.00387
42	.00747	62	.00506	82	.00382
43	.00730	63	.00498	83	.00378
44	.00713	64	.00490	84	.00373
45	.00697	65	.00482	85	.00369
46	.00682	66	.00475	86	.00365
47	.00667	67	.00468	87	.00360
48	.00653	68	.00461	88	.00356
49	.00640	69	.00454	89	.00352
50	.00627	70	.00448	90	.00348
51	.00615	71	.00442		

[a] Source: *ENGINEERING DATA*, Peabody Floway Inc., Fresno, CA.

[b] Overall efficiency as indicated is the input-output efficiency including all losses in the pump unit, pumping 1,000 gallons of clear water one foot total head. Therefore, in determining the kilowatts per 1,000 gallons pumped, it is only necessary to multiply the factor corresponding to the overall efficiency by the number of feet head at which the total dynamic head has been calculated. Example:

Assume an overall efficiency of 65% and a total head of 200 feet.
Kilowatts per 1,000 gallons = .00482 x 200 = .964

use *14.48* percent more energy than a new or reconditioned pump with an O.P.E. of 72 percent. If the O.P.E. drops to 52 percent, then *38.51* percent more energy would be used to pump the same amount of water. When we compare the additional energy used by a worn pump with today's high cost of energy we can readily understand the importance of routine pump performance tests and evaluation. A good operator should be capable of evaluating well pumping plants and determining when it is economically justified to replace a worn pumping unit.

3.176 Pump Electrical

Operators of small water systems, particularly those located in rural areas, should be aware of the common but little understood problem of unbalanced current. Operating a pumping unit with unbalanced current can seriously damage three-phase motors and cause early motor failure. Unbalanced current reduces the starting torque of the motor and can cause overload tripping, excessive heat, vibration, and overall poor performance.

It is common practice for electrical utility companies to furnish power to three-phase customers in open delta or wye configurations. An open delta or wye system is a two-transformer bank that is a suitable configuration where *LIGHTING LOADS ARE LARGE AND THREE-PHASE LOADS ARE LIGHT.* This is the exact opposite of the configuration needed by most pumping facilities where *THREE-PHASE LOADS ARE LARGE.* (Examples of three-transformer banks include Y-delta, delta-Y, and Y-Y.) In most cases, three-phase motors should be fed from three-transformer banks for proper balance. The capacity of a two-transformer bank is only 57 percent of the capacity of a three-transformer bank. The two-transformer configuration can cause one leg of the three-phase current to furnish higher amperage to one leg of the motor, which will greatly shorten its life.

Operators should acquaint themselves with the configuration of their electric power supply. When an open delta or wye configuration is used, operators should calculate the degree of current imbalance existing between legs of their polyphase motors. A small percentage voltage imbalance will result in a much larger percentage current imbalance. If you are unsure about how to determine the configuration of your system or how to calculate the percentage of current imbalance, *ALWAYS* consult a qualified electrician. *CURRENT IMBALANCE BETWEEN LEGS SHOULD NEVER EXCEED 5 PERCENT UNDER NORMAL OPERATING CONDITIONS* (NEMA Standards MGI-14.35).

Another serious consideration for operators is voltage fluctuation caused by neighborhood demands. A pump motor in near perfect balance (for example, 3 percent imbalance) at 9:00 AM could be as much as 17 percent unbalanced by 4:00 PM on a hot day due to the use of air conditioners by customers on the same grid. Also, the hookup of a small market or a new home to the power grid can cause a significant change in the degree of current imbalance in other parts of the power grid. Because energy demands are constantly changing, water system operators should have a qualified electrician check the current balances between legs of their three-phase motors at least once a year.

Do not rely entirely on the power company to detect unbalanced current. Complaints of suspected power problems are frequently met with the explanation that all voltages are within the percentages allowed by law and no mention is made of the percentage of current imbalance which can be a major source of problems with three-phase motors. A little research of your own can pay large benefits. For example, a small water company in Central California configured with an open delta system (and running three-phase imbalances as high as 17 percent as a result) was routinely spending $14,000 a year for energy and burning out a 10-HP motor on the average of every 1½ years (six 10-HP motors in 9 years). After consultation, the local power utility agreed to add a third transformer to each power board to bring the system into better balance. Pump drop leads were then rotated bringing overall current imbalances down to an average of 3 percent, heavy-duty, three-phase capacitors were added to absorb the prevalent voltage surges in the area, and computerized controls were added to the pumps to shut them off when pumping volumes got too low. These modifications resulted in a saving in energy costs the first year alone of $5,500.00!

FORMULAS

Percentage of current unbalance can be calculated by using the following formulas and procedures:

$$\text{Average Current} = \frac{\text{Total of Current Value Measured on Each Leg}}{3}$$

$$\text{\% Current Unbalance} = \frac{\text{Greatest Amp Difference From the Average}}{\text{Average Current}} \times 100\%$$

PROCEDURES

A. Measure and record current readings in amps for each leg (Hookup 1). Disconnect power.

B. Shift or roll the motor leads from left to right so the drop cable lead that was on terminal 1 is now on 2, lead on 2 is now on 3, and lead on 3 is now on 1 (Hookup 2). Rolling the motor leads in this manner will not reverse the motor rotation. Start the motor, measure and record current reading on each leg. Disconnect power.

C. Again shift drop cable leads from left to right so the lead on terminal 1 goes to 2, 2 goes to 3, and 3 to 1 (Hookup 3). Start pump, measure and record current reading on each leg. Disconnect power.

D. Add the values for each hookup.

E. Divide the total by 3 to obtain the average.

F. Compare each single leg reading to the average current amount to obtain the greatest amp difference from the average.

G. Divide this difference by the average to obtain the percentage of unbalance.

H. Use the wiring hookup which provides the lowest percentage of unbalance.

EXAMPLE: CORRECTING THE THREE-PHASE POWER UNBALANCE

Example: Check for current unbalance for a 230-volt, 3-phase, 60-Hz submersible pump motor, 18.6 full load amps.

Solution: Steps 1 to 3 measure and record amps on each motor drop lead for Hookups 1, 2, and 3.

	Step 1 (Hookup 1)	Step 2 (Hookup 2)	Step 3 (Hookup 3)
(T_1)	DL_1 = 25.5 amps	DL_3 = 25 amps	DL_2 = 25.0 amps
(T_2)	DL_2 = 23.0 amps	DL_1 = 24 amps	DL_3 = 24.5 amps
(T_3)	DL_3 = 26.5 amps	DL_2 = 26 amps	DL_1 = 25.5 amps
Step 4	Total = 75 amps	Total = 75 amps	Total = 75 amps

Step 5 Average Current $= \frac{\text{Total Current}}{\text{3 readings}} = \frac{75}{3} = 25$ amps

Step 6	Greatest amp difference from the average:	(Hookup 1) = 25 – 23	= 2
		(Hookup 2) = 26 – 25	= 1
		(Hookup 3) = 25.5 – 25	= .5
Step 7	% Unbalance	(Hookup 1) = 2/25 x 100	= 8
		(Hookup 2) = 1/25 x 100	= 4
		(Hookup 3) = 0.5/25 x 100	= 2

As can be seen, Hookup 3 should be used since it shows the least amount of current unbalance. Therefore, the motor will operate at maximum efficiency and reliability on Hookup 3.

By comparing the current values recorded on each leg, you will note the highest value was always on the same leg, L_3. This indicates the unbalance is in the power source. If the high current values were on a different leg each time the leads were changed, the unbalance would be caused by the motor or a poor connection.

If the current unbalance is greater than 5 percent, contact your power company for help.

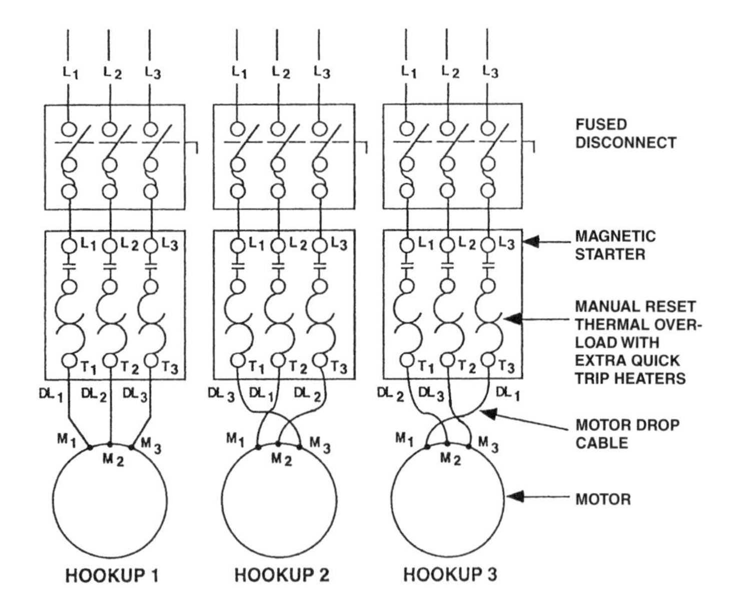

ACKNOWLEDGMENT

Material in this section was provided by James W. Cannell, President, Canyon Meadows Mutual Water Company, Inc., Bodfish, California. His contribution is greatly appreciated.

3.18 ABANDONING AND PLUGGING WELLS

3.180 Reasons for Abandoning and Plugging

Eventually a water well reaches the end of its useful life. Many are simply abandoned and become a safety hazard and a threat to the groundwater supply.

A well that is no longer useful must be properly abandoned and plugged:

1. To ensure that the groundwater supply is protected and preserved for further use,
2. To eliminate the potential physical hazard to people, and
3. To protect nearby wells from contamination.

Private wells that no longer serve any useful purpose should also be abandoned and plugged. When a home is connected to a public water supply, the old well could cause *CROSS CONNECTION*[64] problems if it is not adequately and permanently disconnected from the public water system.

The concept that we all have an interest in our groundwater resources and, therefore, an obligation to protect them for their present and continued use, is a fundamental one. This concept holds true whether the issue is construction or abandoning of a well.

3.181 Objectives

All "abandoned" wells should be properly plugged. A well could be generally classified as abandoned when it has not been used for a period of one year, the pumping unit has been removed, and the well serves no useful purpose. The objective of abandoning and plugging is to restore as nearly as practical, those subsurface conditions which existed prior to the construction of the well. Another important objective is to prevent the possible cross contamination between different aquifers.

3.182 Evaluation

An investigation should be made to determine the condition of the well to be abandoned and plugged, details of its construction, and whether there are obstructions or objects in the well that may hinder the filling and sealing process. Visual inspection using downhole television and photography can be very useful tools in determining the overall condition of a well.

3.183 Permits

In most cases, a permit is required before a well can be abandoned and plugged. This permit may be issued at the local, state, or federal level, and its purpose is to set forth well abandoning and plugging guidelines and inspection to ensure the protection of the groundwater basin. In certain areas, the protection of the groundwater basin is especially important.

As an example, within California alone there are 400 significant groundwater basins; of these, many have been identified as critical basins where the interchange of water between aquifers will result in a significant deterioration of the quality of water in one or more aquifers. Problem areas occur in coastal aquifers that have been invaded by sea water, and in areas where saline water underlies an area at varying depths.

3.184 Abandoning and Plugging Guidelines

Abandoning and plugging procedures and methods will vary according to whether the well is situated in (1) unconsolidated material in an unconfined groundwater zone, (2) creviced or fractured rock, (3) noncreviced consolidated formation, or (4) where several aquifers or formations have been penetrated and an aquifer seal is required to prevent the vertical movement of water between aquifers. Some operators refer to these procedures as well destruction because they destroy the opportunity for aquifers to become contaminated and they prevent the opportunity for people and animals to fall into wells.

The enforcing agency will often establish the procedures to be followed when abandoning and plugging a well. In some cases, the water purveyor will be required to seek the advice of a geologist or a well expert to determine the appropriate methods to use.

Following this paragraph is a brief procedure for abandoning and plugging a well. This procedure reflects the desired end results and may be applicable for many old wells that are likely to be abandoned and plugged. Remember that this is a general guideline only, and that local requirements and ordinances will prevail, plus special conditions that may be encountered due to the construction of the well, and/or the formations penetrated.

1. The well to be abandoned and plugged must be cleaned of foreign debris. In many cases this will require the use of a small drilling rig.
2. Drill out the old well to the estimated depth of its original construction, but in no case less than 100 feet (30 m).
3. With the drilling rig in position, air pump the well for two hours at its maximum capacity or at a minimum rate of 100 GPM (6 liters/sec).
4. Disinfect the well by adding a chlorine compound to produce a chlorine dose of 200 mg/*L*. If a dry chemical is used, it should be mixed with water to form a chlorine solution prior to placing it into the well.
5. Fill the well to within 25 feet (7.5 m) of ground elevation with suitable fill materials such as neat cement, cement grout, concrete, bentonite clay, or clean sand.
6. Excavate a hole around the well casing to a depth of 5 feet (1.5 m) and cut off the well casing 6 inches (150 mm) above the bottom of the excavation.
7. Fill the upper portion of the well with cement grout or concrete and allow to spill over into the excavation to form a cap at least one foot (0.3 m) thick.
8. After the sealing material has set, fill the excavation with native soil.

See Figures 3.63, 3.64, and 3.65 for cross-sectional views of properly abandoned and plugged wells.

[64] *Cross Connection. A connection between a drinking (potable) water system and an unapproved water supply. For example, if you have a pump moving nonpotable water and hook into the drinking water system to supply water for the pump seal, a cross connection or mixing between the two water systems can occur. This mixing may lead to contamination of the drinking water.*

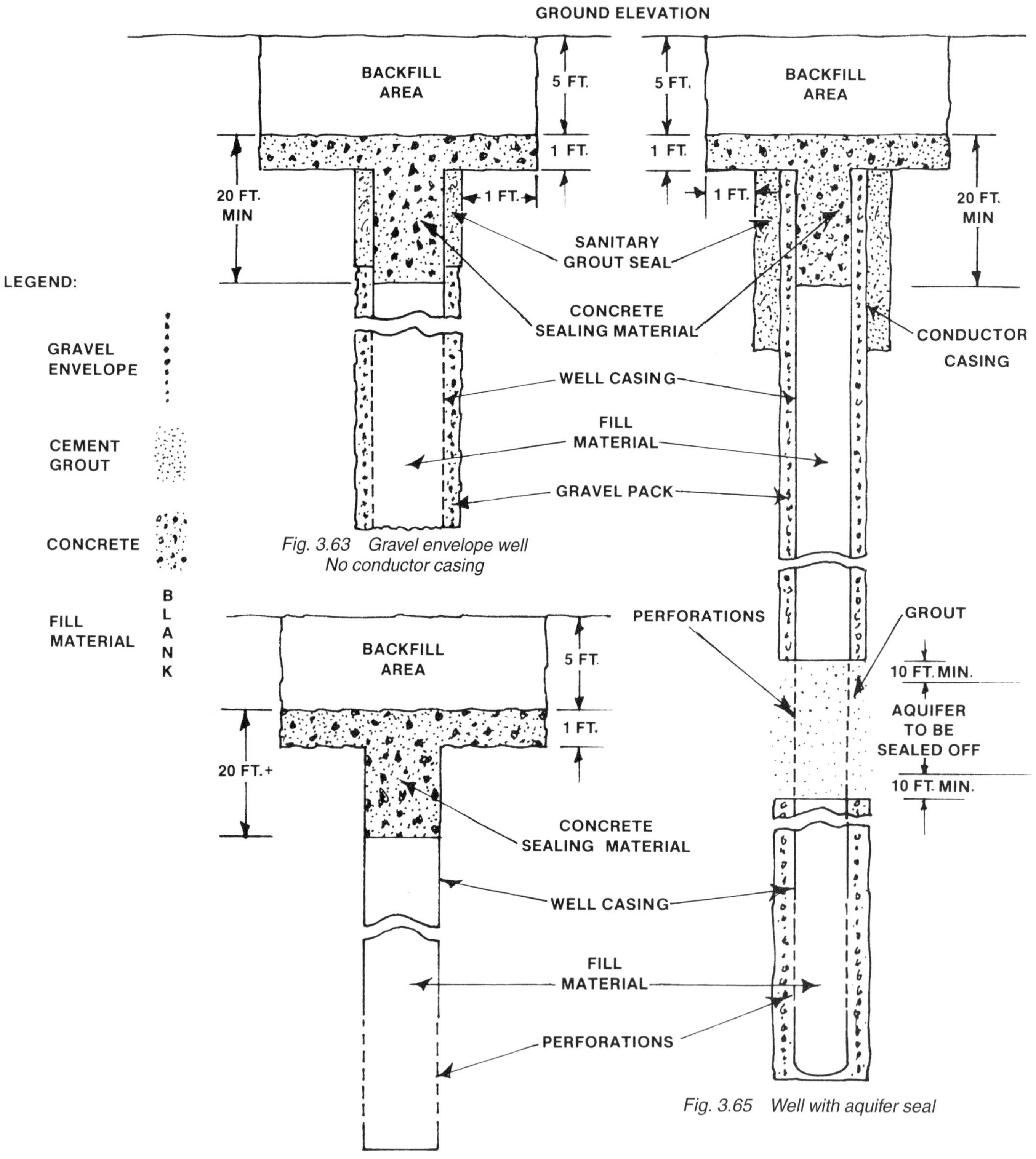

GROUND ELEVATION

BACKFILL AREA

5 FT.

1 FT.

1 FT.

20 FT. MIN

SANITARY GROUT SEAL

CONCRETE SEALING MATERIAL

CONDUCTOR CASING

WELL CASING

FILL MATERIAL

GRAVEL PACK

LEGEND:

GRAVEL ENVELOPE

CEMENT GROUT

CONCRETE

FILL MATERIAL BLANK

Fig. 3.63 Gravel envelope well
No conductor casing

BACKFILL AREA

5 FT.

1 FT.

20 FT.+

CONCRETE SEALING MATERIAL

WELL CASING

FILL MATERIAL

PERFORATIONS

GROUT

10 FT. MIN.

AQUIFER TO BE SEALED OFF

10 FT. MIN.

Fig. 3.64 Old cable tool well
No sanitary grout seal

Fig. 3.65 Well with aquifer seal

CHAPTER 4

SMALL WATER TREATMENT PLANTS

by

Jess Morehouse

TABLE OF CONTENTS

Chapter 4. SMALL WATER TREATMENT PLANTS

OBJECTIVES

Chapter 4. SMALL WATER TREATMENT PLANTS

Following completion of Chapter 4, you should be able to:

1. Explain the importance of small water treatment plants;
2. Operate and maintain treatment processes for
   a. Coagulation,
   b. Flocculation,
   c. Settling,
   d. Filtration (including slow sand filtration),
   e. Disinfection,
   f. Corrosion control,
   g. Solids contact units,
   h. Iron and manganese control, and
   i. Softening; and
3. Safely perform the duties of an operator.

WORDS

Chapter 4. SMALL WATER TREATMENT PLANTS

ABSORPTION (ab-SORP-shun) ABSORPTION

The taking in or soaking up of one substance into the body of another by molecular or chemical action (as tree roots absorb dissolved nutrients in the soil).

ADSORPTION (add-SORP-shun) ADSORPTION

The gathering of a gas, liquid, or dissolved substance on the surface or interface zone of another material.

AERATION (air-A-shun) AERATION

The process of adding air to water. Air can be added to water by either passing air through water or passing water through air.

AEROBIC (AIR-O-bick) AEROBIC

A condition in which atmospheric or dissolved molecular oxygen is present in the aquatic (water) environment.

ALKALINITY (AL-ka-LIN-it-tee) ALKALINITY

The capacity of water to neutralize acids. This capacity is caused by the water's content of carbonate, bicarbonate, hydroxide, and occasionally borate, silicate, and phosphate. Alkalinity is expressed in milligrams per liter of equivalent calcium carbonate. Alkalinity is not the same as pH because water does not have to be strongly basic (high pH) to have a high alkalinity. Alkalinity is a measure of how much acid must be added to a liquid to lower the pH to 4.5.

ANAEROBIC (AN-air-O-bick) ANAEROBIC

A condition in which atmospheric or dissolved molecular oxygen is *NOT* present in the aquatic (water) environment.

BACKWASHING BACKWASHING

The process of reversing the flow of water back through the filter media to remove the entrapped solids.

BREAKPOINT CHLORINATION BREAKPOINT CHLORINATION

Addition of chlorine to water until the chlorine demand has been satisfied. At this point, further additions of chlorine will result in a free chlorine residual that is directly proportional to the amount of chlorine added beyond the breakpoint.

CALCIUM CARBONATE ($CaCO_3$) EQUIVALENT CALCIUM CARBONATE ($CaCO_3$) EQUIVALENT

An expression of the concentration of specified constituents in water in terms of their equivalent value to calcium carbonate. For example, the hardness in water which is caused by calcium, magnesium and other ions is usually described as calcium carbonate equivalent. Alkalinity test results are usually reported as mg/*L* $CaCO_3$ equivalents. To convert chloride to $CaCO_3$ equivalents, multiply the concentration of chloride ions in mg/*L* by 1.41, and for sulfate, multiply by 1.04.

CATHODIC (ca-THOD-ick) PROTECTION CATHODIC PROTECTION

An electrical system for prevention of rust, corrosion, and pitting of metal surfaces which are in contact with water or soil. A low-voltage current is made to flow through a liquid (water) or a soil in contact with the metal in such a manner that the external electromotive force renders the metal structure cathodic. This concentrates corrosion on auxiliary anodic parts which are deliberately allowed to corrode instead of letting the structure corrode.

CHLORINATOR (KLOR-uh-NAY-ter) CHLORINATOR

A metering device which is used to add chlorine to water.

CLEAR WELL CLEAR WELL

A reservoir for the storage of filtered water of sufficient capacity to prevent the need to vary the filtration rate with variations in demand. Also used to provide chlorine contact time for disinfection.

COAGULATION (co-AGG-you-LAY-shun) COAGULATION

The clumping together of very fine particles into larger particles (floc) caused by the use of chemicals (coagulants). The chemicals neutralize the electrical charges of the fine particles, allowing them to come closer and form larger clumps. This clumping together makes it easier to separate the solids from the water by settling, skimming, draining, or filtering.

DIATOMACEOUS (DYE-uh-toe-MAY-shus) EARTH DIATOMACEOUS EARTH

A fine, siliceous (made of silica) "earth" composed mainly of the skeletal remains of diatoms.

DIATOMS (DYE-uh-toms) DIATOMS

Unicellular (single cell), microscopic algae with a rigid (box-like) internal structure consisting mainly of silica.

EFFECTIVE SIZE (E.S.) EFFECTIVE SIZE (E.S.)

The diameter of the particles in a granular sample (filter media) for which 10 percent of the total grains are smaller and 90 percent larger on a weight basis. Effective size is obtained by passing granular material through sieves with varying dimensions of mesh and weighing the material retained by each sieve. The effective size is also approximately the average size of the grains.

ELECTROCHEMICAL REACTION ELECTROCHEMICAL REACTION

Chemical changes produced by electricity (electrolysis) or the production of electricity by chemical changes (galvanic action). In corrosion, a chemical reaction is accompanied by the flow of electrons through a metallic path. The electron flow may come from an external source and cause the reaction, such as electrolysis caused by a D.C. (direct current) electric railway or the electron flow may be caused by a chemical reaction as in the galvanic action of a flashlight dry cell.

FLOCCULATION (FLOCK-you-LAY-shun) FLOCCULATION

The gathering together of fine particles after coagulation to form larger particles by a process of gentle mixing.

FREEBOARD FREEBOARD

(1) The vertical distance from the normal water surface to the top of the confining wall.

(2) The vertical distance from the sand surface to the underside of a trough in a sand filter. This distance is also called AVAILABLE EXPANSION.

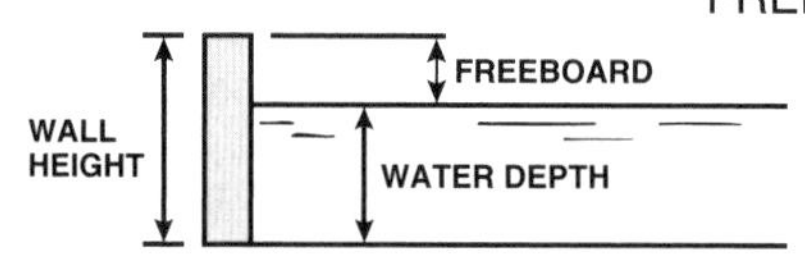

GARNET GARNET

A group of hard, reddish, glassy, mineral sands made up of silicates of base metals (calcium, magnesium, iron and manganese). Garnet has a higher density than sand.

GREENSAND GREENSAND

A mineral (glauconite) material that looks like ordinary filter sand except that it is green in color. Greensand is a natural ion exchange material which is capable of softening water. Greensand which has been treated with potassium permanganate ($KMnO_4$) is called manganese greensand; this product is used to remove iron, manganese and hydrogen sulfide from groundwaters.

HEAD LOSS HEAD LOSS

The head, pressure or energy (they are the same) lost by water flowing in a pipe or channel as a result of turbulence caused by the velocity of the flowing water and the roughness of the pipe, channel walls, or restrictions caused by fittings. Water flowing in a pipe loses head, pressure or energy as a result of friction losses. The head loss through a filter is due to friction losses caused by material building up on the surface or in the top part of a filter.

HYPOCHLORINATORS (HI-poe-KLOR-uh-NAY-tors) HYPOCHLORINATORS

Chlorine pumps, chemical feed pumps or devices used to dispense chlorine solutions made from hypochlorites such as bleach (sodium hypochlorite) or calcium hypochlorite into the water being treated.

MONOMER (MON-o-MER) MONOMER

A molecule of low molecular weight capable of reacting with identical or different monomers to form polymers.

MUDBALLS MUDBALLS

Material that is approximately round in shape and varies from pea-sized up to two or more inches in diameter. This material forms in filters and gradually increases in size when not removed by the backwashing process.

OVERFLOW RATE | OVERFLOW RATE

One of the guidelines for the design of settling tanks and clarifiers in treatment plants. Used by operators to determine if tanks and clarifiers are hydraulically (flow) over- or underloaded. Also called SURFACE LOADING.

$$\text{Overflow Rate, GPD/sq ft} = \frac{\text{Flow, gallons/day}}{\text{Surface Area, sq ft}}$$

OXIDATION (ox-uh-DAY-shun) | OXIDATION

Oxidation is the addition of oxygen, removal of hydrogen, or the removal of electrons from an element or compound. In the environment, organic matter is oxidized to more stable substances. The opposite of REDUCTION.

PARSHALL FLUME | PARSHALL FLUME

A device used to measure the flow in an open channel. The flume narrows to a throat of fixed dimensions and then expands again. The rate of flow can be calculated by measuring the difference in head (pressure) before and at the throat of the flume.

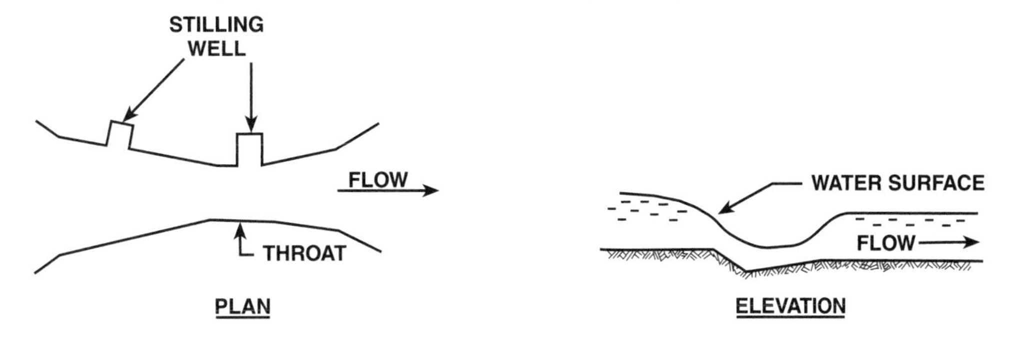

PATHOGENS (PATH-o-jens) | PATHOGENS

Pathogenic or disease-causing organisms.

POLYELECTROLYTE (POLY-ee-LECK-tro-lite) | POLYELECTROLYTE

A high-molecular-weight (relatively heavy) substance having points of positive or negative electrical charges that is formed by either natural or manmade processes. Natural polyelectrolytes may be of biological origin or derived from starch products and cellulose derivatives. Manmade polyelectrolytes consist of simple substances that have been made into complex, high-molecular-weight substances. Used with other chemical coagulants to aid in binding small suspended particles to larger chemical flocs for their removal from water. Often called a POLYMER.

POLYMER (POLY-mer) | POLYMER

A long chain molecule formed by the union of many monomers (molecules of lower molecular weight). Polymers are used with other chemical coagulants to aid in binding small suspended particles to larger chemical flocs for their removal from water.

PRECIPITATE (pre-SIP-uh-TATE) | PRECIPITATE

(1) An insoluble, finely divided substance which is a product of a chemical reaction within a liquid.

(2) The separation from solution of an insoluble substance.

RECARBONATION (re-CAR-bun-NAY-shun) | RECARBONATION

A process in which carbon dioxide is bubbled into the water being treated to lower the pH. The pH may also be lowered by the addition of acid. Recarbonation is the final stage in the lime-soda ash softening process. This process converts carbonate ions to bicarbonate ions and stabilizes the solution against the precipitation of carbonate compounds.

REDUCTION (re-DUCK-shun) | REDUCTION

Reduction is the addition of hydrogen, removal of oxygen, or the addition of electrons to an element or compound. Under anaerobic conditions (no dissolved oxygen present), sulfur compounds are reduced to odor-producing hydrogen sulfide (H_2S) and other compounds. The opposite of OXIDATION.

RELIQUEFACTION (re-LICK-we-FACK-shun) | RELIQUEFACTION

The return of a gas to the liquid state; for example, a condensation of chlorine gas to return it to its liquid form by cooling.

SCHMUTZDECKE (sh-moots-DECK-ee) | SCHMUTZDECKE

A layer of trapped matter at the surface of a slow sand filter in which a dense population of microorganisms develops. These microorganisms within the film or mat feed on and break down incoming organic material trapped in the mat. In doing so the microorganisms both remove organic matter and add mass to the mat, further developing the mat and increasing the physical straining action of the mat.

SHORT-CIRCUITING SHORT-CIRCUITING

A condition that occurs in tanks or basins when some of the water travels faster than the rest of the flowing water. This is usually undesirable since it may result in shorter contact, reaction, or settling times in comparison with the theoretical (calculated) or presumed detention times.

SLUDGE (sluj) SLUDGE

The settleable solids separated from water during processing.

SLURRY (SLUR-e) SLURRY

A watery mixture or suspension of insoluble (not dissolved) matter; a thin, watery mud or any substance resembling it (such as a grit slurry or a lime slurry).

SUPERCHLORINATION (SUE-per-KLOR-uh-NAY-shun) SUPERCHLORINATION

Chlorination with doses that are deliberately selected to produce free or combined residuals so large as to require dechlorination.

SURFACE LOADING SURFACE LOADING

One of the guidelines for the design of settling tanks and clarifiers in treatment plants. Used by operators to determine if tanks and clarifiers are hydraulically (flow) over- or underloaded. Also called OVERFLOW RATE.

$$\text{Surface Loading, GPD/sq ft} = \frac{\text{Flow, gallons/day}}{\text{Surface Area, sq ft}}$$

TRIHALOMETHANES (THMs) (tri-HAL-o-METH-hanes) TRIHALOMETHANES (THMs)

Derivatives of methane, CH_4, in which three halogen atoms (chlorine or bromine) are substituted for three of the hydrogen atoms. Often formed during chlorination by reactions with natural organic materials in the water. The resulting compounds (THMs) are suspected of causing cancer.

TURBIDITY (ter-BID-it-tee) TURBIDITY

The cloudy appearance of water caused by the presence of suspended and colloidal matter. In the waterworks field, a turbidity measurement is used to indicate the clarity of water. Technically, turbidity is an optical property of the water based on the amount of light reflected by suspended particles. Turbidity cannot be directly equated to suspended solids because white particles reflect more light than dark-colored particles and many small particles will reflect more light than an equivalent large particle.

TURBIDITY UNITS (TU) TURBIDITY UNITS (TU)

Turbidity units are a measure of the cloudiness of water. If measured by a nephelometric (deflected light) instrumental procedure, turbidity units are expressed in nephelometric turbidity units (NTU) or simply TU. Those turbidity units obtained by visual methods are expressed in Jackson Turbidity Units (JTU) which are a measure of the cloudiness of water; they are used to indicate the clarity of water. There is no real connection between NTUs and JTUs. The Jackson turbidimeter is a visual method and the nephelometer is an instrumental method based on deflected light.

UNIFORMITY COEFFICIENT (U.C.) UNIFORMITY COEFFICIENT (U.C.)

The ratio of (1) the diameter of a grain (particle) of a size that is barely too large to pass through a sieve that allows 60 percent of the material (by weight) to pass through, to (2) the diameter of a grain (particle) of a size that is barely too large to pass through a sieve that allows 10 percent of the material (by weight) to pass through. The resulting ratio is a measure of the degree of uniformity in a granular material such as filter media.

$$\text{Uniformity Coefficient} = \frac{\text{Particle Diameter}_{60\%}}{\text{Particle Diameter}_{10\%}}$$

WYE STRAINER WYE STRAINER

A screen shaped like the letter Y. The water flows in at the top of the Y and the debris in the water is removed in the top part of the Y.

ZEOLITE ZEOLITE

A type of ion exchange material used to soften water. Natural zeolites are siliceous compounds (made of silica) which remove calcium and magnesium from hard water and replace them with sodium. Synthetic or organic zeolites are ion exchange materials which remove calcium or magnesium and replace them with either sodium or hydrogen. Manganese zeolites are used to remove iron and manganese from water.

CHAPTER 4. SMALL WATER TREATMENT PLANTS

(Lesson 1 of 2 Lessons)

4.0 IMPORTANCE OF SMALL WATER TREATMENT PLANTS

4.00 Need for Effective O & M

Small water treatment plants (Figure 4.1) are much more numerous than large plants. Most water treatment plants are small facilities that supply water to systems ranging in size from a few connections up to several hundred connections. For every large water plant serving a large population in a big service area, there are probably several hundred small plants that serve groups of people in small communities, subdivisions, commercial properties, resorts, and summer camps.

The volume of water produced by these small plants may vary from only fifteen gallons per minute (one liter per second) up to a few hundred gallons per minute. Regardless of the amount of water a plant produces, it must be properly operated to produce safe drinking water. Small plants face many of the same problems as larger ones but often the small plant does not perform as well because of poor design, poor operation, poor maintenance, inadequate budgets, and other causes. However, it is very important that these small plants be operated effectively because many people depend on them for their daily home water supply and many others drink the water on an occasional or intermittent basis.

Small water treatment plants are common because the source of water for many small communities requires treatment. Surface waters must be clarified and disinfected to ensure it is safe and to make it pleasing to the senses. Many groundwater sources require treatment for disinfection or for removal of undesirable minerals and gases in solution. Thus, almost any source of supply is likely to require treatment of some kind. The water supply that needs no treatment whatsoever is increasingly rare today as polluted waters are found everywhere and as standards for domestic water quality become more restrictive.

4.01 Surface Waters

A typical small water treatment system for treating surface waters may include any or all of the following components (see Figure 4.2).

1. *RAW WATER STORAGE*

   Surface water sources are frequently developed from lakes, ponds, or reservoirs. These impoundments provide storage for the raw, untreated water and frequently are helpful to the treatment processes. One important effect of raw water storage is to slow the rate of change in water quality that occurs in surface water sources and to reduce the magnitude of the change. For instance, a rainstorm may cause a sudden, drastic change in the turbidity or the mineral composition of a stream source. If the stream flows into a large reservoir, however, the water quality changes much more slowly and less drastically than in the stream. The overall effect is to make available a raw water supply from the reservoir whose quality is more uniform and consistent. This uniform water quality will make any water treatment plant easier to operate and more effective.

   If the water source is a canal or pipeline that is subject to shutdowns or outages, a raw water reservoir is necessary to keep the treatment plant in operation. Otherwise the plant must stop operating whenever the water source is out of service.

2. *DIVERSION WORKS*

   The diversion works consist of all the facilities used to divert water from the source into the treatment plant.

Fig. 4.1 Small water treatment plants

(Permission of Watermasters, Inc., Burlingame, California)

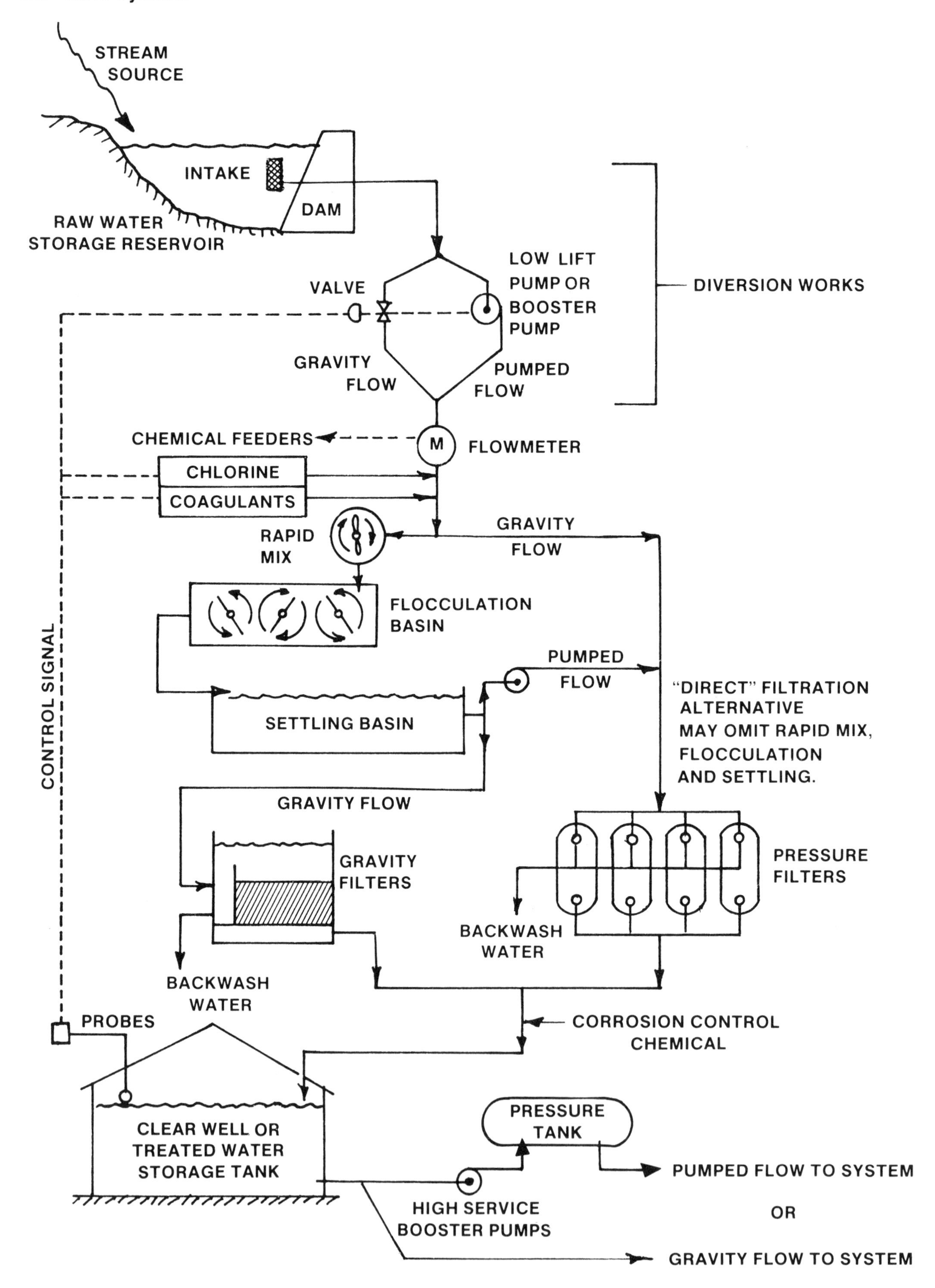

Fig. 4.2 Parts of a typical small water treatment system

These facilities may include a diversion dam, screens to exclude fish and trash, an intake pipe or structure, pumping equipment, piping or conduits to convey the water to the plant, and valves or gates to control the flow of water.

3. *FLOW MEASUREMENT*

   A flowmeter of some type is essential for proper operation of even the smallest water treatment plant. The meter should be the type that indicates both the instantaneous rate of flow as well as the total quantity of water that has flowed through it. The meter must be accurate enough to allow the operator to feed the proper chemical dosages. A fine screen or *WYE STRAINER*[1] should be installed upstream from the meter to keep it from being clogged or damaged by trash or rocks.

4. *DISINFECTION*

   Chlorine is the chemical frequently applied to the water for disinfection. Prechlorination is often used because as the water flows through the treatment units, it receives the maximum contact time before reaching consumers. A chlorine residual in water flowing through the plant also discourages the growths of organisms and helps keep the filter media and other treatment equipment clean and free of organic growths. Prechlorination should not be used if natural organics are in the water and the formation of *TRIHALOMETHANES*[2] is a problem.

5. *COAGULATION*

   COAGULATION[3] facilities include the chemical feeding equipment for injecting coagulants and equipment for providing rapid mixing of the chemicals with the water.

6. *FLOCCULATION*

   Flocculation facilities include equipment for slowly mixing the water to promote growth of floc particles that will settle quickly or be removed by filtration.

7. *SETTLING* (sedimentation)

   Settling allows suspended matter to be separated from the water by gravity. A very large portion of the suspended matter is usually removed in the settling tanks.

8. *FILTRATION*

   Filters remove most of the suspended matter remaining in the water after settling.

9. *CORROSION CONTROL*

   Corrosion control treatment is used to stabilize the chemical nature of the water and make it less aggressive toward the materials used in pipelines, storage reservoirs, and customer appliances.

10. *TREATED WATER STORAGE*

    Storage reservoirs for treated water may be installed either at the treatment plant (*CLEAR WELL*[4]) or out in the distribution system. Such reservoirs allow the treatment plant to be of smaller capacity than would otherwise be necessary because peak water demands can be supplied from storage. Storage also maintains the water supply during outages at the plant or at the source and it provides large amounts of water immediately for firefighting purposes.

11. *HIGH-SERVICE PUMPS*

    Unless the water system can be served entirely by gravity flow from a storage reservoir, high-service pumps are required. The pumps draw water from storage and supply it to the system under pressure. Often one or more hydropneumatic pressure tanks are installed with the pumps.

Many small water treatment facilities are "package plants" (Figure 4.3) which can be purchased as a complete preassembled unit from a single manufacturer. Such package plants are available from a number of manufacturers. These units are most commonly supplied for filtration of turbid waters and for removal of dissolved iron and manganese. Usually the package plant includes all the treatment equipment, pumps, chemical feeders, and controls. As soon as the water pipes and the electric power have been connected, the plant is ready to operate. Thus the package plant is frequently a quick way to provide the needed treatment. Another advantage of the package plant is that the design and the equipment have been proven effective and reliable by experience. All the "bugs" have been eliminated and the purchaser can usually have a high degree of confidence in the performance of the plant.

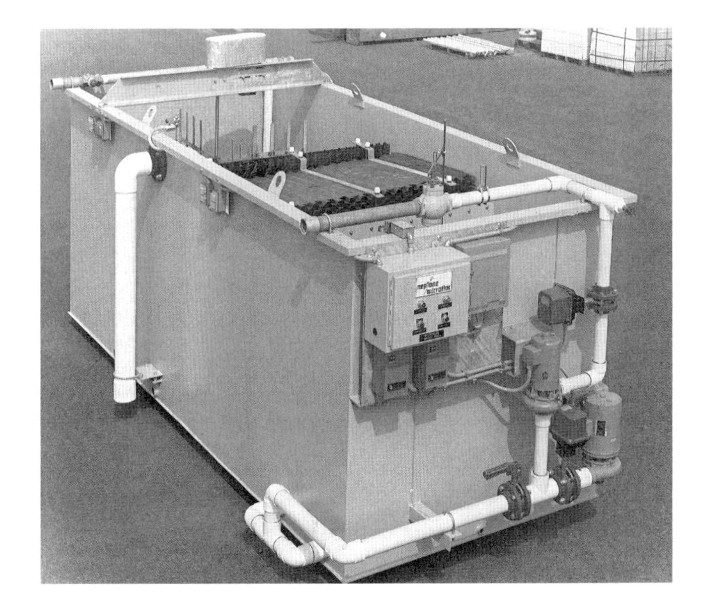

Fig. 4.3 Package water treatment plant
(Permission of Neptune Microfloc)(Now Microfloc Products)

---

[1] *Wye Strainer. A screen shaped like the letter Y. The water flows in at the top of the Y and the debris in the water is removed in the top part of the Y.*

[2] *Trihalomethanes (THMs) (tri-HAL-o-METH-hanes). Derivatives of methane, CH_4, in which three halogen atoms (chlorine or bromine) are substituted for three of the hydrogen atoms. Often formed during chlorination by reactions with natural organic materials in the water. The resulting compounds (THMs) are suspected of causing cancer.*

[3] *Coagulation (co-AGG-you-LAY-shun). The clumping together of very fine particles into larger particles (floc) caused by the use of chemicals (coagulants). The chemicals neutralize the electrical charges of the fine particles, allowing them to come closer and form larger clumps. This clumping together makes it easier to separate the solids from the water by settling, skimming, draining, or filtering.*

[4] *Clear Well. A reservoir for the storage of filtered water of sufficient capacity to prevent the need to vary the filtration rate with variations in demand. Also used to provide chlorine contact time for disinfection.*

When operating a package plant, beware of claims that the plant operates "automatically." Even automatic systems require maintenance, repairs, and occasional process control changes. These plants may be easily upset by sudden changes in source water quality.

The alternative to a purchased package plant is a custom-designed plant which is assembled from equipment and materials supplied by several different manufacturers. This approach is sometimes less expensive than a package plant but the effectiveness of the resulting treatment plant is less predictable and may be unsatisfactory. The success of the plant depends on the ability of the person who designs the equipment and the skill and knowledge of the operator.

4.02 Groundwaters

Many small communities obtain their drinking water from wells. Sometimes the only additional treatment needed is the application of chlorine for disinfection. Other well waters may require treatment to remove iron and manganese or softening to remove excess hardness. In this chapter we will discuss the control of iron and manganese and also hardness at small water treatment plants. The application of chlorine for disinfection follows similar procedures when treating either surface or groundwaters.

1. *IRON AND MANGANESE CONTROL*

   Iron in drinking water can be controlled by converting iron from the liquid ferrous (Fe^{2+}) to the solid ferric (Fe^{3+}) form by the *OXIDATION*[5] process. First an oxidizing agent such as chlorine or potassium permanganate is used. Next oxygen is introduced to the water by the use of sprays, cascades, or trays. Finally, an hour or so (depending on the chemical makeup of the water) is allowed for the oxidation reaction to be completed and the insoluble *PRECIPITATES*[6] of iron and manganese to be formed. The precipitates are removed by sedimentation and filtration or by filtration alone. Sometimes manganese is controlled by the addition of polyphosphates followed by chlorination.

2. *SOFTENING*

   Hardness in water is caused mainly by the presence of calcium and magnesium. Excessive hardness is undesirable to domestic consumers due to difficulties in doing the laundry and washing dishes. Water is softened by either the ion exchange process or by chemical precipitation (lime-soda ash softening). Softened water delivered to consumers usually has a hardness level of around 80 to 90 *mg/L* expressed as *CALCIUM CARBONATE EQUIVALENT.*[7]

4.03 Operator Responsibility

Operators of small water treatment plants probably have the toughest job of anyone in the waterworks field. In many small plants they have to do all of the work themselves. They have full responsibility to produce and deliver a pleasant and potable water. There is no one else to share the responsibility with them. Often these operators receive fairly low pay, support from others is minimal, there never is enough of the right kinds of tools, money for maintenance and repairs is always of concern, and nobody appreciates how hard the operator works. If you are in this position, you can do a good job if you use a lot of imagination and initiative. We hope this manual will provide you with the information to convince your supervisors and consumers that you need more help to do your job.

QUESTIONS

Write your answers in a notebook and then compare your answers with those on page 259.

4.0A Why is a water supply that needs no treatment very rare?

4.0B How does the storage of raw water in lakes, ponds, or reservoirs help the water treatment plant operator?

4.0C What information does an operator obtain from a flowmeter?

4.0D How can a flowmeter be protected?

4.0E Groundwaters may require what types of treatment?

4.1 COAGULATION

Coagulation is the chemical reaction that occurs when a coagulating chemical is added to water. The most common coagulating chemical used is aluminum sulfate (alum) but other chemicals are sometimes used. When one of these coagulants is added to water, it reacts directly with the water and with certain minerals dissolved in the water to form floc particles. These are small, jelly-like, filmy particles that look like snowflakes.

The coagulant also reacts physically with the fine particles of suspended matter in the water. These particles, which cause the water to appear turbid or cloudy, normally have a negative electrical charge on their surface. This charge causes the particles to repel one another and remain suspended in the water rather than clumping together and settling

---

[5] *Oxidation (ox-uh-DAY-shun). Oxidation is the addition of oxygen, removal of hydrogen, or the removal of electrons from an element or compound. In the environment, organic matter is oxidized to more stable substances. The opposite of reduction.*

[6] *Precipitate (pre-SIP-uh-TATE). (1) An insoluble, finely divided substance which is a product of a chemical reaction within a liquid. (2) The separation from solution of an insoluble substance.*

[7] *Calcium Carbonate ($CaCO_3$) Equivalent. An expression of the concentration of specified constituents in water in terms of their equivalent value to calcium carbonate. For example, the hardness in water which is caused by calcium, magnesium and other ions is usually described as calcium carbonate equivalent. Alkalinity test results are usually reported as mg/L $CaCO_3$ equivalents. To convert chloride to $CaCO_3$ equivalents, multiply the concentration of chloride ions in mg/L by 1.41, and for sulfate, multiply by 1.04.*

to the bottom. The effect of the coagulant is to neutralize the negative charge on the suspended particles (destabilize the particles) so they can be brought together into larger clumps that are heavy enough to settle.

Many factors affect the coagulation process. Among the most important are the coagulant used, the coagulant dosage, the pH of the water, the mineral content of the water, the water temperature, and the coagulant injection-mixing arrangement. Under any given set of conditions, a particular water requires a certain optimum dosage of coagulant for coagulation to be successful. If the coagulant dosage is either too high or too low, coagulation will be incomplete and the results of treatment will be unsatisfactory. The pH of the water must be controlled within a narrow range during coagulation to achieve optimum results. The water must have an adequate *ALKALINITY*[8] content to achieve good coagulation with alum. The treatment is even more effective if the water contains abundant amounts of certain minerals, such as calcium and magnesium. Other chemicals (phosphate compounds for example) inhibit coagulation and require larger dosages of coagulant. Any chemical reaction proceeds faster when the water is warm than when it is cold and the method of injecting the coagulant and mixing it with the water has a very significant effect on the results of coagulation treatment. For effective treatment, the chemical must be completely mixed throughout the water being treated.

All these variables make it impossible to calculate what dosage of coagulant will produce the best results under a given set of conditions. Thus, in even the largest water treatment plant, the coagulant dosage must be estimated by performing a *JAR TEST*.[9] A jar test is a laboratory procedure in which varying dosages of coagulant are tested in a series of glass or plastic jars under identical conditions. A jar test apparatus like the one shown in Figure 4.4 is required for the test. The machine consists of a set of six paddle mixers (gang stirrer) all driven by a variable-speed motor. A jar containing the water to be treated is placed under each paddle mixer and varying dosages of coagulant are added to the jars. If the optimum dosage is suspected to be about 20 milligrams per liter of alum, that dosage plus both higher and lower dosages would be tested for comparison. For instance, the following dosages might be tested in one jar test series.

Jar Number	Alum Dosage, mg/*L*
1	14
2	16
3	18
4	20
5	22
6	24

After the proper coagulant dosage is added to each jar, the paddles are run at high speed for a short time to simulate rapid mixing. Then the paddles are adjusted to a very low speed for a longer period of time to simulate conditions in the flocculation basin of the plant. After a period of time, the paddles are stopped and all jars are observed to evaluate the settling results. The evaluation can be made by simple visual observations or by performing further laboratory tests. Ordinarily, the operator will visually observe which coagulant dosage produces the first visible floc, which dosage produces the largest, strongest floc, or which dosage produces the floc that settles the fastest. On this basis the operator can often select the optimum dosage to use in the plant. Other laboratory tests that are sometimes used to evaluate a jar test are (1) the pH of the water, (2) the turbidity of the settled water, and (3) the filterability of the water as determined by filtration through some type of small laboratory filter.

If the characteristics of the source water change rapidly because of storm conditions or other causes, jar tests may have to be conducted very frequently. At other times when the raw water quality is nearly the same from day to day, a jar test may be necessary only on occasion to verify the dosage. Some operators measure the turbidity of the source water and perform the jar test to determine the optimum coagulant dosage. By keeping accurate records they can develop a plot of turbidity as it relates to dosage. Whenever the turbidity changes, they have a good idea of the optimum dosage.

If you do not have a jar test apparatus, collect a water sample in a glass jar directly from the outlet of the rapid mix chamber in the plant. By gently hand stirring the water sample for a few minutes, you can sometimes get an idea of how well the water is coagulating. This method is crude but it is better than doing nothing to monitor coagulation.

Some operators will collect samples from the outlet of the rapid mix chamber and use this water for a jar test. They will add additional doses of coagulant to two jars and dilute a couple of the jars with raw water. Observing the results can indicate whether the dosage should be increased or decreased.

Streaming current meters are a new device used by operators to optimize coagulant doses. The streaming current meter is a continuous on-line measuring instrument. Properly used, the streaming current meter can function as an on-line jar test.

---

[8] *Alkalinity (AL-ka-LIN-it-tee). The capacity of water to neutralize acids. This capacity is caused by the water's content of carbonate, bicarbonate, hydroxide, and occasionally borate, silicate, and phosphate. Alkalinity is expressed in milligrams per liter of equivalent calcium carbonate. Alkalinity is not the same as pH because water does not have to be strongly basic (high pH) to have a high alkalinity. Alkalinity is a measure of how much acid must be added to a liquid to lower the pH to 4.5.*

[9] *Jar Test. See Chapter 11, "Laboratory Procedures," in WATER TREATMENT PLANT OPERATION, Volume I, for additional details on how to perform the Jar Test.*

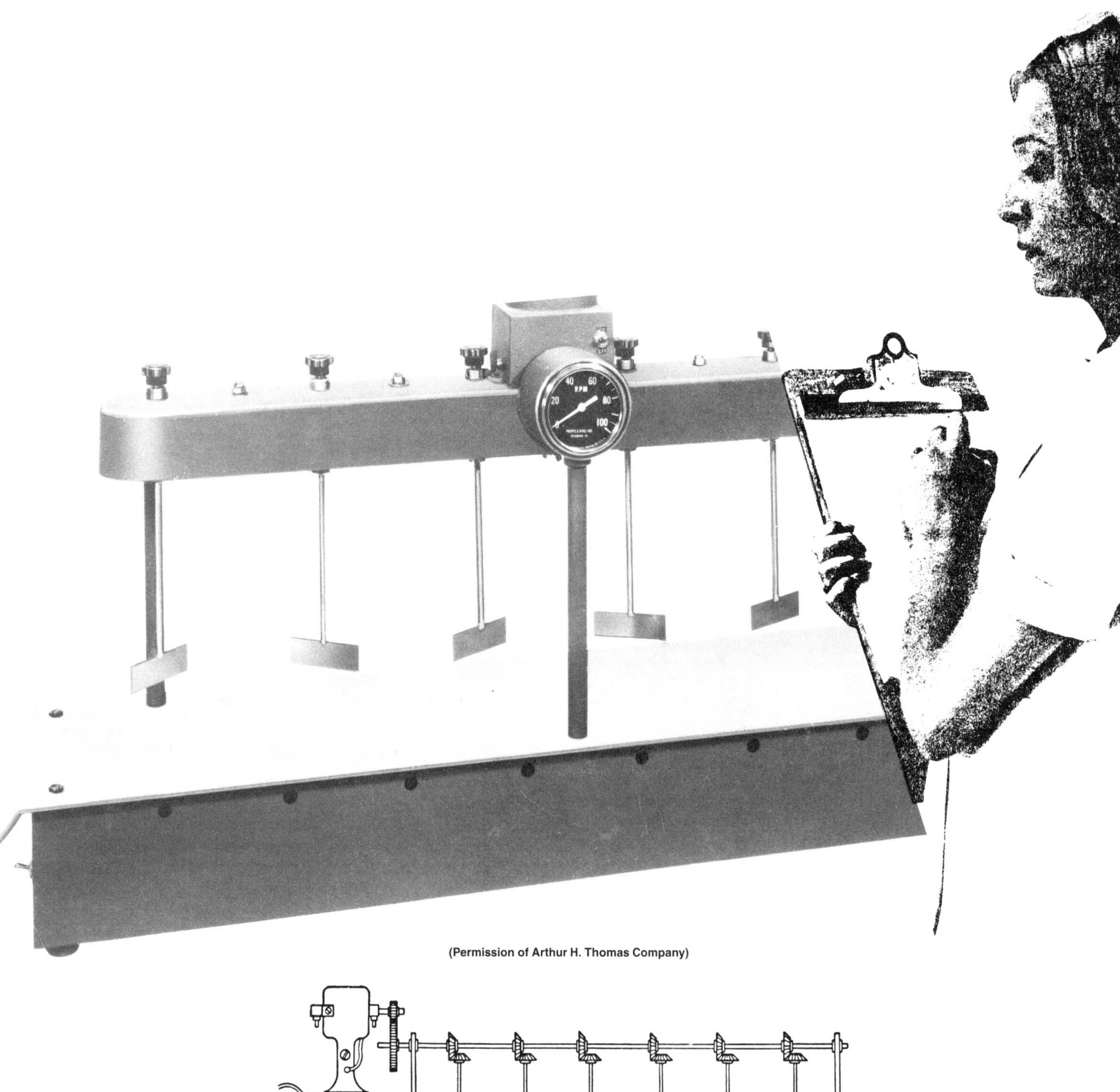

(Permission of Arthur H. Thomas Company)

Laboratory stirring device.

Fig. 4.4 Jar test apparatus

(Reprinted from *HANDBOOK OF APPLIED HYDRAULICS*, by permission. Copyright 1952, McGraw-Hill Book Company)

Most particles in water are anions (negative charge) and most coagulants are cationic (positive charge). The streaming current meter presumes that by bringing the total charge of the water being treated to neutral (zero, 0), the coagulation process has been optimized. Most operators run the charge of their water slightly negative by adjusting the coagulant dose.

The manner in which the coagulant is injected into the water is very important for proper coagulation. The coagulant should be brought into contact with every portion of water and completely mixed with it as fast as possible. Therefore the injection facilities and mixing equipment must be designed to produce good mixing. Mixing must be very rapid and turbulent because the coagulation reactions occur quickly, probably in the first few seconds and certainly within one minute. Experience has shown repeatedly that the optimum injection/rapid mix arrangement can greatly improve the results of coagulation and at the same time significantly decrease the amount of coagulant chemical that is used.

The best type of injector is usually one with multiple-feed orifices that will uniformly distribute the chemical solution throughout the flow of water. Thus a perforated pipe that extends completely across the width of the channel or the diameter of the pipeline is a better injector than one that injects at a single point. Also, locating the injector at a point of turbulent flow, such as a *PARSHALL FLUME*,[10] provides better mixing than locating the injector in an area of smooth, quiet flow. Thorough distribution of the chemical in combination with violent mixing is the most desirable arrangement.

If coagulation is not properly done, the treatment processes that follow (*FLOCCULATION*,[11] settling, and filtration) will be much less effective in purifying the water. Suspended matter will pass through the filters because of poor coagulation and the treated water will be cloudy and turbid. Dissolved alum may also pass through the filters and coagulate later in the clear well reservoir or in distribution system pipelines causing customer complaints about particles in the water and a dirty-tasting water. Therefore, if alum floc appears in the treated water or if alum is present in the treated water in concentrations greater than 0.1 mg/*L*, review the coagulation treatment for proper operation.

If you suspect that the coagulation treatment is substandard or if excessive dosages of coagulant are required, check the items listed below.

1. Perform a series of jar test experiments to simulate conditions in the treatment plant and determine the optimum coagulant dosage needed for the prevailing water quality conditions. Remember, the jar test is an indication of what's happening in your plant. *YOU MUST OBSERVE* the actual coagulation, flocculation, and settling in your plant to determine the optimum chemical dosage.
2. Measure the pH of the water after the coagulant is added. Is the pH the same that normally prevails when good coagulation occurs? A record of daily pH readings is very helpful in determining the optimum pH for coagulation under different conditions. Sometimes jar tests are run at different pH levels to find the optimum pH. Coagulation may not occur unless the pH is very close to optimum.
3. Measure the alkalinity. Is adequate alkalinity present for the coagulation reaction to occur? If not, consider increasing the alkalinity to have at least 30 mg/*L* remaining when the chemical reactions are complete by adding sufficient lime or soda ash prior to coagulation.
4. Is the chemical feeder supplying the correct dosage of coagulant? The feeder may be broken or plugged up. Also, the feeder may be incorrectly adjusted. Measure the amount of chemical actually being fed and calculate the dosage. See Examples 1 and 2 on page 187. If the coagulant chemical is fed in the dry form, the operator can simply weigh the amount of coagulant that the feeder supplies in a certain period of time, say three minutes. A sensitive laboratory scale is needed to weigh the chemical accurately. If the coagulant chemical is fed in a water solution, the simple measuring device illustrated in Figure 4.5 can be used to measure the volume of solution supplied by the feeder pump in a given time. See Examples 3 and 4.
5. Does the chemical feeder inject a steady feed of chemical into the water? If the feeder injects in "pulses" or "slugs," are the "pulses" as frequent as possible? If the feeder injects a "slug" only at rather long intervals, the chemical is not being uniformly distributed throughout the water. This can be corrected by diluting the chemical solution and readjusting the feeder pump to "pulse" more frequently.
6. Is the chemical injector distributing the coagulant completely throughout the flow of water? Injectors with multiple-feed orifices are better than those with a single-feed orifice.
7. Is violent, rapid mixing provided just after the chemical is injected into the water? If not, the operator can install a mechanical mixer or relocate the point of injection to a zone of turbulent flow such as a Parshall flume, a pump intake, or an in-line static mixer.
8. Consider use of a coagulant aid such as an organic *POLYMER*.[12]
9. Examine the source water for a change in quality due to: pollution, storm water runoff, use of a different water source, or other causes.
10. Modify conditions at the intake or diversion to provide a water of more uniform quality to the plant. If large temporary changes occur frequently in the source water (during a storm), if possible turn off the plant and wait until water quality conditions are more stable and uniform.
11. Investigate the suitability of other coagulants. Pure aluminum sulfate is a cheaper and more effective coagulant than potassium alum or ammonium alum. These chemicals can be easily confused by suppliers and operators who are not aware of their differences.

---

[10] *Parshall Flume. A device used to measure the flow in an open channel. The flume narrows to a throat of fixed dimensions and then expands again. The rate of flow can be calculated by measuring the difference in head (pressure) before and at the throat of the flume.*

[11] *Flocculation (FLOCK-you-LAY-shun). The gathering together of fine particles after coagulation to form larger particles by a process of gentle mixing.*

[12] *Polymer (POLY-mer). A long chain molecule formed by the union of many monomers (molecules of lower molecular weight). Polymers are used with other chemical coagulants to aid in binding small suspended particles to larger chemical flocs for their removal from water.*

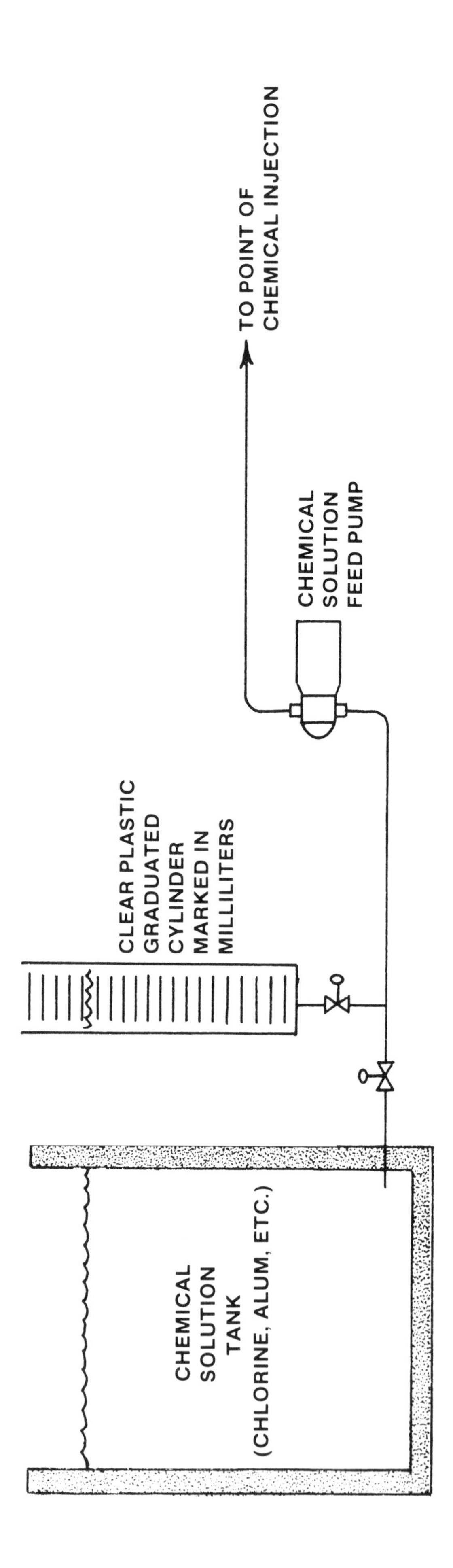

THE FEED RATE OF A CHEMICAL SOLUTION FEED PUMP CAN BE DETERMINED BY MEASURING THE AMOUNT OF SOLUTION WITHDRAWN FROM A GRADUATED CYLINDER IN A GIVEN TIME PERIOD. ALLOW THE CYLINDER TO FILL WITH SOLUTION. THEN CLOSE THE VALVE ON THE LINE FROM THE TANK SO THE FEED PUMP TAKES SUCTION FROM THE CYLINDER ONLY. OBSERVE THE MILLILITERS OF SOLUTION USED IN ONE MINUTE. COMPARE THIS RESULT WITH THE DESIRED FEED RATE AND ADJUST THE FEED PUMP ACCORDINGLY.

Fig. 4.5 Calibration of a chemical feed pump

For additional information on coagulation, see Chapter 4, "Coagulation and Flocculation," in *WATER TREATMENT PLANT OPERATION*, Volume I, in this series of manuals.

FORMULAS

In order to determine the *ACTUAL* feed rate of a dry chemical feeder:

1. Weigh a pan,
2. Place the pan under the feeder and then weigh the pan and chemicals,
3. Record the time the pan collected the chemical, and
4. Calculate the actual chemical feed or dose in pounds per day.

$$\text{Actual Chem Feed, lbs/day} = \frac{(\text{Weight of Chem, lbs})(60\text{ min/hr})(24\text{ hr/day})}{(\text{Time Chemical Collected, min})}$$

$$\text{Weight of Chem, lbs} = \frac{(\text{Weight, grams})}{454\text{ grams/lb}}$$ (Use this formula only if scale weighs in grams.)

In order to determine the *DESIRED* feed rate from a dry chemical feeder:

1. Measure and record the flow of water being treated. If the flow is in gallons per minute (GPM), convert the flow to million gallons per day (MGD).
2. Determine the desired dose in mg/*L* using the jar test.
3. Calculate the desired feed rate in pounds per day.

$$\text{Desired Chem Feed, lbs/day} = (\text{Flow, MGD})(\text{Dose, mg/}L)(8.34\text{ lbs/gal})$$

$$\text{Flow, MGD} = \frac{(\text{Flow, gal/min})(60\text{ min/hr})(24\text{ hr/day})}{1{,}000{,}000/\text{Million}}$$

In Figure 4.5, the chemical solution tank contains a hypochlorite solution containing 0.1 pound of chlorine per gallon of solution or the solution may be given as a percent. The clear plastic graduated cylinder is filled with the hypochlorite solution. In order to determine the actual feed rate of the chemical solution feed pump in pounds per day:

1. Pump a portion of the chemical solution from the graduated cylinder and record the volume pumped in milliliters.
2. Determine the time required to pump the volume in milliliters.
3. Calculate the concentration of the chemical solution in either milligrams per liter or milligrams per milliliter.

$$\text{Chem Conc, mg/}L = \frac{(\text{Chem Conc, lbs/gal})(1{,}000{,}000/\text{Million})}{8.34\text{ lbs/gal}}$$

or

$$\text{Chem Conc, mg/}L = (\text{Chemical Conc, \%})(10{,}000\text{ mg/}L/\%)$$

$$\text{Chem Feed, lbs/day} = \frac{(\text{Chem Conc, mg/}L)(\text{Vol Pumped, m}L)(60\text{ min/hr})(24\text{ hr/day})}{(\text{Time Pumped, min})(1{,}000\text{ m}L/L)(1{,}000\text{ mg/gm})(454\text{ gm/lb})}$$

$$\text{Actual Dose, mg/}L = \frac{\text{Chemical Feed, lbs/day}}{(\text{Flow, MGD})(8.34\text{ lbs/gal})}$$

EXAMPLE 1

A pan is placed under a dry alum feeder for exactly three minutes. The pan and alum are weighed. The alum weighed 45.4 grams. What is the actual alum feed in pounds per day?

Known		Unknown
Time, min	= 3 min	Actual Alum Feed, lbs/day
Alum Weight, gm	= 45.4 gm	

1. Convert alum weight from grams to pounds.

$$\text{Alum Weight, lbs} = \frac{\text{Weight, gm}}{454\text{ gm/lb}}$$

$$= \frac{45.4\text{ gm}}{454\text{ gm/lb}}$$

$$= 0.1\text{ lb}$$

2. Calculate the actual alum feed rate in pounds per day.

$$\text{Actual Alum Feed, lbs/day} = \frac{(\text{Alum Weight, lbs})(60\text{ min/hr})(24\text{ hr/day})}{\text{Time Alum Collected, min}}$$

$$= \frac{(0.1\text{ lb})(60\text{ min/hr})(24\text{ hr/day})}{3\text{ min}}$$

$$= 48\text{ lbs/day}$$

EXAMPLE 2

Jar tests indicate that a water should be dosed with alum at 20 mg/*L*. The flow being treated is 200 GPM. What is the desired alum feed in pounds per day?

Known		Unknown
Alum Dose, mg/*L*	= 20 mg/*L*	Desired Alum Feed, lbs/day
Flow, GPM	= 200 GPM	

1. Convert the flow from gallons per minute (GPM) to million gallons per day (MGD).

$$\text{Flow, MGD} = \frac{(\text{Flow, gal/min})(60\text{ min/hr})(24\text{ hr/day})}{1{,}000{,}000/\text{Million}}$$

$$= \frac{(200\text{ gal/min})(60\text{ min/hr})(24\text{ hr/day})}{1{,}000{,}000/\text{Million}}$$

$$= 0.288\text{ MGD}$$

2. Calculate the desired alum feed rate in pounds per day.

$$\text{Desired Alum Feed, lbs/day} = (\text{Flow, MGD})(\text{Dose, mg/}L)(8.34\text{ lbs/gal})$$

$$= (0.288\text{ MGD})(20\text{ mg/}L)(8.34\text{ lbs/gal})$$

$$= 48\text{ lbs/day}$$

NOTES

1. The actual alum feed in Example 1 of 48 lbs alum per day agrees with the desired alum feed in Example 2 of 48 lbs alum per day. If the actual feed does not agree with the desired feed, then the actual feed rate should be adjusted.
2. Remember that actual performance is what counts. Jar tests are used to help you get close to the desired dose. Fine adjustments must be made on the basis of actual field tests and observations.

EXAMPLE 3

The chemical solution feed pump in Figure 4.5 removes 500 m*L* from the graduated cylinder in 4 minutes. The concentration of the hypochlorite solution is 0.1 pound chlorine per gallon. Determine the chemical feed rate in pounds per day delivered by the pump.

Known		Unknown
Volume Pumped, m*L*	= 500 m*L*	Chemical Feed, lbs/day
Time Pumped, min	= 4 min	
Chem Conc, lbs/gal	= 0.1 lb/gal	

1. Calculate the chemical concentration in the graduated cylinder in milligrams per liter.

$$\text{Chem Conc, mg/}L = \frac{(\text{Chem Conc, lbs/gal})(1{,}000{,}000/\text{Million})}{8.34\text{ lbs/gal}}$$

$$= \frac{(0.1\text{ lb/gal})(1{,}000{,}000/\text{Million})}{8.34\text{ lbs/gal}}$$

$$= \frac{12{,}000\text{ lbs Chlorine}}{1\text{ Million lbs Water}}$$

$$= 12{,}000\text{ mg/}L$$

2. Determine the chemical feed rate of the pump in pounds of chlorine per day.

$$\text{Chem Feed, lbs/day} = \frac{(\text{Chem Conc, mg/}L)(\text{Vol Pumped, m}L)(60\text{ min/hr})(24\text{ hr/day})}{(\text{Time Pumped, min})(1{,}000\text{ m}L/L)(1{,}000\text{ mg/gm})(454\text{ gm/lb})}$$

$$= \frac{(12{,}000\text{ mg/}L)(500\text{ m}L)(60\text{ min/hr})(24\text{ hr/day})}{(4\text{ min})(1{,}000\text{ m}L/L)(1{,}000\text{ mg/gm})(454\text{ gm/lb})}$$

$$= 4.8\text{ lbs/day}$$

NOTE: If the concentration of the hypochlorite solution in Example 3 was given as a 1.2 percent solution, then

$$\text{Chem Conc, mg/}L = (\text{Chemical Conc, \%})(10{,}000\text{ mg/}L/\%)$$

$$= (1.2\%)(10{,}000\text{ mg/}L/\%)$$

$$= 12{,}000\text{ mg/}L$$

EXAMPLE 4

Calculate the actual chlorine dose in milligrams per liter in the water being treated in Example 3. The hypochlorinator (chemical feed pump) delivers 4.8 pounds of chlorine per day and the flow rate of the water being treated is 250,000 gallons per day (0.25 MGD).

Known		Unknown
Chemical Feed, lbs/day	= 4.8 lbs/day	Actual Dose, mg/*L*
Flow, MGD	= 0.25 MGD	

Calculate the actual chlorine dose in milligrams per liter.

$$\text{Actual Dose, mg/}L = \frac{\text{Chemical Feed, lbs/day}}{(\text{Flow, MGD})(8.34\text{ lbs/gal})}$$

$$= \frac{4.8\text{ lbs/day}}{(0.25\text{ MGD})(8.34\text{ lbs/gal})}$$

$$= 2.3\text{ mg/}L$$

QUESTIONS

Write your answers in a notebook and then compare your answers with those on page 259.

4.1A What is the influence of temperature on the coagulation process?

4.1B How are the results of jar tests evaluated?

4.1C How can operators determine whether the coagulant dose should be increased or decreased?

4.1D How can the alkalinity of the water being treated be increased?

4.2 FLOCCULATION

Flocculation is a process of slow, gentle mixing of the water to encourage the tiny floc particles to clump together and grow to a size that will settle quickly. When the particles of floc first form they are too small to be visible to the naked eye. A floc particle of this size is also too small to settle in a reasonable time. Therefore, the process of flocculation or gentle mixing is used to bring the small particles of floc and other suspended matter into contact with each other so they will stick together and form larger particles. By the time the floc particles grow to a size of about 1/16 inch (1.5 mm) diameter or larger, they are usually heavy enough to settle out in a few minutes.

Flocculation can be accomplished either by mechanical mixers or by hydraulic mixing. Mechanical mixers are preferred because their performance can be predicted and controlled more accurately. The typical mechanical flocculator consists of a set of large, slowly rotating paddle wheels. The speed of the second paddle wheel may be slower than the first and the third may be slower than the second. In this manner, the mixing becomes progressively more gentle as the water flows through the flocculation basin. Designers believe that this tapered flocculation encourages the rapid growth of larger floc particles.

Other types of mechanical flocculators exist but the purpose of all flocculators is to provide gentle mixing that will produce a quick settling floc. The mixing must be strong enough to prevent premature settling of floc in the flocculation basin, but the mixing must not be so strong that it breaks apart the floc particles already formed. Mechanical flocculators are preferred because their operation is flexible enough to maintain the proper mixing regardless of the rate of flow through the plant and because the degree of agitation can be adjusted to suit changes in water quality.

Flocculation is sometimes attempted by means of hydraulic mixing. Hydraulic mixing occurs when water flows around obstructions or obstacles such as a series of baffles or through a series of interconnected chambers. Around-the-end baffles are commonly used to create back and forth flow through a basin. Over and under baffles are used to create an up and down rolling motion to the flow of water. Other hydraulic devices such as orifices, weirs, or vortex chambers may also be used to create hydraulic mixing.

Most hydraulic flocculators suffer from serious drawbacks. Mixing is less uniform and controllable, being too violent in one spot and too gentle in another. While hydraulic flocculators may perform well at one rate of flow, they usually give poor results at flows greater or less than that for which they are designed. Thus they have less flexibility of operation than mechanical flocculators because the degree of mixing depends on the flow rate.

The success of flocculation is affected by only a few factors. The primary factor is the degree of mixing. If mixing is too gentle, the suspended particles will not be brought into contact with one another and there will be fewer opportunities for large clumps of floc to form. Conversely, mixing that is too violent will tear apart the floc particles and prevent them from attaining proper size. The time of mixing is also important to proper flocculation. A minimum time of mixing is necessary for flocculation to be completed. Usually a period of 30 to 45 minutes is sufficient. To ensure that all portions of the water are retained in the flocculator for the required time, it is important to limit *SHORT-CIRCUITING.*[13] This can usually be accomplished by proper baffling or by providing several compartments in series (one after another). Three or more compartments in series is recommended. A third factor affecting flocculation is the number of particles. A relatively clear water is harder to flocculate than a turbid water containing a lot of suspended matter. The difference is that the greater number of particles in the turbid water collide with one another more often.

If flocculation is not satisfactory in producing a floc that settles quickly and clarifies the water effectively, take the following corrective steps.

1. Correct any deficiencies in the coagulation process.
2. Check the degree of mixing. Be sure the mixing is neither too gentle nor too violent.
3. Make sure that mixing is provided for an adequate time period. The flow rate through the floc basin may be too high. Are actual flows greater than design flows?
4. See that short-circuiting of flow through the basin is prevented so each portion of the water is retained as long as possible.
5. Adjust the plant to operate on a more continuous basis for longer periods of time. Frequent ON/OFF operation is harmful to flocculation because the floc settles to the bottom when the plant is off and must be resuspended when it starts again. When adjusting flows, consider previous flows, expected demands and available storage.

For additional information on flocculation, see Chapter 4, "Coagulation and Flocculation," in *WATER TREATMENT PLANT OPERATION*, Volume I, in this series of manuals.

QUESTIONS

Write your answers in a notebook and then compare your answers with those on page 259.

4.2A What is flocculation?

4.2B What happens if the flocculation mixing is too strong or too weak?

4.3 SETTLING (SEDIMENTATION)

Settling or sedimentation is the process of holding the water in quiet, low flow conditions so suspended matter and particles can be settled out by gravity to the bottom of the tank and removed as sludge.

The purpose of settling is to remove as much of the floc and other suspended material as possible before the water flows to the filters. Settling is thus the final step of pretreatment prior to filtration. Settling is an economical means of clarifying water; therefore, it is usually practiced whenever the water contains even a moderate amount of suspended matter. Settling may be omitted sometimes when the water contains only a small amount of suspended matter, but settling is required when a lot of suspended matter is present.

To understand the operation of a settling tank, it is helpful to think of it as having four zones. These zones are shown in Figure 4.6. The size of each zone varies and the boundaries between the zones are vague and indefinite rather than sharp and well defined.

In the inlet zone, water entering the tank is distributed across the cross section of the tank and slows to a uniform flow velocity. Next, the water flows into the settling zone which is the main part of the tank. Here the water flows slowly through the tank and the suspended particles settle out. The settled material accumulates in the sludge zone at the bottom of the tank. At the end of the tank, the water enters the outlet zone where it flows to the outlet, collects in suitable channels and leaves the tank.

As the water flows through these zones in the settling tank, various factors have an important effect on the settling of suspended particles. Some of these factors are interrelated.

[13] *Short-Circuiting. A condition that occurs in tanks or basins when some of the water travels faster than the rest of the flowing water. This is usually undesirable since it may result in shorter contact, reaction, or settling times in comparison with the theoretical (calculated) or presumed detention times.*

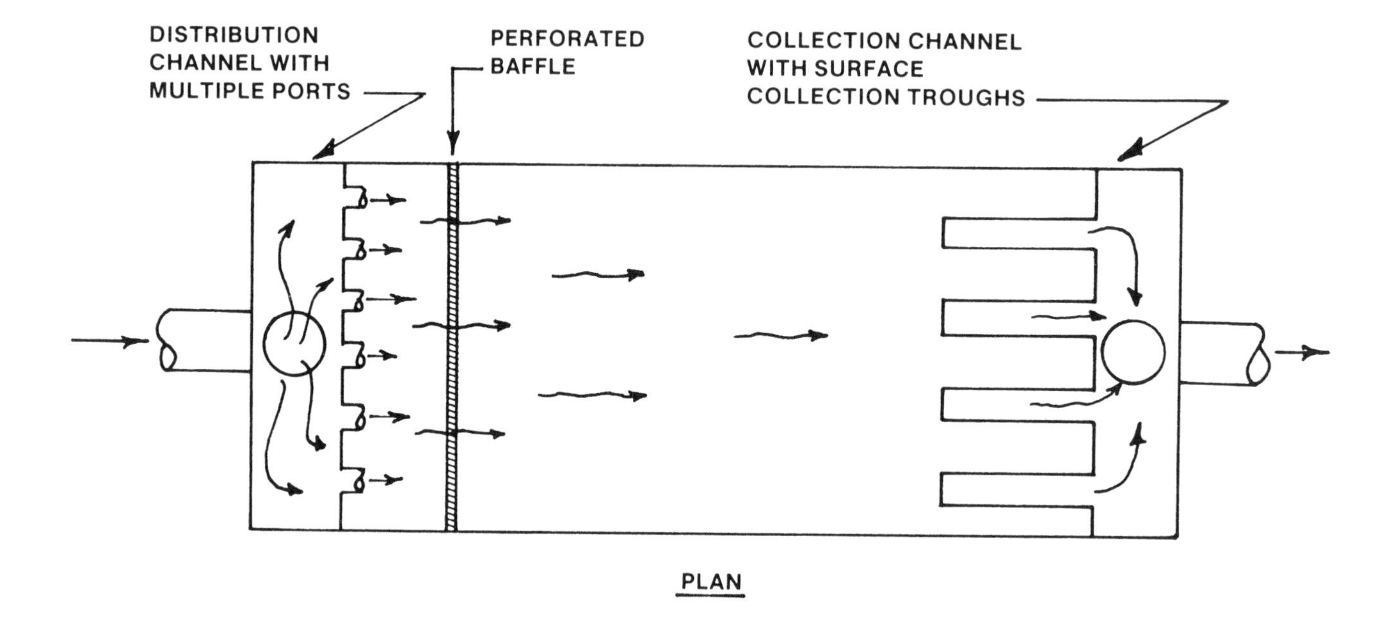

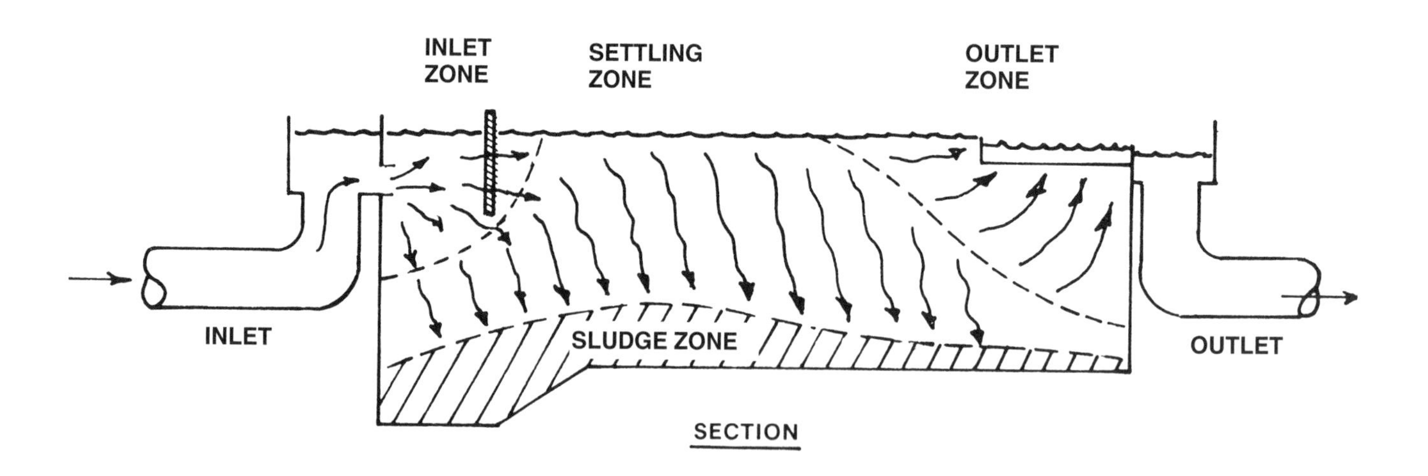

Fig. 4.6 Four zones of a settling tank

1. *TIME PROVIDED FOR SETTLING*

   Commonly, a total settling time of two to four hours is provided. Related to the settling time is the flow velocity through the tank. The faster the water flows, the shorter time a particle has to settle out before the water reaches the outlet of the basin. The theoretical average flow velocity is usually limited to about one to three feet per minute (0.3 to 1 m/min). See Examples 5 and 6 on page 192.

2. *CHARACTERISTICS OF THE SUSPENDED MATTER*

   A dense particle of soil or sand will settle more quickly than a light floc particle. A dense, compact floc containing lots of suspended matter will settle quicker than a light, fluffy particle consisting only of alum floc. The coagulation and flocculation processes determine the settling characteristics of the floc.

3. *DEGREE OF SHORT-CIRCUITING THROUGH THE TANK*

   Short-circuiting occurs when some portions of water pass through the settling tank faster than others. The amount of short-circuiting depends on the shape and dimensions of the tank. For this reason, a long narrow settling tank is more effective than a short, wide tank. Short-circuiting is also affected by the inlet and outlet arrangements of the tank.

4. *TANK INLET AND OUTLET ARRANGEMENTS*

   The inlet arrangement should distribute the incoming water over the full cross section of the settling tank. The water should be distributed evenly both horizontally and vertically, and high velocities and eddy currents should be avoided. This is usually done by using perforated baffles or channels with multiple openings.

   The outlet arrangement should allow the water leaving the tank to be collected near the surface, but uniformly across the width of the tank. The aim is to prevent high flow velocities that will lift already settled particles and carry them out of the tank. Long weir troughs are commonly used to collect the water as well as pipes with multiple ports submerged just below the surface.

   Relatively minor changes in the inlet or outlet arrangement in a settling tank can cause dramatic improvement in its settling efficiency.

5. *TANK OVERFLOW RATE*

   The tank overflow rate, sometimes called the surface loading rate, is determined by dividing the flow rate, Q, by the surface area of the tank. The overflow rate is normally expressed as gallons per day per square foot (GPD/sq ft or cu m/day/sq m). Typical values vary from 400 to 600 GPD/sq ft or 16 to 24 cu m/day/sq m. A lower rate is better than a higher one. See Example 6.

6. *CURRENTS IN THE TANK*

   Currents caused by flow inertia, wind action, temperature differences, and poor design can interfere with efficient settling, resuspend settled particles, and cause short-circuiting. The movement of mechanical sludge scrapers also causes currents that interfere with settling. Many small plants omit sludge scrapers to save costs.

7. *TEMPERATURE OF THE WATER*

   Any particle settles faster in warm water than in cold water because warmer water is less viscous (like warm syrup) and offers less resistance.

8. *WIND*

   High winds can cause surface currents to develop and cause turbulence which will reduce settling.

The water leaving the settling tank should have a turbidity of less than five *TURBIDITY UNITS*[14] and not many floc particles should be carried out of the tank. If good settling is not being obtained, take the following actions to correct it.

1. Check the coagulation and flocculation processes to be sure they are operating as well as possible.
2. Decrease the rate of flow through the tank to lower the overflow rate and the velocity of flow, and to increase the time for settling to occur.
3. Improve inlet conditions to reduce inlet velocity, distribute the flow uniformly, and create uniform flow velocities across the entire cross section of the tank. If more than one tank is used, make sure the flow is equally divided among the tanks as well.
4. Improve outlet conditions to eliminate excessive velocities toward the outlet. Install more weir troughs or outlet ports.
5. Remove accumulated sludge from the bottom of the settling tank.
6. Cover or screen the settling tank to diminish currents caused by changing wind and weather conditions.
7. Recycle sludge to the inlet of the settling tank to increase the number of particles in the water and improve flocculation of the settling particles.

High-rate or tube settlers were developed to increase the settling efficiency of conventional rectangular sedimentation

[14] *Turbidity Units (TU). Turbidity units are a measure of the cloudiness of water. If measured by a nephelometric (deflected light) instrumental procedure, turbidity units are expressed in nephelometric turbidity units (NTU) or simply TU. Those turbidity units obtained by visual methods are expressed in Jackson Turbidity Units (JTU) which are a measure of the cloudiness of water; they are used to indicate the clarity of water. There is no real connection between NTUs and JTUs. The Jackson turbidimeter is a visual method and the nephelometer is an instrumental method based on deflected light.*

basins. They have been installed in circular basins with successful results.

Water enters the inclined settler tubes and is directed upward through the tubes as shown in Figures 4.7, 4.8, and 4.9. Each tube functions as a shallow settling basin. Together, they provide a high ratio of effective settling surface area per unit volume of water. The settled particles can collect on the inside surfaces of the tubes or settle to the bottom of the sedimentation basin.

High-rate settlers are particularly useful for water treatment applications where site area is limited, in package-type water treatment units, and to increase the capacity of existing sedimentation basins. In existing rectangular and circular sedimentation basins, high-rate settler modules can be conveniently installed between the launders. High winds can have an adverse effect on tube settlers.

Water treatment plant sludges are typically alum sludges, with solids concentrations varying from 0.25 to 10 percent when removed from the basin. In gravity flow sludge removal systems, the solids concentration should be limited to 3 percent. If the sludge is to be pumped, solids concentrations as high as 10 percent can be transported.

In horizontal-flow sedimentation basins preceded by coagulation and flocculation, over 50 percent of the floc will settle out in the first third of the basin length. Operationally, this must be considered when establishing the frequency of operation of sludge removal equipment. Also you must consider the volume or amount of sludge to be removed and the sludge storage volume available in the basin.

Sludge may be discharged into sludge basins or ponds for liquid-solids separation. Ultimately the sludge may be disposed of in a landfill.

For additional information on settling, see Chapter 5, "Sedimentation," in *WATER TREATMENT PLANT OPERATION*, Volume I, in this series of manuals.

FORMULAS

To calculate the settling time or detention time in a basin:

1. Determine the basin dimensions and volume, and
2. Measure and record the flow of water being treated.

$$\text{Basin Volume, cu ft} = (\text{Length, ft})(\text{Width, ft})(\text{Depth, ft})$$

$$\text{Basin Volume, gal} = (\text{Basin Volume, cu ft})(7.48\text{ gal/cu ft})$$

$$\text{Detention Time, hr} = \frac{(\text{Basin Volume, gal})(24\text{ hr/day})}{\text{Flow, gal/day}}$$

To calculate the basin overflow rate or surface loading:

1. Determine the basin length and width to determine the surface area, and
2. Measure and record the flow of water being treated.

$$\text{Surface Area, sq ft} = (\text{Length, ft})(\text{Width, ft})$$

$$\text{Overflow Rate, GPD/sq ft} = \frac{\text{Flow, gal/day}}{\text{Surface Area, sq ft}}$$

EXAMPLE 5

A rectangular settling basin 20 feet long, 10 feet wide, and with water 5 feet deep treats a flow of 60,000 gallons per day. Estimate the detention time or settling time in hours for this basin when conveying this flow.

Known		Unknown
Length, ft	= 20 ft	Detention Time, hr
Width, ft	= 10 ft	
Depth, ft	= 5 ft	
Flow, GPD	= 60,000 GPD	

1. Calculate the basin volume in gallons.

$$\begin{aligned}\text{Basin Volume, cu ft} &= (\text{Length, ft})(\text{Width, ft})(\text{Depth, ft})\\ &= (20\text{ ft})(10\text{ ft})(5\text{ ft})\\ &= 1{,}000\text{ cu ft}\end{aligned}$$

$$\begin{aligned}\text{Basin Volume, gal} &= (\text{Basin Volume, cu ft})(7.48\text{ gal/cu ft})\\ &= (1{,}000\text{ cu ft})(7.48\text{ gal/cu ft})\\ &= 7{,}480\text{ gal}\end{aligned}$$

2. Calculate the detention time or settling time in hours.

$$\begin{aligned}\text{Detention Time, hr} &= \frac{(\text{Basin Volume, gal})(24\text{ hr/day})}{\text{Flow, gal/day}}\\ &= \frac{(7{,}480\text{ gal})(24\text{ hr/day})}{60{,}000\text{ gal/day}}\\ &= 3.0\text{ hr}\end{aligned}$$

EXAMPLE 6

A rectangular settling basin 20 feet long, 10 feet wide, and 5 feet deep treats a flow of 60,000 gallons per day. Estimate the overflow rate of the basin in gallons per day per square foot of surface area.

Known		Unknown
Length, ft	= 20 ft	Overflow Rate, GPD/sq ft
Width, ft	= 10 ft	
Depth, ft	= 5 ft	
Flow, GPD	= 60,000 GPD	

1. Calculate the surface area in square feet.

$$\begin{aligned}\text{Surface Area, sq ft} &= (\text{Length, ft})(\text{Width, ft})\\ &= (20\text{ ft})(10\text{ ft})\\ &= 200\text{ sq ft}\end{aligned}$$

2. Determine the basin overflow rate in gallons per day per square foot.

$$\begin{aligned}\text{Overflow Rate, GPD/sq ft} &= \frac{\text{Flow, gal/day}}{\text{Surface Area, sq ft}}\\ &= \frac{60{,}000\text{ gal/day}}{200\text{ sq ft}}\\ &= 300\text{ GPD/sq ft}\end{aligned}$$

QUESTIONS

Write your answers in a notebook and then compare your answers with those on page 259.

4.3A What is the purpose of settling?

4.3B Short-circuiting is influenced by what factors?

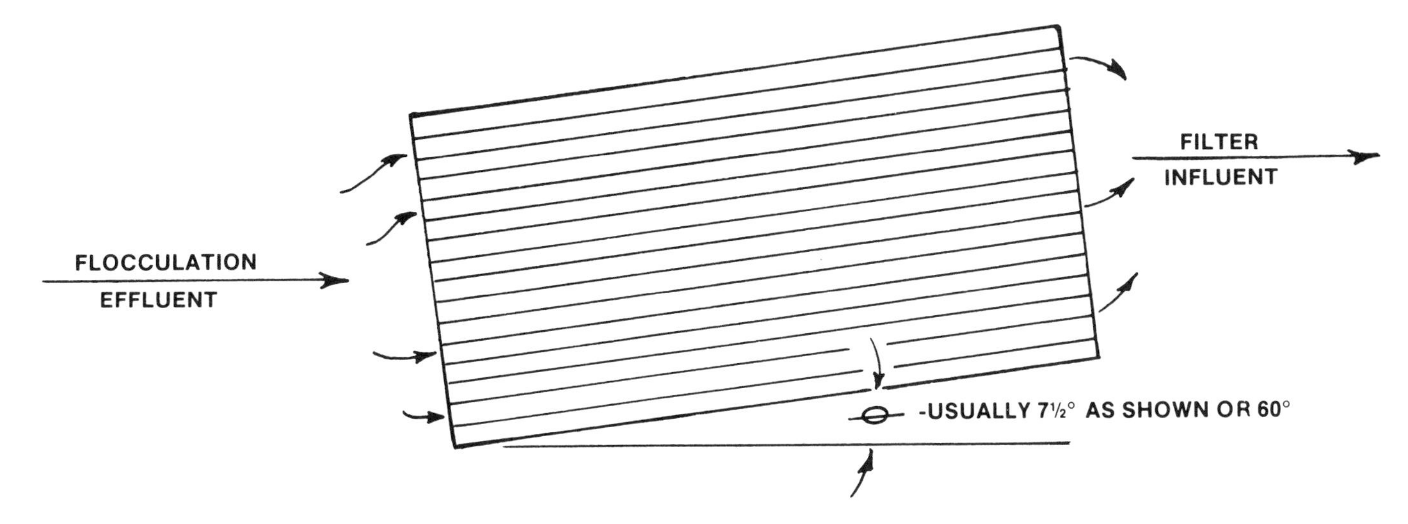

Fig. 4.7 Tube settler (installed in a rectangular or circular sedimentation basin)

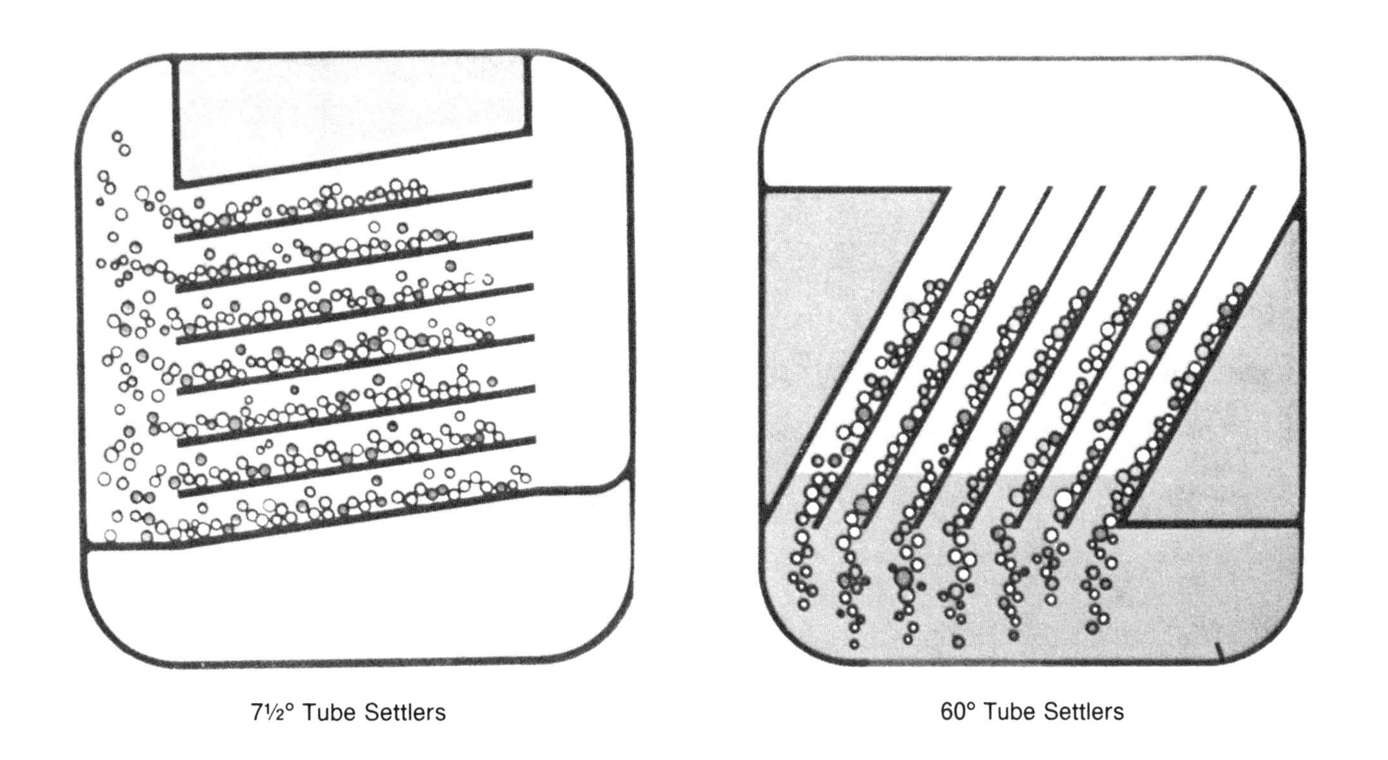

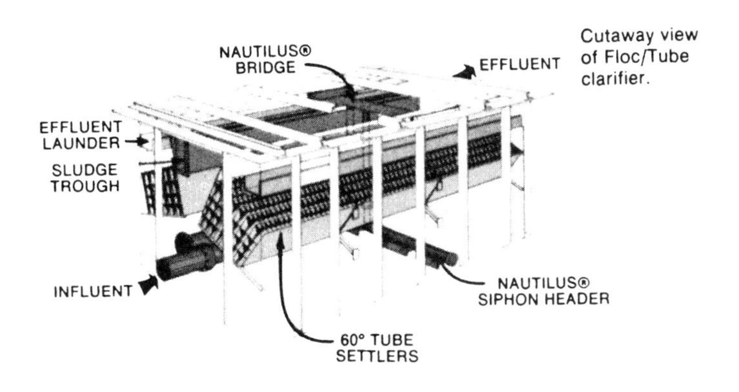

Fig. 4.8 Cutaway view of Floc/Tube clarifier

(Permission of Neptune Microfloc, Inc.)(Now Microfloc Products)

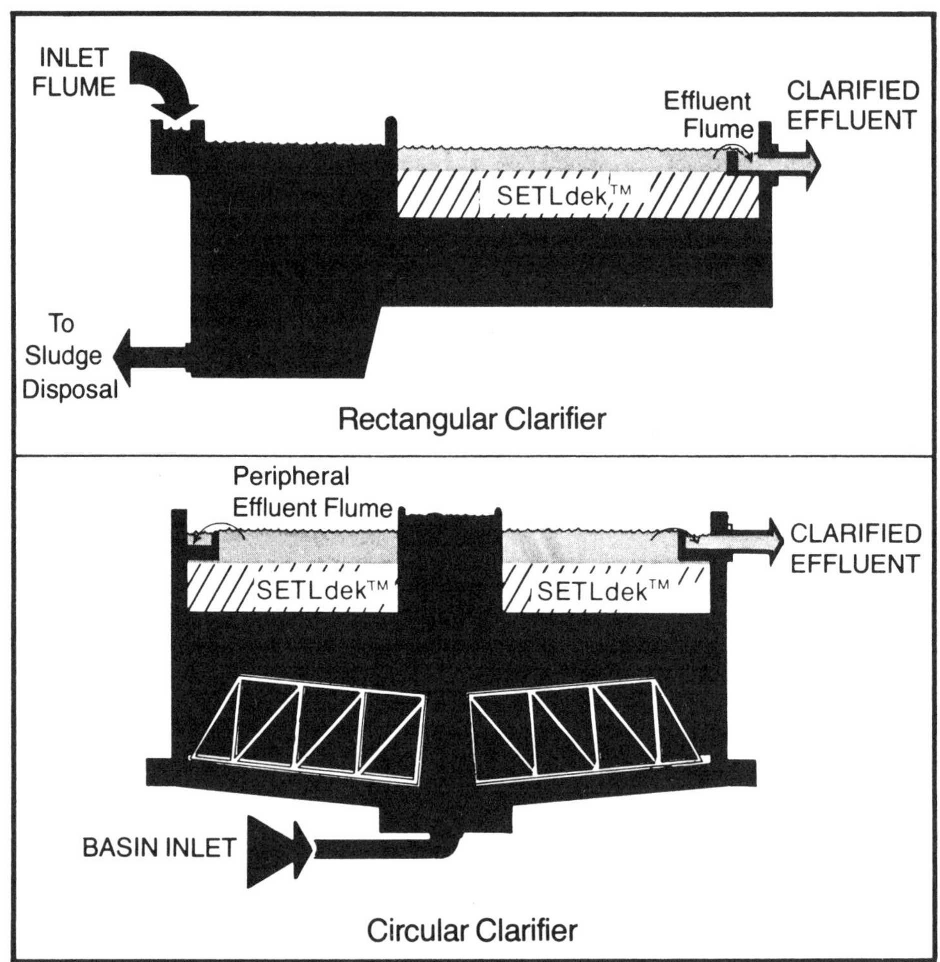

Fig. 4.9 Tube settlers in rectangular and circular clarifiers
(Permission of The Munters Corporation)

4.3C How does temperature influence particle settling?

4.3D Why is sludge recycled to the inlet of the settling tank?

4.4 FILTRATION

Filtration is the process of passing water through a porous bed of fine granular material to remove suspended matter from the water. The suspended matter is mainly particles of floc, soil, and debris; but it also includes living organisms like algae, bacteria, viruses, and protozoa.

There are generally two types of filters used in water treatment. Both the gravity and pressure types are common. The traditional open gravity filter is shown in Figure 4.10. This filter is simply a tank with an open top which contains the water, the filtering media, and other filter equipment. Water enters the filter near the top and flows downward through the filtering media under the force of gravity, hence the name "gravity filter." At the bottom, the filtered water is collected in the underdrain system and flows out of the filter to the filtered water reservoir.

Slow sand filtration is a variation of the gravity filtration process. The physical structure of the filter resembles a standard gravity filter but the slow sand filter uses biological processes as well as physical straining to remove suspended particles from the water. Although slow sand filters have not often been used in the United States, this method of filtration appears to be well suited to meet the needs of small water supply systems. Most small systems are now required to filter their water to comply with the Safe Drinking Water Act (SDWA) of 1986 and the Surface Water Treatment Rule (1989). Because of the growing interest in this technology, slow sand filtration will be discussed fully in Section 4.8 of this chapter.

The second common type of filter is the pressure filter (Figure 4.11). The pressure filter is completely enclosed in a watertight tank and the water is forced through the filter under pressure created by an elevated source, a pump, or other pressure source. When a bed of granular filtering media is used, the water enters the top of the filter tank and flows vertically downward as it does in a gravity filter. The underdrain system is also similar. Some types of pressure filters, called precoat filters (Figure 4.12), use *DIATOMACEOUS EARTH*[15] (D.E.) temporarily deposited on fabric or wire screen elements to accomplish the filtering action. Thus, the interior arrangement of a pressure filter can vary considerably from one manufacturer to another. Pressure filters fre-

[15] *Diatomaceous (DYE-uh-toe-MAY-shus) Earth. A fine, siliceous (made of silica) "earth" composed mainly of the skeletal remains of diatoms.*

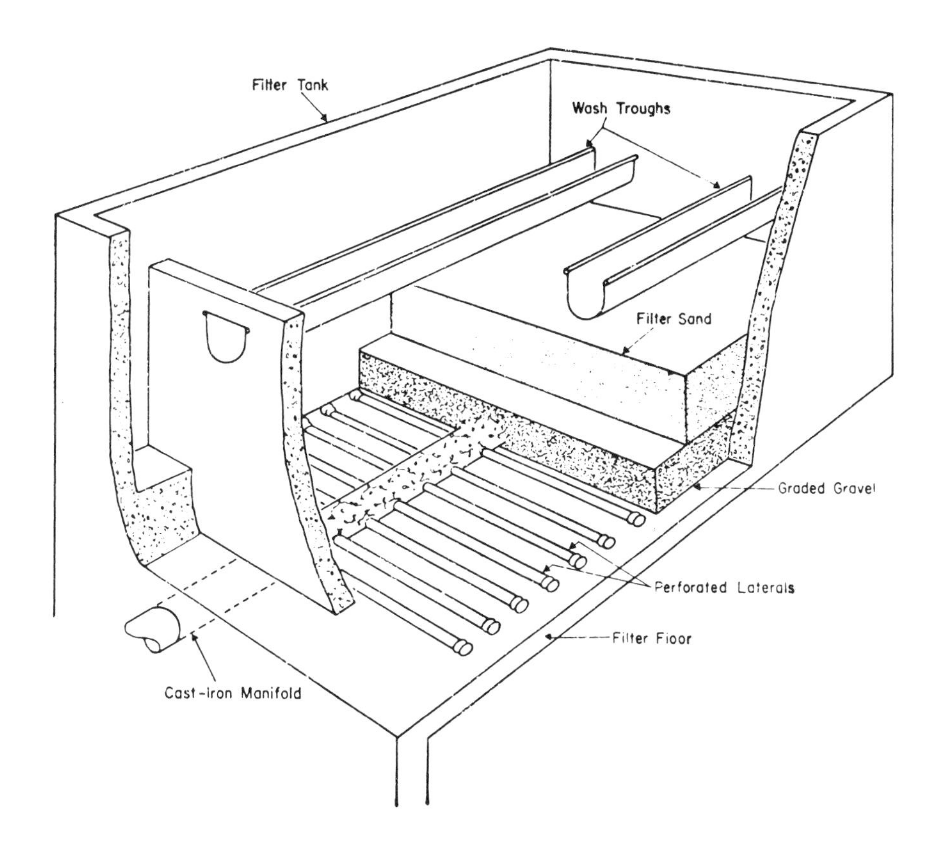

Fig. 4.10 Cutaway view of typical open gravity sand filter

(Reprinted from *WATER TREATMENT PLANT DESIGN*, by permission. Copyright 1969, the American Water Works Association)

quently offer lower installation and operation costs in small filtration plants; *HOWEVER*, they are generally somewhat less reliable than gravity filters (depending upon pressure). Some states do not recommend the use of pressure filters for treating surface waters.

Filtration is the final step in the overall solids removal process which usually includes the pretreatment processes of coagulation, flocculation, and sedimentation. The configuration of these processes, or mode of operation, varies from plant to plant with respect to which steps are included in the process. Three modes (or methods) are common: conventional, direct, and diatomaceous earth. A fourth mode, slow sand filtration, is practiced on a limited scale in the United States, but is also an acceptable method for purposes of meeting regulatory requirements.

CONVENTIONAL FILTRATION: This mode of water treatment is suited for water sources where quality is highly variable or is high in suspended solids, and large volumes of water are required. This process is used in most municipal treatment plants in the United States. A conventional filtration plant commonly operates at filter rates of 2 to 3 GPM/sq ft, but can function up to 10 GPM/sq ft. This mode includes "complete" pretreatment (coagulation, flocculation, and sedimentation), and provides a great amount of flexibility and reliability in plant operations. A chemical application point just prior to filtration permits the application of a filter-aid chemical (such as a nonionic polymer) to assist in the solids removal process, especially during periods of pretreatment process upset, or when operating at high filtration rates.

DIRECT FILTRATION: Direct filtration is the same as conventional filtration except that sedimentation is omitted. Direct filtration is considered a feasible alternative to conventional filtration, particularly when source waters are low in turbidity, color, plankton, and coliform organisms. Filtration rates are usually in the range of 2 to 5 GPM/sq ft. As in conventional filtration, a chemical application point just prior to filtration permits the addition of a filter-aid chemical. Many direct filtration plants provide rapid mix, short detention without agitation (30 to 60 minutes) followed by filtration. Other direct filtration plants practice what is known as in-line filtration, which is the same as direct filtration without separate flocculation. With in-line operation, chemical filter aids are added directly to the filter inlet pipe and are mixed by the flowing water.

DIATOMACEOUS EARTH: In diatomaceous earth (precoat) filtration the filter media is added as a *SLURRY*[16] to the water being treated; it then collects on a septum (a pipe conduit with porous walls) or other appropriate screening device as shown in Figure 4.12. After the initial precoat application, water is filtered by passing it through the coated screen. The coating thickness may be increased during the filtration process by

[16] *Slurry (SLUR-e). A watery mixture or suspension of insoluble (not dissolved) matter; a thin, watery mud or any substance resembling it (such as a grit slurry or a lime slurry).*

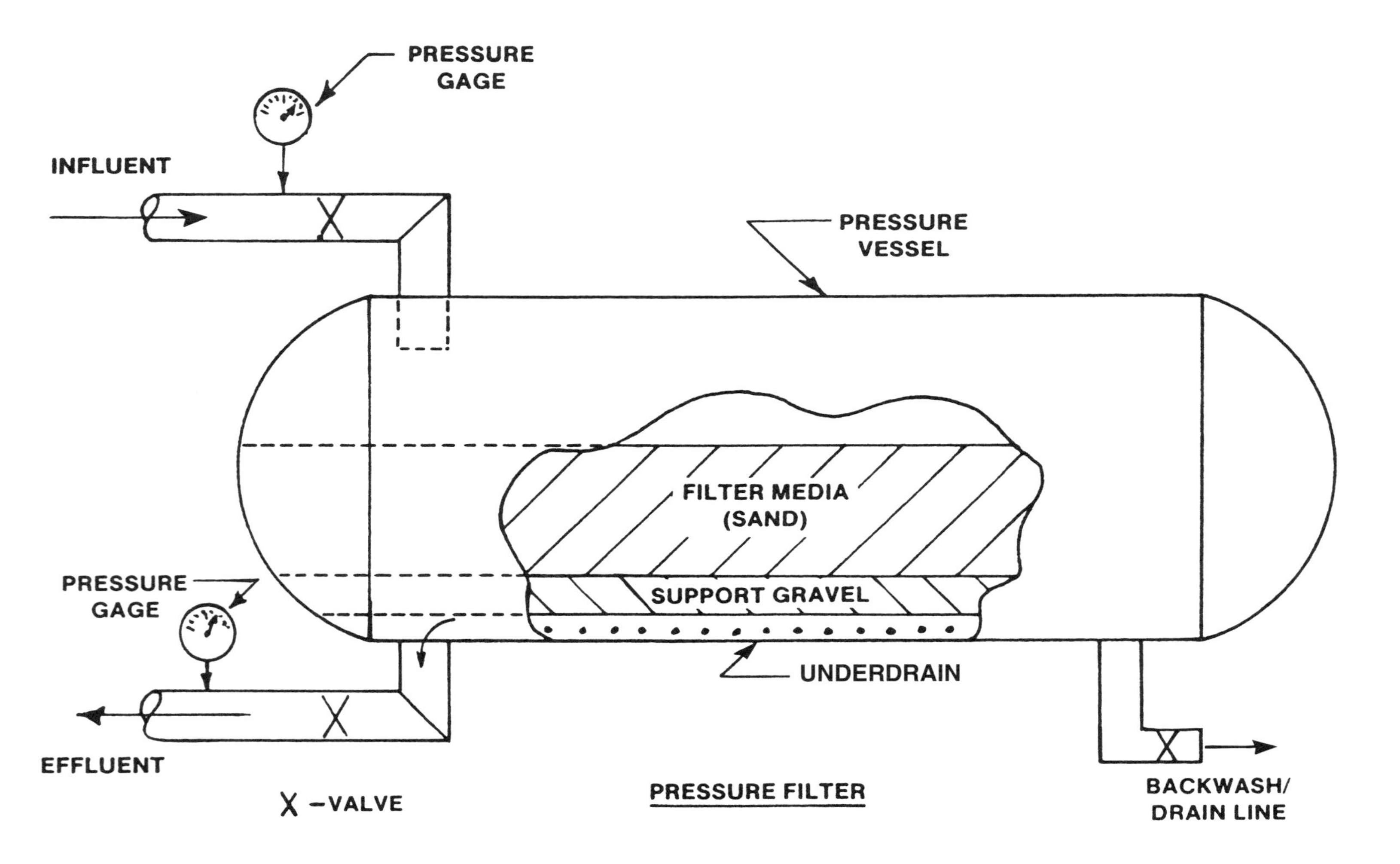

Fig. 4.11 Pressure filter

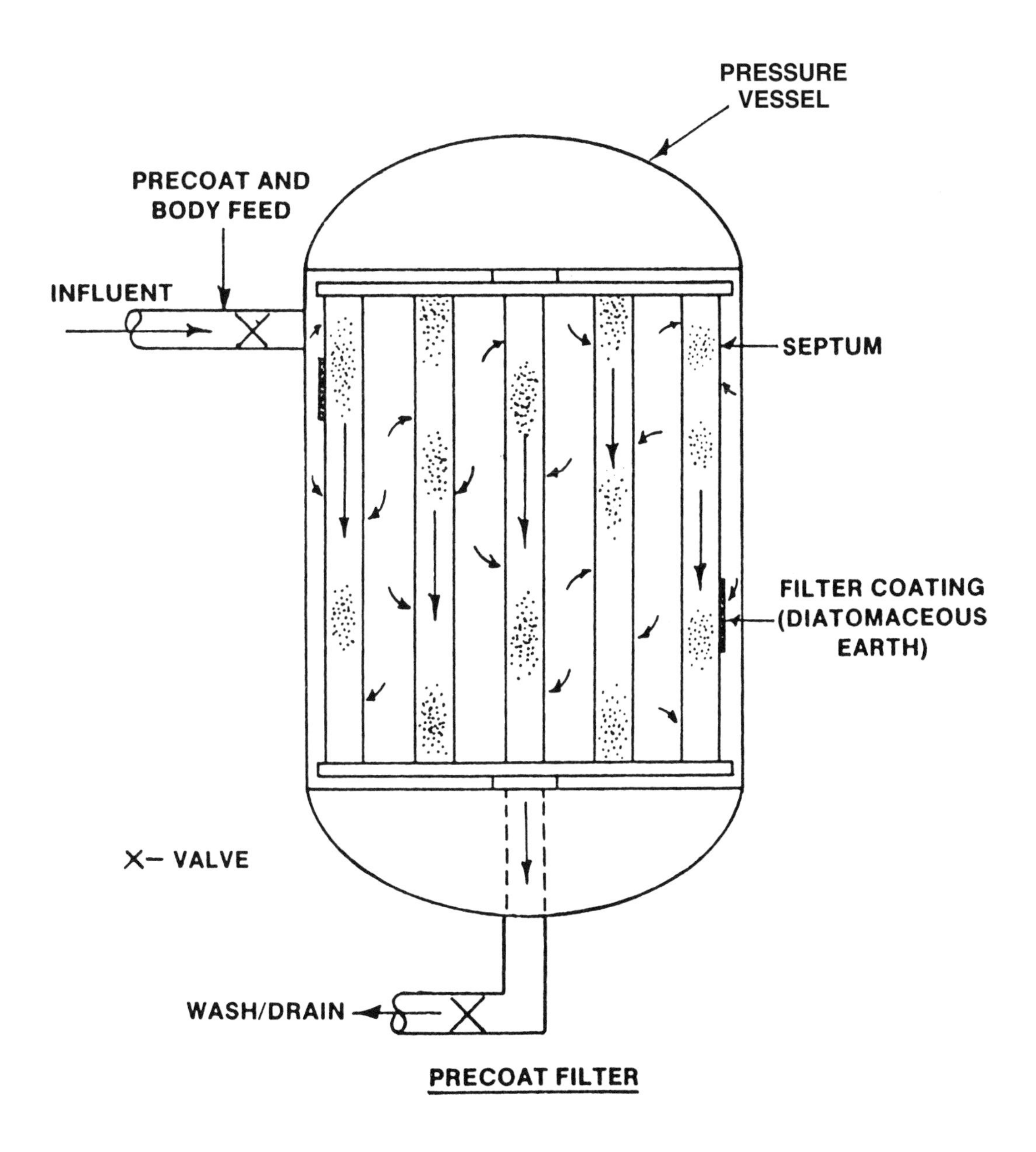

Fig. 4.12 Diatomaceous earth filter

gradually adding more media — a body feed. In most water treatment applications, diatomaceous earth is used for both the precoat and body-feed operations.

Diatomaceous earth filtration is primarily a straining process, and finds wide application where very high particle removal efficiencies (high clarity water) are required, such as in the beverage and food industries. Precoat filters can be operated as gravity, pressure, or vacuum filters. They are also commonly used in swimming pool installations due to their small size, efficiency, ease of operation, and relatively low cost. Use of these filters is limited in larger water treatment plants because of operational considerations such as flow rates and sludge (used diatomaceous earth) disposal requirements.

SLOW SAND: In slow sand filtration, water is drawn downward through the filter media (sand) by gravity as it is in the gravity filtration process. However, this is generally where the similarity between these two filtration processes ends. In the slow sand filtration process, particles are removed by straining, *ADSORPTION*,[17] and biological action. Filtration rates are extremely low (0.015 to 0.15 GPM/sq ft or 0.01 to 0.1 liters per sec/sq m or 0.01 to 0.1 mm/sec). This process usually does not require pretreatment for most surface waters. The majority of the particulate material is removed in the top few inches of sand, so this entire layer must be physically removed, rather than backwashed, when the filter becomes clogged. This filtration process has found limited application due to the large area required and the need to manually clean the filters, but its efficiency in removing or inactivating disease-causing organisms has sparked renewed interest in the process.

CHEMICAL COAGULATION IS ABSOLUTELY ESSENTIAL FOR EFFECTIVE FILTRATION except when using the slow sand process. Filters using a granular filter media will not effectively remove fine suspended matter unless coagulation treatment is provided. Without coagulation, the filter will operate only as a strainer which removes the large, coarse particles while the fine particles of suspended matter will easily flow right between the grains of filter media and pass through the filter without being removed. *ONLY PROPER COAGULATION CAN MAKE A GRANULAR MEDIA FILTER PERFORM EFFECTIVELY TO CLARIFY THE WATER.*

Sand is the most frequently used filtering media for treatment of domestic water. The sand should be a hard material like quartz so it will not erode and crumble or easily dissolve in water. The depth of sand may vary from 20 to 30 inches (50 to 75 cm). A depth of 24 inches (60 cm) is common but some filters of special design may have only 6 to 12 inches (15 to 30 cm) of sand.

Raw sand from just any source is not suitable for filter media. This sand usually is not the right size and it is not properly "graded," that is, the different sizes are not present in the correct proportion. Therefore, filter sand must be prepared by screening the raw sand through a set of sieves and mixing the sand retained on each size sieve in the proper proportion. When this mixing operation is completed, the sand should consist of grains ranging in size from about 0.2 to 1.2 millimeters.

Anthracite is the second most frequently used filtering media. Anthracite is hard coal. Ordinarily anthracite is prepared by crushing the coal and sieving it to achieve the proper gradation. Properly graded anthracite will consist of grains ranging in size from about 0.5 to 3.0 millimeters.

Anthracite may be used as the only filtering media but it is more commonly used along with sand. When the two materials are used together, the filter is called a "dual media" filter. In the dual media filter, about 6 to 10 inches (15 to 25 cm) of sand is placed on the bottom and 18 to 24 inches (45 to 60 cm) of anthracite is placed on top. Since the sand is about two and a half times heavier than water and anthracite is only about one and a half times heavier than water, the two materials stay separated and in the same relative positions (anthracite on top and sand on the bottom) even after backwashing.

Sometimes a layer of crushed *GARNET*[18] media is used at the bottom of the filter in addition to the sand and anthracite. In this case, the filter is called a "mixed media" filter.

Small filters containing replaceable cartridge-type elements are sometimes used in homes or very small treatment plants. When the elements become plugged with dirt they must be discarded and replaced since they cannot be cleaned and reused. The cost of replacement elements is often a significant expense.

The underdrain system of the filter is at the bottom beneath the filtering media. This system has several important functions:

1. To support the filtering media and prevent the media from passing out the bottom of the filter,
2. To collect the filtered water and convey this water out of the filter when the filter is in normal operation, and
3. To uniformly distribute the backwash water across the filter bed and provide uniform upward flow.

Of these three functions, the most important is uniform distribution of the backwash water.

The underdrain system usually consists of two parts, the gravel layers and the water conduits. The gravel layers, which are on top, usually consist of several layers of different size gravel. The uppermost layer of gravel, which is in contact with the sand, may be $^1/_{16}$ to $^1/_8$ inch (1.5 to 3 mm) in diameter and about three inches (75 mm) thick. There may be as many as

---

[17] *Adsorption (add-SORP-shun). The gathering of a gas, liquid, or dissolved substance on the surface or interface zone of another material.*

[18] *Garnet. A group of hard, reddish, glassy, mineral sands made up of silicates of base metals (calcium, magnesium, iron and manganese). Garnet has a higher density than sand.*

six different layers of gravel, each larger in size than the one above. The bottom layer may be as large as 1 1/2 to 2 1/2 inches (35 to 65 mm) in diameter. The total depth of the gravel layers may vary from 12 to 24 inches (30 to 60 cm).

The bottom part of the underdrain system consists of the water conduits for collection of the filtered water and distribution of the backwash water. These conduits may consist of a large central header pipe with numerous small pipe laterals branching off the header. The header-lateral underdrain system is more common in older or smaller filters. Newer and larger filters most often use an underdrain system consisting of a false bottom constructed of porous stone plates or perforated hollow blocks of vitrified clay.

A rate-of-flow control valve orifice is normally placed in the effluent line from each filter. The purpose of this valve is to somewhat equalize the flow among all the filters and to prevent excessive rates of filtration by limiting the rate of flow through each filter to a predetermined value. The rate-of-flow controller also is used to maintain a constant flow through a given filter during the entire filter run. As the head loss through the filter builds up, the controller opens up to maintain a constant flow.

The rate of filtration, usually expressed in gallons per minute per square foot of filter area (GPM/sq ft), is an important characteristic of any filter. The slow sand filters built in the past operated at a filtration rate of about 0.05 GPM/sq ft (0.034 mm/sec or 0.034 liter per sec/sq m). When rapid sand filters were developed, they were usually designed to operate at a filtration rate of two GPM/sq ft (1.4 mm/sec or 1.4 liters per sec/sq m), approximately 40 times faster. Modern filters using dual media and improved design can operate at rates ranging from three to five GPM/sq ft (2 to 3.4 mm/sec or 2 to 3.4 liters per sec/sq m). In a very few cases, filters have been designed and operated successfully at rates ranging from six to eight GPM/sq ft (4 to 5.4 mm/sec) or even as high as 10 GPM/sq ft (6.8 mm/sec or 6.8 liters per sec/sq m), but these filters are rare.

Successful operation of filters at high rates requires the following conditions:

1. Good design by a knowledgeable engineer or consultant,
2. Use of dual or mixed media,
3. Surface wash apparatus to assist cleaning of the media during backwash,
4. Use of polymers or other filter aids,
5. Turbidimeters which continuously monitor the performance of the filters,
6. Competent operators on duty continuously whenever the plant is operating, and
7. Excellent pretreatment (coagulation, flocculation, and settling) to properly condition the water prior to filtration.

Since these conditions usually do not prevail in small water treatment plants, the rate of filtration should usually be restricted in small filters, or ones of poor design, to no more than about two GPM/sq ft (1.4 mm/sec or 1.4 liters per sec/sq m). See Example 7 on page 202.

As a filter operates, it continuously removes suspended matter from the water passing through it. Eventually this matter clogs the openings through the filter and the flow of water is reduced. Another sign of a dirty filter is "breakthrough" of excessive turbidity in the filtered water.

When a filter is operating properly, the head loss through the filter media will build up gradually (Figure 4.13). The actual head loss is measured by a filter head loss gage which is monitored by the operator. The head loss should never be as great as the distance from the water surface to the top of the filter media.

The rate of head loss buildup is an important indicator of filter performance. Sudden increases in head loss might be an indication of surface sealing of the filter media (lack of depth penetration). Early detection of this condition may permit you to make appropriate process changes such as adjustment of the chemical filter-aid feed rate or adjustment of filtration rate.

When the filter becomes dirty (high head loss) it must be cleaned by backwashing. This is a process of reversing the direction of flow through the filter to flush the dirt out of the media. The backwash water enters the underdrain system at the bottom of the filter and flows upward through the filter media. The flow of backwash water is regulated to a rate that will separate the individual grains of media and suspend them in the flow of water. When the media is thus expanded, the grains scrub each other and the dirt particles are flushed out. Backwashing continues until the filter is clean, usually about five to ten minutes are required.

In gravity filters, the backwash water is collected at the top of the filter in troughs that drain it away to waste. In pressure filters, the backwash water is piped to waste by manipulating the proper valves.

The backwash water supply system should provide a maximum upward flow rate of 15 to 20 GPM/sq ft (10 to 13.6 mm/sec or 10 to 13.6 liters per sec/sq m). This is common for sand media. The actual flow rate required for satisfactory backwash will depend on the size and specific gravity of the media and the temperature of the water. Thus a lower flow rate will be necessary for anthracite than for sand. And a lower flow rate will be necessary when the water is cold than when it is warm. In any case, backwash flow must be adequate to lift the individual grains of filter media and expand the filter bed so the foreign matter collected in the bed can be flushed out. Separating the grains of media allows them to scrub each other clean of the coating of floc or dirt. Typically the filter bed will be expanded to 120 to 150 percent of its normal depth. However, the operator must be careful that the flow rate is not so high that the filter media itself is flushed out of the filter and lost during backwashing. See Example 9.

The frequency of backwashing may vary from a few hours to a few days depending on how quickly the filter plugs or on the quality of the filtered water. The filter head loss gages indicate how quickly the filters are plugging and when they need backwashing. Under unfavorable conditions when there is lots of suspended matter in the water being filtered, the filter will plug rapidly and may require backwashing several times a day. At other times when the water to be filtered contains only small amounts of suspended matter, the filter will plug slowly and it may operate from several days up to a week before backwashing is required.

However, backwashing is also required whenever the filtered water quality becomes unacceptable. If the turbidity of the filtered water suddenly increases above the level permitted by standards, backwashing is necessary even though flow through the filter is not seriously affected by plugging.

Operators of small filter plants try to achieve the longest possible filter runs, use the minimum amount of backwash water, and try to produce the best possible filtered water quali-

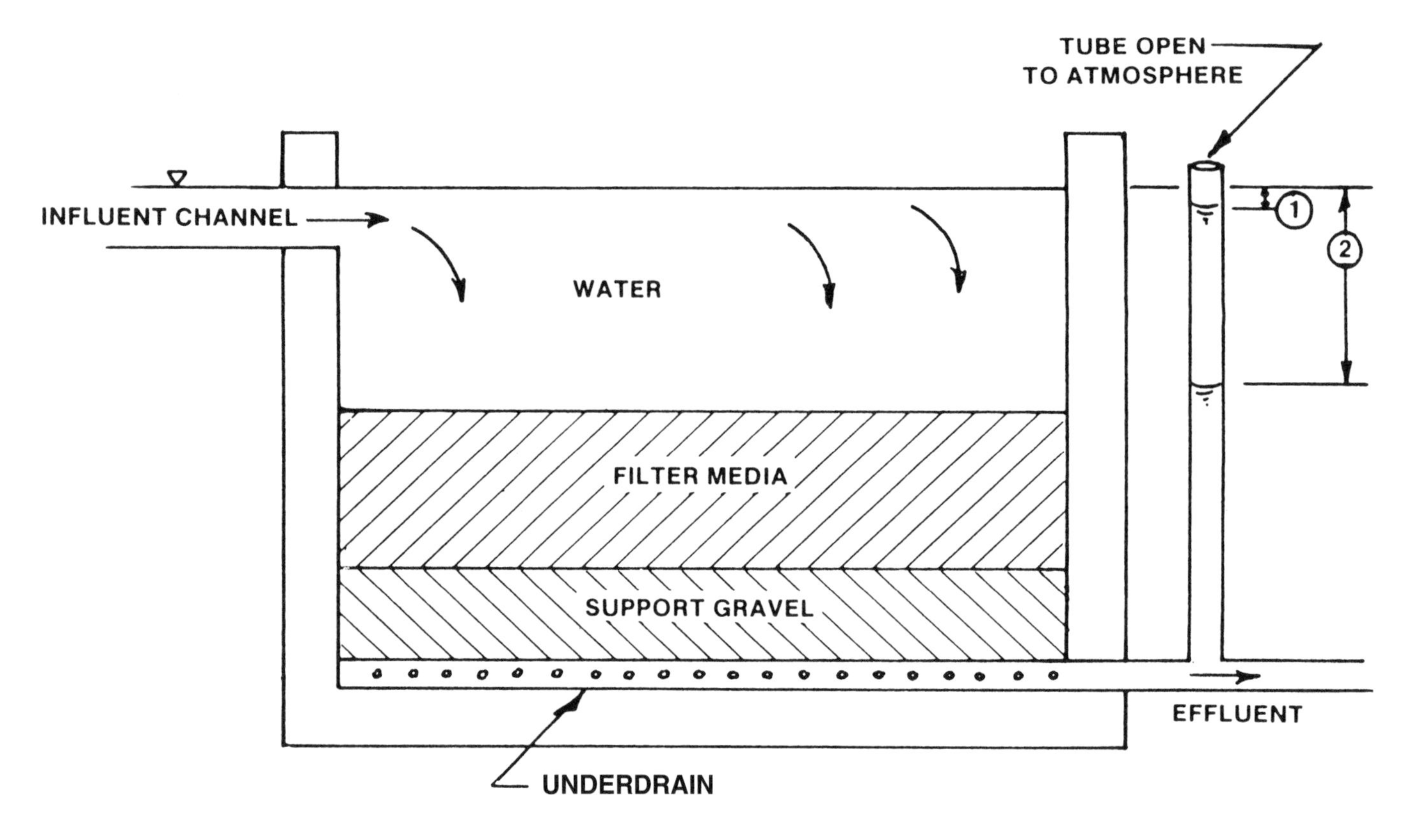
TUBE OPEN TO ATMOSPHERE
INFLUENT CHANNEL
WATER
1
2
FILTER MEDIA
SUPPORT GRAVEL
EFFLUENT
UNDERDRAIN
NOTE: IF A TUBE OPEN TO THE ATMOSPHERE WAS INSTALLED IN THE FILTER EFFLUENT, THEN
1. HEAD LOSS THROUGH FILTER AT START OF RUN, AND
2. HEAD LOSS THROUGH FILTER BEFORE START OF BACKWASH CYCLE.

Fig. 4.13 Head loss through a filter

ty (turbidities of less than 0.1 turbidity unit is a recommended target). For good operation, you should not have to use more than two percent of your filtered water for backwashing. Depending on the quality of the source water, daily backwashing may be often enough. In some cases filters may be operated two or three days without backwashing, but longer periods are not recommended.

Surface wash jets are used on modern filters to improve the efficiency of backwashing. Surface wash is especially good for dual media and mixed media filters. The water jets may be either on a fixed grid or on a rotating arm suspended above the surface of the filtering media. Compressed air is sometimes supplied with the water to the jets. The purpose of the surface wash jets is to provide additional agitation to break up any crust or *MUDBALLS*[19] that may form at the surface of the filter bed and to improve cleaning of the media. Surface washing can be very helpful and should be used whenever available. Therefore, if a permanent system of surface wash jets is not provided, the operator should use a hand-held hose with a fire nozzle to apply a powerful blast of water to the top of the filter media. The blast of water should be applied to the exposed media before and when the backwash water first starts to flow to assist cleaning of the media.

Only fully treated water should be used for filter backwashing. Untreated or partially treated water should not be used because it will contaminate the filter media and underdrain system, and it may leave the filter more dirty after backwashing than it was before. In gravity filter plants, the treated water for backwashing may be supplied by gravity flow from a backwash water storage tank or by special backwash pumps which draw water from the clear well. A third alternative for backwash water supply is gravity flow back from the distribution system itself.

In pressure filter plants, the backwash water is most commonly produced by the filters themselves as it is used. In this mode of operation, the treated water produced by three of the pressure filters is used to backwash the fourth. The filters are backwashed in sequence, one after another, until all are clean. This scheme requires the plant to be out of production while the filters are backwashing, but normally this is not a serious drawback. Figure 4.14 shows the in-service and backwash flows for a pressure filter plant.

The waste backwash water is usually very muddy and it must be disposed of properly. Frequently the backwash water is simply flushed to waste into the nearest stream or drainage

[19] *Mudballs. Material that is approximately round in shape and varies from pea-sized up to two or more inches in diameter. This material forms in filters and gradually increases in size when not removed by the backwashing process.*

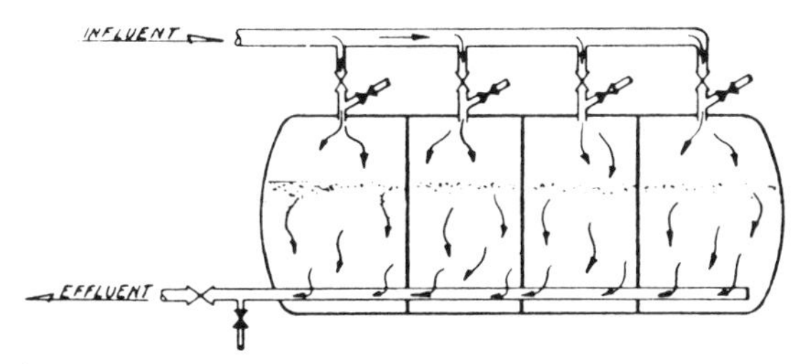

IN SERVICE

The influent water is distributed to each filter compartment, passes evenly through the media, is collected by the underdrain system and leaves the filter through the effluent valve.

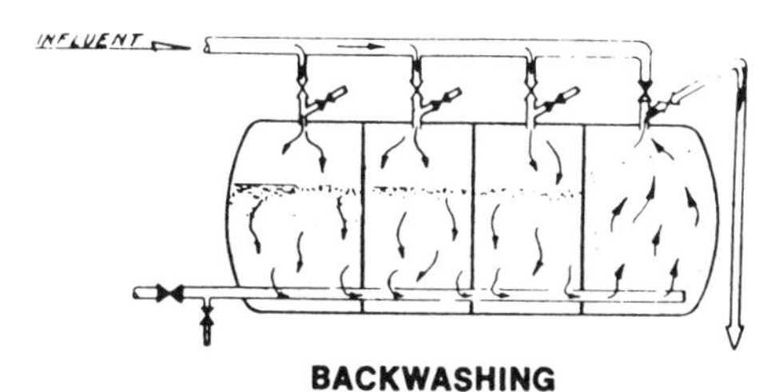

BACKWASHING

The effluent valve is closed. At the same time the influent valve to one compartment is automatically closed. The backwash outlet valve to that compartment is opened. The influent water continues to filter downward through the remaining three compartments and upward through the fourth. This is repeated sequentially until the entire filter has been washed.

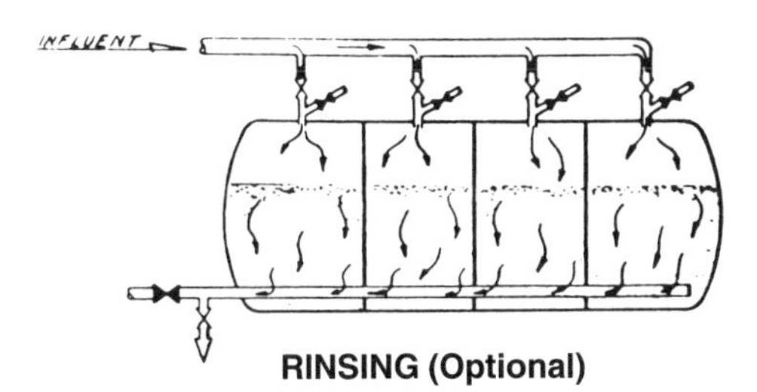

RINSING (Optional)

Before the filter is returned to service a rinse valve is opened allowing rinsing of the entire filter to waste. At the completion of this step the effluent valve is opened as the rinse valve closes, returning the unit to service.

Fig. 4.14 In-service and backwash flows for a four-compartment pressure filter

(Permission of Watermasters, Inc., Burlingame, California)

ditch. This practice is considered a source of pollution and now is no longer recommended. Disposal to a sanitary sewer system is sometimes possible. Many small plants can simply retain the backwash water in a small pond where it will evaporate or seep into the ground. If an evaporation pond is not practical, the backwash water can be contained in a tank or basin until the heavy material settles. Then the clear water can be drawn off and recycled to the inlet of the treatment plant for pretreatment. The settled sludge is periodically removed to a landfill for disposal.

The performance of a filter must be monitored frequently or continuously if possible. Several different manufacturers supply turbidity meters (turbidimeters) that can continuously measure the turbidity of the treated water produced by a filter. Turbidimeters are also available for testing individual water samples collected on a "grab" basis; however, the continuous-reading instruments are much better for monitoring the performance of a filter because they give a rapid indication of any momentary change in quality. This rapid indication feature is indispensable when a polymer is used as a filter aid or when the operator wishes to detect the rapid rise of turbidity in the filtered water which indicates "breakthrough" and the need to backwash.

If a turbidity instrument is not available, the operator should still make an effort to monitor the performance of the filters. A reliable judgment can often be made by simply observing the water in the clear well reservoir which receives the water from the filter. If the bottom of the reservoir is clearly visible through the full depth of water, the turbidity is probably very low. If the water is faintly cloudy and hazy, filter performance is probably substandard. A bright light beam shown through the water is usually a great help in judging the clarity. If the operator routinely observes the water every day, enough experience is soon gained to detect a change for the worse just by visual observation.

A properly operated filter plant should easily produce treated water with a turbidity of less than 1.0 Turbidity Unit and it should usually be less than 0.5 Turbidity Unit. A filtered water turbidity as low as 0.1 Turbidity Unit is achieved regularly by many filter plants which are well designed and operated properly. A filter plant that cannot consistently produce water with a turbidity of less than 1.0 Turbidity Unit is defective either in its design or in its operation, and corrective measures should be undertaken to correct the problem.

Most filters have similar operation and maintenance needs. If the filters are not performing satisfactorily, check the following points and make appropriate corrections.

1. Verify that the coagulation, flocculation, and settling processes are performing as efficiently as possible. The turbidity of the settled water going to the filters should not exceed five Turbidity Units.
2. The rate of filtration should not exceed the design rate. Check the rate of flow controller on each filter to make sure they are preventing excessive filtration rates in any one filter.
3. The filter may need backwashing. Filters must be backwashed frequently, preferably every day. The backwash must effectively clean the filter media. Otherwise mudballs will form in the media and eventually plug the filter completely. Maintain a permanent record of the backwash frequency and duration for each filter. Check the rate of backwashing to make sure it will adequately expand the media and flush out the dirt.
4. The filter must be operated so the rate of flow changes slowly. A rapid increase in the filter rate will very likely cause the filter to produce poorer quality water. For this reason, frequent ON/OFF operation of the filter should be eliminated. In addition, rapid operation of the backwash water valve can cause serious damage to the underdrain system. The backwash valve must always be opened *VERY* slowly.
5. The media must be inspected frequently to determine that it is in good condition. Gravity filters can be inspected easily through the open top, but pressure filters must be opened or taken apart to visually inspect the media. This should be done at least yearly and more often if necessary. Examine the media for loss in depth, mudballs, caking, surface cracks, and mounding or unevenness. Remove the mudballs and replace or replenish the media as required.
6. Determine the condition of the underdrain system by observing the filters during filtering and backwashing. If the gravel layers are upset or the underdrain conduits are broken, the media will escape through the bottom of the filter and appear in the filtered water. Substantial amounts of sand in the filter effluent pipe or filtered water reservoir (clear well) indicates definitely that the underdrain system needs repair.

   If gravel appears near the surface of the media or if the surface of the media is mounded instead of level, the underdrain system should be repaired and new gravel and media installed. Sand "boils" or other evidence of uneven upward flow during backwashing are further evidence of underdrain problems.
7. Accurate gages must be provided to measure the loss of head or the pressure loss between the inlet and the outlet of the filter. These gages give a good idea of how clean the filter is and when it needs backwashing. After a filter has been backwashed, the loss of head through the filter will gradually increase as particles are removed from the water being filtered. These particles will plug the spaces in the filter material and cause the head loss to increase. The operator should maintain a permanent record of the head (pressure) loss in each filter at the beginning and end of each filter run. Variations from past readings indicate a problem.
8. Consider feeding a polymer filter aid at the inlet to the filters. Very small dosages of the proper chemical can often make a startling improvement in the efficiency of filtration.
9. When checking pressure filters with multi-cells, inspect the plates inside the filters for cracks and breaks.

Figure 4.15 shows the installation of the small treatment plant loaded on the pickup shown in Figure 4.1. This plant consists of four vertical pressure filters. Pre- and postchlorination are used for disinfection. Alum is the coagulant. The filters are used to remove turbidity. In some installations this same type of plant is used to remove iron and manganese by changing the media.

For additional information on filtration, see Chapter 6, "Filtration," in *WATER TREATMENT PLANT OPERATION*, Volume I, in this series of operator training manuals.

FORMULAS

To calculate the filtration rate for a rapid sand filter:

1. Determine the dimensions of the filter and surface area, and
2. Measure and record the flow of water being treated.

$$\text{Surface Area, sq ft} = (\text{Length, ft})(\text{Width, ft})$$

$$\text{Flow, gallons/min} = \frac{(\text{Flow, MGD})(1{,}000{,}000/\text{Million})}{(24\ \text{hr/day})(60\ \text{min/hr})}$$

$$\text{Flow, gallons/min} = \frac{(\text{Water Drop, ft})(\text{Surface Area, sq ft})(7.48\ \text{gal/cu ft})}{\text{Time, min}}$$

$$\text{Filtration Rate, GPM/sq ft} = \frac{\text{Flow, gal/min}}{\text{Surface Area, sq ft}}$$

To determine the backwash flow rate for a gravity sand filter:

1. Determine the dimensions of the filter and surface area, and
2. Measure and record the flow rate of the backwash water.

$$\text{Backwash Rate, GPM/sq ft} = \frac{\text{Flow, gal/min}}{\text{Surface Area, sq ft}}$$

To determine the percent of water used for backwashing, divide the gallons of backwash water by the gallons of water filtered and multiply by 100 percent.

$$\text{Backwash, \%} = \frac{(\text{Backwash Water, gal})(100\%)}{\text{Water Filtered, gal}}$$

EXAMPLE 7

A water treatment plant treats a flow of 2 MGD. There are two sand filters and each filter has the dimensions of 20 feet long by 20 feet wide. Determine the filtration rate in gallons per minute per square foot of filter. *NOTE:* Each filter treats a flow of 1 MGD.

Known		Unknown
Length, ft	= 20 ft	Filtration Rate, GPM/sq ft
Width, ft	= 20 ft	
Flow, MGD	= 1 MGD	

Pressure Filters

Chemical Containers

Hydropneumatic Tank

Fig. 4.15 Small pressure filtration plant

(Permission of Watermasters, Inc., Burlingame, California)

1. Calculate the filter surface area in square feet.

$$\text{Surface Area, sq ft} = (\text{Length, ft})(\text{Width, ft})$$
$$= (20 \text{ ft})(20 \text{ ft})$$
$$= 400 \text{ sq ft}$$

2. Convert the flow from MGD to GPM.

$$\text{Flow, gal/min} = \frac{(\text{Flow, MGD})(1{,}000{,}000/\text{Million})}{(24 \text{ hr/day})(60 \text{ min/hr})}$$
$$= \frac{(1 \text{ MGD})(1{,}000{,}000/\text{Million})}{(24 \text{ hr/day})(60 \text{ min/hr})}$$
$$= 694 \text{ gal/min}$$

3. Calculate the filtration rate in gallons per minute per square foot of surface area.

$$\text{Filtration Rate, GPM/sq ft} = \frac{\text{Flow, gal/min}}{\text{Surface Area, sq ft}}$$
$$= \frac{694 \text{ gal/min}}{400 \text{ sq ft}}$$
$$= 1.7 \text{ GPM/sq ft}$$

EXAMPLE 8

To check on the valve position of a controller in a filter, the influent valve to the filter was closed for five minutes. During this time period the water surface in the filter dropped 1.25 feet (1 ft-3 in). The surface area of the filter is 400 square feet. Estimate the flow in gallons per minute.

Known		Unknown
Time, min	= 5 min	Flow, GPM
Water Drop, ft	= 1.25 ft	
Surface Area, sq ft	= 400 sq ft	

Calculate the flow in gallons per minute.

$$\text{Flow, GPM} = \frac{(\text{Water Drop, ft})(\text{Surface Area, sq ft})(7.48 \text{ gal/cu ft})}{\text{Time, min}}$$
$$= \frac{(1.25 \text{ ft})(400 \text{ sq ft})(7.48 \text{ gal/cu ft})}{5 \text{ min}}$$
$$= 748 \text{ GPM}$$

EXAMPLE 9

Estimate the backwash rate for the filter in Example 7 if there are two backwash pumps capable of delivering 4,000 GPM each. The total backwash flow is 8,000 GPM.

Known		Unknown
Surface Area, sq ft	= 400 sq ft	Backwash Rate, GPM/sq ft
Backwash Flow, GPM	= 8,000 GPM	

Calculate the backwash rate in gallons per minute per square foot of surface area.

$$\text{Backwash Rate, GPM/sq ft} = \frac{\text{Backwash Flow, gal/min}}{\text{Surface Area, sq ft}}$$
$$= \frac{8{,}000 \text{ GPM}}{400 \text{ sq ft}}$$
$$= 20 \text{ GPM/sq ft}$$

EXAMPLE 10

During a filter run, the total volume of water filtered was 1.20 million gallons. When the filter was backwashed, 18,000 gallons of water was used. Calculate the percent of filtered water used for backwashing.

Known		Unknown
Water Filtered, gal	= 1,200,000 gal	Backwash, %
Backwash Water, gal	= 18,000 gal	

Calculate the percent of water used for backwashing.

$$\text{Backwash, \%} = \frac{(\text{Backwash Water, gal})(100\%)}{\text{Water Filtered, gal}}$$
$$= \frac{(18{,}000 \text{ gal})(100\%)}{1{,}200{,}000 \text{ gal}}$$
$$= 1.5\%$$

QUESTIONS

Write your answers in a notebook and then compare your answers with those on page 259.

4.4A What is included in the suspended matter removed by filtration?

4.4B Why do anthracite and sand stay separated during and after backwashing?

4.4C What is a "mixed media" filter?

4.4D Under what conditions will mudballs form in filters?

4.5 DISINFECTION

The purpose of disinfection in domestic water treatment is to kill or inactivate any disease-causing organisms that may be present. There are several types of disease-causing organisms and each type has different characteristics that are important to understand.

1. Bacteria

   Bacteria are microscopic organisms. Individual bacteria are usually so small they are invisible to the naked eye. Therefore it is impossible to tell if bacteria are present simply by looking at the water. A glass of water may be sparkling clear and appear to be pure but actually contain millions of bacteria.

   Most bacteria are harmless but a few types cause serious illness and even death in humans. Some bacterial diseases transmitted by drinking contaminated water include typhoid fever, paratyphoid fever, bacillary dysentery, and cholera.

2. Viruses

   Virus agents are very small organisms, much smaller than bacteria, and they too can cause serious illness in humans. Infectious hepatitis and possibly poliomyelitis are two viral diseases that can be transmitted by contaminated drinking water. So-called "sewage poisoning," acute intestinal upset caused by drinking sewage-contaminated water, may also be caused by viral agents.

3. Protozoa

   Protozoa are microscopic animals that are also too small to be detected by the naked eye. Among the waterborne diseases they cause are amoebic dysentery, giardiasis (gee-are-DYE-uh-sis), and cryptosporidiosis.

These three types of organisms (bacteria, viruses, and protozoa) differ in their resistance to chlorine, the most common disinfectant. The bacteria are readily killed by the chlorine residual (0.2 to 0.5 mg/*L*) normally maintained in domestic water from a treatment plant. Most viruses are significantly more resistant to chlorine than are bacteria. Therefore, higher chlorine concentrations and longer contact times are necessary to kill viral agents. The protozoa are the most resistant to chlorine. They normally are not killed by the concentrations of chlorine ordinarily used in water treatment. Therefore, operators must rely on the processes of coagulation, flocculation, settling, and filtration for physical removal of protozoa rather than depend on disinfection to kill the organisms.

The processes of coagulation, flocculation, settling, and filtration can remove a high percentage of the disease organisms from a water supply. However, the operator cannot rely on these processes for 100 percent removal of disease organisms. Disinfection must be applied to all surface water sources to ensure that the water will be entirely safe and free of harmful organisms. Groundwater sources from properly constructed wells and springs may be free of contamination and may not require disinfection.

Nearly all domestic water supplies are disinfected with chlorine, although iodine, ozone, and ultraviolet radiation are also used occasionally. Chlorine is available in several forms. Gaseous chlorine can be purchased as liquid chlorine in steel cylinders. As chlorine gas is removed from the cylinder, the liquid chlorine evaporates and produces more gas. Chlorine can also be purchased in the form of liquid bleach, a solution of sodium hypochlorite in water. Laundry bleach, available from grocery stores, is 5.25 percent chlorine and commercial strength bleach, available from swimming pool suppliers or chemical companies, is usually 12.5 percent chlorine. Chlorine compounds in a solid form, granular or tablets, can also be purchased from swimming pool suppliers and chemical companies. These compounds of chlorine contain calcium hypochlorite. Calcium hypochlorite in solid form contains 65 percent available chlorine. This means that if you add 10 pounds of calcium hypochlorite to water, you are actually adding 6.5 pounds of chlorine (10 lbs x 0.65 = 6.5 lbs). Chlorine dioxide (ClO_2) is another form of chlorine used to disinfect drinking water. All forms of chlorine are hazardous chemicals that can cause serious injury and damage if they are not stored and used properly.

The form of chlorine used at the treatment plant will determine the type of chlorine feeder that is needed. If chlorine gas is used, it must be fed through a *CHLORINATOR*. The chlorinator not only controls and measures the flow of chlorine gas but also dissolves it in water so it can be safely injected into the water supply (Figure 4.16).

If chlorine in the form of liquid bleach (sodium hypochlorite) or a granular compound (calcium hypochlorite) is used, it is mixed with water to make a hypochlorite solution. Then this solution is injected into the water supply by a chemical solution feed pump called a *HYPOCHLORINATOR*.

Small water plants most frequently use hypochlorite forms of chlorine because these are safer to handle and the feeder pumps are relatively inexpensive. Larger plants use chlorine gas because it is less expensive than hypochlorites and because gas chlorinators are usually more reliable and require less attention by the operator than a hypochlorinator.

The following factors have an important effect on disinfection of water with chlorine.

1. *CONCENTRATIONS OF CHLORINE*

   The higher the chlorine concentration, the faster and more complete will be the disinfection.

2. *TIME OF CHLORINE CONTACT*

   The longer that water is in contact with chlorine, the greater will be the degree of disinfection achieved. Short-circuiting of flow through basins and tanks must be avoided because it can drastically shorten the time of chlorine contact. *CLEAR WELLS* which store treated water are used to provide adequate contact time.

3. *THE pH OF THE WATER*

   The pH determines whether the chlorine is in the form of hypochlorous acid, a powerful disinfectant, or in the form of hypochlorite ion, which is a much weaker disinfectant. For best results, the pH should be less than 7.5.

4. *WATER TEMPERATURE*

   Chlorine requires more time to kill organisms in cold water than in warm water. A decrease in the water temperature from summer to winter may require twice the contact time to achieve the same degree of disinfection.

5. *DEGREE OF MIXING*

   Any organisms not exposed to chlorine will not be killed. Therefore, mixing of the chlorine and the water must be rapid and complete to ensure that chlorine comes in contact with all the organisms present.

6. *CLARITY OF WATER*

   Organisms that are encased in a covering of suspended particles, mud, or trash may be so protected from the chlorine that the chlorine never penetrates to the organisms. It is important to remove as much suspended matter as possible so it does not "shield" organisms from chlorine or consume the chlorine before it is effective in killing the organisms.

7. *PRESENCE OF INTERFERING SUBSTANCES*

   Chlorine is a very active chemical and it reacts quickly with many substances. These chemical reactions can consume the chlorine or convert it to forms that are poor disin-

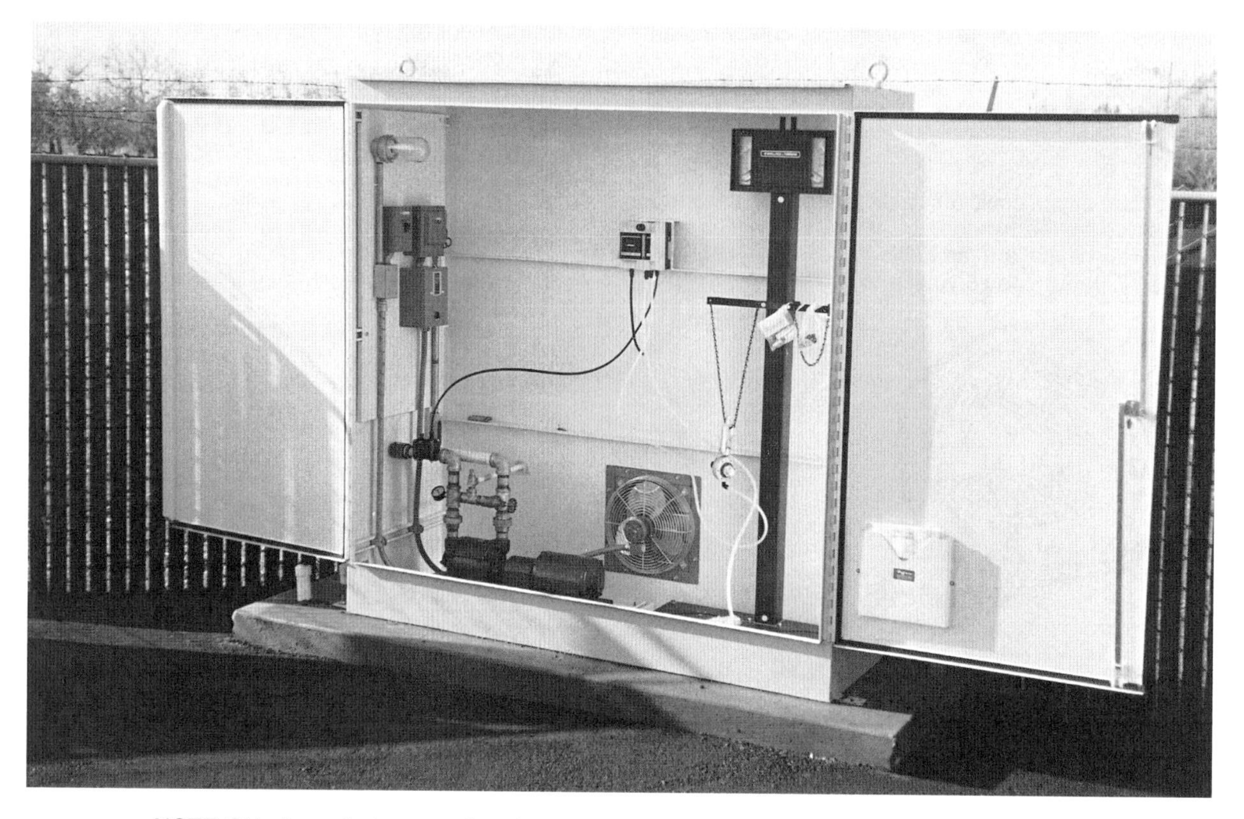

NOTE: Chlorine cylinders are placed on the scales at the right and secured by the chains.

Fig. 4.16 Small chlorination system
(Permission of Watermasters, Inc., Burlingame, California)

fectants. Ammonia is one of the most troublesome substances because it combines readily with chlorine to form chloramine compounds. Chloramines are much weaker disinfectants than free chlorine. Sources of ammonia in water include the effluent from a wastewater treatment plant or seepage from the subsurface leaching system of a septic tank. Other substances which react with chlorine and consume it include dissolved iron and manganese, hydrogen sulfide, and organic matter.

8. *POINTS OF CHLORINE APPLICATION*

   Provisions should be made to feed chlorine at several different locations in the plant. The place of feed may vary from time to time depending on source water quality. If natural organics are not present in the source water, chlorination early in the treatment process can provide effective disinfection and reduce chlorine costs.

In addition to controlling the above factors, the operator must have reliable chlorination equipment and it must be properly maintained and competently operated. An important part of operating the chlorinator is to measure the concentration of chlorine in the water on a daily basis. This measurement is made with a chlorine test kit (Figure 4.17). Many brands of chlorine test kits are available. Most of them consist of a clear sample vial and a set of color standards. The reagent, which produces a color whose intensity is proportional to the chlorine concentration, is added to the vial containing the water sample. The color that forms in the sample is then compared with the known color standards to determine the chlorine concentration. Previously, a chemical solution containing orthotolidine reagent, which produces a yellow color with chlorine, was used in all chlorine test kits. Orthotolidine (O.T.) is still used in the low-cost chlorine test kits that are supplied for testing swimming pool waters. However, for testing domestic water supplies a new reagent called DPD (diethyl-p-phenylene diamine) is much preferred because it is more accurate and versatile. The DPD method is subject to fewer interferences than O.T. and it accurately measures both free and total chlorine residuals.

Operators must be acquainted with the three types of chlorine residuals that can be found in treated water.

1. *FREE*

   Free available chlorine residual includes that chlorine in the form of hypochlorous acid (HOCl) and hypochlorite ion (OCl^-). Hypochlorous acid is the most effective disinfectant form of chlorine. Hypochlorite ion is much less effective as a disinfectant than hypochlorous acid. The pH determines whether the free chlorine is in the form of hypochlorous acid or hypochlorite ion. The lower the pH, the greater the percent of hypochlorous acid.

2. *COMBINED*

   Combined available chlorine residual includes that chlorine in the three chloramine forms (monochloramine, dichloramine, and nitrogen trichloride) which are produced when chlorine reacts with ammonia. Combined chlorine requires up to 100 times the contact time or at least 25 times the chlorine concentration to achieve the same degree of disinfection as free available chlorine.

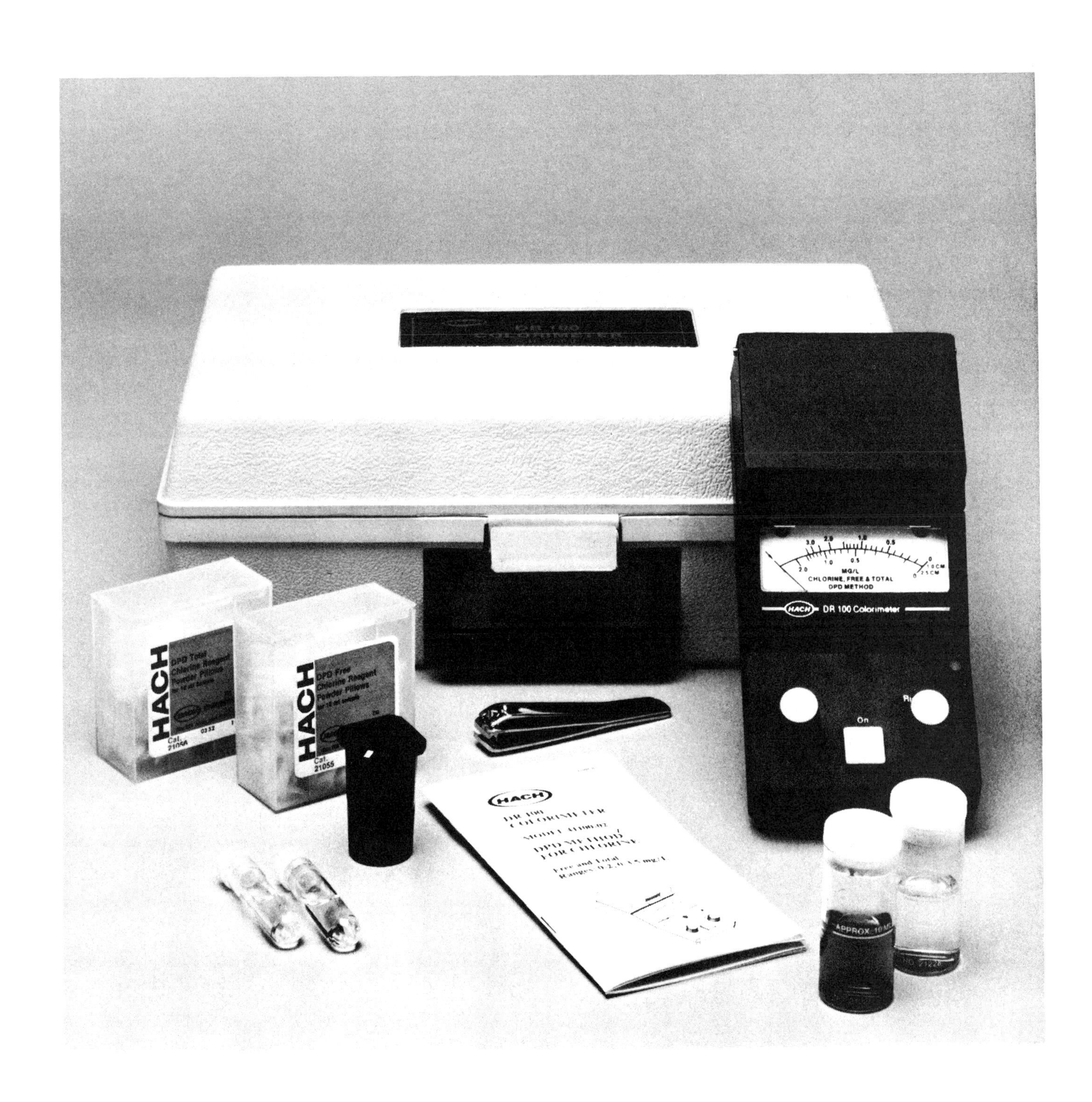

Fig. 4.17 Chlorine test kit

(Permission of HACH Company)

3. *TOTAL*

The total chlorine residual is the sum of the free residual and the combined residual.

Chlorine residual test kits measure the free available chlorine residual and the total chlorine residual. The combined available chlorine residual is the difference. For example, a test kit could measure the free chlorine residual as one mg/*L* and the total chlorine residual as three mg/*L*. The combined available chlorine residual would be two mg/*L* (3 mg/*L* – 1 mg/*L* = 2 mg/*L*).

Operators should strive to maintain a free chlorine residual to accomplish disinfection and to avoid depending on the much weaker combined chlorine residual. Thus, in performing a chlorine residual test, the small system operator should make sure that the test result indicates only the *FREE* chlorine residual concentration.

Operators of small treatment plants frequently ask, "What chlorine residual should I maintain to guarantee that the finished water is adequately disinfected?" This is a very important question and one that must be answered correctly. But it is impossible to give a simple answer covering all situations because, as already explained, the required chlorine concentration is related to the contact time, the pH, and the water temperature.

A reference book, *HANDBOOK OF CHLORINATION AND ALTERNATIVE DISINFECTANTS*[20] by George Clifford White, describes a workable method for calculating the required chlorine concentration under various conditions. The calculation uses the following formula:

$$C \times T = A$$

where C = the free available chlorine residual concentration in milligrams per liter,

T = the chlorine contact time in minutes, and

A = a number which varies with the pH as shown in Table 4.1

TABLE 4.1 RELATIONSHIP BETWEEN pH AND A

pH range	A (for cold water, 0° to 5°C)
7.0 – 7.5	12
7.5 – 8.0	20
8.0 – 8.5	30
8.5 – 9.0	35

For example, suppose in a particular small water plant that the minimum chlorine contact time from the point of chlorine injection to the entrance of the distribution system is 30 minutes. Then for the formula, T = 30. The pH of the water was found to be 7.2 and the water temperature during the winter months is usually just above freezing, say 1° or 2° Celsius. Then from Table 4.1, the value of A in this case is 12. Substituting these values in the formula we obtain:

$$C \times T = A$$

$$C \times 30 = 12$$

Solving for C,

$$C = \frac{A}{T}$$

$$C, mg/L = \frac{12}{30}$$

$$= \frac{4}{10}$$

= 0.4 mg/*L* free chlorine residual

This calculation tells us that good disinfection of the water supply will be accomplished under the existing conditions by maintaining a free chlorine residual of at least 0.4 mg/*L* for the 30-minute contact time. Notice that a stronger chlorine residual would be required if the pH were increased, say to 8.3. Then, A = 30.

$$C \times T = A$$

$$C \times 30 = 30$$

Or

$$C = \frac{30}{30}$$

= 1.0 mg/*L* free chlorine residual

The efficiency of the disinfectant is measured by the time "T" in minutes of the disinfectant's contact in the water and the concentration "C" in mg/*L* of the disinfectant residual measured at the end of the contact time. The product of these two parameters (C × T) provides a measure of the degree of pathogenic inactivation. The required CT value to achieve inactivation depends on the organism in question, pH, and temperature of the water supply. Table 4.2 shows the combinations of disinfectant, pH, and temperature that will produce 99.9 percent *Giardia lamblia* (a disease-causing organism) inactivation.

[20] *HANDBOOK OF CHLORINATION AND ALTERNATIVE DISINFECTANTS, Fourth Edition, by George Clifford White. Obtain from John Wiley & Sons, Inc., Distribution Center, 1 Wiley Drive, Somerset, NJ 08875-1272. ISBN 0-471-29207-9. Price, $199.00, plus $5.00 shipping and handling.*

TABLE 4.2 CT VALUES REQUIRED FOR 99.9% *GIARDIA LAMBLIA* INACTIVATION

Disinfectant	pH	10 °C	15 °C	20 °C	25 °C
Free Chlorine[a]	6	79	53	39	26
	7	112	75	56	37
	8	162	108	81	54
Ozone	6 – 9	1.4	0.95	0.72	0.48
Chloramines	6 – 9	1,850	1,500	1,100	750

[a] with 1 mg/*L* free chlorine residual

Time or "T" is measured from the point of application to the point where "C" is determined. "T" must be based on peak hour flow rate conditions. In pipelines, "T" is calculated by dividing the volume of the pipeline in gallons by the flow rate in gallons per minute (GPM). In reservoirs and basins, dye tracer tests must be used to determine "T." In this case, "T" is the time it takes for 10 percent of the tracer to pass the measuring point.

Sometimes you may increase the chlorine dose setting on your chlorinator and *THE CHLORINE RESIDUAL MAY ACTUALLY DROP* (Figure 4.18, *BREAKPOINT CHLORINATION*)![21] This can happen when ammonia is present. Also under these conditions you may receive complaints that your water tastes like chlorine. The solution to this problem is to increase the chlorine dose even more! For additional information on breakpoint chlorination, see Chapter 5, "Disinfection." You are at the proper setting if:

1. You are maintaining the chlorine residual from the above calculations,
2. An increase in the chlorinator setting will produce a calculated chlorine dose increase of 0.1 mg/*L* and the resulting actual chlorine residual increases by 0.1 mg/*L*,
3. Chlorine residual tests at the far end of the distribution system produce chlorine residuals of at least 0.2 mg/*L*, and
4. Coliform test results from throughout the distribution system are negative.

Another important part of chlorinator operation is keeping adequate records. The records must be legible and kept in an organized format. They should be written down in a permanent manner that can be preserved indefinitely for future reference. Records scribbled with a dull pencil on assorted scraps of paper are worthless and suggest incompetence.

The records should show the date and time the chlorinator was inspected, the flow rate of the water being treated, the total gallons of water treated, the feed rate of the chlorinator, the pounds of chlorine used, the calculated chlorine dosage, the chlorine residual concentration in the water as measured with the test kit, and the operator's name. The operator can conveniently tabulate this vital information on a standard record sheet similar to those in Tables 4.3 and 4.4.

Chlorinators not in service require more maintenance than those in service. The operator must be alert to spot failures and repair the equipment quickly because lack of disinfection is not, of course, acceptable. If possible, the operator should have a spare chlorinator on hand that can replace the failing unit. If not, the treatment plant must be shut down until the chlorinator is repaired. If the plant must operate without a chlorinator, consumers must be notified to boil all drinking water. The operator should save the manufacturer's operation manual for reference and keep repair parts in stock at all times. Contacts should be established with the chlorinator supplier or manufacturer so that service and parts can be obtained quickly. A program of periodic maintenance on chlorination equipment will prevent the vast majority of chlorinator failures. Regular disassembly and cleaning of the chlorinator and replacement of critical parts is recommended insurance against a breakdown.

If a gas chlorinator is not feeding properly the operator should check the items listed below.

1. The chlorine feed rate may not be properly adjusted. Reset it to the desired level.
2. The chlorine gas supply may be interrupted. The gas cylinder may be empty or the gas valve on the cylinder may be closed. Also, the chlorine tubing from the gas cylinder to the chlorinator may be plugged.
3. Are you trying to draw too much chlorine gas from the cylinder? If the withdrawal rate is too high, the liquid chlorine will not evaporate fast enough. High velocities of chlorine gas in the chlorine supply line will remove heat and cause a frost to form on the line. This can cause the chlorine gas to *RELIQUEFY*.[22] This liquid chlorine can plug the supply line and no more chlorine gas will flow. Operators refer to this condition as a "frozen" chlorine supply line: *BE VERY CAREFUL BECAUSE IF YOU DISCONNECT THE SUPPLY*, the liquid chlorine can evaporate again and send liquid chlorine shooting out the end of the disconnected line.
4. The chlorinator may not be drawing a vacuum. Check the water supply to the injector for adequate flow and make sure the injector is not plugged with debris.
5. The internal mechanism of the chlorinator may be malfunctioning. A broken diaphragm, a weak spring, or a leaking gasket will cause a malfunction. Replace the necessary parts and clean the chlorinator thoroughly using the cleaning solvents recommended by the manufacturer.

If a hypochlorinator is not feeding properly, check these points.

1. Adjust the hypochlorinator to feed the desired dosage.
2. The hypochlorite solution may be used up.
3. The intake tubing in the hypochlorite solution tank may be plugged with sediment.
4. The diaphragm in the hypochlorinator may be ruptured.
5. The inlet and outlet poppet valves in the hypochlorinator may be malfunctioning.
6. The drive belt may be broken or slipping.
7. The electrical supply to the hypochlorinator may be interrupted.

[21] *Breakpoint Chlorination. Addition of chlorine to water until the chlorine demand has been satisfied. At this point, further additions of chlorine will result in a free chlorine residual that is directly proportional to the amount of chlorine added beyond the breakpoint.*

[22] *Reliquefaction (re-LICK-we-FACK-shun). The return of a gas to the liquid state; for example, a condensation of chlorine gas to return it to its liquid form by cooling.*

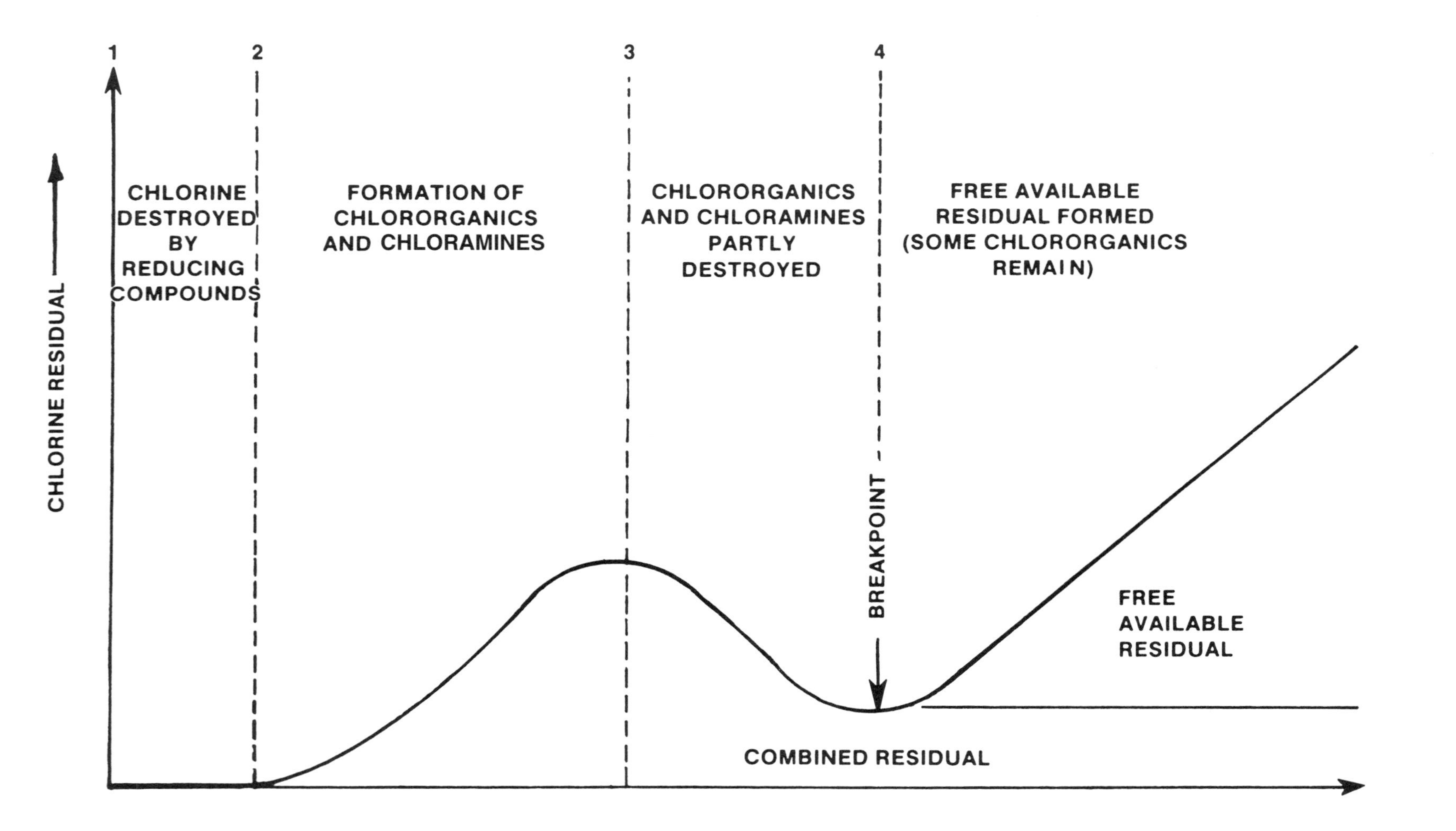

Fig. 4.18 Breakpoint chlorination curve

TABLE 4.3 CHLORINATION RECORD FOR GAS CHLORINATOR

CHLORINATION RECORD
FOR
GAS CHLORINATOR

REPORT NO. ____________

WATER SUPPLIER ____________

SYSTEM NAME ____________ SYSTEM NO. ____________

FOR WEEK OF ____________ THRU ____________ 19____

OPERATOR ____________

DAY	DATE	TIME	CHLORINE RESIDUALS, mg/l			WATER PRODUCTION			CHLORINATION TREATMENT						
			#1	#2	#3	PLANT OPERATING RATE — GPM	WATER METER READING (GALLONS)	GALLONS OF WATER TREATED	CHLORINATOR FEED RATE LBS/24 HRS	WEIGHT OF CHLORINE AND CONTAINER	WEIGHT OF EMPTY CONTAINER	WEIGHT OF CHLORINE REMAINING	POUNDS OF CHLORINE USED	AVG. CHLORINE DOSE FOR PAST DAY (CALCULATED) NOTE 1	INSTANTANEOUS CHLORINE DOSE (CALCULATED) NOTE 2
COLUMN NUMBER			1	2	3	4	5	6	7	8	9	10	11	12	13
SUN.															
SAT.															
FRI.															
THUR.															
WED.															
TUE.															
MON.															
READINGS FORWARD															
WEEKLY TOTALS →															

REPORT SUBMITTED BY:

____
SIGNATURE

REMARKS:

NOTE AVG. CHLORINE DOSE (mg/l) $= \frac{\text{(LBS. OF } CL_2 \text{ USED) (120,000)}}{\text{GAL OF WATER TREATED}} = \frac{\text{(COL II) (120,000)}}{\text{(COL 6)}}$

NOTE 2: INSTANTANEOUS CHLORINE DOSE mg/l $= \frac{\text{(CHLORINATOR FEED RATE)}}{\text{(.012) (PLANT OPERATING RATE — GPM)}} = \frac{\text{(COL. 7)}}{\text{(.012) (COL. 4)}}$

TABLE 4.4 OPERATION RECORD FOR SODIUM HYPOCHLORINATOR

OPERATION RECORD FOR SODIUM HYPOCHLORINATOR

REPORT NO. ____________

WATER SUPPLIER ____________

SYSTEM NAME ____________ SYSTEM NO. ____________

FOR WEEK OF ____________ THRU ____________ 19 ____

ITEM A.	STRENGTH OF BLEACH USED	5.25%	12.5%	1.0%
ITEM B.	POUNDS OF CHLORINE PER GALLON	0.44 LBS.	1.04 LBS.	0.083 LBS.

ITEM C. SIZE OF CHLORINE SOLUTION TANK: ____________ GALLONS

ITEM D. SIDEWALL DEPTH OF SOLUTION TANK: ____________ INCHES

ITEM E. STRENGTH OF SOLUTION: EACH GALLON OF BLEACH IS ADDED TO ____________ GALLONS OF WATER.

ITEM F. GALLONS OF BLEACH IN A FULL TANK OF SOLUTION: ____________ GALLONS.

ITEM G. POUNDS OF CHLORINE IN ONE INCH OF SOLUTION DEPTH: ____________ POUNDS PER INCH DEPTH.

$$\frac{\text{(GAL. BLEACH IN FULL TANK) (POUNDS CHLORINE/GALLON)}}{\text{(SIDEWALL DEPTH OF SOLUTION TANK)}} = \frac{\text{(ITEM F) X (ITEM B)}}{\text{(ITEM D)}}$$

DAY	DATE	TIME	CHLORINE RESIDUALS mg/l			WATER PRODUCTION			CHLORINATION TREATMENT					
			#1	#2	#3	PLANT OPERATING RATE G.P.M.	WATER METER READING (GALLONS)	GALLONS OF WATER TREATED	FEEDER SETTING	INCHES OF SOLUTION IN TANK	INCHES OF SOLUTION USED	POUNDS OF CHLORINE USED (ITEM G) TIMES (COLUMN 8)	AVERAGE CHLORINE DOSE mg/l (NOTE 1)	QUARTS OF BLEACH TO BE ADDED (NOTE 2)
COLUMN NUMBER			1	2	3	4	5	6	7		8	9	10	11
SUN.														
SAT.														
FRI.														
THUR.														
WED.														
TUE.														
MON.														
READINGS FORWARD														
WEEKLY TOTALS →														

REPORT SUBMITTED BY:

____
SIGNATURE

NOTE 1: AVG. CHLORINE DOSE (mg/l) $= \frac{\text{(LBS. OF CHLORINE USED) X (120,000)}}{\text{(GAL OF WATER TREATED)}} = \frac{\text{(COL 9) X (120,000)}}{\text{(COL 6)}}$

NOTE 2: QUARTS OF BLEACH TO BE ADDED $= \frac{\text{(INCHES SOLN. USED) X (4) X (GALLONS OF BLEACH IN A FULL TANK)}}{\text{(SIDEWALL DEPTH OF SOLUTION TANK)}} = \frac{\text{(COL 8) X (4) X (ITEM F)}}{\text{(ITEM D)}}$

8. The feed line from the hypochlorinator to the injection point may be plugged or broken.
9. There may be excessive backpressure at the point of injection which the hypochlorinator cannot overcome.
10. The hypochlorite solution may freeze in the tubing and other small conduits through which it flows. Provide an insulated, heated enclosure for the equipment.

The biggest problem with hypochlorinators is that they can develop calcium deposits if you are using calcium hypochlorite. Hypochlorinators can be cleaned with a weak acid (vinegar or muriatic acid). If you have only one hypochlorinator, turn off the water being disinfected so no unchlorinated water can enter your clear well or distribution system. Pump the acid through the entire pump (hypochlorinator or chemical feed pump, they are the same) and into a waste or drain line. This procedure allows for the cleaning of all feed lines, valves and equipment. If this procedure is followed on a regular or routine basis, the pump will never have to be taken apart for cleaning.

If the operator has problems achieving adequate disinfection of the water or maintaining a chlorine residual, one or more of the following problems could be the cause.

1. The chlorinator is not feeding. Inspect the equipment and determine that the chlorinator is operating properly.
2. The chlorine dosage is inadequate. Readjust the chlorinator to feed the proper dosage. Never mix more than a two- or three-day supply of hypochlorite solution. Hypochlorite solutions lose their strength in time and this fact has the effect of changing the feed rate.
3. The contact time is too short. Relocate the point of chlorine injection to increase the contact time. Eliminate short-circuiting of flow through the plant.
4. Suspended matter in the water is preventing effective disinfection. Improve clarification treatment so that turbidity is minimal.
5. The chlorine is being consumed by organic matter, dissolved gases or minerals, or by sludge in basins and pipelines. Increase the chlorine dosage, clean sludge from basins and storage tanks and flush sediment from the distribution system.
6. The chlorine is being destroyed by exposure to sunlight in an uncovered storage tank. Provide a cover for the storage tank or re-chlorinate the water as it leaves the tank.

For additional information on disinfection, see Chapter 7, "Disinfection," in *WATER TREATMENT PLANT OPERATION*, Volume I, in this series of manuals.

FORMULAS

To estimate the actual average chlorine dose in milligrams per liter for a gas chlorinator:

1. Determine the weight of chlorine used, and
2. Measure and record the amount of water treated.

TABLE 4.5 INSTRUCTIONS FOR COMPLETING TABLE 4.4, "OPERATION RECORD FOR SODIUM HYPOCHLORINATOR"

ITEM A and ITEM B — Circle the strength of chlorine bleach and the corresponding value for pounds of chlorine per gallon that is purchased for mixing the chlorine solution.

ITEM C — Record the capacity of the chlorine solution tank in gallons.

ITEM D — Record the sidewall depth of the solution tank when it is full.

ITEM E — Record the number of gallons of water that are added to each gallon of bleach for preparing chlorine solution.

ITEM F — Using the information in Item E, calculate the gallons of bleach in a full tank of solution.

ITEM G — Calculate the pounds of chlorine in one inch of solution depth in the tank. For example, assume that a 30-gallon solution tank is used and that the chlorine solution is made by mixing 4 gallons of water with each gallon of 5.25% household bleach (Clorox or Purex). Therefore, the full 30-gallon tank will contain 6 gallons of bleach. Since each gallon of 5.25% bleach contains 0.44 pounds of chlorine, 6 gallons of bleach contain $(6) \times (0.44) = 2.64$ pounds of chlorine. If the tank has a sidewall depth of 36 inches, there are $(2.64) \div (36) = 0.073$ pounds of chlorine in each inch of solution depth in the tank.

Columns 1 through 7 in the daily record form are self-explanatory.

COLUMN 8 — Measure the inches of chlorine solution used from the tank and record it in Column 8.

COLUMN 9 — Calculate the pounds of chlorine used by multiplying the inches of chlorine solution used (Column 8) times the pounds of chlorine in one inch of solution depth (Item G). Record the answer in Column 9.

COLUMN 10 — Calculate the average chlorine dose by multiplying the pounds of chlorine used (Column 9) times 120,000 and divide by the gallons of water treated (Column 6). Record the answer in Column 10.

COLUMN 11 — To refill the solution tank proceed as follows. Multiply the inches of solution used from the tank (Column 8) times four. Then multiply that answer times the gallons of bleach that are in the tank when it is full, in this example, 6 gallons. Divide this answer by the total sidewall depth of the solution tank (36") to obtain the number of *QUARTS* of bleach that should be added to the tank. Enter the answer in Column 11. After the bleach is added, fill the tank to the top with clean water. This procedure will maintain the chlorine solution in the tank at a uniform strength.

$$\frac{\text{Inches of Solution Used} \times 4 \times \text{Gallons of Bleach in a Full Tank}}{\text{Sidewall Depth of Solution Tank}} = \frac{\text{Column 8} \times 4 \times \text{Item F}}{\text{Item D}}$$

$$\text{Average Chlorine Dose, mg/}L = \frac{\text{Chlorine Used, lbs/day}}{\text{(Water Treated, MGD)(8.34 lbs/gal)}}$$

$$= \frac{\text{lbs Chlorine}}{\text{M lbs Water}}$$

$$= \frac{\text{mg Chlorine}}{\text{1 M mg Water}}$$

$$= \text{mg Chlorine/Liter Water, or}$$

$$\text{Average Chlorine Dose, mg/}L = \frac{\text{(Chlorine Used, lbs/day)(1,000,000/M)}}{\text{(Water Treated, gal/day)(8.34 lbs/gal)}}$$

$$= \frac{\text{(Chlorine Used, lbs/day)(120,000)}}{\text{Water Treated, gal/day}}$$

$$= \text{mg Chlorine/Liter Water}$$

NOTE: Chlorine used in pounds and water treated in gallons or million gallons must be for the same time interval, but not necessarily for 24 hours. The time period could be 20 hours, 27 hours, or even 2 days.

To estimate the instantaneous chlorine dose in milligrams per liter for a gas chlorinator:

1. Record the chlorinator feed rate setting, and
2. Measure and record the flow rate for the water being treated.

$$\text{Flow Rate, MGD} = \frac{\text{(Flow, gal/min)(60 min/hr)(24 hr/day)}}{\text{1,000,000/M}}$$

$$\text{Instantaneous Chlorine Dose, mg/}L = \frac{\text{Chlorinator Feed Rate, lbs/day}}{\text{(Flow, MGD)(8.34 lbs/gal)}}$$

$$\text{or} = \frac{\text{Chlorinator Feed Rate, lbs/day}}{\text{(Flow, GPM)(0.012)}}$$

To select the chlorinator feed rate setting in pounds of chlorine per 24 hours (lbs/day):

1. Determine the desired chlorine dose, and
2. Measure and record the flow rate for the water being treated.

$$\text{Chlorinator Feed Rate, lbs/day} = \text{(Flow, MGD)(Dose, mg/}L\text{)(8.34 lbs/gal)}$$

To estimate the actual average chlorine dose in milligrams per liter of a hypochlorinator system:

1. Determine the dimensions and volume of the hypochlorite solution container,
2. Determine the number of gallons of water added to each gallon of hypochlorite (bleach) in the solution tank, and
3. Measure and record the amount of water treated.

$$\text{Total Chlorine, lbs} = \text{(Chlorine, lbs/gal)(Bleach, gal)}$$

$$\text{Chlorine Used, lbs} = \frac{\text{(Total Chlorine, lbs)(Chlorine Used, in)}}{\text{Depth of Tank, in}}$$

$$\text{Average Chlorine Dose, mg/}L = \frac{\text{(Chlorine Used, lbs)(1,000,000/Million)}}{\text{(Water Treated, gal)(8.34 lbs/gal)}}$$

$$= \frac{\text{lbs Chlorine}}{\text{M lbs Water}}$$

$$= \frac{\text{mg Chlorine}}{\text{M mg Water}}$$

$$= \text{mg Chlorine/Liter Water}$$

EXAMPLE 11

Estimate the average chlorine dose in mg/*L* for a chlorinator that used 10 pounds of chlorine to treat 500,000 gallons (0.5 million gallons) of water.

Known		**Unknown**
Chlorine Used, lbs	= 10 lbs	Average Chlorine Dose, mg/*L*
Water Treated, M Gal	= 0.5 M Gal	

Calculate the average chlorine dose in mg/*L*.

$$\text{Average Chlorine Dose, mg/}L = \frac{\text{Chlorine Used, lbs}}{\text{(Water Treated, M Gal)(8.34 lbs/gal)}}$$

$$= \frac{\text{10 lbs Chlorine}}{\text{(0.5 M Gal Water)(8.34 lbs/gal)}}$$

$$= \frac{\text{10 lbs Chlorine}}{\text{4.17 M lbs Water}}$$

$$= 2.4 \text{ mg/}L$$

EXAMPLE 12

Estimate the instantaneous chlorine dose in mg/*L* if the chlorinator feed rate is set at 12 pounds per 24 hours and the flow is 300 gallons.

Known		**Unknown**
Chlorine Feed, lbs/day	= 12 lbs/day	Instantaneous Chlorine Dose, mg/*L*
Flow, GPM	= 300 GPM	

1. Convert the flow from GPM to MGD.

$$\text{Flow, MGD} = \frac{\text{(Flow, gal/min)(60 min/hr)(24 hr/day)}}{\text{1,000,000/Million}}$$

$$= \frac{\text{(300 gal/min)(60 min/hr)(24 hr/day)}}{\text{1,000,000/Million}}$$

$$= 0.432 \text{ MGD}$$

2. Calculate the instantaneous chlorine dose in mg/*L*.

$$\text{Instantaneous Chlorine Dose, mg/}L = \frac{\text{Chlorinator Feed Rate, lbs/day}}{\text{(Flow, MGD)(8.34 lbs/gal)}}$$

$$= \frac{\text{12 lbs/day}}{\text{(0.432 MGD)(8.34 lbs/gal)}}$$

$$= 3.3 \text{ mg/}L$$

EXAMPLE 13

Estimate the actual average chlorine dose in milligrams per liter from a hypochlorinator. The depth of hypochlorite solution dropped 10 inches while treating 100,000 gallons of water. The strength of bleach used is 5.25 percent and contains 0.44 lbs of chlorine per gallon of bleach. For every gallon of bleach added to the container, four gallons of water are added. The full mark on the container is at 50 gallons (10 gallons of bleach and 40 gallons of water). When the container is full, there are 50 inches of hypochlorite solution.

Known		Unknown
Chlorine Used, in	= 10 in	Average Chlorine Dose, mg/*L*
Water Treated, gal	= 100,000 gal	
Bleach, lbs/gal	= 0.44 lb/gal	
Bleach Used, gal	= 10 gal	
Depth of Tank, in	= 50 in	

1. Calculate total chlorine in container.

$$\text{Total Chlorine, lbs} = (\text{Chlorine, lbs/gal bleach})(\text{Bleach, gal})$$

$$= (0.44\ \text{lb/gal})(10\ \text{gal})$$

$$= 4.4\ \text{lbs}$$

2. Determine the chlorine used in pounds.

$$\text{Chlorine Used, lbs} = \frac{(\text{Total Chlorine, lbs})(\text{Chlorine Used, in})}{\text{Depth of Tank, in}}$$

$$= \frac{(4.4\ \text{lbs})(10\ \text{in})}{50\ \text{in}}$$

$$= 0.88\ \text{lb}$$

3. Estimate the average chlorine dose in mg/*L*.

$$\text{Average Chlorine Dose, mg/}L = \frac{(\text{Chlorine Used, lbs})(1{,}000{,}000/\text{Million})}{(\text{Water Treated, gal})(8.34\ \text{lbs/gal})}$$

$$= \frac{(0.88\ \text{lb})(1{,}000{,}000/\text{Million})}{(100{,}000\ \text{gal})(8.34\ \text{lbs/gal})}$$

$$= 1.1\ \text{mg/}L$$

QUESTIONS

Write your answers in a notebook and then compare your answers with those on pages 259 and 260.

4.5A List the various methods used to disinfect domestic water supplies.

4.5B List the common forms of chlorine available to disinfect water supplies.

4.5C What does a chlorinator do?

4.5D What substances found in water react with and consume chlorine?

4.5E List the two types of chlorine residuals that can be measured with test kits.

4.5F Calculate the desired free chlorine residual for a water with a pH of 7.6 when the water is 5°C. The contact time is 25 minutes.

End of Lesson 1 of 2 Lessons on SMALL WATER TREATMENT PLANTS

Please answer the discussion and review questions next.

DISCUSSION AND REVIEW QUESTIONS

Chapter 4. SMALL WATER TREATMENT PLANTS

(Lesson 1 of 2 Lessons)

At the end of each lesson in this chapter you will find some discussion and review questions. The purpose of these questions is to indicate to you how well you understand the material in the lesson. Write the answers to these questions in your notebook before continuing.

1. Why do small water treatment plants often not perform satisfactorily?
2. Why might prechlorination be recommended?
3. What happens if the coagulant dosage is either too high or too low?
4. If the cause of turbidity increases in the treated water is a change in the source water, what could have caused changes in the source water?
5. What would you do if a flocculation process is not satisfactorily producing a floc that settles quickly and clarifies the water effectively?
6. List the factors that influence the performance of a settling basin.
7. Successful operation of filters at high rates requires what conditions?
8. If a filter is not performing satisfactorily, what would you do?
9. Why do small water treatment plants frequently use hypochlorite forms instead of chlorine gas?
10. Effectiveness of disinfection of water with chlorine depends on what factors?
11. How is the time or "T" measured to calculate CT values?
12. How would you determine if the chlorinator setting is high enough?
13. How would you try to prevent chlorinator failures?

CHAPTER 4. SMALL WATER TREATMENT PLANTS

(Lesson 2 of 2 Lessons)

4.6 CORROSION CONTROL

Water is frequently corrosive to materials that it contacts, especially metal. Corrosion of metal is an *ELECTROCHEMICAL REACTION*[23] in which materials are dissolved into the water. This process of corrosion may work slowly but over the extended periods of time involved in water systems, it can be very important. Pipelines, storage tanks, and other structures can be seriously damaged by corrosion. Corrosion can decrease the flow capacity of pipelines and significantly increase pumping costs. Also, the quality of water delivered to consumers can be seriously affected by corrosion in the system. Corrosion can cause the water to become discolored or turbid from suspended rust particles and the water may have an unpleasant taste. The water may also contain harmful concentrations of metals like copper, zinc, and lead. Therefore, corrosion control measures are well worthwhile because water system parts that have been damaged by corrosion are very expensive to replace or repair and corrosion can cause the water quality to become unacceptable to consumers.

There are two general methods often used to control corrosion. One method is to use construction materials that resist corrosion. For instance, iron or steel pipe can be lined with cement or coated with coal tar enamel to protect the pipe. A zinc galvanized coating can be applied to small iron plumbing pipes. Copper pipes resist corrosion in most cases and plastic pipes are completely free of corrosion.

The second method of controlling corrosion is to treat the water with chemicals to make it noncorrosive. The chemicals sometimes used include lime, soda ash, caustic soda, sodium silicate, and polyphosphates. Chemical treatment of the water is of limited effectiveness and should be considered only as a supplement to the use of materials that are naturally resistant to corrosion.

The corrosion control chemicals are usually dissolved in water and fed in a solution form. Lime and soda ash can also be fed in the dry granular or powder forms. The dosage of lime, soda ash, or caustic soda is set to adjust the pH of the treated water so a thin film of calcium carbonate is deposited on the pipe surface. This thin coating prevents corrosion of the pipe. Usually the proper pH is in the range of 8.0 to 9.0.

One method to determine the proper pH is to calculate the Langelier Index which takes into account the various factors affecting the corrosiveness of water. The formula and procedures to calculate the Langelier Index are given in Chapter 8, "Corrosion Control," *WATER TREATMENT PLANT OPERATION*, Volume I, in this series of manuals. The dosage of silicate or phosphate is usually determined from past experience with similar types of water. Normally a dosage of several milligrams per liter is necessary. Do not use phosphate compounds if the system has open reservoirs because the phosphate may encourage excessive algae growths.

Corrosion control chemicals are the last ones to be applied after all other treatment has been accomplished. The reason is that the pH required for successful coagulation and disinfection is much lower than the pH required to make the water noncorrosive. Also, a coating of calcium carbonate on the filter media should be avoided because it interferes with effective filtration. The operator must understand that any lime or soda ash applied with the alum to improve coagulation has a minimum impact in controlling corrosion. These materials are all consumed in the chemical reactions that occur during coagulation. Therefore, to make the water noncorrosive, additional lime, soda ash, or caustic soda must be added in the final stage of treatment.

[23] *Electrochemical Reaction. Chemical changes produced by electricity (electrolysis) or the production of electricity by chemical changes (galvanic action). In corrosion, a chemical reaction is accompanied by the flow of electrons through a metallic path. The electron flow may come from an external source and cause the reaction, such as electrolysis caused by a D.C. (direct current) electric railway or the electron flow may be caused by a chemical reaction as in the galvanic action of a flashlight dry cell.*

The pH of the water is frequently used as a day-to-day indicator of the corrosive nature of the water. Water of higher pH is normally less corrosive than water of a lower pH. However, because of the many factors that can affect corrosion, the pH measurement is not totally reliable. The best method of monitoring corrosion is to examine specimens of the actual pipeline material or the surface of the structure involved. This inspection is easy if sections of pipe can be removed periodically. Alternatively, small specimens of the pipeline material can be mounted in the flowing water and periodically removed for inspection.

Waters with a low pH are very aggressive and can dissolve concrete and asbestos cement pipe. Increasing the pH of the water by the use of chemicals can reduce the rate of deterioration of the materials.

For additional information on corrosion control, see Chapter 8, "Corrosion Control," in *WATER TREATMENT PLANT OPERATION*, Volume I, in this series of manuals.

QUESTIONS

Write your answers in a notebook and then compare your answers with those on page 260.

4.6A List the two general methods of controlling corrosion.

4.6B What is the best method of monitoring corrosion in a water main?

4.7 SOLIDS-CONTACT CLARIFICATION
by J.T. Monscvitz

4.70 Process Description

Solids-contact units were first used in the Midwest as a means of handling the large amounts of sludge generated by water softening processes. It quickly became clear that this compact, single-unit process could also be used to remove turbidity from drinking water.

Solids-contact clarifiers (Figures 4.19 and 4.20) go by several names, which may be used interchangeably: solids-contact clarifiers, upflow clarifiers, reactivators, and precipitators. The basic principles of operation are all the same, even though various manufacturers use different terms to describe how the mechanisms remove solids from water. The settled materials from coagulation or settling are referred to as sludge and slurry refers to the suspended floc clumps in the clarifier. Sometimes the terms sludge and slurry are used interchangeably.

The internal mechanism consists of three distinct unit processes that function in the same way as any conventional coagulation-flocculation-sedimentation process chain. Sludge produced by the unit is recycled through the process to act as a coagulant aid, thereby increasing the efficiency of the processes of coagulation, flocculation, and sedimentation. This is the same principle that has been used successfully for many years in the operation of separate coagulation, flocculation, and sedimentation processes.

The advantages of using a solids-contact clarifier over operating the same three processes separately are significant, but are sometimes offset by disadvantages. For example, capital and maintenance costs are greatly reduced because the entire chain of processes is accomplished in a single tank. At the same time, though, operation of the single-unit processes requires a higher level of operator knowledge and skill. The operator must have a thorough understanding of how these processes operate and be able to imagine how all of these processes can occur in a small chamber or clarifier at the same time. Operators often have trouble visualizing what is happening in an upflow clarifier and become discouraged or upset when routine problems associated with solids-contact units occur.

A tremendous advantage in the use of the solids-contact units is the ability to adjust the volume of slurry (sludge blanket). By proper operational control, the operator can increase or decrease the volume of slurry in the clarifier as needed to cope with certain problems. During periods of severe taste and odor problems, for example, the operator can increase the sludge level and add activated carbon. The *ABSORPTIVE*[24] characteristics of activated carbon make it highly effective in treating taste and odor problems. Similarly, when coagulation fails because of increased algal activities, the operator can take advantage of the slurry accumulation to carry the plant through the severe periods of the day when the chemicals will fail to react properly because of changes in the pH, alkalinity, carbonate, and dissolved oxygen. In the conventional plant, the operator cannot respond to this type of breakdown in the coagulation process as well as the operator can with a solids-contact unit. Once algal activities have been determined to be causing the problem (readily checked by pH and DO) the operator can increase the amount of slurry available during good periods of the day and remove it during periods when the coagulation process is not functioning well. By skillfully making these slurry level adjustments, the operator can maintain a high quality effluent from the solids-contact unit.

The most serious limitation of solids-contact units is their instability during rapid changes in flow (throughput), turbidity level, and temperature. The solids-contact unit is most unstable during rapid changes in the flow rate. The operator should identify and keep in mind the design flow for which the unit was constructed. For easier use, convert the design flow to an *OVERFLOW RATE*.[25] This rate may be expressed as the rise rate in inches per minute or feet per minute. For each rise rate there will be an optimum (best) slurry level to be maintained within the unit. A rising flow rate will increase the depth of the slurry without increasing its volume or density. Conversely, a decrease in flow rate will reduce the level of slurry without changing its volume.

Sampling taps can be installed to enable the operator to monitor changes in slurry depth or concentration. The level of

---

[24] *Absorption (ab-SORP-shun). The taking in or soaking up of one substance into the body of another by molecular or chemical action (as tree roots absorb dissolved nutrients in the soil).*

[25] *Overflow Rate. One of the guidelines for the design of settling tanks and clarifiers in treatment plants. Used by operators to determine if tanks and clarifiers are hydraulically (flow) over- or underloaded. Also called surface loading.*

$$\text{Overflow Rate, GPD/sq ft} = \frac{\text{Flow, gallons/day}}{\text{Surface Area, sq ft}}$$

If we divide the overflow rate by 7.48 gallons per cubic foot and also divide by 1,440 minutes per day, we will have converted the overflow rate to a rise rate in feet per minute.

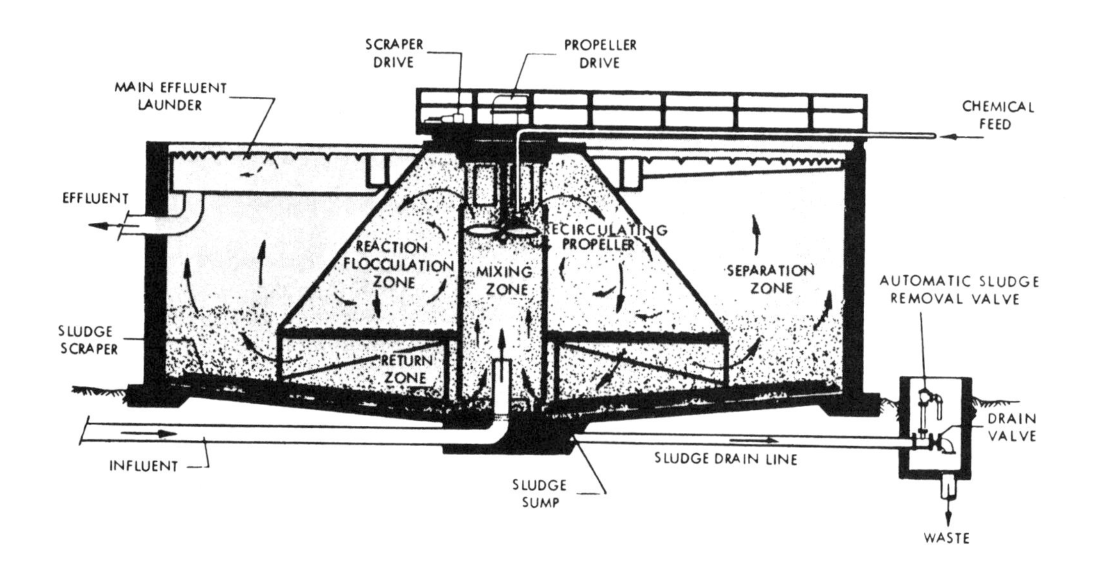

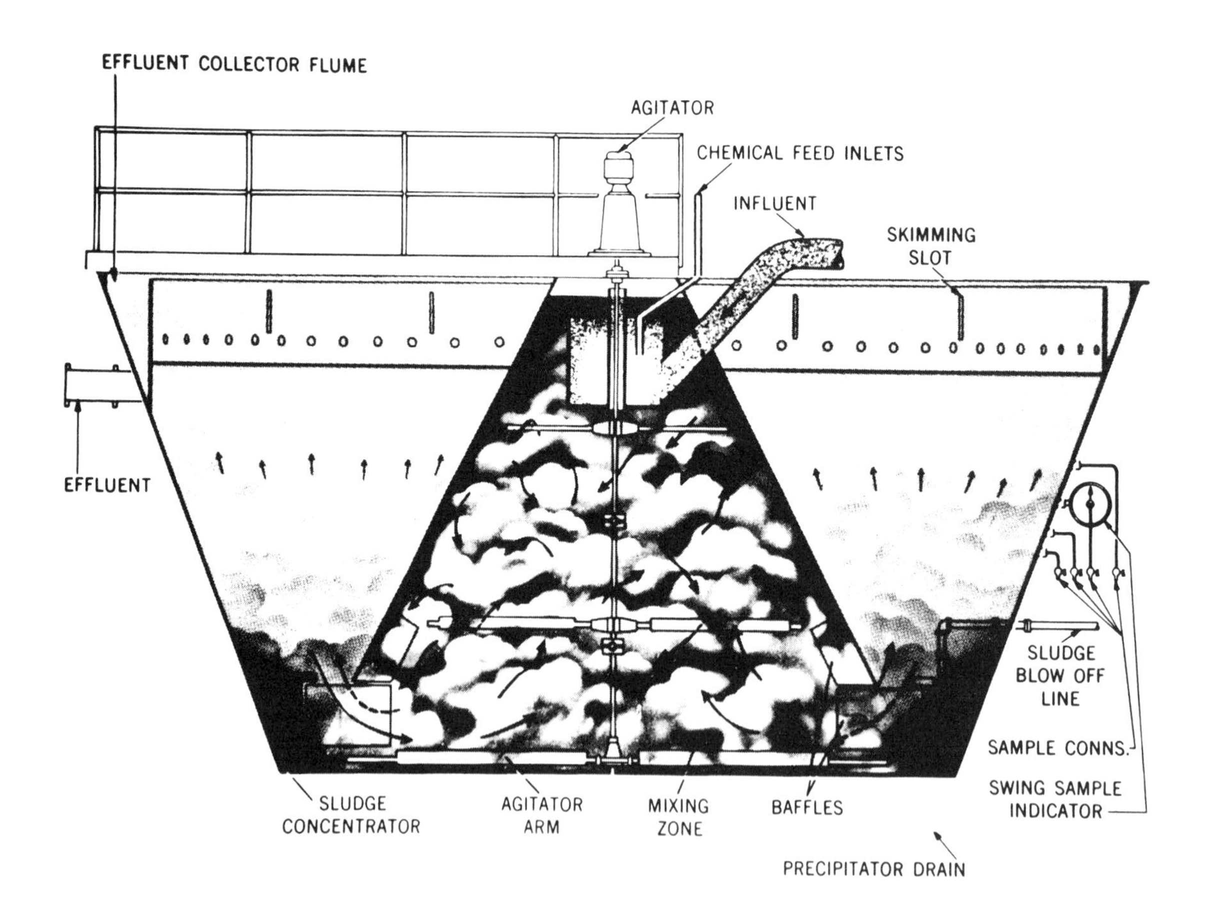

Fig. 4.19 Solids-contact clarifiers
(Permission of Permutit Company)

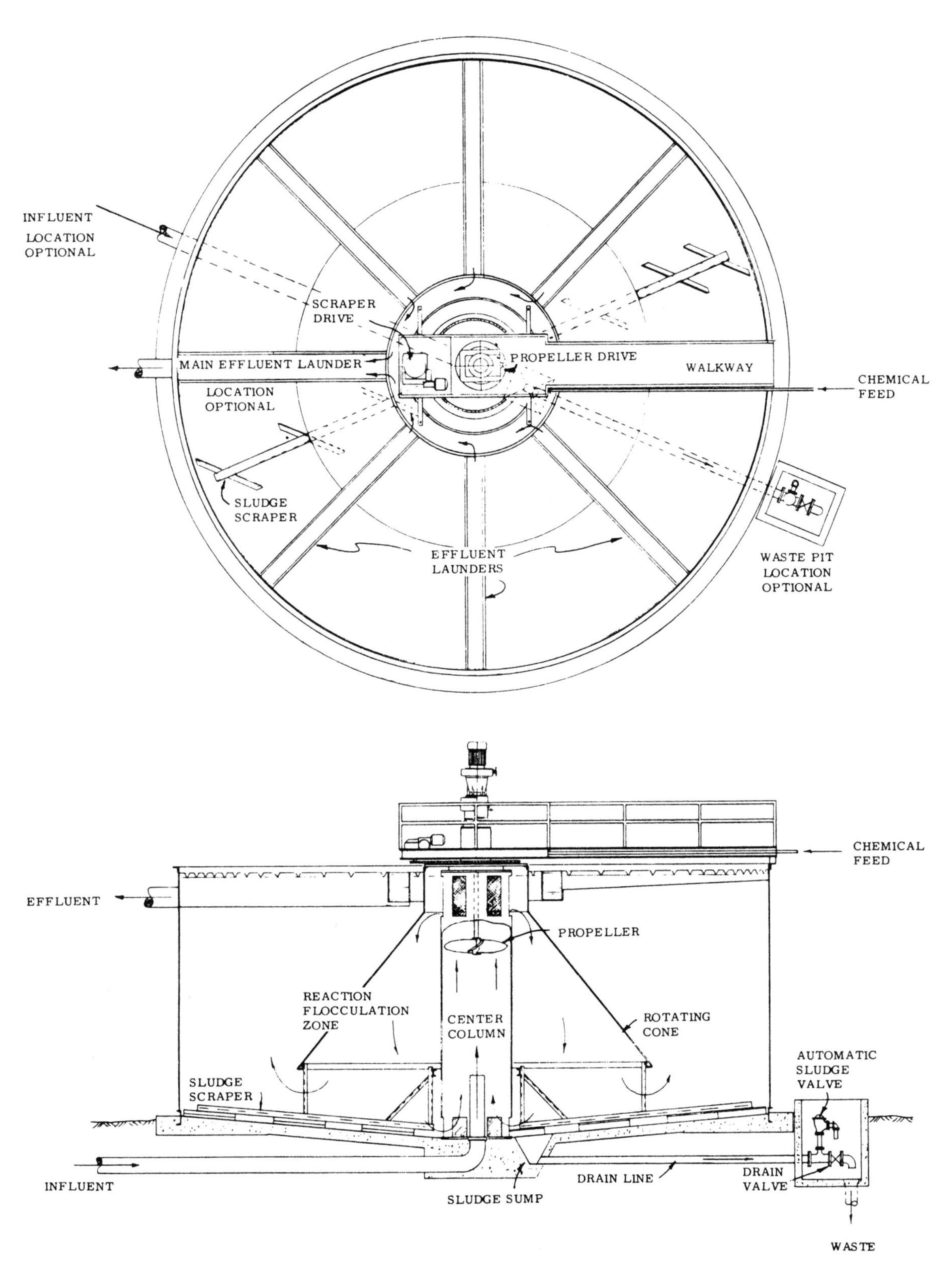

Fig. 4.20 Plan and section views of a solids-contact clarifier

(Permission of General Filter)

the slurry can be identified by placing sampling taps at various depths along the wall of the solids-contact reactor. The taps should penetrate the wall and extend into the slurry zone. It is often necessary to modify existing sampling taps or install additional sampling pipes to accomplish this.

By observing and measuring slurry depths at frequent intervals, the operator can easily monitor the rise or fall of the slurry levels; this will enable the operator to promptly make appropriate adjustments in the recirculation device and/or more tightly control the rate of change in the flow rate. This method of operational control is relatively effective in a gravity flow system when the water demands are moderate and the flow rate can be changed slowly. However, operational control in pressure systems is more difficult. Responding to rapid changes in demand or placing pumps into service at full capacity can easily upset an upflow clarifier immediately. In this case, the operator can witness the crisis occurring by observing the sampling taps, but be next to helpless to respond. The slurry will rise very rapidly in the settling zone, approach the overflow weirs, and spill onto the filters with a complete and total breakdown of the plant process.

Solids-contact clarifiers are also sensitive to severe changes in the turbidity of incoming raw water. The operator must be alert to changes in turbidity and must take immediate action. With experience, the operator will learn to accurately forecast when the turbidity may arrive at the reaction zone and will cope with the problem by increasing the chemical dosage prior to the arrival of excess turbidity. Early application of an increased chemical dosage puts the unit in a mode in which the turbidity can be handled successfully. The control of slurry and its influence on operational control during turbidity changes is discussed in greater detail later in this section.

The third factor which exerts a major influence on operation of a solids-contact unit is temperature. Changes in water temperature will cause changes in the density of the water; changes in density influence the particle settling rate. In extremely cold water, consider using polymers, activated silica, powdered calcium carbonate, or some other weighting agent to aid sedimentation without affecting coagulation. Simple heating by the sun on the wall of the tank or on the flocculant particles within the container will cause a certain amount of carryover of solids to occur. Operators who are not familiar with solids-contact units tend to become upset and overreact because of the potential carryover problem. This phenomenon is not a matter of serious concern because as the position of the sun changes, the convection currents change. The clouds of flocculant particles appear and disappear in response to the currents and there is no real need to control this phenomenon if the overall settled turbidity meets your objective. So long as the major portion of the sludge blanket lies in the settling zone, the few clouds of flocculant particles (which look like billowing clouds) really do not significantly harm the operation of the unit or the quality of the water produced.

Dramatic changes in temperatures and flow rates may sometimes make it impossible to control or prevent process upsets. If the slurry rises to the weirs and is carried over onto the filters, reduce the flow rate. If possible, use weighting agents before changing flow rates in cold water. The use of weighting agents may cause problems with the slurry requiring changes in recirculation rates. However, too high a recirculation rate may also cause the slurry to overflow onto the filters. During a change in temperature (cold water), be very careful in changing the flow rate.

QUESTIONS

Write your answers in a notebook and then compare your answers with those on page 260.

4.7A List at least two advantages of solids-contact units.

4.7B How can the level of the slurry or sludge blanket be determined in solids-contact units?

4.7C What should be done when a rapid change in turbidity is expected?

4.71 Fundamentals of Operation

The operator of a solids-contact unit controls the performance of the unit by adjusting three variables:

1. Chemical dosage,
2. Recirculation rate, and
3. Sludge control.

All three of these variables are interrelated and frequently you may have trouble distinguishing which one is the root of a problem you are encountering. However, if you will use the following analytical techniques, you should be able to separate the three fundamentals into separate groups of symptoms and diagnose the cause of the upset. This is not to say that if you have one problem, it may not co-exist with the other two.

First, let's consider chemical dosage. As in conventional treatment plants, proper consideration must be given to chemical dosage, otherwise the entire system of solids-contact clarification will collapse. There must always be sufficient alkalinity in the raw water to react with the coagulant. Assuming the coagulant used is aluminum sulfate, for every mg/*L* of alum added, 0.45 mg/*L* of bicarbonate alkalinity is required to complete the chemical reaction. To drive the chemical reaction sufficiently to the right (that is, for precipitation to occur), there should be an excess of 20 mg/*L* of alkalinity present. You may have to add sodium hydroxide (caustic soda), calcium hydroxide (lime), or sodium carbonate (soda ash) to cause sufficient alkalinity to be present. For example, if there was only 20 mg/*L* of natural alkalinity present, for every mg/*L* of alum added you should add 0.35 mg/*L* of lime if calcium hydroxide was being used.

All of this information can be verified by jar testing, which is fundamental in determining proper coagulation by chemical dosing. You should never attempt to make changes in solids-contact unit operation without first determining the proper chemical dosage through jar testing. For most solids-contact units, use the chemical dosages that produce floc which gives the lowest turbidity within a five-minute settling period after stopping the jar tester. Using the above criteria, the operator now can set the chemical feeders to dose the raw water entering the solids-contact unit.

The next control mechanism is recirculation. Here most often the plant operator is misled by intuitive judgment. The recirculation rate is established by the speed of the impeller, turbine, pumping unit, or by air injection. Any of these devices causes the slurry to recirculate through the coagulation (reaction) zone. To help you visualize how the slurry should look in the reaction zone, take another look at the lower drawing in Figure 4.19 and note the cloud-like, billowy appearance of the flocculated slurry in this area. Under normal operating conditions, the entire mass of suspended floc clumps billows and flows within the chamber. Its motion is continually being influenced by the mixing of recirculated sludge and incoming raw water. In principle, you are attempting to chemically dose the raw water when it enters the reaction zone and is mixed with the recirculated sludge. Coagulation and flocculation occur in the reaction zone and then the water and sludge pass into the settling zone. Some sludge is recirculated and mixed with incoming raw water and the rest of the sludge settles and is removed from the bottom of the settling area. At the point where water and sludge pass from the reaction zone to the settling zone, approximately one liter of water should rise and one liter of slurry should be returned into the reaction zone.

In order to sort out the effects of chemical dosages, recirculation rate, and sludge control, you should keep a log of the speed (RPM) of the recirculation device. If air is used for mixing, then the cubic feet of air applied per minute should be recorded. There is a direct relationship between the percentage of slurry present and the speed at which the mixing device is traveling.

To control the process, the operator must maintain the correct slurry volume in the reaction zone by exercising control of the rate of recirculation. The percentage of slurry can be determined by performing a volume over volume (V/V) test. The test procedure is as follows: using a 100-m*L* graduated cylinder, collect a sample from the reaction zone. Let the sample sit for five minutes and then determine the volume of slurry accumulated (m*L*) using the formula:

$$\text{V/V, \%} = \frac{(\text{Settled Slurry, m}L)(100\%)}{\text{Total Sample Volume, m}L}$$

At the same time, observe the clarity of the supernatant (settled water) that remains in the graduated cylinder. The clarity of the water above the slurry (the supernatant) will indicate to you how well the chemical reaction is proceeding. The percentage of accumulated solids by volume will indicate whether a proper amount of slurry is in the reaction zone. Customarily such reactors require 5 to 20 percent solids, or a higher percentage in the graduated cylinder at the end of a five- to ten-minute settling period.

Through recordkeeping and experience, you will find an optimum percentage of solids to maintain. You should perform the above analyses hourly and more frequently when the raw water quality is undergoing change. Accurate records must be kept.

The final step in control of the solids-contact unit is the removal of sludge which has accumulated on the bottom of the clarifier (settling zone). There are several means of sludge collection; some devices are located in areas of clarifiers that hold the sludge and are controlled by opening and closing recirculation gates. Others have scrapers that collect the sludge and move it to a discharge sump. In both cases the sludge is removed by hydraulic means (water pressure) through a control valve. The sludge removal mechanisms are generally on a timer, which operates periodically for a time duration set by the operator. The means of making this judgment is quite simple. Once again, use a graduated cylinder to collect a sample from the sludge discharge line. The sludge being discharged should be 90 to 98 percent solids in a V/V test, as indicated above. A five- to ten-minute period should be sufficient to make this determination. When slurries or sludge weaker than 90 percent is pumped, the operator is discharging a considerable amount of water and not leaving enough sludge to be recirculated into the reaction zone. If the percentage is considerably greater than 90 percent, then too much sludge may be accumulating and the recirculation device could become overloaded with too much return slurry.

If you will visualize the above reactions, you can see that with increased speed of the recirculation device, a larger amount of slurry can be retained in the unit. At the same time if this amount becomes too great, it may cause the sludge to rise and ultimately spill over the effluent weirs with the treated water. If the recirculation rate is too low, the solids may settle too soon and, without sufficient recirculation, will not return to the reaction zone. The absence of solids in the reaction zone causes improper coagulation. The net result is a failure of the total solids-contact system.

Considering the above principles and provided with some experience, you should be able to determine an optimum amount of slurry to be present that will satisfy a given recirculation rate, coupled with proper chemical dosage, and a sufficient percentage of solids for recycling. You should always be aware of the amount of solids in the reaction zone and, based upon practical experience, know approximately the percentage required. Some of the obvious difficulties in this judgment will occur as the raw water turbidity changes. For instance, in muddy streams carrying silt, sedimentation may occur very rapidly, thus requiring increased circulation rates to maintain sufficient slurry in the reaction zone (even with proper chemical dosage), and also requiring greater sludge removal rates. As the raw water turbidity becomes lighter, the increased circulation rate may cause the slurry blanket to rise to an uncontrolled depth in the settling zone. Removing too much sludge will also produce this same effect. All of these problems are readily observed in the V/V test for solids determination in the reaction zone; also, this is cause for increased observations of the V/V during water quality changes.

Another problem may be caused by cold water when the recirculation rate may be too high for the densities of the particles present. A set of recirculation speeds for warm-weather operation may be entirely different from those used during cold weather. As a remedy, the operator may select a nonionic polymer as a weighting agent to increase the settling rate in cold waters. Other alternative chemicals are powdered calcium carbonate or the use of activated silica. A note of caution in chemical dosage determination: the reactions in the jar tester should be reasonably rapid to ensure comparable reactions within the solids-contact unit.

Another important point when determining chemical dosage for a solids-contact unit is that a specific set of jar test guidelines will be needed for each plant. For example, you should determine the volume of the reaction zone and the period of detention of the raw water in that reaction zone. This, along with knowledge of the speed of the recirculation device, should allow you to determine the detention time and flocculator speed in the jar tester.

In the real world, this means if the flow rate of the solids-contact unit is 10 minutes in the reaction zone and the speed is two feet per second (0.6 m/sec), then the jar tester mixer should turn at a speed equal to two feet per second with a coagulation period of 10 minutes. You should duplicate in the jar tester, as nearly as possible, those conditions of chemical dosage, detention period, and mixing speeds that occur in the solids-contact unit. Using these guidelines, you should be able to approach approximate real-world conditions in the laboratory and better optimize chemical dosages.

4.72 Maintenance

Solids-contact units, like all waterworks equipment, require at least a minimum of maintenance. The primary consideration is the recirculating device which needs regular inspection of the belt drive, gear boxes, and lubrication. Also if the unit has a sludge collector, then its drive and gear boxes require the same attention. The units should be inspected daily and lubricated in accordance with the manufacturer's recommendations. Also, the contact unit may need to be drained periodically and the sludge collectors inspected for wear and corrosion.

Sludge collector devices are usually constructed of steel within a concrete container. Thus, there is a need to inspect the *CATHODIC PROTECTION*[26] system if one is provided with the unit. Weekly readings should be kept concerning the amperes and voltage supply. Changes in these readings indicate that a problem may be developing. The cathodic protection devices should be inspected and if any defects are detected, they should be corrected as soon as possible.

For additional information on solids-contact units, see Section 5.243, "Arithmetic for Solids-Contact Clarification," page 184, of *WATER TREATMENT PLANT OPERATION*, Volume I, in this series of manuals.

QUESTIONS

Write your answers in a notebook and then compare your answers with those on page 260.

4.7D How is the proper chemical dose selected when operating a solids-contact unit?

4.7E List the devices that may be used to provide recirculation in a solids-contact unit.

4.7F How is the percentage of slurry present in the reaction zone determined?

4.8 SLOW SAND FILTRATION
by Peg Hannah

4.80 Development of Process

The use of slow sand filtration to purify water dates back to 1829 when the process was first used in London. The method was later used widely throughout Europe, but it was not often practiced in the United States. The two most widely used filtration systems in the U.S. today, high rate gravity filters and pressure filters (described in Section 4.4), are modifications of the slow sand filtration process which were the result of efforts to make the process better, faster, and cheaper.

A renewed interest in slow sand filtration developed during the late 1980s, and a growing number of pilot projects were undertaken to study the process. The driving force behind these efforts was the Safe Drinking Water Act of 1974 and its Amendments (1986), in particular the Surface Water Treatment Rule (SWTR) which the Environmental Protection Agency promulgated in 1989. The SWTR is especially significant for small water supply systems. The Rule requires all water systems whose water supply is surface water or groundwater under the influence of surface water[27] to disinfect their water and, under most circumstances, to install filtration equipment. The Rule further specifies which filtration methods are acceptable and the turbidity monitoring frequencies and limits. In 1990, it was estimated that about 2,900 water systems used surface water as a source of supply but did not practice filtration; approximately 90 percent of those systems served populations of fewer than 10,000. The practical result of the SWTR was that most of these small water systems had to begin filtering their source water using one of four modes (or methods) of water treatment by filtration: conventional, direct, diatomaceous earth, or slow sand. The best mode for a particular community will depend upon the characteristics of the supply water, the daily volume of filtered water required to serve the community, and the cost and availability of qualified operators.

[26] *Cathodic (ca-THOD-ick) Protection. An electrical system for prevention of rust, corrosion, and pitting of metal surfaces which are in contact with water or soil. A low-voltage current is made to flow through a liquid (water) or a soil in contact with the metal in such a manner that the external electromotive force renders the metal structure cathodic. This concentrates corrosion on auxiliary anodic parts which are deliberately allowed to corrode instead of letting the structure corrode.*

[27] *States must determine which community and noncommunity groundwater systems are under the direct influence of surface water. The SDWA defines "groundwater under the influence of surface water" as, "Any water beneath the surface of the ground with (i) significant occurrence of insects or other microorganisms, algae, or large diameter pathogens such as Giardia lamblia, or (ii) significant and relatively rapid shifts in water characteristics such as turbidity, conductivity, or pH which closely correlate to climatological or surface water conditions."*

Of the four methods of filtration, slow sand filtration appears to be a simple, cost-effective option for small water systems. The purpose of this section is to describe the slow sand filtration equipment and process, identify the factors that should be considered before installing a slow sand filter, and provide general guidelines for the operation and maintenance of a slow sand filter. (The other three filtration methods (conventional, direct, and diatomaceous earth) were described earlier in Section 4.4, "Filtration," and are also described in *WATER TREATMENT PLANT OPERATION*, Volume I, in this series of manuals.)

Even though the equipment is relatively simple, design guidelines and cost projections are greatly influenced by local conditions. A qualified engineer should assist in the design of a slow sand facility and it is strongly recommended that a pilot project be undertaken prior to construction of a full-scale plant in order to assess the effectiveness of a slow sand filter operating under actual local conditions. Other sources of information on slow sand filters are listed in Section 4.15, "Additional Reading."

4.81 Procedure for Treating Water

Typically, a filter removes suspended particles from water by means of several physical processes. The particles may be trapped in the filter if they are too large to pass through the openings in the filter media; this is the action of a simple strainer. To increase the effectiveness of the straining action, chemicals are usually added to cause the finer particles to clump together (coagulation/flocculation) into larger particles which can more easily be removed. As water moves through the sand (media) bed, suspended particles strike the media (sand grains), their movement is slowed, and they may settle (sedimentation) into the void spaces between sand grains. In addition to this straining and settling action, filters also remove particles which stick (adhere) to the media. The mechanisms which cause adhesion (adsorption) are not fully understood, but may include chemical bonding or electrical forces.

Slow sand filtration is a promising alternative for small water systems even though the filtration rate is 50 to 100 times slower than rapid sand filtration. By combining a simple design, ease of operation requiring minimal staffing, and the ability to remove *Giardia lamblia* cysts as well as *Cryptosporidia*, coliforms, and other microorganisms, slow sand filters meet the needs of many small communities.

In slow sand filters, no filter-aid chemicals are used as they often are in other water filtering systems. However, an additional benefit of the slow sand process is that a biological process assists in the removal of suspended matter. As the layer of trapped matter at the surface of the filter builds up, a dense population of microorganisms develops. This mat is called the "schmutzdecke" (sh-moots-DECK-ee), a German word meaning "dirty skin." The mat itself is only a few centimeters (about an inch) thick but organisms within the film or mat feed on and break down incoming organic material trapped on the schmutzdecke. In doing so they both remove organic matter and add mass to the layer, further developing the schmutzdecke and increasing the physical straining action of the layer.

At the same time other varieties of organisms deeper within the sand media form a biofilm on the sand particles. Organic material passing through the media sticks to the biofilm where the organisms then use the material and thereby remove it or convert it to carbon dioxide and water. The schmutzdecke and biofilm develop or "ripen" over time and the filter is said to be "mature" when the population of organisms becomes well established and the filter produces acceptable filtered water. The filter will continue to operate in this condition until head loss indicates a clogged condition and adequate flow through the filter can no longer be maintained. Filter cycle times are widely variable but usually average at least 1½ to 2 months.

When performance indicators signal the need to clean the filter, the top layer of sand and accumulated material, usually less than one inch thick, is simply scraped off the surface and removed. With time the cleaned sand bed ripens and will again produce design flows of acceptable quality. After several cleanings, new sand is added to bring the media back to its design depth.

The principal removal mechanism in a slow sand filter is physical straining or entrapment of particulates. Both the schmutzdecke and the biofilm contribute to overall particulate removal. There is some speculation that perhaps their relative removal rates vary from filter to filter, depending on factors such as the specific microorganisms which predominate, presence of algae, and maturity of the bed.

4.82 Components (Figure 4.21)

The structure of a slow sand filter is similar to a gravity sand filter. Components include a tank, underdrain piping, gravel media support, sand media bed, water intake and distribution piping, finished water (tailwater) weirs and holding tanks, flow control valves or weirs, and meters to measure head loss and flows. A cover is sometimes provided to prevent freezing of water in the tank, growth of algae, and accumulation of wind-blown debris.

4.820 Filter Tank

Slow sand filter tanks may be constructed at the site using steel-reinforced concrete, poured concrete, or earthen berms with a watertight liner. In small installations, prebuilt tanks made of fiberglass, galvanized steel, or reinforced concrete are sometimes an economical choice. The tanks may be circular, square, or rectangular. The filter tank should consist of two or more equal-sized chambers or cells, each of which can be operated independently of the others so that one can be shut down for cleaning while the others continue to operate. Design questions about the size, shape, and number of filters as well as methods of construction will be greatly influenced by the filter bed area needed to produce adequate treated water.

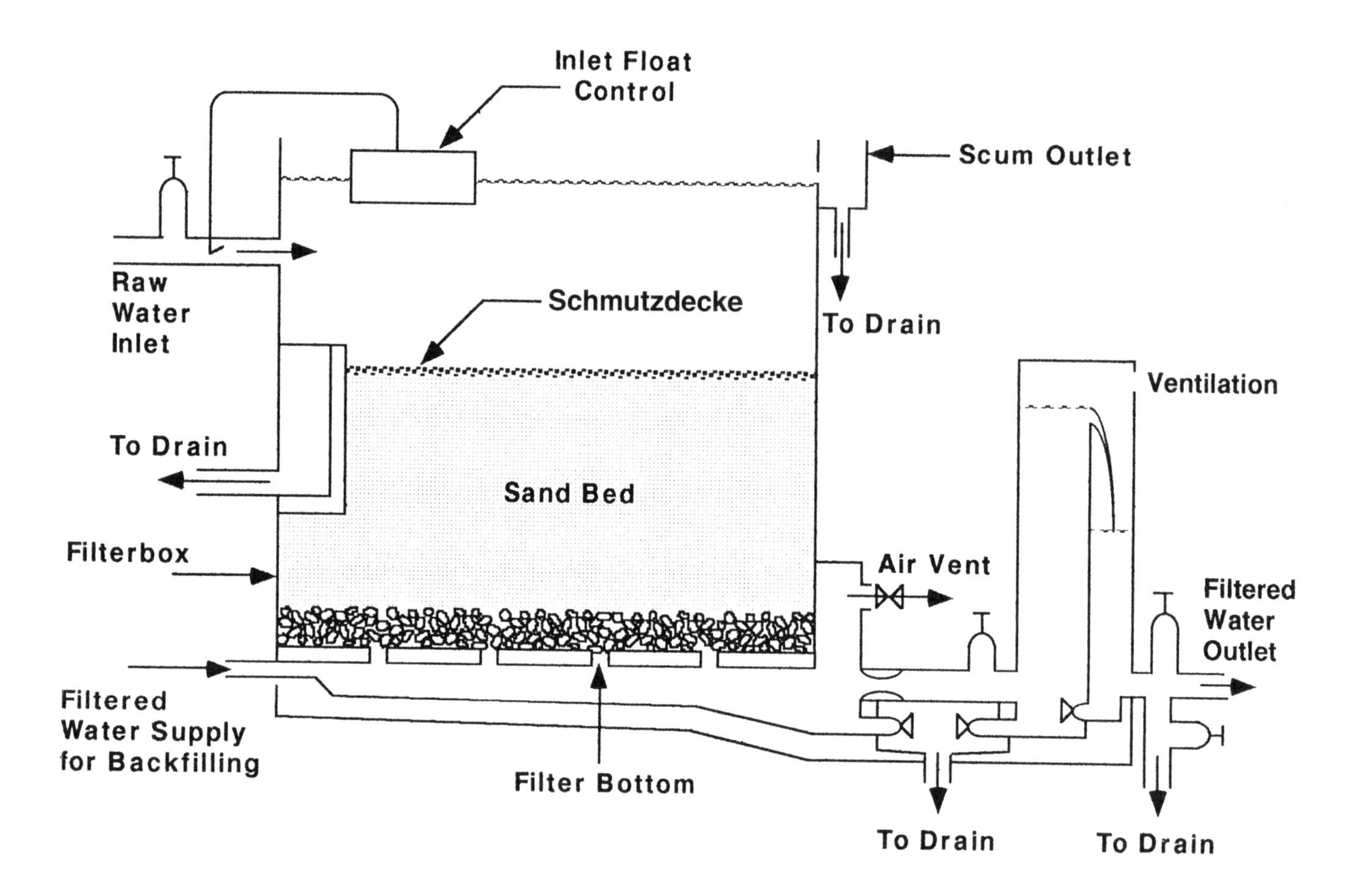

Fig. 4.21 Flow diagram of slow sand filtration system

(Adapted from "What Is Slow Sand Filtration?" *OPFLOW*, Vol. 18, No. 2 (Feb. 1992), by permission. Copyright © 1992, American Water Works Association)

FORMULA

The size of the tank filter surface area can be calculated by dividing the maximum expected flow by the maximum hydraulic loading rate when all but one cell is in service.

$$\text{Filter Surface Area, sq ft} = \frac{\text{Flow Rate, gal/day}}{\text{Hydraulic Loading Rate, GPM/sq ft}}$$

EXAMPLE 14

A slow sand filter has two cells; each is four feet wide and ten feet long. The desired flow is 7,000 gallons of filtered water per day at a hydraulic loading rate of 0.15 gallons per minute per square foot of filter surface area. Is the filter adequately sized to handle this demand?

Known		Unknown
Flow Rate, gal/day	= 7,000 gal/day	1. Surface Area, sq ft
Hydraulic Loading Rate, GPM/sq ft	= 0.15 GPM/sq ft	2. Is surface area adequate?
Length, ft	= 10 ft	
Width, ft	= 4 ft	

1. Calculate the filter surface area needed.

$$\text{Surface Area, sq ft} = \frac{\text{Flow Rate, gal/day}}{\text{Hydraulic Loading Rate, GPM/sq ft}}$$

$$= \frac{7{,}000 \text{ gal/day}}{(0.15 \text{ GPM/sq ft})(60 \text{ min/hr})(24 \text{ hr/day})}$$

$$= \frac{7{,}000}{216 \text{ sq ft}}$$

$$= 32.4 \text{ sq ft}$$

2. Calculate the available surface area for each cell.

$$\text{Surface Area, sq ft} = (\text{L, ft})(\text{W, ft})$$

$$= (10 \text{ ft})(4 \text{ ft})$$

$$= 40 \text{ sq ft}$$

Since the surface area available from one cell is 40 square feet, which is greater than 32.4 square feet, sufficient surface area is available in one cell to meet the flow demand when the other cell is out of service.

Because a slow sand filter must be shut down for cleaning and resanding, it will be necessary to build in enough extra capacity to provide 100 percent of expected flow during shutdowns. In a single tank with two independent chambers, each chamber must accommodate the entire amount. If four tanks or chambers are used, the total capacity of three must equal 100 percent of expected production; the fourth chamber thus represents excess capacity. In Example 14, on the previous page, each of the two cells must have at least 32.4 square feet of filter surface area; therefore, there is 32.4 square feet or more of excess capacity. As the number of chambers or tanks increases, the amount of excess capacity that must be built declines, thereby somewhat reducing initial construction costs. At some point, however, the savings realized by increasing the number of chambers will be offset by the cost of piping and valves to the additional chambers. The designer must balance these factors to achieve the most economical design.

Other factors to be considered in selecting an appropriate tank design include required land area, need for a cover or enclosure, and the filter cleaning method. For example, tank construction for a medium to large plant may be influenced by the need to construct access ramps for cleaning equipment. Access to round tanks for cleaning can be more difficult. Basins formed by earthen berms require more land area than comparable capacity concrete tanks so the type of construction may be limited by the available land area. If the local climate is such that freezing will be a problem, a roof may be needed and the designer must consider the added structural weight in the design of the basin. Small filters may need to be fitted with a removable cover which permits easy access for cleaning. Many such considerations influence tank design, but it is beyond the scope of this discussion to consider them here.

The surface area of the filter will vary from site to site since it is based on the required capacity, but the depth of the tank is somewhat less variable. This dimension is the sum of the depths of the gravel support layer/underdrain system, the sand bed, maximum water depth (headwater) above the filter, and the freeboard. If the filter is covered or enclosed, additional headspace must be provided to permit safe access for cleaning activities and equipment. Figure 4.22 illustrates a cross section of a slow sand filter box. The upper limits shown for filter bed depth and headwater depth in this illustration are slightly higher than typical.

4.821 Underdrain System

The structure and functions of a slow sand filter underdrain system are similar to the underdrain system of the rapid rate gravity filter described in Section 4.4, "Filtration." Perforated or slotted PVC pipe is frequently used for underdrain piping. In the most common configuration, a series of slotted pipes are laid parallel along the bottom of the tank and are connected to a larger manifold or header pipe. Filtered water collects in the perforated pipes, flows through the header, and flows by gravity out of the filter tank to a clear well or reservoir. (In other types of gravity filters the underdrain distributes backwash water to clean the filter media.)

A bed of gravel covers the perforated underdrain piping. As with rapid rate gravity filters, the gravel bed may consist of several layers of different gravel sizes arranged from largest on the bottom in contact with the underdrain piping to smallest on top in contact with the sand. The gravel support layer reduces the velocity of the water, supports the sand media, prevents the sand from being carried into the underdrain pipes, and helps to distribute water evenly when the tank is being filled.

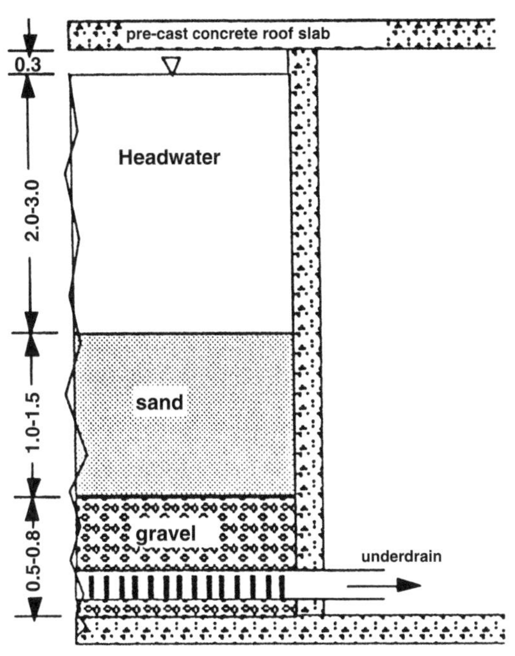

NOTE: To convert the above measurements in meters to feet, multiply the above values by 3.281 feet/meter.

Fig. 4.22 Cross section of slow sand filter showing filter box with underdrain, gravel support, sand, headwater, and freeboard

(Reprinted from *MANUAL OF DESIGN FOR SLOW SAND FILTRATION*, by permission. Copyright © 1991, American Water Works Association)

Before gravel is placed in the filter it should be carefully washed to remove any dirt, clay, or other loose debris. Unless these materials are washed out before the filter is placed into operation, they will wash out during operation and cause increased effluent turbidity.

An important factor which must be considered in the design of a slow sand filter underdrain system is the method that will be used to clean the filter. Allowance must be made for the additional weight of equipment and operators who will be working on the surface of the filter bed during cleaning operations. In a very small plant, this may only be one person with a wheelbarrow, but in larger installations heavier equipment such as tractors may be used.

4.822 Sand Media Bed

The material used as media in a slow sand filter is, as the name implies, sand (generally silica sand). Materials such as anthracite coal, garnet sand, and diatomaceous earth are frequently used in other types of filters (as described in Section 4.4), but long experience has proven silica sand to be an effective and inexpensive filter media for the slow filtration process.

In selecting sand for the media bed of a slow sand filter, the goal is to construct a bed that produces acceptable quality finished water while ensuring that the bulk of the filtered particulate matter accumulates in the upper sand layers without reducing filter cycle time to unreasonably short filter runs. Field studies indicate that this can be accomplished using sand with an *EFFECTIVE SIZE*[28] between 0.15 and 0.35 millimeters. Smaller sand tends to clog and cause short filter runs while larger sands may permit the buildup of filtered particulates deeper in the media. Acceptable *PATHOGEN*[29] rates can be achieved with sand that is larger than the recommended effective size of 0.15 to 0.35 millimeters if the bed is biologically mature.

A second factor that must be considered when selecting sand is the consistency of grain size among sand grains, or *UNIFORMITY COEFFICIENT.*[30] The recommended uniformity coefficient is 1.5 to 3; however, uniformity coefficients up to 5 have been used successfully.

Sand meeting the specifications for effective size and uniformity coefficient can often be purchased locally from building supply companies and is generally classified as masonry or concrete sand. Sand specifically graded for filter use may be purchased from commercial filter media suppliers in 100-pound sacks of a specified size, hardness, and cleanliness.

Sand represents a major capital expense in construction of a slow sand filter. The use of local sand for filter media if it meets size, gradation, and cleanliness standards is preferred for cost considerations. Initially, a slow sand filter may require 30 to 100 times more sand than needed for a high rate filter producing the same amount of finished water. In addition, the dirty sand that is periodically removed during filter cleaning is sometimes discarded and replaced rather than washed and reused. When local sand can be used as filter media, costly shipping expense is avoided. Also, if low-cost, clean local sand is available it may not be cost-effective to wash and reuse dirty sand.

Whatever the source of the sand, it must be clean. Particles of dirt, clay, or organic matter present in the sand when placed in the filter will slowly wash out in the filtered water contributing to increased filtered water turbidity. Since slow sand filters are not backwashed, unacceptably high turbidity could continue for an extended period of time (sometimes as long as one to two years) if the sand is not clean when placed in the filter.

QUESTIONS

Write your answers in a notebook and then compare your answers with those on page 260.

4.8A How does the Surface Water Treatment Rule (SWTR) significantly impact the drinking water treatment processes used by small water supply systems?

4.8B Why is a tank cover sometimes provided on a slow sand filter?

4.8C What factors influence the depth of a slow sand filter?

4.8D What happens if the diameters of sand particles in a slow sand filter are (1) too large, or (2) too small?

4.823 Flow Control Piping, Valves, and Gages

Effective operation of a slow sand filter requires a constant flow of water through the filter. The flow rate through the filter depends on the head (height) of water maintained above the sand filter and the head (friction or energy) losses as the water flows through the sand filter. The head loss increases over time as the schmutzdecke develops and as material is retained in the top layer of the sand. As the head loss increases, a constant flow through the filter can be maintained in either of two ways: (1) by adjusting the raw water influent valve (B. Top of Figure 4.23), or (2) by adjusting the finished water effluent valve (I. Bottom of Figure 4.23) at the point where filtered water flows from the underdrain piping to the outlet chamber.

INFLUENT FLOW CONTROL. Influent flow rate control (Top, Figure 4.23) permits the operator to adjust the filtration rate by controlling the amount of water that flows into the filter. This method is sometimes referred to as the rising water level method because when influent flows at a constant rate, the level of the headwater gradually rises due to increasing head loss through the filter as the run progresses. The filtration rate can be held constant by this method or adjusted as needed by opening or closing the influent valve. When the operator increases the flow into the filter, the headwater depth rises; this increases the head on the filter. Greater head (pressure) increases the filtration rate and thus produces more filtered water.

At the start of a filter run when filter head loss is low, the water surface of an influent-controlled filter is only a short distance above the media. To prevent disturbance and erosion of the sand bed, water entering the filter should first strike a splash plate above the sand surface to reduce the force of the water stream.

The headwater surface level rises as head loss through the filter increases. The rise is gradual at the start of the filter run when head loss increases are gradual. Near the end of the filter run, however, head loss develops more rapidly and the corresponding rise in headwater surface level is easily visible. By monitoring the rising headwater level (both visually and with gages), the operator can anticipate when the filter will need to be shut down and cleaned and can make the necessary preparations.

Filtered water leaves the influent-controlled system over a weir higher than the filter sand surface, but no effluent valve is needed to control the rate of flow.

---

[28] *Effective Size (E.S.). The diameter of the particles in a granular sample (filter media) for which 10 percent of the total grains are smaller and 90 percent are larger on a weight basis. Effective size is obtained by passing granular material through sieves with varying dimensions of mesh and weighing the material retained by each sieve. The effective size is also approximately the average size of the grains.*

[29] *Pathogens (PATH-o-jens). Pathogenic or disease-causing organisms.*

[30] *Uniformity Coefficient (U.C.). The ratio of (1) the diameter of a grain (particle) of a size that is barely too large to pass through a sieve that allows 60 percent of the material (by weight) to pass through, to (2) the diameter of a grain (particle) of a size that is barely too large to pass through a sieve that allows 10 percent of the material (by weight) to pass through. The resulting ratio is a measure of the degree of uniformity in a granular material such as filter media.*

$$\text{Uniformity Coefficient} = \frac{\text{Particle Diameter}_{60\%}}{\text{Particle Diameter}_{10\%}}$$

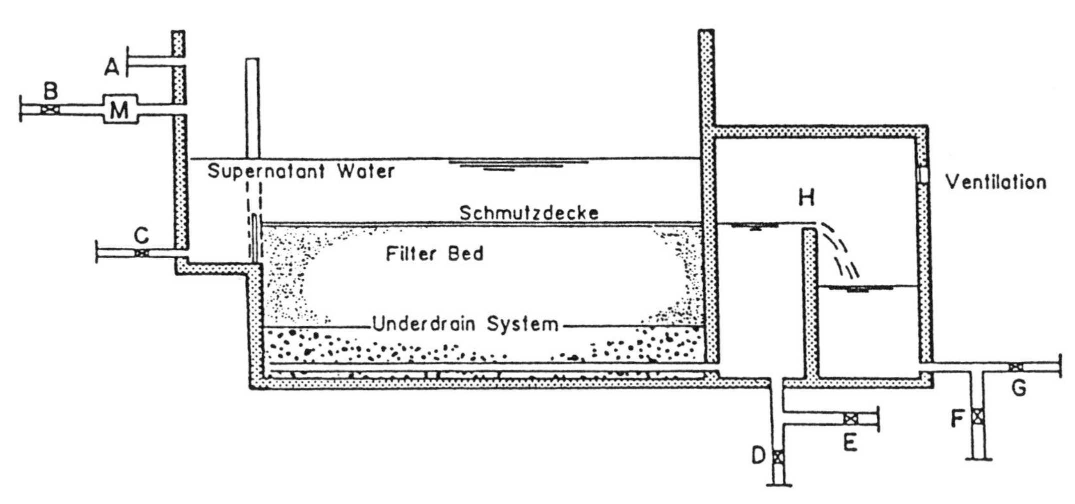

(a) Basic components of a slow sand filter with influent flow control and rising water level.

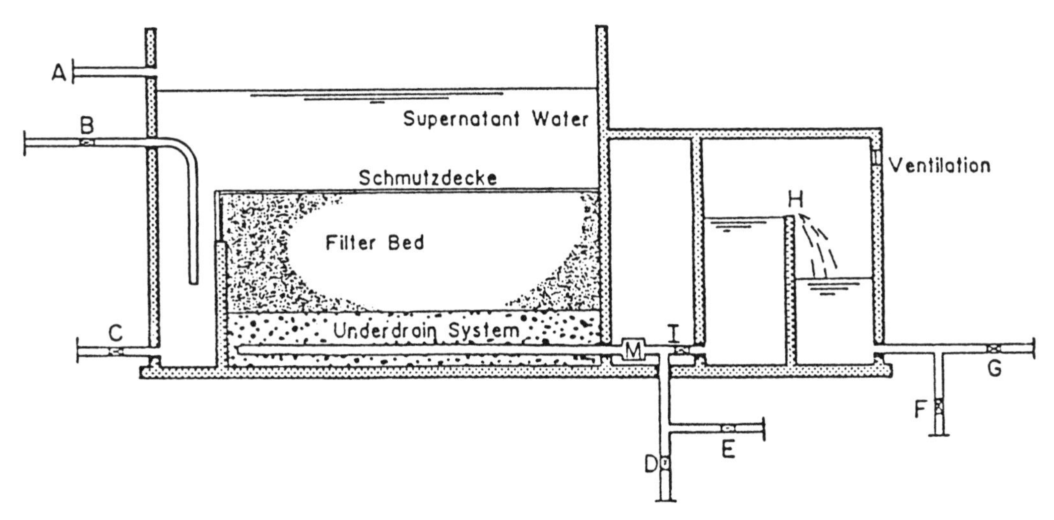

(b) Basic components of a slow sand filter with effluent flow control.

Fig. 4.23 Influent and effluent flow control

(Reprinted from *SLOW SAND FILTRATION*, Logsdon, Gary S., Editor, 1991, by permission of the American Society of Civil Engineers. Copyright © 1991, by the American Society of Civil Engineers)

EFFLUENT FLOW CONTROL. Effluent flow control is another approach to controlling the filtration flow rate to meet flow demands. The effluent flow control valve (I. Bottom of Figure 4.23) can be used to maintain sufficient head to keep the unfiltered water at the desired level above the filter sand. As the head loss increases in the filter during the run, the valve is opened enough to lower the head loss through the valve and achieve the desired flow rate. When the effluent flow control valve is fully opened and the desired flow cannot be sustained or obtained, the filter must be removed from service and cleaned.

Effluent-controlled filters require a float valve or level sensor to regulate the rate of flow into the filter to maintain a constant flow through the filter. The float valve throttles the influent flow to precisely match the effluent flow. However, if the filter is supplied by gravity flow, a slight excess flow can be delivered to the filter (by adjusting an influent valve or weir) and wasted back to the source using the supernatant overflow or filter overflow drain. This condition represents the maximum head possible on the filter and operators should be careful not to exceed design flows or head conditions except during an emergency such as a fire.

4.824 Outlet Chamber

The function of the outlet chamber is to maintain at least three inches (8 cm) of water over the top of the sand bed during operation. This is necessary to prevent accidental dewatering and drying of the filter. It also prevents air from entering and becoming trapped in the media (air binding), a condition that will cause serious operational problems.

An adjustable weir (see Figure 4.23) on the outlet chamber can be raised at the start of the filter run so that the weir crest (tailwater elevation) is about one foot (0.3 m) higher than the surface of the media. As the filter run progresses and filtered material builds up on and in the media, head loss through the filter will gradually increase. As head loss increases, the weir can be lowered gradually until head loss across the sand bed is greater than one foot (0.3 m) and the weir crest is at or slightly above the level of the sand bed surface. Maintaining the tailwater elevation at this level will ensure the sand media is always covered with water, even if the influent flow is interrupted.

If flow measurement is desired, a hook gage can be installed to measure the depth of water flowing over the weir crest. Tables and formulas are available to convert the flow depth to a flow rate in gallons per minute (GPM). The weir should be considered an auxiliary flowmeter and not a substitute for conventional flowmeters.

4.825 Finished Water Holding Facility

The size of the finished water storage facility is based on capacity needed to provide detention time for disinfection, meet peak daily demand and fire protection requirements, and provide emergency reserves expected to be needed in case of power outages or other disruptions in water production. The capability to store enough water to equalize flows over a period of time is very important to the operation of a filter. Slow sand filters operate most efficiently at a steady flow rate; frequent or abrupt changes in flow rate can disrupt the process and may cause a reduction in water quality. A storage capacity equal to 12 hours' production (or more) of finished water would be appropriate at a slow sand filter plant.

Selecting a point of application for disinfection requires consideration of a number of factors. If disinfectant (usually chlorine) is applied before the finished water enters the storage tank, any excess water wasted from the tank will contain chlorine. Chlorinated water can harm wildlife and vegetation and should not be returned to the source water stream or reservoir. Therefore, safe disposal of the chlorinated water may be a problem. At some plants a clear well has been installed between the filter and the storage facility. Water is chlorinated after it leaves the clear well. This permits excess water to be wasted from the clear well before it is chlorinated.

If installing an additional clear well is not feasible, another way to avoid wasting chlorinated water is to apply chlorine as water leaves the storage facility and enters the distribution system. With this arrangement it is important to determine whether sufficient contact time can be provided as water flows through the distribution system. Also, flow past the disinfection equipment will vary with demand so the chlorine must be applied at a rate proportional to the flow.

When calculating CT (**C**oncentration of disinfectant x Contact **T**ime) values (see Section 4.5 for details) to determine the efficiency of disinfection for *Giardia lamblia* inactivation, the time the water is in the holding facility is part of the contact time (T) if the disinfectant is applied before the finished water holding facility. If chlorine is applied as water enters the distribution system, contact time is measured from the point of application of the disinfectant to the point of use by the consumer.

4.826 Hydraulic Controls and Monitoring Devices

In addition to the flow control valves and weirs previously described, several other simple valves, meters, and devices are usually provided to control and monitor the filter operation. As illustrated in Figure 4.23, these devices include:

- Influent valve — used to isolate the filter for maintenance.
- Supernatant drain valve — permits the operator to adjust the level of water in the tank, if necessary, and to dewater the filter rapidly prior to cleaning.
- Filter overflow drain — prevents flooding of the facility if the filter suddenly clogs and the tank fills before the water is shut off.
- Filter bed drain — used to lower the water level below the sand surface prior to cleaning.
- Filter-to-waste valve — permits inadequately treated water to be returned to the influent line or otherwise disposed of during the ripening period after cleaning or resanding.

- Backfilling valve — used to fill a new or newly cleaned filter through the underdrain system.
- Treated water waste valve — permits disposal of excess finished water when stored supply exceeds demand.
- Valve to clear well — used to isolate filter during maintenance.
- Clear well overflow — (not shown in Figure 4.23) permits excess water (from continuous operation) to be sent back through the filter. Placement of the overflow line must not permit a cross connection through which unfiltered water could flow into the clear well and contaminate the finished water.
- Flowmeters — (1) may be installed in raw water intake lines to measure total plant inflow; (2) may be installed to measure flow into each filter tank or chamber to ensure uniform flow distribution and to measure the volumes of water filtered between cleanings; (3) are often installed between the underdrain manifold outlet and the finished water storage facility to measure total production; and (4) are sometimes installed at the clear well outlet to record the total volume of water delivered to the community. In some small plants, residential service meters are used to measure and record flows.
- Head loss gages or piezometers — (not shown in Figure 4.23) installed on each filter unit of effluent-controlled filters to monitor the degree of clogging, which will indicate when the filter needs to be cleaned.
- Rate of flow indicators — (not shown in Figure 4.23) installed on each filter or chamber to indicate actual rate of filtration.
- Staff gages — (not shown in Figure 4.23) may be permanently installed inside each filter cell to permit the operator to monitor the level of sand and/or water in the filters.
- Sight tubes — (not shown in Figure 4.23) sometimes installed in the filter tank and/or clear well to enable the operator to see the level of the supernatant in the filter or water in the clear well without removing the filter cover or opening the clear well.

QUESTIONS

Write your answers in a notebook and then compare your answers with those on page 260.

4.8E The flow rate of water through a slow sand filter depends on what factors?

4.8F List the two methods of controlling flow through a slow sand filter.

4.8G What factors are considered to determine the size of the finished water storage facility?

4.8H What is the purpose of head loss gages or piezometers on slow sand filters?

4.83 Typical Filter Operating Cycle

4.830 Start-Up

When placing a new filter in service or restarting a filter after cleaning, close all outlet valves and fill the tank from the bottom through the underdrain system. Use filtered water and fill slowly to permit air bubbles trapped in the sand to escape to the atmosphere. Continue filling the tank from the bottom until there is enough water (4 to 6 inches) over the sand surface to protect it from disturbance by the force of influent flow from the intake structure (a splash plate below the intake port will help prevent sand erosion by falling water). Once this shallow level of water covers the sand, slowly open the influent flow control valve and continue filling the tank with raw water until the desired supernatant operating level is reached (3 to 5 feet of water above the sand in an effluent-controlled filter).

A new filter, or one that has just been resanded, usually will not produce filtered water of acceptable quality immediately upon start-up. In most cases, the filter must be run continuously for a period of weeks or months before it becomes mature and achieves its full potential to remove turbidity, *Giardia* and other microorganisms. New or resanded filters may produce acceptable turbidity levels (less than 0.1 NTU) during the ripening period shortly after start-up but usually are not capable of removing high levels of *Giardia* until the filter matures. Source water quality and temperature greatly influence the time required for a filter to mature. Operators should therefore rely on the results of turbidity and bacteriological tests to determine when filtered water meets the required standards.

A filter that is being started after scraping will ripen and produce acceptable water much more quickly than a new or resanded filter, often within a few hours or days. This is because only the schmutzdecke and a small amount of sand is removed. The media still contains a mature population of organisms which will quickly recover from the cleaning operation.

During the ripening period, the filtered water that is produced is drained to waste by means of a filter-to-waste valve located between the underdrain manifold and the finished water storage facility. When turbidity and bacteriological tests indicate that the water quality meets desired standards, the ripening period is over. A good target turbidity value is 0.1 NTU. The turbidity level of a filter that has just been cleaned or resanded should at least return to what it was before cleaning. The recommended bacteriological target level is less than 1 coliform per 100 m*L*. When the filter produces acceptable quality water, slowly close the drain-to-waste valve and then slowly open the effluent valve and permit the treated water to flow to storage.

4.831 Daily Operation

The actual day-to-day operation of a properly designed slow sand filter requires very little operator involvement. If the effluent-control design is used, the operator may need to adjust the effluent flow control valve or effluent weir every day or two to offset the slowly increasing resistance (head loss) of the developing filter skin (schmutzdecke). This procedure continues until the valve is fully open or the weir cannot be lowered any more and the desired filtration rate can no longer be maintained, or there is a decline in water quality.

If the filter flow rate is controlled at the influent, the operator must set the rate at start-up and no further adjustments are needed until it is time to shut down the filter for cleaning. For surface waters, a filter rate between 0.04 and 0.08 GPM/sq ft (2.5 and 5.0 MGD/acre) is usually satisfactory. This rate can be increased to a rate as high as 0.16 GPM/sq ft (10 MGD/acre) if necessary for brief periods but steady flow operation is a much safer operating strategy. The operator should occasionally check the flow rate to make sure the flow has not been restricted or increased due to an undetected malfunction of a valve or pump or a plugged water inlet.

An advantage of influent flow control is that the operator can see the effects of increasing head loss. As the schmutzdecke develops and trapped debris begins to accumulate at the filter skin, it offers greater resistance to the flow of water. Since raw water is flowing in at a fixed rate, the increasing resistance of the surface causes a visible increase in water level above the media. Under this situation the water surface would have to be manually skimmed to remove scum and floatables when necessary because the supernatant overflow outlet could not be used. Also, there would be no excess flow to waste, as is sometimes necessary with effluent flow control of a filter.

Other routine daily operator duties include monitoring and recording influent flow, individual filter and total plant production, head loss as measured by headwater and tailwater elevations, and headwater temperature. The operator will also be responsible for collecting daily turbidity samples, inspecting the facility regularly, checking the operation of pumps or any other mechanical equipment and lubricating as necessary, checking the disinfection system, and measuring the disinfectant residual. The finished water storage facility will need to be inspected regularly and cleaned every year or two.

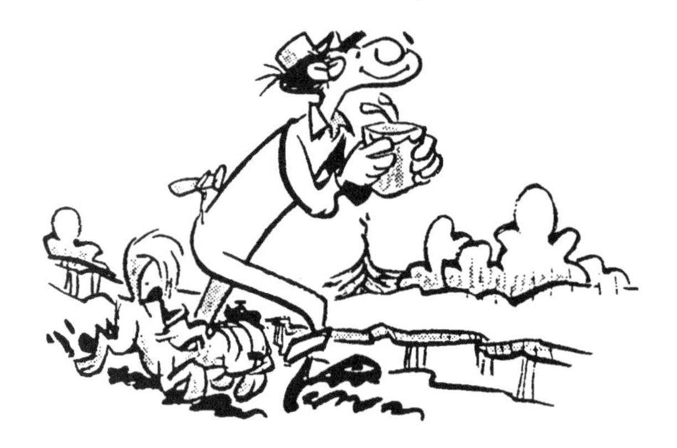

4.832 Cleaning the Filter Media

When head loss reaches the level at which the filter is no longer able to maintain the desired filtration rate, the filter must be shut down and cleaned. Two processes are commonly used to clean slow sand filters: scraping or raking. The goal is to remove the thin schmutzdecke and a small amount of sand just below it, usually totaling about 1 inch (2.5 cm) of material, so that water can once again penetrate the sand media.

The length of time a filter will operate between cleanings is highly variable from plant to plant, ranging from several weeks to several months. Cycle length depends on such factors as filtration rate, available head, media grain size distribution (uniformity coefficient), influent water turbidity, and temperature. Head loss at the beginning of the filter run is usually about 4 inches (10 cm), but it slowly increases to whatever depth is permitted by the height of the filter box, usually no more than 6 feet (2 m). If the raw water is of exceptionally good quality the filter could run for a year or more between cleanings. Filter operating cycles of less than six to eight weeks become uneconomical because of the high labor costs associated with frequent cleaning operations. Scraping a filter is a labor-intensive operation that accounts for a large portion of the operating costs of a filter. On the average, it takes about 4.2 person-hours to scrape 1,000 square feet (100 sq m) of surface area to remove a depth of 1 inch (2.5 cm) of sand.

Prior to cleaning the filter, some operators raise the supernatant level high enough to wash loose material such as leaves and other floating debris out of the filter through the overflow weir or valve. Next, the influent valve is closed and the tank water level is lowered 1 to 2 inches (2 to 5 cm) below the sand surface. Some operators favor rapid dewatering using both the supernatant drain and the waste valve on the underdrain. Others simply close the inlet valve and permit the level in the tank to drop of its own accord to the desired level. Using the slower draining method, it may take several hours for the water surface to drop below the schmutzdecke, but then the water drains rapidly due to lack of resistance from the schmutzdecke.

The filter surface can be scraped and the material removed manually using asphalt rakes, shovels, buckets, and wheelbarrows in small plants. Manual cleaning operations are usually manageable for filters up to about 2,000 square feet (200 sq m) surface area. The sand is scraped into low, parallel windrows (see Figure 4.24) and then shoveled into buckets or wheelbarrows and removed from the filter. The scraping operation should be carried out in a short time period to prevent drying of the media to minimize the impact of cleaning on the microorganisms in the remaining sand media. If it is necessary to return the filter to service quickly, the windrows may be left in place and removed during the next scraping.

Motorized equipment can also be used in the scraping operation. Garden tractors with dump-type carts, all-terrain vehicles, and lawn tractors with flotation tires have been used to remove and haul sand from the filter in small installations. Mechanized scrapers and dump trucks (Figure 4.25) are sometimes used in very large filter installations. The cleaning method to be used must be taken into account during the design of the filter to provide needed access ramps, drains, and adequate structural support.

After scraping has been completed, smooth the surface of the bed by raking or dragging a piece of metal mesh (cyclone fence) across the surface (Figure 4.26). To avoid the possibility of water short-circuiting through the filter at the walls, some operators gently tamp a six-inch strip around the edges of the tank.

Fill the filter through the underdrain system with water from an adjacent filter or from filtered water storage. Use the same procedures that were used for initial start-up, that is, fill slowly (less than 0.6 ft/hr) from the bottom until the sand is covered by several inches of water or water reaches the level of the influent distribution line. Complete the operation by filling to the desired supernatant level with raw influent water. Monitor the water produced by the newly cleaned filter and drain to waste until bacteriological or turbidity tests indicate the ripening period has ended and the filter is producing water of acceptable quality. As previously indicated, a typical turbidity target value is 0.1 NTU or the level at which the filter was operating prior to cleaning.

Fig. 4.24 Operator manually cleans filter by scraping schmutzdecke into parallel windrows for later removal and disposal

(Source: Wickiup Water District, Wickiup, Oregon)

The period during which water is drained to waste is highly variable from plant to plant, but in all cases should be based on the quality of the finished water. Ripening periods of a few days to several weeks are very common. A typical target value for coliforms is less than 1 per 100 m*L*. Some operators simply establish a fixed period of time, usually 24 to 48 hours, during which water from the filter is drained to waste before the filter is placed in service. Use of a fixed period of time is advisable only after experience with a particular filter indicates consistently that water quality standards can be met within that time period.

A controversial process called "raking" is sometimes used to extend the filter run time between scrapings. In this process, an ordinary garden rake is used to disturb the surface of the filter and loosen the schmutzdecke so that water will again pass into the sand layer and be filtered. Some operators lower the water level below the sand surface before raking; others rake the surface while it is still covered with several inches of water. There are reports of a plant that used one to five rakings before the filter had to be shut down for scraping; each successive raking was less effective than the previous one in extending the filter cycle. Raking requires less labor than scraping so it can reduce costs, but there is some evidence that this procedure reduces finished water quality and drives waste material deeper into the media. The short-term savings on labor may thus be offset by the eventual need to remove a deeper layer of sand during scraping.

A less widely practiced cleaning method, wet-harrow cleaning, is used in a few locations as a substitute for scraping. Horizontal and sometimes vertical water streams are used to flush the raked deposits to a surface drain or overflow weir. Resanding is infrequent, but necessary when the depth of sand gets too low and filtered water quality starts to drop.

The process of scraping removes about an inch (2.5 cm) of filter media each time. Eventually the sand must be replaced to restore the original media depth. The process of replacing filter media is called "resanding." In some installations, operators replace scraped sand with clean sand each time the filter is scraped. There is some evidence, however, that this practice causes a buildup of waste material at deeper layers of the media bed which may cause repeated clogging and short filter

Fig. 4.25 Mechanical equipment used for cleaning a large slow sand filter
(Source: Salem, Oregon Water/Wastewater District)

Fig. 4.26 Operator levels the sand surface with a section of chain-link fencing
(Source: Cashmere, Washington Slow Sand Filter, photo by Steve Deems, Washington Department of Health)

runs. More typically, a filter is scraped until overall media depth has dropped 12 inches (30 cm), although for deeper beds, the drop may reach 24 inches (60 cm). Filter performance (particle removal efficiency) is affected by filter bed depth so the bed should never be scraped to less than 1.6 to 2 feet (0.5 to 0.6 m). Thus, if a filter with an initial sand depth of 30 to 36 inches is scraped every two months and one inch (2.5 cm) is removed each time, it will take approximately two years to lower the media surface by one foot, at which time resanding will be necessary.

The most common technique used to resand a slow sand filter is called "throwing over," a process by which clean (new or recycled) sand is added *BELOW* the existing sand to bring the media bed back to its original depth. This method is illustrated in Figure 4.27. To resand a filter, first clean (scrape) the media. Then drain the filter until the water just covers the underdrain gravel support layer (Figure 4.27(a)). Dig a trench along one wall of the tank removing the old sand and placing it on top of the sand along the opposite wall (Figure 4.27(b)). Take care not to disturb or damage the gravel bed and underdrain piping. Fill a portion of the trench with new sand (Figure 4.27(c)). (The depth of the new sand layer can be calculated by the formula: Depth of New Sand = Original Bed Depth – Bed Depth Before Resanding.) Dig another trench along the side of the first one and place the sand from this second trench on top of the new sand in the first trench (Figure 4.27(d)). Partially fill the second trench with new sand (Figure 4.27(e)). Continue in this manner until all of the old sand has been moved to the surface and the original media depth has been restored (Figure. 4.27(f)). Smooth the surface of the bed, fill slowly from below with filtered water as for start-up of a new bed, and drain to waste until the media ripens and produces water of acceptable quality. An alternative start-up procedure for a newly resanded filter involves increasing the filtration rate slowly from one-fourth of the design rate to full flow rate over a period of a few weeks. Filtered water is drained to waste for the first 48 hours (if experience indicates that this time period is adequate) or until laboratory tests indicate the filter is producing water of acceptable quality.

Use of the throwing over technique is thought to minimize the time required for ripening of the resanded bed because the "old" sand in the top of the bed contains a diverse population of microorganisms. This procedure also eliminates any unwanted buildup of organisms and debris deep in the sand bed.

Resanding is approximately 10 times more labor-intensive than scraping. Typically, it takes 53 person-hours to resand 1,000 square feet (50 hr/100 sq m) of surface area.

Replacement sand may be purchased each time the filter requires resanding or the sand removed during scraping and

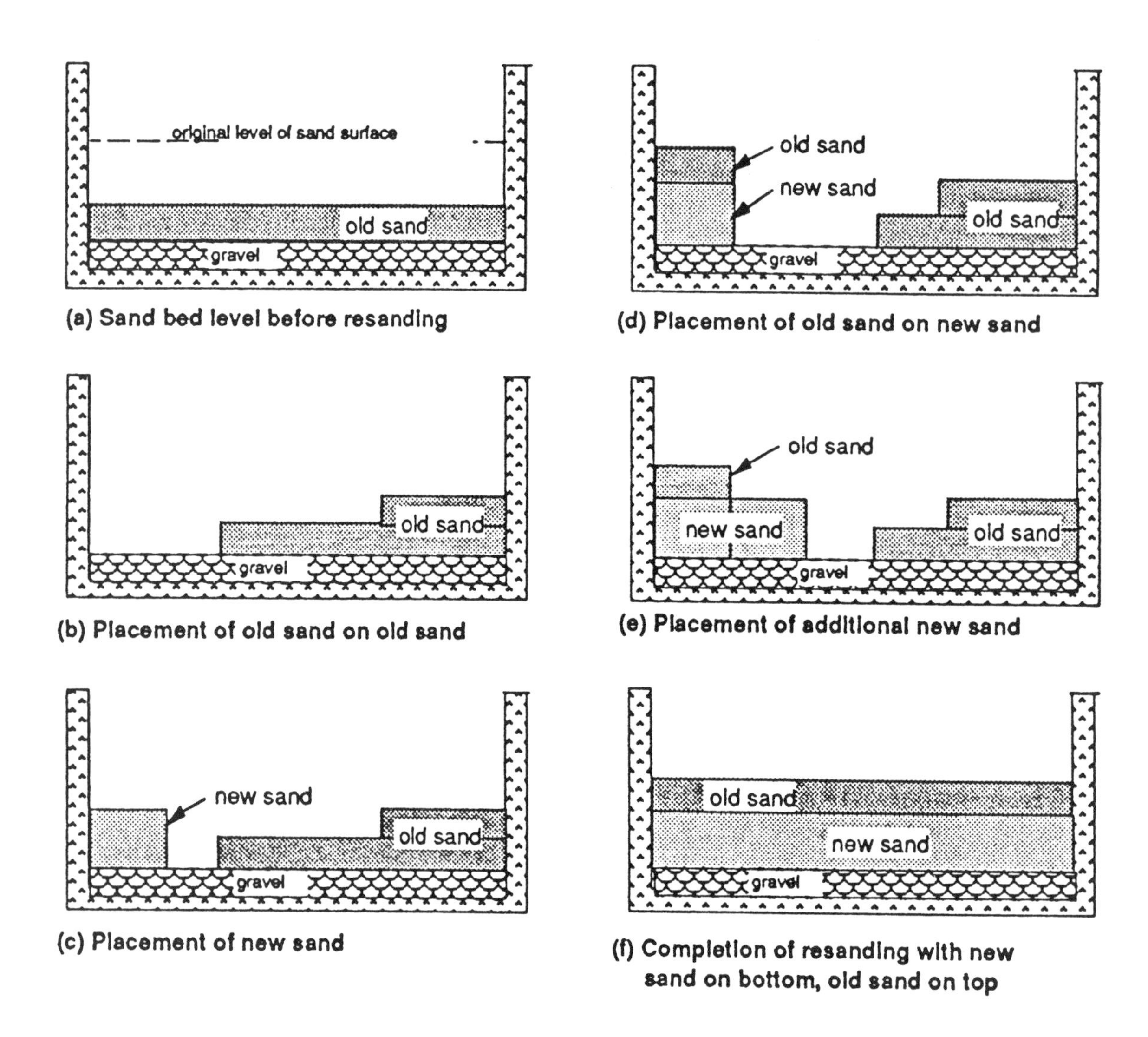

Fig. 4.27 Steps in resanding filter bed

(Reprinted from *MANUAL OF DESIGN FOR SLOW SAND FILTRATION*, by permission. Copyright © 1991, American Water Works Association)

resanding operations can be washed and reused. The decision about which option to use depends mainly on economic considerations. Costs of purchasing new sand include the cost of the sand and transportation, and the cost of disposing of dirty sand when it is removed from the filters. (In one location, discarded sand is reportedly used for road sanding in the winter.) Purchasing new sand is an attractive option when clean sand of the correct grain size and uniformity is available locally.

On the other hand, washing and reusing sand may be a less expensive option if new sand must be hauled a considerable distance, if the filter site can accommodate a sand washing operation, and if there is adequate space to store the washed sand. The cost of labor to wash the sand must also be considered.

Purchased sand should be checked carefully to verify that it is clean and meets the desired specifications for size and uniformity. This precaution is particularly important when buying sand from a new source or resanding a filter with new sand that is different than the sand originally used for the filter.

QUESTIONS

Write your answers in a notebook and then compare your answers with those on page 260.

4.8I How often will a finished water storage facility need to be cleaned?

4.8J When should a slow sand filter be taken out of service and the surface scraped or raked?

4.8K The time between slow sand filter cleanings depends on what factors?

4.8L What is an advantage of using the "throwing over" process to resand a slow sand filter?

4.84 Preventive Maintenance

All mechanical equipment requires periodic inspection, lubrication, and adjustment to keep it functioning as intended. Even though slow sand filters are designed to minimize operator involvement, an established program of preventive maintenance is essential. Reprinted below (with permission) is the suggested general maintenance schedule for the water treatment facility operated by the City of Dover, Idaho.

1. *DAILY*
   a. Check all controls and panels for proper operation.
   b. Visually check all pumps, motors, and valves.
   c. Check turbidimeter and recorder.
   d. Check chlorinator.
2. *WEEKLY*
   a. Wipe down all exposed machinery.
   b. Check operating supplies and order, if necessary.
   c. Lubricate equipment as necessary, per schedule.
   d. Operate all valves and gates; lubricate as necessary.
   e. Change turbidimeter charts.
   f. Mix and refill chlorine solution.
3. *MONTHLY*
   a. Inspect each pump and service as required.
   b. Inspect all safety equipment.
   c. Lubricate rotating equipment per manufacturer's schedule.
   d. Perform general cleanup.
4. *BIANNUALLY*
   a. Inspect intake pipeline.
   b. Calibrate turbidimeter.
5. *ANNUALLY*
   a. Paint equipment and building interiors and exteriors as required.
   b. Dismantle pumps, check impellers, shaft sleeves, shafts, and pump bowls, replace all worn equipment or parts.
   c. Have diver inspect intake structure.

HOUSEKEEPING

The appearance of a treatment facility is generally a good indication of the efficiency of the operator. An operator who takes pride in the appearance of the facility usually is also concerned with the efficient operation of the facility.

4.85 Troubleshooting

A properly designed and constructed slow sand filter should operate reliably with very little operator intervention. Alarm systems are usually installed to alert the operator to existing or developing problems. The following troubleshooting guide was adapted from a guide prepared by Ruen-Yeager & Associates, Inc., for the City of Dover Water Treatment Facility, Coeur D'Alene, Idaho, and is included here with permission of the City of Dover.

Indicator	Corrective Action
Slow Sand Filter Controls	
High water alarm	This alarm will not indicate unless overflow and drain are blocked. Clear the blockage.
Low water alarm	Filtration rate is greater than inflow rate. Decrease filtration rate and allow to restore.
Reservoir Controls	
High water alarm	Reservoir overflowing. Too much water being pumped to reservoir. Allow use to lower level.
Low water alarm	Use greater than supply. Increase filtration rate to pump more water. May indicate if fighting fire which is OK.
Clear Well	
Low water alarm	Pumps are pumping level of clear well down too much. Pump control valve may be malfunctioning.
Intake Pumps	
No flow past flow switch	Check valve could be blocked, intake screen clogged. Clear blockage.
Booster Pumps	
Pump control valve not operating correctly	Valve at reservoir intake is closed; pump has a vapor lock.
Turbidimeter	Indicates an alarm if turbidity is greater than 1.0 NTU. Decrease filtration rate.

Air binding is a problem that sometimes occurs when slow sand filters are filled too quickly during start-up or after cleaning or resanding. If air binding is suspected to be the cause of a sudden drop in water production, try backing filtered water through the underdrain piping and up through the media. If the surface of the water in the filter looks bubbly like a soda bottle that has just been opened, air binding is probably the cause of the problem. To correct the problem, shut down and drain the filter. Then, using the underdrain system, fill the filter slowly enough to allow the trapped air bubbles to rise to the surface of the water and escape to the atmosphere.

4.86 Finished Water Standards

Most of the Drinking Water Standards in the U.S. establish contamination limits in terms of Maximum Contaminant Levels (MCLs) for specific substances; they leave the choice of methods to achieve compliance to the discretion of the water supply or treatment agency. Departing from this regulatory approach, the Surface Water Treatment Rule (SWTR), promulgated in 1989, specifies the treatment techniques that are to be used as well as turbidity and microorganism removal rates. Compliance with the SWTR is measured in terms of how well the treatment plant is operated (removal efficiencies) as determined by laboratory testing of the source water and of the finished water.

The goal of the SWTR is to reduce the contamination of water supplies by disease-causing organisms, which are also called pathogens. The protozoan *Giardia lamblia* is presently the organism most implicated in waterborne disease outbreaks in the United States. These microscopic creatures are found mainly in mountain streams, and have been found to be widespread in small community water systems. Once inside the body, they cause a painful and disabling illness called Giardiasis. The symptoms of Giardiasis are usually severe diarrhea, gas, cramps, nausea, vomiting, and fatigue. Another pathogen found in water supplies, the *legionella* bacterium, causes severe upper respiratory disease.

Viruses, *legionella* bacteria, *Giardia lamblia* cysts, and *Cryptosporidia* are all highly resistant organisms and are difficult and costly to detect. No single method presently available is effective in removing or inactivating all of the various types of pathogens that threaten public health. For this reason the EPA requires most water suppliers to use a combination of treatment techniques (disinfection and filtration) to ensure the safety of drinking water delivered to the public. Disinfection is known to be effective in killing or inactivating bacteria and viruses, which are small enough to pass through most filters. Filtration is known to be effective in removing or inactivating cysts, which are extremely resistant to chemical disinfection methods. The combination of treatment techniques thus ensures removal of both the very smallest and the most resistant types of pathogens.

All public water systems (which may include some privately owned systems serving the public) using surface water or groundwater under the influence of surface water are required by the SWTR to disinfect the water they distribute to the public; no exceptions to this requirement are permitted. Compliance with this rule is confirmed by the presence of at least 0.2 mg/*L* disinfectant (chlorine) residual in the finished water and a detectable residual throughout the distribution system.

In addition to the disinfection requirements, water suppliers may be required to install filtration equipment unless the system can meet stringent guidelines concerning source water quality (turbidity and microbial population), watershed control, and backup disinfection capability. Slow sand filtration is one of the treatment techniques acceptable to the U.S. Environmental Protection Agency for meeting the SWTR requirements. (Conventional filtration, diatomaceous earth filtration, and direct filtration are also acceptable treatment techniques and are described in Section 4.4, "Filtration.")

The SWTR's microbiological standards require that the combination of filtration and disinfection must achieve a 99.9 percent[31] reduction (by removal or inactivation) of *Giardia lamblia* cysts and a 99.99 percent reduction (by removal or inactivation) of viruses. Slow sand filters are particularly notable for their ability to remove *Giardia* cysts and have been shown to achieve the greater than 3-log removal level when the filter bed is mature. Even with new sand, *Giardia* removals remain high.

Since turbidity can interfere with the disinfection process, the SWTR also sets maximum allowable levels for turbidity. The filtration requirement for turbidity of finished water from a slow sand filter is less than one NTU in 95 percent of the samples taken and no sample may exceed 5 NTU, with monitoring to occur at least once per day. Higher effluent turbidity levels (up to 5 NTU) may be allowed by the EPA or state if it can be shown that the higher levels do not interfere with disinfection.

For a complete description of the Safe Drinking Water Act and Surface Water Treatment Rule provisions, see Chapter 22, "Drinking Water Regulations," *WATER TREATMENT PLANT OPERATION*, Volume II, in this series of manuals and consult the poster provided with this manual.

4.87 Factors Affecting Filter Performance

The particle removal efficiency of a slow sand filter depends on many variables such as the hydraulic loading rate, temperature, sand bed depth, sand size, and filter bed maturity, and on interactions between and among these and other variables. Despite having been used since the early 1800s and despite many research efforts to pinpoint the relative importance of various factors, slow sand filtration is still not fully understood. Nonetheless, slow sand filtration has been shown to be an extremely effective, low-cost, low-tech way to remove bacteria, viruses, cysts, spores, silt, and most other suspended organic and inorganic particles from drinking water. However, it is less effective in removing dissolved organic materials and fine clay particles.

Many of the variables which affect filter performance have been described in earlier portions of this section. The following paragraphs describe the effects of raw water quality, cyclic influences, mode of operation, and hydraulic loading rates on slow sand filter performance.

4.870 Source Water Quality

In the most general terms, supported by surveys of operating plants, water with turbidity levels below 10 NTU may be suitable for slow sand filtration. However, it is impossible to predict how well this process will work based solely on turbidity measurements. Finely divided clay particles may not be removed effectively by slow sand filters. Pilot studies should be conducted to assess how well a slow sand filter removes microorganisms and particulate matter from a particular source of water, determine the length of filter runs, and determine the washout time of the new sand.

Ideally, a slow sand filter will produce acceptable quality water for at least a month between cleanings with little operator involvement except for daily monitoring, adjustment, and

[31] *Percentages are often referred to using logarithm (log) numbers. Throughout the SWTR, reduction goals are expressed this way. It is easy to translate percentages into log numbers. Start with 90, and just add nines: 90 is one-log; 99 is two-log; 99.9 is three-log; 99.99 is four-log, and so forth. In this way, log numbers are a form of shorthand: It is much easier to say "three-log" than "99.9 percent." (Footnote from "Surface Water Treatment: The New Rules," by Harry Von Huben, OPFLOW, Volume 16, No. 11, November 1990, by permission of the American Water Works Association. Copyright © 1990 by American Water Works Association.)*

sampling. Cycle length is, of course, directly related to the buildup of filtered material on the schmutzdecke. The mix of filterable mineral sediments, organic detritus, bacteria, cysts, spores, and a variety of other microorganisms varies widely from source to source. Raw water turbidity measurements do *NOT* reliably predict either the rate of filtered particle buildup and resulting head loss that will occur or the microorganism removal rates that can be expected. However, raw water turbidity can serve as a very general starting point for those considering the use of a slow sand filtration system.

4.871 Cyclic Influences

Most slow sand filters are used to purify surface waters which, by their nature, are subject to the seasonal effects of climate changes. For example, during periods of rainfall, the flow of runoff into rivers and streams may temporarily raise turbidity levels as high as 30 to 50 NTU, occasionally peaking at 1,000 NTU. Such an excessive solids load will quickly clog ("blind") the filter media and the filter will have to be cleaned. If high turbidity conditions occur infrequently, the cost and inconvenience of extra scrapings may not be excessive. If high turbidity is more than an occasional problem, installing a settling basin or a sedimentation basin in advance of the filter may reduce turbidity levels enough to permit normal operation of the slow sand filter. If the turbidity is mainly clay, a gravel roughing filter (Figure 4.28) would be more effective than a sedimentation basin. However, installation of additional equipment such as sedimentation basins runs counter to one of the basic aims of slow sand filtration, which is keeping the process simple and inexpensive, with low labor requirements.

Seasonal changes in water temperature and the amount of solar radiation also affect filter performance. Temperature controls the physical viscosity of water and the rate of biological activity. Several studies of particle removal efficiencies in cold water compared to warm water have shown consistently better removal rates in warmer water. The temperature of the filter influent is largely beyond the control of the operator once a plant is built. If low water temperature is anticipated during the design phase, especially if nutrient levels are also low, consideration might be given to alternatives such as below-ground construction or installation of a cover or other enclosure.

Seasonal changes in the amount of solar radiation affect algal growth in the headwater (supernatant) as well as within and immediately below the schmutzdecke in uncovered filters. Under favorable conditions (temperature, nutrients, solar radiation) algae multiply rapidly and algal blooms can occur, causing serious operating problems. Apart from consuming some nutrients, algae do not contribute to the treatment that occurs in a mature sand filter. The decay of algae in the filter skin can impart unpleasant tastes and odors in the filtered water. Also, as algae build up at the surface of the media, head loss will increase rapidly. A roof over the filter or some other form of shading such as shade cloth has been shown to improve filter run lengths in filters experiencing only occasional algal blooms. When algae populations are more consistent, however, shading the filter has produced mixed results and no clear recommendation on the use of shading has emerged from field studies.

Another cyclic factor, diurnal (day/night) fluctuations in solar radiation, also affects the organisms in a slow sand filter, especially if the filter is shut down at night. During daylight hours algae in the headwater and sand media produce oxygen; at night they consume oxygen. All other organisms in the filter consume oxygen at all times, day and night. Therefore, during daylight hours the dissolved oxygen level in the water rises slightly and then drops off at night. If the filter is shut down overnight or operated at very low filtration rates, dissolved oxygen levels drop and there is a possibility of totally depleting the dissolved oxygen. The somewhat longer contact time that accompanies low filtration rates can lead to *ANAEROBIC*[32] conditions and the resulting severe water quality problems. If anaerobic conditions develop, water turns black, gives off foul odors, and has a greatly increased chlorine demand.

Resanding a filter can be considered a cyclic influence on filter performance because it disrupts the balance of microorganisms in a mature filter bed and disturbs the established biofilm. Dewatering the filter for more than a few hours, which might be necessary when resanding medium to large filters, destabilizes the microbial population which will then require time to become reestablished. Additionally, if the new sand added to the bed has not been properly washed to remove organic material, clay, and dirt, turbidity increases in the finished water can be expected to occur while this material washes out of the filter.

One might think that periodic scraping of the filter media would also cause a temporary decrease in filter performance, but removal rates usually are not significantly affected if the sand bed is mature. Field studies in this area suggest that disturbance of the lower layers of the media bed and dewatering of the filter during scraping are more significant factors in reduced removal rates than the removal of the schmutzdecke.

4.872 Mode of Operation

Slow sand filters operate most reliably under continuous steady flow conditions. If the raw water supply is intermittent or uneven, a flow equalization tank can be installed ahead of the filter to dampen fluctuations in the supply to the filter. Intermittent filter operation should be avoided because it has been shown to reduce the bacteriological quality of the finished water within 4 to 5 hours after the filter is restarted.

As previously described, shutting a filter down overnight poses the risk of developing anaerobic conditions; this is a

[32] *Anaerobic (AN-air-O-bick). A condition in which atmospheric or dissolved molecular oxygen is NOT present in the aquatic (water) environment.*

Fig. 4.28 Gravel roughing filters

practice that should be avoided. To reduce production without dropping the filtration rate so low as to cause oxygen depletion, some operators of effluent-controlled filters close off the influent water supply at night (or at other times during the day) and let the filter continue to operate under a declining head of water. Declining rate filtration reportedly produces almost the same quality of finished water as full rate filtration but there is less risk of creating anaerobic conditions in the filter. The dissolved oxygen level that usually signals declining *AEROBIC*[33] bacterial activity due to oxygen starvation is approximately 0.5 mg/*L*.

To ensure that a filter operating in a declining flow rate mode does not run dry (dewater) while operating unattended, the effluent from the filter should be adjusted to empty into the outlet chamber at a level slightly higher than the top of the sand bed.

4.873 Hydraulic Loading Rate

Slow sand filters are operated at hydraulic loading rates from 0.016 GPM/sq ft up to a maximum of 0.16 GPM/sq ft (1 to 10 MGD/acre or 0.04 to 0.4 m/hr). Rates from 0.04 to 0.08 GPM/sq ft (2.5 to 5.0 MGD/acre or 0.1 to 0.2 m/hr) are typical. The rate may vary slightly from season to season, but the daily rate should be steady. Occasional increases in hydraulic loading rates can be tolerated for short periods (for example, to compensate when another filter is shut down for cleaning), but abrupt changes should be avoided.

Field studies have demonstrated that increasing the hydraulic loading rate of a slow sand filter reduces the percent removals of *Giardia* cysts, total coliform bacteria, fecal coliform bacteria, and standard plate count bacteria. However, removals were still high with a hydraulic loading rate of 0.16 GPM/sq ft (10 MGD/acre or 0.40 m/hr).

4.874 Calculation of Hydraulic Loading Rate

Hydraulic loading rates (filtration rates) on slow sand filters may be reported in terms of:

1. Gallons per minute per square foot (GPM/sq ft),
2. Million gallons per day per acre (MGD/ac), and
3. Feet of water fall or drop per hour (ft/hr) or cubic feet of water per hour per square foot of filter surface area (cu ft/hr/sq ft).

EXAMPLE 15

A flow of 20 GPM is applied to a slow sand filter 20 feet long and 20 feet wide. Calculate the hydraulic loading in (1) GPM/sq ft, (2) MGD/ac, and (3) ft/hr.

Known	Unknown
Length, ft = 20 ft	Hydraulic Loading Rate,
Width, ft = 20 ft	1. GPM/sq ft
Flow, GPM = 20 GPM	2. MGD/ac
	3. ft/hr

1. Calculate the filter surface area in square feet.

$$\text{Surface Area, sq ft} = (\text{Length, ft})(\text{Width, ft}) = (20\text{ ft})(20\text{ ft}) = 400\text{ sq ft}$$

2. Convert the flow from GPM to MGD.

$$\text{Flow, MGD} = \frac{\text{Flow, GPM}}{694\text{ GPM/MGD}} = \frac{20\text{ GPM}}{694\text{ GPM/MGD}} = 0.0288\text{ MGD}$$

3. Calculate the hydraulic loading rate in gallons per minute per square foot of surface area.

$$\text{Hydraulic Loading Rate, GPM/sq ft} = \frac{\text{Flow, GPM}}{\text{Surface Area, sq ft}} = \frac{20\text{ GPM}}{400\text{ sq ft}} = 0.05\text{ GPM/sq ft}$$

4. Calculate the hydraulic loading rate in million gallons per day per acre of surface area.

$$\text{Hydraulic Loading Rate, MGD/ac} = \frac{(\text{Flow, MGD})(43{,}560\text{ sq ft/acre})}{\text{Surface Area, sq ft}} = \frac{(0.0288\text{ MGD})(43{,}560\text{ sq ft/acre})}{400\text{ sq ft}} = 3.14\text{ MGD/ac}$$

5. Calculate the hydraulic loading rate in feet per hour.

$$\text{Hydraulic Loading Rate, ft/hr} = \frac{(\text{Flow, GPM})(60\text{ min/hr})}{(\text{Surface Area, sq ft})(7.48\text{ gal/cu ft})} = \frac{(20\text{ GPM})(60\text{ min/hr})}{(400\text{ sq ft})(7.48\text{ gal/cu ft})} = 0.4\text{ ft/hr}$$

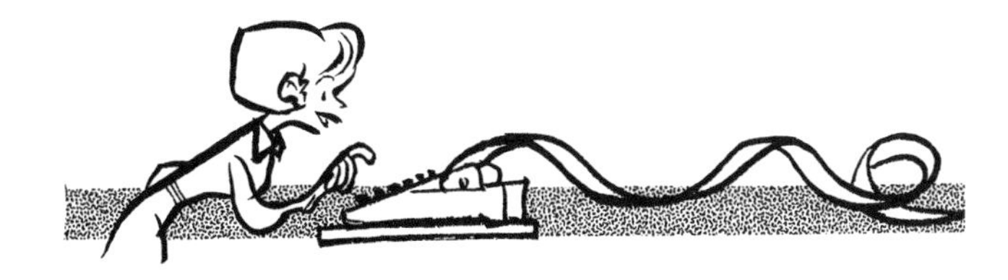

4.88 Recordkeeping

Accurate, up-to-date records which are easily accessible are very important to the proper operation of a water filtration facility. Good records provide a history of past operation and are legal proof of the performance of the facility. Equipment maintenance records which document the performance and repair history of each component are valuable resources for locating and solving operational problems. Accurate water production and cost records enable the plant administrator to prepare realistic budgets and make informed decisions about expenditures.

Recordkeeping does not need to be a complicated, time-consuming activity for the operator of a slow sand filtration system. Only three basic types of records are necessary, the

[33] *Aerobic (AIR-O-bick). A condition in which atmospheric or dissolved molecular oxygen is present in the aquatic (water) environment.*

operator's daily diary, maintenance records, and sampling and monitoring records (Figure 4.29). Information that should be recorded in the operator's daily diary includes equipment repairs, preventive maintenance performed, unusual operating conditions, and all sampling or tests performed. Each entry should indicate the name of the operator as well as the date and time the activity was performed.

The length of time records must be kept depends on the type of record. Maintenance records should be kept at least as long as the equipment is in use and longer if there is a specific reason for keeping them. Federal and state laws govern how long sampling records and laboratory analyses must be retained. Requirements for storage of records vary from state to state and operators are advised to check with their regulatory agencies for exact requirements.

4.89 Process Modifications

The simplicity, effectiveness, and low cost of slow sand filtration is well documented. A variety of modifications have been tried in an effort to adapt the process to less than ideal operating conditions and to extend the range of source waters that can be treated by this process. For example, certain types of roughing filters installed ahead of the filters have reduced raw water suspended solids enough to permit successful treatment by slow sand filtration. Nutrients (primarily nitrate and phosphate) were added to the raw water at one plant to shorten the ripening period of a filter treating low-temperature, nutrient-poor, mountain stream water. Sources of additional information about these and other process modifications are listed in Section 4.15, "Additional Reading," at the end of this chapter.

QUESTIONS

Write your answers in a notebook and then compare your answers with those on page 261.

4.8M How is compliance with the Surface Water Treatment Rule (SWTR) confirmed with regard to disinfection?

4.8N What information can be obtained from pilot studies regarding the performance of slow sand filters?

4.8O What can happen to water quality if anaerobic conditions develop?

4.8P What types of records should be kept at a slow sand filtration plant?

4.8Q Why were nutrients added to the raw water at one slow sand filter plant?

4.810 Summit Lake Slow Sand Filter

This section is presented to illustrate some of the operating considerations at an existing slow sand filter plant. Figure 4.30 shows the layout of the small filter unit installed in 1981 at Summit Lake, Lassen National Park, California.

CONSTRUCTION FEATURES

The filter structure was built with steel-reinforced concrete and the control house at the effluent end of the structure was built of concrete blocks. Some excavation was needed to situate the tanks below the ground surface yet provide easy access to the effluent end of the filter (see Figure 4.31(a)). Wood framing supports the easily removable roof panels. The stilling well at the influent end of each filter measures 3.5 feet by 4 feet, each filter cell measures 8.5 feet by 4 feet, and sidewall depth is 5.5 feet.

The filter media is Monterey sand which is purchased in 100-pound sacks from a vendor in nearby Redding, California. The depth of the sand media is 3 feet 4 inches, effective size is 0.20 to 0.40 millimeters, and uniformity coefficient is 2.0 to 3.0.

Beneath the sand media a 10-inch deep bed of gravel surrounds and covers the 1½-inch perforated PVC underdrain piping. However, the gravel support layer does not entirely cover the bottom of the tank. When the filter was constructed, a one-foot wide band of media sand was placed along the walls and then four layers of gravel graded by size were used to cover the underdrain piping in the center of the tank floor. The purpose of this arrangement was to prevent water from short-circuiting along the walls of the filter. Gravel sizes range from ⅛- to ¼-inch at the top to ¾- to 1½-inch at the bottom.

PVC piping, brass gate valves, and standard residential water meters are used to channel, control, and record flow into and out of the filters. The inlet control valves are located in underground vaults (Figure 4.31(b)) constructed several feet from the main structure.

START-UP

In early May, operators prepare to start the filtration plant by cleaning the finished water storage tank, checking all valves for proper operation, checking the disinfection equipment and supplies, and performing any necessary maintenance. When all equipment has been checked and made ready, the operator begins filling the filter.

The usual procedure for filling filter cells is to fill slowly through the underdrain with finished water until the surface of the media is covered by a few inches of water and then open the inlet valve to finish filling with supply water. In most slow sand filter installations, an adjustable weir in the outlet structure can be used to control tailwater elevation (and thus maintain headwater elevation) until water covers the surface of the media. Because both source water and finished water flow by gravity at the Summit Lake plant, operators there encountered problems filling the filters with finished water by means of the underdrain piping. The finished water storage tank is located about one-quarter of a mile below the slow sand filter so the operators had no way to fill the filter cell using the underdrain except by filling from the other cell next to it. When they tried to do this, however, air entered the underdrain piping and then became trapped in the finished water line to the storage tank when the cell was put back in service after filling the other cell. The operators found that the problems caused by air trapped in the gravity line to the storage tank were more difficult to correct than the relatively minor problems caused by filling the cells with supply water without disturbing the sand surface. The plant has not experienced any adverse effects in either turbidity or coliform levels using this method, but the process of filling the filter cells would be much easier if an outlet chamber with adjustable weir to control tailwater elevation had been installed when the plant was built.

When the headwater level reaches approximately 16 inches, flow through the filter is established by opening the filter-to-waste drain. Filtered water is wasted back to the creek for 24 hours and is then routed to the finished water storage tank. During this relatively short wasting period, some ripening of the media takes place, but the filter doesn't fully mature until late in the summer due largely to the very cold source water. Nonetheless, the plant is able to produce water that meets turbidity and bacteriological standards because of the high quality of the raw water.

CITY OF DOVER
WATER TREATMENT FACILITY

DAILY MONITORING REPORT
MONTH ____________

DAY	TIME	INTAKE PUMPS TOTAL HOURS 1	INTAKE PUMPS TOTAL HOURS 2	BOOSTER PUMPS TOTAL HOURS 1	BOOSTER PUMPS TOTAL HOURS 2	TOT. GAL. PUMPED 1	TOT. GAL. PUMPED 2	SAMPLE
1								
2								
3								
4								
5								
6								
7								
28								
29								
30								
31								

COMMENTS

CITY OF DOVER
WATER TREATMENT FACILITY

MONITORING REPORT
MONTH ____________

DAY	TIME	CHLORINE RESIDUAL (mg/l)	RAW WATER TURBIDITY NTU	EFFLUENT TURBIDITY NTU	SAMPLES TAKEN
1					
2					
3					
4					
5					
6					
7					
28					
29					
30					
31					

COMMENTS

Fig. 4.29 Daily and monthly monitoring report forms for a slow sand filtration plant

(Source: *OPERATION AND MAINTENANCE MANUAL FOR THE CITY OF DOVER WATER TREATMENT FACILITIES*, reprinted with permission of the City of Dover, Idaho)

MONTHLY CHLORINATION REPORT — Month of ______ 19__

SYSTEM NAME

COUNTY

TYPE OF CHLORINATOR

OPERATOR SIGNATURE

IDHW-DEQ
2110 Ironwood Parkway
Coeur d'Alene, ID 83814
(208) 667-3524

Date	Meter Reading	Water Used	Results of Chlorine Residual mg/l Location	Time	Results	Comments
1						
2						
3						
4						
5						
6						
27						
28						
29						
30						
31						

Fig. 4.29 Daily and monthly monitoring report forms for a slow sand filtration plant (continued)

(Source: *OPERATION AND MAINTENANCE MANUAL FOR THE CITY OF DOVER WATER TREATMENT FACILITIES*, reprinted with permission of the City of Dover, Idaho)

OPERATION

The Summit Lake water treatment plant supplies water to Lassen Park visitors at the Summit Lake Campground during the summer season, usually from mid-May through October. At the end of the season the filter is cleaned and shut down for the winter.

Source water of relatively high quality (0.10 to 0.15 NTUs, low in organic matter) flows to the filter by gravity from above the dam on a creek approximately 100 feet above the filter (Figure 4.30). Water is fed continuously at a rate of 7.0 GPM. The supernatant level is maintained at about 16 inches above the sand surface and any excess influent is wasted from the surface of the supernatant in each cell back to the creek.

Filtered water flow is controlled by 3/4-inch (gate) valves (Figures 4.32 and 4.33) on the filter effluent lines. When cells are placed in service, the operator opens the gate valves enough to establish the correct flow rate and head loss across the filters. Each day the operator checks the flowmeters. If the readings indicate that less than the desired amount of water was produced the previous day, the operator opens the effluent valve another half turn. Filtered water flow can be controlled in this manner for approximately two months until head loss readings signal the need to clean the filter.

Finished water flows by gravity from the filter to a storage facility located approximately one-quarter of a mile away. The storage tank is a 10,000-gallon steel tank with a girdered ceiling. As filtered water flows to the tank, sodium hypochlorite is applied for disinfection at a point 30 feet upstream of the storage tank.

Operation of the Summit Lake filter requires very little operator time. An operator visits the facility daily to check and adjust the effluent control valves and chlorination equipment, if necessary. The operator also checks turbidity levels and disinfectant residuals daily and, as necessary, collects samples for coliform analysis by a laboratory in a nearby city.

FILTER CLEANING

The Summit Lake filter operates for about eight weeks between cleanings. When cleaning is necessary, both cells are scraped at the same time since the total surface area is small enough to complete the task in a short time. At this facility, the filters are always cleaned by scraping rather than by raking (which is used at some other plants) in order to avoid the possibility of driving particulate material deeper into the sand bed.

The operator first removes the roof panels and then shuts off the supply water. The filter is allowed to drain until the water level drops to about one foot below the sand surface, just far enough that the sand will support the weight of an operator. When water reaches the correct level, the operator closes the effluent or outlet valves.

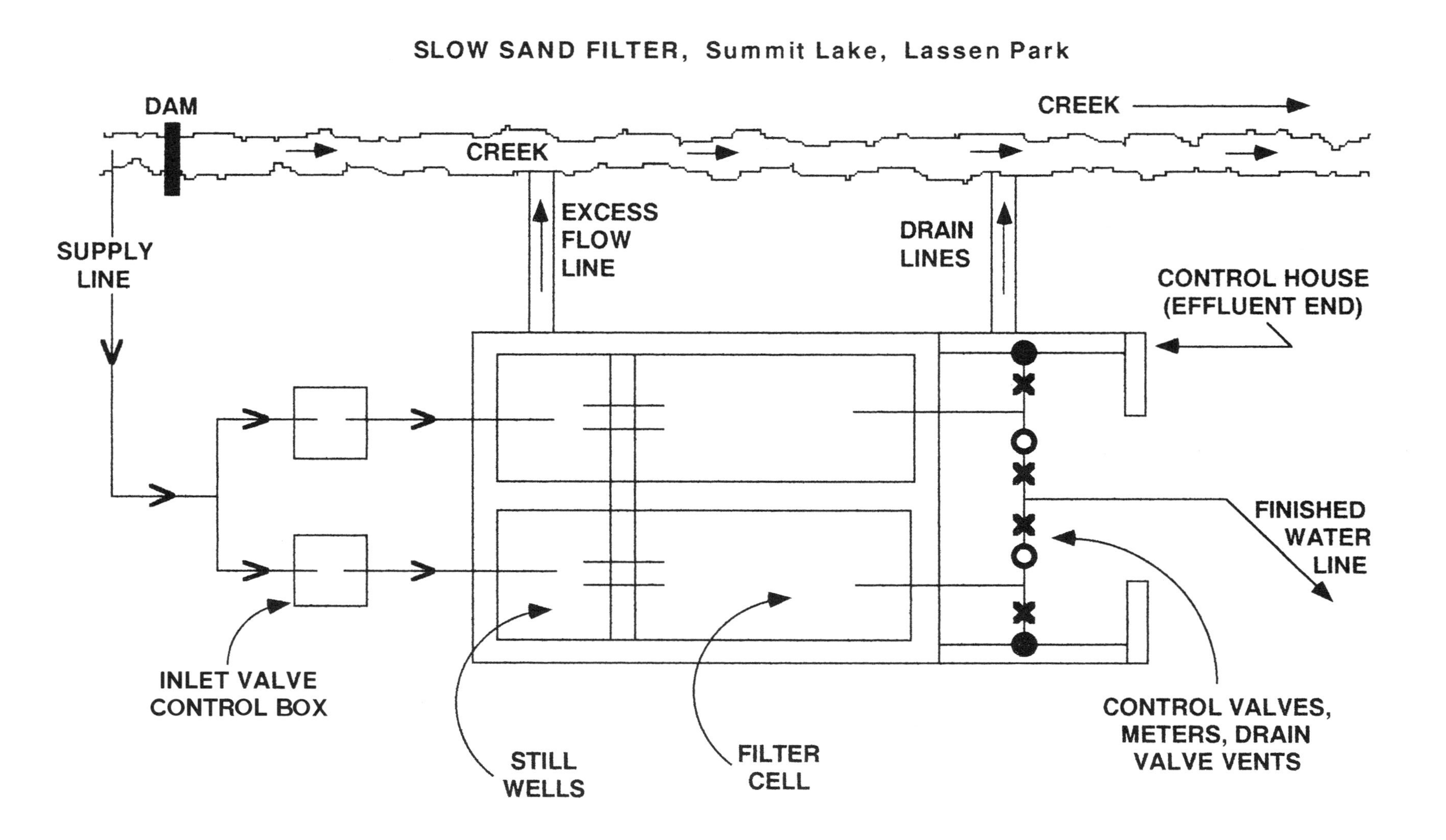

Fig. 4.30 Layout of the slow sand filter at Summit Lake, Lassen National Park, California

(a) Effluent end of filter outside the valve meter area.

(b) Inlet valve boxes (lower right) and cover over filter with two panels opened.

Fig. 4.31 Exterior views of the slow sand filter at Summit Lake, Lassen National Park, California

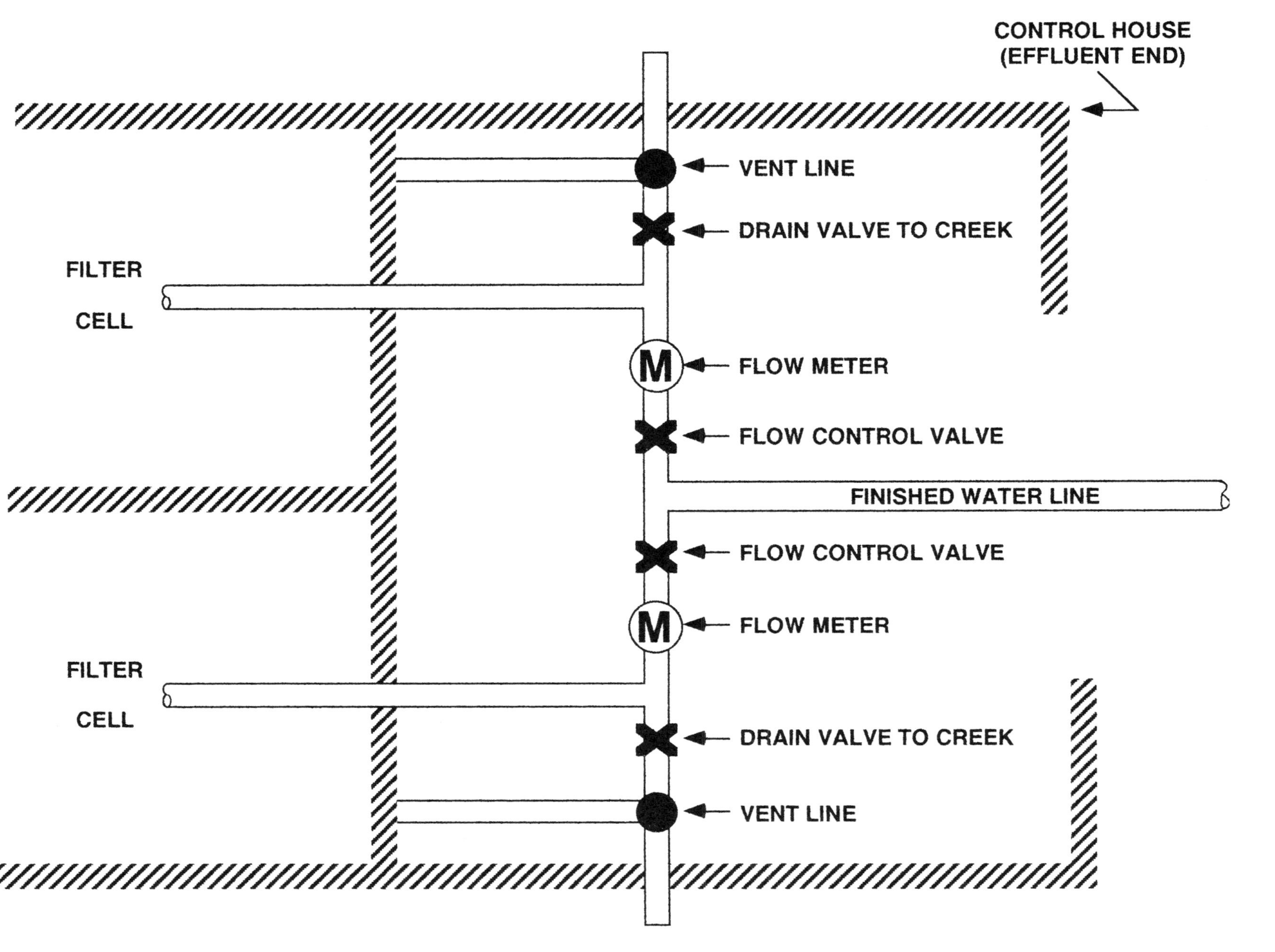

Fig. 4.32 Layout of meters, valves, and vents at effluent end of the Summit Lake filter

1. Finished water line in center.
2. Flow control valves for each cell.
3. Flowmeters for each cell.
4. Drain valves and vent lines for each cell.

Fig. 4.33 Effluent meters and valves

Standing in the center of the filter, an operator uses a flat-edged shovel to remove approximately one inch of material from the surface of the media bed. The operators at Summit Lake report that as the mat begins to dry, it starts to curl away from the sand surface, making it easier to remove. The sand and waste material scraped from the filter is shoveled onto the ground outside the filter. It is later bagged up and hauled to a nearby wastewater treatment plant where it is added to the mounds used for treated wastewater disposal. After scraping the surfaces of both cells, the operator uses a rake to level the surface of the sand. The total cleaning time for one operator is about 30 minutes.

After scraping has been completed, the cells are slowly filled with supply water until the surface of the bed is covered. The inlet valve is then opened to flood the cell to operating depth. When the water reaches a level about one foot above the media surface, the outlet valve is slowly opened to reestablish flow across the filter. Filtered water flows to waste for one hour and then is routed to the finished water storage facility. Samples are collected and tested for coliform and turbidity levels. Experience at this plant has demonstrated that the unusually short (one-hour) filter-to-waste period is a sufficient amount of time for the filtered water quality to return to the levels achieved before cleaning.

SHUTDOWN

At the end of the summer season, the Summit Lake slow sand filter is shut down and scraped in the usual manner. A permanent metal staff gage mounted on an inside wall of each filter permits the operator to easily monitor the depth of the sand bed. In 12 years of operation, new sand was added once to restore the filter media to design depth.

The filtered water storage tank is drained and cleaned at the end of each operating season to prevent freezing.

4.811 Acknowledgment

We wish to thank John Brady, Steve Tanner, Sig Hansen, Graham Dobson, and Mike Lafkas for their review of this material and helpful suggestions.

4.9 IRON AND MANGANESE CONTROL

Excessive iron and manganese in drinking waters are objectionable because they stain clothes and encourage the growth of iron bacteria. These bacteria form thick slimes on the walls of pipes. When these slimes break away from the pipes, the iron causes a rust-colored water and the manganese produces black particles. These slimes also impart foul tastes and odors to the water. For these reasons, iron should not exceed 0.3 mg/*L* and manganese 0.05 mg/*L* in drinking waters.

If a new well is drilled and excessive amounts of iron or manganese are discovered, the best solution may be to drill a new well. Consult with well drillers in the area and also collect and analyze samples from nearby wells. Discuss your situation with the state agency responsible for the regulation of well drilling. Sometimes a new well is cheaper and a lot less trouble than trying to control or remove iron and manganese. In many areas a new well will produce water with just as much iron and manganese. Surface waters also may contain excessive amounts of iron and manganese, especially when you have small, shallow reservoirs. Usually manganese is more of a problem than iron.

If the water contains less than 1.0 mg/*L* iron and less than 0.3 mg/*L* manganese, the use of polyphosphates followed by chlorination can be effective and inexpensive. Any of the three polyphosphates (pyrophosphate, tripolyphosphate, or metaphosphate) can be used, but sodium metaphosphate usually requires a lower dosage than the others.

To determine the best polyphosphate, prepare a series of samples with varying concentrations of polyphosphate. Stir to ensure that the polyphosphate is well mixed. Add enough chlorine to produce a chlorine residual of 0.25 mg/*L* after a five-minute contact time period. Be sure the samples are well mixed. Observe the samples daily against a white background, recording the amount of discoloration. The proper polyphosphate dose is the lowest dose that delays noticeable discoloration for a period of four days.

The addition of either chlorine or potassium permanganate will oxidize iron and manganese to insoluble precipitates which can be removed by filtration. Chlorine will oxidize manganese to insoluble manganese dioxide and iron to insoluble ferric hydroxide. The higher the chlorine residual, the faster the reaction goes. Some plants add chlorine to produce a residual of from 5 to 10 mg chlorine per liter (*SUPERCHLORINATION*[34]). The insoluble precipitates are removed by filtration. The water is dechlorinated by the use of reducing agents such as sulfur dioxide (SO_2), sodium bisulfite ($NaHSO_3$), and sodium sulfite (Na_2SO_3). Bisulfite is commonly used because it is cheaper and more stable than sulfite. A chlorine residual must be maintained in the treated water throughout the distribution system.

Potassium permanganate can be used to accomplish the same result as chlorine. The dose must be exact. Too little permanganate will not oxidize all of the manganese for removal and too much will allow permanganate to enter the distribution system and cause a pink color.

Filtration is used to remove the insoluble precipitates. Sometimes a *GREENSAND*[35] is used which oxidizes iron and manganese to their insoluble oxides. The greensand is capable of both oxidation and filtration. The greensand may be regenerated by the use of potassium permanganate.

Iron can be oxidized by *AERATING*[36] water to form insoluble ferric hydroxide. The higher the pH, the faster the reaction. The oxidation of manganese by aeration is so slow that this process is not used on waters with high manganese concentrations. Aeration is achieved by spraying water into the air, allowing the water to flow over steps, or passing the water over coke trays. After aeration, adequate holding time in a retention basin is required for the oxidation reactions to take place. After the ferric hydroxide is formed, the insoluble precipitate is removed by sedimentation and filtration, or by filtration alone. The main advantage of this method is that no chemicals are required. For information on the operation of pressure filters, see Section 4.4, "Filtration."

Ion exchange can be used to remove both iron and manganese if the water to be treated contains no dissolved oxygen. Usually groundwaters do not contain dissolved oxygen. If dissolved oxygen is present, the exchange resin becomes fouled with iron rust or manganese dioxide. Cleaning the resin is expensive. The procedures for operating and maintaining ion exchange units are outlined in the next section on softening.

For additional information and details on the control of iron and manganese, see Chapter 12, "Iron and Manganese Control," in *WATER TREATMENT PLANT OPERATION*, Volume II, in this series of manuals.

QUESTIONS

Write your answers in a notebook and then compare your answers with those on page 261.

4.9A Why are excessive amounts of iron and manganese in drinking waters objectionable?

4.9B Why is greensand sometimes used to treat iron and manganese?

4.9C What is the main advantage of treating iron by aerating water?

4.9D Why must there be no dissolved oxygen present when using ion exchange units to remove iron and manganese?

4.10 SOFTENING

4.100 Lime-Soda Ash Softening

The exact procedures used to soften water by chemical precipitation using the lime-soda ash process will depend on the hardness and other chemical characteristics of the water being treated (pH, alkalinity, temperature). A series of jar tests is commonly used to determine optimum dosages. In many cases, the feed rates determined by jar tests do not produce the exact same results in an actual plant. This is because of differences in water temperature, size and shape of jar as compared with plant basins, mixing equipment, and influence

[34] *Superchlorination (SUE-per-KLOR-uh-NAY-shun). Chlorination with doses that are deliberately selected to produce free or combined residuals so large as to require dechlorination.*

[35] *Greensand. A mineral (glauconite) material that looks like ordinary filter sand except that it is green in color. Greensand is a natural ion exchange material which is capable of softening water. Greensand which has been treated with potassium permanganate ($KMnO_4$) is called manganese greensand; this product is used to remove iron, manganese and hydrogen sulfide from groundwaters.*

[36] *Aeration (air-A-shun). The process of adding air to water. Air can be added to water by either passing air through water or passing water through air.*

of coagulant (a heavy alum feed will neutralize more of the lime). You must remember that jar test results are a starting point. You may have to make additional adjustments to the chemical feeders in your plant based on actual analyses of the treated water.

Let's set up some jar tests to determine the optimum dosages for lime or lime-soda treatment to remove hardness from well water. To get started, add 10.0 grams of lime to a one-liter container and fill to the one-liter mark with distilled water. Thoroughly mix this stock solution in order to dissolve all of the lime. One m*L* of this solution in a liter of water is the same as a lime dosage of 10 mg/*L* (0.5 m*L* in 500 m*L* is still the same as a 10-mg/*L* lime dose).

Set up a series of hardness tests by adding 5.0 m*L*, 10.0 m*L*, 15.0 m*L*, 20.0 m*L*, 25.0 m*L*, 30.0 m*L*, 35.0 m*L*, and 40.0 m*L* to one-liter (1,000-m*L*) containers or jars. Fill the containers to the 1,000-m*L* mark with the water being tested. Mix thoroughly, allow the precipitate to settle, and measure the hardness remaining in the water above the precipitate. A plot of the hardness remaining against the lime dosage will often reveal the optimum dosage. See Figures 4.34, 4.35, and 4.36.[37]

Examination of Figures 4.34, 4.35, and 4.36 reveals that the water of all three cities responded differently to the increasing lime dosage. City 1 (Figure 4.34) should be providing a lime dosage of 100 mg/*L*. The cost of increasing the dosage to 150 mg/*L* is not worth the slight reduction in hardness from 110 to 100 mg/*L* as $CaCO_3$. Note that an overfeed of lime will actually increase the hardness.

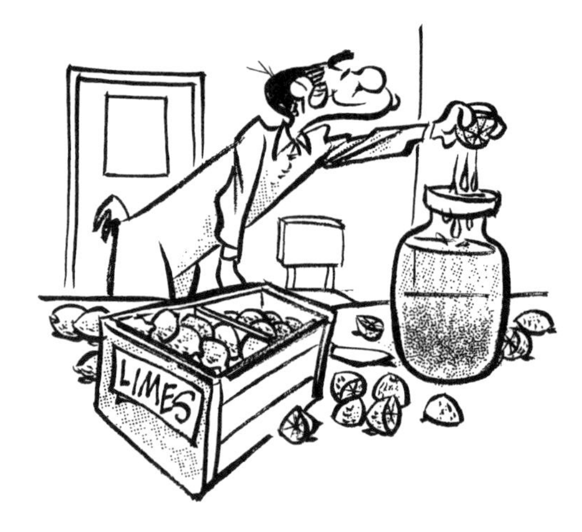

City 2 (Figure 4.35) should be providing a lime dose of 200 mg/*L*. A dose of 300 mg/*L* will reduce hardness, but the increase in lime costs are too great. City 3 (Figure 4.36) should be dosing lime between 200 and 250 mg/*L*. Note that the greater the lime dose, the less the hardness, but the greater the quantities of sludge that must be handled and disposed of.

If lime added to the water does not remove sufficient hardness, select the optimum lime dose and then add varying amounts of soda ash. From Figure 4.36, we found that the optimum lime dose was 200 mg/*L* (300 mg/*L* would have reduced the hardness only slightly more). Let's take six one-liter containers and add 20 m*L* of our lime stock solution (a dosage of 200 mg/*L*). Prepare a stock solution of soda ash similar to our lime solution. Add 10 grams of soda ash to a one-liter container, fill with distilled water and mix thoroughly. Add zero, 2.5 m*L* (2.5 mg/*L* dose), 5 m*L*, 7.5 m*L*, 10 m*L*, 12.5 m*L*, and 15 m*L*. Mix thoroughly, allow the precipitate to settle and measure the hardness remaining in the water above the precipitate. A plot of hardness remaining against the soda ash dosage will reveal the desired dosage. We would like the final hardness to be in the 80- to 90-mg/*L* as $CaCO_3$ range.

To select the optimum doses of lime and soda ash, consider the items discussed below.

1. Optimum dosage of lime was based on increments of 50 mg/*L*. You should refine this test by trying at least two 10 mg/*L* increments above and below the optimum dose. From Figure 4.34 we found that 100 mg/*L* was the optimum dose. Try lime doses of 80, 90, 100, 110, and 120 mg/*L*.
2. Optimum dosage of soda ash can be refined by trying similar increments also.
3. Try slightly increasing the actual lime dose in your plant to see if there is any decrease in the remaining hardness. Is the decrease in hardness worth the increase in lime costs?
4. Try slightly increasing and decreasing both lime and soda ash dosages at your plant one at a time, and evaluate the results.
5. If you are treating well water or a water of constant quality, all you have to do to maintain proper treatment is to make minor adjustments to keep the system fine tuned.
6. If you are treating water from a lake or a river and the water quality (including temperature) changes, you'll have to repeat these procedures whenever the source water quality changes. Water quality changes of concern include raw water hardness, alkalinity, turbidity, and temperature.
7. *REMEMBER*, you do not want to produce water of zero hardness. If you can get the hardness down to around 80 to 90 mg/*L*, that usually will be low enough for most domestic consumers.

After the chemicals have been mixed, the precipitates are allowed to settle out in some type of settling or sedimentation basin. Sludges from the lime or lime-soda softening processes are usually disposed of in sanitary landfills.

The next process is recarbonation. This process introduces carbon dioxide into the treated water to lower the pH and stabilize the water to prevent the precipitation of carbonate compounds on the filter media, in the clear well, and in the pipes in the distribution system. The pH to which the water should be lowered is determined by the Marble test. See Chapter 21, "Advanced Laboratory Procedures," in *WATER TREATMENT PLANT OPERATION*, Volume II, in this series of manuals for details on how to perform the Marble test. For information on how to operate gravity filters, see Section 4.4, "Filtration," and Section 4.8, "Slow Sand Filtration."

[37] *These figures were adapted from an article titled, "Use of Softening Curve for Lime Dosage Control," by Michael D. Curry, P.E., which appeared in THE DIGESTER/OVER THE SPILLWAY, published by the Illinois Environmental Protection Agency.*

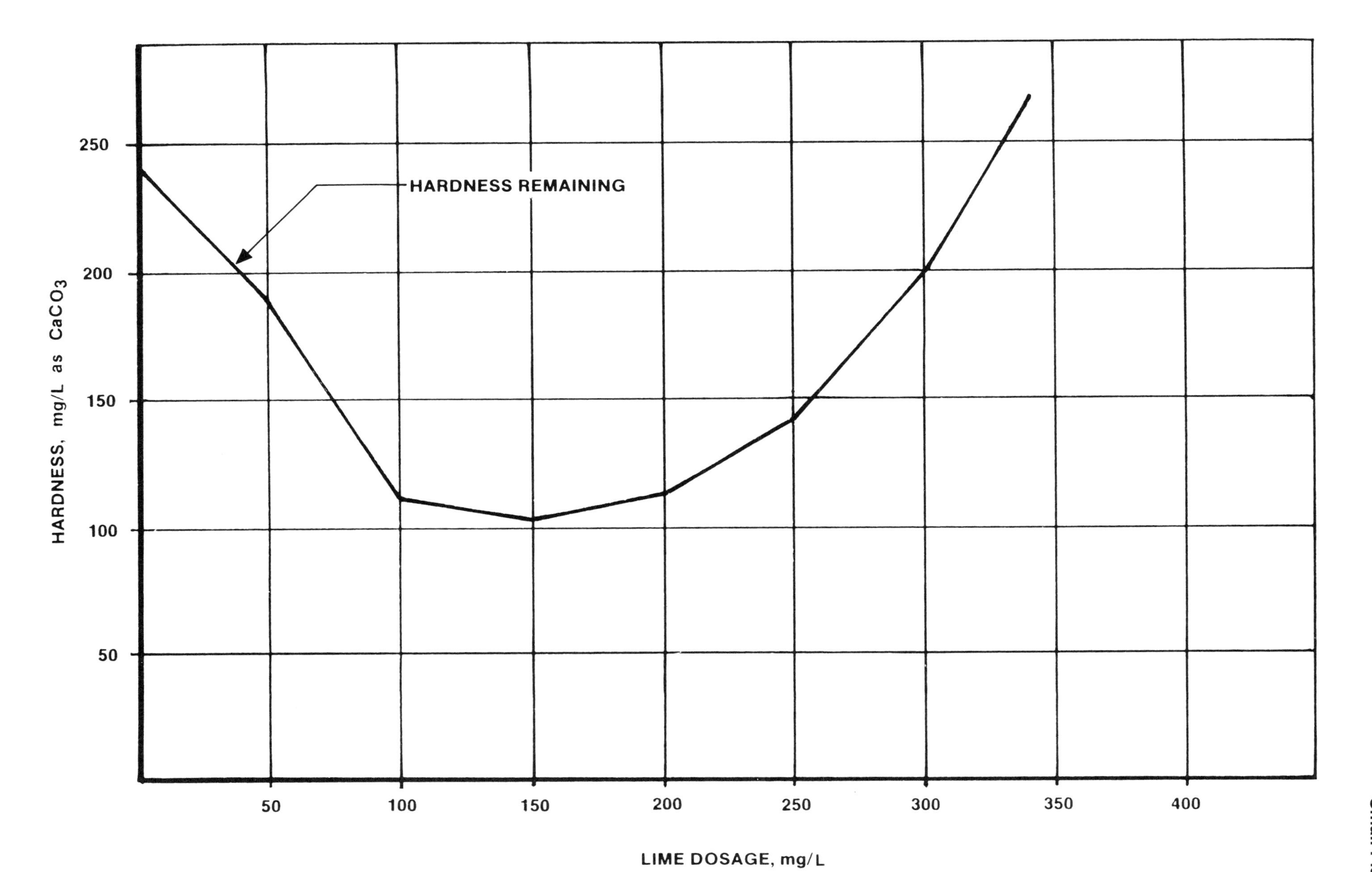

Fig. 4.34 Softening curve for City 1

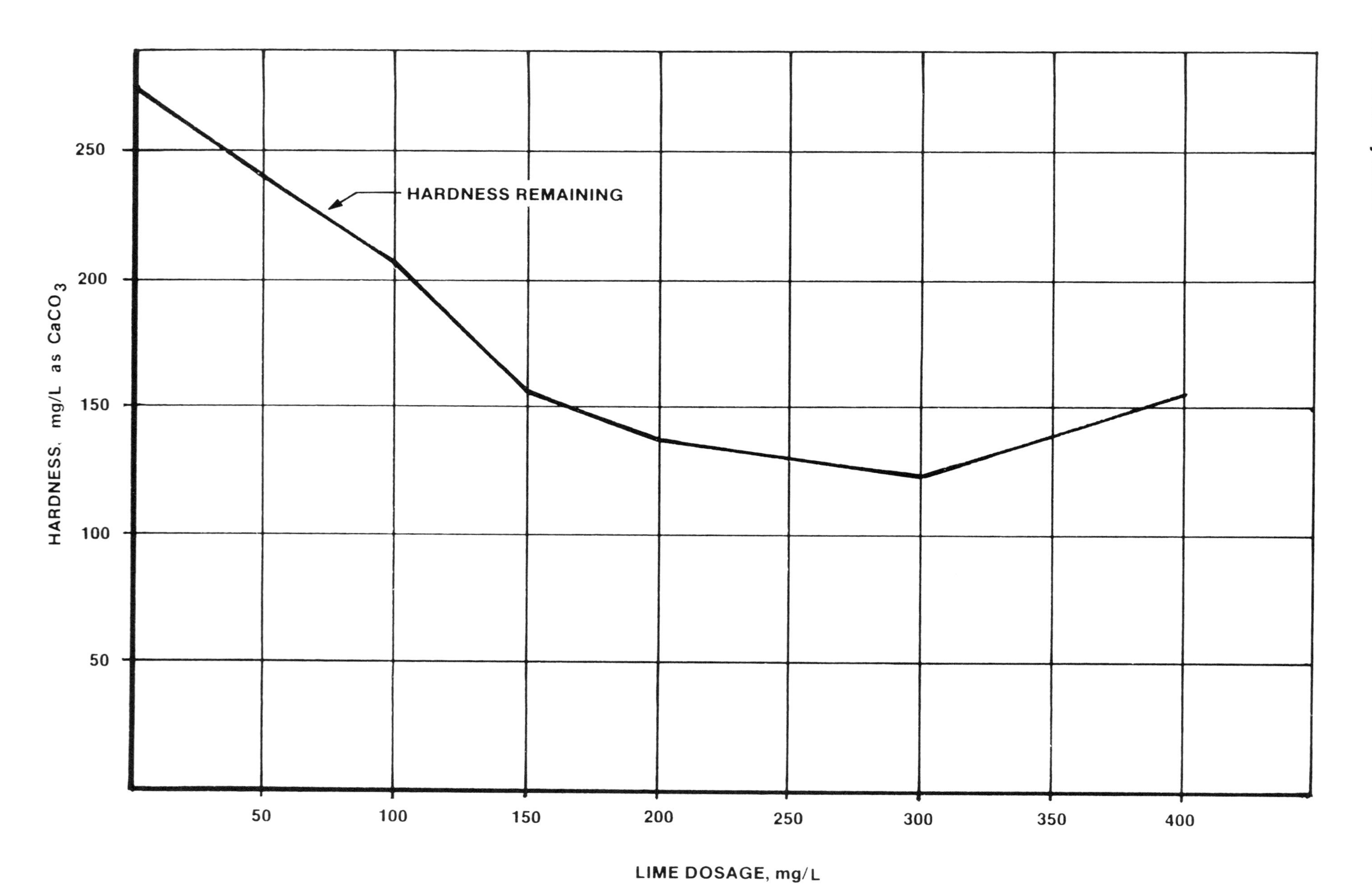

Fig. 4.35 Softening curve for City 2

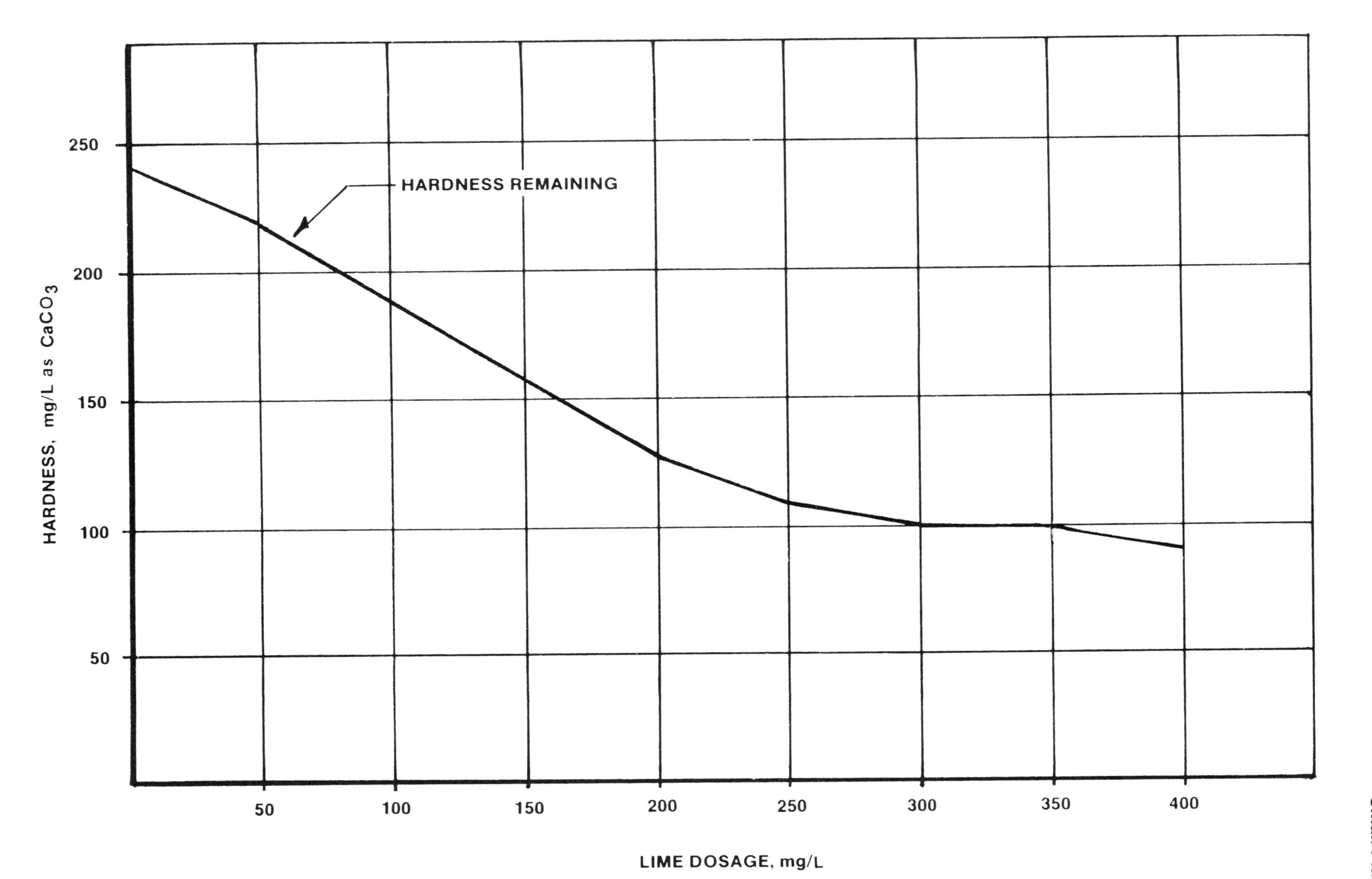

Fig. 4.36 Softening curve for City 3

QUESTIONS

Write your answers in a notebook and then compare your answers with those on page 261.

4.10A The procedures for softening water by chemical precipitation using the lime-soda ash process depend on what factors?

4.10B If lime added to water does not reduce the hardness of a water sufficiently, what would you do?

4.10C Why should softened waters from the lime-soda ash process pass through a recarbonation process before filtration?

4.101 Ion Exchange Softening

Ion exchange water softening is a process in which the hardness-causing ions (mainly calcium and magnesium) are exchanged (or traded) for sodium ions by passing the water through an ion exchange media. *ZEOLITE*[38] softening refers to sodium ion exchangers and will mean the same as ion exchange in this manual.

There are three basic types of ion exchangers in use today. One type is an upflow unit in which the water enters from the bottom and flows up through the resin exchange bed and out the top. Another type operates like a gravity sand filter. The water enters at the top, flows down through the softening media and out the bottom. The most common type is the pressure down-flow ion exchange softener.

Try to operate your ion exchange unit at design flows or less. Measure the hardness of the water entering the unit and the hardness of the water leaving (there should be zero hardness leaving). Monitoring influent hardness is very important, especially for surface waters. Zeolite softeners are designed to remove a specified amount of hardness before they require regeneration. See Example 16.

If design flows through a zeolite softener cannot be achieved or if the head loss through the softener becomes greater than normal, the resin may need backwashing. To backwash an ion exchange softener, take the unit out of service and reverse the flow pattern through the unit. The purpose of backwashing is to expand and clean the media particles, thus freeing any material such as iron and manganese that might have been removed during the softening stage.

Slowly allow the backwash water to enter the softener and gradually increase the backwash rate to the design flow. Periodically during the backwash process, catch a sample of effluent with a glass beaker. A trace amount of media should cause no alarm, but a steady loss of media could indicate a problem in the unit. Try to find the source of the problem as soon as possible and correct the problem. The backwash duration and flow rate will vary depending on the manufacturer and the type of media used.

When the effluent from the zeolite softener no longer has zero hardness or your calculations (Example 16) indicate that the softener is reaching its exchange capacity, the softener must be regenerated. This process is called regeneration or the brine stage. In this stage, the sodium ions present in the brine solution are exchanged with the calcium and magnesium ions on the resin that have been removed during the service or softening stage. If the regeneration process is performed correctly, the result is a bed that is completely recharged with sodium ions and will again soften water when the unit is returned to service.

After the regeneration process is completed, the softener should be rinsed. The flow pattern is very similar to the service or softening stage, with the exception that the softener effluent is going to waste instead of storage. Most rinse stages will last from 20 to 40 minutes depending on the size of the unit and the manufacturer. The rinse must be long enough to remove all traces of the brine from the softener. Near the end of the rinse cycle, taste the effluent to be sure that no salty taste remains. If the effluent tastes OK, the unit is now ready to be returned to service. The effluent from the filter should have zero hardness if the regeneration process was successful.

Zeolite softeners will remove iron and manganese in the soluble or precipitated form. If water high in iron and manganese is applied to the ion exchange media for very long, iron fouling or the loss of exchange capacity could result. If the media becomes coated with iron, the media will develop a rusty or orange appearance and the efficiency of the softener will be greatly reduced. High iron loadings (in the ferric or solid form) can plug the upper layer of the media and force water to channel or short-circuit through the bed. This will result in incomplete or insufficient contact between the water and the media, thus creating hardness leakage and a loss in softening efficiency.

Disposal of spent brine can be a serious problem. The brine is corrosive to material it comes in contact with and is toxic to many living plants and animals in the environment. Each ion exchange treatment plant probably has only one approved method of spent brine disposal which must be carefully followed at all times.

The brine is corrosive to the brine carrying system; therefore, frequent inspection and routine maintenance are important. Immediately repair all leaks in the brine pumping and piping system and also in the brine storage area.

If the bed turns an orange or rusty color, iron fouling is becoming a problem. Try increasing the length of the backwash

[38] *Zeolite. A type of ion exchange material used to soften water. Natural zeolites are siliceous compounds (made of silica) which remove calcium and magnesium from hard water and replace them with sodium. Synthetic or organic zeolites are ion exchange materials which remove calcium or magnesium and replace them with either sodium or hydrogen. Manganese zeolites are used to remove iron and manganese from water.*

stage. A chemical cleaner (sodium bisulfite) can be used to remove heavy iron coatings from the media. The sodium bisulfite can be added to the brine during the regeneration stage or it may be mixed in solution form and poured into the softener when the softener is out of service. The softener should be rinsed before being returned to service.

The water produced by a zeolite softener has near zero hardness and is very corrosive. For these reasons, this water is mixed or blended with other water (untreated or unsoftened water) to produce water of acceptable hardness (80 to 90 mg/*L*) for the consumers. The amount of blend water depends on the desired level of hardness in the finished water.

In most cases ion exchange softening is used by smaller water softening plants instead of the lime-soda softening process due to economics. For additional information on water softening, see Chapter 14, "Softening," in *WATER TREATMENT PLANT OPERATION*, Volume II, in this series of manuals.

FORMULAS

The exchange capacity of ion exchange units is expressed in kilograins. Therefore, if an ion exchange unit has a capacity of 20 kilograins, the unit can remove 20 kilograins of hardness before requiring regeneration.

$$\text{1 grain per gallon or gpg} = 17.1 \text{ mg}/L$$

$$\text{Hardness, grains/gallon} = \frac{\text{Hardness, mg}/L}{17.1 \text{ mg}/L/\text{gpg}}$$

$$\text{Exchange Capacity, grains} = (\text{Media Vol, cu ft})(\text{Removal Capacity, grains/cu ft})$$

$$\text{Water Treated, gal} = \frac{\text{Exchange Capacity, grains}}{\text{Hardness, grains/gallon}}$$

$$\text{Operating Time, hr (Before Regeneration)} = \frac{(\text{Water Treated, gal})(24 \text{ hr/day})}{\text{Avg Daily Flow, gal/day}}$$

EXAMPLE 16

A zeolite softener contains 200 cubic feet of media with a hardness removal capacity of 20 kilograins per cubic foot of media. The water being treated has a hardness of 250 mg/*L* as $CaCO_3$. How much water can be treated before the softener will require regeneration?

Known		Unknown
Media Volume, cu ft	= 200 cu ft	Water Treated, gal
Removal Capacity, gr/cu ft	= 20,000 grains/cu ft	
Hardness, mg/*L*	= 250 mg/*L*	

1. Convert the hardness from mg/*L* to grains per gallon.

$$\text{Hardness, grains/gal} = \frac{\text{Hardness, mg}/L}{17.1 \text{ mg}/L/\text{gpg}}$$

$$= \frac{250 \text{ mg}/L}{17.1 \text{ mg}/L/\text{gpg}}$$

$$= 14.6 \text{ grains per gallon}$$

2. Calculate the exchange capacity of the softener in grains.

$$\text{Exchange Capacity, grains} = (\text{Media Vol, cu ft})(\text{Removal Capacity, grains/cu ft})$$

$$= (200 \text{ cu ft})(20{,}000 \text{ grains/cu ft})$$

$$= 4{,}000{,}000 \text{ grains}$$

3. Calculate the volume of water in gallons that may be treated before regeneration.

$$\text{Water Treated, gal} = \frac{\text{Exchange Capacity, grains}}{\text{Hardness, grains/gallon}}$$

$$= \frac{4{,}000{,}000 \text{ grains}}{14.6 \text{ grains/gallon}}$$

$$= 274{,}000 \text{ gallons}$$

EXAMPLE 17

The zeolite softener described in Example 16 treats an average daily flow of 200,000 gallons of water per day. How often (hours) should the softener be regenerated? From Example 16, the softener can treat 274,000 gallons of water before requiring regeneration.

Known		Unknown
Water Treated, gal	= 274,000 gal	Operating Time, hr
Avg Flow, gal/day	= 200,000 gal/day	

Calculate the number of hours the softener can operate before requiring regeneration.

$$\text{Operating Time, hr} = \frac{(\text{Water Treated, gal})(24 \text{ hr/day})}{\text{Avg Daily Flow, gal/day}}$$

$$= \frac{(274{,}000 \text{ gal})(24 \text{ hr/day})}{200{,}000 \text{ gal/day}}$$

$$= 32.9 \text{ hours}$$

QUESTIONS

Write your answers in a notebook and then compare your answers with those on page 261.

4.10D How does an ion exchange water softening process work?

4.10E When should a zeolite softener be backwashed?

4.10F How can you tell when the rinse cycle of a zeolite softener is adequate?

4.10G What problems does excessive iron cause in a zeolite softener?

4.11 OPERATION

The operators of water treatment plants must be responsible, knowledgeable, and conscientious. A poor operator may not do a good job with even the best equipment, whereas a competent operator, who is conscientious and responsible, can often achieve good results with only ordinary or even inadequate equipment. Over and over again, experience has shown that it is the abilities of the operator that determine whether a water treatment plant produces acceptable quality water. Frequently large sums are spent on the design and construction of a water treatment plant but all too often selection and training of the operator is hardly given a second thought. The need for a capable operator must not be ignored if the goal is to produce good water.

First of all, the operator must be interested in doing a good job. An operator must have the desire to produce the best quality water possible. Next, the operator should receive the necessary training. Prior experience in water treatment is very helpful but additional training is necessary. Many aspects of water treatment and water plant operation involve very technical or specialized subjects. These subjects include mathematics, hydraulics, water chemistry, water quality standards, electricity, plumbing, pumps, motors, bacteriology, recordkeeping, and public relations. Classroom instruction is sometimes available but frequently the operator's only choice is individual study at home. However, when seminars, classes, and meetings are offered, the operator should certainly attend.

The operators of small systems can often receive very helpful training and assistance from the operators of large water treatment plants in the area. Try to visit other plants and become acquainted with the operators. They are usually glad to help another operator.

Operator certification programs are available now in every state. Certification may be granted under a program conducted by the local water works association or by the state government. In either case, the operator should enroll in the program for the training and educational benefits and try to become certified. Operator certification is important because it is evidence of the operator's level of knowledge and proficiency on the job. As a result, operator job descriptions for many water utilities now require certification before hiring. Water treatment plant operators are finding it harder than ever to find work as an operator without being certified.

Following is a list of typical procedures the operator should follow in operating the small treatment plant.

DAILY

1. Inspect the plant daily including weekends and holidays. A trained substitute operator must be available when the regular operator is sick or on vacation.
2. Determine that all the plant equipment is operational.
3. Perform water quality and operational tests like chlorine residual, turbidity, pH, and jar tests.
4. Actually smell and taste the treated water produced to detect any aesthetic problems.
5. Record readings from flowmeters, pressure gages, filter head loss gages, and reservoir level gages.

PERIODICALLY

1. Dismantle, clean, and overhaul chemical feeders like chlorinators and coagulant chemical feed pumps.
2. Test start standby pumps, motors, and generators.
3. Perform "preventive maintenance" to inspect, lubricate, clean, and repair equipment.
4. Replenish chemical stocks.
5. Clean settling basins and dispose of sludge.
6. Overhaul major equipment like filters, clarifiers, and pumps.
7. Calibrate flowmeters and chemical feeders.
8. Repaint piping, equipment, and buildings.
9. Perform electrical efficiency tests on all pumps.
10. Test emergency alarms and control systems.
11. Maintain stock of repair parts and spare equipment.
12. Collect routine bacteriological samples and occasional chemical samples for laboratory analysis.

The operator needs to become acquainted with local contractors, regulatory agencies, and suppliers of chemicals, equipment, and materials. Contractors must be hired occasionally to repair the water system, construct improvements, and perform other services. The operator should be familiar with the capabilities of each contractor and the cost of necessary services. Regulatory agencies, such as the health department, can assist the operator in solving problems that arise. Likewise, suppliers of equipment and materials are often experts in their field and frequently give generously of their time to assist an operator in using their products successfully. These resources should not be overlooked. A lot of good advice is available without charge and the operator should not hesitate to ask for assistance from any of these sources when the need arises.

Records of water system operation are vital and must be kept up to date by the operator. The essential records include the following items.

1. *PLANT OPERATING DATA*

   These data include the date and time that the plant was inspected, the operating flow rate, the amount of water treated, the dosage of chemicals being applied, the amount of chemicals fed, filter head loss, filter backwash time, reservoir levels, and weather conditions.

2. *DAILY LOG OR DIARY*

   Many operators keep a daily diary to supplement and explain information recorded in the log of operating data. The operator's diary usually consists of informal notes and the operator is free to record any important comments, observations, reminders, or explanations. The types of information recorded may include notes about equipment breakdowns, changes in treatment, severe weather, changes in water quality conditions, or visits by repair technicians, health inspectors, customers, or officials. This information is very helpful to the relief operator who must take over operation of the plant suddenly or on occasion.

3. *LABORATORY TEST RESULTS*

   Test results include those for tests performed by the operator as well as by private laboratories. The operator performs jar tests, measures turbidity, pH, chlorine residual, and performs other simple chemical tests such as alkalinity. Complete chemical analyses and routine bacteriological sample testing is usually done by a private analytical laboratory. These test results must usually be reported to the health department and preserved for reference.

4. *SYSTEM RECORDS*

   System records include all available maps and drawings of the distribution system layout. These records are invaluable for locating pipelines, valves, and service connections. The location, size, age, and material of all pipelines should be shown on the plans.

   The operator should carefully preserve the plans and specifications for all structures including the treatment equipment and the storage reservoirs. These documents show the construction details on each piece of equipment or structure and they describe the materials that were used. The operator's files should also contain the technical manual and the maintenance records for each piece of equipment in the system. With this information available for reference, the operator should be able to understand how the system was built and how to keep it operating efficiently.

The backflow of polluted waters or liquids into the domestic water system must be prevented by a program of backflow prevention. Backflow, which is a reversal in the normal direction of flow, can occur in two ways. One way is backflow due to siphonage caused by vacuum conditions in the system. This backsiphonage is caused by a sudden major water demand such as a fire or a major system break. The second way is backflow due to pressure conditions created by a pump or other source of higher pressure.

Backflow from wastewater (sewage) facilities and from industrial or commercial operations is the most hazardous. Many hazardous chemicals and liquids are used in industry and these must be excluded from the water supply to prevent poisoning of consumers. Wastewater contains a variety of germs, often in great numbers, that can cause illness or death.

To limit backflow, the distribution system must be properly constructed and operated. Pipelines must be large enough to handle peak flows without creating vacuum conditions. Also, adequate pressures must be maintained in the system at all times. Pipeline outages and shutdowns should be limited and pipes thoroughly flushed before they are returned to service.

Water service connections to premises where backflow hazards exist should be protected with an appropriate backflow preventer. Typical types of backflow preventers include air gap separations, reduced pressure principle devices, double check valve assemblies, and vacuum breakers. See Chapters 3 and 5 in *WATER DISTRIBUTION SYSTEM OPERATION AND MAINTENANCE* for details on backflow preventers.

For additional information on taste and odor control and plant operation, see Chapter 9, "Taste and Odor Control," and Chapter 10, "Plant Operation," in *WATER TREATMENT PLANT OPERATION*, Volume I, in this series of manuals. For information on fluoridation, see Chapter 13, "Fluoridation," in Volume II.

4.12 MAINTENANCE

4.120 Program

The water plant operator cannot afford to ignore the need for equipment maintenance. The water treatment plant must normally operate every day and equipment outages cannot be tolerated if the treatment is to be effective. Therefore the operator should strive to eliminate equipment breakdowns or at least minimize the time they cause the plant to be out of operation. Frequently, maintenance problems can be avoided by proper design or selection of the right piece of equipment.

Most other problems can be avoided if the operator will regularly inspect, clean, and lubricate the equipment. Maintenance inspections should be performed according to a regular schedule which should be written down so inspections are not overlooked. Unusual noises, vibrations, leaks, and malfunctions should receive prompt attention. Regular cleaning of chemical feeders for chlorine, alum, and lime will prevent many breakdowns in this equipment. Spare parts should be kept on hand so breakdowns can be repaired quickly and so worn parts can be replaced when the feeder is disassembled for cleaning. Preventive maintenance for equipment should be based on manufacturers' recommendations. Other chapters in this manual should be consulted for more specific details on equipment maintenance.

The operator is also responsible for maintaining clean and tidy conditions in the plant grounds. Plant piping, buildings, and tanks should be painted regularly to prevent deterioration and to present a good public appearance. Pipes, plumbing fittings, chemicals, tools, and other materials must be stored in a safe and orderly manner. Junk and clutter should be eliminated.

For additional information on maintenance, see Chapter 18, "Maintenance," and Chapter 19, "Instrumentation," in *WATER TREATMENT PLANT OPERATION*, Volume II, in this series of manuals.

4.121 Tools

Nothing makes a maintenance or repair job tougher than to have the wrong tools or no tools at all. The following list of tools is provided as a guide to help you determine the tools you will need to do your job. This list is not complete. You may need additional tools to do your job and some of the tools you will not need.

Proper care of tools is essential. You must maintain your tools and hand equipment as necessary and keep everything in proper working order. After a job is completed, clean your tools and store them properly so they can be found and used when needed again.

LIST OF OPERATOR'S TOOLS

1. *WRENCHES*
   a. Deep-well socket set, 3/8" to 1 1/4" with 1/2" drive ratchet and "breaker bar"
   b. Pipe wrenches, assorted sizes 6" to 36"
   c. Crescent wrenches, assorted sizes 6" to 18"
   d. Combination open and box end wrench set 1/4" to 3/4"
   e. Allen wrench set
   f. Fire hydrant spanner wrench
   g. Gate valve wrench, 2" nut and 4" nut
   h. Corporation stop wrench
2. *SCREWDRIVERS* — Set, assorted sizes
3. *HAMMERS*
   a. Claw — 16 oz
   b. Ball peen — 16 oz and 8 oz
   c. Brass, rawhide, or soft plastic head
4. *SLEDGEHAMMERS*
   a. 2 pound
   b. 8 pound
5. *PLIERS*
   a. Slip joint common pliers
   b. Groove joint "channel lock" pliers
   c. Electric lineman's pliers
   d. Diagonal cutter pliers
   e. Needle nose pliers
   f. Vise grip pliers
   g. Electric terminal end pliers with wire stripper
   h. Fuse puller pliers
   i. Internal snap-ring pliers
   j. External snap-ring pliers
6. *HACKSAW*
7. *METAL FILES* — set of round and flat
8. *LIGHTS*
   a. 2-cell flashlight
   b. 6-volt battery lantern
9. *MEASURING TAPE*
   a. 10' retractable steel
   b. 100' cloth
10. *SHOVEL*
    a. Round point, heavy duty
    b. Square end
11. *PICK*
12. *PIPE VISE* — Truck mounted
13. *FLARING TOOLS AND CUTTER* — For copper and plastic pipe
14. *TUBING BENDER*
15. *CRIMPING TOOL* — For emergency shutoff of service pipes
16. *POCKET KNIFE* — Folding with blade lock feature
17. *GASKET SCRAPER TOOL*
18. *COLD CHISELS AND PUNCHES* — Set, assorted sizes
19. *PUTTY KNIFE*
20. *MANHOLE COVER PULLER HOOK*
21. *ELECTRIC DRILL* — 1/2" reversible
22. *DRILL BITS* — Set, assorted sizes
23. *TAP AND DIE SET*
24. *"EASY OUTS"* — Set
25. *STUD BOLT PULLER*
26. *PACKING REMOVAL HOOKS* — Flexible, set of sizes
27. *LANTERN RING REMOVAL TOOLS* — Set
28. *PIPE CUTTING TOOLS* — Manual or electric power
29. *WIRE BRUSHES*

30. *STEEL PRY BARS* — 8' long
31. *EXTENSION LADDER* — Aluminum
32. *"COME-A-LONG"* — 2-ton capacity
33. *CHAIN HOIST* — 2-ton capacity
34. *NYLON ROPE* — 1/2" diameter
35. *WATER PRESSURE GAGE* — With connection to hose faucet
36. *FIRE HOSE* — 50'
37. *MUD PUMP* — 3" positive displacement, gasoline powered
38. *CLOTHING*
    a. Rain suit — pants and jacket with hood or hat
    b. Rubber boots — calf length, hip length
    c. Rubber gloves
    d. Hard hat
    e. Face mask, clear plastic
    f. Eye goggles
    g. Leather gloves

39. *ELECTRIC VOLTAGE TESTER* — 110 to 600 volts
40. *VOLTMETER/AMMETER TESTER* — "Amprobe"
41. *WATER TEST KITS*
    a. Chlorine residual
    b. pH
    c. Turbidimeter
42. *AMMONIA SOLUTION* — Industrial strength
43. *CHLORINE BLEACH* — Sodium hypochlorite
44. *CHLORINE TABLETS* — Calcium hypochlorite
45. *BREATHING MASK* — Self-contained, MSA
46. *RESPIRATOR MASK* — With replaceable filters
47. *GREASE GUN* — With graphite cartridges
48. *ELECTRONIC CALCULATOR*
49. *OPERATION MANUALS* — For plant equipment
50. *TOOL BOXES* — Lockable steel
51. *WIPING RAGS*
52. *PLASTIC ELECTRICAL TAPE*
53. *FIRST-AID KIT*
54. *PUSH BROOM* — Heavy duty
55. *PERMATEX NO. 1* — Gasket adhesive

4.13 SAFETY

When operating any type of water treatment facility,

Safety must always come first!

If you are the only operator at a small water treatment plant, you must take care of yourself. There frequently is no one else present to make sure that you follow safe procedures and that there are no safety hazards that can injure you. There may be no one else available to help you or to rescue you if you become injured.

Safety hazards of concern around any water treatment plant include chlorine, electrical shock, and drowning. Studies have shown that the greatest causes of injuries to operators result from muscle and back strains caused by lifting and injuries resulting from slips and falls. These types of injuries can easily be prevented by alert and safety-conscious operators.

Be extremely careful whenever working with chlorine. If you are wearing a self-contained (or air line) breathing apparatus and the respirator falls, you could be overcome by a toxic gas or a lack of oxygen. Under these circumstances, two people should always be standing by for rescue purpose. The following rules are recommended:

1. Two (2) people *SHOULD* be present whenever entrance to a gas chlorine feed station and/or gas chlorine storage area (building or yard) is attempted.
2. Two (2) people *SHALL* be present when any type of maintenance, repair work, and/or work which might create a chlorine leak is attempted.
3. Two (2) approved, self-contained (or air line) supplied air masks *SHALL* be available prior to entering a gas chlorine feed station and/or gas chlorine storage area.

NOTE: *"SHOULD"* indicates that a safety regulation will normally be followed as closely as possible, consistent with good judgment. *"SHALL"* indicates that compliance with a safety regulation is mandatory at all times.

For additional information on safety, see Chapter 6, "Safety," in this manual and also Chapter 20, "Safety," in *WATER TREATMENT PLANT OPERATION*, Volume II, in this series of manuals.

QUESTIONS

Write your answers in a notebook and then compare your answers with those on page 261.

4.11A Why should operators become certified?

4.11B List the essential records that must be kept up to date by the operator.

4.11C Which laboratory tests are commonly performed by operators?

4.12A How can the operator avoid maintenance problems?

4.13A What are the greatest causes of injuries to operators?

4.14 ARITHMETIC ASSIGNMENT

Turn to the Appendix, "How to Solve Small Water System Arithmetic Problems," at the back of this manual and read all of Section A.4, *METRIC SYSTEM*.

In Section A.13, *TYPICAL SMALL WATER SYSTEM PROBLEMS*, read and work the problems in the following sections:

1. A.130, Flows;
2. A.131, Chemical Doses;
3. A.132, Wells; and
4. A.133, Small Water Treatment Plants.

4.15 ADDITIONAL READING

1. *NEW YORK MANUAL*, Chapter 1,* "Purpose of Water Treatment," and Chapter 19,* "Treatment Plant Maintenance and Accident Prevention."
2. *SLOW SAND FILTRATION*. Edited by Gary S. Logsdon. Obtain from the American Society of Civil Engineers (ASCE), Book Orders, PO Box 79404, Baltimore, MD 21279-0404. Stock No. 847. ISBN 0-87262-847-7. Price, $23.00, plus $4.00 shipping and handling.
3. *TEXAS MANUAL*, Chapter 25,* "Water Treatment Plant Waste Disposal."

* Depends on edition.

End of Lesson 2 of 2 Lessons
on
SMALL WATER TREATMENT PLANTS

Please answer the discussion and review questions before continuing with the Objective Test.

DISCUSSION AND REVIEW QUESTIONS

Chapter 4. SMALL WATER TREATMENT PLANTS

(Lesson 2 of 2 Lessons)

Write the answers to these questions in your notebook before continuing with the Objective Test on page 262. The question numbering continues from Lesson 1.

14. How can corrosion be controlled?
15. Why should corrosion control chemicals be applied to water after all other treatment has been accomplished?
16. Sudden changes in what three factors can cause operating problems in solids-contact units?
17. What problems can develop in a solids-contact unit if the recirculation rate is too high or too low?
18. Why are slow sand filters seen as a potentially promising alternative for small water systems?
19. How do microorganisms treat drinking water in the slow sand filtration process?
20. Why will a new filter or one that has just been cleaned or resanded not produce filtered water of acceptable quality immediately upon start-up?
21. How long should a newly cleaned filter operate (filter water) before it is placed back on line?
22. What cyclic factors can influence the performance of slow sand filters?
23. Why must the dose of potassium permanganate for the control of iron and manganese be exact?
24. Why do the results of jar tests often not produce the same results in a water treatment plant using the lime-soda ash softening process?
25. How often should the chemical dosages be checked in a lime-soda ash water softening plant?
26. What is the optimum flow rate through an ion exchange softener?
27. Plant operating data includes what types of information?
28. How can the backflow of polluted waters or liquids into the domestic water system occur?
29. Why is safety especially important for the operator of a small water treatment plant?

SUGGESTED ANSWERS

Chapter 4. SMALL WATER TREATMENT PLANTS

ANSWERS TO QUESTIONS IN LESSON 1

Answers to questions on page 182.

4.0A Water supplies that need no treatment are very rare because of pollution and more restrictive standards for domestic water quality.

4.0B Storage of raw water in lakes, ponds, or reservoirs helps the operator by slowing the rate of change in water quality due to storms and other factors.

4.0C A flowmeter indicates to the operator both the instantaneous rate of flow as well as the total quantity of water that has flowed through it.

4.0D Flowmeters can be protected by a fine screen or wye strainer installed upstream from the meter to prevent clogging or damage by trash or rocks.

4.0E Treatment that may be required by groundwaters includes disinfection, iron and manganese control, and softening.

Answers to questions on page 188.

4.1A The warmer the water, the faster the coagulation chemical reactions. Any chemical reaction proceeds faster when the water is warm than when it is cold.

4.1B The results of jar tests can be evaluated by visual observations or by performing further laboratory tests. Visual observations include which coagulant dosage produces the first visible floc, which produces the largest, strongest floc, or which produces the floc that settles the fastest. Laboratory tests include (1) the pH of the water, (2) the turbidity of the settled water, and (3) the filterability of the water as determined by filtration through some type of small laboratory filter.

4.1C To determine whether the coagulant dose should be increased or decreased, collect a sample from the outlet of the rapid mix chamber and use this water for the jar test. Add additional doses of coagulant to two jars and dilute a couple of jars with raw water. Observing the results can indicate whether the dosage should be increased or decreased.

4.1D The alkalinity of the water being treated can be increased by adding lime or soda ash prior to coagulation.

Answers to questions on page 189.

4.2A Flocculation is a process of slow, gentle mixing of the water to encourage the tiny floc particles to clump together and grow to a size that will settle quickly.

4.2B Flocculation mixing must be strong enough to prevent premature settling of floc in the flocculation basin, but the mixing must not be so strong that it breaks apart the floc particles already formed.

Answers to questions on pages 192 and 194.

4.3A The purpose of settling is to remove as much of the floc and other suspended material as possible before the water flows to the filters.

4.3B Short-circuiting depends on (1) the shape and dimensions of the tank, and (2) the inlet and outlet arrangements of the tank.

4.3C The warmer the water, the faster the particles settle.

4.3D Sludge is recycled to the inlet of the settling tank to increase the number of particles in the water and improve flocculation of the settling particles.

Answers to questions on page 204.

4.4A Suspended matter removed by filtration includes mainly particles of floc, soil, and debris; but it also includes living organisms like algae, bacteria, viruses, and protozoa.

4.4B Anthracite and sand stay separated during and after backwashing because sand is about two and a half times heavier than water and anthracite is only one and a half times heavier than water.

4.4C A mixed media filter contains three layers of media, (1) garnet, (2) sand, and (3) anthracite.

4.4D Mudballs will form in filters if backwashing does not effectively clean the media.

Answers to questions on page 215.

4.5A Domestic water supplies are usually disinfected with chlorine, although iodine, ozone, and ultraviolet radiation are also used occasionally.

4.5B Chlorine is available as (1) gaseous chlorine, (2) liquid chlorine as bleach (sodium hypochlorite), (3) granular or tablet (calcium hypochlorite), and (4) chlorine dioxide.

4.5C A chlorinator controls and measures the flow of chlorine gas and also dissolves the gas in water so it can be safely injected into the water supply.

4.5D Substances found in water which react with and consume chlorine include ammonia, dissolved iron and manganese, hydrogen sulfide, and organic matter.

4.5E The two types of chlorine residuals that can be measured with test kits include (1) free available chlorine, and (2) total chlorine residual.

4.5F Calculate the desired free chlorine residual for a water with a pH of 7.6 when the water is 5°C. The contact time is 25 minutes.

Known		Unknown
Time, min	= 25 min	Chlorine Residual, mg/*L*
pH	= 7.6	
Temp, °C	= 5°C	
A from Table 4.1	= 20	

Calculate the desired free chlorine residual in mg/*L*.

$$\text{Chlorine Residual, mg}/L = \frac{\text{A from Table 4.1}}{\text{Contact Time, min}}$$

$$= \frac{20}{25 \text{ min}}$$

$$= 0.8 \text{ mg}/L \text{ free chlorine residual}$$

ANSWERS TO QUESTIONS IN LESSON 2

Answers to questions on page 217.

4.6A The two general methods of controlling corrosion are (1) use of construction materials that resist corrosion, and (2) treatment of the water with chemicals to make the water noncorrosive.

4.6B The best method of monitoring corrosion in a water main is to examine specimens of the actual pipeline material. This inspection is easy if sections of pipe can be removed periodically. Alternatively, small specimens of the pipeline material can be mounted in the flowing water and periodically removed for inspection.

Answers to questions on page 220.

4.7A Three advantages of solids-contact units include:

1. Reduced capital and maintenance costs,
2. Ability to increase or decrease volume of slurry during periods of severe taste and odor problems, and
3. Ability to use slurry accumulation to carry plant when coagulation fails because of increased algal activities.

4.7B The level of the slurry or sludge blanket can be determined by sampling taps which are placed at various depths on the wall of the solids-contact reactor.

4.7C Operators should attempt to forecast turbidity changes and adjust chemical dosage prior to the arrival of a turbidity change.

Answers to questions on page 222.

4.7D The jar test is used to determine the proper chemical dose when operating a solids-contact unit. Observations and process monitoring are used to "fine tune" the results from jar tests.

4.7E Recirculation in a solids-contact unit may be provided by impellers, turbines, pumping units, or by air injection.

4.7F The percent slurry in the reaction zone is determined by the volume over volume (V/V) test.

Answers to questions on page 226.

4.8A The Surface Water Treatment Rule (SWTR) is especially significant for small water supply systems. The Rule requires all water systems where water supply is surface water or groundwater under the influence of surface water to disinfect their water and, under most circumstances, to install filtration equipment.

4.8B A tank cover is sometimes installed on a slow sand filter to prevent freezing of water in the tank, growth of algae, and accumulation of wind-blown debris.

4.8C The depth of a slow sand filter is the sum of the depths of the gravel support/underdrain system, the sand bed, maximum water (supernatant or headwater) depth above the filter, and freeboard.

4.8D (1) If the diameters of sand particles in a slow sand filter are too small, the filter tends to clog and cause short filter runs while (2) use of larger sands may lead to deep bed plugging which would require removing more sand during cleaning.

Answers to questions on page 229.

4.8E The flow rate of water through a slow sand filter depends on the head (height) of water on the sand filter and the head (friction or energy) losses as the water flows through the sand filter.

4.8F The two methods of controlling flow through a slow sand filter are:

1. Effluent flow control, and
2. Influent flow control.

4.8G The size of the finished water storage facility is based on capacity needed to provide detention time for disinfection, meet peak daily demand, fire protection requirements, and emergency reserves expected to be needed in case of power outages or other disruptions in water production. If a disinfectant is applied before the finished water enters the holding facility, then the size of the facility may be influenced by the length of contact time needed for disinfection.

4.8H The purpose of head loss gages or piezometers is to monitor the degree of clogging, which will indicate when the filter needs to be cleaned.

Answers to questions on page 234.

4.8I Finished water storage facilities need to be cleaned every year or two.

4.8J The surface of a slow sand filter should be scraped or raked when the desired filtration rate can no longer be maintained.

4.8K The time between slow sand filter cleanings depends on such factors as filtration rate, available head, media grain size distribution, influent water quality, and temperature.

4.8L Use of the "throwing over" process of resanding a filter minimizes the time required for ripening of the resanded bed because the "old" sand in the top of the bed contains a diverse population of organisms. This procedure also corrects any unwanted buildup of organisms and wastes deep in the sand bed.

Answers to questions on page 240.

4.8M Compliance with the SWTR is confirmed by the presence of at least 0.2 mg/*L* disinfectant (chlorine) residual in the finished water and a detectable residual throughout the distribution system.

4.8N Pilot studies should be conducted to assess how well a slow sand filter removes microorganisms and particulate matter from a particular source of water, determine the length of filter runs, and determine the washout time for the new sand.

4.8O If anaerobic conditions develop, water turns black, gives off foul odors, and has a greatly increased chlorine demand.

4.8P The basic types of records that should be kept at a slow sand filtration plant include the operator's diary, maintenance records, and sampling and monitoring records.

4.8Q Nutrients (primarily nitrate and phosphate) were added at one plant to shorten the ripening period of a slow sand filter treating low-temperature, nutrient-poor, mountain stream water.

Answers to questions on page 247.

4.9A Excessive amounts of iron and manganese in drinking waters are objectionable because they stain clothes and encourage the growth of iron bacteria. These bacteria form thick slimes on the walls of pipes which break away and cause rust-colored water, black particles, and foul tastes and odors.

4.9B Greensand is sometimes used to treat iron and manganese because it is capable of both oxidation and filtration.

4.9C The main advantage of treating iron by aerating water is that no chemicals are required.

4.9D When using ion exchange units to remove iron and manganese there must be no dissolved oxygen present. Dissolved oxygen causes the resins to become fouled with iron rust or manganese dioxide.

Answers to questions on page 252.

4.10A The procedures for softening water by chemical precipitation using the lime-soda ash process depend on the hardness and other chemical characteristics of the water being treated (pH, alkalinity, temperature).

4.10B If lime added to water does not remove sufficient hardness, select the optimum lime dose and then add enough soda ash to lower the hardness to the 80- to 90-mg/*L* hardness as $CaCO_3$ range.

4.10C Softened waters from the lime-soda ash process must be recarbonated to lower the pH and stabilize the water to prevent the precipitation of carbonate compounds on the filter media, in the clear well, and in the pipes in the distribution system.

Answers to questions on page 253.

4.10D In an ion exchange water softening process, the hardness-causing ions (mainly calcium and magnesium) are exchanged (or traded) for sodium ions by passing the water through an ion exchange media.

4.10E A zeolite softener should be backwashed if design flows cannot be put through the softener or the head loss through the softener becomes greater than normal.

4.10F The rinse cycle of a zeolite softener is adequate when the effluent from the rinse cycle no longer has a salty taste.

4.10G High iron loadings can cause a zeolite softener to lose its exchange capacity. High iron loadings can also plug the media and force water to channel or short-circuit through the bed.

Answers to questions on page 257.

4.11A Operators should become certified because it is evidence of the operator's level of knowledge and proficiency on the job.

4.11B Records that must be kept up to date by the operator include plant operating data, daily log or diary, laboratory test results, and system records.

4.11C Laboratory tests which are commonly performed by operators include jar tests, turbidity, pH, chlorine residual, and alkalinity.

4.12A Maintenance problems can be avoided by regular inspection, cleaning, and lubrication of equipment. Unusual noises, vibrations, leaks, and malfunctions should receive prompt attention.

4.13A The greatest causes of injuries to operators result from muscle and back strains caused by lifting and injuries resulting from slips and falls.

OBJECTIVE TEST

Chapter 4. SMALL WATER TREATMENT PLANTS

Please write your name and mark the correct answers on the answer sheet, as directed at the end of Chapter 1. There may be more than one correct answer to each multiple-choice question.

True-False

1. Prechlorination should be used if natural organics are in the raw water.
   1. True
   2. False
2. Package plants are difficult to upset by sudden changes in source water quality.
   1. True
   2. False
3. Water must have an adequate alkalinity content to achieve good coagulation with alum.
   1. True
   2. False
4. The coagulant dosage must be estimated by performing a jar test.
   1. True
   2. False
5. Direct filtration is the same as conventional filtration except that flocculation is omitted.
   1. True
   2. False
6. Direct filtration is considered a feasible alternative to conventional filtration, particularly when source waters are high in turbidity.
   1. True
   2. False
7. Lower backwash flow rates are necessary when the water is warm than when it is cold.
   1. True
   2. False
8. Only fully treated water should be used for filter backwashing.
   1. True
   2. False
9. All forms of chlorine are hazardous chemicals that can cause serious injury and damage if they are not stored and used properly.
   1. True
   2. False
10. For best results when disinfecting water with chlorine, the pH should be more than 7.5.
    1. True
    2. False
11. Plastic pipes are completely free of corrosion.
    1. True
    2. False
12. Water of higher pH is normally more corrosive than water of a lower pH.
    1. True
    2. False
13. Filter-aid chemicals are used in slow sand filters.
    1. True
    2. False
14. The decay of algae in a slow sand filter skin can impart unpleasant tastes and odors in the filtered water.
    1. True
    2. False
15. Ion exchange resins never need backwashing.
    1. True
    2. False
16. Zeolite softeners will remove iron and manganese in the soluble or precipitated form.
    1. True
    2. False
17. Chloramines are a much stronger disinfectant than free chlorine.
    1. True
    2. False
18. A zeolite water softening plant should deliver water of zero hardness to consumers.
    1. True
    2. False
19. Backflow due to siphonage is caused by vacuum conditions in the water supply system.
    1. True
    2. False
20. The caliber of the person who operates the water treatment plant is more important than any other factor involved in the production of good quality water.
    1. True
    2. False

Best Answer (Select only the closest or best answer.)

21. What is an important advantage of raw water storage?
    1. A raw water supply that is more uniform and consistent
    2. More frequent plant shutdowns and outages
    3. Operators need to watch for more frequent water quality changes
    4. Quicker rates of changes in raw water quality

22. What is the purpose of coagulation facilities?
    1. Allow slow mixing of water
    2. Encourage settling of particles
    3. Feed chemicals to water being treated
    4. Promote growth of particles

23. How can iron be controlled in drinking water?
    1. By converting iron from the liquid ferrous to the solid ferric form
    2. By filtering out the ferrous form
    3. By forming soluble precipitates of iron
    4. By use of a reducing agent

24. Why are mechanical mixers preferred for flocculation?
    1. Degree of agitation does not require adjustment for changes in water quality
    2. Flow rate does not affect mixing time
    3. Mixing allows settling of floc in flocculation basin
    4. Performance can be controlled

25. What is the most important function of a filter underdrain system?
    1. Collect filtered water
    2. Prevent media from passing out bottom of filter
    3. Support filter media
    4. Uniformly distribute backwash water

26. Breakpoint chlorination is the addition of chlorine to water until what occurs?
    1. Chloramines have been formed
    2. Chlorine breaks out of the water
    3. The chlorine demand has been satisfied
    4. The chlorine residual decreases

27. What is the biggest problem with hypochlorinators?
    1. Calcium deposits if using calcium hypochlorite
    2. Chlorine gas leaks can be hazardous
    3. Intake tubing plugged with sediment
    4. Ruptured hypochlorinator diaphragm

28. Why are corrosion control chemicals the last chemicals to be applied after all other treatment has been accomplished?
    1. Chemicals applied to improve coagulation are also effective in controlling corrosion
    2. Most plants add corrosion control processes after corrosion damage is observed
    3. pH for coagulation and disinfection is lower than pH for corrosion control
    4. To provide a coating of calcium carbonate on the filter media

29. When is a slow sand filter "mature"?
    1. When the biofilm develops a "ripe" odor
    2. When the filter produces acceptable filtered water
    3. When the head loss indicates a clogged condition
    4. When the population of organisms in the biofilm starts to die

30. How is chlorine contact time measured?
    1. Flow time as determined by dye tracer studies during low demands
    2. Flow time from point of application to point of entering distribution system
    3. Flow time from point of application to point of use by consumer
    4. Flow time through finished water holding facility

31. How should a slow sand filter be filled with water after cleaning?
    1. Fill quickly through underdrain system to force trapped air bubbles out of sand
    2. Fill the tank from the bottom through the underdrain system with filtered water
    3. Open all outlet valves to allow air bubbles to be pushed out through the underdrain system
    4. Slowly open the influent control valve and fill the sand filter from the sand surface downward

32. What causes air binding in a slow sand filter?
    1. A sudden drop in water production
    2. Filter is filled too quickly after cleaning or resanding
    3. Filter is under excessive vacuum
    4. Filtration rate is too high

33. What determines whether a water treatment plant produces acceptable quality water?
    1. Abilities of the operator
    2. Accurate records
    3. Adequate budget
    4. Construction of plant

34. Why must backflow be prevented?
    1. To avoid changing direction of flow
    2. To keep from causing water to become turbid
    3. To prevent reversal of flows
    4. To protect the public health

35. What should be the basis for equipment preventive maintenance?
    1. Experience
    2. Failure of equipment
    3. Information supplied by other operators
    4. Manufacturers' recommendations

Multiple Choice (Select all correct answers.)

36. What is the purpose of the clear well?
    1. Allows for smaller capacity of treatment plant
    2. Maintains water supply during plant outages
    3. Provides chlorine contact time for disinfection
    4. Provides large amounts of water for firefighting
    5. Stores treated water

37. How can water be softened?
    1. Chemical precipitation
    2. Coagulation
    3. Disinfection
    4. Flocculation
    5. Ion exchange

38. Which factors have an important effect on the coagulation process?
    1. Coagulant dosage
    2. Coagulant used
    3. Mineral content of water
    4. pH of the water
    5. Water temperature

39. Why is short-circuiting undesirable?
    1. It may result in shorter contact times
    2. It may result in shorter electrical circuit times
    3. It may result in shorter laboratory testing times
    4. It may result in shorter reaction times
    5. It may result in shorter settling times

40. Which factors have an important effect on the settling of suspended particles in a settling tank?
    1. Characteristics of suspended matter
    2. Degree of short-circuiting through the tank
    3. Temperature of the water
    4. Time provided for settling
    5. Wind

41. When would a filter-aid chemical (such as a nonionic polymer) be applied just prior to filtration?
    1. Assist in solids removal process
    2. During periods of pretreatment process upset
    3. Precipitate out soluble contaminants
    4. Remove toxic chemicals
    5. When operating at high filtration rates

42. How does the slow sand filtration process remove particles?
    1. Adsorption
    2. Biological action
    3. Oxidation
    4. Precipitation
    5. Straining

43. What conditions are required for successful operation of filters at high rates?
    1. Competent operators on duty whenever plant is operating
    2. Excellent pretreatment to condition water before filtering
    3. Surface wash apparatus to clean media during backwash
    4. Turbidimeters that continuously monitor filter performance
    5. Use of dual or mixed media

44. What factors have an important effect on disinfection of water with chlorine?
    1. Concentration of chlorine
    2. Degree of mixing
    3. pH of the water
    4. Time of chlorine contact
    5. Water temperature

45. What substances react with chlorine and consume it?
    1. Ammonia
    2. Calcium
    3. Dissolved iron
    4. Hydrogen sulfide
    5. Organic matter

46. What kinds of problems can be caused by corrosion?
    1. Damage to pipelines and storage tanks
    2. Decrease in flow capacity of pipelines
    3. Degradation of water quality
    4. Increase in pumping costs
    5. Leaks in storage tanks

47. What variables does an operator adjust to control the performance of a solids-contact unit?
    1. Chemical dosage
    2. Corrosion control
    3. Disinfection
    4. Recirculation rate
    5. Sludge control

48. What factors influence the length of time a slow sand filter will operate between cleanings?
    1. Available head
    2. Filtration rate
    3. Influent water turbidity
    4. Media grain size distribution
    5. Water temperature

49. What basic types of records are necessary for a slow sand filtration system?
    1. Budget records
    2. Maintenance records
    3. Operator's daily diary
    4. Sampling and monitoring records
    5. Work order records

50. Which chemicals can be used to oxidize iron and manganese to insoluble precipitates?
    1. Chlorine
    2. Potassium permanganate
    3. Sodium hydroxide
    4. Sulfur dioxide
    5. Sulfuric acid

51. Which of the following laboratory tests are often performed by operators?
    1. Alkalinity
    2. Chlorine residual
    3. Jar tests
    4. pH
    5. Turbidity

52. What are the greatest causes of injuries to operators?
    1. Confined space entry
    2. Driving in traffic
    3. Handling chlorine
    4. Lifting
    5. Slips and falls

53. Jar tests indicate that water should be dosed with alum at 12 mg/*L*. The flow being treated is 180 GPM. What is the desired alum feed in pounds per day?
    1. 11 lbs/day
    2. 26 lbs/day
    3. 31 lbs/day
    4. 44 lbs/day
    5. 56 lbs/day

54. The optimum dose of liquid alum from the jar tests is 12 mg/*L*. Determine the setting on the liquid alum chemical feeder in gallons per day when the flow is 0.32 MGD. The liquid alum delivered to the plant contains 5.25 pounds of alum per gallon of liquid solution.
    1. 6.1 GPD
    2. 8.1 GPD
    3. 9.0 GPD
    4. 10.7 GPD
    5. 11.9 GPD

55. What is the actual chlorine dose in milligrams per liter if the hypochlorinator delivers 8 pounds of chlorine per day and the flow rate of the water being treated is 600,000 gallons per day (0.6 MGD)?
    1. 0.8 mg/*L*
    2. 1.2 mg/*L*
    3. 1.4 mg/*L*
    4. 1.6 mg/*L*
    5. 2.8 mg/*L*

56. A rectangular settling basin 24 feet long, 8 feet wide, and 6 feet deep treats a flow of 95,000 gallons per day. What is the detention time in the basin for this flow?
    1. 1.0 hr
    2. 2.2 hr
    3. 2.5 hr
    4. 2.9 hr
    5. 3.8 hr

57. A water treatment plant filter treats a flow of 1,000 GPM. The sand filter is 20 feet long by 20 feet wide. What is the filtration rate in gallons per minute per square foot of filter?
    1. 1.5 GPM/sq ft
    2. 1.7 GPM/sq ft
    3. 1.9 GPM/sq ft
    4. 2.1 GPM/sq ft
    5. 2.5 GPM/sq ft

58. To check on the valve position of a controller in a filter, the influent valve to the filter was closed for 8 minutes. During this time period the water surface in the filter dropped 18 inches (1.5 feet). The surface area of the filter is 400 square feet. What is the flow through the filter in gallons per minute?
    1. 468 GPM
    2. 561 GPM
    3. 600 GPM
    4. 750 GPM
    5. 898 GPM

59. During a filter run, the total volume of water filtered was 1.4 million gallons. When the filter was backwashed, 24,000 gallons of water was used. What was the percent of filtered water used for backwashing?
    1. 1.0%
    2. 1.3%
    3. 1.5%
    4. 1.7%
    5. 2.0%

60. A zeolite softener contains 80 cubic feet of media with a hardness removal capacity of 25 kilograins per cubic foot of media. The water being treated has a hardness of 210 mg/*L* as $CaCO_3$. How much water can be treated before the softener will require regeneration?
    1. 125,000 gallons
    2. 136,000 gallons
    3. 145,000 gallons
    4. 154,000 gallons
    5. 164,000 gallons

End of Objective Test

CHAPTER 5

DISINFECTION

by

Tom Ikesaki

NOTICE

Drinking water rules and regulations are continually changing. Two major new laws were signed by the president in December 1998: the Disinfectant/Disinfection By-Products (D/DBP) Rule and the Interim Enhanced Surface Water Treatment Rule (IESWTR). Several other drinking water laws are being developed and are expected to be signed into law over the next two or three years. The regulations described in this chapter and on the poster included with this manual are current as of publication in April 2002.

Keep in contact with your state drinking water agency to obtain the rules and regulations that currently apply to your water utility. For additional information or answers to specific questions about the regulations, phone EPA's toll-free Safe Drinking Water Hotline at (800) 426-4791.

TABLE OF CONTENTS

Chapter 5. DISINFECTION

OBJECTIVES

Chapter 5. DISINFECTION

Following completion of Chapter 5, you should be able to:

1. Disinfect new and existing wells,
2. Disinfect mains and storage tanks,
3. Calculate chlorine dosage,
4. Determine hypochlorinator and chlorinator settings,
5. Operate and maintain hypochlorinators,
6. Operate and maintain chlorinators,
7. Troubleshoot chlorination systems, and
8. Conduct a chlorine safety program.

WORDS

Chapter 5. DISINFECTION

AMPEROMETRIC (am-PURR-o-MET-rick) AMPEROMETRIC

A method of measurement that records electric current flowing or generated, rather than recording voltage. Amperometric titration is a means of measuring concentrations of certain substances in water.

AMPEROMETRIC (am-PURR-o-MET-rick) TITRATION AMPEROMETRIC TITRATION

A means of measuring concentrations of certain substances in water (such as strong oxidizers) based on the electric current that flows during a chemical reaction. Also see TITRATE.

BREAKPOINT CHLORINATION BREAKPOINT CHLORINATION

Addition of chlorine to water until the chlorine demand has been satisfied. At this point, further additions of chlorine will result in a free chlorine residual that is directly proportional to the amount of chlorine added beyond the breakpoint.

BUFFER CAPACITY BUFFER CAPACITY

A measure of the capacity of a solution or liquid to neutralize acids or bases. This is a measure of the capacity of water for offering a resistance to changes in pH.

CARCINOGEN (CAR-sin-o-JEN) CARCINOGEN

Any substance which tends to produce cancer in an organism.

CHLORAMINES (KLOR-uh-means) CHLORAMINES

Compounds formed by the reaction of hypochlorous acid (or aqueous chlorine) with ammonia.

CHLORINATION (KLOR-uh-NAY-shun) CHLORINATION

The application of chlorine to water, generally for the purpose of disinfection, but frequently for accomplishing other biological or chemical results (aiding coagulation and controlling tastes and odors).

CHLORINE DEMAND CHLORINE DEMAND

Chlorine demand is the difference between the amount of chlorine added to water and the amount of residual chlorine remaining after a given contact time. Chlorine demand may change with dosage, time, temperature, pH, and nature and amount of the impurities in the water.

$$\text{Chlorine Demand, mg}/L = \text{Chlorine Applied, mg}/L - \text{Chlorine Residual, mg}/L$$

CHLORINE REQUIREMENT CHLORINE REQUIREMENT

The amount of chlorine which is needed for a particular purpose. Some reasons for adding chlorine are reducing the number of coliform bacteria (Most Probable Number), obtaining a particular chlorine residual, or oxidizing some substance in the water. In each case a definite dosage of chlorine will be necessary. This dosage is the chlorine requirement.

CHLORINE RESIDUAL CHLORINE RESIDUAL

The concentration of chlorine present in water after the chlorine demand has been satisfied. The concentration is expressed in terms of the total chlorine residual, which includes both the free and combined or chemically bound chlorine residuals.

CHLOROPHENOLIC (klor-o-FEE-NO-lick) CHLOROPHENOLIC

Chlorophenolic compounds are phenolic compounds (carbolic acid) combined with chlorine.

CHLORORGANIC (klor-or-GAN-ick) CHLORORGANIC

Organic compounds combined with chlorine. These compounds generally originate from, or are associated with, life processes such as those of algae in water.

COLIFORM (COAL-i-form) COLIFORM

A group of bacteria found in the intestines of warm-blooded animals (including humans) and also in plants, soil, air and water. Fecal coliforms are a specific class of bacteria which only inhabit the intestines of warm-blooded animals. The presence of coliform bacteria is an indication that the water is polluted and may contain pathogenic (disease-causing) organisms.

COLORIMETRIC MEASUREMENT COLORIMETRIC MEASUREMENT

A means of measuring unknown chemical concentrations in water by measuring a sample's color intensity. The specific color of the sample, developed by addition of chemical reagents, is measured with a photoelectric colorimeter or is compared with "color standards" using, or corresponding with, known concentrations of the chemical.

COMBINED AVAILABLE CHLORINE COMBINED AVAILABLE CHLORINE

The total chlorine, present as chloramine or other derivatives, that is present in a water and is still available for disinfection and for oxidation of organic matter. The combined chlorine compounds are more stable than free chlorine forms, but they are somewhat slower in disinfection action.

COMBINED AVAILABLE CHLORINE RESIDUAL COMBINED AVAILABLE CHLORINE RESIDUAL

The concentration of residual chlorine that is combined with ammonia, organic nitrogen, or both in water as a chloramine (or other chloro derivative) and yet is still available to oxidize organic matter and help kill bacteria.

COMBINED CHLORINE COMBINED CHLORINE

The sum of the chlorine species composed of free chlorine and ammonia, including monochloramine, dichloramine, and trichloramine (nitrogen trichloride). Dichloramine is the strongest disinfectant of these chlorine species, but it has less oxidative capacity than free chlorine.

COMBINED RESIDUAL CHLORINATION COMBINED RESIDUAL CHLORINATION

The application of chlorine to water to produce combined available chlorine residual. The residual can be made up of monochloramines, dichloramines, and nitrogen trichloride.

DPD (pronounce as separate letters) DPD

A method of measuring the chlorine residual in water. The residual may be determined by either titrating or comparing a developed color with color standards. DPD stands for N,N-diethyl-p-phenylene-diamine.

DIATOMS (DYE-uh-toms) DIATOMS

Unicellular (single cell), microscopic algae with a rigid (box-like) internal structure consisting mainly of silica.

DISINFECTION (dis-in-FECT-shun) DISINFECTION

The process designed to kill or inactivate most microorganisms in water, including essentially all pathogenic (disease-causing) bacteria. There are several ways to disinfect, with chlorination being the most frequently used in water treatment. Compare with STERILIZATION.

EDUCTOR (e-DUCK-ter) EDUCTOR

A hydraulic device used to create a negative pressure (suction) by forcing a liquid through a restriction, such as a Venturi. An eductor or aspirator (the hydraulic device) may be used in the laboratory in place of a vacuum pump. As an injector, it is used to produce vacuum for chlorinators. Sometimes used instead of a suction pump.

EJECTOR EJECTOR

A device used to disperse a chemical solution into water being treated.

ENTERIC ENTERIC

Of intestinal origin, especially applied to wastes or bacteria.

ENZYMES (EN-zimes) ENZYMES

Organic substances (produced by living organisms) which cause or speed up chemical reactions. Organic catalysts and/or biochemical catalysts.

FREE AVAILABLE RESIDUAL CHLORINE FREE AVAILABLE RESIDUAL CHLORINE

That portion of the total available residual chlorine composed of dissolved chlorine gas (Cl_2), hypochlorous acid (HOCl), and/or hypochlorite ion (OCl^-) remaining in water after chlorination. This does not include chlorine that has combined with ammonia, nitrogen, or other compounds.

HTH (pronounce as separate letters) HTH

High **T**est **H**ypochlorite. Calcium hypochlorite or $Ca(OCl)_2$.

HETEROTROPHIC (HET-er-o-TROF-ick) HETEROTROPHIC

Describes organisms that use organic matter for energy and growth. Animals, fungi and most bacteria are heterotrophs.

HYDROLYSIS (hi-DROLL-uh-sis) HYDROLYSIS

(1) A chemical reaction in which a compound is converted into another compound by taking up water.

(2) Usually a chemical degradation of organic matter.

HYPOCHLORINATION (HI-poe-KLOR-uh-NAY-shun) HYPOCHLORINATION

The application of hypochlorite compounds to water for the purpose of disinfection.

HYPOCHLORITE (HI-poe-KLOR-ite) HYPOCHLORITE

Chemical compounds containing available chlorine; used for disinfection. They are available as liquids (bleach) or solids (powder, granules, and pellets) in barrels, drums, and cans. Salts of hypochlorous acid.

IDLH IDLH

Immediately **D**angerous to **L**ife or **H**ealth. The atmospheric concentration of any toxic, corrosive, or asphyxiant substance that poses an immediate threat to life or would cause irreversible or delayed adverse health effects or would interfere with an individual's ability to escape from a dangerous atmosphere.

MPN (pronounce as separate letters) MPN

MPN is the **M**ost **P**robable **N**umber of coliform-group organisms per unit volume of sample water. Expressed as a density or population of organisms per 100 m*L* of sample water.

NITRIFICATION (NYE-truh-fuh-KAY-shun) NITRIFICATION

An aerobic process in which bacteria reduce the ammonia and organic nitrogen in water into nitrite and then nitrate.

NITROGENOUS (nye-TRAH-jen-us) NITROGENOUS

A term used to describe chemical compounds (usually organic) containing nitrogen in combined forms. Proteins and nitrates are nitrogenous compounds.

ORTHOTOLIDINE (or-tho-TOL-uh-dine) ORTHOTOLIDINE

Orthotolidine is a colorimetric indicator of chlorine residual. If chlorine is present, a yellow-colored compound is produced. This reagent is no longer approved for chemical analysis to determine chlorine residual.

OXIDATION (ox-uh-DAY-shun) OXIDATION

Oxidation is the addition of oxygen, removal of hydrogen, or the removal of electrons from an element or compound. In the environment, organic matter is oxidized to more stable substances. The opposite of REDUCTION.

OXIDIZING AGENT OXIDIZING AGENT

Any substance, such as oxygen (O_2) or chlorine (Cl_2), that will readily add (take on) electrons. The opposite is a REDUCING AGENT.

PALATABLE (PAL-uh-tuh-bull) PALATABLE

Water at a desirable temperature that is free from objectionable tastes, odors, colors, and turbidity. Pleasing to the senses.

PATHOGENIC (PATH-o-JEN-ick) ORGANISMS PATHOGENIC ORGANISMS

Organisms, including bacteria, viruses or cysts, capable of causing diseases (giardiasis, cryptosporidiosis, typhoid, cholera, dysentery) in a host (such as a person). There are many types of organisms which do *NOT* cause disease. These organisms are called non-pathogenic.

PHENOLIC (fee-NO-lick) COMPOUNDS PHENOLIC COMPOUNDS

Organic compounds that are derivatives of benzene.

POTABLE (POE-tuh-bull) WATER POTABLE WATER

Water that does not contain objectionable pollution, contamination, minerals, or infective agents and is considered satisfactory for drinking.

PRECURSOR, THM (pre-CURSE-or) PRECURSOR, THM

Natural organic compounds found in all surface and groundwaters. These compounds *MAY* react with halogens (such as chlorine) to form trihalomethanes (tri-HAL-o-METH-hanes) (THMs); they *MUST* be present in order for THMs to form.

REAGENT (re-A-gent) REAGENT

A pure chemical substance that is used to make new products or is used in chemical tests to measure, detect, or examine other substances.

REDUCING AGENT REDUCING AGENT

Any substance, such as a base metal (iron) or the sulfide ion (S^{2-}), that will readily donate (give up) electrons. The opposite is an OXIDIZING AGENT.

REDUCTION (re-DUCK-shun) REDUCTION

Reduction is the addition of hydrogen, removal of oxygen, or the addition of electrons to an element or compound. Under anaerobic conditions (no dissolved oxygen present), sulfur compounds are reduced to odor-producing hydrogen sulfide (H_2S) and other compounds. The opposite of OXIDATION.

RELIQUEFACTION (re-LICK-we-FACK-shun) RELIQUEFACTION

The return of a gas to the liquid state; for example, a condensation of chlorine gas to return it to its liquid form by cooling.

RESIDUAL CHLORINE RESIDUAL CHLORINE

The concentration of chlorine present in water after the chlorine demand has been satisfied. The concentration is expressed in terms of the total chlorine residual, which includes both the free and combined or chemically bound chlorine residuals.

ROTAMETER (RODE-uh-ME-ter) ROTAMETER

A device used to measure the flow rate of gases and liquids. The gas or liquid being measured flows vertically up a tapered, calibrated tube. Inside the tube is a small ball or bullet-shaped float (it may rotate) that rises or falls depending on the flow rate. The flow rate may be read on a scale behind or on the tube by looking at the middle of the ball or at the widest part or top of the float.

SAPROPHYTES (SAP-row-FIGHTS) SAPROPHYTES

Organisms living on dead or decaying organic matter. They help natural decomposition of organic matter in water.

STERILIZATION (STARE-uh-luh-ZAY-shun) STERILIZATION

The removal or destruction of all microorganisms, including pathogenic and other bacteria, vegetative forms and spores. Compare with DISINFECTION.

TITRATE (TIE-trate) TITRATE

To *TITRATE* a sample, a chemical solution of known strength is added drop by drop until a certain color change, precipitate, or pH change in the sample is observed (end point). Titration is the process of adding the chemical reagent in increments until completion of the reaction, as signaled by the end point.

TOTAL CHLORINE TOTAL CHLORINE

The total concentration of chlorine in water, including the combined chlorine (such as inorganic and organic chloramines) and the free available chlorine.

TOTAL CHLORINE RESIDUAL TOTAL CHLORINE RESIDUAL

The total amount of chlorine residual (value for residual chlorine, including both free chlorine and chemically bound chlorine) present in a water sample after a given contact time.

TRIHALOMETHANES (THMs) (tri-HAL-o-METH-hanes) TRIHALOMETHANES (THMs)

Derivatives of methane, CH_4, in which three halogen atoms (chlorine or bromine) are substituted for three of the hydrogen atoms. Often formed during chlorination by reactions with natural organic materials in the water. The resulting compounds (THMs) are suspected of causing cancer.

TURBIDITY (ter-BID-it-tee) TURBIDITY

The cloudy appearance of water caused by the presence of suspended and colloidal matter. In the waterworks field, a turbidity measurement is used to indicate the clarity of water. Technically, turbidity is an optical property of the water based on the amount of light reflected by suspended particles. Turbidity cannot be directly equated to suspended solids because white particles reflect more light than dark-colored particles and many small particles will reflect more light than an equivalent large particle.

CHAPTER 5. DISINFECTION

(Lesson 1 of 2 Lessons)

5.0 PURPOSE OF DISINFECTION

5.00 Making Water Safe for Consumption

Our single most important natural resource is water. Without water we could not exist. Unfortunately, safe water is becoming very difficult to find. In the past, safe water could be found in remote areas, but with population growth and related pollution of waters, there are very few natural waters left that are safe to drink without treatment of some kind.

Water is the universal solvent and therefore carries all types of dissolved materials. Water also carries biological life forms which can cause diseases. These waterborne pathogenic organisms are listed in Table 5.1. Most of these organisms and the diseases they transmit are no longer a problem in the United States due to proper water protection, treatment, and monitoring. However, many developing regions of the world still experience serious outbreaks of various waterborne diseases.

TABLE 5.1 PATHOGENIC ORGANISMS (DISEASES) TRANSMITTED BY WATER

Bacteria
- Salmonella (salmonellosis)
- Shigella (bacillary dysentery)
- Bacillus typhosus (typhoid fever)
- Salmonella paratyphi (paratyphoid)
- Vibrio cholerae (cholera)

Viruses
- Enterovirus
  - Poliovirus
  - Coxsackie Virus
  - Echo Virus
- Andenovirus
- Reovirus
- Infectious Hepatitis

Intestinal Parasites
- Entamoeba histolytica (amoebic dysentery)
- Giardia lamblia (giardiasis)
- Ascaris lumbricoides (giant roundworm)
- Cryptosporidium (cryptosporidiosis)

One of the cleansing processes in the production of safe water is called disinfection. Disinfection is the selective destruction or inactivation of pathogenic organisms. Don't confuse disinfection with sterilization. Sterilization is the complete destruction of all organisms. Sterilization is not necessary in water treatment and is also quite expensive. (Also note that disinfection does not remove toxic chemicals which could make the water unsafe to drink.)

5.01 Safe Drinking Water Act (SDWA)

In the United States, the U.S. Environmental Protection Agency is responsible for setting drinking water standards and for ensuring their enforcement. This agency sets federal regulations which all state and local agencies must enforce. The 1976 Primary Drinking Water Regulations contain specific maximum allowable levels of substances known to be hazardous to human health. In addition to describing maximum contaminant levels (MCLs), the 1976 Primary Drinking Water Regulations also give detailed instructions on what to do when you exceed the maximum contaminant level for a particular substance. In Table 5.2 you will find an example of the Primary Drinking Water Regulations for *COLIFORM*[1] bacteria which are supposed to be killed by disinfection. Table 5.3 lists the coliform samples required per population served.

The Safe Drinking Water Act (SDWA), originally enacted in 1974, was amended in 1980, 1986, and 1996 to expand and strengthen the protection of drinking water. The 1986 amendments, for example, authorized penalties for tampering with drinking water supplies and mandated complete elimination of lead from drinking water. The 1996 SDWA amendments require EPA to develop several new rules and regulations, including the Disinfectant/Disinfection By-Products (D/DBP) Rule, Enhanced Surface Water Treatment Rule (ESWTR), Ground Water Disinfection Rule (GWDR), Lead and Copper Rule revisions, and regulations for arsenic, radon, and sulfate.

In 1998, the Interim Enhanced Surface Water Treatment Rule (IESWTR) and the Disinfectant/Disinfection By-Products (D/DBP) Rule were signed into law, but further modifications of these two rules are still under development. The goal of the IESWTR is to reduce the occurrence of *Cryptosporidium* and other disease-causing organisms. The new D/DBP Rule was developed to protect the public from harmful concentrations of disinfectants and from trihalomethanes which could form when disinfection by-products combine with organic matter in drinking water. Water systems serving 10,000 or more people have three years (from December 16, 1998) to comply with both the D/DBP and the IESWTR regulations. Small systems (serving fewer than 10,000 people) have five years to comply.

Operators are urged to develop close working relationships with their state regulatory agencies to keep themselves informed of the expected future changes in regulations.

[1] *Coliform (COAL-i-form). A group of bacteria found in the intestines of warm-blooded animals (including humans) and also in plants, soil, air and water. Fecal coliforms are a specific class of bacteria which only inhabit the intestines of warm-blooded animals. The presence of coliform bacteria is an indication that the water is polluted and may contain pathogenic (disease-causing) organisms.*

TABLE 5.2 MICROBIOLOGICAL STANDARDS[a,b]

Maximum Contaminant Level Goal (MCLG): zero

Maximum Contaminant Level (MCL):

1. Compliance based on presence/absence of total coliforms in sample, rather than an estimate of coliform density.
2. MCL for system analyzing at least 40 samples per month: no more than 5.0 percent of the monthly samples may be total coliform-positive.
3. MCL for systems analyzing fewer than 40 samples per month: no more than 1 sample per month may be total coliform-positive.

[a] See Chapter 22, "Drinking Water Regulations," *WATER TREATMENT PLANT OPERATION*, Volume II, and the poster provided with this manual for more details.
[b] See Chapter 7 "Laboratory Procedures," *SMALL WATER SYSTEM OPERATION AND MAINTENANCE*, for details on how to do the coliform bacteria tests (membrane filter (MF), multiple tube fermentation (MPN), presence-absence, Colilert, and Colisure).

MONITORING AND REPEAT SAMPLE FREQUENCY AFTER A TOTAL COLIFORM-POSITIVE ROUTINE SAMPLE

No. Routine Samples/Month	No. Repeat Samples[a]	No. Routine Samples Next Month[b]
1/mo or fewer	4	5/mo
2/mo	3	5/mo
3/mo	3	5/mo
4/mo	3	5/mo
5/mo or more	3	Table 5.3

[a] Number of repeat samples in the same month for each total coliform-positive routine sample.
[b] Except where the state has invalidated the original routine sample, substitutes an on-site evaluation of the problem, or waives the requirement on a case-by-case basis.

QUESTIONS

Write your answers in a notebook and then compare your answers with those on page 330.

5.0A What are pathogenic organisms?

5.0B What is disinfection?

5.0C Drinking water standards are established by what agency of the United States government?

5.0D MCL stands for what words?

5.1 FACTORS INFLUENCING DISINFECTION

5.10 pH

The pH of water being treated can alter the efficiency of disinfectants. Chlorine, for example, disinfects water much faster at a pH around 7.0 than at a pH over 8.0.

5.11 Temperature

Temperature conditions also influence the effectiveness of the disinfectant. The higher the temperature of the water, the more efficiently it can be treated. Water near 70 to 85°F (21 to 29°C) is easier to disinfect than water at 40 to 60°F (4 to 16°C). Longer contact times are required to disinfect water at lower temperatures. To speed up the process, operators often simply use larger amounts of chemicals. Where water is exposed to the atmosphere, the warmer the water temperature the greater the dissipation rate of chlorine into the atmosphere.

5.12 Turbidity

Under normal operating conditions, the turbidity level of water being treated is very low by the time the water reaches the disinfection process. Excessive turbidity will greatly reduce the efficiency of the disinfecting chemical or process. Studies in water treatment plants have shown that when water is filtered to a turbidity of one unit or less, most of the bacteria have been removed.

The suspended matter itself may also change the chemical nature of the water when the disinfectant is added. Some types of suspended solids can create a continuing demand for the chemical, thus changing the effective germicidal (germ killing) properties of the disinfectant.

5.120 Organic Matter

Organics found in the water can consume great amounts of disinfectants while forming unwanted compounds. *TRIHALOMETHANES*[2] are an example of undesirable com-

[2] *Trihalomethanes (THMs) (tri-HAL-o-METH-hanes). Derivatives of methane, CH_4, in which three halogen atoms (chlorine or bromine) are substituted for three of the hydrogen atoms. Often formed during chlorination by reactions with natural organic materials in the water. The resulting compounds (THMs) are suspected of causing cancer.*

TABLE 5.3 TOTAL COLIFORM SAMPLING REQUIREMENTS ACCORDING TO POPULATION SERVED

Population Served	Minimum Number of Routine Samples Per Month[a]	Population Served	Minimum Number of Routine Samples Per Month
25 to 1,000[b]	1[c]	59,001 to 70,000	70
1,001 to 2,500	2	70,001 to 83,000	80
2,501 to 3,300	3	83,001 to 96,000	90
3,301 to 4,100	4	96,001 to 130,000	100
4,101 to 4,900	5	130,001 to 220,000	120
4,901 to 5,800	6	220,001 to 320,000	150
5,801 to 6,700	7	320,001 to 450,000	180
6,701 to 7,600	8	450,001 to 600,000	210
7,601 to 8,500	9	600,001 to 780,000	240
8,501 to 12,900	10	780,001 to 970,000	270
12,901 to 17,200	15	970,001 to 1,230,000	300
17,201 to 21,500	20	1,230,001 to 1,520,000	330
21,501 to 25,000	25	1,520,001 to 1,850,000	360
25,001 to 33,000	30	1,850,001 to 2,270,000	390
33,001 to 41,000	40	2,270,001 to 3,020,000	420
41,001 to 50,000	50	3,020,001 to 3,960,000	450
50,001 to 59,000	60	3,960,001 or more	480

[a] A noncommunity water system using groundwater and serving 1,000 persons or fewer may monitor at a lesser frequency specified by the state until a sanitary survey is conducted and the state reviews the results. Thereafter, noncommunity water systems using groundwater and serving 1,000 persons or fewer must monitor in each calendar quarter during which the system provides water to the public, unless the state determines that some other frequency is more appropriate and notifies the system (in writing). In all cases, noncommunity water systems using groundwater and serving 1,000 persons or fewer must monitor at least once/year.

A noncommunity water system using surface water, or groundwater under the direct influence of surface water, regardless of the number of persons served, must monitor at the same frequency as a like-sized community public system. A noncommunity water system using groundwater and serving more than 1,000 persons during any month must monitor at the same frequency as a like-sized community water system, except that the state may reduce the monitoring frequency for any month the system serves 1,000 persons or fewer.

[b] Includes public water systems which have at least 15 service connections, but serve fewer than 25 persons.

[c] For a community water system serving 25 to 1,000 persons, the state may reduce this sampling frequency if a sanitary survey conducted in the last five years indicates that the water system is supplied solely by a protected groundwater source and is free of sanitary defects. However, in no case may the state reduce the sampling frequency to less than once/quarter.

pounds formed by reactions between chlorine and certain organics. Disinfecting chemicals often react with organics and *REDUCING AGENTS*[3] (Section 5.13). Then, if any of the chemical remains available after this initial reaction, it can act as an effective disinfectant. The reactions with organics and reducing agents, however, will have significantly reduced the amount of chemical available for disinfection.

5.121 Inorganic Matter

Inorganic compounds such as ammonia (NH_3) in the water being treated can create special problems. In the presence of ammonia, some oxidizing chemicals form side compounds causing a partial loss of disinfecting power. Silt can also create a chemical demand. It is clear, then, that the chemical properties of the water being treated can seriously interfere with the effectiveness of disinfecting chemicals.

5.13 Reducing Agents

Chlorine combines with a wide variety of materials, especially reducing agents. Most of the reactions are very rapid, while others are much slower. These side reactions complicate the use of chlorine for disinfection. The demand for chlorine by reducing agents must be satisfied before chlorine becomes available to accomplish disinfection. Examples of inorganic reducing agents present in water which will react with chlorine include hydrogen sulfide (H_2S), ferrous ion

[3] *Reducing Agent. Any substance, such as a base metal (iron) or the sulfide ion (S^{2-}), that will readily donate (give up) electrons. The opposite is an oxidizing agent.*

(Fe^{2+}), manganous ion (Mn^{2+}), and the nitrite ion (NO_2^-). Organic reducing agents in water also will react with chlorine and form chlorinated organic materials of potential health significance.

5.14 Microorganisms

5.140 Number and Types of Microorganisms

Microorganism concentration is important because the higher the number of microorganisms, the greater the demand for a disinfecting chemical. The resistance of microorganisms to specific disinfectants varies greatly. Non-spore-forming bacteria are generally less resistant than spore-forming bacteria. Cysts and viruses can be very resistant to certain types of disinfectants.

5.141 Removal Processes

Pathogenic organisms can be removed from water, killed, or inactivated by various physical and chemical water treatment processes. These processes are:

1. *COAGULATION.* Chemical coagulation followed by sedimentation and filtration will remove 90 to 95 percent of the pathogenic organisms, depending on which chemicals are used. Alum usage can increase virus removals up to 99 percent.
2. *SEDIMENTATION.* Properly designed sedimentation processes can effectively remove 20 to 70 percent of the pathogenic microorganisms. This removal is accomplished by allowing the pathogenic organisms (as well as non-pathogenic organisms) to settle out by gravity, assisted by chemical floc.
3. *FILTRATION.* Filtering water through granular filters is an effective means of removing pathogenic and other organisms from water. The removal rates vary from 20 to 99+ percent depending on the coarseness of the filter media and the type and effectiveness of pretreatment.
4. *DISINFECTION.* Disinfection chemicals such as chlorine are added to water to kill or inactivate pathogenic microorganisms.

In Chapter 4 you have studied the first three processes and were introduced to the disinfection methods used in small plants. This chapter provides an in-depth look at the principles and methods of disinfection.

QUESTIONS

Write your answers in a notebook and then compare your answers with those on page 330.

5.1A How does pH influence the effectiveness of disinfection with chlorine?

5.1B How does the temperature of the water influence disinfection?

5.1C What two factors influence the effectiveness of disinfection on microorganisms?

5.2 PROCESS OF DISINFECTION

5.20 Purpose of Process

The purpose of disinfection is to destroy harmful organisms. This can be accomplished either physically or chemically. Physical methods may (1) physically remove the organisms from the water, or (2) introduce motion that will disrupt the cells' biological activity and kill them.

Chemical methods alter the cell chemistry causing the microorganism to die. The most widely used disinfectant chemical is chlorine. Chlorine is easily obtained, relatively cheap, and most importantly, leaves a *RESIDUAL CHLORINE*[4] that can be measured. Other disinfectants are also used. There has been increased interest in disinfectants other than chlorine because of the *CARCINOGENIC*[5] compounds that chlorine may form (trihalomethanes or THMs).

This chapter will focus primarily on the use of chlorine as a disinfectant. However, let's take a brief look first at other disinfection methods and chemicals. Some of these are being more widely applied today because of the potential adverse side effects of chlorination.

5.21 Agents of Disinfection

5.210 Physical Means of Disinfection

A. *ULTRAVIOLET RAYS* can be used to destroy pathogenic microorganisms. To be effective, the rays must come in contact with each microorganism. The ultraviolet energy disrupts various organic components of the cell causing a biological change that is fatal to the microorganisms.

   This system has not had widespread acceptance because of the lack of measurable residual and the cost of operation. However, advances in UV technology and concern about disinfection by-products (DBPs) produced by other disinfectants have prompted a renewed interest in UV disinfection. Currently, use of ultraviolet rays is limited to small or local systems and industrial applications. Ocean-going ships have used these systems for their water supply.

B. *HEAT* has been used for centuries to disinfect water. Boiling water for about 5 minutes will destroy essentially all mi-

[4] *Residual Chlorine.* *The concentration of chlorine present in water after the chlorine demand has been satisfied. The concentration is expressed in terms of the total chlorine residual, which includes both the free and combined or chemically bound chlorine residuals.*

[5] *Carcinogen (CAR-sin-o-JEN).* *Any substance which tends to produce cancer in an organism.*

croorganisms. This method is very energy intensive and thus very expensive. The only practical application is in the event of a disaster when individual local users are required to boil their water.

C. *ULTRASONIC WAVES* have been used to disinfect water on a very limited scale. Sonic waves destroy the microorganism by vibration. This procedure is not yet practical and is very expensive.

5.211 Chemical Disinfectants (Other Than Chlorine)

A. *IODINE* has been used as a disinfectant in water since 1920, but its use has been limited to emergency treatment of water supplies. Although it has long been recognized as a good disinfectant, iodine's high cost and potential physiological effects (pregnant women can suffer serious side effects) have prevented widespread acceptance. The recommended dosage is two drops of iodine (tincture of iodine which is 7 percent available iodine) in a liter of water.

B. *BROMINE* has been used only on a very limited scale for water treatment because of its handling difficulties. Bromine causes skin burns on contact. Because bromine is a very reactive chemical, residuals are hard to obtain. This also limits its use. Bromine can be purchased at swimming pool supply stores.

C. *BASES* such as sodium hydroxide and lime can be effective disinfectants but the high pH leaves a bitter taste in the finished water. Bases can also cause skin burns when left too long in contact with the skin. Bases effectively kill all microorganisms (they sterilize rather than just disinfect water). Although this method has not been used on a large scale, bases have been used to sterilize water pipes.

D. *OZONE* has been used in the water industry since the early 1900s, particularly in France. In the United States it has been used primarily for taste and odor control. The limited use in the United States has been due to its high costs, lack of residual, difficulty in storing, and maintenance requirements.

Although ozone is effective in disinfecting water, its use is limited by its solubility. The temperature and pressure of water being treated regulate the amount of ozone that can be dissolved in the water. These factors tend to limit the disinfectant strength that can be made available to treat the water.

Many scientists claim that ozone destroys all microorganisms. Unfortunately, significant residual ozone does not guarantee that a water is safe to drink. Organic solids may protect organisms from the disinfecting action of ozone and increase the amount of ozone needed for disinfection. In addition, ozone residuals cannot be maintained in metallic conduits for any period of time because of ozone's reactive nature. The inability of ozone to provide a residual in the distribution system is a major drawback to its use. However, recent information about the formation of trihalomethanes by chlorine compounds has resulted in renewed interest in ozone as an alternative means of disinfection.

QUESTIONS

Write your answers in a notebook and then compare your answers with those on page 330.

5.2A List the physical agents that have been used for disinfection other than chlorine.

5.2B List the chemical agents that have been used for disinfection other than chlorine.

5.2C What is a major limitation to the use of ozone?

5.22 Chlorine (Cl_2)

5.220 Properties of Chlorine

Chlorine is a greenish-yellow gas with a penetrating and distinctive odor. The gas is two-and-a-half times heavier than air. Chlorine has a very high coefficient of expansion. If there is a temperature increase of 50°F (28°C) (from 35°F to 85°F or 2°C to 30°C), the volume will increase by 84 to 89 percent. This expansion could easily rupture a cylinder or a line full of liquid chlorine. For this reason no chlorine containers should be filled to more than 85 percent of their capacity. One liter of liquid chlorine can evaporate and produce 450 liters of chlorine gas.

Chlorine by itself is nonflammable and nonexplosive, but it will support combustion. When the temperature rises, so does the vapor pressure of chlorine. This means that when the temperature increases, the pressure of the chlorine gas inside a chlorine container will increase. This property of chlorine must be considered when:

1. Feeding chlorine gas from a container, and
2. Dealing with a leaking chlorine cylinder.

5.221 Chlorine Disinfection Action

The exact mechanism of chlorine disinfection action is not fully known. One theory holds that chlorine exerts a direct action against the bacterial cell, thus destroying it. Another theory is that the toxic character of chlorine inactivates the *ENZYMES*[6] which enable living microorganisms to use their food supply. As a result, the organisms die of starvation. From the point of view of water treatment, the exact mechanism of chlorine disinfection is less important than its demonstrated effects as a disinfectant.

When chlorine is added to water, several chemical reactions take place. Some involve the molecules of the water itself, and some involve organic and inorganic substances suspended in the water. We will discuss these chemical reactions in more detail in the next few sections of this chapter. First, however, there are some terms associated with chlorine disinfection that you should understand.

[6] *Enzymes (EN-zimes). Organic substances (produced by living organisms) which cause or speed up chemical reactions. Organic catalysts and/or biochemical catalysts.*

When chlorine is added to water containing organic and inorganic materials, it will combine with these materials and form chlorine compounds. If you continue to add chlorine, you will eventually reach a point where the reaction with organic and inorganic materials stops. At this point, you have satisfied what is known as the *"CHLORINE DEMAND."*

The chemical reactions between chlorine and these organic and inorganic substances produce chlorine compounds. Some of these compounds have disinfecting properties; others do not. In a similar fashion, chlorine reacts with the water itself and produces some substances with disinfecting properties. The total of all the compounds with disinfecting properties *PLUS* any remaining free (uncombined) chlorine is known as the *"CHLORINE RESIDUAL."* The presence of this measurable chlorine residual is what indicates to the operator that all possible chemical reactions have taken place and that there is still sufficient *"AVAILABLE RESIDUAL CHLORINE"* to kill the microorganisms present in the water supply.

Now, if you add together the amount of chlorine needed to satisfy the chlorine demand and the amount of chlorine residual needed for disinfection, you will have the *"CHLORINE DOSE."* This is the amount of chlorine you will have to add to the water to disinfect it.

Chlorine Dose, mg/*L* = Chlorine Demand, mg/*L* + Chlorine Residual, mg/*L*

where

Chlorine Demand, mg/*L* = Chlorine Dose, mg/*L* – Chlorine Residual, mg/*L*

and

Chlorine Residual, mg/*L* = Combined Chlorine Forms, mg/*L* + Free Chlorine, mg/*L*

5.222 Reaction With Water

Free chlorine combines with water to form hypochlorous and hydrochloric acids:

Chlorine + Water ⇆ Hypochlorous Acid + Hydrochloric Acid

$$Cl_2 + H_2O \leftrightarrows HOCl + HCl$$

Depending on the pH, hypochlorous acid may be present in the water as the hydrogen ion and hypochlorite ion (Figure 5.1).

Hypochlorous Acid ⇆ Hydrogen Ion + Hypochlorite Ion

$$HOCl \leftrightarrows H^+ + OCl^-$$

In solutions that are dilute (low concentration of chlorine) and have a pH above 4, the formation of HOCl (hypochlorous acid) is most complete and leaves little free chlorine (Cl_2) existing. The hypochlorous acid is a weak acid and hence is poorly dissociated (broken up into ions) at pH levels below 6. Thus any free chlorine or hypochlorite (OCl^-) added to water will immediately form either HOCl or OCl^-, the species formed is thereby controlled by the pH value of the water. This is extremely important since HOCl and OCl^- differ in disinfection ability. HOCl has a much greater disinfection potential than OCl^-. Normally in water with a pH of 7.3, 50 percent of the chlorine present will be in the form of HOCl and 50 percent in the form of OCl^-. The higher the pH level, the greater the percent of OCl^-.

5.223 Reaction With Impurities in Water

Most waters that have been processed still contain some impurities. In this section we will discuss some of the more common impurities that react with chlorine and we will examine the effects of these reactions on the disinfection ability of chlorine.

A. Hydrogen sulfide (H_2S) and ammonia (NH_3) are two inorganic substances that may be found in water when it reaches the disinfection stage of treatment. Their presence can complicate the use of chlorine for disinfection purposes. This is because hydrogen sulfide and ammonia are what is known as *REDUCING AGENTS*. That is, they give up electrons easily. Chlorine reacts rapidly with these particular reducing agents producing some undesirable results.

Hydrogen sulfide produces an odor which smells like rotten eggs. It reacts with chlorine to form sulfuric acid and elemental sulfur (depending on temperature, pH, and hydrogen sulfide concentration). Elemental sulfur is objectionable because it can cause odor problems and will precipitate as finely divided white particles which are sometimes colloidal in nature.

The chemical reactions between hydrogen sulfide and chlorine are as follows:

Hydrogen Sulfide + Chlorine + Oxygen Ion → Elemental Sulfur + Water + Chloride Ions

$$H_2S + Cl_2 + O^{2-} \rightarrow S\downarrow + H_2O + 2\ Cl^-$$

The chlorine required to oxidize hydrogen sulfide to sulfur and water is 2.08 mg/*L* chlorine to 1 mg/*L* hydrogen sulfide. The complete oxidation of hydrogen sulfide to the sulfate form is as follows:

Hydrogen Sulfide + Chlorine + Water → Sulfuric Acid + Hydrochloric Acid

$$H_2S + 4\ Cl_2 + 4\ H_2O \rightarrow H_2SO_4 + 8\ HCl$$

Thus, 8.32 mg/*L* of chlorine are required to oxidize one mg/*L* of hydrogen sulfide to the sulfate form. Note that in both reactions the chlorine is converted to the chloride ion (Cl^- or HCl) which has no disinfecting power and does not produce chlorine residual. In waterworks practice we always chlorinate to produce a chlorine residual; therefore the second reaction (complete oxidation of hydrogen sulfide) occurs before we have any chlorine residual in the water we are treating.

When chlorine is added to water containing ammonia (NH_3), it reacts rapidly with the ammonia and forms *CHLORAMINES.*[7] This means that less chlorine is available to act as a disinfectant. As the concentration of ammonia increases, the disinfectant power of the chlorine drops off at a rapid rate.

[7] *Chloramines (KLOR-uh-means). Compounds formed by the reaction of hypochlorous acid (or aqueous chlorine) with ammonia.*

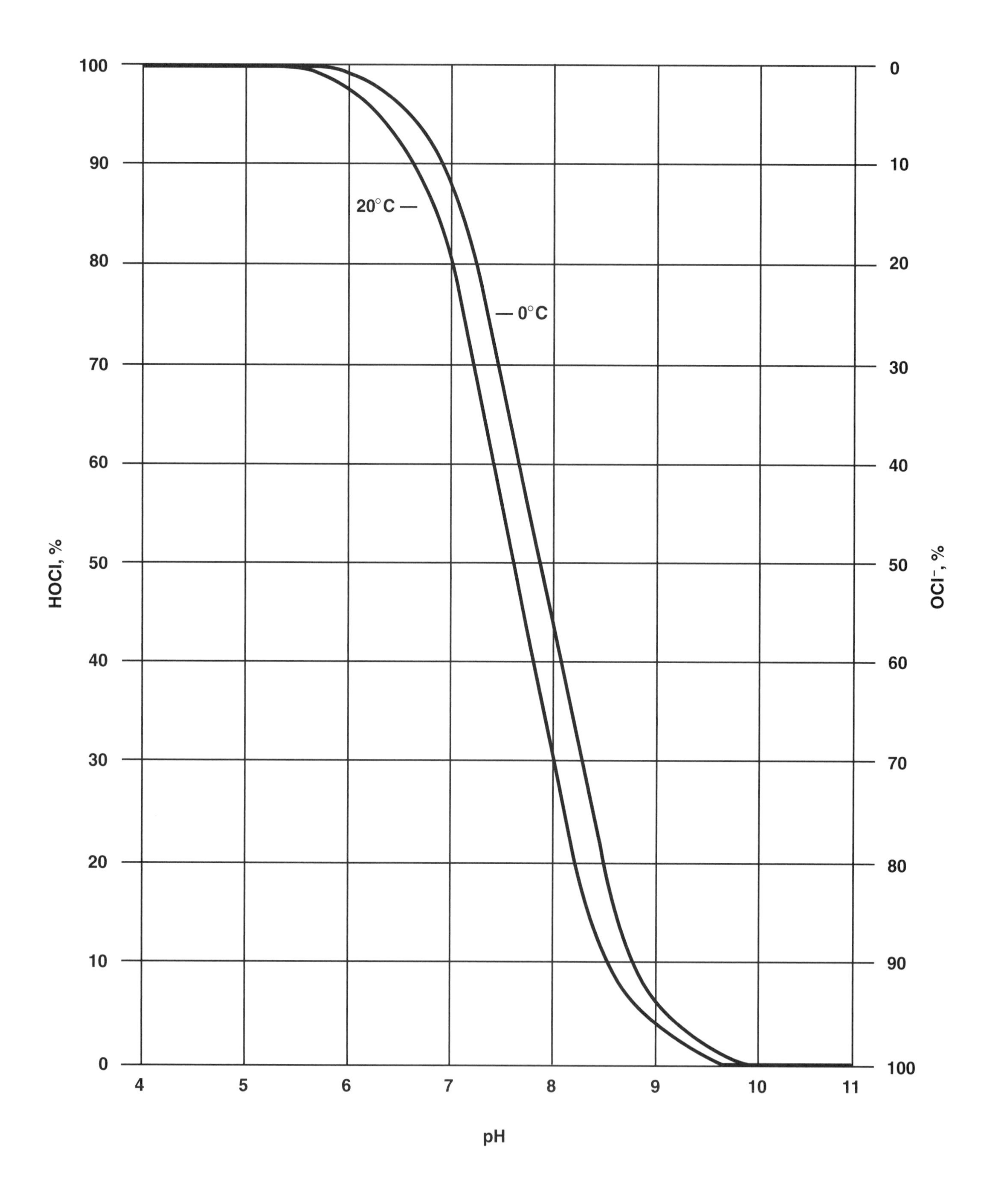

Fig. 5.1 Relationship between hypochlorous acid (HOCl), hypochlorite ion (OCl^-), and pH

B. When organic materials are present in water being disinfected with chlorine, the chemical reactions that take place may produce suspected carcinogenic compounds (trihalomethanes). The formation of these compounds can be prevented by limiting the amount of prechlorination and by removing the organic materials prior to chlorination of the water.

QUESTIONS

Write your answers in a notebook and then compare your answers with those on page 330.

5.2D How is the chlorine dosage determined?

5.2E How is the chlorine demand determined?

5.2F List two inorganic reducing chemicals with which chlorine reacts rapidly.

5.23 Hypochlorite (OCl^-)

5.230 Reactions With Water

The use of hypochlorite to treat potable water achieves the same result as chlorine gas. Hypochlorite may be applied in the form of calcium hypochlorite ($Ca(OCl)_2$) or sodium hypochlorite (NaOCl). The form of calcium hypochlorite most frequently used to disinfect water is known as **H**igh **T**est **H**ypochlorite (HTH). The chemical reactions of hypochlorite in water are similar to those of chlorine gas.

CALCIUM HYPOCHLORITE

Calcium Hypochlorite + Water → Hypochlorous Acid + Calcium Hydroxide

$$Ca(OCl)_2 + 2\ H_2O \rightarrow 2\ HOCl + Ca(OH)_2$$

SODIUM HYPOCHLORITE

Sodium Hypochlorite + Water → Hypochlorous Acid + Sodium Hydroxide

$$NaOCl + H_2O \rightarrow HOCl + NaOH$$

Calcium hypochlorite (HTH) is used by a number of small water supply systems. A problem occurs in these systems when sodium fluoride is injected at the same point as the hypochlorite. A severe crust forms when the calcium and fluoride ions combine.

5.231 Differences Between Chlorine Gas and Hypochlorite Compound Reactions

The only difference between the reactions of the hypochlorite compounds and chlorine gas is the "side" reactions of the end products. The reaction of chlorine gas tends to lower the pH (increases the hydrogen ion (H^+) concentration) by the formation of hydrochloric acid which favors the formation of hypochlorous acid (HOCl). The hypochlorite tends to raise the pH with the formation of hydroxyl ions (OH^-) from the calcium or sodium hydroxide. At a high pH of around 8.5 or higher, the hypochlorous acid (HOCl) is almost completely dissociated to the ineffective hypochlorite ion (OCl^-) (Figure 5.1). This reaction also depends on the *BUFFER CAPACITY*[8] (amount of bicarbonate, HCO_3^-, present) of the water.

Hypochlorous Acid ⇄ Hydrogen Ion + Hypochlorite Ion

$$HOCl \rightleftarrows H^+ + OCl^-$$

5.232 On-Site Chlorine Generation

Small water systems are generating chlorine on site for their water treatment processes. On-site generation (OSG) of process chlorine is attractive due to the lower safety hazards and costs involved. On-site generated chlorine systems produce 0.8 percent sodium hypochlorite. This limited solution strength (about 1/15 the strength of commercial bleach and 1/7 that of household bleach) is below the lower limit deemed a "hazardous liquid," with obvious economic and safety advantages.

Operators' only duties with on-site generation systems are to observe the control panel daily for proper operating guidelines and to dump bags of salt every few weeks. Since the assemblies (which are quite small) include an ion exchange water softener, mineral deposits forming with the electrolytic cell are minimal, with an acid cleaning being necessary only every few months. Cell voltage is controlled at a low value to maximize electrode life, which is about three years. Process brine strength and cell current determine chlorine production at the anode, with hydrogen gas continually vented from the cathode. The units include provisions for storing the chlorine solution in order to deliver chlorine for several days in the event of a power failure or other problems causing equipment failure.

5.24 Chlorine Dioxide (ClO_2)

5.240 Reaction in Water

Chlorine dioxide may be used as a disinfectant. Chlorine dioxide does not form carcinogenic compounds that may be formed by other chlorine compounds. Also it is not affected by ammonia, and is a very effective disinfectant at higher pH levels. In addition, chlorine dioxide reacts with sulfide compounds, thus helping to remove them and eliminate their characteristic odors. Phenolic tastes and odors can be controlled by using chlorine dioxide.

Chlorine dioxide reacts with water to form chlorate and chlorite ions in the following manner:

Chlorine Dioxide + Water → Chlorate Ion + Chlorite Ion + Hydrogen Ions

$$2\ ClO_2 + H_2O \rightarrow ClO_3^- + ClO_2^- + 2\ H^+$$

5.241 Reactions With Impurities in Water

A. *INORGANIC COMPOUNDS*

Chlorine dioxide is an effective *OXIDIZING AGENT*[9] with iron and manganese and does not leave objectionable tastes or odors in the finished water. Because of its oxidizing ability, chlorine dioxide usage must be monitored and the dosage will have to be increased when treating waters with iron and manganese.

---

[8] *Buffer Capacity. A measure of the capacity of a solution or liquid to neutralize acids or bases. This is a measure of the capacity of water for offering a resistance to changes in pH.*

[9] *Oxidizing Agent. Any substance, such as oxygen (O_2) or chlorine (Cl_2), that will readily add (take on) electrons. The opposite is a reducing agent.*

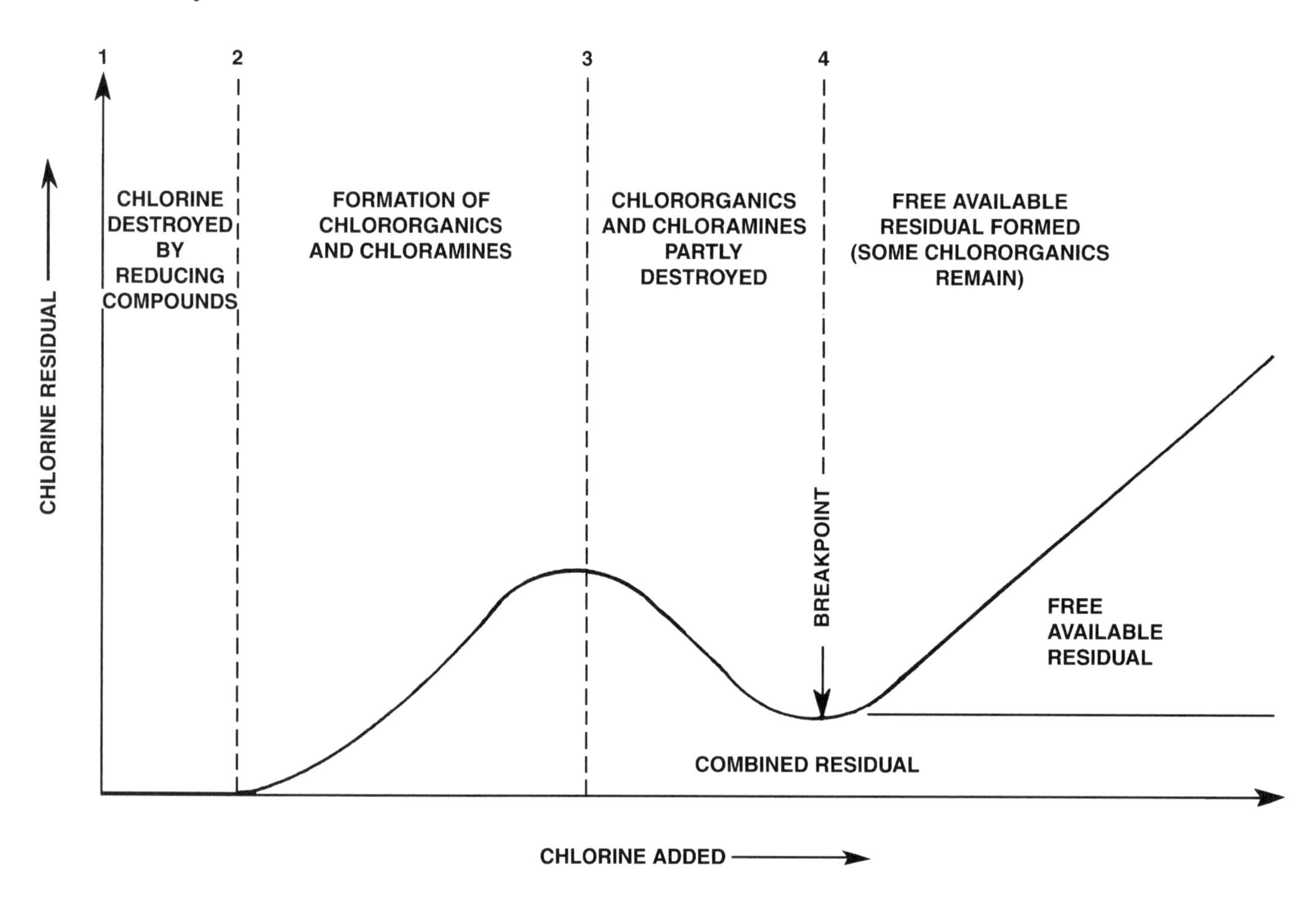

Fig. 5.2 Breakpoint chlorination curve

B. *ORGANIC COMPOUNDS*

Chlorine dioxide does not react with organics in water; therefore, there is little danger of the formation of potentially dangerous trihalomethanes.

QUESTIONS

Write your answers in a notebook and then compare your answers with those on page 330.

5.2G How do chlorine gas and hypochlorite influence pH?

5.2H How does pH influence the relationship between HOCl and OCl^-?

5.25 Breakpoint Chlorination[10]

In determining how much chlorine you will need for disinfection, remember you will be attempting to produce a certain chlorine residual in the form of *FREE AVAILABLE RESIDUAL CHLORINE.*[11] Chlorine in this form has the highest disinfecting ability. *BREAKPOINT CHLORINATION* is the name of this process of adding chlorine to water until the chlorine demand has been satisfied. Further additions of chlorine will result in a chlorine residual that is directly proportional to the amount of chlorine added beyond the breakpoint. Public water supplies are normally chlorinated *PAST THE BREAKPOINT.*

Take a moment here to look at the breakpoint chlorination curve in Figure 5.2. Assume the water being chlorinated contains some manganese, iron, nitrite, organic matter, and ammonia. Now add a small amount of chlorine. The chlorine reacts with (oxidizes) the manganese, iron, and nitrite. That's all that happens — no disinfection and no chlorine residual (Figure 5.2, points 1 to 2). Add a little more chlorine, enough to react with the organics and ammonia; *CHLORORGANICS*[12] and *CHLORAMINES*[13] will form. The chloramines produce a combined chlorine residual — a chlorine residual combined with other substances so it has lost some of its disinfecting strength. Combined residuals have rather poor disinfecting power and may cause tastes and odors (points 2 to 3).

[10] *Breakpoint Chlorination. Addition of chlorine to water until the chlorine demand has been satisfied. At this point, further additions of chlorine will result in a free chlorine residual that is directly proportional to the amount of chlorine added beyond the breakpoint.*

[11] *Free Available Residual Chlorine. That portion of the total available residual chlorine composed of dissolved chlorine gas (Cl_2), hypochlorous acid (HOCl), and/or hypochlorite ion (OCl^-) remaining in water after chlorination. This does not include chlorine that has combined with ammonia, nitrogen, or other compounds.*

[12] *Chlororganic (klor-or-GAN-ick). Organic compounds combined with chlorine. These compounds generally originate from, or are associated with, life processes such as those of algae in water.*

[13] *Chloramines (KLOR-uh-means). Compounds formed by the reaction of hypochlorous acid (or aqueous chlorine) with ammonia.*

With just a little more chlorine the chloramines and some of the chlororganics are destroyed (points 3 to 4). Adding just one last amount of chlorine we get *FREE AVAILABLE RESIDUAL CHLORINE* (beyond point 4) — free in the sense that it has not reacted with anything and available in that it *CAN* and *WILL* react if need be. Free available residual chlorine is the best residual for disinfection. It disinfects faster and without the "swimming pool" odor of combined residual chlorine. Free available residual chlorine begins to form at the breakpoint; the process is called *BREAKPOINT CHLORINATION*. In water treatment plants today it is common practice to go "past the breakpoint." This means that the treated water will have a low chlorine residual, but the residual will be a very effective disinfectant because it is in the form of *FREE AVAILABLE RESIDUAL CHLORINE*.

CAUTION: Ammonia must be present to produce the breakpoint chlorination curve from the addition of chlorine. Sources of ammonia in raw water include fertilizer in agricultural runoff and discharges from wastewater treatment plants. High-quality raw water without any ammonia will not produce a breakpoint curve. Therefore, if there is no ammonia present in the water, and chlorinated water smells and tastes like chlorine, and a chlorine residual is present, DO NOT ADD MORE CHLORINE.

In plants where trihalomethanes (THMs) are not a problem, sufficient chlorine is added to the raw water (prechlorination) to go "past the breakpoint." The chlorine residual will aid coagulation, control algae problems in basins, reduce odor problems in treated water, and provide sufficient chlorine contact time to effectively kill or inactivate pathogenic organisms. Therefore the treated water will have a very low chlorine residual, but the residual will be a very effective disinfectant.

Let's look more closely at some of the chemical reactions that take place during chlorination. When chlorine is added to waters containing ammonia (NH_3), the ammonia reacts with hypochlorous acid (HOCl) to form monochloramine, dichloramine, and trichloramine. The formation of these chloramines depends on the pH of the solution and the initial chlorine-ammonia ratio.

Ammonia + Hypochlorous Acid → Chloramine + Water

$NH_3 + HOCl \rightarrow NH_2Cl + H_2O$ Monochloramine

$NH_2Cl + HOCl \rightarrow NHCl_2 + H_2O$ Dichloramine

$NHCl_2 + HOCl \rightarrow NCl_3 + H_2O$ Trichloramine[14]

As the chlorine to ammonia-nitrogen ratio increases, the ammonia molecule becomes progressively more chlorinated. At Cl_2:NH_3-N weight ratios higher than 7.6:1, all available ammonia is theoretically oxidized to nitrogen gas and chlorine residuals are greatly reduced. The actual Cl_2:NH_3-N ratio for breakpoint for a given source water will usually be greater than 7.6:1 (typically 10:1 for most water), depending on the levels of other substances present in the water (such as nitrite and organic nitrogen). Once this point is reached, additional chlorine dosages yield an equal and proportional increase in free available chlorine.

At the pH levels that are usually found in water (pH 6.5 to 9.5), monochloramine and dichloramine exist together. At pH levels below 5.5, dichloramine exists by itself. Below pH 4.0, trichloramine is the only compound found. The mono- and dichloramine forms have definite disinfection powers and are of interest in the measurement of chlorine residuals. Dichloramine has a more effective disinfecting power than monochloramine. However, dichloramine is not recommended as a disinfectant because of taste and odor problems. Chlorine reacts with *PHENOLIC COMPOUNDS*[15] and salicylic acid (both are leached into water from leaves and blossoms) to form *CHLOROPHENOL*[16] which has an intense medicinal odor. This reaction goes much slower in the presence of monochloramine.

5.26 Critical Factors

Both *CHLORINE RESIDUAL* and *CONTACT TIME* are essential for effective killing or inactivation of pathogenic microorganisms. Complete initial mixing is very important. Changes in pH affect the disinfection ability of chlorine and you must reexamine the best combination of contact time and chlorine residual when the pH fluctuates. Critical factors influencing disinfection are summarized as follows:

1. Effectiveness of upstream treatment processes. The lower the turbidity (suspended solids, organic content, reducing agents) of the water, the better the disinfection.
2. Injection point and method of mixing to get disinfectant in contact with water being disinfected. Depends on whether using prechlorination or postchlorination.
3. Temperature. The higher the temperature, the more rapid the rate of disinfection.
4. Dosage and type of chemical. Usually the higher the dosage, the faster the disinfection rate. The form (chloramines or free chlorine residual) and type of chemical also influence the disinfection rate.
5. pH. The lower the pH, the better the disinfection.
6. Contact time. With good initial mixing, the longer the contact time, the better the disinfection.
7. Chlorine residual.

QUESTIONS

Write your answers in a notebook and then compare your answers with those on pages 330 and 331.

5.2I What is breakpoint chlorination?

5.2J What does chlorine produce when it reacts with organic matter?

5.2K List the critical factors that influence disinfection.

5.27 Chlorine Residual Testing

Many small system operators attempt to maintain a chlorine residual throughout the distribution system. Chlorine is very effective in biological control and especially in elimination of coliform bacteria that might reach water in the distribution system through cross connections or leakage into the system. A chlorine residual also helps to control any microorganisms that could produce slimes, tastes, or odors in the water in the distribution system.

[14] *More commonly called nitrogen trichloride.*

[15] *Phenolic (fee-NO-lick) Compounds. Organic compounds that are derivatives of benzene.*

[16] *Chlorophenolic (klor-o-FEE-NO-lick). Chlorophenolic compounds are phenolic compounds (carbolic acid) combined with chlorine.*

Adequate control of coliform "aftergrowth" is usually obtained only when chlorine residuals are carried to the farthest points of the distribution system. To ensure that this is taking place, make daily chlorine residual tests. A chlorine residual of about 0.2 mg/*L* measured at the extreme ends of the distribution system is usually a good indication that a free chlorine residual is present in all other parts of the system. This small residual can destroy a small amount of contamination, so a lack of chlorine residual could indicate the presence of heavy contamination. If routine checks at a given point show measurable residuals, any sudden absence of a residual at that point should alert the operator to the possibility that a potential problem has arisen which needs prompt investigation. Immediate action that can be taken includes retesting for chlorine residual, then checking chlorination equipment, and finally searching for a source of contamination which could cause an increase in the chlorine demand.

5.28 Chlorine Residual Curve

The chlorine residual curve procedure is a quick and easy way for an operator to estimate the proper chlorine dose, especially when surface water conditions are changing rapidly such as during a storm.

Fill a CLEAN five-gallon bucket from a sample tap located at least two 90-degree elbows (or where chlorine is completely mixed with the water in the pipe) AFTER the chlorine has been injected into the pipe. Immediately measure the chlorine residual and record this value on the "time zero" line of your record sheet (Figure 5.3). This is the initial chlorine residual. At 15-minute intervals, vigorously stir the bucket using an up and down motion. (A large plastic spoon works well for this purpose.) Collect a sample from one or two inches below the water surface and measure the chlorine residual. Record this chlorine residual value on the record sheet. For at least one hour, collect a sample every 15 minutes, measure the chlorine residual, and record the results to indicate the "chlorine demand" of the treated water. Plot these recorded values on a chart or graph paper as shown on Figure 5.3. Connect the plotted points to create a chlorine residual curve. If the chlorine residual after one hour is not correct (about 0.2 mg/*L*), increase or decrease the initial chlorine dose so the final chlorine residual will be approximately at the desired ultimate chlorine residual in the water distribution system. Repeat this procedure until the desired TARGET initial chlorine residual will achieve the desired chlorine residual throughout the distribution system.

Precautions that must be taken when performing this test include being sure the five-gallon plastic test bucket is clean and only used for this purpose. A new bucket does not need to be used for every test, but the bucket should be new when the first test is performed. The stirrer should also be clean. DO NOT USE THE STIRRER FOR THE CHLORINE SOLUTION MIXING AND HOLDING TANK. During the test the bucket should be kept cool so that the chlorine gas does not escape from the water and give false chlorine residual values.

The chlorine demand for groundwater changes slowly, or not at all; therefore, the "initial or target" chlorine residual does not have to be checked more frequently than once a month. Always be sure to measure the chlorine residual in the distribution system on a daily basis. This is also a good check that the chlorination equipment is working properly and that the chlorine stock solution is the correct concentration.

The chlorine demand for surface water can change continuously, especially during storms and the snow melt season. Experience has proven that the required "initial or target" chlorine residual at time zero is directly tied to the turbidity of the finished (treated) water. The higher the finished water turbidity, the higher the "initial" chlorine residual value will have to be to ensure the desired chlorine residual in the distribution system. Careful documentation of this information in your records will greatly reduce the lag time in chlorine addition changes to maintain the desired residual in the distribution system and the delivery of safe drinking water to your consumers. Experience and a review of your records will indicate that for a given turbidity value, you can estimate the desired "initial" chlorine residual, which will require a given chlorinator output level for a given water flow rate.

Acknowledgment

The information in Sections 5.27 and 5.28 was developed by Bill Stokes. His suggestions and procedures are greatly appreciated.

5.29 Chloramination
by David Foust

5.290 Use of Chloramines

Chloramines have been used as an alternative disinfectant by water utilities for over seventy years. An operator's decision to use chloramines depends on several factors, including the quality of the raw water, the ability of the treatment plant to meet various regulations, operational practices, and distribution system characteristics. Chloramines have proven effective in accomplishing the following objectives:

1. Reducing the formation of trihalomethanes (THMs) and other disinfection by-products (DBPs),
2. Maintaining a detectable residual throughout the distribution system,
3. Penetrating the biofilm (the layer of microorganisms on pipeline walls) and reducing the potential for coliform regrowth,
4. Killing or inactivating *HETEROTROPHIC*[17] plate count bacteria, and
5. Reducing taste and odor problems.

5.291 Methods for Producing Chloramines

There are three primary methods by which chloramines are produced: (1) preammoniation followed by later chlorination, (2) addition of chlorine and ammonia at the same time (concurrently), and (3) prechlorination/postammoniation.

1. *PREAMMONIATION FOLLOWED BY LATER CHLORINATION*

   In this method, ammonia is applied at the rapid-mix unit process and chlorine is added downstream at the entrance to the flocculation basins. This approach usually produces

[17] *Heterotrophic (HET-er-o-TROF-ick). Describes organisms that use organic matter for energy and growth. Animals, fungi and most bacteria are heterotrophs.*

Chlorine Detention Rate Curve

ppm
3.0 2.9 2.8 2.7 2.6 2.5 2.4 2.3 2.2 2.1 2.0 1.9 1.8 1.7 1.6 1.5 1.4 1.3 1.2 1.1 1.0 0.9 0.8 0.7 0.6 0.5 0.4 0.3 0.2 0.1 0.0
Time 0 15 30 45 60 75
(Initial) (← Residual →)

Average surface water
(moderate to low turbidity)

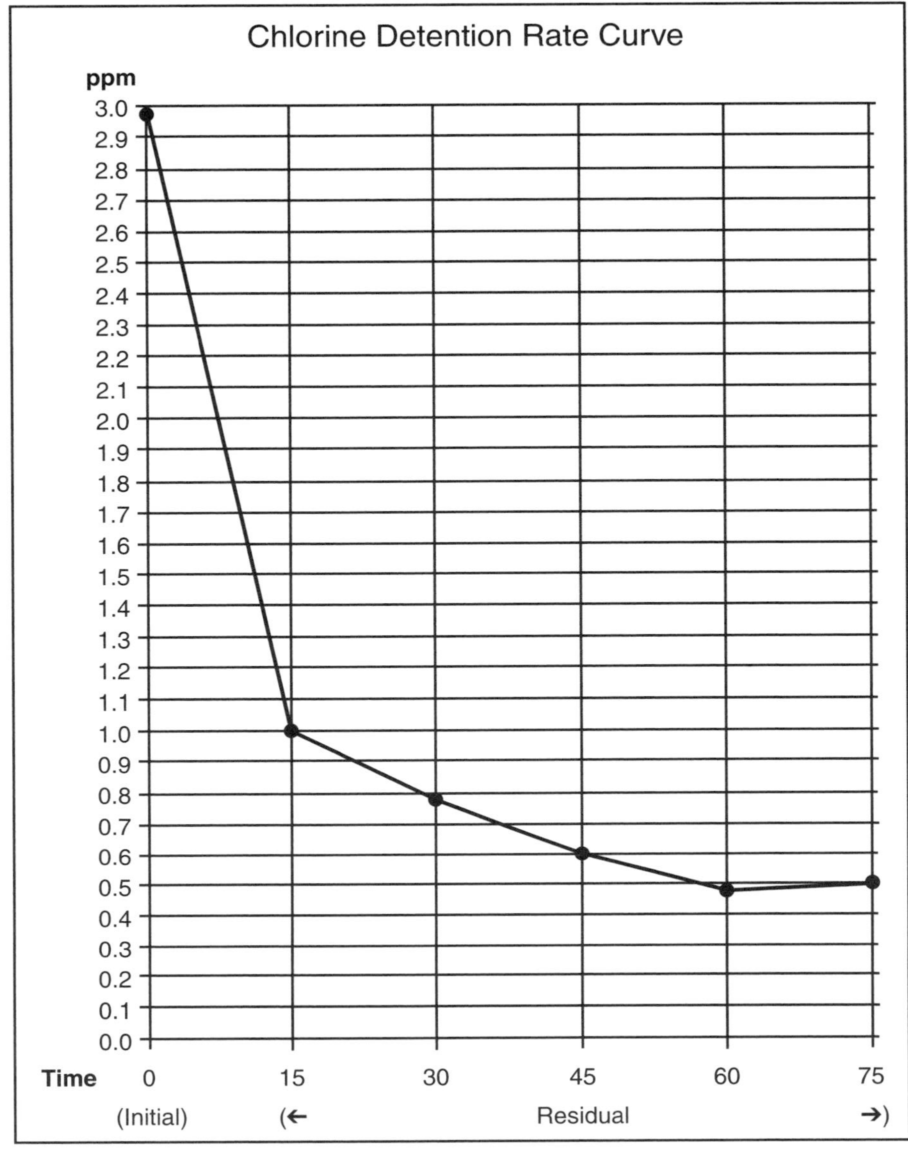

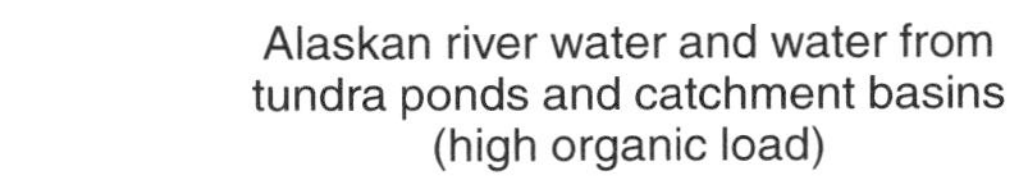

Fig. 5.3 Chlorine detention rate curves

lower THM levels than the postammoniation method. Preammoniation to form chloramines (monochloramine) does not produce phenolic tastes and odors, but this method may not be as effective as postammoniation for controlling tastes and odors associated with *DIATOMS*[18] and anaerobic bacteria in source waters.

2. *CONCURRENT ADDITION OF CHLORINE AND AMMONIA*

   In this method, chlorine is applied to the plant influent and, at the same time or immediately thereafter, ammonia is introduced at the rapid-mix unit process. Concurrent chloramination produces the lowest THM levels of the three methods.

3. *PRECHLORINATION/POSTAMMONIATION*

   In prechlorination/postammoniation, chlorine is applied at the head of the plant and a free chlorine residual is maintained throughout the plant processes. Ammonia is added at the plant effluent to produce chloramines. Because of the longer free chlorine contact time, this application method will result in the formation of more THMs, but it may be necessary to use this method to meet the disinfection requirements of the Surface Water Treatment Rule (SWTR). A major limitation of using chloramine residuals is that chloramines are less effective as a disinfectant than free chlorine residuals.

5.292 Chlorine to Ammonia-Nitrogen Ratios

After a method of chloramine application has been selected, the best ratio of chlorine to ammonia-nitrogen (by weight) and the desired chloramine residual for each system must be determined. A dosage of three parts of chlorine to one part ammonia (3:1) will form monochloramines. This 3:1 ratio provides an excess of ammonia-nitrogen which will be available to react with any chlorine added in the distribution system to boost the chloramine residual.

Higher chlorine to ammonia-nitrogen weight ratios such as 4:1 and 5:1 also have been used successfully by many water agencies. However, the higher the chlorine to ammonia-nitrogen ratio, the less excess ammonia will be available for rechlorination. Some agencies find it necessary to limit the amount of excess available ammonia to prevent incomplete *NITRIFICATION.*[19]

Monochloramines form combined residual chlorine (rising part of curve in Figure 5.4) as the chlorine dose is increased in the presence of ammonia. As the chlorine dose increases, the combined residual increases and excess ammonia decreases. The maximum chlorine to ammonia ratio that can be achieved is 5:1. At a chlorine dose above the 5:1 ratio, the combined residual actually decreases and the total ammonia-nitrogen also begins to decrease as it is oxidized by the additional chlorine. Dichloramines form during this oxidation and may cause tastes and odors. As the chlorine dose is further increased, breakpoint chlorination will eventually occur. Trichloramines are formed past the breakpoint and also may form tastes and odors. As with breakpoint chlorination, further additions of chlorine will result in a chlorine residual that is proportional to the amount of chlorine added beyond the breakpoint.

Calculating the chlorine to ammonia-nitrogen ratio on the basis of actual quantity of chemicals applied can lead to incorrect conclusions regarding the finished water quality. In applications in which chlorine is injected before the ammonia, chlorine demand in the water will reduce the amount of chlorine available to form the combined residual. In such applications the *applied* ratio will be greater than the *actual* ratio of chlorine to ammonia-nitrogen leaving the plant.

As an example, assume that an initial dosage of 5.0 mg/*L* results in a free chlorine residual of 3.5 mg/*L* at the ammonia application point; it can be concluded that a chlorine demand of 1.5 mg/*L* exists. If ammonia-nitrogen is applied at a dose of 1.0 mg/*L*, the applied chlorine to ammonia-nitrogen ratio is 5:1, whereas the actual ratio in water leaving the plant is only 3.5:1.

5.293 Special Water Users

Although chloramines are nontoxic to healthy humans, they can have a weakening effect on individuals with kidney disease who must undergo kidney dialysis. Chloramines must be removed from the water used in the dialysis treatments. Granular activated carbon and ascorbic acid are common substances used to reduce chloramine residuals. All special water users should be notified before chloramines are used as a disinfectant in municipal waters.

Also, chloramines can be deadly to fish. They can damage gill tissue and enter the red blood cells causing a sudden and severe blood disorder. For this reason, all chloramine compounds must be removed from the water prior to any contact with fish.

5.294 Blending Chloraminated Waters

Care must be taken when blending chloraminated water with water that has been disinfected with free chlorine. Depending on the ratio of the blend, these two different disinfectants can cancel each other out resulting in very low disinfectant residuals. When chlorinated water is blended with chloraminated water, the chloramine residual will decrease after the excess ammonia has been combined (Figure 5.4). Knowing the amount of uncombined ammonia available is important in determining how much chlorinated water can be blended with a particular chloraminated water without significantly affecting the monochloramine residual. Knowing how much uncombined ammonia-nitrogen is available is also important before you make any attempt to boost the chloramine residual by adding chlorine.

5.295 Chloramine Residuals

When measuring combined chlorine residuals (chloramines) in the field, analyze for total chlorine. No free chlorine should be present at chlorine to ammonia-nitrogen ratios of 3:1 to 5:1. Care must be taken when attempting to measure free chlorine with chloraminated water because the chloramine residual will interfere with the DPD method for measuring free chlorine. (See Section 5.5, "Measurement of Chlorine Residual," for more specific information about the methods commonly used to measure chlorine residuals, including the DPD method.)

---

[18] *Diatoms (DYE-uh-toms). Unicellular (single cell), microscopic algae with a rigid (box-like) internal structure consisting mainly of silica.*

[19] *Nitrification (NYE-truh-fuh-KAY-shun). An aerobic process in which bacteria reduce the ammonia and organic nitrogen in water into nitrite and then nitrate.*

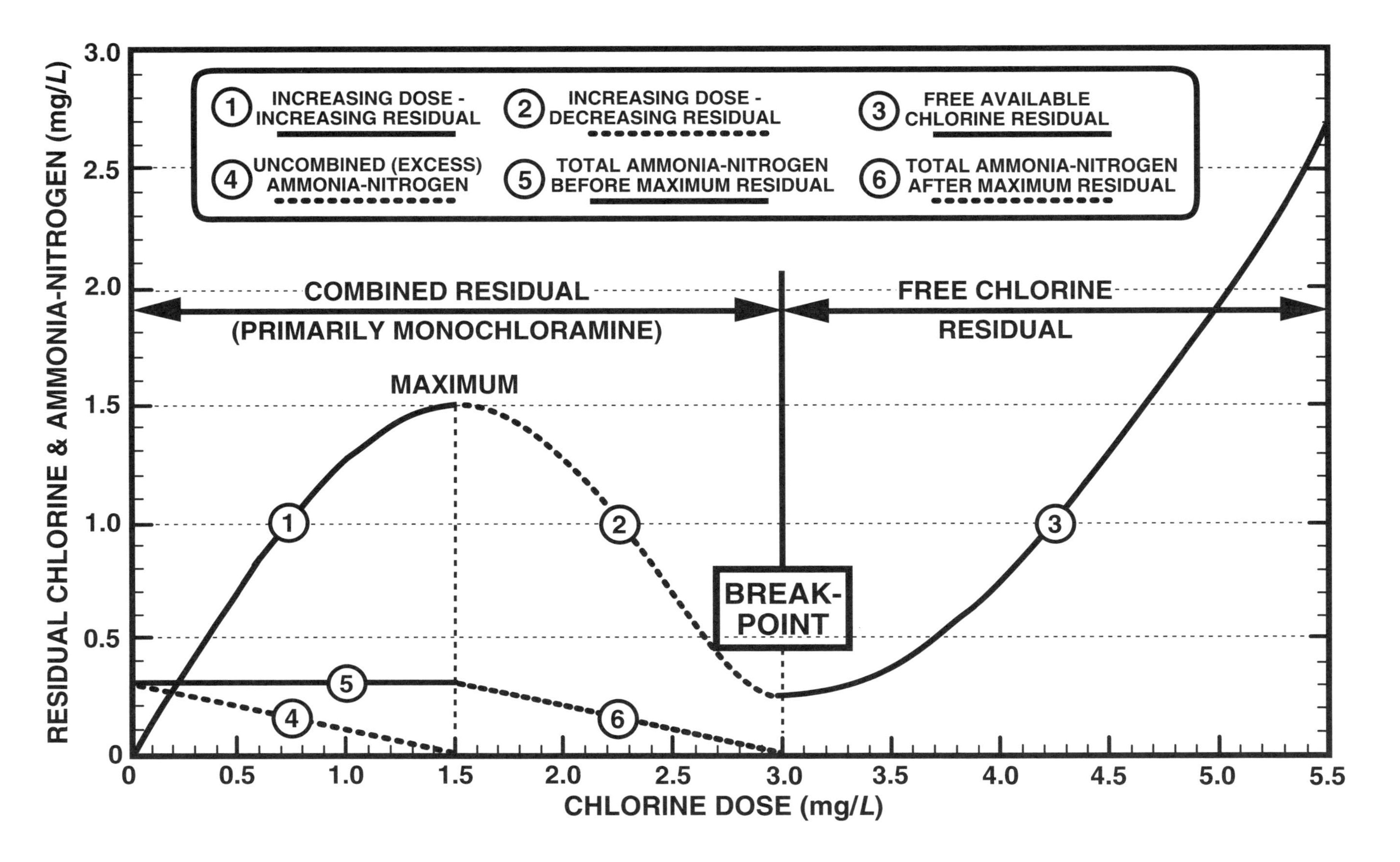

Fig. 5.4 Typical chloramination dose-residual curve

5.296 Nitrification

Nitrification is an important and effective microbial process in the oxidation of ammonia in both land and water environments. Two groups of organisms are involved in the nitrification process: ammonia-oxidizing bacteria (AOB) (see Figure 5.5) and nitrite-oxidizing bacteria. Nitrification has been well recognized as a beneficial treatment for the removal of ammonia in municipal wastewater.

When nitrification occurs in chloraminated drinking water, however, the process may lower the water quality unless the nitrification process reaches completion. Incomplete or partial nitrification causes the production of nitrite from the growth of AOB (ammonia-oxidizing bacteria, see Figure 5.5). This nitrite, in turn, rapidly reduces free chlorine and can interfere with the measurement of free chlorine. The end result may be a loss of total chlorine and ammonia and an increase in the concentration of heterotrophic plate count bacteria.

Factors influencing nitrification include the water temperature, the detention time in the reservoir or distribution system, excess ammonia in the water system, and the chloramine concentration used. The conditions most likely to lead to nitrification when using chloramines are a pH of 7.5 to 8.5, a water temperature of 25 to 30°C, a free ammonia concentration in the water, and a dark environment. The danger in allowing nitrification episodes to occur is that you may be left with very low or no total chlorine residual.

5.297 Nitrification Prevention and Control

When using chloramines for disinfection, an early warning system should be developed to detect the signs that nitrification is beginning to occur so that you can prevent or at least control the nitrification process. The best way to do this is to set up a regularly scheduled monitoring program. The warning signs to watch for include decreases in ammonia level, total chlorine level and pH, increases in nitrite level, and an increase in heterotrophic plate count bacteria. In addition, action response levels should be established for chloraminated distribution systems and reservoirs. Normal background levels of nitrite should be measured and then alert levels should be established so that increasing nitrite levels will not be overlooked.

An inexpensive way to help keep nitrite levels low is to reduce the detention times through the reservoirs and the distribution system, especially during warmer weather. Adding more chlorine to reservoir inlets and increasing the chlorine to ammonia-nitrogen ratio from 3:1 up to 5:1 at the treatment plant effluent will further control nitrification by decreasing the amount of uncombined ammonia in the distribution system. However, at a chlorine and ammonia-nitrogen ratio of 5:1, it is critical that the chlorine and ammonia feed systems operate accurately and reliably because an overdose of chlorine can reduce the chloramine residual.

Other strategies for controlling nitrification include establishing a flushing program and increasing the chloramine residual. A uniform flushing program should be a key component of any nitrification control program. Flushing reduces the detention time in low-flow areas, increases the water velocity within pipelines to remove sediments and biofilm that would harbor nitrifying bacteria, and draws higher disinfectant residuals into problem areas. Increasing the chloramine residual in the distribution system to greater than 2.0 mg/*L* is also effective in preventing the onset of nitrification.

QUESTIONS

Write your answers in a notebook and then compare your answers with those on page 331.

5.2L An operator's decision to use chloramines depends on what factors?

5.2M What are the three primary methods by which chloramines are produced?

5.2N Why is the *applied* chlorine to ammonia-nitrogen ratio usually greater than the *actual* chlorine to nitrogen ratio leaving the plant?

5.2O Incomplete nitrification causes the production of nitrite which produces what problems in disinfection of water?

5.3 POINTS OF APPLICATION

5.30 Wells and Pumps[20]

5.300 Disinfecting New and Existing Wells

When a new well is completed, it is necessary to disinfect the well, pump and screen. A 50 mg/*L* residual of free chlorine in contact with all surfaces of the well, screen, pump and piping is recommended. Consideration must be given to the fact that the water aquifer could have been contaminated during the drilling process. Disinfection is usually accomplished by using a chlorine solution applied into the well and the aquifer around the well. To accomplish this process the use of sodium hypochlorite is recommended. As previously mentioned, 50 mg/*L* is needed and 24 hours of contact time is recommended. At the end of this period, pump the well until all evidence of a chlorine residual is gone. Then take a sample and test for *TOTAL COLIFORMS*[21] to determine the effectiveness of the chlorine dosage.

Since organic matter such as oil may be used during drilling, the chlorine solution should have an initial concentration of about 1,000 mg/*L* if it is being injected into the well through the vent pipe. If the chlorine solution is applied to the well by injecting the chlorine through the pump column pipe, a chlorine dosage of 100 mg/*L* is acceptable. By using the column pipe, the oil on the water surface is avoided and the large amount of chlorine is not needed.

---

[20] *Also see Section 3.5, "Disinfection of Wells and Pumps," in Chapter 3.*

[21] *Total Coliforms. See procedures for Coliform Test in Chapter 7, "Laboratory Procedures."*

Nitrification is a biological process caused by naturally occurring ammonia-oxidizing bacteria. These bacteria feed on free ammonia and convert it to nitrite and then nitrate. They thrive in covered reservoirs during warm summer months and are very resistant to chloramine disinfection. The by-products of their biological breakdown can support the growth of coliform bacteria.

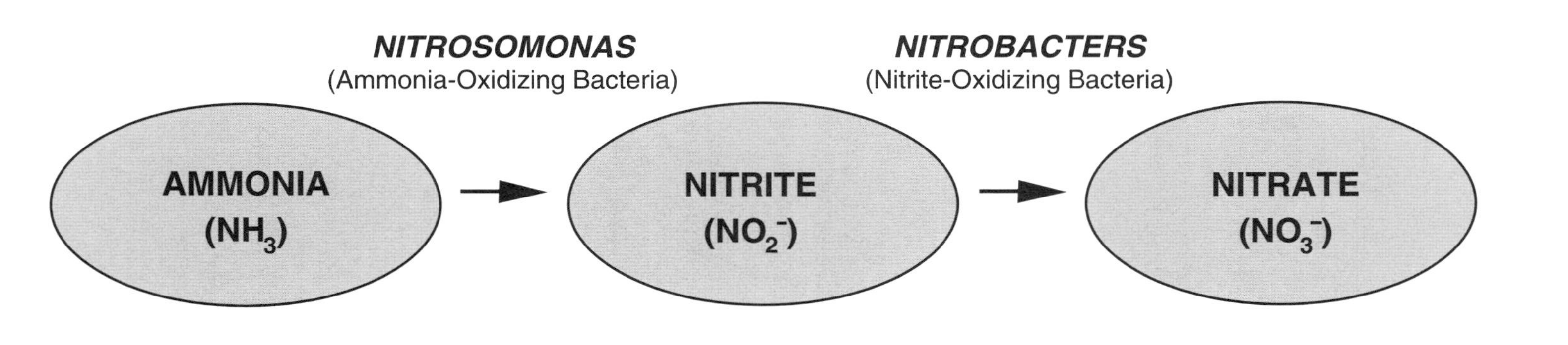

<u>Factors Influencing</u>

Water Temperature
Detention Time
Excess Ammonia
Chloramine Concentration

<u>Early Warning Signs</u>

Decrease in Ammonia
Decrease in Chlorine
Decrease in pH
Increase in Nitrite
Increase in Plate Count Bacteria

<u>Optimum Conditions</u>

pH 7.5 - 8.5
Temp. 25 - 30°C
Free Ammonia
Dark Environment

<u>Prevention and Control</u>

Decrease Detention Times
Decrease Free Ammonia
Increase Disinfection Dosage Ratio
Breakpoint Chlorinate
Establish a Flushing Program

Fig. 5.5 The nitrification process

5.301 Sample Problem

FORMULAS

To disinfect a well, the first step is to determine the volume of water in the well. There are three approaches to calculating the volume of water in a well.

1. Calculate the volume of water using the diameter of the casing and the depth of water.

$$\text{Volume, gal} = \frac{(0.785)(\text{Diameter, in})^2(\text{Depth, ft})(7.48\ \text{gal/cu ft})}{144\ \text{sq in/sq ft}}$$

2. When the depth of water in a well changes from time to time, many operators calculate a constant for their well in terms of gallons of water per foot of water in the well.

$$\text{Volume, gal/ft} = \frac{(0.785)(\text{Diameter, in})^2(7.48\ \text{gal/cu ft})}{144\ \text{sq in/sq ft}}$$

3. Sometimes tables are available which allow you to find the volume in gallons per foot from the table if you know the diameter of the casing.

To find the pounds of chlorine needed to disinfect a well, we need to know the volume of water in gallons and the desired chlorine dose in milligrams per liter. Convert the volume in gallons to million gallons by dividing by 1,000,000.

$$\text{Volume, M Gal} = \frac{\text{Volume, Gal}}{1{,}000{,}000/\text{Million}}$$

$$\boxed{\text{Chlorine, lbs} = (\text{Volume, M Gal})(\text{Dose, mg}/L)(8.34\ \text{lbs/gal})}$$

We can "prove" the above formula if we know that one liter of water weighs 1,000,000 mg.

$$\text{Chlorine, lbs} = (\text{Volume, M Gal})\left(\frac{\text{Dose, mg}}{1{,}000{,}000\ \text{mg}}\right)(8.34\ \text{lbs/gal})$$

Note that the dose is now in milligrams per million milligrams, or parts per million. Now change the milligrams to pounds so we will have dose, pounds chlorine per million pounds of water.

$$\text{Chlorine, lbs} = (\text{Volume, M Gal})\left(\frac{\text{Dose, lbs}}{\text{M lbs}}\right)(8.34\ \text{lbs/gal})$$

To calculate the gallons of sodium hypochlorite needed to disinfect a well, we need to know the pounds of chlorine needed and the percent available chlorine in the hypochlorite.

$$\text{Sodium Hypochlorite Solution, gallons} = \frac{(\text{Chlorine, lbs})(100\%)}{(8.34\ \text{lbs/gal})(\text{Hypochlorite, \%})}$$

EXAMPLE 1

A new 200-foot deep well is to be disinfected with a five percent sodium hypochlorite solution. The top 150 feet of the well has a 12-inch casing and the bottom 50 feet has an 8-inch casing and well screen. The water level in the well is at 80 feet from the top of the well. Since this is a new well, the desired chlorine concentration in the initial dose should be 100 mg/*L*. How many gallons of five percent sodium hypochlorite will be needed?

Known		Unknown
Depth of Well, ft	= 200 ft	5% Hypochlorite, gallons
Hypochlorite, %	= 5%	
Chlorine Dose, mg/*L*	= 100 mg/*L*	
Top 150 ft	= 12-in casing	
Bottom 50 ft	= 8-in casing and screen	
Water Level, ft	= 80 ft from top	
or	= 120 ft from bottom	

1. Find the volume of water in the well. Use Table 5.4.

TABLE 5.4 VOLUME OF WATER IN WELL PER FOOT OF DEPTH

Casing Size, in	Volume, gallons per foot of depth
4	0.65
5	1.02
6	1.47
8	2.61
10	4.08
12	5.87
14	7.99
16	10.44
18	13.21
20	16.31

a. From 80 feet from top to 150 feet from top we have 70 feet (150 ft – 80 ft) of 12-inch casing.

From Table 5.4, 12-inch casing has 5.87 gallons of water per foot.

b. From 150 feet from top to 200 feet from top we have 50 feet (200 ft – 150 ft) of 8-inch casing.

From Table 5.4, 8-inch casing has 2.61 gallons of water per foot.

$$\begin{aligned}\text{Total Volume of Water, gallons} &= (\text{Length, ft})(\text{Volume, gal/ft})\\ &= (70\ \text{ft})(5.87\ \text{gal/ft}) + (50\ \text{ft})(2.61\ \text{gal/ft})\\ &= 411\ \text{gal} + 131\ \text{gal}\\ &= 542\ \text{gallons of Water}\end{aligned}$$

If Table 5.4 was not available, we could calculate the approximate volume in gallons per foot of depth and the total volume of water in gallons.

a. For the 12-inch casing, calculate the volume in gallons per foot of depth.

$$\text{Volume, gal} = (\text{Area, sq ft})(\text{Depth, ft})(7.48\ \text{gal/cu ft})$$

$$\begin{aligned}\text{Volume, gal/ft} &= (\text{Area, sq ft})(7.48\ \text{gal/cu ft})\\ &= \frac{(0.785)(12\ \text{in})^2(7.48\ \text{gal/cu ft})}{144\ \text{sq in/sq ft}}\\ &= 5.87\ \text{gal/ft}\end{aligned}$$

b. For the 8-inch casing, calculate the volume in gallons per foot of depth.

$$\text{Volume, gal/ft} = (\text{Area, sq ft})(7.48 \text{ gal/cu ft})$$

$$= \frac{(0.785)(8 \text{ in})^2(7.48 \text{ gal/cu ft})}{144 \text{ sq in/sq ft}}$$

$$= 2.61 \text{ gal/ft}$$

$$\text{Total Volume of Water, gallons} = (\text{Length, ft})(\text{Volume, gal/ft})$$

$$= (70 \text{ ft})(5.87 \text{ gal/ft}) + (50 \text{ ft})(2.61 \text{ gal/ft})$$

$$= 411 \text{ gal} + 131 \text{ gal}$$

$$= 542 \text{ gallons of Water}$$

This answer is the same as the 542 gallons we obtained using the values in Table 5.4.

2. Find the pounds of chlorine needed.

$$\text{Chlorine, lbs} = (\text{Vol Water, M Gal})(\text{Chlorine Dose, mg}/L)(8.34 \text{ lbs/gal})$$

$$= (0.000542 \text{ M Gal})(100 \text{ mg}/L)(8.34 \text{ lbs/gal})$$

$$= 0.45 \text{ lbs Chlorine}$$

3. Calculate the gallons of 5 percent sodium hypochlorite solution needed.

$$\text{Sodium Hypochlorite Solution, gallons} = \frac{(\text{Chlorine, lbs})(100\%)}{(8.34 \text{ lbs/gal})(\text{Hypochlorite, }\%)}$$

$$= \frac{(0.45 \text{ lbs})(100\%)}{(8.34 \text{ lbs/gal})(5\%)}$$

$$= 1.08 \text{ gallons}$$

A little over one gallon of 5 percent solution sodium hypochlorite should do the job.

The gravel in a gravel envelope well must be disinfected as the gravel is added to the well. This is accomplished by adding half a pound (227 gm) of five gram *HTH*[22] granules per ton of gravel.

When disinfecting a well that draws from more than one aquifer, remember that there is nearly always flow from one aquifer to another. Also if water is flowing through a well, the procedures described in this chapter may not work very well because the flowing water will carry away the chlorine.

Once water from a well starts producing positive coliform test results, it is almost impossible to correct the problem with chlorine treatment. However, some wells will clear up after long periods of pumping (sometimes two years are required).

Disinfection of the well may not be accomplished the first time so this procedure may have to be repeated a second and even possibly a third time.

5.302 Procedures

Procedures described in this section are one method for disinfecting a well; however, other methods are available and may be used

a. Calculate volume of water in well using guidelines in Example 1.

b. Wash the pump column or drop pipe (pipe attached to the pump) with chlorine solution as it is lowered into well. The chlorine solution can be fed through a hose to wash the pipe.

c. After the chlorine solution has been fed into the well, operate the pump so as to thoroughly mix the disinfectant with the water in the well. Pump until the water discharged has the odor of chlorine. Then shut down pump and let water surge back down into well. Repeat this procedure several times over an hour.

d. Allow the well to stand for 24 hours.

e. Pump well to waste until all traces of chlorine are gone.

f. Take a bacteriological sample for analysis. Use the 24-hour membrane filter method for the quickest results.

g. If results of bacteriological analysis for total coliforms indicate unsafe conditions, repeat disinfection procedure. Conditions are considered unsafe when the test results are positive (there are coliforms present). Stated another way, a well has been successfully disinfected if the results of a bacteriological analysis for total coliforms are negative (no coliforms).

5.303 Continuous Disinfecting of Wells

Normally wells are treated by a single application of disinfecting solution, but since more and more groundwater sources are becoming contaminated, it is becoming a standard practice to disinfect wells on a continuous basis.

The type of disinfection installation to be used will depend on which type of chemical is to be used. From a safety standpoint the use of hypochlorite is the safest. Normally there are fewer hazards associated with hypochlorite compounds than with liquid or gaseous chlorine. However, liquid or gaseous chlorine is cheaper to use than hypochlorite, especially if large volumes of chlorine are required.

[22] *HTH (pronounce as separate letters).* ***H****igh* ***T****est* ***H****ypochlorite. Calcium hypochlorite or* $Ca(OCl)_2$.

QUESTIONS

Write your answers in a notebook and then compare your answers with those on page 331.

5.3A When disinfecting a well, why should the well, pump, screen, and aquifer around the well all be disinfected?

5.3B What are the advantages of disinfecting a well by applying chlorine through the pump column pipe rather than through the vent pipe?

5.3C How would you determine if a well has been successfully disinfected?

5.3D Why is it becoming standard practice to disinfect wells on a continuous basis?

5.31 Mains

5.310 Procedures

A. Preventive Measures

One of the most effective steps in disinfecting water mains is to do everything possible to prevent the mains from becoming contaminated. Keep outside material such as dirt, construction materials, animals, rodents, and dirty water out of mains being installed or repaired. Inspect the interior of all pipes for cleanliness as the main is laid in the trench. Keep the trenches dry or dewatered. Install water-

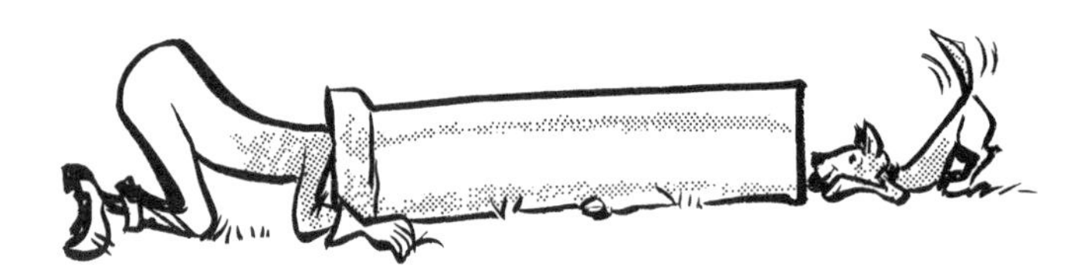

tight plugs in all open-end joints whenever the trench is unattended for any length of time. An exception to the use of watertight plugs in the open ends of joints is when groundwater could cause the pipe to float. When a long pipeline is to be laid, a dry trench may not always be possible during the night when crews are not on duty.

B. Preliminary Measures

Flush the mains before attempting to disinfect them. Flushing is not an alternative to cleanliness when laying or repairing mains. Flushing cannot be expected to remove debris caked on joints, crevices and other parts of the system. Flush the mains using water flowing with a velocity of at least 2.5 ft/sec (0.75 m/sec). See Table 5.5 for flows required in gallons per minute (GPM) to flush various diameter mains with 2.5 ft/sec and 5.0 ft/sec flushing velocities.

Special care must be exercised to be sure fittings and valves are clean before disinfecting a main. Also, all air pockets or other conditions which would prevent proper disinfection should be eliminated. Many water mains contain mechanical joints. These joints have spaces that are difficult to chlorinate once they become filled with water. This is one reason why initial flushing may be ineffective. Some operators crush calcium hypochlorite tablets and place the crushed tablets in joints to improve the disinfection process.

TABLE 5.5 FLOWS REQUIRED FOR VARIOUS FLUSHING VELOCITIES

Line Size, inches	Flow Required, GPM[a] Velocity, 2.5 ft/sec	Velocity, 5 ft/sec
4	100	200
6	220	440
8	390	780
10	610	1,220
12	880	1,760
14	1,200	2,400
16	1,570	3,140

[a] Line should be flushed for at least 30 minutes.

If you wish to calculate the flow required in gallons per minute of any pipe diameter and flushing velocity not shown in Table 5.5, or if you wish to verify any numbers in Table 5.5, use the following procedure.

1. Calculate the cross-sectional area of the pipe in square feet. Use a 10-inch diameter pipe.

$$\text{Area, sq ft} = \frac{(0.785)(\text{Diameter, in})^2}{144 \text{ sq in/sq ft}}$$

$$= \frac{(0.785)(10 \text{ in})^2}{144 \text{ sq in/sq ft}}$$

$$= 0.545 \text{ sq ft}$$

2. Calculate the flow rate in cubic feet per second (CFS). Use a velocity of five feet per second and Q = AV.

$$\text{Flow Rate, CFS} = (\text{Area, sq ft})(\text{Velocity, ft/sec})$$

$$= (0.545 \text{ sq ft})(5 \text{ ft/sec})$$

$$= 2.725 \text{ CFS}$$

3. Convert the flow from cubic feet per second to gallons per minute.

$$\text{Flow Rate, GPM} = (\text{Flow, cu ft/sec})(7.48 \text{ gal/cu ft})(60 \text{ sec/min})$$

$$= (2.725 \text{ cu ft/sec})(7.48 \text{ gal/cu ft})(60 \text{ sec/min})$$

$$= 1{,}223 \text{ GPM}$$

NOTE: Table 5.5 shows a flow of 1,220 GPM because this is as close as you can read most flowmeters, as well as regulate the flow.

C. Disinfection Alternatives

Three forms of chlorine commonly used for disinfection are chlorine gas, calcium hypochlorite, and sodium hypochlorite. If chlorine gas is used, a trained operator is required to operate the solution-feed chlorinator in combination with a booster pump. Applying gas directly from a cylinder is dangerous and, if proper mixing of chlorine and water is not obtained, a highly corrosive condition could develop.

Another danger is that water could be drawn back into the chlorine cylinder. A mixture of water and liquid chlorine will form a very concentrated hydrochloric acid solution.

This acid solution could "eat" a hole in the wall of the cylinder and allow liquid or gaseous chlorine to escape. For these reasons, *NEVER* use water on a chlorine leak. The corrosive action of chlorine and water will *ALWAYS* make a leak worse.

Calcium hypochlorite is available in powder, granular, or tablet form at 65 percent available chlorine. Calcium hypochlorite is relatively soluble in water and can be applied by the use of a solution feeder. When using calcium hypochlorite (chlorine) tablets, the tablets will not dissolve readily if the water temperature is below 41°F (5°C), which will reduce the chlorine concentration in the water being disinfected. Temperature control of water is difficult, but you can control contact time. Therefore, if you are disinfecting when the water temperature is low, increase the contact time to achieve effective disinfection.

Calcium hypochlorite requires special storage to avoid contact with organic material. When organic material and calcium hypochlorite come in contact, the resulting chemical reactions can generate enough heat and oxygen to start and support a fire. When calcium hypochlorite is mixed with water, heat is given off. To adequately disperse the heat generated, the dry calcium hypochlorite should be added to the correct volume of water, rather than adding water to the calcium hypochlorite.

WARNING

Do not use HTH powder in pipes with solvent welded plastic or screwed-joint steel pipe because reaction between the joint compounds and the calcium hypochlorite could cause a fire or an explosion.

Sodium hypochlorite is available in liquid form at five to fifteen percent available chlorine and can be fed by the use of a hypochlorinator. Sodium hypochlorite can lose from two to four percent of its available chlorine content per month at room temperatures. Therefore, manufacturers recommend a maximum storage period of 60 to 90 days.

See Section 5.4, "Operation of Chlorination Equipment," for procedures on how to operate equipment.

Three common methods of disinfecting mains are summarized in Table 5.6, "Chlorination Methods for Disinfecting Water Mains," and two methods are illustrated in Figures 5.6 and 5.7.

1. *NEW MAINS*

All pipes, fittings, valves, and other items which will not be disinfected by the filled line must be precleaned and disinfected. Valve bonnets and other high spots where cleaning may not be effective due to air pockets should be precleaned and chlorinated. Calcium hypochlorite tablets may be crushed and placed in joints and hydrant branches to assist in the total disinfection process.

When applying chlorine by either the continuous or slug method, be sure that the chlorine fed is well mixed with the water used to fill the pipe. Solution feeders can be used to inject chlorine into the water (continuous or slug method) used to disinfect the main. If possible, recycle the flows from the continuous method to minimize the use of chlorine and any problems that might be encountered from disposal of water with a high chlorine residual.

TABLE 5.6 CHLORINATION METHODS FOR DISINFECTING WATER MAINS

Chlorination Method Used	Maximum Chlorine Dose, mg/*L*[a]	Minimum Contact Time, hr	Minimum Chlorine Residual, mg/*L*
Continuous	50	24	25
Slug	500	3	300
Tablet[b]	50	24	25

[a] *NOTE:* AWWA Standard C651-99 recommends the following doses: Continuous, 25 mg/*L*; Slug, 100 mg/*L*; and Tablet, 25 mg/*L*. The minimum chlorine dose depends on whether you are disinfecting an existing main (high dose, 500 mg/*L*, and short contact time (5 minutes)), or a new main (use a continuous minimum residual of 25 mg/*L* for 24 hours). Use whatever dose you need that will produce no positive coliform test results.

[b] Tablets must be placed at inside top of pipe when the pipe is being laid. Also, two tablets must be placed at all joints on both sides of the pipe at the half-full location. Place one ounce (28 gm) of HTH powder per inch (25 mm) of pipe diameter in the first length of pipe and again after each 500 feet (150 m) of pipe. This ensures that the first water entering the spaces at joints will have a high chlorine residual. Fill the pipe with water at velocities of less than one ft/sec.

Care must be exercised when disposing of all water with a high chlorine residual (greater than 1 mg/*L*). Possible means of disposal include sanitary sewers, storm sewers, or on land. If sanitary sewers are used, there should be adequate dilution and travel time so there will be no chlorine residual when the water reaches the wastewater treatment plant. Be sure to notify the plant operator in advance. If a storm sewer is used, be sure there is no chlorine residual remaining when the water reaches the receiving waters (creek, river, or lake). Chlorine is toxic to fish and other aquatic life. Land disposal may be acceptable if percolation rates are high and there are no nearby wells pumping groundwater.

2. *MAIN REPAIRS*

   a. If repairs are made with the line continuously full of water and under pressure, no disinfection is required.

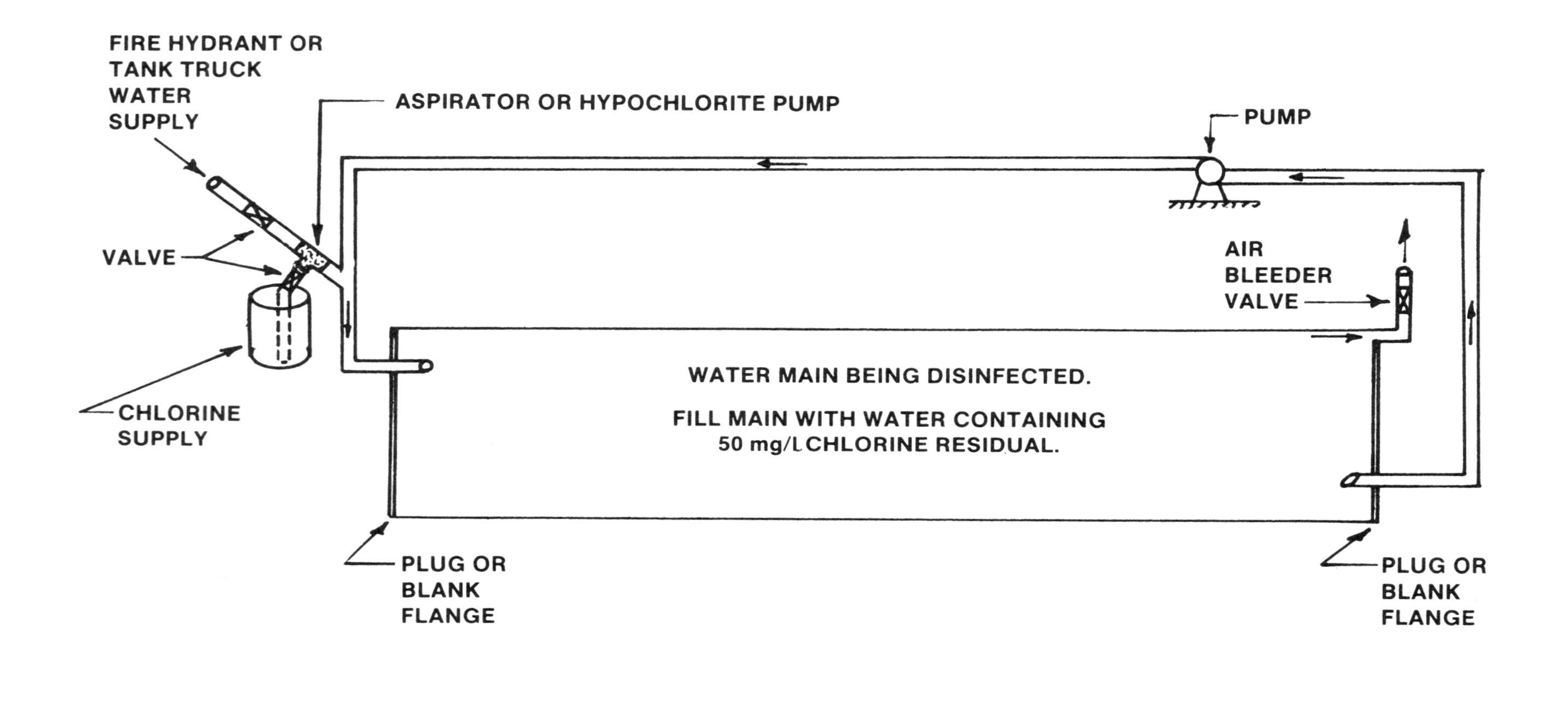
FIRE HYDRANT OR
TANK TRUCK
WATER
SUPPLY
ASPIRATOR OR HYPOCHLORITE PUMP
PUMP
VALVE
AIR
BLEEDER
VALVE
CHLORINE
SUPPLY
WATER MAIN BEING DISINFECTED.
FILL MAIN WITH WATER CONTAINING
50 mg/L CHLORINE RESIDUAL.
PLUG OR
BLANK
FLANGE
PLUG OR
BLANK
FLANGE
CHLORINE SUPPLY
CAN USE HYPOCHLORITE
AND PREMIX SOLUTION
PRIOR TO INJECTION.
FILLING MAIN
WATER AND CHLORINE SOLUTION
SHOULD BE ADDED UNTIL MAIN
IS COMPLETELY FULL. ALLOW AIR
TO ESCAPE THROUGH AIR
BLEEDER VALVE.
EMPTYING MAIN
USE CAUTION WHEN DISPOSING
OF DISINFECTING SOLUTION.
IF DISCHARGING TO SEWER
SYSTEM BE SURE THAT
THE PLANT OPERATOR IS
NOTIFIED. DISCHARGE
WATER TO SEWER VERY
SLOWLY OVER A PERIOD OF
TIME RATHER THAN IN
ONE LARGE SLUG.

Fig. 5.6 Disinfection of a water main by the continuous method

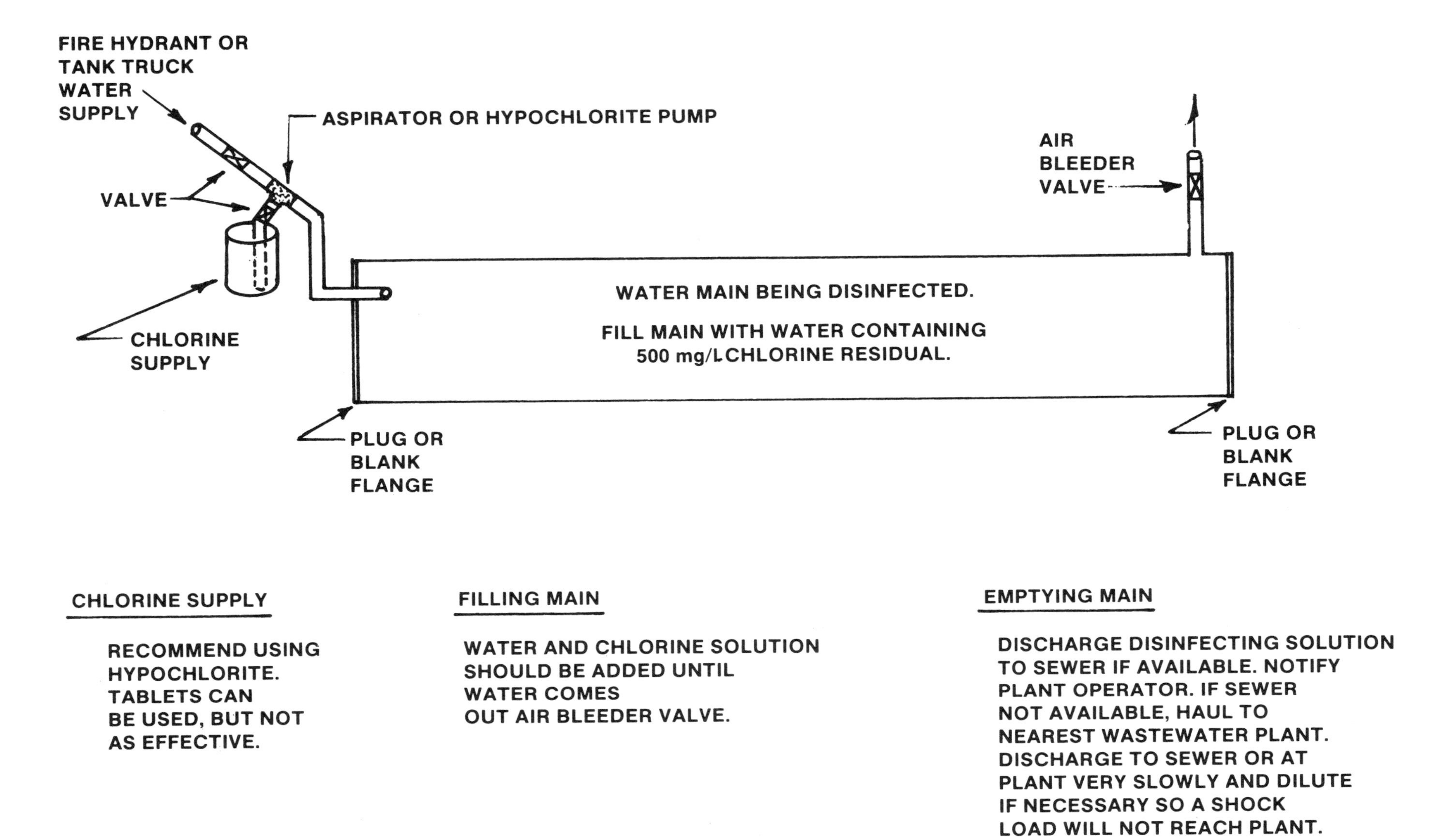
FIRE HYDRANT OR TANK TRUCK WATER SUPPLY
ASPIRATOR OR HYPOCHLORITE PUMP
VALVE
CHLORINE SUPPLY
AIR BLEEDER VALVE
WATER MAIN BEING DISINFECTED.
FILL MAIN WITH WATER CONTAINING 500 mg/L CHLORINE RESIDUAL.
PLUG OR BLANK FLANGE
PLUG OR BLANK FLANGE
CHLORINE SUPPLY
RECOMMEND USING HYPOCHLORITE. TABLETS CAN BE USED, BUT NOT AS EFFECTIVE.
FILLING MAIN
WATER AND CHLORINE SOLUTION SHOULD BE ADDED UNTIL WATER COMES OUT AIR BLEEDER VALVE.
EMPTYING MAIN
DISCHARGE DISINFECTING SOLUTION TO SEWER IF AVAILABLE. NOTIFY PLANT OPERATOR. IF SEWER NOT AVAILABLE, HAUL TO NEAREST WASTEWATER PLANT. DISCHARGE TO SEWER OR AT PLANT VERY SLOWLY AND DILUTE IF NECESSARY SO A SHOCK LOAD WILL NOT REACH PLANT.

Fig. 5.7 Disinfection of a water main by slug method

b. Where lines are opened:

(1) Dewater open trench areas using trash pumps,

(2) Flush and swab all portions of all pipe, fittings, and materials used in repairs, and which will be in contact with the water supply, with a 5 percent hypochlorite solution,

(3) Flush system, and

(4) Disinfect the main. Slug disinfection is recommended, if practical, using a 500 mg/*L* dosage and 30-minute minimum contact time.

D. Flushing After Disinfection

Flush lines after disinfection under all alternative procedures until residual chlorine is less than 1 mg/*L*. Velocity of flushing is not critical if the preventive and preliminary procedures described above were adequately performed.

E. Testing After Disinfection

After disinfection and prior to placing line in service, bacteriological tests (24-hour membrane filter) are required as follows:

1. In a chlorinated water system, test at least one sample for each section disinfected.
2. Test at least two samples for each section disinfected in an unchlorinated water system.
3. For long lines, test samples along line. Line in excess of 2,500 feet (750 meters) is considered a long line.

If bacteriological tests are unsatisfactory, disinfection must be repeated. Repeat of the tablet method, of course, is impossible and an alternative procedure must be used. A suggested procedure is to flush the main again and then take additional samples. Be sure the sample tap is satisfactory because many unsatisfactory samples from new mains are the result of using poor sample taps such as fire hydrants and blow-off valves.

If samples are still unsatisfactory, dewater the main as completely as possible. Use air compressors to blow out all remaining water. Refill the empty main with water containing a chlorine residual between 50 and 100 mg/*L*. Allow this water to stand in the main for 48 hours. Flush the main again and resample. Also collect samples from the water entering the main to be sure that this water is not the source of the positive coliform test results.

5.311 Emergency or Maintenance Disinfection

Mains may be disinfected by spraying a high concentration of chlorine on the insides of the mains. This method is often used for disinfecting sections of a water main where a break is repaired or where a crew has cut into an existing main, or in the very short (less than 100 feet or 30 meters) extension of an existing main. This method is frequently used when the main must be quickly returned to service and you can't wait for a 24-hour contact time.

Spray with a solution of one to five percent (10,000 to 50,000 mg/*L*) chlorine at a pressure of 100 psi (690 kPa or 7 kg/sq cm). An ordinary pressure type stainless-steel fire extinguisher will do the job. Spray the complete interior of each section of pipe or fitting as it is lowered into the trench. The stainless-steel extinguisher will not corrode as rapidly from contact with the chlorine solution if the interior of the extinguisher tank is lined with a protective coating, such as a chlorinated rubber or an epoxy coating.

5.312 Disinfection Specifications

If water mains are to be disinfected by others as part of a contract, the following should be included in the specifications:

1. Specify AWWA C651-99 as the standard for disinfection procedures to be followed,
2. Specify where flushing may be done, rates of flushing, and location of suitable drainage facilities,
3. Form of chlorine to be used and method of application,
4. The type, number, and frequency of samples for bacteriological tests,
5. The method of taking samples, and
6. Who is responsible for testing and use of a certified laboratory.

5.313 Areas of Disinfection Problems

When disinfecting mains, there are certain aspects which are most likely to cause problems. Joints and connections must be thoroughly cleaned before the mains are disinfected. Whenever anyone connects a service line into a main, they must use clean tools and materials. Each service line should be disinfected before being connected to a water main.

Some water supply systems may be lax with their disinfection practices. Personnel working for these systems are sometimes inadequately trained, underpaid, and understaffed. They use the excuse that they have neither the time nor the budget to adequately disinfect new mains and existing mains after repairs.

THERE IS NO EXCUSE OR REASON FOR EVER PLACING A LINE IN SERVICE THAT HAS NOT BEEN ADEQUATELY DISINFECTED AND CONFIRMED BY THE 24-HOUR MEMBRANE FILTER TEST FOR COLIFORMS.

5.314 Sample Problem

FORMULAS

Use the same formulas that are used to disinfect a well.

EXAMPLE 2

A new 6-inch water main 500 feet long needs to be disinfected. An initial chlorine dose of 400 mg/*L* is expected to maintain a chlorine residual of over 300 mg/*L* during the three-hour disinfection period. How many gallons of 5 percent sodium hypochlorite solution will be needed?

Known		Unknown
Diameter of Pipe, in	= 6 in	5% Hypochlorite, gallons
Length of Pipe, ft	= 500 ft	
Chlorine Dose, mg/*L*	= 400 mg/*L*	
Hypochlorite, %	= 5%	

1. Calculate the volume of water in the pipe in gallons.

$$\text{Pipe Volume, gallons} = \frac{(0.785)(\text{Diameter, in})^2(\text{Length, ft})(7.48 \text{ gal/cu ft})}{144 \text{ sq in/sq ft}}$$

$$= \frac{(0.785)(6 \text{ in})^2(500 \text{ ft})(7.48 \text{ gal/cu ft})}{144 \text{ sq in/sq ft}}$$

$$= 734 \text{ gallons of water}$$

2. Determine the pounds of chlorine needed.

$$\text{Chlorine, lbs} = (\text{Volume, M Gal})(\text{Dose, mg}/L)(8.34 \text{ lbs/gal})$$

$$= (0.000734 \text{ M Gal})(400 \text{ mg}/L)(8.34 \text{ lbs/gal})$$

$$= 2.45 \text{ lbs chlorine}$$

3. Calculate the gallons of 5 percent sodium hypochlorite solution needed.

$$\text{Sodium Hypochlorite Solution, gallons} = \frac{(\text{Chlorine, lbs})(100\%)}{(8.34 \text{ lbs/gal})(\text{Hypochlorite, \%})}$$

$$= \frac{(2.45 \text{ lbs})(100\%)}{(8.34 \text{ lbs/gal})(5\%)}$$

$$= 5.9 \text{ gallons}$$

Six gallons of 5 percent sodium hypochlorite solution should do the job.

5.32 Tanks

5.320 Procedures

Procedures for disinfecting tanks are similar to those used for mains. Thoroughly clean the tank after construction, maintenance, or repairs. Add chlorine to the water used to fill the tank during the disinfection process and mix thoroughly. Maintain a chlorine residual of at least 50 mg/*L* for at least six hours and preferably for 24 hours. When the disinfection procedure is completed, carefully dispose of the disinfection water using the same procedures as for disposal of water used to disinfect mains.

When disinfecting large tanks which hold more than one million gallons, it may not be practical or economical to fill and drain the tank with a disinfecting solution with a high chlorine residual. A solution to this problem is to fill the tank, increase the chlorine residual slightly, collect samples, and run bacteriological tests while keeping the tank full of water (see SPECIAL NOTE, Section 5.34). If results from the 24-hour membrane filter tests are acceptable, the tank may be placed in service. If necessary, dilute the tank contents as the water flows into the distribution system.

Another approach to disinfecting large tanks is to spray the walls with a jet of water containing a chlorine residual of 200 mg/*L*. Thoroughly spraying the walls with water with a high chlorine residual is an effective means of disinfecting large tanks. After the tank has been sprayed, allow the tank to stand unused for 30 minutes before filling. Fill the tank with distribution system water that has been treated with chlorine to provide a residual of 3 mg/*L*. Let the water in the tank stand for 3 to 6 hours. Operators doing the spraying should wear a self-contained breathing apparatus because the chlorine fumes from the spray water are very unpleasant and could be hazardous to your health. Be sure to provide plenty of ventilation.

See Section 5.82, "Disinfection of Storage Facilities," in *WATER DISTRIBUTION SYSTEM OPERATION AND MAINTENANCE*, for a more detailed discussion of three different methods of disinfecting water storage tanks. Typically, only one method is used for a particular storage facility disinfection, but combinations of the three methods' chlorine concentrations and contact times may be used.

5.321 Sample Problem

FORMULAS

Use the same formulas used to disinfect wells.

EXAMPLE 3

A new service storage reservoir needs to be disinfected before being placed in service. The tank is 8 feet high and 20 feet in diameter. An initial chlorine dose of 100 mg/*L* is expected to maintain a chlorine residual of over 50 mg/*L* during the 24-hour disinfection period. How many gallons of five percent sodium hypochlorite solution will be needed?

Known		**Unknown**
Diameter of Tank, ft	= 20 ft	5% Hypochlorite, gallons
Height of Tank, ft	= 8 ft	
Chlorine Dose, mg/*L*	= 100 mg/*L*	
Hypochlorite, %	= 5%	

1. Calculate the volume of water in the tank in gallons.

$$\text{Tank Volume, gallons} = (0.785)(\text{Diameter, ft})^2(\text{Height, ft})(7.48 \text{ gal/cu ft})$$

$$= (0.785)(20 \text{ ft})^2(8 \text{ ft})(7.48 \text{ gal/cu ft})$$

$$= 18{,}800 \text{ gallons}$$

2. Determine the pounds of chlorine needed.

$$\text{Chlorine, lbs} = (\text{Vol Water, M Gal})(\text{Chlorine Dose, mg}/L)(8.34 \text{ lbs/gal})$$

$$= (0.0188 \text{ M Gal})(100 \text{ mg}/L)(8.34 \text{ lbs/gal})$$

$$= 15.68 \text{ lbs chlorine}$$

3. Calculate the gallons of 5 percent sodium hypochlorite solution needed.

$$\text{Sodium Hypochlorite Solution, gallons} = \frac{(\text{Chlorine, lbs})(100\%)}{(8.34 \text{ lbs/gal})(\text{Hypochlorite, \%})}$$

$$= \frac{(15.68 \text{ lbs})(100\%)}{(8.34 \text{ lbs/gal})(5\%)}$$

$$= 37.6 \text{ gallons}$$

Thirty-eight gallons of five percent sodium hypochlorite solution should do the job.

5.33 Water Treatment Plants

For additional information on disinfection at water treatment plants, see 4, "Small Water Treatment Plants," Section 4.5, "Disinfection," in this manual and Chapter 7, "Disinfection," in *WATER TREATMENT PLANT OPERATION*, Volume I, in this series of manuals.

5.34 Sampling

SPECIAL NOTE

Whenever you collect a sample for a bacteriological test (coliforms), be sure to use a sterile plastic or glass bottle. If the sample contains any chlorine residual, sufficient sodium thiosulfate should be added to neutralize all of the chlorine residual. Usually 0.1 m*L*, 10 percent sodium thiosulfate per 4-ounce bottle before sterilization is sufficient, unless you are disinfecting mains or storage tanks. If the chlorine residual in the sample is greater than 15 mg/*L*, more "thio" is required to neutralize the chlorine.

QUESTIONS

Write your answers in a notebook and then compare your answers with those on page 331.

5.3E How can water mains be kept clean during construction or repair?

5.3F Before a water main is disinfected, what flushing velocity should be used and for how long?

5.3G What areas in a water main require extra effort for successful disinfection?

5.3H List three forms of chlorine used for disinfection.

5.3I What precautions should be taken when disposing of water with a high chlorine residual?

End of Lesson 1 of 2 Lessons on DISINFECTION

Please answer the discussion and review questions next.

DISCUSSION AND REVIEW QUESTIONS

Chapter 5. DISINFECTION

(Lesson 1 of 2 Lessons)

At the end of each lesson in this chapter you will find some discussion and review questions. The purpose of these questions is to indicate to you how well you understand the material in the lesson. Write the answers to these questions in your notebook before continuing.

1. Why is drinking water disinfected?
2. If a water is disinfected, will it be safe to drink?
3. What is the chlorine demand of a water?
4. How would you determine the chlorine dose for water?
5. Why should a chlorine residual be maintained in a water distribution system?
6. Why would you consider recycling water used to disinfect a water main when using the continuous chlorination method?
7. How would you dispose of the water used to disinfect a water main if it has a high chlorine residual (greater than 1 mg/*L*)?

CHAPTER 5. DISINFECTION

(Lesson 2 of 2 Lessons)

5.4 OPERATION OF CHLORINATION EQUIPMENT

5.40 Description of Various Units

5.400 Field Equipment

1. *HYPOCHLORINATORS* (equipment that feeds liquid chlorine (bleach) solutions)

Hypochlorinators used on small water systems are very simple and relatively easy to install. Typical installations are shown in Figures 5.8 and 5.9. Hypochlorinator systems usually consist of a chemical solution tank for the hypochlorite, diaphragm-type pump (Figure 5.10), power supply, water pump, pressure switch, and water storage tank.

There are two methods of feeding the hypochlorite solution into the water being disinfected. The hypochlorite solution may be pumped directly into the water (Figure 5.11). In the other method, the hypochlorite solution is pumped through an *EJECTOR*[23] (also called an eductor or injector) which draws in additional water for dilution of the hypochlorite solution (Figure 5.12).

2. *CHLORINATORS* (equipment that feeds gaseous chlorine)

Disinfection by means of gaseous chlorine is typically accomplished in small systems with the equipment shown in Figure 5.13. For small water treatment systems, small chlorinators which are mounted directly on a chlorine container (as shown in Figures 5.13, 5.14, and 5.15) have proven to be safer, easier to operate and maintain, and less expensive to install than larger in-place chlorination systems, yet they provide the same reliable service.

A direct-mounted chlorinator meters prescribed (preset, or selected) doses of chlorine gas from a chlorine cylinder, conveys it under a vacuum, and injects it into the water supply. Direct cylinder mounting is the safest and simplest way to connect the chlorinator to the chlorine cylinder. The valves on the cylinder and chlorinator inlet are connected by a positive metallic yoke, which is sealed by a single lead or fiber gasket.

CHLORINATOR PARTS AND THEIR PURPOSE

THE EJECTOR: The ejector, fitted with a Venturi nozzle, creates the vacuum that moves the chlorine gas. Water supplied by a pump moves across the Venturi nozzle creating a differential pressure which establishes the vacuum. The gas chlorinator is able to transport the chlorine gas to the water supply by reducing the gas pressure from the chlorine cylinder to less than the atmospheric pressure (vacuum). Figure 5.13 illustrates such an arrangement. The flow diagrams in Figure 5.14 are cutaway views of the ejector and check valve assembly.

In the past it was not uncommon to find the ejector and the vacuum regulator mounted inside some type of cabinet. However, it makes better sense to locate the ejector at the site where the chlorine is to be applied, eliminating the necessity of pumping the chlorine over long distances and the associated problems inherent with gas pressure lines. Also, by placing the ejector at the application point, any tubing break will cause the chlorinator to shut down. This halting of operation stops the flow of gas and any damage that could result from a chlorine solution leak.

CHECK VALVE ASSEMBLY: The vacuum created by the ejector moves through the check valve assembly. This assembly prevents water from back-feeding, that is, entering the vacuum-regulator portion of the chlorinator (Figure 5.14).

RATE VALVE: The rate valve controls the flow rate at which chlorine gas enters the chlorinator. The rate valve controls the vacuum level and thus directly affects the action of the diaphragm assembly in the vacuum regulator. A reduction in vacuum lets the diaphragm close, causing the needle valve to reduce the inlet opening which restricts chlorine gas flow to the chlorinator. An increase in the rate valve setting applies more vacuum to the diaphragm assembly, pulling the needle valve back away from the inlet opening and permitting an increased chlorine gas flow rate.

DIAPHRAGM ASSEMBLY: This assembly connects directly to the inlet valve of the vacuum regulator, as described above. A vacuum (of at least 20 inches (508 mm) of water column) exists on one side of the diaphragm; the other side is open to atmospheric pressure through the vent. This differential in pressure causes the diaphragm to open the chlorine inlet valve allowing the gas to move (under vacuum) through the *ROTAMETER*,[24] past the rate valve and through the tubing to the check valve assembly, into the ejector nozzle area, and then to the point of application. If for some reason the vacuum is lost, the diaphragm will seat the needle valve on the inlet, stopping chlorine gas flow to the chlorinator.

INTERCONNECTION MANIFOLD: If several gas cylinders provide the chlorine gas, direct cylinder mounting is not possible. An interconnection manifold made of seamless steel pipe and flexible connectors of cadmium-plated copper fitted with isolation valves must be used as the bridge between the chlorinator and the various cylinders.

[23] *Ejector. A device used to disperse a chemical solution into water being treated.*

[24] *Rotameter (RODE-uh-ME-ter). A device used to measure the flow rate of gases and liquids. The gas or liquid being measured flows vertically up a tapered, calibrated tube. Inside the tube is a small ball or bullet-shaped float (it may rotate) that rises or falls depending on the flow rate. The flow rate may be read on a scale behind or on the tube by looking at the middle of the ball or at the widest part or top of the float.*

TYPICAL INSTALLATION

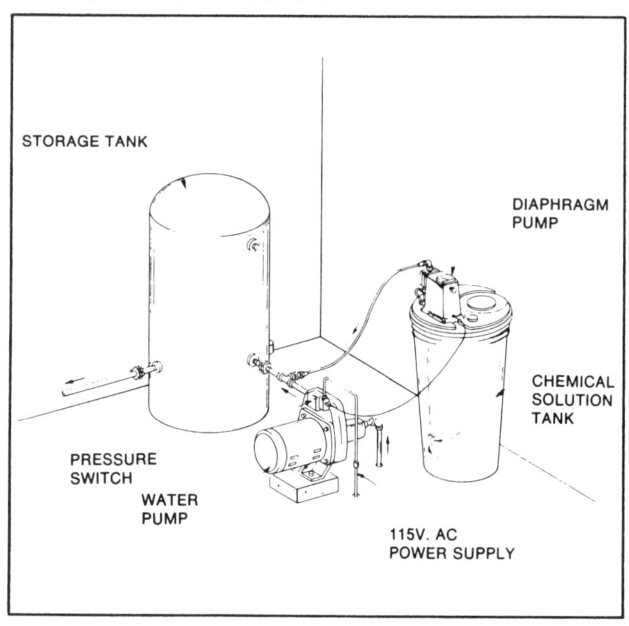

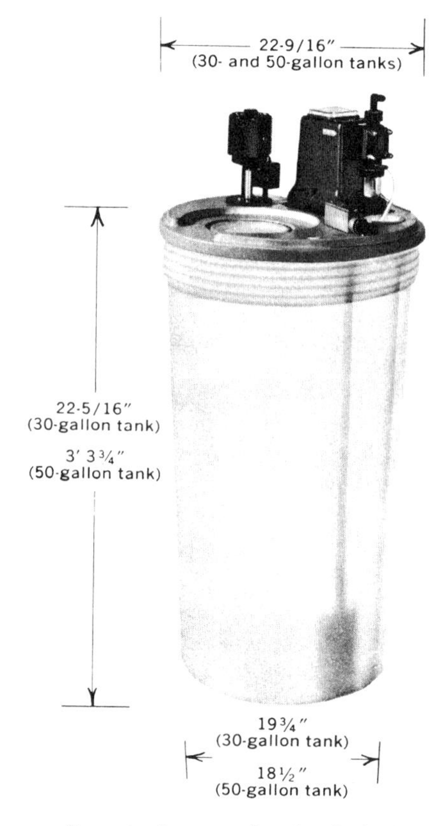

Pump-tank system for chemical mixing and metering. Cover supports pump, impeller-type mixer, and liquid-level switch.

Fig. 5.8 Typical hypochlorinator installation
(Permission of Wallace & Tiernan Division, Pennwalt Corporation)

The steel gas manifold with chlorine valve is mounted to the chlorinator. The flexible connector links the rigid manifold and the chlorine cylinder. The isolation valve between the flexible connector and the cylinder valve provides a way to close off the flexible connector when a new gas cylinder must be attached. This limits the amount of moisture that enters the system. Moisture in the system will combine with the chlorine gas to produce hydrochloric acid and cause corrosion. *CORROSION CAN CAUSE THE MANIFOLD TO FAIL.*

The chlorine is usually injected directly into the water supply pipe and there may not be contact chambers or mixing units. The location of the injection point is important. The injection should never be on the intake side of the pump as it will cause corrosion problems. There should be a check valve and a meter to monitor the chlorine dose.

On most well applications a chlorine booster pump is needed to overcome the higher water pump discharge pressures. The low volume-high pressure booster pump shown in Figure 5.13 must have extremely small clearance between the impeller and the casing. If the well produces sand in the water, this pump will wear rapidly and become unreliable. In this situation, the chlorine solution should be introduced down the well through a polyethylene tube.

The polyethylene tube (½ inch or 12 mm) must be installed in the well so as to discharge a few inches below the suction screen. The chlorinator should operate *ONLY* when the pump is running. The chlorine solution flowing through the polyethylene tube is extremely corrosive. If the tube does not discharge into flowing water, the effect of the solution touching the metal surface can be disastrous. Wells have been destroyed by corrosion from chlorine.

5.401 Chlorine Containers

1. *HYPOCHLORINATORS*

Plastic containers are sufficient (Figures 5.8 and 5.9). Size depends on usage. Plastic containers should be large enough to hold a two or three days' supply of hypochlorite solution. The solution should be prepared every two or three days. If a larger amount of solution is mixed, the solution may lose its strength and thus affect the chlorine feed rate. Normally a week's supply of hypochlorite should be in storage and available for preparing hypochlorite solutions. Store the hypochlorite in a cool, dark place. Sodium hypochlorite can lose from two to four percent of its available chlorine content per month at room temperature. Therefore, manufacturers recommend a maximum shelf life of 60 to 90 days.

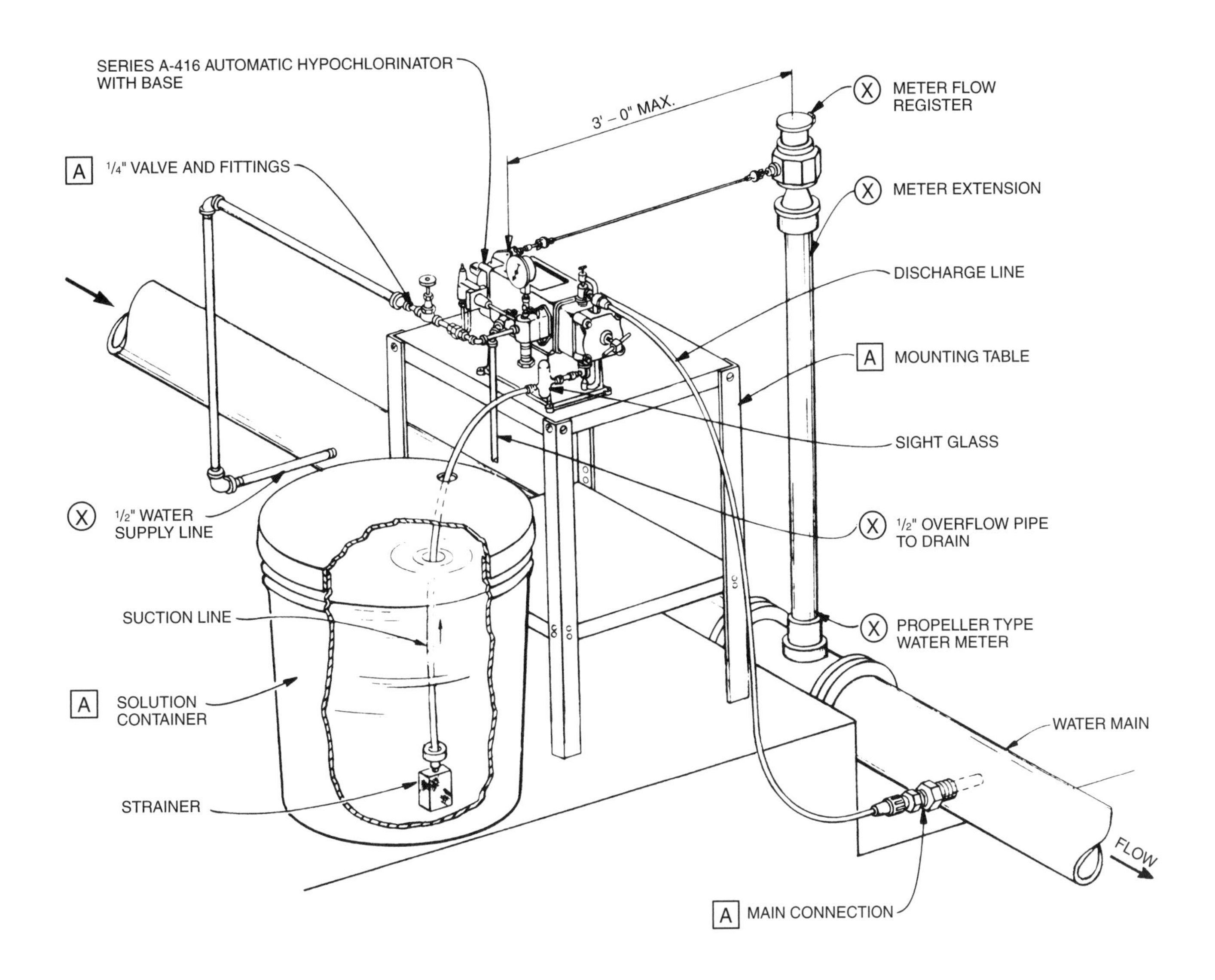

(X) NOT FURNISHED BY W & T.

[A] ACCESSORY ITEM FURNISHED ONLY IF SPECIFICALLY LISTED IN QUOTATION AND AS CHECKED ON THIS DRAWING.

NOTE: Hypochlorinator paced by a propeller-type water meter.

Fig. 5.9 Typical hypochlorinator installation

(Permission of Wallace & Tiernan Division, Pennwalt Corporation)

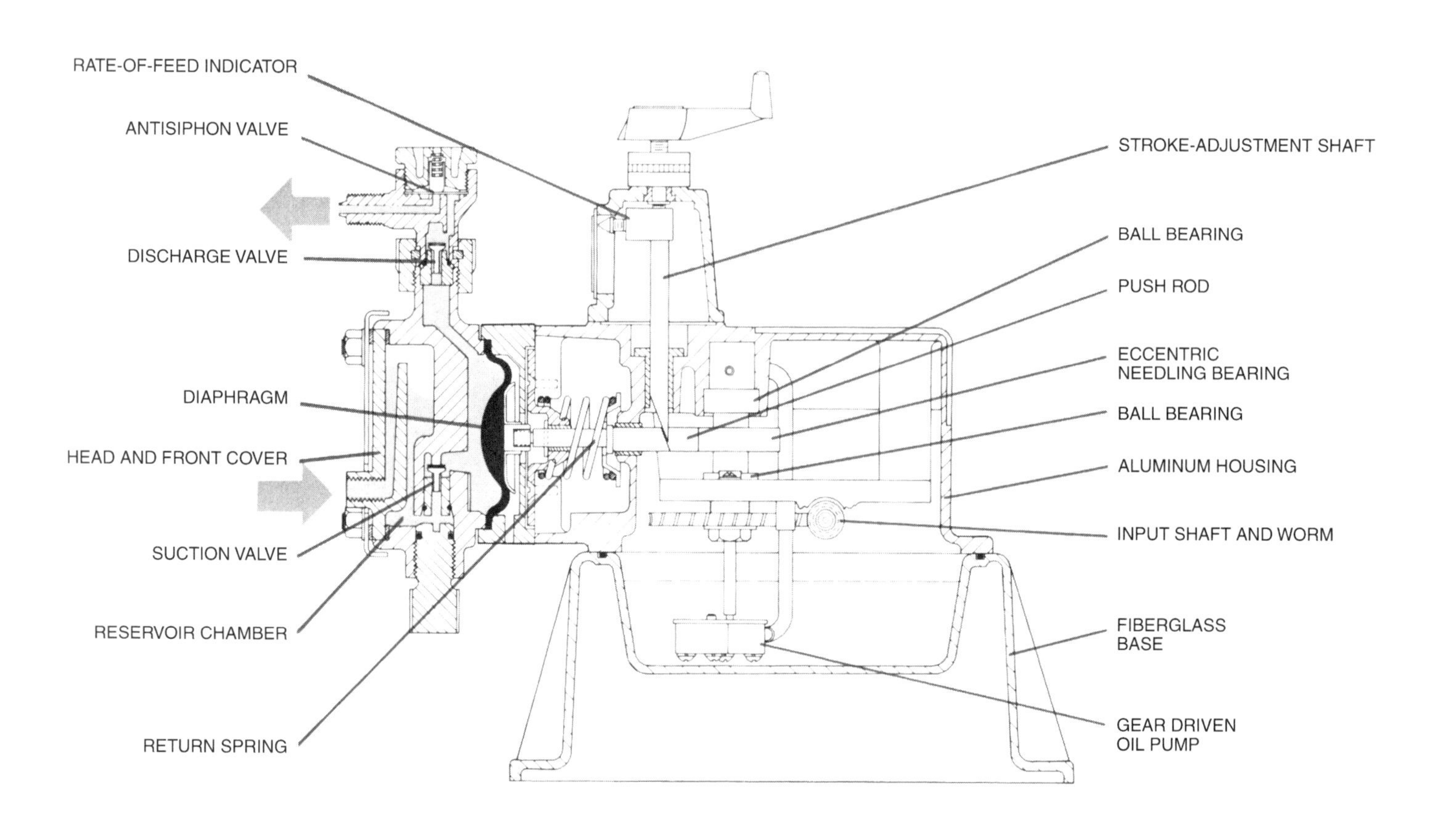

Fig. 5.10 Diaphragm-type pump
(Permission of Wallace & Tiernan Division, Pennwalt Corporation)

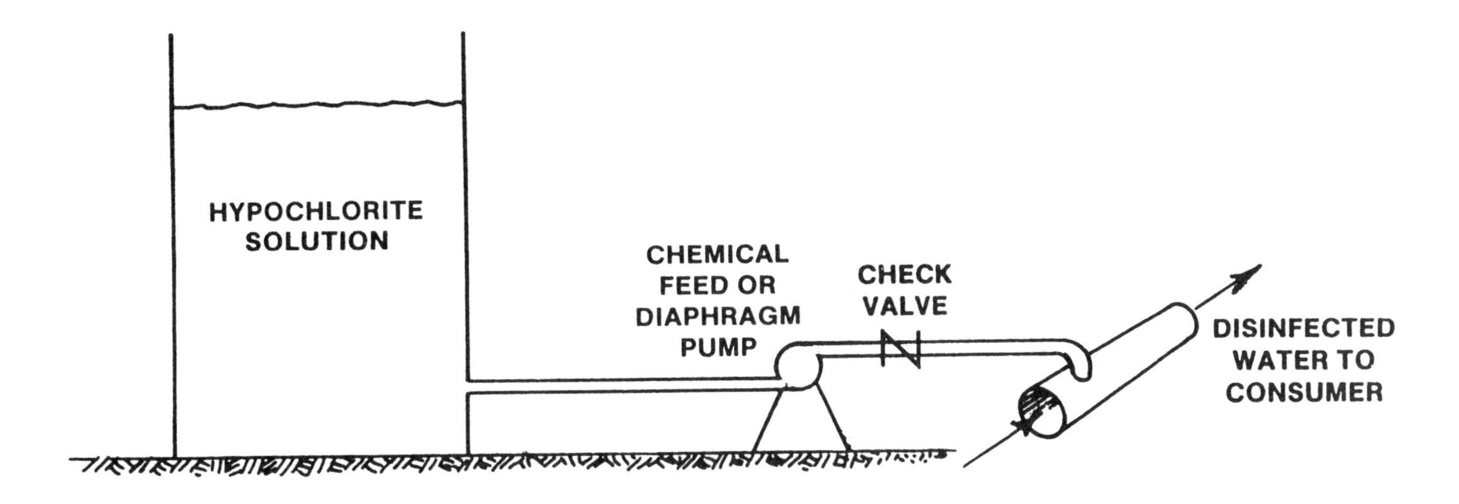

Fig. 5.11 Hypochlorinator direct pumping system

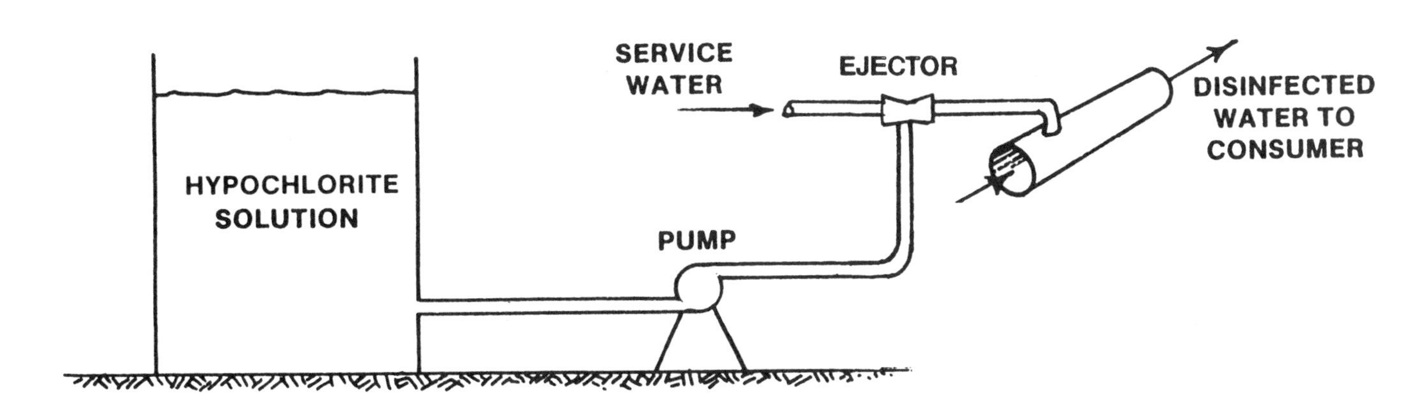

NOTE: Pump is chemical feed or diaphragm pump.

Fig. 5.12 Hypochlorinator injector feed system

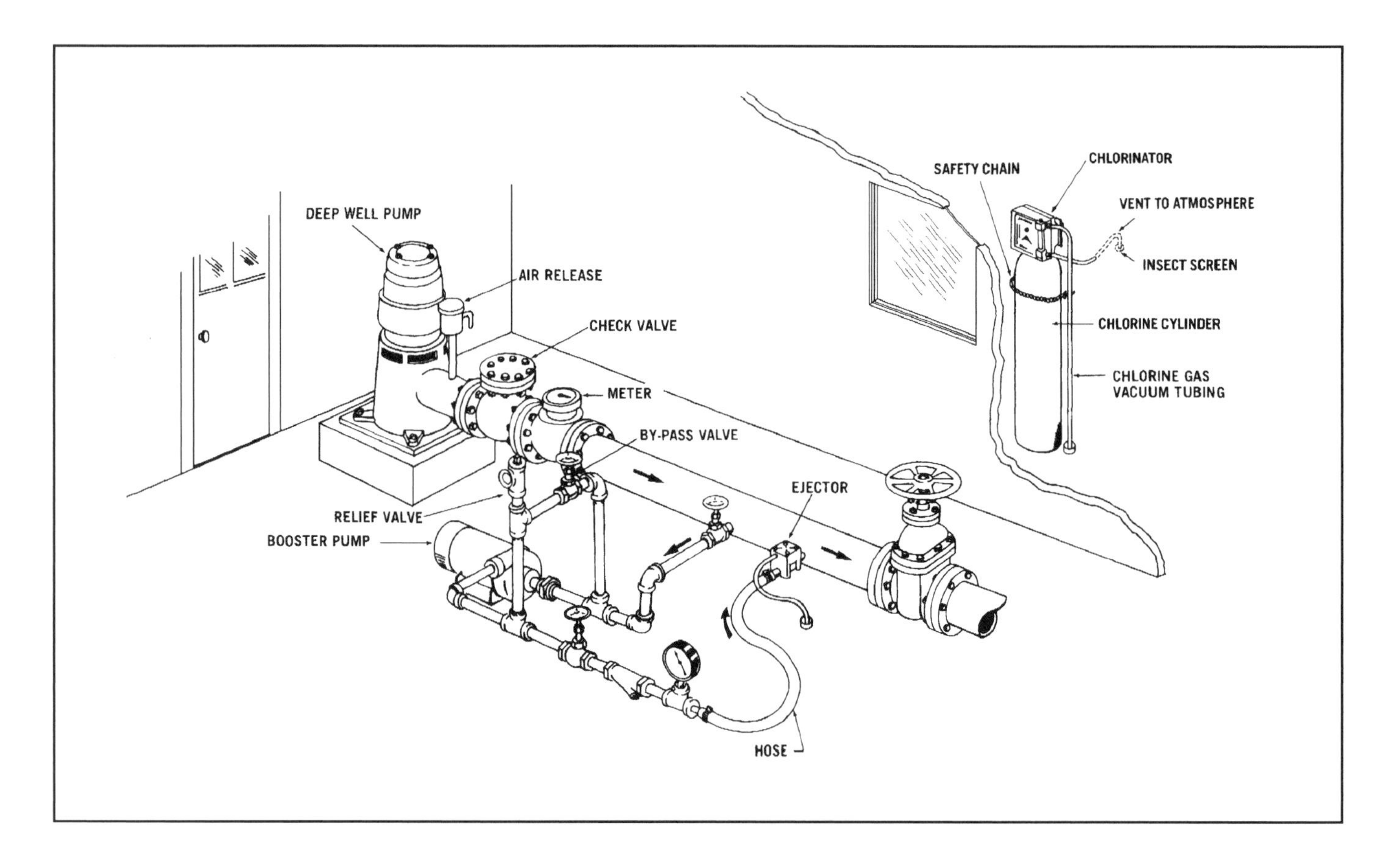

Fig. 5.13 Typical deep well chlorination system

(Permission of Capital Controls Company, Colmar, PA)

2. *GAS CHLORINATORS*

Chlorine is delivered for use by chlorinators in 100- and 150-pound (45- to 68-kg) cylinders (Figures 5.16 and 5.17), one-ton (900-kg) tanks or chlorine tank cars in sizes from 16 to 90 tons (14,500 to 81,800 kg). The 100- and 150-pound (45- to 68-kg) cylinders will be discussed in this section.

These cylinders are usually made of seamless carbon steel. A fusible plug is placed in the valve below the valve seat (Figure 5.18). This plug is a safety device. The fusible metal softens or melts at 158 to 165°F (70 to 74°C) to prevent buildup of excessive pressures and the possibility of rupture due to fire or high surrounding temperatures.

The maximum rate of chlorine removal from a 150-pound (18-kg) cylinder is 40 pounds (18 kg) of chlorine per day. If the rate of removal is greater, "freezing" can occur and less chlorine will be delivered.

WARNING

When frost appears on valves and flex connectors, the chlorine gas may condense to liquid (reliquify). The liquid chlorine may plug the chlorine supply lines (sometimes this is referred to as chlorine ice or frozen chlorine). If you disconnect the chlorine supply line to unplug it, *BE VERY CAREFUL.* The liquid chlorine in the line could reevaporate, expand as a gas, build up pressure in the line, and cause liquid chlorine to come shooting out the open end of a disconnected chlorine supply line.

5.41 Chlorine Handling

Cylinders containing 100 to 150 pounds (45 to 68 kg) of chlorine are convenient for very small plants with capacities less than 0.5 MGD (1,890 cu m/day or 1.89 ML/day).[25]

[25] *ML/day. Megaliters or million liters per day.*

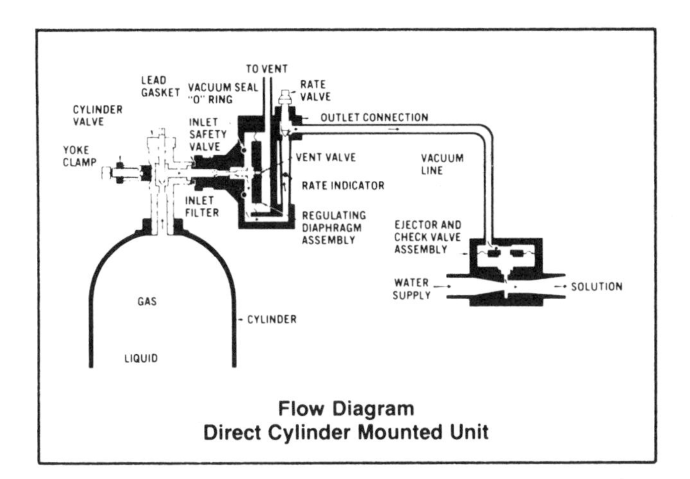

Flow Diagram
Direct Cylinder Mounted Unit

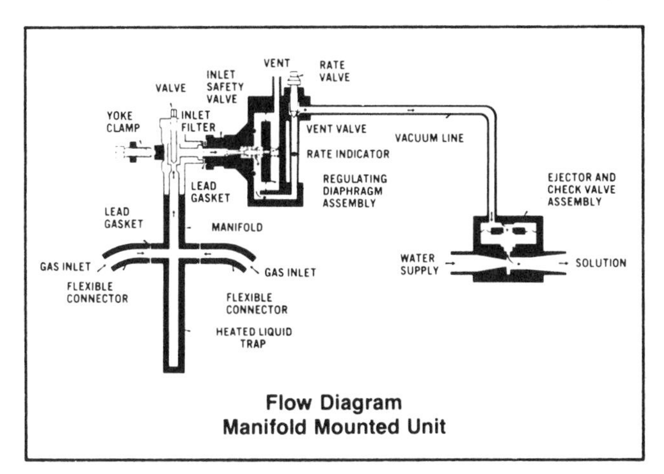

Flow Diagram
Manifold Mounted Unit

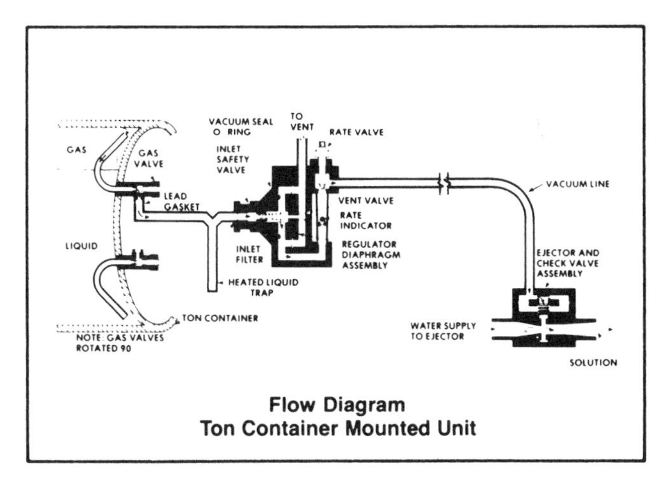

Flow Diagram
Ton Container Mounted Unit

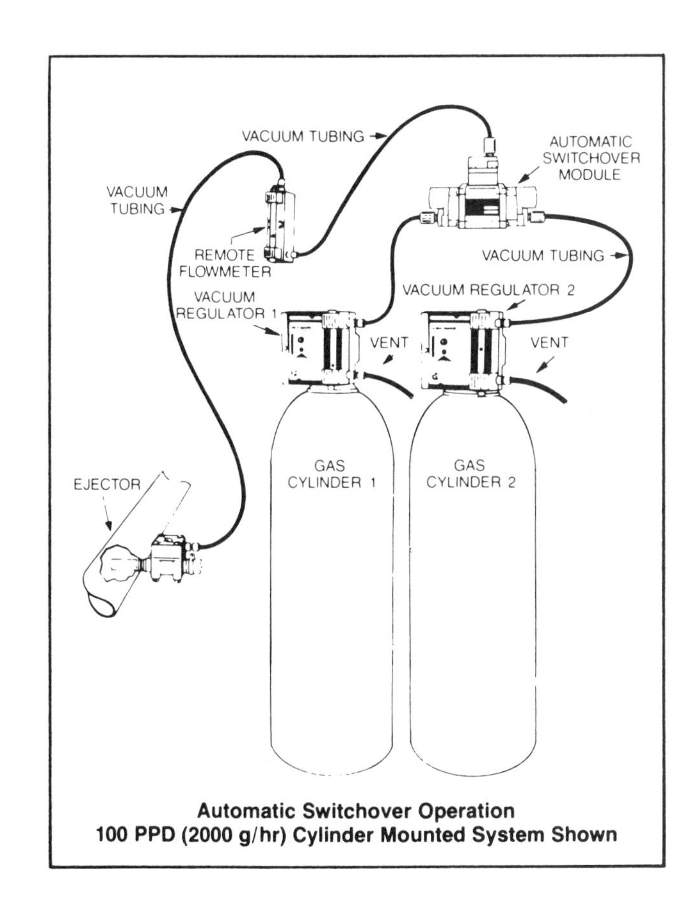

Automatic Switchover Operation
100 PPD (2000 g/hr) Cylinder Mounted System Shown

Fig. 5.14 Typical chlorinator flow diagrams

(Permission of Capital Controls Company, Colmar, PA)

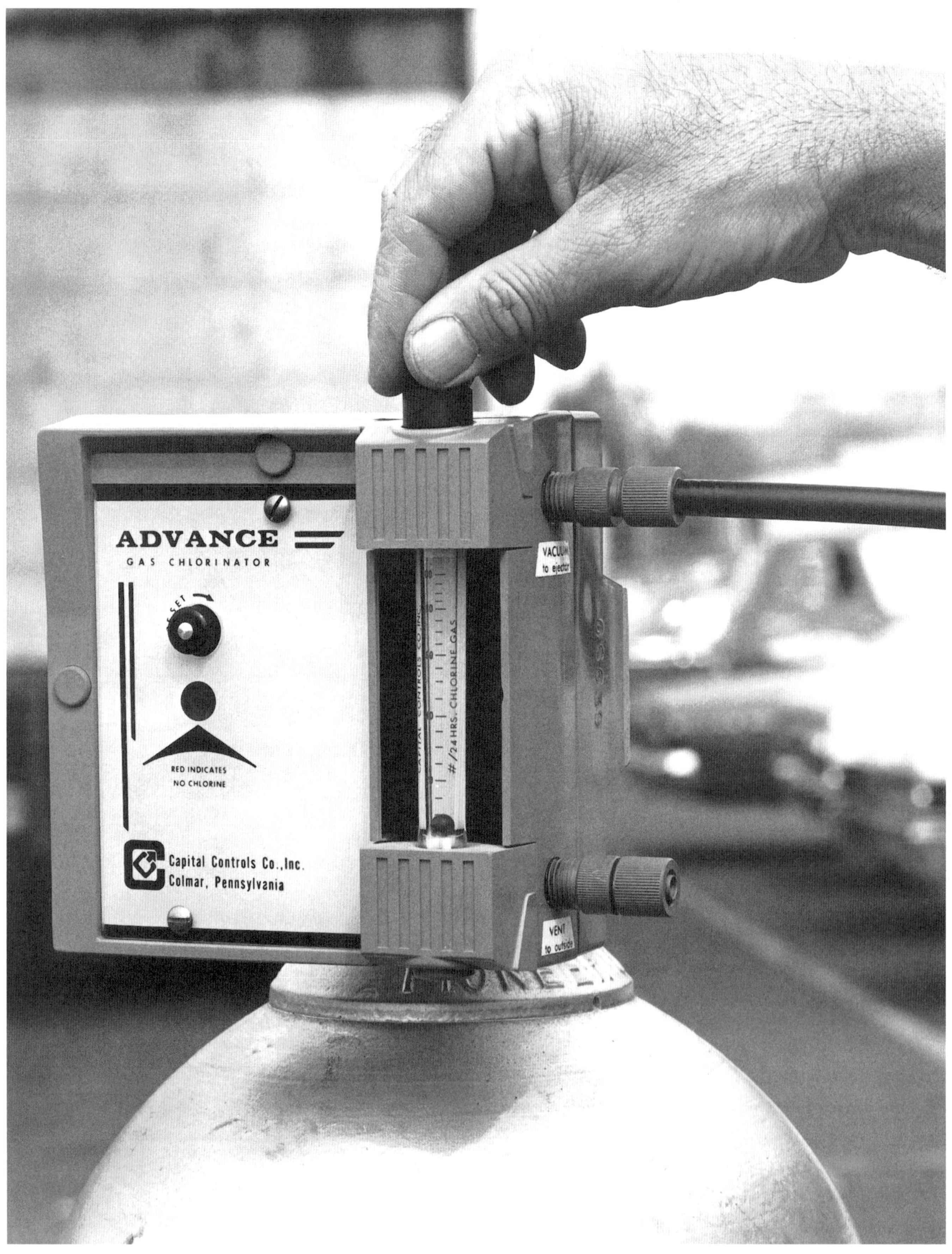

Fig. 5.15 Chlorinator with rotameter showing feed rate in #/24 hr (lbs/day) chlorine gas

(Permission of Capital Controls Company, Colmar, PA)

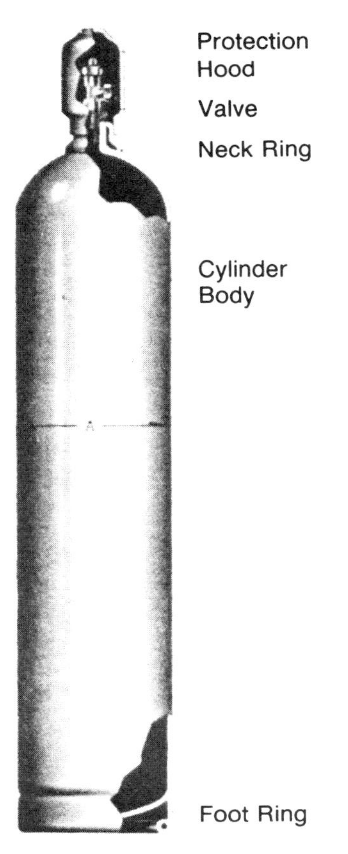

Net Cylinder Contents	Approx. Tare, Lbs.*	Dimensions, Inches	
		A	B
100 Lbs.	73	8 1/4	54 1/2
150 Lbs.	92	10 1/4	54 1/2

*Stamped tare weight on cylinder shoulder does not include valve protection hood.

Fig. 5.16 Chlorine cylinder
(Courtesy of PPG Industries)

NOTES:
1. Scale for weighing chlorine cylinders and chlorine.
2 Flexible tubing (pigtail).
3. Cylinders chained to wall.

Fig. 5.17 Typical chlorine cylinder station for water treatment
(Courtesy of PPG Industries)

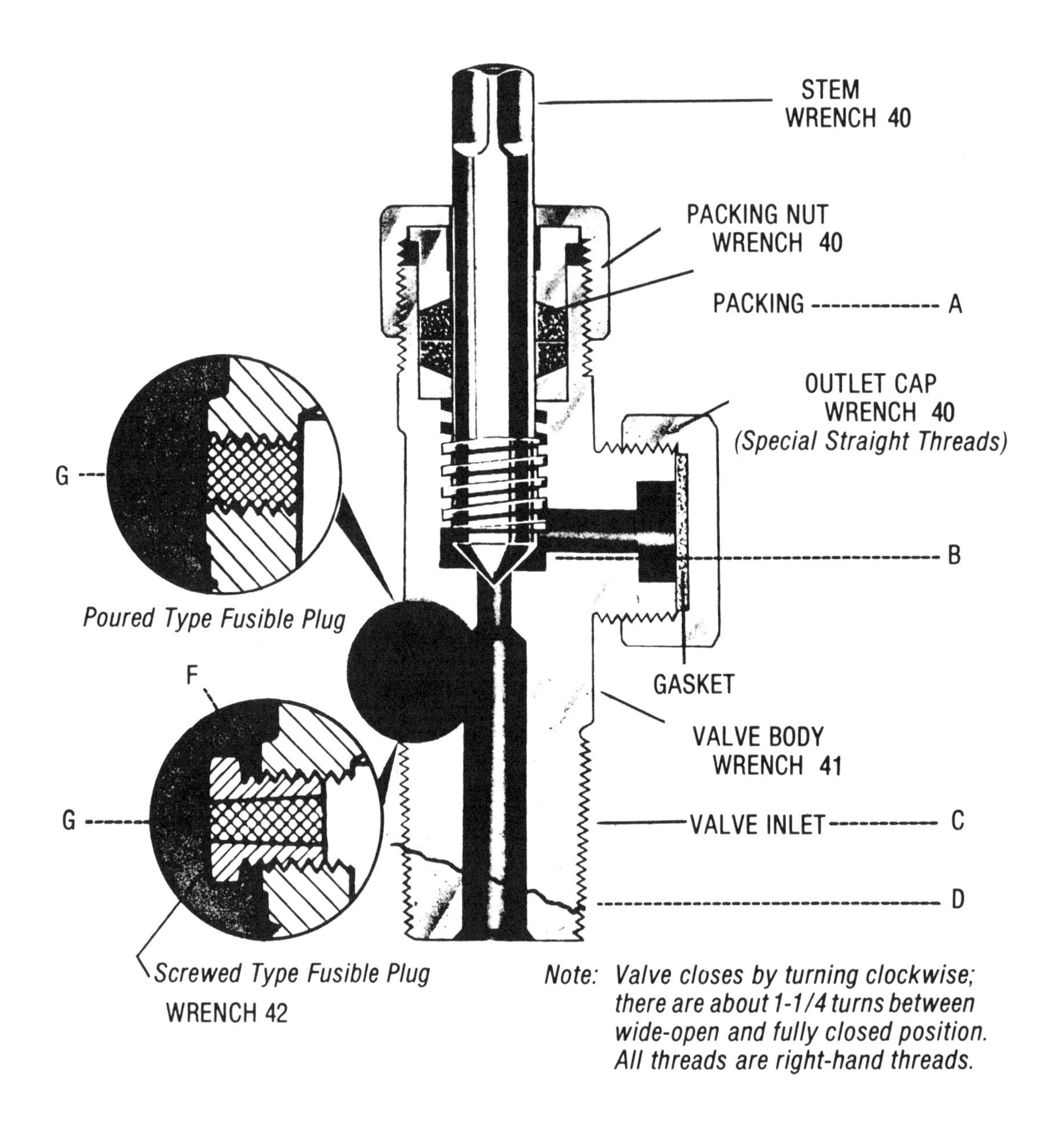

TYPICAL VALVE LEAKS OCCUR THROUGH . . .

A - VALVE PACKING GLAND

B - VALVE SEAT

C - VALVE INLET THREADS

D - BROKEN OFF VALVE

E - VALVE BLOWN OUT

F - FUSIBLE PLUG THREADS

G - FUSIBLE METAL OF PLUG

H - VALVE STEM BLOWN OUT

Fig. 5.18 Standard chlorine cylinder valve

(Permission of Chlorine Specialties, Inc.)

The following are procedures for safely handling chlorine cylinders.

1. Move cylinders with a properly balanced hand truck with clamp supports that fasten at least two-thirds of the way up the cylinder (Figure 5.19).

Fig. 5.19 Hand truck for chlorine cylinder
(Courtesy of PPG Industries)

2. 100- and 150-pound (45- to 68-kg) cylinders can be rolled in a vertical position. Avoid lifting these cylinders except with approved equipment. Use a lifting clamp, cradle, or carrier. Never lift with chains, rope slings, or magnetic hoists. Never roll, push, or drop cylinders off the back of trucks or loading docks.
3. Always replace the protective cap when moving a cylinder.
4. Keep cylinders away from direct heat (steam pipes or radiators) and direct sun, especially in warm climates.
5. Transport and store cylinders in an upright position.
6. Firmly secure cylinders to an immovable object.
7. Store empty cylinders separately from full cylinders. All empty chlorine cylinders must be tagged as empty. (*NOTE:* Never store chlorine cylinders near turpentine, ether, anhydrous ammonia, finely divided metals, hydrocarbons, or other materials that are flammable in air or will react violently with chlorine.)
8. Remove the outlet cap from the cylinder and inspect the threads on the outlet. Cylinders having outlet threads that are corroded, worn, cross-threaded, broken, or missing should be rejected and returned to the supplier.
9. The specifications and regulations of the U.S. Interstate Commerce Commission require that chlorine cylinders be tested at 800 psi (5,516 kPa or 56.24 kg/sq cm) every five years. The date of testing is stamped on the dome of the cylinder. Cylinders which have not been tested within that period of time should be rejected and returned to the supplier.

QUESTIONS

Write your answers in a notebook and then compare your answers with those on page 331.

5.4A List the two major types of chlorine feeders.

5.4B Why are chlorine booster pumps needed on most well applications?

5.4C What is the maximum rate of chlorine removal from a 150-pound cylinder?

5.42 Performance of Chlorination Units

5.420 Hypochlorinators

1. *HYPOCHLORINATOR START-UP*
   a. Solution. Chemical solutions have to be made up. Most agencies buy commercial or industrial hypochlorite at around 12 to 15 percent chlorine. This solution is usually diluted down to a two percent solution. If using commercially prepared solutions, dosage rates will have to be calculated.
   b. Electrical. Lock out the circuit while making an inspection of an electrical circuit. Normally no adjustments are needed. Look for frayed wires. Turn the power back on. Leave the solution switch off.
   c. Turn the chemical pump on. Make any necessary adjustments while the pump is running. Never adjust the pump while it is off because damage to the pump will occur.
   d. Make sure solution is being fed into the system. Measure the chlorine residual just downstream from where solution is being fed into the system. You may have a target residual you wish to maintain at the beginning of the system, such as 2.0 mg/*L*.
   e. Check the chlorine residual in the system. The residual should be measured at the most remote test location within the distribution system and should be at least 0.2 mg/*L* free residual chlorine. This chlorine residual is necessary to protect the treated water from any recontamination. Adjust the chemical feed as needed.
2. *HYPOCHLORINATOR SHUTDOWN*
   a. Short Duration
      (1) Turn the water supply pump off. You do not want to pump any unchlorinated water and possibly contaminate the rest of the system.
      (2) Turn the hypochlorinator off.
      (3) When making any repairs, lock out the circuit or pull the plug from an electric socket.
   b. Long Duration
      (1) Obtain another hypochlorinator as a replacement.
3. *NORMAL OPERATION OF HYPOCHLORINATOR*

Normal operation of the hypochlorination process requires routine observation and preventive maintenance.

DAILY

a. Inspect the building to make sure only authorized personnel have been there.

b. Read and record the level of the solution tank at the same time every day.

c. Read the meters and record the amount of water pumped.

d. Check the chlorine residual (0.2 mg/*L*) in the system and adjust the chlorine feed rate as necessary. Try to maintain a chlorine residual of 0.2 mg/*L* at the most remote point in the distribution system. The suggested free chlorine residual for treated water or well water is 0.5 mg/*L* at the point of chlorine application provided the 0.2 mg/*L* residual is maintained throughout the distribution system and coliform test results are negative.

e. Check the chemical feed pump operation. Most hypochlorinators have a dial with a range from 0 to 10 which indicates the chlorine feed rate. Initially set the pointer on the dial to approximately 6 or 7 on the dial and use a two percent hypochlorite solution. The pump should be operated in the upper ranges of the dial. This will require the frequency of the strokes or pulses from the pump to be frequent enough so that the chlorine will be fed continuously to the water being treated. Adjust the feed rate after testing chlorine residual levels.

WEEKLY

a. Clean the building.

b. Replace the chemicals and wash the chemical storage tank. Try to have a 15- to 30-day supply of chlorine in storage for future needs. When preparing hypochlorite solutions, prepare only enough for a two- or three-day supply.

MONTHLY

a. Check the operation of the check valve.

b. Perform any required preventive maintenance.

c. Cleaning

Commercial sodium hypochlorite solutions (such as Clorox) contain an excess of caustic (sodium hydroxide or NaOH). When this solution is diluted with water containing calcium and also carbonate alkalinity, the resulting solution becomes supersaturated with calcium carbonate. This calcium carbonate tends to form a coating on the poppet valves in the solution feeder. The coated valves will not seal properly and the feeder will fail to feed properly.

Use the following procedure to remove the carbonate scale:

(1) Fill a one-quart (one-liter) Mason jar half-full of tap water.

(2) Place one fluid ounce (20 m*L*) of 30 to 37 percent hydrochloric acid (swimming pool acid) in the jar. *ALWAYS ADD ACID TO WATER, NEVER THE REVERSE.*

(3) Fill the jar with tap water.

(4) Place the suction hose of the hypochlorinator in the jar and pump the entire contents of the jar through the system.

(5) Return the suction hose to the hypochlorite solution tank and resume normal operation.

You can prevent the formation of the calcium carbonate coatings by obtaining the dilution water from an ordinary home water softener.

NORMAL OPERATION CHECKLIST

a. Check chemical usage. Record solution level and the water pump meter reading or number of hours of pump operation. Calculate the amount of chemical solution used and compare with the desired feed rate. See Example 4.

b. Determine if every piece of equipment is operating.

c. Inspect the lubrication of the equipment.

d. Check the building for any possible problems.

e. Clean up the area.

FORMULAS

To determine the actual chlorine dose of water being treated by either a chlorinator or a hypochlorinator, we need to know the gallons of water treated and the pounds of chlorine used to disinfect the water.

1. Calculate the pounds of water disinfected.

$$\text{Water, lbs} = (\text{Water Pumped, gallons})(8.34\ \text{lbs/gal})$$

2. To calculate the volume of hypochlorite used from a container, we need to know the dimensions of the container.

$$\text{Volume, gallons} = \frac{(0.785)(\text{Diameter, in})^2(\text{Depth, ft})(7.48\ \text{gal/cu ft})}{144\ \text{sq in/sq ft}}$$

3. To calculate the pounds of chlorine used to disinfect water, we need to know the gallons of hypochlorite used and the percent available chlorine in the hypochlorite solution.

$$\text{Chlorine, lbs} = (\text{Hypochlorite, gal})(8.34\ \text{lbs/gal})\left(\frac{\text{Hypochlorite, \%}}{100\%}\right)$$

4. To determine the actual chlorine dose in milligrams per liter, we divide the pounds of chlorine used by the millions of pounds of water treated. Pounds of chlorine per million pounds of water is the same as parts per million or milligrams per liter.

$$\text{Chlorine Dose, mg}/L = \frac{\text{Chlorine Used, lbs}}{\text{Water Treated, Million lbs}}$$

EXAMPLE 4

Water pumped from a well is disinfected by a hypochlorinator. A chlorine dosage of 1.2 mg/*L* is necessary to maintain an adequate chlorine residual throughout the system. During a one-week time period, the water meter indicated that 2,289,000 gallons of water were pumped. A two-percent sodium hypochlorite solution is stored in a three-foot diameter plastic tank. During this one-week period, the level of hypochlorite in the tank dropped 2 feet, 8 inches (2.67 feet). Does the chlorine feed rate appear to be too high, too low, or about right?

Known		Unknown
Desired Chlorine Dose, mg/*L*	= 1.2 mg/*L*	1. Actual Chlorine Dose, mg/*L*
Water Pumped, gal	= 2,289,000 gal	2. Is Actual Dose OK?
Hypochlorite, %	= 2%	
Chemical Tank Diameter, ft	= 3 ft	
Chemical Drop in Tank, ft	= 2.67 ft	

1. Calculate the pounds of water disinfected.

$$\text{Water, lbs} = (\text{Water Pumped, gallons})(8.34\ \text{lbs/gal})$$
$$= (2{,}289{,}000\ \text{gal})(8.34\ \text{lbs/gal})$$
$$= 19{,}090{,}000\ \text{lbs}$$
$$\text{or} = 19.09\ \text{Million lbs}$$

2. Calculate the volume of 2 percent sodium hypochlorite used in gallons.

$$\text{Hypochlorite, gallons} = (0.785)(\text{Diameter, ft})^2(\text{Depth, ft})(7.48\ \text{gal/cu ft})$$
$$= (0.785)(3\ \text{ft})^2(2.67\ \text{ft})(7.48\ \text{gal/cu ft})$$
$$= 141.1\ \text{gallons}$$

3. Determine the pounds of chlorine used to disinfect the water.

$$\text{Chlorine, lbs} = (\text{Hypochlorite, gal})(8.34\ \text{lbs/gal})\left(\frac{\text{Hypochlorite, \%}}{100\%}\right)$$
$$= (141.1\ \text{gal})(8.34\ \text{lbs/gal})\left(\frac{2\%}{100\%}\right)$$
$$= 23.5\ \text{lbs Chlorine}$$

4. Calculate the chlorine dosage in mg/*L*.

$$\text{Chlorine Dose, mg}/L = \frac{\text{Chlorine Used, lbs}}{\text{Water Treated, Million lbs}}$$
$$= \frac{23.5\ \text{lbs chlorine}}{19.09\ \text{Million lbs water}}$$
$$= 1.23\ \text{mg}/L$$

Since the actual chlorine dose (1.23 mg/*L*) was slightly greater than the desired dose of 1.2 mg/*L*, the chlorine feed rate appears OK.

4. *ABNORMAL OPERATION OF HYPOCHLORINATOR*

   a. Inform your supervisor of the problem.

   b. If the hypochlorinator malfunctions, it should be repaired immediately. See the shutdown operation (Step 2 in this section).

   c. Solution tank level.

      (1) If too low: Check the adjustment of the pump.
      Check the hour meter of the water pump.

      (2) If too high: Check the chemical pump.
      Check the hour meter of the water pump.

   d. Determine if the chemical pump is not operating.

      TROUBLESHOOTING GUIDELINES

      (1) Check the electrical connection.
      (2) Check the circuit breaker.
      (3) Check for stoppages in the flow lines.

      CORRECTIVE MEASURES

      (1) Shut off the water pump so that no contaminated water is pumped into the system.
      (2) Check for a blockage in the solution tank.
      (3) Check the operation of the check valve.
      (4) Check the electrical circuits.
      (5) Replace the chemical feed pump with another pump while repairing the defective unit.

   e. The solution is not being pumped into the water line.

      TROUBLESHOOTING GUIDELINES

      (1) Check the solution level.
      (2) Check for blockages in the solution line.

5. *MAINTENANCE OF HYPOCHLORINATORS*

Hypochlorinators on small systems are normally small, sealed systems that cannot be repaired so replacement of the entire unit is the only solution. Maintenance requirements are normally minor such as changing the oil and lubricating the moving parts. Review the manufacturer's specifications for maintenance requirements.

QUESTIONS

Write your answers in a notebook and then compare your answers with those on page 331.

5.4D What is the basis for adjusting the chemical feed of a hypochlorinator?

5.4E When should the level of the hypochlorite solution tank be read?

5.4F What is the basis for determining the strength of the hypochlorite solution in the solution tank?

5.4G What maintenance is usually required on hypochlorinators?

5.421 Chlorinators

1. *SAFETY EQUIPMENT REQUIRED AND AVAILABLE OUTSIDE THE CHLORINATOR ROOM*

   a. Protective clothing

      (1) Gloves
      (2) Rubber suit

   b. Self-contained pressure-demand air supply system (Figure 5.20)

   c. Chlorine leak detector/warning device should be located outside the room storing chlorine and should have a battery backup in case of a power failure. The chlorine sensor unit should be in the chlorine room and connected to the leak detector/warning device which is located outside the chlorine storage room.

2. *START-UP OF CHLORINATORS*

Work in pairs. Never work alone when hooking up a chlorine system.

   a. Inspect the chlorine container for leaks. Position the container or chlorine cylinder in its location for connection. Install safety chains or locking devices to prevent cylinder or container movement. Remove the valve protective hood on cylinders, or the valve bonnet on one-ton containers.

   b. Inspect the chlorination equipment.

   c. Use new gaskets for connection of the chlorinator to the chlorine cylinder. The chlorine supplier usually attaches two fiber gaskets to the cylinder valve for connection. Most agencies prefer to use lead gaskets because they seat better, permitting fewer connection leaks. Fiber gaskets tend to leave a deposit of fiber material on the faces of both the cylinder valve outlet and the chlorinator inlet. This deposited material must be scraped or wire brushed off both faces to obtain a leakproof connection.

   d. Prior to the hookup, inspect for moisture and foreign substances in the lines.

   e. Have an ammonia bottle readily available to detect any leaks in the system. Dip a rag on the end of a stick in the ammonia[26] bottle and place the rag near the location of suspected leaks. A white cloud will reveal the location of a chlorine leak.

   f. Have safety equipment available which may be used in case of a leak.

   g. Hook up the chlorinator to the chlorine container with the chlorine valve turned off. Use the gas side (not liquid side) if using a ton tank. Remove the chlorine cylinder valve outlet cap and check the valve outlet face for burrs, deep scratches, or debris left from former connections; clean or wire brush if necessary. If the valve face is smooth, clean, and free of deep cuts or corrosion, proceed with hooking up the cylinder. If there is any evidence of damage to the valve face, replace the outlet cap and protection hood, and ask the supplier to pick up the cylinder and replace the damaged valve. Check the inlet face of the chlorinator and clean if necessary. Place a *NEW LEAD GASKET* on the chlorinator inlet, place the chlorinator on the cylinder valve, install the yoke clamp, and slowly tighten the yoke until the two faces are against the lead gasket. Continue to slowly tighten the yoke, compressing the gasket connection one-half to three-quarters of a turn. *DO NOT OVERTIGHTEN;* this can damage the yoke or the chlorinator and cause a leak.

   h. Open the cylinder valve one-quarter of a turn and check for leaks. *IF THE VALVE IS DIFFICULT TO OPEN, RETURN THE CYLINDER TO YOUR SUPPLIER.* Some agencies require suppliers to torque the valve stems to no more than 35 psi; this permits operation of the valve stem with the correct valve wrench, without having to strike the wrench with the palm of your hand to unseat the valve stem. Striking the wrench usually results in the cylinder rotating because the safety chain usually does not securely hold the cylinder. Jarring the cylinder in this way may result in a gas leak, particularly if two cylinders are used with a manifold and pigtails. Remember — don't leave the valve in this position for a long time because it will plug up. Commercial chlorine usually contains small amounts of chlorinated organic compounds which are not volatile. Trace amounts of these compounds can be carried over in the chlorine vapor. A slightly opened valve, with the resulting change in velocity, is an ideal place for these materials to collect, build up, and cause the valve to plug up. Open the valve *ONE TURN;* this provides sufficient gas flow to prevent plugging and permits quick shutoff in case of a leak.

   i. Adjust the chlorinator to the proper setting. The desired chlorine residual should be maintained throughout the distribution system. Coliform test results should be negative when the chlorination system is operating properly.

   j. Check the system for leaks by applying a concentrated ammonia solution (28 to 30 percent ammonia as NH_3) *VAPOR* from a "squeeze bottle" to the chlorine cylinder valve and the chlorinator. Any leaks will be detected by the presence of a white smoke. *IF A LEAK IS PRESENT, CLOSE THE GAS CYLINDER VALVE.*

---

[26] *Use a concentrated ammonia solution containing 28 to 30 percent ammonia as NH_3 (this is the same as 58 percent ammonium hydroxide, NH_4OH, or commercial 26° Baumé).*

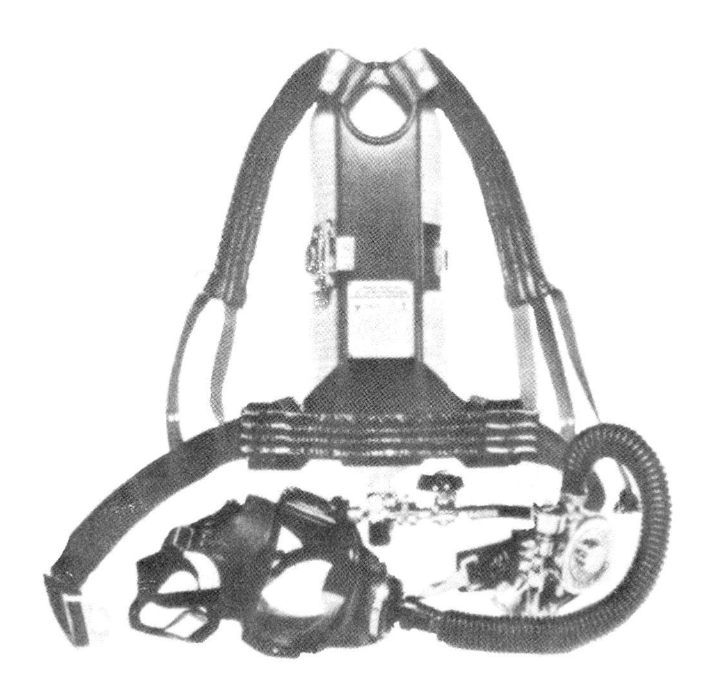

Fig. 5.20 Self-contained breathing apparatus
(Courtesy of PPG Industries)

- Make sure the packing nut on the chlorine cylinder valve stem is tight.
- Check the lead gasket. If the lead gasket is distorted, clean the connection and refit with a new gasket. Then repeat the above procedure for checking the system for leaks.

3. *CHLORINATOR SHUTDOWN*

Work in pairs. A plan should be used where both people are not exposed to the chlorine at the same time.

a. Have safety equipment available in the event of a chlorine leak.
b. Shut off the chlorine valve from the supply source.
c. Allow sufficient time for the chlorine to purge out of the line.
d. Turn the chlorinator off.
e. Leave the discharge line open. If chlorine is trapped between two valves, the chlorine gas can expand when heated by sunlight and develop high pressures in the line. Leave the discharge line open to prevent this hazard from developing.

4. *NORMAL OPERATION OF CHLORINATORS*

Normal operation of a chlorinator requires routine observation and preventive maintenance.

DAILY

a. Inspect the building to make sure that only authorized personnel have been there.
b. Read the chlorinator rotameter.
c. Record the reading, time, and date and initial the entries.
d. Read the meters and record the number of gallons of water pumped.
e. Check the chlorine residual. If the residual is below 0.2 mg/*L* in the distribution system, increase the feed rate by adjusting the rotameter. If the residual is too high, lower the feed rate by adjusting the rotameter.
f. Calculate the chlorine usage. Refer to Examples 6 and 7.

WEEKLY

a. Clean the equipment and the building.
b. Perform preventive maintenance on the equipment.
c. Calculate chlorine usage so that replacement supply containers can be ordered and constant chlorination can be maintained. Refer to Example 7. Try to have a 15- to 30-day supply of chlorine in storage.

NORMAL OPERATION CHECKLIST

a. Chemical usage in pounds.
b. Meter readings for water usage in gallons.
c. Equipment log.
d. Lubrication inspection log.
e. Building inspection.

5. *ABNORMAL OPERATION*

a. Inform supervisor of the problem.
b. If the chlorinator malfunctions, repair the unit immediately.
c. If repairs cannot be completed quickly, shut off the water supply so that unchlorinated or contaminated water will not be delivered to the consumers.
d. Follow standard chlorinator shutdown procedures.

ABNORMAL OPERATION, TROUBLESHOOTING

TABLE 5.7 DIRECT-MOUNT CHLORINATOR TROUBLESHOOTING GUIDE

Operating Symptoms	Probable Cause	Remedy
1. Water in the chlorine metering tube.	Check valve failure, deposits on seat of check valve, or check valve seat distorted by high pressure.	Clean deposits from check ball and seat with dilute muriatic acid. Badly distorted check valve may have to be replaced.
2. Water venting to atmosphere.	Excess water pressure in the vacuum regulator.	Remove vacuum regulator from chlorine cylinder and allow chlorinator to pull air until dry.
3. No indication on flowmeter when vacuum is present.	Vacuum leak due to bad or brittle vacuum tubing, connections, rate valve o-rings, or gasket on top of flowmeter.	Check the vacuum tubing, rate valve o-rings and flowmeter gasket for vacuum leaks. Replace bad tubing connectors, o-rings, or gasket.
4. Indication on flowmeter but air present, not chlorine gas.	Connection below meter tube gasket leaks.	Check connections and replace damaged elements.

6. *MAINTENANCE*

Most direct-mounted chlorinators are simple units and are more easily replaced than repaired on line. Remove and repair in shop or have repaired by others who are qualified to repair these chlorinators.

FORMULAS

To determine the setting on a chlorinator in pounds per day, multiply the flow in MGD times the dose in mg/*L* times 8.34 lbs/gal.

Chlorinator Setting, lbs/day = (Flow, MGD)(Dose, mg/*L*)(8.34 lbs/gal)

To calculate the number of chlorine cylinders used per month, determine the pounds of chlorine used per month and divide by the pounds of chlorine per cylinder.

$$\text{Cylinders Used, number/month} = \frac{\text{Chlorine Used, lbs/mo}}{\text{Chlorine Cylinders, lbs/cylinder}}$$

EXAMPLE 5

A deep well turbine pump delivers approximately 200 GPM against typical operating heads. If the desired chlorine dose is 2 mg/*L*, what should be the setting on the rotameter for the chlorinator (lbs chlorine per 24 hours)?

Known		Unknown
Pump Flow, GPM	= 200 GPM	Rotameter Setting, lbs chlorine/24 hours
Chlorine Dose, mg/*L*	= 2 mg/*L*	

1. Convert pump flow to million gallons per day (MGD).

$$\text{Flow, MGD} = \frac{(200\text{ GPM})(60\text{ min/hr})(24\text{ hr/day})}{1{,}000{,}000/\text{Million}}$$

$$= 0.288\text{ MGD}$$

2. Calculate the rotameter setting in pounds of chlorine per 24 hours.

$$\text{Rotameter Setting, lbs/day} = (\text{Flow, MGD})(\text{Dose, mg/}L)(8.34\text{ lbs/gal})$$

$$= (0.288\text{ M Gal/day})(2\text{ mg/}L)(8.34\text{ lbs/gal})$$

$$= 4.8\text{ lbs/day}$$

$$= 4.8\text{ lbs/24 hours}$$

EXAMPLE 6

Using the results from Example 5 (a chlorinator setting of 4.8 lbs/day), how many pounds of chlorine would be used during one week if the pump hour meter showed 100 hours of pump operation? If the chlorine cylinder contained 78 pounds of chlorine at the start of the week, how many pounds of chlorine should be remaining at the end of the week?

Known		Unknown
Chlorinator Setting, lbs/day	= 4.8 lbs/day	1. Chlorine Used, lbs/week
Time, hr/week	= 100 hr/week	2. Chlorine Remaining, lbs
Chlorine Cylinder, lbs	= 78 lbs	

1. Calculate the chlorine used in pounds per week.

$$\text{Chlorine Used, lbs/week} = (\text{Chlorinator Setting, lbs/day})(\text{Time, hr/week})$$

$$= (4.8\text{ lbs/day})\left(\frac{100\text{ hr/wk}}{24\text{ hr/day}}\right)$$

$$= 20\text{ lbs chlorine/week}$$

2. Determine the amount of chlorine that should be remaining in the cylinder at the end of the week.

$$\text{Chlorine Remaining, lbs} = \text{Chlorine at Start, lbs} - \text{Chlorine Used, lbs}$$

$$= 78\text{ lbs} - 20\text{ lbs}$$

$$= 58\text{ lbs chlorine remaining at end of week}$$

EXAMPLE 7

Given the pumping rate and chlorination system in Examples 5 and 6, if 20 pounds of chlorine are used during an average week, how many 150-pound chlorine cylinders will be used per month (assume 30 days per month)?

Known		Unknown
Chlorine Use, lbs/week	= 20 lbs/week	1. Amount of Chlorine Used per Month, lbs 2. Number of 150-lb Cylinders Used per Month

1. Calculate the amount of chlorine used in pounds of chlorine per month.

$$\text{Chlorine Used, lbs/month} = (\text{Chlorine Use, lbs/week})(\text{Number Weeks/mo})$$

$$= (20\text{ lbs/week})\left(\frac{(1\text{ week})(30\text{ days})}{(7\text{ days})(1\text{ month})}\right)$$

$$= 85.7\text{ lbs/month}$$

2. Determine the number of 150-pound chlorine cylinders used per month.

$$\text{Cylinders Used, number/month} = \frac{\text{Chlorine Used, lbs/mo}}{\text{Chlorine Cylinders, lbs/cylinder}}$$

$$= \frac{85.7\text{ lbs/mo}}{150\text{ lbs/cylinder}}$$

$$= 0.57\text{ cylinders/month}$$

This installation requires less than one 150-pound chlorine cylinder per month.

5.43 Laboratory Tests

1. Chlorine Residual in System

   a. Chlorine residual tests using the *DPD*[27] *METHOD*[28] should be taken daily at various locations in the system. A remote tap is ideal for one sampling location. Take the test sample from a tap as close to the main as possible. Allow the water to run at least 5 minutes before sampling to ensure a representative sample from the main.

   Operators using the DPD colorimetric method to test water for a free chlorine residual need to be aware of a potential error that may occur. If the DPD test is run on water containing a combined chlorine residual, a precipitate may form during the test. The particles of precipitated material will give the sample a turbid appearance or the appearance of having color. This turbidity can produce a positive test result for free chlorine residual when there is actually no chlorine present. Operators call this error a "false positive" chlorine residual reading.

   b. Chlorine residual test kits are available for small systems.

[27] *DPD (pronounce as separate letters). A method of measuring the chlorine residual in water. The residual may be determined by either titrating or comparing a developed color with color standards. DPD stands for N,N-diethyl-p-phenylene-diamine.*

[28] *See Section 5.5, "Measurement of Chlorine Residual," for details on how to perform the DPD test for measuring chlorine residual.*

2. Bacteriological Analysis (Coliform Tests)

Samples should be taken routinely in accordance with health department requirements. Take samples according to approved procedures.[29] Be sure to use a sterile plastic or glass bottle. If the sample contains any chlorine residual, sufficient sodium thiosulfate should be added to neutralize all of the chlorine residual. Usually 0.1 milliliter of 10 percent sodium thiosulfate in a 120-m*L* (4 oz) bottle is sufficient for distribution systems. The "thio" should be added to the sample bottle before sterilization.

5.44 Troubleshooting

TABLE 5.8 DISINFECTION TROUBLESHOOTING GUIDE

Operating Symptoms	Probable Cause	Remedy
1. Increase in coliform level	Low chlorine residual	Raise chlorine dose
2. Drop in chlorine level	a. Increase in chlorine demand b. Drop in chlorine feed rate	Raise chlorine dose and find out why chlorine demand increased or chlorine feed rate dropped

5.45 Chlorination System Failure

IF YOUR CHLORINATION SYSTEM FAILS, DO NOT ALLOW UNCHLORINATED WATER TO ENTER THE DISTRIBUTION SYSTEM. Never allow unchlorinated water to be delivered to your consumers. If your chlorination system fails and cannot be repaired within a reasonable time period, notify your supervisor and officials of the health department. To prevent this problem from occurring, your plant should have backup or standby chlorination facilities.

5.46 Emergency Disinfection Plan

All water treatment plants should have an emergency disinfection plan. The plan must be ready to be implemented any time a disinfection failure occurs in order to prevent the delivery to the distribution system of any water that is not disinfected or is inadequately disinfected. The plan should be posted in the plant or at a place that is readily available to the plant operator or an emergency crew.

The emergency disinfection plan should include information outlining the corrective actions that must be taken until the disinfection problem is properly corrected. The plan should include a description of the existing disinfection facilities and the operating and monitoring procedures of these facilities. Emergency telephone numbers must be listed for the appropriate health department officials and also of the operators available and the equipment suppliers needed to make the repairs. Review your emergency disinfection plan at least once a year and check to be sure all phone numbers are still current.

Procedures should be outlined for the emergency response if the disinfection system failed, but no inadequately disinfected water entered the water distribution system. These procedures should include how to immediately shut down the water treatment plant if possible. Use of alternative water sources, if available, should be explained and procedures for implementation of water conservation measures should be outlined. If a backup chlorinator or a chemical feeder from a less critical treatment process is available, this equipment should be used. If no backup equipment is available, procedures for manual disinfection at the plant and also the distribution reservoirs need to be outlined. Also increased monitoring of bacteriological quality and chlorine residual levels of the water being delivered to and within the distribution system must be performed.

If inadequately disinfected water entered the water distribution system, additional procedures must be implemented. Health department officials must be notified immediately. The distribution system must be flushed to remove inadequately disinfected water using properly disinfected water and properly disinfected water must be distributed within the system as quickly as possible.

5.47 CT Values

The purpose of the Surface Water Treatment Rule (SWTR) is to ensure that pathogenic organisms are removed and/or inactivated by the treatment process. To meet this goal, all systems are required to disinfect their water supplies. For some water systems using very clean source water and meeting the other criteria to avoid filtration, disinfection alone can achieve the 3-log (99.9%) *Giardia* and 4-log (99.99%) virus inactivation levels required by the SWTR. (For procedures to calculate log removals, see Section A.16, "Calculation of Log Removals," in the Arithmetic Appendix at the end of this manual.)

Several methods of disinfection are in common use, including free chlorination, chloramination, use of chlorine dioxide, and application of ozone. The concentration of chemical needed and the length of contact time needed to ensure disinfection are different for each disinfectant. Therefore, the efficiency of the disinfectant is measured by the time "T" in minutes of the disinfectant's contact in the water and the concentration "C" of the disinfectant residual in mg/*L* measured at the end of the contact time. The product of these two factors (CxT) provides a measure of the degree of pathogenic inactivation. The required CT value to achieve inactivation is dependent upon the organism in question, type of disinfectant, pH, and temperature of the water supply.

Time or "T" is measured from point of application to the point where "C" is determined. "T" must be based on peak hour flow rate conditions. In pipelines, "T" is calculated by dividing the volume of the pipeline in gallons by the flow rate in gallons per minute (GPM). In reservoirs and basins, dye tracer

[29] *See Chapter 7, "Laboratory Procedures," for proper procedures for collecting and analyzing samples for chlorine residuals and coliform tests.*

tests must be used to determine "T." In this case "T" is the time it takes for 10 percent of the tracer to pass the measuring point.

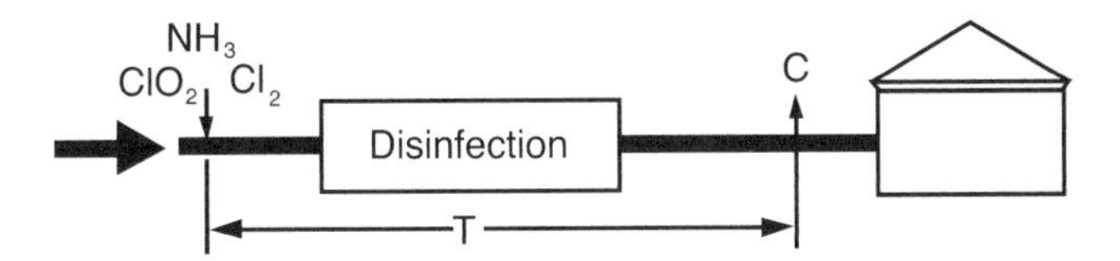

A properly operated filtration system can achieve limited removal or inactivation of microorganisms. Because of this, systems that are required to filter their water are permitted to apply a factor that represents the microorganism removal value of filtration when calculating CT values to meet the disinfection requirements. The factor (removal credit) varies with the type of filtration system. Its purpose is to take into account the combined effect of both disinfection and filtration in meeting the SWTR microbial standards.

Please refer to the Arithmetic Appendix at the end of *WATER TREATMENT PLANT OPERATION*, Volume I, in this series of manuals (Section A.16, "Calculation of CT Values") for instructions on how to perform these calculations for a water treatment plant. Also see Volume II for more information on CT values.

For more detailed information about the requirements and application of the Surface Water Treatment Rule, you may wish to order a copy of the publication, *GUIDANCE MANUAL FOR COMPLIANCE WITH THE FILTRATION AND DISINFECTION REQUIREMENTS FOR PUBLIC WATER SYSTEMS USING SURFACE WATER SOURCES*. It is available from American Water Works Association (AWWA), Bookstore, 6666 West Quincy Avenue, Denver, CO 80235. Order No. 20271. ISBN 0-89867-558-8. Price to members, $50.50; nonmembers, $72.50; price includes cost of shipping and handling.

5.48 Acknowledgment

Some of the material in this section on gas chlorinators was prepared by Joe Habraken, Treatment Supervisor, City of Tampa, Florida. His contribution is greatly appreciated.

QUESTIONS

Write your answers in a notebook and then compare your answers with those on page 331.

5.4H What personal safety equipment should be available before attempting to locate and repair a chlorine gas leak?

5.4I How is ammonia used to detect a chlorine leak?

5.4J What would you do if you could not repair a broken chlorinator quickly?

5.4K What two water quality tests are run on samples of water from a water supply system?

5.5 MEASUREMENT OF CHLORINE RESIDUAL

5.50 Methods of Measuring Chlorine Residual

AMPEROMETRIC TITRATION[30] provides for the most convenient and most repeatable chlorine residual results. However, amperometric titration equipment is more expensive than equipment for other methods. DPD tests can be used and are less expensive than other methods, but this method requires the operator to match the color of a sample with the colors on a comparator. See Chapter 7, "Laboratory Procedures," for detailed information on these tests.

Residual chlorine measurements of treated water should be taken at least three times per day on small systems and once every two hours on large systems to ensure that the treated water is being adequately disinfected. A free chlorine residual of at least 0.5 mg/*L* in the treated water at the point of application is usually recommended.

ALL surface water systems and groundwater systems under the influence of surface water must provide disinfection. Systems are required to monitor the disinfectant residual leaving the plant and at various points in the distribution system. The water leaving the plant must have at least 0.2 mg/*L* of the disinfectant, and the samples taken in the distribution system must have a detectable residual. Certain guidelines must be followed to ensure that there is enough contact time between the disinfectant and the water so that the microorganisms are inactivated.

If at any time the disinfectant residual leaving the plant is less than 0.2 mg/*L*, the system is allowed up to four hours to correct the problem. If the problem is corrected within this time, it is not considered a violation but the regulatory agency must be notified. The disinfectant residual must be measured continuously. For systems serving fewer than 3,300 people, this may be reduced to once per day.

The disinfectant in the distribution system must be measured at the same frequency and location as the total coliform samples. Measurements for heterotrophic plate count (HPC) bacteria may be substituted for disinfectant residual measurements. If the HPC is less than 500 colonies per m*L*, then the sample is considered equivalent to a detectable disinfectant residual. For systems serving fewer than 500 people, the regulatory agency may determine the adequacy of the disinfectant residual in place of monitoring.

5.51 Amperometric Titration for Free Chlorine Residual

1. Place a 200-m*L* sample of water in the titrator.
2. Start the agitator.
3. Add 1 m*L* of pH 7 buffer.
4. Titrate with 0.00564 *N* phenylarsene oxide solution.
5. End point is reached when further additions (drops) will not cause a deflection on the microammeter.
6. m*L* of phenylarsene oxide used in titration is equal to mg/*L* of free chlorine residual.

5.52 DPD Colorimetric Method for Free Chlorine Residual (Figures 5.21 and 5.22)

This procedure is for the use of prepared powder pillows.

1. Collect a 100-m*L* sample.
2. Add color reagent.
3. Match color sample with a color on the comparator to obtain the chlorine residual in mg/*L*.

---

[30] *Amperometric (am-PURR-o-MET-rick) Titration. A means of measuring concentrations of certain substances in water (such as strong oxidizers) based on the electric current that flows during a chemical reaction.*

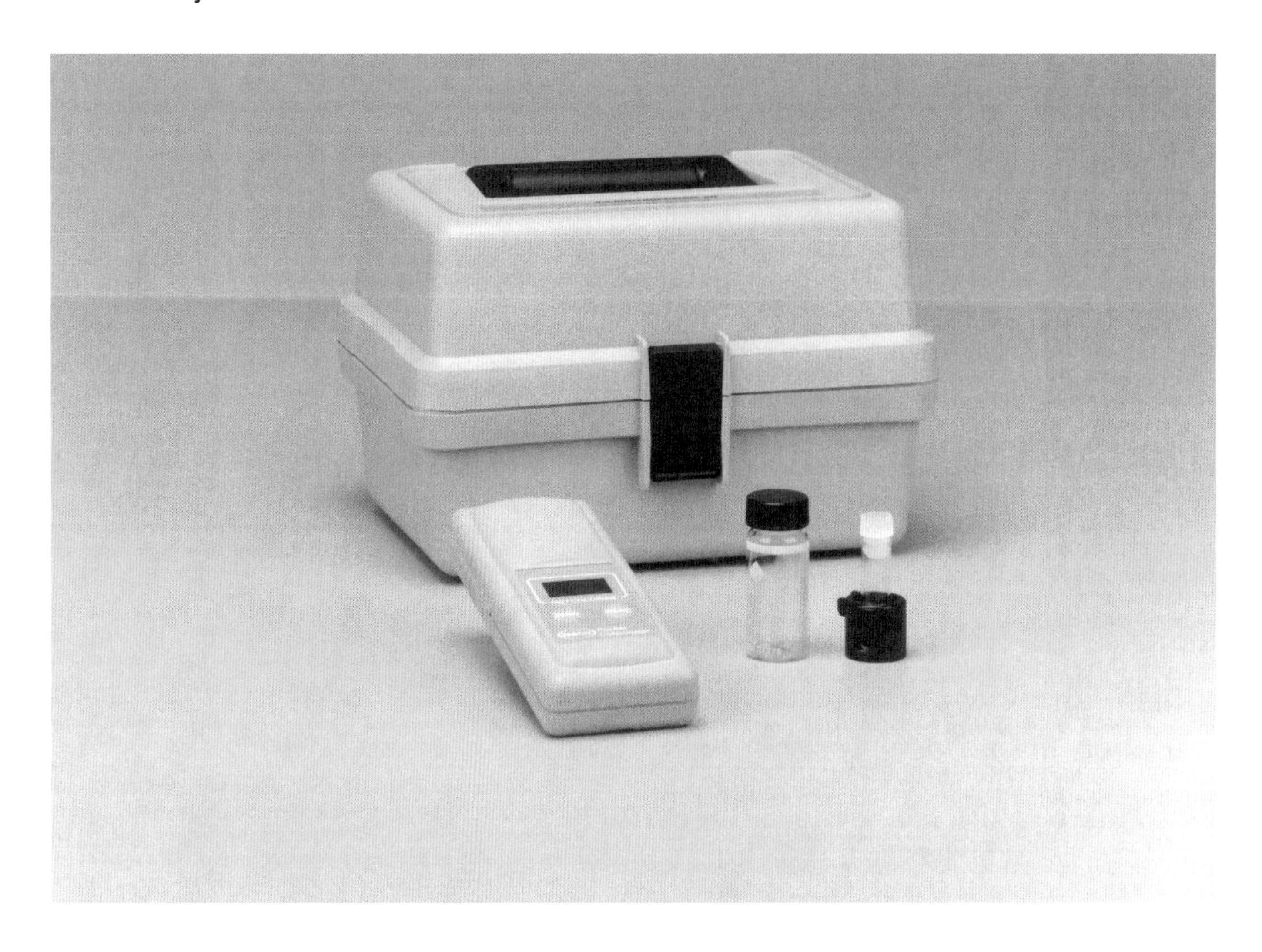

Fig. 5.21 Direct reading colorimeter for free chlorine residuals
(Permission of the HACH Company)

Operators using the DPD colorimetric method to test water for a free chlorine residual need to be aware of a potential error that may occur. If the DPD test is run on water containing a combined chlorine residual, a precipitate may form during the test. The particles of precipitated material will give the sample a turbid appearance or the appearance of having color. This turbidity can produce a positive test result for free chlorine residual when there is actually no chlorine present. Operators call this error a "false positive" chlorine residual reading.

5.53 ORP Probes

ORP (Oxidation-Reduction Potential) probes are being used to optimize chlorination processes in water treatment plants. ORP (also called the redox potential) is a direct measure of the effectiveness of a chlorine residual in disinfecting the water being treated. Chlorine forms that are toxic to microorganisms (including coliforms) are missing one or more electrons in their molecular structure. They satisfy their need for electrons by taking electrons from any organic substances or microorganisms present in the water being treated. When microorganisms lose electrons they become inactivated and can no longer transmit a disease or reproduce.

The ability of chlorine to take electrons (the electrical attraction or electrical potential) is the ORP and is measurable in millivolts. The strength of the millivoltage (or the redox measurement) is directly proportional to the oxidative disinfection

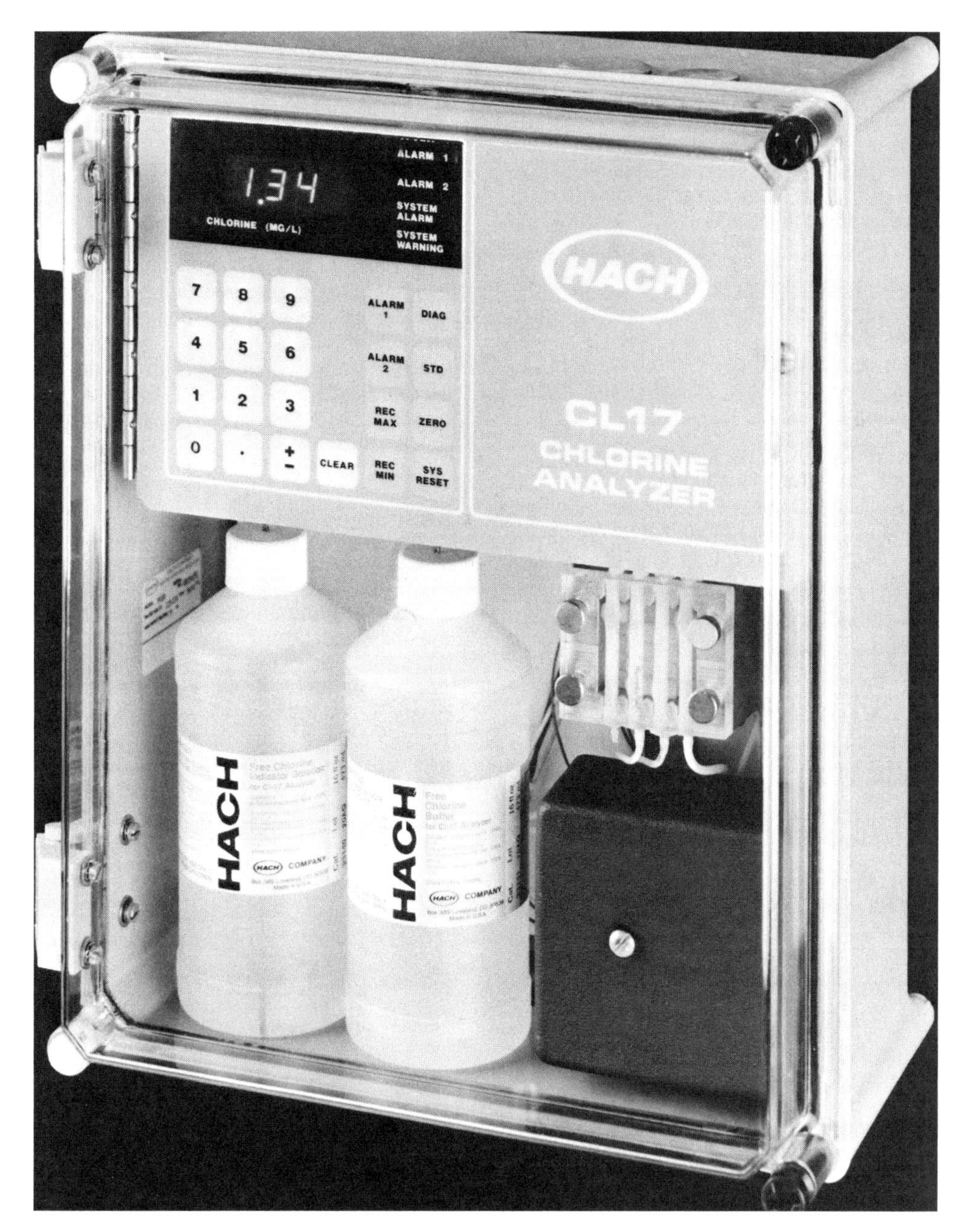

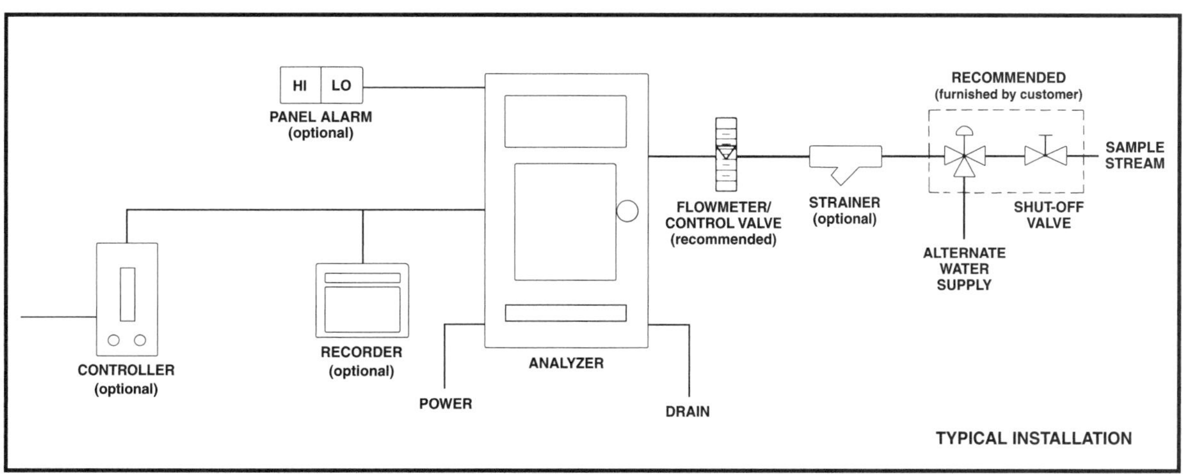

Fig. 5.22 Continuous on-line free chlorine residual analyzer

(Permission of the HACH Company)

strength of the chlorine in the treatment system. The higher the concentration of chlorine disinfectant, the higher the measured ORP voltage. Conversely, the higher the concentration of organics (chlorine demanding substances), the lower the measured ORP voltage. The redox sensing unit (ORP probe) measures the voltage present in the water being treated and thus provides a direct measure of the disinfecting power of the disinfectant present in the water.

In a typical installation, a High Resolution Redox (HRR) chlorine controller monitors the chlorine residual using a redox (ORP) probe suspended in the chlorine contact chamber approximately 6.5 minutes downstream from the chlorine injection point. The controller converts the redox signal to a 4 to 20 milliamp (mA) signal that automatically adjusts the chlorine feed rate from the chlorinator.

The High Resolution Redox (HRR) units control the chlorination chemical feed rates according to actual demand in the treatment processes. These HRR units automatically treat the water with the chlorine dosages required to maintain chemical residuals in the ideal ranges, regardless of changes in the chemical demand or water flow.

Maintenance for the chlorine ORP probe consists of cleaning the unit's sensor once a month.

QUESTIONS

Write your answers in a notebook and then compare your answers with those on pages 331 and 332.

5.5A What three methods are used to measure chlorine residual in treated water?

5.5B How often should treated water residual chlorine measurements be made?

5.5C What does an ORP probe measure in a disinfection system?

5.5D What happens to a microorganism when it loses an electron?

5.5E What maintenance is required on ORP probes?

5.6 CHLORINE SAFETY PROGRAM

Every good safety program begins with cooperation between the employee and the employer. The employee must take an active part in the overall program. The employee must be responsible and should take all necessary steps to prevent accidents. This begins with the attitude that as good an effort as possible must be made by everyone. Safety is everyone's priority. The employer also must take an active part by supporting safety programs. There must be funding to purchase equipment and to enforce safety regulations required by OSHA and state industrial safety programs. The following items should be included in all safety programs.

1. Establishment of a formal safety program.
2. Written rules.
3. Periodic hands-on training using safety equipment.
   a. Leak-detection equipment
   b. Self-contained breathing apparatus (Figure 5.20)
   c. Atmospheric monitoring devices
4. Establishment of emergency procedures for chlorine leaks and first aid.
5. Establishment of a maintenance and calibration program for safety devices and equipment.
6. Provide police and fire departments with tours of facilities to locate hazardous areas and provide chlorine safety information.

All persons handling chlorine should be thoroughly aware of its hazardous properties. Personnel should know the location and use of the various pieces of protective equipment and be instructed in safety procedures. In addition, an emergency procedure should be established and each individual should be instructed how to follow the procedures. An emergency checklist also should be developed and available. For additional information on this topic, see the Chlorine Institute's *CHLORINE MANUAL*.[31] Also see Chapter 6, "Safety."

5.60 Chlorine Hazards

Chlorine is a gas that is 2.5 times heavier than air, extremely toxic, and corrosive in moist atmospheres. Dry chlorine gas can be safely handled in steel containers and piping, but with moisture must be handled in corrosion-resistant materials such as silver, glass, Teflon, and certain other plastics. Chlorine gas at container pressure should never be piped in silver, glass, Teflon, or any other material that cannot handle the pressure. Even in dry atmospheres chlorine combines with the moisture in the mucous membranes of the nose and throat, and with the fluids in the eyes and lungs; a very small percentage in the air can be very irritating and can cause severe coughing. Heavy exposure can be fatal (see Table 5.9).

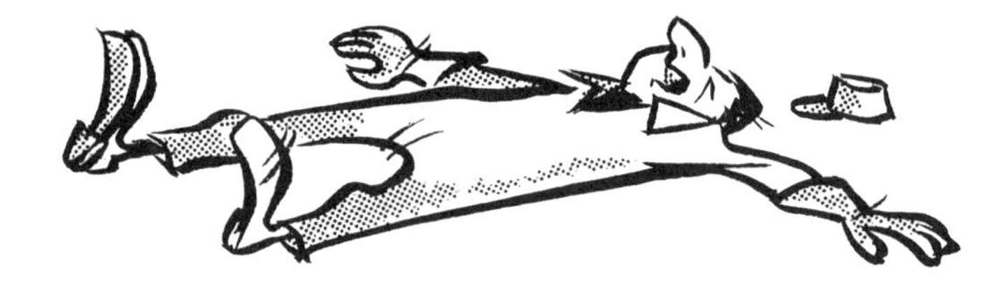

WARNING

WHEN ENTERING A ROOM THAT MAY CONTAIN CHLORINE GAS, OPEN THE DOOR SLIGHTLY AND CHECK FOR THE SMELL OF CHLORINE. **NEVER** GO INTO A ROOM CONTAINING CHLORINE GAS WITH HARMFUL CONCENTRATIONS IN THE AIR WITHOUT A SELF-CONTAINED AIR SUPPLY, PROTECTIVE CLOTHING, AND HELP STANDING BY. HELP MAY BE OBTAINED FROM YOUR CHLORINE SUPPLIER AND YOUR LOCAL FIRE DEPARTMENT.

Most people can usually detect concentrations of chlorine gas above 0.3 ppm and you should not be exposed to concentrations greater than 1 ppm. However, chlorine gas can deaden your sense of smell and cause a false sense of security. **NEVER** rely on your sense of smell to protect you from chlorine because **YOUR** sense of smell might not be able to detect harmful levels of chlorine.

[31] *Write to: The Chlorine Institute, Inc., 2001 L Street, NW, Suite 506, Washington, DC 20036. Pamphlet 1. Price to members, $15.00; nonmembers, $30.00; plus 10 percent of order total for shipping and handling.*

TABLE 5.9 PHYSIOLOGICAL RESPONSE TO CONCENTRATIONS OF CHLORINE GAS[a]

Effect	Parts of Chlorine Gas Per Million Parts of Air by Volume (ppm)
Slight symptoms after several hours' exposure	1[b]
Detectable odor	0.3 to 3.5
Noxiousness	5
Throat irritation	15
Coughing	30
Dangerous from one-half to one hour	40
Death after a few deep breaths	1,000

[a] Adapted from data in U.S. Bureau of Mines *TECHNICAL PAPER 248* (1955).
[b] OSHA regulations specify that exposure to chlorine shall at *NO* time exceed 1 ppm.

5.61 Why Chlorine Must Be Handled With Care

You must always remember that chlorine is a hazardous chemical and must be handled with respect. Concentrations of chlorine gas in excess of 1,000 ppm (0.1% by volume in air) may be fatal after a few breaths.

Because the characteristic sharp odor of chlorine can be noticeable even when the amount of chlorine in the air is small, it is possible to get out of the gas area before serious harm is suffered. This feature of chlorine gas helps reduce potential operator exposure as compared with other hazardous gases such as carbon monoxide, which is odorless and gives no warning of exposure, and hydrogen sulfide, which impairs your sense of smell in a short time.

Inhaling chlorine causes general restlessness, panic, severe irritation of the throat, sneezing, and production of much saliva. These symptoms are followed by coughing, retching and vomiting, and difficulty in breathing. Chlorine is particularly irritating to persons suffering from asthma and certain types of chronic bronchitis. Liquid chlorine causes severe irritation and blistering on contact with the skin. Regular exposure to chlorine can produce chronic (long-term) effects such as permanent damage to lung tissue.

5.62 Protect Yourself From Chlorine

Every person working with chlorine should know the proper ways to handle it, should be trained in the use of self-contained breathing apparatus (SCBA), and should know what to do in case of emergencies. Wear a SCBA which protects your face, eyes and nose. The clothing of persons exposed to chlorine can be saturated with chlorine which can irritate the skin if exposed to moisture or sweat. These people should not enter confined spaces before their clothing is purged of chlorine (stand out in the open air for a while). This is particularly applicable to police and fire department personnel who leave the scene of a chlorine leak and ride back to their stations in closed vehicles. Suitable protective clothing for working in an atmosphere containing chlorine includes a disposable rainsuit with hood to protect your body, head, and limbs, and rubber boots to protect your feet.

WARNING

CANISTER TYPE **'GAS MASKS'** ARE USUALLY **INADEQUATE** AND **INEFFECTIVE** IN SITUATIONS WHERE CHLORINE LEAKS OCCUR AND ARE THEREFORE NOT RECOMMENDED FOR USE UNDER ANY CIRCUMSTANCES. **SELF-CONTAINED AIR OR OXYGEN SUPPLY TYPE BREATHING APPARATUS ARE RECOMMENDED.** OPERATORS SERVING ON "EMERGENCY CHLORINE TEAMS" MUST BE CAREFULLY SELECTED AND RECEIVE REGULAR APPROVED TRAINING. THEY MUST BE PROVIDED THE PROPER EQUIPMENT WHICH RECEIVES REGULAR MAINTENANCE AND IS READY FOR USE AT ALL TIMES.

Self-contained air supply and positive pressure/demand breathing equipment must fit properly and be used properly. Pressure demand and rebreather kits may be safer. Pressure demand units use more air from the air bottle which reduces the time a person may work on a leak.

The 1991 Uniform Fire Code requires proper ventilation of chlorine storage rooms and rooms where chlorine is used. Mechanical exhaust systems must draw air from the room at a point no higher than 12 inches (30.5 cm) above the floor at a rate of not less than one cubic foot of air per minute per square foot (0.00508 cu m/sec/sq m) of floor area in the storage area. (The system should not draw air through the fan itself because chlorine gas can damage the fan motor.) Normally ventilated air from chlorine storage rooms is discharged to the atmosphere. When a chlorine leak occurs, the ventilated air containing the chlorine must be treated to reduce the chlorine concentration. A caustic scrubbing system can be used. The treatment must reduce the chlorine concentration to one-half the *IDLH*[32] (**I**mmediately **D**angerous to **L**ife or **H**ealth) level at the point of discharge to the atmosphere. The IDLH level for chlorine is 10 ppm. A secondary standby source of power is also required for the chlorine detection, alarm, ventilation and treatment systems.

Before entering an area with a chlorine leak, wear protective clothing. Gloves and a rubber suit will prevent chlorine from contacting the sweat on your body and forming hydrochloric acid. Rubber suits are very cumbersome, but should be worn when the weather is hot and humid and the chlorine concentration is high. Otherwise, use your own judgment regarding whether or not to wear protective clothing.

The best protection that one can have when dealing with chlorine is to respect it. Each individual should practice rules of safe handling and good PREVENTIVE MAINTENANCE.

PREVENTION IS THE BEST EMERGENCY TOOL YOU HAVE. PLAN AHEAD:

1. Have your fire department tour the area so that they know where the facilities are located. Give them a clearly marked map indicating the location of the chlorine storage area, chlorinators, and gas masks.

[32] *IDLH. **I**mmediately **D**angerous to **L**ife or **H**ealth. The atmospheric concentration of any toxic, corrosive, or asphyxiant substance that poses an immediate threat to life or would cause irreversible or delayed adverse health effects or would interfere with an individual's ability to escape from a dangerous atmosphere.*

2. Have emergency drills using chlorine gas masks and chlorine repair kits.
3. Have a supply of ammonia available to detect chlorine leaks.
4. Write emergency procedures:

   Prepare a CHLORINE EMERGENCY LIST of names or companies and phone numbers of persons to call during an emergency. This list should include:

   a. Fire department,
   b. Chlorine emergency personnel, and
   c. Chlorine supplier.
5. Follow established procedures during all emergencies.
   a. Never work alone during chlorine emergencies.
   b. Obtain help immediately and quickly repair the problem. *PROBLEMS DO NOT GET BETTER.*
   c. Only authorized and properly trained persons with adequate equipment should be allowed in the danger area to correct the problem.
   d. If you are caught in a chlorine atmosphere without a gas mask, shallow breathing is safer than breathing deeply. Recovery depends upon the duration and amount of chlorine inhaled, so it is important to keep that amount as small as possible.
   e. If you discover a chlorine leak, leave the area immediately unless it is a very minor leak. Small leaks can be found by using a rag soaked with ammonia. A white gas will form near the leak so it can be located and corrected.
   f. Use approved respiratory protection and wear disposable clothing when repairing a chlorine leak.
   g. Notify your police department that you need help if it becomes necessary to stop traffic on roads and to evacuate persons in the vicinity of the chlorine leak.
6. Develop emergency evacuation procedures for use during a serious chlorine leak. Coordinate these procedures with your police department and other officials.
7. Post emergency procedures in all operating areas.
8. Inspect equipment and routinely make any necessary repairs.
9. At least twice weekly, inspect area where chlorine is stored and where chlorinators are located. Remove all obstructions from the area.
10. Schedule routine maintenance on *ALL* chlorine equipment at least once every six months or more frequently.
11. Have health appraisal for employees on chlorine emergency duty. No one with heart and respiratory problems should be allowed on emergency teams.

REMEMBER: Small amounts of chlorine cause large problems. Leaks never get better.

5.63 First-Aid Measures

MILD CASES

Whenever you have a mild case of chlorine exposure (which does happen from time to time around chlorination equipment), you should first leave the contaminated area. Move slowly, breathe lightly without exertion, remain calm, keep warm, and resist coughing. Notify other operators and have them repair the leak immediately.

If clothing has been contaminated, remove as soon as possible. Otherwise the clothing will continue to give off chlorine gas which will irritate the body even after leaving the contaminated area. Immediately wash the area affected by chlorine. Shower and put on clean clothes.

If victim has slight throat irritation, immediate relief can be accomplished by drinking milk. Drinking spirits of peppermint also will help reduce throat irritation. See a physician.

EXTREME CASES

1. Follow established emergency procedures.
2. Always use proper safety equipment. Do not enter area without a self-contained breathing apparatus.
3. Remove patient from affected area immediately. Call a physician and begin appropriate treatment immediately.
4. First aid:
   a. Remove contaminated clothes to prevent clothing giving off chlorine gas which will irritate the body.
   b. Keep patient warm and cover with blankets if necessary.
   c. Place patient in a comfortable position on back.
   d. If breathing is difficult, administer oxygen if equipment and trained personnel are available.
   e. If breathing seems to have stopped, begin artificial respiration immediately. Mouth-to-mouth resuscitation or any of the approved methods may be used.
   f. EYES! If even a small amount of chlorine gets into the eyes, they should be flushed immediately with large amounts of lukewarm water so that all traces of chlorine are flushed from the eyes (at least 15 minutes). Hold the eyelids apart forcibly to ensure complete washing of all eye and lid tissues.
5. See a physician.

5.64 Hypochlorite Safety

Hypochlorite does not present the hazards that gaseous chlorine does and therefore is safer to handle. When spills occur, wash with large volumes of water. The solution is messy to handle. Hypochlorite causes damage to your eyes and skin upon contact. Immediately wash affected areas thoroughly with water. Consult a physician if the area appears burned. Hypochlorite solutions are very corrosive. Hypochlorite compounds are nonflammable; however, they can cause fires when they come in contact with organics or other easily oxidizable substances.

5.65 Chlorine Dioxide Safety

Chlorine dioxide is generated in much the same manner as chlorine and should be handled with the same care. Of special concern is the use of sodium chlorite to generate chlorine dioxide. Sodium chlorite is very combustible around organic compounds. Whenever spills occur, sodium chlorite must be neutralized with anhydrous sodium sulfite. Combustible materials (including combustible gloves) should not be worn when handling sodium chlorite. If sodium chlorite comes in contact with clothing, the clothes should be removed immediately and soaked in water to remove all traces of sodium chlorite or the clothes should be burned immediately.

5.66 Operator Safety Training

Training is a concern to everyone, especially when your safety and perhaps your life is involved. Every utility agency should have an operator chlorine safety training program that introduces new operators to the program and updates previously trained operators. As soon as a training session ends, obsolescence begins. People will forget what they have learned if they don't use and practice their knowledge and skills. Operator turnover can dilute a well-trained staff. New equipment and also new techniques and procedures can dilute the readiness of trained operators. An ongoing training program could include a monthly luncheon seminar, a monthly safety bulletin that is to be read by every operator, and outside speakers who reinforce and refresh specific elements of safety training.

5.67 CHEMTREC (800) 424-9300

Safely handling chemicals used in daily water treatment is an operator's responsibility. However, if the situation ever gets out of hand, there are emergency teams that will respond with help anywhere there is an emergency. If an emergency does develop in your plant and you need assistance, call CHEMTREC (Chemical Transportation Emergency Center) for assistance. CHEMTREC will provide immediate advice for those at the scene of an emergency and then quickly alert experts whose products are involved for more detailed assistance and appropriate follow-up.

CHEMTREC'S EMERGENCY TOLL-FREE TELEPHONE NUMBER IS (800) 424-9300.

QUESTIONS

Write your answers in a notebook and then compare your answers with those on page 332.

5.6A What properties make chlorine gas so hazardous?

5.6B What type of breathing apparatus is recommended when repairing a chlorine leak?

5.6C What first-aid measures should be taken if a person comes in contact with chlorine gas?

5.6D What would you do if hypochlorite came in contact with your hand?

5.7 CHLORINATION ARITHMETIC

All calculations in this section can be performed by addition, subtraction, multiplication and division on a pocket electronic calculator.

FORMULAS

There are two approaches to calculating chlorine doses in milligrams per liter. They both give the same results, but have a slightly different form. From the basic equation,

$$\text{Chlorine, lbs} = (\text{Volume, M Gal})(\text{Dose, mg}/L)(8.34\text{ lbs/gal}),$$

we can rearrange the equation and solve for the dose in milligrams per liter.

$$\text{Chlorine Dose, mg}/L = \frac{\text{Chlorine, lbs}}{(\text{Volume, M Gal})(8.34\text{ lbs/day})}$$

If the basic equation is expressed as a chemical feeder setting in pounds per day, then the flow would be in million gallons per day (MGD).

$$\text{Chlorine Dose, mg}/L = \frac{\text{Chlorine, lbs/day}}{(\text{Flow, MGD})(8.34\text{ lbs/gal})}$$

Both of the above equations are also expressed in terms of pounds or pounds per day of chlorine per million pounds or million pounds per day of water.

$$\text{Chlorine Dose, mg}/L = \frac{\text{Chlorine, lbs/day}}{\text{Water, Million lbs/day}}$$

5.70 Disinfection of Facilities

5.700 Wells and Pumps

EXAMPLE 8

How many gallons of 5.25 percent sodium hypochlorite will be needed to disinfect a well with an 18-inch diameter casing and well screen? The well is 300 feet deep and there is 200 feet of water in the well. Use an initial chlorine dose of 100 mg/*L*.

Known		Unknown
Hypochlorite, %	= 5.25%	5.25% Hypochlorite, gal
Chlorine Dose, mg/*L*	= 100 mg/*L*	
Diameter, in	= 18 in	
Water Depth, ft	= 200 ft	

1. Find the volume of water in the well in gallons.

$$\text{Water Vol, gal} = \frac{(0.785)(\text{Diameter, in})^2(\text{Water Depth, ft})(7.48\text{ gal/cu ft})}{144\text{ sq in/sq ft}}$$

$$= \frac{(0.785)(18\text{ in})^2(200\text{ ft})(7.48\text{ gal/cu ft})}{144\text{ sq in/sq ft}}$$

$$= 2{,}642\text{ gal}$$

2. Determine the pounds of chlorine needed.

$$\text{Chlorine, lbs} = (\text{Volume, M Gal})(\text{Dose, mg}/L)(8.34\text{ lbs/gal})$$

$$= (0.002642\text{ M Gal})(100\text{ mg}/L)(8.34\text{ lbs/gal})$$

$$= 2.2\text{ lbs chlorine}$$

3. Calculate the gallons of 5.25 percent sodium hypochlorite solution needed.

$$\text{Sodium Hypochlorite Solution, gallons} = \frac{(\text{Chlorine, lbs})(100\%)}{(8.34\ \text{lbs/gal})(\text{Hypochlorite, \%})}$$

$$= \frac{(2.2\ \text{lbs})(100\%)}{(8.34\ \text{lbs/gal})(5.25\%)}$$

$$= 5.0\ \text{gallons}$$

Five gallons of 5.25 percent sodium hypochlorite should do the job.

5.701 Mains

EXAMPLE 9

A section of an old 8-inch water main has been replaced and a 350-foot section of pipe needs to be disinfected. An initial chlorine dose of 400 mg/*L* is expected to maintain a chlorine residual of over 300 mg/*L* during the three-hour disinfection period. How many gallons of 5.25 percent sodium hypochlorite solution will be needed?

Known		Unknown
Diameter of Pipe, in	= 8 in	5.25% Hypochlorite, gallons
or 8 in/12 in/ft	= 0.67 ft	
Length of Pipe, ft	= 350 ft	
Chlorine Dose, mg/*L*	= 400 mg/*L*	
Hypochlorite, %	= 5.25%	

1. Calculate the volume of water in the pipe in gallons.

$$\text{Pipe Volume, gallons} = (0.785)(\text{Diameter, ft})^2(\text{Length, ft})(7.48\ \text{gal/cu ft})$$

$$= (0.785)(0.67\ \text{ft})^2(350\ \text{ft})(7.48\ \text{gal/cu ft})$$

$$= 923\ \text{gallons of water}$$

2. Determine the pounds of chlorine needed.

$$\text{Chlorine, lbs} = (\text{Volume, M Gal})(\text{Dose, mg/}L)(8.34\ \text{lbs/gal})$$

$$= (0.000923\ \text{M Gal})(400\ \text{mg/}L)(8.34\ \text{lbs/gal})$$

$$= 3.08\ \text{lbs chlorine}$$

3. Calculate the gallons of 5.25 percent sodium hypochlorite solution needed.

$$\text{Sodium Hypochlorite Solution, gallons} = \frac{(\text{Chlorine, lbs})(100\%)}{(8.34\ \text{lbs/gal})(\text{Hypochlorite, \%})}$$

$$= \frac{(3.08\ \text{lbs})(100\%)}{(8.34\ \text{lbs/gal})(5.25\%)}$$

$$= 7.0\ \text{gallons}$$

Seven gallons of 5.25 percent solution of sodium hypochlorite should do the job.

5.702 Tanks

EXAMPLE 10

An existing service storage reservoir has been taken out of service for inspection, maintenance and repairs. The reservoir needs to be disinfected before being placed back on line. The reservoir is 6 feet deep, 10 feet wide, and 25 feet long. An initial chlorine dose of 100 mg/*L* is expected to maintain a chlorine residual of over 50 mg/*L* during the 24-hour disinfection period. How many gallons of 5.25 percent sodium hypochlorite solution will be needed?

Known		Unknown
Tank Depth, ft	= 6 ft	5.25% Hypochlorite, gal
Tank Width, ft	= 10 ft	
Tank Length, ft	= 25 ft	
Chlorine Dose, mg/*L*	= 100 mg/*L*	
Hypochlorite, %	= 5.25%	

1. Calculate the volume of water in the tank in gallons.

$$\text{Tank Volume, gallons} = (\text{Length, ft})(\text{Width, ft})(\text{Depth, ft})(7.48\ \text{gal/cu ft})$$

$$= (25\ \text{ft})(10\ \text{ft})(6\ \text{ft})(7.48\ \text{gal/cu ft})$$

$$= 11{,}220\ \text{gallons}$$

2. Determine the pounds of chlorine needed.

$$\text{Chlorine, lbs} = (\text{Vol water, M Gal})(\text{Chlorine Dose, mg/}L)(8.34\ \text{lbs/gal})$$

$$= (0.01122\ \text{M Gal})(100\ \text{mg/}L)(8.34\ \text{lbs/gal})$$

$$= 9.36\ \text{lbs chlorine}$$

3. Calculate the gallons of 5.25 percent sodium hypochlorite solution needed.

$$\text{Sodium Hypochlorite Solution, gallons} = \frac{(\text{Chlorine, lbs})(100\%)}{(8.34\ \text{lbs/gal})(\text{Hypochlorite, \%})}$$

$$= \frac{(9.36\ \text{lbs})(100\%)}{(8.34\ \text{lbs/gal})(5.25\%)}$$

$$= 21.4\ \text{gallons}$$

Twenty-two gallons of 5.25 percent sodium hypochlorite solution should do the job.

5.71 Disinfection of Water From Wells

5.710 Chlorine Dose

EXAMPLE 11

A chlorine demand test from a well water sample produced a result of 1.2 mg/*L*. The water supplier would like to maintain a chlorine residual of 0.2 mg/*L* throughout the system. What should be the chlorine dose in mg/*L* from either a chlorinator or hypochlorinator?

Known		Unknown
Chlorine Demand, mg/*L*	= 1.2 mg/*L*	Chlorine Dose, mg/*L*
Chlorine Residual, mg/*L*	= 0.2 mg/*L*	

Calculate the chlorine dose in mg/*L*.

$$\text{Chlorine Dose, mg/}L = \text{Chlorine Demand, mg/}L + \text{Chlorine Residual, mg/}L$$

$$= 1.2\ \text{mg/}L + 0.2\ \text{mg/}L$$

$$= 1.4\ \text{mg/}L$$

NOTE: Be sure to check the chlorine residual regularly throughout the system. If the residual is low or there are coliforms present in the test results, then the residual should be increased.

5.711 Chlorinator

EXAMPLE 12

A deep well turbine pump is connected to a hydropneumatic tank. Under normal operating heads, the pump delivers 500 GPM. If the desired chlorine dosage is 3.5 mg/*L*, what should be the setting on the rotameter for the chlorinator (lbs chlorine per 24 hours)?

Known		**Unknown**
Pump Flow, GPM	= 500 GPM	Rotameter Setting, lbs chlorine/24 hours
Chlorine Dose, mg/*L*	= 3.5 mg/*L*	

1. Convert pump flow to million gallons per day (MGD).

$$\text{Flow, MGD} = \frac{(500\text{ GPM})(60\text{ min/hr})(24\text{ hr/day})}{1{,}000{,}000/\text{Million}}$$

$$= 0.72\text{ MGD}$$

2. Calculate the rotameter setting in pounds of chlorine per 24 hours.

$$\text{Rotameter Setting, lbs/day} = (\text{Flow, MGD})(\text{Dose, mg}/L)(8.34\text{ lbs/gal})$$

$$= (0.72\text{ MGD})(3.5\text{ mg}/L)(8.34\text{ lbs/gal})$$

$$= 21.0\text{ lbs chlorine/day}$$

$$= 21.0\text{ lbs chlorine/24 hours}$$

EXAMPLE 13

Using the results from Example 12 (a chlorinator setting of 21 lbs per 24 hours), how many pounds of chlorine would be used in one month if the pump hour meter shows the pump operates an average of 20 hours per day? The chlorinator operates only when the pump operates. How many 150-pound cylinders will be needed per month?

Known		**Unknown**
Chlorinator Setting, lbs/day	= 21 lbs/day	1. Chlorine Used, lbs/mo
Pump Operation, hr/day	= 20 hr/day	2. Cylinders Needed, no/mo
Chlorine Cylinders, lbs/cyl	= 150 lbs/cyl	

1. Calculate the chlorine used in pounds per month.

$$\text{Chlorine Used, lbs/mo} = \frac{(\text{Cl Setting, lbs/day})(\text{Operation, hr/day})(30\text{ days/mo})}{24\text{ hr/day}}$$

$$= \frac{(21\text{ lbs/day})(20\text{ hr/day})(30\text{ days/mo})}{24\text{ hr/day}}$$

$$= 525\text{ lbs/mo}$$

2. Determine the number of 150-pound cylinders needed per month.

$$\text{Cylinders Needed, no/mo} = \frac{\text{Chlorine Used, lbs/mo}}{\text{Chlorine Cylinders, lbs/cyl}}$$

$$= \frac{525\text{ lbs Cl/mo}}{150\text{ lbs Cl/cylinder}}$$

$$= 3.5\text{ Cylinders/month}$$

EXAMPLE 14

A deep well turbine pump delivers 400 GPM throughout a 24-hour period. The weight of chlorine in a 150-pound cylinder was 123 pounds at the start of the time period and 109 pounds at the end of the 24 hours. What was the chlorine dose rate in mg/*L*?

Known		**Unknown**
Pump Flow, GPM	= 400 GPM	Chlorine Dose, mg/*L*
Time Period, hr	= 24 hr	
Chlorine Wt at Start, lbs	= 123 lbs	
Chlorine Wt at End, lbs	= 109 lbs	

1. Convert flow of 400 GPM to MGD.

$$\text{Flow, MGD} = \frac{(400\text{ gal/min})(60\text{ min/hr})(24\text{ hr/day})}{1{,}000{,}000/\text{Million}}$$

$$= 0.576\text{ MGD}$$

2. Calculate the chlorine dose rate in mg/*L*.

$$\text{Chlorine Dose, mg}/L = \frac{\text{Chlorine Used, lbs/day}}{(\text{Flow, MGD})(8.34\text{ lbs/gal})}$$

$$= \frac{(123\text{ lbs} - 109\text{ lbs})/1\text{ day}}{(0.576\text{ MGD})(8.34\text{ lbs/gal})}$$

$$= \frac{14\text{ lbs chlorine/day}}{(0.576\text{ MGD})(8.34\text{ lbs/gal})}$$

$$= \frac{2.9\text{ lbs Chlorine}}{1\text{ M lbs Water}}$$

$$= 2.9\text{ mg}/L$$

5.712 Hypochlorinator

EXAMPLE 15

Water from a well is being treated by a hypochlorinator. If the hypochlorinator is set at a pumping rate of 50 gallons per day (GPD) and uses a 3 percent available hypochlorite solution, what is the chlorine dose rate in mg/*L* if the pump delivers 350 GPM?

Known		**Unknown**
Hypochlorinator, GPD	= 50 GPD	Chlorine Dose, mg/*L*
Hypochlorite, %	= 3%	
Pump, GPM	= 350 GPM	

1. Convert the pumping rate to MGD.

$$\text{Pumping Rate, MGD} = \frac{(350\text{ GPM})(60\text{ min/hr})(24\text{ hr/day})}{1{,}000{,}000/\text{Million}}$$

$$= 0.50\text{ MGD}$$

2. Calculate the chlorine dose rate in pounds per day.

$$\text{Chlorine Dose, lbs/day} = \frac{(\text{Flow, gal/day})(\text{Hypochlorite, \%})(8.34\text{ lbs/gal})}{100\%}$$

$$= \frac{(50\text{ gal/day})(3\%)(8.34\text{ lbs/gal})}{100\%}$$

$$= 12.5\text{ lbs/day}$$

3. Calculate the chlorine dose in mg/*L*.

$$\text{Chlorine Dose, mg/}L = \frac{\text{Chlorine Dose, lbs/day}}{\text{(Flow, MGD)(8.34 lbs/gal)}}$$

$$= \frac{\text{12.5 lbs chlorine/day}}{\text{(0.50 M Gal/day)(8.34 lbs/gal)}}$$

$$= \text{3 lbs chlorine/M lbs water}$$

$$= 3 \text{ mg/}L$$

EXAMPLE 16

Water pumped from a well is disinfected by a hypochlorinator. During a one-week time period, the water meter indicated that 1,098,000 gallons of water were pumped. A 2.0 percent sodium hypochlorite solution is stored in a 2.5-foot diameter plastic tank. During this one-week time period, the level of hypochlorite in the tank dropped 18 inches (1.50 ft). What was the chlorine dose in mg/*L*?

Known		Unknown
Water Treated, M Gal	= 1.098 M Gal	Chlorine Dose, mg/*L*
Hypochlorite, %	= 2.0%	
Hypochlorite Tank D, ft	= 2.5 ft	
Hypochlorite Used, ft	= 1.5 ft	

1. Calculate the pounds of water disinfected.

$$\text{Water, lbs} = \text{(Water Treated, M Gal)(8.34 lbs/gal)}$$

$$= \text{(1.098 M Gal)(8.34 lbs/gal)}$$

$$= \text{9.16 M lbs water}$$

2. Calculate the volume of hypochlorite solution used in gallons.

$$\text{Hypochlorite, gal} = \text{(0.785)(Diameter, ft)}^2\text{(Drop, ft)(7.48 gal/cu ft)}$$

$$= \text{(0.785)(2.5 ft)}^2\text{(1.5 ft)(7.48 gal/cu ft)}$$

$$= \text{55.0 gallons}$$

3. Determine the pounds of chlorine used to treat the water.

$$\text{Chlorine, lbs} = \text{(Hypochlorite, gal)}\left(\frac{\text{Hypochlorite, \%}}{100\%}\right)\text{(8.34 lbs/gal)}$$

$$= \text{(55.0 gal)}\left(\frac{2.0\%}{100\%}\right)\text{(8.34 lbs/gal)}$$

$$= \text{9.17 lbs chlorine}$$

4. Calculate the chlorine dose in mg/*L*.

$$\text{Chlorine Dose, mg/}L = \frac{\text{Chlorine Used, lbs}}{\text{Water Treated, Million lbs}}$$

$$= \frac{\text{9.17 lbs Chlorine}}{\text{9.16 M lbs Water}}$$

$$= \frac{\text{1.0 lbs Chlorine}}{\text{1 M lbs Water}}$$

$$= 1.0 \text{ mg/}L$$

EXAMPLE 17

Estimate the required concentration of a hypochlorite solution (%) if a pump delivers 600 GPM from a well. The hypochlorinator can deliver a maximum of 120 GPD and the desired chlorine dose is 1.8 mg/*L*.

Known		Unknown
Pump Flow, GPM	= 600 GPM	Hypochlorite Strength, %
Hypochl Flow, GPD	= 120 GPD	
Chlorine Dose, mg/*L*	= 1.8 mg/*L*	

1. Calculate the flow of water treated in million gallons per day.

$$\text{Water Treated, M Gal/day} = \frac{\text{(600 GPM)(60 min/hr)(24 hr/day)}}{\text{1,000,000/Million}}$$

$$= \text{0.864 MGD}$$

2. Determine the pounds of chlorine required per day.

$$\text{Chlorine Required, lbs/day} = \text{(Flow, MGD)(Dose, mg/}L\text{)(8.34 lbs/gal)}$$

$$= \text{(0.864 MGD)(1.8 mg/}L\text{)(8.34 lbs/gal)}$$

$$= \text{13.0 lbs chlorine/day}$$

3. Calculate the hypochlorite solution strength as a percent.

$$\text{Hypochlorite Strength, \%} = \frac{\text{(Chlorine Required, lbs/day)(100\%)}}{\text{(Hypochlorinator Flow, GPD)(8.34 lbs/gal)}}$$

$$= \frac{\text{(13.0 lbs/day)(100\%)}}{\text{(120 GPD)(8.34 lbs/gal)}}$$

$$= 1.3\%$$

EXAMPLE 18

A hypochlorite solution for a hypochlorinator is being prepared in a 55-gallon drum. If 10 gallons of 5 percent hypochlorite is added to the drum, how much water should be added to the drum to produce a 1.3 percent hypochlorite solution?

Known		Unknown
Drum Capacity, gal	= 55 gal	Water Added, gal
Hypochlorite, gal	= 10 gal	
Actual Hypo, %	= 5%	
Desired Hypo, %	= 1.3%	

or

$$\text{Desired Hypo, \%} = \frac{\text{(Hypo, gal)(Hypo, \%)}}{\text{Hypo, gal + Water Added, gal}}$$

Rearrange the terms in the equation.

$$\text{(Desired Hypo, \%)(Hypo, gal + Water Added, gal)} = \text{(Hypo, gal)(Hypo, \%)}$$

$$\text{(Desired Hypo, \%)(Hypo, gal)} + \text{(Desired Hypo, \%)(Water Added, gal)} = \text{(Hypo, gal)(Hypo, \%)}$$

$$\text{(Desired Hypo, \%)(Water Added, gal)} = \text{(Hypo, gal)(Hypo, \%)} - \text{(Desired Hypo, \%)(Hypo, gal)}$$

Calculate the volume of water to be added in gallons.

$$\text{Water Added, gal} = \frac{(\text{Hypo, gal})(\text{Actual Hypo, \%}) - (\text{Desired Hypo, \%})(\text{Hypo, gal})}{\text{Desired Hypo, \%}}$$

$$= \frac{(10\text{ gal})(5\%) - (1.3\%)(10\text{ gal})}{1.3\%}$$

$$= \frac{50 - 13}{1.3}$$

$$= 28.5\text{ gallons of water}$$

Add 28.5 gallons of water to the 10 gallons of 5 percent hypochlorite in the drum.

QUESTIONS

Write your answers in a notebook and then compare your answers with those on page 332.

5.7A A section of 12-inch water main has been repaired and a 400-ft section of pipe needs to be disinfected. An initial chlorine dose of 450 mg/*L* is expected to maintain a chlorine residual of over 300 mg/*L* during the three-hour disinfection period. How many gallons of 5 percent sodium hypochlorite solution will be needed?

5.7B Estimate the chlorine demand in milligrams per liter of a water that is dosed at 2.0 mg/*L*. The chlorine residual is 0.2 mg/*L* after a 30-minute contact period.

5.7C What should be the setting on a chlorinator (lbs chlorine per 24 hours) if a pump usually delivers 600 GPM and the desired chlorine dosage is 4.0 mg/*L*?

5.7D Water from a well is being disinfected by a hypochlorinator. If the hypochlorinator is set at a pumping rate of 60 gallons per day (GPD) and uses a 2 percent available chlorine solution, what is the chlorine dose rate in mg/*L*? The pump delivers 400 GPM.

5.8 ARITHMETIC ASSIGNMENT

Turn to the Appendix, "How to Solve Small Water System Arithmetic Problems," at the back of this manual and read Section A.9, *STEPS IN SOLVING PROBLEMS*. Also work the example problems and check the arithmetic using your calculator.

In Section A.13, *TYPICAL SMALL WATER SYSTEM PROBLEMS*, read and work the problems in Section A.134, Disinfection.

5.9 ADDITIONAL READING

1. *NEW YORK MANUAL*, Chapter 10,* "Chlorination."
2. *TEXAS MANUAL*, Chapter 10,* "Disinfection of Water."
3. *CHLORINE MANUAL*, Sixth Edition. Obtain from the Chlorine Institute, Inc., 2001 L Street, NW, Suite 506, Washington, DC 20036. Pamphlet 1. Price to members, $15.00; nonmembers, $30.00; plus 10 percent of order total for shipping and handling.
4. *AWWA STANDARD FOR DISINFECTING WATER MAINS*, C651-99. Obtain from American Water Works Association (AWWA), Bookstore, 6666 West Quincy Avenue, Denver, CO 80235. Order No. 43651. Price to members, $28.50; nonmembers, $41.50; price includes cost of shipping and handling.
5. *AWWA STANDARD FOR DISINFECTION OF WATER-STORAGE FACILITIES*, C652-92. Obtain from American Water Works Association (AWWA), Bookstore, 6666 West Quincy Avenue, Denver, CO 80235. Order No. 43652. Price to members, $28.50; nonmembers, $41.50; price includes cost of shipping and handling.
6. *AWWA STANDARD FOR DISINFECTION OF WATER TREATMENT PLANTS*, C653-97. Obtain from American Water Works Association (AWWA), Bookstore, 6666 West Quincy Avenue, Denver, CO 80235. Order No. 43653. Price to members, $28.50; nonmembers, $41.50; price includes cost of shipping and handling.
7. *AWWA STANDARD FOR DISINFECTION OF WELLS*, C654-97. Obtain from American Water Works Association (AWWA), Bookstore, 6666 West Quincy Avenue, Denver, CO 80235. Order No. 43654. Price to members, $28.50; nonmembers, $41.50; price includes cost of shipping and handling.

* Depends on edition.

5.10 ACKNOWLEDGMENT

Malcolm P. Dalton, General Manager, and James F. Brace, Staff Assistant, Navajo Tribal Utility Authority, provided many helpful contributions to all aspects of safety throughout this entire manual.

End of Lesson 2 of 2 Lessons on DISINFECTION

Please answer the discussion and review questions before continuing with the Objective Test.

DISCUSSION AND REVIEW QUESTIONS

Chapter 5. DISINFECTION

(Lesson 2 of 2 Lessons)

Write the answers to these questions in your notebook before continuing with the Objective Test on page 333. The question numbering continues from Lesson 1.

8. What procedures would you follow to safely handle chlorine cylinders?
9. What precautions would you take when shutting down a hypochlorinator for a short duration to make repairs?
10. How would you determine whether or not the chlorine feed rate of a chlorinator needs adjustment?
11. How would you determine the desired strength of a hypochlorite solution in the solution tank for effective operation of a hypochlorinator?
12. Why is safety equipment needed when repairing a chlorine leak?
13. How would you detect a chlorine leak?
14. What would you do if you could not repair a broken chlorinator quickly?
15. Why should clothing be removed from a person who has been in an area contaminated with liquid or gaseous chlorine?

SUGGESTED ANSWERS

Chapter 5. DISINFECTION

ANSWERS TO QUESTIONS IN LESSON 1

Answers to questions on page 277.

5.0A Pathogenic organisms are disease-producing organisms.

5.0B Disinfection is the selective destruction or inactivation of pathogenic organisms.

5.0C The U.S. Environmental Protection Agency establishes drinking water standards.

5.0D MCL stands for **M**aximum **C**ontaminant **L**evel.

Answers to questions on page 279.

5.1A Chlorine disinfects water much faster at a pH around 7.0 than at a pH over 8.0.

5.1B Relatively cold water requires longer disinfection time or greater quantities of disinfectants.

5.1C The number and type of organisms present in water influence the effectiveness of disinfection on microorganisms.

Answers to questions on page 280.

5.2A Physical agents that have been used for disinfection include (1) ultraviolet rays, (2) heat, and (3) ultrasonic waves.

5.2B Chemical agents that have been used for disinfection other than chlorine include (1) iodine, (2) bromine, (3) bases (sodium hydroxide and lime), and (4) ozone.

5.2C A major limitation to the use of ozone is the inability of ozone to provide a residual in the distribution system.

Answers to questions on page 283.

5.2D
$$\text{Chlorine Dose, mg}/L = \text{Chlorine Demand, mg}/L + \text{Chlorine Residual, mg}/L$$

5.2E
$$\text{Chlorine Demand, mg}/L = \text{Chlorine Dose, mg}/L - \text{Chlorine Residual, mg}/L$$

5.2F Hydrogen sulfide and ammonia are two inorganic reducing chemicals with which chlorine reacts rapidly.

Answers to questions on page 284.

5.2G Chlorine gas tends to lower the pH while hypochlorite tends to increase the pH.

5.2H The higher the pH level, the greater the percent of OCl^-.

Answers to questions on page 285.

5.2I Breakpoint chlorination is the addition of chlorine to water until the chlorine demand has been satisfied and further additions of chlorine result in a free available residual chlorine that is directly proportional to the amount of chlorine added beyond the breakpoint.

5.2J Chlorine reacts with organic matter to form chlororganic compounds and chloramines. Also, suspected carcinogenic compounds (trihalomethanes) may be formed.

5.2K Critical factors that influence disinfection include:

1. Effectiveness of upstream treatment processes,
2. Injection point and method of mixing,
3. Temperature,
4. Type of chemical and dosage,
5. pH,
6. Contact time, and
7. Chlorine residual.

Answers to questions on page 290.

5.2L An operator's decision to use chloramines depends on the ability to meet various regulations, the quality of the raw water, operational practices, and distribution system characteristics.

5.2M The three primary methods by which chloramines are produced are: (1) preammoniation followed by chlorination, (2) concurrent addition of chlorine and ammonia, and (3) prechlorination/postammoniation.

5.2N The *applied* chlorine to ammonia-nitrogen ratio is usually greater than the *actual* chlorine to nitrogen ratio leaving the plant because of the chlorine demand of the water.

5.2O Production of nitrite rapidly reduces free chlorine and can interfere with the measurement of free chlorine. The end result of incomplete nitrification may be a loss of total chlorine and ammonia and an increase in the concentration of heterotrophic plate count bacteria.

Answers to questions on page 294.

5.3A When disinfecting a well, the well, pump, screen and aquifer around the well all should be disinfected because they all could have been contaminated during construction and are potential sources of contamination.

5.3B The advantages of applying chlorine solution to the well by injecting the chlorine down the pump column pipe rather than injecting it through the vent pipe are the lower dosage required and the fact that use of the column pipe avoids contact with oil on the water surface.

5.3C A well has been successfully disinfected if the results of a bacteriological analysis for total coliforms are negative (no coliforms).

5.3D It is becoming standard practice to disinfect wells on a continuous basis because more and more groundwater sources are becoming contaminated.

Answers to questions on page 300.

5.3E Water mains can be kept clean during construction and repair by keeping material such as dirt, construction materials, animals, rodents and dirty water out of the mains. Inspecting the pipes when laid, keeping trenches dry and installing watertight plugs all help to keep pipes clean.

5.3F Before a water main is disinfected, the main should be flushed for at least 30 minutes with a flushing velocity of 2.5 ft/sec.

5.3G Areas in a water main that require an extra effort for successful disinfection include fittings, valves and air pockets where chlorine might not come in contact with the surface.

5.3H Three forms of chlorine used for disinfection include:

1. Chlorine gas (liquid chlorine in cylinders),
2. Calcium hypochlorite, and
3. Sodium hypochlorite.

5.3I Water used for disinfection that has a high chlorine residual may be disposed of in sanitary sewers, storm sewers, or on land, but should not be disposed of in a manner that will cause an adverse impact on the environment.

ANSWERS TO QUESTIONS IN LESSON 2

Answers to questions on page 311.

5.4A The two major types of chlorine feeders are (1) hypochlorinators, and (2) chlorinators.

5.4B Chlorine booster pumps are needed on most well applications to overcome the higher water pump discharge pressures.

5.4C The maximum rate of chlorine removal from a 150-pound cylinder is 40 pounds of chlorine per day.

Answers to questions on page 314.

5.4D The chemical feed of a hypochlorinator is adjusted until an adequate chlorine residual (0.2 mg/*L*) exists throughout the system and coliform test results are negative.

5.4E The level of the hypochlorite solution tank should be read at the same time every day.

5.4F The strength of the hypochlorite solution in the solution tank is adjusted so that the frequency of the strokes or pulses from the solution feed pump will be close together. This ensures that chlorine will be fed continuously to the water treated.

5.4G Maintenance usually required of hypochlorinators includes oil changes and lubrication.

Answers to questions on page 319.

5.4H Before attempting to locate and repair a chlorine gas leak, you should have protective clothing (gloves and rubber suit) and a self-contained pressure-demand air supply.

5.4I A chlorine leak can be detected by holding an ammonia-soaked rag near suspected leaks. A white cloud will reveal the location of the leak.

5.4J If a chlorinator cannot be repaired quickly, shut off the water supply so that unchlorinated or contaminated water will not be delivered to consumers. Follow standard procedures to shut down and repair the chlorinator.

5.4K The two water quality tests run on samples of water from a water supply system are (1) chlorine residual, and (2) coliform tests.

Answers to questions on page 322.

5.5A Chlorine residual is measured in treated water by the use of (1) amperometric titration, (2) DPD colorimetric method, and (3) ORP probes.

5.5B Residual chlorine measurements of treated water should be taken three times per day on small systems and once every two hours on large systems.

5.5C In a disinfection system, ORP is a direct measure of the effectiveness of a chlorine residual in disinfecting the water being treated.

5.5D When a microorganism loses an electron, it becomes inactivated and can no longer transmit a disease or reproduce.

5.5E Maintenance for the chlorine ORP probe consists of cleaning the unit's sensor once a month.

Answers to questions on page 325.

5.6A Chlorine gas is extremely toxic and corrosive in moist atmospheres.

5.6B A properly fitting self-contained air or oxygen supply type of breathing apparatus, pressure demand, or a rebreather kit is recommended when repairing a chlorine leak.

5.6C First-aid measures depend on the severity of the contact. Move the victim out of the gas area and remove contaminated clothing. Flush skin/eyes as needed and keep the victim warm and quiet. Call a doctor and the fire department immediately.

5.6D Whenever hypochlorite comes in contact with your hand, immediately wash the hypochlorite off and thoroughly wash your hand. Consult a physician if the area appears burned.

Answers to questions on page 329.

5.7A

Known		Unknown
Diameter of Pipe, in	= 12 in	5% Hypochlorite, gallons
or ft	= 1.0 ft	
Length of Pipe, ft	= 400 ft	
Chlorine Dose, mg/*L*	= 450 mg/*L*	
Hypochlorite, %	= 5%	

1. Calculate the volume of water in the pipe in gallons.

$$\text{Pipe Volume, gallons} = (0.785)(\text{Diameter, ft})^2(\text{Length, ft})(7.48 \text{ gal/cu ft})$$
$$= (0.785)(1.0 \text{ ft})^2(400 \text{ ft})(7.48 \text{ gal/cu ft})$$
$$= 2{,}349 \text{ gallons}$$

2. Determine the pounds of chlorine needed.

$$\text{Chlorine, lbs} = (\text{Vol water, M Gal})(\text{Chlorine Dose, mg}/L)(8.34 \text{ lbs/gal})$$
$$= (0.002349 \text{ M Gal})(450 \text{ mg}/L)(8.34 \text{ lbs/gal})$$
$$= 8.82 \text{ lbs chlorine}$$

3. Calculate the gallons of 5 percent sodium hypochlorite solution needed.

$$\text{Sodium Hypochlorite Solution, gallons} = \frac{(\text{Chlorine, lbs})(100\%)}{(8.34 \text{ lbs/gal})(\text{Hypochlorite, }\%)}$$
$$= \frac{(8.82 \text{ lbs Chlorine})(100\%)}{(8.34 \text{ lbs/gal})(5\%)}$$
$$= 21.2 \text{ gallons}$$

A little over 21 gallons of hypochlorite should do the job.

5.7B

Known		Unknown
Chlorine Dose, mg/*L*	= 2.0 mg/*L*	Chlorine Demand, mg/*L*
Chlorine Residual, mg/*L*	= 0.2 mg/*L*	

Calculate the chlorine demand in mg/*L*.

$$\text{Chlorine Demand, mg}/L = \text{Chlorine Dose, mg}/L - \text{Chlorine Residual, mg}/L$$
$$= 2.0 \text{ mg}/L - 0.2 \text{ mg}/L$$
$$= 1.8 \text{ mg}/L$$

5.7C

Known		Unknown
Pump Flow, GPM	= 600 GPM	Chlorinator Setting, lbs Chlorine/24 hr
Chlorine Dose, mg/*L*	= 4.0 mg/*L*	

1. Convert pump flow to million gallons per day (MGD).

$$\text{Flow, MGD} = \frac{(600 \text{ GPM})(60 \text{ min/hr})(24 \text{ hr/day})}{1{,}000{,}000/\text{Million}}$$
$$= 0.864 \text{ MGD}$$

2. Calculate the chlorinator setting in pounds of chlorine per 24 hours.

$$\text{Chlorinator Setting, lbs/24 hr} = (\text{Flow, MGD})(\text{Dose, mg}/L)(8.34 \text{ lbs/gal})$$
$$= (0.864 \text{ MGD})(4.0 \text{ mg}/L)(8.34 \text{ lbs/gal})$$
$$= 28.8 \text{ lbs chlorine/day}$$
$$= 28.8 \text{ lbs chlorine/24 hours}$$

5.7D

Known		Unknown
Hypochlorinator, GPD	= 60 GPD	Chlorine Dose, mg/*L*
Hypochlorite, %	= 2%	
Pump, GPM	= 400 GPM	

1. Convert the pumping rate to MGD.

$$\text{Pumping Rate, MGD} = \frac{(400 \text{ GPM})(60 \text{ min/hr})(24 \text{ hr/day})}{1{,}000{,}000/\text{Million}}$$
$$= 0.58 \text{ MGD}$$

2. Calculate the chlorine dose rate in pounds per day.

$$\text{Chlorine Dose, lbs/day} = \frac{(\text{Flow, gal/day})(\text{Hypochlorite, }\%)(8.34 \text{ lbs/gal})}{100\%}$$
$$= \frac{(60 \text{ GPD})(2\%)(8.34 \text{ lbs/gal})}{100\%}$$
$$= 10.0 \text{ lbs Chlorine/day}$$

3. Calculate the chlorine dose in mg/*L*.

$$\text{Chlorine Dose, mg}/L = \frac{\text{Chlorine Dose, lbs/day}}{(\text{Flow, MGD})(8.34 \text{ lbs/gal})}$$
$$= \frac{10.0 \text{ lbs Chlorine/day}}{(0.58 \text{ MGD})(8.34 \text{ lbs/gal})}$$
$$= 2.1 \text{ lbs chlorine/M lbs water}$$
$$= 2.1 \text{ mg}/L$$

OBJECTIVE TEST

Chapter 5. DISINFECTION

Please write your name and mark the correct answers on the answer sheet, as directed at the end of Chapter 1. There may be more than one correct answer to each multiple-choice question.

True-False

1. Sterilization is necessary when treating water for drinking purposes.
   1. True
   2. False
2. Longer contact times are required to disinfect water at lower temperatures.
   1. True
   2. False
3. Organics found in water can consume great amounts of disinfectants while forming unwanted compounds.
   1. True
   2. False
4. The demand for chlorine by reducing agents is satisfied after chlorine accomplishes disinfection.
   1. True
   2. False
5. Hypochlorite has a much greater disinfection potential than hypochlorous acid.
   1. True
   2. False
6. Chlorine dioxide reacts with organics in water.
   1. True
   2. False
7. Public water supplies are normally chlorinated precisely to the breakpoint.
   1. True
   2. False
8. Always add dry calcium hypochlorite to water, *NEVER* the reverse.
   1. True
   2. False
9. Each service line should be disinfected before being connected to a water main.
   1. True
   2. False
10. Never adjust the hypochlorinator pump while it is off because damage to the pump will occur.
    1. True
    2. False
11. Hypochlorinators on small systems are normally small, sealed systems that can be easily repaired.
    1. True
    2. False
12. When shutting down a chlorinator, be sure to close the chlorine discharge line.
    1. True
    2. False
13. The concentration of chemical and the length of contact time needed to ensure disinfection are the same for each disinfectant.
    1. True
    2. False
14. The employer must take an active part by supporting safety programs.
    1. True
    2. False
15. Hypochlorite presents the same hazards as presented by gaseous chlorine.
    1. True
    2. False

Best Answer (Select only the closest or best answer.)

16. What is purpose of disinfection of drinking water?
    1. The cleansing of organisms from drinking water
    2. The complete destruction of all organisms
    3. The removal of coliform organisms
    4. The selective destruction of pathogenic organisms
17. What is the chlorine dose?
    1. Chlorine demand minus chlorine residual
    2. Chlorine demand plus chlorine residual
    3. Chlorine demand plus free chlorine
    4. Chlorine residual minus chlorine demand
18. What problem occurs when calcium and fluoride ions combine in water supply systems?
    1. A corrosive solution develops
    2. A severe crust forms
    3. Calcium prevents fluoride uptake in teeth
    4. Disinfection power of chlorine is reduced
19. What is the only difference between the reactions of the hypochlorite compounds and chlorine gas?
    1. The form of hypochlorous acid produced
    2. The magnitude of the disinfecting power of chlorine
    3. The reactions with reducing agents
    4. The "side" reactions of the end products

20. After a chlorine solution has been fed into a well, how is the disinfectant mixed with the water in the well?
    1. By inducing a hydraulic jump
    2. By lowering a propeller mixer into the well
    3. By operating the pump on and off
    4. By pumping a nearby well to create turbulence

21. Why do operators never use water on a chlorine leak?
    1. The hydrochloric acid formed will make the leak worse
    2. The mixture of chlorine and water will reduce the disinfection strength of the chlorine
    3. The reaction between water and chlorine will cause a false sealing crust
    4. Water entering the cylinder will cause an explosion

22. HTH calcium hypochlorite powder is *NOT* used in pipes with solvent-welded plastic or screwed-joint steel pipe because the reaction between joint compounds and HTH could cause what problem?
    1. Blockage in the pipe
    2. Fire or explosion
    3. Leaks in joints
    4. Tastes and odors

23. How do direct-mounted chlorinators deliver chlorine to the water supply?
    1. Convey chlorine under a vacuum and inject it into the water supply
    2. Dissolve chlorine gas into a water solution and mix it into the water supply
    3. Feed chlorine solution through an ejector which draws additional dilution water
    4. Pump chlorine from a chemical solution tank into the water supply

24. What happens when moisture combines with chlorine gas?
    1. Chlorine will dry up the moisture
    2. Corrosive hydrochloric acid will be produced
    3. The chlorine and water will produce a salty scale
    4. The mixture will disinfect the system

25. Why should chlorine never be injected on the intake side of the pump?
    1. Chlorine will cause pump corrosion problems
    2. Chlorine will come out of solution when passing through the pump
    3. Insufficient contact time will occur in the pump
    4. Pump impeller will cause excessive chlorine mixing

26. What is the most important task when shutting down a hypochlorinator?
    1. Check the chlorine residual throughout the distribution system
    2. Exhaust hypochlorite solution so it won't deteriorate
    3. Perform necessary maintenance while pump is off
    4. Turn off water supply pump to avoid contaminating system

27. How can an operator prevent the formation of calcium carbonate coatings on the poppet valves in a hypochlorite solution feeder?
    1. By exercising the valves when performing scheduled maintenance
    2. By obtaining dilution water from an ordinary home water softener
    3. By switching to calcium hypochlorite
    4. By wiping the poppet valves clean every day

28. What is a "false positive" chlorine residual reading?
    1. Turbidity causing a higher coliform MPN than actually present
    2. Turbidity causing a positive coliform test when no coliforms are present
    3. Turbidity producing a higher chlorine residual than actually present
    4. Turbidity producing a positive test result for free chlorine residual when none is present

29. Which maintenance task is required for a chlorine ORP probe?
    1. Clean the unit's sensor once a month
    2. Lubricate all moving parts on a weekly basis
    3. Measure the chlorine strength daily
    4. Run a voltage test on the meter daily

30. Under what conditions can hypochlorite compounds cause fires?
    1. When in contact with electricity
    2. When in contact with gaseous chlorine
    3. When in contact with organics
    4. When in contact with oxygen

Multiple Choice (Select all correct answers.)

31. Which diseases are transmitted by water?
    1. Colds
    2. Cryptosporidiosis
    3. Dysentery
    4. Giardiasis
    5. Salmonellosis

32. What factors influence the effectiveness of chlorine disinfection?
    1. Organic matter
    2. pH
    3. Reducing agents
    4. Temperature
    5. Turbidity

33. Which water treatment processes are used to remove or kill pathogenic organisms?
    1. Coagulation
    2. Comminution
    3. Disinfection
    4. Filtration
    5. Sedimentation

34. Chlorine reacts with what impurities in water?
    1. Ammonia
    2. Chloride
    3. Hydrogen sulfide
    4. Organic materials
    5. Sulfate

35. What problems may be caused by combined chlorine residuals?
    1. Corrosive waters
    2. Excessive free available chlorine
    3. Formation of crusts
    4. Poor disinfecting power
    5. Tastes and odors

36. What problems may be caused by chloramines?
    1. Blending chlorinated waters can lower disinfectant residuals
    2. Can be deadly to fish
    3. Denitrification may occur
    4. Nitrification may occur
    5. Weakening effect on individuals with kidney disease undergoing kidney dialysis
37. What are the critical factors for effective chlorine disinfection?
    1. Chloride dose
    2. Chlorine residual
    3. Contact time
    4. Water pH
    5. Water temperature
38. Which of the following tasks are performed daily during normal operation of a hypochlorinator?
    1. Check the chemical feed pump operation
    2. Check the chlorine residual in the system
    3. Make any necessary adjustments in the feed rate after testing chlorine residuals
    4. Read and record level of solution tank
    5. Read the meters and record amount of water pumped
39. Which of the following tasks are performed when starting up a chlorinator?
    1. Have an ammonia bottle readily available to detect any leaks in the system
    2. Have safety equipment available that may be used in case of a leak
    3. Inspect the chlorine container for leaks
    4. Prior to the hookup, inspect for moisture and foreign substances in the lines
    5. Use new gaskets for connection of the chlorinator to the chlorine cylinder
40. What procedures should be implemented immediately if inadequately disinfected water enters a water distribution system?
    1. Distribute properly disinfected water as soon as possible
    2. Flush distribution system to remove inadequately disinfected water
    3. Implement purchase order procedures for replacement equipment
    4. Notify health department officials immediately
    5. Order additional chlorine supplies
41. What factors influence the required CT value to achieve pathogenic inactivation?
    1. Analytical coliform procedure
    2. pH of water
    3. Temperature of water
    4. Type of disinfectant
    5. Type of organism
42. Which conditions must be satisfied before an operator can safely enter a room containing high concentrations of chlorine gas?
    1. A self-contained air supply
    2. Help standing by
    3. Notify proper authorities
    4. Notify supervisor
    5. Wear protective clothing
43. What should be the chlorine dose of a water that has a chlorine demand of 2.3 mg/*L* if a residual of 0.5 mg/*L* is desired?
    1. 1.8 mg/*L*
    2. 2.3 mg/*L*
    3. 2.8 mg/*L*
    4. 3.3 mg/*L*
    5. 3.6 mg/*L*
44. What should be the setting on a chlorinator if a pump usually delivers approximately 595 GPM (assume 0.85 MGD) and the desired chlorine dose is 2.8 mg/*L*?
    1. 20 lbs chlorine/24 hours
    2. 22 lbs chlorine/24 hours
    3. 25 lbs chlorine/24 hours
    4. 28 lbs chlorine/24 hours
    5. 30 lbs chlorine/24 hours
45. Water from a well is being disinfected by a hypochlorinator. If the hypochlorinator is set at a pumping rate of 60 GPD and uses a 2.5 percent available chlorine solution, what is the chlorine dose rate in mg/*L*? The water pump delivers 500 GPM.
    1. 2.1 mg/*L*
    2. 2.4 mg/*L*
    3. 2.6 mg/*L*
    4. 2.8 mg/*L*
    5. 3.0 mg/*L*
46. A 10-inch water main 300 feet long needs to be disinfected. An initial chlorine dose of 450 mg/*L* is expected to maintain a chlorine residual of over 300 mg/*L* during the three-hour disinfection period. How many gallons of 5.25 percent sodium hypochlorite solution will be needed?
    1. 10.5 gallons
    2. 14.0 gallons
    3. 15.1 gallons
    4. 17.5 gallons
    5. 20.0 gallons
47. How many gallons of 15 percent sodium hypochlorite solution will be needed to disinfect a reservoir 20 feet in diameter and 6 feet deep? An initial chlorine dose of 65 mg/*L* is expected to maintain a chlorine residual of over 50 mg/*L* during the 24-hour disinfection period.
    1. 5.9 gallons
    2. 6.1 gallons
    3. 8.1 gallons
    4. 9.5 gallons
    5. 12.7 gallons
48. How many gallons of hypochlorite were pumped by a hypochlorinator if the hypochlorite solution was in a container with a diameter of 42 inches and the hypochlorite level drops 15 inches in 24 hours?
    1. 26.5 gallons
    2. 80.0 gallons
    3. 82.5 gallons
    4. 90.0 gallons
    5. 100.0 gallons

49. What is the desired strength (as a percent chlorine) of a hypochlorite solution being pumped by a hypochlorinator that delivers 60 gallons per day. The water being treated requires a chlorine feed rate of 15 pounds of chlorine per day.
    1. 2.5%
    2. 3.0%
    3. 3.3%
    4. 4.0%
    5. 5.2%

50. A hypochlorite solution for a hypochlorinator is being prepared in a 55-gallon drum. If 8 gallons of 5 percent hypochlorite is added to the drum, how much water should be added to produce a 1.5 percent hypochlorite solution.
    1. 18.7 gallons of water
    2. 21.5 gallons of water
    3. 24.7 gallons of water
    4. 25.3 gallons of water
    5. 28.5 gallons of water

Review Questions (Select all correct answers.)

51. What is the jar test commonly used to measure?
    1. Alkalinity
    2. Filterability
    3. Optimum coagulant dosages
    4. pH
    5. Turbidity

52. How can short-circuiting in flocculators be corrected?
    1. Adjust pH
    2. Increase coagulant dosages
    3. Increase flows
    4. Install proper baffling
    5. Provide several compartments in series

53. Which materials are suitable for use as filter media?
    1. Activated carbon
    2. Anthracite
    3. Diatomaceous earth
    4. Loam
    5. Sand

54. Which of the following devices are typical types of backflow preventers?
    1. Air gap separators
    2. Double check valve assemblies
    3. Gate valves
    4. Reduced-pressure principle devices
    5. Vacuum breakers

CHAPTER 6

SAFETY

by

Dan Saenz

and

Russ Armstrong

TABLE OF CONTENTS

Chapter 6. SAFETY

SAFETY ASSIGNMENT: DO AS I SAY... NOT AS I DO!!

OBJECTIVES

Chapter 6. SAFETY

Following completion of Chapter 6, you should be able to:

1. *THINK SAFETY,*
2. Develop a safety program for a water utility agency,
3. Prepare and conduct tailgate safety sessions,
4. Safely operate and maintain pumps and wells, with attention to the safety of operators and consumers,
5. Work safely in streets,
6. Protect the motoring public and pedestrians from work areas in streets and sidewalks, and
7. Conduct a safety inspection of waterworks facilities.

WORDS

Chapter 6. SAFETY

ACUTE HEALTH EFFECT ACUTE HEALTH EFFECT

An adverse effect on a human or animal body, with symptoms developing rapidly.

CAUTION CAUTION

This word warns against potential hazards or cautions against unsafe practices. Also see DANGER, NOTICE, and WARNING.

CHRONIC HEALTH EFFECT CHRONIC HEALTH EFFECT

An adverse effect on a human or animal body with symptoms that develop slowly over a long period of time or that recur frequently.

COMPETENT PERSON COMPETENT PERSON

A competent person is defined by OSHA as a person capable of identifying existing and predictable hazards in the surroundings, or working conditions which are unsanitary, hazardous or dangerous to employees, and who has authorization to take prompt corrective measures to eliminate the hazards.

CONFINED SPACE CONFINED SPACE

Confined space means a space that:

A. Is large enough and so configured that an employee can bodily enter and perform assigned work; and

B. Has limited or restricted means for entry or exit (for example, tanks, vessels, silos, storage bins, hoppers, vaults, and pits are spaces that may have limited means of entry); and

C. Is not designed for continuous employee occupancy.

(Definition from the Code of Federal Regulations (CFR) Title 29 Part 1910.146.)

CONFINED SPACE, NON-PERMIT CONFINED SPACE, NON-PERMIT

A non-permit confined space is a confined space that does not contain or, with respect to atmospheric hazards, have the potential to contain any hazard capable of causing death or serious physical harm.

CONFINED SPACE, PERMIT-REQUIRED (PERMIT SPACE) CONFINED SPACE, PERMIT-REQUIRED (PERMIT SPACE)

A confined space that has one or more of the following characteristics:

- Contains or has a potential to contain a hazardous atmosphere,
- Contains a material that has the potential for engulfing an entrant,
- Has an internal configuration such that an entrant could be trapped or asphyxiated by inwardly converging walls or by a floor which slopes downward and tapers to a smaller cross section, or
- Contains any other recognized serious safety or health hazard.

(Definition from the Code of Federal Regulations (CFR) Title 29 Part 1910.146.)

DANGER DANGER

The word *DANGER* is used where an immediate hazard presents a threat of death or serious injury to employees. Also see CAUTION, NOTICE, and WARNING.

DANGEROUS AIR CONTAMINATION DANGEROUS AIR CONTAMINATION

An atmosphere presenting a threat of causing death, injury, acute illness, or disablement due to the presence of flammable and/or explosive, toxic or otherwise injurious or incapacitating substances.

A. Dangerous air contamination due to the flammability of a gas or vapor is defined as an atmosphere containing the gas or vapor at a concentration greater than 10 percent of its lower explosive (lower flammable) limit.

B. Dangerous air contamination due to a combustible particulate is defined as a concentration greater than 10 percent of the minimum explosive concentration of the particulate.

C. Dangerous air contamination due to the toxicity of a substance is defined as the atmospheric concentration immediately hazardous to life or health.

DECIBEL (DES-uh-bull) DECIBEL

A unit for expressing the relative intensity of sounds on a scale from zero for the average least perceptible sound to about 130 for the average level at which sound causes pain to humans. Abbreviated dB.

MATERIAL SAFETY DATA SHEET (MSDS) MATERIAL SAFETY DATA SHEET (MSDS)

A document which provides pertinent information and a profile of a particular hazardous substance or mixture. An MSDS is normally developed by the manufacturer or formulator of the hazardous substance or mixture. The MSDS is required to be made available to employees and operators whenever there is the likelihood of the hazardous substance or mixture being introduced into the workplace. Some manufacturers are preparing MSDSs for products that are not considered to be hazardous to show that the product or substance is *NOT* hazardous.

NOTICE NOTICE

This word calls attention to information that is especially significant in understanding and operating equipment or processes safely. Also see CAUTION, DANGER, and WARNING.

OSHA (O-shuh) OSHA

The Williams-Steiger **O**ccupational **S**afety and **H**ealth **A**ct of 1970 (OSHA) is a federal law designed to protect the health and safety of industrial workers and also the operators of water supply systems and treatment plants. The Act regulates the design, construction, operation and maintenance of water supply systems and water treatment plants. OSHA also refers to the federal and state agencies which administer the OSHA regulations.

OXYGEN DEFICIENCY OXYGEN DEFICIENCY

An atmosphere containing oxygen at a concentration of less than 19.5 percent by volume.

OXYGEN ENRICHMENT OXYGEN ENRICHMENT

An atmosphere containing oxygen at a concentration of more than 23.5 percent by volume.

PERMIT-REQUIRED CONFINED SPACE (PERMIT SPACE) PERMIT-REQUIRED CONFINED SPACE (PERMIT SPACE)

See CONFINED SPACE, PERMIT-REQUIRED (PERMIT SPACE).

TAILGATE SAFETY MEETING TAILGATE SAFETY MEETING

Brief (10 to 20 minutes) safety meetings held every 7 to 10 working days. The term *TAILGATE* comes from the safety meetings regularly held by the construction industry around the tailgate of a truck.

WARNING WARNING

The word *WARNING* is used to indicate a hazard level between *CAUTION* and *DANGER*. Also see CAUTION, DANGER, and NOTICE.

CHAPTER 6. SAFETY[1]

(Lesson 1 of 2 Lessons)

6.0 IMPORTANCE OF SAFETY

6.00 Think Safety

LET'S START THINKING SAFETY NOW! In this section we are going to provide you with some ideas for your safety program. The remaining sections of this chapter will discuss how to perform specific jobs safely. Two very important aspects of your safety program are:

1. Making people aware of unsafe acts, and
2. Conducting and/or participating in regular safety training programs.

6.01 What Is Safety?

Webster's dictionary states that safety is the following:

> "The condition of being safe; freedom from exposure to danger; exemption from hurt, injury or loss; to protect against failure, breakage or other accidents; knowledge of or skill in methods of avoiding accident or disease."

Safety is more than words. Safety is the action of Webster's definition. Safety is using one's knowledge or skill to avoid accidents or to protect oneself and others from accidents. Safety is a form of preventive maintenance which includes equipment and machinery and its proper handling. Who is responsible for safety? Everyone should be responsible, from top management to all employees.

Safety is a program for everybody, not only on the job but also at home. Unfortunately, people miss more time at work from off-the-job accidents than from on-the-job accidents!

Management has its responsibilities for safety; its main function is to set the tone and provide the training and funding for an effective safety program. Management should be responsible for the following:

1. Establish a safety policy;
2. Assign responsibility for accident prevention, which includes description of the duties and responsibilities of a safety officer, department heads, line supervisors, safety committees, and operators;
3. Appoint a safety officer or coordinator;
4. Establish realistic goals and periodically revise them to ensure continuous and maximum effort; and
5. Evaluate the results of the program.

While management has its responsibilities, the operators also have their particular responsibilities. Operators should:

1. Perform their jobs in accordance with established safe procedures,
2. Recognize the responsibility for their own safety and that of fellow operators,
3. Report all injuries,
4. Report all observed hazards, and
5. Actively participate in the safety program.

QUESTIONS

Write your answers in a notebook and then compare your answers with those on page 390.

6.0A List two very important aspects of a safety program.

6.0B What are management's responsibilities for safety?

6.0C What are the operator's responsibilities for safety?

6.1 SAFETY PROGRAM

6.10 Objective of Safety Program

A SAFETY PROGRAM HAS ONE OBJECTIVE: TO PREVENT ACCIDENTS. Accidents do not happen, they are caused. They may be caused by unsafe acts of operators or result from hazardous conditions or may be a combination of both. An analysis of accident statistics reveals that operator negligence and carelessness are the causes of most accidents. We know how to do our job safely, but we just don't do it safely.

Accidents reduce efficiency and effectiveness. An accident is that occurrence in a sequence of events that usually produc-

[1] *Portions of this chapter include safety material developed and distributed by the Department of Water and Power, City of Los Angeles.*

es unintended injury, death, or property damage. Accidents affect the lives and morale of operators. They raise the costs not only to management but also to operators.

6.11 Unsafe Acts

Safety authorities tell us that nine out of ten injuries are the result of unsafe acts of either the person injured or someone else. Here are some of the principal reasons for unsafe acts.

IGNORANCE. This may be due to lack of experience or training, or to a temporary condition that prevents the recognition of a hazard.

INDIFFERENCE. Some people know better but don't care. They take unnecessary risks and disregard the rules or instructions.

POOR WORK HABITS. Some people either don't learn the right way of doing things, or they develop a wrong way. Supervisors or fellow operators who see things done unsafely must speak up.

LAZINESS. Laziness affects speed and quality of work. Laziness also affects safety because safety requires an effort. In most jobs you cannot reduce your "safety effort" and still maintain the same level of safety, even if you slow down or lower your quality standards.

HASTE. When we rush, we work too fast to think about what we are doing; we take dangerous shortcuts, and we are more likely to be injured.

POOR PHYSICAL CONDITION. Some of us just won't take reasonable care of ourselves. We ignore our bodily needs in regard to exercise, rest, and diet, lessening our endurance and alertness.

TEMPER. Impatience and anger cause many accidents. Again, our thinking is interfered with and the way is prepared for an accident to occur.

Therefore, every one of these unsafe acts could be considered due to operator negligence or carelessness.

SAFETY QUOTE: THE MORE YOU TALK ABOUT SAFETY, THE LESS YOU HEAR ABOUT ACCIDENTS.

6.12 Driving Safety

Defensive drivers are safe drivers. Good drivers check and maintain vehicles properly, use the safety equipment, are courteous to other drivers, use proper signals in advance of any directional changes, and know and observe all traffic regulations. Defensive drivers always use seat safety belts and make sure that all lights — headlights, taillights and directional lights — are operating correctly. Good drivers check to see that the tires are inflated to the proper pressure, that there are no bald spots, and that the tires are not wearing unevenly. Brakes should be operating correctly. They should be able to stop your vehicle without grabbing or pulling to one side. Routine oil changes, observation of adequate coolant levels, and tune-ups should be performed periodically. The vehicle should be operated following the manufacturer's recommendations, observing your water utility's safe driving policies, and observing all state and local driving regulations.

QUESTIONS

Write your answers in a notebook and then compare your answers with those on page 390.

6.1A What is an accident?

6.1B List the principal reasons for unsafe acts.

6.1C What problems are caused by laziness?

6.1D How can you determine if the brakes on a vehicle are working properly?

6.13 Towing a Trailer

Frequently it is necessary to tow trailer-mounted equipment such as generators, compressors, and pumps. Towing a trailer requires experience, knowledge, and a great amount of common sense. The job of towing a trailer seems to be a simple task; however, there are many pitfalls if good judgment is not exercised, or if the safety of others is not considered.

Many accidents related to trailer towing occur while simply coupling the trailer to the vehicle. Back strains and cuts and bruises to the hands and fingers are the types of injuries that often occur before starting to tow a trailer.

The trailer and towing vehicle should be inspected and checked for the following:

1. Rear, turn signal, and brake lights,
2. Tire tread and inflation,
3. Wheel attachment (vandals may have loosened lug nuts),
4. Trailer brake operation if equipped with brakes,
5. Trailer hitch and tongue, and
6. Safety chain. After the trailer is coupled, the safety chain should be securely attached to the frame of the towing vehicle with enough slack to allow jackknifing, but not enough to drag on the ground.

Once underway with a trailer in tow, the three items below are examples of when good judgment should be exercised:

1. Maximum speed for all vehicles is 55 MPH — remember that some trailers will not handle properly even at 45 MPH.
2. Braking or stopping distances will increase if the trailer itself is not equipped with brakes.
3. Trailers make your vehicle much longer. Allow extra room when making turns, changing lanes, or passing.

6.14 How To Charge a Battery

At one time or another you may have given a battery a boost or may have seen it done. There is a correct procedure to follow to eliminate damage to electrical components and to prevent a battery explosion (Figure 6.1).

To boost the battery of a disabled vehicle from that of another vehicle, follow this procedure. First put out all ciga-

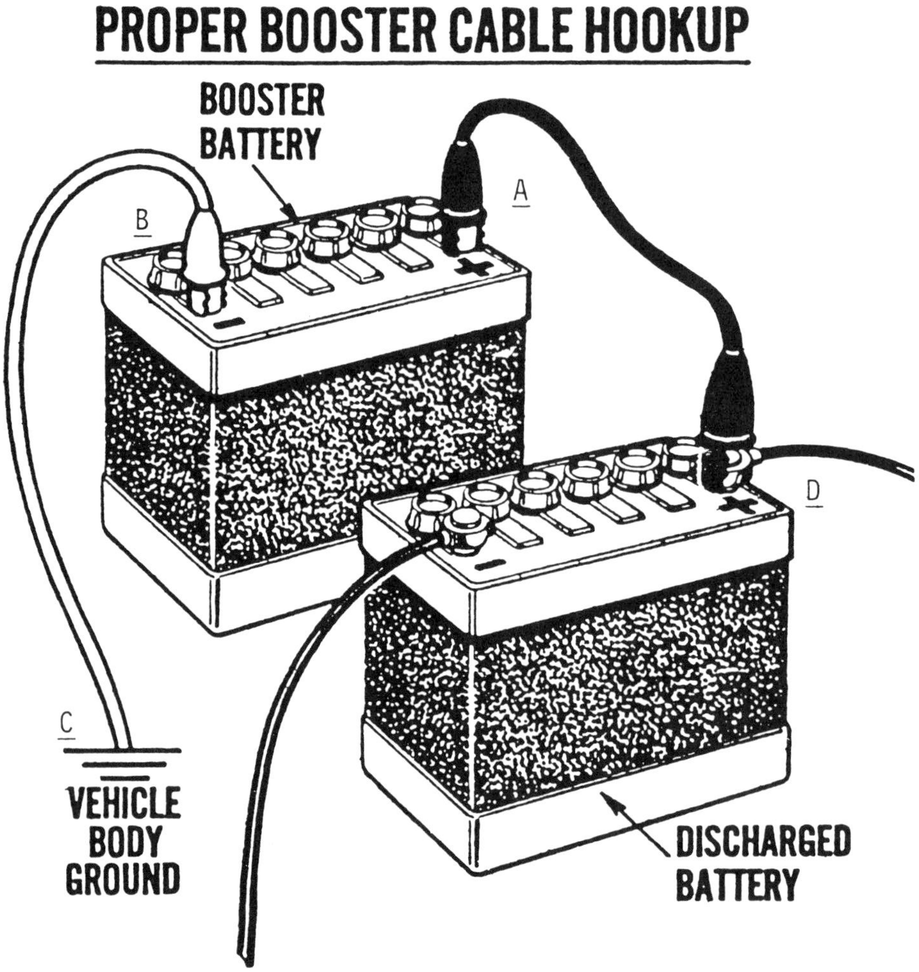

Fig. 6.1 Proper booster cable hookup

rettes and flames. A spark can ignite hydrogen gas from the battery fluid. Next take off the battery caps, if removable, and add distilled water if it is needed. Check for ice in the battery fluid. Never jump-start a frozen battery! Replace the caps.

Next, park the auto with the "live" battery close enough so the cables will reach between the batteries of the two autos. The cars can be parked close, but do not allow them to touch. If they touch, a dangerous "arcing" situation could occur. Now set each car's brake. Be sure that an automatic transmission is set in park; put a manual-shift transmission in neutral. Make sure your headlights, heater, and all other electrical accessories are off (you don't want to sap electricity away from your dead battery while you are trying to start your car). If the two batteries have vent caps, remove them and lay a cloth over the open holes. This will reduce the risk of explosion (relieves pressure within the battery).

Attach one end of the jumper cable to the positive terminal of the booster battery (A) (that's the good battery in the other car) and the other end to the positive terminal of your battery (D). The positive terminal is identified by a + sign, a red color, or a "P" on the battery in your car. Each of the two booster cables has an alligator clip at each end (not shown in Figure 6.1). To attach, you simply squeeze the clip, place it over the terminal, then let it shut. Now attach one end of the remaining booster cable to the negative terminal of the booster battery (B). The negative terminal is marked with a – sign, a black color, or the letter "N." Attach the other end of the cable to a metal part on the engine of your car (C). Many mechanics simply attach it to the negative post of the battery. This is not recommended because a resulting arc could ignite hydrogen gas present at the battery surface and cause an explosion. Be sure the cables do not interfere with the fan blades or belts. The engine in the other car should be running, although it is not an absolute necessity.

Get in the disabled car and start the engine in a normal manner. After it starts, remove the booster cables. Removal is the exact reverse of installation. Remove the black cable attached to your engine, then remove it from the negative terminal of the booster battery. Remove the remaining cable from the positive terminal of the "dead" battery and then from the booster battery. Remove the cloth cover from the fill-holes, replace the vent caps and you are done. Have the battery and/or charging system of the car checked by a mechanic to correct any problems.

For maximum eye safety, everyone working with car batteries or standing nearby should wear protective goggles to keep flying battery fragments and chemicals out of the eyes.

Should battery acid get into the eyes, immediately flush them with water continuously for 15 minutes. Then see a doctor.

The recommended procedure for jump-starting is listed, step by step, on a 4-inch by 8-inch yellow vinyl sticker which has a permanent adhesive. The sticker can be affixed to any clean, dry surface under the hood or kept inside the car's glove compartment. Stickers are available — no charge for the first three, $24.00 per 100 after that — by writing to Prevent Blindness America, 500 East Remington Road, Schaumburg, IL 60173.

6.15 Boat Safety

Boats are commonly used in the operation of a water utility. They are used for sampling, surveying, or inspecting reservoirs. Boats may be either motor operated (inboard or outboard) or rowboats. All boats should be inspected annually. The U.S. Coast Guard is usually the inspecting agency. They also recommend the safety practices to be used for boat safety. All boats should be equipped with a safety vest for each occupant or some type of approved flotation cushion. There should be oarlocks for the oars and oars should be of a suitable length for the respective boat and maintained in good operable condition. No boat should be loaded in excess of its loading capacity.

Motorized boats should meet the same standards as rowboats. In addition, the motors (inboard and outboard) should be maintained in good operating condition. Periodic tune-ups and safety checks should be made following your utility's maintenance procedures or the manufacturer's recommendations. All motorized boats should have a set of oars to be used in an emergency and a fire extinguisher. At least two people should go out on a boat. If it is necessary that only one person use a boat, someone on shore should be aware that the person is out on the boat alone. Following safety practices when using boats could save someone from a potential drowning accident.

6.16 Corrosive Chemicals

A corrosive chemical is any chemical which may weaken, burn, and/or destroy a person's skin or eyes. A corrosive chemical may be either acid or base (alkali). The pH scale from 0 to 14 is an indication of the strength of a solution. The "acids" with the low pH values are the most corrosive. Hydrochloric (HCl), nitric (HNO_3), and sulfuric (H_2SO_4) are among the strongest acids. A "base" with the highest pH values of around 13 is the most corrosive. Sodium hydroxide (NaOH) is a strong base. Some common corrosive chemicals are:

1. Ammonium hydroxide,
2. Calcium hypochlorite,
3. Chlorine,
4. Ferric chloride,
5. Hydrochloric acid,
6. Potassium permanganate,
7. Sodium chlorite,
8. Sodium hydroxide,
9. Sodium hypochlorite,
10. Sodium nitrate,
11. Sulfuric acid, and
12. Zinc orthophosphate.

This list is by no means complete.

Safety procedures must be followed in the handling and use of all corrosive chemicals. Use safety protection equipment to protect your eyes and skin (see Section 6.21, "Protective Clothing"). Safety showers and eye/face wash stations should be installed at a location close to where any corrosive chemical is being used. Test the shower and eye/face wash station at least once a week. Run the water using the emergency levers (the pull chains, hand-actuated valves, or treadle foot-actuated valves). Let the water run for approximately three to five minutes or until the water runs clear. Rust may be in the water lines and may increase the problem of washing chemicals from your eyes.

6.17 Tailgate Safety Sessions

Tailgate safety sessions are so called because the session consists of a small group of operators gathered around the tailgate of a pickup or truck to discuss safety. These sessions are usually held near the work site and new and old safety hazards and safer approaches or techniques to deal with the problems of the day or week are discussed. The amount or degree of organization or formality of the sessions depends on the types of safety problems and what you think is the most effective way of educating your crew. Table 6.1 is a typical topic for a tailgate safety session.

TABLE 6.1 TAILGATE SAFETY TOPIC

Most injuries that happen on water utility jobs are caused directly by the injured person. Only a small percent are caused by defective equipment or devices. Because of this, you must be primarily responsible for your own safety.

Managers and supervisors are usually regarded as being the ones responsible for safety. Without proper interest on the part of management and supervisory staff, a total safety program can't be effective. But you must realize that you, more than anyone else, are responsible for not only your own safety but for the safety of your fellow operators. In other words, you must be your "brother's keeper."

Here's an example of what we're trying to say. Management can purchase new trucks and equip them with all the known safety devices and maintain them in perfect operating condition. But a truck has to be operated by a driver (you); you alone are responsible for its safe operation so that neither you nor any of your fellow operators will get hurt.

Here's another example. Take the simple wooden ladder. The ladder may be built to the best safety specifications; it may have been properly stored and frequently inspected for defects. Somebody, however, must place the ladder in position and somebody must use it. If it's not properly placed, if the footing is insecure, if it hasn't been secured to the building by some individual, it is entirely likely that the person using it or some other operator will get hurt. And how can either management or a supervisor be held responsible for such accidents? You must realize that you are the most vital factor in the control of accidents.

Some operators seem to think that the safety officer is responsible for preventing accidents. Even though the safety officer frequently makes inspections and counsels the operators and supervisors, the safety operator can't be in all places at all times. And the safety officer cannot, therefore, be blamed or held responsible whenever an accident occurs. So let's bear in mind that we ourselves, as individuals, must constantly be alert to the hazards around us. If we can't remove a hazard ourselves, we should call it to the attention of those who have the authority to do so.

6.18 Employee "Right-To-Know" Laws

Employee "Right-To-Know" legislation has been implemented to require employers to inform employees (operators) of the possible health effects resulting from contact with hazardous substances. At locations where this legislation is in force, employers must meet certain requirements. Employers must provide operators with information regarding any hazardous substances which the operators might be exposed to under either normal work conditions or reasonably foreseeable emergency conditions resulting from workplace conditions. Information regarding a hazardous substances is available from the manufacturer in the form of a *MATERIAL SAFETY DATA SHEET (MSDS)*.[2] Employers must provide operators with a copy of MSDSs upon request and also must train operators to work safely with the hazardous substances that are encountered in the workplace.

As an operator you have the right to ask your employer if you are working with any hazardous substances. If you work with any hazardous substances, your employer must provide you with information on the health implications resulting from contact with the substances, the Material Safety Data Sheets, and training for working safely with the hazardous substances.

QUESTIONS

Write your answers in a notebook and then compare your answers with those on page 390.

6.1E List the items that should be inspected before towing a trailer.

6.1F Why should all cigarettes and flames be put out before attaching cables to charge a battery?

6.1G List the safety practices for using boats.

6.1H What is a corrosive chemical?

6.1I What is a tailgate safety session?

6.2 PERSONAL SAFETY

Personal safety goes hand-in-hand with a good, effective safety program. You are the key individual in an effective safety program.

Don't you, as an operator, become a statistic.

Table 6.2 is a tabulation of the types of accidents encountered in the water utility field. This chapter will provide you with safe procedures which will help you avoid becoming an accident statistic.

TABLE 6.2 ANALYSIS OF ACCIDENT TYPES[a]

	Percent
1. Sprains/strains in lifting, pulling, or pushing objects	28
2. Sprains/strains due to awkward position or sudden twist or slip	16
3. Struck by falling or flying objects	10
4. Struck against stationary or moving objects	9
5. Falls to different level from platform, ladder, stairs	7
6. Caught in, under, or between objects	6
7. Falls on same level to working surface	5
8. Contact with radiation, caustics, toxic, and noxious substances	3
9. Animal or insect bites	1
10. Contact with temperature extremes	1
11. Rubbed or abraded	1
12. Contact with electric current	0
13. Miscellaneous	11

[a] AWWA Safety Bulletin

6.20 Monitoring Equipment

There are many locations in a water utility where it is necessary to monitor the atmosphere before entering. Some of these specific locations may be underground regulator vaults, solution vaults, manholes, tanks, trenches, and other *CONFINED SPACES*[3] (also see Section 6.63, "Confined Spaces"). A variety of devices are used to monitor (check) the available oxygen and also combustible and toxic gases. Regardless of what type of monitoring device or devices are used, you should always test for *DANGEROUS AIR CONTAMINATION*,[4] *OXYGEN DEFICIENCY*,[5] and/or *OXYGEN ENRICHMENT*[6] with an approved device immediately prior to an operator en-

[2] *Material Safety Data Sheet (MSDS). A document which provides pertinent information and a profile of a particular hazardous substance or mixture. An MSDS is normally developed by the manufacturer or formulator of the hazardous substance or mixture. The MSDS is required to be made available to employees and operators whenever there is the likelihood of the hazardous substance or mixture being introduced into the workplace. Some manufacturers are preparing MSDSs for products that are not considered to be hazardous to show that the product or substance is NOT hazardous.*

[3] *Confined Space. Confined space means a space that:*
A. Is large enough and so configured that an employee can bodily enter and perform assigned work; and
B. Has limited or restricted means for entry or exit (for example, tanks, vessels, silos, storage bins, hoppers, vaults, and pits are spaces that may have limited means of entry); and
C. Is not designed for continuous employee occupancy.
(Definition from the Code of Federal Regulations (CFR) Title 29 Part 1910.146.)

[4] *Dangerous Air Contamination. An atmosphere presenting a threat of causing death, injury, acute illness, or disablement due to the presence of flammable and/or explosive, toxic or otherwise injurious or incapacitating substances.*
A. Dangerous air contamination due to the flammability of a gas or vapor is defined as an atmosphere containing the gas or vapor at a concentration greater than 10 percent of its lower explosive (lower flammable) limit.
B. Dangerous air contamination due to a combustible particulate is defined as a concentration greater than 10 percent of the minimum explosive concentration of the particulate.
C. Dangerous air contamination due to the toxicity of a substance is defined as the atmospheric concentration immediately hazardous to life of health.

[5] *Oxygen Deficiency. An atmosphere containing oxygen at a concentration of less than 19.5 percent by volume.*

[6] *Oxygen Enrichment. An atmosphere containing oxygen at a concentration of more than 23.5 percent by volume.*

tering a confined space and at intervals frequent enough to ensure a safe atmosphere during the time an operator is in the structure. Dangerous air contamination includes explosive conditions and toxic gases. Toxic gases include hydrogen sulfide and carbon monoxide. *A RECORD OF THE TEST MUST BE KEPT AT THE JOB SITE* for the duration of the work. *YOUR LIFE* and that of *YOUR FELLOW OPERATORS* is in jeopardy if the air in a confined space is not tested before entering. The following five steps should be performed before entering a confined space:

1. *CALIBRATE* the gas detection device,
2. *BARELY OPEN* the confined space,
3. *TEST* for combustible and toxic gases and oxygen deficiency/enrichment using a probe or tube to collect the sample,
4. *RECORD* the results of these tests in your gas log, and
5. *VENTILATE* the confined space.

If a hazardous atmospheric condition is discovered, you must repeat the tests for combustible gases, toxic gases, and oxygen deficiency/enrichment while ventilating and again *RECORD* the results.

On rare occasions, unusual conditions may exist or be created even though the required testing procedures have been followed. Air currents through duct lines can easily change if other vaults in the same system are opened. Toxic or explosive gases from broken gas lines or decayed vegetation may then flow through the ducts into a previously gas-free vault. Therefore it is important that *ADEQUATE VENTILATION BE MAINTAINED AND THAT THE SPACE BE RECHECKED FOR HAZARDOUS ATMOSPHERES PERIODICALLY WHILE OPERATORS ARE WORKING IN SUCH LOCATIONS.* Use portable fans or ventilators to provide fresh air.

At least one backup person must stay outside the confined work area with another person standing by. This backup person should check continuously on the status of personnel working in the confined space. Should a person working in the confined space be rendered unconscious due to asphyxiation or lack of oxygen, the backup person at the surface could descend into the space after:

1. Securing the help of an additional backup rescue person, and
2. Putting on the correct type of respiratory protective equipment. (Self-contained breathing equipment should be nearby and used as necessary.)

Follow the procedures outlined in the *SAFETY NOTICE* whenever anyone must enter a confined space.

SAFETY NOTICE

Before anyone ever enters a tank for any reason, these safety procedures must be followed:

1. Test atmosphere in the tank for toxic and explosive gases and for adequate or inadequate oxygen. Contact your local safety equipment supplier for the proper types of atmospheric testing devices. These devices should have alarms which are activated whenever an unsafe atmosphere is encountered;
2. Provide adequate ventilation, especially when painting. A self-contained, positive-pressure breathing apparatus may be necessary when painting;
3. All persons entering a tank must wear a safety harness; and
4. One person must be stationed at the tank entrance to observe the actions of all people in the tank. An additional person must be readily available to help the person at the tank entrance with any rescue operation.

Table 6.3 lists some common dangerous gases which may be encountered by the operators of water supply systems and water treatment plants.

Not every mixture of gas with air will explode or burn. The "proper" mixture of oxygen with a gas is necessary to create explosive or flammable conditions. If there is not enough gas present (too much oxygen), the mixture is too lean and will not burn or explode. Also, if there is too much gas (not enough oxygen), the mixture is too rich and will not burn or explode. The range of explosive or flammable gas mixtures is defined by the Lower Explosive Limit (LEL) and Upper Explosive Limit (UEL). The objective of gas detection equipment is to warn us when we encounter potentially explosive conditions. Any time a gas mixture exceeds 10 percent of the Lower Explosive Limit, this is considered a hazardous condition.

Figure 6.2 shows the relationship between the Lower Explosive Limit (LEL) and the Upper Explosive Limit (UEL) of a mixture of air and gas. Note in Table 6.3 that the percent by volume of a gas in air for the Lower Explosive Limit to occur is different for each gas. Also note in Figure 6.2 that if you have all gas (100% gas and 0% oxygen), the mixture is too rich and will not explode. If a mixture of gas and oxygen from the air is greater than 10 percent of the Lower Explosive Limit, the mixture is considered hazardous.

Monitoring equipment is essential to protect you from hazardous atmospheres when entering and working in enclosed spaces. This type of monitoring equipment is available in three different types:

1. Portable and attached to your clothing or your body,
2. Portable and carried by you or placed near your work site, and
3. Permanently installed in locations where hazardous atmospheres could develop.

TABLE 6.3 COMMON DANGEROUS GASES ENCOUNTERED IN WATER SUPPLY SYSTEMS AND AT WATER TREATMENT PLANTS[a]

Name of Gas	Chemical Formula	Specific Gravity or Vapor Density[b] (Air = 1)	Explosive Range (% by volume in air) Lower Limit	Upper Limit	Common Properties (Percentages below are percent in air by volume)	Physiological Effects (Percentages below are percent in air by volume)	Most Common Sources in Sewers	Simplest and Cheapest Safe Method of Testing[c]
Oxygen (in Air)	O_2	1.11	Not flammable		Colorless, odorless, tasteless, nonpoisonous gas. Supports combustion.	Normal air contains 20.93% of O_2. If O_2 is less than 19.5%, do not enter space without respiratory protection.	Oxygen depletion from poor ventilation and absorption or chemical consumption of available O_2.	Oxygen deficiency indicator.
Gasoline Vapor	C_5H_{12} to C_9H_{20}	3.0 to 4.0	1.3	7.0	Colorless, odor noticeable in 0.03%. Flammable. Explosive.	Anesthetic effects when inhaled. 2.43% rapidly fatal. 1.1% to 2.2% dangerous for even short exposure.	Leaking storage tanks, discharges from garages, and commercial or home dry-cleaning operations.	1. Combustible gas indicator. 2. Oxygen deficiency indicator for concentrations over 0.3%.
Carbon Monoxide	CO	0.97	12.5	74.2	Colorless, odorless, nonirritating, tasteless. Flammable. Explosive.	Hemoglobin of blood has strong affinity for gas, causing oxygen starvation. 0.2 to 0.25% causes unconsciousness in 30 minutes.	Manufactured fuel gas.	CO ampoules.
Hydrogen	H_2	0.07	4.0	74.2	Colorless, odorless, tasteless, nonpoisonous, flammable. Explosive. Propagates flame rapidly; very dangerous.	Acts mechanically to deprive tissues of oxygen. Does not support life. A simple asphyxiant.	Manufactured fuel gas.	Combustible gas indicator.
Methane	CH_4	0.55	5.0	15.0	Colorless, tasteless, odorless, nonpoisonous. Flammable. Explosive.	See hydrogen.	Natural gas, marsh gas, mfg. fuel gas, gas found in sewers.	1. Combustible gas indicator. 2. Oxygen deficiency indicator.
Hydrogen Sulfide	H_2S	1.19	4.3	46.0	Rotten egg odor in small concentrations, but sense of smell rapidly impaired. Odor not evident at high concentrations. Colorless. Flammable. Explosive. Poisonous.	Death in a few minutes at 0.2%. Paralyzes respiratory center.	Petroleum fumes, from blasting, gas found in sewers.	1. H_2S analyzer. 2. H_2S ampoules.
Carbon Dioxide	CO_2	1.53	Not flammable		Colorless, odorless, nonflammable. Not generally present in dangerous amounts unless there is already a deficiency of oxygen.	10% cannot be endured for more than a few minutes. Acts on nerves of respiration.	Issues from carbonaceous strata. Gas found in sewers.	Oxygen deficiency indicator.
Nitrogen	N_2	0.97	Not flammable		Colorless, tasteless, odorless. Nonflammable. Nonpoisonous. Principal constituent of air (about 79%).	See hydrogen.	Issues from some rock strata. Gas found in sewers.	Oxygen deficiency indicator.
Ethane	C_2H_4	1.05	3.1	15.0	Colorless, tasteless, odorless, nonpoisonous. Flammable. Explosive.	See hydrogen.	Natural gas.	Combustible gas indicator.
Chlorine	Cl_2	2.5	Not flammable Not explosive		Greenish yellow gas, or amber color liquid under pressure. Highly irritating and penetrating odor. Highly corrosive in presence of moisture.	Respiratory irritant, irritating to eyes and mucous membranes. 30 ppm causes coughing. 40-60 ppm dangerous in 30 minutes. 1,000 ppm apt to be fatal in a few breaths.	Leaking pipe connections. Overdosage.	Chlorine detector. Odor. Strong ammonia on swab gives off white fumes.

[a] Originally printed in Water and Sewage Works, August 1953. Adapted from "Manual of Instruction for Sewage Treatment Plant Operators," State of New York.
[b] Gases with a specific gravity less than 1.0 are lighter than air; those more than 1.0 heavier than air.
[c] The first method given is the preferable testing procedure.

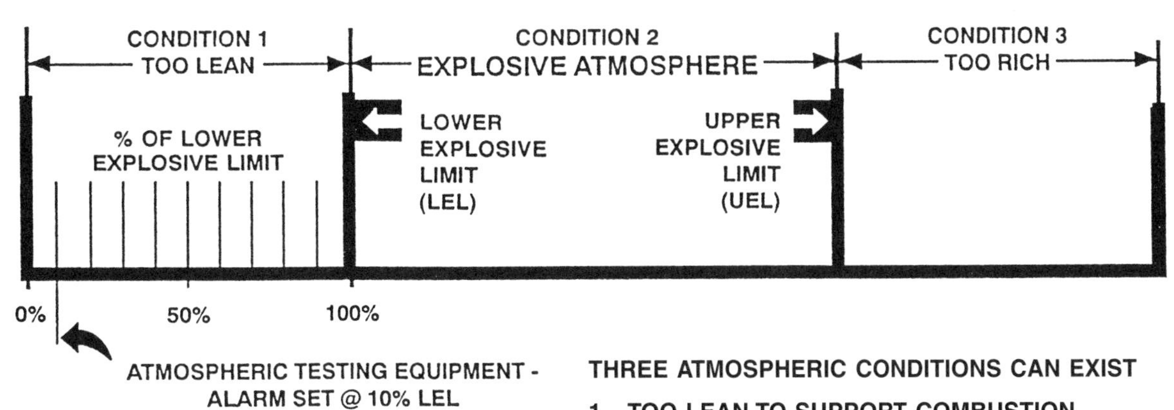

Fig. 6.2 Relationship between the lower explosive limit or level (LEL) and the upper explosive limit (UEL) of a mixture of air and gas

You may encounter three types of hazardous atmospheres:

1. Toxic gases such as chlorine, carbon monoxide, or hydrogen sulfide,
2. Explosive or flammable conditions, and
3. Oxygen deficiency/enrichment.

Regardless of the type of detection devices you are using to monitor hazardous atmospheres, regularly calibrate the detection devices according to the manufacturer's recommendations. These calibration procedures must be scheduled and performed as part of your preventive maintenance program. Remember, the detection devices are sensitive instruments. Handle them gently and keep them accurate if you wish to stay alive.

The recommended warning devices should be an audible noise and/or a flashing light that should start whenever a hazardous condition is detected. An operator should not be expected to read a meter while working and decide whether or not a hazard exists.

Sometimes operators working in small communities or for small agencies experience problems convincing their supervisors of the need for gas detection devices, portable ventilators, safety harnesses, and qualified people standing by when they must enter tanks, vaults, and manholes. *OSHA*[7] regulations allow operators to refuse to perform work under unsafe conditions. This is a very difficult situation. We recommend that operators contact their local or state OSHA officials or other industrial safety officials. Obtain copies of the rules and regulations for entering tanks and confined spaces (see Section 6.63, "Confined Spaces"). These regulations usually specify the conditions which must be monitored. More important, however, is that these regulations will specify the penalties imposed on supervisors and employers for failing to provide operators with the proper safety equipment, procedures, and training. These penalties include stiff fines and jail sentences. Responsible employers and supervisors would rather comply with safety regulations than pay fines and spend time in jail.

OPERATORS HAVE THE RIGHT TO REFUSE TO DO ANYTHING THAT ENDANGERS THEIR LIVES.

QUESTIONS

Write your answers in a notebook and then compare your answers with those on page 390.

6.2A List the typical locations in a water utility where it is necessary to monitor the atmosphere before entering.

6.2B Why should a backup person stand outside the confined work area?

6.2C At what level is a mixture of air and gas considered hazardous?

---

[7] *OSHA (O-shuh). The Williams-Steiger **O**ccupational **S**afety and **H**ealth **A**ct of 1970 (OSHA) is a federal law designed to protect the health and safety of industrial workers and also the operators of water supply systems and treatment plants. The Act regulates the design, construction, operation and maintenance of water supply systems and water treatment plants. OSHA also refers to the federal and state agencies which administer the OSHA regulations.*

6.21 Protective Clothing (Figure 6.3)

Protective clothing is required for the protection of operators. Some clothing may be the responsibility of each operator and other protective equipment may be supplied by the water utility. Typical protective equipment supplied by the operators may be:

1. Safety toe (hard toe) shoes, and
2. Safety prescription glasses.

Protective equipment which may be required or supplied by the utility includes:

1. Safety (hard) hats,
2. Earplugs,
3. Safety glasses (nonprescription),
4. Safety goggles,
5. Face shields,
6. Respirators and masks,
7. Canvas gloves,
8. Rubber gloves for use with corrosive chemicals,
9. Safety aprons,
10. Leggings and knee pads,
11. Safety toe caps,
12. Protective clothing (lab coats or rubber suits), and
13. Safety belts or harnesses.

Protective clothing must be worn and equipment used (as appropriate) whenever *YOU* are exposed to safety hazards as you do your job.

6.22 Slips and Falls

Slips and falls are common accidents which can be a major problem in a water utility. All floors should be level and kept as slip resistant as possible. Floors should be free of all debris. Material that drips or spills can be collected in drip pans and appropriate gutters, or splash guards may be used to deflect drips. If liquids do get on floors, nonflammable absorbent materials should be available for cleaning up.

CAUTION: Sawdust should never be used as an absorbent because it is combustible.

When cracks, splinters, ruts, or breaks occur in floors they should be repaired as soon as possible. Good housekeeping and maintenance go hand in hand to prevent slips and falls. If water and chemicals spill on the floor, they should be wiped up as soon as possible. Catwalks or safety tread may be necessary on floors where water or chemicals are commonly spilled. The catwalk or safety tread will help prevent unnecessary slips and falls.

All stairways and elevated catwalks should have safety railings. Approved caution signs should be used where necessary to call attention to potential hazards or remind operators to use caution or to use safety equipment. (Turn back to the "WORDS" section at the beginning of this chapter and read the definitions of the warning terms *CAUTION, DANGER, NOTICE*, and *WARNING*.)

SAFETY QUOTE: Falls cause more accidents than summers, winters, or springs.

6.23 Handling and Lifting

Injuries caused by handling and lifting are many and varied. Approximately 28 percent of operator injuries can be directly attributed to the handling of objects. A large portion of strains and sprains, fractures and bruises, back injuries, and hernias are the result of common handling injuries. These injuries are caused primarily by unsafe work practices such as:

1. Improper lifting,
2. Carrying too heavy a load,
3. Incorrect gripping,
4. Failing to observe proper foot or hand clearance, and
5. Failing to use or wear proper equipment.

Many of us do not think much of lifting but if we are lifting an object incorrectly, we may suffer bad results such as pulled muscles, disc lesions, or painful hernias. Remember Figure 6.4 whenever you have a lifting job.

Here are seven steps to safe lifting:

1. Keep feet parted — one along the side of and one behind the object,
2. Keep back straight — nearly vertical,
3. Tuck your chin in,
4. Grip the object with the whole hand,
5. Tuck elbows and arms in,
6. Keep body weight directly over feet, and
7. Lift slowly.

Safety equipment should be used to help protect your body. To protect your hands, use gloves; to protect your feet, use safety shoes, instep protectors, and ankle guards. Estimate the weight of objects you must lift. Never lift more than you can safely carry. Use handles or holders that attach to the objects such as handles for moving auto batteries or baskets for carrying laboratory samples. If the object to be moved is awkward or too large or heavy for one operator, have two operators move the object in question. When more than one operator is used, the object to be lifted must be lifted in unison (all at once) and everyone must know how the object is to be lifted and handled. Only one operator is assigned the task of giving orders such as when to start, lift, carry, and set down the object. All parties concerned should lift correctly with the legs bearing the bulk of the physical effort.

If the object to be lifted and handled is too heavy for a person or persons to move safely, then use hand trucks, dollies, or powered hand trucks to move and/or lift objects.

Remember, 28 percent of injuries are caused by improper lifting and handling. Your health and safety depend on you using correct lifting and handling procedures.

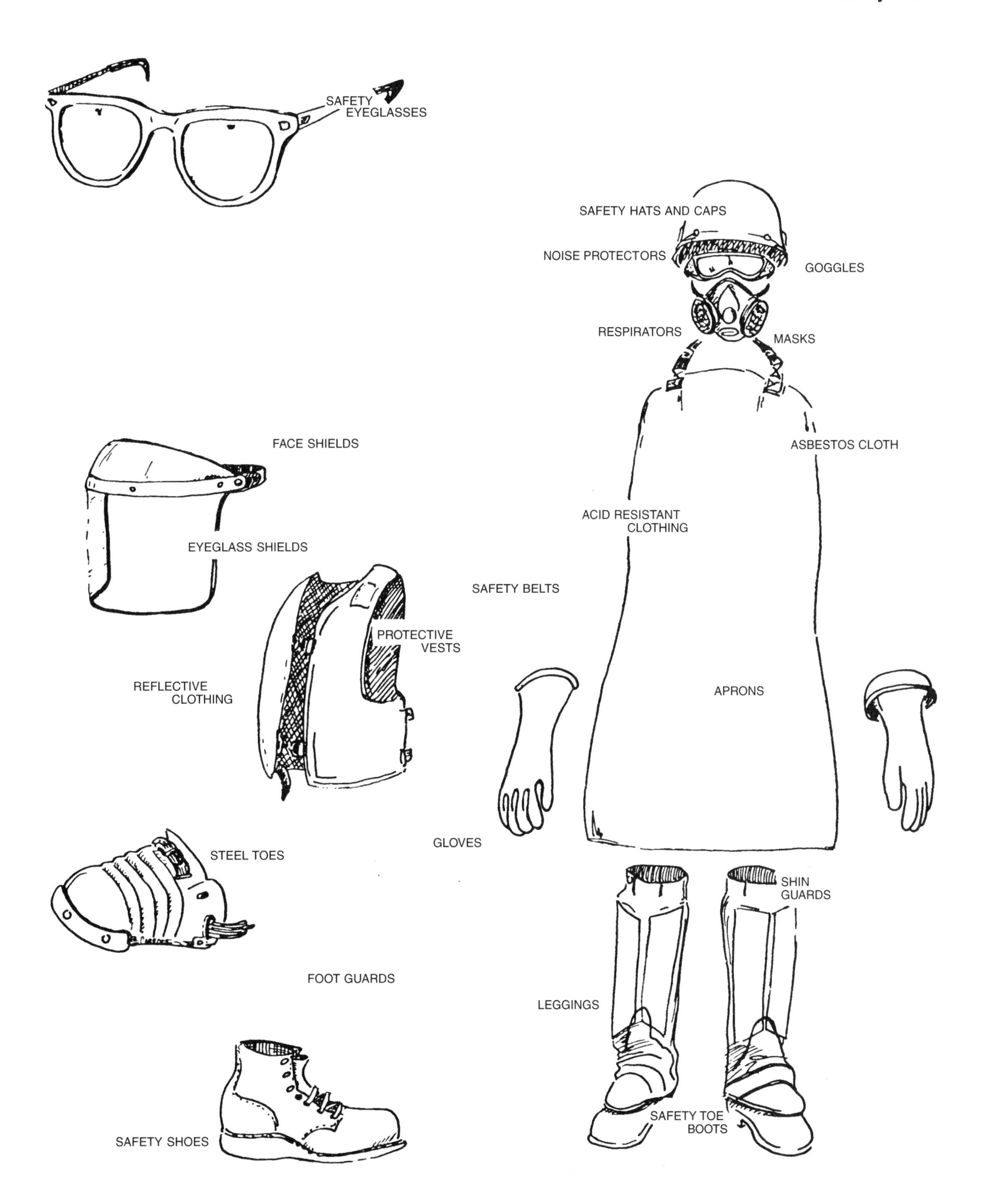

Fig. 6.3 Protect all of you

BE CAREFUL...

THE BACK YOU SAVE MAY BE YOUR OWN!

1. Size up the load.
2. Bend your knees.
3. Get a firm grip.
4. Lift with your legs... gradually.
5. Keep the load close to you.

IT'S HOW YOU LIFT as well as WHAT YOU LIFT.

Fig. 6.4 Lifting guidelines

6.24 Electrical Safety

Electrical safety is very important to your life. Whenever you must work on electrical equipment or equipment that is run by electric motors, be sure to lock out and tag all of the electric switches (see Section 6.43, "Lockout/Tagout Procedure"). Only you should ever be allowed to remove a lockout device and tag that you have installed. Also you should never remove a tag or lockout device installed by another operator.

Water is a good conductor of electricity. Do not stand in water when working on electrical equipment.

Many water supply and water treatment monitoring, measuring, and control systems are electrical. Always be very careful when working around or troubleshooting these instrumentation systems.

QUESTIONS

Write your answers in a notebook and then compare your answers with those on pages 390 and 391.

6.2D List as many items of protective clothing for operators as you can recall.

6.2E How can liquids spilled on floors be cleaned up?

6.2F List the steps to safe lifting.

6.2G What precautions should be taken before working on electric motors?

6.3 SAFETY AROUND WELLS

6.30 Importance of Well Safety

Safety around wells is important for two reasons:

1. Safety control around the well site is necessary to prevent contamination or pollution of the well which could affect the health and safety of the consumer, and
2. Safety housekeeping of the site will help prevent an accident to operators or employees working near or around the well.

The well operator has the responsibility to preserve the quality of the well through preventive maintenance of the area where the well is located.

6.31 Location of Well Site

Selection of a well site requires serious consideration of several factors. The well should be located a suitable distance from any potential pollution source which may affect the groundwater. Some potential sources of pollution may be septic tanks, subsurface leaching systems, mining operations, solid and hazardous waste disposal sites, and wastewater treatment facilities.

Some wells have been located in areas where initially the environment was correct for the site. The following are examples of well sites which were contaminated after a well had been in operation for some time.

EXAMPLE 1

A Girl Scout Camp leased a camp in the Angeles National Forest. This camp had two wells which supplied all the drinking water. The two wells were located in a small, narrow, secluded valley and for several years the bacteriological samples were negative. Camp officials decided to purchase horses to provide a special camping session where the camper could specialize in equestrian events. After a few years the well water bacteriological samples began to turn up positive. This water even had positive coliform samples with chlorination. After an investigation it was determined that the well was being contaminated because the horse stables were located above the well. During the rainy season the rain washed the horse manure down into the wells located below the stables.

EXAMPLE 2

A utility began to cut personnel because of fiscal problems. Some of the first cuts affected gardening personnel who cleaned up around the well site. After a few months the area around the well site became a dumping site. Old lumber, furniture, sofas, chairs, and garden clippings began to pile up around the well site. This debris attracted rodents and reptiles (rats, ground squirrels, and snakes) close to the well site. This caused a potential problem to the employees and to the utility's customers.

6.32 New Wells

No sewers should be permitted within 50 feet (15 m) of any well. If the well must be located within 100 feet (30 m) of a sanitary or storm sewer under gravity flow, special pipe and fittings must be used[8] (see Chapter 3, "Wells," Section 3.12, "Well Site Selection"). This is necessary to prevent intrusion of leaking wastewater into the well.

6.33 Sanitary Seal

The sanitary seal on a well is the most important factor to prevent contamination of any type of well. This seal will prevent contaminated water or debris from entering around the well casing and into the water (review Chapter 3 Appendix, "Wells," Section 3.14, "Well Structure and Components").

---

[8] *Check with your local health department for distances which apply to your well.*

When applying disinfectant or using well cleaning agents, the operator must protect the well against the entrance of surface water or foreign matter.

6.34 Surface Portion of Well

All surface openings installed on top of a well must be protected against the entrance of surface water or foreign matter. Well casing vents should be constructed so that openings are in a vertical downward position. They should be a minimum of 36 inches (90 centimeters) in height above the finished surface of the well lot (ground surface) and should be covered with a fine mesh or device to exclude small insects from the inside of the well.

Gravel tubes must be capped at the top end, kept tightly sealed, and elevated above ground level to prevent flood waters from entering the well.

The sounding tube must be kept tightly sealed. All pipe vents, sounding lines, and gravel fill pipes must be continuous conduit through the concrete pedestal. Also, all conduits which penetrate the casing must be provided with a continuous watertight weld at the point of entry into the well interior. A pump motor base seal must be watertight. All these safety requirements are necessary to ensure that contaminants do not get into the well's interior. When these safety features are incorporated in a well's construction, the probability is lessened that the well will be contaminated by surface debris or waters. Review Chapter 3, "Wells," Section 3.1, "Surface Features of a Well."

6.35 Tank Coatings

Hydropneumatic pressure tanks incorporated with wells are normally painted on the interior surface with a protective coating to extend the life of the tank. Special consideration must be used in selecting the proper type of coating.

The coating should be approved by the AWWA or the Additives Evaluation Branch of EPA and not be a cause of pollution or contamination of the drinking water. The application procedures are most important and the curing time is essential for proper results.

> **CAUTION**
>
> When applying a coating, use forced ventilation and/or a protective mask with an air line to provide an adequate air supply that is safe to work in. Coatings used in tanks are *VOLATILE* and can affect the respiratory system.

When applying a coating, at least two operators should work together — one inside the tank and the other close by. The operator outside checks on the status of the person working inside and checks periodically on the forced ventilation system and air pressure for the air mask, if used. A third person must be readily available to assist the person outside with any rescue operations. See the *SAFETY NOTICE* on page 349.

QUESTIONS

Write your answers in a notebook and then compare your answers with those on page 391.

6.3A What are the safety responsibilities of the well operator?

6.3B How is contaminated water or debris kept from entering around the well casing and reaching the groundwater?

6.3C List the surface openings on top of a well that must be protected against the entrance of surface water or debris into the well.

6.36 Well Chemicals

6.360 Acid Cleaning

An acid treatment may be necessary to loosen incrustations from a well casing and well. Normally hydrochloric acid or sulfamic acid is used. Hydrochloric (HCl) acid should be used at full strength. Hydrochloric acid can be introduced into a well by means of a wide-mouthed funnel and a 3/4- or 1-inch (16- or 25-mm) place pipe. Extend the place pipe into the well and attach the funnel to the end outside of the well. Pour the acid into the funnel and allow it to run into the well. If hydrochloric acid (muriatic acid — industrial name) is used, be careful! The pH of hydrochloric acid is approximately 1.0. This acid is very corrosive.

Inhalation of hydrochloric acid fumes can cause coughing, choking, and inflammation of the entire respiratory tract.

Swallowing even a small amount of this acid can cause corrosion of mucous membranes, esophagus, and stomach; difficulty swallowing; nausea; vomiting; intense thirst; and diarrhea. Circulatory collapse and death may occur.

Hydrochloric acid causes severe skin burns. Spills, splashes, or fumes may affect the eyes and could cause permanent eye damage.

Safety apparatus must be used when working with hydrochloric acid. Protective clothing includes rubber gloves, preferably long for hands — approximately 15 to 18 inches (38 to 45 cm), rubber apron for clothes protection, boots for foot protection in case of spillage, and face shield for protection against splashes and fumes.

Extreme care should be used in transporting acid. Acid should be carried in cases (gallons) or in carboys for protection and must be placed securely in a vehicle whenever transported.

Sulfamic acid has advantages over hydrochloric acid. Sulfamic acid may be used in granular form or as an acid solution mixed on site. Granular sulfamic acid is nonirritating to dry skin and its solution gives off no fumes except when reacting with incrusting materials. Spillage, therefore, presents no hazard and handling is easier, cheaper, and safer.

Although sulfamic acid is nonirritating to dry skin and it gives off no fumes, rubber gloves should still be used because hands may become wet from perspiration and the acid will cause burns on the skin. Safety goggles or glasses, preferably a face shield, should be used in case granules are blown into the face and/or the eyes. If sulfamic acid is mixed on site, care should be used in mixing.

CAUTION

Acid should always be added slowly to the water. Use an acid-proof crock for mixing. Mix with a wooden paddle or electric mixer. Again, always use rubber gloves, rubber boots, a rubber apron, and a face shield when mixing sulfamic acid.

NEVER ADD WATER TO ACID. ACID COULD SPLASH ALL OVER YOU.

When using any *CORROSIVE* or *CAUSTIC* chemical, an eye/face wash station must be available in case of an emergency. There are a variety of portable eye/face wash units commercially available. The unit should be located at a site which has easy access in case it is needed, preferably within 20 feet (6 m) of where the acid is being used. All personnel concerned should know the location of the eye/face wash station and be instructed on how to use it.

If anyone spills acid on their skin, the exposed area should be washed with water as soon as possible. Also, they should change their clothes and shower if necessary. If acid is splashed on the face or in the eyes, immediately wash face/eyes with eye wash for at least 15 minutes. Make sure the eyes are thoroughly flushed with water. If irritation continues, resume washing eyes with water. Keep eyes moist with a wet towel and have a physician examine them for further treatment.

QUESTIONS

Write your answers in a notebook and then compare your answers with those on page 391.

6.3D Why must hydrochloric acid be handled carefully?

6.3E What are the advantages of using sulfamic acid over hydrochloric acid to acid clean a well?

6.3F What safety precautions should be used when working with sulfamic acid in the granular form?

6.361 Chlorine Treatment

When a well requires treatment with chlorine, either granular calcium hypochlorite or liquid sodium hypochlorite is commonly used. Regardless of the type of hypochlorite used, safety equipment should be used. Again, use rubber gloves, preferably 15 to 18 inches (38 to 45 cm) long, a rubber apron, rubber boots, and a face shield to protect against splashes. A portable eye/face wash station should be available in case of splashes. Wear a "nuisance mask" when working with granular or powdered chemicals because their dusts can be harmful.

Use caution in the transfer of hypochlorite compounds. They should be transferred in their original containers and handled with care. Operators must be trained in the use of calcium and sodium hypochlorite products. Follow manufacturers' recommendations or your utility's safety policy.

CAUTION

When using acid or chlorine products, be mindful of the dangerous potential that fumes may have. Therefore, whenever using an acid or a chlorine product, use them where there is adequate ventilation. If necessary, use a forced ventilation system in pump houses or confined spaces.

When working with chemicals around a well, keep all personnel out of pits or depressions around the well. During treatment some of the toxic gases, such as hydrogen sulfide, may rise from the well and settle in the lowest areas nearby because these gases are usually heavier than air.

6.362 Polyphosphate

Sodium polymetaphosphate is a chemical sometimes added to wells. This chemical may be sold commercially as Graham's salt, "Sodium hexametaphosphate," glassy sodium metaphosphate, or Hy-Phos $(NaPO_3)x$. The chemical is clear, hygroscopic (attracts moisture from the air), and soluble in water, but it dissolves slowly in water. The chemical is supplied in the form of a powder, flake, or as small, broken glass-like particles which possess dispersing and deflocculating properties.

CAUTION

Glassy phosphate consists of broken glass-like particles which are very sharp and can cut your skin. Use heavy-duty gloves when handling. Respiration equipment or a face shield should be used.

An eye/face wash station should be available if solution is splashed into the face or eyes. Wash face continuously for 15 minutes. Get medical attention.

6.363 Disinfection of Wells and Pumps

Care should be taken during the construction of new wells. All pipes used in the construction of a well should be as clean as possible. Regardless of how clean the pipe for the well casing is kept, a new well pump should be disinfected. This disinfection is necessary because contamination could be introduced into the well from the drilling tools and mud, makeup water, topsoil falling into or sticking to tools, and from the gravel itself.

Disinfect a well or well pump with a chlorine solution strong enough to produce a chlorine concentration of 50 mg/*L* in the well casing. Use caution when disinfecting wells and pumps. Follow safety procedures listed under chlorine treatment in this chapter. Also refer to Chapter 3, Section 3.5, "Disinfection of Wells and Pumps."

6.37 Working Around Electrical Units

Electricity is supplied as an alternating current (A.C.) at 120, 240, or 480 volts. Most wells or large pumps will be energized by 240 or 480 volts. Extremely large pumps may be energized by even higher voltages. Place nonconducting rubber mats on the floor in front of all power panels and motor control centers.

Care and caution should be used when working around equipment that is hooked up to any electric current.

When maintenance is to be done on wells or pumps, lock out the electric current to the equipment. Open the breakers so that electric current is turned off.[9] Place a tag (Figure 6.5) with your name on it on opened breakers and locks to indicate that you turned off the breakers and locks. No one should remove the tag or close the breakers or locks but you. This will allow the equipment to be pulled and worked on without the fear of being shocked or burned. Personnel working around high-voltage equipment should be trained and have respect for electric current.

After work is completed on the equipment, personnel should stand clear when the equipment is energized (electric current turned on). Make sure that the equipment is properly grounded. Personnel working with electrical equipment should have a good knowledge of electric circuits and circuit testing. They should get clearance (approval) to work on the equipment only if they are qualified. If no one is knowledgeable about electric circuits, the water utility should contact an electrician from a local electrical firm.

Safe procedures that must be used when working around electrical equipment include:

1. Only qualified persons can work on electrical equipment;
2. Electrical installations must be maintained in a safe condition;
3. Electrical equipment and wiring must be protected from mechanical damage and environmental deterioration;
4. Covers or barriers must be installed on boxes, fittings, and enclosures to prevent accidental contact with live parts;

DANGER

OPERATOR WORKING ON LINE

DO NOT CLOSE THIS SWITCH WHILE THIS TAG IS DISPLAYED

TIME OFF: ____________________

DATE: ____________________

SIGNATURE: ____________________

This is the ONLY person authorized to remove this tag.

INDUSTRIAL INDEMNITY/INDUSTRIAL UNDERWRITERS/ INSURANCE COMPANIES

4E210—R66

Fig. 6.5 Typical lockout warning tag

(Source: Industrial Indemnity/Industrial Underwriters/Insurance Companies)

[9] *When an electrician "closes" an electric circuit, the circuit is connected together and electricity will flow. Closing a circuit is similar to opening a valve in a pipeline. The reverse is also true. Opening an electric circuit is similar to closing a valve on a pipeline.*

5. An acceptable service pole must be used;
6. Equipment must be suitably grounded;
7. Provisions must be made for suitable overcurrent protection;
8. Machinery must be locked out during cleaning, servicing, or adjusting;
9. Machinery must be de-energized, locked, or blocked to prevent movement if exposed parts are dangerous to personnel; and
10. If a switch or circuit breaker is tagged and locked out, only the person placing the tag should remove it.

6.38 Abandoning and Plugging Wells

Wells that are no longer useful should be abandoned and plugged for the following reasons:

1. To ensure that groundwater is protected and preserved for further use,
2. To eliminate the potential physical hazard to people, and
3. To protect nearby wells from contamination.

Obtain the appropriate permits from the proper agency. Depending on the locality, this may be a local, state, or federal agency. Follow the guidelines issued by the permitting agency and use experts (such as a geologist) where necessary. Review Chapter 3 Appendix, Section 3.18, "Abandoning and Plugging Wells."

6.39 Safety Inspection

Safety around wells is the responsibility of the well operator. A periodic inspection should be performed at each well site whether the well is in service or out of service. Keep a record of each monthly inspection and use the remarks column to make special notations. Some things that you should inspect for are:

1. Cleanliness of site,
2. Locks on electrical panels,
3. Condition of fencing around site, and
4. Caution/warning signs in place where necessary.

Any problems or hazards, safety or otherwise, should be corrected as soon as possible. If you cannot take care of the problem, notify your supervisor about the nature of the problem and what type of support you will need to correct the problem.

QUESTIONS

Write your answers in a notebook and then compare your answers with those on page 391.

6.3G What protective clothing should be worn when handling hypochlorite?

6.3H How should glassy phosphate be handled?

6.3I If no one working for a water utility is knowledgeable about electric circuits, how can a water utility have a malfunctioning electrical system repaired?

6.3J List the items that should be included in the safety inspection of a well site.

Please answer the discussion and review questions next.

DISCUSSION AND REVIEW QUESTIONS

Chapter 6. SAFETY

(Lesson 1 of 2 Lessons)

At the end of each lesson in this chapter you will find some discussion and review questions. The purpose of these questions is to indicate to you how well you understand the material in the lesson. Write the answers to these questions in your notebook before continuing.

1. What is safety?
2. Who is responsible for safety?
3. What is the one objective of a safety program?
4. What are the causes of accidents?
5. How would you test a safety shower and eye/face wash station?
6. What should be done before entering a confined space?
7. Why should hazardous gas detection devices be calibrated regularly?
8. How can an object be moved if it is too heavy for operators to lift?
9. Why is safety around wells important?
10. How should hydrochloric acid be transported?
11. What would you do if acid splashed in your eyes?

CHAPTER 6. SAFETY

(Lesson 2 of 2 Lessons)

6.4 PUMP SAFETY

6.40 Uses of a Pump

When we think of pumps, we commonly think of pumping water. Actually, pumps can do many things as is illustrated by the following definitions of a pump:

1. An apparatus or machine which forces liquids, air, or gas into or out of things.
2. A machine which increases the static pressure of fluids (air and water).
3. Pumping is the addition of energy to a fluid which is used mainly for the purpose of moving the fluid from one point to another.

6.41 Guards Over Moving Parts

A pump has moving parts. Where there are moving parts there is potential danger. All mechanical action or motion is hazardous, but in varying degrees. *ROTATION MEMBERS, RECIPROCATING ARMS, MOVING PARTS,* and *MESHING GEARS* are some examples of action and motion requiring protection on pumps.

Rotating, reciprocating, and transverse (crosswise) motions (Figure 6.6) create hazards in two general areas:

1. At the point of operation where work is done, and
2. At the points where power or motion is being transmitted from one part of a mechanical linkage to another.

Any rotating device (pump shaft) is dangerous. Whatever is rotating can grip clothing or hair and possibly force an arm or hand into a dangerous position. While accidents due to contact with rotating objects are not frequent, the severity of injury is very high.

Whenever hazardous machine actions or motions are identified, a means for providing protection for operators is essential. Enclosed guards are used to protect operators from rotating parts on pumps. The enclosed guard prevents access to dangerous parts at all times by enclosing the hazardous operation completely. The basic requirement for an enclosed guard is that it must prevent hands, arms, or any other part of an operator's body from making contact with dangerous moving parts. A good guarding system eliminates the possibility of the operator or another worker placing their hands near hazardous moving parts. Figures 6.7 and 6.8 are pictures showing examples of good, effective guarding systems.

QUESTIONS

Write your answers in a notebook and then compare your answers with those on page 391.

6.4A What are the uses of pumps?

6.4B List the moving parts on a pump which require guards.

6.4C How can rotating devices injure you?

6.4D What is the basic requirement for an enclosed guard?

6.42 Maintenance and Repair

Pumps, like any other equipment, need preventive maintenance and periodic repair. A proper maintenance program and routine inspections will allow you to obtain satisfactory service from your pumps and minimize any potential safety hazards. Figure 6.9 identifies danger areas on pumps. These danger areas could be safety hazards to operators or areas where the pump could fail.

CAUTION

Do not operate a pump against a closed discharge valve, otherwise overheating will occur.

Whenever maintenance is to be performed on pumps or any other equipment, the work must be carefully planned and coordinated with other operators and the operation of the facilities. Good planning includes figuring out how to do the job from start to finish. Planning includes identifying potential safety hazards *BEFORE* starting a job and deciding how they can be avoided. To do a job safely, use the proper tools and equipment and follow the lockout/tagout procedure described in Section 6.43.

If a stationary pump or motor must be moved, use a lifting sling. Attach the sling properly to the equipment and lift the equipment up off the base. Set the pump or motor down on a cart or pallet and transfer the pump or motor to a workshop or another location as necessary for repair.

NEVER LEAVE ANY EQUIPMENT HANGING FROM A SLING IN MIDAIR. NEVER WORK ON EQUIPMENT WHEN IT IS HANGING FROM A SLING. ALWAYS LOWER EQUIPMENT TO THE FLOOR, SET IT ON A CART, OR PLACE IT ON A WORKBENCH BEFORE STARTING MAINTENANCE.

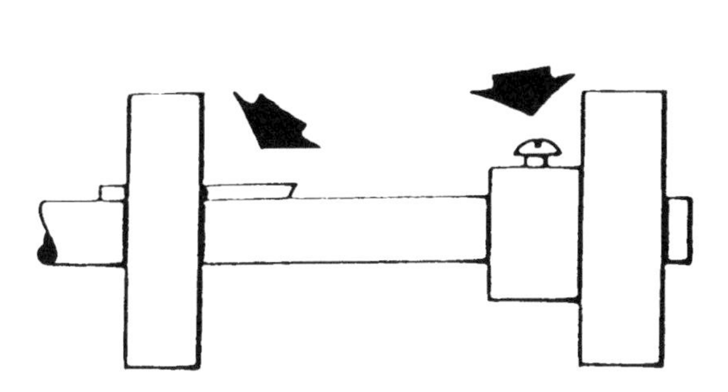

Rotating shaft and pulleys with projecting key and set screw

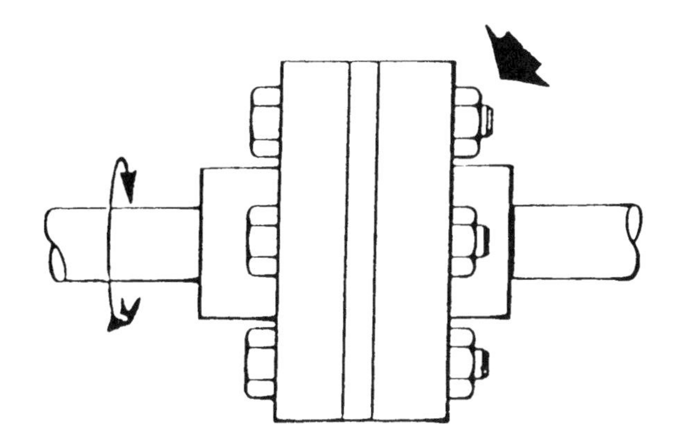

Rotating coupling with projecting bolt heads

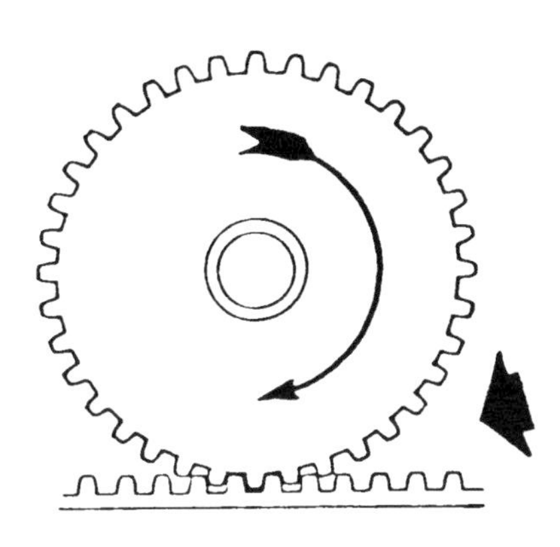

Rack and gear

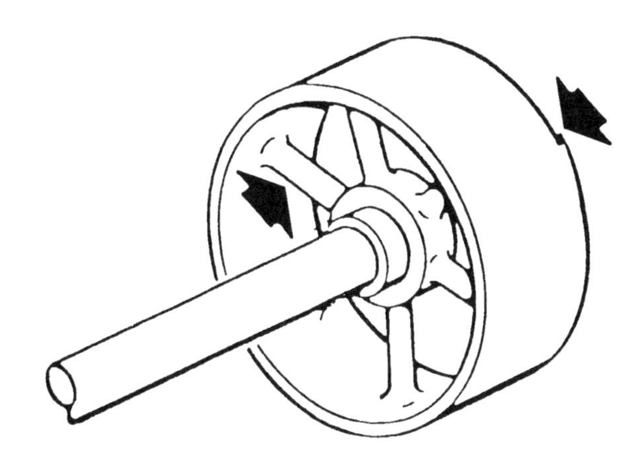

Rotating pulley with spokes and projecting burr on face of pulley

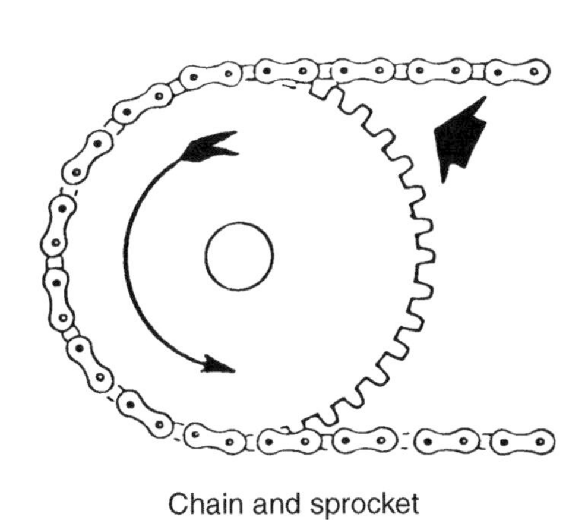

Chain and sprocket

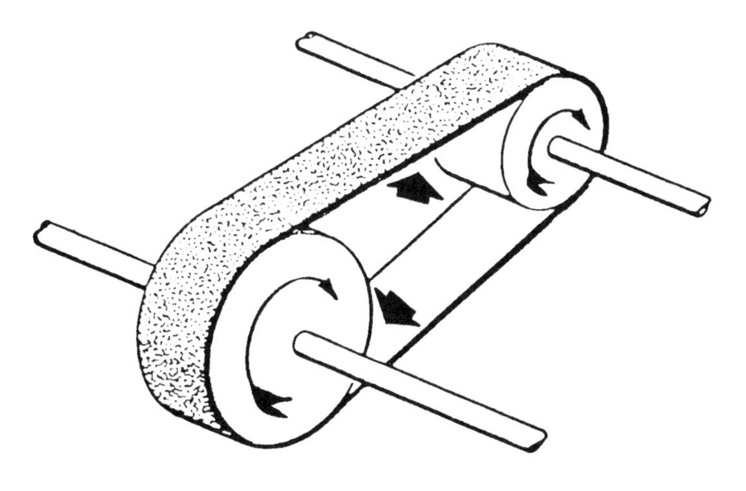

Belt and pulley

NOTE: Short, thick arrows indicate danger points

Fig. 6.6 Examples of typical rotating, reciprocating, and transverse mechanisms

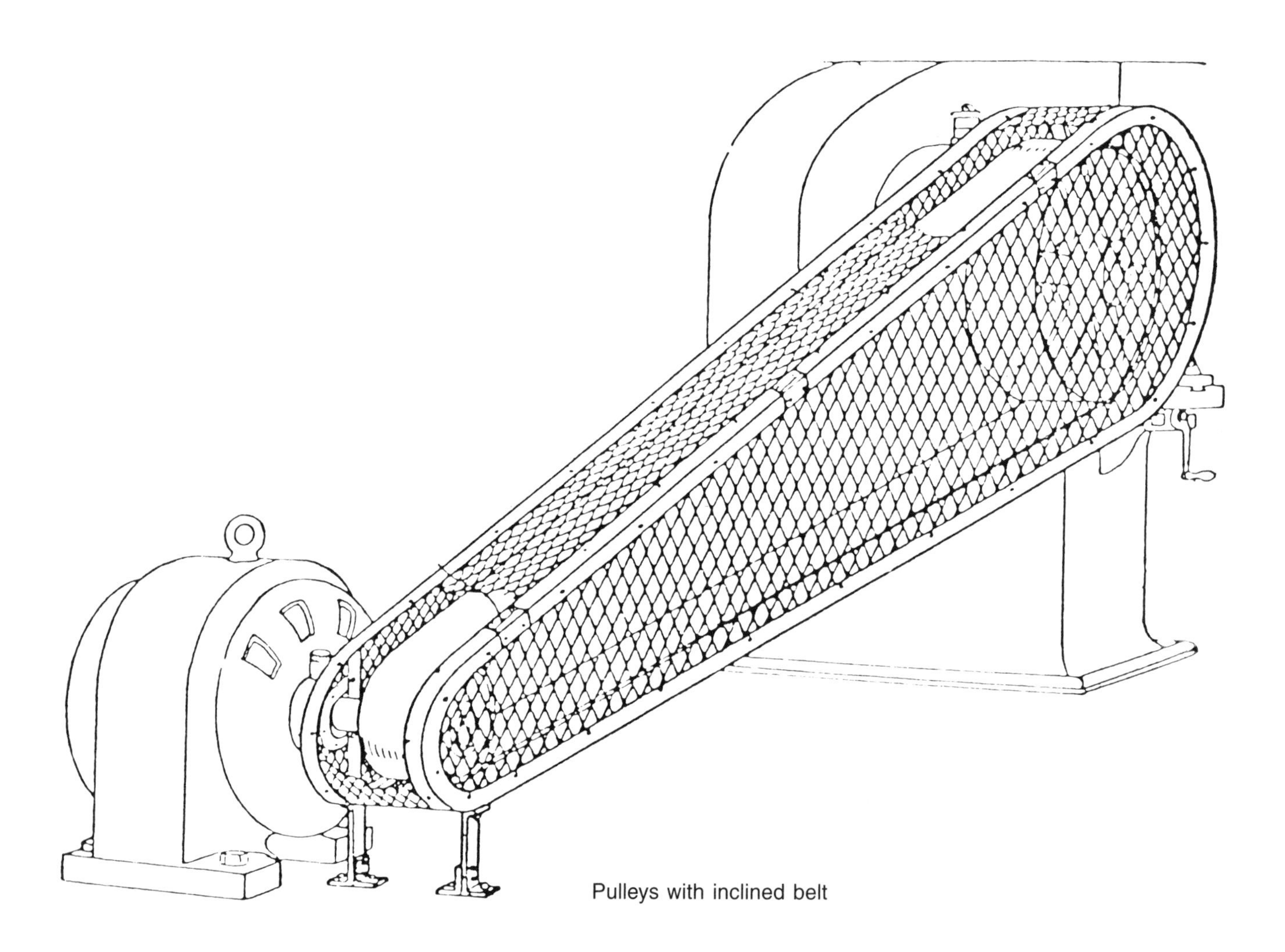

Fig. 6.7 Enclosed guard protecting operators from moving pulleys and belt

Guard over belt on gas engine

Guard over coupling

Guard on gas engine and pump

Guard on hypochlorinator

Fig. 6.8 Effective guarding systems

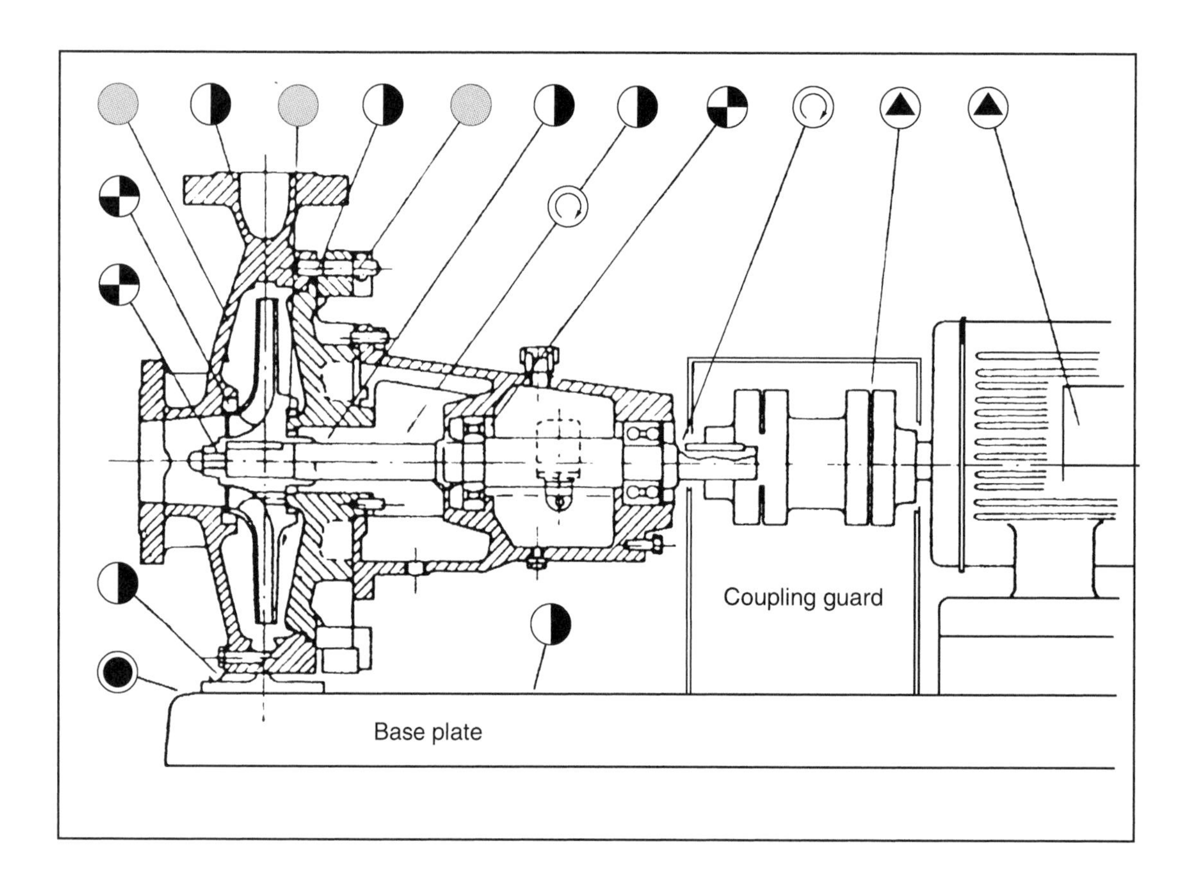

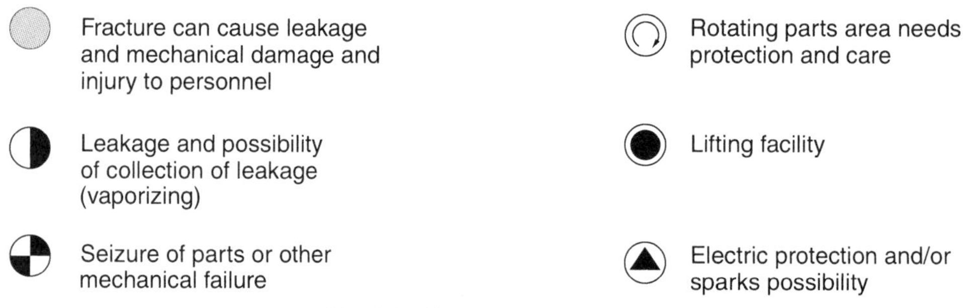

Fig. 6.9 Danger areas on pumps

When transferring pumps or equipment from one location to another, use a sturdy cart or truck, or transfer the equipment on a pallet by using a forklift. Many operators have seriously injured their backs attempting to lift and carry heavy pumps, motors, and other equipment.

6.43 Lockout/Tagout Procedure

OSHA standards require that all equipment that could unexpectedly start up or release stored energy must be locked out and tagged whenever it is being worked on. Common forms of stored energy are electrical energy, spring-loaded equipment, hydraulic pressure, and compressed gases which are stored under pressure.

The operator who will be performing the work on the equipment is the person who installs a lock (either key or combination) on the energy isolating device (switch, valve) to positively lock the switch or valve in a safe position, preventing the equipment from starting or moving. At the same time, a tag such as the one shown in Figure 6.5 is installed at the lock indicating why the equipment is locked out and who is working on it. NO OTHER PERSON IS AUTHORIZED TO REMOVE THE TAG OR LOCKOUT DEVICE unless the employer has provided specific procedures and training for removal by others.

Even after equipment is locked out and tagged, it still may not be safe to work on. Stored energy in gas, air, water, steam, and hydraulic systems (such as water pumping systems) must be drained or bled down to release the stored energy. Elevated machine members, flywheels, and springs should be physically blocked and/or secured in place to prevent movement.

Before bleeding down pressurized systems, think about the pressures involved and what will be discharged to the atmosphere and work area (toxic or explosive gases, corrosive chemicals). Will other safety precautions have to be taken? What volume of bleed down material will there be and where will it go? If dealing with chemicals, greases, or lubricants, how can they be cleaned up or contained? Many times an operator has removed a pump volute cleanout only to find that the discharge check valve wasn't seated or one of the isolation valves was not fully seated. Once the pump was open, however, the pump intake structure drained into the pump room through the opened pump.

All rotating mechanical equipment must have guards installed to protect operators from becoming entangled or caught up in belts, pulleys, drive lines, flywheels, and couplings. Keep the guards installed even when no one is working on the equipment.

Various pieces of machinery are equipped with travel limit switches, pressure sensors, pressure reliefs, shear pins, and stall or torque switches to ensure proper and safe operation of the equipment. Never disconnect a device, or install larger shear pins than those specified on the original design, or modify pressure or temperature settings.

The basic elements of a proper lockout/tagout procedure are as follows:

1. Notify all affected employees that a lockout or tagged system is going to be used and the reason why. The authorized employee shall know the type and level of energy that the equipment uses and shall understand the hazard presented by the equipment.
2. If the equipment is operating, shut it down using the procedures established for the equipment.
3. Operate the switch, valve, or other energy isolating device(s) so that the equipment is isolated from its energy source(s). Stored energy such as that in springs, elevated machine parts, rotating flywheels, hydraulic systems, and air, gas, steam, or water pressure must be released or restrained by methods such as repositioning, blocking, or bleeding down.
4. Lock out and/or tag out the energy isolating device with your assigned individual lock or tag.
5. After ensuring that no personnel are exposed, and as a check that the energy source is disconnected, operate the pushbutton or other normal operating controls to make certain the equipment will not operate. A common problem when motor control centers (MCCs) contain many breakers is to lock out the wrong equipment (locking pump #3 and thinking it is #2). Always confirm the dead circuit. CAUTION! RETURN OPERATING CONTROLS TO THE NEUTRAL OR OFF POSITION AFTER THE TEST.
6. The equipment is now locked out or tagged out and work on the equipment may begin.
7. After the work on the equipment is complete, all tools have been removed, guards have been reinstalled, and employees are in the clear, remove all lockout or tagout devices. Operate the energy isolating devices to restore energy to the equipment.
8. Notify affected employees that the lockout or tagout device(s) has been removed before starting the equipment.

6.44 Storage of Lubricants and Fuel

Lubricants and fuels must be stored in a separate facility from other storage areas because of the potential fire hazard. Clean up all spills immediately and keep the area neat and clean. A fire extinguisher must be readily available near the storage facility, but never inside the facility.

QUESTIONS

Write your answers in a notebook and then compare your answers with those on page 391.

6.4E What safety planning should be done before beginning maintenance on a pump?

6.4F Why should you never work on equipment hanging from a sling in midair?

6.4G Why is it necessary to bleed down pressurized systems before making repairs?

6.4H Why should lubricants and fuels be stored in a separate facility from other storage areas?

6.5 WORKING IN STREETS

6.50 Definition of Terms

Before we discuss how to work safely in streets, let's review some of the terms we'll use.

WORK SITE — general place where a construction or maintenance operation is taking place.

WORK SPACE — space set apart which is identified for use by workers and equipment performing work which is protected, marked, or signed to exclude vehicular and pedestrian traffic.

TRAFFIC CONTROL — the process of advising motorists and pedestrians as to detailed requirements or conditions affecting road use at specific places and times in order that proper action may be taken and accidents or delays avoided.

Work site traffic control applies to maintenance and construction activities or other special temporary conditions affecting road use at specific places and times.

TRAFFIC CONTROL DEVICE — all signs, signals, markings, and devices placed on or adjacent to a street or highway by a public body or official having jurisdiction to regulate, warn, or guide traffic. Traffic control devices are used to slow and guide traffic and to warn motorists of changes or possible changes in conditions.

TRAFFIC CONTROL ZONE — the entire area of the roadway which encompasses all traffic control devices used to warn, regulate, or guide drivers through or around a work site.

ROADWAY CLOSURE — the taking of some portion of the roadway for the exclusive use of workers, materials, or equipment during a construction or maintenance operation. Traffic is excluded from the closed portion and must use the remaining portion of the roadway if any exists. The option exists to close a portion of a lane or one or more travel lanes.

6.51 Work Area Protection

All work done in streets and highways must be performed in accordance with city, local, and state regulations regarding work in roadways. In many locations a roadway encroachment permit must be obtained *BEFORE* starting work.

Most work area accidents on streets and highways can be attributed to the lack of proper advance warning, signing, and guidance. Speed of traffic and the increased maintenance of utility lines under our streets have created hazards for the motorists. We must channel traffic flow through a work area safely with the following items considered.

1. PREWARN
Give the motorist time to THINK and REACT to avoid construction hazards.

2. SIGN PROPERLY
Use the standard shapes, sizes, and colors designated by your state highway or transportation agency.

3. REGULATE SPEED
The motorists' speed must be controlled so they may proceed safely through the construction area with a minimum of inconvenience.

4. GUIDE PROPERLY
The driver must be directed through the work area by clearly indicated instructions and traffic control devices (traffic cones).

The person installing traffic control devices should drive through the area as another motorist would. This is a final check to determine if the work area protection system is actually effective.

6.52 Setting Up Equipment in Streets

The most common cause of worker fatality is an accident involving moving vehicles. Personnel of a water utility are exposed to this potential hazard because of work necessary to install, repair, and/or operate valves in the street. Operators must work in manholes, open trenches, regulation vaults, and on roadway shoulders. Any type of work in the street, roadways, or within a highway right-of-way increases the potential for hazard not only because of the nature of the job, but also because of the added complication of motorists and pedestrians.

Each job has its particular hazard(s) which should be reviewed in a *TAILGATE SAFETY MEETING.*[10] All aspects of the job should be reviewed and someone should be responsible for each particular part of the job in order to make the job safe. Making work in streets safe requires good planning, adherence to safety standards, adequate supervision, and frequent inspection.

Traffic control systems are used on street work sites to protect the:

1. Work force,
2. Motorist,
3. Pedestrian,
4. Equipment, and
5. Facility.

Traffic control procedures are used at work sites to:

1. Warn motorists and pedestrians of hazards involved and to advise them of the proper manner to travel through the area,
2. Inform the drivers of changes in regulations or additional regulations that apply to traffic through the area,
3. Guide traffic through and around the work site, and
4. Identify areas where traffic should not operate.

Whenever workers (personnel working on the street) are exposed to vehicular traffic, they should wear a high-visibility shirt, poncho, vest, or coat over their normal clothing. This safety clothing should be colored red or safety orange. A safety hat (hard hat) should be worn whenever you are near traffic.

Some jobs will require a flagger (Figure 6.10) to help control the work space. Flaggers are required where:

1. Workers or equipment intermittently block a traffic lane,
2. One lane must be used for two directions of traffic (a flagger is required for each direction of traffic), and
3. The safety of the public and/or workers requires it.

[10] *Tailgate Safety Meeting. Brief (10 to 20 minutes) safety meetings held every 7 to 10 working days. The term TAILGATE comes from the safety meetings regularly held by the construction industry around the tailgate of a truck.*

Fig. 6.10 Alerting and directing traffic

(Courtesy of the Bureau of Contract Administration, City of Los Angeles)

All flaggers must wear an orange jacket or vest for daytime use and a reflectorized belt and suspender harness for use at night. During the day, flaggers use a double-sided sign paddle with SLOW printed on one side and STOP on the other. At night a flashlight with red lens is used.

The flagger should be positioned at least 100 feet (30 m) in front of the work space and a sign reading FLAGGER AHEAD should be placed as far ahead of the flaggers as practical (500 feet or 150 meter minimum). Figure 6.11 shows flaggers and hand signaling motions and devices. Flaggers are expected to regulate and control the flow of traffic in a safe and orderly fashion. Flaggers and utilities have been sued because improper actions by flaggers caused an accident.

Portable arrows such as the ones shown in Figure 6.12 can also be used to reduce drivers' confusion, and effective placement of signs, barricades, and traffic cones (Figure 6.13) can make work sites in streets safe places to work. You must follow local safety regulations regarding the locations and types of signs to protect your agency from lawsuits if an accident occurs. However, you should study each work site and make any necessary improvements to ensure a safe area to work.

Accidents associated with work on the street are a prime indicator of a breakdown in the traffic control system. Any accident that occurs at the work site should be analyzed to determine whether adjustments to the traffic control zone are needed. The time you spend to properly protect the work space with traffic controls may save your life. Figure 6.14 illustrates some good practices in work area protection.

Suggested distances for the spacing of delineators (traffic cones) and warning devices or signs are summarized in Table 6.4.

TABLE 6.4 SUGGESTED DISTANCES FOR PLACEMENT OF WARNING DEVICES[a]

TRAFFIC CONES

Approach Speed	Number of Cones	Cone Spacing
Up to 30 MPH	7	30 feet
30 to 40 MPH	7 to 9	30 to 40 feet
45 to 55 MPH	13 to 21	45 to 50 feet

HIGH-LEVEL WARNING DEVICE

Approach Speed	Distance
25 MPH or below	150 feet
35 MPH	250 feet
45 MPH	550 feet
55 MPH	1,000 feet

[a] Contact your state highway department or transportation agency for your regulations.

TO *SLOW* DAYTIME TRAFFIC
Short up-and-down motion with extended right hand.

TO *STOP* DAYTIME TRAFFIC
STOP paddle high enough to be above garment.

TRAFFIC PROCEED
Never use a flag or paddle as a signal to move traffic.

Fig. 6.11 Flaggers and hand signaling motion and devices

Fig. 6.12 Portable arrows reduce driver confusion in construction area
(Permission of Safety Tech, Inc.)

Fig. 6.13 Signs, barricades, and traffic cones placed to warn traffic and protect workers

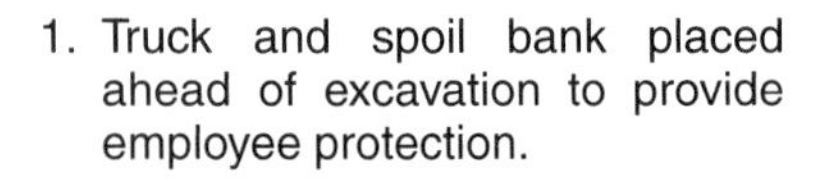

1. Truck and spoil bank placed ahead of excavation to provide employee protection.
2. Cone pattern arranged with gentle curves — traffic adjusts smoothly.
3. Pipe blocked to prevent rolling into street. Barricades provide pedestrian warning.
4. Material neatly stacked.
5. High-level warning or barricades of solid material to give audible warning of vehicles entering work area.
6. Pedestrian bridge over excavation.
7. Left side of truck protected by cone pattern, work area entirely outlined.
8. Tools out of way of pedestrians, tools not in use replaced in truck.
9. Pickup parked in work area or on street away from work area.

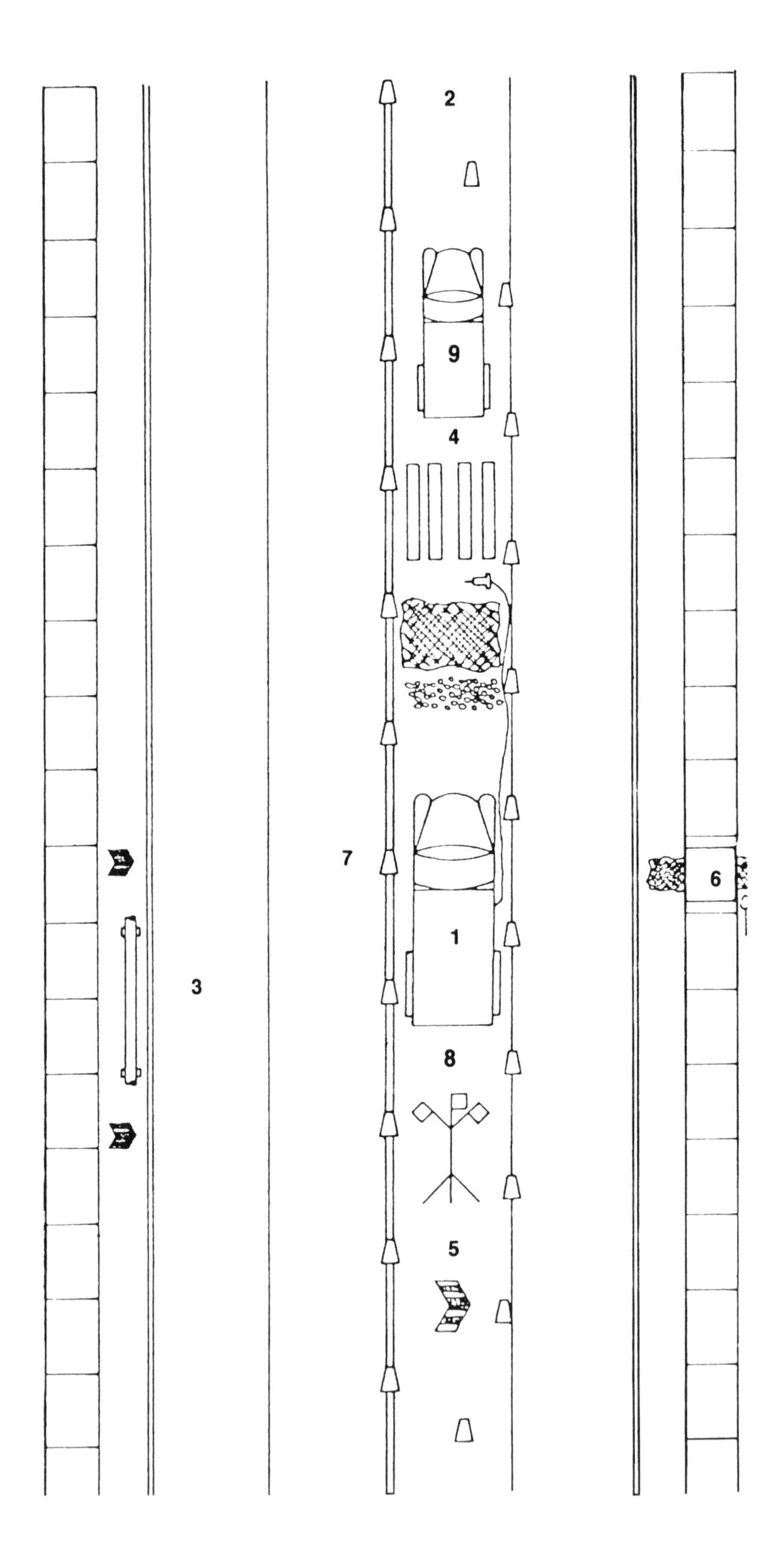

Fig. 6.14 Good practices in work area protection

This sign, though not strictly a work area protection device, helps promote customer acceptance and understanding of the necessity for important construction work.

Figures 6.15, 6.16, and 6.17 show some typical setups for working in the street.

For additional information about setting up a safe traffic control zone, see *OPERATION AND MAINTENANCE OF WASTEWATER COLLECTION SYSTEMS*, Volume I, Section 4.3, "Routing Traffic Around the Job Site," in this series of operator training manuals.

6.53 Tailgate Topic on Street Warning Devices

Each job in a roadway presents problems in methods of giving motorists adequate warning of our presence. Whenever there are new mains to be laid, mains repaired, or mains relocated because of freeway construction, you may be required to work in traffic. All this construction has a bad effect on the motorist in one respect; drivers pay less attention to work going on in the street, and they do not slow down in many cases where they pass our jobs.

Our only protection is to give the motorist sufficient warning by use of barricades, cones, and flaggers. Placement of signs requires careful thought. Some job locations can present real problems when ensuring safety for all concerned. We all know most of the established rules for placing warning devices. A review of some of these rules may remind us of a few points we forget to consider.

1. Whenever possible, use a flagger when setting up or breaking down the work area.
2. Use more, rather than too little, warning equipment at the work area.
3. Anticipate the unexpected and irrational acts of pedestrians and drivers.
4. Do not use the same setups on streets with fast-moving traffic as would be used on slow-traffic streets.
5. Keep all tools and materials within the confines of the protected work area.

These are just a few points to consider. Every job is different. Use good safety judgment and past experience as a guide. Remember, we not only want to set up the job so we are obeying the law, we want to go even further and set up so an accident can't happen.

QUESTIONS

Write your answers in a notebook and then compare your answers with those on pages 391 and 392.

6.5A What items should be considered when channeling traffic flow through a work area?

6.5B What should persons who must work on streets wear?

6.5C List the hand signaling motions commonly used by flaggers.

6.5D List the rules that should be considered when setting up street warning devices.

6.54 Excavations[11]

6.540 Safety Rules for Excavations

Excavations are necessary to install or repair water lines in streets or roadways. Regulations and requirements governing excavations may be found in your state construction safety orders or other similar documents. A permit may be required for any excavations four feet (1.2 meters) or more in depth in which personnel must work. Shoring is required in excavations five feet or more in depth and in excavations less than five feet deep if a *COMPETENT PERSON*[12] determines that there is a potential for a cave-in.

General safety rules include:

1. Employer must inspect any excavation for hazards from possible moving ground before employees may work in or adjacent to the excavation.
2. The location of underground installations (such as utility lines) must be determined before excavation begins.

---

[11] *Also see Chapter 3, Distribution System Facilities, Section 3.653, "Excavation and Shoring," WATER DISTRIBUTION SYSTEM OPERATION AND MAINTENANCE, in this series of manuals.*

[12] *Competent Person. A competent person is defined by OSHA as a person capable of identifying existing and predictable hazards in the surroundings, or working conditions which are unsanitary, hazardous or dangerous to employees, and who has authorization to take prompt corrective measures to eliminate the hazards.*

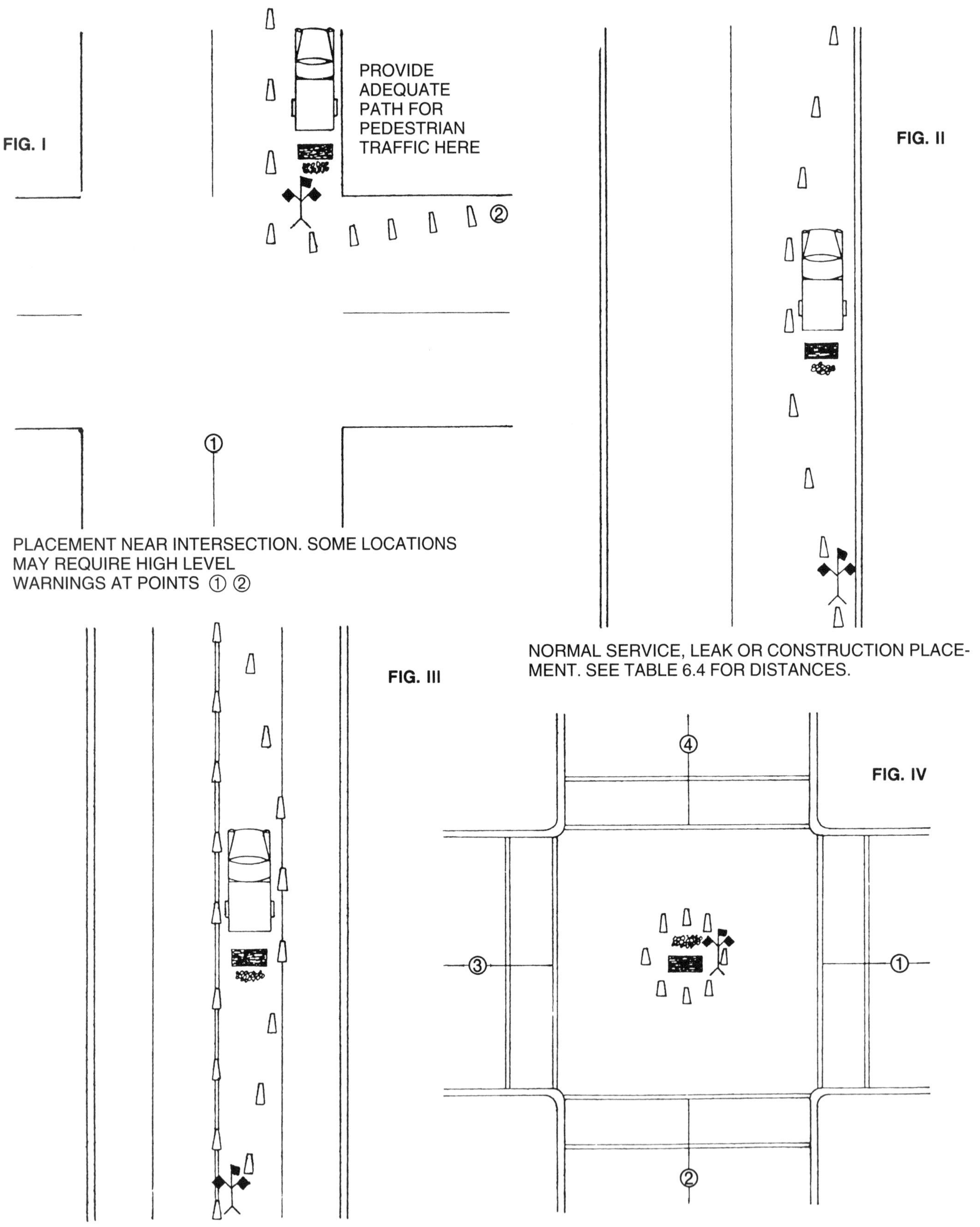
FIG. I
PROVIDE ADEQUATE PATH FOR PEDESTRIAN TRAFFIC HERE
①
②
PLACEMENT NEAR INTERSECTION. SOME LOCATIONS MAY REQUIRE HIGH LEVEL WARNINGS AT POINTS ① ②
FIG. II
NORMAL SERVICE, LEAK OR CONSTRUCTION PLACEMENT. SEE TABLE 6.4 FOR DISTANCES.
FIG. III
MULTI-LANE HIGHWAY PLACEMENT. PLACE HI-LEVEL WARNING IN SAME LANE AS OBSTRUCTION. SEE TABLE 6.4 FOR DISTANCES.
FIG. IV
①
②
③
④
PLACEMENT AT MAJOR TRAFFIC SIGNAL CONTROLLED INTERSECTION WHERE CONGESTION IS EXTREME. SOME LOCATIONS MAY PERMIT WARNINGS AT POINTS 1 2 3 4.

Fig. 6.15 Placement of traffic cones and signs

TYPICAL HIGH-LEVEL WARNING

FIG. V

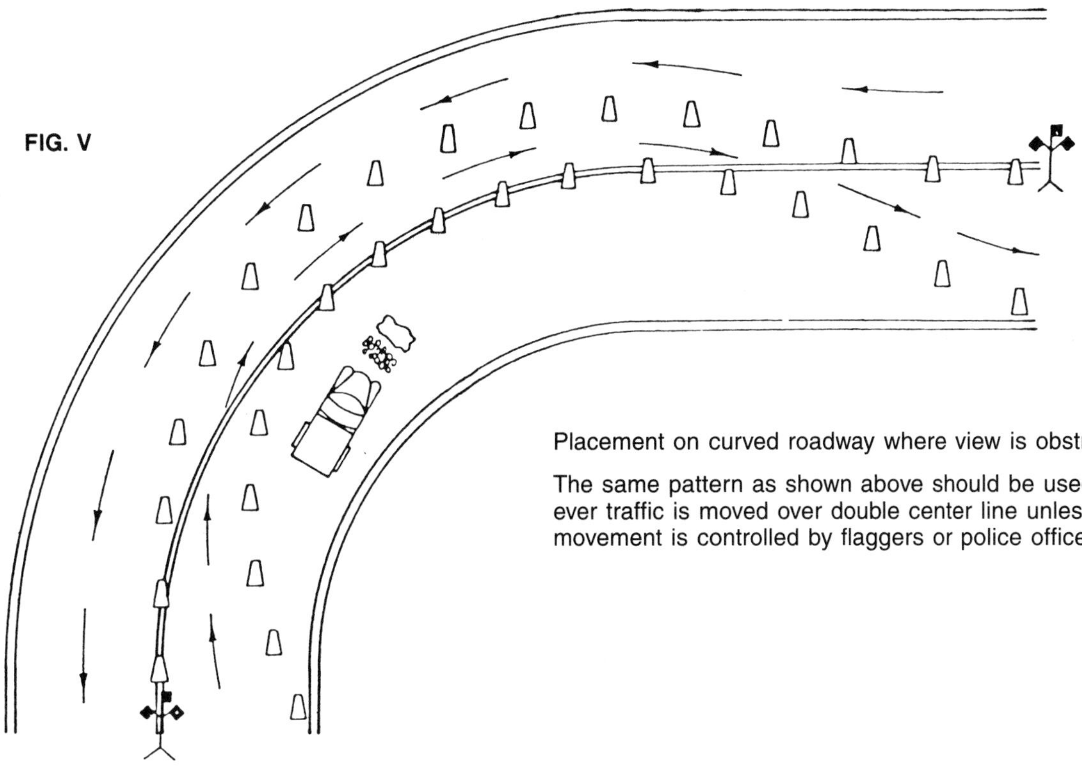

Placement on curved roadway where view is obstructed.

The same pattern as shown above should be used whenever traffic is moved over double center line unless traffic movement is controlled by flaggers or police officer.

FIG. VI

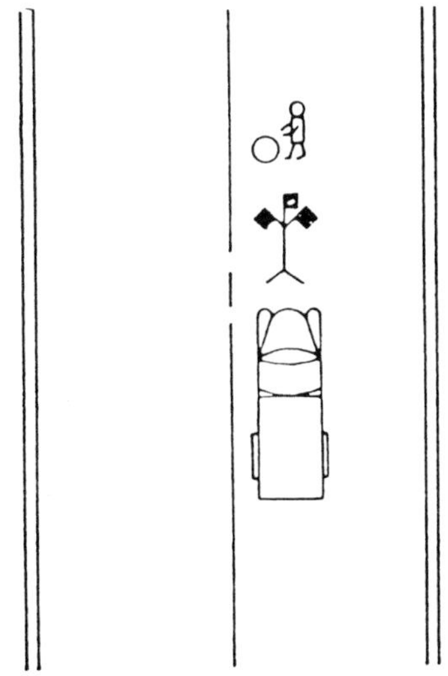

FIG. VII

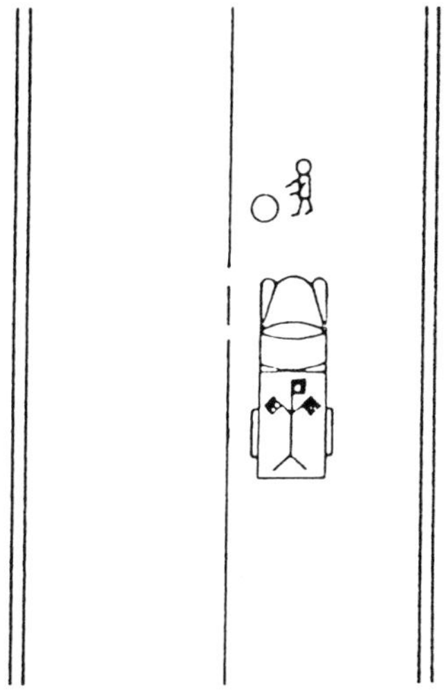

Placement for gate operation or other jobs of short duration. Employee must wear high-visibility red jacket.

Alternate method for operation described at left. High-level warning is mounted on rear of vehicle that is parked in advance of work location. Employee must wear high-visibility red jacket.

Fig. 6.15 Placement of traffic cones and signs (continued)

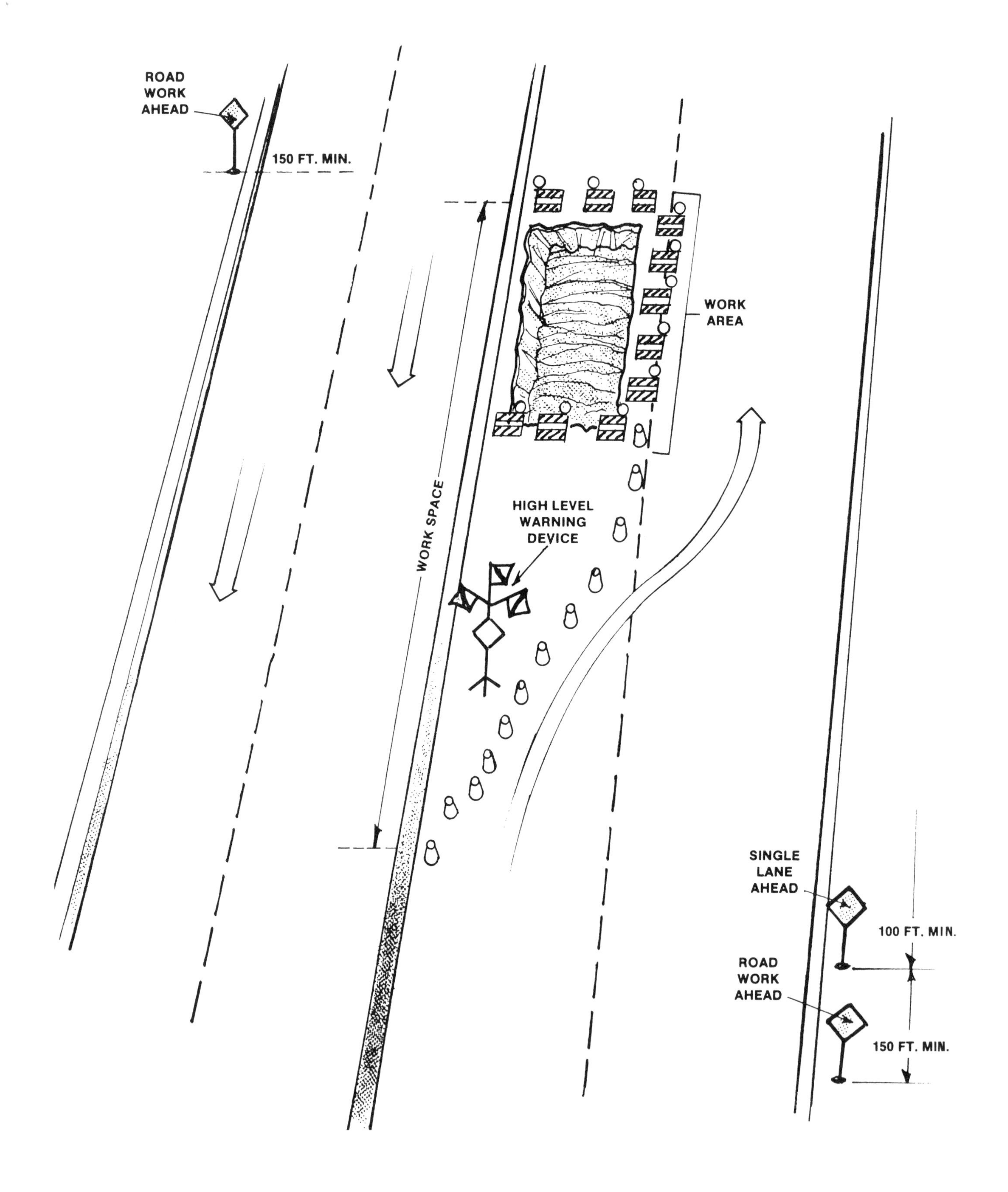

Fig. 6.16 Closing of left lane

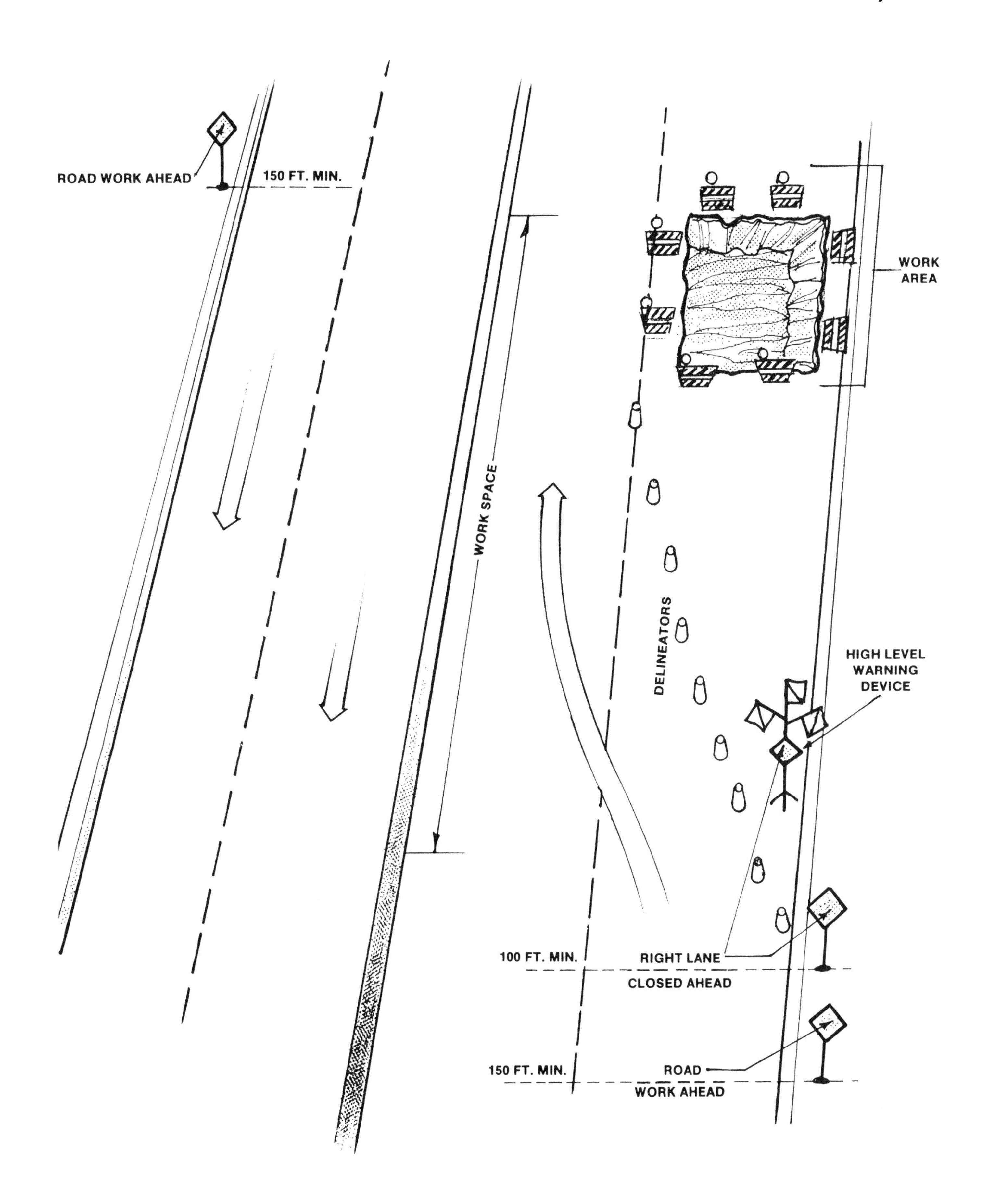

Fig. 6.17 Closing of right lane

3. Excavations greater than four feet deep must be tested as often as necessary for oxygen deficiency/enrichment and hazardous atmospheres, and appropriate precautions must be taken to protect workers.
4. Excavations must be inspected daily by a competent person and, if a problem exists, workers must be removed until precautions have been taken to ensure their safety.
5. Properly qualified supervisors must be on site at all times during excavations.
6. Safety provisions must be taken to protect workers installing or removing shoring systems.
7. Spoil (material removed from excavation) must be kept back at least two feet (0.6 m) from the edge of all excavations.
8. Safe and convenient access to an excavation must be provided.
9. Effective barriers must be installed at excavations adjacent to mobile equipment.
10. Water must not be allowed to accumulate in any excavation.

6.541 Trenches

A trench is an excavation in which the average depth exceeds the width and the width is 15 feet (4.5 meters) or less at the bottom. The same general requirements which apply to excavations apply to trenches. Trenching is more common than excavations in water utility work. Exits must be provided at 25-foot (7.4-meter) intervals for all occupied trenches four or more feet (1.2 m) deep. Walkways must also be provided where operators must cross the trench. If the trench is six feet or more in depth at the crossing area, guardrails must be provided on the walkway.

Covers placed over trenches in roadways must be secured against displacement. Covers are usually placed over a trench at the end of the day. This allows opening of the street for peak traffic flow. Occasionally excavations or trenches must cross sidewalks or other pedestrian rights of way. When this is necessary, a safe, temporary right of way must be provided or foot traffic should be rerouted.

Figure 6.18 illustrates various types of trench shoring conditions.

6.542 Cave-Ins

Improper shoring can contribute to cave-ins. Many cave-ins can be avoided by using the proper shoring where warranted by soil type and condition. Also, proper shoring procedures can avoid the embarrassing situation of finding oneself in a position of helplessness while installing the bottom brace or screw jack first simply because it is the path of least resistance. That is the easy way, but not the safe way.

Another hazard is the excavated spoil bank. Usually the spoil bank is set back far enough to prevent the material from falling back into the trench and also far enough to provide a walkway or tool-rest for shovels and crowbars. Of great significance, however, is the fact that the proper setback of the spoil bank will keep the weight away from the edge of the trench. If the spoil is too heavy and too close to the edge, the wall of the trench could cave in on top of you.

The following procedures are recommended in order to prevent violations of the safety rules, to prevent cave-ins, and to help keep us from getting hurt.

1. In excavations five feet (1.5 meters) or deeper (four feet in some regulations), shoring is required and the spoil bank is to be kept back a minimum of two feet (0.6 meter) from the edge of the trench. Increase this distance consistent with the type and condition of the soil.
2. When installing braces or screw jacks, start with the top one first and progressively work downward. Removal is in reverse order, starting with the bottom brace and working upward.
3. Eight feet (2.4 meters) is the maximum spacing of uprights under the ideal conditions of hard, compact soil excavated to a depth not exceeding 10 feet (3 meters). Spacing of the uprights shall be decreased and solid sheathing may be required when the depth exceeds 10 feet (3 meters) or if unstable soil is encountered.
4. When sources of external vibration (trucks, railroads) are nearby, special safety precautions may be necessary. Additional bracing or other effective means are recommended.
5. In trenches four feet (1.2 meters) or deeper, ladders or ramps should be provided for access and should be located so that workers need not move farther than 25 feet (7.5 meters). The ladders should extend a minimum of 30 inches (75 cm) above the top of the excavations.

Field inspections and injury reports have indicated that the items listed above are frequently overlooked. Attention to these items is necessary if we are to reduce accidents in this area. If we continue to ignore safety regulations, the pain and suffering of accidents will continue.

The economic impact of shoring requirements is a known factor. Provision of the proper materials, equipment, and the time to install and remove the shoring are required expenses. The installation of a substandard shoring system is just as expensive as one properly installed. Given the alternative, it makes sense to install the shoring system properly!

These requirements are minimal and, on occasion, the shoring and methods may have to be improved. In large excavations or where complex shoring systems deviate from regular shoring procedures, alternate shoring systems must be designed by a registered civil engineer and submitted to the appropriate authorities for approval.

RUNNING MATERIAL

Sheet Piling or equivalent solid sheeting is required for trenches four feet or more deep.

Longitudinal-stringer dimensions depend upon the strut braces, the stringer spacing, and the depth of stringer below the ground surface.

Greater loads are encountered as the depth increases, so more or stronger stringers and struts are required near the trench bottom.

HARD COMPACT GROUND

Trenches 5 feet or more deep *and* over 8 feet long must be braced at intervals of 8 feet or less.

SATURATED, FILLED OR UNSTABLE GROUND

Sheeting must be provided, and must be sufficient to hold the material in place.

Longitudinal-stringer dimensions depend upon the strut and stringer spacing and upon the degree of instability encountered.

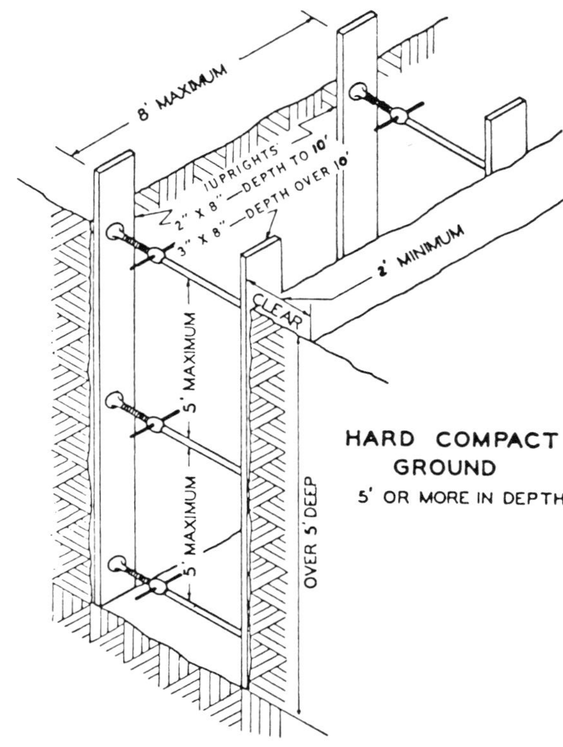

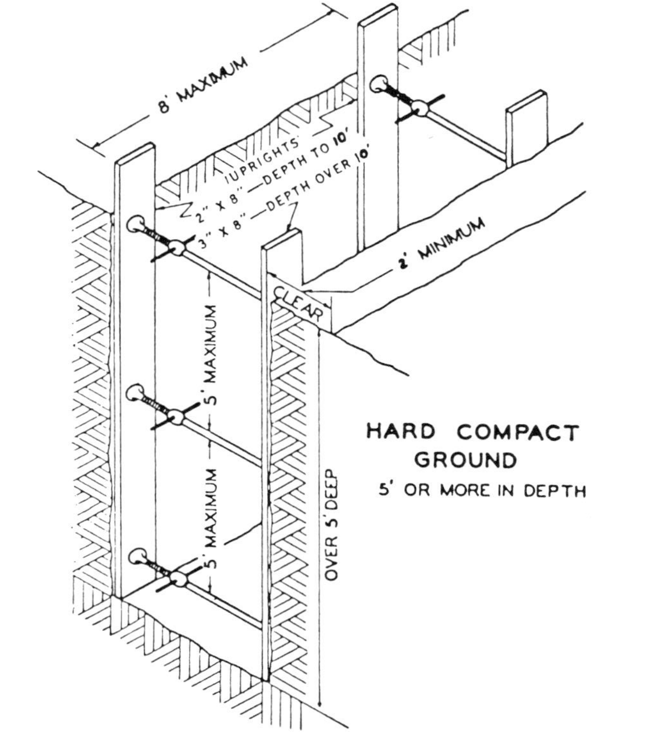

Fig. 6.18 Various types of trench shoring conditions

6.543 *Ladders*

Ladders must be placed in excavations so that any worker is never more than 25 feet (7.5 meters) from the ladder. For example, one ladder is required for each 50 feet (15 meters) of trench if the ladder is placed in the center of the trench. If the ladder is placed near one end of the trench, two ladders would be required. The ladders must extend 2½ feet (75 centimeters) above the surface of the excavation. Ladders are required in all excavations where a problem exists in entering or exiting the excavation. All ladders must be kept clean and in good repair.

6.544 *WARNING: Look Out for Other Underground Utilities*

Before you do any excavating, determine the location of underground utilities and advise the utility owners of the proposed work. Some underground installations that could be damaged by excavations include underground cables, electrical services, gas mains, and occasionally storm water and sewer lines that may be adjacent to proposed excavations. *NEVER* disturb a thrust block supporting an underground utility. Many phone books list an "Underground Service Alert" or "One-Call" phone number which you should call *BEFORE* you do any underground digging or drilling. Figures 6.19 and 6.20 give some typical telephone company warnings and indicators of underground cable lines.

6.55 Protect the Public

When the work area crosses the sidewalks, walkways, or crosswalk areas, special consideration must be given to pedestrian safety. Every effort must be made to separate the pedestrian from the work area. Protective barricades, fencing, handrails, and bridges must be used with warning devices and signs to direct foot traffic to areas where pedestrians may pass safely. Walkways in construction areas should be at least four feet (1.2 meters) wide and free of abrupt changes in grade. Where traffic must change grade, an incline should be used. Special consideration should be given to the visually impaired and handicapped. There should be a minimum vertical clearance of seven feet (2.1 meters) and illumination should be provided during hours of darkness. Occasionally, a walkway must be closed completely. When this is necessary, foot traffic should be diverted at the closest crosswalks with the appropriate signs and barricades (Figure 6.21).

When welding is necessary on the job, suitable screens must be placed to protect the public from arc welding flash or sparks from cutting operations.

When equipment is stored on the job site, such as pipe, it should be securely blocked so it cannot be displaced.

When work is to be done in a residential area, it is good public relations to advise the public what is to be done, how long it is expected to take, that all safety regulations will be observed, and that every effort will be taken not to disrupt the routine of the neighborhood.

QUESTIONS

Write your answers in a notebook and then compare your answers with those on page 392.

6.5E List the general safety rules for an excavation.

6.5F What type of shoring material should be used with trenches in running material?

6.5G List the underground utilities that could be damaged by water utility personnel searching for a leaking underground water main.

6.5H What would you tell the public when excavation work is planned in a residential neighborhood?

6.6 SAFETY AROUND WATER STORAGE FACILITIES

6.60 Slips and Falls

Industrial accident reports indicate that approximately 25 percent of all accidents are caused by slips and falls. The water utility industry is no different than other industries. Operators suffer from too many accidents due to slippery, wet surfaces. Slippery surfaces are common around storage facilities. They are caused by water or chemical spills on the floor. When cleaning a storage facility, one cannot avoid working on a wet surface.

When you must work inside a water tank, wear boots which have a special tread to help prevent a person from slipping. The area around a storage facility may be wet and could be the cause of slimes and growths which create slippery surfaces. Since rubber boots with special anti-slip tread are not always available, it may be necessary to treat the wet, slippery surface. There are many anti-slip coatings commercially available. Many of these anti-slip coatings may be applied with a brush, rollers, or sprayed. These coatings usually have an abrasive (carbide crystals or ground walnut shells) substance suspended in a resin and use of them can change a slippery surface to a safe, nonslip surface.

In addition to coatings, there are other products which you might consider. These are specially treated strips of tape-like materials which can be affixed to areas where foot traffic is heavy. This type of tape material is often affixed every 6 to 12 inches (15 to 30 cm) along a slippery walkway.

WARNING
UNDERGROUND
CABLES
BEFORE
DIGGING
FOR FREE CABLE
LOCATING SERVICE
call collect
AREA CODE 213
621-3111
24 HOUR SERVICE
PACIFIC TELEPHONE

WARNING
UNDERGROUND CABLE
BELL
SYSTEM
BEFORE DIGGING
PLEASE CALL COLLECT
AREA CODE 213
621-3111
24 HOUR SERVICE

Fig. 6.19 Typical identification and warning signs used by a phone company to identify facilities

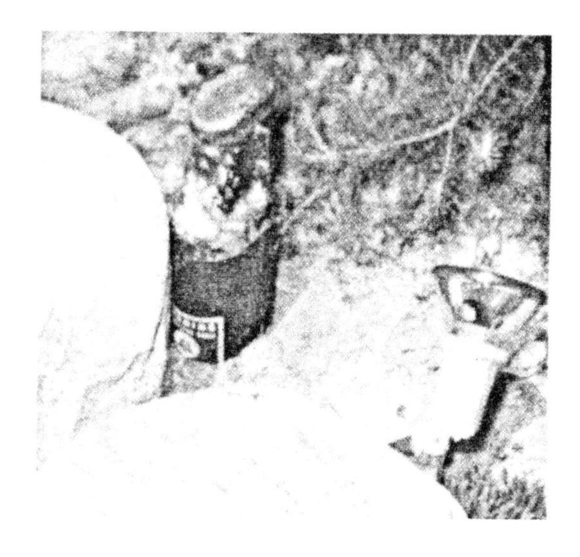

Fig. 6.20 Typical identification that can be found near areas where the telephone company has installations

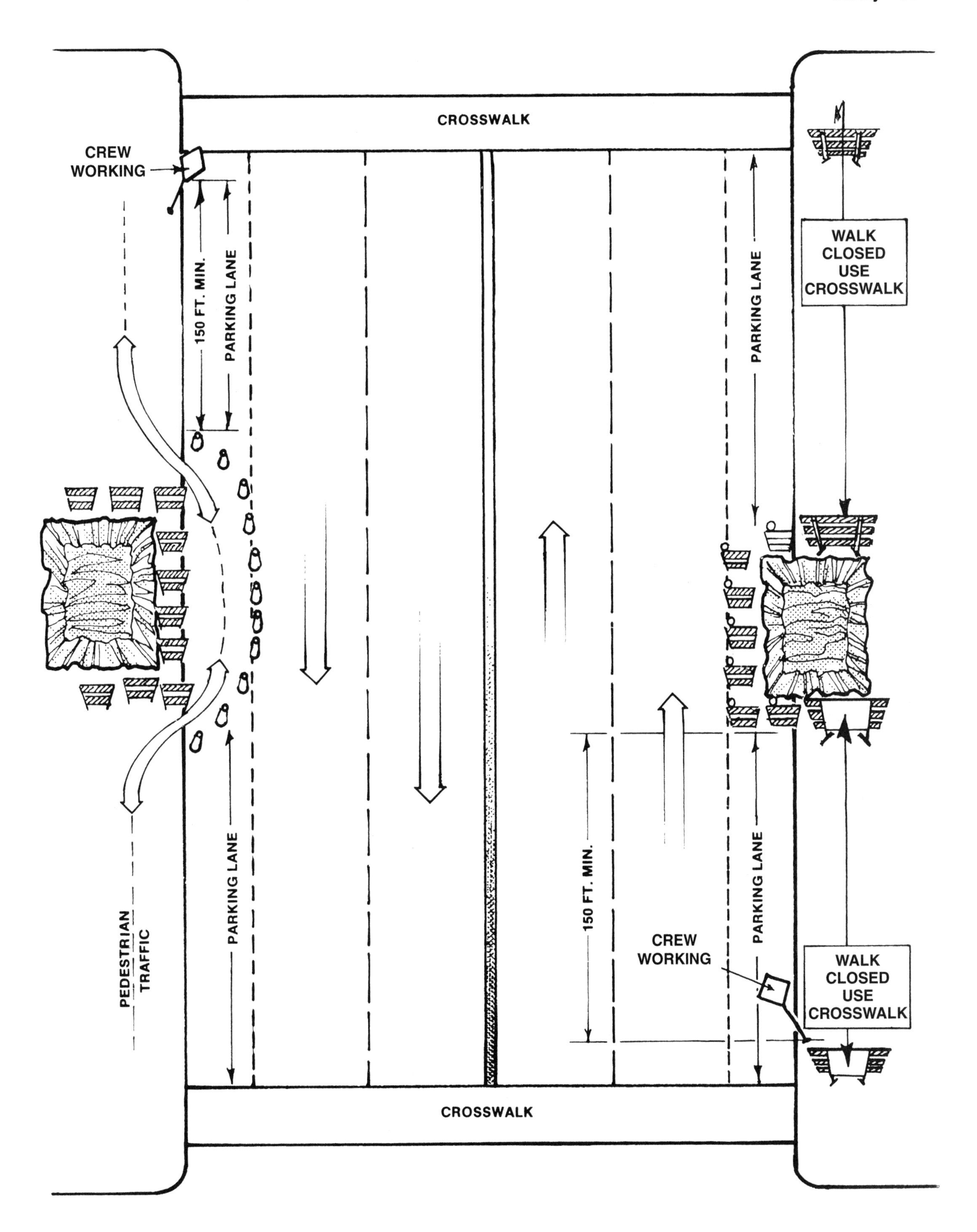

Fig. 6.21 Pedestrian control

6.61 Ladders

6.610 Fixed Ladders (*Figure 6.22*)

Most storage facilities are equipped with a fixed ladder which provides the only means of access to the top of an elevated facility (tank). A fixed ladder usually consists of individual rungs which are attached to the tank. A recommended type of fixed ladder should have provisions for use of a safety belt equipped with hooks and collar. The major hazard in the use of a fixed ladder is free fall. Other hazards include falls from carrying loads, running up or down, jumping from a ladder, and reaching too far out to the side while working from the ladder. The major hazard may be eliminated by the installation and proper use of available safety climbing devices.

6.611 Climbing Safety Devices (*Figure 6.22*)

Climbing safety devices are intended to prevent free fall of a climber if, for any reason, the climber should lose a grip or footing. Two basic types of safety devices are available:

1. A rail or cable attached to the ladder on which a sleeve or collar travels, and
2. A sleeve or collar which is fastened to the climber's safety belt by hooks and short lengths of chain.

   The safety sleeve should be of a type which can be operated entirely by the person using the device to ascend or descend without having to continuously manipulate the safety sleeve. At normal climbing speed, the sleeve or collar slides up and down without hindrance. If the climber falls, a locking trigger or friction brake is activated and automatically stops the climber's fall.

Many hazards may be minimized by training personnel in how to go up and down a ladder.

Potential ladder users should be:

1. Physically capable of exertion required, and
2. Without a previous history of heart ailment, dizzy or fainting spells, or other physical impairment which would make this climbing dangerous.

6.612 Training

Personnel should be properly trained in the use of ladders. The safe use of ladders requires two hands for the job. When climbing a ladder, use the following procedures:

1. Face the ladder and use both hands to grip the rungs or side rails firmly.
2. Place feet firmly on each rung before transferring full weight of body to each foot.
3. Climb deliberately, but without haste. Never run up or down and never slide down a ladder.
4. Never jump from a ladder. Check footing before stepping off a ladder.
5. Wear shoes with heels.
6. Always give your full attention when climbing. Be mindful that you are responsible for your own safety.

6.613 Fixed Ladder Inspection

All fixed ladders and climbing devices should be inspected at least once a year. This inspection should be done by a person who is familiar with ladder inspection. Give prompt attention to correcting any defect which may be the potential cause of an accident. Some items to check for during a routine inspection include:

1. Loose, worn, and damaged rungs or side rails;
2. Corroded or damaged parts of the security door;
3. Corroded bolts or rivet heads where ladder is affixed to tank; and
4. Defects in climbing devices, including loose or damaged carrier rails or cables.

NEVER LEAVE A DEFECT UNCORRECTED. A ladder or safety device is only as safe as the person who is inspecting or using the ladder and climbing device.

6.62 Application of Coatings

The application of interior coatings to a storage facility (water tank) is necessary to preserve a tank interior in operating order. The application of a coating to a tank interior should not be taken lightly. Coordination and safety equipment are necessary to complete the job effectively. After the tank has been cleaned, it is recommended that it be sandblasted before applying any type of coating.

After the tank has been declared safe to enter (safe atmospheric conditions are discussed in Section 6.63, "Confined Spaces"), sandblasting may take place. The person operating the nozzle end of the sandblaster must wear protective clothing. This protection should include some type of clothing which has long sleeves — preferably a coverall type of clothing; a respirator and goggles or face shield should be used for face protection; and earplugs may be helpful because of noise.

Make certain that sufficient light is available and that sufficient air is available for the person doing the sandblasting. A respirator may not be sufficient. You may need to provide air by means of a portable air cylinder and an adequate length of air hose to a mask worn by the person doing the work.

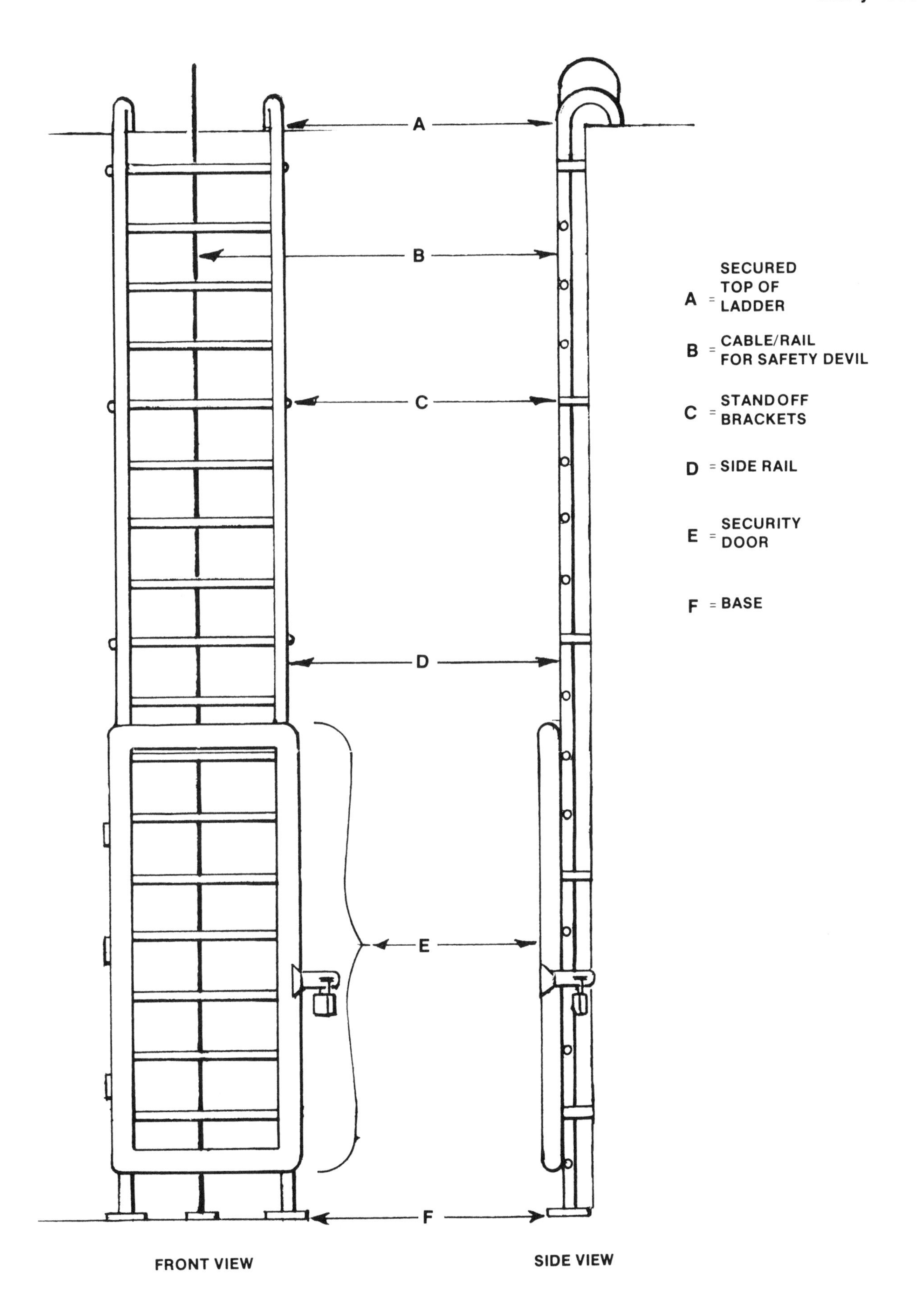

Fig. 6.22 Fixed ladder with climbing safety device

After sandblasting has been completed, remove sand and paint debris from the tank. Make certain the tank has adequate air available for anyone working inside. After sand and debris are removed, scaffolding may be necessary for painters to apply the coating. Make certain that scaffolding meets your local safety requirements. If a mobile ladder stand is used instead of scaffolds, make certain that it meets safety requirements.

While painters are applying the coating, make certain that they have air masks supplied with air and that someone is close by to check on their status periodically. Painters or any other person who must work in an enclosed area (tank) must have a safety harness with a safety line secured at the tank's entrance.

6.63 Confined Spaces

Confined space means a space that:

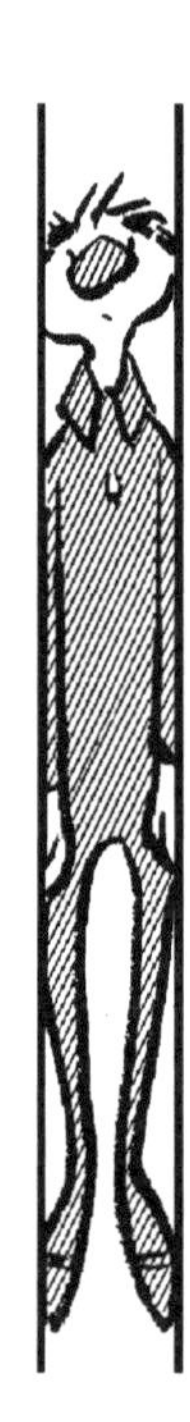

- Is large enough and so configured that an operator can bodily enter and perform assigned work
- Has limited or restricted means for entry or exit, and
- Is not designed for continuous operator occupancy.

In addition, if the space has or may have one or more of the following conditions, it is a permit-required confined space:

- A hazardous atmosphere,
- The potential for engulfing an operator. This means the space could contain a material, such as water or granular or powdered chemicals that might flow down and around an operator thereby surrounding and trapping the operator,
- An internal configuration (shape) such that an operator could be trapped or asphyxiated (suffocated) by inwardly converging walls or by a floor which slopes downward and tapers to a smaller cross section, and/or
- Any other recognized serious safety or health hazard.

Let's apply some common sense to simplify the definition of a confined space. First, we can get into the space. Second, even though we can get into the space, it has limited means of getting into it and getting back out, which is important in case rescue is required. Third, the space has the potential for the existence of a toxic or explosive atmosphere or an oxygen deficiency/enrichment that could kill or injure us. Finally, there is a potential for engulfment by a material (such as water).

The federal regulations regarding confined spaces identify both permit-required confined spaces and non-permit-required confined spaces. All spaces must be considered permit-required confined spaces until a competent person following prescribed pre-entry procedures determines otherwise. Clearly, any operator who is either making this type of evaluation or will be entering the confined space must have completed minimum confined space entry training requirements.

A non-permit confined space means a confined space that does not contain or, with respect to atmospheric hazards, have the potential to contain any hazard capable of causing death or serious physical harm.

A written copy of Operating and Rescue Procedures that are required by the confined space entry procedure must be at the work site during any work. The checklist includes the control of atmospheric and engulfment hazards, surveillance of the surrounding area to avoid hazards such as drifting vapors, and atmospheric testing for oxygen deficiency/enrichment and lower explosive limit.

Confined spaces may be entered without the need for a written permit (Figure 6.23) or an additional person standing by provided that the only hazard posed by the space is an actual or potential hazardous atmosphere and continuous forced-air ventilation alone can maintain the space safe for entry.

Covered water storage facilities are considered a confined space. This includes elevated tanks, facilities with roofs or covers, and underground facilities. Hazardous atmospheric conditions in confined spaces are a threat to the health of anyone who attempts to enter the facility. Other potential confined spaces encountered by operators include manholes, valve vaults, and any empty covered tank such as a dewatered filter or clear well.

When entering any confined space, the greatest consideration must be given to *DANGEROUS AIR CONTAMINATION* and *OXYGEN DEFICIENCY/ENRICHMENT.* Many operators have lost their lives as a result of these conditions. Use portable ventilation fans and *ALWAYS PROVIDE ADEQUATE VENTILATION.* Adequate ventilation will remove hazardous gases and replenish the oxygen. *BEFORE* entering a confined space, test the atmosphere with gas detection instruments capable of detecting and measuring explosive gases, toxic conditions, and/or oxygen deficiency/enrichment (see Section 6.20, "Monitoring Equipment"). These instruments should sound an alarm and/or produce a flashing light whenever hazardous atmospheric conditions are detected. Operators working in confined spaces should not be expected to read meters and determine whether or not a hazard exists. Operators who work around these conditions should be trained in applying cardiopulmonary resuscitation (CPR).

Use the buddy system. Wear a safety harness and be certain your buddy is trained and knows what to do in the event that you get into trouble.

Safety is a must — never cut corners no matter how cumbersome the safety equipment may be — its use may save your life. Always review the *SAFETY NOTICE* before entering a confined space.

Confined Space Pre-Entry Checklist/Confined Space Entry Permit

Date and Time Issued: ________ Date and Time Expires: ________ Job Site/Space I.D.: ________

Job Supervisor: ________ Equipment to be worked on: ________ Work to be performed: ________

Standby personnel: ________ ________ ________

1. Atmospheric Checks: Time ________ Oxygen ________ % Toxic ________ ppm
   Explosive ________ % LEL Carbon Monoxide ________ ppm

2. Tester's signature: ________
3. Source isolation: (No Entry) N/A Yes No
   Pumps or lines blinded, disconnected, or blocked ( ) ( ) ( )
4. Ventilation Modification: N/A Yes No
   Mechanical ( ) ( ) ( )
   Natural ventilation only ( ) ( ) ( )
5. Atmospheric check after isolation and ventilation: Time ________
   Oxygen ________ % > 19.5% < 23.5% Toxic ________ ppm < 10 ppm H_2S
   Explosive ________ % LEL < 10% Carbon Monoxide ________ ppm < 35 ppm CO

Tester's signature: ________

6. Communication procedures: ________
7. Rescue procedures: ________

8. Entry, standby, and backup persons Yes No
   Successfully completed required training? ( ) ( )
   Is training current? ( ) ( )
9. Equipment: N/A Yes No
   Direct reading gas monitor tested ( ) ( ) ( )
   Safety harnesses and lifelines for entry and standby persons ( ) ( ) ( )
   Hoisting equipment ( ) ( ) ( )
   Powered communications ( ) ( ) ( )
   SCBAs for entry and standby persons ( ) ( ) ( )
   Protective clothing ( ) ( ) ( )
   All electric equipment listed for Class I, Division I, Groups A, B, C, and D, and nonsparking tools ( ) ( ) ( )
10. Periodic atmospheric tests:
    Oxygen: ____% Time ____; ____% Time ____; ____% Time ____; ____% Time ____;
    Explosive: ____% Time ____; ____% Time ____; ____% Time ____; ____% Time ____;
    Toxic: ____ppm Time ____; ____ppm Time ____; ____ppm Time ____; ____ppm Time ____;
    Carbon Monoxide: ____ppm Time ____; ____ppm Time ____; ____ppm Time ____; ____ppm Time ____;

We have reviewed the work authorized by this permit and the information contained herein. Written instructions and safety procedures have been received and are understood. Entry cannot be approved if any brackets ( ) are marked in the "No" column. This permit is not valid unless all appropriate items are completed.

Permit Prepared By: (Supervisor) ________ Approved By: (Unit Supervisor) ________

Reviewed By: (CS Operations Personnel) ________
(Entrant) (Attendant) (Entry Supervisor)

This permit to be kept at job site. Return job site copy to Safety Office following job completion.

Fig. 6.23 Confined space pre-entry checklist/permit

> **SAFETY NOTICE**
>
> Before anyone ever enters a tank for any reason, these safety procedures must be followed:
>
> 1. Test atmosphere in the tank for toxic and explosive gases and for adequate or inadequate oxygen. Contact your local safety equipment supplier for the proper types of atmospheric testing devices. These devices should have alarms which are activated whenever an unsafe atmosphere is encountered;
> 2. Provide adequate ventilation, especially when painting. A self-contained, positive-pressure breathing apparatus may be necessary when painting;
> 3. All persons entering a tank must wear a safety harness; and
> 4. One person must be stationed at the tank entrance to observe the actions of all people in the tank. An additional person must be readily available to help the person at the tank entrance with any rescue operation.

QUESTIONS

Write your answers in a notebook and then compare your answers with those on page 392.

6.6A How do anti-slip coatings prevent slippery surfaces?

6.6B What are the two basic types of safety devices used by operators when climbing fixed ladders?

6.6C What are the physical requirements for operators who must climb fixed ladders?

6.6D A safety inspection of a fixed ladder should include what items?

6.6E Why do storage tanks need interior coatings?

6.6F What atmospheric hazards may be encountered in a confined space?

6.7 WORKING NEAR NOISE

Operators must be protected against the effects of exposure to noise when the sound levels exceed the limits listed in Table 6.5. Engineering methods of controlling noise must be initiated whenever possible. If noise levels cannot be controlled within acceptable limits, operators must be provided and use approved earplugs, muffs, and/or personal protective equipment.

TABLE 6.5 PERMISSIBLE NOISE EXPOSURES[a]

1. When employees are exposed to loud or extended noise, earplugs and/or protective devices shall be provided and employees shall wear them accordingly.
2. Protection against the effects of noise exposure shall be provided when sound levels exceed those shown in rule number 7. These measurements shall be made on the A-scale of a standard sound level meter at a slow response.
3. When employees are subjected to sound levels exceeding those listed in rule number 7, feasible administrative and/or engineering controls shall be utilized. If such controls fail to reduce employee exposure to within the permissible noise levels listed in this table, personal protective equipment shall be provided and used.
4. Plain cotton shall not be considered acceptable as an earplug.
5. Machinery creating excessive noise shall be equipped with mufflers.
6. Exposure to impulsive or impact noise should not exceed 140 dB peak sound pressure level.
7. Employees shall not be exposed to noise which exceeds the levels listed below for a time period not to exceed those listed below.

DURATION (Hours Per Day)	SOUND LEVEL (dB[b] Slow Response)
8	90
6	92
4	95
3	97
2	100
1½	102
1	105
½ (thirty minutes)	110
one-fourth (¼) hour, or less	115

[a] Occupational Safety and Health Standards, Subpart G, Paragraph 1910.94, Noise Exposure

[b] Decibel (DES-uh-bull). A unit for expressing the relative intensity of sounds on a scale from zero for the average least perceptible sound to about 130 for the average level at which sound causes pain to humans. Abbreviated dB.

6.8 SAFETY INSPECTIONS

Regular safety inspections are a good way to be sure that equipment and facilities receive a regular safety review. To ensure that important items are not overlooked or forgotten, inspection forms or reports are very helpful. These forms also provide a record of who inspected what items and when. Figures 6.24 and 6.25 are typical safety inspection reports.

SAFETY INSPECTION REPORT
SANITARY ENGINEERING DIVISION

Location______________________________

Inspec. By - Date						
Housekeeping						
Material Handling						
Material Storage						
Aisles & Walkways						
*Ladders & Stairs						
Floors & Railings						
Exits						
Lighting						
Ventilation						
Hand Tools						
Electric Equipment						
Machinery						
Safety Guards						
Safety Devices						
Clothing & Equipment						
Dusts & Fumes						
Fire Hazards						
Expolsion Hazards						
Chemical Hazards						
Fire Equipment						
Unsafe Practices						
Other						

Reviewed By - Date						

Remarks:

*Note ladders

Fig. 6.24 Safety inspection report used by operators to inspect facilities

SAFETY INSPECTION REPORT

LOCATION INSPECTED Palos Verdes Chlorination Station DATE April 28, 1998

DIVISION Sanitary Engineering SECTION Water Treatment

SUPERVISOR Danny Saenz TITLE

REASON FOR CALL	ROUTINE	REQUEST	SPECIAL	ANNUAL	CALIF. D.I.S.
				X	

ROUTING	SUPERVISOR	SECTION	DIVISION	SYSTEM
	X			

INSPECTED BY Clarence Butrum DISCUSSED WITH Danny Saenz

CHECK LIST

NUMERIC RATING VALUE: 0 POOR OR DEFICIENT, 2 FAIR OR AVERAGE, 3 GOOD, 4 EXCELLENT

ITEMS MARKED X WERE NOT INSPECTED ON THIS VISIT

1. 4 HOUSEKEEPING
2. 4 MATERIAL HANDLING
3. 4 MATERIAL STORAGE
4. 4 AISLES AND WALKWAYS
5. 2 LADDERS AND STAIRS
6. 4 FLOORS, PLATFORMS AND RAILINGS
7. 4 EXITS
8. 4 LIGHTING
9. 4 VENTILATION
10. – HAND TOOLS
11. 4 ELECTRIC TOOLS AND EQUIPMENT
12. – MACHINERY AND EQUIPMENT
13. – GUARDS AND SAFETY DEVICES
14. – WELDING EQUIPMENT
15. 4 PROTECTIVE CLOTHING AND EQUIPMENT
16. – PERSONAL TOOLS AND EQUIPMENT
17. 4 DUSTS, FUMES, MISTS, GASES, AND VAPORS
18. – FIRE HAZARDS
19. – EXPLOSION HAZARDS
20. 4 CHEMICAL HAZARDS
21. – HAND AND POWER TRUCKS
22. 4 FIRE FIGHTING EQUIPMENT
23. – VEHICLES
24. – UNSAFE PRACTICES
25. – HORSEPLAY
26. – OTHER

TO SUPERVISOR: ITEMS FOUND TO BE DEFICIENT ARE DESCRIBED IN DETAIL BELOW.

#5. Ladder needs 1998 inspection.

J. K. JACOBS
MAY 0 4 1998

J. L. SAENZ
MAY 4 1998

SIGNED Clarence Butrum – Safety

Fig. 6.25 Safety inspection report used by safety section personnel

QUESTIONS

Write your answers in a notebook and then compare your answers with those on page 392.

6.7A How can operators protect themselves from excessive noise?

6.8A Why are safety inspection forms helpful?

6.9 ARITHMETIC ASSIGNMENT

There is no arithmetic assignment for this chapter.

6.10 ADDITIONAL READING

1. *NEW YORK MANUAL*, Chapter 19,* "Treatment Plant Maintenance and Accident Prevention."
2. *TEXAS MANUAL*, Chapter 21,* "Safety."
3. Chapter 20, "Safety," in *WATER TREATMENT PLANT OPERATION*, Volume II. Available from the Office of Water Programs, California State University, Sacramento, 6000 J Street, Sacramento, CA 95819-6025. Price $33.00.
4. *SAFETY BASICS FOR WATER UTILITIES.* Video, 27 minutes. Obtain from American Water Works Association (AWWA), Bookstore, 6666 West Quincy Avenue, Denver, CO 80235. Order No. 65119. Price to members, $186.00; nonmembers, $280.50; price includes cost of shipping and handling.
5. *LET'S TALK SAFETY—2002 SAFETY TALKS.* A series of 52 lectures on common water utility safety practices. Obtain from American Water Works Association (AWWA), Bookstore, 6666 West Quincy Avenue, Denver, CO 80235. Order No. 10123. ISBN 1-58321-145-4. Price to members, $33.50; nonmembers, $47.50; price includes cost of shipping and handling.
6. The American Red Cross is another source of up-to-date information on safety and first aid. Contact your local Area Chapter for a catalog of materials.
7. Read the *SAFETY* chapters in the other operator manuals in this series.

* Depends on edition.

End of Lesson 2 of 2 Lessons on SAFETY

Please answer the discussion and review questions before continuing with the Objective Test.

DISCUSSION AND REVIEW QUESTIONS

Chapter 6. SAFETY

(Lesson 2 of 2 Lessons)

Write the answers to these questions in your notebook before continuing with the Objective Test on page 393. The question numbering continues from Lesson 1.

12. Where do rotating, reciprocating, and transverse (crosswise) motions on pumps create hazards?
13. Why must water utility operators work in streets?
14. Where are flaggers required?
15. How can cave-ins be prevented?
16. How can the public be protected from excavation hazards?
17. How can you reduce the chances of operators slipping and falling on slippery surfaces?

SUGGESTED ANSWERS

Chapter 6. SAFETY

ANSWERS TO QUESTIONS IN LESSON 1

Answers to questions on page 344.

6.0A Two very important aspects of a safety program are:

1. Making people aware of unsafe acts, and
2. Conducting and/or participating in regular safety training programs.

6.0B Management's responsibilities for safety include:

1. Establishing a safety policy,
2. Assigning responsibility for accident prevention,
3. Appointing a safety officer,
4. Establishing realistic goals and periodically revising them, and
5. Evaluating the results of the program.

6.0C Operators' responsibilities for safety include:

1. Performing their jobs in accordance with established safe procedures,
2. Recognizing the responsibility for their own safety and that of fellow operators,
3. Reporting all injuries,
4. Reporting all observed hazards, and
5. Actively participating in the safety program.

Answers to questions on page 345.

6.1A An accident is that occurrence in a sequence of events that usually produces unintended injury, death, or property damage.

6.1B The principal reasons for unsafe acts include ignorance, indifference, poor work habits, laziness, haste, poor physical condition, and temper.

6.1C Laziness affects the speed and quality of work. Laziness also affects safety because safety requires an effort. In most jobs you cannot reduce your "safety effort" and still maintain the same level of safety, even if you slow down or lower your quality standards.

6.1D The brakes on a vehicle are working properly when they are able to stop the vehicle without grabbing or pulling to one side.

Answers to questions on page 348.

6.1E Before towing a trailer, inspect:

1. Rear, turn signal, and brake lights.
2. Tire tread and inflation,
3. Wheel attachment,
4. Trailer brake operation if equipped with brakes,
5. Trailer hitch and tongue, and
6. Safety chain.

6.1F All cigarettes and flames should be put out before attaching cables to charge a battery because a spark can ignite the hydrogen gas from the battery fluid.

6.1G Good safety practices for using boats include:

1. An annual safety inspection,
2. A safety vest or flotation device for each occupant,
3. Not overloading a boat, and
4. At least two people should go out in a boat.

6.1H A corrosive chemical is any chemical which may weaken, burn, and/or destroy a person's skin or eyes.

6.1I A tailgate safety session is usually held near the work site and new and old safety hazards and safer approaches and techniques to deal with the problems of the day or week are discussed.

Answers to questions on page 351.

6.2A Some of the typical locations in a water utility where it is necessary to monitor the atmosphere before entering include underground regulator vaults, solution vaults, manholes, tanks, trenches, and other confined spaces.

6.2B The backup person outside the confined area should check continuously on the status of personnel working in the confined space. Should a person working in the confined space be rendered unconscious due to asphyxiation or lack of oxygen, the backup person at the surface could enter into the space after:

1. Securing the help of an additional backup rescue person, and
2. Putting on the correct type of respiratory protective equipment.

6.2C A mixture of air and gas is considered hazardous when the mixture exceeds 10 percent of the Lower Explosive Limit (LEL).

Answers to questions on page 355.

6.2D Items of protective clothing for operators include:

1. Safety toe (hard toe) shoes,
2. Safety prescription glasses,
3. Safety (hard) hats,
4. Earplugs,
5. Safety glasses (nonprescription),
6. Safety goggles,
7. Face shields,
8. Respirators and masks,
9. Canvas gloves,
10. Rubber gloves for use with corrosive chemicals,
11. Safety aprons,
12. Leggings or knee pads,
13. Safety toe caps,
14. Protective clothing (lab coats or rubber suits), and
15. Safety belts or harnesses.

6.2E Liquids spilled on floors can be cleaned up by using nonflammable absorbent materials.

6.2F The steps to safe lifting include:

1. Keep feet parted — one along the side of and one behind the object,
2. Keep back straight — nearly vertical,
3. Tuck in your chin,
4. Grip the object with the whole hand,
5. Tuck elbows and arms in,
6. Keep body weight directly over feet, and
7. Lift slowly.

6.2G Before attempting to work on electric motors, be sure to lock out and tag all of the electric switches.

Answers to questions on page 356.

6.3A The well operator has the responsibility to preserve the quality of the well through preventive maintenance of the area where the well is located. This responsibility also extends to good housekeeping to prevent the development of safety hazards to operators working around the well site.

6.3B A sanitary seal on a well is used to keep contaminated water and debris from entering around the well casing and reaching the groundwater.

6.3C Surface openings on top of a well that must be protected against the entrance of surface water or debris into the well include:

1. Well casing vents,
2. Gravel tubes,
3. Sounding tube,
4. Conduits, and
5. Pump motor base seal.

Answers to questions on page 357.

6.3D Hydrochloric acid must be handled carefully because it is extremely corrosive. Hydrochloric acid can burn your eyes and skin, if inhaled it will cause coughing and choking, and if swallowed will cause corrosion of your insides (including your circulatory system).

6.3E Sulfamic acid has advantages over hydrochloric acid for acid cleaning a well. Granular sulfamic acid is non-irritating to dry skin and its solution gives off no fumes except when reacting with incrusting materials. Spillage, therefore, presents no hazard and handling is easier, cheaper, and safer.

6.3F Although the granular form of sulfamic acid is non-irritating to dry skin and it gives off no fumes, rubber gloves should still be used because hands may become wet from perspiration and the acid will cause burns on the skin. Safety goggles or glasses, preferably a face shield, should be used in case granules are blown into the face and/or eyes. Acid should always be added slowly to the water.

Answers to questions on page 359.

6.3G When handling hypochlorite use rubber gloves, a rubber apron, rubber boots, and a face shield.

6.3H Glassy phosphate should be handled very carefully because the glass-like particles are very sharp and can cut your skin. Use heavy-duty gloves when handling. Respiration equipment or a face shield should be worn.

6.3I If no one working for a water utility is knowledgeable about electric circuits, the utility should contact an electrician from a local electrical firm.

6.3J Items that should be included in a safety inspection of a well site include:

1. Cleanliness of site,
2. Locks on electrical panels,
3. Fencing around site, and
4. Caution/warning signs in place.

ANSWERS TO QUESTIONS IN LESSON 2

Answers to questions on page 360.

6.4A Pumps are used to move liquids, air, or gas. They are also used to increase pressures and energy of fluids.

6.4B Moving parts on a pump which require guards include rotation members, reciprocating arms, moving parts, and meshing gears. This includes rotating shafts, couplings, belts, pulleys, chains, and sprockets.

6.4C Rotating devices can injure you by gripping your clothing or hair and possibly forcing an arm or hand into a dangerous position.

6.4D The basic requirement for an enclosed guard is that it must prevent hands, arms, or any other part of an operator's body from making contact with dangerous moving parts. A good guarding system eliminates the possibility of the operator or another worker placing their hands near hazardous moving parts.

Answers to questions on page 365.

6.4E Safety planning includes figuring out how to do the job from start to finish as well as identifying potential safety hazards *BEFORE* starting a job and deciding how they can be avoided.

6.4F Never work on any equipment when it is hanging from a sling in midair because the equipment could slip from the sling and fall on you.

6.4G Stored energy must be bled down to prevent its sudden release, which could injure an operator working on the system or equipment.

6.4H Lubricants and fuels must be stored in a separate facility from other storage areas because of the potential fire hazard.

Answers to questions on page 371.

6.5A When channeling traffic flow through a work area, consider the following items:

1. Prewarn,
2. Sign properly,
3. Regulate speed, and
4. Guide properly.

6.5B Persons who must work in streets should wear their normal clothing plus a high-visibility shirt, poncho, vest, or coat. This safety clothing should be colored red or safety orange. A safety hat (hard hat) should be worn whenever you are near traffic.

6.5C Hand signaling motions commonly used by flaggers include (1) slow, (2) stop, and (3) proceed.

6.5D The following rules should be considered when setting up street warning devices:

1. Whenever possible, use a flagger when setting up or breaking down the work area,
2. Use more, rather than too little, warning equipment at the work area,
3. Anticipate the unexpected and irrational acts of pedestrians and drivers,
4. Do not use the same setup on streets with fast-moving traffic as would be used on slow-traffic streets, and
5. Keep all tools and materials within the confines of the protected area.

Answers to questions on page 378.

6.5E General safety rules for an excavation include:

1. Employer must inspect excavations for hazards before employees may work in or adjacent to the excavation.
2. The location of underground installations (such as utility lines) must be determined before excavation begins.
3. Excavations greater than four feet deep must be tested as often as necessary for oxygen deficiency/enrichment and hazardous atmospheres, and appropriate precautions must be taken to protect workers.
4. Excavations must be inspected daily by a competent person and appropriate safety precautions must be taken.
5. Properly qualified supervisors must be on site at all times during excavations.
6. Safety provisions must be taken to protect workers installing or removing shoring systems.
7. Spoil must be kept back at least two feet from the edge of all excavations.
8. Safe and convenient access to an excavation must be provided.
9. Effective barriers must be installed at excavations adjacent to mobile equipment.
10. Water must not be allowed to accumulate in any excavation.

6.5F Trenches in running material should have sheet piling or equivalent solid sheeting, longitudinal stringers, and cross braces.

6.5G Underground utilities that could be damaged by water utility personnel searching for a leaking underground water main include telephone cables, television cables, electrical services, gas mains, and occasionally storm water and sewer lines.

6.5H When work is to be done in a residential neighborhood, it is good public relations to advise the public what is to be done, how long it is expected to take, that all safety regulations will be observed, and that every effort will be taken not to disrupt the routine of the neighborhood.

Answers to questions on page 386.

6.6A Anti-slip coatings usually have an abrasive (carbide crystals or ground walnut shells) substance suspended in a resin which can change a slippery surface to a safe, nonslip surface.

6.6B The two basic types of safety devices used by operators when climbing fixed ladders are:

1. A rail or cable, which is fixed to the ladder, on which a sleeve or collar travels, and
2. A sleeve or collar which is fastened to the climber's safety belt by hooks and short lengths of chain.

6.6C Physical requirements for operators who must climb fixed ladders include:

1. Physically capable of the exertion required, and
2. Without a previous history of heart ailment, dizzy or fainting spells, or other physical impairment which would make this climbing dangerous.

6.6D A safety inspection of a fixed ladder should include checking for:

1. Loose, worn, and damaged rungs or side rails;
2. Corroded or damaged parts of the security door;
3. Corroded bolts or rivet heads where ladder is affixed to tank; and
4. Defects in climbing devices, including loose or damaged carrier rails or cables.

6.6E Storage tanks need interior coatings to preserve a tank interior in operating order.

6.6F Atmospheric hazards that may be encountered in a confined space include explosive conditions, toxic gases, and oxygen deficiency and/or oxygen enrichment.

Answers to questions on page 389.

6.7A Operators can protect themselves from excessive noise by using approved earplugs, muffs, and/or personal protective equipment.

6.8A Safety inspection forms are helpful because (1) they help to ensure that important safety items are not overlooked or forgotten during regular safety inspections, and (2) they provide a record of who inspected what items and when.

OBJECTIVE TEST

Chapter 6. SAFETY

Please write your name and mark the correct answers on the answer sheet, as directed at the end of Chapter 1. There may be more than one correct answer to each multiple-choice question.

True-False

1. People miss more time at work from on-the-job accidents than from off-the-job accidents.
   1. True
   2. False
2. Make sure headlights, heater, and other electrical accessories are on when attempting to charge a dead battery on a vehicle.
   1. True
   2. False
3. All motorized boats should have a fire extinguisher.
   1. True
   2. False
4. OSHA regulations allow operators to refuse to perform work under unsafe conditions.
   1. True
   2. False
5. Opening an electric circuit is similar to closing a valve on a pipeline.
   1. True
   2. False
6. Always isolate equipment from the driving force before starting any work.
   1. True
   2. False
7. Always store a fire extinguisher inside a fuel storage facility.
   1. True
   2. False
8. Tops of ladders should extend a minimum of 30 inches (75 cm) above the top of trench excavations.
   1. True
   2. False
9. Slippery surfaces are caused by water or chemical spills on the floor.
   1. True
   2. False
10. Never run up or down and never slide down a ladder.
    1. True
    2. False
11. Plain cotton is considered acceptable as an earplug.
    1. True
    2. False

Best Answer (Select only the closest or best answer.)

12. Who is responsible for safety?
    1. Everyone
    2. Operators
    3. Safety officer
    4. Top management
13. What is the one objective of a safety program?
    1. To ensure adequate funding
    2. To involve everyone
    3. To prevent accidents
    4. To provide safety training
14. What would you do if you were unable to remove a safety hazard?
    1. Bring hazard to the attention of those who have proper authority
    2. Leave work area immediately
    3. Prevent others from encountering the safety hazard
    4. Try to finish the job safely
15. What is the purpose of employee "Right-to-Know" laws? To require employers to inform operators of possible health effects resulting from
    1. Hazardous substances.
    2. Lack of training.
    3. Negligence or carelessness.
    4. Reckless drivers.
16. Why should sawdust never be used as an absorbent? Because it is
    1. Combustible.
    2. Dusty.
    3. Expensive.
    4. Slippery.
17. What should be done before working on electrical equipment or equipment that is run by electric motors?
    1. Inform other operators that equipment is turned off
    2. Inspect all electric circuits
    3. Lock out and tag all electric switches
    4. Turn all equipment off
18. Why should you *NEVER ADD WATER TO ACID*?
    1. Acid could splash all over you
    2. Oxygen depletion will occur
    3. Poor mixing will result
    4. The pH will increase rapidly

19. What will happen if a centrifugal pump is operated against a closed discharge valve?
    1. Consumers will complain of excess pressures
    2. Meters will cease to work properly
    3. Overheating will occur
    4. Pipes will burst

20. What is the cause of most work area accidents on streets and highways?
    1. Drivers under the influence of alcohol
    2. Excessive speed
    3. Failure to plan work in advance
    4. Lack of proper advance warning, signing, and guidance

21. Why must the spoil bank be kept away from the edge of a trench?
    1. To keep spoil from falling into trench
    2. To keep spoil in proper location
    3. To keep spoil out of roadway
    4. To keep wall of trench from caving in on top of you

22. Why should the location of underground utilities be determined before performing any excavation work?
    1. To allow other utilities to perform maintenance and repair work at the same time
    2. To avoid having work crews interfere with each other's activities
    3. To coordinate repaving of roadways
    4. To prevent damage to underground utilities

Multiple Choice (Select all correct answers.)

23. Management should be responsible for which aspects of safety?
    1. Appoint a safety officer
    2. Assign responsibility for accident prevention
    3. Establish a safety policy
    4. Establish realistic goals
    5. Evaluate results of program

24. Operators should be responsible for which aspects of safety?
    1. Actively participate in the safety program
    2. Perform jobs in accordance with established safe procedures
    3. Recognize the responsibility for their own safety
    4. Report all injuries
    5. Report all observed hazards

25. Which of the following are principal reasons for unsafe acts?
    1. Haste
    2. Ignorance
    3. Indifference
    4. Laziness
    5. Temper

26. Which gases could cause dangerous air contamination?
    1. Carbon monoxide gas
    2. Hydrogen sulfide gas
    3. Methane gas
    4. Oxygen
    5. Toxic gases

27. What are the recommended warning devices for gas detection equipment?
    1. Audible noise
    2. Flashing light
    3. Red area on meter scale
    4. Vibrating meter
    5. Warning odor

28. What safety procedures must be followed before anyone ever enters a tank for any reason?
    1. All persons entering a tank must wear a safety harness
    2. An additional person must be readily available to help the person at the tank entrance with any rescue operations
    3. One person must be at the tank entrance and observe the actions of all people in the tank
    4. Provide adequate ventilation, especially when painting
    5. Test atmosphere in the tank for toxic and explosive gases and for oxygen deficiency/enrichment

29. Machinery must be locked out and tagged during what activities?
    1. Adjusting
    2. Cleaning
    3. Operating
    4. Servicing
    5. Testing

30. What are traffic control systems on street work sites used to protect?
    1. Animals
    2. Equipment
    3. Motorists
    4. Pedestrians
    5. Work force

31. When utility work is to be done in a residential area, what information should be provided to the public?
    1. A list of the cost to each residence
    2. Every effort will be made not to disrupt the routine of the neighborhood
    3. How long the work is expected to take
    4. That all safety regulations will be observed
    5. What work is to be done

32. Which items should be checked during a routine inspection of a fixed ladder?
    1. Corroded bolts or rivet heads where ladder is affixed to tank
    2. Corroded or damaged parts of the security door
    3. Defects in climbing devices
    4. Loose, worn, and damaged rungs or side rails
    5. Moving parts properly lubricated

33. What types of confined spaces could be encountered by operators?
    1. Cabs of pickup trucks
    2. Dewatered clear wells
    3. Elevated tanks
    4. Manholes
    5. Valve vaults

Review Questions (Select all correct answers.)

34. The maximum rate of chlorine removal from a 150-pound cylinder is __________ pounds of chlorine per 24 hours.
    1. 20
    2. 40
    3. 60
    4. 80
    5. 100

35. Which of the following precautions should be taken when disposing of water with a high chlorine residual?
    1. Ensure enough dilution and travel time to remove residual chlorine before water reaches the treatment plant
    2. Dispose on land only where percolation rates are low
    3. Do not allow water with chlorine residual to reach receiving waters
    4. Do not dispose on land near wells pumping groundwater
    5. Notify the treatment plant operator in advance

End of Objective Test

CHAPTER 7

LABORATORY PROCEDURES

by

Jim Sequeira

TABLE OF CONTENTS

Chapter 7. LABORATORY PROCEDURES

OBJECTIVES

Chapter 7. LABORATORY PROCEDURES

Following completion of Chapter 7, you should be able to:

1. Work safely in a laboratory,
2. Operate laboratory equipment,
3. Collect representative samples and also preserve and transport the samples,
4. Prepare samples for analysis,
5. Describe the limitations of lab tests,
6. Recognize precautions to be taken for lab tests,
7. Record laboratory test results, and
8. Perform the following field or laboratory tests — alkalinity, chlorine residual, chlorine demand, coliform, hardness, jar test, pH, temperature, and turbidity.

WORDS

Chapter 7. LABORATORY PROCEDURES

ACIDIC (uh-SID-ick) ACIDIC

The condition of water or soil which contains a sufficient amount of acid substances to lower the pH below 7.0.

ALIQUOT (AL-li-kwot) ALIQUOT

Portion of a sample. Often an equally divided portion of a sample.

ALKALI (AL-ka-lie) ALKALI

Any of certain soluble salts, principally of sodium, potassium, magnesium, and calcium, that have the property of combining with acids to form neutral salts and may be used in chemical water treatment processes.

ALKALINE (AL-ka-LINE) ALKALINE

The condition of water or soil which contains a sufficient amount of alkali substances to raise the pH above 7.0.

AMBIENT (AM-bee-ent) TEMPERATURE AMBIENT TEMPERATURE

Temperature of the surrounding air (or other medium). For example, temperature of the room where a gas chlorinator is installed.

AMPEROMETRIC (am-PURR-o-MET-rick) AMPEROMETRIC

A method of measurement that records electric current flowing or generated, rather than recording voltage. Amperometric titration is a means of measuring concentrations of certain substances in water.

AMPEROMETRIC (am-PURR-o-MET-rick) TITRATION AMPEROMETRIC TITRATION

A means of measuring concentrations of certain substances in water (such as strong oxidizers) based on the electric current that flows during a chemical reaction. Also see TITRATE.

ASEPTIC (a-SEP-tick) ASEPTIC

Free from the living germs of disease, fermentation, or putrefaction. Sterile.

BACTERIA (back-TEAR-e-ah) BACTERIA

Bacteria are living organisms, microscopic in size, which usually consist of a single cell. Most bacteria use organic matter for their food and produce waste products as a result of their life processes.

BLANK BLANK

A bottle containing only dilution water or distilled water; the sample being tested is not added. Tests are frequently run on a *SAMPLE* and a *BLANK* and the differences are compared.

BUFFER BUFFER

A solution or liquid whose chemical makeup neutralizes acids or bases without a great change in pH.

BUFFER CAPACITY BUFFER CAPACITY

A measure of the capacity of a solution or liquid to neutralize acids or bases. This is a measure of the capacity of water for offering a resistance to changes in pH.

CALCIUM CARBONATE ($CaCO_3$) EQUIVALENT CALCIUM CARBONATE ($CaCO_3$) EQUIVALENT

An expression of the concentration of specified constituents in water in terms of their equivalent value to calcium carbonate. For example, the hardness in water which is caused by calcium, magnesium and other ions is usually described as calcium carbonate equivalent. Alkalinity test results are usually reported as mg/*L* $CaCO_3$ equivalents. To convert chloride to $CaCO_3$ equivalents, multiply the concentration of chloride ions in mg/*L* by 1.41, and for sulfate, multiply by 1.04.

CARCINOGEN (CAR-sin-o-JEN) CARCINOGEN

Any substance which tends to produce cancer in an organism.

CHLORORGANIC (klor-or-GAN-ick) CHLORORGANIC

Organic compounds combined with chlorine. These compounds generally originate from, or are associated with, life processes such as those of algae in water.

COLORIMETRIC MEASUREMENT COLORIMETRIC MEASUREMENT

A means of measuring unknown chemical concentrations in water by measuring a sample's color intensity. The specific color of the sample, developed by addition of chemical reagents, is measured with a photoelectric colorimeter or is compared with "color standards" using, or corresponding with, known concentrations of the chemical.

COMPOSITE (come-PAH-zit) (PROPORTIONAL) SAMPLE COMPOSITE (PROPORTIONAL) SAMPLE

A composite sample is a collection of individual samples obtained at regular intervals, usually every one or two hours during a 24-hour time span. Each individual sample is combined with the others in proportion to the rate of flow when the sample was collected. The resulting mixture (composite sample) forms a representative sample and is analyzed to determine the average conditions during the sampling period.

COMPOUND COMPOUND

A pure substance composed of two or more elements whose composition is constant. For example, table salt (sodium chloride, NaCl) is a compound.

DPD (pronounce as separate letters) DPD

A method of measuring the chlorine residual in water. The residual may be determined by either titrating or comparing a developed color with color standards. DPD stands for N,N-diethyl-p-phenylene-diamine.

DESICCATOR (DESS-uh-KAY-tor) DESICCATOR

A closed container into which heated weighing or drying dishes are placed to cool in a dry environment in preparation for weighing. The dishes may be empty or they may contain a sample. Desiccators contain a substance, such as anhydrous calcium chloride, which absorbs moisture and keeps the relative humidity near zero so that the dish or sample will not gain weight from absorbed moisture.

DISINFECTION (dis-in-FECT-shun) DISINFECTION

The process designed to kill or inactivate most microorganisms in water, including essentially all pathogenic (disease-causing) bacteria. There are several ways to disinfect, with chlorination being the most frequently used in water treatment. Compare with STERILIZATION.

ELEMENT ELEMENT

A substance which cannot be separated into its constituent parts and still retain its chemical identity. For example, sodium (Na) is an element.

END POINT END POINT

Samples are titrated to the end point. This means that a chemical is added, drop by drop, to a sample until a certain color change (blue to clear, for example) occurs. This is called the *END POINT* of the titration. In addition to a color change, an end point may be reached by the formation of a precipitate or the reaching of a specified pH. An end point may be detected by the use of an electronic device such as a pH meter. The completion of a desired chemical reaction.

FACULTATIVE (FACK-ul-TAY-tive) FACULTATIVE

Facultative bacteria can use either dissolved molecular oxygen or oxygen obtained from food materials such as sulfate or nitrate ions. In other words, facultative bacteria can live under aerobic or anaerobic conditions.

FLAME POLISHED FLAME POLISHED

Melted by a flame to smooth out irregularities. Sharp or broken edges of glass (such as the end of a glass tube) are rotated in a flame until the edge melts slightly and becomes smooth.

GRAB SAMPLE GRAB SAMPLE

A single sample of water collected at a particular time and place which represents the composition of the water only at that time and place.

GRAVIMETRIC GRAVIMETRIC

A means of measuring unknown concentrations of water quality indicators in a sample by *WEIGHING* a precipitate or residue of the sample.

INDICATOR (CHEMICAL) INDICATOR (CHEMICAL)

A substance that gives a visible change, usually of color, at a desired point in a chemical reaction, generally at a specified end point.

INORGANIC INORGANIC

Material such as sand, salt, iron, calcium salts and other mineral materials. Inorganic substances are of mineral origin, whereas organic substances are usually of animal or plant origin. Also see ORGANIC.

INORGANIC WASTE INORGANIC WASTE

Waste material such as sand, salt, iron, calcium, and other mineral materials which are only slightly affected by the action of organisms. Inorganic wastes are chemical substances of mineral origin; whereas organic wastes are chemical substances of an animal or plant origin.

M or MOLAR ***M*** or MOLAR

A molar solution consists of one gram molecular weight of a compound dissolved in enough water to make one liter of solution. A gram molecular weight is the molecular weight of a compound in grams. For example, the molecular weight of sulfuric acid (H_2SO_4) is 98. A one *M* solution of sulfuric acid would consist of 98 grams of H_2SO_4 dissolved in enough distilled water to make one liter of solution.

MPN (pronounce as separate letters) MPN

MPN is the **M**ost **P**robable **N**umber of coliform-group organisms per unit volume of sample water. Expressed as a density or population of organisms per 100 m*L* of sample water.

MENISCUS (meh-NIS-cuss) MENISCUS

The curved surface of a column of liquid (water, oil, mercury) in a small tube. When the liquid wets the sides of the container (as with water), the curve forms a valley. When the confining sides are not wetted (as with mercury), the curve forms a hill or upward bulge.

MILLIGRAMS PER LITER, mg/*L* MILLIGRAMS PER LITER, mg/*L*

A measure of the concentration by weight of a substance per unit volume. For practical purposes, one mg/*L* of a substance in fresh water is equal to one part per million parts (ppm). Thus a liter of water with a specific gravity of 1.0 weighs one million milligrams. If water contains 10 milligrams of calcium, the concentration is 10 milligrams per million milligrams, or 10 milligrams per liter (10 mg/*L*), or 10 parts of calcium per million parts of water, or 10 parts per million (10 ppm).

MOLE MOLE

The molecular weight of a substance, usually expressed in grams.

MOLECULAR WEIGHT MOLECULAR WEIGHT

The molecular weight of a compound in grams is the sum of the atomic weights of the elements in the compound. The molecular weight of sulfuric acid (H_2SO_4) in grams is 98.

Element	Atomic Weight	Number of Atoms	Molecular Weight
H	1	2	2
S	32	1	32
O	16	4	64
			98

MOLECULE (MOLL-uh-KULE) MOLECULE

The smallest division of a compound that still retains or exhibits all of the properties of the substance.

N or NORMAL — *N* or NORMAL

A normal solution contains one gram equivalent weight of reactant (compound) per liter of solution. The equivalent weight of an acid is that weight which contains one gram atom of ionizable hydrogen or its chemical equivalent. For example, the equivalent weight of sulfuric acid (H_2SO_4) is 49 (98 divided by 2 because there are two replaceable hydrogen ions). A one *N* solution of sulfuric acid would consist of 49 grams of H_2SO_4 dissolved in enough water to make one liter of solution.

NEPHELOMETRIC (NEFF-el-o-MET-rick) — NEPHELOMETRIC

A means of measuring turbidity in a sample by using an instrument called a nephelometer. A nephelometer passes light through a sample and the amount of light deflected (usually at a 90-degree angle) is then measured.

NONVOLATILE MATTER — NONVOLATILE MATTER

Material such as sand, salt, iron, calcium, and other mineral materials which are only slightly affected by the actions of organisms and are not lost on ignition of the dry solids at 550°C. Volatile materials are chemical substances usually of animal or plant origin. Also see INORGANIC WASTE and VOLATILE MATTER or VOLATILE SOLIDS.

OSHA (O-shuh) — OSHA

The Williams-Steiger **O**ccupational **S**afety and **H**ealth **A**ct of 1970 (OSHA) is a federal law designed to protect the health and safety of industrial workers and also the operators of water supply systems and treatment plants. The Act regulates the design, construction, operation and maintenance of water supply systems and water treatment plants. OSHA also refers to the federal and state agencies which administer the OSHA regulations.

ORGANIC — ORGANIC

Substances that come from animal or plant sources. Organic substances always contain carbon. (Inorganic materials are chemical substances of mineral origin.) Also see INORGANIC.

ORGANISM — ORGANISM

Any form of animal or plant life. Also see BACTERIA.

OXIDATION (ox-uh-DAY-shun) — OXIDATION

Oxidation is the addition of oxygen, removal of hydrogen, or the removal of electrons from an element or compound. In the environment, organic matter is oxidized to more stable substances. The opposite of REDUCTION.

OXIDATION-REDUCTION POTENTIAL (ORP) — OXIDATION-REDUCTION POTENTIAL (ORP)

The electrical potential required to transfer electrons from one compound or element (the oxidant) to another compound or element (the reductant); used as a qualitative measure of the state of oxidation in water treatment systems. ORP is measured in millivolts, with negative values indicating a tendency to reduce compounds or elements and positive values indicating a tendency to oxidize compounds or elements.

PARTS PER MILLION (PPM) — PARTS PER MILLION (PPM)

Parts per million parts, a measurement of concentration on a weight or volume basis. This term is equivalent to milligrams per liter (mg/*L*) which is the preferred term.

PATHOGENIC (PATH-o-JEN-ick) ORGANISMS — PATHOGENIC ORGANISMS

Organisms, including bacteria, viruses or cysts, capable of causing diseases (giardiasis, cryptosporidiosis, typhoid, cholera, dysentery) in a host (such as a person). There are many types of organisms which do *NOT* cause disease. These organisms are called non-pathogenic.

PATHOGENS (PATH-o-jens) — PATHOGENS

Pathogenic or disease-causing organisms.

PERCENT SATURATION — PERCENT SATURATION

The amount of a substance that is dissolved in a solution compared with the amount dissolved in the solution at saturation, expressed as a percent.

$$\text{Percent Saturation, \%} = \frac{\text{Amount of Substance That Is Dissolved} \times 100\%}{\text{Amount Dissolved in Solution at Saturation}}$$

pH (pronounce as separate letters) pH

pH is an expression of the intensity of the basic or acidic condition of a liquid. Mathematically, pH is the logarithm (base 10) of the reciprocal of the hydrogen ion activity.

$$pH = Log \frac{1}{[H^+]}$$

The pH may range from 0 to 14, where 0 is most acidic, 14 most basic, and 7 neutral. Natural waters usually have a pH between 6.5 and 8.5.

POTABLE (POE-tuh-bull) WATER POTABLE WATER

Water that does not contain objectionable pollution, contamination, minerals, or infective agents and is considered satisfactory for drinking.

PRECIPITATE (pre-SIP-uh-TATE) PRECIPITATE

(1) An insoluble, finely divided substance which is a product of a chemical reaction within a liquid.

(2) The separation from solution of an insoluble substance.

REAGENT (re-A-gent) REAGENT

A pure chemical substance that is used to make new products or is used in chemical tests to measure, detect, or examine other substances.

REDUCTION (re-DUCK-shun) REDUCTION

Reduction is the addition of hydrogen, removal of oxygen, or the addition of electrons to an element or compound. Under anaerobic conditions (no dissolved oxygen present), sulfur compounds are reduced to odor-producing hydrogen sulfide (H_2S) and other compounds. The opposite of OXIDATION.

REPRESENTATIVE SAMPLE REPRESENTATIVE SAMPLE

A sample portion of material or water that is as nearly identical in content and consistency as possible to that in the larger body of material or water being sampled.

SOLUTION SOLUTION

A liquid mixture of dissolved substances. In a solution it is impossible to see all the separate parts.

STANDARD METHODS STANDARD METHODS

STANDARD METHODS FOR THE EXAMINATION OF WATER AND WASTEWATER, 20th Edition. A joint publication of the American Public Health Association (APHA), American Water Works Association (AWWA), and the Water Environment Federation (WEF) which outlines the accepted laboratory procedures used to analyze the impurities in water and wastewater. Available from American Water Works Association Bookstore, 6666 West Quincy Avenue, Denver, CO 80235. Order No. 10079. Price to members, $166.00; nonmembers, $211.00; price includes cost of shipping and handling.

STANDARD SOLUTION STANDARD SOLUTION

A solution in which the exact concentration of a chemical or compound is known.

STANDARDIZE STANDARDIZE

To compare with a standard.

(1) In wet chemistry, to find out the exact strength of a solution by comparing it with a standard of known strength. This information is used to adjust the strength by adding more water or more of the substance dissolved.

(2) To set up an instrument or device to read a standard. This allows you to adjust the instrument so that it reads accurately, or enables you to apply a correction factor to the readings.

STERILIZATION (STARE-uh-luh-ZAY-shun) STERILIZATION

The removal or destruction of all microorganisms, including pathogenic and other bacteria, vegetative forms and spores. Compare with DISINFECTION.

SUPERNATANT (sue-per-NAY-tent) SUPERNATANT

Liquid removed from settled sludge. Supernatant commonly refers to the liquid between the sludge on the bottom and the scum on the water surface of a basin or container.

SURFACTANT (sir-FAC-tent) SURFACTANT

Abbreviation for surface-active agent. The active agent in detergents that possesses a high cleaning ability.

TITRATE (TIE-trate) TITRATE

To *TITRATE* a sample, a chemical solution of known strength is added drop by drop until a certain color change, precipitate, or pH change in the sample is observed (end point). Titration is the process of adding the chemical reagent in increments until completion of the reaction, as signaled by the end point.

TURBIDITY UNITS (TU) TURBIDITY UNITS (TU)

Turbidity units are a measure of the cloudiness of water. If measured by a nephelometric (deflected light) instrumental procedure, turbidity units are expressed in nephelometric turbidity units (NTU) or simply TU. Those turbidity units obtained by visual methods are expressed in Jackson Turbidity Units (JTU) which are a measure of the cloudiness of water; they are used to indicate the clarity of water. There is no real connection between NTUs and JTUs. The Jackson turbidimeter is a visual method and the nephelometer is an instrumental method based on deflected light.

VOLATILE (VOL-uh-tull) VOLATILE

(1) A volatile substance is one that is capable of being evaporated or changed to a vapor at relatively low temperatures. Volatile substances also can be partially removed by air stripping.

(2) In terms of solids analysis, volatile refers to materials lost (including most organic matter) upon ignition in a muffle furnace for 60 minutes at 550°C. Natural volatile materials are chemical substances usually of animal or plant origin. Manufactured or synthetic volatile materials such as ether, acetone, and carbon tetrachloride are highly volatile and not of plant or animal origin. Also see NONVOLATILE MATTER.

VOLATILE ACIDS VOLATILE ACIDS

Fatty acids produced during digestion which are soluble in water and can be steam-distilled at atmospheric pressure. Also called organic acids. Volatile acids are commonly reported as equivalent to acetic acid.

VOLATILE LIQUIDS VOLATILE LIQUIDS

Liquids which easily vaporize or evaporate at room temperature.

VOLATILE MATTER VOLATILE MATTER

Matter in water, wastewater, or other liquids that is lost on ignition of the dry solids at 550°C.

VOLATILE SOLIDS VOLATILE SOLIDS

Those solids in water or other liquids that are lost on ignition of the dry solids at 550°C.

VOLUMETRIC VOLUMETRIC

A measurement based on the volume of some factor. Volumetric titration is a means of measuring unknown concentrations of water quality indicators in a sample *BY DETERMINING THE VOLUME* of titrant or liquid reagent needed to complete particular reactions.

CHAPTER 7. LABORATORY PROCEDURES

(Lesson 1 of 5 Lessons)

7.0 BASIC WATER LABORATORY PROCEDURES

7.00 Importance of Laboratory Procedures

Water treatment processes cannot be controlled effectively unless the operator has some means to check and evaluate the quality of water being treated and produced. Laboratory quality control tests provide the necessary information to monitor the treatment processes and to ensure a safe and pleasant-tasting drinking water for all who use it. By relating laboratory results to treatment operations, the water treatment or supply system operator can first select the most effective operational procedures, then determine the efficiency of the treatment processes, and identify potential problems before they affect finished water quality. For these reasons, a clear understanding of laboratory procedures is a must for every waterworks operator.

NOTICE

THE COLLECTION OF A BAD SAMPLE OR A BAD LABORATORY RESULT IS ABOUT AS USEFUL AS NO RESULTS. TO PREVENT BAD RESULTS REQUIRES (1) CONSTANT MAINTENANCE AND CALIBRATION OF LABORATORY EQUIPMENT, AND (2) USE OF CORRECT LAB PROCEDURES. ALSO RESULTS OF LAB TESTS ARE OF NO VALUE TO ANYONE UNLESS THEY ARE USED.

7.01 Metric System

The metric system is used in the laboratory to express units of length, volume, weight (mass), concentration, and temperature. The metric system is based on the decimal system. All units of length, volume, and weight use factors of 10 to express larger or smaller quantities of these units. Below is a summary of metric and English unit names and their abbreviations.

Type of Measurement	English System	Metric Name	Metric Abbreviation
Length	inch foot yard	meter	m
Temperature	Fahrenheit	Celsius	°C
Volume	quart gallon	liter	*L*
Weight	ounce pound	gram	gm
Concentration	lbs/gal strength, %	milligrams per liter	mg/*L*

Many times in the water laboratory we use smaller amounts than a meter, a liter, or a gram. To express these smaller amounts, prefixes are added to the names of the base metric unit. There are many prefixes in use; however, we commonly use two or three prefixes more than any others in the laboratory.

Prefix	Abbreviation	Meaning
centi-	c	1/100 of; or 0.01 times
milli-	m	1/1,000 of; or 0.001 times
micro-	μ	1/1,000,000 of; or 0.000001 times

One centimeter (cm) is 1/100 (one hundredth) of a meter, one milliliter (m*L*) is 1/1,000 (one thousandth) of a liter, and likewise, one microgram (μgm) is 1/1,000,000 (one millionth) of a gram.

EXAMPLES:

(1) Convert 3 grams into milligrams.

1 milligram = 1 mg = 1/1,000 grams

therefore, 1 gram = 1,000 milligrams

(3 grams)(1,000 mg/gram) = 3,000 mg

(2) Convert 750 milliliters (m*L*) to liters.

1 m*L* = 1/1,000 liter

therefore, 1 liter = 1,000 m*L*

(750 m*L*)(1 liter/1,000 m*L*) = 0.750 liters

(3) Convert 50 micrograms (μgm) to grams.

1 μgm = 1/1,000,000 gram

therefore, 1 gram = 1,000,000 μgm

50 μgm x 1 gram/1,000,000 μgm = 0.00005 grams

Larger amounts than a meter, liter, or gram can be expressed using such prefixes as kilo- meaning 1,000. A kilogram is 1,000 grams.

The Celsius (or centigrade) temperature scale is used in the water laboratory rather than the more familiar Fahrenheit scale.

	Fahrenheit (°F)	Celsius (°C)
Freezing point of water	32	0
Boiling point of water	212	100

To convert Fahrenheit to Celsius, you can use the following formula:

Temperature, °C = 5/9(°F – 32°F)

EXAMPLE: Convert 68°F to °C

$$\begin{aligned} \text{Temperature, °C} &= 5/9(\text{°F} - 32\text{°F}) \\ &= 5/9(68\text{°F} - 32\text{°F}) \\ &= 5/9(36) \\ &= 20\text{°C} \end{aligned}$$

To convert Celsius to Fahrenheit, the following formula can be used:

$$\text{Temperature, °F} = 9/5(\text{°C}) + 32\text{°F}$$

EXAMPLE: Convert 35°C to °F

$$\begin{aligned} \text{Temperature, °F} &= 9/5(\text{°C}) + 32\text{°F} \\ &= 9/5(35\text{°C}) + 32\text{°F} \\ &= 63 + 32 \\ &= 95\text{°F} \end{aligned}$$

7.02 Chemical Names and Formulas

In the laboratory, chemical symbols are used as "shorthand" for the names of the elements. The names and symbols for some of these elements are listed below.

Chemical Name	Symbol
Calcium	Ca
Carbon	C
Chlorine	Cl
Copper	Cu
Fluorine	F
Hydrogen	H
Iron	Fe
Lead	Pb
Magnesium	Mg
Manganese	Mn
Nitrogen	N
Oxygen	O
Sodium	Na
Sulfur	S

A compound is a substance composed of two or more different elements and whose composition (proportion of elements) is constant. Generally, all chemical compounds can be divided into two main groups, organic and inorganic. Organic compounds are those which contain the element carbon (C). There are, however, a few simple substances containing carbon which are considered to belong to the realm of inorganic chemistry. These include carbon dioxide (CO_2), carbon monoxide (CO), bicarbonate (HCO_3^-) and carbonate (CO_3^{2-}) as in calcium carbonate ($CaCO_3$).

Many different compounds can be made from the same two or three elements. Therefore, you must carefully read the formula and name to prevent errors and accidents. A chemical formula is a "shorthand" or abbreviated way to write the name of a chemical compound. For example, the name sodium chloride (common table salt) can be written "NaCl." Table 7.1 lists commonly used chemical compounds found in the water laboratory.

TABLE 7.1 NAMES AND FORMULAS OF CHEMICALS COMMONLY USED IN WATER ANALYSES

Chemical Name	Chemical Formula
Acetic Acid	CH_3COOH
Aluminum Sulfate (alum)	$Al_2(SO_4)_3 \cdot 14.3\ H_2O$ [a]
Ammonium Hydroxide	NH_4OH
Calcium Carbonate	$CaCO_3$
Chloroform	$CHCl_3$
Copper Sulfate	$CuSO_4$
Ferric Chloride	$FeCl_3$
Nitric Acid	HNO_3
Phenylarsine Oxide	C_6H_5AsO
Potassium Iodide	KI
Sodium Bicarbonate	$NaHCO_3$
Sodium Hydroxide	NaOH
Sulfuric Acid	H_2SO_4

[a] 14.3 H_2O. Alum in the dry form based on 17% Al_2O_3.

Poor results and safety hazards are often caused by using a chemical from the shelf that is *NOT* exactly the same chemical called for in a particular procedure. The mistake usually occurs when the chemicals are not properly labeled or have similar names or formulas. This problem can be eliminated if you use *BOTH* the chemical name and formula as a double check. The spellings of many chemical names are quite similar. These slight differences are critical because the chemicals do not behave alike. For example, the chemicals potassium nitr*A*te (KNO_3) and potassium nitr*I*te (KNO_2) are just as different in meaning chemically as the words f*A*t and f*I*t are to your doctor.

7.03 Helpful References

1. *METHODS FOR CHEMICAL ANALYSIS OF WATER AND WASTES.* Obtain from National Technical Information Service (NTIS), 5285 Port Royal Road, Springfield, VA 22161. Order No. PB84-128677. EPA No. 600-4-79-020. Price, $101.00, plus $5.00 shipping and handling per order.
2. *SIMPLIFIED PROCEDURES FOR WATER EXAMINATION* (M12). Obtain from American Water Works Association (AWWA), Bookstore, 6666 West Quincy Avenue, Denver, CO 80235. Order No. 30012. ISBN 0-89867-914-1. Price to members, $62.50; nonmembers, $92.50; price includes cost of shipping and handling.
3. *STANDARD METHODS FOR THE EXAMINATION OF WATER AND WASTEWATER,* 20th Edition, 1998. Obtain from American Water Works Association (AWWA), Bookstore, 6666 West Quincy Avenue, Denver, CO 80235. Order No. 10079. Price to members, $166.00; nonmembers, $211.00; price includes cost of shipping and handling.
4. *HANDBOOK FOR ANALYTICAL QUALITY CONTROL IN WATER AND WASTEWATER LABORATORIES.* Obtain from National Technical Information Service (NTIS), 5285 Port Royal Road, Springfield, VA 22161. Order No. PB-297451/7. EPA No. 600-4-79-019. Price, $45.00, plus $5.00 shipping and handling per order.

5. *MICROBIOLOGICAL METHODS FOR MONITORING THE ENVIRONMENT—WATER AND WASTES*, U.S. Environmental Protection Agency, December 1978. Obtain from National Technical Information Service (NTIS), 5285 Port Royal Road, Springfield, VA 22161. Order No. PB-290329/2. EPA No. 600-8-78-017. Price, $75.00, plus $5.00 shipping and handling per order.
6. *WATER QUALITY.* Obtain from American Water Works Association (AWWA), Bookstore, 6666 West Quincy Avenue, Denver, CO 80235. Order No. 1958. ISBN 0-89867-804-8. Price to members, $67.50; nonmembers, $97.50; price includes cost of shipping and handling.
7. *WATER ANALYSIS HANDBOOK.* Available from HACH Company, 5600 Lindbergh Drive, PO Box 389, Loveland, CO 80539-0389. Request literature No. 8374. No charge.

QUESTIONS

Write your answers in a notebook and then compare your answers with those on page 469.

7.0A Why are laboratory quality control tests important?

7.0B What does the prefix milli- mean?

7.0C What's the proper name of the chemical compound, $CaCO_3$?

7.1 LABORATORY EQUIPMENT AND TECHNIQUES

7.10 Water Laboratory Equipment

The items of equipment in a water laboratory are the operator's "tools-of-the-trade." In any laboratory there are certain basic pieces of equipment that are used routinely to perform water analysis tests. The following is a brief description of several of the more common items of glassware and pieces of equipment used in the analysis of water.

Volumetric glassware (graduated cylinders and pipets) is calibrated either "to contain" (TC) or "to deliver" (TD). Glassware designed "to deliver" will do so accurately only when the inner surface is so scrupulously clean that water wets the surface immediately and forms a uniform film on the surface upon emptying.

BEAKERS. Beakers are the most common pieces of laboratory equipment. They come in sizes from 1 m*L* to 4,000 m*L*. They are used mainly for mixing chemicals and to measure approximate volumes.

Beaker

GRADUATED CYLINDERS. Graduated cylinders also are basic to any laboratory and come in sizes from 5 m*L* to 4,000 m*L*. They are used to measure volumes more accurately than beakers.

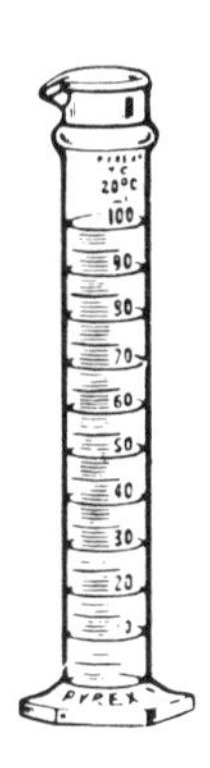

Cylinder, Graduated

PIPETS. Pipets are used to deliver accurate volumes and range in size from 0.1 m*L* to 100 m*L*.

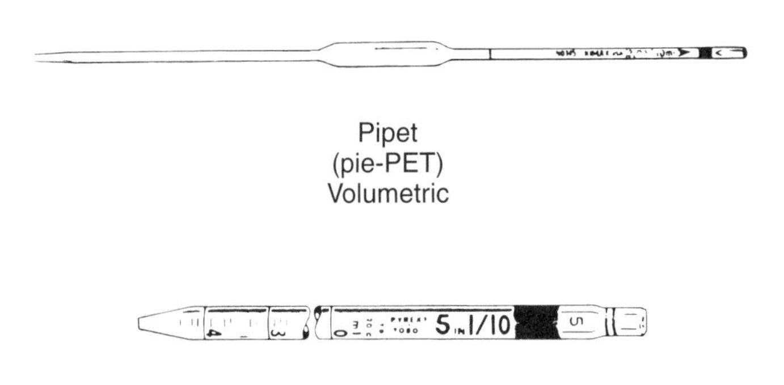

Pipet (pie-PET) Volumetric

Pipet, Serological

BURETS. Burets are also used to deliver accurate volumes. They are especially useful in a procedure called "titration." Burets come in sizes from 10 to 1,000 m*L*.

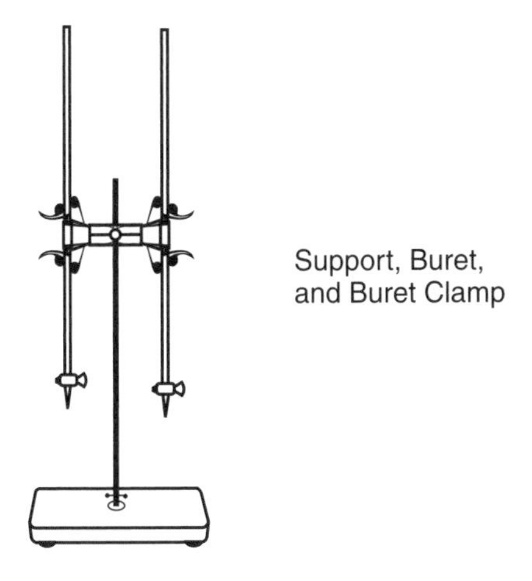

Support, Buret, and Buret Clamp

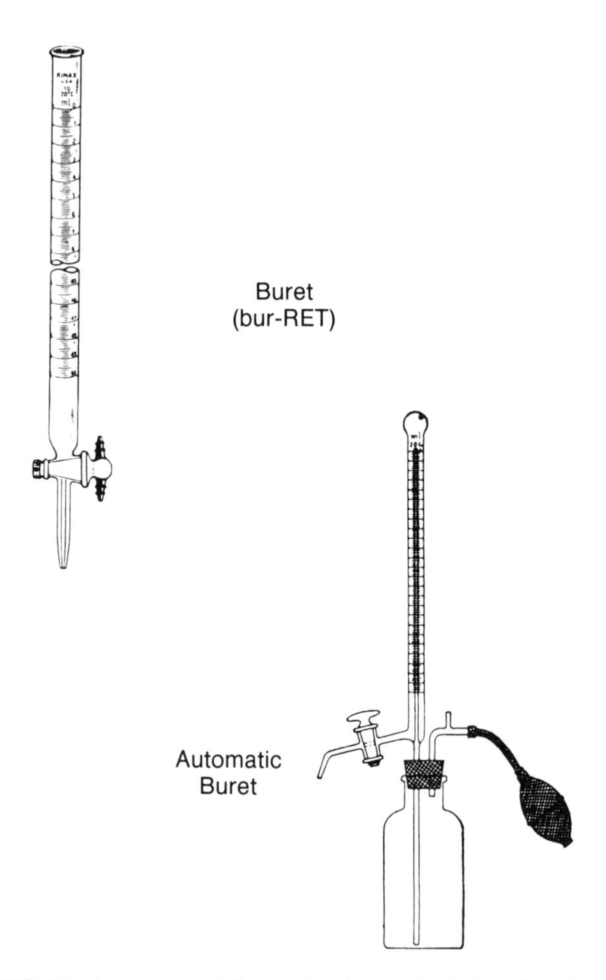

Buret
(bur-RET)

Automatic
Buret

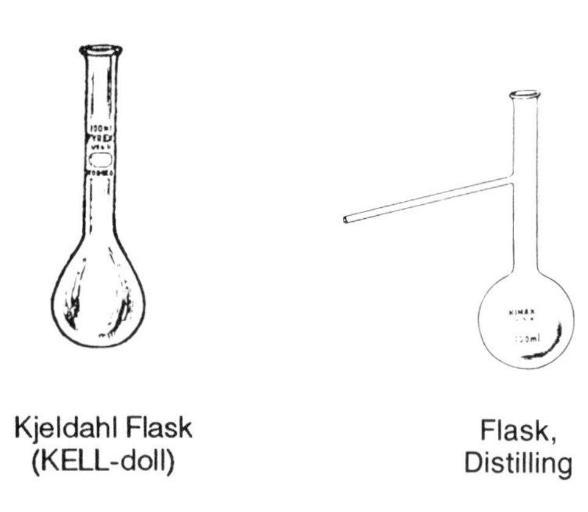

Kjeldahl Flask
(KELL-doll)

Flask,
Distilling

BOTTLES. Bottles are used to store chemicals, to collect samples for testing purposes, and to dispense liquids.

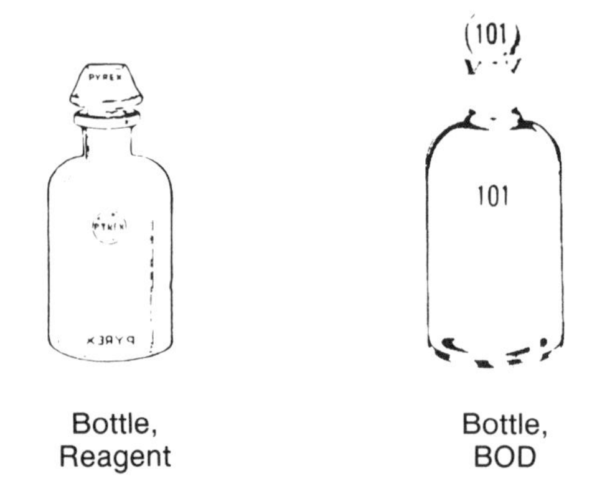

Bottle,
Reagent

Bottle,
BOD

FLASKS. Flasks are used for containing and mixing chemicals. There are many different sizes and shapes.

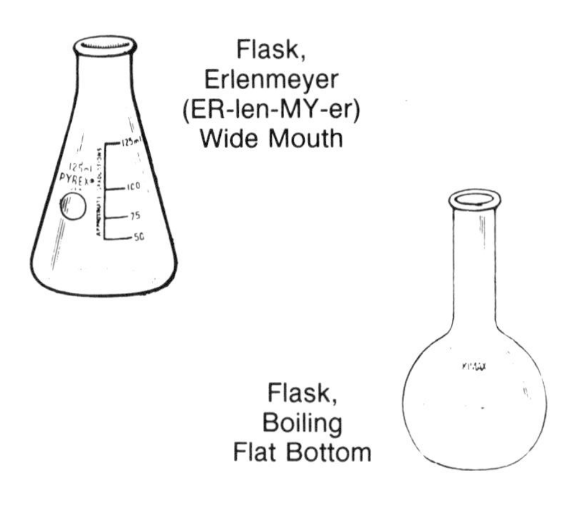

Flask,
Erlenmeyer
(ER-len-MY-er)
Wide Mouth

Flask,
Boiling
Flat Bottom

Flask,
Boiling
Round Bottom
Short Neck

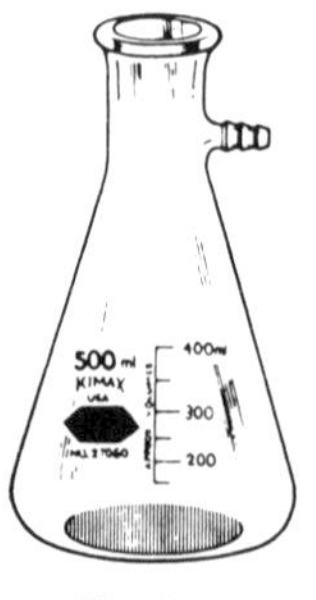

Flask,
Filtering

FUNNELS. A funnel is used for pouring solutions or transferring solid chemicals. This funnel also can be used with filter paper to remove solids from a solution.

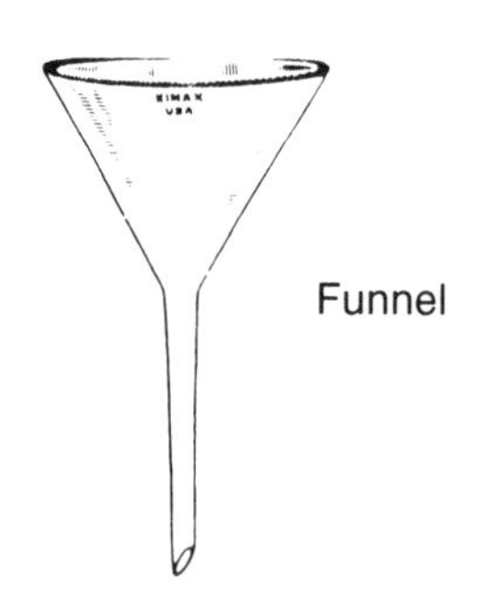

Funnel

A Buchner funnel is used to separate solids from a mixture. It is used with a filter flask and a vacuum.

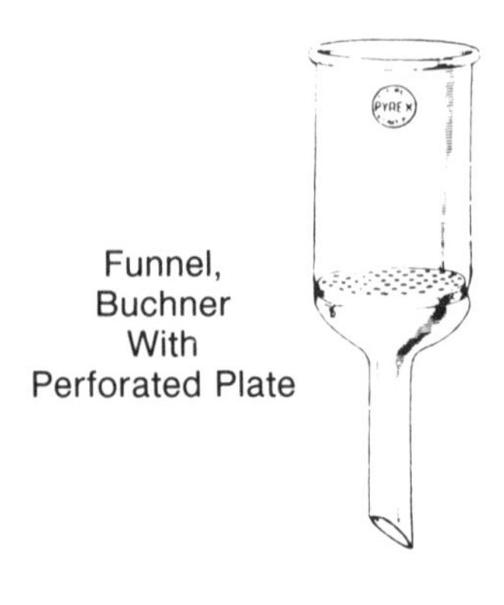

Funnel,
Buchner
With
Perforated Plate

Separatory funnels are used to separate one chemical mixture from another. The separated chemical usually is dissolved in one or two layers of liquid.

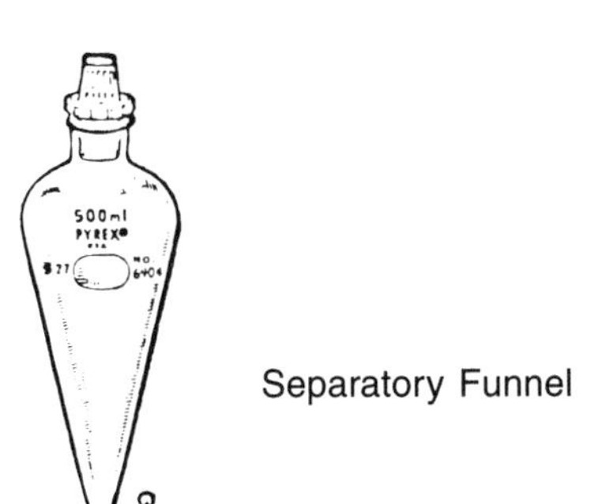

Separatory Funnel

TUBES. Test tubes are used for mixing small quantities of chemicals. They are also used as containers for bacterial testing (culture tubes).

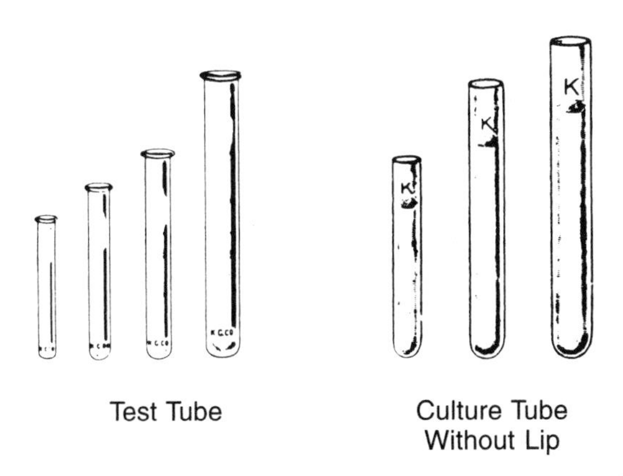

Test Tube

Culture Tube Without Lip

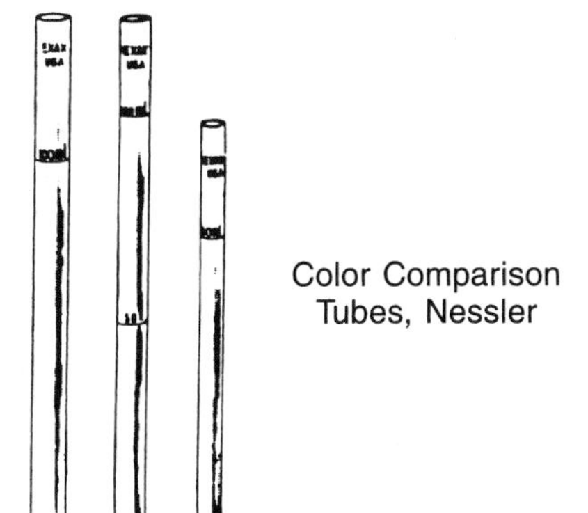

Color Comparison Tubes, Nessler

OTHER LABWARE AND EQUIPMENT.

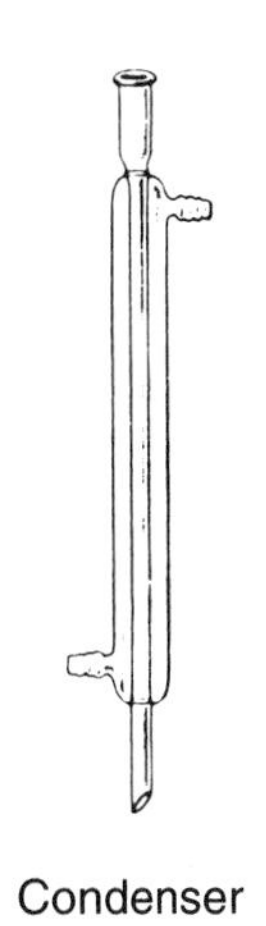

Condenser

Dish, Petri

Desiccator (DESS-uh-KAY-tor)

Thermometer, Dial

Oven, Mechanical Convection

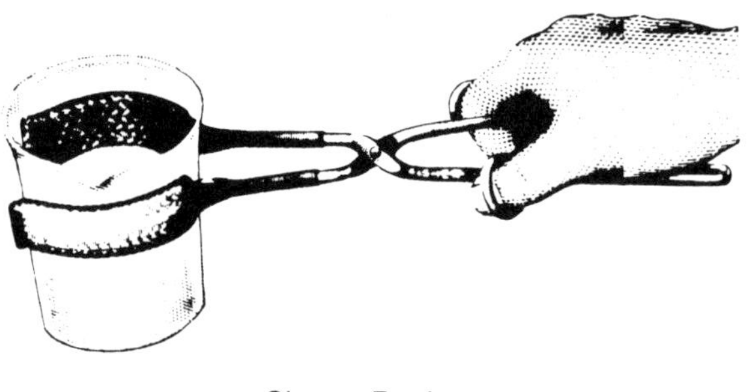

Clamp, Beaker,
Safety Tongs

Clamp, Dish,
Safety Tongs

Hot Plate

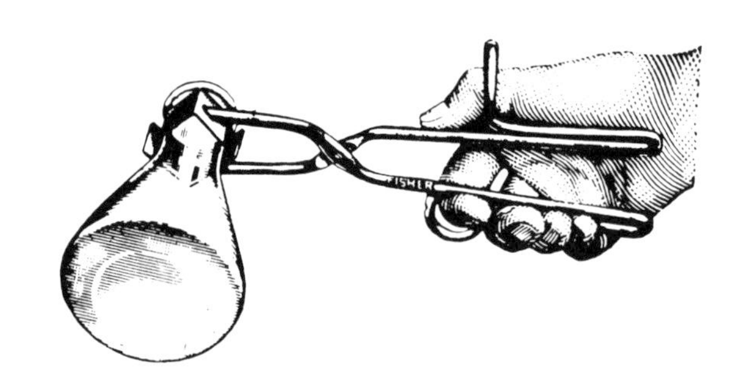

Clamp, Flask,
Safety Tongs

Clamp, Test Tube

Muffle Furnace, Electric

Clamp Holder

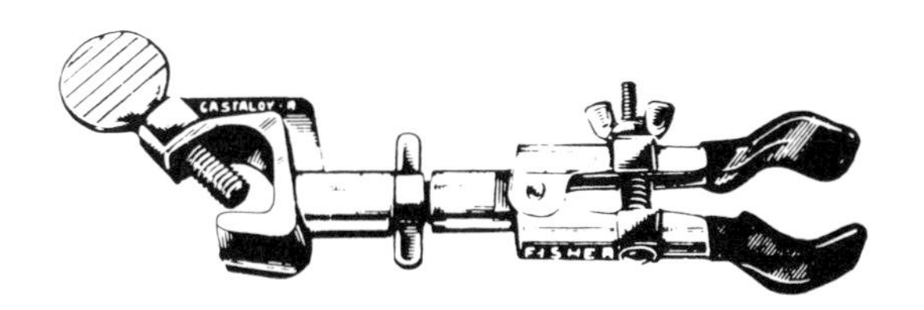

Clamp, Utility

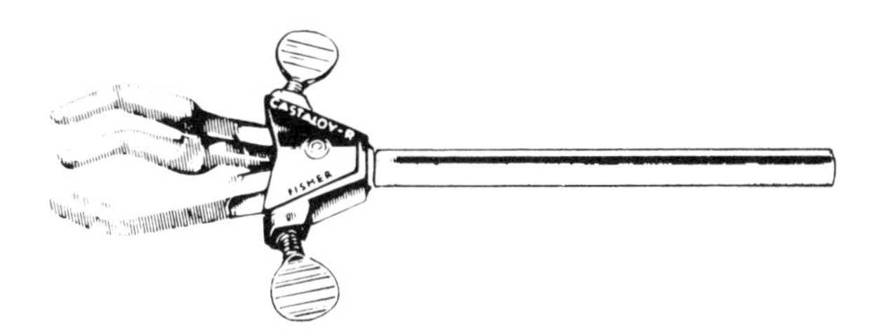

Clamp

Tripod, Concentric Ring

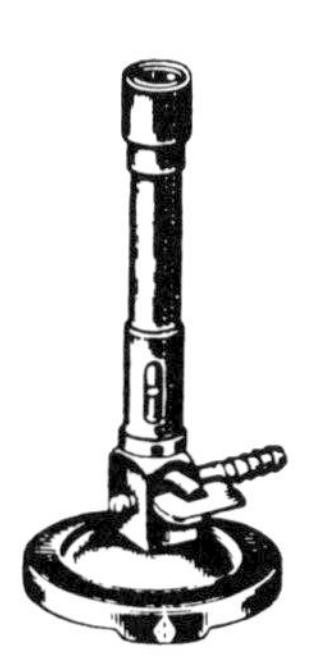

Burner, Bunsen

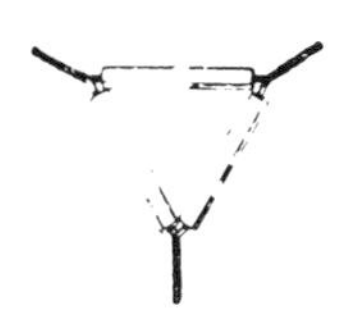

Triangle, Fused

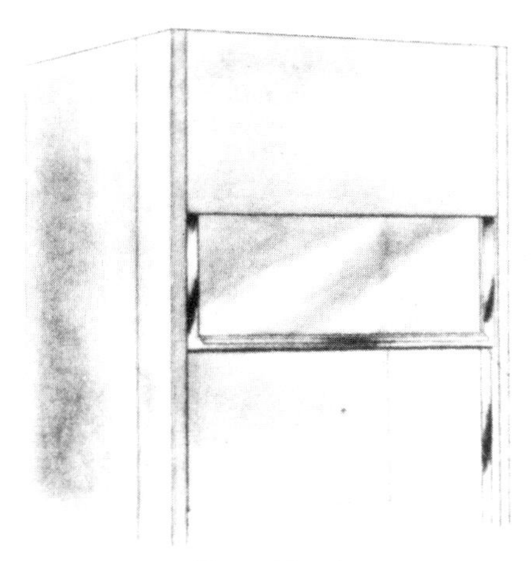

Fume Hood

Portable Dissolved Oxygen Meter (with computer docking station)
(Courtesy of HACH Company)

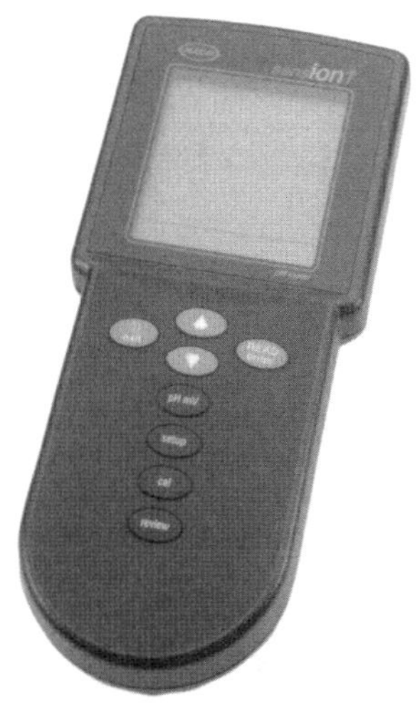

Portable pH Meter
(Courtesy of HACH Company)

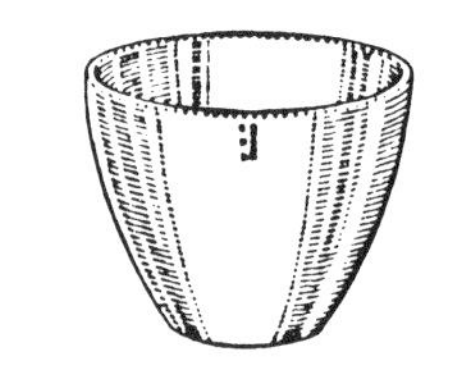

Crucible (CREW-suh-bull), Porcelain

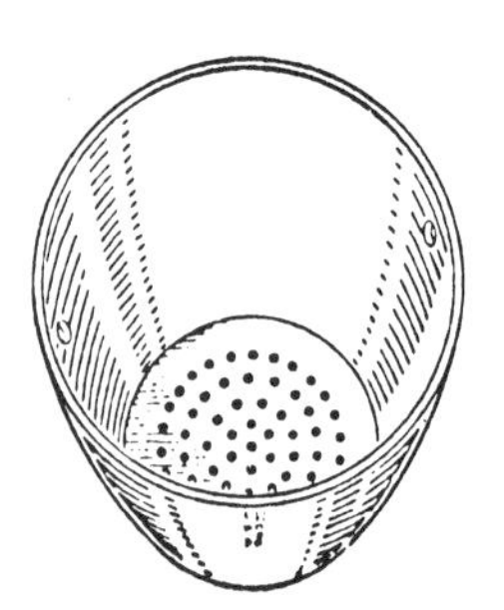

Crucible, Gooch (GOO-ch) Porcelain

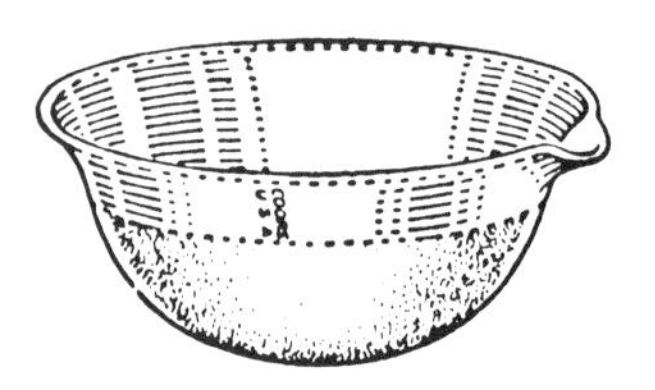

Dish, Evaporating

Test Paper, pH 1-11

Pump, Air Pressure and Vacuum

Pipet Bulb

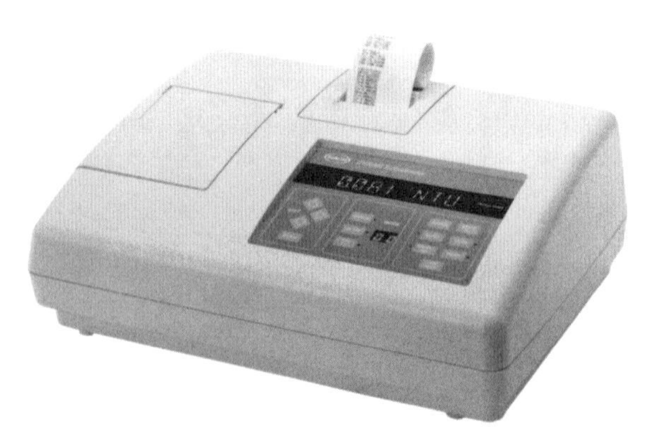

Laboratory Turbidimeter
(ratio mode — ON/OFF — is keypad selectable)
(Courtesy of HACH Company)

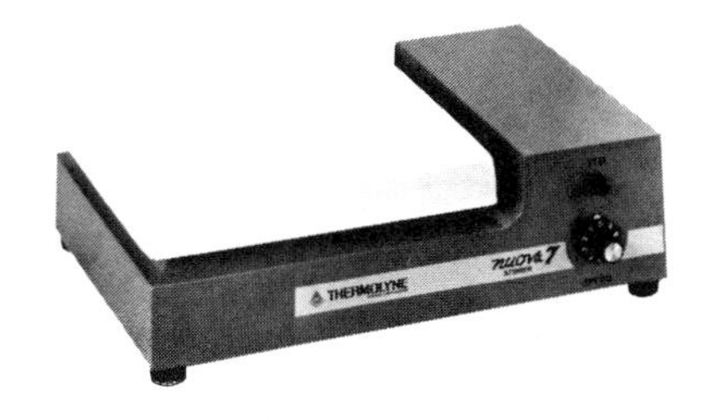

Magnetic Stirrer
(Permission of Thermolyne)

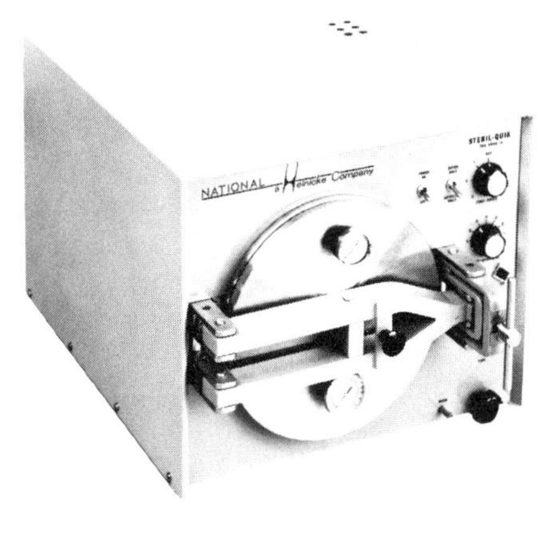

Autoclave
(Permission of Napco)

Incubator
(Permission of Blue M Electric)

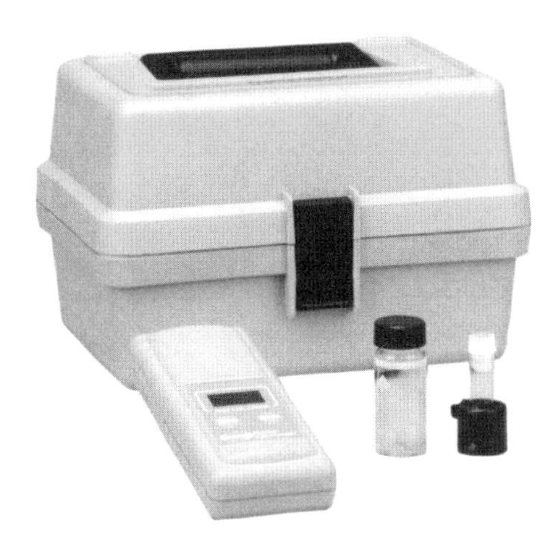

Direct-Reading Colorimeter
(free chlorine residual)
(Courtesy of HACH Company)

Portable Spectrophotometer
(Courtesy of HACH Company)

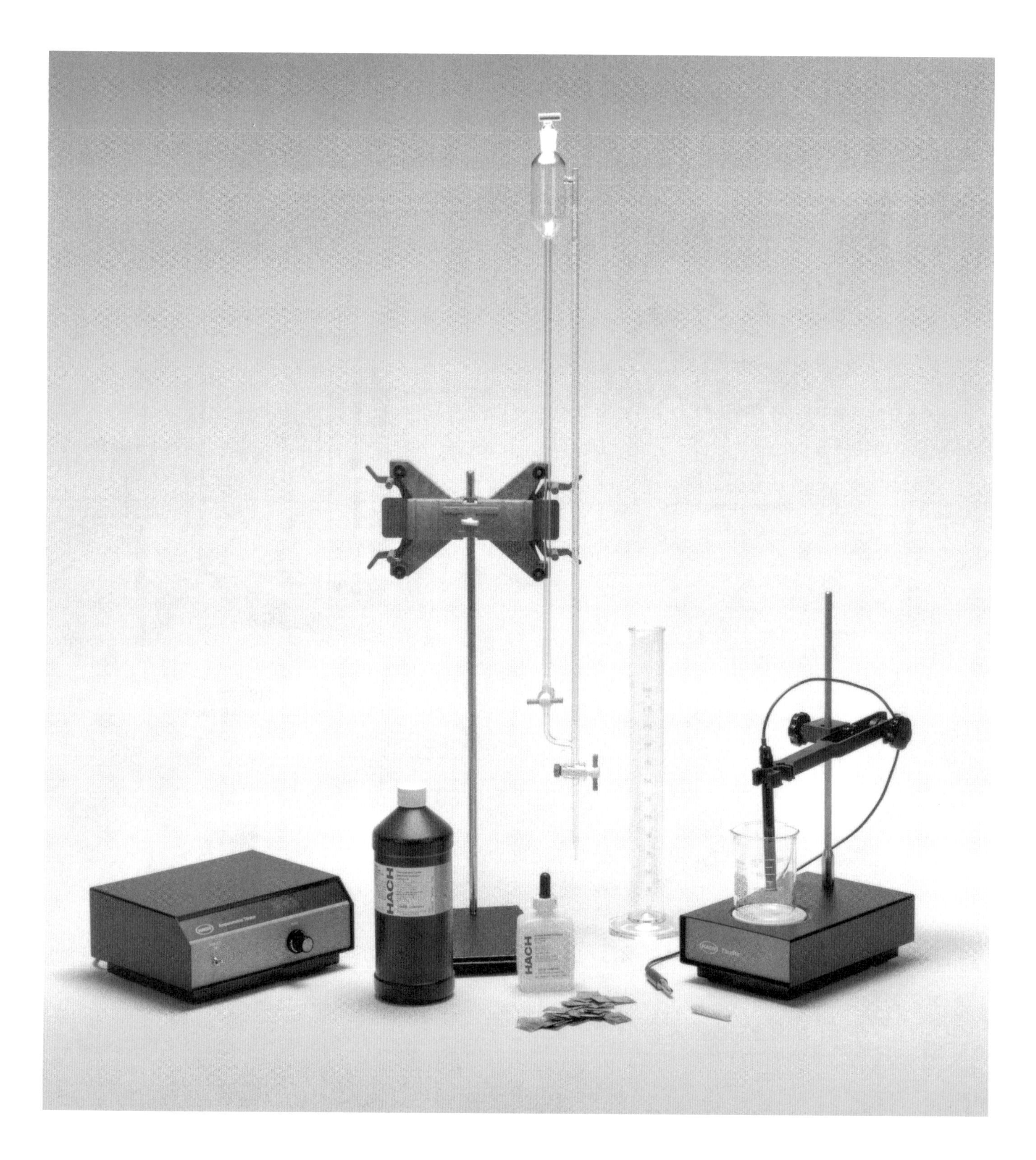

Amperometric Titrator
(Courtesy of HACH Company)

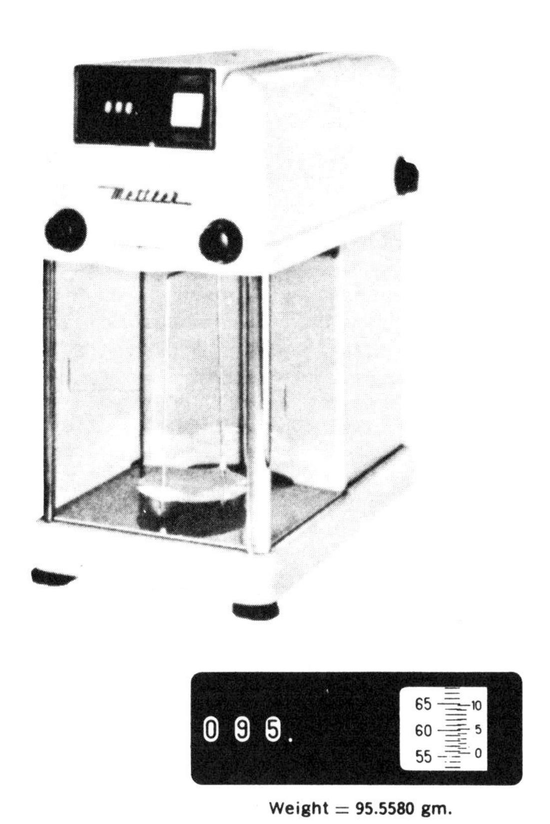

Balance, Analytical
(Permission of Mettler)

7.11 Use of Laboratory Glassware

BURETS

A buret is used to give accurate measurements of liquid volumes. The stopcock controls the amount of liquid which will flow from the buret. A glass stopcock must be lubricated (stopcock grease) and should not be used with alkaline solutions. A teflon stopcock never needs to be lubricated.

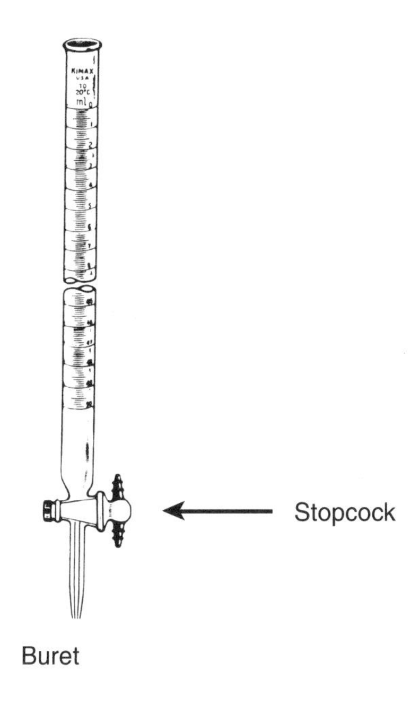

Burets come in several sizes, with those holding 10 to 25 milliliters used most frequently.

When a buret is filled with liquid, the surface of the liquid is curved. This curve of the surface is called the meniscus (meh-NIS-cuss). Depending on the liquid, the curve forms a valley, as with water, or forms a hill, as with mercury. Since most solutions used in the laboratory are water based, always read the bottom of the meniscus with your eye at the same level (Figure 7.1). If you have the meniscus at eye level, the closest marks that go all the way around the buret will appear as straight lines, not circles.

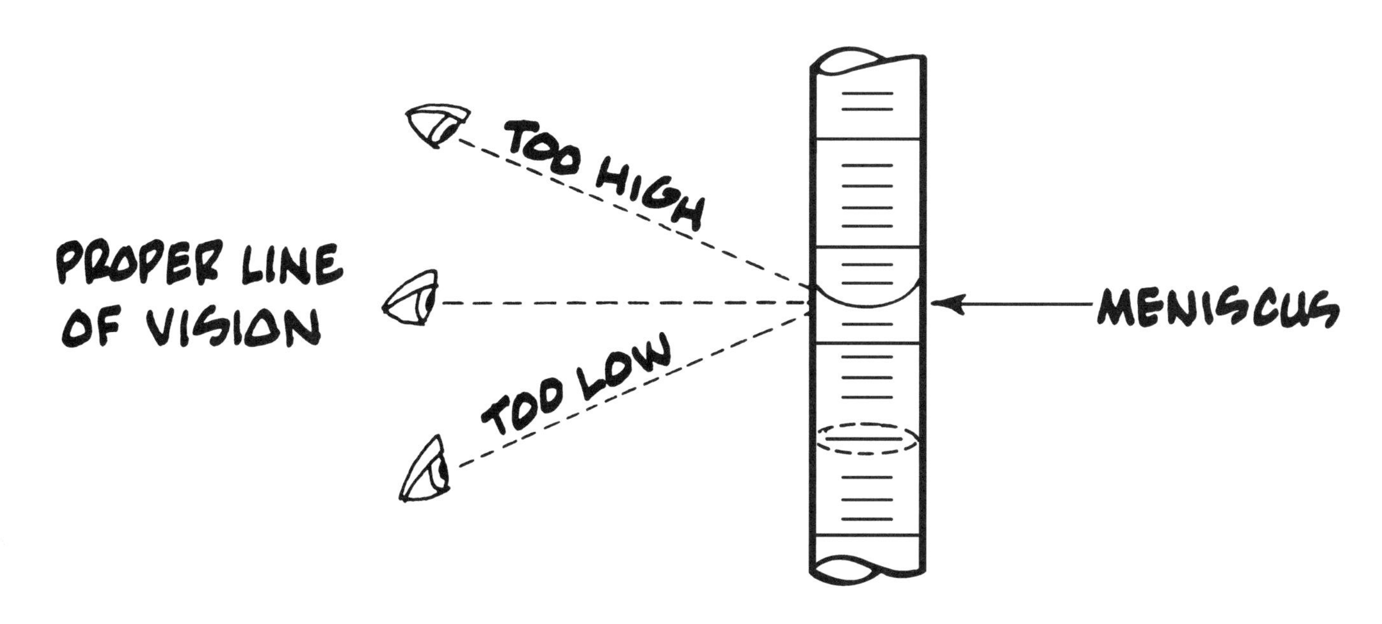

Fig. 7.1 How to read meniscus

GRADUATED CYLINDERS

The graduated cylinder or "graduate" is one of the most often used pieces of laboratory equipment. This cylinder is made either of glass or of plastic and ranges in sizes from 10 m*L* to 4 liters. The graduate is used to measure volumes of liquid with an accuracy *LESS* than burets but *GREATER* than beakers or flasks. Graduated cylinders should never be heated in an open flame because they will break.

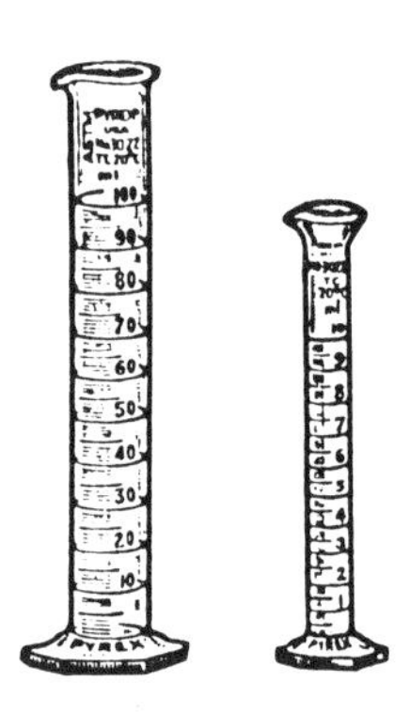

FLASKS AND BEAKERS

Beakers and flasks are used for mixing, heating, and weighing chemicals. Most beakers and flasks are *NOT* calibrated with exact volume lines; however, they are sometimes marked with approximate volumes and can be used to estimate volumes.

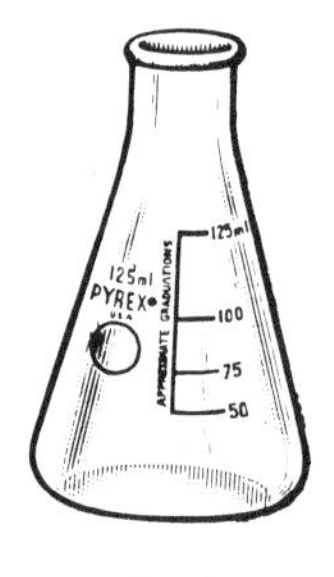

Flask

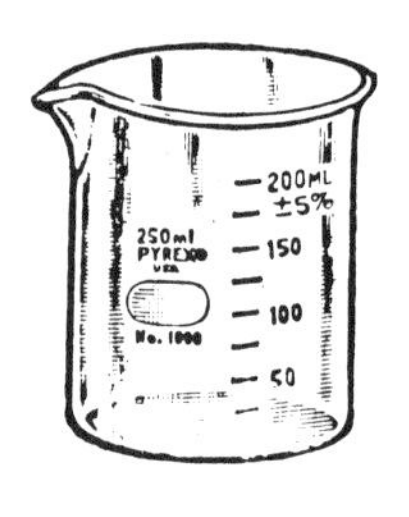

Beaker

VOLUMETRIC FLASKS

Volumetric flasks are used to prepare solutions; they come in sizes from 10 to 2,000 m*L*. Volumetric flasks should *NEVER* be heated. Rather than store liquid chemicals in volumetric flasks, the chemicals should be transferred to a storage bottle.

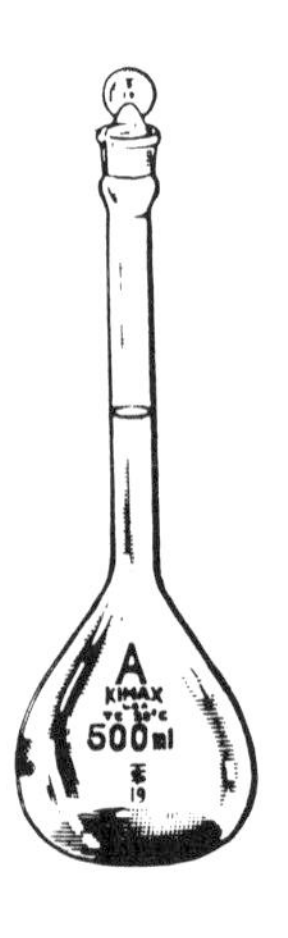

PIPETS

Pipets are used for accurate volume measurements and transfer. There are three types of pipets commonly used in the laboratory — volumetric pipets, graduated (measuring) or Mohr pipets, and serological pipets.

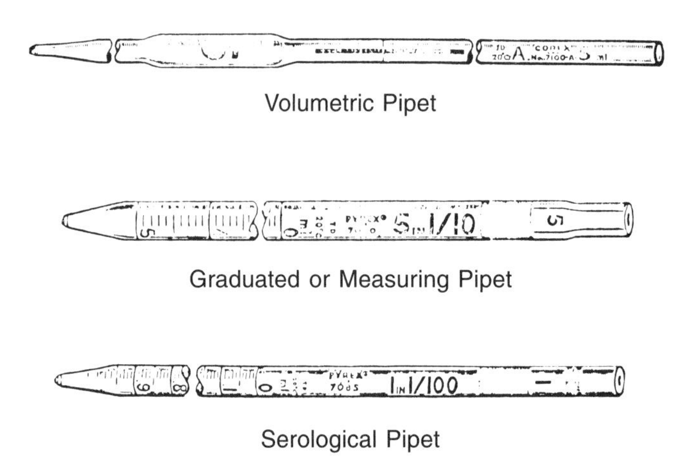

Volumetric Pipet

Graduated or Measuring Pipet

Serological Pipet

Volumetric pipets are available in sizes such as 1, 10, 25, 50, and 100 m*L*. They are used to deliver a single volume. Measuring and serological pipets, however, will deliver fractions of the total volume indicated on the pipet.

Volumetric pipets should be held in a vertical position when emptying and the outflow should be unrestricted. The tip should be touched to the wet surface of the receiving vessel and kept in contact with it until the emptying is complete. Under no circumstance should the small amount remaining in the tip be blown out.

Measuring and serological pipets should be held in the vertical position. After outflow has stopped, the tip should be touched to the wet surface of the receiving vessel. No drainage period is allowed. When the small amount remaining in the tip is to be blown out and added, this will be indicated by a frosted band near the top of the pipet.

Use of a pipet filler or pipet bulb (page 414) is recommended to draw the sample into a pipet. Never pipet chemical reagent solutions or unknown water samples by mouth. Use the following techniques for best results.

1. Draw liquid up into the pipet past the calibration mark.
2. Quickly remove the bulb and place dry fingertip over the top end of the pipet.
3. Wipe excess liquid from the tip of the pipet using laboratory tissue paper.
4. Lift finger and allow desired amount, or all, of liquid to drain. Pipets can be drained without removing the pipet bulb.

NOTE: There are pipet bulbs with valves that can control the flow of liquid from the pipet without removing the bulb.

ACKNOWLEDGMENTS

Pictures of laboratory glassware and equipment in this manual are reproduced with the permission of VWR Scientific, San Francisco, California, with exceptions noted.

QUESTIONS

Write your answers in a notebook and then compare your answers with those on page 469.

7.1A For each type of glassware listed below, describe the item and its use or purpose.

1. Beaker
2. Graduated cylinders
3. Pipets
4. Burets

7.1B What is a meniscus?

7.1C Why should graduated cylinders never be heated in an open flame?

7.12 Chemical Solutions

Many laboratory procedures do not give the concentrations of standard solutions in grams/liter or milligrams/liter. Instead, the concentrations are usually given as *NORMALITY (N)*,[1] which is the standard designation for solution strengths in chemistry.

EXAMPLES:

0.025 $N\ H_2SO_4$ means a 0.025 normal solution of sulfuric acid

2 N NaOH means that the normality of a sodium hydroxide solution is 2

The *LARGER* the number in front of the *N*, the *MORE* concentrated the solution. For example, 1 *N* NaOH solution is more concentrated than a 0.2 *N* NaOH solution.

Another method of specifying the concentration of solutions uses the "a + b system." This means that "a" volumes of concentrated reagent are diluted with "b" volumes of distilled water to form the required solution.

EXAMPLES:

1 + 1 HCl means 1 volume of concentrated HCl is diluted with 1 volume of distilled water

1 + 5 H_2SO_4 means 1 volume of concentrated sulfuric acid is diluted with 5 volumes of distilled water

When the exact concentration of a prepared chemical solution is known, it is referred to as a "standard solution." Many times standard solutions can be ordered already prepared from chemical supply companies. Once a standard has been prepared, it can then be used to standardize other laboratory solutions. To standardize a solution means to determine and adjust its concentration accurately, thereby making it a standard solution. "Standardization" is the process of using one solution of known concentration to determine the concentration of another solution. This action often involves a procedure called a "titration."

When preparing standard solutions or reagents, the directions may say to weigh out 7.6992 grams of a chemical and dilute to one liter with distilled water. To weigh out 7.6992 grams of a chemical, determine the weight of a weighing dish and add this weight to the weight of the chemical. Place the weighing dish on the weighing platform of an analytical balance. (See page 417. Some balances have the weighing platform on top for the weighing dish.) Gently add the chemical to the weighing dish until you are slightly below the desired weight. The weighing mechanism should be off while the chemical is being added and then turned on to determine the exact weight. When you get close to the exact weight, place some of the chemical on a spatula. Gently tap the spatula to add very small amounts of chemical to the weighing dish. Continue this procedure until you've reached the exact weight. If you add too much chemical, remove some of the chemical with the spatula and again repeat the procedure until you reach the exact weight.

Another procedure is to place approximately the desired weight in the weighing dish. Weigh this amount exactly. Then add a proportionate amount of distilled water.

EXAMPLE 1

The directions for preparing a standard reagent indicate that you should weigh out 7.6992 grams and dilute to one liter. You weigh out 7.5371 grams. How much water should be added to produce the desired concentration or normality of the standard reagent?

Known		Unknown
Desired Weight, gm	= 7.6992 gm	Water, m*L*
Actual Weight, gm	= 7.5371 gm	

The chemical should be diluted to how many milliliters?

$$\text{Dilute to m}L = \frac{(\text{Actual Weight, gm})(1{,}000\ \text{m}L)}{\text{Desired Weight, gm}}$$

$$= \frac{(7.5371\ \text{gm})(1{,}000\ \text{m}L)}{7.6992\ \text{gm}}$$

$$= 979\ \text{m}L$$

The 7.5371 grams of chemical should be diluted to 979 m*L*.

7.13 Titrations

A titration involves the measured addition of a standardized solution, which is usually in a buret, to another solution in a flask or beaker. The solution in the buret is referred to as the "titrant" and is added to the other solution until there is a measurable change in the test solution in the flask or beaker. This change is frequently a color change as a result of the addition of another chemical called an "indicator" to the solution in the flask before the titration begins. The solution in the buret is added slowly to the flask until the change, which is called the "end point," is reached. The entire process is the "titration."

---

[1] ***N** or Normal. A normal solution contains one gram equivalent weight of reactant (compound) per liter of solution. The equivalent weight of an acid is that weight which contains one gram atom of ionizable hydrogen or its chemical equivalent. For example, the equivalent weight of sulfuric acid (H_2SO_4) is 49 (98 divided by 2 because there are two replaceable hydrogen ions). A one N solution of sulfuric acid would consist of 49 grams of H_2SO_4 dissolved in enough water to make one liter of solution.*

Figure 7.2 illustrates the four general steps used during a chemical titration.

7.14 Data Recording and Recordkeeping

The use of a laboratory notebook and worksheets are a must for laboratory analysts, water supply system and treatment plant operators. Notebooks and worksheets help you record data in an orderly manner. Too often, hours of work are wasted when test results and other data (such as a sample volume) are written down on a scrap of paper only to be misplaced or thrown away by mistake. Notebooks and worksheets help prevent errors and provide a record of your work. The routine use of laboratory worksheets and notebooks is the only way an operator or a lab person can be sure that all important information is properly recorded.

There is no standard laboratory form. Most operators usually develop their own data sheets for recording test results and other important data. These data sheets should be prepared in a manner that makes it easy for you to record results, review them, and recover these results when it is necessary. Each treatment plant will have different needs for collecting and recording data and may require several different data or worksheets. Figures 7.3 and 7.4 illustrate two typical laboratory worksheets.

7.15 Laboratory Quality Control

Having good equipment and using the correct methods are not enough to ensure correct analytical results. Each operator must be constantly alert to factors in the water treatment process which can lead to poor quality of data. Such factors include sloppy laboratory technique, deteriorated reagents and standards, poorly operating instruments, and calculation mistakes. One of the best ways to ensure quality control in your laboratory is to analyze reference-type samples to provide independent checks on your analysis. These reference-type samples are available from the U.S. Environmental Protection Agency and from commercial sources. From time to time, it is also a good idea to split a sample with one of your fellow operators or another laboratory and compare analytical results. In addition, frequent self-appraisal and evaluation — from sampling to reporting results — can help you gain full confidence in your results.

QUESTIONS

Write your answers in a notebook and then compare your answers with those on page 469.

7.1D What is a "standard solution"?

7.1E What is the primary purpose of laboratory notebooks and worksheets?

7.1F List three sources or causes of poor quality of analytical data.

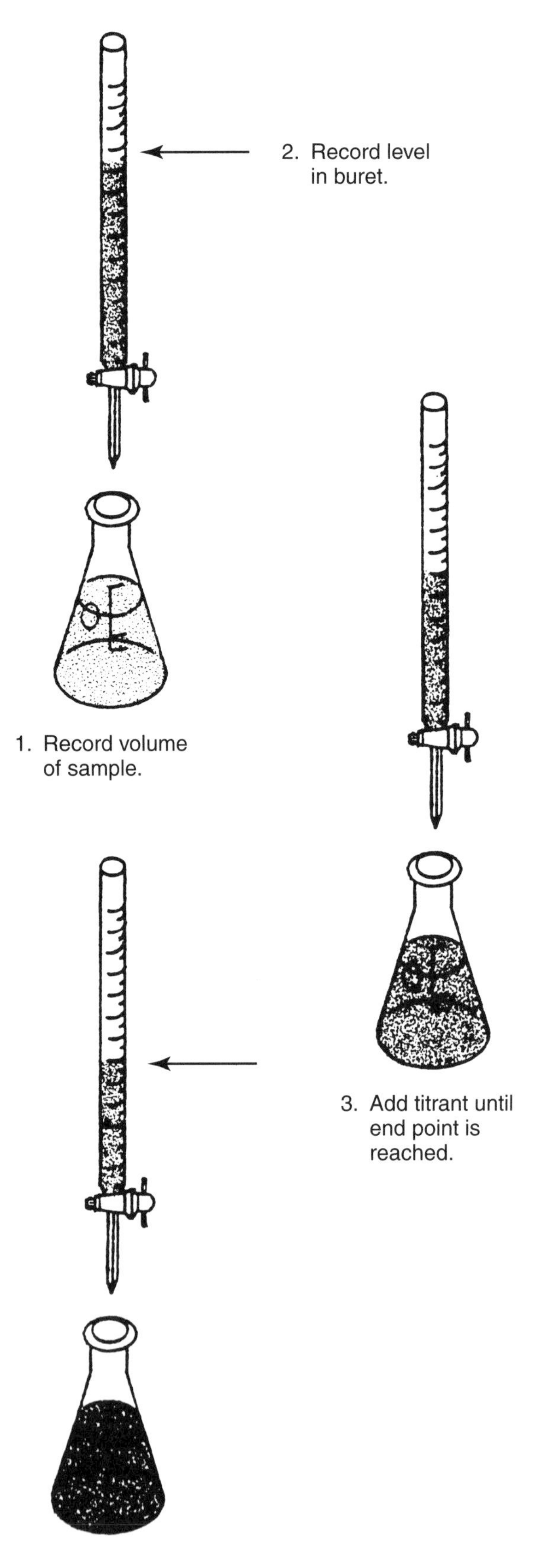
2. Record level in buret.
1. Record volume of sample.
3. Add titrant until end point is reached.
4. Record level of titrant at end point.

Fig. 7.2 Titration steps

WATER ANALYSES REPORT SHEET

(All results in mg/*L* except as noted)

SAMPLE SOURCE:			
DATE SAMPLED:	BY:		
ANALYST:	DATE COMPLETED:		
A. PHYSICAL PARAMETERS	ADDITIONAL INFORMATION		
color (units)	Water Tr., Vol. II, pg. 471		
odor (units)	Water Tr., Vol. II, pg. 492		
turbidity (NTU)	This Manual, pg. 466		
B. GENERAL MINERAL	ADDITIONAL INFORMATION		
pH (units)	This Manual, pg. 464		
total alkalinity ($CaCO_3$)	This Manual, pg. 434		
specific conductance, μmhos/cm	Water Tr., Vol. II, pg. 489		
total dissolved solids	Water Tr., Vol. II, pg. 497		
total hardness ($CaCO_3$)	This Manual, pg. 460		
calcium	Water Tr., Vol. II, pg. 468		
fluoride	Water Tr., Vol. II, pg. 475		
iron	Water Tr., Vol. II, pg. 479		
manganese	Water Tr., Vol. II, pg. 481		
chloride	Water Tr., Vol. II, pg. 469		
sulfate	Water Tr., Vol. II, pg. 490		

Fig. 7.3 Typical laboratory worksheet

7.16 Laboratory Safety

Safety is just as important in the laboratory as it is outside the lab. State laws and the Occupational Safety and Health Act (OSHA) demand that proper safety procedures be exercised in the laboratory at all times. OSHA specifically deals with "safety at the place of work." The Act requires that "each employer has the general duty to furnish all employees with employment free from recognized hazards causing, or likely to cause, death or serious physical harm."

Personnel working in the water industry must realize that a number of hazardous materials and conditions can exist. Always be alert and careful. Be aware of potential dangers at all times. Safe practice in the laboratory and any time while working around chemicals requires hardly any more effort than unsafe practices, with the important benefits from prevention of injury to you and your fellow operators.

On specific questions of safety, consult your state's General Industrial Safety Orders or OSHA regulations.

7.160 Laboratory Hazards

Working with chemicals and other materials in the water laboratory can be dangerous. Laboratory hazards include:

1. Hazardous materials,
2. Explosions,
3. Cuts and bruises,
4. Electric shock,
5. Fire, and
6. Burns (heat and chemical).

Hazardous materials include (1) corrosive, (2) toxic, and (3) explosive or flammable materials.

1. Corrosive Materials

   ACIDS

   a. Examples: Hydrochloric or muriatic (HCl), hydrofluoric (HF), glacial acetic (CH_3COOH), nitric (NHO_3), and sulfuric (H_2SO_4).

   b. Acids can be extremely corrosive and hazardous to human tissue, metals, clothing, cement, stone, and concrete.

Sheet No. __________

Day __________

Date __________

AMERICAN RIVER WATER TREATMENT PLANT

WATER QUALITY

Washwater used M.G.	
Gals. Alum used	
Lbs. Cl_2 used	
Lbs. Lime used	

Time	Turbidity		Hydrogen Ion Conc. (pH)			Alkalinity			Cl_2 Residual			Chlorine		Chemical Dose			TEMP. °C	
	RAW	Filter Effluent	RAW	Res. Influent	TAP Res.	RAW T-Alk	TAP T-Alk	TAP P-alk	Pre-Chlor	Res. Influent	TAP Res	Chlorine Pre	Chlorine Post	Alum	Lime		Raw / Tap	
0001																		
100																		
200																		
300																		
400																		
500																		
600																		
700																		
800																		
900																		
1000																		
1100																		
1200																		
1300																		
1400																		
1500																		
1600																		
1700																		
1800																		
1900																		
2000																		
2100																		
2200																		
2300																		
2400																		
Average																		

Fig. 7.4 Typical operational worksheet

c. Commercially available spill cleanup kits should be kept on hand to neutralize the acid in the event of an accidental spill. Baking soda (bicarbonate, *NOT* laundry soda) effectively neutralizes acids. Baking soda can be used on lab and human surfaces without worrying about toxicity.

BASES (Caustics)

a. Examples: Sodium hydroxide (caustic soda or lye, NaOH), quicklime (CaO), hydrated lime ($Ca(OH)_2$), and alkaline iodine-azide solution (used in dissolved oxygen test).

b. Bases are extremely corrosive to skin, clothing, and leather. Caustics can quickly and permanently cloud vision if not immediately flushed out of eyes. Determine location of safety showers and eyewash station *BEFORE* starting to work with dangerous chemicals.

c. Commercially available spill cleanup materials should be kept on hand for use in the event of an accidental spill. A jug of ordinary vinegar can be kept on hand to neutralize bases and it will not harm your skin.

MISCELLANEOUS CHEMICALS

a. Examples: Alum, chlorine, ferric salts (ferric chloride), and other strong oxidants.

2. Toxic Materials

   Examples:

   a. Solids: Cyanide compounds, chromium, orthotolidine, cadmium, mercury, and other heavy metals.

   b. Liquids: Chloroform and other organic solvents.

   c. Gases: Chlorine, ammonia, sulfur dioxide, and chlorine dioxide.

3. Explosive or Flammable Materials

   Examples:

   a. Liquids: Acetone, ethers, and gasoline.

   b. Gases: Propane and hydrogen.

7.161 Personal Safety and Hygiene

Laboratory work can be quite dangerous if proper precautions are not taken. *ALWAYS* follow these basic rules:

1. *NEVER* work alone in the laboratory. Someone should always be available to help in case you should have an accident which blinds you, leaves you unconscious, or starts a fire you can't handle. If necessary, have someone check on you regularly to be sure you are OK.

2. Wear protective goggles or eyeglasses at all times in the laboratory. Contact lenses should not be worn, even under safety goggles, because fumes can seep between the lens and the eyeball and irritate the eye.

Safety Glasses

DON'T PIPET HAZARDOUS LIQUIDS BY MOUTH.

3. Never pipet hazardous materials by your mouth.

4. Always wear a lab coat or apron in the laboratory to protect your skin and clothes.

5. Wear insulated gloves when handling hot objects. If there is a danger of hot liquid erupting from a container, wear a face shield, too.

6. Don't keep food in a refrigerator that is used for chemical and/or sample storage.

7. Good housekeeping is an effective way to prevent accidents.

7.162 Prevention of Laboratory Accidents

7.1620 CHEMICAL STORAGE

An adequate chemical storeroom is essential for safety in the water laboratory. The storeroom should be properly ventilated and lighted and laid out to segregate incompatible chemicals. Order and cleanliness must be maintained. Clearly label and date all chemicals and bottles of reagents.

Store heavy items on or as near to the floor as possible. *VOLATILE LIQUIDS*[2] which may escape as a gas, such as ether, must be kept away from heat sources, sunlight, and electric switches.

Cap and secure cylinders of gas in storage to prevent rolling or tipping. They should also be placed away from any possible sources of heat or open flames.

[2] *Volatile Liquids. Liquids which easily vaporize or evaporate at room temperature.*

CLAMPS, RAISED SHELF EDGES, AND PROPER ARRANGEMENT PREVENT STOCKROOM FALLOUT.

Follow usual common sense rules of storage. Good housekeeping is a most significant contribution toward an active safety campaign.

7.1621 MOVEMENT OF CHEMICALS

The next area of concern is the transfer of chemicals, apparatus, gases, or other hazardous materials from the storeroom to the laboratory for use. Use cradles or tilters to facilitate handling carboys or other large chemical vessels.

Drum Tilter

In transporting cylinders of compressed gases, use a trussed hand truck. Never roll a cylinder by its valve. Immediately after they are positioned for use, cylinders should be clamped securely into place to prevent shifting or toppling.

Carry flammable liquids in safety cans or, in the case of reagent-grade chemicals, protect the bottle with a carrier. Always wear protective gloves, safety shoes, and rubber aprons in case of accidental spilling of chemical containers.

7.1622 PROPER LABORATORY TECHNIQUES

Faulty technique is one of the chief causes of accidents and, because it involves the human element, is one of the most difficult to correct.

Because of their nature and prevalence in the laboratory, acids and other corrosive materials constitute a series of hazards ranging from poisoning, burning, and gassing through explosion. Always flush the outsides of acid bottles with water before opening them. Don't lay the stopper down on the countertop where a person might lay a hand or rest an arm on it. Keep all acids tightly stoppered when not in use and make sure no spilled acid remains on the floor, table, or bottle after use. To avoid splashing of acid, don't pour water into acid; *ALWAYS POUR ACID INTO WATER.*

Mercury requires special care. Even a small amount in the bottom of a drawer can poison the atmosphere in a room. After an accident involving mercury, go over the entire area carefully until there are no globules remaining. Keep all mercury containers tightly stoppered.

7.1623 ACCIDENT PREVENTION

ELECTRIC SHOCK. Wherever there are electrical outlets, plugs, and wiring connections, there is a danger of electric shock. The usual "do's" and "don'ts" of protection against shock in the home are equally applicable in the laboratory. Don't use worn or frayed wires. Replace connections when there is any sign of thinning insulation. Ground all apparatus using three-prong plugs. Don't continue to run a motor after liquid has spilled on it. Turn it off immediately and clean and dry the inside thoroughly before attempting to use it again.

Electrical units which are operated in an area exposed to flammable vapors should be explosion-proof. All permanent wiring should be installed by an electrician with proper conduit or BX cable to eliminate any danger of circuit overloading.

CUTS. Some of the pieces of glass used in the laboratory, such as glass tubing, thermometers, and funnels, must be inserted through rubber stoppers. If the glass is forced through the hole in the stopper by applying a lot of pressure, the glass usually breaks. This is one of the most common sources of cuts in the laboratory.

Use care in making rubber-to-glass connections. Lengths of glass tubing should be supported while they are being inserted into rubber. The ends of the glass should be *FLAME POLISHED*[3] and either wetted or covered with a lubricating jelly for ease in joining connections. Never use oil or grease. Wear gloves when making such connections, and hold the tubing as close to the end being inserted as possible to prevent bending or breaking. Also, never try to force rubber tubing or stoppers from glassware. Cut off the rubber or materials.

A FIRST-AID kit must be available in the laboratory.

BURNS. All glassware and porcelain look cold after the red from heating has disappeared. The red is gone in seconds but the glass is hot enough to burn for several minutes. After heating a piece of glass, put it out of the way until cold.

Spattering from acids, caustic materials, and strong oxidizing solutions should be washed off immediately with large quantities of water. Every worker in the water laboratory

[3] *Flame Polished. Melted by a flame to smooth out irregularities. Sharp or broken edges of glass (such as the end of a glass tube) are rotated in a flame until the edge melts slightly and becomes smooth.*

should have access to a sink and an emergency deluge shower. Keep vinegar and soda handy to neutralize acids and bases (caustic materials).

Many safeguards against burns are available. Gloves, safety tongs, aprons, and emergency deluge showers are but a few examples. Never decide it is too much trouble to put on a pair of gloves or use a pair of tongs to handle a dish or flask that has been heated.

USE TONGS — DON'T JUGGLE HOT CONTAINERS.

Perhaps the most harmful and painful chemical burn occurs when small objects, chemicals, or fumes get into your eyes. Immediately flood your eyes with water or a special "eye wash" solution from a safety kit or from an eyewash station or fountain.

TOXIC FUMES. Use a ventilated laboratory fume hood for routine reagent preparation. Select a hood that has adequate air displacement and expels harmful vapors and gases at their source. An annual check should be made of the entire laboratory building. Sometimes noxious fumes are spread by the heating and cooling system of the building.

When working with chlorine and other toxic substances, always wear a self-contained breathing apparatus. If possible, try to clear the atmosphere with adequate ventilation *BEFORE* entry.

WASTE DISPOSAL. A good safety program requires constant care in disposal of laboratory waste. Corrosive materials should never be poured down an ordinary sink or drain. These substances can corrode away the drain pipe and/or trap. Corrosive acids should be neutralized and poured down corrosion-resistant sinks and sewers using large quantities of water to dilute and flush the acid.

To protect maintenance personnel, use separate covered containers to dispose of broken glass.

DON'T POUR VOLATILE LIQUIDS INTO THE SINK.

FIRE. The laboratory should be equipped with a fire blanket. The fire blanket is used to smother clothing fires. Small fires which occur in an evaporating dish or beaker may be put out by covering the container with a glass plate, wet towel, or wet blanket. For larger fires, or ones which may spread rapidly, promptly use a fire extinguisher. Do not use a fire extinguisher on small beaker fires because the force of the spray will knock over the beaker and spread the fire. Take time to become familiar with the operation and use of your fire extinguishers.

The use of the proper type of extinguisher for each class of fire will give the best control of the situation and avoid compounding the problem. For example, water should not be poured on grease, electrical fires, or metal fires because water could increase the hazards, such as splattering of the fire and electric shock. The classes of fires given here are based on the type of material being consumed.

Fire classifications are important for determining the type of fire extinguisher needed to control the fire. Classifications also aid in recordkeeping. Fires are classified as A, B, C, or D fires based on the type of material being consumed: A, ordinary combustibles; B, flammable liquids and vapors; C, energized electrical equipment; and D, combustible metals. Fire extinguishers are also classified as A, B, C, or D to correspond with the class of fire each will extinguish.

Class A fires: ordinary combustibles such as wood, paper, cloth, rubber, many plastics, dried grass, hay, and stubble. Use foam, water, soda-acid, carbon dioxide gas, or almost any type of extinguisher.

Class B fires: flammable and combustible liquids such as gasoline, oil, grease, tar, oil-based paint, lacquer, and solvents, and also flammable gases. Use foam, carbon dioxide, or dry chemical extinguishers.

Class C fires: energized electrical equipment such as starters, breakers, and motors. Use carbon dioxide or dry chemical extinguishers to smother the fire; both types are nonconductors of electricity.

Class D fires: combustible metals such as magnesium, sodium, zinc, and potassium. Operators rarely encounter this type of fire. Use a Class D extinguisher or use fine dry soda ash, sand, or graphite to smother the fire. Consult with your local fire department about the best methods to use for specific hazards that exist at your facility.

Multipurpose extinguishers are also available, such as a Class BC carbon dioxide extinguisher that can be used to smother Class B and Class C fires. A multipurpose ABC carbon dioxide extinguisher will handle most laboratory fire situations. (When using carbon dioxide extinguishers, remember that the carbon dioxide can displace oxygen—take appropriate precautions.)

There is no single type of fire extinguisher that is effective for all fires so it is important that you understand the class of fire you are trying to control. You must be trained in the use of the different types of extinguishers, and the proper type should be located near the area where that class of fire may occur.

7.163 Acknowledgments

Portions of this section were taken from material written by A.E. Greenberg, "Safety and Hygiene," which appeared in the California Water Pollution Control Association's *OPERATORS' LABORATORY MANUAL*. Some of the ideas and material also came from the *FISHER SAFETY MANUAL*.

7.164 Additional Reading

1. *FISHER SCIENTIFIC SAFETY PRODUCTS REFERENCE MANUAL*. Obtain from Fisher Scientific Company, Safety Division, 9999 Veterans Memorial Drive, Houston, TX 77038.
2. *GENERAL INDUSTRY, OSHA. SAFETY AND HEALTH STANDARDS* (CFR, Title 29, Labor Pt.1900-1910 (most recent edition)). Obtain from the U.S. Government Printing Office, Superintendent of Documents, PO Box 371954, Pittsburgh, PA 15250-7954. Order No. 869-044-00104-7. Price, $55.00.
3. See *SAFETY*, page 1-38, *STANDARD METHODS*, 20th Edition, 1998.

QUESTIONS

Write your answers in a notebook and then compare your answers with those on page 469.

7.1G List five laboratory hazards.

7.1H Why should you not work alone in a laboratory?

7.1I True or False? You may *ADD ACID TO WATER*, but never water to acid.

7.1J How would you dispose of a corrosive acid?

End of Lesson 1 of 5 Lessons on LABORATORY PROCEDURES

Please answer the discussion and review questions next.

DISCUSSION AND REVIEW QUESTIONS

Chapter 7. LABORATORY PROCEDURES

(Lesson 1 of 5 Lessons)

At the end of each lesson in this chapter you will find some discussion and review questions. The purpose of these questions is to indicate to you how well you understand the material in the lesson. Write the answers to these questions in your notebook before continuing.

1. How do the operators of water treatment plants use the results from laboratory tests?
2. Why must chemicals be properly labeled?
3. How are pipets emptied or drained?
4. How would you titrate a test solution?
5. List as many of the seven basic rules for working in a laboratory as you can remember.
6. Why should work with certain chemicals be conducted under a ventilated laboratory fume hood?
7. Why should water not be poured on certain types of fires?

CHAPTER 7. LABORATORY PROCEDURES

(Lesson 2 of 5 Lessons)

7.2 SAMPLING

7.20 Importance of Sampling

Sampling is a vital part of studying the quality of water in a water treatment process, distribution system, or source of water supply. The major source of error in the whole process of obtaining water quality information often occurs during sampling. Proper sampling procedures are essential in order to obtain an accurate description of the material or water being sampled and tested. This fact is not well enough recognized and cannot be overemphasized.

In any type of testing program where only small samples (a liter or two) are withdrawn from perhaps millions of gallons of water under examination, there is potential uncertainty because of possible sampling errors. Water treatment decisions based upon incorrect data may be made if sampling is performed in a careless and thoughtless manner. Obtaining good results will depend to a great extent upon the following factors:

1. Ensuring that the sample taken is truly representative of the water under consideration,
2. Using proper sampling techniques, and
3. Protecting and preserving the samples until they are analyzed.

The greatest errors in laboratory tests are usually caused by improper sampling, poor preservation, or lack of enough mixing during testing. The accuracy of your analysis is only as good as the care that was taken in obtaining a representative sample.[4]

7.21 Representative Sampling

7.210 Importance of Representative Sampling

A representative sample must be collected in order for test results to have any significant meaning. Without a representative sample, the test results will not reflect actual water conditions.

The sampling of a tank or a lake that is completely mixed is a simple matter. Unfortunately, most bodies of water are not well mixed and obtaining samples that are truly representative of the whole body depends to a great degree upon sampling technique. A sample that is properly mixed (integrated) by taking small portions of the water at points distributed over the whole body represents the material better than a sample collected from a single point. The more portions taken, the more nearly the sample represents the original. The sample error would reach zero when the size of the sample became equal to the original volume of material being sampled, but for obvious reasons this method of decreasing sample error is not practical. The size of sample depends on which water quality indicators are being tested and how many. Every precaution must be taken to ensure that the sample collected is as representative of the water source or process being examined as is feasible.

7.211 Source Water Sampling

RIVERS. To adequately determine the composition of a flowing stream, each sample (or set of samples taken at the same time) must be representative of the entire flow at the sampling point at that instant. Furthermore, the sampling process must be repeated frequently enough to show significant changes of water quality that may occur over time in the water passing the sampling point.

On small or medium-sized streams, it is usually possible to find a sampling point at which the composition of the water is presumably uniform at all depths and across the stream. Obtaining representative samples in these streams is relatively simple. For larger streams, more than one sample may be required. A portable conductivity meter is very useful in selecting good sample sites.

RESERVOIRS AND LAKES. Water stored in reservoirs and lakes is usually poorly mixed. Thermal stratification and associated depth changes in water composition (such as dissolved oxygen) are among the most frequently observed effects. Single samples can therefore be assumed to represent only the spot of water from which the sample came. Therefore, a number of samples must be collected at different depths and from different areas of the impoundment to accurately sample reservoirs and lakes.

GROUNDWATER. Most of the physical factors which promote mixing in surface waters are absent or much less effective in groundwater systems. Wells usually draw water from a considerable thickness of saturated rock and often from sever-

---

[4] *Representative Sample. A sample portion of material or water that is as nearly identical in content and consistency as possible to that in the larger body of material or water being sampled.*

al different strata. These water components are mixed by the turbulent flow of water in the well before they reach the surface and become available for sampling. Most techniques for well sampling and exploration are usable only in unfinished or nonoperating wells. Usually the only means of sampling the water tapped by a well is the collection of a pumped sample. The operator is cautioned to remember that well pumps and casings can contribute to sample contamination. If a pump has not run for an extended period of time prior to sampling, the water collected may not be representative of the normal water quality.

7.212 In-Plant Sampling

Collection of representative samples within the water treatment plants is really no different from sample collection in a stream or river. The operator simply wants to be sure the water sampled is representative of the water passing that sample point. In many water plants, money is spent to purchase sample pumps and piping only to sample from a point that is not representative of the passing water. A sample tap in a dead area of a reservoir or on the floor of a process basin serves no purpose in helping the plant operator with control of water quality. The operator is urged to find each and every sample point and ensure it is located to provide a useful and representative sample. If the sampling point is not properly located, plan to move the piping to a better location.

7.213 Distribution System Sampling

Representative sampling in the distribution system is a true indication of system water quality. Results of sampling should show if there are quality changes in the entire, or parts of, the system and may point to the source of a problem (such as tastes and/or odors). Sampling points should be selected, in part, to trace the course from the finished water source (at the well or plant), through the transmission mains, and then through the major and minor arteries of the system. A sampling point on a major artery, or on an active main directly connected to it, would be representative of the water quality being furnished to a subdivision of this network. Generally, these primary points are used as "official" sample points in evaluating prevailing water quality.

Obtaining a representative sample from the distribution system is not as easy as it might seem. One would think almost any faucet would do, but experience has shown otherwise. Local conditions at the tap and in its connection to the main can easily make the point unrepresentative of water being furnished to your customers.

The truest evaluation of water in a distribution system can be obtained from samples drawn directly from the main. You might think that samples taken from a fire hydrant would prove satisfactory, but this is usually not the case. The problem with fire hydrants as sampling points is that they give erratic (uneven) results due to the way they are constructed and their lack of use. In general, an ideal sample station is one that has a short, direct connection with the main and is made of corrosion-resistant material.

In most smaller water systems, special sample taps are not available. Therefore, customer's faucets must be used to collect samples. The best sample points are front yard faucets on homes supplied by short service lines (homes with short service lines are located on the same side of the street as the water main).

If the customer is home, you should contact the person in the home and obtain permission to collect the sample. Disconnect the hose from the faucet if one is attached and don't forget to reconnect the hose when finished collecting the sample. Open the faucet to a convenient flow for sampling (usually about half a gallon per minute). Allow the water to flow until the water in the service line has been replaced twice. Since 50 feet (15 m) of three-quarter inch (18 mm) pipe contains over one gallon (3.8 liters), four or five minutes will be required to replace the water in the line twice. Collect the sample. Be sure the sample container does not touch the faucet.

Do not try to save time by turning the faucet handle to wide open to flush the service line. This will disturb sediment and incrustations in the line which must be flushed out before the sample can be collected.

FORMULAS

To estimate the flow from a faucet, use a gallon jug and a watch. If you want a flow of half a gallon per minute, then the jug should be half full in one minute or completely full in two minutes.

$$\text{Flow, GPM} = \frac{\text{Volume, gallons}}{\text{Time, minutes}}$$

To calculate the volume of a service line, multiply the area of the pipe in square feet times the length of the pipe in feet to obtain cubic feet. The diameter of a pipe is given in inches, so this value must be divided by 12 inches per foot to obtain a volume in cubic feet. Multiply cubic feet by 7.48 gallons per cubic foot to obtain the volume in gallons.

$$\text{Pipe Volume, cu ft} = \frac{\text{(Pipe Area, sq in)(Pipe Length, ft)}}{\text{144 sq in/sq ft}}$$

$$\text{Pipe Volume, gal} = \text{(Pipe Volume, cu ft)(7.48 gal/cu ft)}$$

To determine the time to allow water to flow from a faucet to flush a service line twice, divide the pipe volume in gallons by the flow in gallons per minute. Then multiply the result by two so the line will be flushed twice.

$$\text{Flushing Time, min} = \frac{\text{(Pipe Volume, gal)(2)}}{\text{Flow, gal/min}}$$

EXAMPLE 2

How long should a three-quarter inch service line 80 feet long be flushed if the flow is 0.5 GPM?

Known		**Unknown**
Diameter, in	= $^3/_4$ in	Flushing Time, min
	= 0.75 in	
Length, ft	= 80 ft	
Flow, GPM	= 0.5 GPM	

Calculate the pipe volume in cubic feet and then in gallons.

$$\text{Pipe Volume, cu ft} = \frac{(\text{Pipe Area, sq in})(\text{Pipe Length, ft})}{144 \text{ sq in/sq ft}}$$

$$= \frac{(0.785)(0.75 \text{ in})^2(80 \text{ ft})}{144 \text{ sq in/sq ft}}$$

$$= 0.245 \text{ cu ft}$$

$$\text{Pipe Volume, gal} = (\text{Pipe Volume, cu ft})(7.48 \text{ gal/cu ft})$$

$$= (0.245 \text{ cu ft})(7.48 \text{ gal/cu ft})$$

$$= 1.833 \text{ gal}$$

Calculate the flushing time for the service line in minutes.

$$\text{Flushing Time, min} = \frac{(\text{Pipe Volume, gal})(2)}{\text{Flow, gal/min}}$$

$$= \frac{(1.833 \text{ gal})(2)}{0.5 \text{ GPM}}$$

$$= 7.3 \text{ min}$$

$$\text{or} = 7 \text{ min} + (0.3 \text{ min})(60 \text{ sec/min})$$

$$= 7 \text{ min and } 18 \text{ sec}$$

QUESTIONS

Write your answers in a notebook and then compare your answers with those on page 469.

7.2A What are frequently the greatest causes of errors in laboratory tests?

7.2B Why must a representative sample be collected?

7.2C How are sampling points selected in a distribution system?

7.22 Types of Samples

There are generally two types of samples collected by waterworks operators, and either type may be obtained manually or automatically. The two types are grab samples and composite samples.

7.220 Grab Samples

A grab sample is a single water sample collected at no specific time. Grab samples will show the water characteristics at the time the sample was taken. A grab sample may be preferred over a composite sample when:

1. The water to be sampled does not flow continuously,
2. The water's characteristics are relatively constant, and
3. The water is to be analyzed for water quality indicators that may change with time, such as dissolved gases, coliform bacteria, residual chlorine, temperature, and pH.

7.221 Composite Samples

In many processes, the water quality is changing from moment to moment or hour to hour. A continuous sampler-analyzer would give the most accurate results in these cases. However, since operators themselves are often the sampler-analyzer, continuous analysis would leave little time for anything but sampling and testing. Except for tests which cannot wait due to rapid physical, chemical, or biological changes of the sample (such as tests for dissolved oxygen, pH, and temperature) a fair compromise may be reached by taking samples throughout the day at hourly or two-hour intervals. Each sample should be refrigerated immediately after it is collected. At the end of 24 hours, a portion of each sample is mixed with the other samples. The size of the portion is in direct proportion to the flow when the sample was collected and the total size of sample needed for testing. For example, if hourly samples were collected when the flow was 1.2 MGD, use a 12-m*L* portion sample, and when the flow was 1.5 MGD, use a 15-m*L* portion sample. The resulting mixture of portions of samples is called a *COMPOSITE SAMPLE.* In no case, however, should a composite sample be collected for bacteriological examination.

When the samples are taken, they can either be set aside to be combined later or combined as they are collected. In both cases, they should be stored at a temperature of 4°C until they are analyzed.

7.23 Sampling Devices

Automatic sampling devices are wonderful time-savers but are expensive. As with anything automatic, problems do arise and the operator should be on the lookout for potential difficulties.

Manual sampling equipment includes dippers, weighted bottles, hand-operated pumps, and cross-section samplers. Dippers consist of wide-mouth, corrosion-resistant containers (such as cans or jars) on long handles that collect a sample for testing. A weighted bottle is a collection container which is lowered to a desired depth. At this depth a cord or wire removes the bottle stopper so the bottle can be filled (see Figure 7.5).

Some water treatment facilities use sample pumps to collect the sample and transport it to a central location. The pump and its associated piping should be corrosion-resistant and sized to deliver the sample at a high enough velocity to prevent sedimentation in the sample line.

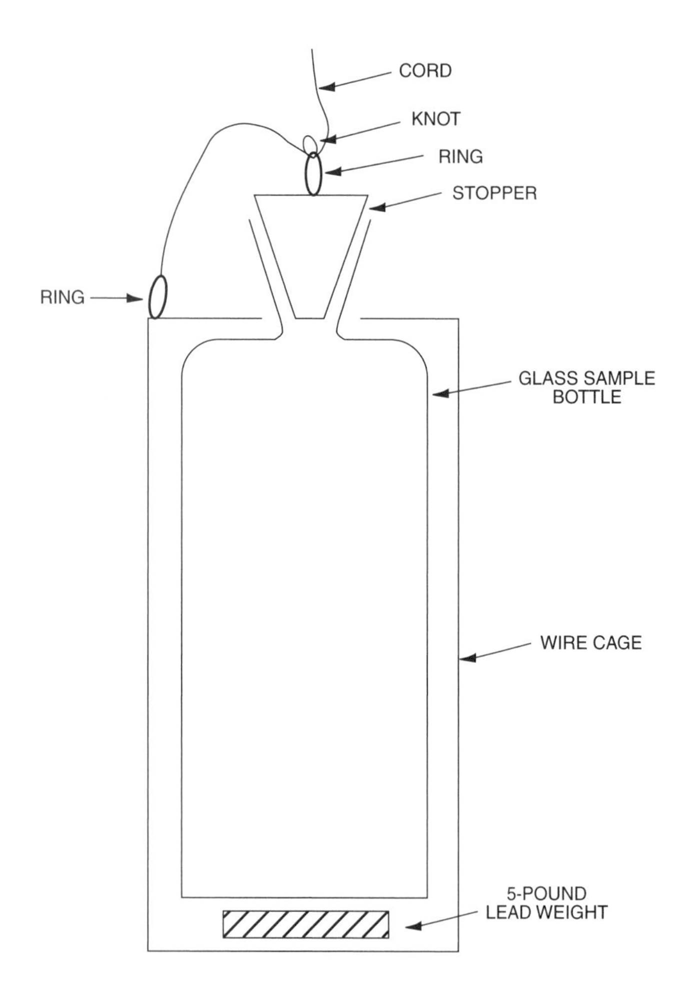

Fig. 7.5 Sectional view of homemade depth sampler

Fig. 7.6 Distribution system sampling station

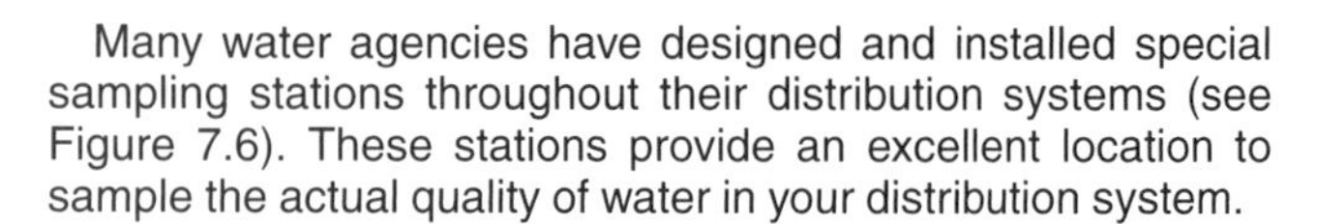

Many water agencies have designed and installed special sampling stations throughout their distribution systems (see Figure 7.6). These stations provide an excellent location to sample the actual quality of water in your distribution system.

7.24 Sampling Techniques

7.240 Surface Sampling

A surface sample is obtained by grasping the sample container at the base with one hand and plunging the bottle mouth down into the water to avoid introducing any material floating on the surface. Position the mouth of the bottle into the current and away from the hand of the collector (see Figure 7.7). If the water is not flowing, then an artificial current can be created by moving the bottle horizontally in the direction it is pointed and away from the sampler. Tip the bottle slightly upward to allow air to exit so the bottle can fill. Tightly stopper and label the bottle.

Another technique for collecting a surface sample is to place the bottle in a weighted frame that holds the bottle securely when sampling from a walkway or other structure above a body of water. Remove the stopper or lid and lower the device to the water surface. A nylon rope which does not absorb water and will not rot is recommended. Face the bottle mouth upstream by swinging the sampling device downstream and then allow it to drop into the water, without slack in the rope. Pull the sample device rapidly upstream and out of the water simulating (imitating) the scooping motion of the hand-sampling described previously. Take care not to dislodge dirt or other debris that might fall into the open sample container from above. Be sure to label the container when sampling is completed.

7.241 Depth Sampling

Several additional pieces of equipment are needed for collection of depth samples from basins, tanks, lakes, and reservoirs. These depth samplers require lowering the sample device and container to the desired water depth, then opening, filling, and closing the container, and returning the device to the surface. Although depth measurements are best made with a pre-marked steel cable, the sample depths can be determined by pre-measuring and marking a nylon rope at intervals with a non-smearing ink, paint, or fingernail polish. One of the most common commercial devices is called a Kemmerer Sampler (see Figure 7.8). This type of depth sampler consists of a cylindrical tube that contains a rubber stopper or valve at each end. The device is lowered into the water in the open position and the water sample is trapped in the cylinder when the valves are closed by the dropped messenger.

Figures 7.5 and 7.9 show typical depth samplers. These samplers are lowered to the desired depth. A jerk on the cord will remove the stopper and allow the bottle in the depth sampler to fill. Good samples can be collected in depths of water up to 40 feet (12 m).

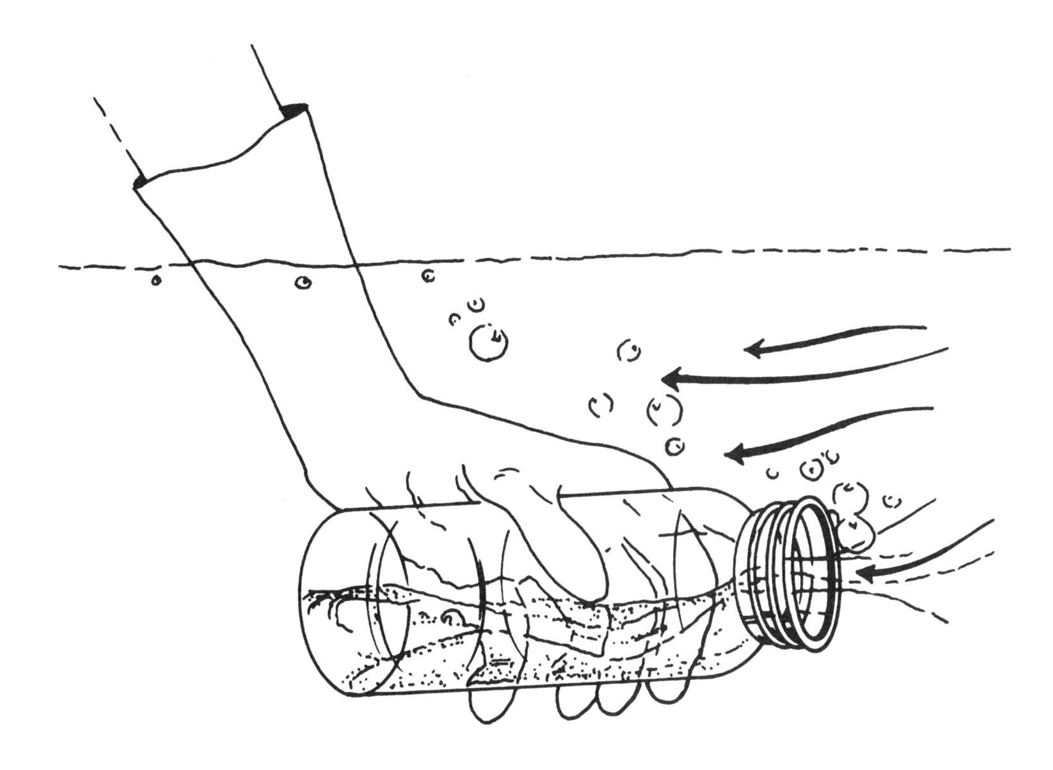

Fig. 7.7 Demonstration of technique used in grab sampling of surface waters

(Source: US EPA "Microbiological Methods For Monitoring the Environment," December 1978)

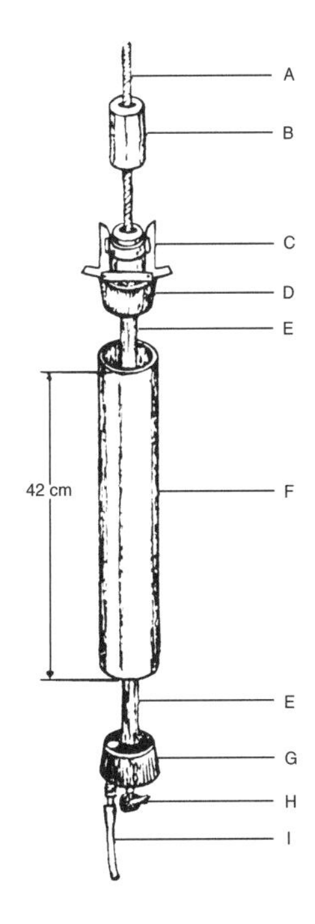

Fig. 7.8 Kemmerer depth sampler. (A) nylon line, (B) messenger, (C) catch set so that the sampler is open, (D) top rubber valve, (E) connecting rod between the valves, (F) tube body, (G) bottom rubber valve, (H) knot at the bottom of the suspension line, and (I) rubber tubing attached to the spring-loaded check valve.

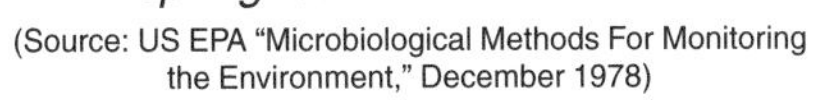

(Source: US EPA "Microbiological Methods For Monitoring the Environment," December 1978)

Fig. 7.9 Depth sampler

(Permission of HACH Company)

7.242 Water Tap Sampling

To collect samples from water main connections, first flush the service line for a brief period of time. Samples should not be taken from drinking fountains, restrooms, or taps that have aerators. Aerators can change water quality indicators such as pH and dissolved oxygen, and can harbor bacteria under some conditions. Do not sample from taps surrounded by excessive foliage (leaves, flowers) or from taps that are dirty, corroded, or are leaking. Never collect a sample from a hose or any other attachment fastened to a faucet. Care must be taken to be sure that the sample collector does not come in contact with the faucet.

7.243 First-Draw Samples

The Lead and Copper Rule calls for first-draw or first-flush samples. These are water samples taken at the customer's tap after the water stands motionless in the plumbing pipes for at least six hours. This usually means taking a sample early in the day before water is used in the kitchen or bathroom. (For information about the Lead and Copper Rule, see Section 8.6, "The Lead and Copper Rule," in *WATER TREATMENT PLANT OPERATION*, Volume I, in this series of operator training manuals.

7.25 Sampling Containers and Preservation of Samples

The shorter the time that elapses between the actual collection of the sample and the analysis, the more reliable your results will be. Samples should be preserved if they are not going to be analyzed immediately due to remoteness of the laboratory or workload. Preservation of some types of samples is essential to prevent deterioration of the sample. Some water quality indicators, such as residual chlorine and temperature, require immediate analysis, while others can be preserved and transported to the laboratory. A summary of acceptable sample containers, preservative, and maximum time between sampling and analysis is shown on Table 7.2.

Whatever type of container you use, clearly identify the sample location, date and time of collection, name of collector, and any other pertinent information.

7.26 Reporting

The water system owner (water utility agency) is responsible for reporting lab results at regular frequencies to the regulatory agency as required by the Safe Drinking Water Act.

7.27 Additional Reading

1. See page 1-27, *COLLECTION AND PRESERVATION OF SAMPLES, STANDARD METHODS*, 20th Edition.

QUESTIONS

Write your answers in a notebook and then compare your answers with those on pages 469 and 470.

7.2D What are the two general types of samples collected by water treatment personnel?

7.2E List three water quality indicators that are usually measured with a grab sample.

7.2F How would you collect a depth sample from a lake?

7.2G Samples should not be collected from water taps under what conditions?

7.2H What information should be recorded when a sample is collected?

End of Lesson 2 of 5 Lessons
on
LABORATORY
PROCEDURES

Please answer the discussion and review questions next.

DISCUSSION AND REVIEW QUESTIONS

Chapter 7. LABORATORY PROCEDURES

(Lesson 2 of 5 Lessons)

Write the answers to these questions in your notebook before continuing. The question numbering continues from Lesson 1.

8. Why are proper sampling procedures important?
9. What is meant by a representative sample?
10. Generally speaking, how would you obtain a representative sample?
11. Under what conditions and why would you preserve a sample?

TABLE 7.2 RECOMMENDATION FOR SAMPLING AND PRESERVATION OF SAMPLES ACCORDING TO MEASUREMENT [a]

Measurement	Vol. Req. (m*L*)	Container [b]	Preservative	Max. Holding Time [c]
PHYSICAL PROPERTIES				
Color	500	P,G	Cool, 4°C	48 hours
Conductance	500	P,G	Cool, 4°C	28 days
Hardness [d]	100	P,G	HNO_3 to pH <2, H_2SO_4 to pH <2	6 months
pH [d]	25	P,G	Det. on site	Immediately
Residue, Filterable	100	P,G	Cool, 4°C	7 days
Temperature	1,000	P,G	Det. on site	Immediately
Turbidity	100	P,G	Cool, 4°C	48 hours
METALS (Fe, Mn)				
Dissolved or Suspended	200	P,G	Filter on site, HNO_3 to pH <2	6 months
Total	100	P,G	Filter on site, HNO_3 to pH <2	6 months
INORGANICS, NONMETALLICS				
Acidity	100	P,G	Cool, 4°C	14 days
Alkalinity	200	P,G	Cool, 4°C	14 days
Bromide	100	P,G	None Req.	28 days
Chloride	50	P,G	None Req.	28 days
Chlorine, Total Residual	500	P,G	None Req.	Immediately
Cyanide, Total and Amenable to Chlorination	500	P,G	Cool, 4°C, NaOH to pH >12, 0.6 gm ascorbic acid [e]	14 days
Fluoride	300	P	None Req.	28 days
Nitrogen				
Ammonia	500	P,G	Cool, 4°C, H_2SO_4 to pH <2	28 days
Kjeldahl and Organic	500	P,G	Cool, 4°C, H_2SO_4 to pH <2	28 days
Nitrate-Nitrite	200	P,G	Cool, 4°C, H_2SO_4 to pH <2	28 days
Nitrate	100	P,G	Cool, 4°C	48 hours
Nitrite	100	P,G	Cool, 4°C	48 hours
Dissolved Oxygen				
Probe	300	G with top	Det. on site	Immediately
Winkler	300	G with top	Fix on site, store in dark	8 hours
Phosphorus				
Orthophosphate	50	P,G	Filter on site, Cool, 4°C	48 hours
Elemental	50	G	Cool, 4°C	48 hours
Total	50	P,G	Cool, 4°C, H_2SO_4 to pH <2	28 days
Silica	50	P	Cool, 4°C	28 days
Sulfate	100	P,G	Cool, 4°C	28 days
Sulfide	100	P,G	Cool, 4°C, add zinc acetate plus H_2SO_4 to pH >9	7 days
Sulfite	50	P,G	Det. on site	Immediately

[a] "Required Containers, Preservation Techniques, and Holding Times." *CODE OF FEDERAL REGULATIONS*, Protection of the Environment, 40, Parts 136-149, 2001. This publication is available from the U.S. Government Printing Office, Superintendent of Documents, PO Box 371954, Pittsburgh, PA 15250-7954. Order No. 869-044-00152-7. Price, $55.00.

[b] Polyethylene (P) or Glass (G). For metals, polyethylene with a polypropylene cap (no liner) is preferred.

[c] Holding times listed above are recommended for properly preserved samples based on currently available data. It is recognized that for some sample types, extension of these times may be possible while for other types, these times may be too long. Where shipping regulations prevent the use of the proper preservation technique or the holding time is exceeded, such as the case of a 24-hour composite, the final reported data for these samples should indicate the specific variance.

[d] Hardness and pH are usually considered chemical properties of water rather than physical properties.

[e] Use ascorbic acid only if residual chlorine is present.

SPECIAL NOTE: Whenever you collect a sample for a bacteriological test (coliforms), be sure to use a sterile plastic or glass bottle. If the sample contains any chlorine residual, sufficient sodium thiosulfate should be added to neutralize all of the chlorine residual. Usually two drops (0.1 m*L*) of ten percent sodium thiosulfate for every 100 m*L* of sample is sufficient, unless you are disinfecting mains or storage tanks.

CHAPTER 7. LABORATORY PROCEDURES

(Lesson 3 of 5 Lessons)

7.3 WATER LABORATORY TESTS

1. Alkalinity

A. Discussion

The alkalinity of a water sample is a measure of the water's capacity to neutralize acids. In natural and treated waters, alkalinity is the result of bicarbonates, carbonates, and hydroxides of the metals of calcium, magnesium, and sodium.

Many of the chemicals used in water treatment, such as alum, chlorine, or lime, cause changes in alkalinity. The alkalinity determination is needed when calculating chemical dosages used in coagulation and water softening. Alkalinity must also be known to calculate corrosivity and to estimate the carbonate hardness of water. Alkalinity is usually expressed in terms of *CALCIUM CARBONATE ($CaCO_3$) EQUIVALENT.*[5]

There are five alkalinity conditions possible in a water sample: (1) bicarbonate alone, (2) bicarbonate and carbonate, (3) carbonate alone, (4) carbonate and hydroxide, and (5) hydroxide alone. These five conditions may be distinguished and quantities determined from the results of acid titrations by the method given below.

B. What Is Tested

Sample	Common Range, mg/*L*
Raw and Treated Surface Water	20 - 300
Well Water	80 - 500

C. Apparatus Required

pH meter	graduated cylinder (100 m*L*)
reference electrode	buret (25 m*L*)
glass electrode	buret support
magnetic stirrer	beaker (250 m*L*)
magnetic stir-bar	wash bottle
analytical balance	desiccator

D. Reagents

NOTE: Standardized solutions are commercially available for most reagents. Refer to *STANDARD METHODS* if you wish to prepare your own reagents.

1. Sodium carbonate (Na_2CO_3) solution, approximately 0.05 *N*.
2. Sulfuric acid (H_2SO_4), 0.1 *N*.
3. Standard sulfuric acid, 0.02 *N*: Dilute 200 m*L* 0.10 *N* standard acid to 1 liter using a volumetric flask. To determine the volume to be diluted, use the following formula:

$$\text{Volume Diluted, m}L = \frac{(\text{Standard, } 0.02\ N) \times (1{,}000\ \text{m}L)}{(\text{Calculated Normality, } 0.10\ N)}$$

$$= \frac{(0.02\ N) \times (1{,}000\ \text{m}L)}{(0.10\ N)}$$

$$= 200\ \text{m}L$$

E. Procedure

Total and Phenolphthalein Alkalinity

1. Take a clean beaker and add 100 m*L* of sample (or other sample volume that will give a titration volume of less than 50 m*L* of acid titrant).
2. Place electrodes of pH meter into beaker containing sample.
3. Stir sample slowly (with a magnetic stirrer if possible).
4. Check pH of sample. If pH is 8.3 or below, then there is no phenolphthalein alkalinity present and you can go to step 6.
5. If the pH is greater than 8.3, titrate very carefully to a pH of 8.3 with 0.02 *N* H_2SO_4. Record the amount of acid used to this point.
6. Continue to titrate to pH 4.5 with 0.02 *N* H_2SO_4.[6] Record the total amount of acid used from starting point to finish.
7. Calculate Total and Phenolphthalein (if present) Alkalinities.

F. Example

Results from alkalinity titrations on a finished water sample were as follows:

Sample size = 100 m*L*

m*L* titrant used to pH 8.3, A = 0.5 m*L*

total m*L* of titrant used, B = 6.8 m*L*

Acid Normality, *N* = 0.02 *N* H_2SO_4

---

[5] *Calcium Carbonate ($CaCO_3$) Equivalent. An expression of the concentration of specified constituents in water in terms of their equivalent value to calcium carbonate. For example, the hardness in water which is caused by calcium, magnesium and other ions is usually described as calcium carbonate equivalent. Alkalinity test results are usually reported as mg/L $CaCO_3$ equivalents. To convert chloride to $CaCO_3$ equivalents, multiply the concentration of chloride ions in mg/L by 1.41, and for sulfate, multiply by 1.04.*

[6] *STANDARD METHODS, 20th Edition, page 2-27, recommends titrating to a pH of 4.5 for routine or automated analyses. However, other pH levels are suggested for various levels of alkalinity.*

(Alkalinity)

OUTLINE OF PROCEDURE FOR ALKALINITY

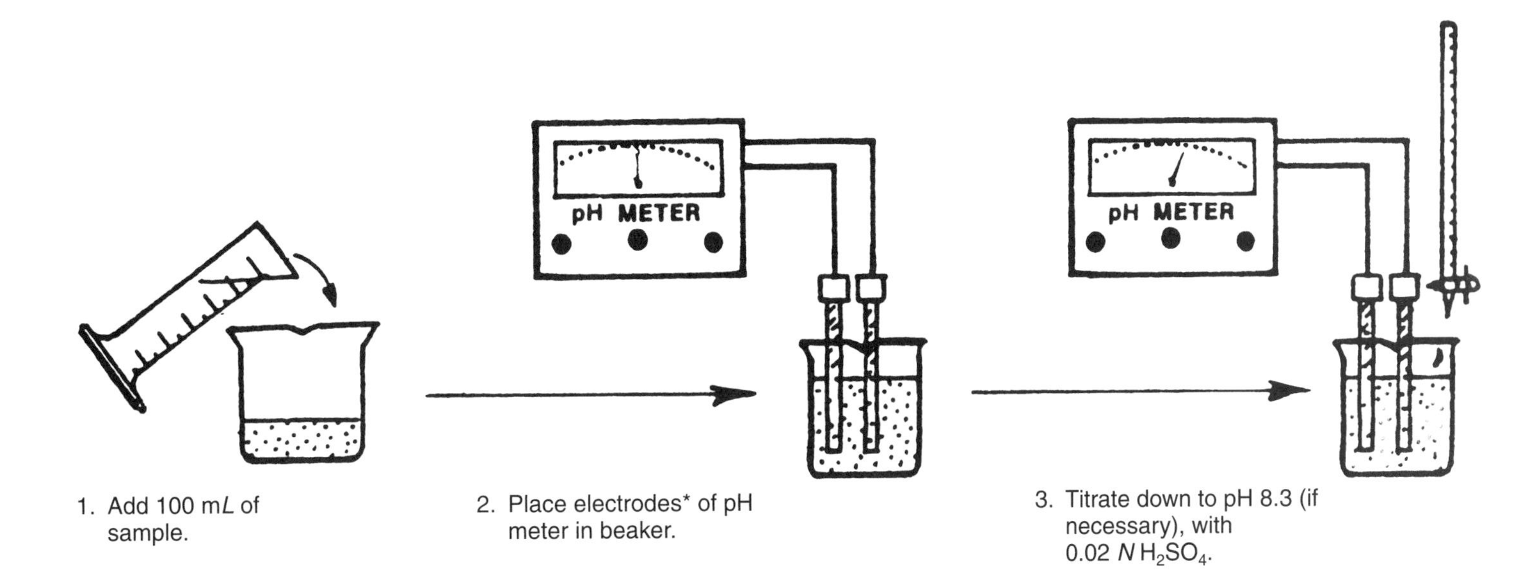

1. Add 100 m*L* of sample.

2. Place electrodes* of pH meter in beaker.

3. Titrate down to pH 8.3 (if necessary), with 0.02 *N* H_2SO_4.

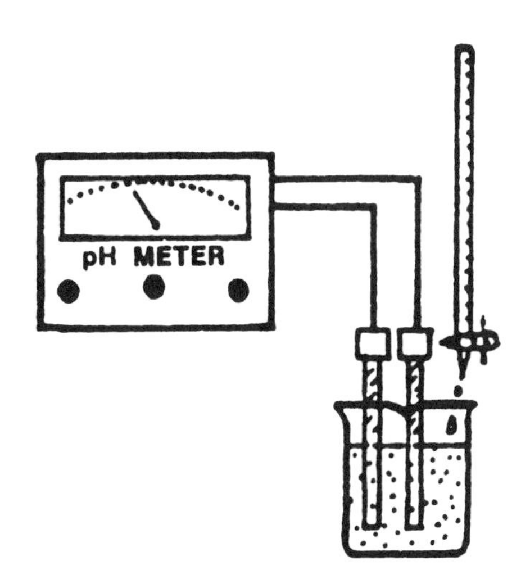

4. Continue to titrate to pH 4.5.

* Some pH meters have a single "combination" electrode.

(Chlorine Residual)

G. Calculations

Phenolphthalein Alkalinity, mg/L as $CaCO_3$

$$= \frac{A \times N \times 50{,}000}{mL \text{ of sample}}$$

$$= \frac{(0.5\ mL) \times (0.02\ N) \times 50{,}000}{100\ mL}$$

$$= 5\ mg/L$$

Total Alkalinity, mg/L as $CaCO_3$

$$= \frac{B \times N \times 50{,}000}{mL \text{ of sample}}$$

$$= \frac{(6.8\ mL) \times (0.02\ N) \times 50{,}000}{100\ mL}$$

$$= 68\ mg/L$$

H. Interpretation of Results

From the test results and the information given below, the different types of alkalinity contained in a water sample can be determined.

Titration Result	Alkalinity, mg/L as $CaCO_3$ Bicarbonate	Carbonate	Hydroxide
P = 0	T	0	0
P is less than 1/2 T	T – 2P	2P	0
P = 1/2 T	0	2P	0
P is greater than 1/2 T	0	2T – 2P	2P – T
P = T	0	0	T

where P = phenolphthalein alkalinity
T = total alkalinity

Example: The example in "G" above gave the following results:

phenolphthalein alkalinity = 5 mg/L

total alkalinity = 68 mg/L

Since the phenolphthalein alkalinity (5 mg/L) is less than one half of the total alkalinity (68 mg/L) from the table, then there is bicarbonate and carbonate alkalinity in the water.

The bicarbonate alkalinity in this case is equal to T – 2P or 68 mg/L – (2 x 5 mg/L) = 58 mg/L as $CaCO_3$.

The carbonate alkalinity is equal to 2P or 2 x 5 mg/L = 10 mg/L as $CaCO_3$.

I. Precautions

1. The sample should be analyzed as soon as possible, at least within a few hours after collection.
2. The sample should not be agitated, warmed, filtered, diluted, concentrated, or altered in any way.

J. Reference

1. See page 2-26, *STANDARD METHODS*, 20th Edition.
2. See page 310-1-1 *METHODS FOR CHEMICAL ANALYSIS OF WATER AND WASTES*, March 1979.

QUESTIONS

Write your answers in a notebook and then compare your answers with those on page 470.

7.3A What chemicals used in water treatment will cause changes in alkalinity?

7.3B Why is it important to know the alkalinity of a water sample?

2. Chlorine Residual

A. Discussion

Chlorine is not only an excellent disinfectant but also serves to react with iron, manganese, protein substances, sulfide, and many taste- and odor-producing compounds to help improve the quality of treated water. In addition, chlorine helps to control microorganisms that might interfere with coagulation and flocculation, keeps filter media free of slime growths, and helps bleach out undesirable color.

There are two general types of residual chlorine produced in chlorinated water. They are (1) free residual chlorine and (2) combined residual chlorine. Free residual chlorine refers to chlorine (Cl_2), hypochlorus acid (HOCl), and the hypochlorite ion (OCl^-). Combined residual chlorine generally refers to the chlorine-ammonia compounds of monochloramine (NH_2Cl), dichloramine ($NHCl_2$), and trichloramine (NCl_3 or nitrogen trichloride). Both types of residuals act as disinfectants, but differ in their capacity to produce a germ-free water supply during the same contact time.

In addition to all the positive aspects of chlorination, there may be some adverse effects. Potentially carcinogenic chlororganic compounds such as chloroform and other THMs may be formed during the chlorination process. To minimize any adverse effects, the operator should be familiar with the concentrations of free and combined residual chlorine produced in a water supply following chlorination. Both residuals are extremely important in producing a potable water that is not only safe to drink but is also free of objectionable tastes and odors.

B. What Is Tested

Source	Common Range Residual Chlorine, mg/L Free	Total
Chlorinated Raw Surface Water (Prechlorination)	0.3 to 3	0.5 to 5
Chlorinated Finished Surface Water (Post-chlorination)	0.2 to 1	0.3 to 1.5
Well Water	0.2 to 1	0.2 to 1

C. Methods

There are eight methods listed for measuring residual chlorine in the 20th Edition of *STANDARD METHODS*. Selection of the most practical and appropriate procedure in any particular instance generally depends upon the characteristics of the water being examined. The *AMPEROMETRIC*[7] titration method is a

[7] *Amperometric (am-PURR-o-MET-rick). A method of measurement that records electric current flowing or generated, rather than recording voltage. Amperometric titration is a means of measuring concentrations of certain substances in water.*

standard of comparison for determining free or combined chlorine residual. This method is relatively free of interferences but does require greater operator skill to obtain good results. In addition, the titration instrument is expensive.

The *DPD*[8] methods are simpler to perform than amperometric titration but are subject to interferences due to manganese. Field comparator kits are available from several suppliers such as Orbeco-Hellige, Wallace & Tiernan, and HACH.

D. Apparatus Required

1. Amperometric Titration Method

   See page 4-58, *STANDARD METHODS*, 20th Edition and amperometric titrator's instruction manual.

2. DPD Colorimetric Method (Field Comparator Kit)

   Field Comparator
   Sample cells

3. DPD Titrimetric Method[9]

   Graduated cylinder (100 m*L*)
   Pipets (1 and 10 m*L*)
   Flask, Erlenmeyer (250 m*L*)
   Buret (10 m*L*)
   Magnetic stirrer
   Magnetic stir-bar
   Balance, analytical

E. Reagents

   NOTE: Prepared reagents may be purchased from laboratory chemical supply houses.

Amperometric Titration Method

1. Standard phenylarsine oxide (PAO) solution, 0.00564 *N* *CAUTION:* Toxic — avoid ingestion.
2. Acetate buffer solution, pH 4.
3. Phosphate buffer solution, pH 7.
4. Potassium iodide solution.

   Store in brown glass, stoppered bottle, preferably in the refrigerator. Discard when solution becomes yellow.

(Chlorine Residual)

DPD Colorimetric Method (Field Comparator Kit)

Use reagents supplied by kit manufacturer.

DPD Titrimetric Method

1. 1 + 3 H_2SO_4. *CAREFULLY* add 10 m*L* concentrated sulfuric acid to 30 m*L* distilled water. Cool.
2. Phosphate Buffer Solution.
3. DPD Indicator Solution.
4. Standard Ferrous Ammonium Sulfate (FAS) Titrant, 0.00282 *N*.
5. Potassium iodide, KI, crystals.

F. Procedure

Amperometric Titration Method

Follow manufacturer's instructions.

DPD Colorimetric Method (Field Comparator Kit)

To measure chlorine residuals you should follow the directions provided by the manufacturer of the equipment or instrument you are using.

If you are disinfecting clear wells, distribution reservoirs, or mains and very high chlorine residuals must be measured, a drop-dilution technique can be used to estimate the chlorine residual. The procedure is as follows:

1. Add 10 m*L* of distilled water and one powder pillow of DPD reagent (or 0.5 m*L* of DPD solution) to the sample tube of the test kit.
2. Add a sample of the water being tested drop by drop to the sample tube until a color is produced.
3. Record the number of drops added to the sample tube. Assume one drop equals 0.05 m*L*.
4. Determine the chlorine residual in the sample as a result of the color produced and record the residual in milligrams per liter.

EXAMPLE 3

The recorded chlorine residual is 0.3 mg/*L*. Two drops of sample produced a chlorine residual of 0.3 mg/*L* in 10 m*L* of distilled water. Assume 0.05 m*L* per drop.

Known		Unknown
Chlorine Residual, mg/*L*	= 0.3 mg/*L*	Actual Chlorine Residual, mg/*L*
Sample Volume, drops	= 2 drops	
Distilled Water, m*L*	= 10 m*L*	

Calculate the actual residual in milligrams per liter.

$$\text{Actual Chlorine Residual, mg}/L = \frac{(\text{Chlorine Residual, mg}/L)(\text{Distilled Water, m}L)}{(\text{Sample Volume, drops})(0.05\text{ m}L/\text{drop})}$$

$$= \frac{(0.3\text{ mg}/L)(10\text{ m}L)}{(2\text{ drops})(0.05\text{ m}L/\text{drop})}$$

$$= 30\text{ mg}/L$$

[8] *DPD (pronounce as separate letters). A method of measuring the chlorine residual in water. The residual may be determined by either titrating or comparing a developed color with color standards. DPD stands for N,N-diethyl-p-phenylene-diamine.*

[9] *Some regulatory agencies require the use of the DPD Titrimetric Method if chlorine residual testing is used in place of some of the coliform tests.*

(Chlorine Residual)

DPD Titrimetric Method

FOR FREE RESIDUAL CHLORINE

1. Place 5 m*L* each of buffer reagent and DPD indicator in a 250-m*L* flask and mix.
2. Add 100 m*L* of sample and mix.
3. Titrate rapidly with standard FAS titrant until red color is discharged (disappears).
4. Record amount of FAS used (Reading A). If combined residual chlorine fractions are desired, continue to next step.

FOR COMBINED RESIDUAL CHLORINE

Monochloramine

5. Add one very small crystal of KI and mix. If monochloramine is present, the red color will reappear.
6. Continue titrating carefully until red color again disappears.
7. Record reading of FAS in the buret used to this point (this includes amount used above). This is Reading B.

Dichloramine

8. Add several crystals KI (about 1 gm) and mix until dissolved.
9. Let stand 2 minutes.
10. Continue titrating until red color again disappears. Record amount of FAS used to this point. This includes amounts used in two previous titrations for free and monochloramine. This is Reading C.

NOTE: Manufacturers of laboratory equipment are continually developing faster and more accurate ways to measure chlorine residual. If you have new equipment, follow the manufacturer's procedures.

G. Precautions

1. For accurate results, careful pH control is essential. The pH of the sample, buffer, and DPD indicator together should be between 6.2 and 6.5.
2. If the sample contains oxidized manganese, an inhibitor must be used.
3. Samples should be analyzed as soon as possible after collection.
4. If nitrogen trichloride or chlorine dioxide are present, special procedures are necessary.
5. Operators using the DPD colorimetric method to test water for a free chlorine residual need to be aware of a potential error that may occur. If the DPD test is run on water containing a combined chlorine residual, a precipitate may form during the test. The particles of precipitated material will give the sample a turbid appearance or the appearance of having color. This turbidity can produce a positive test result for free chlorine residual when there is actually no chlorine present. Operators call this error a "false positive" chlorine residual reading.

H. Example

A sample taken after prechlorination at a filtration plant was tested for residual chlorine using the DPD Titrimetric Method.

m*L* of sample = 100 m*L*
Reading A, m*L* = 1.4 m*L*
Reading B, m*L* = 1.6 m*L*
Reading C, m*L* = 2.7 m*L*

I. Calculation

READING	*CHLORINE RESIDUAL*
A	= mg/*L* free residual chlorine
B – A	= mg/*L* monochloramine
C – B	= mg/*L* dichloramine
C – A	= mg/*L* combined available chlorine
C	= mg/*L* total residual chlorine

Example: The concentrations of the different types of residual chlorine present can be calculated from the information given in (H).

READING		*CHLORINE RESIDUAL*
A	= 1.4 m*L*	= 1.4 mg/*L* free residual chlorine
B – A	= 1.6 – 1.4	= 0.2 mg/*L* monochloramine
C – B	= 2.7 – 1.6	= 1.1 mg/*L* dichloramine
C – A	= 2.7 – 1.4	= 1.3 mg/*L* combined available chlorine
C	= 2.7 m*L*	= 2.7 mg/*L* total residual chlorine

J. Interpretation of Results

1. Any "chlorine" taste and odor that may result from chlorination would generally be from the dichloramine or nitrogen trichloride nuisance residuals.
2. Free residual chlorination produces the best results when the free residual makes up more than 80 percent of the total residual. However, this will not be the case if ammonia is added to the water being treated to form chloramines in order to prevent the formation of THMs.

K. Reference

See page 4-53, *STANDARD METHODS*, 20th Edition.

QUESTIONS

Write your answers in a notebook and then compare your answers with those on page 470.

7.3C List some of the important benefits of chlorinating water.

7.3D What is a potential adverse effect from chlorination?

OUTLINE OF PROCEDURE FOR CHLORINE RESIDUAL

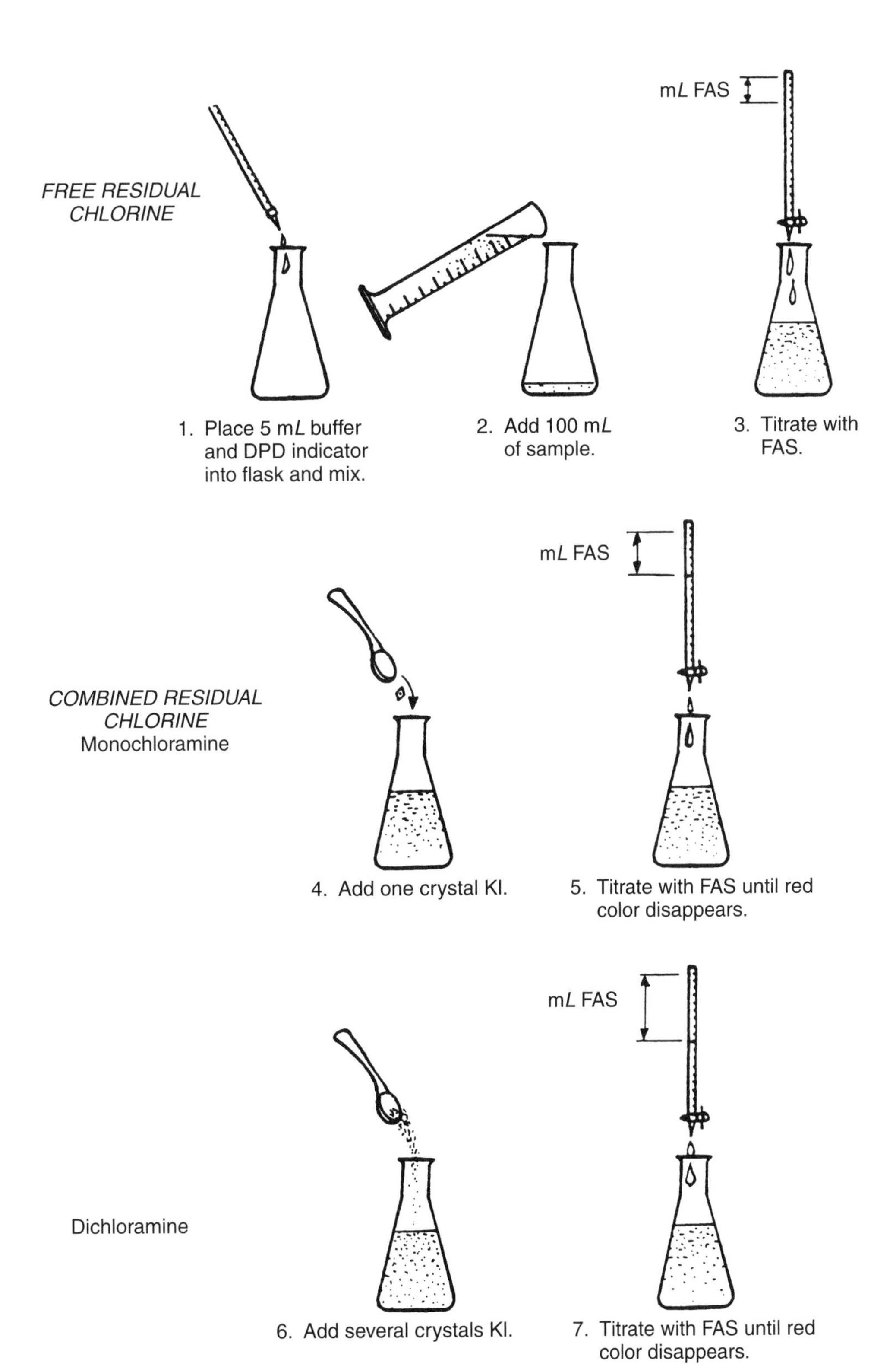
mL FAS
FREE RESIDUAL CHLORINE
1. Place 5 mL buffer and DPD indicator into flask and mix.
2. Add 100 mL of sample.
3. Titrate with FAS.
mL FAS
COMBINED RESIDUAL CHLORINE
Monochloramine
4. Add one crystal KI.
5. Titrate with FAS until red color disappears.
mL FAS
Dichloramine
6. Add several crystals KI.
7. Titrate with FAS until red color disappears.

(Chlorine Demand)

3. Chlorine Demand

A. Discussion

The chlorine demand of water is the difference between the amount of chlorine applied (or dosed) and the amount of free, combined, or total residual chlorine remaining at the end of a specific contact period. The chlorine demand varies with the amount of chlorine applied, length of contact time, pH, and temperature. Also the presence of organics and reducing agents in water will influence the chlorine demand. The chlorine demand test should be conducted with chlorine gas or with granular hypochlorite, depending upon which form you usually use for chlorination.

The chlorine demand test can be used to determine the best chlorine dosage to achieve specific chlorination objectives. The measurement of chlorine demand is performed by treating a series of water samples in question with known but varying amounts of chlorine or hypochlorite. After the desired contact time, calculation of residual chlorine in the samples will demonstrate which dosage satisfied the requirements of the chlorine demand in terms of the residual desired.

B. What Is Tested

	Common Range Chlorine Demand, mg/*L*
Surface Water	0.5 to 5
Well Water	0.1 to 1.3

C. Apparatus

In addition to the apparatus described under one of the methods for chlorine residual, the following items are required:

Flasks, Erlenmeyer (1,000 m*L*)

Pipets (5 and 10 m*L*)

Graduated cylinder (500 m*L*)

Flask, volumetric (1,000 m*L*)

Flask, Erlenmeyer (250 m*L*)

Buret (25 m*L*)

D. Reagents

In addition to the reagents described under the method you will use to determine chlorine residual, the following items are also needed:

1. Stock chlorine solution. Obtain a suitable solution from a chlorinator solution line or by purchasing a bottle of household bleach ("Clorox" or similar product). Store in a dark, cool place to maintain chemical strength. Household bleach products usually contain approximately five percent available chlorine which is about 50,000 mg/*L*.

2. Chlorine dosing solution. If using household bleach, carefully pipet about 10 m*L* of the bleach into a 1,000 m*L* volumetric flask. Fill to the mark with chlorine demand-free water and standardize. If using a stock chlorine solution obtained from a chlorinator solution line, simply standardize this solution directly.

   Standardization:

   a. Place 2 m*L* acetic acid and 20 m*L* chlorine demand-free water in a 250-m*L* flask.

   b. Add about 1 gm KI crystals.

   c. Measure into the flask a suitable volume of chlorine dosing solution. If using household bleach as your stock solution, add 25 m*L* of the dosing solution. Note: In measuring the volume of the dosing solution, notice that 1 m*L* of the 0.025 *N* thiosulfate titrant is equal to 0.9 mg chlorine.

   d. Titrate with standardized 0.025 *N* thiosulfate titrant until the yellow iodine color almost disappears.

   e. Add 1 to 2 m*L* starch indicator solution.

   f. Continue to titrate until blue color disappears.

$$\text{mg}/L \text{ Cl as } Cl_2/\text{m}L = \frac{(\text{m}L \text{ thiosulfate used}) \times N \times 35.45}{\text{m}L \text{ of dosing solution}}$$

where N = normality of thiosulfate titrant

3. Acetic acid, concentrated (glacial).

4. Potassium Iodide, KI, crystals.

5. Standard sodium thiosulfate 0.025 *N*.

6. Chlorine demand-free water. Prepare chlorine demand-free water from good quality distilled or deionized water by adding sufficient chlorine to give 5 mg/*L* free chlorine residual. After standing 2 days, this solution should contain at least 2 mg/*L* free residual chlorine. If not, discard and obtain better quality water. Remove remaining free chlorine by placing the solution in the direct sunlight. After several hours, measure total chlorine residual. Do not use until last trace of chlorine has been removed.

7. Starch indicator.

E. Procedure

1. Measure a 500-m*L* sample into each of five to ten 1,000-m*L* flasks or bottles.

2. To the first flask, add an amount of chlorine that leaves no residual at the end of the contact time. Mix.

3. Add increasing amounts of chlorine to successive portions of the sample and mix.

4. Measure residual chlorine after the specific contact time. Record results.

5. On graph paper, plot the residual chlorine (or the amount of chlorine consumed) versus chlorine dosage.

F. Precautions

1. Dose sample portions at time intervals that will leave you enough time for chlorine residual testing at predetermined contact times.

2. Conduct test over the desired contact time. If test objective is to duplicate your plant contact time, then match plant detention time as closely as possible.

3. Keep samples in the dark, protected from sunlight.

4. Keep temperature as constant as possible.

5. Sterilize all glassware if test has a bacteriologic objective.

G. Example

A raw water sample was collected from a river to determine chlorine demand.

Contact time = 30 minutes (plant detention time)
pH = 7.6
Temperature = 15°C

(Chlorine Demand)

OUTLINE OF PROCEDURE FOR CHLORINE DEMAND

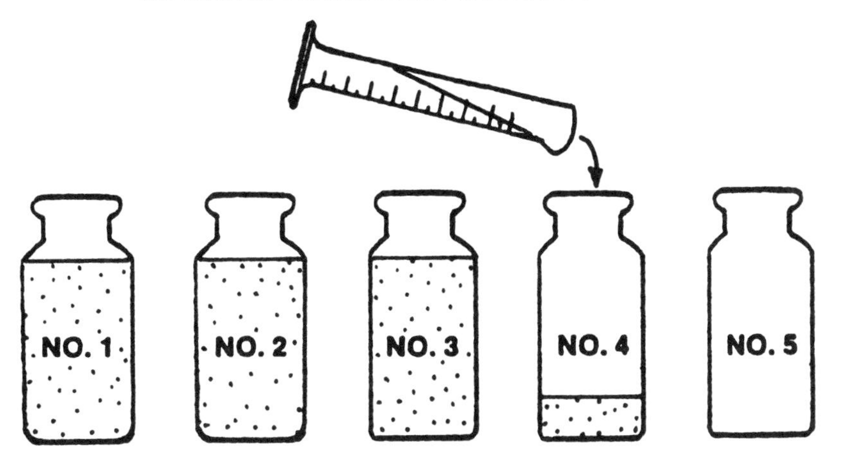

1. Measure 500 m*L* water sample into each container.

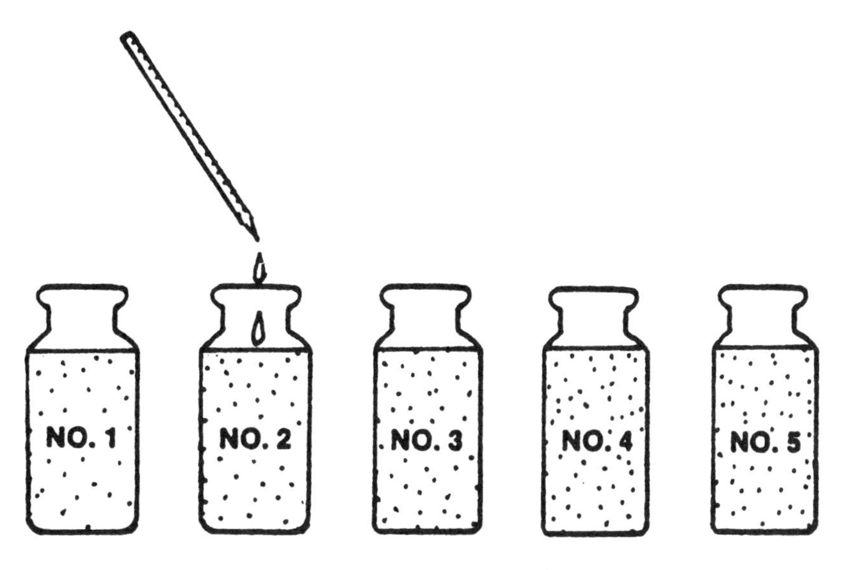

2. Add desired dosage of chlorine to each container.

3. After end of desired contact time, measure chlorine residual. Plot results on graph paper.

(Chlorine Demand)

Results from the chlorine demand test were as follows:

Flask No.	Chlorine Added, mg/L	Total Residual Chlorine after 30 min, mg/L
1	0.5	0.36
2	1.0	0.82
3	1.5	1.14
4	2.0	0.60
5	2.5	0.75
6	3.0	1.25

H. Calculation

Calculate the chlorine demand by using the formula

$$\text{Chlorine Demand, mg/}L = \text{Chlorine Added, mg/}L - \text{Total Residual Chlorine, mg/}L$$

Flask No.	Chlorine Added, mg/L	–	Total Residual Chlorine, mg/L	=	Chlorine Demand, mg/L
1	0.5	–	0.36	=	0.14
2	1.0	–	0.82	=	0.18
3	1.5	–	1.14	=	0.36
4	2.0	–	0.60	=	1.40
5	2.5	–	0.75	=	1.75
6	3.0	–	1.25	=	1.75

Valuable knowledge can be gained by plotting the data from test results. Figure 7.10 is a plot of the chlorine added vs. the free chlorine residual after 30 minutes. By drawing a smooth line between the plotting points, a typical breakpoint chlorination curve is produced.

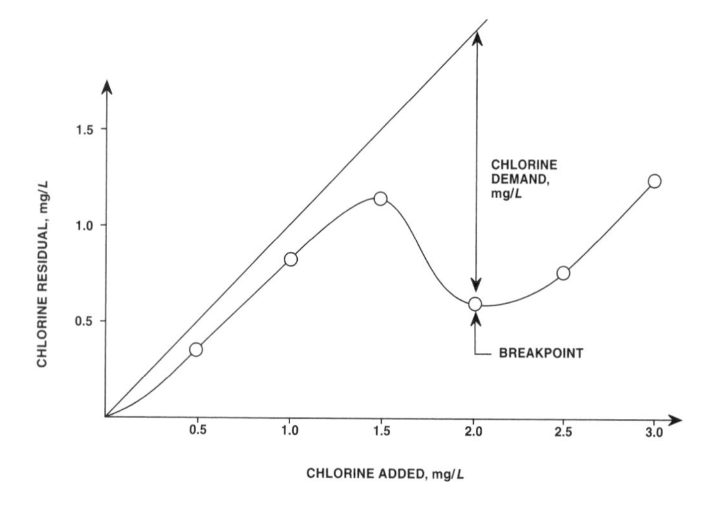

Fig. 7.10 Breakpoint chlorination curve

Figure 7.11 is a plot of the chlorine demand curve. Note that the far right end of the curve is flat which indicates that the chlorine demand has been satisfied. The curve reveals that the chlorine demand will increase as chlorine is added to water until you have gone past the breakpoint.

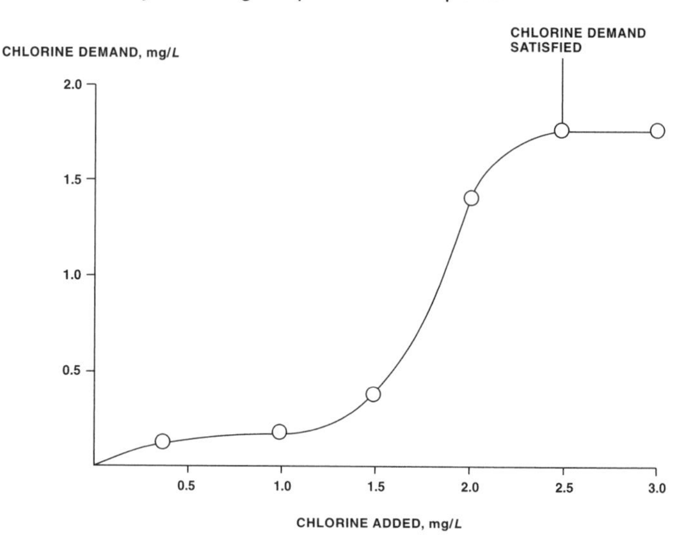

Fig. 7.11 Plot of chlorine demand on graph paper

I. Reference

See page 2-40, *STANDARD METHODS*, 20th Edition.

QUESTIONS

Write your answers in a notebook and then compare your answers with those on page 470.

7.3E What conditions can cause variations in the chlorine demand?

7.3F How does the operator use the results from the chlorine demand test?

7.3G Calculate the chlorine demand for a raw water sample if chlorine is added at 2.0 mg/L and the free residual chlorine after 30 minutes was 0.4 mg/L.

End of Lesson 3 of 5 Lessons on LABORATORY PROCEDURES

Please answer the discussion and review questions next.

DISCUSSION AND REVIEW QUESTIONS

Chapter 7. LABORATORY PROCEDURES

(Lesson 3 of 5 Lessons)

Write the answers to these questions in your notebook before continuing. The question numbering continues from Lesson 2.

12. What precautions must be taken with the sample when conducting the alkalinity test?

13. What could be the causes of tastes and odors that might result from chlorination?

14. What precautions should be exercised when performing a chlorine demand test?

CHAPTER 7. LABORATORY PROCEDURES

(Lesson 4 of 5 Lessons)

4. Coliform Bacteria

A. Discussion

An improperly treated or unprotected water supply may contain microorganisms that are pathogenic, that is, capable of producing disease. Testing for specific pathogenic microorganisms (pathogens which cause diseases such as typhoid, dysentery, cryptosporidiosis, or giardiasis) is very time-consuming and requires special techniques and equipment. So instead of testing for specific pathogens, water is generally analyzed for the presence of an "indicator organism," the coliform group of bacteria.

The presence or absence of coliform bacteria is a good index of the degree of bacteriologic safety of a water supply. In general, coliform bacteria can be divided into fecal or non-fecal groups. Fecal coliform bacteria occur normally in the intestines of humans and other warm-blooded animals. They are discharged in great numbers in human and animal wastes. Coliforms are generally more hardy than true pathogenic bacteria and their absence from water is thus a good indication that the water is bacteriologically safe for human consumption. The presence of coliforms, however, indicates the potential presence of pathogenic organisms that may have entered the water with them, and suggests that the water is unsafe.

The coliform group of bacteria includes all the aerobic and *FACULTATIVE*[10] anaerobic gram-negative, nonspore-forming, rod-shaped bacteria that ferment lactose (a sugar) within 48 hours at 35°C (human body temperature). The fecal coliform can grow at a higher temperature (45°C) than the non-fecal coliform.

The bacteriological quality of water supplies is subject to control by federal, state, and local agencies, all of which are governed by the rules and regulations contained in the Safe Drinking Water Act. This law stipulates the methods to be used, the number of samples required, and the maximum levels allowed for coliform organisms in drinking water supplies. The number of samples required is generally based on population served by the water system.

The primary standards (MCLs) for coliform bacteria have been established to indicate the likely presence of disease-causing bacteria. The Total Coliform Rule uses a presence-absence approach (rather than an estimation of how many coliforms are present) to determine compliance with the standards. The Maximum Contaminant Level Goal (MCLG) for coliform is zero.

Compliance under the Total Coliform Rule is determined on a monthly basis. In general, coliform bacteria must be absent in at least 95 percent of the samples for those larger systems that collect more than 40 samples per month. Smaller systems that collect fewer than 40 samples cannot have coliform-positive results in more than one sample per month.

Whenever a routine coliform sample is coliform-positive, the regulation calls for determination of the presence of fecal coliform bacteria (*E. coli*) and for repeat sampling. Whenever fecal coliform bacteria (*E. coli*) are present in the routine sample and the repeat samples are coliform-positive, there is a violation of the MCL, additional repeat sampling is required, and notification of both the state and the public is required.

Also see *WATER TREATMENT PLANT OPERATION*, Volume II, Sections 22.22, "Microbial Standards," and 22.220, "Total Coliform Rule."

B. Test Methods

Approved water testing procedures for total coliform bacteria are: (1) the Multiple Tube Fermentation Method (sometimes called the Most Probable Number or MPN procedure), (2) the Membrane Filter (MF) Method, (3) Presence-Absence Method (P-A) (see page 455), (4) Colilert™ Method (see page 457), and (5) Colisure Method (see page 459).

1. Multiple Tube Fermentation Method (MPN)

The multiple tube coliform test has been a standard method for determining the coliform group of bacteria since 1936. In this procedure tubes of lauryl tryptose broth are inoculated with a water sample. Lauryl tryptose broth contains lactose which is the source of carbohydrates (sugar). The coliform density is then calculated from statistical probability formulas that predict the most probable number (MPN) of coliforms in a 100-m*L* sample necessary to produce certain combinations of "gas-positive" (gas forming) and gas-negative tubes in the series of inoculated tubes.

There are three distinct test states for coliform testing using the Multiple Tube Fermentation Method — the Presumptive Test, the Confirmed Test, and the Completed Test. Each test makes the coliform test more valid and specific. These tests are described in detail in the following paragraphs.

2. Membrane Filter (MF) Method

This method was introduced as a tentative method in 1955 and has been an approved test for coliform bacteria since 1960. The basic procedure involves filtering a known volume of water through a membrane filter of optimum pore size for full coliform bacteria retention. As the water passes through the microscopic pores, bacteria are entrapped on the upper surface of the filter. The membrane filter is then placed in contact with either a paper pad saturated with liquid medium or directly over an agar (gelatin-like) medium to provide proper nutrients for bacterial growth. Following incubation under pre-

[10] *Facultative (FACK-ul-TAY-tive). Facultative bacteria can use either dissolved molecular oxygen or oxygen obtained from food materials such as sulfate or nitrate ions. In other words, facultative bacteria can live under aerobic or anaerobic conditions.*

(Coliform)

scribed conditions of time, temperature, and humidity, the cultures are examined for coliform colonies that are counted directly and recorded as a density of coliforms per 100 m*L* of water sample.

There are certain important limitations to membrane filter methods. Some types of samples cannot be filtered because of excessive turbidity, high non-coliform bacterial densities, or heavy metal (bactericidal) compounds. In addition, coliforms contained in chlorinated supplies sometimes do not give characteristic reactions on the media and hence special procedures must then be used.

C. What Is Tested

	Common Range Total Coliforms per 100 m*L*
Surface waters	50 to 1,000,000
Treated water supplies	0
Well water	0 to 50

D. Materials Required

1. Sampling Bottles

Bottles of glass or other material which are watertight, resistant to the solvent action of water, and capable of being sterilized may be used for bacteriologic sampling. Plastic bottles made of nontoxic materials are also satisfactory and eliminate the possibility of breakage during transport. The bottles should hold a sufficient volume of sample for all tests, permit proper washing, and maintain the samples uncontaminated until examinations are complete.

Before sterilization by autoclave, add 0.1 m*L* 10 percent sodium thiosulfate per 4-ounce bottle (120 m*L*). This will neutralize a sample containing about 15 mg/*L* residual chlorine. If the residual chlorine is not neutralized, it would continue to be toxic to the coliform organisms remaining in the sample and give false results.

When filling bottles with sample, do not flush out the sodium thiosulfate or contaminate the bottle or sample. Fill bottles approximately three-quarters full and start the test in the laboratory within six hours. If the samples cannot be processed within one hour, they should be held below 10°C for not longer than six hours.

2. Glassware

All glassware must be thoroughly cleansed using a suitable detergent and hot water (160°F or 71°C), rinsed with hot water (180°F or 82°C) to remove all traces of residual detergent, and finally rinsed with distilled or deionized water.

3. Water

Only distilled water or demineralized water which has been tested and found free from traces of dissolved metals and bactericidal and inhibitory compounds may be used for preparation of culture media.

4. Buffered[11] Dilution Water

Prepare stock solution by dissolving 34 grams of KH_2PO_4 in 500 m*L* distilled water, adjusting the pH to 7.2 with 1 *N* NaOH and dilute to one liter. Prepare dilution water by adding 1.25 m*L* of the stock solution and 5.0 m*L* magnesium sulfate (50 grams $MgSO_4 \cdot 7\ H_2O$ dissolved in one liter of water) to 1 liter distilled water. This solution can be dispersed into various size dilution blanks or used as a sterile rinse for the membrane filter test.

5. Media Preparation

Careful media preparation is necessary for meaningful bacteriological testing. Attention must be given to the quality, mixing, and sterilization of the ingredients. The purpose of this care is to ensure that if the bacteria being tested for are indeed present in the sample, every opportunity is presented for the development and ultimate identification. Bacteriological identification is often done by noting changes in the medium; consequently, the composition of the medium must be standardized. Much of the tedium of media preparation can be avoided by purchase of dehydrated media (Difco, BBL, or equivalent) from local scientific supply houses. The operator is advised to make use of these products, and, if only a limited amount of testing is to be done, consider using tubed, prepared media.

MEDIA — MPN (TOTAL COLIFORM)

a. Lauryl Tryptose Broth

For the Presumptive Coliform Test, dissolve the recommended amount of the dehydrated media in distilled water. Dispense solution into fermentation tubes containing an inverted glass vial (see illustration of the tube with vial on page 447). For 10-m*L* water portions from samples, double-strength media is required while all other inoculations require single strength. Directions for preparation are given on the media bottle label.

b. Brilliant Green Bile (BGB) Broth

For the Confirmed Coliform Test, dissolve 40 grams of the dehydrated media in one liter of distilled water. Dispense and sterilize as with Lauryl Tryptose Broth.

MEDIA — MEMBRANE FILTER METHOD (TOTAL COLIFORM)

a. M-Endo Broth

Prepare this media by dissolving 48 grams of the dehydrated product in one liter of distilled water which contains 20 m*L* of ethyl alcohol[12] per liter. Heat solution to boiling only — *DO NOT AUTOCLAVE.* Promptly remove solution from heat and cool. Prepared media should be stored in a refrigerator and used within 96 hours.

b. LES Endo Agar

Prepare this media, used for the two-step procedure, following the instructions on the bottle.

6. Media Storage

Culture media should be prepared in batches of such size that the entire batch will be used in less than one week.

---

[11] *Buffer. A solution or liquid whose chemical makeup neutralizes acids or bases without a great change in pH.*

[12] *In some states the ethyl alcohol used in bacteriological media preparation CANNOT be the specially denatured alcohol sold by supply houses to people without an alcohol permit. One way to obtain suitable ethanol under these circumstances is to buy a brand of ethanol sold for human consumption.*

7. Autoclaving

Steam autoclaves are used for the sterilization of the liquid media and associated apparatus. They sterilize (kill all organisms) at a relatively low temperature of 121°C within 15 minutes using moist heat.

Components of the media, particularly sugars such as lactose, may decompose at higher temperatures or longer heating times. For this reason adherence to time and temperature schedules is vital. The maximum elapsed time for exposure of the media to any heat (from the time the autoclave door is closed to unloading) is 45 minutes. Preheating the autoclave can reduce total heating time.

Autoclaves operate in a manner similar to the familiar kitchen pressure cooker:

a. Water is heated in a boiler to produce steam.

b. The steam is vented to drive out air.

c. The steam vent is closed when the air is gone.

d. Continued heat raises the pressure to 15 lbs/sq in (103.4 kPa or 1.05 kg/sq cm); (at this pressure, pure steam has a temperature of 121°C at sea level only).

e. The heat and pressure are maintained for 15 minutes.

f. The heat is turned off.

g. The steam vent is opened slowly to vent steam until atmospheric pressure is reached. (Fast venting will cause the liquids to boil and overflow tubes.)

h. Sterile material is removed to cool.

In autoclaving fermentation tubes, a vacuum is formed in the inner tubes. As the tubes cool, the inner tubes are filled with sterile medium. Capture of gas in this inner tube from the culture of bacteria is the evidence of fermentation and is recorded as a *POSITIVE TEST.*

E. Procedure for Testing Total Coliform Bacteria

Multiple Tube Fermentation Method

1. General Discussion

Coliform bacteria are detected in water by placing portions of a sample of the water in lauryl tryptose broth. Lauryl tryptose broth is a standard bacteriological media containing lactose (milk) sugar in tryptose broth. The coliform bacteria are those which can grow in this media at 35°C temperature and are able to ferment and produce gas from the lactose within 48 hours. Thus, to detect these bacteria the operator need only inspect the fermentation tubes for gas. A schematic of the coliform test procedure is shown in Figure 7.12.

2. Materials Needed

FOR UNTREATED WATER SAMPLES

a. Fifteen sterile tubes containing 10 m*L* of lauryl tryptose broth are needed for each sample. Use five tubes for each dilution.

b. Dilution tubes or blanks containing 99 m*L* of sterile buffered distilled water.

c. A quantity of 1-m*L* and 10-m*L* serological pipets. The 1-m*L* pipets should be graduated in 0.1-m*L* increments.

d. Incubator set at 35° ± 0.5°C.

e. Thermometer verified to be accurate by comparison with a National Bureau of Standards (NBS) certified thermometer.

FOR DRINKING WATER SAMPLES

a. Five sterile tubes of 10-m*L*, double-strength lauryl tryptose broth are needed if 10 m*L* of sample is added to each tube. Ten m*L* of lauryl tryptose broth is required in all tubes containing one m*L* or less of sample.

b. Sterile 10-m*L* pipet for each sample.

c. Incubator set at 35° ± 0.5°C.

d. Thermometer verified to be accurate by comparison with a National Bureau of Standards (NBS) certified thermometer, or equivalent.

3. Technique for Inoculation of Sample (Figures 7.13 and 7.14)

All inoculations and dilutions of water samples must be accurate and made so that no contaminants from the air, equipment, clothes, or fingers reach the sample, either directly or by way of a contaminated pipet. Clean, sterile pipets must be used for each separate sample.

FOR UNTREATED WATER SAMPLES

a. Shake the sample bottle vigorously 20 times before removing sample volumes.

b. Pipet 10 m*L* of sample directly into each of the first five tubes. Each tube must contain 10 m*L* lauryl tryptose broth (double strength).

NOTE: You must realize that the sample volume applied to the first five tubes will depend upon the type of water being tested. The sample volume applied to each tube can vary from 10 m*L* for high-quality waters to as low as 0.00001 m*L* (applied as 1 m*L* of a diluted sample) for highly polluted raw water samples.

NOTE: When delivering the sample into the culture medium, deliver sample portions of 1 m*L* or less down into the culture tube near the surface of the medium. *DO NOT* deliver small sample volumes at the top of the tube and allow them to run down inside the tube; too much of the sample will fail to reach the culture medium.

NOTE: Use 10-m*L* pipets for 10-m*L* sample portions, and 1-m*L* pipets for portions of 1 m*L* or less. Handle

(Coliform)

OUTLINE OF PROCEDURE FOR TOTAL COLIFORM

1. Presumptive Test

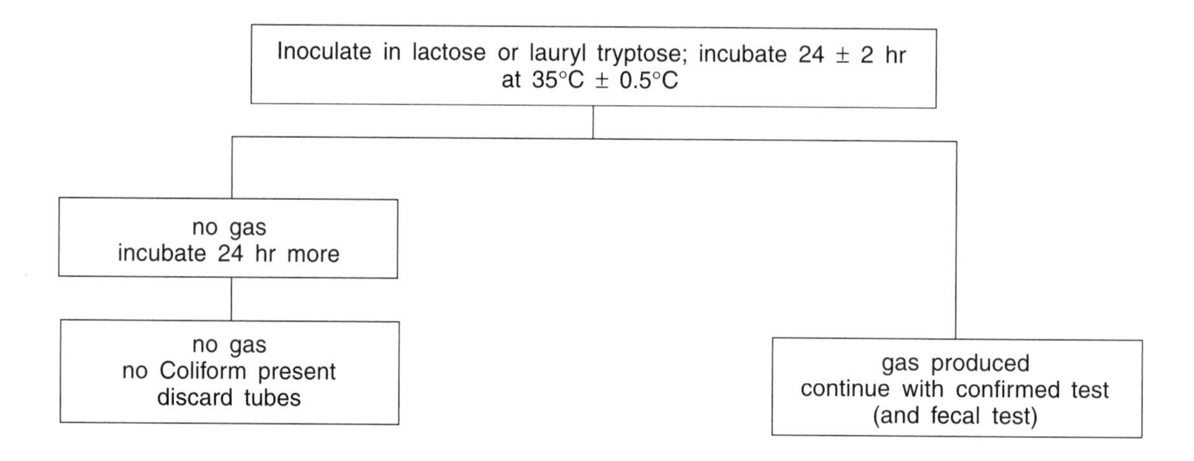

2. Confirmed Test

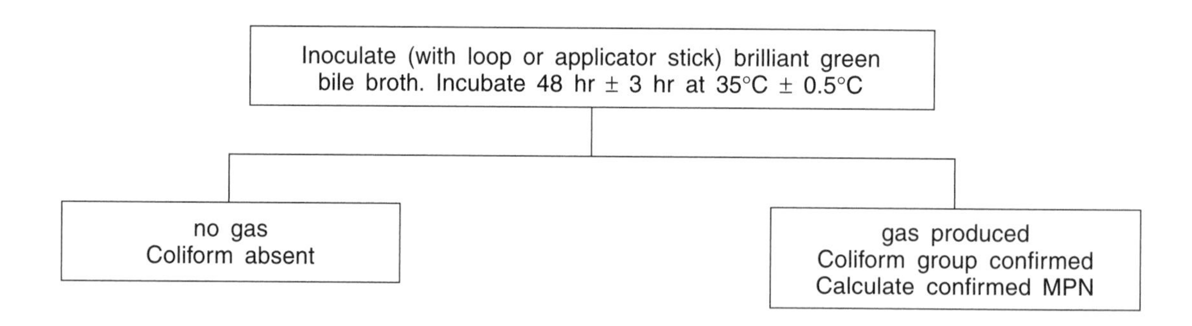

Fig. 7.12 Schematic outline of test procedure for Total Coliform — Multiple Tube Fermentation Method

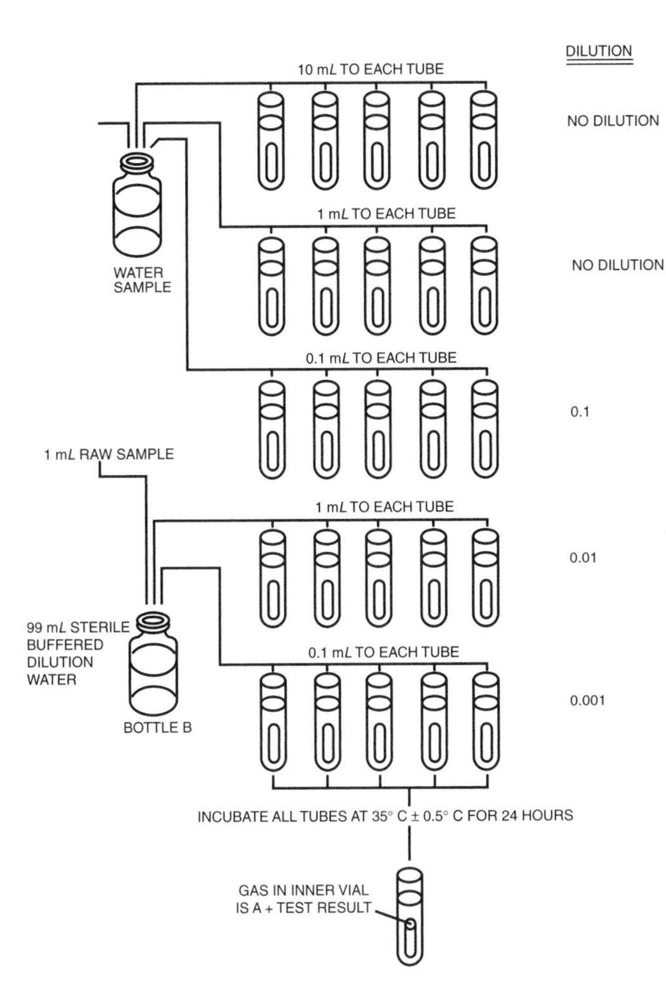

Fig. 7.13 Coliform bacteria test — raw water

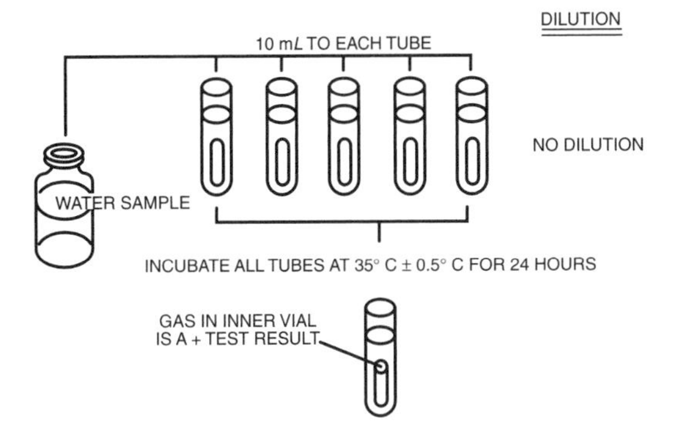

Fig. 7.14 Coliform bacteria test — drinking water

sterile pipet only near the mouthpiece, and protect the delivery end from external contamination.

c. Pipet 1 m*L* of water sample into each of the next five lauryl tryptose broth (single strength) tubes.

d. Pipet 1/10 m*L* (0.1 m*L*) of water sample into each of the next five lauryl tryptose broth (single strength) tubes. This makes a 0.1 (1 to 10) dilution.

At this point you have 15 tubes inoculated and can place these three sets of tubes in the incubator; however, your sample specimen may show gas production in all fermentation tubes. This means your sample was not diluted enough and you have no usable results.

e. To make a 1/100 (0.01) dilution, add 1 m*L* of well-mixed water sample to 99 m*L* of sterile buffered dilution water. Mix thoroughly by shaking. Add 1 m*L* from this bottle directly into each of five more lauryl tryptose broth tubes.

f. To make a 1/1,000 (0.001) dilution, add 0.1 m*L* from the 1/100 dilution bottle directly into each of five more lauryl tryptose broth tubes.

The first time a sample is analyzed, 25 tubes of lauryl tryptose broth should be prepared. Once you find out what dilutions give usable results for determining the MPN index, you will only need to prepare 15 tubes to analyze subsequent samples from the same source.

FOR DRINKING WATER SAMPLES

a. Shake the sample bottle vigorously 20 times before removing a sample volume.

b. Pipet 10 m*L* of sample directly into each of five tubes containing 10 m*L* of double-strength lauryl tryptose broth.

4. Incubation (Total Coliform)

a. 24-Hour Lauryl Tryptose (LT) Broth Presumptive Test

Place all inoculated LT broth tubes in 35°C ± 0.5°C incubator. After 24 ± 2 hours have elapsed, examine each tube for gas formation in inverted vial (inner tube). Mark plus (+) on report form such as shown in Figure 7.15 for all tubes that show presence of gas. Mark minus (–) for all tubes showing no gas formation. Immediately perform Confirmation Test on all positive (+) tubes (see paragraph c. below, "24-Hour Brilliant Green Bile (BGB) Confirmation Test"). The negative (–) tubes must be reincubated for an additional 24 hours.

b. 48-Hour Lauryl Tryptose Broth Presumptive Test

Record both positive and negative tubes at the end of 48 ± 3 hours. Immediately perform Total Coliform Confirmation Test on all new positive tubes.

c. 24-Hour Brilliant Green Bile (BGB) Confirmation Test

Confirm all presumptive tubes that show gas at 24 or 48 hours. Transfer, with the aid of a sterile 3-mm platinum wire loop (sterile wood applicator or disposable loops may be used also), one loop-full of the broth from the lauryl tryptose broth tubes showing gas, and inoculate a corresponding tube of BGB broth by mixing the loop of broth in the BGB broth. Discard all positive lauryl tryptose broth tubes after transferring is completed.

Always sterilize inoculation loops and needles in flame immediately before transfer of culture; do not lay loop down or touch it to any nonsterile object before making the transfer. After sterilization in a flame, allow sufficient time for cooling, in the air, to prevent the heat of the loop from killing the bacterial cells being transferred. Sterile wood applicator sticks also are used to transfer cultures, especially in the field where a flame is not available for sterilization. If using hardwood applicators, sterilize by autoclaving before use and discard after each transfer.

(Coliform)

FOR UNTREATED WATER

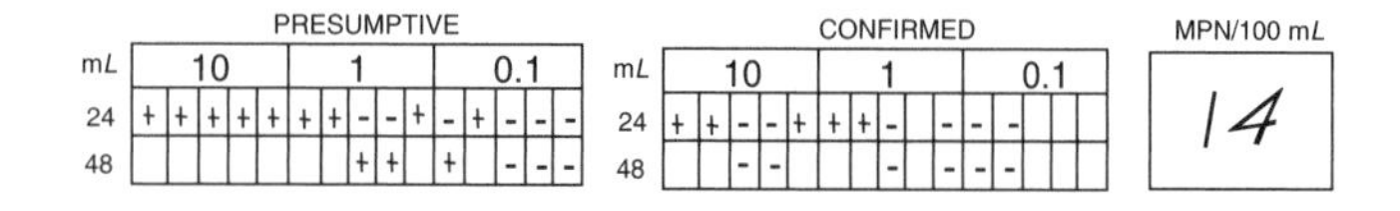

mL = mL OF SAMPLE: 10 mL, 1 mL, AND 0.1 mL
24 = RESULTS AFTER 24 HOURS OF INCUBATION
48 = RESULTS AFTER 48 HOURS OF INCUBATION

FOR DRINKING WATER

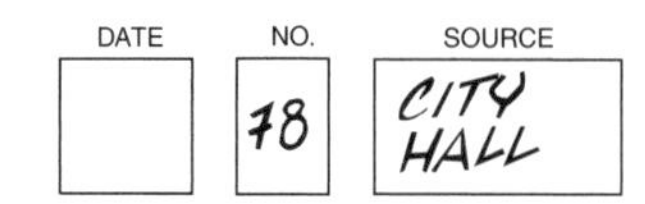

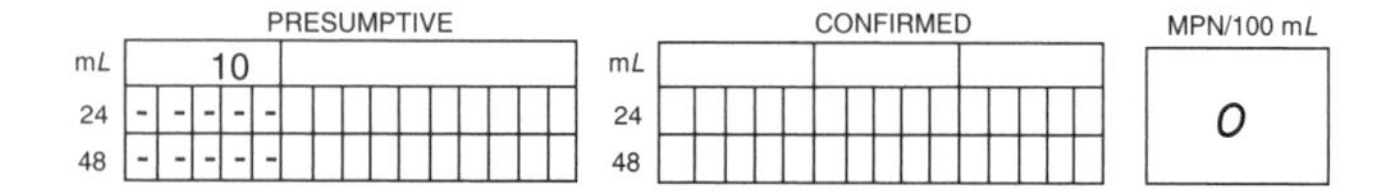

Fig. 7.15 Recorded coliform test results

After 24 hours have elapsed, inspect each of the BGB tubes for gas formation. Those with any amount of gas are considered positive and are so recorded on the data sheet. Negative BGB tubes are reincubated for an additional 24 hours.

d. 48-Hour Brilliant Green Bile Confirmation Test

(1) Examine tubes for gas at the end of the 48 ± 3 hour period. Record both positive and negative tubes.
(2) Complete reports by determining MPN Index and recording MPN on worksheets.

5. Recording Results

Results should be recorded on data sheets prepared especially for this test. Examples are shown in Figure 7.15.

6. Method of Calculating the Most Probable Number (MPN)

Table 7.3 is used to estimate the Most Probable Number (MPN) of coliforms per 100 mL in a sample. The numbers are calculated using a statistical probability formula. *NOTE:* Table 7.3 is extracted from the most recent edition of "Standard Methods." Some facilities may be obligated to use the MPN Index provided in a previous edition. Check with your local regulatory agency for the correct edition.

To estimate the MPN, select the highest dilution or inoculation with all positive tubes, before a negative tube occurs, plus the next two dilutions.

EXAMPLE 1 (UNTREATED WATER) — Select the underlined inoculations.

mL of sample	10	1	0.1	0.01	0.001
dilutions	0	0	–1	–2	–3
positive tubes	5	1	0	0	0

Read MPN as 30 per 100 mL from Table 7.3
Report MPN as 30 per 100 mL

EXAMPLE 2 (UNTREATED WATER) — Select the underlined inoculations (see Figure 7.16).

mL of sample	10	1	0.1	0.01	0.001
dilutions	0	0	–1	–2	–3
positive tubes	5	5	2	0	0

Read MPN as 50 per 100 mL from Table 7.3
Report results as 500/100 mL

We added one zero to 50 because we started with the 1 mL sample and Table 7.3 begins with one dilution column to the left.

For Example No. 2

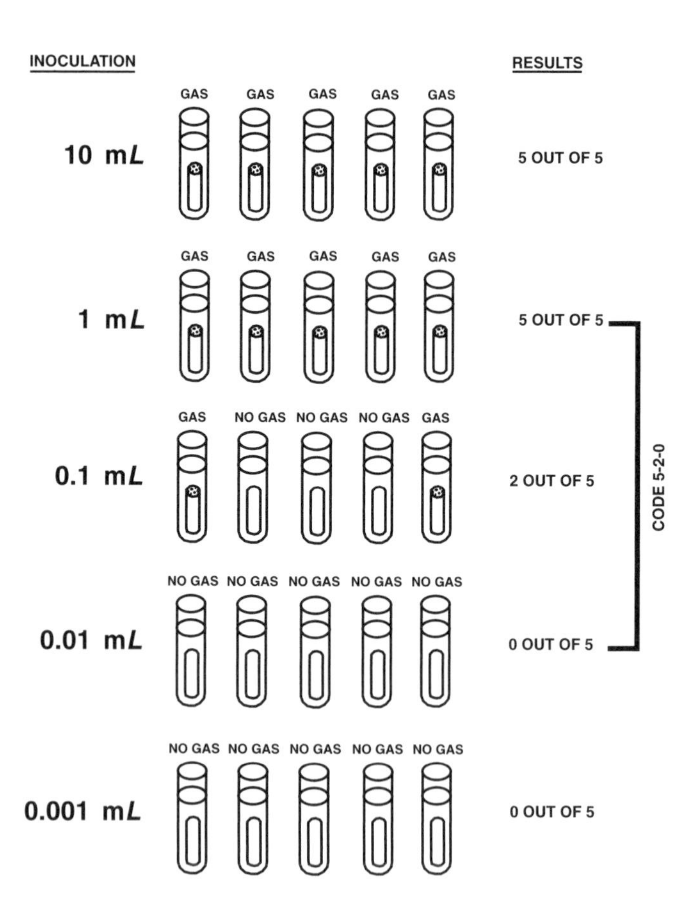

Fig. 7.16 Results of coliform test — untreated water

(Coliform)

TABLE 7.3 MPN INDEX FOR VARIOUS COMBINATIONS OF POSITIVE AND NEGATIVE RESULTS IN A PLANTING SERIES OF FIVE 10-mL, FIVE 1-mL, AND FIVE 0.1-mL PORTIONS OF SAMPLE

Number of tubes giving positive reaction out of			MPN Index
Five 10-mL portions	Five 1-mL portions	Five 0.1-mL portions	(organisms) per 100 mL
0	0	0	<2
0	0	1	2
0	0	2	4
0	1	0	2
0	1	1	4
0	1	2	5
0	2	0	4
0	2	1	6
0	3	0	6
1	0	0	2
1	0	1	4
1	0	2	6
1	0	3	8
1	1	0	4
1	1	1	6
1	1	2	8
1	2	0	6
1	2	1	8
1	2	2	10
1	3	0	8
1	3	1	10
1	4	0	11
2	0	0	4
2	0	1	7
2	0	2	9
2	0	3	12
2	1	0	7
2	1	1	9
2	1	2	12
2	2	0	9
2	2	1	12
2	2	2	14
2	3	0	12
2	3	1	14
2	4	0	15
3	0	0	8
3	0	1	11
3	0	2	13
3	1	0	11
3	1	1	14
3	1	2	17
3	1	3	20
3	2	0	14
3	2	1	17
3	2	2	20

(Coliform)

TABLE 7.3 MPN INDEX FOR VARIOUS COMBINATIONS OF POSITIVE AND NEGATIVE RESULTS IN A PLANTING SERIES OF FIVE 10-m*L*, FIVE 1-m*L*, AND FIVE 0.1-m*L* PORTIONS OF SAMPLE (continued)

Number of tubes giving positive reaction out of			MPN Index
Five 10-m*L* portions	Five 1-m*L* portions	Five 0.1-m*L* portions	(organisms) per 100 m*L*
3	3	0	17
3	3	1	21
3	4	0	21
3	4	1	24
3	5	0	25
4	0	0	13
4	0	1	17
4	0	2	21
4	0	3	25
4	1	0	17
4	1	1	21
4	1	2	26
4	2	0	22
4	2	1	26
4	2	2	32
4	3	0	27
4	3	1	33
4	3	2	39
4	4	0	34
4	4	1	40
4	5	0	41
4	5	1	48
5	0	0	23
5	0	1	30
5	0	2	40
5	0	3	60
5	0	4	80
5	1	0	30
5	1	1	50
5	1	2	60
5	1	3	80
5	2	0	50
5	2	1	70
5	2	2	90
5	2	3	130
5	2	4	150
5	2	5	180
5	3	0	80
5	3	1	110
5	3	2	140
5	3	3	170
5	3	4	210
5	3	5	250
5	4	0	130
5	4	1	170
5	4	2	220
5	4	3	280

(Coliform)

TABLE 7.3 MPN INDEX FOR VARIOUS COMBINATIONS OF POSITIVE AND NEGATIVE RESULTS IN A PLANTING SERIES OF FIVE 10-m*L*, FIVE 1-m*L*, AND FIVE 0.1-m*L* PORTIONS OF SAMPLE (continued)

Number of tubes giving positive reaction out of			MPN Index
Five 10-m*L* portions	Five 1-m*L* portions	Five 0.1-m*L* portions	(organisms) per 100 m*L*
5	4	4	350
5	4	5	430
5	5	0	240
5	5	1	300
5	5	2	500
5	5	3	900
5	5	4	1,600
5	5	5	≥1,600

EXAMPLE 3 (DRINKING WATER SAMPLE) — (see Figure 7.17).

m*L* of sample 10
positive tubes 1

Read MPN as 2.2 per 100 m*L* from Table 7.4

For Example No. 3

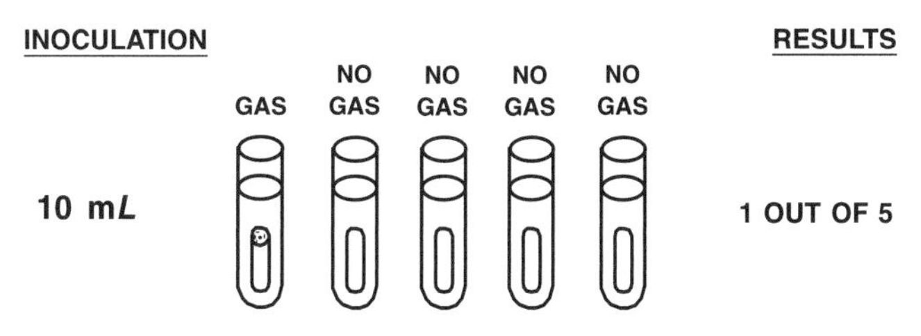

Fig. 7.17 Results of coliform test — drinking water

TABLE 7.4 MPN INDEX FOR VARIOUS COMBINATIONS OF POSITIVE AND NEGATIVE RESULTS WHEN FIVE 10-m*L* PORTIONS ARE USED

Number of tubes giving positive reaction out of five 10-m*L* portions	MPN Index per 100 m*L*
0	less than 2.2
1	2.2
2	5.1
3	9.2
4	16
5	greater than 16

When you wish to summarize with a single MPN value the results from a series of samples, the arithmetic mean, the median, or the geometric mean may be used (see following sections for *FORMULAS* and *EXAMPLES*).

FORMULAS

The MPN for combinations not appearing in the table given or for other combinations of tubes or dilutions may be estimated by the formula:

$$\text{MPN/100 m}L = \frac{(\text{No. of positive tubes})(100)}{\sqrt{\left(\text{m}L\text{ sample in negative tubes}\right) \times \left(\text{m}L\text{ sample in all tubes}\right)}}$$

Coliform results may be summarized with a single MPN value by using the arithmetic mean, median, or geometric mean. The arithmetic mean is often called the average.

$$\text{Mean} = \frac{\text{Sum of Items of Values}}{\text{Number of Items or Values}}$$

The median is the middle value in a set or group of data. There are just as many values larger than the median as there are smaller than the median. To determine the median, the data should be written in ascending (increasing) or descending (decreasing) order and the middle value identified.

Median = Middle Value of a Group of Data

The median and geometric mean are used when a sample contains a few very large values. This frequently happens when measuring the MPN of raw water. If **X** is a measurement (MPN) and **n** is the number of measurements, then

$$\text{Geometric Mean} = [(X_1)(X_2)(X_3) \ldots\ldots (X_n)]^{1/n}$$

This calculation can be easily performed on many electronic calculators.

EXAMPLE 4

The results from an MPN test using five fermentation tubes for each dilution on a sample of water were as follows:

Sample Size, m*L*	10	1.0	0.1
Number of Positive Tubes Out of Five Tubes	3+	1+	0+

(Coliform)

Determine the information necessary to solve the formula:

$$\text{MPN/100 mL} = \frac{(\text{No. of positive tubes})(100)}{\sqrt{\begin{pmatrix}\text{mL sample in}\\ \text{negative tubes}\end{pmatrix} \times \begin{pmatrix}\text{mL sample in}\\ \text{all tubes}\end{pmatrix}}}$$

1. Number of positive tubes. There were 3 positive tubes with 10 m*L* of sample and 1 positive tube with 1 m*L* of sample. Therefore, 3 + 1 = 4 positive tubes.

2. Determine the m*L* of sample in negative tubes.

Sample Size, m*L*	Number of Negative Tubes	m*L* of Sample in Negative Tubes		
10 m*L*	2–	(10 m*L*)(2)	=	20 m*L*
1.0 m*L*	4–	(1.0 m*L*)(4)	=	4 m*L*
0.1 m*L*	5–	(0.1 m*L*)(5)	=	0.5 m*L*
		Total	=	24.5 m*L*

3. Determine the m*L* of sample in all tubes.

Sample Size, m*L*	Number of Tubes	m*L* of Sample in All Tubes		
10 m*L*	5	(10 m*L*)(5)	=	50 m*L*
1.0 m*L*	5	(1.0 m*L*)(5)	=	5 m*L*
0.1 m*L*	5	(0.1 m*L*)(5)	=	0.5 m*L*
		Total	=	55.5 m*L*

4. Estimate the MPN/100 m*L*.

$$\text{MPN/100 mL} = \frac{(\text{No. of positive tubes})(100)}{\sqrt{\begin{pmatrix}\text{mL sample in}\\ \text{negative tubes}\end{pmatrix} \times \begin{pmatrix}\text{mL sample in}\\ \text{all tubes}\end{pmatrix}}}$$

$$= \frac{(4)(100)}{\sqrt{(24.5\text{ mL})(55.5\text{ mL})}}$$

$$= \frac{400}{36.87}$$

$$= 10.8\text{ MPN Coliforms/100 mL}$$

$$= 11\text{ MPN Coliforms/100 mL}$$

EXAMPLE 5

Results from MPN tests during one week were as follows:

Day	S	M	T	W	TH	F	S
MPN/100 m*L*	2	8	14	6	10	26	4

Estimate the (1) mean, (2) median, and (3) geometric mean of the data in MPN/100 m*L*.

1. Calculate the mean.

$$\text{Mean, MPN/100 mL} = \frac{\text{Sum of All MPNs}}{\text{Number of MPNs}}$$

$$= \frac{2 + 8 + 14 + 6 + 10 + 26 + 4}{7}$$

$$= \frac{70}{7}$$

$$= 10\text{ MPN/100 mL}$$

2. To determine the median, rearrange the data in ascending (increasing) order and select the middle value (three will be smaller and three will be larger in this example).

Order	1	2	3	4	5	6	7
MPN/100 m*L*	2	4	6	8 ↑	10	14	26

Median, MPN/100 m*L* = Middle value of a group of data

= 8 MPN/100 m*L*

3. Calculate the geometric mean for the given data.

$$\text{Geometric Mean, MPN/100 mL} = [(X_1)(X_2)(X_3)(X_4)(X_5)(X_6)(X_7)]^{1/7}$$

$$= [(2)(8)(14)(6)(10)(26)(4)]^{1/7}$$

$$= [1{,}397{,}760]^{1/7}$$

$$= 7.5\text{ MPN/100 mL}$$

4. SUMMARY

   1. Mean = 10 MPN/100 m*L*
   2. Median = 8 MPN/100 m*L*
   3. Geometric Mean = 7.5 MPN/100 m*L*

As you can see from the summary, the geometric mean more nearly describes most of the MPNs. This is the reason why the geometric mean is sometimes used to describe the results of MPN tests when there are a few very large values.

Membrane Filter Method

1. General Discussion

In addition to the fermentation tube test for coliform bacteria, another test is used for these same bacteria in water analysis. This test uses a cellulose ester filter, called a membrane filter, the pore size of which can be manufactured to close tolerances. Not only can the pore size be made to selectively trap bacteria from water filtered through the membrane, but nutrients can be diffused (from an enriched pad) through the membrane to grow these bacteria into colonies. These colonies are recognizable as coliform because the nutrients include fuchsin dye which peculiarly colors the colony. Knowing the number of colonies and the volume of water filtered, the operator can then compare the water tested with water quality standards.

A two-step pre-enrichment technique is included at the end of this section for samples which have been chlorinated. Chlorinated bacteria which are still living have had their enzyme systems damaged and require a 2-hour enrichment media before contact with the selective M-Endo Media.

2. Materials Needed

   a. One sterile membrane filter having a 0.45μ pore size.
   b. One sterile 47-mm petri dish with lid.
   c. One sterile funnel and support stand.
   d. Two sterile pads.

e. One receiving flask (side-arm, 1,000 m*L*).

f. Vacuum pump, trap, suction or vacuum gage, connection sections of plastic tubing, glass "T" hose clamp to adjust pressure bypass.

g. Forceps (round-tipped tweezers), alcohol, Bunsen burner, grease pencil.

h. Sterile, buffered, distilled water for rinsing.

i. M-Endo Media.

j. Sterile pipets — two 5-m*L* graduated pipets and one 1-m*L* pipet for sample or one 10-m*L* pipet for larger sample. Quantity of 1-m*L* pipets if dilution of sample is necessary. Also, quantity of dilution water blanks if dilution of sample is necessary.

k. One moist incubator at 35°C; auxiliary incubator dish with cover.

l. Enrichment media — lauryl tryptose broth (for pre-enrichment technique).

m. A binocular, wide-field, dissecting microscope is recommended for counting. The light source should be a cool white, fluorescent lamp.

3. Selection of Sample Size

Size of the sample will be governed by the expected bacterial density. An ideal quantity will result in the growth of 20 to 80 coliform colonies, but not more than 200 bacterial colonies of all types. The table below lists suggested sample volumes for MF total coliform testing.

	Quantities Filtered (m*L*)				
	100	10	1	0.1	0.01
Well Water	x				
Drinking Water	x				
Lakes	x	x	x		
Rivers		x	x	x	x

When less than 20 m*L* of sample is to be filtered, a small amount of sterile dilution water should be added to the funnel before filtration. This increase in water volume aids in uniform dispersion of the sample over the membrane filter.

4. Preparation of Petri Dish for Membrane Filter

a. Sterilize forceps by dipping in alcohol and passing quickly through Bunsen burner flame to burn off the alcohol. An alcohol burner may be used also.

b. Place sterile absorbent pad into sterile petri dish.

c. Add 1.8 to 2.0 m*L* M-Endo Medium to absorbent pad using a sterile pipet. Remove excess media.

5. Procedure for Filtration of Unchlorinated Samples

All filtrations and dilutions of water specimens must be accurate and should be made so that no contaminants from the air, equipment, clothes, or fingers reach the specimen either directly or by way of the contaminated pipet.

a. Secure tubing from pump and bypass to receiving flask. Place palm of hand on flask opening and start pump. Adjust suction to 1/4 atmosphere with hose clamp on pressure bypass. Turn pump switch to OFF.

b. Set sterile filter support stand and funnel on receiving flask. Loosen wrapper. Rotate funnel counterclockwise to disengage pin. Secure wrapper.

c. Place petri dish on bench with lid up. Write identification on lid with grease pencil.

d. Unwrap sterile pad container. Light Bunsen burner.

e. Unwrap membrane filter container.

f. Sterilize forceps by dipping in alcohol and passing quickly through Bunsen burner to burn off the alcohol.

g. Center membrane filter on filter stand with forceps after lifting funnel. Membrane filter with printed grid should show grid uppermost (Fig. I, next page).

h. Replace funnel and lock against pin (Fig. II).

i. Shake sample or diluted sample. Measure proper *ALIQUOT*[13] with sterile pipet and add to funnel.

j. Add a small amount of the sterile dilution water to funnel. This will help check for leakage and also aid in dispersing small volumes (Fig. III).

k. Now start vacuum pump.

l. Rinse filter with three 20- to 30-m*L* portions of sterile dilution water.

m. When membrane filter appears barely moist, switch pump to OFF.

n. Sterilize forceps as before.

o. Remove membrane filter with forceps after first removing funnel as before (Fig. I).

p. Center membrane filter on pad containing M-Endo Media with a rolling motion to ensure water seal. Inspect membrane to ensure no captured air bubbles are present (Fig. IV).

q. Place *INVERTED* petri dish in incubator for 22 ± 2 hours. Incubate at 35°C.

6. Procedure for Counting Membrane Filter Colonies

a. Remove petri dish from incubator.

b. Remove lid from petri dish.

c. Turn so that your back is to window.

d. Tilt membrane filter in base of petri dish so that green and yellow-green colonies are most apparent. Direct sunlight has too much red to facilitate counting.

e. Count individual colonies using an overhead fluorescent light. Use a low-power (10 to 15 magnifications), binocular, wide-field, dissecting microscope or other similar optical device. The typical colony has a pink to dark red color with a metallic surface sheen. The

[13] *Aliquot (AL-li-kwot). Portion of a sample. Often an equally divided portion of a sample.*

(Coliform)

OUTLINE OF PROCEDURE FOR INOCULATION OF MEMBRANE FILTER

Fig. I

1. Center membrane filter on filter holder. Handle membrane only on outer 3/16 inch with forceps sterilized before use in ethyl or methyl alcohol and passed lightly through a flame.

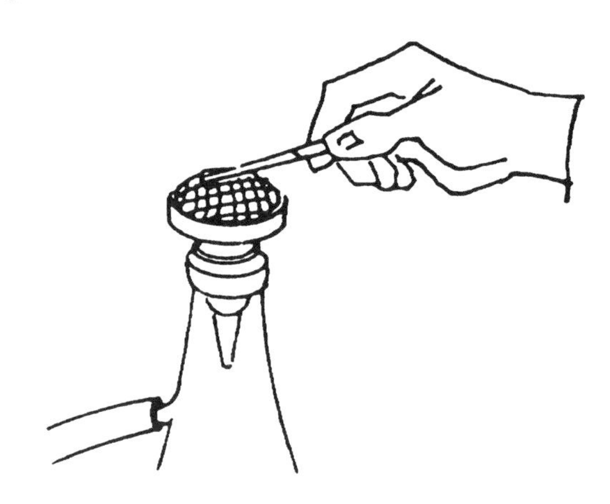

Fig. II

2. Place funnel onto filter holder.

Fig. III

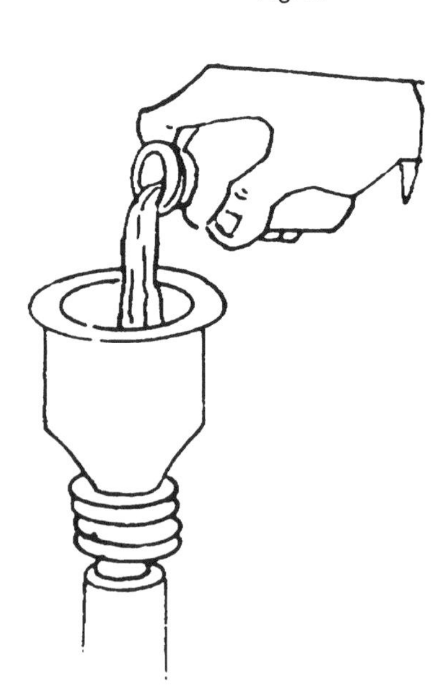

3. Pour or pipet sample aliquot into funnel. Avoid spattering. After suction is applied, rinse four times with sterile, buffered, distilled water.

Fig. IV

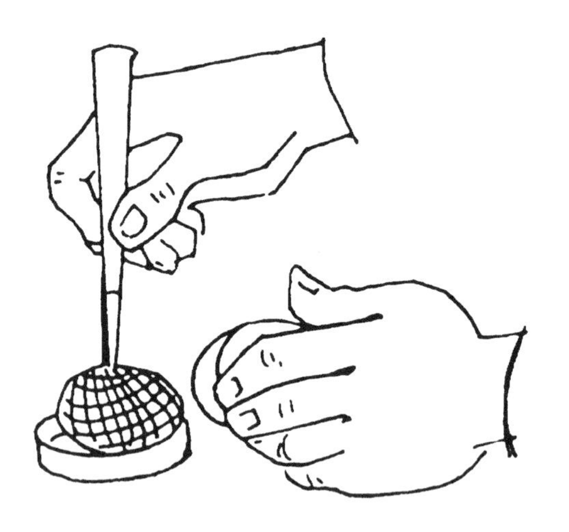

4. Remove membrane filter from filter holder with sterile forceps. Place membrane on pad. Cover with petri top.

Fig. V

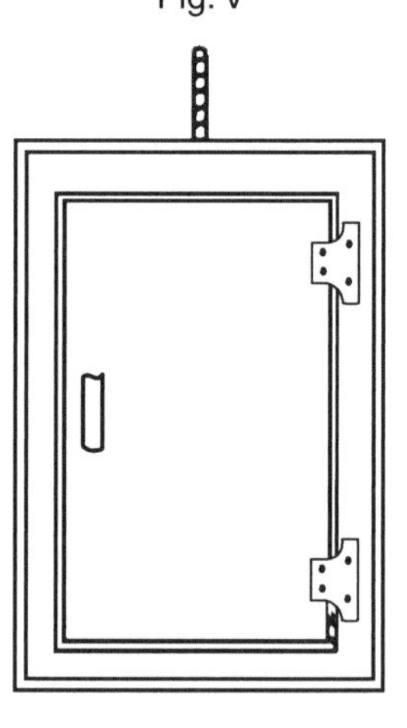

5. Incubate in inverted position for 22 ± 2 hours.
6. Count coliform-appearing colonies on membrane.

sheen area may vary from a small pin-head size to complete coverage of the colony surface. Only those showing this sheen should be counted.

f. Report total number of "coliform colonies" on worksheet. Results are often reported as "colony formation units" or CFUs. Use the membranes that show from 20 to 90 colonies and do not have more than 200 colonies of all types (including non-sheen or, in other words, non-coliforms).

EXAMPLE:

A total of 42 colonies grew after filtering a 10-m*L* sample.

$$\text{Bacteria/100 m}L = \frac{\text{No. of Colonies Counted x 100 m}L}{\text{Sample Volume Filtered, m}L\text{ x 100 m}L}$$

$$= \frac{(42\text{ colonies})(100\text{ m}L)}{(10\text{ m}L)(100\text{ m}L)}$$

$$= \frac{(4.2)(100\text{ m}L)}{100\text{ m}L}$$

$$= 420\text{ per 100 m}L$$

SPECIAL NOTE:

Inexperienced persons often have great difficulty with connected colonies, with mirror reflections of fluorescent tubes (which are confused with metallic sheen), and with water condensate and particulate matter which are occasionally mistaken for colonies. Thus there is a tendency for inexperienced persons to make errors on the high side in MF counts. At least five apparent coliform colonies should be transferred to lauryl tryptose broth tubes for verification as coliform organisms.

7. Procedure for Filtration of Chlorinated Samples Using Enrichment Technique

a. Place a sterile absorbent pad in the upper half of a sterile petri dish and pipet 1.8 to 2.0 m*L* sterile lauryl tryptose broth. Carefully remove any surplus liquid.

b. *ASEPTICALLY*[14] place the membrane filter through which the sample has been passed on the pad.

c. Incubate the filter, without inverting the dish, for 1 1/2 to 2 hours at 35°C in an atmosphere of 90 percent humidity (damp paper towels added to a plastic container with a snap-on lid can be used to produce the humidity).

d. Remove the enrichment culture from the incubator. Place a fresh, sterile, absorbent pad in the bottom half of the petri dish and saturate with 1.8 to 2.0 m*L* M-Endo Broth.

e. Transfer the membrane filter to the new pad. The used pad of lauryl tryptose may be discarded.

f. Invert the dish and incubate for 20 to 22 hours at 35° ± 0.5°C.

g. Count colonies as in previous method.

8. Procedure for Fecal Coliform

For drinking water samples, all positive total coliform colonies must be tested for fecal coliform. When the membrane filter method is used, growth from the positive total coliform colony should be transferred to a tube of EC media to determine the presence of fecal coliform.

a. Transfer, with the aid of a sterile 3-mm platinum wire loop, growth from each positive total coliform colony to a corresponding tube of EC broth.

b. Incubate inoculated EC broth tube in a water bath at 44.5° ± 0.2°C for 24 hours.

c. Gas production in an EC broth culture within 24 hours or less is considered a positive fecal coliform reaction.

d. Calculate total number of "fecal coliform colonies."

EXAMPLE:

For 100 m*L* of sample, 2 colonies gave a positive total coliform result. Only 1 of the 2 gave positive results when growth was transferred to EC broth.

Fecal Coliform/100 m*L* = 1 Coliform/100 m*L*.

Presence-Absence Method

1. General Discussion

The presence-absence (P-A) test for the coliform group in drinking water is a simple modification of the multiple tube procedure described above. One large test portion (100 m*L*) in a single culture bottle is used to obtain qualitative information on the presence or absence of coliforms. The media used is a mixture of lactose and lauryl tryptose broths with bromocresol purple added to indicate pH. Following incubation, gas and/or acid (yellow color of media) is produced if coliforms are present. A schematic of the P-A test procedure is shown in Figure 7.18.

2. Materials Needed

a. P-A broth. Prepare according to the instructions on the bottle.

b. 250-m*L* screw-cap, milk dilution bottle with 50 m*L* sterile triple-strength P-A broth.

c. Autoclave for sterilization of media and glassware.

d. Incubator set at 35°C ± 0.5°C.

e. Sterile, 100-m*L* graduated cylinder.

f. Thermometer verified to be accurate by comparison with a National Bureau of Standards (NBS) certified thermometer.

g. Brilliant Green Bile Broth. Prepare as instructed under MPN method (page 444).

h. Water bath set at 44.5°C ± 0.2°C.

i. EC Broth. Prepare according to instructions on bottle.

---

[14] *Aseptic (a-SEP-tick). Free from the living germs of disease, fermentation, or putrefaction. Sterile.*

(Coliform)

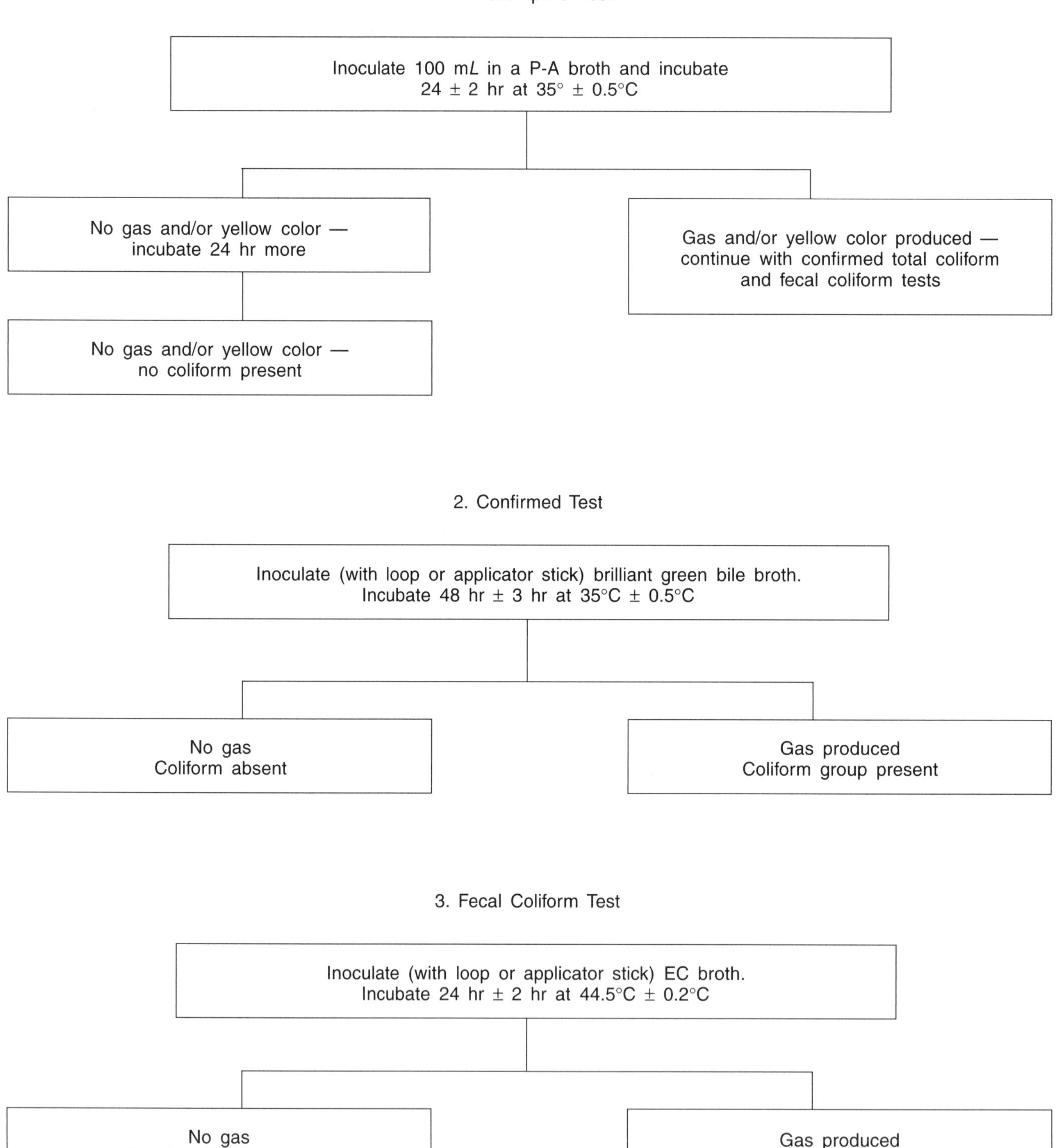
OUTLINE OF PROCEDURE FOR TOTAL COLIFORM
1. Presumptive Test
Inoculate 100 mL in a P-A broth and incubate 24 ± 2 hr at 35° ± 0.5°C
No gas and/or yellow color — incubate 24 hr more
Gas and/or yellow color produced — continue with confirmed total coliform and fecal coliform tests
No gas and/or yellow color — no coliform present
2. Confirmed Test
Inoculate (with loop or applicator stick) brilliant green bile broth. Incubate 48 hr ± 3 hr at 35°C ± 0.5°C
No gas Coliform absent
Gas produced Coliform group present
3. Fecal Coliform Test
Inoculate (with loop or applicator stick) EC broth. Incubate 24 hr ± 2 hr at 44.5°C ± 0.2°C
No gas Fecal Coliform absent
Gas produced Fecal Coliform present

Fig. 7.18 Schematic outline of test procedure for total coliform — Presence-Absence Method

(Coliform)

FOR DRINKING WATER SAMPLES

a. Shake sample approximately 25 times.

b. Inoculate 100 m*L* into P-A culture bottle and mix thoroughly.

c. Incubate at 35°C ± 0.5°C and inspect after 24 and 48 hours for acid reactions (yellow color). Record both positive and negative culture bottles.

d. A distinct yellow color forms in the culture when acid conditions exist. If gas is also being produced, gentle shaking of the bottle will result in a foaming action. Any amount of gas and/or acid constitutes a positive presumptive test requiring confirmation for total coliform in addition to further testing for fecal coliform.

e. 24-Hour Brilliant Green Bile Confirmation Test

Confirm all presumptive culture bottles that show gas and/or acid at 24 or 48 hours. Transfer liquid, with aid of a sterile platinum wire loop or sterile wood applicator, from the P-A broth showing gas and/or acid to a tube of brilliant green bile (BGB) broth.

After 24 hours, inspect the tube for gas formation. Any amount of gas formation is considered positive and so recorded on the data sheet. Negative BGB tubes are reincubated for an additional 24 hours.

f. 48-Hour Brilliant Green Bile Confirmation Test

Examine tubes for gas at the end of the 48 ± 3 hour period. Record both positive and negative tubes.

g. Recording Results

Gas production in the BGB broth culture within 48 hours confirms the presence of coliform bacteria. Report result as presence-absence test positive or negative for total coliforms in 100 m*L* of sample.

EXAMPLE 6 (DRINKING WATER SAMPLE) (see Figure 7.19).

m*L* of sample, 100 m*L*; Presence-Absence Method

Colilert™ Method

1. General Discussion

Colilert provides a media that contains specific indicator nutrients for total coliform and *E. coli*. As these nutrients are metabolized, yellow color and fluorescence are released confirming the presence of total coliform and *E. coli* respectively. Non-coliform bacteria are chemically suppressed and cannot metabolize the indicator nutrients. Consequently, they do not interfere with the identification of the target microbes. Total coliforms and *E. coli* are specifically and simultaneously detected and identified in 24 hours or less.

Although the Colilert method can yield presence-absence results within 24 hours and is easier to perform than the Membrane Filtration (MF) method, operators should be aware of the limitations of these tests in evaluating samples for regulatory purposes. The results for the Colilert and MF tests are not always comparable, which may be due to:

- Interferences in the sample that may suppress or mask bacterial growth,
- Greater sensitivity of the Colilert media,
- Added stress to organisms related to filtering, or
- The fact that different media may obtain better growth for some bacteria.

Both test methods are approved by the EPA for reporting under the Safe Drinking Water Program. When these tests produce conflicting results, however, the safest course of action is to increase monitoring and treatment efforts until the results for *both* tests are negative.

2. Materials Needed

a. 10 culture tubes each containing Colilert reagent for 10 m*L* of sample (available from IDEXX Laboratories, 1 IDEXX Dr., Westbrook, ME 04092, 800-321-0207).

b. Sterile 10-m*L* pipets.

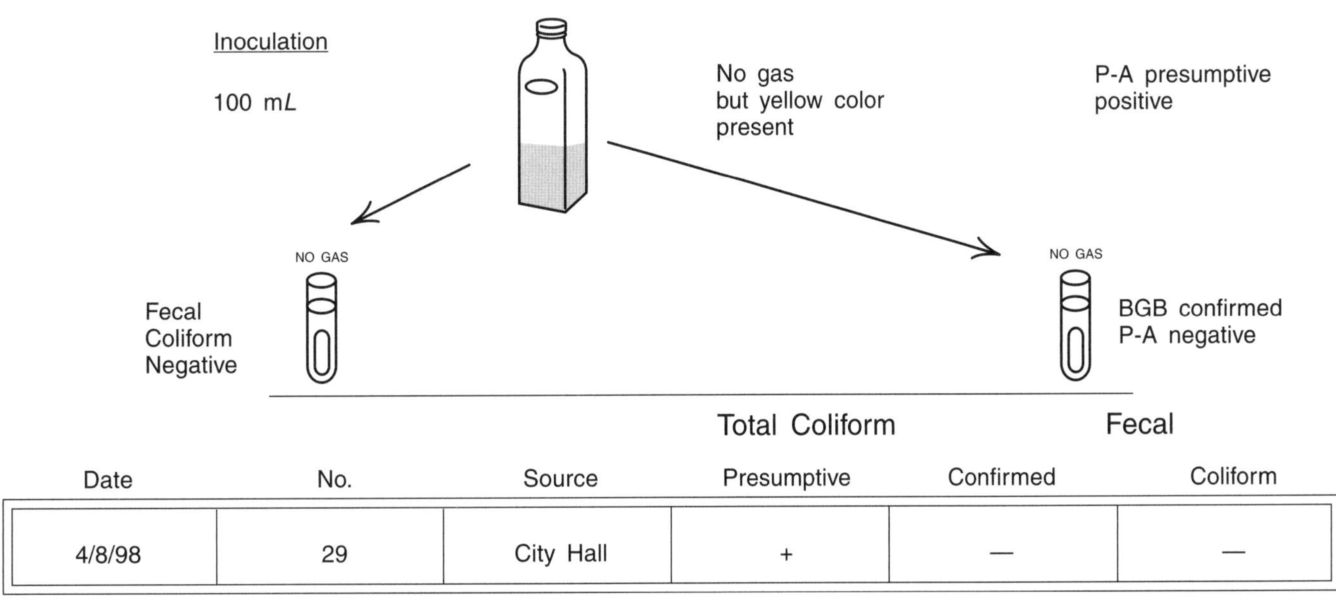

Date	No.	Source	Total Coliform Presumptive	Total Coliform Confirmed	Fecal Coliform
4/8/98	29	City Hall	+	—	—

Fig. 7.19 Record Presence-Absence coliform test results

(Coliform)

c. Incubator at 35°C.

d. Long wavelength ultraviolet lamp.

e. Color and fluorescence comparator.

3. Test Procedure

a. Remove the front panel of the Colilert tube box along the perforation line to allow easy access and removal of tube carriers.

b. Remove a 10-tube carrier from the kit and label each 10-tube MPN test as indicated on the carrier.

c. Crack the carrier along the front perforation and bend the carrier along the back crease line. Stand the carrier on the table top.

d. Aseptically fill each Colilert tube with sample water to the level of the back of the tube carrier (10 m*L*).

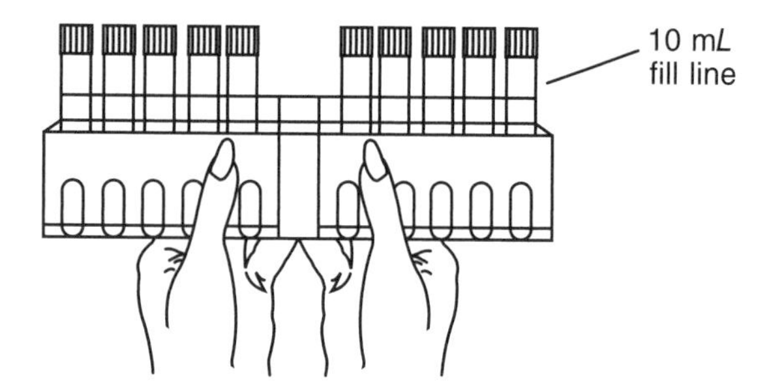

e. Cap the tubes tightly.

f. Mix vigorously to dissolve the reagent by repeated inversion of carrier.

NOTE: High calcium salt concentrations in certain waters may cause precipitate. This will not affect test results.

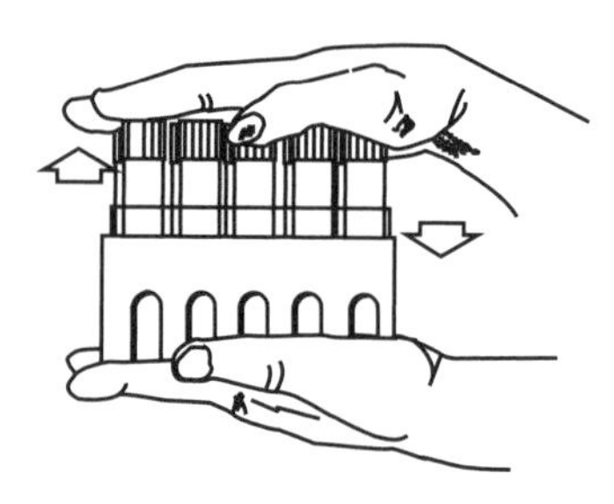

g. Incubate inoculated reagent tubes in the carrier at 35°C for 24 hours. Incubation should begin within 30 minutes of inoculation. (Incubation exceeding 28 hours should be avoided.)

h. Read tubes in the carrier at 24 hours. If yellow color is seen, check for fluorescence. Color should be uniform throughout the tube. If not, mix by inversion before reading.

4. Test Results and Interpretation

a. Compare each tube against the color comparator. If the inoculated reagent has a yellow color greater or equal to the comparator, the presence of total coliforms is confirmed.

b. If any of the tubes are yellow in color, observe each tube for fluorescence by placing carrier 2 to 5 inches from the long wavelength, ultraviolet lamp. If the fluorescence of the tube(s) is greater or equal to the comparator, the presence of *E. coli* is specifically confirmed.

c. Samples are negative for total coliforms if no color is observed at 24 hours. Should a sample be so lightly yellow after 24 hours' incubation that you cannot definitively read it relative to the positive comparator tube, you may incubate it up to an additional 4 hours. If the sample is coliform positive, the color will intensify. If it does not intensify, consider the sample negative.

d. To find the concentration of total coliforms or *E. coli* per 100 m*L*, compare the number of positive reaction tubes per sample set (10 tubes) to the standard MPN (Most Probable Number) chart shown on Table 7.5 or in *STANDARD METHODS.*

TABLE 7.5 MPN INDEX FOR VARIOUS COMBINATIONS OF POSITIVE AND NEGATIVE RESULTS WHEN TEN 10-m*L* PORTIONS ARE USED

Number of tubes giving positive reaction out of ten 10-m*L* portions	MPN Index per 100 m*L*
0	<1.1
1	1.1
2	2.2
3	3.6
4	5.1
5	6.9
6	9.2
7	12.0
8	16.1
9	23.0
10	>23.0

5. Recording Results

Record results on data sheets prepared especially for this test. See example in Figure 7.20.

DATE	NO.	SOURCE	24 HRS		MPN/100 m*L* TOTAL COLIFORM	MPN/100 m*L* *E. COLI*
4/16/98	X 29	1701 Main St	---- + ---- + ----	Yellow	2.2	<1.1
			— —	Fluorescence		

Fig. 7.20 Record coliform test results using Colilert Method

6. Method of Calculating the Most Probable Number (MPN)

EXAMPLE 7 (DRINKING WATER SAMPLE) — (see Figure 7.21).

m*L* of sample	100 m*L*
positive tubes	2
yellow color	yes
fluorescence	no

Read MPN as 2.2 per 100 m*L* for total coliform and <1.1 for *E. coli* from Table 7.5

Inoculation	Results
10 m*L*	2 of 10 - yellow 0 of 10 - fluorescence

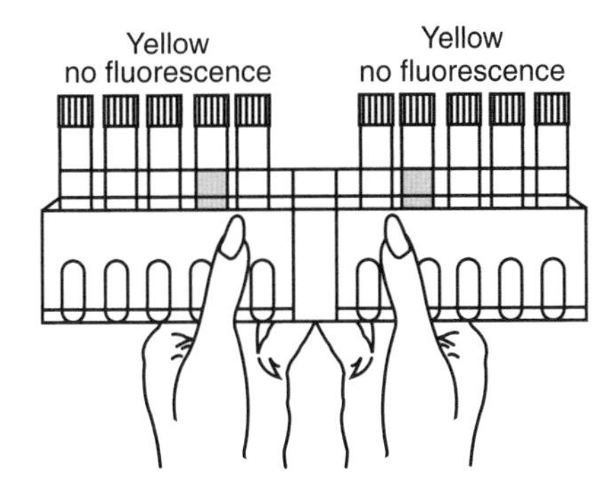

Fig. 7.21 Results of Colilert test — drinking water

Colisure Method

The Colisure test is a presence-absence test for coliform bacteria. A sample of water is added to a dehydrated medium and, after 28 to 48 hours, the medium is examined for the presence of coliforms. (The Colisure test is available from IDEXX Laboratories, 1 IDEXX Drive, Westbrook, ME 04092.)

QUESTIONS

Write your answers in a notebook and then compare your answers with those on page 470.

7.3H Why are drinking waters not tested for specific pathogens (such as *Cryptosporidium* or *Giardia*)?

7.3I To perform coliform tests, how would you decide how many samples to test?

7.3J Why should sodium thiosulfate be added to coliform sample bottles?

7.3K Steam autoclaves sterilize (kill all organisms) at a pressure of ____ psi and temperature of ____ °C during a ____ minute time period (at sea level).

7.3L Estimate the Most Probable Number (MPN) of coliform group bacteria for a raw water sample from the following test results:

m*L* of sample	1	0.1	0.01	0.001
Dilutions	0	–1	–2	–3
Readings (+ tubes)	5	5	3	1

7.3M How is the number of coliforms estimated by the membrane filter method?

7.3N What are the incubation conditions for chlorinated samples when using the membrane filter method (enrichment technique)?

Please answer the discussion and review questions next.

DISCUSSION AND REVIEW QUESTIONS

Chapter 7. LABORATORY PROCEDURES

(Lesson 4 of 5 Lessons)

Write the answers to these questions in your notebook before continuing. The question numbering continues from Lesson 3.

15. What is the purpose of the coliform group bacteria test?
16. How would you determine the number of dilutions for an MPN test?
17. What factors can cause errors when counting colonies on membrane filters?
18. How can questionable colonies on membrane filters be verified as coliform colonies?

CHAPTER 7. LABORATORY PROCEDURES

(Lesson 5 of 5 Lessons)

5. Hardness

A. Discussion

Hardness is caused principally by the calcium and magnesium ions commonly present in water. Hardness may also be caused by iron, manganese, aluminum, strontium, and zinc if present in significant amounts. Because only calcium and magnesium are present in significant concentrations in most waters, hardness can be defined as the total concentration of calcium and magnesium ions expressed as the calcium carbonate ($CaCO_3$) equivalent.

There are two types or classifications of water hardness: carbonate and noncarbonate. Carbonate hardness is due to calcium/magnesium bicarbonate or carbonate. Hardness that is due to calcium/magnesium sulfate, chloride, or nitrate is called noncarbonate hardness.

Hard water can cause incrustations (scale) when the water evaporates, or when heated in household hot water heaters and piping. Hardness-producing substances in water also combine with soap to form insoluble precipitates. The common method of minimizing these and other problems due to hardness is water supply softening. This procedure is discussed in Chapter 14, "Softening," in Volume II of *WATER TREATMENT PLANT OPERATION.*

B. What Is Tested

	Common Range mg/*L* as $CaCO_3$
Surface Water	30 to 500*
Well Water	80 to 500*

* Levels of hardness depend on local conditions.

C. Apparatus

Buret (25 m*L*)
Buret support
Graduated cylinder (100 m*L*)

Beaker (250 m*L*)
Magnetic stirrer
Magnetic stir-bar
Flask, Erlenmeyer (500 m*L*)
Funnel
Hot plate
Flask, volumetric (1,000 m*L*)

D. Reagents

NOTE: Standardized solutions are available already prepared from laboratory chemical supply companies.

1. Buffer solution.
2. Standard EDTA or CDTA titrant. EDTA is disodium ethylene-diaminetetraacetate dihydrate, also called (ethylenedinitrilo)-tetraacetic acid disodium salt. CDTA is disodium-CDTA (1,2 cyclohexanediaminetetraacetic acid).
3. Indicator solution.

OUTLINE OF PROCEDURE

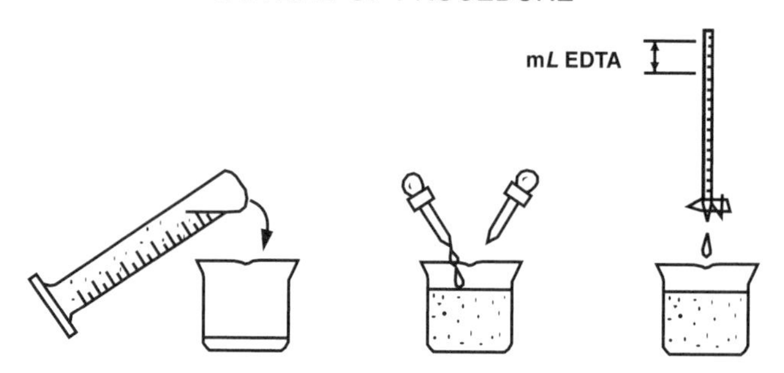
mL EDTA
1. Add 50 mL of sample to clean beaker.
2. Add 2 mL of buffer and 2 drops of indicator.
3. Titrate with EDTA to blue end point.

E. Procedure

1. Take a clean beaker and add 50 m*L* of sample.
2. Add 2 m*L* of buffer solution.
3. Add 2 drops indicator solution.
4. Titrate with standard EDTA solution until the last reddish tinge disappears from the solution. The solution is pure blue when the end point is reached.
5. Calculate total hardness.

F. Example

Results from water hardness testing of a well water sample were as follows:

Sample size = 50 m*L*

m*L* of EDTA titrant used, A = 10.5 m*L*

(Jar Test)

G. Calculation

$$\text{Hardness, mg/}L \text{ as } CaCO_3 = \frac{A \times 1{,}000}{mL \text{ of sample}}$$

$$= \frac{10.5\ mL \times 1{,}000}{50\ mL \text{ of sample}}$$

$$= 210\ mg/L$$

H. Precautions

1. Some metal ions interfere with this procedure by causing fading or indistinct end points. In these cases, an inhibitor reagent should be used. You may titrate with either standard CDTA solution or EDTA solution.
2. The titration should be completed within five minutes to minimize $CaCO_3$ precipitation.
3. A sample volume should be selected that requires less than 15 m*L* of EDTA titrant to be used.
4. For titrations of samples containing low hardness concentrations (less than 150 mg/*L*, as $CaCO_3$) a larger sample volume should be used.

I. Reference

See page 2-36, *STANDARD METHODS*, 20th Edition.

QUESTIONS

Write your answers in a notebook and then compare your answers with those on page 470.

7.3O What are the principal hardness-causing ions in water?

7.3P What problems are caused by hardness in water?

6. Jar Test

A. Discussion

Jar tests are tests designed to show the effectiveness of chemical treatment in a water treatment facility. Many of the chemicals we add to water can be evaluated on a small laboratory scale by the use of a jar test. The most important of these chemicals are those used for coagulation, such as alum and polymers. Using the jar test, the operator can approximate the correct coagulant dosage for plant use when varying amounts of turbidity, color, or other factors indicate raw water quality changes. The jar test is also a very useful tool in evaluating new coagulants or polymers being considered for use on a plant scale.

B. What Is Tested

Raw water, for optimum coagulant dose, which varies depending on coagulant(s) used and water quality.

C. Apparatus

1. A stirring machine with six paddles capable of variable speeds from 0 to 100 revolutions per minute (RPM).
2. An illuminator located underneath the stirring mechanism (optional).
3. Beakers (1,000 m*L*)
4. Pipets (10 m*L*)
5. Flask, volumetric (1,000 m*L*)
6. Balance, analytical

D. Reagents

1. Stock Coagulant Solution
   a. Dry alum, $Al_2(SO_4)_3 \cdot 14.3\ H_2O$.[15] Dissolve 10.0 gm dry alum (17 percent) in 600 m*L* distilled water contained in a 1,000 m*L* volumetric flask. Fill to mark. This solution contains 10,000 mg/*L* or 10 mg/m*L*.
   b. Liquid alum, $Al_2(SO_4)_3 \cdot 49.6\ H_2O$.[15] The operator should verify the strength of the alum with a hydrometer. Liquid alum is usually shipped as 8.0 to 8.5 percent Al_2O_3 and contains about 5.36 pounds of dry aluminum sulfate (17 percent dry) per gallon (specific gravity 1.325). This converts to 624,336 mg/*L*. Therefore, add 15.6 m*L* liquid alum to a 1,000-m*L* volumetric flask and fill to mark. This solution contains 10,000 mg/*L* or 10 mg/m*L*.
   c. Table 7.6 indicates the strengths of stock solutions for various dosages.

TABLE 7.6 DRY CHEMICAL CONCENTRATIONS USED FOR JAR TESTING [a]

Approx. Dosage Required, mg/*L* [b]	Grams/Liter to Prepare [c]	1 mL Added to 1 Liter Sample Equals	Stock Solution Conc., mg/*L* (%)
1-10 mg/*L*	1 gm/*L*	1 mg/*L*	1,000 mg/*L* (0.1%)
10-50 mg/*L*	10 gm/*L*	10 mg/*L*	10,000 mg/*L* (1.0%)
50-500 mg/*L*	100 gm/*L*	100 mg/*L*	100,000 mg/*L* (10.0%)

[a] From *JAR TEST* by E. E. Arasmith, Linn-Benton Community College, Albany, OR.
[b] Use this column which indicates the approximate dosage required by raw water to determine the trial dosages to be used in the jar test.
[c] This column indicates the grams of dry chemical that should be used when preparing the stock solution. The stock solution consists of the chemical plus enough water to make a one-liter solution.

E. Procedure

1. Collect a two-gallon (8-liter) sample of the water to be tested.
2. Immediately measure six 1,000-m*L* quantities and place into each of six 1,000-m*L* beakers.
3. Place all six beakers on stirring apparatus.
4. With a measuring pipet, add increasing dosages of coagulant solution to the beakers as rapidly as possible. Select a series of dosages so that the first beaker will represent an underdose and the last an overdose.
5. With stirring paddles lowered into the beakers, start stirring apparatus and operate it for one minute at a speed of 80 RPM.[16]

[15] *Values of 14.3 H_2O and 49.6 H_2O were obtained from ALUMINUM SULFATE published by Stauffer Chemicals. Actual values for commercial alum purchased by your water treatment plant may vary slightly.*
[16] *Use stirring speeds and times which are similar to actual conditions in your water treatment plant.*

(Jar Test)

6. Reduce the stirring speed for the next 30 minutes to 20 RPM.[16]

OUTLINE OF PROCEDURE

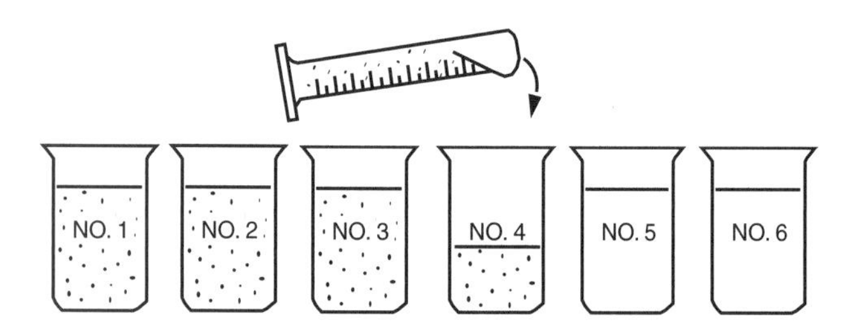

1. Add 1,000 m*L* to each of 6 beakers.

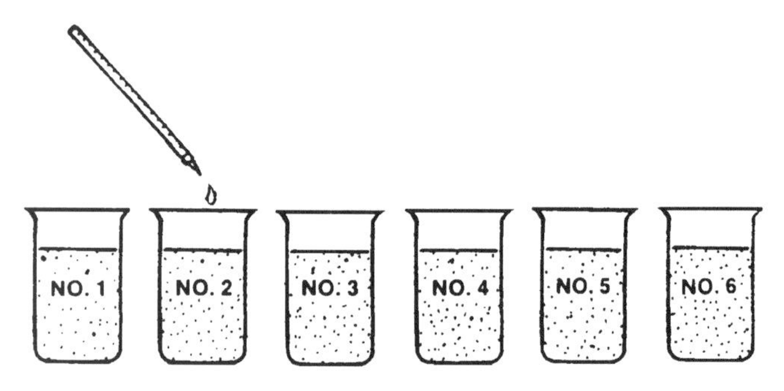

2. Add increasing dosages of coagulant.

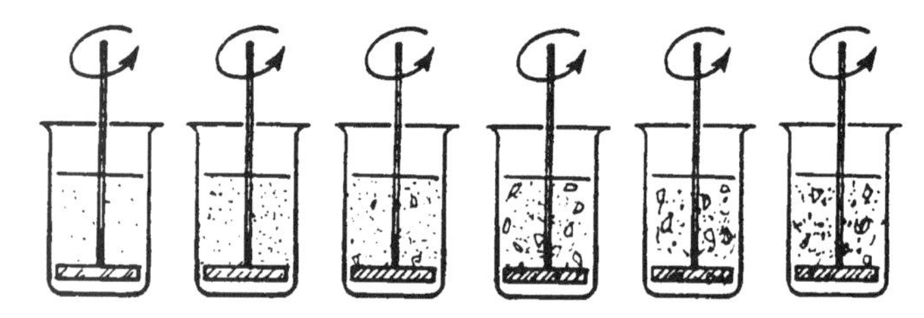

3. Stir for appropriate time period. Evaluate floc quality.

7. Observe and evaluate each beaker as to that specific dosage's floc quality. Record results.

8. Stop the stirring apparatus and allow samples in beakers to settle for 30 minutes.[16] Observe the floc settling characteristics. A hazy sample indicates poor coagulation. Properly coagulated water contains floc particles that are well-formed and dense, with the liquid between the particles clear. Describe results as poor, fair, good, or excellent.

F. Precautions

Without going to extreme measures, it is very difficult to duplicate in the jar test exactly what is happening on a plant scale. The jar test, therefore, should be used as an indication of what is to be expected on a larger scale in a water treatment plant.

There are a number of tests that can be performed to improve the jar test and the interpretation of the results. These tests include:

1. Alkalinity (before and after),
2. pH (before and after),
3. Turbidity of *SUPERNATANT*[17] (before and after), and
4. Filtered turbidity of supernatant.

See Figure 7.22 for a helpful jar test data sheet.

After estimating the optimum coagulant dosage, run another jar test with the optimum coagulant dosage constant, but vary the pH. These results will give you the optimum pH.

Alkalinity must be monitored very carefully before and after the jar test. Alkalinity must always be *AT LEAST* half of the coagulant dose. For example, if the optimum coagulant dose is 50 mg/*L*, the total alkalinity must be at least 25 mg/*L*. If the natural alkalinity is less than 25 mg/*L*, adjust the total alkalinity up to 25 mg/*L* by adding lime. For additional information on how to calculate the amount of lime needed to increase the alkalinity, see Section 5.243, "Arithmetic for Solids-Contact Clarification," *EXAMPLE 6*, in *WATER TREATMENT PLANT OPERATION,* Volume I.

G. Example

A sample of river water was collected for jar test analysis to determine the "optimum" alum dosage for effective coagulation.

The following series of alum dosages were added:

Beaker No.	m*L* Alum Solution Added	Alum Added, mg/*L*
1	1.0	10
2	1.2	12
3	1.4	14
4	1.6	16
5	1.8	18
6	2.0	20

H. Interpretation of Results

Results of the above jar testing were recorded as follows:

Beaker No.	Alum Dose, mg/*L*	Floc Quality
1	10	poor
2	12	fair
3	14	good
4	16	excellent
5	18	excellent
6	20	good

The above results seem to indicate that a dose of 16 or 18 mg/*L* would be optimum. The operator should, however, verify this result with visual observation of what is actually happening in the flocculation basin.

[16] *Use stirring speeds and times which are similar to actual conditions in your water treatment plant.*

[17] *Supernatant (sue-per-NAY-tent). Liquid removed from settled sludge. Supernatant commonly refers to the liquid between the sludge on the bottom and the scum on the water surface of a basin or container.*

JAR TEST DATA SHEET

ANALYST:

SOURCE: DATE:
TIME: COAGULANT AID:
COAGULANT: pH ADJUST:

JAR NUMBER	SAMPLE SIZE	TEMP START/FINISH	TURB START/FINISH	pH START/FINISH	ALKA START/FINISH	FILTERED TURBIDITY	COAG DOSE	COAG-AID DOSE	ADJUSTED pH	DESCRIPTION OF FLOC IN JARS: 1ST APPEAR	5 MIN	10 MIN	15 MIN	20 MIN	25 MIN	30 MIN	45 MIN	HARDNESS START/FINISH	COMMENTS
1																			
2																			
3																			
4																			
5																			
6																			

CODE: VS = VERY SMALL; S = SMALL; M = MEDIUM
D = DENSE; VD = VERY DENSE; SP = SPARCE
L = LARGE; VL = VERY LARGE

Fig. 7.22 Typical jar test data sheet

QUESTIONS

Write your answers in a notebook and then compare your answers with those on page 470.

7.3Q What is the purpose of the jar test?

7.3R What stirring speeds are used during the jar tests for optimum alum dosage?

7. pH

A. Discussion

The pH of a water indicates the intensity of its acidic or basic strength. The pH scale runs from 0 to 14. Water having a pH of 7 is at the midpoint of the scale and is considered neutral. Such a water is neither acidic nor basic. A pH of greater than 7 indicates basic water. The stronger the basic intensity, the greater the pH. The opposite is true of the acidity. The stronger the intensity of the acidity, the lower will be the pH.

pH SCALE

0 ← INCREASING ACID – – – – 7 – – – – INCREASING BASE → 14

1 ← 2 ← 3 ← 4 ← 5 ← 6 /\ 8 → 9 → 10 → 11 → 12 → 13

Neutral

Mathematically, pH is the logarithm (base 10) of the reciprocal of the hydrogen ion activity, or the negative logarithm of the hydrogen ion activity.

$$pH = Log \frac{1}{[H^+]} = -Log\ [H^+]$$

For example, if a water has a pH of 1, then the hydrogen ion activity $[H^+] = 10^{-1} = 0.1$. If the pH is 7, then $[H^+] = 10^{-7} = 0.0000001$. A change in the pH of one unit is caused by changing the hydrogen ion level by a factor of 10 (10 times).

In a solution, both hydrogen ions $[H^+]$ and the hydroxyl ions $[OH^-]$ are always present. At a pH of 7, the activity of both hydrogen and hydroxyl ions equals 10^{-7} *MOLES*[18] per liter. When the pH is less than 7, the activity of hydrogen ions is greater than the hydroxyl ions.

pH plays an important role in the water treatment processes such as disinfection, coagulation, softening, and corrosion control. The pH test also indicates changes in raw and finished water quality.

B. What Is Tested

	Common Range
Surface water	6.5 to 8.5
Finished water	7.5 to 9.0
Well water	6.5 to 8.0

C. Apparatus

1. pH meter
2. Glass electrode
3. Reference electrode
4. Magnetic stirrer
5. Magnetic stir-bar

D. Reagents

1. Buffer tablets for various pH value solutions (available from laboratory chemical supply houses).
2. Distilled water.

E. Procedure

1. Due to the difference between the various makes and models of pH meters commercially available, specific instructions cannot be provided for the correct operation of all instruments. In each case, follow the manufacturer's instructions for preparing the electrodes and operating the instrument.
2. *STANDARDIZE THE INSTRUMENT (pH METER) AGAINST A BUFFER SOLUTION WITH A pH CLOSE TO THAT OF THE SAMPLE.*
3. Rinse electrodes thoroughly with distilled water after removal from buffer solution.
4. Place electrodes in sample and measure pH.
5. Remove electrodes from sample, rinse thoroughly with distilled water.
6. Immerse electrode ends in beaker of pH 7 buffer storage solution.
7. Turn meter to "standby" (or OFF).

F. Precautions

1. To avoid faulty instrument calibration, prepare fresh buffer solutions as needed, and at least once per week (from commercially available buffer tablets).
2. pH meter, buffer solution, and samples should all be near the same temperature because temperature variations will give somewhat erroneous results. Allow a few minutes for the probes to adjust to the buffers before calibrating a pH meter to ensure accurate pH readings.
3. Watch for erratic results arising from electrodes, faulty connections, or fouling of electrodes with oily precipitated matter. Films may be removed from electrodes by placing isopropanol on a tissue or a Q-tip and cleaning the probe.

[18] *Mole. The molecular weight of a substance, usually expressed in grams.*

4. The temperature compensator on the pH meter adjusts the meter for changes in electrode response with temperature. However, the pH of water also changes with temperature and the pH meter cannot compensate for this change.

5. We recommend standardizing the pH meter using a pH buffer solution close to the pH of the sample. However, if you use another buffer solution with a different pH to determine the calibration of the pH meter for a range of pH values and the pH meter does not give the pH of the second buffer, follow the manufacturer's directions to adjust the pH meter. This procedure is called adjusting the "slope" of the pH meter.

6. If you're measuring pH in colored samples or samples with high solids content, or if you're taking measurements that need to be reported to the US EPA, you should use a pH electrode and meter instead of a colorimetric method or test papers. pH meters are capable of providing ±0.1 pH accuracy in most applications. In contrast, colorimetric tests provide ±0.1 pH accuracy only in a limited range. pH papers provide even less accuracy.

G. Interpretation of Results

The pH of water has a very important influence on the effectiveness of chlorine disinfection. Chlorination is a chemical reaction in which chlorine is an oxidizing agent. Chlorine is a more effective oxidant at lower pH values. Simply stated, a chlorine residual of 0.2 mg/*L* at a pH of 7 is just as effective as 1 mg/*L* at a pH of 10. Therefore, five times as much chlorine is required to do the same disinfecting job at pH 10 as it does at pH 7.

The finished water of some treatment facilities is adjusted with lime or caustic soda to a slightly basic pH for the purpose of minimizing corrosion in the distribution system.

H. Reference

See page 4-86, *STANDARD METHODS*, 20th Edition.

QUESTIONS

Write your answers in a notebook and then compare your answers with those on pages 470 and 471.

7.3S What does the pH of a water indicate?

7.3T What precautions should be exercised when using a pH meter?

8. Temperature

A. Discussion

Temperature is one of the most frequently taken tests in the water industry. Accurate water temperature readings are important not only for historical purposes but also because of its influence on chemical reaction rates, biological growth, dissolved gas concentrations, and water stability with respect to calcium carbonate, in addition to its acceptability by consumers for drinking.

(Temperature)

B. What Is Tested

	Common Range, °C
Raw and Treated Surface Water	5 to 25
Well Water	10 to 20

C. Apparatus

1. One NBS (National Bureau of Standards) thermometer for calibration of the other thermometers.
2. One Fahrenheit, mercury-filled, 1° subdivided thermometer.
3. One Celsius (formerly called Centigrade), mercury-filled, 1° subdivided thermometer.
4. One metal case to fit each thermometer.

NOTE: There are three types of thermometers and two scales.

SCALES

1. Fahrenheit, marked °F.
2. Celsius, marked °C (formerly Centigrade).

THERMOMETER STYLES

1. Total immersion. This type of thermometer must be totally immersed when read. Readings with this type of thermometer will change most rapidly when removed from the liquid to be recorded.
2. Partial immersion. This type thermometer will have a solid line (water-level indicator) around the stem below the point where the scale starts.
3. Dial. This type has a dial that can be easily read while the thermometer is still immersed. Dial thermometer readings should be checked (calibrated) against the NBS thermometer. Some dial thermometers can be recalibrated (adjusted) to read at a set temperature against the NBS thermometer.

D. Reagents

None required.

E. Procedure

1. Collect as large a volume of sample as is practical. The temperature will have less chance to change in a large volume than in a small container.
2. Immerse the thermometer to the proper depth. Do not touch the bottom or sides of the sample container with the thermometer.
3. Record temperature to the nearest fraction of a degree which can be estimated from the thermometer available.
4. When measuring the temperature of well water samples, allow the water to continuously overflow a small container (a polystyrene coffee cup is ideal). Place the thermometer in the cup. After there has been no change in the temperature reading for one minute, record the temperature. The temperature of water samples collected from a distribution system mainly depend on the soil temperature at the depth of the water main.

(Turbidity)

F. Precautions

To avoid breaking or damaging a glass thermometer, store it in a shielded metal case. Check your thermometer's accuracy against the NBS-certified thermometer by measuring the temperature of a sample with both thermometers simultaneously. Some of the poorer quality thermometers are substantially inaccurate (off as much as 6°F or 3°C).

G. Example

To measure the temperature of treated water, a sample was obtained in a gallon bottle, the thermometer immediately immersed, and a temperature of 15°C recorded after the reading became constant.

H. Calculation

Normally we measure and record temperatures using a thermometer with the proper scale. We could, however, measure a temperature in °C and convert to °F. The following formulas are used to convert temperatures from one scale to the other.

1. If we measure in °F and want °C,

$$°C = \frac{5}{9}(°F - 32°F)$$

2. If we measure in °C and want °F,

$$°F = \frac{9}{5}(°C) + 32°F$$

3. Sample Calculation

The measured treated water temperature was 15°C. What is the temperature in °F?

$$\begin{aligned} °F &= 9/5(°C) + 32°F \\ &= 9/5(15°C) + 32°F \\ &= 27 + 32 \\ &= 59°F \end{aligned}$$

I. Reference

See page 2-60, *STANDARD METHODS*, 20th Edition.

QUESTIONS

Write your answers in a notebook and then compare your answers with those on page 471.

7.3U Why are temperature readings important?

7.3V Why should the thermometer remain immersed in the liquid while being read?

7.3W Why should thermometers be calibrated against an accurate NBS-certified thermometer?

9. Turbidity

A. Discussion

The term turbidity is simply an expression of the physical cloudiness of water. Turbidity is caused by the presence of suspended matter such as silt, finely divided organic and inorganic matter, and microscopic organisms such as algae.

The accepted method used to measure turbidity is called the nephelometric method. The nephelometric turbidimeter or nephelometer (Figure 7.23) is designed to measure particle-reflected light at an angle of 90 degrees to the source beam. The greater the intensity of scattered light, the higher the turbidity.

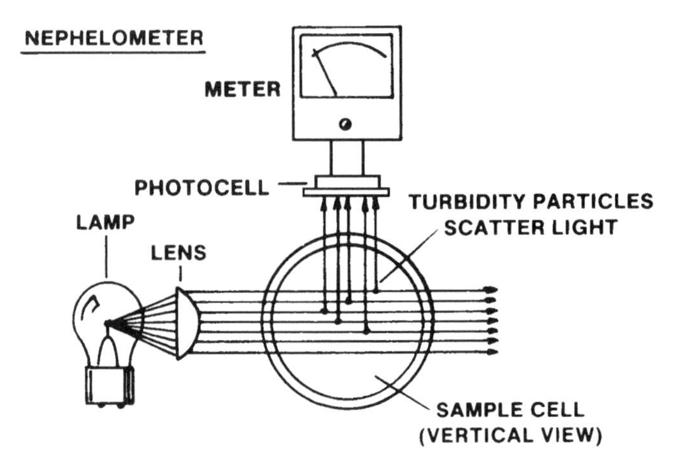

Fig. 7.23 Nephelometer

(Permission of HACH Company)

The *TURBIDITY UNIT (NTU)*[19] is an empirical quantity which is based on the amount of light that is scattered by particles of a polymer reference standard called formazin which produces particles that scatter light in a reproducible manner. Formazin, the primary turbidity standard, is an aqueous (water-based) suspension of an insoluble polymer formed by the condensation reaction between hydrazine sulfate and hexamethylenetetramine.

Secondary turbidity standards are suspensions of various materials formulated to match the primary formazin solutions. These secondary standards are generally used because of their convenience and the instability of dilute formazin primary standard solutions. Examples of these secondary standards include "standards" that are supplied by the turbidimeter manufacturer with the instrument. Periodic checks of these secondary standards against the primary formazin standard are a must and will provide assurance of measurement accuracy.

The turbidity measurement is one of the most important tests the plant operator performs. The Safe Drinking Water Act stipulates specific monitoring requirements for turbidity. Turbidity of treated surface water must be measured continuously or determined daily and the results reported to the appropriate authority. Turbidity testing is the most critical tool in recognizing changes in raw water quality, detecting problems in coagulation and sedimentation, and troubleshooting filtration problems.

The maximum contaminant level (MCL) for turbidity is one TU with five TU allowed under certain circumstances. Water

[19] *Turbidity Units (TU). Turbidity units are a measure of the cloudiness of water. If measured by a nephelometric (deflected light) instrumental procedure, turbidity units are expressed in nephelometric turbidity units (NTU) or simply TU. Those turbidity units obtained by visual methods are expressed in Jackson Turbidity Units (JTU) which are a measure of the cloudiness of water; they are used to indicate the clarity of water. There is no real connection between NTUs and JTUs. The Jackson turbidimeter is a visual method and the nephelometer is an instrumental method based on deflected light.*

(Turbidity)

treatment plant operators should strive to produce a finished water with a turbidity of 0.1 TU or less.

B. What Is Tested

	Common Range, NTU
Untreated Surface Water	1 to 300
Filtered Water	0.03 to 0.50
Well Water	0.05 to 1.0+

C. Apparatus

1. Turbidimeter: To minimize differences in turbidity measurements, rigorous specifications for turbidimeters are necessary. The turbidimeter should have the following important characteristics:
   a. The turbidimeter should consist of a nephelometer with a light source illuminating the sample, and one or more photoelectric detectors to indicate the intensity of scattered light at a 90° angle with a "readout" device.
   b. The light source should be an intense tungsten filament lamp.
   c. The total distance traveled by the light through the sample should be less than about 5 centimeters.
   d. The instrument should have several measurement ranges. The instrument should be able to measure from 0 to 100 turbidity units, with sufficient sensitivity (on the lowest scale) to detect differences of 0.02 or less in filtered waters having turbidities of less than one unit.
2. Sample tubes. These are usually provided with the instrument.

D. Reagents

1. Turbidity-free water: Pass distilled water through a membrane filter having a pore size no greater than 0.2 microns (available from laboratory supply houses). Discard the first 200 m*L* collected. If filtration does not reduce turbidity of the distilled water, use unfiltered distilled water.
2. Stock formazin turbidity suspension:[20]
   a. Solution I — Dissolve 1.000 gm hydrazine sulfate in distilled water and dilute to 100 m*L* in a volumetric flask.
   b. Solution II — Dissolve 10.00 gm hexamethylenetetramine in distilled water and dilute to 100 m*L* in a volumetric flask.
   c. In a 100-m*L* volumetric flask, add (using 5-m*L* volumetric pipets) 5.0 m*L* Solution I and 5.0 m*L* of Solution II. Mix and allow to stand 24 hours at 25°C. Then dilute to the mark and mix. The turbidity of this suspension is considered 400 NTU exactly.
   d. Prepare solutions and suspensions monthly.
3. Standard turbidity suspensions: Dilute 10.00 m*L* stock turbidity suspension to 100 m*L* with turbidity-free water. Prepare weekly. The turbidity of this suspension is defined as 40 NTU.
4. Dilute turbidity standards: Dilute portions of the standard turbidity suspension with turbidity-free water as required. Prepare weekly.

E. Procedure

1. Turbidimeter calibration: The manufacturer's operating instructions should be followed. Measure your standard solutions on the turbidimeter covering the range of interest. If the instrument is already calibrated in standard turbidity units, this procedure will check the accuracy of the calibration scales. At least one standard should be run in each instrument range to be used. Some instruments permit adjustments of sensitivity so that scale values will correspond to turbidities. Reliance on an instrument manufacturer's scattering standards for calibrating the instrument is not an acceptable practice unless they are in acceptably close agreement with prepared standards. If a precalibrated scale is not supplied, then calibration curves should be prepared for each usable range of the instrument.
2. Turbidities less than 40 units: Shake the sample to thoroughly disperse the suspended solids. Wait until air bubbles disappear, then pour the sample into the turbidimeter tube. Read the turbidity directly from the instrument scale or from the appropriate calibration curve.
3. Turbidities exceeding 40 units: Dilute the sample with one or more volumes of turbidity-free water until the turbidity falls below 40 units. The turbidity of the original sample is then computed from the turbidity of the diluted sample and the dilution factor.

F. Example

A reservoir sample was collected and the turbidity was found to be greater than 40 units when first checked. 30 m*L* of this sample was then diluted with 60 m*L* of turbidity-free water. This diluted sample showed a turbidity of 25 units.

NTU found in diluted sample, A	= 25 NTU
m*L* of dilution water used, B	= 60 m*L*
m*L* of sample volume taken for dilution, C	= 30 m*L*

G. Calculation

$$\text{Nephelometric Turbidity Units (NTU)} = \frac{(A)x(B+C)}{C}$$

$$= \frac{(25\text{ NTU})x(60\text{ m}L + 30\text{ m}L)}{30\text{ m}L}$$

$$= 75\text{ NTU}$$

H. Interpretation of Results

Report turbidity results as follows:

Turbidity Reading	Record to Nearest
0.0 to 1.0	0.05
1 to 10	0.1
10 to 40	1
40 to 100	5
100 to 1,000	10
>1,000	100

[20] *Stock secondary standard turbidity suspensions that require no preparation are available from commercial suppliers and approved for use.*

(Turbidity)

I. Notes and Precautions

1. Sample tubes must be kept scrupulously clean, both inside and out. Discard them when they become scratched or etched. Never handle them where the light strikes them.
2. Fill the tubes with samples and standards that have been agitated thoroughly, and allow sufficient time for bubbles to escape.
3. Turbidity should be measured in a sample as soon as possible to obtain accurate results. The turbidity of a sample can change after the sample is collected. Shaking the sample will not re-create the original turbidity.

J. Reference

See page 2-8, *STANDARD METHODS*, 20th Edition.

QUESTIONS

Write your answers in a notebook and then compare your answers with those on page 471.

7.3X What are the causes of turbidity in water?

7.3Y How is turbidity measured?

7.4 ARITHMETIC ASSIGNMENT

Turn to the Appendix, "How to Solve Small Water System Arithmetic Problems," at the back of this manual and read Section A.135, Laboratory Procedures. Check all of the arithmetic in this section using an electronic pocket calculator. You should be able to get the same answers.

7.5 ADDITIONAL READING

1. *NEW YORK MANUAL*, Chapter 4,* "Water Chemistry," Chapter 5,* "Microbiology," and Chapter 21,* "Laboratory Examinations of Water."
2. *TEXAS MANUAL*, Chapter 6,* "Water Chemistry," and Chapter 12,* "Laboratory Examinations."

* Depends on edition.

7.6 WATER LABORATORY TESTS IN *WATER TREATMENT PLANT OPERATION*, VOLUME II

Laboratory procedures for the following tests are provided in *WATER TREATMENT PLANT OPERATION*, Volume II, Chapter 21, "Advanced Laboratory Procedures."

1. Algae Counts
2. Calcium
3. Chloride
4. Color
5. Dissolved Oxygen
6. Fluoride
7. Iron (Total)
8. Manganese
9. Marble Test
10. Metals
11. Nitrate
12. pH
13. Specific Conductance
14. Sulfate
14. Taste and Odor
16. Trihalomethanes
17. Total Dissolved Solids

End of Lesson 5 of 5 Lessons on LABORATORY PROCEDURES

Please answer the discussion and review questions before continuing with the Objective Test.

DISCUSSION AND REVIEW QUESTIONS

Chapter 7. LABORATORY PROCEDURES

(Lesson 5 of 5 Lessons)

Write the answers to these questions in your notebook before continuing with the Objective Test on page 471. The question numbering continues from Lesson 4.

19. What precautions should be considered when performing the hardness determination on a water sample?
20. How could you estimate the most effective dose of alum or a polymer in a water treatment process?
21. What precautions should be exercised when taking temperature measurements?
22. Why should turbidity be measured in a sample as soon as possible?

SUGGESTED ANSWERS

Chapter 7. LABORATORY PROCEDURES

ANSWERS TO QUESTIONS IN LESSON 1

Answers to questions on page 409.

7.0A Laboratory quality control tests are important because they provide the necessary information to monitor the treatment processes and ensure a safe and pleasant-tasting drinking water.

7.0B The prefix milli- means 1/1,000 (0.001) of the unit following.

7.0C The proper name of the chemical compound $CaCO_3$ is calcium carbonate.

Answers to questions on page 419.

7.1A Descriptions of laboratory glassware and their use or purpose.

Item	Description	Use or Purpose
1. Beakers	Short, wide cylinders in sizes from 1 m*L* to 4,000 m*L*	Mixing chemicals
2. Graduated cylinders	Long, narrow cylinders in sizes from 5 m*L* to 4,000 m*L*	Measuring volumes
3. Pipets	Small-diameter, graduated tubes, with a pointed tip, in sizes from 1.0 m*L* to 100 m*L*	Delivering accurate volumes
4. Burets	Long tubes with graduated walls and a stopcock in sizes from 10 m*L* to 1,000 m*L*	Delivering and measuring accurate volumes used in "titrations"

7.1B A meniscus is the curve of the surface of a liquid in a small tube.

7.1C Never heat graduated cylinders in an open flame because they may break.

Answers to questions on page 420.

7.1D A "standard solution" is a solution in which the exact concentration of a chemical or compound in a solution is known.

7.1E Laboratory notebooks and worksheets help record data in an orderly manner.

7.1F Some sources or causes of poor quality of analytical data are:

1. Sloppy laboratory technique,
2. Deteriorated reagents and standards,
3. Poorly operating instruments, and
4. Calculation mistakes.

Answers to questions on page 426.

7.1G Six laboratory hazards are:

1. Hazardous materials,
2. Explosions,
3. Cuts and bruises,
4. Electric shock,
5. Fire, and
6. Burns (heat and chemical).

7.1H *NEVER* work alone in the laboratory. Someone should always be available to help in case you should have an accident which blinds you, leaves you unconscious, or starts a fire you can't handle. If necessary, have someone check on you regularly to be sure you are OK.

7.1I True. You may *ADD ACID TO WATER*, but never the reverse.

7.1J Dispose of small amounts of corrosive acids by pouring the neutralized acid down corrosion-resistant sinks and sewers using large quantities of water to dilute and flush the acid.

ANSWERS TO QUESTIONS IN LESSON 2

Answers to questions on page 429.

7.2A The greatest errors in laboratory tests are frequently caused by (1) improper sampling, (2) poor sample preservation, and (3) lack of enough mixing during testing.

7.2B A representative sample must be collected in order for test results to have any significant meaning. Without a representative sample, the test results will not reflect actual water conditions.

7.2C Sampling points in a distribution system should be selected in order to trace the course from the finished water source (at the well or plant), through the transmission mains, and then through the major and minor arteries of the system.

Answers to questions on page 432.

7.2D The two general types of samples collected by water treatment personnel are (1) grab samples, and (2) composite samples.

7.2E Water quality indicators that are usually measured with a grab sample include (1) dissolved gases, (2) coliform bacteria, (3) residual chlorine, (4) temperature, and (5) pH.

7.2F Depth samples are collected by the use of a Kemmerer Sampler or similar device. The sampling device and container are lowered to the desired depth, then opened, filled, closed, and returned to the surface.

7.2G Samples should not be collected from taps surrounded by excessive foliage (leaves, flowers) or from taps that are dirty, corroded, or are leaking.

7.2H When collecting a sample, record the sample location, date and time of collection, name of collector, and any other pertinent information.

ANSWERS TO QUESTIONS IN LESSON 3

Answers to questions on page 436.

7.3A Chemicals used in water treatment that will cause changes in alkalinity include alum, chlorine, and lime.

7.3B Alkalinity determination is needed when calculating chemical dosages for coagulation and water softening. Alkalinity must also be known to calculate corrosivity and to estimate carbonate hardness of water.

Answers to questions on page 438.

7.3C The benefits of chlorinating water include:

1. Disinfection;
2. Improved quality of water due to chlorine reacting with iron, manganese, protein substances, sulfide, and many taste- and odor-producing compounds;
3. Control of microorganisms that might interfere with coagulation and flocculation;
4. Keeps filter media free of slime growths; and
5. Helps bleach out undesirable color.

7.3D A potential adverse effect from chlorination is the possibility of the formation of carcinogenic chlororganic compounds such as chloroform and other THMs.

Answers to questions on page 442.

7.3E Conditions that can cause variations in the chlorine demand of water include (1) the amount of chlorine applied, (2) length of contact time, (3) pH, (4) temperature, (5) organics, and (6) reducing agents.

7.3F The operator uses the chlorine demand test to determine the best chlorine dosage to achieve specific chlorination objectives.

7.3G
$$\text{Chlorine Demand, mg}/L = \text{Chlorine Added, mg}/L - \text{Free Residual Chlorine, mg}/L$$
$$= 2.0 \text{ mg}/L - 0.4 \text{ mg}/L$$
$$= 1.6 \text{ mg}/L$$

ANSWERS TO QUESTIONS IN LESSON 4

Answers to questions on page 459.

7.3H Drinking waters are not tested for specific pathogens because the tests are very time-consuming and require special techniques and equipment.

7.3I The number of samples required for coliform tests is generally based on the population served by the water system.

7.3J Sodium thiosulfate should be added to sample bottles for coliform tests to neutralize any residual chlorine in the sample. If residual chlorine is not neutralized, it will continue to be toxic to the coliform organisms remaining in the sample and give false (negative) results.

7.3K Steam autoclaves sterilize (kill all organisms) at a pressure of *15* psi and temperature of *121°C* during a *15*-minute time period (at sea level).

7.3L

m*L* of sample	0.1	0.01	0.001
Dilutions	–1	–2	–3
Readings (+ tubes)	5	3	1

MPN = 11,000/100 m*L*

7.3M The number of coliforms is determined by counting the number of coliform-appearing colonies grown on the membrane filter.

7.3N Incubate the filter, without inverting the dish, for 1 ½ to 2 hours at 35°C in an atmosphere of 90 percent humidity. Transfer the membrane filter to a new pad enriched with M-Endo Broth. Invert the dish and incubate for 20 to 22 hours at 35° ± 0.5°C.

ANSWERS TO QUESTIONS IN LESSON 5

Answers to questions on page 461.

7.3O The principal hardness-causing ions in water are calcium and magnesium.

7.3P Problems caused by hardness in water include (1) incrustations when water evaporates or scale when heated, and (2) formation of precipitates when combined with soap.

Answers to questions on page 464.

7.3Q The jar test is used to (1) determine the correct coagulant dosage, and (2) evaluate new coagulants or polymers.

7.3R Speeds used during the jar test are as follows:

1. 80 RPM for the first minute,
2. 20 RPM for the next 30 minutes, and
3. Stop stirring (0 RPM) for the next 30 minutes.

These speeds and times should be adjusted if necessary to be similar to actual conditions in the water treatment plant.

Answers to questions on page 465.

7.3S The pH of a water indicates the intensity of its basic or acidic strength.

7.3T Precautions to be exercised when using a pH meter include:

1. Prepare fresh buffer solution weekly for calibration purposes;
2. Have pH meter, samples, and buffer solutions all near the same temperature;
3. Watch for erratic results arising from electrodes, faulty connections, or fouling of electrodes with interfering matter;
4. The pH of water changes with temperature; and
5. If testing waters with a range of pH values, the "slope" of the pH meter may require adjusting.

Answers to questions on page 466.

7.3U Temperature readings are important because temperature influences chemical reaction rates, biological growth, dissolved gas concentrations, and water stability with respect to calcium carbonate. Also consumers are sensitive to the temperature of the water they drink.

7.3V The thermometer should remain immersed in the liquid while being read for accurate results. When removed from the liquid, the reading will change.

7.3W All thermometers should be calibrated against an accurate National Bureau of Standards thermometer because some poorer quality thermometers are substantially inaccurate (off as much as 6°F or 3°C).

Answers to questions on page 468.

7.3X Turbidity in water can be caused by the presence of suspended matter such as silt, finely divided organic and inorganic matter, and microscopic organisms such as algae.

7.3Y Turbidity is measured by the nephelometric method.

OBJECTIVE TEST

Chapter 7. LABORATORY PROCEDURES

Please write your name and mark the correct answers on the answer sheet, as directed at the end of Chapter 1. There may be more than one correct answer to each multiple-choice question.

True-False

1. The collection of a bad sample or a bad laboratory result is about as useful as no results.
   1. True
   2. False

2. Results of lab tests are of no value to anyone unless they are used.
   1. True
   2. False

3. Always keep Teflon stopcocks in burets properly lubricated.
   1. True
   2. False

4. Completely empty volumetric pipets by blowing out the small amount remaining in the tip.
   1. True
   2. False

5. Never pipet chemical reagent solutions or unknown water samples by mouth.
   1. True
   2. False

6. Never work alone in the laboratory.
   1. True
   2. False

7. Food may be kept in a refrigerator that is used for chemical and/or sample storage.
   1. True
   2. False

8. Never use oil or grease as a lubricant when making rubber-to-glass connections.
   1. True
   2. False

9. Well pumps and casings can contribute to sample contamination.
   1. True
   2. False

10. pH meters should be standardized against a buffer solution with a pH close to that of the sample.
    1. True
    2. False

Best Answer (Select only the closest or best answer.)

11. Generally, all chemical compounds can be divided into what two main groups?
    1. Acids and bases
    2. Elements and symbols
    3. Liquids and solids
    4. Organics and inorganics

12. Which piece of laboratory equipment is the most accurate way to measure a volume of liquid?
    1. Beaker
    2. Flask
    3. Graduated cylinder
    4. Pipet

13. Which piece of laboratory equipment is used to titrate a chemical reagent?
    1. Buchner funnel
    2. Buret
    3. Graduated cylinder
    4. Pipet

14. What does the alkalinity of a water sample measure?
    1. Acidity
    2. Amount of carbon in sample
    3. Capacity to neutralize acids
    4. pH

15. What is the purpose of the coliform bacteria test? Test for
    1. Aerobic and facultative anaerobic gram-negative bacteria.
    2. Bacteria that ferment lactose.
    3. *Cryptosporidium*.
    4. Potential presence of pathogens.

16. How are the results of the presence-absence (P-A) test for the coliform group in drinking water reported?
    1. Colony formation units (CFUs)
    2. MPN/100 m*L*
    3. Qualitative
    4. Quantitative

17. What is the purpose of a jar test?
    1. To determine the need to treat coliform bacteria
    2. To evaluate mixing techniques using jars
    3. To prepare dilutions for MPN tests
    4. To show the effectiveness of chemical treatment

18. Why are water temperature measurements important to consumers?
    1. Acceptability of water for drinking
    2. Concentration of dissolved gases in water
    3. Effectiveness of disinfection process
    4. Minimization of cost of chemicals

19. Which style of thermometer will change most rapidly when removed from the liquid sample?
    1. Dial
    2. Digital
    3. Partial immersion
    4. Total immersion

20. What is the accepted method used to measure turbidity?
    1. Candle
    2. Jackson
    3. Light
    4. Nephelometric

Multiple Choice (Select all correct answers.)

21. Which factors can lead to poor quality laboratory data?
    1. Calculation mistakes
    2. Deteriorated reagents
    3. Neatly recorded results in lab notebooks
    4. Poorly operating instruments
    5. Sloppy laboratory technique

22. Which of the following hazards could be encountered in a laboratory?
    1. Burns
    2. Drowning
    3. Electric shock
    4. Explosions
    5. Hazardous materials

23. Which types of hazardous materials could be encountered in a laboratory?
    1. Caustic materials
    2. Corrosive materials
    3. Flammable materials
    4. Saline materials
    5. Toxic materials

24. What are the characteristics of an ideal water distribution sampling station?
    1. Drinking fountains in parks
    2. Fire hydrants
    3. Made of corrosion-resistant material
    4. Short, direct connection to main
    5. Well sampling station

25. Grab samples should be used to measure which of the following water quality indicators?
    1. Chlorine residual
    2. Coliform bacteria
    3. Dissolved gases
    4. pH
    5. Temperature

26. What information must be identified on a sample container?
    1. Date of collection
    2. Location of collection
    3. Name of collector
    4. Results of collection
    5. Time of collection

27. Which of the following chemicals used in water treatment cause changes in alkalinity?
    1. Alum
    2. Calcium
    3. Chlorine
    4. Lime
    5. Sodium

28. An alkalinity determination is needed when calculating chemical dosages used in what water treatment processes?
    1. Coagulation
    2. Filtration
    3. Flocculation
    4. Settling
    5. Softening

29. Why is chlorine added to drinking water?
    1. Disinfection
    2. Helps bleach out undesirable color
    3. Keeps filter media free of slime growths
    4. Kills corrosion-causing bacteria
    5. Reacts with many taste- and odor-producing compounds

30. What factors can cause a variation in chlorine demand?
    1. Length of contact time
    2. pH of water
    3. Presence of organics
    4. Presence of reducing agents
    5. Temperature of water

31. Hardness is caused in water by which ions?
    1. Alkalinity
    2. Calcium
    3. Carbonate
    4. Magnesium
    5. Sodium

32. What tests can be performed to improve the jar test and interpretation of the results?
    1. Alkalinity
    2. Coliform bacteria
    3. Filtered turbidity of supernatant
    4. pH
    5. Turbidity of supernatant

33. pH plays an important role in which water treatment processes?
    1. Coagulation
    2. Corrosion control
    3. Disinfection
    4. Filtration
    5. Settling

34. What causes turbidity?
    1. Algae
    2. Dissolved solids
    3. Finely divided organic matter
    4. Microscopic organisms
    5. Silt

35. Results from a coliform bacteria multiple fermentation tube test of raw water are as follows:

m*L* of sample	10	1	0.1	0.01	0.001
Dilutions	0	0	–1	–2	–3
Positive tubes	5	5	5	3	1

What is the most probable number of coliforms per 100 m*L* in the sample? Use Table 7.3 on page 449.

1. 110 coliforms per 100 m*L*
2. 900 coliforms per 100 m*L*
3. 1,100 coliforms per 100 m*L*
4. 9,000 coliforms per 100 m*L*
5. 11,000 coliforms per 100 m*L*

36. Results from a coliform bacteria Colilert™ Method test from a 100-m*L* sample indicated that there were three positive tubes with a yellow color. How would you report the results? Use Table 7.5 on page 458.
    1. 1.1 total coliform and 2.2 *E. coli* MPN per 100 m*L*
    2. 2.2 total coliform and 1.1 *E. coli* MPN per 100 m*L*
    3. 3.3 total coliform MPN per 100 m*L*
    4. 3.6 *E. coli* MPN per 100 m*L*
    5. 3.6 total coliform MPN per 100 m*L*

Data to answer questions 37 and 38.

Results from MPN tests during one week were as follows:

Day	S	M	T	W	T	F	S
MPN/100 m*L*	4	7	14	8	9	16	5

37. What is the mean MPN per 100 m*L* from the data?
    1. 6 MPN/100 m*L*
    2. 7 MPN/100 m*L*
    3. 8 MPN/100 m*L*
    4. 9 MPN/100 m*L*
    5. 10 MPN/100 m*L*

38. What is the median MPN per 100 m*L* from the data?
    1. 6 MPN/100 m*L*
    2. 7 MPN/100 m*L*
    3. 8 MPN/100 m*L*
    4. 9 MPN/100 m*L*
    5. 10 MPN/100 m*L*

39. A temperature of 77°F is equivalent to how many degrees Celsius?
    1. 17°C
    2. 25°C
    3. 62°C
    4. 77°C
    5. 81°C

40. How long should a ¾-inch service line be flushed before sampling if the line is 75 feet long and the flow is 0.5 GPM? Flush out twice the pipe volume.
    1. 2 min 26 sec
    2. 3 min 15 sec
    3. 4 min 30 sec
    4. 4 min 58 sec
    5. 6 min 53 sec

End of Objective Test

CHAPTER 8

SETTING WATER RATES FOR SMALL WATER UTILITIES

by

James D. Beard

TABLE OF CONTENTS

Chapter 8. SETTING WATER RATES FOR SMALL WATER UTILITIES

HOW MUCH?

OBJECTIVES

Chapter 8. SETTING WATER RATES FOR SMALL WATER UTILITIES

Following completion of Chapter 8, you should be able to:

1. Determine revenue needs for a small water utility,
2. Itemize various system expenses,
3. Identify possible funding sources for capital improvements,
4. Relate costs to level of service,
5. Distribute costs to customers,
6. Develop water rates for a small water utility,
7. Describe the financial strength of the utility,
8. Prepare a Consumer Confidence Report for the utility's customers,
9. Plan for financial stability, and
10. Keep accurate records of costs and sources of revenues.

WORDS

Chapter 8. SETTING WATER RATES FOR SMALL WATER UTILITIES

AMERICAN WATER WORKS ASSOCIATION — AMERICAN WATER WORKS ASSOCIATION

A professional organization for all persons working in the water utility field. This organization develops and recommends goals, procedures and standards for water utility agencies to help them improve their performance and effectiveness. For information on AWWA membership and publications, contact AWWA, 6666 W. Quincy Avenue, Denver, CO 80235. Phone (303) 794-7711.

BASE-EXTRA CAPACITY METHOD — BASE-EXTRA CAPACITY METHOD

A cost allocation method used by water utilities to determine water rates for various water user groups. This method considers base costs (O & M expenses and capital costs), extra capacity costs (additional costs for maximum day and maximum hour demands), customer costs (meter maintenance and reading, billing, collection, accounting) and fire protection costs.

BOND — BOND

(1) A written promise to pay a specified sum of money (called the face value) at a fixed time in the future (called the date of maturity). A bond also carries interest at a fixed rate, payable periodically. The difference between a note and a bond is that a bond usually runs for a longer period of time and requires greater formality. Utility agencies use bonds as a means of obtaining large amounts of money for capital improvements.

(2) A warranty by an underwriting organization, such as an insurance company, guaranteeing honesty, performance, or payment by a contractor.

CALL DATE — CALL DATE

First date a bond can be paid off.

COMMODITY-DEMAND METHOD — COMMODITY-DEMAND METHOD

A cost allocation method used by water utilities to determine water rates for the various water user groups. This method considers the commodity costs (water, chemicals, power, amount of water use), demand costs (treatment, storage, distribution), customer costs (meter maintenance and reading, billing, collection, accounting) and fire protection costs.

CONSUMER CONFIDENCE REPORTS — CONSUMER CONFIDENCE REPORTS

An annual report prepared by a water utility to communicate with its consumers. The report provides consumers with information on the source and quality of their drinking water. The report is an opportunity for positive communication with consumers and to convey the importance of paying for good quality drinking water.

COVERAGE RATIO — COVERAGE RATIO

The coverage ratio is a measure of the ability of the utility to pay the principal and interest on loans and bonds (this is known as "debt service") in addition to any unexpected expenses.

DEBT SERVICE — DEBT SERVICE

The amount of money required annually to pay the (1) interest on outstanding debts; or (2) funds due on a maturing bonded debt or the redemption of bonds.

DEPRECIATION — DEPRECIATION

The gradual loss in service value of a facility or piece of equipment due to all the factors causing the ultimate retirement of the facility or equipment. This loss can be caused by sudden physical damage, wearing out due to age, obsolescence, inadequacy or availability of a newer, more efficient facility or equipment. The value cannot be restored by maintenance.

EQUITY — EQUITY

The value of an investment in a facility.

FIXED COSTS — FIXED COSTS

Costs that a utility must cover or pay even if there is no demand for water or no water to sell to customers. Also see VARIABLE COSTS.

OPERATING RATIO OPERATING RATIO

The operating ratio is a measure of the total revenues divided by the total operating expenses.

OVERHEAD OVERHEAD

Indirect costs necessary for a water utility to function properly. These costs are not related to the actual production and delivery of water to consumers, but include the costs of rent, lights, office supplies, management and administration.

PRESENT WORTH PRESENT WORTH

The value of a long-term project expressed in today's dollars. Present worth is calculated by converting (discounting) all future benefits and costs over the life of the project to a single economic value at the start of the project. Calculating the present worth of alternative projects makes it possible to compare them and select the one with the largest positive (beneficial) present worth or minimum present cost.

RATE OF RETURN RATE OF RETURN

A value which indicates the return of funds received on the basis of the total equity capital used to finance physical facilities. Similar to the interest rate on savings accounts or loans.

VARIABLE COSTS VARIABLE COSTS

Costs that a utility must cover or pay that are associated with the production and delivery of water. The costs vary or fluctuate on the basis of the volume of water treated and delivered to customers (water production). Also see FIXED COSTS.

CHAPTER 8. SETTING WATER RATES FOR SMALL WATER UTILITIES

8.0 DEVELOPING WATER RATES

8.00 General Rate Setting Philosophy

The goal established by the *AMERICAN WATER WORKS ASSOCIATION (AWWA)*[1] for setting water rates is that each utility develop a rate setting methodology that fairly allocates costs to various classes of water users, and then recovers these costs through rates which will sustain the enterprise and not unduly discriminate against any particular class of users. To put this goal in simpler terms — rates should be "cost based" and "nondiscriminatory."

As water utility managers, we must develop rate setting practices and procedures which fairly allocate costs to the classes of users that benefit most from our services. Again, the burden of water costs should not be unfairly thrust upon any individual class of customers whether they be small or large, new or old, domestic or commercial/industrial water users.

A second goal in rate setting is stability of rates. Careful analysis of past, projected, and actual costs will help you avoid frequent, unexpected rate increases. No matter how fair and equitable the rate structure is, sudden large increases upset consumers and raise questions about the competence of the utility manager. A rate structure that is fair to all customers reflects well on your integrity, and a rate structure that is stable and predictable over a period of years reflects well on your professional ability.

The rate setting practices and procedures described in this chapter can be readily used by small water utilities. We will use the AWWA definition of a small utility — one having fewer than 5,000 customers (active service connections). The procedures and ideas could be applied by larger utilities as well.

It would be helpful if a simple formula or model could be used to establish water rates — but this is seldom the case. Each water utility is unique and this yields special sets of circumstances which need to be fully considered in setting fair water rates.

Since a simple formula isn't available, cost-of-service or rate studies are usually conducted to establish fair and equitable rates. The results of these studies are highly dependent on the accuracy and reliability of the utility's information and record-keeping systems. For larger utilities more comprehensive rate studies may be needed but for smaller utilities it is generally advisable to keep rate studies simple. However, even in the small utility, full or partial rate studies may be required when the utility has a substantial industrial base or other special water sale categories or customers.

The procedures and practices described in this chapter should be used as guidelines to establish rates and charges. Each water utility is unique and will have its own special considerations to deal with. By using these procedures, the small utility should be able to develop a cost-based, fair, and reasonable set of rates and charges which accurately reflect the cost of service for each class of consumer.

8.01 General Plan of Attack

The first step in setting water rates is to develop a general plan of attack. That is to say, we need to outline the specific steps we'll take to reach our goal. Fortunately, many of the resources that we will need to use are readily available in the form of budgets, budget reports showing conformance to the spending plan, and customer-related records. Hopefully, the customer records will include such essential information as water usage and customer-related operation and maintenance expenses.

Armed with a good working knowledge of our past financial and operating history, we can begin to develop a schedule of water rates and charges by using the following procedures:

1. Determine total annual revenue requirements,
2. Allocate revenue requirements to cost components,
3. Distribute component costs to customer classes, and
4. Design rates that will recover these costs from each class of customer.

The following sections will describe each of these procedures in enough detail that you can apply them to your own

[1] *American Water Works Association. A professional organization for all persons working in the water utility field. This organization develops and recommends goals, procedures and standards for water utility agencies to help them improve their performance and effectiveness. For information on AWWA membership and publications, contact AWWA, 6666 W. Quincy Avenue, Denver, CO 80235. Phone (303) 794-7711.*

water utility. Figure 8.1 outlines the plan of attack or procedures developed in this chapter for an operator to determine water rates for a small utility agency. As you begin this procedure, keep in mind one basic difference between budget planning for a utility and budgeting your own personal expenses. You probably begin with a fixed annual income and try to adjust your expenses to fit within that amount. In budgeting for a utility, you will begin by figuring out all present and anticipated expenses and then set fair and equitable rates that will produce the amount of annual income needed.

8.02 Let's Develop a Hypothetical Water District

To better illustrate the application of the rate setting procedures in this chapter, a hypothetical water district will be used as a model of a typical water utility.

Assume that the XYZ Water District has 4,000 customers (services) and serves a predominantly residential community with a population of approximately 11,200 people. It has two large industrial users and several small commercial users. Water-related statistics for this water district are given in Table 8.1.

TABLE 8.1 SERVICE AREA STATISTICS

SERVICE AREA SIZE		= 8 sq miles
SOURCES OF SUPPLY		
Surface Water		= 25%
Groundwater		= 75%
CUSTOMERS (SERVICES)		= 4,000
POPULATION		
(4,000 Services)(2.8 Persons/Household)		= 11,200
AVERAGE DAILY CONSUMPTION		= 1.71 MGD
WATER USE (From Meter Records)		
Residential	(11,200)(125 GPD)	= 1.40 MGD
Industrial:	1 @ 100,000 GPD	
	1 @ 75,000 GPD	
	175,000 GPD	= 0.18 MGD
Commercial		= 0.05 MGD
Miscellaneous		= 0.08 MGD
	Average Daily Use	= 1.71 MGD
MAXIMUM DAILY CONSUMPTION		
(1.40 MGD)(1.8 Residential Peak Factor) + 0.31 MGD		= 2.83 MGD
SYSTEM DESIGN FLOW CAPACITY		
(1.71 MGD)(1.25 Capacity Factor)		= 2.14 MGD

Given the average daily consumption of 1.71 MGD, the annual water use for this community would be about 625 million gallons or 1,915 acre-feet. However, the treatment plant, storage, and distribution facilities must be sized to deliver maximum day demand with some margin of safety (25 percent is used in this example).

QUESTIONS

Write your answers in a notebook and then compare your answers with those on page 503.

8.0A What are the goals of a fair water rate setting program?

8.0B The accuracy and reliability of rate studies depend on what factors?

8.1 REVENUE REQUIREMENTS

8.10 Determining Revenue Needs

Developing a statement of revenue requirements which accurately reflects the cash needs of our utility is the first step in our plan of attack for setting water rates. If we have developed a good budget and spending plan, this should merely be an extension of existing budgeting and accounting practices.

Our objective is to determine the cash needs and timing of expenditures (cash flow) for the utility. Examples of typical revenue needs are:

— O & M expenses

— Debt service on borrowed funds

— Rehabilitation and replacement costs

— Administrative expenses related to the operation of our system

— Cash reserves

To avoid cash flow shortages during the rate period, we need to be sure to project our revenue needs far enough into the future and allow some cash reserve for emergencies and contingencies (unplanned events). Sometimes it is best to project several years into the future even though the rate period is typically only one year.

We also need to allow for unexpected decreases in water demand, inflation, and system growth since these factors play a significant role in setting stable water rates. A good historical database from which to project future needs will greatly assist in setting long-term rates.

PLAN OF ATTACK

**1. DETERMINE REVENUE NEEDS (TABLE 8.3)
IDENTIFY EXPECTED COSTS FOR**

a. O & M expenses, including administration
b. Debt service on borrowed funds
c. Rehabilitation, replacement, and expansion costs
d. Cash reserve (10% of O & M)

2. SELECT A METHOD FOR A REVENUE BASE

a. Cash basis
b. Utility basis

3. ALLOCATE COSTS

a. Commodity-demand method
b. Base-extra capacity method

4. ALLOCATE O & M EXPENSES (TABLE 8.5)

5. ALLOCATE PLANT VALUE AND OTHER CAPITAL FACILITIES (TABLE 8.6)

6. PREPARE OVERALL ALLOCATION OF COSTS OF SERVICE (TABLE 8.7)

7. DETERMINE RATES PER CUSTOMER ON BASIS OF METER SIZE

a. Fixed costs (meter service charge and fire protection)
b. Volumetric charge

Fig. 8.1 Procedure to determine water rates

Revenue requirements may be broadly divided into two major categories — Operating Expenses and Capital Costs. A description of these costs is given in Table 8.2.

TABLE 8.2 MAJOR REVENUE REQUIREMENTS

OPERATING EXPENSES (O & M)

Costs of operating the plant, reservoir, groundwater, and distribution systems

Costs of maintaining facilities and equipment

Administration

CAPITAL COSTS

Annual costs associated with plant construction investment

Debt service expenses (principal and interest on bonds or other loans)

Facilities replacement costs

Facilities extension (expansion) costs

Facilities improvement costs, including new facilities

Cash reserves

8.11 Forecasting Expenditures

As previously mentioned, it is necessary to forecast expenses carefully in establishing a stable system of rates and charges. *A stable system is one in which rates are adequate to cover all anticipated expenses and still maintain a positive cash position at the end of the rate period.* This avoids "bumping rates" up in the middle of the rate period, and allows the utility to build a modest cash reserve which can be called upon in times of serious financial need. A good rule of thumb for a small water system is to establish a general fund reserve which is a minimum of 10 percent of one year's O & M expenses. Special reserve funds may also need to be established for capital program needs.

In forecasting expenditures, the following items should be considered:

- Unexpected decreases in water demand
- System growth rate
- Anticipated inflation
- Long-range capital improvement program needs
- Long-range financing requirements

8.110 Declining Demand for Water

A major concern when projecting revenue needs is the sensitivity of water demand projections and the ability of revenues from water sales to meet projected target values. A drop in planned water sales by only ten percent in a below-normal water year could have a significant impact on projected revenue from water sales for a small water utility.

To minimize the financial impact of variable water sales, a water utility might consider establishing a dry-year contingency reserve whereby a small surcharge is assessed each year (on a monthly or twice monthly billing basis). These funds are set aside for use during dry years or when there is an unexpected drop in demand. Since dry periods typically occur only in one or two years out of every ten in most regions, building up a reserve is a good method for making up reduced revenue during periods when water is not available due to drought or the demand drops due to a shift in economic conditions.

Another method of guarding against an unexpected drop in revenue during dry years is to make sure that the *FIXED COSTS*[2] are always recovered by water rates. Since a high percentage of the revenue requirement can be assigned to fixed cost items, a rate structure can be established that is not too dependent on the commodity or volume charge component. If a fluctuating or unstable demand is a concern in a cost analysis, consider shifting some of the *VARIABLE COSTS*[3] or volumetric cost burden to the customer service charge component. Always try to design water rates in a manner that will recover base (fixed) costs.

8.111 System Growth Rate

The growth rate of your community may have a significant impact on water demand, capital facility, and staffing needs. If your community is growing rapidly, you will need a long-range as well as a short-range capital improvement program designed to meet system demands as they occur over time. A short-range program is generally forecast for one to five years; whereas a long-range program may estimate system demands and facility needs for 25 years or more. Obviously, short-range planning is much better focused than long-range programs. We need to review our long-range planning program periodically and revise these plans as appropriate.

Traditionally, *BONDS*[4] have been used to finance major capital improvements to the water system. In recent years, however, there has been increasing pressure to include capital costs which are general in nature when calculating the general water rate. Other capital costs incurred, for such uses as system expansion required by new development, might better be met by new-user capital contributions in the form of connection fees or other related charges.

Connection fees are generally "upfront" fees paid by new users connecting to the system to cover new capital costs incurred to provide service to their property. Or, new users are sometimes required to pay a "buy-in" fee at the time of connection to the system to pay back an appropriate portion of capital already invested in the water system. User-fee-based financing systems are often referred to as "pay-as-you-go" financing. The advantages of this method of assessment are that it avoids payment of interest expense on long-term debt and does not put all of the burden of system expansion on current rate payers.

[2] *Fixed Costs. Costs that a utility must cover or pay even if there is no demand for water or no water to sell to customers. Also see VARIABLE COSTS.*

[3] *Variable Costs. Costs that a utility must cover or pay that are associated with the production and delivery of water. The costs vary or fluctuate on the basis of the volume of water treated and delivered to customers (water production). Also see FIXED COSTS.*

[4] *Bond. A written promise to pay a specified sum of money (called the face value) at a fixed time in the future (called the date of maturity). A bond also carries interest at a fixed rate, payable periodically. The difference between a note and a bond is that a bond usually runs for a longer period of time and requires greater formality. Utility agencies use bonds as a means of obtaining large amounts of money for capital improvements.*

8.112 Inflation

Inflation is another important consideration in setting stable water rates. There are two general components of inflation which will affect rates and charges — Consumer Price Index or CPI, and Construction Cost Index or CCI. The CPI is established by the federal government and is published for most areas of the United States. The CPI should be applied to general O & M expense items such as materials and supplies as well as labor. The CCI published by *ENGINEERING NEWS RECORD* magazine is generally used for estimating construction-related capital improvement costs. It is advisable to use both of these indices since inflation rates may affect these two major sectors of the economy differently.

8.113 Capital Improvement Program

A capital improvement fund must be a part of the utility budget. Be sure that everyone, your governing body and the public, understands the capital improvement fund is not a profit for the utility but a replacement fund to keep the utility operating in the future.

Capital planning starts with a look at changes in the community. Where are the areas of growth in the community, where are the areas of decline, and what are the anticipated changes in industry within the community? After identifying the changing needs in the community, you should examine the existing utility structure. Identify your weak spots (in the distribution system or with in-plant processes). Make a list of the areas that will be experiencing growth, weak spots in the system, and anticipated new regulatory requirements. The list should include expected capital improvements that will need to be made over the next year, two years, five years, and ten years. You can use the information in your annual reports and other operational logs to help compile the list.

Once you have compiled this information, prioritize the list and make a timetable for improving each of the areas. Starting at the top of the priority list, estimate the costs for improvements and incorporate these costs into your capital improvement budget.

You may find that some of your capital improvement needs could be met in more than one way. How do you decide which of several options is most cost-effective? How do you compare fundamentally different solutions? For example, assume your community's population is growing rapidly and you will need to increase treated water production by 20 percent by the end of the next ten years. Your existing wells cannot provide that amount of additional water and your filtration equipment is operating at 90 percent of design capacity. Possible solutions might include a combination of the following options, some of which might be implemented immediately while others might be brought on line in five or ten years:

- Rehabilitate some declining wells,
- Drill additional new wells,
- Develop an available surface water source,
- Install another filtration unit, and/or
- Install additional distribution system storage reservoirs.

To compare alternative plans you will need to calculate the present value (or *PRESENT WORTH*[5]) of each plan; that is, the costs and benefits of each plan in today's dollars. This is done by identifying all the costs and benefits of each alternative plan over the same time period or time horizon. Costs should include not only the initial purchase price or construction costs, but also financing costs over the life of the loans or bonds and all operation and maintenance costs. Benefits include all of the revenue that would be produced by this facility or equipment, including connection and user fees. With the help of an experienced accountant, apply standard inflation, depreciation, and other economic discount factors to calculate the present value of all the benefits and costs of each plan during the same planning period. This will give you the cost of each plan in the equivalent of today's dollars.

Remember to involve all of your local officials and the public in the development of this capital improvement budget so they understand what will be needed.

Long-term capital improvements, such as a new plant or a new treatment process, are usually anticipated in your 10-year or 20-year projection. These long-term capital improvements usually require some additional financing. The basic ways for a utility to finance capital improvements are through general obligation bonds, revenue bonds, or loan funding programs.

General obligation bonds or *ad valorem* (based on value) taxes are assessed based on property taxes. These bonds usually have a lower interest rate and longer payback time, but the total bond limit is determined for the entire community. This means that the water utility will have available only a portion of the total bond capacity of the community. These bonds are not often used for funding water utility improvements today.

The second type of bond, the revenue bond, is commonly used to fund utility improvements. This bond has no limit on the amount of funds available and the user charges provide repayment on the bond. To qualify for these bonds, the utility must show sound financial management and the ability to repay the bond. Learn as much as you can about the provisions of the bond. Be sure the bond has a call date, which is the first date when you can pay off the bond. The common practice is for a 20-year bond to have a 10-year call date and for a 15-year bond to have an 8-year call date. The bond will also have a call premium, which is the amount of extra funds

---

[5] *Present Worth. The value of a long-term project expressed in today's dollars. Present worth is calculated by converting (discounting) all future benefits and costs over the life of the project to a single economic value at the start of the project. Calculating the present worth of alternative projects makes it possible to compare them and select the one with the largest positive (beneficial) present worth or minimum present cost.*

needed to pay off the debt on the call date. You should try to get your bonds a call premium of no more than 102 percent par. This means that for a debt of $200,000 on the call date, the total payoff would be $204,000, which includes the extra two percent for the call premium. You will need to get help from a financial advisor to prepare for and issue the bonds. These advisors will help you negotiate the best bond structure for your community.

Special assessment bonds may be used to extend services into specific areas. The direct users pay the capital costs and the assessment is usually based on frontage or area of real estate. These special assessments carry a greater risk to investors but may be the best way to extend service to some areas.

The most common way to finance water supply system improvements in the past has been federal and state grant programs. The Block Grants from HUD are still available for some projects and Rural Utilities Service (RUS) loans may also be used as a funding source. In addition, state revolving fund (SRF) programs provide loans (but not direct grants) for improvements. The SRF program has already been implemented with wastewater improvements and the new Safe Drinking Water Regulations include an SRF program for funding water treatment improvements. These SRF programs will be very competitive and utilities must provide evidence of sound financial management to qualify for these loans. You should contact your state regulatory agency to find out more about the SRF program in your state.

8.114 Financial Assistance

Many small water treatment and distribution systems need additional funds to repair and upgrade their systems. Potential funding sources include loans and grants from federal and state agencies, banks, foundations, and other sources. Some of the federal funding programs for small public utility systems include:

- Appalachian Regional Commission (ARC),
- Department of Housing and Urban Development (HUD) (provides Community Development Block Grants),
- Economic Development Administration (EDA),
- Indian Health Service (IHS), and
- Rural Utilities Service (RUS)(formerly Farmer's Home Administration (FmHA) and Rural Development Administration (RDA)).

For additional information, see "Financing Assistance Available for Small Public Water Systems," by Susan Campbell, Benjamin W. Lykins, Jr., and James A. Goodrich, *Journal American Water Works Association*, June 1993, pages 47-53.

Another valuable contact is the Environmental Financing Information Network which provides information on financing alternatives for state and local environmental programs and projects in the form of abstracts of publications, case studies, and contacts. Contact Environmental Financing Information Network, U.S. Environmental Protection Agency (EPA), EFIN (mail code 2731R), Ariel Rios Building, 1200 Pennsylvania Avenue, NW, Washington, DC 20460. Phone (202) 564-4994 and Fax (202) 565-2587.

Also many states have one or more special financing mechanisms for small public utility systems. These funds may be in the form of grants, loans, bonds, or revolving loan funds. Contact your state drinking water agency for more information.

QUESTIONS

Write your answers in a notebook and then compare your answers with those on page 503.

8.1A What items should be considered when forecasting expenditures?

8.1B What is a connection fee?

8.1C What are the two general components of inflation which will affect setting of rates and charges?

8.1D What is a revenue bond?

8.12 Itemizing System Expenses

In order to develop a complete statement of our revenue needs, we need to look more closely at our plant and facility expenses. As you read through this section, look at Table 8.3 to see how each expense item is listed in an overall statement of projected revenue requirements.

Let's assume that the surface water source of supply for our water district is contracted for from another water agency and this accounts for about 25 percent of our needs, or 480 acre-feet per year. Further, let's assume that we take this water from a service connection in a "joint-use aqueduct," and that raw water storage is accomplished in a common upstream reservoir. Raw water transmission and storage costs are included in the $150 per acre-foot water charge. In order to treat this water, we own and operate a small 1 MGD water treatment plant. So, costs associated with the surface water *source of supply and treatment* would include the following:

O & M EXPENSES

- Source of supply ($150/ac-ft)(480 ac-ft) $ 72,000
- Chemicals ($5/ac-ft)(480 ac-ft) 2,400
- Power ($12/ac-ft)(480 ac-ft) 5,760
- Labor 100,000

DEBT SERVICE EXPENSES

- Water treatment plant costs 120,000

REPAIR, REPLACEMENT, EXTENSIONS, AND IMPROVEMENT EXPENSES

- Surface water system 35,000

The groundwater source of supply for our water district provides about 75 percent of our needs, or about 1,435 acre-feet per year. Production facilities which are owned and operated by the district to meet this need include five 350-GPM wells. Costs associated with *GROUNDWATER PRODUCTION* include:

O & M EXPENSES

- Source of supply (recharge and replenishment) ($50/ac-ft)(1,435 ac-ft) $ 71,750
- Chemicals ($3/ac-ft)(1,435 ac-ft) 4,305
- Power ($25/ac-ft)(1,435 ac-ft) 35,875

DEBT SERVICE EXPENSES

- Wells 16,000

REPAIR, REPLACEMENT, EXTENSION, AND IMPROVEMENT EXPENSES

- Well sites 7,000

TABLE 8.3 PROJECTED REVENUE REQUIREMENTS FOR XYZ WATER DISTRICT

Expense Item	Unit Price ($/ac-ft)	Expense Revenue Requirements ($)
O & M EXPENSES		
Source of Supply		
Surface water (water cost, transmission, and storage)	150	72,000
Groundwater (recharge and replenishment)	50	71,750
Power		
Surface water treatment plant	12	5,760
Groundwater pumping	25	35,875
Distribution system pumping		5,000
Water Treatment		
Surface water treatment chemicals	5	2,400
Groundwater treatment chemicals	3	4,305
Labor		100,000
Transmission and Distribution		
Storage reservoirs		5,000
Mains		30,000
Meters		20,000
Valves		5,000
Service lines		10,000
Fire hydrants		3,000
Administration and general		
Meter reading		25,000
Billing and collection		40,000
General administration		50,000
Fringe benefits		35,000
Cash reserves[a]		—
DEBT SERVICE EXPENSES[b]		
Water treatment plant		120,000
Wells		16,000
Transmission and distribution system		60,000
Other capital costs (service lines, meters, hydrants, etc.)		24,000
MAJOR REPAIR, REPLACEMENT, EXTENSION, AND IMPROVEMENT EXPENSES[c]		
Surface water treatment plant		35,000
Well sites		7,000
Transmission and distribution system		28,000
Total Revenue Requirement		$ 810,090

[a] A reserve of 10% of one year's O & M costs is suggested. No reserve fund has been budgeted in this example.

[b] Debt service is equal annual payments spread over 25 years (to retire bonds/loans).

[c] Repair and replacements calculated at 2 percent of plant's value.

Joint surface water/groundwater transmission, storage, and distribution facilities include five treated-water storage reservoirs totaling 8.5 MG of operational storage, 50 miles of transmission mains, valve structures, a booster pumping station, and a public fire protection system. Costs of the *distribution system* include:

O & M EXPENSES

- Storage reservoirs $ 5,000
- Distribution
  - Mains 30,000
  - Meters 20,000
  - Valves 5,000
  - Service lines 10,000
  - Fire hydrants 3,000
- Power for pumping 5,000

DEBT SERVICE EXPENSES

- Transmission and distribution system 60,000
- Service lines, meters, and hydrants 24,000

MAJOR REPAIR, REPLACEMENT, EXTENSION, AND IMPROVEMENT EXPENSES

- Transmission and distribution system 28,000

Costs of *administration and general expenses* items include:

ADMINISTRATION AND GENERAL EXPENSES

- Meter reading $ 25,000
- Billing and collection 40,000
- General administration 50,000
- Fringe benefits 35,000

These administration and general expenses are usually included as a separate cost category under O & M expenses.

8.13 Two Different Methods of Establishing Revenue Base

There are two basic accounting methods for establishing the revenue base: (1) the cash basis, and (2) the utility basis. Using the cash basis, the utility must raise enough money to cover all cash needs including *DEBT SERVICE*.[6] The cash basis is generally used by publicly owned utilities and is, therefore, the focus of this chapter. It is essentially an extension of the cash-oriented budgeting and accounting system used by many governmental agencies.

CASH BASIS

Revenue base requirements (as illustrated in Table 8.3) include:

- O & M expenses
- Administration and general expenses
- Debt service expenses
- Capital expenses
- Cash reserves

[6] *Debt Service. The amount of money required annually to pay the (1) interest on outstanding debts; or (2) funds due on a maturing bonded debt or the redemption of bonds.*

Operation and maintenance expenses are usually based on actual expenditures and are adjusted to reflect conditions expected to occur during the rate period. Historical spending patterns are normally used as a starting point for developing cost estimates. System operation and maintenance expense items include:

- Purchased water
- Water treatment
- Salaries and fringe benefits (for O & M)
- Power costs
- Water transmission and distribution
- Outside services (contracts)
- Equipment rental
- Materials and supplies

Administration and general expenses include:

- Salaries and fringe benefits (administrators, secretarial, janitorial)
- Building rental
- Meter reading
- Billing and collection

These administration and general expenses are usually listed under operation and maintenance expenses.

Debt service expenses on bonds or other loans include:

- Principal and interest

Capital expense items include:

- Replacements (equipment)
- Main extensions and improvements
- Major rehabilitation of facilities
- Cash reserves

Most major capital projects are financed by bonds and/or developer-related fees and charges in order to extend the repayment period over the useful life of the facilities. This helps to avoid putting the burden of cost for these facilities on current users when facilities are designed to meet current as well as future system demands.

To adequately plan for the future, every utility should have a repair/replacement fund. The purpose of this fund is to generate additional revenue to pay for the routine, minor repairs and replacement of capital equipment as the equipment wears out. When a treatment plant is new, the balance in the repair/replacement fund should be increasing each year. As the plant gets older, the funds will have to be used and the balance may get dangerously low as equipment breakdowns occur. However, it is important to maintain a positive balance in this account with the understanding that this account is not meant to generate a "profit" for the utility but rather to plan for future equipment needs. In water treatment facilities construction, providing an adequate repair/replacement fund is very important; if this repair/replacement fund hasn't been reviewed annually, it must be updated.

To set up a repair/replacement fund for your utility, first put together a list of the equipment (called an asset inventory) required for each process in your utility. Once you have this list, estimate the life expectancy of the equipment and the replacement cost. From this list you can predict the amount of money you should set aside each year so that when each piece of equipment wears out, you will have enough money to replace that piece of equipment.

UTILITY BASIS

The utility basis for determining revenue requirements is used primarily by investor-owned utilities, although some publicly owned utilities may also use this method to establish rates for customers outside the utility's service area. The major difference in this accounting method is that it entitles the utility owner to a reasonable rate of return (profit) on *EQUITY*[7] in the system. To accommodate this additional cost item, capital costs are additionally broken down into two other components: (1) depreciation expenses, and (2) return on rate base.

Depreciation expense is related to the gradual loss in service value of facilities and equipment over time due to age (wear and tear, obsolescence). Depreciation expense essentially provides for recovery of the capital investment over the useful life of the facility and equipment. The rate-of-return component is designed to pay the cost of debt service and to provide a fair rate of return for total equity capital used to finance physical facilities.

The utility method requires development of a "rate base" (value of assets on which the utility is entitled to earn a return). The rate base is usually based on the value of the utility's plant and property actually used to serve the utility's customers.

In contrast to the cash basis accounting method, capital improvement costs in the utility basis accounting method are recovered through depreciation (wear and tear) and rate of return (debt service), and no provision is made for reserves. Since these cost items tend to be offsetting, the revenue requirement could actually be less under the utility basis for capital items. This possibility should be thoroughly investigated if you are considering using the utility basis for a public or municipal utility.

---

[7] *Equity. The value of an investment in a facility.*

The total revenue requirement for investor-owned utilities includes:

- O & M expenses
- Administration and general expenses
- Federal and State income taxes
- Other taxes (such as property taxes)
- Depreciation expense
- Return on rate base

QUESTIONS

Write your answers in a notebook and then compare your answers with those on page 503.

8.1E List the O & M costs associated with a water distribution system.

8.1F List the two basic accounting methods for establishing a revenue base.

8.1G What is depreciation?

8.2 COST ALLOCATION METHODS

8.20 Relate Costs to Level of Service

If all water users made similar demands on the water system, this section would not be necessary. We could simply divide the total revenue requirements of the utility by the average water use and thus allocate costs evenly across the board. This is usually not the case, however, so costs need to be allocated to customers on the basis of their particular service requirements.

Water service to customers has two distinct components: (1) average demand, and (2) peak flow demand. These demands differ between customer classes and require the utility to provide different levels of service (plant and facilities) to meet these demands. As an illustration, a class of customers with a very high peak rate of use (flow) as compared with the average use may require larger capacity pumps, larger sized mains, and additional reservoir storage capacity. The water rate for this class of customer should reflect their proportionate share of users' *peak flow demand*, as well as the average usage. A customer with a high peak rate of use is sometimes referred to as having a "poor load factor."

In the following sections we will review two popular cost allocation methods used to assign costs to appropriate levels of service. The two methods are called:

1. Commodity-demand method, and
2. Base-extra capacity method.

Cost allocation methods which require the use of significant amounts of personal judgment should be avoided.

8.21 Commodity-Demand Method

The commodity-demand cost allocation method is most appropriate for use by small utilities since it is less complicated to administer, but it should be noted that this method may not be as accurate as the base-extra capacity method in some instances.

To distribute costs of service to user groups by the commodity-demand cost allocation method, we must first identify the utility costs related to the following service categories: (1) commodity (water) charges, (2) demand (peak flow), (3) customer service costs, and (4) public fire protection.

1. Commodity costs tend to vary as a function of *average* water use. These costs include:
   - Purchased water,
   - Chemicals,
   - Power costs, and
   - Other costs directly related to the amounts of water use.
2. Demand costs are associated with providing facilities to meet *peak rates* of use and include:
   - Transmission and distribution system pumping capacities,
   - Treatment plant flow capacity,
   - Transmission and distribution mains,
   - Storage facilities sized to meet peak demands, and
   - Source of supply costs such as "peaking wells" (wells used only during periods of high demand).

   These cost components may be further broken down to assign specific demands to maximum day and maximum hour users.
3. Customer service costs are primarily those related to providing service to customers regardless of the amount of water used. Customer costs include:
   - Meter and service line maintenance,
   - Meter reading and billing,
   - Customer accounting, and
   - Collection costs.

4. Public fire protection costs are primarily demand-related due to the nature of this use. These costs can be allocated based on cost-of-service studies that seem to fit your specific needs or you can estimate these costs based on standard curves such as the one given in AWWA Manual M-1 (for ordering information, see Section 8.8, "Additional Reading," reference 1, at the end of this chapter). A summary of the AWWA allocation method is given in Table 8.4. These costs can be distributed to all public fire protection system users. To use Table 8.4 you must know the number of customers served. For example, if the number of customers served is 1,000, then Table 8.4 indicates that 28 percent of the total revenue can be allocated to fire protection costs.

TABLE 8.4 ALLOCATION OF FIRE PROTECTION COSTS[a]

Number of Customers	Percentage of Total Revenue
500	32.0
1,000	28.0
2,000	24.0
3,000	21.5
4,000	20.0
5,000	18.5

[a] Adapted from *AWWA WATER RATES* Manual No. M-1.

8.22 Base-Extra Capacity Method

The base-extra capacity method of cost allocation has four major cost components:

1. Base costs,
2. Extra capacity costs,
3. Customer costs, and
4. Direct fire protection costs.

1. Base costs tend to vary as a function of total water quantity used and include O & M and capital costs associated with service to customers under average load conditions (excludes peak demand elements). Base costs include:
   - O & M expenses of supply, treatment, pumping, and distribution facilities, and
   - Capital costs related to plant investment to serve average system flow.
2. Extra capacity costs are those required to meet demands in excess of average use. These include O & M and capital costs for system capacity over and above the average flow rate. Extra capacity costs may be further subdivided into additional costs to meet:
   - Max-day extra demand costs,
   - Max-hour extra demand costs, and
   - Other extra demand-related costs.
3. Customer costs are allocated in the same way as they were in the commodity-demand cost allocation method described in Section 8.21.
4. Direct fire protection costs are allocated in the same way as they were in the commodity-demand cost allocation method described in Section 8.21.

To split costs between base and extra capacity we usually use an "allocation factor." For example, assume the maximum daily use exceeds average daily use by a factor of two. This factor can be used to allocate applicable cost components to users responsible for the maximum daily use. In a similar way, a factor based on actual or estimated flow measurements can be used to allocate maximum hourly and other peak demands.

When allocating costs for portions of certain facilities required to accommodate peak demands, be sure to use good judgment in developing allocation factors which are fair to all users and based on all the facts available to you.

The greatest advantage of this cost allocation method is its ability to establish the minimum cost of service under ideal (average) conditions. By using this minimum cost figure to establish base minimum costs, you will avoid setting rates that could result in selling water below cost.

8.23 Examples of Cost Allocations

To illustrate the allocation of costs to various user groups, we will use the commodity-demand method described in Section 8.21 and apply it to our hypothetical water district.

The first step in distributing costs is the allocation of operation and maintenance expenses, as shown in Table 8.5. The figures used in Table 8.5 are those that we projected to be the revenue needs of our water district in Table 8.3.

Cost allocations to the commodity component are primarily expenses which relate to the volume of water delivered such as water cost, power and chemical costs, and the bulk of labor for operating and maintaining the water system.

Costs allocated to the demand component include expenses associated with peak operation of the water system, including a portion of the system operation and maintenance labor.

Customer costs are those not related to volume of water or peak flows. In other words, these costs would be incurred regardless of the amount of water used by the customer.

In this illustration we used a split of 70% commodity and 30% demand allocation for some of the expense items.[8] However, in actual applications these allocation percentages should be based on actual water use data or reasonable estimates of who uses how much water. The general administration and fringe benefits expense items were spread to all categories based on the overall weighted average subtotal of all other expenses. This is a generally acceptable method of allocating these costs. In this example the following procedure was used with the information in Table 8.5 to develop the allocation percentages for administration and fringe benefits.

Cost Components In Table 8.5	Sum of Expense Items Except Admin. and Fringe	Allocation Percentage
Commodity	$ 209,960	$\frac{(209{,}960)(100\%)}{435{,}090} = 48\%$
Demand	127,130	$\frac{(127{,}130)(100\%)}{435{,}090} = 29\%$
Customer	95,000	$\frac{(95{,}000)(100\%)}{435{,}090} = 22\%$
Direct fire protection	3,000	$\frac{(3{,}000)(100\%)}{435{,}090} = 1\%$
TOTALS	$ 435,090	100%

[8] *From Table 8.1 the average daily residential consumption was 1.4 MGD and the maximum daily consumption was (1.4 MGD)(1.8) = 2.52 MGD. Therefore the percent commodity demand would be (100%)(1.4 MGD) / (2.52 MGD) = 68%. For ease of calculations use a split of 70% commodity and 30% demand allocation.*

TABLE 8.5 ALLOCATION OF O & M EXPENSES

Expense Item	Total,[a] $/yr	Commodity, $/yr	Demand, $/yr	Customer, $/yr	Direct Fire Protection, $/yr
Source of Supply:					
— Surface Water	72,000[b]	50,400	21,600	0	0
— Groundwater	71,750[b]	50,225	21,525	0	0
Power:					
— Surface Water T.P.	5,760[b]	4,030	1,730	0	0
— Groundwater Pumping	35,875[b]	25,100	10,775	0	0
— Dist. System Pumping	5,000[b]	3,500	1,500	0	0
Water Treatment:					
— Plant Treat. Chemicals	2,400	2,400	0	0	0
— Well Treat. Chemicals	4,305	4,305	0	0	0
— Labor	100,000[b]	70,000	30,000	0	0
Transmission and Distribution:					
— Storage Reservoirs	5,000	0	5,000	0	0
— Mains	30,000	0	30,000	0	0
— Meters	20,000	0	0	20,000	0
— Valves	5,000	0	5,000	0	0
— Service Connections	10,000	0	0	10,000	0
— Fire Hydrants	3,000	0	0	0	3,000
Admin. and General:					
— Meter Reading	25,000	0	0	25,000	0
— Billing and Collection	40,000	0	0	40,000	0
(Subtotals for calculating allocation percentages)	435,090	209,960	127,130	95,000	3,000
— General Administration Labor	50,000[c]	24,000	14,500	11,000	500
— Fringe Benefits	35,000[c]	16,800	10,150	7,700	350
TOTALS	520,090[d]	250,760	151,780	113,700	3,850

[a] From Table 8.3
[b] Allocation — 70% Commodity; 30% Demand.
[c] Based on other allocation totals.
[d] This figure is normally the basis for calculating the suggested 10% reserve account expense.

It should be noted that if you compare cost figures for the commodity-demand method and the base-extra capacity method, you may see major differences in the percentage allocations between individual water systems for "commodity" and "demand" categories. This occurs because the commodity and demand categories are defined by individual system configuration and water users' demands.

The next step is to allocate capital expense items, but before we can do this we need to develop an overall allocation schedule of plant and facility values. This allocation is shown in Table 8.6; these values are the basis of the long-term debt service expenses shown in Table 8.3. The annual revenue requirements are determined on the basis of the period of debt payment (25 years) and the interest rate (6%). With this information you can use a table found in engineering economic textbooks which will give the compound interest rate factor (0.08) which is multiplied times the "total plant value" to give the "annual revenue requirements." Table 8.6 also includes other capital expense items such as service connections, meters, and hydrants.

Again, the major differences in this schedule are found in the commodity and demand components. In the commodity-demand method of cost allocation, major plant facilities, including those designed to supply peak-day or peak-hour demands, are assigned to the demand component. In the base-extra capacity method, these costs would be broken down between base costs and extra-capacity costs according to the demands of the users. While the latter method is more complex, it also provides a more accurate allocation of expenses if system demands vary considerably during the year.

The last step in cost allocation is to combine these costs into a summary table which reflects the overall costs of service. Such an allocation summary for our hypothetical water district is given in Table 8.7, which includes capital-related expenses for debt service and major repair, replacements, extension, and improvement expenses. Note that we used the cost allocation values given in Table 8.4 to offset fire protection costs because these will be paid as a separate cost item. If this amount proves to be too high or too low, then a cost-of-service study or other analysis should be conducted to establish a more appropriate value.

TABLE 8.6 ALLOCATION OF PLANT VALUE AND OTHER CAPITAL FACILITIES

Capital Item	Total, $/yr	Commodity, $/yr	Demand, $/yr	Customer, $/yr	Direct Fire Protection, $/yr
Water Treatment Plant	1,500,000	0	1,500,000	0	0
Wells	200,000	0	200,000	0	0
Trans. and Dist. System	750,000	0	750,000	0	0
Other Capital					
— Service Connections	150,000	0	0	150,000	0
— Meters	90,000	0	0	90,000	0
— Hydrants	60,000	0	0	0	60,000
Total "Physical Plant" Value	2,750,000	0	2,450,000	240,000	60,000
Annual Revenue Requirements[a]	220,000	0	196,000	19,200	4,800

[a] Based on equal debt payments over 25 years and a 6% interest rate (compound interest factor 0.08). Multiply "Total Physical Plant Value" by 0.08 to obtain "Annual Revenue Requirements."

TABLE 8.7 OVERALL ALLOCATION OF COSTS OF SERVICE

Item	Total, $/yr	Commodity, $/yr	Demand, $/yr	Customer, $/yr	Direct Fire Protection, $/yr
O & M Expenses[a]	520,090	250,760	151,780	113,700	3,850
Capital Expenses					
— Debt Service[b]	220,000	0	196,000	19,200	4,800
— Major Repair and Replacement[c]	70,000	0	70,000	0	0
Total Cost of Service[d]	810,090	250,760	417,780	132,900	8,650
Less Fire Protection Revenue[e]	<162,000>	—	<153,350>	—	<8,650>
General Cost of Water Service	648,090	250,760	264,430	132,900	0

[a] From Table 8.5
[b] Based on Annual Revenue Requirements given in Table 8.6
[c] From Table 8.3
[d] This amount does not include the suggested 10% O & M reserve funds ($52,009).
[e] 20% of Revenue based on Table 8.4. Values on this line subtracted because they will be paid as a separate cost item.

QUESTIONS

Write your answers in a notebook and then compare your answers with those on page 504.

8.2A List the utility service cost categories used to distribute costs in the commodity-demand method.

8.2B What is a commodity cost?

8.2C What is a demand cost?

8.2D What is a customer cost?

8.3 DISTRIBUTION OF COSTS TO CUSTOMERS

8.30 Establishing Customer Classes

The next step in our planning effort is to establish different classes of water service by customer groups. In setting up customer classes you must be sure to create sufficient classes and broad enough category definitions to cover all user groups. Always keep in mind that the overall intent of establishing customer classes is to group customers into categories of similar *use patterns.* Thus, classes should separate users by load demand and peaking characteristics.

While the process of establishing customer classes will vary depending on a system's unique mix of customers, the following general customer classes are frequently used by many water utilities:

1. Residential,
2. Commercial,
3. Industrial,
4. Agricultural (irrigation), and
5. Other (public authority, government, etc.).

Residential customers are typically single-family residences. In some instances, this class includes multiple-family dwelling units. Commercial customers are nonindustrial, business, and service-type enterprises. Industrial customers are usually manufacturing and processing enterprises. Agricultural customers may be heavy irrigation users or livestock or poultry feeding operations. The "other" category may be used for schools, governmental uses, parks, and recreation facilities.

8.31 Special Considerations

In distributing costs to customers you will need to give special consideration to such items as:

- Unusual service characteristics,
- Demand patterns,
- Wholesale versus retail use,
- Interruptible service options, and
- Inside versus outside service area.

Unusual service characteristics are unique factors which require special consideration for a particular customer or class of users. For example, demand patterns could influence rates for heavy irrigation users or other seasonal users such as canneries. In addition, if a system has both wholesale and retail customers, recognition should be given to the wholesalers during periods when they operate at less than 100 percent capacity. You may want to establish a lower rate as an incentive for customers to contract for interruptible service. This type of service contract allows the utility to cut off service during periods of high system demand or under drought conditions. For customers who reside outside of district boundaries, you may want to set a slightly higher rate to provide a return on capital investment within the system. In some cases, special assessments may be charged to customers at higher elevations in the form of a pressure zone surcharge to cover additional pumping and storage costs. In any event, the intent of distributing costs to different classes of water users is to recognize the differing demands on the water system and thereby set water rates which accurately reflect the actual cost of service to each customer class.

8.4 RATE DESIGN

8.40 General

The final step in our plan of attack is to design water rates which will recover appropriate costs from each class of customer. In this section we will develop a typical water bill for our hypothetical water district.

Simplified methods of rate design can be used in small utilities when the cost of preparing a comprehensive rate study are prohibitive. However, when a utility, regardless of size, has special circumstances such as a large industrial base, a comprehensive or partial rate study may be in order.

While rates should always be based on the cost of service, good judgment must also be used in rate setting. Remember that the goal is to develop a rate structure which attains the maximum degree of equity among all customer classes.

Before we proceed we need to touch on a somewhat thorny issue in some parts of the country — metered vs. unmetered customers. It has generally been shown that unmetered customers tend to use more water than equivalent metered customers. Therefore, providing service to unmetered customers when the metering option is available is discouraged. In this section we will assume that all customers are metered.

8.41 Information and Data Requirements

In order to set accurate rates that differentiate between customer classes, we need a good historical database from which information on the various user classes can be taken. General information and data requirements include the following:

1. Revenue requirements for the projected rate period in sufficient detail to permit allocation to cost categories,
2. A bill-frequency analysis of metered water use by customer class and meter size, and
3. A record of the number and size of meters for each customer class.

The basic revenue requirement for our hypothetical water district was developed in Section 8.13, and is illustrated in Table 8.3. A further breakdown and allocation of the revenue requirement is given in Tables 8.5 and 8.7.

A bill-frequency analysis can be prepared manually by summarizing metered water use data for a given period of time, say the last three to five years. With the assistance of a personal computer, this database can be created easily on a spreadsheet and updated routinely on a monthly basis.

Records of the number and size of meters installed in the system can be manually determined from customer records. Again, the use of a personal computer can greatly reduce the effort required to produce this information in a readily usable format.

By analyzing the above data we can determine if water use patterns differ significantly from the average and, if so, which customer classes are responsible for these deviations. If the makeup of the water system is predominantly residential, you may not find any major deviations. But if significant differences are noted in a given customer's use pattern, then consider establishing a separate rate for this water user. In many cases, a uniform rate for the commodity-demand cost components may prove to be quite satisfactory.

8.42 Rate Components

The overall water rate for our typical water district will be divided into four major components:

1. Service charge,
2. Volume charge,
3. Fire protection charge, and
4. Miscellaneous charge.

The service charge is designed to recover customer-related costs such as meter-reading expenses. These costs are incurred regardless of the amount of water used, and are distributed over all customer classes. The service charge is based on meter size. However, if revenue from your water sales fluctuates widely, consider shifting more of your costs to the service charge to avoid revenue shortages.

The volume charge should reflect the costs associated with actual water use. In a true commodity-demand system, this charge would be broken down into the two components of commodity and demand volumes. To do this on an ongoing basis, however, demand water meters would be required and these meters are generally not cost-effective, especially in residential applications. Thus, if water use patterns dictate that different rates should be established for the commodity and demand components, historical water use patterns and good judgment should be used in setting these rates for the small water utility.

The fire protection cost component includes costs for fire hydrants, service lines, and other facilities required to provide an adequate water supply in the event of a fire. In public systems the simplest method of recovering these costs is to spread them equally among all who benefit from the public fire protection system. In the case of private fire protection systems, costs should be paid by the individual user in much the same way as initial service connection fees.

Miscellaneous charges might include a fee for establishing a new account or other one-time costs which should be segregated from the other rate components.

Special rate schedules may also need to be established for certain customers such as those located outside district boundaries. As mentioned in a previous section, you may want to assess a surcharge to these customers to recover a portion of the system capital costs above the actual investment in the form of a rate of return.

In the past, multiple-block rate structures have been established for quantity users who have a small demand factor or who place lower excess-capacity requirements on the water system. This rate structure has generally been applied to large industrial users in the form of volume discounts. However, in view of the current trend to encourage water conservation, this rate structure is finding less acceptance in the water industry, and uniform rate structures are finding wider acceptance. Uniform rate structures make it easier to promote conservation of water when every customer pays the same for each gallon of water.

8.43 Rate Design Example

Let's go through an example of developing a water bill for our typical residential customer to put everything we've learned about rate setting into perspective.

For purposes of illustration, let's assume that we have a uniform rate structure (no differentiation between commodity and demand volume use) and the average monthly water use is 11,000 gallons for our typical residential user. Volume of use is frequently expressed as cubic feet or hundred cubic feet (ccf), so 11,000 gallons is equal to 14.7 ccf (1 cubic foot = 7.48 gallons).

We calculate the service charge component on the basis of meter size. In this example, we will assume the use of a 5/8-inch residential meter. Before we can assign a service charge, however, we must determine the total number of equivalent 5/8-inch meters in the system. The total number of equivalent meters for our hypothetical water district is 4,231 as shown in Table 8.8. The annual customer cost of service component is $132,900 as given in Table 8.7. To establish the monthly service charge for a 5/8-inch residential meter, divide the annual cost of service by the equivalent number of 5/8-inch meters, then divide this result by 12 as follows:

$$\begin{matrix}\text{5/8-Inch Meter} \\ \text{Service Charge}\end{matrix} = \frac{\$132{,}900/\text{yr}}{(4{,}231 \text{ meters})(12 \text{ mo/yr})} = \begin{matrix}\$2.62 \text{ per month} \\ \text{per meter}\end{matrix}$$

For a 1-inch meter we would multiply the monthly service charge of $2.62 by 1.4 (from Table 8.8), yielding an equivalent 1-inch meter service charge of $3.67 per month.

TABLE 8.8 DETERMINATION OF EQUIVALENT 5/8-INCH METERS[a]

Meter Size (In)	Number of Meters	Service[b] Ratio	Equivalent 5/8-Inch Meters
5/8	3,550	1.0	3,550
1	350	1.4	490
1 1/2	90	1.8	162
2	10	2.9	29
TOTAL	4,000		4,231

[a] Adapted from *AWWA Manual M-1.*
[b] Service ratio based on relative cost of meter and service investment.

The volume charge component is designed to recover commodity- and demand-related costs as given in Table 8.7. This charge is calculated by dividing the revenues to be recovered by total water sales as follows:

Commodity cost	$ 250,760
Demand cost	264,430
Total volumetric cost	$ 515,190

Annual water use = 1,915 acre-feet

= (1,915 ac-ft)(43,560 sq ft/ac)(7.48 gal/cu ft)

= 624,000,000 gallons

= 624,000 • 1,000 gallons

= 624,000 thousand gallons

$$\text{Volumetric Rate} = \frac{\$515{,}190}{624{,}000 \text{ thous. gal}} = \$0.83/\text{thousand gal}$$

For our typical residential user who consumed 11,000 gallons of water for the month, this volume charge would be

$$\frac{(\$0.83)(11{,}000 \text{ gal/mo})}{1{,}000 \text{ gal}} = \$9.13 \text{ per month.}$$

The public fire protection charge component is designed to recover the demand and direct fire protection costs as given in Table 8.7. The fire protection charge is calculated in a manner similar to that of the service charge based on the annual demand-related costs ($153,350) and direct fire protection costs ($8,650) given in Table 8.7, as follows:

$$\text{5/8-Inch Meter Fire Protection Charge} = \frac{\$153{,}350/\text{yr} + \$8{,}650/\text{yr}}{(4{,}231 \text{ meters})(12 \text{ mo/yr})} = \$3.19 \text{ per month per meter}$$

In this example we will assume that there are no miscellaneous charges, so this completes our water bill. Bills for commercial or industrial customers would be calculated in a similar fashion. In summary, our typical residential customer would pay the following monthly water bill:

Service Charge =	$ 2.62
Volume Charge =	9.13
Fire Protection Charge =	3.19
Total Monthly Charge	$ 14.94

On the basis of these calculations, the water rates for various sized residential meters can be established as follows:

Meter Size (inches)	Service Ratio (From Table 8.8)	Meter Service Charge ($2.62/mo) (Ratio)	Fire Protection Charge ($3.19/mo) (Ratio)	Total Fixed Charge ($/mo)
5/8	1.0	$2.62	$ 3.19	$ 5.81
1	1.4	3.67	4.47	8.14
1 1/2	1.8	4.72	5.74	10.46
2	2.9	7.60	9.25	16.85

Therefore, monthly residential charges would be the total fixed charge based on meter size plus a volumetric charge of $0.83 per thousand gallons used.

In our example water district (Table 8.1) there are two industrial water users and the number of commercial and miscellaneous customers is not indicated. If the industrial and commercial users have demands similar to the demands of residential customers, the fixed charges should be based on meter size and the volumetric charge the same as for residential customers.

8.44 Typical Rate Structure

Most water districts use a rate structure similar to the one developed in the previous section with a fixed charge plus a volumetric charge. Very often the charges are rounded upward to the nearest 10 cents, 25 cents, 50 cents, or a dollar for ease of computation and to generate some surplus revenue for a reserve fund. Some water districts have a minimum charge and the first 1,500 gallons or 200 cubic feet of water used per month are included in this charge with an additional volumetric charge for consumption above the minimum level.

Typical charges for a small water district are listed below by type of water user.

Number of Gallons	$ Rate per 1,000 Gallons	Base Charge
RESIDENTIAL		
(5/8-inch meter)		
First 1,000 gallons		$ 6.00
Additional gallons	$ 0.80	
CLINIC		
First 5,000 gallons		30.00
Additional gallons	1.20	
ELEMENTARY SCHOOL		
First 10,000 gallons		45.00
Additional gallons	1.20	
COMMERCIAL		
Restaurant		
First 5,000 gallons		30.00
Additional gallons	1.20	
Other Commercial		
First 1,000 gallons		
1.0 inch or smaller		10.00
1.01 to 2.0 inch		13.25
3.0 inch		16.50
4.0 inch		26.50
5.0 inch		34.75
6.0 inch		46.50
Additional gallons	0.80	

In addition to water usage fees, water districts usually also have a list of other important fees or charges to cover the costs of managing the water supply system. Typical other charges are listed below:

GUARANTEE DEPOSIT	
Single-family residences	$ 75.00
Commercial accounts	150.00
MISCELLANEOUS CHARGES	
Transfer fee	25.00
Service call (after hours)	20.00 per call
Turn-on fee (after shutoff for delinquent account)	35.00
WATER CONNECTION CHARGES (New Installation)	
Capital improvement fee	$1,000.00
Water main front footage charge	5.00 per foot
Water service meter charge	Cost of meter plus 15%
Water main tap-in charge	100.00

8.45 Very Small System Example

To illustrate the application of the concepts presented in this chapter, a rate structure will be developed for a very small water supply system.

1. Determine the service area statistics similar to Table 8.1.

Sources of Supply	2 wells with hydropneumatic tanks and one storage tank
Customers (services)	200
Population	500 people
Average Daily Consumption	50,000 gallons per day

2. Estimate the annual revenue needs based on current and expected expenses (Table 8.3).

O & M EXPENSES	
Power (pumping, lights)	$ 1,000.00
Chemicals (chlorine)	700.00
Salaries	25,000.00
Materials and supplies	5,000.00
Vehicles (gas, oil, tires)	1,000.00
Administration (secretarial, bookkeeping)	20,000.00
Utilities (phone, heat)	500.00
Office rent	1,800.00
DEBT SERVICE (from bookkeeper)	5,000.00
REPAIR, REPLACEMENT, RESERVE	6,000.00
Total Revenue Requirement =	$66,000.00

3. The next step is to allocate the costs (Table 8.5, Allocation of O & M Expenses, Table 8.6, Allocation of Plant Value and Other Capital Facilities, and Table 8.7, Overall Allocation of Cost of Service). Assign the last three O & M expense items to the customer (administration, $20,000, + utilities, $500, + rent, $1,800 = $22,300) because these are not costs directly related to the production of water.

Item	Total	Commodity and Demand	Customer
O & M Expenses	$55,000	$32,700	$22,300
Debt Service	5,000	5,000	—
Repair, Replacement, Reserve	6,000	6,000	—
Total Costs	$66,000	$43,700	$22,300

4. Determine the monthly meter service charge.

$$\text{Meter Service Charge, \$/mo} = \frac{\text{Customer Charge, \$/yr}}{\text{(Number of Meters)(12 mo/yr)}}$$

$$= \frac{\$22{,}300\text{/yr}}{\text{(200 meters)(12 mo/yr)}}$$

$$= \$9.29\text{/mo}$$

5. Determine the monthly volumetric water charge.

$$\text{Total Annual Water Use, 1,000 gal/yr} = \text{(Average Daily Use, GPD)(365 days/yr)}$$

$$= \text{(50,000 gal/day)(365 days/yr)}$$

$$= 18{,}250{,}000 \text{ gallons/yr}$$

$$= 18{,}250 \text{ thousand gallons/yr}$$

$$\text{Volumetric Water Charge, \$/1,000 gal} = \frac{\text{Total Commodity and Demand Cost, \$/yr}}{\text{Total Water Use, 1,000 gal/yr}}$$

$$= \frac{\$43{,}700\text{/yr}}{18{,}250 \text{ thousand gallons/yr}}$$

$$= \$2.40 \text{ per thousand gallons}$$

On the basis of these calculations, the customers for this very small water district will pay $9.29 per month per meter plus $2.40 per thousand gallons of water used.

QUESTIONS

Write your answers in a notebook and then compare your answers with those on page 504.

8.3A What is the basis for establishing customer classes?

8.3B Who would be included in a commercial customer class?

8.4A Who uses more water — metered or unmetered customers?

8.4B List the four major components of a typical water rate structure.

8.5 ADMINISTRATION OF RATES AND CHARGES

8.50 Presenting Rates to the Rate Payer

If you have followed the procedures properly up to this point, then presenting your rate needs to the rate payer should not be a problem, but a rather routine matter.

An important aspect in presenting revenue needs to the rate payer is having the ability to justify your system's revenue needs. This is best accomplished by practicing sound financial management and being able to back up your needs with factual records.

In making rate increase presentations to an elected Board of Directors or other elected officials at a public hearing, it is helpful to prepare illustrations of the utility's needs in the form of charts and graphs which state the following information clearly and simply:

- Recent cost performance of utility (budget comparisons),
- Current revenue requirements vs. projected needs,
- Effect of proposed rate increases on typical user groups, and
- Comparison of proposed rates with other similar water utilities in your area.

It is good practice to consider potential rate increases on a regular basis, usually annually. This will promote a stable rate adjustment program and will avoid increasing rates by 15 to 20 percent or more during periods of high inflation or greater demand on utility funds. The latter sometimes occurs when rates are not adjusted for multiple-year periods. The consumer expects utility costs to rise moderately over time, but significant one-time bumps in rates every three or four years can be judged by the consumer to reflect poor fiscal management.

8.51 Consumer Confidence Reports (CCRs)

EPA has developed regulations requiring every public water system to prepare and distribute an annual Consumer Confidence Report. The reports are an opportunity for positive communication with consumers and a means to convince consumers of the importance of paying for good quality water.

Consumer Confidence Reports are an effective way for a water utility to communicate with consumers that their water is safe to drink. These reports also provide an opportunity to inform rate payers of the need for sufficient funds to properly operate and maintain the water supply, treatment, and distribution systems. If higher rates are necessary to fund a capital improvement program, the reports can explain the importance of having sufficient water with adequate pressure and high quality.

Items that should be covered in Consumer Confidence Reports include:

1. Information on source of drinking water supply,
2. Brief definition of terms,
3. List of contaminants detected including level detected, MCL (maximum contaminant level), and MCLG (maximum contaminant level goal),
4. For MCLs violated, information on health effects, and
5. Information on any unregulated contaminants.

Some utility agencies try to have an article published in the local newspaper explaining the report before it is made available to consumers. This advance information helps the consumers understand the report. CCRs should be a short, concise letter report of one or two pages that can be mailed directly to consumers or mailed in the envelope with the utility bill. All utility staff should be familiar with the contents of the report because consumers who know individual staff members frequently will ask questions about the report. Water utilities should emphasize the good job they are doing as guardians of health in these Consumer Confidence Reports.

Contact your local or state drinking water supply agency for suggestions and details on how to prepare a Consumer Confidence Report. The regulations required the first CCR to be delivered to customers by October 19, 1999 and the second one by July 1, 2000. The third and all subsequent CCRs must be issued by July 1 each year.

QUESTIONS

Write your answers in a notebook and then compare your answers with those on page 504.

8.5A What information should be presented to officials when making a rate increase presentation?

8.5B Why should all utility staff be familiar with the contents of the Consumer Confidence Report?

8.6 PLANNING FOR FINANCIAL STABILITY

Financial management for a utility should include providing financial stability for the utility, careful budgeting, and providing capital improvement funds for future utility expansion. These three areas must be examined on a routine basis to ensure the continued operation of the utility. They may be formally reviewed on an annual basis or more frequently when the utility is changing rapidly. The utility manager should understand what is required for each of the three areas and be able to develop record systems that keep the utility on track and financially prepared for the future.

8.60 Measuring Stability

How do you measure financial stability for a utility? Two very simple calculations can be used to help you determine how healthy and stable the finances are for the utility. These two calculations are the OPERATING RATIO and the COVERAGE RATIO. The operating ratio is a measure of the total revenues divided by the total operating expenses. The coverage ratio is a measure of the ability of the utility to pay the principal and interest on loans and bonds (this is known as debt service) in addition to any unexpected expenses. A utility that is in good financial shape will have an operating ratio and coverage ratio above 1.0. In fact, most bonds and loans require the utility to have a coverage ratio of at least 1.25. As state and federal funds for utility improvements have become much more difficult to obtain, these financial indicators have become more important for utilities. Being able to show and document the financial stability of the utility is an important part of getting funding for more capital improvements.

The operating ratio is perhaps the simplest measure of a utility's financial stability. In essence, the utility must be generating enough revenue to pay its operating expenses. The actual ratio is usually computed on a yearly basis, since many utilities may have monthly variations that do not reflect the overall performance. The total revenue is calculated by adding up all revenue generated by user fees, hook-up charges, taxes or assessments, interest income, and special income. Next determine the total operating expenses by adding up the expenses of the utility, including administrative costs, salaries, benefits, energy costs, chemicals, supplies, fuel, equipment costs, equipment replacement fund, principal and interest payments, and other miscellaneous expenses.

EXAMPLE 1

The total revenues for a utility are $1,686,000 and the operating expenses for the utility are $1,278,899. The debt service expenses are $560,000. What is the operating ratio? What is the coverage ratio?

Known		Unknown
Total Revenue, $	= $1,686,000	Operating Ratio
Operating Expenses, $	= $1,278,899	Coverage Ratio
Debt Service Expenses, $	= $560,000	

1. Calculate operating ratio.

$$\text{Operating Ratio} = \frac{\text{Total Revenue, \$}}{\text{Operating Expenses, \$}}$$

$$= \frac{\$1{,}686{,}000}{\$1{,}278{,}899}$$

$$= 1.32$$

2. Calculate non-debt expenses.

$$\text{Non-Debt Expenses, \$} = \text{Operating Exp, \$} - \text{Debt Service Exp, \$}$$

$$= \$1{,}278{,}899 - \$560{,}000$$

$$= \$718{,}899$$

3. Calculate coverage ratio.

$$\text{Coverage Ratio} = \frac{\text{Total Revenue, \$} - \text{Non-Debt Expenses, \$}}{\text{Debt Service Expenses, \$}}$$

$$= \frac{\$1{,}686{,}000 - \$718{,}899}{\$560{,}000}$$

$$= 1.73$$

These calculations provide a good starting point for looking at the financial strength of the utility. Both of these calculations use the total revenue for the utility, which is an important component for any utility budgeting. As managers we often focus on the expense side and forget to look carefully at the revenue side of utility management. The fees collected by the utility, including hook-up fees and user fees, must accurately reflect the cost of providing service. These fees must be reviewed annually and they must be increased as expenses rise to maintain financial stability. Some other areas to examine on the revenue side include how often and how well user fees are collected, the number of delinquent accounts, and the accuracy of meters. Some small communities have found they can cut their administrative costs significantly by switching to a quarterly billing cycle. The utility must have the support of the community to determine and collect user fees, and the utility must keep track of revenue generation as carefully as resource spending.

8.61 Budgeting

Budgeting for the utility is perhaps the most unpleasant task of the year for many managers. The list of needs usually is much larger than the possible revenue for the utility. The only way for the manager to prepare a good budget is to have good records from the year before. A system of recording or filing purchase orders (see "Procurement Records" in Section 8.622) or a requisition records system must be in place to keep track of expenses and prevent spending money that is not in the budget.

To budget effectively, a manager needs to understand how the money has been spent over the last year, the needs of the utility, and how the needs should be prioritized. The manager also must take into account cost increases that cannot be controlled while trying to minimize the expenses as much as possible. The following problem is an example of the types of decisions a manager must make to keep the budget in line while also improving service from the utility.

EXAMPLE 2

A pump which has been in operation for 25 years pumps a constant 600 GPM through 47 feet of dynamic head. The pump uses 6,071 kilowatt-hours of electricity per month, at a cost of $0.085 per kilowatt-hr. The old pump efficiency has dropped to 63 percent. Assuming a new pump that operates at 86 percent efficiency is available for $9,730.00, how long would it take to pay for replacing the old pump?

Known		Unknown
Electricity, kW-hr/mo	= 6,071 kW-hr/mo	New Pump Payback Time, yr
Electricity Cost, $/kW-hr	= $0.085/kW-hr	
Old Pump Efficiency, %	= 63%	
New Pump Efficiency, %	= 86%	
New Pump Cost, $	= $9,730	

1. Calculate old pump operating costs in dollars per month.

$$\text{Old Pump Operating Costs, \$/mo} = (\text{Electricity, kW-hr/mo})(\text{Electricity Cost, \$/kW-hr})$$

$$= (6{,}071 \text{ kW-hr/mo})(\$0.085/\text{kW-hr})$$

$$= \$516.04/\text{mo}$$

2. Calculate new pump operating electricity requirements.

$$\text{New Pump Electricity, kW-hr/mo} = (\text{Old Pump Electricity, kW-hr/mo})\frac{(\text{Old Pump Eff, \%})}{(\text{New Pump Eff, \%})}$$

$$= (6{,}071 \text{ kW-hr/mo})\frac{(63\%)}{(86\%)}$$

$$= 4{,}447 \text{ kW-hr/mo}$$

3. Calculate new pump operating costs in dollars per month.

$$\text{New Pump Operating Costs, \$/mo} = (\text{Electricity, kW-hr/mo})(\text{Electricity Cost, \$/kW-hr})$$

$$= (4{,}447 \text{ kW-hr/mo})(\$0.085/\text{kW-hr})$$

$$= \$378.03/\text{mo}$$

4. Calculate annual cost savings of new pump.

$$\text{Cost Savings, \$/yr} = (\text{Old Costs, \$/mo} - \text{New Costs, \$/mo})(12 \text{ mo/yr})$$

$$= (\$516.04/\text{mo} - \$378.03/\text{mo})(12 \text{ mo/yr})$$

$$= \$1{,}656.12/\text{yr}$$

5. Calculate the new pump payback time in years.

$$\text{Payback Time, yr} = \frac{\text{Initial Cost, \$}}{\text{Savings, \$/yr}}$$

$$= \frac{\$9{,}730.00}{\$1{,}656.12/\text{yr}}$$

$$= 5.9 \text{ years}$$

In this example a payback time of 5.9 years is acceptable and would probably justify the expense for a new pump. This calculation was a simple payback calculation which did not take into account the maintenance on each pump, depreciation, and inflation. Many excellent references are available from EPA to help utility managers make more complex decisions about purchasing new equipment (see Section 8.8, "Additional Reading").

The annual report should be used to help develop the budget so that long-term planning will have its place in the budgeting process. The utility manager must track revenue generation and expenses with adequate records to budget effectively. The manager must also get input from other personnel in the utility as well as community leaders as the budgeting process proceeds. This input from others is invaluable to gain support for the budget and to keep the budget on track once adopted.

8.62 Recordkeeping

8.620 Purpose of Records

As described in Section 8.41, good recordkeeping is the basis for setting accurate water rates. A good historical database of customers' water use will provide the basis for allocation of costs to appropriate users. Accurate records also contribute to effective utility management, satisfy legal requirements, and provide valuable operations and maintenance information that will save time when problems develop. When accurately kept, records provide a sound basis for design of future changes or expansions of the treatment plant. If legal questions or problems occur in connection with the treatment processes or the operation of the plant, accurate and complete records will provide evidence of what actually occurred and what procedures were followed.

8.621 Computer Recordkeeping Systems

Until fairly recently, water supply system recordkeeping has been done manually. The current availability of low-cost personal computer systems puts automation of many manual bookkeeping functions within the means of all water utilities.

To automate your recordkeeping functions as they relate to customer billing, you will need to develop a simple database management system that will create tables similar to those illustrated in this chapter. This can be readily accomplished by use of standard spreadsheet software programs which are available in the marketplace at a cost of $300 to $400. Hardware including a personal computer, data storage system, and a printer can be purchased for under $5,000.

Excellent computer software packages are being developed and offered to assist utility managers. SURF (Small Utility Rates and Finances) has been developed by the American Water Works Association (AWWA). SURF is a self-guided, interactive spreadsheet application designed to assist small drinking water systems in developing budgets, setting user rates, and tracking expenses. SURF requires very little computer or software knowledge and can be used by system operators, bookkeepers, and managers to improve the financial management practices of their utilities. SURF can print out three separate modules: (1) system budget, (2) user rate(s), and (3) system expenses.

SURF hardware and software requirements are modest.

Hardware Requirements:	IBM Compatible PC with at least 4 MB RAM (8 MB recommended). Color monitor. Ink jet or laser jet printer.
Software Requirements:	DOS (Version 5 or newer), Windows 95 or higher and Microsoft Excel 97.

The SURF software and an excellent user's manual are available free from the American Water Works Association. They can be obtained by calling the AWWA Small Systems Program at (800) 366-0107 or by downloading the program from the AWWA website (www.AWWA.org).

Two computer programs available for water and wastewater utility managers for rate setting, impact fees, and financial planning are RateMod Pro ($1,495) and RateMod XP ($1,795). For details and assistance in selecting the most appropriate program for your utility, contact RateMod Associates at (202) 237-2455.

8.622 Types of Records

Many different types of records are required for effective management and operation of water supply, treatment, and distribution system facilities. The following paragraphs describe some of the most important types of records that should be kept and Section 8.623 discusses how long records need to be kept.

Equipment and Maintenance Records

A good plant maintenance effort depends heavily on good recordkeeping. You will need to keep accurate records to monitor the operation and maintenance of each piece of plant equipment. Equipment control cards and work orders can be used to:

- Record important equipment data such as make, model, serial number, and date purchased,
- Record maintenance and repair work performed to date,
- Anticipate preventive maintenance needs, and
- Schedule future maintenance work.

Whenever a piece of equipment is changed, repaired, or tested, the work performed should be recorded on an equipment history card of some type. Complete, up-to-date equipment records will enable the plant operators to evaluate the reliability of equipment and will provide the basis for a realistic preventive maintenance program.

Plant Operations Data

Plant operations logs can be as different as the treatment plants whose information they record. The differences in amount, nature, and format of data are so significant that any attempt to prepare a "typical" log would be very difficult. For detailed information and example recordkeeping forms for a water supply utility, see *WATER TREATMENT PLANT OPERATION*, Volume I, Chapter 10, "Plant Operation," Section 10.6, "Operating Records and Reports."

Procurement Records

Ordering repair parts and supplies usually is done when the on-hand quantity of a stocked part or chemical falls below the reorder point, a new item is added to stock, or an item has been requested that is not stocked. Most organizations require employees to submit a requisition (Figure 8.2) when they need to purchase equipment or supplies. When the requisition has been approved by the authorized person (a supervisor or purchasing agent, in most cases) the items are ordered using a form called a purchase order. A purchase order contains a number of important items, including: (1) the date, (2) a complete description of each item and quantity needed, (3) prices, (4) the name of the vendor, and (5) a purchase order number.

A copy of the purchase order should be retained in a suspense file or on a clipboard until the ordered items arrive. This procedure helps keep track of the items that have been ordered but have not yet been received.

All supplies should be processed through the storeroom immediately upon arrival. When an item is received, it should be so recorded on an inventory card. The inventory card will keep track of the numbers of an item in stock, when last ordered, cost, and other information. Furthermore, by always logging in supplies immediately upon receipt, you are in a position to reject defective or damaged shipments and control shortages or errors in billing. Many utilities now use personal computers to keep track of orders and deliveries.

Inventory Records

An inventory consists of the supplies the treatment plant needs to keep on hand to operate the facility. These maintenance supplies may include repair parts, spare valves, electrical supplies, tools, and lubricants. The purpose of maintaining an inventory is to provide needed parts and supplies quickly, thereby reducing equipment downtime and work delays.

In deciding what supplies to stock, keep in mind the economics involved in buying and stocking an item as opposed to depending on outside availability to provide needed supplies. Is the item critical to continued plant or process operation? Should certain frequently used repair parts be kept on hand? Does the item have a shelf life?

Inventory costs can be held to a minimum by keeping on hand only those parts and supplies for which a definite need exists or which would take too long to obtain from an outside vendor. A "definite need" for an item is usually demonstrated by a history of regular use. Some items may be infrequently used but may be vital in the event of an emergency; these items should also be stocked. Take care to exclude any parts and supplies that may become obsolete, and do not stock parts for equipment scheduled for replacement.

Personnel Records

Documentation of all aspects of personnel management provides an important measure of legal protection for the utility, even a small water utility with only a few employees. If an employee files a lawsuit alleging discriminatory hiring practices or treatment, harassment, breach of contract, or other grievance, the utility's ability to defend its practices and procedures will depend almost entirely on complete and accurate records. Similarly, if an employee is injured on the job, written records can help establish whether the utility was responsible for the accident.

Each personnel action should be fully documented in writing and filed. Even verbal discussions of a supervisor with an employee about job performance should be summarized in writing upon completion of the conversation. Also file copies of all written warnings and disciplinary actions.

An employee's personnel file should also contain a complete record of accomplishments, certificates earned, commendations received, and formal performance reviews. Personnel records often contain sensitive, confidential information; therefore, access to these records should be closely controlled.

8.623 Disposition of Utility Records

An important question is how long records should be kept. As a general rule, records should be kept as long as they may be useful or as long as legally required. Some information will become useless after a short time, while other data may be valuable for many years. Data that might be used for future design or expansion should be kept indefinitely. Laboratory data will always be useful and should be kept indefinitely. Regulatory agencies may require you to keep certain water quality analyses (bacteriological test results) and customer complaint records on file for specified time periods (10 years for chemical analyses and bacteriological tests).

Even if old records are not consulted every day, this does not lessen their potential value. For orderly records handling and storage, set up a schedule to periodically review old records and to dispose of those records that are no longer needed. A decision can be made when a record is established regarding the time period for which it must be retained.

8.7 ACKNOWLEDGMENT

Portions of the material in this chapter were provided by Lorene Lindsay and the chapter was reviewed by Leonard Ainsworth, Mike Cherniak, and Dave Davidson. Their contributions are greatly appreciated.

P-1

CITY OF SACRAMENTO
REQUISITION

1 BID NO.

DUE DATE

2 PURCHASE ORDER NO.

DATE

3 DELIVER TO

4 FUND GP.	5 FUND	6 ORGAN.	7 REV. COST CTR	8 OBJECT	

9 REQUISITION NO

DATE

10 REFER QUESTIONS TO:

VENDORS: USE THESE COLS. ONLY

11 ITEM NO.	12 COMMODITY CODE NO.	13 DESCRIPTION	14 QUANTITY / UNIT	15 PRICE / UNIT	16 BRAND OFFERED	17 LOT PRICE FOR EACH ITEM

18 DEPT.

19 ORIG BY

TEL. NO.

THIS REQUISITION WILL BE REPRODUCED TO CREATE A PURCHASE ORDER QUICKLY AND PROVIDE FASTER DELIVERY OF THESE ITEMS TO YOUR DEPARTMENT. PLEASE FOLLOW INSTRUCTIONS CAREFULLY.

SUB TOTAL

SALES TAX

INVOICE AMOUNT

20 EST. OF COST

22 SPECIAL REQUIREMENTS OR INSTRUCTIONS - PREVIOUS P.O. NO. OR BID NO., OR RECOMMENDED VENDORS

21

NUMBER OF REQUISITION ATTACHMENT SHEETS

23 CERTIFICATION IS HEREBY MADE THAT THE ABOVE IS A LEGAL CHARGE AGAINST THE APPROPRIATION INDICATED

DEPT. HEAD ________________ BY ________________

____
OFFICIAL TITLE

24

VENDOR ________________ (A)

ADDRESS ________________ (B)

CITY, STATE AND ZIP ________________ (C)

TERMS ________________ TEL. NO. ________________ (D)

DELIVERY ________________ FOB DESTINATION (E)

BY ________________ TITLE ________________

Fig. 8.2 Requisition/purchase order form

8.8 ADDITIONAL READING

Setting rates for a small water utility is often one of the main responsibilities of the utility manager. For a more detailed discussion of a manager's role and responsibilities, see *UTILITY MANAGEMENT* in this series of operator training manuals. It is available from the Office of Water Programs, California State University, Sacramento, 6000 J Street, Sacramento, CA 95819-6025. Price, $11.00.

Another helpful source of information on financial management of a utility is *A WATER AND WASTEWATER MANAGER'S GUIDE FOR STAYING FINANCIALLY HEALTHY*, July 1989. Obtain from National Technical Information Service (NTIS), 5285 Port Royal Road, Springfield, VA 22161. Order No. PB90-114455. EPA No. 430-9-89-004. Price, $28.50, plus $5.00 shipping and handling per order.

The following American Water Works Association publications may be obtained by contacting AWWA, Bookstore, 6666 West Quincy Avenue, Denver, CO 80235. Prices include cost of shipping and handling.

1. *PRINCIPLES OF WATER RATES, FEES, AND CHARGES* (M1), Fifth Edition. Order No. 30001. ISBN 1-58321-069-5. Price to members, $77.50; nonmembers, $114.00.
2. *WATER UTILITY CAPITAL FINANCING* (M29). Order No. 30029. ISBN 0-89867-957-5. Price to members, $50.50; nonmembers, $72.50.
3. *BASIC MANAGEMENT PRINCIPLES FOR SMALL WATER SYSTEMS*. Order No. 20462. This volume is currently being revised and should be available in late 2002.
4. *WATER UTILITY MANAGEMENT* (M5). Order No. 30005. ISBN 0-89867-063-2. Price to members, $62.50; nonmembers, $92.50.

QUESTIONS

Write your answers in a notebook and then compare your answers with those on page 504.

8.6A How is a utility's operating ratio calculated?

8.6B Why is it important for a manager to consult with other utility personnel and with community leaders during the budget process?

8.6C What are some of the benefits of keeping complete, up-to-date records?

8.6D What is the purpose of maintaining an inventory?

8.6E As a general rule, how long should utility records be kept?

Please answer the discussion and review questions before continuing with the Objective Test.

DISCUSSION AND REVIEW QUESTIONS

Chapter 8. SETTING WATER RATES FOR SMALL WATER UTILITIES

Write the answers to these questions in your notebook before continuing with the Objective Test on page 505. The purpose of these questions is to indicate to you how well you understand the material in the chapter.

1. How can water rates be set which accurately reflect the cost of service for each customer?
2. How can future needs be estimated when setting water rates?
3. What are the basic ways for a utility to finance long-term capital improvements?
4. What is the difference between establishing a revenue base on a cash basis or on a utility basis?
5. What is the difference between the "commodity-demand method" and the "base-extra capacity method" of allocating costs to water users?
6. What is the greatest advantage of the base-extra capacity method of cost allocation?
7. What factors would you consider in determining the customer classes for water rates?
8. What is the difference between multiple-block rate structures and uniform rate structures?
9. Why is it thought to be a good practice to consider rate increases on a regular basis, usually annually?
10. What items should be covered in a Consumer Confidence Report?
11. How do you measure financial stability for a utility?
12. List four major types of records a utility should maintain.

SUGGESTED ANSWERS

Chapter 8. SETTING WATER RATES FOR SMALL WATER UTILITIES

Answers to questions on page 482.

8.0A The goals of a fair water rate setting program are to determine rates on a "cost basis" and be "nondiscriminatory" to all users.

8.0B The accuracy and reliability of rate studies depend on the accuracy and reliability of the utility's information and recordkeeping systems.

Answers to questions on page 486.

8.1A In forecasting expenditures, the following items should be considered:

1. Unexpected decreases in water demand,
2. System growth rate,
3. Anticipated inflation,
4. Long-range capital improvement needs, and
5. Long-range financing requirements.

8.1B A connection fee is an upfront fee paid by new users connecting to the system to cover new capital costs to provide service to their property.

8.1C The two general components of inflation which will affect setting of rates and charges are the Consumer Price Index (CPI) and the Construction Cost Index (CCI). The CPI should be applied to general O & M expense items such as labor, materials, and supplies. The CCI is used for estimating construction-related capital improvement costs.

8.1D A revenue bond is a debt incurred by the community, often to finance utility improvements. User charges provide repayment on the bond.

Answers to questions on page 489.

8.1E The O & M costs associated with a water distribution system include storage reservoirs, distribution system mains, meters, valves, hydrants, service lines, and power for pumping.

8.1F The two basic accounting methods for establishing a revenue base are the cash basis and the utility basis.

8.1G Depreciation is the gradual loss in service value of facilities and equipment over time due to age (wear and tear, obsolescence).

Answers to questions on page 492.

8.2A Utility service cost categories used to distribute costs using the commodity-demand method include (1) commodity, (2) demand, (3) customer, and (4) public fire protection.

8.2B A commodity cost is an expense directly related to the amount of water use. Examples of commodity costs are water costs, power and chemical costs, and labor costs for operating and maintaining the water system.

8.2C A demand cost is an expense associated with providing facilities to meet peak rates of use. Examples of demand costs are storage facilities sized to meet peak demands, transmission and distribution mains, pumping, peaking wells, and some treatment costs.

8.2D A customer cost is the cost of providing service to customers regardless of the amount of water used. Customer costs include meter maintenance, meter reading, customer accounting, and collection costs.

Answers to questions on page 496.

8.3A When establishing customer classes, customers should be grouped into categories of similar use patterns. Thus, classes should separate users by load demand and peaking characteristics.

8.3B A commercial customer class includes nonindustrial, business, and service-type enterprises.

8.4A Unmetered customers tend to use more water than metered customers.

8.4B The four major components of a typical water rate structure are:

1. Service charge,
2. Volume charge,
3. Fire protection charge, and
4. Miscellaneous charge.

Answers to questions on page 497.

8.5A When making a rate increase presentation, illustrate the utility's needs in the form of charts and graphs which simply state the following:

1. Recent cost performance of utility (budget comparisons),
2. Current revenue requirements versus projected needs,
3. Effect of proposed rate increase on typical user groups, and
4. Comparison of proposed rates with other comparable water utilities in your area.

8.5B All utility staff should be familiar with the contents of the CCR because consumers who know individual staff members frequently will ask questions about the report.

Answers to questions on page 502.

8.6A The operating ratio for a utility is calculated by dividing total revenues by total operating expenses.

8.6B It is important for a manager to get input from other personnel in the utility as well as community leaders as the budgeting process proceeds in order to gain support for the budget and to keep the budget on track once adopted.

8.6C Keeping complete, up-to-date records contributes to more effective utility management, helps to satisfy legal requirements, provides valuable operations and maintenance information, and assists in preparing budget requests. Accurate records provide a sound basis for design of future changes or expansions of the treatment plant. If legal questions or problems occur in connection with the treatment processes or the operation of the plant, accurate and complete records will provide evidence of what actually occurred and what procedures were followed.

8.6D The purpose of maintaining an inventory is to provide needed parts and supplies quickly, thereby reducing equipment downtime and work delays.

8.6E As a general rule, utility records should be kept for as long as they may be useful or as long as legally required.

OBJECTIVE TEST

Chapter 8. SETTING WATER RATES FOR SMALL WATER UTILITIES

Please write your name and mark the correct answers on the answer sheet, as directed at the end of Chapter 1. There may be more than one correct answer to each multiple-choice question.

True-False

1. Water rates need to be designed so that they will recover appropriate costs from each class of customer.
   1. True
   2. False
2. Cost allocation methods that require the use of significant amounts of personal judgment should be avoided.
   1. True
   2. False
3. Commodity costs tend to vary as a function of the peak water rate.
   1. True
   2. False
4. Customer costs are those related to volume of water or peak flows.
   1. True
   2. False
5. An important aspect in presenting revenue needs to the rate payer is having the ability to justify the system's revenue needs.
   1. True
   2. False

Best Answer (Select only the closest or best answer.)

6. How can stability of rates be achieved?
   1. By careful analysis of past projected and actual costs
   2. By maintaining constant revenue
   3. By scheduling work on basis of funds received
   4. By spending only revenue received
7. What is the purpose of a cost-of-service or rate study?
   1. To determine the cost of operating a utility
   2. To establish fair and equitable rates
   3. To investigate need for increasing rates
   4. To study rate structures of similar utilities
8. What is the major difference between (1) the cash basis and (2) the utility basis accounting methods for establishing a revenue base? The utility basis allows for the use of
   1. A bond financing program for long-range improvements.
   2. A debt service program to fund equipment replacement.
   3. A reasonable rate of return (profit) on equity in the system.
   4. An establishment of credit to borrow funds.
9. Why is the commodity-demand cost allocation method most appropriate for use by small utilities?
   1. Easier to understand by rate payers
   2. Less complicated to administer
   3. More accurate than other methods
   4. Requires less information
10. Which costs are assigned or allocated to the customer cost category?
    1. Costs associated with the delivery of water to the customers
    2. Costs related to volume of water and peak flows
    3. Costs required to obtain source water for customers
    4. Costs incurred regardless of the amount of water used by the customers
11. What type of rate structure promotes water conservation?
    1. Multiple-block rate structure
    2. Pay-as-you-go structure
    3. Rate-of-return structure
    4. Uniform rate structure

Multiple Choice (Select all correct answers.)

12. Which of the following are revenue needs for a typical small water utility?
    1. Administrative expenses
    2. Cash reserves
    3. Debt service
    4. O & M expenses
    5. Rehabilitation and replacement costs
13. Capital costs that represent major revenue requirements include which of the following items?
    1. Administration
    2. Cash reserves
    3. Costs of maintaining facilities and equipment
    4. Debt service expenses
    5. Facilities replacement costs

14. When forecasting expenditures, which of the following items should be considered?
    1. Anticipated inflation
    2. Long-range capital improvement program needs
    3. Long-range financing requirements
    4. System growth rate
    5. Unexpected decreases in water demand

15. How can a water utility minimize the financial impact of variable water sales due to droughts or economic conditions?
    1. Adjust rate charges every month to cover costs
    2. Always recover fixed costs
    3. Establish a dry-year contingency reserve
    4. Increase variable cost component of the rates
    5. Invest reserves in stocks sensitive to market fluctuations

16. What are the basic ways for a utility to finance long-term capital improvements?
    1. General obligation bonds
    2. Hook-up fees
    3. Loan funding programs
    4. Revenue bonds
    5. User service charges

17. What is debt service?
    1. Debts owed contractors for work completed
    2. Funds due on a maturing bonded debt
    3. Funds due on redemption of bonds
    4. Interest on outstanding debts
    5. Unpaid debts owed by customers

18. What are the distinct components of water service to customers?
    1. Average flow demand
    2. Dry weather flow demand
    3. Minimum flow demand
    4. Peak flow demand
    5. Wet weather flow demand

19. What are the popular cost allocation methods used to assign costs to appropriate levels of service?
    1. Base-extra capacity method
    2. Cash-basis method
    3. Commodity-demand method
    4. Debt-service method
    5. Utility-basis method

20. The commodity-demand cost allocation method uses utility costs related to which of the following service categories?
    1. Administration
    2. Commodity (water) charges
    3. Customer service costs
    4. Demand (peak flow)
    5. Public fire protection

21. Commodity costs include which of the following items?
    1. Chemicals
    2. Power costs
    3. Purchased water
    4. System pumping capacities
    5. Treatment plant flow capacity

22. What are the major cost components of the base-extra capacity method?
    1. Base costs
    2. Customer costs
    3. Direct fire protection costs
    4. Extra capacity costs
    5. O & M costs

23. What factors should be considered when establishing customer classes?
    1. Class of customer (user)
    2. Time user has been a customer
    3. User financial status
    4. User load demand
    5. User peaking characteristics

24. What is the purpose of Consumer Confidence Reports (CCRs)?
    1. Chance to gain confidence in reporting needs
    2. Effective means of communicating with consumers that water is safe
    3. Method of informing public of need for more water
    4. Opportunity to inform rate payers of need for sufficient funds
    5. To publicly inform consumers about the utility's safety record

25. What information should be included in a Consumer Confidence Report (CCR)?
    1. Definition of terms
    2. Information on any unregulated contaminants
    3. Information on health effects of MCLs violated
    4. List of contaminants detected
    5. Source of drinking water supply

26. What are the main activities involved in financial management of a utility?
    1. Careful budgeting
    2. Collecting unpaid bills
    3. Providing capital improvement funds
    4. Providing financial stability
    5. Stopping services for delinquent accounts

27. How can the financial stability of a utility be measured?
    1. Consumer Price Index
    2. Coverage ratio
    3. Depreciation expenses
    4. Operating ratio
    5. Payback time.

28. How long should utility records be kept?
    1. As long as auditors require
    2. As long as legally required
    3. As long as space is available
    4. As long as they may be useful
    5. Forever

End of Objective Test

APPENDIX

SMALL WATER SYSTEM OPERATION AND MAINTENANCE

FINAL EXAMINATION

This final examination was prepared *TO HELP YOU REVIEW* the material in this manual. The questions are divided into five types:

1. True-False,
2. Best Answer,
3. Multiple Choice,
4. Short Answer, and
5. Problems.

To work this examination:

1. Write the answer to each question in your notebook,
2. After you have worked a group of questions (you decide how many), check your answers with the suggested answers at the end of this exam, and
3. If you missed a question and don't understand why, reread the material in the manual.

You may wish to use this examination for review purposes when preparing for civil service and certification examinations.

Since you have already completed this course, you do not have to send your answers to California State University, Sacramento.

True-False

1. Operators are responsible for the drinking water of their community.
   1. True
   2. False
2. Clear water is always safe to drink.
   1. True
   2. False
3. The specific disease-producing organisms present in water are easily isolated and identified.
   1. True
   2. False
4. Conditions for growth and survival of bacteria and viruses in groundwater are generally favorable when compared with surface waters.
   1. True
   2. False
5. Flowmeters should be calibrated in place to ensure accurate flow measurements.
   1. True
   2. False
6. A pump will shut off when the cutoff pressure setting is too high.
   1. True
   2. False
7. Water must have an adequate alkalinity content to achieve good coagulation with alum.
   1. True
   2. False
8. The coagulant dosage must be estimated by performing a jar test.
   1. True
   2. False
9. Ion exchange resins never need backwashing.
   1. True
   2. False
10. Chloramines are a much stronger disinfectant than free chlorine.
    1. True
    2. False
11. Sterilization is necessary when treating water for drinking purposes.
    1. True
    2. False
12. Hypochlorite has a much greater disinfection potential than hypochlorous acid.
    1. True
    2. False
13. Hypochlorinators on small systems are normally small, sealed systems that can be easily repaired.
    1. True
    2. False
14. When shutting down a chlorinator, leave the chlorine discharge line closed.
    1. True
    2. False
15. Make sure headlights, heater, and other electrical accessories are on when attempting to charge a dead battery on a vehicle.
    1. True
    2. False
16. Always store a fire extinguisher inside a fuel storage facility.
    1. True
    2. False

17. Tops of ladders should extend a minimum of 30 inches (75 cm) above the top of trench excavations.
    1. True
    2. False

18. Plain cotton is considered acceptable as an earplug.
    1. True
    2. False

19. The collection of a bad sample or a bad laboratory result is about as useful as no results.
    1. True
    2. False

20. Completely empty volumetric pipets by blowing out the small amount remaining in the tip.
    1. True
    2. False

21. Never pipet chemical reagent solutions or unknown water samples by mouth.
    1. True
    2. False

22. Water rates need to be designed so that they will recover appropriate costs from each class of customer.
    1. True
    2. False

23. Commodity costs tend to vary as a function of the peak water rate.
    1. True
    2. False

24. Customer costs are those related to volume of water or peak flows.
    1. True
    2. False

Best Answer (Select only the closest or best answer.)

1. What is the name of storage facilities for treated water at water treatment plants?
    1. Basins
    2. Clear wells
    3. Reservoirs
    4. Tanks

2. Why do operators take certification examinations?
    1. To certify that their water is safe to drink
    2. To certify that they have an operator's license
    3. To gain experience taking certification examinations
    4. To indicate a level of professional competence

3. How can the delivery of a constant, safe drinking water to consumers be ensured?
    1. Attention to the operation and maintenance of the distribution system
    2. Recommend consumers boil water
    3. Suggest bottled water be used for drinking purposes
    4. Use of clean delivery facilities

4. What is a sanitary survey?
    1. A detailed evaluation of a source of water supply and all water facilities
    2. A detailed survey of the boundaries of a watershed
    3. A survey of consumers' sanitary disposal practices
    4. A survey of sanitary landfills located over an aquifer

5. What is the most widely used test to indicate the bacteriological quality of water?
    1. Coliform
    2. *Cryptosporidium*
    3. *Giardia*
    4. Hepatitis

6. An aquifer is a natural underground layer of what?
    1. Nonporous, nonwater-bearing materials usually capable of yielding very little water
    2. Porous, water-bearing materials usually capable of yielding a large amount of water
    3. Porous, water-bearing materials usually capable of yielding a small amount of water
    4. Water which is directly related to the rate of evapotranspiration

7. What type of drain field problem can occur when garbage disposals are used in septic tank systems?
    1. Adsorption
    2. Clogging
    3. Leaching
    4. Nitrification

8. When should existing wells be disinfected?
    1. After an intense, long-duration storm
    2. After well or pump repairs
    3. Every year
    4. Never. Existing wells don't need to be disinfected

9. What could cause a pump to start too frequently?
    1. Cutoff pressure setting too high
    2. Impellers worn
    3. Leaking foot valve
    4. Motor bearings worn

10. How can iron be controlled in drinking water?
    1. By converting iron from the liquid ferrous to the solid ferric form
    2. By filtering out the ferrous form
    3. By forming soluble precipitates of iron
    4. By use of a reducing agent

11. Why are mechanical mixers preferred for flocculation?
    1. Degree of agitation does not require adjustment for changes in water quality
    2. Flow rate does not affect mixing time
    3. Mixing allows settling of floc in flocculation basin
    4. Performance can be controlled

12. How is chlorine contact time measured?
    1. Flow time as determined by dye tracer studies during low demands
    2. Flow time from point of application to point of entering distribution system
    3. Flow time from point of application to point of use by consumer
    4. Flow time through finished water holding facility

13. What is the chlorine dose?
    1. Chlorine demand minus chlorine residual
    2. Chlorine demand plus chlorine residual
    3. Chlorine demand plus free chlorine
    4. Chlorine residual minus chlorine demand

14. Why do operators never use water on a chlorine leak?
    1. The hydrochloric acid formed will make the leak worse
    2. The mixture of chlorine and water will reduce the disinfection strength of the chlorine
    3. The reaction between water and chlorine will cause a false sealing crust
    4. Water entering the cylinder will cause an explosion
15. Why should chlorine never be injected on the intake side of the pump?
    1. Chlorine will cause pump corrosion problems
    2. Chlorine will come out of solution when passing through the pump
    3. Insufficient contact time will occur in the pump
    4. Pump impeller will cause excessive chlorine mixing
16. What is a "false positive" chlorine residual reading?
    1. Turbidity causing a higher coliform MPN than actually present
    2. Turbidity causing a positive coliform test when no coliforms are present
    3. Turbidity producing a higher chlorine residual than actually present
    4. Turbidity producing a positive test result for free chlorine residual when none is present
17. Who is responsible for safety?
    1. Everyone
    2. Operators
    3. Safety officer
    4. Top management
18. What would you do if you were unable to remove a safety hazard?
    1. Bring hazard to the attention of those who have proper authority
    2. Leave work area immediately
    3. Prevent others from encountering the safety hazard
    4. Try to finish the job safely
19. What should be done before working on electrical equipment or equipment that is run by electric motors?
    1. Inform other operators that equipment is turned off
    2. Inspect all electric circuits
    3. Lock out and tag all electric switches
    4. Turn all equipment off
20. Why must the spoil bank be kept away from the edge of a trench?
    1. To keep spoil from falling into trench
    2. To keep spoil in proper location
    3. To keep spoil out of roadway
    4. To keep wall of trench from caving in on top of you
21. Which piece of laboratory equipment is the most accurate way to measure a volume of liquid?
    1. Beaker
    2. Flask
    3. Graduated cylinder
    4. Pipet
22. Which piece of laboratory equipment is used to titrate a chemical reagent?
    1. Buchner funnel
    2. Buret
    3. Graduated cylinder
    4. Pipet
23. Which style of thermometer will change most rapidly when removed from the liquid sample?
    1. Dial
    2. Digital
    3. Partial immersion
    4. Total immersion
24. What is the accepted method used to measure turbidity?
    1. Candle
    2. Jackson
    3. Light
    4. Nephelometric
25. How can stability of water rates be achieved?
    1. By careful analysis of past projected and actual costs
    2. By maintaining constant revenue
    3. By scheduling work on basis of funds received
    4. By spending only revenue received
26. Why is the commodity-demand cost allocation method most appropriate for use by small utilities?
    1. Easier to understand by rate payers
    2. Less complicated to administer
    3. More accurate than other methods
    4. Requires less information

Multiple Choice (Select all correct answers.)

1. What do consumers expect from the operators of water systems?
    1. Adequate water pressure
    2. High-cost water
    3. Pleasant water
    4. Safe water
    5. Sufficient water
2. What safety hazards are encountered on the job by small water system operators?
    1. Drinking treated water
    2. Explosive conditions in elevated tanks
    3. Lack of ventilation in elevated tanks
    4. Work in excavations
    5. Work in traffic
3. What are potential sources of groundwater contamination?
    1. Agricultural drainage systems
    2. Evapotranspiration from ponds
    3. Improper disposal of hazardous wastes
    4. Sea water intrusion
    5. Seepage from septic tank leaching systems
4. What undesirable results can occur if wells pump water from an aquifer at a rate that will deplete the aquifer?
    1. Artesian flows
    2. Land subsidence
    3. Replenishment of aquifers
    4. Sea water intrusion
    5. Sustained yields
5. Which treatment processes are in operation in the soil to treat contaminants?
    1. Adsorption
    2. Biological decay
    3. Filtration
    4. Ion exchange
    5. Oxidation

6. What water quality tests should be performed on groundwater samples if a study indicates that septic tank system failures may have occurred near shallow wells?
    1. Coliform
    2. Nitrate
    3. Pesticides
    4. Petroleum
    5. Saline
7. What causes water hammer?
    1. Pressure relief valves
    2. Quick closing of valves
    3. Quick opening of valves
    4. Shutdown of pumps
    5. Start-up of pumps
8. What problems are associated with sand in water?
    1. Consumer complaints
    2. Damage to water meters
    3. Damage to well pumping facilities
    4. Decreasing friction loss in pipes
    5. Reducing pipe carrying capacity
9. How can water be softened?
    1. Chemical precipitation
    2. Coagulation
    3. Disinfection
    4. Flocculation
    5. Ion exchange
10. Which factors have an important effect on the settling of suspended particles in a settling tank?
    1. Characteristics of suspended matter
    2. Degree of short-circuiting through the tank
    3. Temperature of the water
    4. Time provided for settling
    5. Wind
11. What variables does an operator adjust to control the performance of a solids-contact unit?
    1. Chemical dosage
    2. Corrosion control
    3. Disinfection
    4. Recirculation rate
    5. Sludge control
12. What factors influence the effectiveness of chlorine disinfection?
    1. Organic matter
    2. pH
    3. Reducing agents
    4. Temperature
    5. Turbidity
13. Which water treatment processes are used to remove or kill pathogenic organisms?
    1. Coagulation
    2. Comminution
    3. Disinfection
    4. Filtration
    5. Sedimentation
14. Which of the following tasks are performed when starting up a chlorinator?
    1. Have an ammonia bottle readily available to detect any leaks in the system
    2. Have safety equipment available that may be used in case of a leak
    3. Inspect the chlorine container for leaks
    4. Prior to the hookup, inspect for moisture and foreign substances in the lines
    5. Use new gaskets for connection of the chlorinator to the chlorine cylinder
15. What procedures should be implemented immediately if inadequately disinfected water enters a water distribution system?
    1. Distribute properly disinfected water as soon as possible
    2. Flush distribution system to remove inadequately disinfected water
    3. Implement purchase order procedures for replacement equipment
    4. Notify health department officials immediately
    5. Order additional chlorine supplies
16. Which gases could cause dangerous air contamination?
    1. Carbon monoxide gas
    2. Hydrogen sulfide gas
    3. Methane gas
    4. Oxygen
    5. Toxic gases
17. What safety procedures must be followed before anyone ever enters a tank for any reason?
    1. All persons entering a tank must wear a safety harness
    2. An additional person must be readily available to help the person at the tank entrance with any rescue operations
    3. One person must be at the tank entrance and observe the actions of all people in the tank
    4. Provide adequate ventilation, especially when painting
    5. Test atmosphere in the tank for toxic and explosive gases and for oxygen deficiency/enrichment
18. What types of confined spaces could be encountered by operators?
    1. Cabs of pickup trucks
    2. Dewatered clear wells
    3. Elevated tanks
    4. Manholes
    5. Valve vaults
19. Which factors can lead to poor quality laboratory data?
    1. Calculation mistakes
    2. Deteriorated reagents
    3. Neatly recorded results in lab notebooks
    4. Poorly operating instruments
    5. Sloppy laboratory technique

20. Which of the following hazards could be encountered in a laboratory?
    1. Burns
    2. Drowning
    3. Electric shock
    4. Explosions
    5. Hazardous materials

21. Grab samples should be used to measure which of the following water quality indicators?
    1. Chlorine residual
    2. Coliform bacteria
    3. Dissolved gases
    4. pH
    5. Temperature

22. An alkalinity determination is needed when calculating chemical dosages used in what water treatment processes?
    1. Coagulation
    2. Filtration
    3. Flocculation
    4. Settling
    5. Softening

23. When forecasting expenditures, which of the following items should be considered?
    1. Anticipated inflation
    2. Long-range capital improvement program needs
    3. Long-range financing requirements
    4. System growth rate
    5. Unexpected decreases in water demand

24. What are the popular cost allocation methods used to assign costs to appropriate levels of service?
    1. Base-extra capacity method
    2. Cash-basis method
    3. Commodity-demand method
    4. Debt-service method
    5. Utility-basis method

25. What factors should be considered when establishing customer classes?
    1. Class of customer (user)
    2. Time user has been a customer
    3. User financial status
    4. User load demand
    5. User peaking characteristics

26. What is the purpose of Consumer Confidence Reports (CCRs)?
    1. Chance to gain confidence in reporting needs
    2. Effective means of communicating with consumers that water is safe
    3. Method of informing public of need for more water
    4. Opportunity to inform rate payers of need for sufficient funds
    5. To publicly inform consumers about the utility's safety record

27. How can the financial stability of a utility be measured?
    1. Consumer Price Index
    2. Coverage ratio
    3. Depreciation expenses
    4. Operating ratio
    5. Payback time

Short Answer

1. Why should surface waters be treated?
2. Why are adequate and reliable records very important?
3. What is the hydrologic cycle?
4. List the common physical characteristics of water.
5. What causes turbidity in water?
6. Who must comply with the Safe Drinking Water Act (SDWA) regulations?
7. What problems can be caused by the overdraft of a groundwater supply?
8. How would you determine the distance down to the water level in a well?
9. What is the term used to describe the water volume that can move through the pores in a rock and is affected by gravitational forces?
10. What options are available for a community whose groundwater supply has been contaminated?
11. Under what circumstances might a booster pump be installed?
12. What are the advantages of ON/OFF pump controls?
13. Why do small water treatment plants often not perform satisfactorily?
14. What happens if the coagulant dosage is either too high or too low?
15. Why should corrosion control chemicals be applied to water after all other treatment has been accomplished?
16. Why are slow sand filters seen as a potentially promising alternative for small water systems?
17. Why do the results of jar tests often not produce the same results in a water treatment plant using the lime-soda ash softening process?
18. MCL stands for what words?
19. How does the temperature of the water influence disinfection?
20. When disinfecting a well, why should the well, pump, screen, and aquifer around the well all be disinfected?
21. What is the chlorine demand of a water?
22. How would you determine the chlorine dose for water?
23. What is the maximum rate of chlorine removal from a 150-pound cylinder?
24. How would you detect a chlorine leak?
25. What would you do if you could not repair a broken chlorinator quickly?
26. What first-aid measures should be taken if a person comes in contact with chlorine gas?
27. What are the causes of accidents?
28. How can you determine if the brakes on a vehicle are working properly?

29. How would you test a safety shower and eye/face wash station?

30. What is a tailgate safety session?

31. List the typical locations in a water utility where it is necessary to monitor the atmosphere before entering.

32. At what level is a mixture of air and gas considered hazardous?

33. How is contaminated water or debris kept from entering around the well casing and reaching the groundwater?

34. Why is it necessary to bleed down pressurized systems before making repairs?

35. How can the public be protected from excavation hazards?

36. What atmospheric hazards may be encountered in a confined space?

37. What is a meniscus?

38. How would you titrate a test solution?

39. True or False? You may *ADD ACID TO WATER*, but never water to acid.

40. Why should water not be poured on certain types of fires?

41. Why are proper sampling procedures important?

42. What is meant by a representative sample?

43. What is a potential adverse effect from chlorination?

44. Why are drinking waters not tested for specific pathogens (such as *Cryptosporidium* or *Giardia*)?

45. What are the principal hardness-causing ions in water?

46. Why should turbidity be measured in a sample as soon as possible?

47. The accuracy and reliability of rate studies depend on what factors?

48. What are the basic ways for a utility to finance long-term capital improvements?

49. What is a revenue bond?

50. What is a commodity cost?

51. What factors would you consider in determining the customer classes for water rates?

52. Why should all utility staff be familiar with the contents of the Consumer Confidence Report?

53. How do you measure financial stability for a utility?

54. How is a utility's operating ratio calculated?

Problems

1. What is the surface area of a rectangular settling basin 40 feet long and 8 feet wide?
   1. 230 square feet
   2. 320 square feet
   3. 360 square feet
   4. 460 square feet
   5. 540 square feet

2. Calculate the volume in cubic feet of a rectangular settling basin 30 feet long, 10 feet wide, and 8 feet deep.
   1. 240 cu ft
   2. 2,400 cu ft
   3. 3,000 cu ft
   4. 3,200 cu ft
   5. 4,800 cu ft

3. Calculate the gallons of sodium hypochlorite required to disinfect a well if the well casing contains 1,200 gallons of water. The chlorine dose is 50 mg/*L*. Sodium hypochlorite is 5.25 percent or 52,500 mg/*L* chlorine. Select the closest answer.
   1. 0.60 gallon
   2. 1.14 gallons
   3. 2.28 gallons
   4. 3.36 gallons
   5. 6.0 gallons

4. Jar tests indicate that water should be dosed with alum at 12 mg/*L*. The flow being treated is 180 GPM. What is the desired alum feed in pounds per day?
   1. 11 lbs/day
   2. 26 lbs/day
   3. 31 lbs/day
   4. 44 lbs/day
   5. 56 lbs/day

5. What is the actual chlorine dose in milligrams per liter if the hypochlorinator delivers 8 pounds of chlorine per day and the flow rate of the water being treated is 600,000 gallons per day (0.6 MGD)?
   1. 0.8 mg/*L*
   2. 1.2 mg/*L*
   3. 1.4 mg/*L*
   4. 1.6 mg/*L*
   5. 2.8 mg/*L*

6. During a filter run, the total volume of water filtered was 1.4 million gallons. When the filter was backwashed, 24,000 gallons of water was used. What was the percent of filtered water used for backwashing?
   1. 1.0%
   2. 1.3%
   3. 1.5%
   4. 1.7%
   5. 2.0%

7. What should be the chemical dose of a water that has a chlorine demand of 2.3 mg/*L* if a residual of 0.5 mg/*L* is desired?
   1. 1.8 mg/*L*
   2. 2.3 mg/*L*
   3. 2.8 mg/*L*
   4. 3.3 mg/*L*
   5. 3.6 mg/*L*

8. Water from a well is being disinfected by a hypochlorinator. If the hypochlorinator is set at a pumping rate of 60 GPD and uses a 2.5 percent available chlorine solution, what is the chlorine dose rate in mg/*L*? The water pump delivers 500 GPM.
   1. 2.1 mg/*L*
   2. 2.4 mg/*L*
   3. 2.6 mg/*L*
   4. 2.8 mg/*L*
   5. 3.0 mg/*L*

9. How many gallons of hypochlorite were pumped by a hypochlorinator if the hypochlorite solution was in a container with a diameter of 42 inches and the hypochlorite level drops 15 inches in 24 hours?
   1. 26.5 gallons
   2. 80.0 gallons
   3. 82.5 gallons
   4. 90.0 gallons
   5. 100.0 gallons

10. A hypochlorite solution for a hypochlorinator is being prepared in a 55-gallon drum. If 8 gallons of 5 percent hypochlorite is added to the drum, how much water should be added to produce a 1.5 percent hypochlorite solution?
   1. 18.7 gallons of water
   2. 21.5 gallons of water
   3. 24.7 gallons of water
   4. 25.3 gallons of water
   5. 28.5 gallons of water

11. Results from a coliform bacteria multiple fermentation tube test of raw water are as follows:

m*L* of sample	10	1	0.1	0.01	0.001
Dilutions	0	0	–1	–2	–3
Positive tubes	5	5	5	3	1

What is the most probable number of coliforms per 100 m*L* in the sample? Use Table 7.3 on page 449.
   1. 110 coliforms per 100 m*L*
   2. 900 coliforms per 100 m*L*
   3. 1,100 coliforms per 100 m*L*
   4. 9,000 coliforms per 100 m*L*
   5. 11,000 coliforms per 100 m*L*

Data to answer questions 12 and 13.

Results from MPN tests during one week were as follows:

Day	S	M	T	W	T	F	S
MPN/100 m*L*	4	7	14	8	9	16	5

12. What is the mean MPN per 100 m*L* from the data?
   1. 6 MPN/100 m*L*
   2. 7 MPN/100 m*L*
   3. 8 MPN/100 m*L*
   4. 9 MPN/100 m*L*
   5. 10 MPN/100 m*L*

13. What is the median MPN per 100 m*L* from the data?
   1. 6 MPN/100 m*L*
   2. 7 MPN/100 m*L*
   3. 8 MPN/100 m*L*
   4. 9 MPN/100 m*L*
   5. 10 MPN/100 m*L*

SUGGESTED ANSWERS FOR FINAL EXAMINATION

True-False

1. True — Operators are responsible for the drinking water of their community.
2. False — Clear water is *NOT* always safe to drink.
3. False — Specific disease-producing organisms in water are *NOT* easily identified.
4. False — Conditions for growth and survival of bacteria and viruses in groundwater are unfavorable when compared with surface waters.
5. True — Flowmeters should be calibrated in place to ensure accurate flow measurements.
6. False — A pump will *NOT* shut off when the cutoff pressure setting is too high.
7. True — Water must have adequate alkalinity to achieve good coagulation with alum.
8. True — Coagulant dosage must be estimated by performing a jar test.
9. False — Ion exchange resins may need backwashing if head loss is greater than normal.
10. False — Free chlorine is a much stronger disinfectant than chloramines.
11. False — Sterilization is *NOT* necessary when treating water for drinking purposes.
12. False — Hypochlorous acid has a much greater disinfection potential than hypochlorite.
13. False — Sealed hypochlorinators cannot be repaired, they must be replaced.
14. False — Leave chlorine discharge line open to prevent high chlorine pressures from developing.
15. False — Make sure all electrical accessories are *OFF* when attempting to charge battery.
16. False — Always store fire extinguisher near a fuel storage facility, but never inside.
17. True — Ladders should extend a minimum of 30 inches above top of trench excavations.
18. False — Plain cotton is *NOT* considered acceptable as an earplug.
19. True — Bad samples or bad lab results are as useful as no results.
20. False — Under no circumstance should the small amount remaining in tip be blown out.

21. True — Never pipet chemical reagent solutions or unknown water samples by mouth.

22. True — Design rates to recover appropriate costs from each class of customer.

23. False — Commodity costs tend to vary as a function of average water use.

24. False — Customer costs are those *NOT* related to volume of water or peak flows.

Best Answer

1. 2 — Storage facilities for treated water at plants are called clear wells.

2. 4 — Operators take certification exams to indicate level of competence.

3. 1 — Distribution system O & M can ensure delivery of safe drinking water.

4. 1 — A sanitary survey is a detailed evaluation of a source and facilities.

5. 1 — Coliform is the most widely used bacteriological quality test.

6. 2 — An aquifer consists of porous, water-bearing materials that yield large amounts of water.

7. 2 — Garbage disposals can cause drain field clogging.

8. 2 — Disinfect existing wells after well or pump repairs.

9. 3 — Leaking foot valve could cause pump to start too frequently.

10. 1 — Control iron by converting it from the liquid ferrous to solid ferric form.

11. 4 — Mechanical mixers are preferred because performance can be controlled.

12. 3 — Contact time is time from point of application to point of use.

13. 2 — Chlorine dose is chlorine demand plus chlorine residual.

14. 1 — The hydrochloric acid formed will make the leak worse.

15. 1 — Chlorine will cause pump corrosion problems if injected on intake side of pump.

16. 4 — False positive is when turbidity produces free chlorine residual when none is present.

17. 1 — Everyone is responsible for safety.

18. 1 — Bring hazard to attention of those who have proper authority.

19. 3 — Lock out and tag all electric switches before working on equipment.

20. 4 — Keep spoil back to prevent wall of trench from caving in.

21. 4 — Pipets are the most accurate way to measure a volume of liquid.

22. 2 — Burets are used to titrate chemical reagents.

23. 4 — Total immersion thermometers change most rapidly when removed from liquid.

24. 4 — The nephelometric method is the accepted method to measure turbidity.

25. 1 — Stable water rates can be achieved by analysis of projected and actual costs.

26. 2 — Commodity-demand cost allocation is less complicated to administer.

Multiple Choice

1. 1, 3, 4, 5 — Consumers expect operators to supply sufficient water with adequate pressure that is safe and pleasant to drink.

2. 2, 3, 4, 5 — Explosive conditions and/or lack of ventilation in elevated tanks are safety hazards encountered on the job by small water system operators. In addition, work in excavations and in traffic can be hazardous to operators.

3. 1, 3, 4, 5 — Potential sources of groundwater contamination include agricultural drainage systems, improper disposal of hazardous wastes, sea water intrusion, and seepage from septic tank leaching systems.

4. 2, 4 — Undesirable results that occur when wells deplete an aquifer include land subsidence and sea water intrusion.

5. 1, 2, 3, 4, 5 — Treatment processes operating in the soil to treat contaminants include adsorption, biological decay, filtration, ion exchange, and oxidation.

6. 1, 2 — Test groundwater samples for coliform and nitrate if septic tank systems fail near shallow wells.

7. 2, 3, 4, 5 — Water hammer can be caused by quick opening and closing of valves, and start-up and shutdown of pumps.

8. 1, 2, 3, 5 — Problems associated with sand in water are consumer complaints, damage to water meters and well pumping facilities, and reduction of pipe carrying capacity.

9. 1, 5 — Chemical precipitation (lime-soda ash) and ion exchange are used to soften water.

10. 1, 2, 3, 4, 5 — Factors that have an important effect on the settling of suspended particles in a settling tank include the characteristics of the suspended matter, the degree of short-circuiting through the tank, the temperature of the water, the time provided for settling, and wind.

11. 1, 4, 5 — Variables an operator adjusts to control the performance of a solids-contact unit include chemical dosages, recirculation rates, and sludge control.

12. 1, 2, 3, 4, 5 — Organic matter, pH, reducing agents, temperature, and turbidity all influence the effectiveness of chlorine disinfection.

13. 1, 3, 4, 5 — Water treatment processes used to remove or kill pathogenic organisms include coagulation, disinfection, filtration, and sedimentation.

14. 1, 2, 3, 4, 5 — When starting up a chlorinator, have an ammonia bottle and safety equipment readily available, inspect the chlorine container for leaks, inspect for moisture and foreign substances in the lines prior to hookup, and use new gaskets for connection of the chlorinator to the chlorine cylinder.

15. 1, 2, 4 — If inadequately disinfected water enters a water distribution system, immediately notify health department officials, flush the distribution system, and distribute properly disinfected water.

16. 1, 2, 3, 4, 5 — Gases that could cause dangerous air contamination include carbon monoxide, hydrogen sulfide, methane, oxygen in concentrations above 23.5%, and toxic gases.

17. 1, 2, 3, 4, 5 — All the safety procedures listed must be followed before anyone ever enters a tank for any reason: all persons entering must wear a safety harness; one person must be at the entrance for observation; an additional person must be readily available to assist the person at the entrance; adequate ventilation must be provided; and the atmosphere in the tank must be tested for toxic and explosive gases and oxygen levels.

18. 2, 3, 4, 5 — Confined spaces that could be encountered by operators include dewatered clear wells, elevated tanks, manholes, and valve vaults.

19. 1, 2, 4, 5 — Poor quality laboratory data will result from calculation mistakes, deteriorated reagents, poorly operating instruments, and sloppy laboratory technique.

20. 1, 3, 4, 5 — Hazards in the lab include burns, electric shock, explosions, and exposure to hazardous materials.

21. 1, 2, 3, 4, 5 — Grab samples are used to measure chlorine residuals, coliform bacteria, dissolved gases, pH, and temperature.

22. 1, 5 — An alkalinity determination is needed when calculating chemical dosages for coagulation and softening.

23. 1, 2, 3, 4, 5 — When forecasting expenditures, a utility must consider anticipated inflation, long-range capital improvement and financing requirements, system growth rate, and unexpected decreases in water demand.

24. 1, 3 — Popular cost allocation methods used to assign costs to appropriate levels of service are base-extra capacity and commodity-demand.

25. 1, 4, 5 — Important factors to consider in establishing customer classes include the class of customer — residential, commercial, agricultural; user load demand; and user peaking characteristics.

26. 2, 4 — Consumer Confidence Reports (CCRs) are opportunities to communicate to consumers that water is safe, and to inform rate payers of the need for funds.

27. 2, 4 — Coverage ratio and operating ratio are measures of a utility's financial stability.

Short Answer

1. Surface water is usually treated to remove suspended and dissolved materials and to kill disease-causing organisms.

2. Adequate and reliable records are very important to document the effectiveness of your operation.

3. The hydrologic cycle is the cycle or path water follows from evaporation from oceans, to formation of clouds, to precipitation, to runoff, to evaporation and transpiration back to the atmosphere, and eventually back to the ocean.

4. The common physical characteristics of water are observable color, turbidity, temperature, taste, and odor.

5. Turbidity in water is caused by the presence of suspended material such as clay, silt, finely divided organic material, plankton, and other inorganic material.

6. All public water systems must comply with the Safe Drinking Water Act (SDWA) regulations.

7. Problems that can be caused by overdraft include:
   1. Permanent damage to the water storage and transmitting properties of aquifers, and
   2. Subsidence of land.

8. The water level in a well is determined by inserting a measuring tape into the sounding tube, lowering it down the tube to the water level, and recording the distance. Or, air pressure in a sounding line may be used; the pressure necessary to force water out of the line is recorded and converted to feet. Then,

   $$\text{Distance Down to Water Surface, ft} = \text{Length of Line, ft} - \text{Water Depth in Line, ft}$$

9. Specific yield describes the water volume that can move through the pores in the rock and is affected by gravitational forces.

10. Options available to a community whose groundwater supply has been contaminated include:
    1. Contain the contaminants to prevent their migration from their source,
    2. Withdraw the pollutants from the aquifer,
    3. Treat the groundwater where it is withdrawn or at its point of use,
    4. Rehabilitate the aquifer by either immobilizing or detoxifying the contaminants while they are still in the aquifer, and
    5. Abandon the use of the aquifer and find alternative sources of water.

11. Booster pumps are installed to pump a smaller amount of water to a pressure zone that operates at a much higher pressure or elevation than the main well pump system.

12. The advantages of ON/OFF pump controls include (1) low cost, (2) few parts, and (3) usually very reliable.

13. Small water treatment plants often do not perform satisfactorily due to poor design, poor operation, poor maintenance, inadequate budgets and other causes.

14. If the coagulant dosage is either too high or too low, coagulation will be incomplete and the results of treatment will be unsatisfactory. Too high a dose increases chemical costs unnecessarily.

15. Corrosion control chemicals should be applied to water after all other treatment has been accomplished because the pH required for successful coagulation and disinfection is much lower than the pH required to make the water noncorrosive. Also, a coating of calcium carbonate on the filter media should be avoided because it interferes with effective filtration.

16. The reason slow sand filters are seen as a potentially promising alternative for small water systems is a combination of simple design, ease of operation requiring minimal staffing, and their ability to remove *Giardia lamblia* cysts as well as *Cryptosporidia*, coliforms, and other microorganisms.

17. In many cases the feed rates determined by jar tests do not produce the exact same results in an actual plant. This is because of differences in water temperature, size and shape of jar as compared with plant basins, mixing equipment, and influence of coagulant (a heavy alum feed will neutralize more of the lime).

18. MCL stands for **M**aximum **C**ontaminant **L**evel.

19. Relatively cold water requires longer disinfection time or greater quantities of disinfectants.

20. When disinfecting a well, the well, pump, screen, and aquifer around the well all should be disinfected because they all could have been contaminated during construction and are potential sources of contamination.

21. The chlorine demand of water is defined as the difference between the amount of chlorine added to water and the amount of residual chlorine remaining after all chemical reactions with organic and inorganic substances have been completed.

22. The chlorine dose is determined by adding the amount of chlorine needed to meet the chlorine demand and the amount of chlorine residual needed for disinfection. The chlorine dose may be expressed in mg/*L* or as a chlorinator setting in pounds per 24 hours.

23. The maximum rate of chlorine removal from a 150-pound cylinder is 40 pounds of chlorine per day.

24. A chlorine leak may be detected by the use of a rag on the end of a stick dipped into a solution of ammonia water. Place the rag near the location of a suspected chlorine leak. A white cloud will reveal the location of the leak.

25. If a broken chlorinator cannot be repaired quickly, shut off the water supply so that unchlorinated or contaminated water will not be delivered to consumers. Follow standard procedures to shut down the chlorinator.

26. First-aid measures depend on the severity of the contact. Move the victim out of the gas area and remove contaminated clothing. Flush skin/eyes as needed and keep the victim warm and quiet. Call a doctor and the fire department immediately.

27. Accidents may be caused by unsafe acts of operators or result from hazardous conditions or may be a combination of both. An analysis of accident statistics reveals that operator negligence and carelessness are the causes of most accidents.

28. The brakes on a vehicle are working properly when they are able to stop the vehicle without grabbing or pulling to one side.

29. To test a safety shower and eye/face wash station, run the water using the emergency levers (the pull chains, hand-actuated valves or treadle foot-actuated valves). Let the water run for approximately three to five minutes or until the water runs clear. Rust may be in the water lines and may increase the problem of washing chemicals from your eyes.

30. A tailgate safety session is usually held near the work site and new and old safety hazards and safer approaches and techniques to deal with the problems of the day or week are discussed.

31. Some of the typical locations in a water utility where it is necessary to monitor the atmosphere before entering include underground regulator vaults, solution vaults, manholes, tanks, trenches, and other confined spaces.

32. A mixture of air and gas is considered hazardous when the mixture exceeds 10 percent of the Lower Explosive Limit (LEL).

33. A sanitary seal on a well is used to keep contaminated water and debris from entering around the well casing and reaching the groundwater.

34. Stored energy must be bled down to prevent its sudden release, which could injure an operator working on the system or equipment.

35. The public can be protected from excavation hazards by providing protective barricades, fencing, handrails, bridges, warning devices, and signs to direct the public to areas where they may pass safely.

36. Atmospheric hazards that may be encountered in a confined space include explosive conditions, toxic gases, and oxygen deficiency and/or oxygen enrichment.

37. A meniscus is the curve of the surface of a liquid in a small tube.

38. Titration involves the addition of a standardized solution, which is usually in a buret, to another solution in a flask or beaker. Slowly add the solution in the buret to the test solution in the flask until the change, which is called the "end point," is reached.

39. True. You may *ADD ACID TO WATER*, but never the reverse.

40. Water should not be poured on grease, electrical fires, or metal fires because water could increase the hazards, such as splattering of the fire and electrical shock.

41. Proper sampling procedures are essential in order to obtain an accurate description of the material or water being sampled and tested.

42. A representative sample is a sample portion of material or water that is as nearly identical in content and consistency as possible to that in the larger body of material or water being sampled.

43. A potential adverse effect from chlorination is the possibility of the formation of carcinogenic chlororganic compounds such as chloroform and other THMs.

44. Drinking waters are not tested for specific pathogens because the tests are very time-consuming and require special techniques and equipment.

45. The principal hardness-causing ions in water are calcium and magnesium.

46. Turbidity should be measured in a sample as soon as possible to obtain accurate results. The turbidity of a sample can change after the sample is collected.

47. The accuracy and reliability of rate studies depend on the accuracy and reliability of the utility's information and recordkeeping systems.

48. The basic ways for a utility to finance long-term capital improvements are through general obligation bonds, revenue bonds, or loan funding programs.

49. A revenue bond is a debt incurred by the community, often to finance utility improvements. User charges provide repayment on the bond.

50. A commodity cost is an expense directly related to the amount of water use. Examples of commodity costs are water costs, power and chemical costs, and labor costs for operating and maintaining the water system.

51. When determining customer classes, be sure to use sufficient classes and broad enough category definitions to cover all user groups. Customer classes should be separated into categories of similar use patterns. Classes should also segregate users by load demand and peaking characteristics.

52. All utility staff should be familiar with the contents of the CCR because consumers who know individual staff members frequently will ask questions about the report.

53. Financial stability is measured by calculating (1) the operating ratio, and (2) the coverage ratio. Both of these ratios should be greater than 1.0.

54. The operating ratio for a utility is calculated by dividing total revenues by total operating expenses.

Problems

1. What is the surface area of a rectangular settling basin 40 feet long and 8 feet wide?

Known	Unknown
Length, ft = 40 ft	Surface Area, sq ft
Width, ft = 8 ft	

Calculate the surface area of the basin in square feet.

Surface Area, sq ft = (Length, ft)(Width, ft)

= (40 ft)(8 ft)

= 320 sq ft

2. Calculate the volume in cubic feet of a rectangular settling basin 30 feet long, 10 feet wide, and 8 feet deep.

Known	Unknown
Length, ft = 30 ft	Volume, cubic feet
Width, ft = 10 ft	
Depth, ft = 8 ft	

Calculate the volume in cubic feet.

Volume, cu ft = (Length, ft)(Width, ft)(Depth, ft)

= (30 ft)(10 ft)(8 ft)

= 2,400 cu ft

3. Calculate the gallons of sodium hypochlorite required to disinfect a well if the well casing contains 1,200 gallons of water. The chlorine dose is 50 mg/*L*. Sodium hypochlorite is 5.25 percent or 52,500 mg/*L* chlorine. Select the closest answer.

Known		Unknown
Casing Volume, gal	= 1,200 gal	Chlorine Required, gal
Chlorine Dose, mg/*L*	= 50 mg/*L*	
Chlorine Compound, mg/*L* (Sodium Hypochlorite)	= 52,500 mg/*L*	

Calculate the required gallons of sodium hypochlorite.

$$\text{Chlorine Required, gal} = \frac{(\text{Casing Volume, gal})(\text{Desired Dose, mg}/L)}{\text{Chlorine Compound Solution, mg}/L}$$

$$= \frac{(1{,}200 \text{ gal})(50 \text{ mg}/L)}{52{,}500 \text{ mg}/L}$$

= 1.14 gallons

4. Jar tests indicate that water should be dosed with alum at 12 mg/*L*. The flow being treated is 180 GPM. What is the desired alum feed in pounds per day?

Known		Unknown
Alum Dose, mg/*L*	= 12 mg/*L*	Desired Alum Feed, lbs/day
Flow, GPM	= 180 GPM	

1. Convert the flow from gallons per minute (GPM) to million gallons per day (MGD).

$$\text{Flow, MGD} = \frac{(\text{Flow, gal/min})(60 \text{ min/hr})(24 \text{ hr/day})}{1{,}000{,}000/\text{Million}}$$

$$= \frac{(180 \text{ gal/min})(60 \text{ min/hr})(24 \text{ hr/day})}{1{,}000{,}000/\text{Million}}$$

= 0.26 MGD

2. Calculate the desired alum feed rate in pounds per day.

Desired Alum Feed, lbs/day = (Flow, MGD)(Dose, mg/*L*)(8.34 lbs/gal)

= (0.26 MGD)(12 mg/*L*)(8.34 lbs/gal)

= 26 lbs/day

5. What is the actual chlorine dose in milligrams per liter if the hypochlorinator delivers 8 pounds of chlorine per day and the flow rate of the water being treated is 600,000 gallons per day (0.6 MGD)?

Known		Unknown
Chemical Feed, lbs/day	= 8 lbs/day	Actual Dose, mg/L
Flow, MGD	= 0.6 MGD	

Calculate the actual chlorine dose in milligrams per liter.

$$\text{Actual Dose, mg}/L = \frac{\text{Chemical Feed, lbs/day}}{(\text{Flow, MGD})(8.34\ \text{lbs/gal})}$$

$$= \frac{8\ \text{lbs/day}}{(0.6\ \text{MGD})(8.34\ \text{lbs/gal})}$$

$$= 1.6\ \text{mg}/L$$

6. During a filter run, the total volume of water filtered was 1.4 million gallons. When the filter was backwashed, 24,000 gallons of water was used. What was the percent of filtered water used for backwashing?

Known		Unknown
Water Filtered, gal	= 1,400,000 gal	Backwash, %
Backwash Water, gal	= 24,000 gal	

Calculate the percent of filtered water used for backwashing.

$$\text{Backwash, \%} = \frac{(\text{Backwash Water, gal})(100\%)}{\text{Water Filtered, gal}}$$

$$= \frac{(24{,}000\ \text{gal})(100\%)}{1{,}400{,}000\ \text{gal}}$$

$$= 1.7\%$$

7. What should be the chemical dose of a water that has a chlorine demand of 2.3 mg/*L* if a residual of 0.5 mg/*L* is desired?

Known		Unknown
Chlorine Demand, mg/L	= 2.3 mg/L	Chlorine Dose, mg/L
Chlorine Residual, mg/L	= 0.5 mg/L	

Calculate the chlorine dose in mg/*L*.

$$\text{Chlorine Dose, mg}/L = \text{Chlorine Demand, mg}/L + \text{Chlorine Residual, mg}/L$$

$$= 2.3\ \text{mg}/L + 0.5\ \text{mg}/L$$

$$= 2.8\ \text{mg}/L$$

8. Water from a well is being disinfected by a hypochlorinator. If the hypochlorinator is set at a pumping rate of 60 GPD and uses a 2.5 percent available chlorine solution, what is the chlorine dose rate in mg/*L*? The water pump delivers 500 GPM.

Known		Unknown
Hypochlorinator, GPD	= 60 GPD	Chlorine Dose, mg/L
Hypochlorite, %	= 2.5%	
Pump, GPM	= 500 GPM	

1. Convert the pumping rate to MGD.

$$\text{Pumping Rate, MGD} = \frac{(500\ \text{GPM})(60\ \text{min/hr})(24\ \text{hr/day})}{1{,}000{,}000/\text{Million}}$$

$$= 0.72\ \text{MGD}$$

2. Calculate the chlorine feed rate in pounds per day.

$$\text{Chlorine Feed, lbs/day} = \frac{(\text{Flow, gal/day})(\text{Hypochlorite, \%})(8.34\ \text{lbs/gal})}{100\%}$$

$$= \frac{(60\ \text{GPD})(2.5\%)(8.34\ \text{lbs/gal})}{100\%}$$

$$= 12.5\ \text{lbs chlorine/day}$$

3. Calculate the chlorine dose in mg/*L*.

$$\text{Chlorine Dose, mg}/L = \frac{\text{Chlorine Feed, lbs/day}}{(\text{Flow, MGD})(8.34\ \text{lbs/gal})}$$

$$= \frac{12.5\ \text{lbs chlorine/day}}{(0.72\ \text{MGD})(8.34\ \text{lbs/gal})}$$

$$= 2.1\ \text{lbs chlorine/M lbs water}$$

$$= 2.1\ \text{mg}/L$$

9. How many gallons of hypochlorite were pumped by a hypochlorinator if the hypochlorite solution was in a container with a diameter of 42 inches and the hypochlorite level drops 15 inches in 24 hours?

Known		Unknown
Diameter, in	= 42 in	Hypochlorite Pumped, gal
Drop, in	= 15 in	
Time, hr	= 24 hr*	

Estimate the gallons of hypochlorite pumped.

$$\text{Hypochlorite, gal} = (\text{Container Area, sq in})(\text{Drop, in})(7.48\ \text{gal/cu ft})$$

$$= \frac{(0.785)(42\ \text{in})^2(15\ \text{in})(7.48\ \text{gal/cu ft})}{(144\ \text{sq in/sq ft})(12\ \text{in/ft})}$$

$$= 90.0\ \text{gal}$$

* The time, 24 hr, is not needed to calculate the gallons of hypochlorite pumped.

10. A hypochlorite solution for a hypochlorinator is being prepared in a 55-gallon drum. If 8 gallons of 5 percent hypochlorite is added to the drum, how much water should be added to produce a 1.5 percent hypochlorite solution?

Known		Unknown
Drum Capacity, gal	= 55 gal	Water Added, gal
Hypochlorite, gal	= 8 gal	
Actual Hypochlorite, %	= 5%	
Desired Hypochlorite, %	= 1.5%	

Calculate the volume of water to be added in gallons.

$$\text{Water Added, gal} = \frac{(\text{Hypo, gal})(\text{Actual Hypo, \%}) - (\text{Desired Hypo, \%})(\text{Hypo, gal})}{\text{Desired Hypo, \%}}$$

$$= \frac{(8\text{ gal})(5\%) - (1.5\%)(8\text{ gal})}{1.5\%}$$

$$= \frac{40 - 12}{1.5}$$

$$= 18.7\text{ gallons}$$

11. Results from a coliform bacteria multiple fermentation tube test of raw water are as follows:

m*L* of sample	10	1	0.1	0.01	0.001
Dilutions	0	0	–1	–2	–3
Positive tubes	5	5	5	3	1

What is the most probable number of coliforms per 100 m*L* in the sample? Use Table 7.3 on page 449.

Select the underlined inoculations.

m*L* of sample	10	1	0.1	0.01	0.001
Dilutions	0	0	–1	–2	–3
Positive tubes	5	5	<u>5</u>	<u>3</u>	<u>1</u>

Read MPN as 110 from Table 7.3.
Report results as 11,000 coliforms per 100 m*L*.

We added two zeros to 110 because we started with the 0.1 m*L* sample and Table 7.3 begins with two dilution columns to the left (from 0.1 m*L* to 1 m*L* to 10 m*L* or two dilution columns to the left).

Data to answer questions 12 and 13.

Results from MPN tests during one week were as follows:

Day	S	M	T	W	T	F	S
MPN/100 m*L*	4	7	14	8	9	16	5

12. What is the mean MPN per 100 m*L* from the data?

Calculate the mean MPN/100 m*L* from the data.

$$\text{Mean, MPN/100 mL} = \frac{\text{Sum of All MPNs}}{\text{Number of MPNs}}$$

$$= \frac{4 + 7 + 14 + 8 + 9 + 16 + 5}{7}$$

$$= \frac{63}{7}$$

$$= 9\text{ MPN/100 mL}$$

13. What is the median MPN per 100 m*L* from the data?

Rearrange the data in ascending (increasing) order and select the middle value (three will be smaller and three will be larger).

Order	1	2	3	4	5	6	7
MPN/100 m*L*	4	5	7	8 ↑	9	14	16

$$\text{Median, MPN/100 mL} = \text{Middle value of a group of data}$$

$$= 8\text{ MPN/100 mL}$$

APPENDIX

HOW TO SOLVE SMALL WATER SYSTEM ARITHMETIC PROBLEMS

by

Ken Kerri

TABLE OF CONTENTS

HOW TO SOLVE SMALL WATER SYSTEM ARITHMETIC PROBLEMS

OBJECTIVES

HOW TO SOLVE SMALL WATER SYSTEM ARITHMETIC PROBLEMS

After completion of this Appendix, you should be able to:

1. Add, subtract, multiply, and divide;
2. List from memory basic conversion factors and formulas; and
3. Solve small water system arithmetic problems.

APPENDIX

HOW TO SOLVE SMALL WATER SYSTEM ARITHMETIC PROBLEMS

A.0 HOW TO STUDY THIS APPENDIX

This appendix may be worked early in your training program to help you gain the greatest benefit from your efforts. Whether to start this appendix early or wait until later is your decision. The chapters in this manual were written in a manner requiring very little background in arithmetic. You may wish to concentrate your efforts on the chapters and refer to this appendix when you need help. Some operators prefer to complete this appendix early so they will not have to worry about how to do the arithmetic when they are studying the chapters. You may try to work this appendix early or refer to it while studying the other chapters.

The intent of this appendix is to provide you with a quick review of the addition, subtraction, multiplication, and division needed to work the arithmetic problems in this manual. This appendix is not intended to be a math textbook. There are no fractions because you don't need fractions to work the problems in this manual. Some operators will be able to skip over the review of addition, subtraction, multiplication, and division. Others may need more help in these areas. If you need help in solving problems, read Section A.9, "Steps in Solving Problems." Basic arithmetic textbooks are available at every local library or bookstore and should be referred to if needed. Most instructional or operating manuals for pocket electronic calculators contain sufficient information on how to add, subtract, multiply, and divide.

After you have worked a problem involving your job, you should check your calculations, examine your answer to see if it appears reasonable, and if possible have another operator check your work before making any decisions or changes.

A.1 BASIC ARITHMETIC

In this section we provide you with basic arithmetic problems involving addition, subtraction, multiplication, and division. You may work the problems "by hand" if you wish, but we recommend you use an electronic pocket calculator. The operating or instructional manual for your calculator should outline the step-by-step procedures to follow. All calculators use similar procedures, but most of them are slightly different from others.

We will start with very basic, simple problems. Try working the problems and then comparing your answers with the given answers. If you can work these problems, you should be able to work the more difficult problems in the text of this training manual by using the same procedures.

A.10 Addition

2 3 — 5	6.2 8.5 — 14.7	16.7 38.9 — 55.6	6.12 38.39 — 44.51	43 39 34 38 39 37 29 — 259
2.12 9.80 — 11.92	0.12 2.0 — 2.12	63 32 — 95	120 60 — 180	
4 7 2 — 13	23 79 31 — 133	16.2 43.5 67.8 — 127.5	45.98 28.09 114.00 — 188.07	70 50 40 80 — 240

A.11 Subtraction

7 – 5 — 2	12 – 3 — 9	25 – 5 — 20	78 – 30 — 48	83 – 69 — 14
61 – 37 — 24	485 – 296 — 189	4.3 – 0.8 — 3.5	3.5 – 0.7 — 2.8	123 – 109 — 14
8.6 – 8.22 — 0.38	11.92 – 3.70 — 8.22	27.32 – 12.96 — 14.36	3.574 – 0.042 — 3.532	75.132 – 49.876 — 25.256

A.12 Multiplication

(3)(2)* = 6
(10)(5) = 50
(2)(22.99) = 45.98
(16)(17.1) = 273.6
(40)(2.31) = 92.4

(4)(7) = 28
(10)(1.3) = 13
(6)(19.5) = 117
(50)(20,000) = 1,000,000
(80)(0.433) = 34.64

(40)(20)(6) = 4,800
(4,800)(7.48) = 35,904
(1.6)(2.3)(8.34) = 30.6912
(0.001)(200)(8.34) = 1.668
(0.785)(7.48)(60) = 352.308
(12,000)(500)(60)(24) = 8,640,000,000 or 8.64×10^9
(4)(1,000)(1,000)(454) = 1,816,000,000 or 1.816×10^9

NOTE: The term, $\times 10^9$, means that the number is multiplied by 10^9 or 1,000,000,000. Therefore $8.64 \times 10^9 = 8.64 \times 1{,}000{,}000{,}000 = 8{,}640{,}000{,}000$.

* (3)(2) is the same as 3 x 2 = 6.

A.13 Division

$$\frac{6}{3} = 2 \qquad \frac{48}{12} = 4$$

$$\frac{50}{25} = 2 \qquad \frac{300}{20} = 15$$

$$\frac{20}{7.1} = 2.8 \qquad \frac{11{,}400}{188} = 60.6$$

$$\frac{1{,}000{,}000}{17.5} = 57{,}143 \qquad \frac{861{,}429}{30{,}000} = 28.7$$

$$\frac{4{,}000{,}000}{74{,}880} = 53.4 \qquad \frac{1.67}{8.34} = 0.20$$

$$\frac{80}{2.31} = 34.6 \qquad \frac{62}{454} = 0.137$$

$$\frac{250}{17.1} = 14.6 \qquad \frac{4{,}000{,}000}{14.6} = 273{,}973$$

NOTE: When we divide $1/3 = 0.3333$, we get a long row of 3s. Instead of the row of 3s, we "round off" our answer so $1/3 = 0.33$. For a discussion of rounding off numbers, see Section A.95, "Significant Figures."

A.14 Rules for Solving Equations

Most of the arithmetic problems we work in the waterworks field require us to plug numbers into formulas and calculate the answer. There are a few basic rules which apply to solving formulas. These rules are:

1. Work from left to right.
2. Do all the multiplication and division above the line (in the numerator) and below the line (in the denominator); then do the addition and subtraction above and below the line.
3. Perform the division (divide the numerator by the denominator).

Parentheses ( ) are used in formulas to identify separate parts of a problem. A fourth rule tells us how to handle numbers within parentheses.

4. Work the arithmetic within the parentheses before working outside the parentheses. Use the same order stated in rules 1, 2, and 3: work left to right, above and below the line, then divide the top number by the bottom number.

Let's look at an example problem to see how these rules apply. This year one of the responsibilities of the operators at our plant is to paint both sides of the wooden fence across the front of the facility. The fence is 145 feet long and 9 feet high. The steel access gate, which does not need painting, measures 14 feet wide by 9 feet high. Each gallon of paint will cover 150 square feet of surface area. How many gallons of paint should be purchased?

STEP 1: Identify the correct formula.

$$\text{Paint Req, gal} = \frac{\text{Total Area, sq ft}}{\text{Coverage, sq ft/gal}}$$

or

$$\text{Paint Req, gal} = \frac{(\text{Fence L, ft x H, ft x No. Sides}) - (\text{Gate L, ft x H, ft x No. Sides})}{\text{Coverage, sq ft/gal}}$$

STEP 2: Plug numbers into the formula.

$$\text{Paint Req, gal} = \frac{(145 \text{ ft x } 9 \text{ ft x } 2) - (14 \text{ ft x } 9 \text{ ft x } 2)}{150 \text{ sq ft/gal}}$$

STEP 3: Work the multiplication within parentheses.

$$\text{Paint Req, gal} = \frac{(2{,}610 \text{ sq ft}) - (252 \text{ sq ft})}{150 \text{ sq ft/gal}}$$

STEP 4: Work the subtraction above the line.

$$\text{Paint Req, gal} = \frac{2{,}358 \text{ sq ft}}{150 \text{ sq ft/gal}}$$

STEP 5: Divide the numerator by the denominator.

Paint Req, gal = 15.72 gal
or 16 gallons of paint will be needed.

Instructions for your electronic calculator can provide you with the detailed procedures for working the practice problems below.

$$\frac{(3)(4)}{2} = 6 \qquad \frac{64}{(8)(4)} = 2$$

$$\frac{(2+3)(4)}{5} = 4 \qquad \frac{54}{(4+2)(3)} = 3$$

$$\frac{(7-2)(8)}{4} = 10 \qquad \frac{48}{(8-3)(4)} = 2.4$$

$$\frac{(0.1)(60)(24)}{3} = 48$$

$$\frac{(12{,}000)(500)(60)(24)}{(4)(1{,}000)(1{,}000)(454)} = 4.76$$

$$\frac{12}{(0.432)(8.34)} = 3.3$$

$$\frac{(274{,}000)(24)}{200{,}000} = 32.88$$

A.15 Actual Problems

Let's look at the last four problems in the previous Section A.14, "Rules for Solving Equations," as they might be encountered by an operator.

1. To determine the actual chemical feed rate from an alum feeder, an operator collects the alum from the feeder in a bucket for three minutes. The alum in the bucket weighs 0.1 pounds.

Known		Unknown
Weight of Alum, lbs	= 0.1 lbs	Actual Alum Feed, lbs/day
Time, min	= 3 min	

Calculate the actual alum feed rate in pounds per day.

$$\text{Actual Alum Feed Rate, lbs/day} = \frac{(\text{Alum Wt, lbs})(60 \text{ min/hr})(24 \text{ hr/day})}{\text{Time Alum Collected, min}}$$

$$= \frac{(0.1 \text{ lb})(60 \text{ min/hr})(24 \text{ hr/day})}{3 \text{ min}}$$

$$= 48 \text{ lbs/day}$$

2. A solution chemical feeder is calibrated by measuring the time to feed 500 milliliters of chemical solution. The test calibration run required four minutes. The chemical concentration in the solution is 12,000 mg/*L* or 1.2%. Determine the chemical feed in pounds per day.

Known		Unknown
Volume Pumped, m*L*	= 500 m*L*	Chemical Feed, lbs/day
Time Pumped, min	= 4 min	
Chemical Conc, mg/*L*	= 12,000 mg/*L*	

Estimate the chemical feed rate in pounds per day.

$$\text{Chemical Feed, lbs/day} = \frac{(\text{Chem Conc, mg}/L)(\text{Vol Pumped, m}L)(60\text{ min/hr})(24\text{ hr/day})}{(\text{Time Pumped, min})(1{,}000\text{ m}L/L)(1{,}000\text{ mg/gm})(454\text{ gm/lb})}$$

$$= \frac{(12{,}000\text{ mg}/L)(500\text{ m}L)(60\text{ min/hr})(24\text{ hr/day})}{(4\text{ min})(1{,}000\text{ m}L/L)(1{,}000\text{ mg/gm})(454\text{ gm/lb})}$$

$$= 4.76\text{ lbs/day}$$

3. A chlorinator is set to feed 12 pounds of chlorine per day to a flow of 300 gallons per minute (0.432 million gallons per day). What is the chlorine dose in milligrams per liter?

Known		Unknown
Chlorinator Feed, lbs/day	= 12 lbs/day	Chlorine Dose, mg/*L*
Flow, MGD	= 0.432 MGD	

Determine the chlorine dose in milligrams per liter.

$$\text{Chlorine Dose, mg}/L = \frac{\text{Chlorinator Feed Rate, lbs/day}}{(\text{Flow, MGD})(8.34\text{ lbs/gal})}$$

$$= \frac{12\text{ lbs/day}}{(0.432\text{ MGD})(8.34\text{ lbs/gal})}$$

$$= 3.3\text{ mg}/L$$

4. Estimate the operating time of a water softening ion exchange unit before the unit needs regeneration. The unit can treat 274,000 gallons of water before the exchange capacity is exhausted. The average daily flow is 200,000 gallons per day.

Known		Unknown
Water Treated, gal	= 274,000 gal	Operating Time, hr
Avg Daily Flow, gal/day	= 200,000 gal/day	

Estimate the operating time of the ion exchange unit in hours.

$$\text{Operating Time, hr} = \frac{(\text{Water Treated, gal})(24\text{ hr/day})}{\text{Avg Daily Flow, gal/day}}$$

$$= \frac{(274{,}000\text{ gal})(24\text{ hr/day})}{200{,}000\text{ gal/day}}$$

$$= 32.9^*\text{ hours}$$

* We rounded off 32.88 hours to 32.9 hours.

A.2 AREAS

A.20 Units

Areas are measured in two dimensions or in square units. In the English system of measurement the most common units are square inches, square feet, square yards, and square miles. In the metric system the units are square millimeters, square centimeters, square meters, and square kilometers.

A.21 Rectangle

The area of a rectangle is equal to its length (L) multiplied by its width (W).

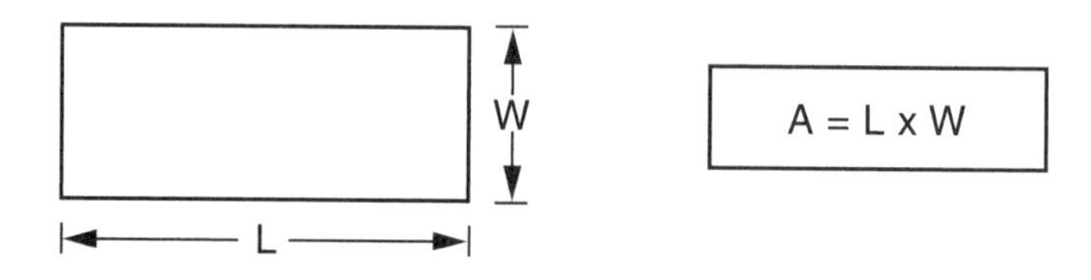

EXAMPLE: Find the area of a rectangle if the length is 5 feet and the width is 3.5 feet.

$$\text{Area, sq ft} = \text{Length, ft} \times \text{Width, ft}$$

$$= 5\text{ ft} \times 3.5\text{ ft}$$

$$= 17.5\text{ ft}^2$$

$$\text{or} = 17.5\text{ sq ft}$$

EXAMPLE: The surface area of a settling basin is 330 square feet. One side measures 15 feet. How long is the other side?

$$A = L \times W$$

$$330\text{ sq ft} = L\text{, ft} \times 15\text{ ft}$$

$$\frac{L\text{, ft} \times 15\text{ ft}}{15\text{ ft}} = \frac{330\text{ sq ft}}{15\text{ ft}}$$ Divide both sides of equation by 15 ft.

$$L\text{, ft} = \frac{330\text{ sq ft}}{15\text{ ft}}$$

$$= 22\text{ ft}$$

A.22 Triangle

The area of a triangle is equal to one half the base multiplied by the height. This is true for any triangle.

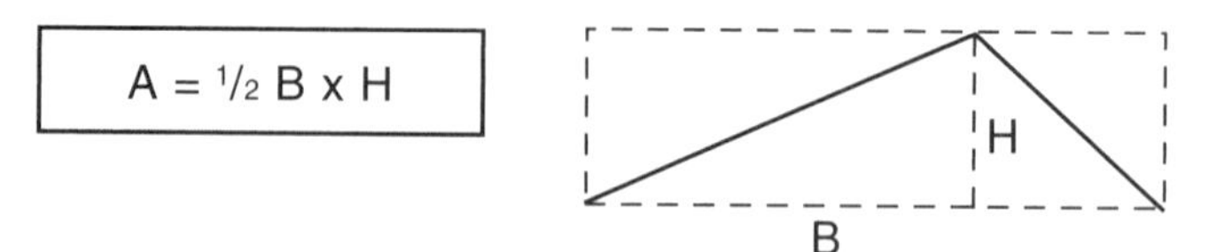

NOTE: The area of *ANY* triangle is equal to 1/2 the area of the rectangle that can be drawn around it. The area of the rectangle is B x H. The area of the triangle is 1/2 B x H.

EXAMPLE: Find the area of triangle ABC.

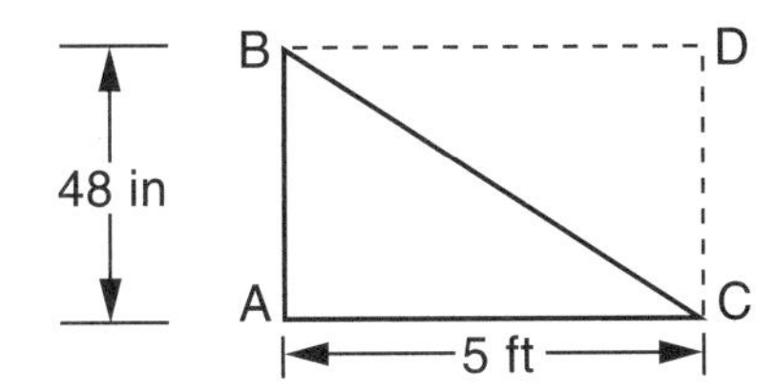

The first step in the solution is to make all the units the same. In this case, it is easier to change inches to feet.

$$48 \text{ in} = 48 \not{\text{in}} \times \frac{1 \text{ ft}}{12 \not{\text{in}}} = \frac{48}{12} \text{ ft} = 4 \text{ ft}$$

NOTE: All conversions should be calculated in the above manner. Since 1 ft/12 in is equal to unity, or one, multiplying by this factor changes the form of the answer but not its value.

$$\text{Area, sq ft} = \tfrac{1}{2}(\text{Base, ft})(\text{Height, ft})$$

$$= \tfrac{1}{2} \times 5 \text{ ft} \times 4 \text{ ft}$$

$$= \frac{20}{2} \text{ sq ft}$$

$$= 10 \text{ sq ft}$$

NOTE: Triangle ABC is one half the area of rectangle ABCD. The triangle is a special form called a *RIGHT TRIANGLE* since it contains a 90° angle at point A.

A.23 Circle

A square with sides of 2R can be drawn around a circle with a radius of R.

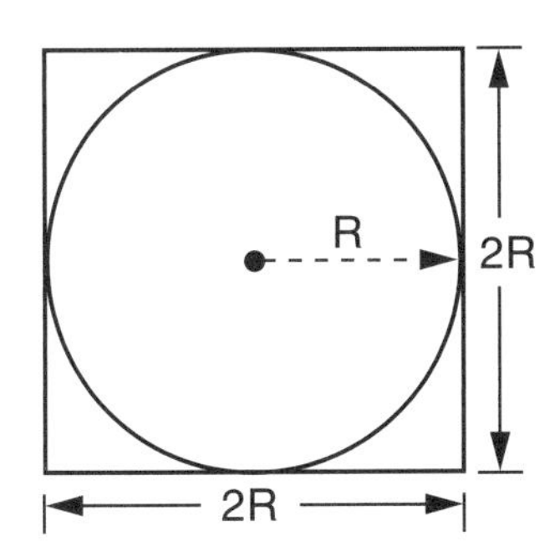

The area of the square is: $A = 2R \times 2R = 4R^2$.

It has been found that the area of any circle inscribed within a square is slightly more than 3/4 of the area of the square. More precisely, the area of the preceding circle is:

$$A \text{ circle} = 3\tfrac{1}{7} R^2 = 3.14 R^2$$

The formula for the area of a circle is usually written:

$$\boxed{A = \pi R^2}$$

The Greek letter π (pronounced pie) merely substitutes for the value 3.1416.

Since the diameter of any circle is equal to twice the radius, the formula for the area of a circle can be rewritten as follows:

$$A = \pi R^2 = \pi \times R \times R = \pi \times \frac{D}{2} \times \frac{D}{2} = \frac{\pi D^2}{4} = \frac{3.14}{4} D^2 = \boxed{0.785 D^2}$$

The type of problem and the magnitude of the numbers in a problem will determine which of the two formulas will provide a simpler solution. All of these formulas will give the same results if you use the same number of digits to the right of the decimal point.

EXAMPLE: What is the area of a circle with a diameter of 20 centimeters?

In this case, the formula using a radius is more convenient since it takes advantage of multiplying by 10.

$$\text{Area, sq cm} = \pi(\text{R, cm})^2$$

$$= 3.14 \times 10 \text{ cm} \times 10 \text{ cm}$$

$$= 314 \text{ sq cm}$$

EXAMPLE: What is the area of a clarifier with a 50-foot radius?

In this case, the formula using diameter is more convenient.

$$\text{Area, sq ft} = (0.785)(\text{Diameter, ft})^2$$

$$= 0.785 \times 100 \text{ ft} \times 100 \text{ ft}$$

$$= 7{,}850 \text{ sq ft}$$

Occasionally the operator may be confronted with a problem giving the area and requesting the radius or diameter. This presents the special problem of finding the square root of the number.

EXAMPLE: The surface area of a circular clarifier is approximately 5,000 square feet. What is the diameter?

$A = 0.785 D^2$, or

$\text{Area, sq ft} = (0.785)(\text{Diameter, ft})^2$

$5{,}000 \text{ sq ft} = 0.785 D^2$ To solve, substitute given values in equation.

$\dfrac{0.785 D^2}{0.785} = \dfrac{5{,}000 \text{ sq ft}}{0.785}$ Divide both sides by 0.785 to find D^2.

$$D^2 = \frac{5{,}000 \text{ sq ft}}{0.785}$$

$$= 6{,}369 \text{ sq ft. Therefore,}$$

D = square root of 6,369 sq ft, or

$$\text{Diameter, ft} = \sqrt{6{,}369 \text{ sq ft}}$$

Press the $\sqrt{\ }$ sign on your calculator and get D, ft = 79.8 ft.

Sometimes a trial-and-error method can be used to find square roots. Since 80 x 80 = 6,400, we know the answer is close to 80 feet.

Try 79 x 79 = 6,241

Try 79.5 x 79.5 = 6,320.25

Try 79.8 x 79.8 = 6,368.04

The diameter is 79.8 ft, or approximately 80 feet.

A.24 Cylinder

With the formulas presented thus far, it would be a simple matter to find the number of square feet in a room that was to be painted. The length of each wall would be added together and then multiplied by the height of the wall. This would give the surface area of the walls (minus any area for doors and windows). The ceiling area would be found by multiplying length times width and the result added to the wall area gives the total area.

The surface area of a circular cylinder, however, has not been discussed. If we wanted to know how many square feet of surface area are in a tank with a diameter of 60 feet and a height of 20 feet, we could start with the top and bottom.

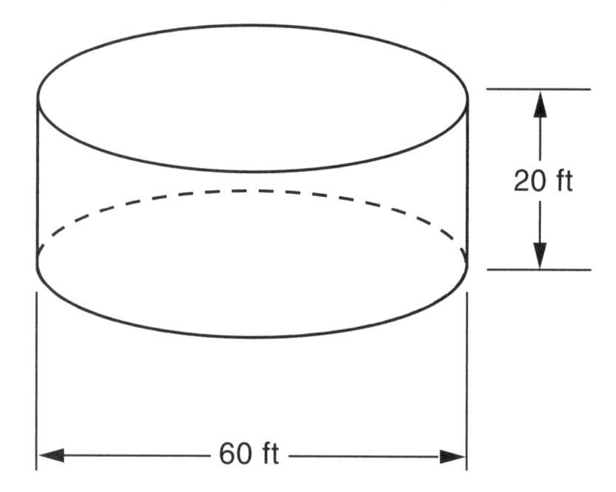

The area of the top and bottom ends are both $\pi \times R^2$.

$$\text{Area, sq ft} = (2\text{ ends})(\pi)(\text{Radius, ft})^2$$
$$= 2 \times \pi \times (30\text{ ft})^2$$
$$= 5{,}652\text{ sq ft}$$

The surface area of the wall must now be calculated. If we made a vertical cut in the wall and unrolled it, the straightened wall would be the same length as the circumference of the floor and ceiling.

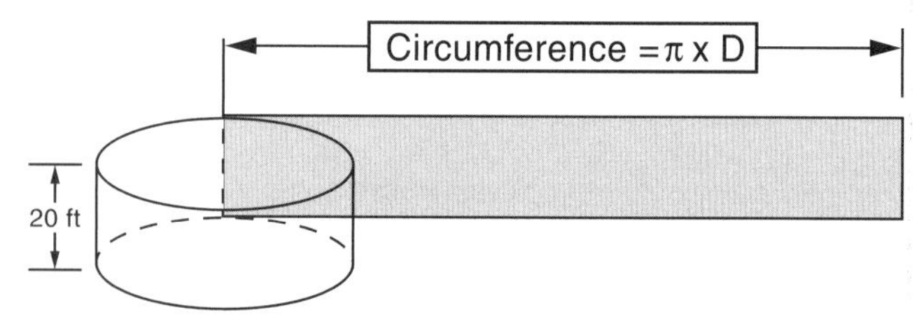

This length has been found to always be $\pi \times D$. In the case of the tank, the length of the wall would be:

$$\text{Length, ft} = (\pi)(\text{Diameter, ft})$$
$$= 3.14 \times 60\text{ ft}$$
$$= 188.4\text{ ft}$$

Area would be:

$$A_w\text{, sq ft} = \text{Length, ft} \times \text{Height, ft}$$
$$= 188.4\text{ ft} \times 20\text{ ft}$$
$$= 3{,}768\text{ sq ft}$$

$$\text{Outside Surface Area to Paint, sq ft} = \text{Area of Top and Bottom, sq ft} + \text{Area of Wall, sq ft}$$
$$= 5{,}652\text{ sq ft} + 3{,}768\text{ sq ft}$$
$$= 9{,}420\text{ sq ft}$$

A container has inside and outside surfaces and you may need to paint both of them.

A.25 Cone

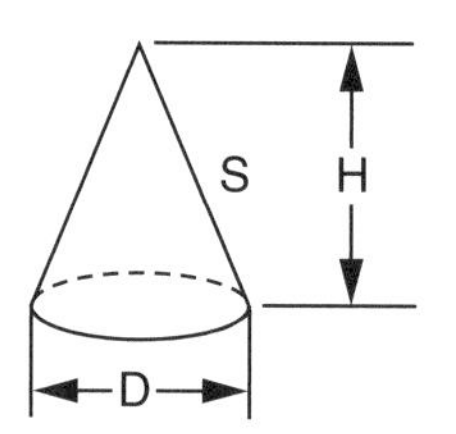

The lateral area of a cone is equal to 1/2 of the slant height (S) multiplied by the circumference of the base.

$$A_L = \tfrac{1}{2} S \times \pi \times D = \pi \times S \times R$$

In this case the slant height is not given, but it may be calculated by:

$$S = \sqrt{R^2 + H^2}$$

EXAMPLE: Find the entire outside area of a cone with a diameter of 30 inches and a height of 20 inches.

$$\text{Slant Height, in} = \sqrt{(\text{Radius, in})^2 + (\text{Height, in})^2}$$
$$= \sqrt{(15\text{ in})^2 + (20\text{ in})^2}$$
$$= \sqrt{225\text{ sq in} + 400\text{ sq in}}$$
$$= \sqrt{625\text{ sq in}}$$
$$= 25\text{ in}$$

$$\text{Lateral Area of Cone, sq in} = \pi(\text{Slant Height, in})(\text{Radius, in})$$
$$= 3.14 \times 25\text{ in} \times 15\text{ in}$$
$$= 1{,}177.5\text{ sq in}$$

Since the entire area was asked for, the area of the base must be added.

$$\text{Area, sq in} = 0.785(\text{Diameter, in})^2$$
$$= 0.785 \times 30\text{ in} \times 30\text{ in}$$
$$= 706.5\text{ sq in}$$

$$\text{Total Area, sq in} = \text{Area of Cone, sq in} + \text{Area of Bottom, sq in}$$
$$= 1{,}177.5\text{ sq in} + 706.5\text{ sq in}$$
$$= 1{,}884\text{ sq in}$$

A.26 Sphere

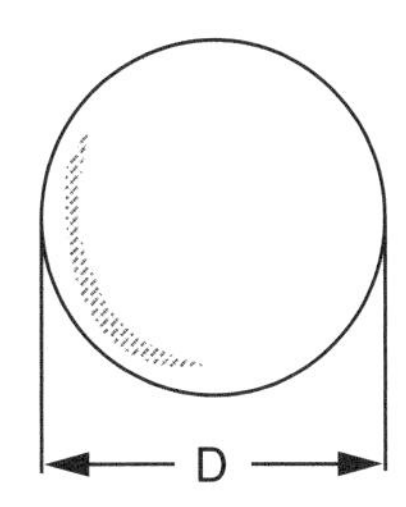

The surface area of a sphere or ball is equal to π multiplied by the diameter squared which is four times the cross-sectional area.

$$A_s = \pi D^2$$

If the radius is used, the formula becomes:

$$A_s = \pi D^2 = \pi \times 2R \times 2R = 4\pi R^2$$

EXAMPLE: What is the surface area of a sphere-shaped water tank 20 feet in diameter?

Area, sq ft = π(Diameter, ft)^2

= 3.14 x 20 ft x 20 ft

= 1,256 sq ft

A.3 VOLUMES

A.30 Rectangle

Volumes are measured in three dimensions or in cubic units. To calculate the volume of a rectangle, the area of the base is calculated in square units and then multiplied by the height. The formula then becomes:

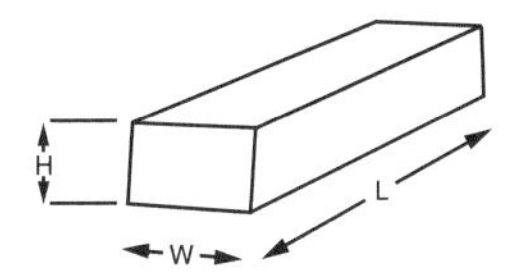

$$V = L \times W \times H$$

EXAMPLE: The length of a box is 2 feet, the width is 15 inches, and the height is 18 inches. Find its volume.

Volume, cu ft = Length, ft x Width, ft x Height, ft

$= 2 \text{ ft} \times \frac{15 \text{ in}}{12 \text{ in/ft}} \times \frac{18 \text{ in}}{12 \text{ in/ft}}$

= 2 ft x 1.25 ft x 1.5 ft

= 3.75 cu ft

A.31 Prism

The same general rule that applies to the volume of a rectangle also applies to a prism.

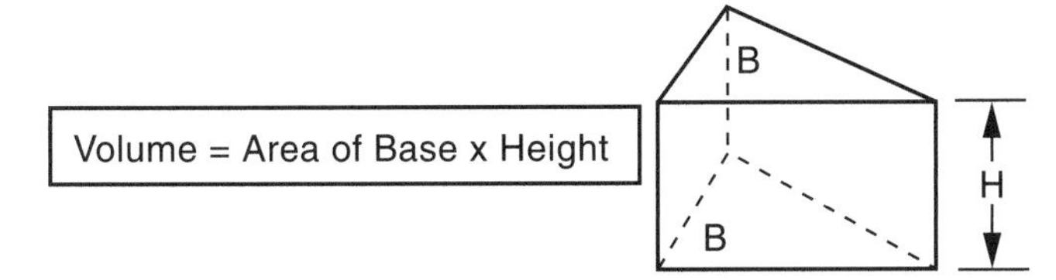

EXAMPLE: Find the volume of a prism with a base area of 10 square feet and a height of 5 feet. (Note that the base of a prism is triangular in shape.)

Volume, cu ft = Area of Base, sq ft x Height, ft

= 10 sq ft x 5 ft

= 50 cu ft

A.32 Cylinder

The volume of a cylinder is equal to the area of the base multiplied by the height.

$$V = \pi R^2 \times H = 0.785\ D^2 \times H$$

EXAMPLE: A tank has a diameter of 100 feet and a depth of 12 feet. Find the volume.

Volume, cu ft = 0.785 x (Diameter, ft)^2 x Height, ft

= 0.785 x 100 ft x 100 ft x 12 ft

= 94,200 cu ft

A.33 Cone

The volume of a cone is equal to 1/3 the volume of a circular cylinder of the same height and diameter.

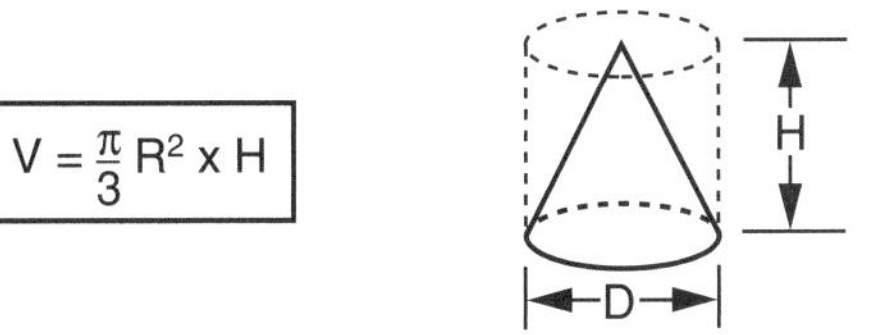

$$V = \frac{\pi}{3} R^2 \times H$$

EXAMPLE: Calculate the volume of a cone if the height at the center is 4 feet and the diameter is 100 feet (radius is 50 feet).

Volume, cu ft = $\frac{\pi}{3}$ x (Radius)^2 x Height, ft

= $\frac{\pi}{3}$ x 50 ft x 50 ft x 4 ft

= 10,500 cu ft

A.34 Sphere

The volume of a sphere is equal to π/6 times the diameter cubed.

$$V = \frac{\pi}{6} \times D^3$$

EXAMPLE: How much gas can be stored in a sphere with a diameter of 12 feet? (Assume atmospheric pressure.)

Volume, cu ft = $\frac{\pi}{6}$ x (Diameter, ft)^3

= $\frac{\pi}{\cancel{6}} \times \overset{2}{\cancel{12}}$ ft x 12 ft x 12 ft

= 904.32 cubic feet

A.4 METRIC SYSTEM

The two most common systems of weights and measures are the English system and the metric system (Le Système International d' Unités, SI). Of these two, the metric system is more popular with most of the nations of the world. The reason for this is that the metric system is based on a system of tens and is therefore easier to remember and easier to use than the English system. Even though the basic system in the United States is the English system, the scientific community uses the metric system almost exclusively. Many organizations have urged, for good reason, that the United States switch to the metric system. Today the metric system is gradually becoming the standard system of measurement in the United States.

As the United States changes from the English to the metric system, some confusion and controversy has developed. For example, which is the correct spelling of the following words:

1. Liter or litre?
2. Meter or metre?

The U.S. National Bureau of Standards, the Water Environment Federation, and the American Water Works Association use litre and metre. The U.S. Government uses liter and meter and accepts no deviations. Some people argue that METRE should be used to measure LENGTH and that METER should be used to measure FLOW RATES (like a water or electric meter). Liter and meter are used in this manual because this is most consistent with spelling in the United States.

One of the most frequent arguments heard against the U.S. switching to the metric system was that the costs of switching manufacturing processes would be excessive. Pipe manufacturers have agreed upon the use of a "soft" metric conversion system during the conversion to the metric system. Past practice in the U.S. has identified some types of pipe by external (outside) diameter while other types are classified by nominal (existing only in name, not real or actual) bore. This means that a six-inch pipe does not have a six-inch inside diameter. With the strict or "hard" metric system, a six-inch pipe would be a 152.4-mm (6 in x 25.4 mm/in) pipe. In the "soft" metric system a six-inch pipe is a 150-mm (6 in x 25 mm/in) pipe. Typical customary and "soft" metric pipe-size designations are shown below:

PIPE-SIZE DESIGNATIONS

Customary, in	2	4	6	8	10	12	15	18
"Soft" Metric, mm	50	100	150	200	250	300	375	450
Customary, in	24	30	36	42	48	60	72	84
"Soft" Metric, mm	600	750	900	1050	1200	1500	1800	2100

In order to study the metric system, you must know the meanings of the terminology used. Following is a list of Greek and Latin prefixes used in the metric system.

PREFIXES USED IN THE METRIC SYSTEM

Prefix	Symbol	Meaning
Micro	μ	1/1 000 000 or 0.000 001
Milli	m	1/1000 or 0.001
Centi	c	1/100 or 0.01
Deci	d	1/10 or 0.1
Unit		1
Deka	da	10
Hecto	h	100
Kilo	k	1000
Mega	M	1 000 000

A.40 Measures of Length

The basic measure of length is the meter.

1 kilometer (km) = 1,000 meters (m)
1 meter (m) = 100 centimeters (cm)
1 centimeter (cm) = 10 millimeters (mm)

Kilometers are usually used in place of miles, meters are used in place of feet and yards, centimeters are used in place of inches and millimeters are used for inches and fractions of an inch.

LENGTH EQUIVALENTS

1 kilometer	= 0.621 mile	1 mile	= 1.61 kilometers
1 meter	= 3.28 feet	1 foot	= 0.305 meter
1 meter	= 39.37 inches	1 inch	= 0.0254 meter
1 centimeter	= 0.3937 inch	1 inch	= 2.54 centimeters
1 millimeter	= 0.0394 inch	1 inch	= 25.4 millimeters

NOTE: The above equivalents are reciprocals. If one equivalent is given, the reverse can be obtained by division. For instance, if one meter equals 3.28 feet, one foot equals 1/3.28 meter, or 0.305 meter.

A.41 Measures of Capacity or Volume

The basic measure of capacity in the metric system is the liter. For measurement of large quantities the cubic meter is sometimes used.

1 kiloliter (k*L*) = 1,000 liters (*L*) = 1 cu meter (cu m)
1 liter (*L*) = 1,000 milliliters (m*L*)

Kiloliters, or cubic meters, are used to measure capacity of large storage tanks or reservoirs in place of cubic feet or gallons. Liters are used in place of gallons or quarts. Milliliters are used in place of quarts, pints, or ounces.

CAPACITY EQUIVALENTS

1 kiloliter	= 264.2 gallons	1 gallon	= 0.003785 kiloliter
1 liter	= 1.057 quarts	1 quart	= 0.946 liter
1 liter	= 0.2642 gallon	1 gallon	= 3.785 liters
1 milliliter	= 0.0353 ounce	1 ounce	= 29.57 milliliters

A.42 Measures of Weight

The basic unit of weight in the metric system is the gram. One cubic centimeter of water at maximum density weighs one gram, and thus there is a direct, simple relation between volume of water and weight in the metric system.

1 kilogram (kg) = 1,000 grams (gm)
1 gram (gm) = 1,000 milligrams (mg)
1 milligram (mg) = 1,000 micrograms (μg)

Grams are usually used in place of ounces, and kilograms are used in place of pounds.

WEIGHT EQUIVALENTS

1 kilogram	= 2.205 pounds	1 pound	= 0.4536 kilogram
1 gram	= 0.0022 pound	1 pound	= 453.6 grams
1 gram	= 0.0353 ounce	1 ounce	= 28.35 grams
1 gram	= 15.43 grains	1 grain	= 0.0648 gram

A.43 Temperature

Just as the operator should become familiar with the metric system, you should also become familiar with the centigrade (Celsius) scale for measuring temperature. There is nothing magical about the centigrade scale — it is simply a different size than the Fahrenheit scale. The two scales compare as follows:

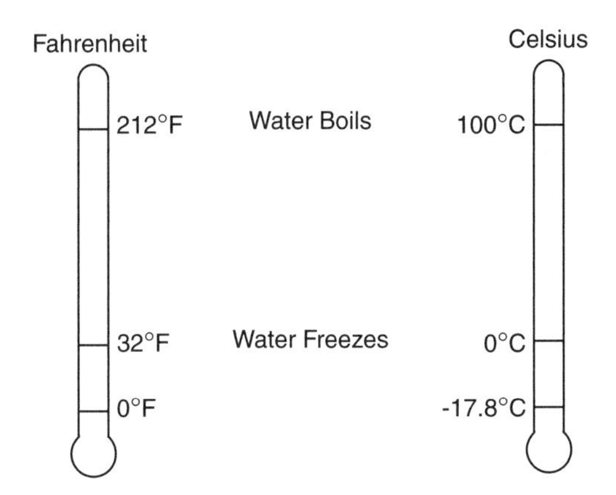

The two scales are related in the following manner:

Fahrenheit	= (°C x 9/5) + 32°
Celsius	= (°F – 32°) x 5/9

EXAMPLE: Convert 20° Celsius to degrees Fahrenheit.

$$°F = (°C \times 9/5) + 32°$$

$$°F = (20° \times 9/5) + 32°$$

$$°F = \frac{180°}{5} + 32°$$

$$= 36° + 32°$$

$$= 68°F$$

EXAMPLE: Convert –10°C to °F.

$$°F = (-10° \times 9/5) + 32°$$

$$°F = -90°/5 + 32°$$

$$= -18° + 32°$$

$$= 14°F$$

EXAMPLE: Convert –13°F to °C.

$$°C = (°F - 32°) \times \frac{5}{9}$$

$$°C = (-13° - 32°) \times \frac{5}{9}$$

$$= -45° \times \frac{5}{9}$$

$$= -5° \times 5$$

$$= -25°C$$

A.44 Milligrams Per Liter

Milligrams per liter (mg/*L*) is a unit of measurement used in laboratory and scientific work to indicate very small concentrations of dilutions. Since water contains small concentrations of dissolved substances and solids, and since small amounts of chemical compounds are sometimes used in water treatment processes, the term milligrams per liter is also common in treatment plants. It is a weight/volume relationship.

As previously discussed:

1,000 liters = 1 cubic meter = 1,000,000 cubic centimeters.

Therefore,

1 liter = 1,000 cubic centimeters.

Since one cubic centimeter of water weighs one gram,

1 liter of water = 1,000 grams or 1,000,000 milligrams.

$$\frac{1 \text{ milligram}}{\text{liter}} = \frac{1 \text{ milligram}}{1{,}000{,}000 \text{ milligrams}} = \frac{1 \text{ part}}{\text{million parts}} = \text{1 part per million (ppm)}$$

Milligrams per liter and parts per million (parts) may be used interchangeably as long as the liquid density is 1.0 gm/cu cm or 62.43 lb/cu ft. A concentration of 1 milligram/liter (mg/*L*) or 1 ppm means that there is 1 part of substance by weight for every 1 million parts of water. A concentration of 10 mg/*L* would mean 10 parts of substance per million parts of water.

To get an idea of how small 1 mg/*L* is, divide the numerator and denominator of the fraction by 10,000. This, of course, does not change its value since, 10,000 ÷ 10,000 is equal to one.

$$1\,\frac{mg}{L} = \frac{1 \text{ mg}}{1{,}000{,}000 \text{ mg}} = \frac{1/10{,}000 \text{ mg}}{1{,}000{,}000/10{,}000 \text{ mg}} = \frac{0.0001 \text{ mg}}{100 \text{ mg}} = 0.0001\%$$

Therefore, 1 mg/*L* is equal to one ten-thousandth of a percent, or

1% is equal to 10,000 mg/*L*.

To convert mg/*L* to %, move the decimal point four places or numbers to the left.

Working problems using milligrams per liter or parts per million is a part of everyday operation in most water treatment plants.

A.45 Example Problems

EXAMPLE: Raw water flowing into a plant at a rate of five million pounds per day is prechlorinated at 5 mg/*L*. How many pounds of chlorine are used per day?

$$5 \text{ mg}/L = \frac{5 \text{ lbs chlorine}}{\text{million lbs water}}$$

$$\text{Chlorine Feed, lbs/day} = \text{Conc, lbs/M lbs} \times \text{Flow, lbs/day}$$

$$= \frac{5 \text{ lbs}}{\cancel{\text{million lbs}}} \times \frac{5 \cancel{\text{ million lbs}}}{\text{day}}$$

$$= 25 \text{ lbs/day}$$

There is one thing that is unusual about the above problem and that is the flow is reported in pounds per day. In most treatment plants, flow is reported in terms of gallons per minute or gallons per day. To convert these flow figures to weight, an additional conversion factor is needed. One gallon of water weighs 8.34 pounds. Using this factor, it is possible to convert flow in gallons per day to flow in pounds per day.

EXAMPLE: A well pump with a flow of 3.5 million gallons per day (MGD) chlorinates the water with 2.0 mg/*L* chlorine. How many pounds of chlorine are used per day?

$$\text{Flow, lbs/day} = \text{Flow, } \frac{\text{M gal}}{\text{day}} \times \frac{8.34 \text{ lbs}}{\text{gal}}$$

$$= \frac{3.5 \text{ million } \cancel{\text{gal}}}{\text{day}} \times \frac{8.34 \text{ lbs}}{\cancel{\text{gal}}}$$

$$= 29.19 \text{ million lbs/day}$$

$$\text{Chlorine Feed, lbs/day} = \text{Level, mg}/L \times \text{Flow, M lbs/day}$$

$$= \frac{2.0 \cancel{\text{ mg}}^*}{\cancel{\text{million mg}}} \times \frac{29.19 \cancel{\text{ million}} \text{ lbs}}{\text{day}}$$

$$= 58.38 \text{ lbs/day}$$

* Remember that $\frac{1 \text{ mg}}{\text{M mg}} = \frac{1 \text{ lb}}{\text{M lb}}$. They are identical ratios.

In solving the above problem, a relation was used that is most important to understand and commit to memory.

Feed, lbs/day = Flow, MGD x Dose, mg/*L* x 8.34 lbs/gal

EXAMPLE: A chlorinator is set to feed 50 pounds of chlorine per day to a flow of 0.8 MGD. What is the chlorine dose in mg/*L*?

$$\text{Conc or Dose, mg}/L = \frac{\text{lbs/day}}{\text{MGD} \times 8.34 \text{ lbs/gal}}$$

$$= \frac{50 \text{ lbs/day}}{0.80 \text{ MG/day} \times 8.34 \text{ lbs/gal}}$$

$$= \frac{50 \text{ lbs}}{6.672 \text{ M lbs}}$$

$$= 7.5 \text{ mg}/L\text{, or } 7.5 \text{ ppm}$$

EXAMPLE: A pump delivers 500 gallons per minute to a water treatment plant. Alum is added at 10 mg/*L*. How much alum is used in pounds per day?

$$\text{Flow, MGD} = \text{Flow, GPM} \times 60 \text{ min/hr} \times 24 \text{ hr/day}$$

$$= \frac{500 \text{ gal}}{\cancel{\text{min}}} \times \frac{60 \cancel{\text{ min}}}{\cancel{\text{hr}}} \times \frac{24 \cancel{\text{ hr}}}{\text{day}}$$

$$= 720{,}000 \text{ gal/day}$$

$$= 0.72 \text{ MGD}$$

$$\text{Alum Feed, lbs/day} = \text{Flow, MGD} \times \text{Dose, mg}/L \times 8.34 \text{ lbs/ga}$$

$$= \frac{0.72 \cancel{\text{ M gal}}}{\text{day}} \times \frac{10 \cancel{\text{ mg}}}{\cancel{\text{M mg}}} \times \frac{8.34 \text{ lbs}}{\cancel{\text{gal}}}$$

$$= 60.048 \text{ lbs/day or about } 60 \text{ lbs/day}$$

A.5 WEIGHT — VOLUME RELATIONS

Another factor for the operator to remember, in addition to the weight of a gallon of water, is the weight of a cubic foot of water. One cubic foot of water weighs 62.4 lbs. If these two weights are divided, it is possible to determine the number of gallons in a cubic foot.

$$\frac{62.4 \cancel{\text{ pounds}}\text{/cu ft}}{8.34 \cancel{\text{ pounds}}\text{/gal}} = 7.48 \text{ gal/cu ft}$$

Thus we have another very important relationship to commit to memory.

8.34 lbs/gal x 7.48 gal/cu ft = 62.4 lbs/cu ft

It is only necessary to remember two of the above items since the third may be found by calculation. For most problems, $8\frac{1}{3}$ lbs/gal and $7\frac{1}{2}$ gal/cu ft will provide sufficient accuracy.

EXAMPLE: Change 1,000 cu ft of water to gallons.

1,000 cu ft x 7.48 gal/cu ft = 7,480 gallons

EXAMPLE: What is the weight of three cubic feet of water?

62.4 lbs/cu ft x 3 cu ft = 187.2 lbs

EXAMPLE: The net weight of a tank of water is 750 lbs. How many gallons does it contain?

$$\frac{750 \cancel{\text{ lbs}}}{8.34 \cancel{\text{ lbs}}\text{/gal}} = 90 \text{ gal}$$

A.6 FORCE, PRESSURE, AND HEAD

In order to study the forces and pressures involved in fluid flow, it is first necessary to define the terms used.

FORCE: The push exerted by water on any surface being used to confine it. Force is usually expressed in pounds, tons, grams, or kilograms.

PRESSURE: The force per unit area. Pressure can be expressed in many ways, but the most common term is pounds per square inch (psi).

HEAD: Vertical distance from the water surface to a reference point below the surface. Usually expressed in feet or meters.

An *EXAMPLE* should serve to illustrate these terms.

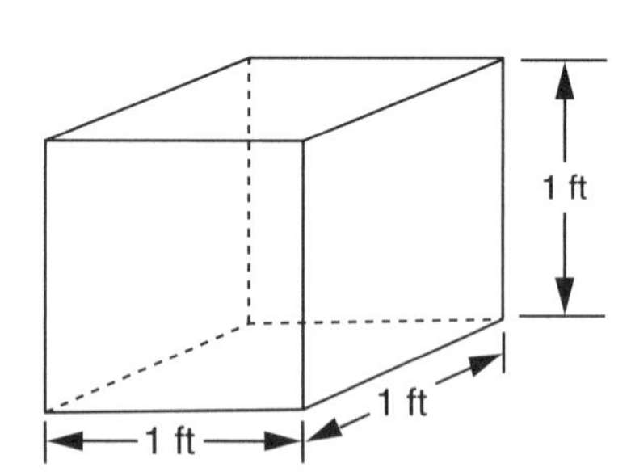

If water were poured into a one-foot cubical container, the *FORCE* acting on the bottom of the container would be 62.4 pounds.

The *PRESSURE* acting on the bottom would be 62.4 pounds per square foot. The area of the bottom is also 12 in x 12 in = 144 sq in. Therefore, the pressure may also be expressed as:

$$\text{Pressure, psi} = \frac{62.4 \text{ lbs}}{\text{sq ft}} = \frac{62.4 \text{ lbs/sq ft}}{144 \text{ sq in/sq ft}}$$

$$= 0.433 \text{ lb/sq in}$$

$$= 0.433 \text{ psi}$$

Since the height of the container is one foot, the *HEAD* would be one foot.

The pressure in any vessel at one foot of depth or one foot of head is 0.433 psi acting in any direction.

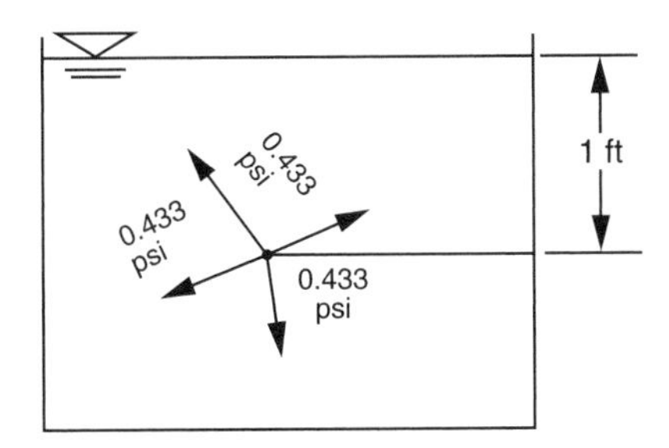

If the depth of water in the previous example were increased to two feet, the pressure would be:

$$p = \frac{2(62.4 \text{ lbs})}{144 \text{ sq in}} = \frac{124.8 \text{ lbs}}{144 \text{ sq in}} = 0.866 \text{ psi}$$

Therefore we can see that for every foot of head, the pressure increases by 0.433 psi. Thus, the general formula for pressure becomes:

p, psi = 0.433 (H, ft)

H = feet of head

p = pounds per square *INCH* of pressure

P, lbs/sq ft = 62.4 (H, ft)

H = feet of head

P = pounds per square *FOOT* of pressure

We can now draw a diagram of the pressure acting on the side of a tank. Assume a 4-foot deep tank. The pressures shown on the tank are gage pressures. These pressures do not include the atmospheric pressure acting on the surface of the water.

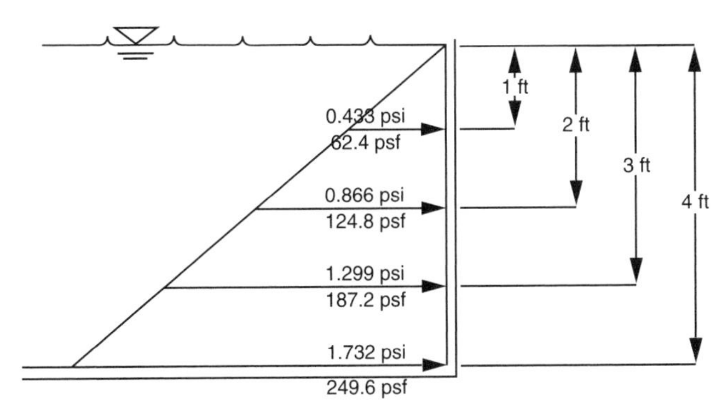

$p_0 = 0.433 \times 0 = 0.0$ psi
$p_1 = 0.433 \times 1 = 0.433$ psi
$p_2 = 0.433 \times 2 = 0.866$ psi
$p_3 = 0.433 \times 3 = 1.299$ psi
$p_4 = 0.433 \times 4 = 1.732$ psi

$P_0 = 62.4 \times 0 = 0.0$ lb/sq ft
$P_1 = 62.4 \times 1 = 62.4$ lbs/sq ft
$P_2 = 62.4 \times 2 = 124.8$ lbs/sq ft
$P_3 = 62.4 \times 3 = 187.2$ lbs/sq ft
$P_4 = 62.4 \times 4 = 249.6$ lbs/sq ft

The average *PRESSURE* acting on the tank wall is 1.732 psi/2 = 0.866 psi, or 249.2 psf/2 = 124.8 psf. We divided by two to obtain the average pressure because there is zero pressure at the top and 1.732 psi pressure on the bottom of the wall.

If the wall were 5 feet long, the pressure would be acting over the entire 20-square-foot (5 ft x 4 ft) area of the wall. The total force acting to push the wall would be:

$$\text{Force, lbs} = (\text{Pressure, lbs/sq ft})(\text{Area, sq ft})$$
$$= 124.8 \text{ lbs/sq ft} \times 20 \text{ sq ft}$$
$$= 2{,}496 \text{ lbs}$$

If the pressure in psi were used, the problem would be similar:

$$\text{Force, lbs} = (\text{Pressure, lbs/sq in})(\text{Area, sq in})$$
$$= 0.866 \text{ psi} \times 48 \text{ in} \times 60 \text{ in}$$
$$= 2{,}494 \text{ lbs*}$$

* Difference in answer due to rounding off of decimal points.

The general formula, then, for finding the total force acting on a side wall of a tank is:

$F = 31.2 \times H^2 \times L$

F = force in pounds

H = head in feet

L = length of wall in feet

31.2 = constant with units of lbs/cu ft and considers the fact that the force results from H/2 or half the depth of the water which is the average depth. The force is exerted at H/3 from the bottom.

EXAMPLE: Find the force acting on a 5-foot long wall in a 4-foot deep tank.

$$\text{Force, lbs} = 31.2 (\text{Head, ft})^2(\text{Length, ft})$$
$$= 31.2 \text{ lbs/cu ft} \times (4 \text{ ft})^2 \times 5 \text{ ft}$$
$$= 2{,}496 \text{ lbs}$$

Occasionally an operator is warned: *NEVER EMPTY A TANK DURING PERIODS OF HIGH GROUNDWATER.* Why? The pressure on the bottom of the tank caused by the water surrounding the tank will tend to float the tank like a cork if the upward force of the water is greater than the weight of the tank.

F = 62.4 x H x A

F = upward force in pounds

H = head of water on tank bottom in feet

A = area of bottom of tank in square feet

62.4 = a constant with units of lbs/cu ft

This formula is approximately true if the tank doesn't crack, leak, or start to float.

EXAMPLE: Find the upward force on the bottom of an empty tank caused by a groundwater depth of 8 feet above the tank bottom. The tank is 20 ft wide and 40 ft long.

$$\text{Force, lbs} = 62.4 \text{ (Head, ft)(Area, sq ft)}$$
$$= 62.4 \text{ lbs/cu ft} \times 8 \text{ ft} \times 20 \text{ ft} \times 40 \text{ ft}$$
$$= 399{,}400 \text{ lbs}$$

A.7 VELOCITY AND FLOW RATE

A.70 Velocity

The velocity of a particle or substance is the speed at which it is moving. It is expressed by indicating the length of travel and how long it takes to cover the distance. Velocity can be expressed in almost any distance and time units. For instance, a car may be traveling at a rate of 280 miles per five hours. However, it is normal to express the distance traveled per unit time. The above example would then become:

$$\text{Velocity, mi/hr} = \frac{280 \text{ miles}}{5 \text{ hours}}$$
$$= 56 \text{ miles/hour}$$

The velocity of water in a channel, pipe, or other conduit can be expressed in the same way. If the particle of water travels 600 feet in five minutes, the velocity is:

$$\text{Velocity, ft/min} = \frac{\text{Distance, ft}}{\text{Time, minutes}}$$
$$= \frac{600 \text{ ft}}{5 \text{ min}}$$
$$= 120 \text{ ft/min}$$

If you wish to express the velocity in feet per second, multiply by 1 min/60 seconds.

NOTE: Multiplying by $\frac{1 \text{ minute}}{60 \text{ seconds}}$ is like multiplying by $\frac{1}{1}$; it does not change the relative value of the answer. It only changes the form of the answer.

$$\text{Velocity, ft/sec} = \text{(Velocity, ft/min)(1 min/60 sec)}$$
$$= \frac{120 \text{ ft}}{\cancel{\text{min}}} \times \frac{1 \cancel{\text{min}}}{60 \text{ sec}}$$
$$= \frac{120 \text{ ft}}{60 \text{ sec}}$$
$$= 2 \text{ ft/sec}$$

A.71 Flow Rate

If water in a one-foot wide channel is one foot deep, then the cross-sectional area of the channel is 1 ft x 1 ft = 1 sq ft.

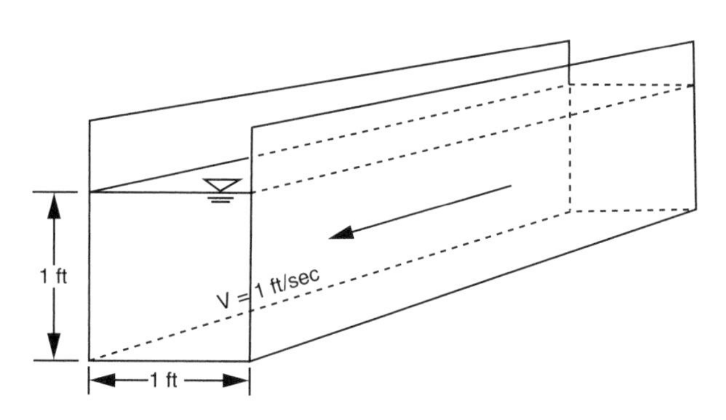

If the velocity in this channel is 1 ft per second, then each second a body of water 1 sq ft in area and 1 ft long will pass a given point. The volume of this body of water would be 1 cubic foot. Since 1 cubic foot of water would pass by every second, the flow rate would be equal to 1 cubic foot per second, or 1 CFS.

To obtain the flow rate in the above example the velocity was multiplied by the cross-sectional area. This is another important general formula.

$$\boxed{Q = V \times A}$$

Q = flow rate, CFS or cu ft/sec

V = velocity, ft/sec

A = area, sq ft

EXAMPLE: A rectangular channel 3 feet wide contains water 2 feet deep and flowing at a velocity of 1.5 feet per second. What is the flow rate in CFS?

$$Q = V \times A$$
$$\text{Flow Rate, CFS} = \text{Velocity, ft/sec} \times \text{Area, sq ft}$$
$$= 1.5 \text{ ft/sec} \times 3 \text{ ft} \times 2 \text{ ft}$$
$$= 9 \text{ cu ft/sec}$$

EXAMPLE: Flow in a 2.5-foot wide channel is 1.4 ft deep and measures 11.2 CFS. What is the average velocity?

In this problem we want to find the velocity. Therefore, we must rearrange the general formula to solve for velocity.

$$V = \frac{Q}{A}$$
$$\text{Velocity, ft/sec} = \frac{\text{Flow Rate, cu ft/sec}}{\text{Area, sq ft}}$$
$$= \frac{11.2 \text{ cu ft/sec}}{2.5 \text{ ft} \times 1.4 \text{ ft}}$$
$$= \frac{11.2 \text{ cu ft/sec}}{3.5 \text{ sq ft}}$$
$$= 3.2 \text{ ft/sec}$$

EXAMPLE: Flow in an 8-inch pipe is 500 GPM. What is the average velocity?

$$\text{Area, sq ft} = 0.785(\text{Diameter, ft})^2$$
$$= 0.785(8/12\text{ ft})^2$$
$$= 0.785(0.67\text{ ft})^2$$
$$= 0.785(0.67\text{ ft})(0.67\text{ ft})$$
$$= 0.785(0.45\text{ sq ft})$$
$$= 0.35\text{ sq ft}$$

$$\text{Flow, CFS} = \text{Flow, gal/min} \times \frac{\text{cu ft}}{7.48\text{ gal}} \times \frac{1\text{ min}}{60\text{ sec}}$$
$$= \frac{500\text{ gal}}{\text{min}} \times \frac{\text{cu ft}}{7.48\text{ gal}} \times \frac{1\text{ min}}{60\text{ sec}}$$
$$= \frac{500\text{ cu ft}}{448.8\text{ sec}}$$
$$= 1.114\text{ CFS}$$

$$\text{Velocity, ft/sec} = \frac{\text{Flow, cu ft/sec}}{\text{Area, sq ft}}$$
$$= \frac{1.114\text{ cu ft/sec}}{0.35\text{ sq ft}}$$
$$= 3.18\text{ ft/sec}$$

A.8 PUMPS

A.80 Pressure

Atmospheric pressure at sea level is approximately 14.7 psi. This pressure acts in all directions and on all objects. If a tube is placed upside down in a basin of water and a 1 psi partial vacuum is drawn on the tube, the water in the tube will rise 2.31 feet.

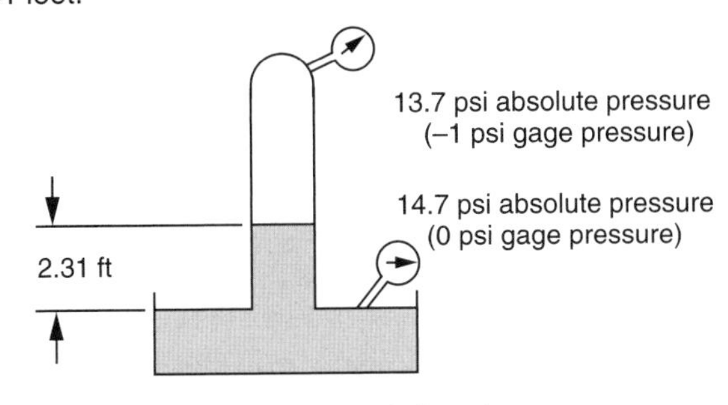

NOTE: 1 ft of water = 0.433 psi; therefore,

$$1\text{ psi} = \frac{1}{0.433}\text{ ft} = 2.31\text{ ft of water}$$

The action of the partial vacuum is what gets water out of a sump or well and up to a pump. It is not sucked up, but it is pushed up by atmospheric pressure on the water surface in the sump. If a complete vacuum could be drawn, the water would rise 2.31 x 14.7 = 33.9 feet; but this is impossible to achieve. The practical limit of the suction lift of a positive displacement pump is about 22 feet, and that of a centrifugal pump is 15 feet.

A.81 Work

Work can be expressed as lifting a weight a certain vertical distance. It is usually defined in terms of foot-pounds.

EXAMPLE: A 165-pound man runs up a flight of stairs 20 feet high. How much work did he do?

$$\text{Work, ft-lbs} = \text{Weight, lbs} \times \text{Height, ft}$$
$$= 165\text{ lbs} \times 20\text{ ft}$$
$$= 3{,}300\text{ ft-lbs}$$

A.82 Power

Power is a rate of doing work and is usually expressed in foot-pounds per minute.

EXAMPLE: If the man in the above example runs up the stairs in three seconds, how much power has he exerted?

$$\text{Power, ft-lbs/sec} = \frac{\text{Work, ft-lbs}}{\text{Time, sec}}$$
$$= \frac{3{,}300\text{ ft-lbs}}{3\text{ }\cancel{\text{sec}}} \times \frac{60\text{ }\cancel{\text{sec}}}{\text{minute}}$$
$$= 66{,}000\text{ ft-lbs/min}$$

A.83 Horsepower

Horsepower is also a unit of power. One horsepower is defined as 33,000 ft-lbs per minute or 746 watts.

EXAMPLE: How much horsepower has the man in the previous example exerted as he climbs the stairs?

$$\text{Horsepower, HP} = (\text{Power, ft-lbs/min})\left(\frac{\text{HP}}{33{,}000\text{ ft-lbs/min}}\right)$$
$$= 66{,}000\text{ ft-lbs/min} \times \frac{\text{Horsepower}}{33{,}000\text{ ft-lbs/min}}$$
$$= 2\text{ HP}$$

Work is also done by lifting water. If the flow from a pump is converted to a weight of water and multiplied by the vertical distance it is lifted, the amount of work or power can be obtained.

$$\text{Horsepower, HP} = \frac{\text{Flow, gal}}{\text{min}} \times \text{Lift, ft} \times \frac{8.34\text{ lbs}}{\text{gal}} \times \frac{\text{Horsepower}}{33{,}000\text{ ft-lbs/min}}$$

Solving the above relation, the amount of horsepower necessary to lift the water is obtained. This is called water horsepower.

$$\text{Water, HP} = \frac{(\text{Flow, GPM})(\text{H, ft})}{3{,}960^*}$$

$$^*\frac{8.34\text{ lbs}}{\text{gal}} \times \frac{\text{HP}}{33{,}000\text{ ft-lbs/min}} = \frac{1}{3{,}960}$$

1 gallon weighs 8.34 pounds and 1 horsepower is the same as 33,000 ft-lbs/min.

H or Head in feet is the same as Lift in feet.

However, since pumps are not 100% efficient (they cannot transmit all the power put into them), the horsepower supplied to a pump is greater than the water horsepower. Horsepower supplied to the pump is called brake horsepower.

$$\text{Brake, HP} = \frac{\text{Flow, GPM} \times \text{H, ft}}{3{,}960 \times E_p}$$

E_p = Efficiency of Pump (Usual range 50-85%, depending on type and size of pump)

Motors are also not 100% efficient; therefore, the power supplied to the motor is greater than the motor transmits.

$$\text{Motor, HP} = \frac{\text{Flow, GPM} \times \text{H, ft}}{3{,}960 \times E_p \times E_m}$$

E_m = Efficiency of Motor (Usual range 80-95%, depending on type and size of motor)

The above formulas have been developed for the pumping of water and wastewater which have a specific gravity of 1.0. If other liquids are to be pumped, the formulas must be multiplied by the specific gravity of the liquid.

EXAMPLE: A flow of 500 GPM of water is to be pumped against a total head of 100 feet by a pump with an efficiency of 70%. What is the pump horsepower?

$$\text{Brake, HP} = \frac{\text{Flow, GPM} \times \text{H, ft}}{3{,}960 \times E_p}$$

$$= \frac{500 \times 100}{3{,}960 \times 0.70}$$

$$= 18 \text{ HP}$$

EXAMPLE: Find the horsepower required to pump gasoline (specific gravity = 0.75) in the previous problem.

$$\text{Brake, HP} = \frac{500 \times 100 \times 0.75}{3{,}960 \times 0.70}$$

$$= 13.5 \text{ HP}$$ (gasoline is lighter and requires less horsepower)

A.84 Head

Basically, the head that a pump must work against is determined by measuring the vertical distance between the two water surfaces, or the distance the water must be lifted. This is called the static head. Two typical conditions for lifting water are shown below.

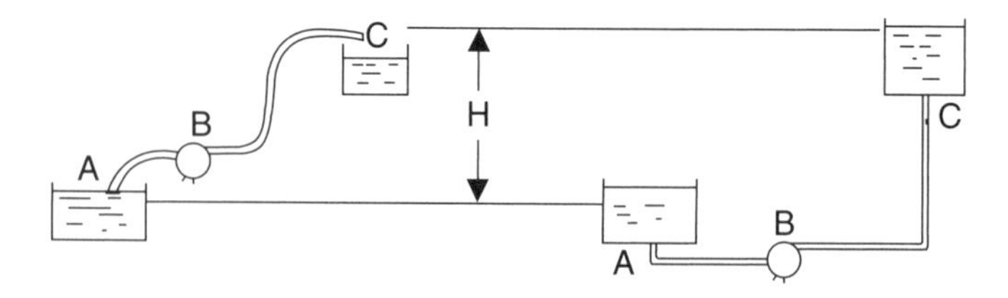

If a pump were designed in the above examples to pump only against head H, the water would never reach the intended point. The reason for this is that the water encounters friction in the pipelines. Friction depends on the roughness and length of pipe, the pipe diameter, and the flow velocity. The turbulence caused at the pipe entrance (point A); the pump (point B); the pipe exit (point C); and at each elbow, bend, or transition also adds to these friction losses. Tables and charts are available in Section A.88 for calculation of these friction losses so they may be added to the measured or static head to obtain the total head. For short runs of pipe which do not have high velocities, the friction losses are generally less than 10 percent of the static head.

EXAMPLE: A pump is to be located 8 feet above a wet well and must lift 1.8 MGD another 50 feet to a storage reservoir. If the pump has an efficiency of 75% and the motor an efficiency of 90%, what is the cost of the power consumed if one kilowatt hour costs 4 cents?

Since we are not given the length or size of pipe and the number of elbows or bends, we will assume friction to be 10% of static head.

$$\text{Static Head, ft} = \text{Suction Lift, ft} + \text{Discharge Head, ft}$$

$$= 8 \text{ ft} + 50 \text{ ft}$$

$$= 58 \text{ ft}$$

$$\text{Friction Losses, ft} = 0.1 \text{ (Static Head, ft)}$$

$$= 0.1 \text{ (58 ft)}$$

$$= 5.8 \text{ ft}$$

$$\text{Total Dynamic Head, ft} = \text{Static Head, ft} + \text{Friction Losses, ft}$$

$$= 58 \text{ ft} + 5.8 \text{ ft}$$

$$= 63.8 \text{ ft}$$

$$\text{Flow, GPM} = \frac{1{,}800{,}000 \text{ gal}}{\cancel{\text{day}}} \times \frac{\cancel{\text{day}}}{24 \cancel{\text{hr}}} \times \frac{1 \cancel{\text{hr}}}{60 \text{ min}}$$

$$= 1{,}250 \text{ GPM}$$ (assuming pump runs 24 hours per day)

$$\text{Motor, HP} = \frac{\text{Flow, GPM} \times \text{H, ft}}{3{,}960 \times E_p \times E_m}$$

$$= \frac{1{,}250 \times 63.8}{3{,}960 \times 0.75 \times 0.9}$$

$$= 30 \text{ HP}$$

$$\text{Kilowatt-hr} = 30 \cancel{\text{HP}} \times 24 \text{ hr/day} \times 0.746 \text{ kW/}\cancel{\text{HP}}^{*}$$

$$= 537 \text{ kilowatt-hr/day}$$

$$\text{Cost} = \text{kWh} \times \$0.04/\text{kWh}$$

$$= 537 \times 0.04$$

$$= \$21.48/\text{day}$$

* See Section A.10, "Basic Conversion Factors," *POWER*, page 546.

A.85 Pump Characteristics

The discharge of a centrifugal pump, unlike a positive displacement pump, can be made to vary from zero to a maximum capacity which depends on the speed, head, power, and specific impeller design. The interrelation of capacity, efficiency, head, and power is known as the characteristics of the pump.

The first relation normally looked at when selecting a pump is the head vs. capacity. The head of a centrifugal pump normally rises as the capacity is reduced. If the values are plotted on a graph they appear as follows:

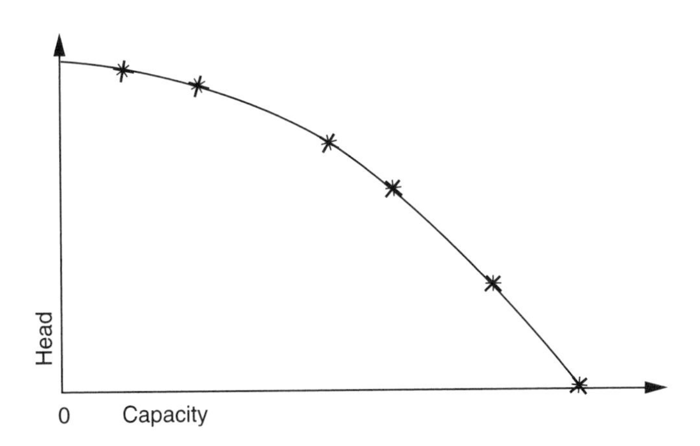

Another important characteristic is the pump efficiency. It begins from zero at no discharge, increases to a maximum, and then drops as the capacity is increased. Following is a graph of efficiency vs. capacity:

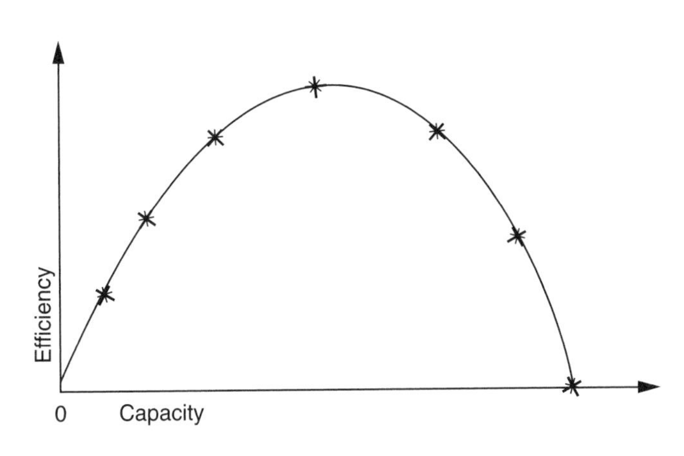

The last important characteristic is the brake horsepower or the power input to the pump. The brake horsepower usually increases with increasing capacity until it reaches a maximum, then it normally reduces slightly.

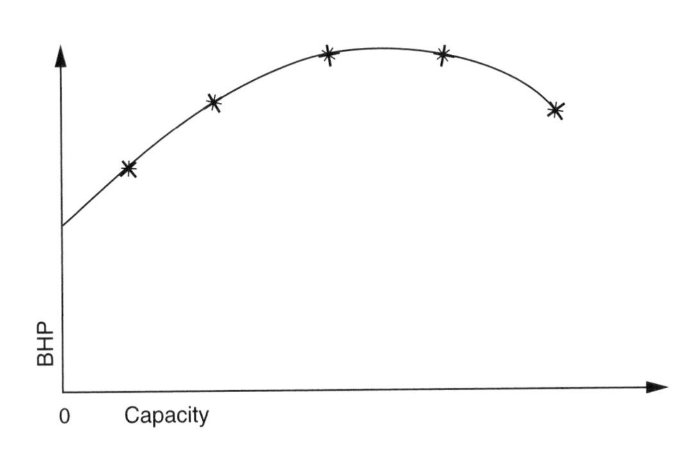

These pump characteristic curves are quite important. Pump sizes are normally picked from these curves rather than calculations. For ease of reading, the three characteristic curves are normally plotted together. A typical graph of pump characteristics is shown as follows:

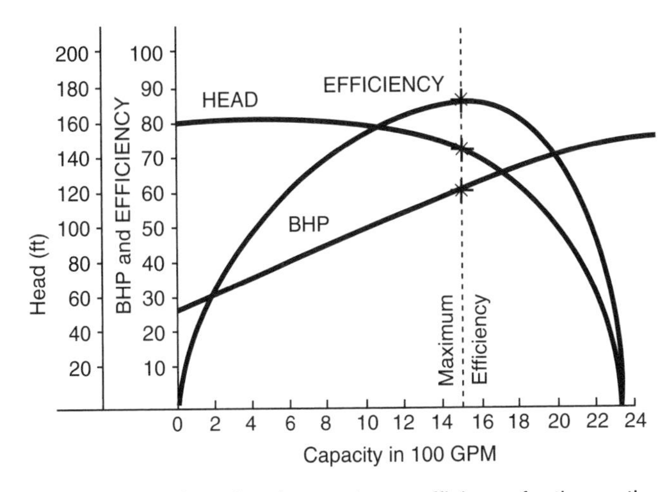

The curves show that the maximum efficiency for the particular pump in question occurs at approximately 1,475 GPM, a head of 132 feet, and a brake horsepower of 58. Operating at this point the pump has an efficiency of approximately 85%. This can be verified by calculation:

$$\text{BHP} = \frac{\text{Flow, GPM} \times \text{H, ft}}{3{,}960 \times \text{E}}$$

As previously explained, a number can be written over one without changing its value:

$$\frac{\text{BHP}}{1} = \frac{\text{GPM} \times \text{H}}{3{,}960 \times \text{E}}$$

Since the formula is now in ratio form, it can be cross multiplied.

$$\text{BHP} \times 3{,}960 \times \text{E} = \text{GPM} \times \text{H} \times 1$$

Solving for E,

$$\text{E} = \frac{\text{GPM} \times \text{H}}{3{,}960 \times \text{BHP}}$$

$$\text{E} = \frac{1{,}475 \text{ GPM} \times 132 \text{ ft}}{3{,}960 \times 58 \text{ HP}}$$

$$= 0.85 \text{ or } 85\% \text{ (Check)}$$

The preceding is only a brief description of pumps to familiarize the operator with their characteristics. The operator does not normally specify the type and size of pump needed at a plant. If a pump is needed, the operator should be able to supply the information necessary for a pump supplier to provide the best possible pump for the lowest cost. Some of the information needed includes:

1. Flow range desired;
2. Head conditions:
   a. Suction head or lift,
   b. Pipe and fitting friction head, and
   c. Discharge head;
3. Type of fluid pumped and temperature; and
4. Pump location.

A.86 Evaluation of Pump Performance

1. Capacity

Sometimes it is necessary to determine the capacity of a pump. This can be accomplished by determining the time it takes a pump to fill or empty a portion of a wet well or diversion box when all inflow is blocked off.

EXAMPLE:

a. Measure the size of the wet well.

Length = 10 ft

Width = 10 ft

Depth = 5 ft (We will measure the time it takes to lower the well a distance of 5 feet.)

$$\text{Volume, cu ft} = \text{L, ft} \times \text{W, ft} \times \text{D, ft}$$
$$= 10\text{ ft} \times 10\text{ ft} \times 5\text{ ft}$$
$$= 500\text{ cu ft}$$

b. Record time for water to drop 5 feet in wet well.

Time = 10 minutes 30 seconds

= 10.5 minutes

c. Calculate pumping rate or capacity.

$$\text{Pumping Rate, GPM} = \frac{\text{Volume, gallons}}{\text{Time, minutes}}$$
$$= \frac{(500\text{ cu ft})(7.5\text{ gal/cu ft})}{10.5\text{ min}}$$
$$= \frac{3{,}750}{10.5}$$
$$= 357\text{ GPM}$$

If you know the total dynamic head and have the pump's performance curves, you can determine if the pump is delivering at design capacity. If not, try to determine the cause (see Chapter 18, "Maintenance," in *WATER TREATMENT PLANT OPERATION*, Volume II). After a pump overhaul, the pump's actual performance (flow, head, power, and efficiency) should be compared with the pump manufacturer's performance curves. This procedure for calculating the rate of filling or emptying of a wet well or diversion box can be used to calibrate flowmeters.

2. Efficiency

To estimate the efficiency of the pump in the previous example, the total head must be known. This head may be estimated by measuring the suction and discharge pressures. Assume these were measured as follows:

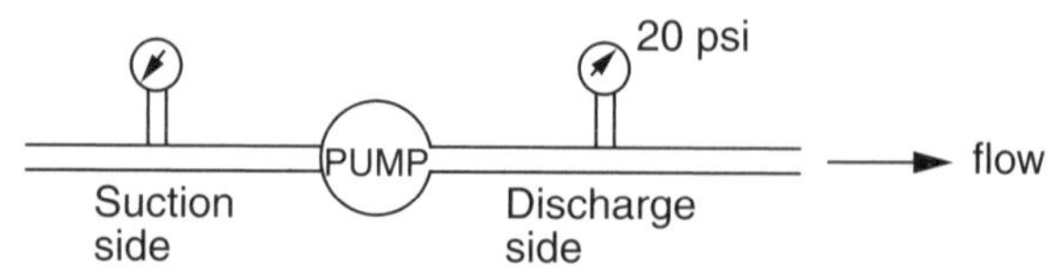

No additional information is necessary if we assume the pressure gages are at the same height and the pipe diameters are the same. Both pressure readings must be converted to feet.

$$\text{Suction Lift, ft} = 2\text{ in Mercury} \times \frac{1.133\text{ ft water*}}{1\text{ in Mercury}}$$
$$= 2.27\text{ ft}$$

$$\text{Discharge Head, ft} = 20\text{ psi} \times 2.31\text{ ft/psi*}$$
$$= 46.20\text{ ft}$$

$$\text{Total Head, ft} = \text{Suction Lift, ft} + \text{Discharge Head, ft}$$
$$= 2.27\text{ ft} + 46.20\text{ ft}$$
$$= 48.47\text{ ft}$$

* See Section A.10, "Basic Conversion Factors," *PRESSURE*, page 546.

Calculate the power output of the pump or water horsepower:

$$\text{Water Horsepower, HP} = \frac{(\text{Flow, GPM})(\text{Head, ft})}{3{,}960}$$
$$= \frac{(357\text{ GPM})(48.47\text{ ft})}{3{,}960}$$
$$= 4.4\text{ HP}$$

To estimate the efficiency of the pump, measure the kilowatts drawn by the pump motor. Assume the meter indicates 8,000 watts or 8 kilowatts. The manufacturer claims the electric motor is 80% efficient.

$$\text{Brake Horsepower, HP} = (\text{Power to Elec. Motor})(\text{Motor Eff.})$$
$$= \frac{(8\text{ kW})(0.80)}{0.746\text{ kW/HP}}$$
$$= 8.6\text{ HP}$$

$$\text{Pump Efficiency, \%} = \frac{\text{Water Horsepower, HP} \times 100\%}{\text{Brake Horsepower, HP}}$$
$$= \frac{4.4\text{ HP} \times 100\%}{8.6\text{ HP}}$$
$$= 51\%$$

The following diagram may clarify the previous problem:

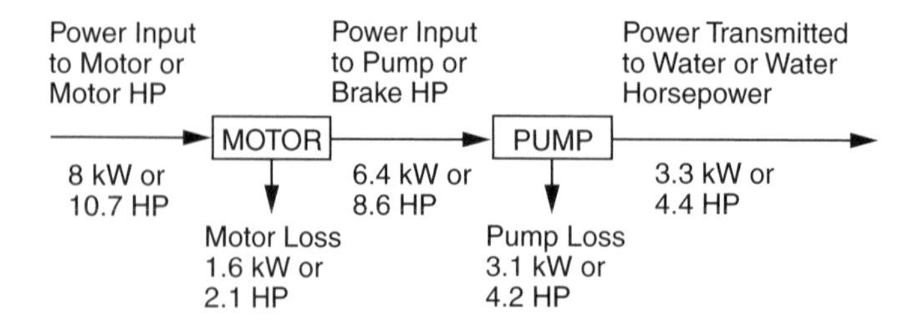

The wire-to-water efficiency is the efficiency of the power input to produce water horsepower.

$$\text{Wire-to-Water Efficiency, \%} = \frac{\text{Water Horsepower, HP}}{\text{Power Input, HP}} \times 100\%$$

$$= \frac{4.4 \text{ HP}}{10.7 \text{ HP}} \times 100\%$$

$$= 41\%$$

The wire-to-water efficiency of a pumping system (pump and electric motor) can be calculated by using the following formula:

$$\text{Efficiency, \%} = \frac{(\text{Flow, GPM})(\text{TDH, ft})(100\%)}{(\text{Voltage, volts})(\text{Current, amps})(5.308)}$$

$$= \frac{(357 \text{ GPM})(48.47 \text{ ft})(100\%)}{(220 \text{ volts})(36 \text{ amps})(5.308)}$$

$$= 41\%$$

A.87 Pump Speed — Performance Relationships

Changing the velocity of a centrifugal pump will change its operating characteristics. If the speed of a pump is changed, the flow, head developed, and power requirements will change. The operating characteristics of the pump will change with speed approximately as follows:

$$\text{Flow, } Q_n = \left[\frac{N_n}{N_r}\right] Q_r$$

$$\text{Head, } H_n = \left[\frac{N_n}{N_r}\right]^2 H_r$$

$$\text{Power, } P_n = \left[\frac{N_n}{N_r}\right]^3 P_r$$

r = rated

n = now

N = pump speed

Actually, pump efficiency does vary with speed; therefore, these formulas are not quite correct. If speeds do not vary by more than a factor of two (if the speeds are not doubled or cut in half), the results are close enough. Other factors contributing to changes in pump characteristic curves include impeller wear and roughness in pipes.

EXAMPLE: To illustrate these relationships, assume a pump has a rated capacity of 600 GPM, develops 100 ft of head, and has a power requirement of 15 HP when operating at 1,500 RPM. If the efficiency remains constant, what will be the operating characteristics if the speed drops to 1,200 RPM?

Calculate new flow rate or capacity:

$$\text{Flow, } Q_n = \left[\frac{N_n}{N_r}\right] Q_r$$

$$= \left[\frac{1{,}200 \text{ RPM}}{1{,}500 \text{ RPM}}\right](600 \text{ GPM})$$

$$= \left(\frac{4}{5}\right)(600 \text{ GPM})$$

$$= (4)(120 \text{ GPM})$$

$$= 480 \text{ GPM}$$

Calculate new head:

$$\text{Head, } H_n = \left[\frac{N_n}{N_r}\right]^2 H_r$$

$$= \left[\frac{1{,}200 \text{ RPM}}{1{,}500 \text{ RPM}}\right]^2 (100 \text{ ft})$$

$$= \left(\frac{4}{5}\right)^2 (100 \text{ ft})$$

$$= \left(\frac{16}{25}\right)(100 \text{ ft})$$

$$= (16)(4 \text{ ft})$$

$$= 64 \text{ ft}$$

Calculate new power requirement:

$$\text{Power, } P_n = \left[\frac{N_n}{N_r}\right]^3 P_r$$

$$= \left[\frac{1{,}200 \text{ RPM}}{1{,}500 \text{ RPM}}\right]^3 (15 \text{ HP})$$

$$= \left(\frac{4}{5}\right)^3 (15 \text{ HP})$$

$$= \left(\frac{64}{125}\right)(15 \text{ HP})$$

$$= \left(\frac{64}{25}\right)(3 \text{ HP})$$

$$= 7.7 \text{ HP}$$

A.88 Friction or Energy Losses

Whenever water flows through pipes, valves, and fittings, energy is lost due to pipe friction (resistance), friction in valves and fittings, and the turbulence resulting from the flowing water changing its direction. Figure A.1 can be used to convert the friction losses through valves and fittings to lengths of straight pipe that would produce the same amount of friction losses. To estimate the friction or energy losses resulting from water flowing in a pipe system, we need to know:

1. Water flow rate,
2. Pipe size or diameter and length, and
3. Number, size, and type of valve fittings.

An easy way to estimate friction or energy losses is to follow these steps:

1. Determine the flow rate,
2. Determine the diameter and length of pipe,
3. Convert all valves and fittings to equivalent lengths of straight pipe (see Figure A.1),
4. Add up total length of equivalent straight pipe, and
5. Estimate friction or energy losses by using Figure A.2. With the flow in GPM and diameter of pipe, find the friction loss per 100 feet of pipe. Multiply this value by equivalent length of straight pipe.

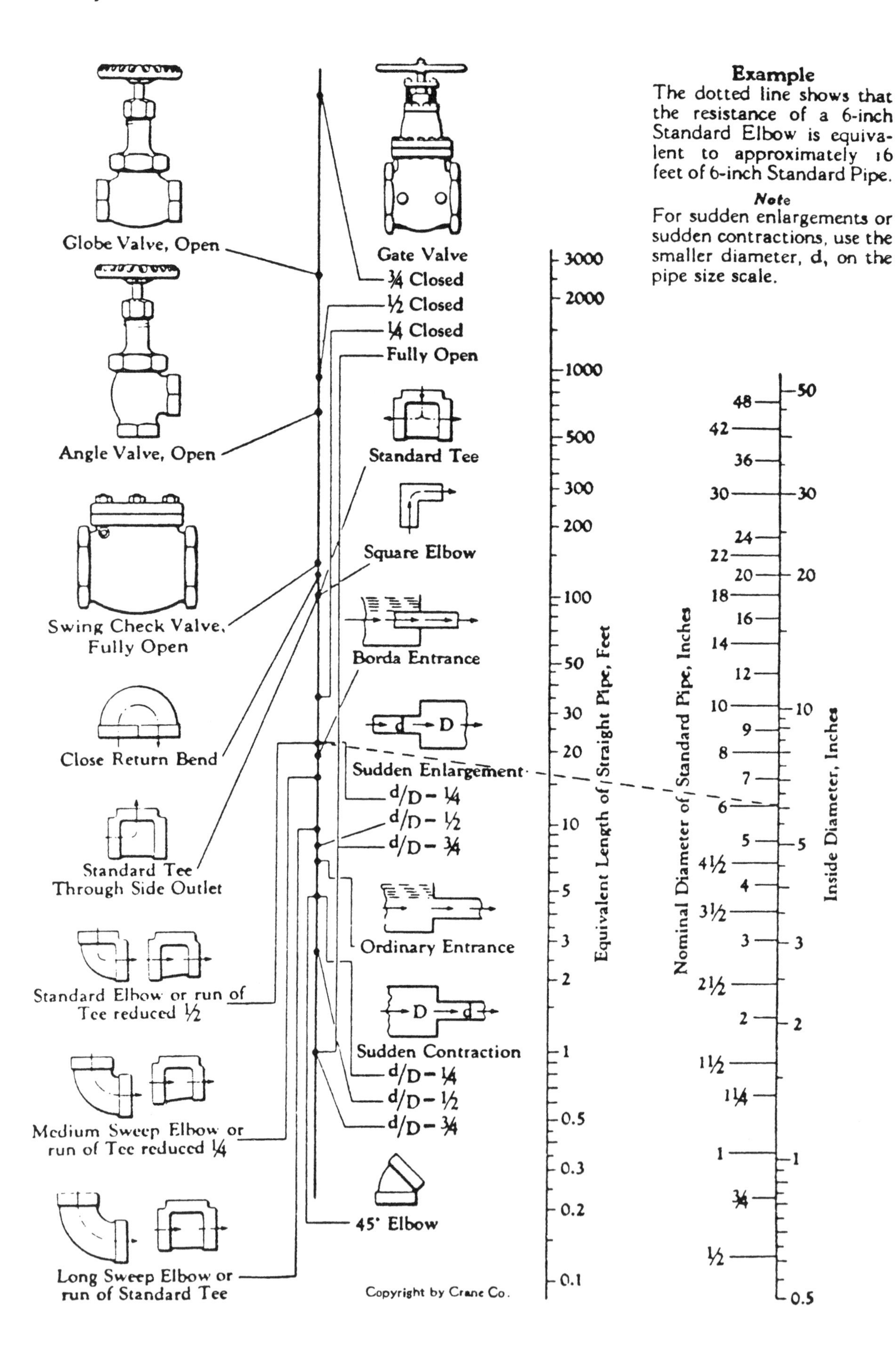
Example
The dotted line shows that the resistance of a 6-inch Standard Elbow is equivalent to approximately 16 feet of 6-inch Standard Pipe.
Note
For sudden enlargements or sudden contractions, use the smaller diameter, d, on the pipe size scale.
Globe Valve, Open
Angle Valve, Open
Swing Check Valve, Fully Open
Close Return Bend
Standard Tee Through Side Outlet
Standard Elbow or run of Tee reduced ½
Medium Sweep Elbow or run of Tee reduced ¼
Long Sweep Elbow or run of Standard Tee
Gate Valve
¾ Closed
½ Closed
¼ Closed
Fully Open
Standard Tee
Square Elbow
Borda Entrance
Sudden Enlargement
d/D = ¼
d/D = ½
d/D = ¾
Ordinary Entrance
Sudden Contraction
d/D = ¼
d/D = ½
d/D = ¾
45° Elbow
Copyright by Crane Co.
Equivalent Length of Straight Pipe, Feet
3000
2000
1000
500
300
200
100
50
30
20
10
5
3
2
1
0.5
0.3
0.2
0.1
Nominal Diameter of Standard Pipe, Inches
48
42
36
30
24
22
20
18
16
14
12
10
9
8
7
6
5
4½
4
3½
3
2½
2
1½
1¼
1
¾
½
Inside Diameter, Inches
50
30
20
10
5
3
2
1
0.5

Fig. A.1 Resistance of valves and fittings to flow of water

(Reprinted by permission of Crane Co.)

U.S. GPM	0.5 in.		0.75 in.		1 in.		1.25 in.		1.5 in.		2 in.		2.5 in.	
	Vel.	Frict.	Vel.	Frict.	Vel.	Frict.	Vel.	Frict.	Vel.	Frict.	Vel.	Frict.	Vel.	Frict.
10	10.56	95.9	6.02	23.0	3.71	6.86	2.15	1.77	1.58	.83	.96	.25	.67	.11
20	....	....	12.0	86.1	7.42	25.1	4.29	6.34	3.15	2.94	1.91	.87	1.34	.36
30	....	....	....	....	11.1	54.6	6.44	13.6	4.73	6.26	2.87	1.82	2.01	.75
40	....	....	....	....	14.8	95.0	8.58	23.5	6.30	10.79	3.82	3.10	2.68	1.28
50	....	....	....	....	....	....	10.7	36.0	7.88	16.4	4.78	4.67	3.35	1.94
60	....	....	....	....	....	....	12.9	51.0	9.46	23.2	5.74	6.59	4.02	2.72
70	....	....	....	....	....	....	15.0	68.8	11.03	31.3	6.69	8.86	4.69	3.63
80	....	....	....	....	....	....	17.2	89.2	12.6	40.5	7.65	11.4	5.36	4.66
90	....	....	....	....	....	....	....	....	14.2	51.0	8.60	14.2	6.03	5.82
100	....	....	....	....	....	....	....	....	15.8	62.2	9.56	17.4	6.70	7.11
120	....	....	....	....	....	....	....	....	18.9	88.3	11.5	24.7	8.04	10.0
140	....	....	....	....	....	....	....	....		....	13.4	33.2	9.38	13.5
160	....	....	....	....	....	....	....	....		....	15.3	43.0	10.7	17.4
180	....	....	....	....	....	....	....	....		....	17.2	54.1	12.1	21.9
200	....	....	....	....	....	....	....	....		....	19.1	66.3	13.4	26.7
220	....	....	....	....	....	....	....	....		....	21.0	80.0	14.7	32.2
240	....	....	....	....	....	....	....	....		....	22.9	95.0	16.1	38.1
260	....	....	....	....	....	....	....	....		....	....	....	17.4	44.5
280	....	....	....	....	....	....	....	....		....	....	....	18.8	51.3
300	....	....	....	....	....	....	....	....		....	....	....	20.1	58.5
350	....	....	....	....	....	....	....	....		....	....	....	23.5	79.2

U.S. GPM	3 in.		4 in.		5 in.		6 in.		8 in.		10 in.		12 in.		14 in.		16 in.		18 in.		20 in.	
	Vel.	Frict.	Vel.	Frict.	Vel.	Frict.	Vel.	Frict.	Vel.	Frict.	Vel.	Frict.	Vel.	Frict.	Vel.	Frict.	Vel.	Frict.	Vel.	Frict.	Vel.	Frict.
20	.91	.15																				
40	1.82	.55	1.02	.13																		
50	2.72	1.17	1.53	.28	.96	.08																
80	3.63	2.02	2.04	.48	1.28	.14	.91	.06														
100	4.54	3.10	2.55	.73	1.60	.20	1.13	.10														
120	5.45	4.40	3.06	1.03	1.92	.29	1.36	.13														
140	6.35	5.93	3.57	1.38	2.25	.38	1.59	.18														
160	7.26	7.71	4.08	1.78	2.57	.49	1.82	.23														
180	8.17	9.73	4.60	2.24	2.89	.61	2.04	.28														
200	9.08	11.9	5.11	2.74	3.21	.74	2.27	.35														
220	9.98	14.3	5.62	3.28	3.53	.88	2.50	.42	1.40	.10												
240	10.9	17.0	6.13	3.88	3.85	1.04	2.72	.49	1.53	.12												
260	11.8	19.8	6.64	4.54	4.17	1.20	2.95	.57	1.66	.14												
280	12.7	22.8	7.15	5.25	4.49	1.38	3.18	.66	1.79	.16												
300	13.6	26.1	7.66	6.03	4.81	1.58	3.40	.75	1.91	.18												
350	....	....	8.94	8.22	5.61	2.11	3.97	1.01	2.24	.24												
400	....	....	10.20	10.7	6.41	2.72	4.54	1.30	2.55	.30												
460	....	....	11.45	13.4	7.22	3.41	5.11	1.64	2.87	.38	1.84	.12										
500	....	....	12.8	16.6	8.02	4.16	5.67	2.02	3.19	.46	2.04	.15	1.42	.06								
550	....	....	14.0	19.9	8.82	4.98	6.24	2.42	3.51	.56	2.25	.18	1.56	.07								
600	....	....	....	....	9.62	5.88	6.81	2.84	3.83	.66	2.45	.21	1.70	.08	1.25	.04						
700	....	....	....	....	11.2	7.93	7.94	3.87	4.47	.88	2.86	.29	1.99	.12	1.46	.05						
800	....	....	....	....	12.8	10.22	9.08	5.06	5.11	1.14	3.27	.37	2.27	.15	1.67	.07						
900	....	....	....	....	14.4	12.9	10.2	6.34	5.74	1.44	3.68	.46	2.55	.18	1.88	.09						
1000	....	....	....	....	....	....	11.3	7.73	6.38	1.76	4.09	.57	2.84	.22	2.08	.11						
1100	....	....	....	....	....	....	12.5	9.80	7.02	2.14	4.49	.68	3.12	.27	2.29	.13						
1200	....	....	....	....	....	....	13.6	11.2	7.66	2.53	4.90	.81	3.40	.32	2.50	.15	1.91	.08				
1300	....	....	....	....	....	....	14.7	13.0	8.30	2.94	6.31	.95	3.69	.37	2.71	.17	2.07	.09				
1400	....	....	....	....	....	....	....	....	8.93	3.40	5.72	1.09	3.97	.43	2.92	.20	2.23	.10				
1500	....	....	....	....	....	....	....	....	9.57	3.91	6.13	1.25	4.26	.49	3.13	.23	2.34	.12				
1600	....	....	....	....	....	....	....	....	10.2	4.45	6.54	1.42	4.54	.55	3.33	.25	2.55	.13	2.02	.07		
1700	....	....	....	....	....	....	....	....	10.8	5.00	6.94	1.60	4.87	.62	3.54	.29	2.71	.15	2.15	.08		
1800	....	....	....	....	....	....	....	....	11.5	5.58	7.35	1.78	5.11	.70	3.75	.32	2.87	.16	2.27	.09		
1900	....	....	....	....	....	....	....	....	12.1	6.19	7.76	1.97	5.39	.77	3.96	.35	3.03	.18	2.40	.10		
2000	....	....	....	....	....	....	....	....	12.8	6.84	8.17	2.17	5.67	.86	4.17	.39	3.19	.20	2.52	.11		
2500	....	....	....	....	....	....	....	....	....	....	10.2	3.38	7.10	1.33	5.21	.60	3.99	.31	3.15	.17		
3000	....	....	....	....	....	....	....	....	....	....	12.3	4.79	8.51	1.88	6.25	.86	4.79	.44	3.78	.24	3.06	.14
3500	....	....	....	....	....	....	....	....	....	....	14.3	6.55	9.93	2.56	7.29	1.16	5.58	.58	4.41	.32	3.57	.19
4000	....	....	....	....	....	....	....	....	....	....	....	....	11.3	3.31	8.34	1.50	6.38	.75	5.04	.42	4.08	.24
4500	....	....	....	....	....	....	....	....	....	....	....	....	12.8	4.18	9.38	1.88	7.18	.95	5.67	.53	4.59	.31
5000	....	....	....	....	....	....	....	....	....	....	....	....	14.7	5.13	10.4	2.30	7.98	1.17	6.30	.65	5.11	.38
6000	....	....	....	....	....	....	....	....	....	....	....	....	....	....	12.5	3.31	9.57	1.66	7.56	.92	6.13	.53
7000	....	....	....	....	....	....	....	....	....	....	....	....	....	....	14.6	4.50	11.2	2.26	8.83	1.24	7.15	.72
8000	....	....	....	....	....	....	....	....	....	....	....	....	....	....	....	....	12.8	2.96	10.09	1.61	8.17	.94
9000	....	....	....	....	....	....	....	....	....	....	....	....	....	....	....	....	14.4	3.73	11.3	2.02	9.19	1.18
10000	....	....	....	....	....	....	....	....	....	....	....	....	....	....	....	....	....	....	12.6	2.48	10.2	1.45

No allowance has been made for age, differences in diameter, or any other abnormal condition of interior surface. Any Factor of Safety must be estimated from the local conditions and the requirements of each particular installation. For general purposes, 15% is a responsible Factor of Safety.

Fig. A.2 Friction Loss for Water in Feet Per 100 Feet of Pipe

(Reprinted from the 10th Edition of the Standards of the Hydraulic Institute, 122 East 42nd Street, New York)

The procedure for using Figure A.1 is very easy. Locate the type of valve or fitting you wish to convert to an equivalent pipe length; find its diameter on the right-hand scale; and draw a straight line between these two points to locate the equivalent length of straight pipe.

EXAMPLE: Estimate the friction losses in the piping system of a pump station when the flow is 1,000 GPM. The 8-inch suction line is 10 feet long and contains a 90-degree bend (long sweep elbow), a gate valve and an 8-inch by 6-inch reducer at the inlet to the pump. The 6-inch discharge line is 30 feet long and contains a check valve, a gate valve, and three 90-degree bends (medium sweep elbows):

SUCTION LINE (8-inch diameter)

Item	Equivalent Length, ft
1. Length of pipe	10
2. 90-degree bend	14
3. Gate valve	4
4. 8-inch by 6-inch reducer	3
5. Ordinary entrance	12
Total equivalent length	43 feet

Friction loss (Figure A.2) = 1.76 ft/100 ft of pipe

DISCHARGE LINE (6-inch diameter)

Item	Equivalent Length, ft
1. Length of pipe	30
2. Check valve	38
3. Gate valve	4
4. Three 90-degree bends (3)(14)	42
Total equivalent length	114 feet

Friction loss (Figure A.2) = 7.73 ft/100 ft of pipe

Estimate the total friction losses in pumping system for a flow of 1,000 GPM.

SUCTION

Loss = (1.76 ft/100 ft)(43 ft) = 0.8 ft

DISCHARGE

Loss = (7.73 ft/100 ft)(114 ft) = 8.8 ft

Total Friction Losses, ft = 9.6 ft

A.9 STEPS IN SOLVING PROBLEMS

A.90 Identification of Problem

To solve any problem, you have to identify the problem, determine what kind of answer is needed, and collect the information needed to solve the problem. A good approach to this type of problem is to examine the problem and make a list of *KNOWN* and *UNKNOWN* information.

EXAMPLE: Find the theoretical detention time in a rectangular sedimentation tank 8 feet deep, 30 feet wide, and 60 feet long when the flow is 1.4 MGD.

Known	Unknown
Depth, ft = 8 ft	Detention Time, hours
Width, ft = 30 ft	
Length, ft = 60 ft	
Flow, MGD = 1.4 MGD	

Sometimes a drawing or sketch will help to illustrate a problem and indicate the knowns, unknowns, and possibly additional information needed.

A.91 Selection of Formula

Most problems involving mathematics in water treatment plant operation can be solved by selecting the proper formula, inserting the known information, and calculating the unknown. In our example, we could look in Chapter 4, "Small Water Treatment Plants," or in Section A.11 of this chapter, "Basic Formulas," to find a formula for calculating detention time.

From Section A.11:

$$\text{Detention Time, hr} = \frac{(\text{Tank Volume, cu ft})(7.48\text{ gal/cu ft})(24\text{ hr/day})}{\text{Flow, gal/day}}$$

To convert the known information to fit the terms in a formula sometimes requires extra calculations. The next step is to find the values of any terms in the formula that are not in the list of known values.

$$\text{Flow, gal/day} = 1.4\text{ MGD}$$

$$= 1{,}400{,}000\text{ gal/day}$$

From Section A.30:

$$\text{Tank Volume, cu ft} = (\text{Length, ft})(\text{Width, ft})(\text{Height, ft})$$

$$= 60\text{ ft} \times 30\text{ ft} \times 8\text{ ft}$$

$$= 14{,}400\text{ cu ft}$$

Solution of Problem:

$$\text{Detention Time, hr} = \frac{(\text{Tank Volume, cu ft})(7.48\text{ gal/cu ft})(24\text{ hr/day})}{\text{Flow, gal/day}}$$

$$= \frac{(14{,}400\text{ cu ft})(7.48\text{ gal/cu ft})(24\text{ hr/day})}{1{,}400{,}000\text{ gal/day}}$$

$$= 1.85\text{ hr}$$

The remainder of this section discusses the details that must be considered in solving this problem.

A.92 Arrangement of Formula

Once the proper formula is selected, you may have to rearrange the terms to solve for the unknown term. From Section A.71, "Flow Rate," we can develop the formula:

$$\text{Velocity, ft/sec} = \frac{\text{Flow Rate, cu ft/sec}}{\text{Cross-Sectional Area, sq ft}}$$

or $$V = \frac{Q}{A}$$

In this equation if Q and A were given, the equation could be solved for V. If V and A were known, the equation would have to be rearranged to solve for Q. To move terms from one side of an equation to another, use the following rule:

When moving a term or number from one side of an equation to the other, move the numerator (top) of one side to the denominator (bottom) of the other; or from the denominator (bottom) of one side to the numerator (top) of the other.

$$V = \frac{Q}{A} \text{ or } Q = AV \text{ or } A = \frac{Q}{V}$$

If the volume of a sedimentation tank and the desired detention time were given, the detention time formula could be rearranged to calculate the design flow.

$$\text{Detention Time, hr} = \frac{(\text{Tank Vol, cu ft})(7.48 \text{ gal/cu ft})(24 \text{ hr/day})}{\text{Flow, gal/day}}$$

By rearranging the terms

$$\text{Flow, gal/day} = \frac{(\text{Tank Vol, cu ft})(7.48 \text{ gal/cu ft})(24 \text{ hr/day})}{\text{Detention Time, hr}}$$

A.93 Unit Conversions

Each term in a formula or mathematical calculation must be of the correct units. The area of a rectangular clarifier (Area, sq ft = Length, ft x Width, ft) can't be calculated in square feet if the width is given as 246 inches or 20 feet 6 inches. The width must be converted to 20.5 feet. In the example problem, if the tank volume were given in gallons, then the 7.48 gal/cu ft would not be needed. *THE UNITS IN A FORMULA MUST ALWAYS BE CHECKED BEFORE ANY CALCULATIONS ARE PERFORMED TO AVOID TIME-CONSUMING MISTAKES.*

$$\text{Detention Time, hr} = \frac{(\text{Tank Volume, cu ft})(7.48 \text{ gal/cu ft})(24 \text{ hr/day})}{\text{Flow, gal/day}}$$

$$= \frac{\cancel{\text{cu ft}}}{} \times \frac{\cancel{\text{gal}}}{\cancel{\text{cu ft}}} \times \frac{\text{hr}}{\cancel{\text{day}}} \times \frac{\cancel{\text{day}}}{\cancel{\text{gal}}}$$

$$= \text{hr (all other units cancel)}$$

NOTE: We have hours = hr. One should note that the hour unit on both sides of the equation can be cancelled out and nothing would remain. This is one more check that we have the correct units. By rearranging the detention time formula, other unknowns could be determined.

If the design detention time and design flow were known, the required capacity of the tank could be calculated.

$$\text{Tank Volume, cu ft} = \frac{(\text{Detention Time, hr})(\text{Flow, gal/day})}{(7.48 \text{ gal/cu ft})(24 \text{ hr/day})}$$

If the tank volume and design detention time were known, the design flow could be calculated.

$$\text{Flow, gal/day} = \frac{(\text{Tank Volume, cu ft})(7.48 \text{ gal/cu ft})(24 \text{ hr/day})}{\text{Detention Time, hr}}$$

Rearrangement of the detention time formula to find other unknowns illustrates the need to always use the correct units.

A.94 Calculations

Sections A.12, "Multiplication," and A.13, "Division," outline the steps to follow in mathematical calculations. In general, do the calculations inside parentheses ( ) first and brackets [ ] next. Calculations should be done left to right above and below the division line before dividing.

$$\text{Detention Time, hr} = \frac{[(\text{Tank Volume, cu ft})(7.48 \text{ gal/cu ft})(24 \text{ hr/day})]}{\text{Flow, gal/day}}$$

$$= \frac{[(14{,}400 \text{ cu ft})(7.48 \text{ gal/cu ft})(24 \text{ hr/day})]}{1{,}400{,}000 \text{ gal/day}}$$

$$= \frac{2{,}585{,}088 \text{ gal-hr/day}}{1{,}400{,}000 \text{ gal/day}}$$

$$= 1.85, \text{ or}$$

$$= 1.9 \text{ hr}$$

A.95 Significant Figures

In calculating the detention time in the previous section, the answer is given as 1.9 hr. The answer could have been calculated:

$$\text{Detention Time, hr} = \frac{2{,}585{,}088 \text{ gal-hr/day}}{1{,}400{,}000 \text{ gal/day}}$$

$$= 1.846491429 \ldots \text{ hours}$$

How does one know when to stop dividing? Common sense and significant figures both help.

First, consider the meaning of detention time and the measurements that were taken to determine the knowns in the formula. Detention time in a tank is a theoretical value and assumes that all particles of water throughout the tank move through the tank at the same velocity. This assumption is not correct; therefore, detention time can only be a representative time for some of the water particles.

Will the flow of 1.4 MGD be constant throughout the 1.9 hours, and is the flow exactly 1.4 MGD, or could it be 1.35 MGD or 1.428 MGD? A carefully calibrated flowmeter may give a reading within 2% of the actual flow rate. Flows into a tank fluctuate and flowmeters do not measure flows extremely accurately; so the detention time again appears to be a representative or typical detention time.

Tank dimensions are probably satisfactory within 0.1 ft. A flowmeter reading of 1.4 MGD is less precise and it could be 1.3 or 1.5 MGD. A 0.1 MGD flowmeter error when the flow is 1.4 MGD is (0.1/1.4) x 100% = 7% error. A detention time of 1.9 hours, based on a flowmeter reading error of plus or minus 7%, also could have the same error or more, even if the flow was constant. Therefore, the detention time error could be 1.9 hours x 0.07 = ±0.13 hour.

In most of the calculations in the operation of water treatment plants, the operator uses measurements determined in the lab or read from charts, scales, or meters. The accuracy of every measurement depends on the sample being measured, the equipment doing the measuring, and the operator reading or measuring the results. Your estimate is no better than the least precise measurement. Do not retain more than one doubtful number.

To determine how many figures or numbers mean anything in an answer, the approach called "significant figures" is used. In the example the flow was given in two significant figures (1.4 MGD), and the tank dimensions could be considered accurate to the nearest tenth of a foot (depth = 9.0 ft) or two significant figures. Since all measurements and the constants contained two significant figures, the results should be reported as two significant figures or 1.9 hours. The calculations are normally carried out to three significant figures (1.85 hours) and rounded off to two significant figures (1.9 hours).

Decimal points require special attention when determining the number of significant figures in a measurement.

Measurement	Significant Figures
0.00325	3
11.078	5
21,000.	2

EXAMPLE: The distance between two points was divided into three sections, and each section was measured by a different group. What is the distance between the two points if each group reported the distance it measured as follows?

Group	Distance, ft	Significant Figures
A	11,300.	3
B	2,438.9	5
C	87.62	4
Total Distance	13,826.52	

Group A reported the length of the section it measured to three significant figures; therefore, the distance between the two points should be reported as 13,800 feet (3 significant figures).

When adding, subtracting, multiplying, or dividing, the number of significant figures in the answer should not be more than the term in the calculations with the least number of significant figures.

A.96 Check Your Results

After completing your calculations, you should carefully examine your calculations and answer. Does the answer seem reasonable? If possible, have another operator check your calculations before making any operational changes.

A.10 BASIC CONVERSION FACTORS (ENGLISH SYSTEM)

UNITS			
1,000,000	=	1 Million	1,000,000/1 Million
LENGTH			
12 in	=	1 ft	12 in/ft
3 ft	=	1 yd	3 ft/yd
5,280 ft	=	1 mi	5,280 ft/mi
AREA			
144 sq in	=	1 sq ft	144 sq in/sq ft
43,560 sq ft	=	1 acre	43,560 sq ft/ac
VOLUME			
7.48 gal	=	1 cu ft	7.48 gal/cu ft
1,000 m*L*	=	1 liter	1,000 m*L*/*L*
3.785 *L*	=	1 gal	3.785 *L*/gal
231 cu in	=	1 gal	231 cu in/gal
WEIGHT			
1,000 mg	=	1 gm	1,000 mg/gm
1,000 gm	=	1 kg	1,000 gm/kg
454 gm	=	1 lb	454 gm/lb
2.2 lbs	=	1 kg	2.2 lbs/kg
POWER			
0.746 kW	=	1 HP	0.746 kW/HP
DENSITY			
8.34 lbs	=	1 gal	8.34 lbs/gal
62.4 lbs	=	1 cu ft	62.4 lbs/cu ft
DOSAGE			
17.1 mg/*L*	=	1 grain/gal	17.1 mg/*L*/gpg
64.7 mg	=	1 grain	64.7 mg/grain
PRESSURE			
2.31 ft water	=	1 psi	2.31 ft water/psi
0.433 psi	=	1 ft water	0.433 psi/ft water
1.133 ft water	=	1 in Mercury	1.133 ft water/in Mercury
FLOW			
694 GPM	=	1 MGD	694 GPM/MGD
1.55 CFS	=	1 MGD	1.55 CFS/MGD
TIME			
60 sec	=	1 min	60 sec/min
60 min	=	1 hr	60 min/hr
24 hr	=	1 day	24 hr/day

NOTE: In our equations the values in the right-hand column may be written either as 24 hr/day or 1 day/24 hours depending on which units we wish to convert to obtain our desired results.

A.11 BASIC FORMULAS

FLOWS

1. $$\text{Flow, MGD} = \frac{(\text{Flow, GPM})(60\ \text{min/hr})(24\ \text{hr/day})}{1{,}000{,}000/\text{Million}}$$

   or

   $$\text{Flow, GPM} = \frac{(\text{Flow, MGD})(1{,}000{,}000/\text{Million})}{(60\ \text{min/hr})(24\ \text{hr/day})}$$

CHEMICAL DOSES

2. $\text{Chemical Feed, lbs/day} = (\text{Flow, MGD})(\text{Dose, mg}/L)(8.34\text{ lbs/gal})$

Calibration of a Dry Chemical Feeder

3. $\text{Chemical Feed, lbs/day} = \dfrac{\text{Chemical Applied, lbs}}{\text{Length of Application, day}}$

Calibration of a Solution Chemical Feeder (Chemical Feed Pump or a Hypochlorinator)

4. $\text{Chemical Feed, lbs/day} = \dfrac{(\text{Chem Conc, mg}/L)(\text{Vol Pumped, m}L)(60\text{ min/hr})(24\text{ hr/day})}{(\text{Time Pumped, min})(1{,}000\text{ m}L/L)(1{,}000\text{ mg/gm})(454\text{ gm/lb})}$

WELLS

5. $\text{Chlorine Required, gal} = \dfrac{(\text{Casing Volume, gal})(\text{Desired Dose, mg}/L)}{\text{Chlorine Solution, mg}/L}$

SMALL WATER TREATMENT PLANTS

6. $\text{Chemical Feeder Setting, GPD} = \dfrac{(\text{Flow, MGD})(\text{Alum Dose, mg}/L)(8.34\text{ lbs/gal})}{\text{Liquid Alum, lbs/gal}}$

7. $\text{Detention Time, hr} = \dfrac{(\text{Volume, gal})(24\text{ hr/day})}{(\text{Flow, gal/day})}$

   $\text{Detention Time, hr} = \dfrac{(\text{Volume, cu ft})(7.48\text{ gal/cu ft})(24\text{ hr/day})}{\text{Flow, gal/day}}$

8. $\text{Overflow Rate, GPD/sq ft} = \dfrac{\text{Flow, GPD}}{\text{Surface Area, sq ft}}$

9. $\text{Weir Loading, GPM/ft} = \dfrac{\text{Flow, GPM}}{\text{Weir Length, ft}}$

10. $\text{Filtration Rate, GPM/sq ft} = \dfrac{\text{Flow, GPM}}{\text{Surface Area, sq ft}}$

11. $\text{Backwash Rate, GPM/sq ft} = \dfrac{\text{Backwash Flow, GPM}}{\text{Surface Area, sq ft}}$

12. $\text{Ion Exchange Capacity, grains} = (\text{Media Vol, cu ft})(\text{Removal Capacity, gr/cu ft})$

DISINFECTION

13. $\text{Chlorine, lbs} = (\text{Volume, M Gal})(\text{Dose, mg}/L)(8.34\text{ lbs/gal})$

13a. $\text{Sodium Hypochlorite Solution, gal} = \dfrac{(\text{Chlorine, lbs})(100\%)}{(8.34\text{ lbs/gal})(\text{Hypochlorite, \%})}$

14. To disinfect a main use the same formulas as in 13 to disinfect a well.

15. To disinfect a storage tank use the same formulas as in 13 to disinfect a well.

16. $\text{Actual Dose, mg}/L = \dfrac{\text{Chlorine, lbs}}{(\text{Volume of Water, M Gal})(8.34\text{ lbs/gal})}$

17. $\text{Chlorine Feed, lbs/day} = (\text{Flow, MGD})(\text{Dose, mg}/L)(8.34\text{ lbs/gal})$

18. $\text{Chlorine Demand, mg}/L = \text{Chlorine Dose, mg}/L - \text{Chlorine Residual, mg}/L$

19. $\text{Hypochlorite, gal} = (\text{Container Area, sq ft})(\text{Drawdown, ft})(7.48\text{ gal/cu ft})$

20. $\text{Hypochlorite Strength, \%} = \dfrac{(\text{Chlorine Required, lbs/day})(100\%)}{(\text{Hypochlorinator Flow, gal/day})(8.34\text{ lbs/gal})}$

21. $\text{Water Added, gal (to Hypochlorite Solution)} = \dfrac{(\text{Hypo, gal})(\text{Hypo, \%}) - (\text{Hypo, gal})(\text{Desired Hypo, \%})}{\text{Desired Hypo, \%}}$

LABORATORY PROCEDURES

22. $\text{Temperature, °C} = (°\text{F} - 32°\text{F}) \times \dfrac{5}{9}$

23. $\text{Temperature, °F} = \left(\dfrac{9}{5} \times °\text{C}\right) + 32°\text{F}$

A.12 HOW TO USE THE BASIC FORMULAS

One clever way of using the basic formulas is to use the Davidson* Pie Method. To apply this method to the basic formula for chemical doses,

1. $\text{Chemical Feed, lbs/day} = (\text{Flow, MGD})(\text{Dose, mg}/L)(8.34\text{ lbs/gal})$

   (a) Draw a circle and draw a horizontal line through the middle of the circle;

   (b) Write the Chemical Feed, lbs/day in the top half;

   (c) Divide the bottom half into three parts; and

   (d) Write Flow, MGD; Dose, mg/*L*; and 8.34 lbs/gal in the other three parts.

   (e) The line across the middle of the circle represents the line of the equation. Items above the line stay above the line and those below the line stay below the line.

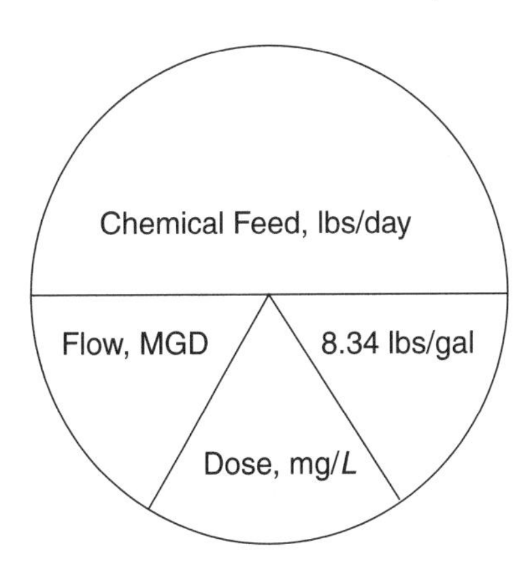

* Gerald Davidson, Manager, Clear Lake Oaks Water District, Clear Lake Oaks, California.

If you want to find the Chemical Feed, lbs/day, cover up the Chemical Feed, lbs/day, and what is left uncovered will give you the correct formula.

2. $\text{Chemical Feed, lbs/day} = (\text{Flow, MGD})(\text{Dose, mg}/L)(8.34\text{ lbs/gal})$

If you know the chlorinator setting in pounds per day and the flow in MGD and would like to know the dose in mg/*L*, cover up the Dose, mg/*L*, and what is left uncovered will give you the correct formula.

3. $\text{Dose, mg}/L = \dfrac{\text{Chemical Feed, lbs/day}}{(\text{Flow, MGD})(8.34\text{ lbs/gal})}$

Another approach to using the basic formulas is to memorize the basic formula, for example the detention time formula.

4. Detention Time, hr $= \frac{\text{(Tank Volume, gal)(24 hr/day)}}{\text{Flow, gal/day}}$

This formula works fine to solve for the detention time when the Tank Volume, gal, and Flow, gal/day, are given.

If you wish to determine the Flow, gal/day, when the Detention Time, hr, and Tank Volume, gal, are given, you must change the basic formula. You want the Flow, gal/day, on the left of the equal sign and everything else on the right of the equal sign. This is done by moving the terms diagonally (from top to bottom or from bottom to top) past the equal sign.

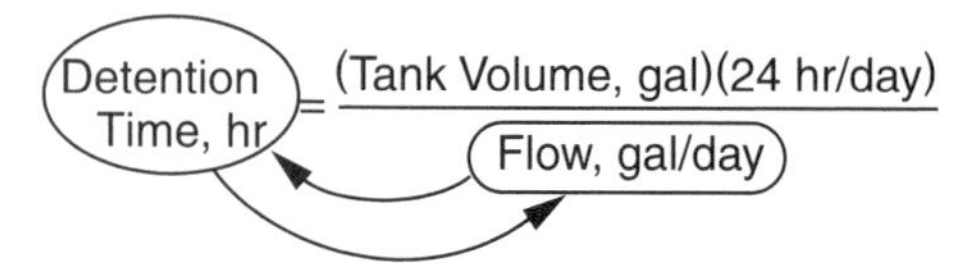

or

$$\text{Flow, gal/day} = \frac{\text{(Tank Volume, gal)(24 hr/day)}}{\text{Detention Time, hr}}$$

This same approach can be used if the Tank Volume, gal, was unknown and the Detention Time, hr, and Flow, gal/day, were given. We want Tank Volume, gal, on one side of the equation and everything else on the other side.

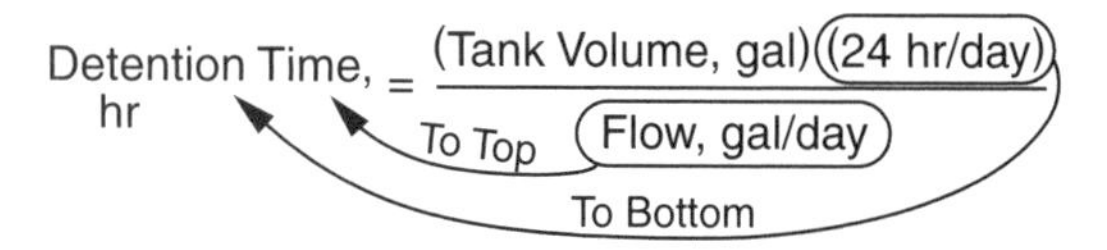

or

$$\frac{\text{(Detention Time, hr)(Flow, gal/day)}}{\text{24 hr/day}} = \text{Tank Volume, gal}$$

or

$$\text{Tank Volume, gal} = \frac{\text{(Detention Time, hr)(Flow, gal/day)}}{\text{24 hr/day}}$$

One more check is to be sure the units in the rearranged formula cancel out correctly.

For additional information on the use of the basic formulas, refer to Sections:

A.91, "Selection of Formula,"

A.92, "Arrangement of Formula,"

A.93, "Unit Conversions," and

A.94, "Calculations."

A.13 TYPICAL SMALL WATER SYSTEM PROBLEMS (ENGLISH SYSTEM)

A.130 Flows

EXAMPLE 1

Convert a flow of 450 gallons per minute to million gallons per day.

Known	Unknown
Flow, GPM = 450 GPM	Flow, MGD

Convert flow from 450 GPM to MGD.

$$\text{Flow, MGD} = \frac{\text{(Flow, GPM)(60 min/hr)(24 hr/day)}}{\text{1,000,000/Million}}$$

$$= \frac{\text{(450 GPM)(60 min/hr)(24 hr/day)}}{\text{1,000,000/Million}}$$

$$= 0.648 \text{ MGD}$$

A.131 Chemical Doses

EXAMPLE 2

Determine the chlorinator setting in pounds per 24 hours if a well pump delivers 300 GPM and the desired chlorine dose is 2.0 mg/*L*.

Known		Unknown
Flow, GPM	= 300 GPM	Chlorinator Setting, lbs/24 hours
Chlorine Dose, mg/*L*	= 2.0 mg/*L*	

1. Convert flow from gallons per minute to million gallons per day.

$$\text{Flow, MGD} = \frac{\text{(Flow, gal/min)(60 min/hr)(24 hr/day)}}{\text{1,000,000/Million}}$$

$$= \frac{\text{(300 gal/min)(60 min/hr)(24 hr/day)}}{\text{1,000,000/Million}}$$

$$= 0.432 \text{ MGD}$$

NOTE: When we divide an equation by 1,000,000/Million we do not change anything except the units. This is just like dividing an equation by 12 inches/foot or 60 min/hr; all we are doing is changing units.

2. Determine the chlorinator setting in pounds per 24 hours or pounds per day.

$$\text{Chemical Feed, lbs/day} = \text{(Flow, MGD)(Dose, mg/}L\text{)(8.34 lbs/gal)}$$

$$= \text{(0.432 MGD)(2.0 mg/}L\text{)(8.34 lbs/gal)}$$

$$= 7.2 \text{ lbs/day}$$

EXAMPLE 3

Determine the actual chemical dose or chemical feed in pounds per day from a dry chemical feeder. A bucket placed under the chemical feeder weighed 0.2 pound empty and 2.6 pounds after 30 minutes.

Known		Unknown
Empty Bucket, lbs	= 0.2 lb	Chemical Feed, lbs/day
Full Bucket, lbs	= 2.6 lbs	
Time to Fill, min	= 30 min	

Determine the chemical feed in pounds of chemical per day.

$$\text{Chemical Feed, lbs/day} = \frac{\text{Chemical Applied, lbs}}{\text{Length of Application, days}}$$

$$= \frac{(2.6\text{ lbs} - 0.2\text{ lb})(60\text{ min/hr})(24\text{ hr/day})}{30\text{ min}}$$

$$= 115\text{ lbs/day}$$

EXAMPLE 4

Determine the chemical feed in pounds of chlorine per day from a hypochlorinator. The hypochlorite solution is 1.4 percent or 14,000 mg chlorine per liter. During a test run the hypochlorinator delivers 400 m*L* during 12 minutes.

Known		Unknown
Cl Solution, %	= 1.4%	Chemical Feed, lbs/day
Cl Conc, mg/*L*	= 14,000 mg/*L*	
Vol Pumped, m*L*	= 400 m*L*	
Time Pumped, min	= 12 min	

Calculate the chlorine fed by the hypochlorinator in pounds of chlorine per day.

$$\text{Chlorine Feed, lbs/day} = \frac{(\text{Cl Conc, mg/}L)(\text{Vol Pumped, m}L)(60\text{ min/hr})(24\text{ hr/day})}{(\text{Time Pumped, min})(1{,}000\text{ m}L/L)(1{,}000\text{ mg/gm})(454\text{ gm/lb})}$$

$$= \frac{(14{,}000\text{ mg/}L)(400\text{ m}L)(60\text{ min/hr})(24\text{ hr/day})}{(12\text{ min})(1{,}000\text{ m}L/L)(1{,}000\text{ mg/gm})(454\text{ gm/lb})}$$

$$= 1.48\text{ lbs/day}$$

$$\text{or} = 1.5\text{ lbs chlorine/day}$$

A.132 Wells

EXAMPLE 5

How much sodium hypochlorite is required to dose a well at 100 mg/*L*? The casing diameter is 18 inches (1.5 ft) and the length of water-filled casing is 80 feet. Sodium hypochlorite is 5.25 percent or 52,500 mg/*L* chlorine.

Known		Unknown
Casing Diameter, ft	= 1.5 ft	Chlorine Required, gal
Casing Length, ft	= 80 ft	
Chlorine Dose, mg/*L*	= 100 mg/*L*	
Chlorine Solution, mg/*L*	= 52,500 mg/*L*	

1. Calculate the water-filled casing volume in gallons.

$$\text{Casing Volume, gal} = (0.785)(\text{Casing Diam, ft})^2(\text{Length, ft})(7.48\text{ gal/cu ft})$$

$$= (0.785)(1.5\text{ ft})^2(80\text{ ft})(7.48\text{ gal/cu ft})$$

$$= 1{,}057\text{ gal}$$

2. Calculate the required gallons of sodium hypochlorite.

$$\text{Chlorine Required, gal} = \frac{(\text{Casing Volume, gal})(\text{Desired Dose, mg/}L)}{\text{Chlorine Solution, mg/}L}$$

$$= \frac{(1{,}057\text{ gal})(100\text{ mg/}L)}{52{,}500\text{ mg/}L}$$

$$= 2.0\text{ gallons}$$

A.133 Small Water Treatment Plants

EXAMPLE 6

The optimum dose of liquid alum from the jar tests is 13 mg/*L*. Determine the setting on the liquid alum chemical feeder in gallons per day when the flow is 1.1 MGD. The liquid alum delivered to the plant contains 5.36 pounds of alum per gallon of liquid solution.

Known		Unknown
Alum Dose, mg/*L*	= 13 mg/*L*	Chemical Feeder Setting, GPD
Flow, MGD	= 1.1 MGD	
Liquid Alum, lbs/gal	= 5.36 lbs/gal	

Calculate the chemical feeder setting in gallons per day (GPD).

$$\text{Chemical Feeder Setting, GPD} = \frac{(\text{Flow, MGD})(\text{Alum Dose, mg/}L)(8.34\text{ lbs/gal})}{\text{Liquid Alum, lbs/gal}}$$

$$= \frac{(1.1\text{ MGD})(13\text{ mg/}L)(8.34\text{ lbs/gal})}{5.36\text{ lbs/gal}}$$

$$= 22.2\text{ GPD}$$

EXAMPLE 7

A rectangular settling basin that is 18 feet long, 8 feet wide, and the water is 4 feet deep treats a flow of 50,000 gallons per day. Estimate the detention time or settling time in hours in the basin for this flow.

Known		Unknown
Length, ft	= 18 ft	Detention Time, hr
Width, ft	= 8 ft	
Depth, ft	= 4 ft	
Flow, GPD	= 50,000 GPD	

1. Calculate the basin volume in gallons.

$$\text{Basin Volume, cu ft} = (\text{Length, ft})(\text{Width, ft})(\text{Depth, ft})$$

$$= (18\text{ ft})(8\text{ ft})(4\text{ ft})$$

$$= 576\text{ cu ft}$$

$$\text{Basin Volume, gal} = (\text{Basin Volume, cu ft})(7.48\text{ gal/cu ft})$$

$$= (576\text{ cu ft})(7.48\text{ gal/cu ft})$$

$$= 4{,}308\text{ gal}$$

2. Calculate the detention time in hours.

$$\text{Detention Time, hr} = \frac{(\text{Basin Volume, gal})(24\ \text{hr/day})}{\text{Flow, gal/day}}$$

$$= \frac{(4{,}308\ \text{gal})(24\ \text{hr/day})}{50{,}000\ \text{gal/day}}$$

$$= 2.07\ \text{hr}$$

$$\text{or} = 2.1\ \text{hr}$$

EXAMPLE 8

Estimate the surface overflow rate in gallons per day per square foot for a 20-foot diameter clarifier that treats a flow of 70,000 gallons per day.

Known	Unknown
Diameter, ft = 20 ft	Overflow Rate, GPD/sq ft
Flow, GPD = 70,000 GPD	

1. Calculate the surface area of the clarifier in square feet.

$$\text{Surface Area, sq ft} = (0.785)(\text{Diameter, ft})^2$$

$$= (0.785)(20\ \text{ft})^2$$

$$= 314\ \text{sq ft}$$

2. Determine the basin overflow rate in gallons per day per square foot of surface area.

$$\text{Overflow Rate, GPD/sq ft} = \frac{\text{Flow, gal/day}}{\text{Surface Area, sq ft}}$$

$$= \frac{70{,}000\ \text{GPD}}{314\ \text{sq ft}}$$

$$= 223\ \text{GPD/sq ft}$$

EXAMPLE 9

A 20-foot diameter circular clarifier treats a flow of 0.1 MGD. Calculate the weir loading in gallons per minute per foot of weir length if the weir is on the outside edge of the clarifier.

Known	Unknown
Weir Diameter, ft = 20 ft	Weir Loading, GPM/ft
Flow, MGD = 0.1 MGD	

1. Calculate length of weir.

$$\text{Weir Length, ft} = \pi(\text{Weir Diameter, ft})$$

$$= (3.14)(20\ \text{ft})$$

$$= 62.8\ \text{ft}$$

2. Convert flow from million gallons per day to gallons per minute.

$$\text{Flow, GPM} = \frac{(\text{Flow, M Gal/day})(1{,}000{,}000/\text{Million})}{(24\ \text{hr/day})(60\ \text{min/hr})}$$

$$= \frac{(0.1\ \text{MGD})(1{,}000{,}000/\text{Million})}{(24\ \text{hr/day})(60\ \text{min/hr})}$$

$$= 69.4\ \text{GPM}$$

3. Estimate the weir loading in gallons per minute per foot.

$$\text{Weir Loading, GPM/ft} = \frac{\text{Flow, GPM}}{\text{Weir Length, ft}}$$

$$= \frac{69.4\ \text{GPM}}{62.8\ \text{ft}}$$

$$= 1.1\ \text{GPM/ft of weir length}$$

EXAMPLE 10

A 15-foot wide by 20-foot long rapid sand filter treats a flow of 720 gallons per minute. Calculate the filtration rate in gallons per minute per square foot of filter surface area.

Known	Unknown
Width, ft = 15 ft	Filtration Rate, GPM/sq ft
Length, ft = 20 ft	
Flow, GPM = 720 GPM	

1. Calculate the surface area of the filter.

$$\text{Surface Area, sq ft} = (\text{Length, ft})(\text{Width, ft})$$

$$= (20\ \text{ft})(15\ \text{ft})$$

$$= 300\ \text{sq ft}$$

2. Determine the filtration rate in gallons per minute per square foot of filter surface area.

$$\text{Filtration Rate, GPM/sq ft} = \frac{\text{Flow, GPM}}{\text{Surface Area, sq ft}}$$

$$= \frac{720\ \text{GPM}}{300\ \text{sq ft}}$$

$$= 2.4\ \text{GPM/sq ft}$$

EXAMPLE 11

Calculate the backwash flow required in gallons per minute to backwash a 10-foot wide by 15-foot long filter if the desired backwash flow rate is 20 gallons per minute per square foot.

Known	Unknown
Width, ft = 10 ft	Backwash Flow, GPM
Length, ft = 15 ft	
Backwash Rate, GPM/sq ft = 20 GPM/sq ft	

1. Calculate the surface area of the filter.

$$\text{Surface Area, sq ft} = (\text{Length, ft})(\text{Width, ft})$$

$$= (15\ \text{ft})(10\ \text{ft})$$

$$= 150\ \text{sq ft}$$

2. Calculate the backwash flow in gallons per minute.

BASIC FORMULA

$$\text{Backwash Rate, GPM/sq ft} = \frac{\text{Backwash Flow, GPM}}{\text{Surface Area, sq ft}}$$

$$\text{Backwash Flow, GPM} = (\text{Backwash Rate, GPM/sq ft})(\text{Surface Area, sq ft})$$

$$= (20\ \text{GPM/sq ft})(150\ \text{sq ft})$$

$$= 3{,}000\ \text{GPM}$$

EXAMPLE 12

Estimate the ion exchange capacity in grains of hardness for a zeolite softener containing 150 cubic feet of media with a hardness removal capacity of 25 kilograins per cubic foot of media.

Known		Unknown
Media Vol, cu ft	= 150 cu ft	Ion Exchange Capacity, grains
Removal Capacity, gr/cu ft	= 25,000 grains/cu ft	

Calculate the ion exchange capacity in grains of hardness.

Ion Exchange Capacity, grains = (Media Vol, cu ft)(Removal Capacity, grains/cu ft)

= (150 cu ft)(25,000 grains/cu ft)

= 3,750,000 grains

A.134 Disinfection

EXAMPLE 13

How many gallons of 5 percent sodium hypochlorite will be needed to disinfect a well with a 15-inch diameter casing and well screen? The well is 180 feet deep and there are 75 feet of water in the well. Use an initial chlorine dose of 100 mg/*L*.

Known		Unknown
Hypochlorite, %	= 5%	5% Hypochlorite, gal
Chlorine Dose, mg/*L*	= 100 mg/*L*	
Diameter, in	= 15 in	
Water Depth, ft	= 75 ft	

1. Calculate the volume of water in the well in gallons.

$$\text{Water Vol, gal} = \frac{(0.785)(\text{Diameter, in})^2(\text{Water Depth, ft})(7.48\text{ gal/cu ft})}{144\text{ sq in/sq ft}}$$

$$= \frac{(0.785)(15\text{ in})^2(75\text{ ft})(7.48\text{ gal/cu ft})}{144\text{ sq in/sq ft}}$$

$$= 688\text{ gal}$$

2. Determine the pounds of chlorine needed.

Chlorine, lbs = (Volume, M Gal)(Dose, mg/*L*)(8.34 lbs/gal)

= (0.000688 M Gal)(100 mg/*L*)(8.34 lbs/gal)

= 0.57 lbs chlorine

3. Calculate the gallons of 5 percent sodium hypochlorite solution needed.

$$\text{Sodium Hypochlorite Solution, gal} = \frac{(\text{Chlorine, lbs})(100\%)}{(8.34\text{ lbs/gal})(\text{Hypochlorite, }\%)}$$

$$= \frac{(0.57\text{ lbs})(100\%)}{(8.34\text{ lbs/gal})(5\%)}$$

$$= 1.37\text{ gallons}$$

Use 1.4 gallons of 5 percent sodium hypochlorite to disinfect the well.

EXAMPLE 14

A new 10-inch diameter water main 650 feet long needs to be disinfected. An initial chlorine dose of 400 mg/*L* is expected to maintain a chlorine residual of over 300 mg/*L* during the three-hour disinfection period. How many gallons of 5.25 percent sodium hypochlorite solution will be needed?

Known		Unknown
Diameter of Pipe, in	= 10 in	5.25% Hypochlorite, gal
Length of Pipe, ft	= 650 ft	
Chlorine Dose, mg/*L*	= 400 mg/*L*	
Hypochlorite, %	= 5.25%	

1. Calculate the volume of water in the pipe in gallons.

$$\text{Pipe Volume, gal} = \frac{(0.785)(\text{Diameter, in})^2(\text{Length, ft})(7.48\text{ gal/cu ft})}{144\text{ sq in/sq ft}}$$

$$= \frac{(0.785)(10\text{ in})^2(650\text{ ft})(7.48\text{ gal/cu ft})}{144\text{ sq in/sq ft}}$$

$$= 2{,}650\text{ gal}$$

2. Determine the pounds of chlorine needed.

Chlorine, lbs = (Volume, M Gal)(Dose, mg/*L*)(8.34 lbs/gal)

= (0.00265 M Gal)(400 mg/*L*)(8.34 lbs/gal)

= 8.84 lbs chlorine

3. Calculate the gallons of 5.25 percent sodium hypochlorite solution needed.

$$\text{Sodium Hypochlorite Solution, gallons} = \frac{(\text{Chlorine, lbs})(100\%)}{(8.34\text{ lbs/gal})(\text{Hypochlorite, }\%)}$$

$$= \frac{(8.84\text{ lbs})(100\%)}{(8.34\text{ lbs/gal})(5.25\%)}$$

$$= 20.2\text{ gallons}$$

Twenty gallons of 5.25 percent sodium hypochlorite solution should do the job.

EXAMPLE 15

A service storage reservoir has been taken out of service for inspection, maintenance, and repairs. The reservoir needs to be disinfected before being placed back on line. The reservoir is 30 feet in diameter and 8 feet deep. An initial chlorine dose of 100 mg/*L* is expected to maintain a chlorine residual of over 50 mg/*L* during the 24-hour disinfection period. How many gallons of 15 percent sodium hypochlorite solution will be needed?

Known		Unknown
Diameter, ft	= 30 ft	15% Hypochlorite, gal
Tank Depth, ft	= 8 ft	
Chlorine Dose, mg/*L*	= 100 mg/*L*	
Hypochlorite, %	= 15%	

1. Calculate the volume of water in the tank in gallons.

$$\text{Tank Volume, gal} = (0.785)(\text{Diameter, ft})^2(\text{Depth, ft})(7.48\text{ gal/cu ft})$$

$$= (0.785)(30\text{ ft})^2(8\text{ ft})(7.48\text{ gal/cu ft})$$

$$= 42{,}277\text{ gal}$$

2. Determine the pounds of chlorine needed.

Chlorine, lbs = (Volume, M Gal)(Dose, mg/*L*)(8.34 lbs/gal)

= (0.042277 M Gal)(100 mg/*L*)(8.34 lbs/gal)

= 35.3 lbs chlorine

3. Calculate the gallons of 15 percent sodium hypochlorite solution needed.

$$\text{Sodium Hypochlorite Solution, gallons} = \frac{(\text{Chlorine, lbs})(100\%)}{(8.34\text{ lbs/gal})(\text{Hypochlorite, }\%)}$$

$$= \frac{(35.3\text{ lbs})(100\%)}{(8.34\text{ lbs/gal})(15\%)}$$

$$= 28.2\text{ gallons}$$

Twenty-eight gallons of 15 percent sodium hypochlorite solution should do the job.

EXAMPLE 16

Calculate the actual chlorine dose in milligrams per liter if 200 gallons of a two percent sodium hypochlorite solution were used to treat two million gallons of water.

Known		Unknown
Volume Hypocl, gal	= 200 gal	Chlorine Dose, mg/*L*
Volume Water, M Gal	= 2 M Gal	
Hypochlorite, %	= 2%	

1. Calculate the pounds of chlorine used.

$$\text{Chlorine, lbs} = \frac{(\text{Hypochlorite, gal})(8.34\text{ lbs/gal})(\text{Hypochlorite, }\%)}{100\%}$$

$$= \frac{(200\text{ gal})(8.34\text{ lbs/gal})(2\%)}{100\%}$$

$$= 33.36\text{ lbs chlorine}$$

2. Calculate the actual chlorine dose in milligrams per liter.

$$\text{Actual Dose, mg}/L = \frac{\text{Chlorine, lbs}}{(\text{Volume Water, M Gal})(8.34\text{ lbs/gal})}$$

$$= \frac{33.36\text{ lbs chlorine}}{(2\text{ M Gal})(8.34\text{ lbs/gal})}$$

$$= 2.0\text{ mg}/L$$

EXAMPLE 17

A deep well turbine pump delivers 250 GPM against typical operating heads. If the desired chlorine dose is 2.5 mg/*L*, what should be the chlorine feed rate in pounds per day?

Known		Unknown
Flow, GPM	= 250 GPM	Chlorine Feed, lbs/day
Dose, mg/*L*	= 2.5 mg/*L*	

1. Convert flow from gallons per minute to million gallons per day.

$$\text{Flow, MGD} = \frac{(\text{Flow, GPM})(60\text{ min/hr})(24\text{ hr/day})}{1{,}000{,}000/\text{Million}}$$

$$= \frac{(250\text{ GPM})(60\text{ min/hr})(24\text{ hr/day})}{1{,}000{,}000/\text{Million}}$$

$$= 0.36\text{ MGD}$$

2. Calculate the chlorine feed rate in pounds of chlorine per day.

$$\text{Chlorine Feed, lbs/day} = (\text{Flow, MGD})(\text{Dose, mg}/L)(8.34\text{ lbs/gal})$$

$$= (0.36\text{ MGD})(2.5\text{ mg}/L)(8.34\text{ lbs/gal})$$

$$= 7.5\text{ lbs/day}$$

EXAMPLE 18

Estimate the chlorine demand for a water in milligrams per liter if the chlorine dose is 2.6 mg/*L* and the chlorine residual is 0.4 mg/*L*.

Known		Unknown
Chlorine Dose, mg/*L*	= 2.6 mg/*L*	Chlorine Demand, mg/*L*
Chlorine Residual, mg/*L*	= 0.4 mg/*L*	

Estimate the chlorine demand in milligrams per liter.

$$\text{Chlorine Demand, mg}/L = \text{Chlorine Dose, mg}/L - \text{Chlorine Residual, mg}/L$$

$$= 2.6\text{ mg}/L - 0.4\text{ mg}/L$$

$$= 2.2\text{ mg}/L$$

EXAMPLE 19

Estimate the gallons of hypochlorite pumped by a hypochlorinator if the hypochlorite solution is in a container with a diameter of 3 feet and the hypochlorite level drops 18 inches (1.5 feet) during a specific time period.

Known		Unknown
Diameter, ft	= 3 ft	Hypochlorite Pumped, gal
Drop, ft	= 1.5 ft	

Estimate the gallons of hypochlorite pumped.

$$\text{Hypochlorite, gal} = (\text{Container Area, sq ft})(\text{Drop, ft})(7.48\text{ gal/cu ft})$$

$$= (0.785)(3\text{ ft})^2(1.5\text{ ft})(7.48\text{ gal/cu ft})$$

$$= 79.3\text{ gallons}$$

EXAMPLE 20

Estimate the desired strength (as a percent chlorine) of a hypochlorite solution being pumped by a hypochlorinator that delivers 80 gallons per day. The water being treated requires a chlorine feed rate of 10 pounds of chlorine per day.

Known		Unknown
Hypochlorinator Flow, GPD	= 80 GPD	Hypochlorite Strength, %
Chlorine Required, lbs/day	= 10 lbs/day	

Estimate the desired hypochlorite strength as a percent chlorine.

$$\text{Hypochlorite Strength, \%} = \frac{(\text{Chlorine Required, lbs/day})(100\%)}{(\text{Hypochlorinator Flow, gal/day})(8.34\text{ lbs/gal})}$$

$$= \frac{(10\text{ lbs/day})(100\%)}{(80\text{ GPD})(8.34\text{ lbs/gal})}$$

$$= 1.5\%$$

EXAMPLE 21

How many gallons of water must be added to 10 gallons of a 5 percent hypochlorite solution to produce a 1.5 percent hypochlorite solution?

Known		Unknown
Hypochlorite, gal	= 10 gal	Water Added, gal
Desired Hypochlorite, %	= 1.5%	
Actual Hypochlorite, %	= 5%	

Calculate the gallons of water that must be added to produce a 1.5 percent hypochlorite solution.

$$\text{Water Added, gal (to hypochlorite solution)} = \frac{(\text{Hypo, gal})(\text{Hypo, \%}) - (\text{Hypo, gal})(\text{Desired Hypo, \%})}{\text{Desired Hypo, \%}}$$

$$= \frac{(10\text{ gal})(5\%) - (10\text{ gal})(1.5\%)}{1.5\%}$$

$$= \frac{50 - 15}{1.5}$$

$$= 23.3\text{ gallons}$$

A.135 Laboratory Procedures

EXAMPLE 22

Convert a temperature of 59°F to degrees Celsius.

Known	Unknown
Temp, °F = 59°F	Temp, °C

Change 59°F to degrees Celsius.

$$\text{Temperature, °C} = \frac{5}{9}(\text{°F} - 32\text{°F})$$

$$= \frac{5}{9}(59\text{°F} - 32\text{°F})$$

$$= 15\text{°C}$$

EXAMPLE 23

Convert a temperature of 5°C to degrees Fahrenheit.

Known	Unknown
Temp, °C = 5°C	Temp, °F

Change 5°C to degrees Fahrenheit.

$$\text{Temperature, °F} = \frac{9}{5}(\text{°C}) + 32\text{°F}$$

$$= \frac{9}{5}(5\text{°C}) + 32\text{°F}$$

$$= 41\text{°F}$$

A.14 BASIC CONVERSION FACTORS (METRIC SYSTEM)

LENGTH		
100 cm	= 1 m	100 cm/m
3.281 ft	= 1 m	3.281 ft/m
AREA		
2.4711 ac	= 1 ha*	2.4711 ac/ha
10,000 sq m	= 1 ha	10,000 sq m/ha
VOLUME		
1,000 m*L*	= 1 liter	1,000 m*L*/*L*
1,000 *L*	= 1 cu m	1,000 *L*/cu m
3.785 *L*	= 1 gal	3.785 *L*/gal
WEIGHT		
1,000 mg	= 1 gm	1,000 mg/gm
1,000 gm	= 1 kg	1,000 gm/kg
DENSITY		
1 kg	= 1 liter	1 kg/*L*
PRESSURE		
10.015 m	= 1 kg/sq cm	10.015 m/kg/sq cm
1 Pascal	= 1 N/sq m	1 Pa/N/sq m
1 psi	= 6,895 Pa	1 psi/6,895 Pa
FLOW		
3,785 cu m/day	= 1 MGD	3,785 cu m/day/MGD
3.785 M*L*/day	= 1 MGD	3.785 M*L*/day/MGD

* hectare

A.15 TYPICAL SMALL WATER SYSTEM PROBLEMS (METRIC SYSTEM)

A.150 Flows

EXAMPLE 1

Convert a flow of 500 gallons per minute to liters per second and cubic meters per day.

Known	Unknown
Flow, GPM = 500 GPM	1. Flow, liters/sec 2. Flow, cu m/day

1. Convert the flow from 500 GPM to liters per second.

$$\text{Flow, liters/sec} = \frac{(\text{Flow, gal/min})(3.785 \text{ liters/gal})}{60 \text{ sec/min}}$$

$$= \frac{(500 \text{ gal/min})(3.785 \text{ liters/gal})}{60 \text{ sec/min}}$$

$$= 31.5 \text{ liters/sec}$$

2. Convert the flow from 500 GPM to cubic meters per day.

$$\text{Flow, cu m/day} = \frac{(\text{Flow, gal/min})(3.785\ L/\text{gal})(60 \text{ min/hr})(24 \text{ hr/day})}{1{,}000\ L/\text{cu m}}$$

$$= \frac{(500 \text{ gal/min})(3.785\ L/\text{gal})(60 \text{ min/hr})(24 \text{ hr/day})}{1{,}000\ L/\text{cu m}}$$

$$= 2{,}725 \text{ cu m/day}$$

A.151 Chemical Doses

EXAMPLE 2

Determine the chlorinator setting in kilograms per 24 hours if 4,000 cubic meters of water per day are to be treated with a desired chlorine dose of 2.5 mg/*L*.

Known		Unknown
Flow, cu m/day	= 4,000 cu m/day	Chlorinator Setting, kg/24 hours
Chlorine Dose, mg/*L*	= 2.5 mg/*L*	

Determine the chlorinator setting in kilograms per 24 hours.

$$\text{Chlorinator Setting, kg/day} = \frac{(\text{Flow, cu m/day})(\text{Dose, mg}/L)(1{,}000\ L/\text{cu m})}{(1{,}000 \text{ mg/gm})(1{,}000 \text{ gm/kg})}$$

$$= \frac{(4{,}000 \text{ cu m/day})(2.5 \text{ mg}/L)(1{,}000\ L/\text{cu m})}{(1{,}000 \text{ mg/gm})(1{,}000 \text{ gm/kg})}$$

$$= 10 \text{ kg/day}$$

EXAMPLE 3

Determine the actual chemical dose or chemical feed in kilograms per day from a dry chemical feeder. A bucket placed under the chemical feeder weighed 100 grams empty and 1,400 grams after 30 minutes.

Known		Unknown
Empty Bucket, gm	= 100 gm	Chemical Feed, kg/day
Full Bucket, gm	= 1,400 gm	
Time to fill, min	= 30 min	

Determine the chemical feed in kilograms of chemical per day.

$$\text{Chemical Feed, kg/day} = \frac{\text{Chemical Applied, kg}}{\text{Length of Application, days}}$$

$$= \frac{(1{,}400 \text{ gm} - 100 \text{ gm})(60 \text{ min/hr})(24 \text{ hr/day})}{(1{,}000 \text{ gm/kg})(30 \text{ min})}$$

$$= 62.4 \text{ kg/day}$$

EXAMPLE 4

Determine the chemical feed in kilograms of chlorine per day from a hypochlorinator. The hypochlorite solution is 1.4 percent or 14,000 mg chlorine per liter. During a test run the hypochlorinator delivered 400 m*L* during 12 minutes.

Known		Unknown
Cl Solution, %	= 1.4%	Chemical Feed, kg/day
Cl Conc, mg/*L*	= 14,000 mg/*L*	
Vol Pumped, m*L*	= 400 m*L*	
Time Pumped, min	= 12 min	

Calculate the chlorine fed by the hypochlorinator in kilograms of chlorine per day.

$$\text{Chlorine Feed, kg/day} = \frac{(\text{Cl Conc, mg}/L)(\text{Vol Pumped, m}L)(60 \text{ min/hr})(24 \text{ hr/day})}{(\text{Time Pumped, min})(1{,}000 \text{ m}L/L)(1{,}000 \text{ mg/gm})(1{,}000 \text{ gm/kg})}$$

$$= \frac{(14{,}000 \text{ mg}/L)(400 \text{ m}L)(60 \text{ min/hr})(24 \text{ hr/day})}{(12 \text{ min})(1{,}000 \text{ m}L/L)(1{,}000 \text{ mg/gm})(1{,}000 \text{ gm/kg})}$$

$$= 0.67 \text{ kg/day}$$

A.152 Wells

EXAMPLE 5

How much sodium hypochlorite is required to dose a well at 100 mg/*L*? The casing diameter is 50 cm (0.5 m) and the length of the water-filled casing is 24 meters. Sodium hypochlorite is 5.25 percent or 52,500 mg/*L* chlorine.

Known		Unknown
Casing Diameter, m	= 0.5 m	Chlorine Required, *L*
Casing Length, m	= 24 m	
Chlorine Dose, mg/*L*	= 100 mg/*L*	
Chlorine Solution, mg/*L*	= 52,500 mg/*L*	

1. Calculate the volume of the water-filled casing in liters.

$$\text{Casing Volume, } L = (0.785)(\text{Diam, m})^2(\text{Length, m})(1{,}000\ L/\text{cu m})$$

$$= (0.785)(0.5 \text{ m})^2(24 \text{ m})(1{,}000\ L/\text{cu m})$$

$$= 4{,}710 \text{ liters}$$

2. Calculate the required liters of sodium hypochlorite.

$$\text{Chlorine Required, } L = \frac{(\text{Casing Volume, } L)(\text{Desired Dose, mg}/L)}{\text{Chlorine Solution, mg}/L}$$

$$= \frac{(4{,}710\ L)(100 \text{ mg}/L)}{52{,}500 \text{ mg}/L}$$

$$= 9.0 \text{ liters}$$

A.153 Small Water Treatment Plants

EXAMPLE 6

The optimum dose of liquid alum from the jar tests is 13 mg/*L*. Determine the setting on the liquid alum chemical feeder in liters per day and milliliters per minute when the flow is 4 M*L*D (mega or million liters per day). The liquid alum delivered to the plant contains 642.3 milligrams of alum per milliliter of liquid solution.

Known		Unknown
Alum Dose, mg/*L*	= 13 mg/*L*	1. Chemical Feeder Setting, *L*/day
Flow, M*L*D	= 4 M*L*D	2. Chemical Feeder Setting, m*L*/min
Liquid Alum	= 642.3 mg/m*L*	

1. Calculate the chemical feeder setting in liters per day.

$$\text{Chemical Feeder Setting, } LPD = \frac{(\text{Flow, M}L\text{D})(\text{Alum Dose, mg}/L)(1{,}000{,}000/\text{Million})}{(\text{Liquid Alum, mg/m}L)(1{,}000\text{ m}L/L)}$$

$$= \frac{(4\text{ M}L\text{D})(13\text{ mg}/L)(1{,}000{,}000/\text{Million})}{(642.3\text{ mg/m}L)(1{,}000\text{ m}L/L)}$$

$$= 81\ LPD$$

2. Calculate the chemical feeder setting in milliliters per minute.

$$\text{Chemical Feeder Setting, m}L\text{/min} = \frac{(\text{Flow, M}L\text{D})(\text{Alum Dose, mg}/L)(1{,}000{,}000/\text{Million})}{(\text{Liquid Alum, mg/m}L)(24\text{ hr/day})(60\text{ min/hr})}$$

$$= \frac{(4\text{ M}L\text{D})(13\text{ mg}/L)(1{,}000{,}000/\text{Million})}{(642.3\text{ mg/m}L)(24\text{ hr/day})(60\text{ min/hr})}$$

$$= 56\text{ m}L\text{/min}$$

EXAMPLE 7

A rectangular settling basin that is 9 meters long, 5 meters wide, and the water is 2 meters deep treats a flow of 800 cubic meters per day. Estimate the detention time or settling time in hours in the basin for this flow.

Known		Unknown
Length, m	= 9 m	Detention Time, hr
Width, m	= 5 m	
Depth, m	= 2 m	
Flow, cu m/day	= 800 cu m/day	

1. Calculate the basin volume in cubic meters.

$$\text{Basin Volume, cu m} = (\text{Length, m})(\text{Width, m})(\text{Depth, m})$$

$$= (9\text{ m})(5\text{ m})(2\text{ m})$$

$$= 90\text{ cu m}$$

2. Calculate the detention time in hours.

$$\text{Detention Time, hr} = \frac{(\text{Basin Volume, cu m})(24\text{ hr/day})}{\text{Flow, cu m/day}}$$

$$= \frac{(90\text{ cu m})(24\text{ hr/day})}{800\text{ cu m/day}}$$

$$= 2.7\text{ hr}$$

EXAMPLE 8

Estimate the surface overflow rate in cubic meters per day per square meter for a 7-meter diameter clarifier that treats a flow of 300 cubic meters per day.

Known		Unknown
Diameter, m	= 7 m	Overflow Rate, cu m/day/sq m
Flow, cu m/day	= 300 cu m/day	

1. Calculate the surface area of the clarifier in square meters.

$$\text{Surface Area, sq m} = (0.785)(\text{Diameter, m})^2$$

$$= (0.785)(7\text{ m})^2$$

$$= 38.5\text{ sq m}$$

2. Determine the basin overflow rate in cubic meters per day per square meter of surface area.

$$\text{Overflow Rate, cu m/day/sq m} = \frac{\text{Flow, cu m/day}}{\text{Surface Area, sq m}}$$

$$= \frac{300\text{ cu m/day}}{38.5\text{ sq m}}$$

$$= 7.8\text{ cu m/day/sq m}$$

EXAMPLE 9

A 7-meter diameter circular clarifier treats a flow of 400 cubic meters per day. Calculate the weir loading in cubic meters per day per meter of weir length if the weir is on the outside edge of the clarifier.

Known		Unknown
Weir Diameter, m	= 7 m	Weir Loading, cu m/day/m
Flow, cu m/day	= 400 cu m/day	

1. Calculate the length of weir.

$$\text{Weir Length, m} = \pi(\text{Weir Diameter, m})$$

$$= (3.14)(7\text{ m})$$

$$= 22\text{ m}$$

2. Estimate the weir loading in cubic meters per day per meter.

$$\text{Weir Loading, cu m/day/m} = \frac{\text{Flow, cu m/day}}{\text{Weir Length, m}}$$

$$= \frac{400\text{ cu m/day}}{22\text{ m}}$$

$$= 18\text{ cu m/day/m of weir length}$$

EXAMPLE 10

A 5-meter wide by 7-meter long rapid sand filter treats a flow of 4,000 cubic meters per day. Calculate the filtration rate in liters per second per square meter of filter surface area and also in millimeters per second.

Known		Unknown
Width, m	= 5 m	1. Filtration Rate, *L*/sec/sq m
Length, m	= 7 m	2. Filtration Rate, mm/sec
Flow, cu m/day	= 4,000 cu m/day	

1. Calculate the surface area of the filter.

$$\text{Surface Area, sq m} = (\text{Length, m})(\text{Width, m})$$
$$= (7 \text{ m})(5 \text{ m})$$
$$= 35 \text{ sq m}$$

2. Convert the flow from cubic meters per day to liters per second.

$$\text{Flow, } L\text{/sec} = \frac{(\text{Flow, cu m/day})(1{,}000\ L\text{/cu m})}{(24 \text{ hr/day})(60 \text{ min/hr})(60 \text{ sec/min})}$$
$$= \frac{(4{,}000 \text{ cu m/day})(1{,}000\ L\text{/cu m})}{(24 \text{ hr/day})(60 \text{ min/hr})(60 \text{ sec/min})}$$
$$= 46.3\ L\text{/sec}$$

3. Determine the filtration rate in liters per second per square meter of filter surface area.

$$\text{Filtration Rate, } L\text{/sec/sq m} = \frac{\text{Flow, liters/sec}}{\text{Surface Area, sq m}}$$
$$= \frac{46.3\ L\text{/sec}}{35 \text{ sq m}}$$
$$= 1.32\ L\text{/sec/sq m}$$

4. Calculate the filtration rate in millimeters per second.

$$\text{Filtration Rate, mm/sec} = \frac{(\text{Flow, liters/sec})(1{,}000 \text{ mm/m})}{(\text{Surface Area, sq m})(1{,}000\ L\text{/cu m})}$$
$$= \frac{(46.3\ L\text{/sec})(1{,}000 \text{ mm/m})}{(35 \text{ sq m})(1{,}000\ L\text{/cu m})}$$
$$= 1.32 \text{ mm/sec}$$

EXAMPLE 11

Calculate the backwash flow required in cubic meters per second to backwash a 3-meter wide by 5-meter long filter if the desired backwash flow rate is 15 liters per second per square meter.

Known		Unknown
Width, m	= 3 m	Backwash Flow, cu m/sec
Length, m	= 5 m	
Backwash Rate, *L*/sec/sq m	= 15 *L*/sec/sq m	

1. Calculate the surface area of the filter.

$$\text{Surface Area, sq m} = (\text{Length, m})(\text{Width, m})$$
$$= (5 \text{ m})(3 \text{ m})$$
$$= 15 \text{ sq m}$$

2. Calculate the backwash flow in cubic meters per day.

$$\text{Backwash Flow cu m/sec} = \frac{(\text{Backwash Rate, } L\text{/sec/sq m})(\text{Area, sq m})}{1{,}000\ L\text{/cu m}}$$
$$= \frac{(15\ L\text{/sec/sq m})(15 \text{ sq m})}{1{,}000\ L\text{/cu m}}$$
$$= 0.225 \text{ cu m/sec}$$

EXAMPLE 12

Estimate the ion exchange capacity in milligrams* of hardness for a zeolite softener containing 5 cubic meters of media with a hardness removal capacity of 70 kilograms per cubic meter of media.

Known		Unknown
Media Vol, cu m	= 5 cu m	Ion Exchange Capacity, grams
Removal Capacity gm/cu m	= 70,000 gm/cu m	

Calculate the ion exchange capacity in grams of hardness.

$$\text{Ion Exchange Capacity, grams} = (\text{Media Vol, cu m})(\text{Removal Capacity, gm/cu m})$$
$$= (5 \text{ cu m})(70{,}000 \text{ gm/cu m})$$
$$= 350{,}000 \text{ grams}$$

* Grams are the metric units of hardness; grains are the English units.

A.154 Disinfection

EXAMPLE 13

How many liters of 5 percent sodium hypochlorite will be needed to disinfect a well with a 0.5-meter diameter casing and well screen? The well is 60 meters deep and there are 25 meters of water in the well. Use an initial chlorine dose of 100 mg/*L*.

Known		Unknown
Hypochlorite, %	= 5%	5% Hypochlorite, *L*
Chlorine Dose, mg/*L*	= 100 mg/*L*	
Diameter, m	= 0.5 m	
Water Depth, m	= 25 m	

1. Calculate the volume of water in the well in liters.

$$\text{Water Vol, liters} = (0.785)(\text{Diameter, m})^2(\text{Depth, m})(1{,}000\ L/\text{cu m})$$
$$= (0.785)(0.5\ \text{m})^2(25\ \text{m})(1{,}000\ L/\text{cu m})$$
$$= 4{,}906\ \text{liters}$$

2. Determine the grams of chlorine needed.

$$\text{Chlorine, gm} = \frac{(\text{Volume, }L)(\text{Dose, mg}/L)}{1{,}000\ \text{mg/gm}}$$
$$= \frac{(4{,}906\ L)(100\ \text{mg}/L)}{1{,}000\ \text{mg/gm}}$$
$$= 491\ \text{gm}$$

3. Calculate the liters of 5 percent sodium hypochlorite solution needed. One liter of water weighs 1,000 grams.

$$\text{Sodium Hypochlorite Solution, liters} = \frac{(\text{Chlorine, gm})(100\%)}{(1{,}000\ \text{gm}/L)(\text{Hypochlorite, }\%)}$$
$$= \frac{(491\ \text{gm})(100\%)}{(1{,}000\ \text{gm}/L)(5\%)}$$
$$= 9.8\ \text{liters}$$

Ten liters should do the job.

4. *ALTERNATE SOLUTION*

Calculate the liters of 5 percent sodium hypochlorite solution needed. A 5 percent solution is the same as 50,000 mg per liter.

$$\text{Sodium Hypochlorite Solution, liters} = \frac{(\text{Volume, }L)(\text{Dose, mg}/L)}{\text{Hypochlorite Solution, mg}/L}$$
$$= \frac{(4{,}906\ L)(100\ \text{mg}/L)}{50{,}000\ \text{mg}/L}$$
$$= 9.8\ \text{liters}$$

EXAMPLE 14

A new 250-mm diameter water main 200 meters long needs to be disinfected. An initial chlorine dose of 400 mg/*L* is expected to maintain a chlorine residual of over 300 mg/*L* during the three-hour disinfection period. How many liters of 5.25 percent (52,500 mg/*L*) sodium hypochlorite solution will be needed?

Known		Unknown
Diameter of Pipe, m	= 0.25 m	5.25% Hypochlorite, *L*
Length of Pipe, m	= 200 m	
Chlorine Dose, mg/*L*	= 400 mg/*L*	
Hypochlorite, %	= 5.25%	
Hypochlorite, mg/*L*	= 52,500 mg/*L*	

1. Calculate the volume of water in the pipe in liters.

$$\text{Pipe Vol, }L = (0.785)(\text{Diameter, m})^2(\text{Length, m})(1{,}000\ L/\text{cu m})$$
$$= (0.785)(0.25\ \text{m})^2(200\ \text{m})(1{,}000\ L/\text{cu m})$$
$$= 9{,}813\ L$$

2. Calculate the liters of 5.25 percent sodium hypochlorite solution needed.

$$\text{Sodium Hypochlorite Solution, }L = \frac{(\text{Volume, }L)(\text{Dose, mg}/L)}{\text{Hypochlorite Solution, mg}/L}$$
$$= \frac{(9{,}813\ L)(400\ \text{mg}/L)}{52{,}500\ \text{mg}/L}$$
$$= 75\ \text{liters}$$

EXAMPLE 15

A service storage reservoir has been taken out of service for inspection, maintenance, and repairs. The reservoir needs to be disinfected before being placed back on line. The reservoir is 12 meters in diameter and 3 meters deep. An initial chlorine dose of 100 mg/*L* is expected to maintain a chlorine residual of over 50 mg/*L* during the 24-hour disinfection period. How many liters of 15 percent sodium hypochlorite solution will be needed?

Known		Unknown
Diameter, m	= 12 m	15% Hypochlorite, *L*
Tank Depth, m	= 3 m	
Chlorine Dose, mg/*L*	= 100 mg/*L*	
Hypochlorite, %	= 15%	
Hypochlorite, mg/*L*	= 150,000 mg/*L*	

1. Calculate the volume of water in the tank in liters.

$$\text{Tank Volume, }L = (0.785)(\text{Diameter, m})^2(\text{Depth, m})(1{,}000\ L/\text{cu m})$$
$$= (0.785)(12\ \text{m})^2(3\ \text{m})(1{,}000\ L/\text{cu m})$$
$$= 339{,}120\ L$$

2. Calculate the liters of 15 percent sodium hypochlorite solution needed.

$$\text{Sodium Hypochlorite Solution, }L = \frac{(\text{Volume, }L)(\text{Dose, mg}/L)}{\text{Hypochlorite Solution, mg}/L}$$
$$= \frac{(339{,}120\ L)(100\ \text{mg}/L)}{150{,}000\ \text{mg}/L}$$
$$= 226\ \text{liters}$$

EXAMPLE 16

Calculate the actual chlorine dose in milligrams per liter if 800 liters of a 2 percent sodium hypochlorite solution were used to treat 8 megaliters of water.

Known		Unknown
Volume Hypocl, *L*	= 800 *L*	Chlorine Dose, mg/*L*
Volume Water, M*L*	= 8 M*L*	
Hypochlorite, %	= 2%	
Hypochlorite, mg/*L*	= 20,000 mg/*L*	

Calculate the actual chlorine dose in milligrams per liter.

$$\text{Actual Dose, mg}/L = \frac{(\text{Hypochlorite, mg}/L)(\text{Volume Hypocl, }L)}{(\text{Volume Water, M}L)(1{,}000{,}000/\text{Million})}$$
$$= \frac{(20{,}000\ \text{mg}/L)(800\ L)}{(8\ \text{M}L)(1{,}000{,}000/\text{Million})}$$
$$= 2.0\ \text{mg}/L$$

EXAMPLE 17

A deep well turbine pump delivers 15 liters per second against typical operating heads. If the desired chlorine dose is 2.5 mg/*L*, what should be the chlorine feed rate in kilograms per day and milligrams per second?

Known	Unknown
Flow, *L*/sec = 15 *L*/sec	1. Chlorine Feed, kg/day
Dose, mg/*L* = 2.5 mg/*L*	2. Chlorine Feed, mg/sec

1. Calculate the chlorine feed rate in kilograms per day.

$$\text{Chlorine Feed, kg/day} = \frac{(\text{Flow, } L\text{/sec})(\text{Dose, mg/}L)(60\text{ sec/min})(60\text{ min/hr})(24\text{ hr/day})}{(1{,}000\text{ mg/gm})(1{,}000\text{ gm/kg})}$$

$$= \frac{(15\ L\text{/sec})(2.5\text{ mg/}L)(60\text{ sec/min})(60\text{ min/hr})(24\text{ hr/day})}{(1{,}000\text{ mg/gm})(1{,}000\text{ gm/kg})}$$

$$= 3.24\text{ kg/day}$$

2. Calculate the chlorine feed in milligrams per second.

$$\text{Chlorine Feed, mg/sec} = (\text{Flow, } L\text{/sec})(\text{Dose, mg/}L)$$

$$= (15\ L\text{/sec})(2.5\text{ mg/}L)$$

$$= 37.5\text{ mg/sec}$$

EXAMPLE 18

Estimate the chlorine demand for a water in milligrams per liter if the chlorine dose is 2.6 mg/*L* and the chlorine residual is 0.4 mg/*L*.

Known	Unknown
Chlorine Dose, mg/*L* = 2.6 mg/*L*	Chlorine Demand, mg/*L*
Chlorine Residual, mg/*L* = 0.4 mg/*L*	

Estimate the chlorine demand in milligrams per liter.

$$\text{Chlorine Demand, mg/}L = \text{Chlorine Dose, mg/}L - \text{Chlorine Residual, mg/}L$$

$$= 2.6\text{ mg/}L - 0.4\text{ mg/}L$$

$$= 2.2\text{ mg/}L$$

EXAMPLE 19

Estimate the liters of hypochlorite pumped by a hypochlorinator if the hypochlorite solution is in a container with a diameter of one meter and the hypochlorite level drops 45 centimeters during a specific time period.

Known	Unknown
Diameter, m = 1 m	Hypochlorite Pumped, liters
Drop, cm = 45 cm	

Estimate the liters of hypochlorite pumped.

$$\text{Hypochlorite, liters} = (\text{Container Area, sq m})(\text{Drop, m})(1{,}000\ L\text{/cu m})$$

$$= \frac{(0.785)(1\text{ m})^2(45\text{ cm})(1{,}000\ L\text{/cu m})}{(100\text{ cm/m})}$$

$$= 353\text{ liters}$$

EXAMPLE 20

Estimate the desired strength (as a percent chlorine) of a hypochlorite solution which is pumped by a hypochlorinator that delivers 300 liters per day. The water being treated requires a chlorine dose of 5 kilograms of chlorine per day.

Known	Unknown
Hypochlorinator Flow, *L*/day = 300 *L*/day	Hypochlorite Strength, %
Chlorine Required, kg/day = 5 kg/day	

Estimate the desired hypochlorite strength as a percent chlorine.

$$\text{Hypochlorite Strength, \%} = \frac{(\text{Chlorine Required, kg/day})(100\%)}{(\text{Hypochlorinator Flow, } L\text{/day})(1\text{ kg/}L)}$$

$$= \frac{(5\text{ kg/day})(100\%)}{(300\ L\text{/day})(1\text{ kg/}L)}$$

$$= 1.67\%$$

EXAMPLE 21

How many liters of water must be added to 40 liters of a 5 percent hypochlorite solution to produce a 1.67 percent hypochlorite solution?

Known	Unknown
Hypochlorite, *L* = 40 *L*	Water Added, *L*
Desired Hypo, % = 1.67%	
Actual Hypo, % = 5%	

Calculate the liters of water that must be added to produce a 1.67 percent hypochlorite solution.

$$\text{Water Added, } L \text{ (to hypochlorite solution)} = \frac{(\text{Hypo, } L)(\text{Hypo, \%}) - (\text{Hypo, } L)(\text{Desired Hypo, \%})}{\text{Desired Hypo, \%}}$$

$$= \frac{(40\ L)(5\%) - (40\ L)(1.67\%)}{1.67\%}$$

$$= \frac{200 - 66.8}{1.67}$$

$$= 80\text{ liters}$$

A.155 Laboratory Procedures

EXAMPLE 22

Convert a temperature of 77°F to degrees Celsius.

Known	Unknown
Temp, °F = 77°F	Temp, °C

Change 77°F to degrees Celsius.

$$\text{Temperature, °C} = \frac{5}{9}(\text{°F} - 32\text{°F})$$

$$= \frac{5}{9}(77\text{°F} - 32\text{°F})$$

$$= 25\text{°C}$$

EXAMPLE 23

Convert a temperature of 15°C to degrees Fahrenheit.

Known	Unknown
Temp, °C = 15°C	Temp, °F

Change 15°C to degrees Fahrenheit.

$$\text{Temperature, °F} = \frac{9}{5}(\text{°C}) + 32\text{°F}$$

$$= \frac{9}{5}(15\text{°C}) + 32\text{°F}$$

$$= 59\text{°F}$$

A.16 CALCULATION OF LOG REMOVALS

Regulations may require the calculation of log removals for inactivation of *Giardia* cysts, viruses, or particle counts. How are log removals calculated? The following example illustrates two methods of calculating log removals.

EXAMPLE 24

Calculate the log removal of 5- to 15-micron particles per milliliter if the influent particle count to a water treatment filter reported 2,100 particles in the 5- to 15-microns range per milliliter of water and the filter effluent reported 30 particles in the 5- to 15-microns range per milliliter of filtered water.

Known	Unknown
Influent, particles/m*L* = 2,100 particles/m*L*	Log Removal, particles/m*L*
Effluent, particles/m*L* = 30 particles/m*L*	

Calculate the log removal of particles in the 5- to 15-microns range per m*L* by the filter.

PROCEDURE 1

$$\text{Log Removal, particles/m}L = \text{Log Influent, particles/m}L - \text{Log Effluent, particles/m}L$$

$$= \text{Log 2,100 particles/m}L - \text{Log 30 particles/m}L$$

$$= 3.3 - 1.5$$

$$= 1.8$$

PROCEDURE 2

$$\text{Log Removal, particles/m}L = \text{Log}\left[\frac{\text{Influent, particles/m}L}{\text{Effluent, particles/m}L}\right]$$

$$= \text{Log}\left[\frac{\text{2,100 particles/m}L}{\text{30 particles/m}L}\right]$$

$$= \text{Log } 70$$

$$= 1.8$$

WATER ABBREVIATIONS

ac	acre
ac-ft	acre-feet
af	acre feet
amp	ampere
°C	degrees Celsius
CFM	cubic feet per minute
CFS	cubic feet per second
Ci	Curie
cm	centimeter
cu ft	cubic feet
cu in	cubic inch
cu m	cubic meter
cu yd	cubic yard
°F	degrees Fahrenheit
ft	feet or foot
ft-lb/min	foot-pounds per minute
g	gravity
gal	gallon
gal/day	gallons per day
gm	gram
GPD	gallons per day
gpg	grains per gallon
GPM	gallons per minute
gr	grain
ha	hectare
HP	horsepower
hr	hour
in	inch
k	kilo
kg	kilogram
km	kilometer
kN	kilonewton
kW	kilowatt
kWh	kilowatt-hour
L	liter
lb	pound
lbs/sq in	pounds per square inch
m	meter
M	mega
M	million
mg	milligram
MGD	million gallons per day
mg/*L*	milligram per liter
min	minute
m*L*	milliliter
mm	millimeter
N	Newton
ohm	ohm
Pa	Pascal
pCi	picoCurie
ppb	parts per billion
ppm	parts per million
psf	pounds per square foot
psi	pounds per square inch
psig	pounds per square inch gage
RPM	revolutions per minute
sec	second
sq ft	square feet
sq in	square inches
W	watt

WATER WORDS

A Summary of the Words Defined

in

SMALL WATER SYSTEM OPERATION AND MAINTENANCE,

WATER DISTRIBUTION SYSTEM OPERATION AND MAINTENANCE,

and

WATER TREATMENT PLANT OPERATION

PROJECT PRONUNCIATION KEY

by Warren L. Prentice

The Project Pronunciation Key is designed to aid you in the pronunciation of new words. While this key is based primarily on familiar sounds, it does not attempt to follow any particular pronunciation guide. This key is designed solely to aid operators in this program.

You may find it helpful to refer to other available sources for pronunciation help. Each current standard dictionary contains a guide to its own pronunciation key. Each key will be different from each other and from this key. Examples of the difference between the key used in this program and the *WEBSTER'S NEW WORLD COLLEGE DICTIONARY*[1] "Key" are shown below.

In using this key, you should accent (say louder) the syllable that appears in capital letters. The following chart is presented to give examples of how to pronounce words using the Project Key.

	SYLLABLE				
WORD	1st	2nd	3rd	4th	5th
acid	AS	id			
coliform	COAL	i	form		
biological	BUY	o	LODGE	ik	cull

The first word, *ACID*, has its first syllable accented. The second word, *COLIFORM*, has its first syllable accented. The third word, *BIOLOGICAL*, has its first and third syllables accented.

We hope you will find the key useful in unlocking the pronunciation of any new word.

Term
Project Key
Webster Key
acid
AS-id
aś id
coliform
COAL-i-form
kō′ lə fôrm
biological
BUY-o-LODGE-ik-cull
bī ə läj′ i kəl

[1] *The WEBSTER'S NEW WORLD COLLEGE DICTIONARY, Fourth Edition, 1999, was chosen rather than an unabridged dictionary because of its availability to the operator. Other editions may be slightly different.*

WATER WORDS

A

ABC ABC

See **A**SSOCIATION OF **B**OARDS OF **C**ERTIFICATION.

ACEOPS ACEOPS

See **A**LLIANCE OF **C**ERTIFIED **O**PERATOR**S**, LAB ANALYSTS, INSPECTORS, AND SPECIALISTS (ACEOPS).

atm atm

The abbreviation for atmosphere. One atmosphere is equal to 14.7 psi or 100 kPa.

AWWA AWWA

See **A**MERICAN **W**ATER **W**ORKS **A**SSOCIATION.

ABSORPTION (ab-SORP-shun) ABSORPTION

The taking in or soaking up of one substance into the body of another by molecular or chemical action (as tree roots absorb dissolved nutrients in the soil).

ACCOUNTABILITY ACCOUNTABILITY

When a manager gives power/responsibility to an employee, the employee ensures that the manager is informed of results or events.

ACCURACY ACCURACY

How closely an instrument measures the true or actual value of the process variable being measured or sensed.

ACID RAIN ACID RAIN

Precipitation which has been rendered (made) acidic by airborne pollutants.

ACIDIC (uh-SID-ick) ACIDIC

The condition of water or soil which contains a sufficient amount of acid substances to lower the pH below 7.0.

ACIDIFIED (uh-SID-uh-FIE-d) ACIDIFIED

The addition of an acid (usually nitric or sulfuric) to a sample to lower the pH below 2.0. The purpose of acidification is to "fix" a sample so it won't change until it is analyzed.

ACRE-FOOT ACRE-FOOT

A volume of water that covers one acre to a depth of one foot, or 43,560 cubic feet (1,233.5 cubic meters).

ACTIVATED CARBON ACTIVATED CARBON

Adsorptive particles or granules of carbon usually obtained by heating carbon (such as wood). These particles or granules have a high capacity to selectively remove certain trace and soluble materials from water.

ACUTE HEALTH EFFECT ACUTE HEALTH EFFECT

An adverse effect on a human or animal body, with symptoms developing rapidly.

ADSORBATE (add-SORE-bait) ADSORBATE

The material being removed by the adsorption process.

ADSORBENT (add-SORE-bent) ADSORBENT

The material (activated carbon) that is responsible for removing the undesirable substance in the adsorption process.

ADSORPTION (add-SORP-shun) ADSORPTION

The gathering of a gas, liquid, or dissolved substance on the surface or interface zone of another material.

AERATION (air-A-shun) AERATION

The process of adding air to water. Air can be added to water by either passing air through water or passing water through air.

AEROBIC (AIR-O-bick) AEROBIC

A condition in which atmospheric or dissolved molecular oxygen is present in the aquatic (water) environment.

AESTHETIC (es-THET-ick) AESTHETIC

Attractive or appealing.

AGE TANK AGE TANK

A tank used to store a known concentration of chemical solution for feed to a chemical feeder. Also called a DAY TANK.

AIR BINDING AIR BINDING

The clogging of a filter, pipe or pump due to the presence of air released from water. Air entering the filter media is harmful to both the filtration and backwash processes. Air can prevent the passage of water during the filtration process and can cause the loss of filter media during the backwash process.

AIR GAP AIR GAP

An open vertical drop, or vertical empty space, that separates a drinking (potable) water supply to be protected from another water system in a water treatment plant or other location. This open gap prevents the contamination of drinking water by backsiphonage or backflow because there is no way raw water or any other water can reach the drinking water supply.

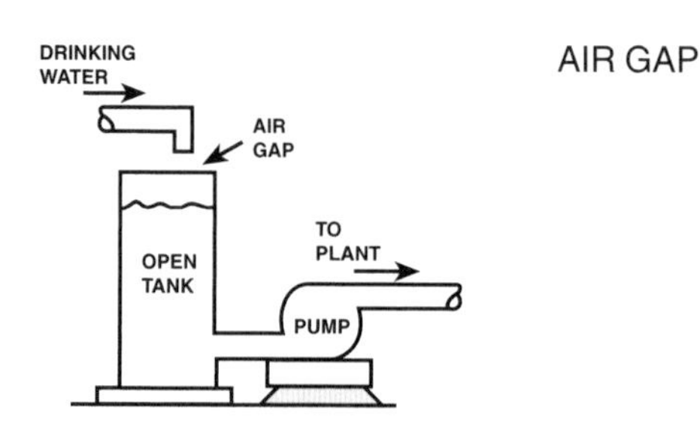

AIR PADDING AIR PADDING

Pumping dry air (dew point –40°F) into a container to assist with the withdrawal of a liquid or to force a liquified gas such as chlorine out of a container.

AIR STRIPPING AIR STRIPPING

A treatment process used to remove dissolved gases and volatile substances from water. Large volumes of air are bubbled through the water being treated to remove (strip out) the dissolved gases and volatile substances.

ALARM CONTACT ALARM CONTACT

A switch that operates when some preset low, high or abnormal condition exists.

ALGAE (AL-gee) ALGAE

Microscopic plants which contain chlorophyll and live floating or suspended in water. They also may be attached to structures, rocks or other submerged surfaces. Excess algal growths can impart tastes and odors to potable water. Algae produce oxygen during sunlight hours and use oxygen during the night hours. Their biological activities appreciably affect the pH, alkalinity, and dissolved oxygen of the water.

ALGAL (AL-gull) BLOOM ALGAL BLOOM

Sudden, massive growths of microscopic and macroscopic plant life, such as green or blue-green algae, which develop in lakes and reservoirs.

ALGICIDE (AL-juh-SIDE) ALGICIDE

Any substance or chemical specifically formulated to kill or control algae.

ALIPHATIC (AL-uh-FAT-ick) HYDROXY ACIDS ALIPHATIC HYDROXY ACIDS

Organic acids with carbon atoms arranged in branched or unbranched open chains rather than in rings.

ALIQUOT (AL-li-kwot) ALIQUOT

Portion of a sample. Often an equally divided portion of a sample.

ALKALI (AL-ka-lie) ALKALI

Any of certain soluble salts, principally of sodium, potassium, magnesium, and calcium, that have the property of combining with acids to form neutral salts and may be used in chemical water treatment processes.

ALKALINE (AL-ka-LINE) ALKALINE

The condition of water or soil which contains a sufficient amount of alkali substances to raise the pH above 7.0.

ALKALINITY (AL-ka-LIN-it-tee) ALKALINITY

The capacity of water to neutralize acids. This capacity is caused by the water's content of carbonate, bicarbonate, hydroxide, and occasionally borate, silicate, and phosphate. Alkalinity is expressed in milligrams per liter of equivalent calcium carbonate. Alkalinity is not the same as pH because water does not have to be strongly basic (high pH) to have a high alkalinity. Alkalinity is a measure of how much acid must be added to a liquid to lower the pH to 4.5.

ALLIANCE OF CERTIFIED OPERATORS, LAB ANALYSTS, INSPECTORS, AND SPECIALISTS (ACEOPS) ALLIANCE OF CERTIFIED OPERATORS, LAB ANALYSTS, INSPECTORS, AND SPECIALISTS (ACEOPS)

A professional organization for operators, lab analysts, inspectors, and specialists dedicated to improving professionalism; expanding training, certification, and job opportunities; increasing information exchange; and advocating the importance of certified operators, lab analysts, inspectors, and specialists. For information on membership, contact ACEOPS, 1810 Bel Air Drive, Ames, IA 50010, phone (515) 663-4128 or e-mail: ACEOPS@aol.com.

ALLUVIAL (uh-LOU-vee-ul) ALLUVIAL

Relating to mud and/or sand deposited by flowing water. Alluvial deposits may occur after a heavy rainstorm.

ALTERNATING CURRENT (A.C.) ALTERNATING CURRENT (A.C.)

An electric current that reverses its direction (positive/negative values) at regular intervals.

ALTITUDE VALVE ALTITUDE VALVE

A valve that automatically shuts off the flow into an elevated tank when the water level in the tank reaches a predetermined level. The valve automatically opens when the pressure in the distribution system drops below the pressure in the tank.

AMBIENT (AM-bee-ent) TEMPERATURE AMBIENT TEMPERATURE

Temperature of the surrounding air (or other medium). For example, temperature of the room where a gas chlorinator is installed.

AMERICAN WATER WORKS ASSOCIATION AMERICAN WATER WORKS ASSOCIATION

A professional organization for all persons working in the water utility field. This organization develops and recommends goals, procedures and standards for water utility agencies to help them improve their performance and effectiveness. For information on AWWA membership and publications, contact AWWA, 6666 W. Quincy Avenue, Denver, CO 80235. Phone (303) 794-7711.

AMPERAGE (AM-purr-age) AMPERAGE

The strength of an electric current measured in amperes. The amount of electric current flow, similar to the flow of water in gallons per minute.

AMPERE (AM-peer) AMPERE

The unit used to measure current strength. The current produced by an electromotive force of one volt acting through a resistance of one ohm.

AMPEROMETRIC (am-PURR-o-MET-rick) AMPEROMETRIC

A method of measurement that records electric current flowing or generated, rather than recording voltage. Amperometric titration is a means of measuring concentrations of certain substances in water.

AMPEROMETRIC (am-PURR-o-MET-rick) TITRATION AMPEROMETRIC TITRATION

A means of measuring concentrations of certain substances in water (such as strong oxidizers) based on the electric current that flows during a chemical reaction. Also see TITRATE.

AMPLITUDE AMPLITUDE

The maximum strength of an alternating current during its cycle, as distinguished from the mean or effective strength.

ANAEROBIC (AN-air-O-bick) ANAEROBIC

A condition in which atmospheric or dissolved molecular oxygen is *NOT* present in the aquatic (water) environment.

ANALOG ANALOG

The readout of an instrument by a pointer (or other indicating means) against a dial or scale.

ANALYZER ANALYZER

A device which conducts periodic or continuous measurement of some factor such as chlorine, fluoride or turbidity. Analyzers operate by any of several methods including photocells, conductivity or complex instrumentation.

ANGSTROM (ANG-strem) ANGSTROM

A unit of length equal to one-tenth of a nanometer or one-tenbillionth of a meter (1 Angstrom = 0.000 000 000 1 meter). One Angstrom is the approximate diameter of an atom.

ANION (AN-EYE-en) ANION

A negatively charged ion in an electrolyte solution, attracted to the anode under the influence of a difference in electrical potential. Chloride ion (Cl^-) is an anion.

ANIONIC (AN-eye-ON-ick) POLYMER ANIONIC POLYMER

A polymer having negatively charged groups of ions; often used as a filter aid and for dewatering sludges.

ANNULAR (AN-you-ler) SPACE ANNULAR SPACE

A ring-shaped space located between two circular objects, such as two pipes.

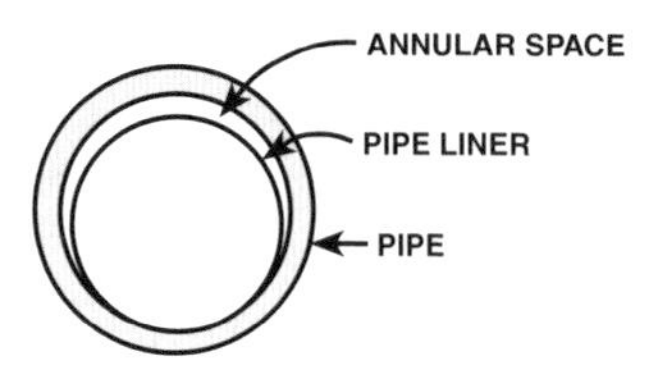

ANODE (an-O-d) ANODE

The positive pole or electrode of an electrolytic system, such as a battery. The anode attracts negatively charged particles or ions (anions).

APPARENT COLOR APPARENT COLOR

Color of the water that includes not only the color due to substances in the water but suspended matter as well.

APPROPRIATIVE APPROPRIATIVE

Water rights to or ownership of a water supply which is acquired for the beneficial use of water by following a specific legal procedure.

APPURTENANCE (uh-PURR-ten-nans) APPURTENANCE

Machinery, appliances, structures and other parts of the main structure necessary to allow it to operate as intended, but not considered part of the main structure.

AQUEOUS (A-kwee-us) AQUEOUS

Something made up of, similar to, or containing water; watery.

AQUIFER (ACK-wi-fer) AQUIFER

A natural underground layer of porous, water-bearing materials (sand, gravel) usually capable of yielding a large amount or supply of water.

ARCH ARCH

(1) The curved top of a sewer pipe or conduit.

(2) A bridge or arch of hardened or caked chemical which will prevent the flow of the chemical.

ARTESIAN (are-TEE-zhun) ARTESIAN

Pertaining to groundwater, a well, or underground basin where the water is under a pressure greater than atmospheric and will rise above the level of its upper confining surface if given an opportunity to do so.

ASEPTIC (a-SEP-tick) ASEPTIC

Free from the living germs of disease, fermentation, or putrefaction. Sterile.

ASSOCIATION OF BOARDS OF CERTIFICATION (ABC) ASSOCIATION OF BOARDS OF CERTIFICATION (ABC)

An international organization representing over 150 boards which certify the operators of waterworks and wastewater facilities. For information on ABC publications regarding the preparation of and how to study for operator certification examinations, contact ABC, 208 Fifth Street, Ames, IA 50010-6259. Phone (515) 232-3623.

ASYMMETRIC (A-see-MET-rick) ASYMMETRIC

Not similar in size, shape, form or arrangement of parts on opposite sides of a line, point or plane.

ATOM ATOM

The smallest unit of a chemical element; composed of protons, neutrons and electrons.

AUDIT, WATER AUDIT, WATER

A thorough examination of the accuracy of water agency records or accounts (volumes of water) and system control equipment. Water managers can use audits to determine their water distribution system efficiency. The overall goal is to identify and verify water and revenue losses in a water system.

AUTHORITY AUTHORITY

The power and resources to do a specific job or to get that job done.

AVAILABLE CHLORINE AVAILABLE CHLORINE

A measure of the amount of chlorine available in chlorinated lime, hypochlorite compounds, and other materials that are used as a source of chlorine when compared with that of elemental (liquid or gaseous) chlorine.

AVAILABLE EXPANSION AVAILABLE EXPANSION

The vertical distance from the sand surface to the underside of a trough in a sand filter. This distance is also called FREEBOARD.

AVERAGE AVERAGE

A number obtained by adding quantities or measurements and dividing the sum or total by the number of quantities or measurements. Also called the arithmetic mean.

$$\text{Average} = \frac{\text{Sum of Measurements}}{\text{Number of Measurements}}$$

AVERAGE DEMAND AVERAGE DEMAND

The total demand for water during a period of time divided by the number of days in that time period. This is also called the average daily demand.

AXIAL TO IMPELLER AXIAL TO IMPELLER

The direction in which material being pumped flows around the impeller or flows parallel to the impeller shaft.

AXIS OF IMPELLER AXIS OF IMPELLER

An imaginary line running along the center of a shaft (such as an impeller shaft).

B

BOD (pronounce as separate letters) BOD

Biochemical **O**xygen **D**emand. The rate at which organisms use the oxygen in water while stabilizing decomposable organic matter under aerobic conditions. In decomposition, organic matter serves as food for the bacteria and energy results from its oxidation. BOD measurements are used as a measure of the organic strength of wastes in water.

BACK PRESSURE BACK PRESSURE

A pressure that can cause water to backflow into the water supply when a user's water system is at a higher pressure than the public water system.

BACKFLOW BACKFLOW

A reverse flow condition, created by a difference in water pressures, which causes water to flow back into the distribution pipes of a potable water supply from any source or sources other than an intended source. Also see BACKSIPHONAGE.

BACKSIPHONAGE BACKSIPHONAGE

A form of backflow caused by a negative or below atmospheric pressure within a water system. Also see BACKFLOW.

BACKWASHING BACKWASHING

The process of reversing the flow of water back through the filter media to remove the entrapped solids.

BACTERIA (back-TEAR-e-ah) BACTERIA

Bacteria are living organisms, microscopic in size, which usually consist of a single cell. Most bacteria use organic matter for their food and produce waste products as a result of their life processes.

BAFFLE BAFFLE

A flat board or plate, deflector, guide or similar device constructed or placed in flowing water or slurry systems to cause more uniform flow velocities, to absorb energy, and to divert, guide, or agitate liquids (water, chemical solutions, slurry).

BAILER (BAY-ler) BAILER

A 10- to 20-foot-long pipe equipped with a valve at the lower end. A bailer is used to remove slurry from the bottom or the side of a well as it is being drilled.

BASE-EXTRA CAPACITY METHOD BASE-EXTRA CAPACITY METHOD

A cost allocation method used by water utilities to determine water rates for various water user groups. This method considers base costs (O & M expenses and capital costs), extra capacity costs (additional costs for maximum day and maximum hour demands), customer costs (meter maintenance and reading, billing, collection, accounting) and fire protection costs.

BASE METAL BASE METAL

A metal (such as iron) which reacts with dilute hydrochloric acid to form hydrogen. Also see NOBLE METAL.

BATCH PROCESS BATCH PROCESS

A treatment process in which a tank or reactor is filled, the water is treated or a chemical solution is prepared, and the tank is emptied. The tank may then be filled and the process repeated.

BENCH SCALE TESTS BENCH SCALE TESTS

A method of studying different ways or chemical doses for treating water on a small scale in a laboratory.

BIOCHEMICAL OXYGEN DEMAND (BOD) BIOCHEMICAL OXYGEN DEMAND (BOD)

The rate at which organisms use the oxygen in water while stabilizing decomposable organic matter under aerobic conditions. In decomposition, organic matter serves as food for the bacteria and energy results from its oxidation. BOD measurements are used as a measure of the organic strength of wastes in water.

BIOLOGICAL GROWTH BIOLOGICAL GROWTH

The activity and growth of any and all living organisms.

BLANK BLANK

A bottle containing only dilution water or distilled water; the sample being tested is not added. Tests are frequently run on a *SAMPLE* and a *BLANK* and the differences are compared.

BOND BOND

(1) A written promise to pay a specified sum of money (called the face value) at a fixed time in the future (called the date of maturity). A bond also carries interest at a fixed rate, payable periodically. The difference between a note and a bond is that a bond usually runs for a longer period of time and requires greater formality. Utility agencies use bonds as a means of obtaining large amounts of money for capital improvements.

(2) A warranty by an underwriting organization, such as an insurance company, guaranteeing honesty, performance, or payment by a contractor.

BONNET (BON-it) BONNET

The cover on a gate valve.

BOWL, PUMP — BOWL, PUMP

The submerged pumping unit in a well, including the shaft, impellers and housing.

BRAKE HORSEPOWER — BRAKE HORSEPOWER

(1) The horsepower required at the top or end of a pump shaft (input to a pump).

(2) The energy provided by a motor or other power source.

BREAKPOINT CHLORINATION — BREAKPOINT CHLORINATION

Addition of chlorine to water until the chlorine demand has been satisfied. At this point, further additions of chlorine will result in a free chlorine residual that is directly proportional to the amount of chlorine added beyond the breakpoint.

BREAKTHROUGH — BREAKTHROUGH

A crack or break in a filter bed allowing the passage of floc or particulate matter through a filter. This will cause an increase in filter effluent turbidity. A breakthrough can occur (1) when a filter is first placed in service, (2) when the effluent valve suddenly opens or closes, and (3) during periods of excessive head loss through the filter (including when the filter is exposed to negative heads).

BRINELLING (bruh-NEL-ing) — BRINELLING

Tiny indentations (dents) high on the shoulder of the bearing race or bearing. A type of bearing failure.

BUFFER — BUFFER

A solution or liquid whose chemical makeup neutralizes acids or bases without a great change in pH.

BUFFER CAPACITY — BUFFER CAPACITY

A measure of the capacity of a solution or liquid to neutralize acids or bases. This is a measure of the capacity of water for offering a resistance to changes in pH.

C

C FACTOR — C FACTOR

A factor or value used to indicate the smoothness of the interior of a pipe. The higher the C Factor, the smoother the pipe, the greater the carrying capacity, and the smaller the friction or energy losses from water flowing in the pipe. To calculate the C Factor, measure the flow, pipe diameter, distance between two pressure gages, and the friction or energy loss of the water between the gages.

$$\text{C Factor} = \frac{\text{Flow, GPM}}{193.75(\text{Diameter, ft})^{2.63}(\text{Slope})^{0.54}}$$

CT VALUE — CT VALUE

Residual concentration of a given disinfectant in mg/*L* times the disinfectant's contact time in minutes.

CAISSON (KAY-sawn) — CAISSON

A structure or chamber which is usually sunk or lowered by digging from the inside. Used to gain access to the bottom of a stream or other body of water.

CALCIUM CARBONATE EQUILIBRIUM — CALCIUM CARBONATE EQUILIBRIUM

A water is considered stable when it is just saturated with calcium carbonate. In this condition the water will neither dissolve nor deposit calcium carbonate. Thus, in this water the calcium carbonate is in equilibrium with the hydrogen ion concentration.

CALCIUM CARBONATE ($CaCO_3$) EQUIVALENT — CALCIUM CARBONATE ($CaCO_3$) EQUIVALENT

An expression of the concentration of specified constituents in water in terms of their equivalent value to calcium carbonate. For example, the hardness in water which is caused by calcium, magnesium and other ions is usually described as calcium carbonate equivalent. Alkalinity test results are usually reported as mg/*L* $CaCO_3$ equivalents. To convert chloride to $CaCO_3$ equivalents, multiply the concentration of chloride ions in mg/*L* by 1.41, and for sulfate, multiply by 1.04.

CALIBRATION — CALIBRATION

A procedure which checks or adjusts an instrument's accuracy by comparison with a standard or reference.

CALL DATE — CALL DATE

First date a bond can be paid off.

CAPILLARY ACTION — CAPILLARY ACTION

The movement of water through very small spaces due to molecular forces.

CAPILLARY FORCES — CAPILLARY FORCES

The molecular forces which cause the movement of water through very small spaces.

CAPILLARY FRINGE — CAPILLARY FRINGE

The porous material just above the water table which may hold water by capillarity (a property of surface tension that draws water upward) in the smaller void spaces.

CARCINOGEN (CAR-sin-o-JEN) — CARCINOGEN

Any substance which tends to produce cancer in an organism.

CATALYST (CAT-uh-LIST) — CATALYST

A substance that changes the speed or yield of a chemical reaction without being consumed or chemically changed by the chemical reaction.

CATALYZE (CAT-uh-LIZE) — CATALYZE

To act as a catalyst. Or, to speed up a chemical reaction.

CATALYZED (CAT-uh-LIZED) — CATALYZED

To be acted upon by a catalyst.

CATHODE (KA-thow-d) — CATHODE

The negative pole or electrode of an electrolytic cell or system. The cathode attracts positively charged particles or ions (cations).

CATHODIC (ca-THOD-ick) PROTECTION — CATHODIC PROTECTION

An electrical system for prevention of rust, corrosion, and pitting of metal surfaces which are in contact with water or soil. A low-voltage current is made to flow through a liquid (water) or a soil in contact with the metal in such a manner that the external electromotive force renders the metal structure cathodic. This concentrates corrosion on auxiliary anodic parts which are deliberately allowed to corrode instead of letting the structure corrode.

CATION (CAT-EYE-en) — CATION

A positively charged ion in an electrolyte solution, attracted to the cathode under the influence of a difference in electrical potential. Sodium ion (Na^+) is a cation.

CATIONIC POLYMER — CATIONIC POLYMER

A polymer having positively charged groups of ions; often used as a coagulant aid.

CAUTION — CAUTION

This word warns against potential hazards or cautions against unsafe practices. Also see DANGER, NOTICE, and WARNING.

CAVITATION (CAV-uh-TAY-shun) — CAVITATION

The formation and collapse of a gas pocket or bubble on the blade of an impeller or the gate of a valve. The collapse of this gas pocket or bubble drives water into the impeller or gate with a terrific force that can cause pitting on the impeller or gate surface. Cavitation is accompanied by loud noises that sound like someone is pounding on the impeller or gate with a hammer.

CENTRATE — CENTRATE

The water leaving a centrifuge after most of the solids have been removed.

CENTRIFUGAL (sen-TRIF-uh-gull) PUMP — CENTRIFUGAL PUMP

A pump consisting of an impeller fixed on a rotating shaft that is enclosed in a casing, and having an inlet and discharge connection. As the rotating impeller whirls the liquid around, centrifugal force builds up enough pressure to force the water through the discharge outlet.

CENTRIFUGE CENTRIFUGE

A mechanical device that uses centrifugal or rotational forces to separate solids from liquids.

CERTIFICATION EXAMINATION CERTIFICATION EXAMINATION

An examination administered by a state agency that small water supply system operators take to indicate a level of professional competence. In the United States, certification of small water system operators is mandatory.

CHARGE CHEMISTRY CHARGE CHEMISTRY

A branch of chemistry in which the destabilization and neutralization reactions occur between stable negatively charged and stable positively charged particles.

CHECK SAMPLING CHECK SAMPLING

Whenever an initial or routine sample analysis indicates that a Maximum Contaminant Level (MCL) has been exceeded, *CHECK SAMPLING* is required to confirm the routine sampling results. Check sampling is in addition to the routine sampling program.

CHECK VALVE CHECK VALVE

A special valve with a hinged disc or flap that opens in the direction of normal flow and is forced shut when flows attempt to go in the reverse or opposite direction of normal flows.

CHELATING (key-LAY-ting) AGENT CHELATING AGENT

A chemical used to prevent the precipitation of metals (such as copper).

CHELATION (key-LAY-shun) CHELATION

A chemical complexing (forming or joining together) of metallic cations (such as copper) with certain organic compounds, such as EDTA (ethylene diamine tetracetic acid). Chelation is used to prevent the precipitation of metals (copper). Also see SEQUESTRATION.

CHLORAMINATION (KLOR-ah-min-NAY-shun) CHLORAMINATION

The application of chlorine and ammonia to water to form chloramines for the purpose of disinfection.

CHLORAMINES (KLOR-uh-means) CHLORAMINES

Compounds formed by the reaction of hypochlorous acid (or aqueous chlorine) with ammonia.

CHLORINATION (KLOR-uh-NAY-shun) CHLORINATION

The application of chlorine to water, generally for the purpose of disinfection, but frequently for accomplishing other biological or chemical results (aiding coagulation and controlling tastes and odors).

CHLORINATOR (KLOR-uh-NAY-ter) CHLORINATOR

A metering device which is used to add chlorine to water.

CHLORINE DEMAND CHLORINE DEMAND

Chlorine demand is the difference between the amount of chlorine added to water and the amount of residual chlorine remaining after a given contact time. Chlorine demand may change with dosage, time, temperature, pH, and nature and amount of the impurities in the water.

$$\text{Chlorine Demand, mg}/L = \text{Chlorine Applied, mg}/L - \text{Chlorine Residual, mg}/L$$

CHLORINE REQUIREMENT CHLORINE REQUIREMENT

The amount of chlorine which is needed for a particular purpose. Some reasons for adding chlorine are reducing the number of coliform bacteria (Most Probable Number), obtaining a particular chlorine residual, or oxidizing some substance in the water. In each case a definite dosage of chlorine will be necessary. This dosage is the chlorine requirement.

CHLORINE RESIDUAL CHLORINE RESIDUAL

The concentration of chlorine present in water after the chlorine demand has been satisfied. The concentration is expressed in terms of the total chlorine residual, which includes both the free and combined or chemically bound chlorine residuals.

CHLOROPHENOLIC (klor-o-FEE-NO-lick) CHLOROPHENOLIC

Chlorophenolic compounds are phenolic compounds (carbolic acid) combined with chlorine.

CHLOROPHENOXY (KLOR-o-fuh-KNOX-ee) CHLOROPHENOXY

A class of herbicides that may be found in domestic water supplies and cause adverse health effects. Two widely used chlorophenoxy herbicides are 2,4-D (2,4-Dichlorophenoxy acetic acid) and 2,4,5-TP (2,4,5-Trichlorophenoxy propionic acid (silvex)).

CHLORORGANIC (klor-or-GAN-ick) CHLORORGANIC

Organic compounds combined with chlorine. These compounds generally originate from, or are associated with, life processes such as those of algae in water.

CHRONIC HEALTH EFFECT CHRONIC HEALTH EFFECT

An adverse effect on a human or animal body with symptoms that develop slowly over a long period of time or that recur frequently.

CIRCLE OF INFLUENCE CIRCLE OF INFLUENCE

The circular outer edge of a depression produced in the water table by the pumping of water from a well. Also see CONE OF INFLUENCE and CONE OF DEPRESSION.

[SEE DRAWING ON PAGE 573]

CIRCUIT CIRCUIT

The complete path of an electric current, including the generating apparatus or other source; or, a specific segment or section of the complete path.

CIRCUIT BREAKER CIRCUIT BREAKER

A safety device in an electric circuit that automatically shuts off the circuit when it becomes overloaded. The device can be manually reset.

CISTERN (SIS-turn) CISTERN

A small tank (usually covered) or a storage facility used to store water for a home or farm. Often used to store rainwater.

CLARIFIER (KLAIR-uh-fire) CLARIFIER

A large circular or rectangular tank or basin in which water is held for a period of time during which the heavier suspended solids settle to the bottom. Clarifiers are also called settling basins and sedimentation basins.

CLASS, PIPE AND FITTINGS CLASS, PIPE AND FITTINGS

The working pressure rating, including allowances for surges, of a specific pipe for use in water distribution systems. The term is used for cast iron, ductile iron, asbestos cement and some plastic pipe.

CLEAR WELL CLEAR WELL

A reservoir for the storage of filtered water of sufficient capacity to prevent the need to vary the filtration rate with variations in demand. Also used to provide chlorine contact time for disinfection.

COAGULANT AID COAGULANT AID

Any chemical or substance used to assist or modify coagulation.

COAGULANTS (co-AGG-you-lents) COAGULANTS

Chemicals that cause very fine particles to clump (floc) together into larger particles. This makes it easier to separate the solids from the water by settling, skimming, draining or filtering.

COAGULATION (co-AGG-you-LAY-shun) COAGULATION

The clumping together of very fine particles into larger particles (floc) caused by the use of chemicals (coagulants). The chemicals neutralize the electrical charges of the fine particles, allowing them to come closer and form larger clumps. This clumping together makes it easier to separate the solids from the water by settling, skimming, draining or filtering.

CODE OF FEDERAL REGULATIONS (CFR) CODE OF FEDERAL REGULATIONS (CFR)

A publication of the United States Government which contains all of the proposed and finalized federal regulations, including environmental regulations.

COLIFORM (COAL-i-form) COLIFORM

A group of bacteria found in the intestines of warm-blooded animals (including humans) and also in plants, soil, air and water. Fecal coliforms are a specific class of bacteria which only inhabit the intestines of warm-blooded animals. The presence of coliform bacteria is an indication that the water is polluted and may contain pathogenic (disease-causing) organisms.

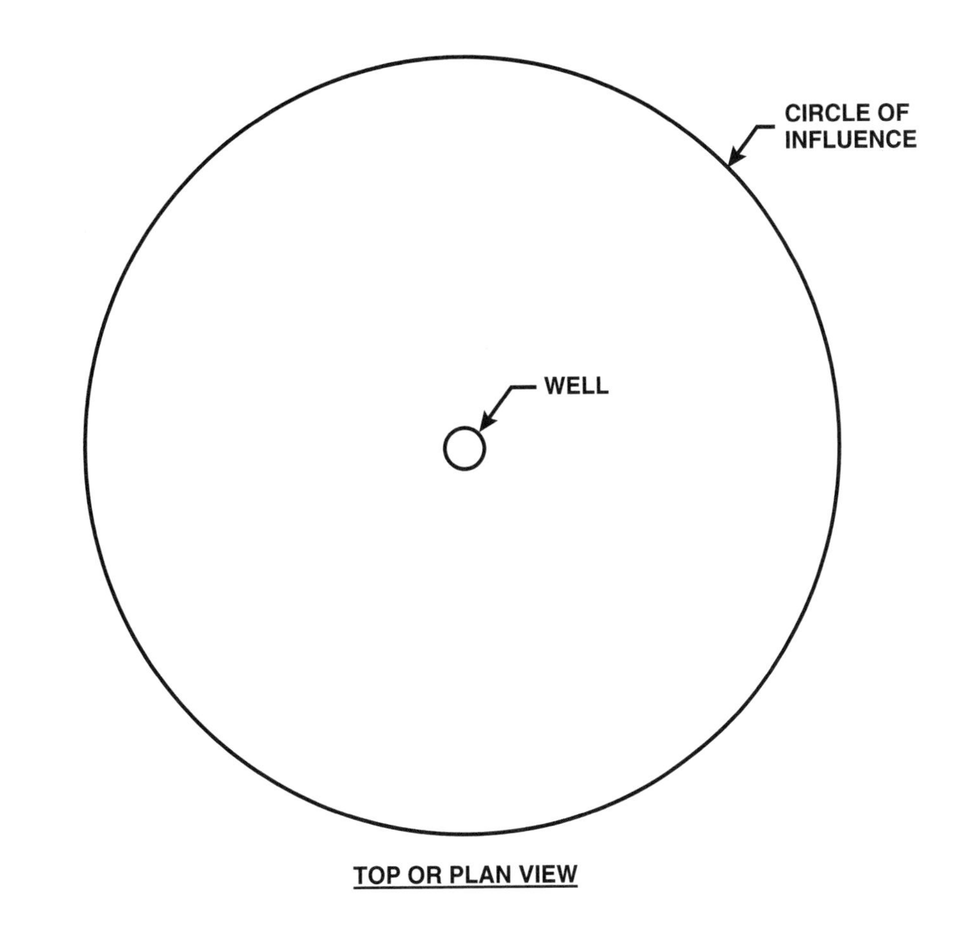

TOP OR PLAN VIEW

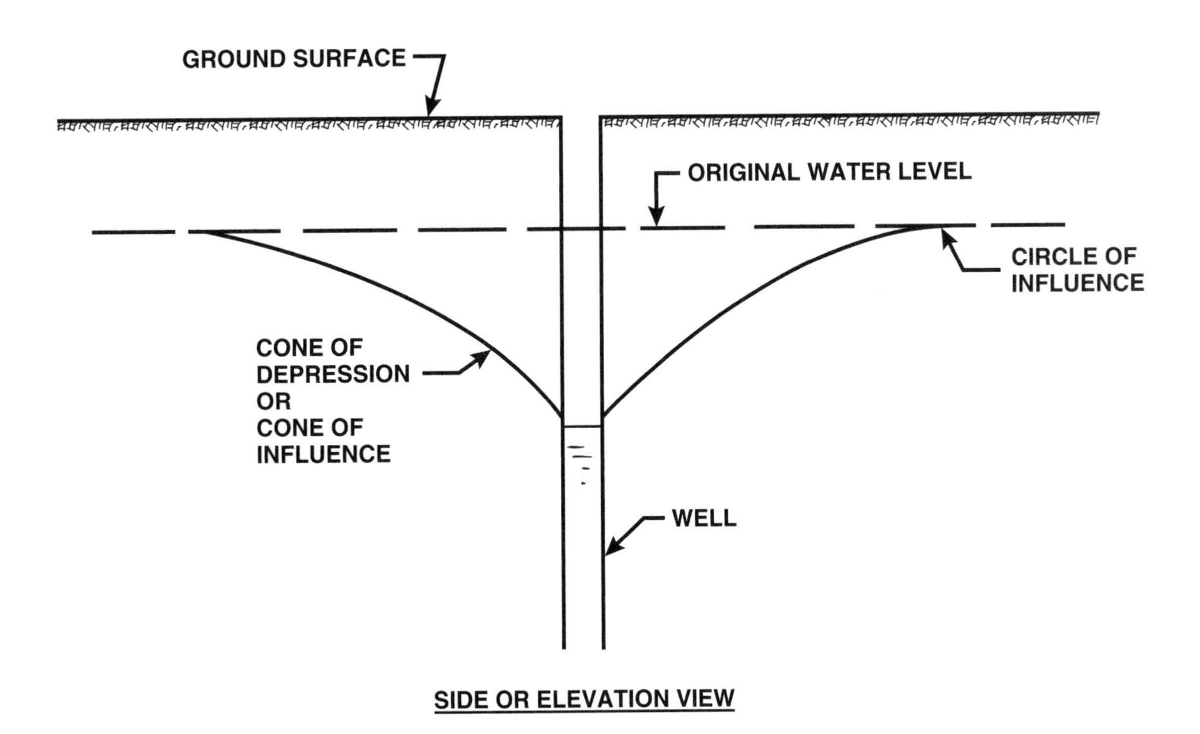

SIDE OR ELEVATION VIEW

CIRCLE OF INFLUENCE and CONE OF DEPRESSION/CONE OF INFLUENCE

COLLOIDS (CALL-loids) COLLOIDS

Very small, finely divided solids (particles that do not dissolve) that remain dispersed in a liquid for a long time due to their small size and electrical charge. When most of the particles in water have a negative electrical charge, they tend to repel each other. This repulsion prevents the particles from clumping together, becoming heavier, and settling out.

COLOR COLOR

The substances in water that impart a yellowish-brown color to the water. These substances are the result of iron and manganese ions, humus and peat materials, plankton, aquatic weeds, and industrial waste present in the water. Also see TRUE COLOR.

COLORIMETRIC MEASUREMENT COLORIMETRIC MEASUREMENT

A means of measuring unknown chemical concentrations in water by measuring a sample's color intensity. The specific color of the sample, developed by addition of chemical reagents, is measured with a photoelectric colorimeter or is compared with "color standards" using, or corresponding with, known concentrations of the chemical.

COMBINED AVAILABLE CHLORINE COMBINED AVAILABLE CHLORINE

The total chlorine, present as chloramine or other derivatives, that is present in a water and is still available for disinfection and for oxidation of organic matter. The combined chlorine compounds are more stable than free chlorine forms, but they are somewhat slower in disinfection action.

COMBINED AVAILABLE CHLORINE RESIDUAL COMBINED AVAILABLE CHLORINE RESIDUAL

The concentration of residual chlorine that is combined with ammonia, organic nitrogen, or both in water as a chloramine (or other chloro derivative) and yet is still available to oxidize organic matter and help kill bacteria.

COMBINED CHLORINE COMBINED CHLORINE

The sum of the chlorine species composed of free chlorine and ammonia, including monochloramine, dichloramine, and trichloramine (nitrogen trichloride). Dichloramine is the strongest disinfectant of these chlorine species, but it has less oxidative capacity than free chlorine.

COMBINED RESIDUAL CHLORINATION COMBINED RESIDUAL CHLORINATION

The application of chlorine to water to produce combined available chlorine residual. This residual can be made up of monochloramines, dichloramines, and nitrogen trichloride.

COMMODITY-DEMAND METHOD COMMODITY-DEMAND METHOD

A cost allocation method used by water utilities to determine water rates for the various water user groups. This method considers the commodity costs (water, chemicals, power, amount of water use), demand costs (treatment, storage, distribution), customer costs (meter maintenance and reading, billing, collection, accounting) and fire protection costs.

COMPETENT PERSON COMPETENT PERSON

A competent person is defined by OSHA as a person capable of identifying existing and predictable hazards in the surroundings, or working conditions which are unsanitary, hazardous or dangerous to employees, and who has authorization to take prompt corrective measures to eliminate the hazards.

COMPLETE TREATMENT COMPLETE TREATMENT

A method of treating water which consists of the addition of coagulant chemicals, flash mixing, coagulation-flocculation, sedimentation and filtration. Also called CONVENTIONAL FILTRATION.

COMPOSITE (come-PAH-zit) (PROPORTIONAL) SAMPLE COMPOSITE (PROPORTIONAL) SAMPLE

A composite sample is a collection of individual samples obtained at regular intervals, usually every one or two hours during a 24-hour time span. Each individual sample is combined with the others in proportion to the rate of flow when the sample was collected. The resulting mixture (composite sample) forms a representative sample and is analyzed to determine the average conditions during the sampling period.

COMPOUND COMPOUND

A pure substance composed of two or more elements whose composition is constant. For example, table salt (sodium chloride, NaCl) is a compound.

CONCENTRATION POLARIZATION CONCENTRATION POLARIZATION

(1) A buildup of retained particles on the membrane surface due to dewatering of the feed closest to the membrane. The thickness of the concentration polarization layer is controlled by the flow velocity across the membrane.
(2) Used in corrosion studies to indicate a depletion of ions near an electrode.
(3) The basis for chemical analysis by a polarograph.

CONDITIONING CONDITIONING

Pretreatment of sludge to facilitate removal of water in subsequent treatment processes.

CONDUCTANCE CONDUCTANCE

A rapid method of estimating the dissolved solids content of a water supply. The measurement indicates the capacity of a sample of water to carry an electrical current, which is related to the concentration of ionized substances in the water. Also called SPECIFIC CONDUCTANCE.

CONDUCTIVITY CONDUCTIVITY

A measure of the ability of a solution (water) to carry an electric current.

CONDUCTOR CONDUCTOR

A substance, body, device or wire that readily conducts or carries electric current.

CONDUCTOR CASING CONDUCTOR CASING

The outer casing of a well. The purpose of this casing is to prevent contaminants from surface waters or shallow groundwaters from entering a well.

CONE OF DEPRESSION CONE OF DEPRESSION

The depression, roughly conical in shape, produced in the water table by the pumping of water from a well. Also called the CONE OF INFLUENCE. Also see CIRCLE OF INFLUENCE.

[SEE DRAWING ON PAGE 573]

CONE OF INFLUENCE CONE OF INFLUENCE

The depression, roughly conical in shape, produced in the water table by the pumping of water from a well. Also called the CONE OF DEPRESSION. Also see CIRCLE OF INFLUENCE.

[SEE DRAWING ON PAGE 573]

CONFINED SPACE CONFINED SPACE

Confined space means a space that:

A. Is large enough and so configured that an employee can bodily enter and perform assigned work; and

B. Has limited or restricted means for entry or exit (for example, tanks, vessels, silos, storage bins, hoppers, vaults, and pits are spaces that may have limited means of entry); and

C. Is not designed for continuous employee occupancy.

(Definition from the Code of Federal Regulations (CFR) Title 29 Part 1910.146.)

CONFINED SPACE, CLASS "A" CONFINED SPACE, CLASS "A"

A confined space that presents a situation that is immediately dangerous to life or health (IDLH). These include but are not limited to oxygen deficiency, explosive or flammable atmospheres, and/or concentrations of toxic substances.

(Definition from NIOSH, "Criteria for a Recommended Standard: Working in Confined Spaces.")

CONFINED SPACE, CLASS "B" CONFINED SPACE, CLASS "B"

A confined space that has the potential for causing injury and illness, if preventive measures are not used, but not immediately dangerous to life and health.

(Definition from NIOSH, "Criteria for a Recommended Standard: Working in Confined Spaces.")

CONFINED SPACE, CLASS "C" CONFINED SPACE, CLASS "C"

A confined space in which the potential hazard would not require any special modification of the work procedure.

(Definition from NIOSH, "Criteria for a Recommended Standard: Working in Confined Spaces.")

CONFINED SPACE, NON-PERMIT CONFINED SPACE, NON-PERMIT

A non-permit confined space is a confined space that does not contain or, with respect to atmospheric hazards, have the potential to contain any hazard capable of causing death or serious physical harm.

CONFINED SPACE, PERMIT-REQUIRED (PERMIT SPACE)

CONFINED SPACE, PERMIT-REQUIRED (PERMIT SPACE)

A confined space that has one or more of the following characteristics:

- Contains or has a potential to contain a hazardous atmosphere,
- Contains a material that has the potential for engulfing an entrant,
- Has an internal configuration such that an entrant could be trapped or asphyxiated by inwardly converging walls or by a floor which slopes downward and tapers to a smaller cross section, or
- Contains any other recognized serious safety or health hazard.

(Definition from the Code of Federal Regulations (CFR) Title 29 Part 1910.146.)

CONFINING UNIT

CONFINING UNIT

A layer of rock or soil of very low hydraulic conductivity that hampers the movement of groundwater in and out of an aquifer.

CONSOLIDATED FORMATION

CONSOLIDATED FORMATION

A geologic material whose particles are stratified (layered), cemented or firmly packed together (hard rock); usually occurring at a depth below the ground surface. Also see UNCONSOLIDATED FORMATION.

CONSUMER CONFIDENCE REPORTS

CONSUMER CONFIDENCE REPORTS

An annual report prepared by a water utility to communicate with its consumers. The report provides consumers with information on the source and quality of their drinking water. The report is an opportunity for positive communication with consumers and to convey the importance of paying for good quality drinking water.

CONTACTOR

CONTACTOR

An electric switch, usually magnetically operated.

CONTAMINATION

CONTAMINATION

The introduction into water of microorganisms, chemicals, toxic substances, wastes, or wastewater in a concentration that makes the water unfit for its next intended use.

CONTINUOUS SAMPLE

CONTINUOUS SAMPLE

A flow of water from a particular place in a plant to the location where samples are collected for testing. This continuous stream may be used to obtain grab or composite samples. Frequently, several taps (faucets) will flow continuously in the laboratory to provide test samples from various places in a water treatment plant.

CONTROL LOOP

CONTROL LOOP

The path through the control system between the sensor, which measures a process variable, and the controller, which controls or adjusts the process variable.

CONTROL SYSTEM

CONTROL SYSTEM

An instrumentation system which senses and controls its own operation on a close, continuous basis in what is called proportional (or modulating) control.

CONTROLLER

CONTROLLER

A device which controls the starting, stopping, or operation of a device or piece of equipment.

CONVENTIONAL FILTRATION

CONVENTIONAL FILTRATION

A method of treating water which consists of the addition of coagulant chemicals, flash mixing, coagulation-flocculation, sedimentation and filtration. Also called COMPLETE TREATMENT. Also see DIRECT FILTRATION and IN-LINE FILTRATION.

CONVENTIONAL TREATMENT

CONVENTIONAL TREATMENT

See CONVENTIONAL FILTRATION. Also called COMPLETE TREATMENT.

CORPORATION STOP

CORPORATION STOP

A water service shutoff valve located at a street water main. This valve cannot be operated from the ground surface because it is buried and there is no valve box. Also called a corporation cock.

CORROSION | CORROSION

The gradual decomposition or destruction of a material by chemical action, often due to an electrochemical reaction. Corrosion may be caused by (1) stray current electrolysis, (2) galvanic corrosion caused by dissimilar metals, or (3) differential-concentration cells. Corrosion starts at the surface of a material and moves inward.

CORROSION INHIBITORS | CORROSION INHIBITORS

Substances that slow the rate of corrosion.

CORROSIVE GASES | CORROSIVE GASES

In water, dissolved oxygen reacts readily with metals at the anode of a corrosion cell, accelerating the rate of corrosion until a film of oxidation products such as rust forms. At the cathode where hydrogen gas may form a coating on the cathode and slow the corrosion rate, oxygen reacts rapidly with hydrogen gas forming water, and again increases the rate of corrosion.

CORROSIVITY | CORROSIVITY

An indication of the corrosiveness of a water. The corrosiveness of a water is described by the water's pH, alkalinity, hardness, temperature, total dissolved solids, dissolved oxygen concentration, and the Langelier Index.

COULOMB (COO-lahm) | COULOMB

A measurement of the amount of electrical charge carried by an electric current of one ampere in one second. One coulomb equals about 6.25×10^{18} electrons (6,250,000,000,000,000,000 electrons).

COUPON | COUPON

A steel specimen inserted into water to measure the corrosiveness of water. The rate of corrosion is measured as the loss of weight of the coupon (in milligrams) per surface area (in square decimeters) exposed to the water per day. 10 decimeters = 1 meter = 100 centimeters.

COVERAGE RATIO | COVERAGE RATIO

The coverage ratio is a measure of the ability of the utility to pay the principal and interest on loans and bonds (this is known as "debt service") in addition to any unexpected expenses.

CROSS CONNECTION | CROSS CONNECTION

A connection between a drinking (potable) water system and an unapproved water supply. For example, if you have a pump moving nonpotable water and hook into the drinking water system to supply water for the pump seal, a cross connection or mixing between the two water systems can occur. This mixing may lead to contamination of the drinking water.

CRYPTOSPORIDIUM (CRIP-toe-spo-RID-ee-um) | CRYPTOSPORIDIUM

A waterborne intestinal parasite that causes a disease called cryptosporidiosis (CRIP-toe-spo-rid-ee-O-sis) in infected humans. Symptoms of the disease include diarrhea, cramps, and weight loss. *Cryptosporidium* contamination is found in most surface waters and some groundwaters. Commonly referred to as "crypto."

CURB STOP | CURB STOP

A water service shutoff valve located in a water service pipe near the curb and between the water main and the building. This valve is usually operated by a wrench or valve key and is used to start or stop flows in the water service line to a building. Also called a curb cock.

CURIE | CURIE

A measure of radioactivity. One Curie of radioactivity is equivalent to 3.7×10^{10} or 37,000,000,000 nuclear disintegrations per second.

CURRENT | CURRENT

A movement or flow of electricity. Water flowing in a pipe is measured in gallons per second past a certain point, not by the number of water molecules going past a point. Electric current is measured by the number of coulombs per second flowing past a certain point in a conductor. A coulomb is equal to about 6.25×10^{18} electrons (6,250,000,000,000,000,000 electrons). A flow of one coulomb per second is called one ampere, the unit of the rate of flow of current.

CYCLE | CYCLE

A complete alternation of voltage and/or current in an alternating current (A.C.) circuit.

D

DBP DBP

See **D**ISINFECTION **B**Y-**P**RODUCT.

DPD (pronounce as separate letters) DPD

A method of measuring the chlorine residual in water. The residual may be determined by either titrating or comparing a developed color with color standards. DPD stands for N,N-diethyl-p-phenylene-diamine.

DANGER DANGER

The word *DANGER* is used where an immediate hazard presents a threat of death or serious injury to employees. Also see CAUTION, NOTICE, and WARNING.

DANGEROUS AIR CONTAMINATION DANGEROUS AIR CONTAMINATION

An atmosphere presenting a threat of causing death, injury, acute illness, or disablement due to the presence of flammable and/or explosive, toxic or otherwise injurious or incapacitating substances.

A. Dangerous air contamination due to the flammability of a gas or vapor is defined as an atmosphere containing the gas or vapor at a concentration greater than 10 percent of its lower explosive (lower flammable) limit.

B. Dangerous air contamination due to a combustible particulate is defined as a concentration greater than 10 percent of the minimum explosive concentration of the particulate.

C. Dangerous air contamination due to the toxicity of a substance is defined as the atmospheric concentration immediately hazardous to life or health.

DATEOMETER (day-TOM-uh-ter) DATEOMETER

A small calendar disc attached to motors and equipment to indicate the year in which the last maintenance service was performed.

DATUM LINE DATUM LINE

A line from which heights and depths are calculated or measured. Also called a datum plane or a datum level.

DAY TANK DAY TANK

A tank used to store a chemical solution of known concentration for feed to a chemical feeder. A day tank usually stores sufficient chemical solution to properly treat the water being treated for at least one day. Also called an AGE TANK.

DEAD END DEAD END

The end of a water main which is not connected to other parts of the distribution system by means of a connecting loop of pipe.

DEBT SERVICE DEBT SERVICE

The amount of money required annually to pay the (1) interest on outstanding debts; or (2) funds due on a maturing bonded debt or the redemption of bonds.

DECANT DECANT

To draw off the upper layer of liquid (water) after the heavier material (a solid or another liquid) has settled.

DECANT WATER DECANT WATER

Water that has separated from sludge and is removed from the layer of water above the sludge.

DECHLORINATION (dee-KLOR-uh-NAY-shun) DECHLORINATION

The deliberate removal of chlorine from water. The partial or complete reduction of residual chlorine by any chemical or physical process.

DECIBEL (DES-uh-bull) DECIBEL

A unit for expressing the relative intensity of sounds on a scale from zero for the average least perceptible sound to about 130 for the average level at which sound causes pain to humans. Abbreviated dB.

DECOMPOSITION, DECAY DECOMPOSITION, DECAY

The conversion of chemically unstable materials to more stable forms by chemical or biological action. If organic matter decays when there is no oxygen present (anaerobic conditions or putrefaction), undesirable tastes and odors are produced. Decay of organic matter when oxygen is present (aerobic conditions) tends to produce much less objectionable tastes and odors.

DEFLUORIDATION (de-FLOOR-uh-DAY-shun) DEFLUORIDATION

The removal of excess fluoride in drinking water to prevent the mottling (brown stains) of teeth.

DEGASIFICATION (DEE-GAS-if-uh-KAY-shun) DEGASIFICATION

A water treatment process which removes dissolved gases from the water. The gases may be removed by either mechanical or chemical treatment methods or a combination of both.

DELEGATION DELEGATION

The act in which power is given to another person in the organization to accomplish a specific job.

DEMINERALIZATION (DEE-MIN-er-al-uh-ZAY-shun) DEMINERALIZATION

A treatment process which removes dissolved minerals (salts) from water.

DENSITY (DEN-sit-tee) DENSITY

A measure of how heavy a substance (solid, liquid or gas) is for its size. Density is expressed in terms of weight per unit volume, that is, grams per cubic centimeter or pounds per cubic foot. The density of water (at 4°C or 39°F) is 1.0 gram per cubic centimeter or about 62.4 pounds per cubic foot.

DEPOLARIZATION DEPOLARIZATION

The removal or depletion of ions in the thin boundary layer adjacent to a membrane or pipe wall.

DEPRECIATION DEPRECIATION

The gradual loss in service value of a facility or piece of equipment due to all the factors causing the ultimate retirement of the facility or equipment. This loss can be caused by sudden physical damage, wearing out due to age, obsolescence, inadequacy or availability of a newer, more efficient facility or equipment. The value cannot be restored by maintenance.

DESALINIZATION (DEE-SAY-leen-uh-ZAY-shun) DESALINIZATION

The removal of dissolved salts (such as sodium chloride, NaCl) from water by natural means (leaching) or by specific water treatment processes.

DESICCANT (DESS-uh-kant) DESICCANT

A drying agent which is capable of removing or absorbing moisture from the atmosphere in a small enclosure.

DESICCATION (DESS-uh-KAY-shun) DESICCATION

A process used to thoroughly dry air; to remove virtually all moisture from air.

DESICCATOR (DESS-uh-KAY-tor) DESICCATOR

A closed container into which heated weighing or drying dishes are placed to cool in a dry environment in preparation for weighing. The dishes may be empty or they may contain a sample. Desiccators contain a substance, such as anhydrous calcium chloride, which absorbs moisture and keeps the relative humidity near zero so that the dish or sample will not gain weight from absorbed moisture.

DESTRATIFICATION (de-STRAT-uh-fuh-KAY-shun) DESTRATIFICATION

The development of vertical mixing within a lake or reservoir to eliminate (either totally or partially) separate layers of temperature, plant, or animal life. This vertical mixing can be caused by mechanical means (pumps) or through the use of forced air diffusers which release air into the lower layers of the reservoir.

DETECTION LAG DETECTION LAG

The time period between the moment a process change is made and the moment when such a change is finally sensed by the associated measuring instrument.

DETENTION TIME DETENTION TIME

(1) The theoretical (calculated) time required for a small amount of water to pass through a tank at a given rate of flow.

(2) The actual time in hours, minutes or seconds that a small amount of water is in a settling basin, flocculating basin or rapid-mix chamber. In storage reservoirs, detention time is the length of time entering water will be held before being drafted for use (several weeks to years, several months being typical).

$$\text{Detention Time, hr} = \frac{(\text{Basin Volume, gal})(24\text{ hr/day})}{\text{Flow, gal/day}}$$

DEW POINT DEW POINT

The temperature to which air with a given quantity of water vapor must be cooled to cause condensation of the vapor in the air.

DEWATER DEWATER

(1) To remove or separate a portion of the water present in a sludge or slurry. To dry sludge so it can be handled and disposed of.

(2) To remove or drain the water from a tank or a trench.

DIATOMACEOUS (DYE-uh-toe-MAY-shus) EARTH DIATOMACEOUS EARTH

A fine, siliceous (made of silica) "earth" composed mainly of the skeletal remains of diatoms.

DIATOMS (DYE-uh-toms) DIATOMS

Unicellular (single cell), microscopic algae with a rigid (box-like) internal structure consisting mainly of silica.

DIELECTRIC (DIE-ee-LECK-trick) DIELECTRIC

Does not conduct an electric current. An insulator or nonconducting substance.

DIGITAL READOUT DIGITAL READOUT

Use of numbers to indicate the value or measurement of a variable. The readout of an instrument by a direct, numerical reading of the measured value. The signal sent to such readouts is usually an analog signal.

DILUTE SOLUTION DILUTE SOLUTION

A solution that has been made weaker usually by the addition of water.

DIMICTIC (die-MICK-tick) DIMICTIC

Lakes and reservoirs which freeze over and normally go through two stratification and two mixing cycles within a year.

DIRECT CURRENT (D.C.) DIRECT CURRENT (D.C.)

Electric current flowing in one direction only and essentially free from pulsation.

DIRECT FILTRATION DIRECT FILTRATION

A method of treating water which consists of the addition of coagulant chemicals, flash mixing, coagulation, minimal flocculation, and filtration. The flocculation facilities may be omitted, but the physical-chemical reactions will occur to some extent. The sedimentation process is omitted. Also see CONVENTIONAL FILTRATION and IN-LINE FILTRATION.

DIRECT RUNOFF DIRECT RUNOFF

Water that flows over the ground surface or through the ground directly into streams, rivers, or lakes.

DISCHARGE HEAD DISCHARGE HEAD

The pressure (in pounds per square inch or psi) measured at the centerline of a pump discharge and very close to the discharge flange, converted into feet. The pressure is measured from the centerline of the pump to the hydraulic grade line of the water in the discharge pipe.

Discharge Head, ft = (Discharge Pressure, psi)(2.31 ft/psi)

DISINFECTION (dis-in-FECT-shun) DISINFECTION

The process designed to kill or inactivate most microorganisms in water, including essentially all pathogenic (disease-causing) bacteria. There are several ways to disinfect, with chlorination being the most frequently used in water treatment. Compare with STERILIZATION.

DISINFECTION BY-PRODUCT (DBP) DISINFECTION BY-PRODUCT (DBP)

A contaminant formed by the reaction of disinfection chemicals (such as chlorine) with other substances in the water being disinfected.

DISTILLATE (DIS-tuh-late) DISTILLATE

In the distillation of a sample, a portion is evaporated; the part that is condensed afterward is the distillate.

DIVALENT (die-VAY-lent) DIVALENT

Having a valence of two, such as the ferrous ion, Fe^{2+}. Also called bivalent.

DIVERSION DIVERSION

Use of part of a stream flow as a water supply.

DRAFT DRAFT

(1) The act of drawing or removing water from a tank or reservoir.

(2) The water which is drawn or removed from a tank or reservoir.

DRAWDOWN DRAWDOWN

(1) The drop in the water table or level of water in the ground when water is being pumped from a well.

(2) The amount of water used from a tank or reservoir.

(3) The drop in the water level of a tank or reservoir.

DRIFT DRIFT

The difference between the actual value and the desired value (or set point); characteristic of proportional controllers that do not incorporate reset action. Also called OFFSET.

DYNAMIC PRESSURE DYNAMIC PRESSURE

When a pump is operating, the vertical distance (in feet) from a reference point (such as a pump centerline) to the hydraulic grade line is the dynamic head. Also see ENERGY GRADE LINE, STATIC HEAD, STATIC PRESSURE, and TOTAL DYNAMIC HEAD.

$$\text{Dynamic Pressure, psi} = (\text{Dynamic Head, ft})(0.433\ \text{psi/ft})$$

E

EPA EPA

United States **E**nvironmental **P**rotection **A**gency. A regulatory agency established by the U.S. Congress to administer the nation's environmental laws. Also called the U.S. EPA.

EDUCTOR (e-DUCK-ter) EDUCTOR

A hydraulic device used to create a negative pressure (suction) by forcing a liquid through a restriction, such as a Venturi. An eductor or aspirator (the hydraulic device) may be used in the laboratory in place of a vacuum pump. As an injector, it is used to produce vacuum for chlorinators. Sometimes used instead of a suction pump.

EFFECTIVE RANGE EFFECTIVE RANGE

That portion of the design range (usually from 10 to 90+) in which an instrument has acceptable accuracy. Also see RANGE and SPAN.

EFFECTIVE SIZE (E.S.) EFFECTIVE SIZE (E.S.)

The diameter of the particles in a granular sample (filter media) for which 10 percent of the total grains are smaller and 90 percent larger on a weight basis. Effective size is obtained by passing granular material through sieves with varying dimensions of mesh and weighing the material retained by each sieve. The effective size is also approximately the average size of the grains.

EFFLUENT EFFLUENT

Water or other liquid—raw (untreated), partially or completely treated—flowing *FROM* a reservoir, basin, treatment process, or treatment plant.

EJECTOR EJECTOR

A device used to disperse a chemical solution into water being treated.

ELECTROCHEMICAL REACTION ELECTROCHEMICAL REACTION

Chemical changes produced by electricity (electrolysis) or the production of electricity by chemical changes (galvanic action). In corrosion, a chemical reaction is accompanied by the flow of electrons through a metallic path. The electron flow may come from an external source and cause the reaction, such as electrolysis caused by a D.C. (direct current) electric railway or the electron flow may be caused by a chemical reaction as in the galvanic action of a flashlight dry cell.

ELECTROCHEMICAL SERIES ELECTROCHEMICAL SERIES

A list of metals with the standard electrode potentials given in volts. The size and sign of the electrode potential indicates how easily these elements will take on or give up electrons, or corrode. Hydrogen is conventionally assigned a value of zero.

ELECTROLYSIS (ee-leck-TRAWL-uh-sis) ELECTROLYSIS

The decomposition of material by an outside electric current.

ELECTROLYTE (ee-LECK-tro-LITE) ELECTROLYTE

A substance which dissociates (separates) into two or more ions when it is dissolved in water.

ELECTROLYTIC (ee-LECK-tro-LIT-ick) CELL ELECTROLYTIC CELL

A device in which the chemical decomposition of material causes an electric current to flow. Also, a device in which a chemical reaction occurs as a result of the flow of electric current. Chlorine and caustic (NaOH) are made from salt (NaCl) in electrolytic cells.

ELECTROMOTIVE FORCE (E.M.F.) ELECTROMOTIVE FORCE (E.M.F.)

The electrical pressure available to cause a flow of current (amperage) when an electric circuit is closed. Also called VOLTAGE.

ELECTROMOTIVE SERIES ELECTROMOTIVE SERIES

A list of metals and alloys presented in the order of their tendency to corrode (or go into solution). Also called the GALVANIC SERIES. This is a practical application of the theoretical ELECTROCHEMICAL SERIES.

ELECTRON ELECTRON

(1) A very small, negatively charged particle which is practically weightless. According to the electron theory, all electrical and electronic effects are caused either by the movement of electrons from place to place or because there is an excess or lack of electrons at a particular place.

(2) The part of an atom that determines its chemical properties.

ELEMENT ELEMENT

A substance which cannot be separated into its constituent parts and still retain its chemical identity. For example, sodium (Na) is an element.

END BELLS END BELLS

Devices used to hold the rotor and stator of a motor in position.

END POINT END POINT

Samples are titrated to the end point. This means that a chemical is added, drop by drop, to a sample until a certain color change (blue to clear, for example) occurs. This is called the *END POINT* of the titration. In addition to a color change, an end point may be reached by the formation of a precipitate or the reaching of a specified pH. An end point may be detected by the use of an electronic device such as a pH meter. The completion of a desired chemical reaction.

ENDEMIC (en-DEM-ick) ENDEMIC

Something peculiar to a particular people or locality, such as a disease which is always present in the population.

ENDRIN (EN-drin) ENDRIN

A pesticide toxic to freshwater and marine aquatic life that produces adverse health effects in domestic water supplies.

ENERGY GRADE LINE (EGL) ENERGY GRADE LINE (EGL)

A line that represents the elevation of energy head (in feet) of water flowing in a pipe, conduit or channel. The line is drawn above the hydraulic grade line (gradient) a distance equal to the velocity head ($V^2/2g$) of the water flowing at each section or point along the pipe or channel. Also see HYDRAULIC GRADE LINE.

[SEE DRAWING ON PAGE 583]

ENTERIC ENTERIC

Of intestinal origin, especially applied to wastes or bacteria.

ENTRAIN ENTRAIN

To trap bubbles in water either mechanically through turbulence or chemically through a reaction.

ENZYMES (EN-zimes) ENZYMES

Organic substances (produced by living organisms) which cause or speed up chemical reactions. Organic catalysts and/or biochemical catalysts.

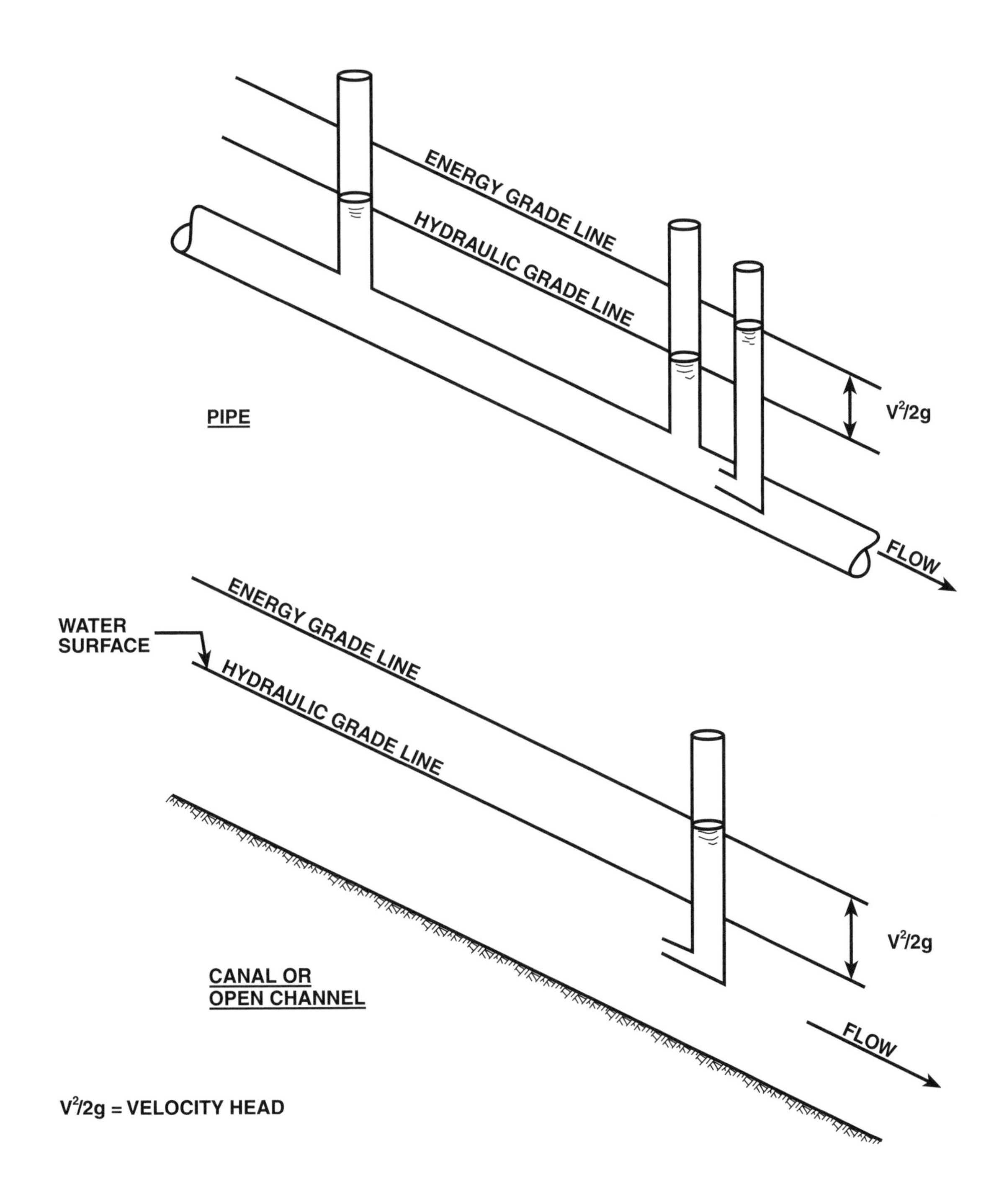

ENERGY GRADE LINE and HYDRAULIC GRADE LINE

EPIDEMIC (EP-uh-DEM-ick) EPIDEMIC

A disease that occurs in a large number of people in a locality at the same time and spreads from person to person.

EPIDEMIOLOGY (EP-uh-DE-me-ALL-o-gee) EPIDEMIOLOGY

A branch of medicine which studies epidemics (diseases which affect significant numbers of people during the same time period in the same locality). The objective of epidemiology is to determine the factors that cause epidemic diseases and how to prevent them.

EPILIMNION (EP-uh-LIM-knee-on) EPILIMNION

The upper layer of water in a thermally stratified lake or reservoir. This layer consists of the warmest water and has a fairly uniform (constant) temperature. The layer is readily mixed by wind action.

EQUILIBRIUM, CALCIUM CARBONATE EQUILIBRIUM, CALCIUM CARBONATE

A water is considered stable when it is just saturated with calcium carbonate. In this condition the water will neither dissolve nor deposit calcium carbonate. Thus, in this water the calcium carbonate is in equilibrium with the hydrogen ion concentration.

EQUITY EQUITY

The value of an investment in a facility.

EQUIVALENT WEIGHT EQUIVALENT WEIGHT

That weight which will react with, displace or is equivalent to one gram atom of hydrogen.

ESTER ESTER

A compound formed by the reaction between an acid and an alcohol with the elimination of a molecule of water.

EUTROPHIC (you-TRO-fick) EUTROPHIC

Reservoirs and lakes which are rich in nutrients and very productive in terms of aquatic animal and plant life.

EUTROPHICATION (you-TRO-fi-KAY-shun) EUTROPHICATION

The increase in the nutrient levels of a lake or other body of water; this usually causes an increase in the growth of aquatic animal and plant life.

EVAPORATION EVAPORATION

The process by which water or other liquid becomes a gas (water vapor or ammonia vapor).

EVAPOTRANSPIRATION (ee-VAP-o-TRANS-purr-A-shun) EVAPOTRANSPIRATION

(1) The process by which water vapor passes into the atmosphere from living plants. Also called TRANSPIRATION.

(2) The total water removed from an area by transpiration (plants) and by evaporation from soil, snow and water surfaces.

F

FACULTATIVE (FACK-ul-TAY-tive) FACULTATIVE

Facultative bacteria can use either dissolved molecular oxygen or oxygen obtained from food materials such as sulfate or nitrate ions. In other words, facultative bacteria can live under aerobic or anaerobic conditions.

FEEDBACK FEEDBACK

The circulating action between a sensor measuring a process variable and the controller which controls or adjusts the process variable.

FEEDWATER FEEDWATER

The water that is fed to a treatment process; the water that is going to be treated.

FINISHED WATER FINISHED WATER

Water that has passed through a water treatment plant; all the treatment processes are completed or “finished.” This water is ready to be delivered to consumers. Also called PRODUCT WATER.

FIX, SAMPLE — FIX, SAMPLE

A sample is "fixed" in the field by adding chemicals that prevent the water quality indicators of interest in the sample from changing before final measurements are performed later in the lab.

FIXED COSTS — FIXED COSTS

Costs that a utility must cover or pay even if there is no demand for water or no water to sell to customers. Also see VARIABLE COSTS.

FLAGELLATES (FLAJ-el-LATES) — FLAGELLATES

Microorganisms that move by the action of tail-like projections.

FLAME POLISHED — FLAME POLISHED

Melted by a flame to smooth out irregularities. Sharp or broken edges of glass (such as the end of a glass tube) are rotated in a flame until the edge melts slightly and becomes smooth.

FLOAT ON SYSTEM — FLOAT ON SYSTEM

A method of operating a water storage facility. Daily flow into the facility is approximately equal to the average daily demand for water. When consumer demands for water are low, the storage facility will be filling. During periods of high demand, the facility will be emptying.

FLOC — FLOC

Clumps of bacteria and particulate impurities that have come together and formed a cluster. Found in flocculation tanks and settling or sedimentation basins.

FLOCCULATION (FLOCK-you-LAY-shun) — FLOCCULATION

The gathering together of fine particles after coagulation to form larger particles by a process of gentle mixing.

FLUIDIZED (FLEW-id-I-zd) — FLUIDIZED

A mass of solid particles that is made to flow like a liquid by injection of water or gas is said to have been fluidized. In water treatment, a bed of filter media is fluidized by backwashing water through the filter.

FLUORIDATION (FLOOR-uh-DAY-shun) — FLUORIDATION

The addition of a chemical to increase the concentration of fluoride ions in drinking water to a predetermined optimum limit to reduce the incidence (number) of dental caries (tooth decay) in children. Defluoridation is the removal of excess fluoride in drinking water to prevent the mottling (brown stains) of teeth.

FLUSHING — FLUSHING

A method used to clean water distribution lines. Hydrants are opened and water with a high velocity flows through the pipes, removes deposits from the pipes, and flows out the hydrants.

FLUX — FLUX

A flowing or flow.

FOOT VALVE — FOOT VALVE

A special type of check valve located at the bottom end of the suction pipe on a pump. This valve opens when the pump operates to allow water to enter the suction pipe but closes when the pump shuts off to prevent water from flowing out of the suction pipe.

FREE AVAILABLE RESIDUAL CHLORINE — FREE AVAILABLE RESIDUAL CHLORINE

That portion of the total available residual chlorine composed of dissolved chlorine gas (Cl_2), hypochlorous acid (HOCl), and/or hypochlorite ion (OCl^-) remaining in water after chlorination. This does not include chlorine that has combined with ammonia, nitrogen, or other compounds.

FREE RESIDUAL CHLORINATION — FREE RESIDUAL CHLORINATION

The application of chlorine to water to produce a free available chlorine residual equal to at least 80 percent of the total residual chlorine (sum of free and combined available chlorine residual).

FREEBOARD

FREEBOARD

(1) The vertical distance from the normal water surface to the top of the confining wall.

(2) The vertical distance from the sand surface to the underside of a trough in a sand filter. This distance is also called AVAILABLE EXPANSION.

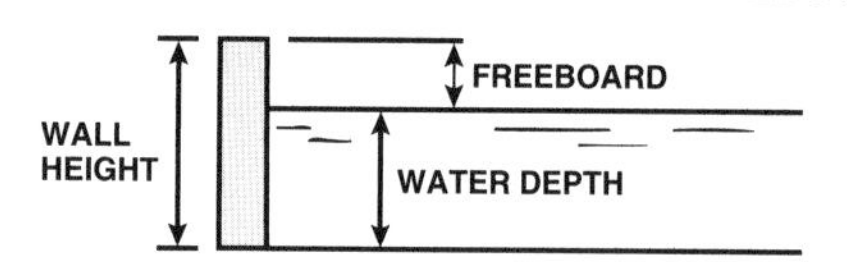

FRICTION LOSSES

FRICTION LOSSES

The head, pressure or energy (they are the same) lost by water flowing in a pipe or channel as a result of turbulence caused by the velocity of the flowing water and the roughness of the pipe, channel walls, or restrictions caused by fittings. Water flowing in a pipe loses head, pressure or energy as a result of friction losses. Also see HEAD LOSS.

FUNGI (FUN-ji)

FUNGI

Mushrooms, molds, mildews, rusts, and smuts that are small non-chlorophyll-bearing plants lacking roots, stems and leaves. They occur in natural waters and grow best in the absence of light. Their decomposition may cause objectionable tastes and odors in water.

FUSE

FUSE

A protective device having a strip or wire of fusible metal which, when placed in a circuit, will melt and break the electric circuit if heated too much. High temperatures will develop in the fuse when a current flows through the fuse in excess of that which the circuit will carry safely.

G

GIS

GIS

Geographic **I**nformation **S**ystem. A computer program that combines mapping with detailed information about the physical locations of structures such as pipes, valves, and manholes within geographic areas. The system is used to help operators and maintenance personnel locate utility system features or structures and to assist with the scheduling and performance of maintenance activities.

GAGE PRESSURE

GAGE PRESSURE

The pressure within a closed container or pipe as measured with a gage. In contrast, absolute pressure is the sum of atmospheric pressure (14.7 lbs/sq in) *PLUS* pressure within a vessel (as measured with a gage). Most pressure gages read in "gage pressure" or psig (**p**ounds per **s**quare **i**nch **g**age pressure).

GALVANIC CELL

GALVANIC CELL

An electrolytic cell capable of producing electric energy by electrochemical action. The decomposition of materials in the cell causes an electric (electron) current to flow from cathode to anode.

GALVANIC SERIES

GALVANIC SERIES

A list of metals and alloys presented in the order of their tendency to corrode (or go into solution). Also called the ELECTROMOTIVE SERIES. This is a practical application of the theoretical ELECTROCHEMICAL SERIES.

GALVANIZE

GALVANIZE

To coat a metal (especially iron or steel) with zinc. Galvanization is the process of coating a metal with zinc.

GARNET

GARNET

A group of hard, reddish, glassy, mineral sands made up of silicates of base metals (calcium, magnesium, iron and manganese). Garnet has a higher density than sand.

GAUGE, PIPE

GAUGE, PIPE

A number that defines the thickness of the sheet used to make steel pipe. The larger the number, the thinner the pipe wall.

GEOGRAPHIC INFORMATION SYSTEM (GIS)

GEOGRAPHIC INFORMATION SYSTEM (GIS)

A computer program that combines mapping with detailed information about the physical locations of structures such as pipes, valves, and manholes within geographic areas. The system is used to help operators and maintenance personnel locate utility system features or structures and to assist with the scheduling and performance of maintenance activities.

GEOLOGICAL LOG

GEOLOGICAL LOG

A detailed description of all underground features discovered during the drilling of a well (depth, thickness and type of formations).

GEOPHYSICAL LOG GEOPHYSICAL LOG

A record of the structure and composition of the earth encountered when drilling a well or similar type of test hole or boring.

GERMICIDE (GERM-uh-SIDE) GERMICIDE

A substance formulated to kill germs or microorganisms. The germicidal properties of chlorine make it an effective disinfectant.

GIARDIA (GEE-ARE-dee-ah) GIARDIA

A waterborne intestinal parasite that causes a disease called giardiasis (GEE-are-DIE-uh-sis) in infected humans. Symptoms of the disease include diarrhea, cramps, and weight loss. *Giardia* contamination is found in most surface waters and some groundwaters.

GIARDIASIS (GEE-are-DIE-uh-sis) GIARDIASIS

Intestinal disease caused by an infestation of *Giardia* flagellates.

GRAB SAMPLE GRAB SAMPLE

A single sample of water collected at a particular time and place which represents the composition of the water only at that time and place.

GRADE GRADE

(1) The elevation of the invert (or bottom) of a pipeline, canal, culvert, or similar conduit.

(2) The inclination or slope of a pipeline, conduit, stream channel, or natural ground surface; usually expressed in terms of the ratio or percentage of number of units of vertical rise or fall per unit of horizontal distance. A 0.5 percent grade would be a drop of one-half foot per hundred feet of pipe.

GRAVIMETRIC GRAVIMETRIC

A means of measuring unknown concentrations of water quality indicators in a sample by *WEIGHING* a precipitate or residue of the sample.

GRAVIMETRIC FEEDER GRAVIMETRIC FEEDER

A dry chemical feeder which delivers a measured weight of chemical during a specific time period.

GREENSAND GREENSAND

A mineral (glauconite) material that looks like ordinary filter sand except that it is green in color. Greensand is a natural ion exchange material which is capable of softening water. Greensand which has been treated with potassium permanganate ($KMnO_4$) is called manganese greensand; this product is used to remove iron, manganese and hydrogen sulfide from groundwaters.

GROUND GROUND

An expression representing an electrical connection to earth or a large conductor which is at the earth's potential or neutral voltage.

H

HTH (pronounce as separate letters) HTH

High **T**est **H**ypochlorite. Calcium hypochlorite or $Ca(OCl)_2$.

HARD WATER HARD WATER

Water having a high concentration of calcium and magnesium ions. A water may be considered hard if it has a hardness greater than the typical hardness of water from the region. Some textbooks define hard water as water with a hardness of more than 100 mg/*L* as calcium carbonate.

HARDNESS, WATER HARDNESS, WATER

A characteristic of water caused mainly by the salts of calcium and magnesium, such as bicarbonate, carbonate, sulfate, chloride and nitrate. Excessive hardness in water is undesirable because it causes the formation of soap curds, increased use of soap, deposition of scale in boilers, damage in some industrial processes, and sometimes causes objectionable tastes in drinking water.

HEAD HEAD

The vertical distance (in feet) equal to the pressure (in psi) at a specific point. The pressure head is equal to the pressure in psi times 2.31 ft/psi.

HEAD LOSS

HEAD LOSS

The head, pressure or energy (they are the same) lost by water flowing in a pipe or channel as a result of turbulence caused by the velocity of the flowing water and the roughness of the pipe, channel walls, or restrictions caused by fittings. Water flowing in a pipe loses head, pressure or energy as a result of friction losses. The head loss through a filter is due to friction losses caused by material building up on the surface or in the top part of a filter. Also see FRICTION LOSSES.

HEADER

HEADER

A large pipe to which the ends of a series of smaller pipes are connected. Also called a MANIFOLD.

HEAT SENSOR

HEAT SENSOR

A device that opens and closes a switch in response to changes in the temperature. This device might be a metal contact, or a thermocouple which generates a minute electric current proportional to the difference in heat, or a variable resistor whose value changes in response to changes in temperature. Also called a TEMPERATURE SENSOR.

HECTARE (HECK-tar)

HECTARE

A measure of area in the metric system similar to an acre. One hectare is equal to 10,000 square meters and 2.4711 acres.

HEPATITIS (HEP-uh-TIE-tis)

HEPATITIS

Hepatitis is an inflammation of the liver usually caused by an acute viral infection. Yellow jaundice is one symptom of hepatitis.

HERBICIDE (HERB-uh-SIDE)

HERBICIDE

A compound, usually a manmade organic chemical, used to kill or control plant growth.

HERTZ

HERTZ

The number of complete electromagnetic cycles or waves in one second of an electric or electronic circuit. Also called the frequency of the current. Abbreviated Hz.

HETEROTROPHIC (HET-er-o-TROF-ick)

HETEROTROPHIC

Describes organisms that use organic matter for energy and growth. Animals, fungi and most bacteria are heterotrophs.

HIGH-LINE JUMPERS

HIGH-LINE JUMPERS

Pipes or hoses connected to fire hydrants and laid on top of the ground to provide emergency water service for an isolated portion of a distribution system.

HOSE BIB

HOSE BIB

Faucet. A location in a water line where a hose is connected.

HYDRATED LIME

HYDRATED LIME

Limestone that has been "burned" and treated with water under controlled conditions until the calcium oxide portion has been converted to calcium hydroxide ($Ca(OH)_2$). Hydrated lime is quicklime combined with water. $CaO + H_2O \rightarrow Ca(OH)_2$. Also called slaked lime. Also see QUICKLIME.

HYDRAULIC CONDUCTIVITY (K)

HYDRAULIC CONDUCTIVITY (K)

A coefficient describing the relative ease with which groundwater can move through a permeable layer of rock or soil. Typical units of hydraulic conductivity are feet per day, gallons per day per square foot, or meters per day (depending on the unit chosen for the total discharge and the cross-sectional area).

HYDRAULIC GRADE LINE (HGL)

HYDRAULIC GRADE LINE (HGL)

The surface or profile of water flowing in an open channel or a pipe flowing partially full. If a pipe is under pressure, the hydraulic grade line is at the level water would rise to in a small vertical tube connected to the pipe. Also see ENERGY GRADE LINE.

[SEE DRAWING ON PAGE 583]

HYDRAULIC GRADIENT

HYDRAULIC GRADIENT

The slope of the hydraulic grade line. This is the slope of the water surface in an open channel, the slope of the water surface of the groundwater table, or the slope of the water pressure for pipes under pressure.

HYDROGEOLOGIST (HI-dro-gee-ALL-uh-gist)

HYDROGEOLOGIST

A person who studies and works with groundwater.

HYDROLOGIC (HI-dro-LOJ-ick) CYCLE — HYDROLOGIC CYCLE

The process of evaporation of water into the air and its return to earth by precipitation (rain or snow). This process also includes transpiration from plants, groundwater movement, and runoff into rivers, streams and the ocean. Also called the WATER CYCLE.

HYDROLYSIS (hi-DROLL-uh-sis) — HYDROLYSIS

(1) A chemical reaction in which a compound is converted into another compound by taking up water.

(2) Usually a chemical degradation of organic matter.

HYDROPHILIC (HI-dro-FILL-ick) — HYDROPHILIC

Having a strong affinity (liking) for water. The opposite of HYDROPHOBIC.

HYDROPHOBIC (HI-dro-FOE-bick) — HYDROPHOBIC

Having a strong aversion (dislike) for water. The opposite of HYDROPHILIC.

HYDROPNEUMATIC (HI-dro-new-MAT-ick) — HYDROPNEUMATIC

A water system, usually small, in which a water pump is automatically controlled (started and stopped) by the air pressure in a compressed-air tank.

HYDROSTATIC (HI-dro-STAT-ick) PRESSURE — HYDROSTATIC PRESSURE

(1) The pressure at a specific elevation exerted by a body of water at rest, or

(2) In the case of groundwater, the pressure at a specific elevation due to the weight of water at higher levels in the same zone of saturation.

HYGROSCOPIC (HI-grow-SKOP-ick) — HYGROSCOPIC

Absorbing or attracting moisture from the air.

HYPOCHLORINATION (HI-poe-KLOR-uh-NAY-shun) — HYPOCHLORINATION

The application of hypochlorite compounds to water for the purpose of disinfection.

HYPOCHLORINATORS (HI-poe-KLOR-uh-NAY-tors) — HYPOCHLORINATORS

Chlorine pumps, chemical feed pumps or devices used to dispense chlorine solutions made from hypochlorites such as bleach (sodium hypochlorite) or calcium hypochlorite into the water being treated.

HYPOCHLORITE (HI-poe-KLOR-ite) — HYPOCHLORITE

Chemical compounds containing available chlorine; used for disinfection. They are available as liquids (bleach) or solids (powder, granules, and pellets) in barrels, drums, and cans. Salts of hypochlorous acid.

HYPOLIMNION (HI-poe-LIM-knee-on) — HYPOLIMNION

The lowest layer in a thermally stratified lake or reservoir. This layer consists of colder, more dense water, has a constant temperature and no mixing occurs.

I

ICR — ICR

The **I**nformation **C**ollection **R**ule (ICR) specifies the requirements for monitoring microbial contaminants and disinfection by-products (DBPs) by large public water systems (PWSs). It also requires large PWSs to conduct either bench- or pilot-scale testing of advanced treatment techniques.

IDLH — IDLH

Immediately **D**angerous to **L**ife or **H**ealth. The atmospheric concentration of any toxic, corrosive or asphyxiant substance that poses an immediate threat to life or would cause irreversible or delayed adverse health effects or would interfere with an individual's ability to escape from a dangerous atmosphere.

IMHOFF CONE — IMHOFF CONE

A clear, cone-shaped container marked with graduations. The cone is used to measure the volume of settleable solids in a specific volume (usually one liter) of water.

IMPELLER IMPELLER

A rotating set of vanes in a pump or compressor designed to pump or move water or air.

IMPERMEABLE (im-PURR-me-uh-BULL) IMPERMEABLE

Not easily penetrated. The property of a material or soil that does not allow, or allows only with great difficulty, the movement or passage of water.

INDICATOR (CHEMICAL) INDICATOR (CHEMICAL)

A substance that gives a visible change, usually of color, at a desired point in a chemical reaction, generally at a specified end point.

INDICATOR (INSTRUMENT) INDICATOR (INSTRUMENT)

A device which indicates the result of a measurement. Most indicators in the water utility field use either a fixed scale and movable indicator (pointer) such as a pressure gage or a movable scale and movable indicator like those used on a circular flow-recording chart. Also called a RECEIVER.

INFILTRATION (IN-fill-TRAY-shun) INFILTRATION

The seepage of groundwater into a sewer system, including service connections. Seepage frequently occurs through defective or cracked pipes, pipe joints, connections or manhole walls.

INFLUENT INFLUENT

Water or other liquid—raw (untreated) or partially treated—flowing *INTO* a reservoir, basin, treatment process, or treatment plant.

INFORMATION COLLECTION RULE (ICR) INFORMATION COLLECTION RULE (ICR)

The Information Collection Rule (ICR) specifies the requirements for monitoring microbial contaminants and disinfection by-products (DBPs) by large public water systems (PWSs). It also requires large PWSs to conduct either bench- or pilot-scale testing of advanced treatment techniques.

INITIAL SAMPLING INITIAL SAMPLING

The very first sampling conducted under the Safe Drinking Water Act (SDWA) for each of the applicable contaminant categories.

INJECTOR WATER INJECTOR WATER

Service water in which chlorine is added (injected) to form a chlorine solution.

IN-LINE FILTRATION IN-LINE FILTRATION

The addition of chemical coagulants directly to the filter inlet pipe. The chemicals are mixed by the flowing water. Flocculation and sedimentation facilities are eliminated. This pretreatment method is commonly used in pressure filter installations. Also see CONVENTIONAL FILTRATION and DIRECT FILTRATION.

INORGANIC INORGANIC

Material such as sand, salt, iron, calcium salts and other mineral materials. Inorganic substances are of mineral origin, whereas organic substances are usually of animal or plant origin. Also see ORGANIC.

INORGANIC WASTE INORGANIC WASTE

Waste material such as sand, salt, iron, calcium, and other mineral materials which are only slightly affected by the action of organisms. Inorganic wastes are chemical substances of mineral origin; whereas organic wastes are chemical substances of an animal or plant origin.

INPUT HORSEPOWER INPUT HORSEPOWER

The total power used in operating a pump and motor.

$$\text{Input Horsepower, HP} = \frac{(\text{Brake Horsepower, HP})(100\%)}{\text{Motor Efficiency, \%}}$$

INSECTICIDE INSECTICIDE

Any substance or chemical formulated to kill or control insects.

INSOLUBLE (in-SAWL-you-bull) INSOLUBLE

Something that cannot be dissolved.

INTEGRATOR — INTEGRATOR

A device or meter that continuously measures and calculates (adds) a process rate variable in cumulative fashion; for example, total flows displayed in gallons, million gallons, cubic feet, or some other unit of volume measurement. Also called a TOTALIZER.

INTERFACE — INTERFACE

The common boundary layer between two substances such as water and a solid (metal); or between two fluids such as water and a gas (air); or between a liquid (water) and another liquid (oil).

INTERLOCK — INTERLOCK

An electric switch, usually magnetically operated. Used to interrupt all (local) power to a panel or device when the door is opened or the circuit is exposed to service.

INTERNAL FRICTION — INTERNAL FRICTION

Friction within a fluid (water) due to cohesive forces.

INTERSTICE (in-TUR-stuhz) — INTERSTICE

A very small open space in a rock or granular material. Also called a PORE, VOID, or void space. Also see VOID.

INVERT (IN-vert) — INVERT

The lowest point of the channel inside a pipe, conduit, or canal.

ION — ION

An electrically charged atom, radical (such as SO_4^{2-}), or molecule formed by the loss or gain of one or more electrons.

ION EXCHANGE — ION EXCHANGE

A water treatment process involving the reversible interchange (switching) of ions between the water being treated and the solid resin. Undesirable ions in the water are switched with acceptable ions on the resin.

ION EXCHANGE RESINS — ION EXCHANGE RESINS

Insoluble polymers, used in water treatment, that are capable of exchanging (switching or giving) acceptable cations or anions to the water being treated for less desirable ions.

IONIC CONCENTRATION — IONIC CONCENTRATION

The concentration of any ion in solution, usually expressed in moles per liter.

IONIZATION (EYE-on-uh-ZAY-shun) — IONIZATION

The splitting or dissociation (separation) of molecules into negatively and positively charged ions.

J

JAR TEST — JAR TEST

A laboratory procedure that simulates a water treatment plant's coagulation/flocculation units with differing chemical doses and also energy of rapid mix, energy of slow mix, and settling time. The purpose of this procedure is to *ESTIMATE* the minimum or ideal coagulant dose required to achieve certain water quality goals. Samples of water to be treated are commonly placed in six jars. Various amounts of chemicals are added to each jar, stirred and the settling of solids is observed. The dose of chemicals that provides satisfactory settling, removal of turbidity and/or color is the dose used to treat the water being taken into the plant at that time. When evaluating the results of a jar test, the operator should also consider the floc quality in the flocculation area and the floc loading on the filter.

JOGGING — JOGGING

The frequent starting and stopping of an electric motor.

JOULE (jewel) — JOULE

A measure of energy, work or quantity of heat. One joule is the work done when the point of application of a force of one newton is displaced a distance of one meter in the direction of the force. Approximately equal to 0.7375 ft-lbs (0.1022 m-kg).

K

KELLY — KELLY

The square section of a rod which causes the rotation of the drill bit. Torque from a drive table is applied to the square rod to cause the rotary motion. The drive table is chain or gear driven by an engine.

KILO — KILO

(1) Kilogram.

(2) Kilometer.

(3) A prefix meaning "thousand" used in the metric system and other scientific systems of measurement.

KINETIC ENERGY — KINETIC ENERGY

Energy possessed by a moving body of matter, such as water, as a result of its motion.

KJELDAHL (KELL-doll) NITROGEN — KJELDAHL NITROGEN

Nitrogen in the form of organic proteins or their decomposition product ammonia, as measured by the Kjeldahl Method.

L

LANGELIER INDEX (L.I.) — LANGELIER INDEX (L.I.)

An index reflecting the equilibrium pH of a water with respect to calcium and alkalinity. This index is used in stabilizing water to control both corrosion and the deposition of scale.

$$\text{Langelier Index} = pH - pH_S$$

where pH = actual pH of the water, and

pH_S = pH at which water having the same alkalinity and calcium content is just saturated with calcium carbonate.

LAUNDERING WEIR (LAWN-der-ing weer) — LAUNDERING WEIR

Sedimentation basin overflow weir. A plate with V-notches along the top to ensure a uniform flow rate and avoid short-circuiting.

LAUNDERS (LAWN-ders) — LAUNDERS

Sedimentation basin and filter discharge channels consisting of overflow weir plates (in sedimentation basins) and conveying troughs.

LEAD (LEE-d) — LEAD

A wire or conductor that can carry electric current.

LEATHERS — LEATHERS

O-rings or gaskets used with piston pumps to provide a seal between the piston and the side wall.

LEVEL CONTROL — LEVEL CONTROL

A float device (or pressure switch) which senses changes in a measured variable and opens or closes a switch in response to that change. In its simplest form, this control might be a floating ball connected mechanically to a switch or valve such as is used to stop water flow into a toilet when the tank is full.

LINDANE (LYNN-dane) — LINDANE

A pesticide that causes adverse health effects in domestic water supplies and also is toxic to freshwater and marine aquatic life.

LINEARITY (LYNN-ee-AIR-it-ee) — LINEARITY

How closely an instrument measures actual values of a variable through its effective range; a measure used to determine the accuracy of an instrument.

LITTORAL (LIT-or-al) ZONE — LITTORAL ZONE

(1) That portion of a body of fresh water extending from the shoreline lakeward to the limit of occupancy of rooted plants.

(2) The strip of land along the shoreline between the high and low water levels.

LOGARITHM (LOG-a-rith-m) LOGARITHM

The exponent that indicates the power to which a number must be raised to produce a given number. For example: if $B^2 = N$, the 2 is the logarithm of N (to the base B), or $10^2 = 100$ and $\log_{10} 100 = 2$. Also abbreviated to "log."

LOGGING, ELECTRICAL LOGGING, ELECTRICAL

A procedure used to determine the porosity (spaces or voids) of formations in search of water-bearing formations (aquifers). Electrical probes are lowered into wells, an electric current is induced at various depths, and the resistance measured of various formations indicates the porosity of the material.

LOWER EXPLOSIVE LIMIT (LEL) LOWER EXPLOSIVE LIMIT (LEL)

The lowest concentration of gas or vapor (percent by volume in air) that explodes if an ignition source is present at ambient temperature. At temperatures above 250°F the LEL decreases because explosibility increases with higher temperature.

M

M or MOLAR ***M*** or MOLAR

A molar solution consists of one gram molecular weight of a compound dissolved in enough water to make one liter of solution. A gram molecular weight is the molecular weight of a compound in grams. For example, the molecular weight of sulfuric acid (H_2SO_4) is 98. A one *M* solution of sulfuric acid would consist of 98 grams of H_2SO_4 dissolved in enough distilled water to make one liter of solution.

MBAS MBAS

Methylene - **B**lue - **A**ctive **S**ubstances. These substances are used in surfactants or detergents.

MCL MCL

Maximum **C**ontaminant **L**evel. The largest allowable amount. MCLs for various water quality indicators are specified in the National Primary Drinking Water Regulations (NPDWR).

MCLG MCLG

Maximum **C**ontaminant **L**evel **G**oal. MCLGs are health goals based entirely on health effects. They are a preliminary standard set but not enforced by EPA. MCLs consider health effects, but also take into consideration the feasibility and cost of analysis and treatment of the regulated MCL. Although often less stringent than the corresponding MCLG, the MCL is set to protect health.

mg/*L* mg/*L*

See MILLIGRAMS PER LITER, mg/*L*.

MPN (pronounce as separate letters) MPN

MPN is the **M**ost **P**robable **N**umber of coliform-group organisms per unit volume of sample water. Expressed as a density or population of organisms per 100 m*L* of sample water.

MSDS MSDS

See **M**ATERIAL **S**AFETY **D**ATA **S**HEET.

MACROSCOPIC (MACK-row-SKAWP-ick) ORGANISMS MACROSCOPIC ORGANISMS

Organisms big enough to be seen by the eye without the aid of a microscope.

MANDREL (MAN-drill) MANDREL

A special tool used to push bearings in or to pull sleeves out.

MANIFOLD MANIFOLD

A large pipe to which the ends of a series of smaller pipes are connected. Also called a HEADER.

MANOMETER (man-NAH-mut-ter) MANOMETER

An instrument for measuring pressure. Usually, a manometer is a glass tube filled with a liquid that is used to measure the difference in pressure across a flow measuring device such as an orifice or a Venturi meter. The instrument used to measure blood pressure is a type of manometer.

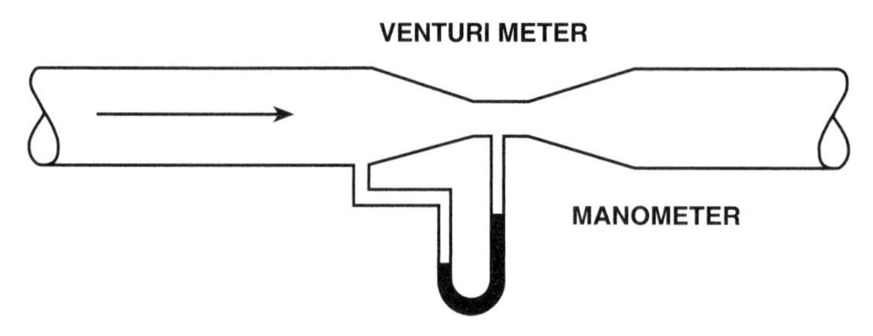

MATERIAL SAFETY DATA SHEET (MSDS) MATERIAL SAFETY DATA SHEET (MSDS)

A document which provides pertinent information and a profile of a particular hazardous substance or mixture. An MSDS is normally developed by the manufacturer or formulator of the hazardous substance or mixture. The MSDS is required to be made available to employees and operators whenever there is the likelihood of the hazardous substance or mixture being introduced into the workplace. Some manufacturers are preparing MSDSs for products that are not considered to be hazardous to show that the product or substance is *NOT* hazardous.

MAXIMUM CONTAMINANT LEVEL (MCL) MAXIMUM CONTAMINANT LEVEL (MCL)

See MCL.

MEASURED VARIABLE MEASURED VARIABLE

A characteristic or component part that is sensed and quantified (reduced to a reading of some kind) by a primary element or sensor.

MECHANICAL JOINT MECHANICAL JOINT

A flexible device that joins pipes or fittings together by the use of lugs and bolts.

MEG MEG

A procedure used for checking the insulation resistance on motors, feeders, bus bar systems, grounds, and branch circuit wiring. Also see MEGGER.

MEGGER (from megohm) MEGGER

An instrument used for checking the insulation resistance on motors, feeders, bus bar systems, grounds, and branch circuit wiring. Also see MEG.

MEGOHM MEGOHM

Meg means one million, so 5 megohms means 5 million ohms. A megger reads in millions of ohms.

MENISCUS (meh-NIS-cuss) MENISCUS

The curved surface of a column of liquid (water, oil, mercury) in a small tube. When the liquid wets the sides of the container (as with water), the curve forms a valley. When the confining sides are not wetted (as with mercury), the curve forms a hill or upward bulge.

MESH MESH

One of the openings or spaces in a screen or woven fabric. The value of the mesh is usually given as the number of openings per inch. This value does not consider the diameter of the wire or fabric; therefore, the mesh number does not always have a definite relationship to the size of the hole.

MESOTROPHIC (MESS-o-TRO-fick) MESOTROPHIC

Reservoirs and lakes which contain moderate quantities of nutrients and are moderately productive in terms of aquatic animal and plant life.

METABOLISM (meh-TAB-uh-LIZ-um) METABOLISM

(1) The biochemical processes in which food is used and wastes are formed by living organisms.

(2) All biochemical reactions involved in cell formation and growth.

METALIMNION (MET-uh-LIM-knee-on) METALIMNION

The middle layer in a thermally stratified lake or reservoir. In this layer there is a rapid decrease in temperature with depth. Also called the THERMOCLINE.

METHOXYCHLOR (meth-OXY-klor) METHOXYCHLOR

A pesticide which causes adverse health effects in domestic water supplies and is also toxic to freshwater and marine aquatic life. The chemical name for methoxychlor is 2,2-bis(p-methoxyphenol)-1,1,1-trichloroethane.

METHYL ORANGE ALKALINITY METHYL ORANGE ALKALINITY

A measure of the total alkalinity in a water sample. The alkalinity is measured by the amount of standard sulfuric acid required to lower the pH of the water to a pH level of 4.5, as indicated by the change in color of methyl orange from orange to pink. Methyl orange alkalinity is expressed as milligrams per liter equivalent calcium carbonate.

MICROBIAL (my-KROW-bee-ul) GROWTH MICROBIAL GROWTH

The activity and growth of microorganisms such as bacteria, algae, diatoms, plankton and fungi.

MICRON (MY-kron) MICRON

μm, Micrometer or Micron. A unit of length. One millionth of a meter or one thousandth of a millimeter. One micron equals 0.00004 of an inch.

MICROORGANISMS (MY-crow-OR-gan-IS-zums) MICROORGANISMS

Living organisms that can be seen individually only with the aid of a microscope.

MIL MIL

A unit of length equal to 0.001 of an inch. The diameter of wires and tubing is measured in mils, as is the thickness of plastic sheeting.

MILLIGRAMS PER LITER, mg/*L* MILLIGRAMS PER LITER, mg/*L*

A measure of the concentration by weight of a substance per unit volume. For practical purposes, one mg/*L* of a substance in fresh water is equal to one part per million parts (ppm). Thus a liter of water with a specific gravity of 1.0 weighs one million milligrams. If water contains 10 milligrams of calcium, the concentration is 10 milligrams per million milligrams, or 10 milligrams per liter (10 mg/*L*), or 10 parts of calcium per million parts of water, or 10 parts per million (10 ppm).

MILLIMICRON (MILL-uh-MY-kron) MILLIMICRON

A unit of length equal to $10^{-3}\mu$ (one thousandth of a micron), 10^{-6} millimeters, or 10^{-9} meters; correctly called a nanometer, nm.

MOLAR MOLAR

See ***M*** for MOLAR.

MOLE MOLE

The molecular weight of a substance, usually expressed in grams.

MOLECULAR WEIGHT MOLECULAR WEIGHT

The molecular weight of a compound in grams is the sum of the atomic weights of the elements in the compound. The molecular weight of sulfuric acid (H_2SO_4) in grams is 98.

Element	Atomic Weight	Number of Atoms	Molecular Weight
H	1	2	2
S	32	1	32
O	16	4	64
			98

MOLECULE (MOLL-uh-KULE) MOLECULE

The smallest division of a compound that still retains or exhibits all the properties of the substance.

MONOMER (MON-o-MER) MONOMER

A molecule of low molecular weight capable of reacting with identical or different monomers to form polymers.

MONOMICTIC (mo-no-MICK-tick) MONOMICTIC

Lakes and reservoirs which are relatively deep, do not freeze over during the winter months, and undergo a single stratification and mixing cycle during the year. These lakes and reservoirs usually become destratified during the mixing cycle, usually in the fall of the year.

MONOVALENT MONOVALENT

Having a valence of one, such as the cuprous (copper) ion, Cu^+.

MOST PROBABLE NUMBER (MPN) MOST PROBABLE NUMBER (MPN)

See MPN.

MOTILE (MO-till) MOTILE

Capable of self-propelled movement. A term that is sometimes used to distinguish between certain types of organisms found in water.

MOTOR EFFICIENCY MOTOR EFFICIENCY

The ratio of energy delivered by a motor to the energy supplied to it during a fixed period or cycle. Motor efficiency ratings will vary depending upon motor manufacturer and usually will be near 90.0 percent.

MUDBALLS MUDBALLS

Material that is approximately round in shape and varies from pea-sized up to two or more inches in diameter. This material forms in filters and gradually increases in size when not removed by the backwashing process.

MULTI-STAGE PUMP MULTI-STAGE PUMP

A pump that has more than one impeller. A single-stage pump has one impeller.

N

N or NORMAL ***N*** or NORMAL

A normal solution contains one gram equivalent weight of reactant (compound) per liter of solution. The equivalent weight of an acid is that weight which contains one gram atom of ionizable hydrogen or its chemical equivalent. For example, the equivalent weight of sulfuric acid (H_2SO_4) is 49 (98 divided by 2 because there are two replaceable hydrogen ions). A one *N* solution of sulfuric acid would consist of 49 grams of H_2SO_4 dissolved in enough water to make one liter of solution.

NETA NETA

See **N**ATIONAL **E**NVIRONMENTAL **T**RAINING **A**SSOCIATION.

NIOSH (NYE-osh) NIOSH

The **N**ational **I**nstitute of **O**ccupational **S**afety and **H**ealth is an organization that tests and approves safety equipment for particular applications. NIOSH is the primary federal agency engaged in research in the national effort to eliminate on-the-job hazards to the health and safety of working people. The NIOSH Publications Catalog, Sixth Edition, NIOSH Pub. No. 84-118, lists the NIOSH publications concerning industrial hygiene and occupational health. To obtain a copy of the catalog, write to National Technical Information Service (NTIS), 5285 Port Royal Road, Springfield, VA 22161. NTIS Stock No. PB-86-116-787, price, $103.50, plus $5.00 shipping and handling per order.

NOM (NATURAL ORGANIC MATTER) NOM (NATURAL ORGANIC MATTER)

Humic substances composed of humic and fulvic acids that come from decayed vegetation.

NPDES PERMIT NPDES PERMIT

National **P**ollutant **D**ischarge **E**limination **S**ystem permit is the regulatory agency document issued by either a federal or state agency which is designed to control all discharges of pollutants from point sources and storm water runoff into U.S. waterways. NPDES permits regulate discharges into navigable waters from all point sources of pollution, including industries, municipal wastewater treatment plants, sanitary landfills, large agricultural feedlots and return irrigation flows.

NPDWR NPDWR

National **P**rimary **D**rinking **W**ater **R**egulations.

NSDWR NSDWR

National **S**econdary **D**rinking **W**ater **R**egulations.

NTU NTU

Nephelometric **T**urbidity **U**nits. See TURBIDITY UNITS (TU).

NAMEPLATE NAMEPLATE

A durable metal plate found on equipment which lists critical operating conditions for the equipment.

NATIONAL ENVIRONMENTAL TRAINING ASSOCIATION (NETA) NATIONAL ENVIRONMENTAL TRAINING ASSOCIATION (NETA)

A professional organization devoted to serving the environmental trainer and promoting better operation of waterworks and pollution control facilities. For information on NETA membership and publications, contact NETA, 5320 North 16th Street, Suite 114, Phoenix, AZ 85016-3241. Phone (602) 956-6099.

NATIONAL INSTITUTE OF OCCUPATIONAL SAFETY AND HEALTH NATIONAL INSTITUTE OF OCCUPATIONAL SAFETY AND HEALTH

See NIOSH.

NATIONAL PRIMARY DRINKING WATER REGULATIONS NATIONAL PRIMARY DRINKING WATER REGULATIONS

Commonly referred to as NPDWR.

NATIONAL SECONDARY DRINKING WATER REGULATIONS NATIONAL SECONDARY DRINKING WATER REGULATIONS

Commonly referred to as NSDWR.

NEPHELOMETRIC (NEFF-el-o-MET-rick) NEPHELOMETRIC

A means of measuring turbidity in a sample by using an instrument called a nephelometer. A nephelometer passes light through a sample and the amount of light deflected (usually at a 90-degree angle) is then measured.

NEWTON NEWTON

A force which, when applied to a body having a mass of one kilogram, gives it an acceleration of one meter per second per second.

NITRIFICATION (NYE-truh-fuh-KAY-shun) NITRIFICATION

An aerobic process in which bacteria reduce the ammonia and organic nitrogen in water into nitrite and then nitrate.

NITROGENOUS (nye-TRAH-jen-us) NITROGENOUS

A term used to describe chemical compounds (usually organic) containing nitrogen in combined forms. Proteins and nitrate are nitrogenous compounds.

NOBLE METAL NOBLE METAL

A chemically inactive metal (such as gold). A metal that does not corrode easily and is much scarcer (and more valuable) than the so-called useful or base metals. Also see BASE METAL.

NOMINAL DIAMETER NOMINAL DIAMETER

An approximate measurement of the diameter of a pipe. Although the nominal diameter is used to describe the size or diameter of a pipe, it is usually not the exact inside diameter of the pipe.

NONIONIC (NON-eye-ON-ick) POLYMER NONIONIC POLYMER

A polymer that has no net electrical charge.

NON-PERMIT CONFINED SPACE NON-PERMIT CONFINED SPACE

See CONFINED SPACE, NON-PERMIT.

NONPOINT SOURCE NONPOINT SOURCE

A runoff or discharge from a field or similar source. A point source refers to a discharge that comes out the end of a pipe.

NONPOTABLE (non-POE-tuh-bull) NONPOTABLE

Water that may contain objectionable pollution, contamination, minerals, or infective agents and is considered unsafe and/or unpalatable for drinking.

NONVOLATILE MATTER NONVOLATILE MATTER

Material such as sand, salt, iron, calcium, and other mineral materials which are only slightly affected by the actions of organisms and are not lost on ignition of the dry solids at 550°C. Volatile materials are chemical substances usually of animal or plant origin. Also see INORGANIC WASTE and VOLATILE MATTER or VOLATILE SOLIDS.

NORMAL NORMAL

See ***N*** for NORMAL.

NOTICE NOTICE

This word calls attention to information that is especially significant in understanding and operating equipment or processes safely. Also see CAUTION, DANGER, and WARNING.

NUTRIENT NUTRIENT

Any substance that is assimilated (taken in) by organisms and promotes growth. Nitrogen and phosphorus are nutrients which promote the growth of algae. There are other essential and trace elements which are also considered nutrients.

O

ORP (pronounce as separate letters) ORP

Oxidation-**R**eduction **P**otential. The electrical potential required to transfer electrons from one compound or element (the oxidant) to another compound or element (the reductant); used as a qualitative measure of the state of oxidation in water treatment systems. ORP is measured in millivolts, with negative values indicating a tendency to reduce compounds or elements and positive values indicating a tendency to oxidize compounds or elements.

OSHA (O-shuh) OSHA

The Williams-Steiger **O**ccupational **S**afety and **H**ealth **A**ct of 1970 (OSHA) is a federal law designed to protect the health and safety of industrial workers and also the operators of water supply systems and treatment plants. The Act regulates the design, construction, operation and maintenance of water supply systems and water treatment plants. OSHA also refers to the federal and state agencies which administer the OSHA regulations.

OCCUPATIONAL SAFETY AND HEALTH ACT OF 1970 OCCUPATIONAL SAFETY AND HEALTH ACT OF 1970

See OSHA.

ODOR THRESHOLD ODOR THRESHOLD

The minimum odor of a water sample that can just be detected after successive dilutions with odorless water. Also called THRESHOLD ODOR.

OFFSET OFFSET

The difference between the actual value and the desired value (or set point); characteristic of proportional controllers that do not incorporate reset action. Also called DRIFT.

OHM OHM

The unit of electrical resistance. The resistance of a conductor in which one volt produces a current of one ampere.

OLFACTORY (ol-FAK-tore-ee) FATIGUE OLFACTORY FATIGUE

A condition in which a person's nose, after exposure to certain odors, is no longer able to detect the odor.

OLIGOTROPHIC (AH-lig-o-TRO-fick) OLIGOTROPHIC

Reservoirs and lakes which are nutrient poor and contain little aquatic plant or animal life.

OPERATING PRESSURE DIFFERENTIAL OPERATING PRESSURE DIFFERENTIAL

The operating pressure range for a hydropneumatic system. For example, when the pressure drops below 40 psi the pump will come on and stay on until the pressure builds up to 60 psi. When the pressure reaches 60 psi the pump will shut off.

OPERATING RATIO OPERATING RATIO

The operating ratio is a measure of the total revenues divided by the total operating expenses.

ORGANIC ORGANIC

Substances that come from animal or plant sources. Organic substances always contain carbon. (Inorganic materials are chemical substances of mineral origin.) Also see INORGANIC.

ORGANICS ORGANICS

(1) A term used to refer to chemical compounds made from carbon molecules. These compounds may be natural materials (such as animal or plant sources) or manmade materials (such as synthetic organics). Also see ORGANIC.
(2) Any form of animal or plant life. Also see BACTERIA.

ORGANISM ORGANISM

Any form of animal or plant life. Also see BACTERIA.

ORGANIZING ORGANIZING

Deciding who does what work and delegating authority to the appropriate persons.

ORIFICE (OR-uh-fiss) ORIFICE

An opening (hole) in a plate, wall, or partition. An orifice flange or plate placed in a pipe consists of a slot or a calibrated circular hole smaller than the pipe diameter. The difference in pressure in the pipe above and at the orifice may be used to determine the flow in the pipe.

ORTHOTOLIDINE (or-tho-TOL-uh-dine) ORTHOTOLIDINE

Orthotolidine is a colorimetric indicator of chlorine residual. If chlorine is present, a yellow-colored compound is produced. This reagent is no longer approved for chemical analysis to determine chlorine residual.

OSMOSIS (oz-MOE-sis) OSMOSIS

The passage of a liquid from a weak solution to a more concentrated solution across a semipermeable membrane. The membrane allows the passage of the water (solvent) but not the dissolved solids (solutes). This process tends to equalize the conditions on either side of the membrane.

OUCH PRINCIPLE OUCH PRINCIPLE

This principle says that as a manager when you delegate job tasks you must be **O**bjective, **U**niform in your treatment of employees, **C**onsistent with utility policies, and **H**ave job relatedness.

OVERALL EFFICIENCY, PUMP OVERALL EFFICIENCY, PUMP

The combined efficiency of a pump and motor together. Also called the WIRE-TO-WATER EFFICIENCY.

OVERDRAFT OVERDRAFT

The pumping of water from a groundwater basin or aquifer in excess of the supply flowing into the basin. This pumping results in a depletion or “mining” of the groundwater in the basin.

OVERFLOW RATE OVERFLOW RATE

One of the guidelines for the design of settling tanks and clarifiers in treatment plants. Used by operators to determine if tanks and clarifiers are hydraulically (flow) over- or underloaded. Also called SURFACE LOADING.

$$\text{Overflow Rate, GPD/sq ft} = \frac{\text{Flow, gallons/day}}{\text{Surface Area, sq ft}}$$

OVERHEAD OVERHEAD

Indirect costs necessary for a water utility to function properly. These costs are not related to the actual production and delivery of water to consumers, but include the costs of rent, lights, office supplies, management and administration.

OVERTURN OVERTURN

The almost spontaneous mixing of all layers of water in a reservoir or lake when the water temperature becomes similar from top to bottom. This may occur in the fall/winter when the surface waters cool to the same temperature as the bottom waters and also in the spring when the surface waters warm after the ice melts.

OXIDATION (ox-uh-DAY-shun) OXIDATION

Oxidation is the addition of oxygen, removal of hydrogen, or the removal of electrons from an element or compound. In the environment, organic matter is oxidized to more stable substances. The opposite of REDUCTION.

OXIDATION-REDUCTION POTENTIAL (ORP) OXIDATION-REDUCTION POTENTIAL (ORP)

The electrical potential required to transfer electrons from one compound or element (the oxidant) to another compound or element (the reductant); used as a qualitative measure of the state of oxidation in water treatment systems. ORP is measured in millivolts, with negative values indicating a tendency to reduce compounds or elements and positive values indicating a tendency to oxidize compounds or elements.

OXIDIZING AGENT OXIDIZING AGENT

Any substance, such as oxygen (O_2) or chlorine (Cl_2), that will readily add (take on) electrons. The opposite is a REDUCING AGENT.

OXYGEN DEFICIENCY OXYGEN DEFICIENCY

An atmosphere containing oxygen at a concentration of less than 19.5 percent by volume.

OXYGEN ENRICHMENT OXYGEN ENRICHMENT

An atmosphere containing oxygen at a concentration of more than 23.5 percent by volume.

OZONATION (O-zoe-NAY-shun) OZONATION

The application of ozone to water for disinfection or for taste and odor control.

P

PCBs PCBs

See **P**OLY**C**HLORINATED **B**IPHENYLS.

pCi/*L* pCi/*L*

picoCurie per liter. A picoCurie is a measure of radioactivity. One picoCurie of radioactivity is equivalent to 0.037 nuclear disintegrations per second.

pcu (PLATINUM COBALT UNITS) pcu (PLATINUM COBALT UNITS)

Platinum cobalt units are a measure of color using platinum cobalt standards by visual comparison.

PMCL PMCL

Primary **M**aximum **C**ontaminant **L**evel. Primary MCLs for various water quality indicators are established to protect public health.

PPM PPM

See **P**ARTS **P**ER **M**ILLION.

PSIG PSIG

Pounds per **S**quare **I**nch **G**age pressure. The pressure within a closed container or pipe measured with a gage in pounds per square inch. See GAGE PRESSURE.

PACKER ASSEMBLY PACKER ASSEMBLY

An inflatable device used to seal the tremie pipe inside the well casing to prevent the grout from entering the inside of the conductor casing.

PALATABLE (PAL-uh-tuh-bull) PALATABLE

Water at a desirable temperature that is free from objectionable tastes, odors, colors, and turbidity. Pleasing to the senses.

PARSHALL FLUME PARSHALL FLUME

A device used to measure the flow in an open channel. The flume narrows to a throat of fixed dimensions and then expands again. The rate of flow can be calculated by measuring the difference in head (pressure) before and at the throat of the flume.

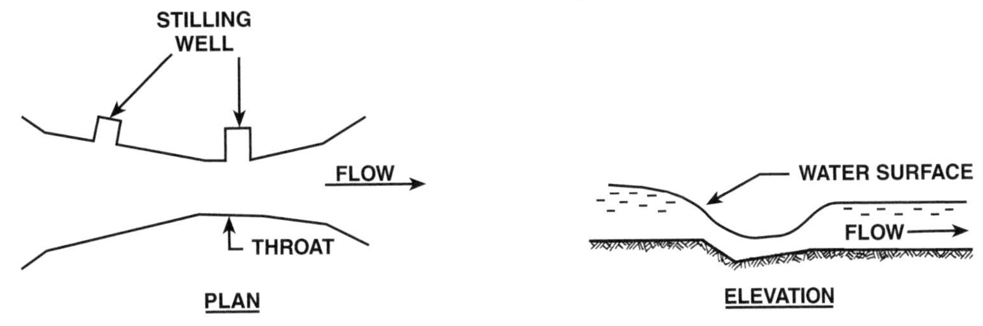

PARTICLE COUNT PARTICLE COUNT

The results of a microscopic examination of treated water with a special particle counter which classifies suspended particles by number and size.

PARTICLE COUNTER PARTICLE COUNTER

A device which counts and measures the size of individual particles in water. Particles are divided into size ranges and the number of particles is counted in each of these ranges. The results are reported in terms of the number of particles in different particle diameter size ranges per milliliter of water sampled.

PARTICLE COUNTING PARTICLE COUNTING

A procedure for counting and measuring the size of individual particles in water. Particles are divided into size ranges and the number of particles is counted in each of these ranges. The results are reported in terms of the number of particles in different particle diameter size ranges per milliliter of water sampled.

PARTICULATE (par-TICK-you-let) PARTICULATE

A very small solid suspended in water which can vary widely in size, shape, density, and electrical charge. Colloidal and dispersed particulates are artificially gathered together by the processes of coagulation and flocculation.

PARTS PER MILLION (PPM) PARTS PER MILLION (PPM)

Parts per million parts, a measurement of concentration on a weight or volume basis. This term is equivalent to milligrams per liter (mg/*L*) which is the preferred term.

PASCAL PASCAL

The pressure or stress of one newton per square meter. Abbreviated Pa.

1 psi = 6,895 Pa = 6.895 kN/sq m = 0.0703 kg/sq cm

PATHOGENIC (PATH-o-JEN-ick) ORGANISMS PATHOGENIC ORGANISMS

Organisms, including bacteria, viruses or cysts, capable of causing diseases (giardiasis, cryptosporidiosis, typhoid, cholera, dysentery) in a host (such as a person). There are many types of organisms which do *NOT* cause disease. These organisms are called non-pathogenic.

PATHOGENS (PATH-o-jens) PATHOGENS

Pathogenic or disease-causing organisms.

PEAK DEMAND PEAK DEMAND

The maximum momentary load placed on a water treatment plant, pumping station or distribution system. This demand is usually the maximum average load in one hour or less, but may be specified as the instantaneous load or the load during some other short time period.

PERCENT SATURATION PERCENT SATURATION

The amount of a substance that is dissolved in a solution compared with the amount dissolved in the solution at saturation, expressed as a percent.

$$\text{Percent Saturation, \%} = \frac{\text{Amount of Substance That Is Dissolved x 100\%}}{\text{Amount Dissolved in Solution at Saturation}}$$

PERCOLATING (PURR-co-LAY-ting) WATER — PERCOLATING WATER

Water that passes through soil or rocks under the force of gravity.

PERCOLATION (PURR-co-LAY-shun) — PERCOLATION

The slow passage of water through a filter medium; or, the gradual penetration of soil and rocks by water.

PERIPHYTON (pair-e-FI-tawn) — PERIPHYTON

Microscopic plants and animals that are firmly attached to solid surfaces under water such as rocks, logs, pilings and other structures.

PERMEABILITY (PURR-me-uh-BILL-uh-tee) — PERMEABILITY

The property of a material or soil that permits considerable movement of water through it when it is saturated.

PERMEATE (PURR-me-ate) — PERMEATE

(1) To penetrate and pass through, as water penetrates and passes through soil and other porous materials.

(2) The liquid (demineralized water) produced from the reverse osmosis process that contains a *LOW* concentration of dissolved solids.

PERMIT-REQUIRED CONFINED SPACE (PERMIT SPACE) — PERMIT-REQUIRED CONFINED SPACE (PERMIT SPACE)

See CONFINED SPACE, PERMIT-REQUIRED (PERMIT SPACE).

PESTICIDE — PESTICIDE

Any substance or chemical designed or formulated to kill or control animal pests. Also see INSECTICIDE and RODENTICIDE.

PET COCK — PET COCK

A small valve or faucet used to drain a cylinder or fitting.

pH (pronounce as separate letters) — pH

pH is an expression of the intensity of the basic or acidic condition of a liquid. Mathematically, pH is the logarithm (base 10) of the reciprocal of the hydrogen ion activity.

$$pH = Log \frac{1}{[H^+]}$$

The pH may range from 0 to 14, where 0 is most acidic, 14 most basic, and 7 neutral. Natural waters usually have a pH between 6.5 and 8.5.

PHENOLIC (fee-NO-lick) COMPOUNDS — PHENOLIC COMPOUNDS

Organic compounds that are derivatives of benzene.

PHENOLPHTHALEIN (FEE-nol-THAY-leen) ALKALINITY — PHENOLPHTHALEIN ALKALINITY

The alkalinity in a water sample measured by the amount of standard acid required to lower the pH to a level of 8.3, as indicated by the change in color of phenolphthalein from pink to clear. Phenolphthalein alkalinity is expressed as milligrams per liter of equivalent calcium carbonate.

PHOTOSYNTHESIS (foe-toe-SIN-thuh-sis) — PHOTOSYNTHESIS

A process in which organisms, with the aid of chlorophyll (green plant enzyme), convert carbon dioxide and inorganic substances into oxygen and additional plant material, using sunlight for energy. All green plants grow by this process.

PHYTOPLANKTON (FI-tow-PLANK-ton) — PHYTOPLANKTON

Small, usually microscopic plants (such as algae), found in lakes, reservoirs, and other bodies of water.

PICO — PICO

A prefix used in the metric system and other scientific systems of measurement which means 10^{-12} or 0.000 000 000 001.

PICOCURIE — PICOCURIE

A measure of radioactivity. One picoCurie of radioactivity is equivalent to 0.037 nuclear disintegrations per second.

PITLESS ADAPTER — PITLESS ADAPTER

A fitting which allows the well casing to be extended above ground while having a discharge connection located below the frost line. Advantages of using a pitless adapter include the elimination of the need for a pit or pump house and it is a watertight design, which helps maintain a sanitary water supply.

PLAN VIEW — PLAN VIEW

A diagram or photo showing a facility as it would appear when looking down on top of it.

PLANKTON — PLANKTON

(1) Small, usually microscopic, plants (phytoplankton) and animals (zooplankton) in aquatic systems.

(2) All of the smaller floating, suspended or self-propelled organisms in a body of water.

PLANNING — PLANNING

Management of utilities to build the resources and financial capability to provide for future needs.

PLUG FLOW — PLUG FLOW

A type of flow that occurs in tanks, basins or reactors when a slug of water moves through a tank without ever dispersing or mixing with the rest of the water flowing through the tank.

POINT SOURCE — POINT SOURCE

A discharge that comes out the end of a pipe. A nonpoint source refers to runoff or a discharge from a field or similar source.

POLARIZATION — POLARIZATION

The concentration of ions in the thin boundary layer adjacent to a membrane or pipe wall.

POLE SHADER — POLE SHADER

A copper bar circling the laminated iron core inside the coil of a magnetic starter.

POLLUTION — POLLUTION

The impairment (reduction) of water quality by agricultural, domestic, or industrial wastes (including thermal and radioactive wastes) to a degree that has an adverse effect on any beneficial use of water.

POLYANIONIC (poly-AN-eye-ON-ick) — POLYANIONIC

Characterized by many active negative charges especially active on the surface of particles.

POLYCHLORINATED BIPHENYLS — POLYCHLORINATED BIPHENYLS

A class of organic compounds that cause adverse health effects in domestic water supplies.

POLYELECTROLYTE (POLY-ee-LECK-tro-lite) — POLYELECTROLYTE

A high-molecular-weight (relatively heavy) substance having points of positive or negative electrical charges that is formed by either natural or manmade processes. Natural polyelectrolytes may be of biological origin or derived from starch products and cellulose derivatives. Manmade polyelectrolytes consist of simple substances that have been made into complex, high-molecular-weight substances. Used with other chemical coagulants to aid in binding small suspended particles to larger chemical flocs for their removal from water. Often called a POLYMER.

POLYMER (POLY-mer) — POLYMER

A long chain molecule formed by the union of many monomers (molecules of lower molecular weight). Polymers are used with other chemical coagulants to aid in binding small suspended particles to larger chemical flocs for their removal from water.

PORE — PORE

A very small open space in a rock or granular material. Also called an INTERSTICE, VOID, or void space. Also see VOID.

POROSITY — POROSITY

(1) A measure of the spaces or voids in a material or aquifer.

(2) The ratio of the volume of spaces in a rock or soil to the total volume. This ratio is usually expressed as a percentage.

$$\text{Porosity, \%} = \frac{(\text{Volume of Spaces})(100\%)}{\text{Total Volume}}$$

POSITIVE BACTERIOLOGICAL SAMPLE　　POSITIVE BACTERIOLOGICAL SAMPLE

A water sample in which gas is produced by coliform organisms during incubation in the multiple tube fermentation test. See Chapter 7, Laboratory Procedures, "Coliform Bacteria," for details.

POSITIVE DISPLACEMENT PUMP　　POSITIVE DISPLACEMENT PUMP

A type of piston, diaphragm, gear or screw pump that delivers a constant volume with each stroke. Positive displacement pumps are used as chemical solution feeders.

POSTCHLORINATION　　POSTCHLORINATION

The addition of chlorine to the plant effluent, *FOLLOWING* plant treatment, for disinfection purposes.

POTABLE (POE-tuh-bull) WATER　　POTABLE WATER

Water that does not contain objectionable pollution, contamination, minerals, or infective agents and is considered satisfactory for drinking.

POWER FACTOR　　POWER FACTOR

The ratio of the true power passing through an electric circuit to the product of the voltage and amperage in the circuit. This is a measure of the lag or lead of the current with respect to the voltage. In alternating current the voltage and amperes are not always in phase; therefore, the true power may be slightly less than that determined by the direct product.

PRECHLORINATION　　PRECHLORINATION

The addition of chlorine at the headworks of the plant *PRIOR TO* other treatment processes mainly for disinfection and control of tastes, odors and aquatic growths. Also applied to aid in coagulation and settling.

PRECIPITATE (pre-SIP-uh-TATE)　　PRECIPITATE

(1) An insoluble, finely divided substance which is a product of a chemical reaction within a liquid.

(2) The separation from solution of an insoluble substance.

PRECIPITATION (pre-SIP-uh-TAY-shun)　　PRECIPITATION

(1) The process by which atmospheric moisture falls onto a land or water surface as rain, snow, hail, or other forms of moisture.

(2) The chemical transformation of a substance in solution into an insoluble form (precipitate).

PRECISION　　PRECISION

The ability of an instrument to measure a process variable and repeatedly obtain the same result. The ability of an instrument to reproduce the same results.

PRECURSOR, THM (pre-CURSE-or)　　PRECURSOR, THM

Natural organic compounds found in all surface and groundwaters. These compounds *MAY* react with halogens (such as chlorine) to form trihalomethanes (tri-HAL-o-METH-hanes) (THMs); they *MUST* be present in order for THMs to form.

PRESCRIPTIVE (pre-SKRIP-tive)　　PRESCRIPTIVE

Water rights which are acquired by diverting water and putting it to use in accordance with specified procedures. These procedures include filing a request (with a state agency) to use unused water in a stream, river or lake.

PRESENT WORTH　　PRESENT WORTH

The value of a long-term project expressed in today's dollars. Present worth is calculated by converting (discounting) all future benefits and costs over the life of the project to a single economic value at the start of the project. Calculating the present worth of alternative projects makes it possible to compare them and select the one with the largest positive (beneficial) present worth or minimum present cost.

PRESSURE CONTROL　　PRESSURE CONTROL

A switch which operates on changes in pressure. Usually this is a diaphragm pressing against a spring. When the force on the diaphragm overcomes the spring pressure, the switch is actuated (activated).

PRESSURE HEAD　　PRESSURE HEAD

The vertical distance (in feet) equal to the pressure (in psi) at a specific point. The pressure head is equal to the pressure in psi times 2.31 ft/psi.

PRESTRESSED — PRESTRESSED

A prestressed pipe has been reinforced with wire strands (which are under tension) to give the pipe an active resistance to loads or pressures on it.

PREVENTIVE MAINTENANCE UNITS — PREVENTIVE MAINTENANCE UNITS

Crews assigned the task of cleaning sewers (for example, balling or high-velocity cleaning crews) to prevent stoppages and odor complaints. Preventive maintenance is performing the most effective cleaning procedure, in the area where it is most needed, at the proper time in order to prevent failures and emergency situations.

PRIMARY ELEMENT — PRIMARY ELEMENT

(1) A device that measures (senses) a physical condition or variable of interest. Floats and thermocouples are examples of primary elements. Also called a SENSOR.

(2) The hydraulic structure used to measure flows. In open channels, weirs and flumes are primary elements or devices. Venturi meters and orifice plates are the primary elements in pipes or pressure conduits.

PRIME — PRIME

The action of filling a pump casing with water to remove the air. Most pumps must be primed before start-up or they will not pump any water.

PROCESS VARIABLE — PROCESS VARIABLE

A physical or chemical quantity which is usually measured and controlled in the operation of a water treatment plant or an industrial plant.

PRODUCT WATER — PRODUCT WATER

Water that has passed through a water treatment plant. All the treatment processes are completed or finished. This water is the product from the water treatment plant and is ready to be delivered to the consumers. Also called FINISHED WATER.

PROFILE — PROFILE

A drawing showing elevation plotted against distance, such as the vertical section or *SIDE* view of a pipeline.

PRUSSIAN BLUE — PRUSSIAN BLUE

A blue paste or liquid (often on a paper like carbon paper) used to show a contact area. Used to determine if gate valve seats fit properly.

PUMP BOWL — PUMP BOWL

The submerged pumping unit in a well, including the shaft, impellers and housing.

PUMPING WATER LEVEL — PUMPING WATER LEVEL

The vertical distance in feet from the centerline of the pump discharge to the level of the free pool while water is being drawn from the pool.

PURVEYOR (purr-VAY-or), WATER — PURVEYOR, WATER

An agency or person that supplies water (usually potable water).

PUTREFACTION (PEW-truh-FACK-shun) — PUTREFACTION

Biological decomposition of organic matter, with the production of foul-smelling and -tasting products, associated with anaerobic (no oxygen present) conditions.

Q

QUICKLIME — QUICKLIME

A material that is mostly calcium oxide (CaO) or calcium oxide in natural association with a lesser amount of magnesium oxide. Quicklime is capable of combining with water, that is, becoming slaked. Also see HYDRATED LIME.

R

RADIAL TO IMPELLER RADIAL TO IMPELLER

Perpendicular to the impeller shaft. Material being pumped flows at a right angle to the impeller.

RADICAL RADICAL

A group of atoms that is capable of remaining unchanged during a series of chemical reactions. Such combinations (radicals) exist in the molecules of many organic compounds; sulfate (SO_4^{2-}) is an inorganic radical.

RANGE RANGE

The spread from minimum to maximum values that an instrument is designed to measure. Also see EFFECTIVE RANGE and SPAN.

RANNEY COLLECTOR RANNEY COLLECTOR

This water collector is constructed as a dug well from 12 to 16 feet (3.5 to 5 m) in diameter that has been sunk as a caisson near the bank of a river or lake. Screens are driven radially and approximately horizontally from this well into the sand and the gravel deposits underlying the river.

[SEE DRAWING ON PAGE 607]

RATE OF RETURN RATE OF RETURN

A value which indicates the return of funds received on the basis of the total equity capital used to finance physical facilities. Similar to the interest rate on savings accounts or loans.

RAW WATER RAW WATER

(1) Water in its natural state, prior to any treatment.

(2) Usually the water entering the first treatment process of a water treatment plant.

REAERATION (RE-air-A-shun) REAERATION

The introduction of air through forced air diffusers into the lower layers of the reservoir. As the air bubbles form and rise through the water, oxygen from the air dissolves into the water and replenishes the dissolved oxygen. The rising bubbles also cause the lower waters to rise to the surface where oxygen from the atmosphere is transferred to the water. This is sometimes called surface reaeration.

REAGENT (re-A-gent) REAGENT

A pure chemical substance that is used to make new products or is used in chemical tests to measure, detect, or examine other substances.

RECARBONATION (re-CAR-bun-NAY-shun) RECARBONATION

A process in which carbon dioxide is bubbled into the water being treated to lower the pH. The pH may also be lowered by the addition of acid. Recarbonation is the final stage in the lime-soda ash softening process. This process converts carbonate ions to bicarbonate ions and stabilizes the solution against the precipitation of carbonate compounds.

RECEIVER RECEIVER

A device which indicates the result of a measurement. Most receivers in the water utility field use either a fixed scale and movable indicator (pointer) such as a pressure gage or a movable scale and movable indicator like those used on a circular flow-recording chart. Also called an INDICATOR.

RECORDER RECORDER

A device that creates a permanent record, on a paper chart or magnetic tape, of the changes in a measured variable.

REDUCING AGENT REDUCING AGENT

Any substance, such as base metal (iron) or the sulfide ion (S^{2-}), that will readily donate (give up) electrons. The opposite is an OXIDIZING AGENT.

REDUCTION (re-DUCK-shun) REDUCTION

Reduction is the addition of hydrogen, removal of oxygen, or the addition of electrons to an element or compound. Under anaerobic conditions (no dissolved oxygen present), sulfur compounds are reduced to odor-producing hydrogen sulfide (H_2S) and other compounds. The opposite of OXIDATION.

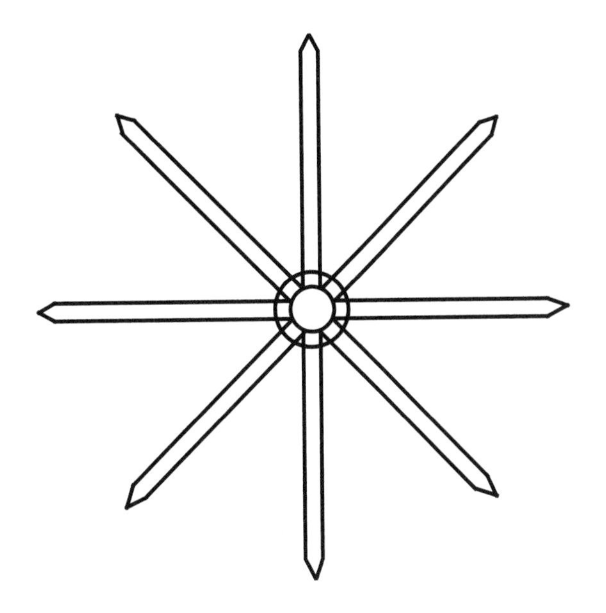

PLAN VIEW OF COLLECTOR PIPES

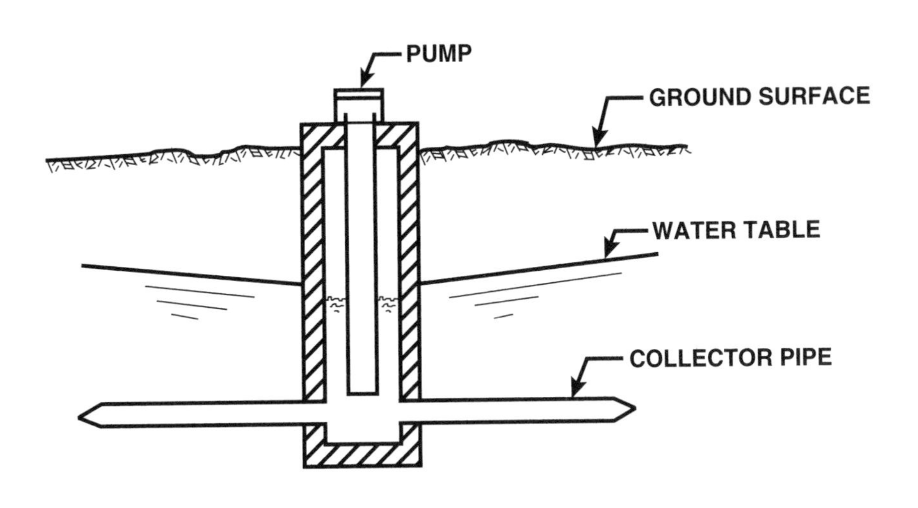

ELEVATION VIEW

RANNEY COLLECTOR

REFERENCE REFERENCE

A physical or chemical quantity whose value is known exactly, and thus is used to calibrate instruments or standardize measurements. Also called a STANDARD.

REGULATORY NEGOTIATION REGULATORY NEGOTIATION

A process whereby the U.S. Environmental Protection Agency acts on an equal basis with outside parties to reach consensus on the content of a proposed rule. If the group reaches consensus, the US EPA commits to propose the rule with the agreed upon content.

RELIQUEFACTION (re-LICK-we-FACK-shun) RELIQUEFACTION

The return of a gas to the liquid state; for example, a condensation of chlorine gas to return it to its liquid form by cooling.

REPRESENTATIVE SAMPLE REPRESENTATIVE SAMPLE

A sample portion of material or water that is as nearly identical in content and consistency as possible to that in the larger body of material or water being sampled.

RESIDUAL CHLORINE RESIDUAL CHLORINE

The concentration of chlorine present in water after the chlorine demand has been satisfied. The concentration is expressed in terms of the total chlorine residual, which includes both the free and combined or chemically bound chlorine residuals.

RESIDUE RESIDUE

The dry solids remaining after the evaporation of a sample of water or sludge. Also see TOTAL DISSOLVED SOLIDS.

RESINS RESINS

See ION EXCHANGE RESINS.

RESISTANCE RESISTANCE

That property of a conductor or wire that opposes the passage of a current, thus causing electric energy to be transformed into heat.

RESPIRATION RESPIRATION

The process in which an organism uses oxygen for its life processes and gives off carbon dioxide.

RESPONSIBILITY RESPONSIBILITY

Answering to those above in the chain of command to explain how and why you have used your authority.

REVERSE OSMOSIS (oz-MOE-sis) REVERSE OSMOSIS

The application of pressure to a concentrated solution which causes the passage of a liquid from the concentrated solution to a weaker solution across a semipermeable membrane. The membrane allows the passage of the water (solvent) but not the dissolved solids (solutes). The liquid produced is a demineralized water. Also see OSMOSIS.

RIPARIAN (ri-PAIR-ee-an) RIPARIAN

Water rights which are acquired together with title to the land bordering a source of surface water. The right to put to beneficial use surface water adjacent to your land.

RODENTICIDE (row-DENT-uh-SIDE) RODENTICIDE

Any substance or chemical used to kill or control rodents.

ROTAMETER (RODE-uh-ME-ter) ROTAMETER

A device used to measure the flow rate of gases and liquids. The gas or liquid being measured flows vertically up a tapered, calibrated tube. Inside the tube is a small ball or bullet-shaped float (it may rotate) that rises or falls depending on the flow rate. The flow rate may be read on a scale behind or on the tube by looking at the middle of the ball or at the widest part or top of the float.

ROTOR ROTOR

The rotating part of a machine. The rotor is surrounded by the stationary (non-moving) parts (stator) of the machine.

ROUTINE SAMPLING ROUTINE SAMPLING

Sampling repeated on a regular basis.

S

SCADA (ss-KAY-dah) SYSTEM | SCADA SYSTEM

Supervisory **C**ontrol **A**nd **D**ata **A**cquisition system. A computer-monitored alarm, response, control and data acquisition system used by drinking water facilities to monitor their operations.

SCFM | SCFM

Cubic **F**eet of air per **M**inute at **S**tandard conditions of temperature, pressure, and humidity (0°C, 14.7 psia, and 50% relative humidity).

SDWA | SDWA

See **S**AFE **D**RINKING **W**ATER **A**CT.

SMCL | SMCL

Secondary **M**aximum **C**ontaminant **L**evel. Secondary MCLs for various water quality indicators are established to protect public welfare.

SNARL | SNARL

Suggested **N**o **A**dverse **R**esponse **L**evel. The concentration of a chemical in water that is expected not to cause an adverse health effect.

SACRIFICIAL ANODE | SACRIFICIAL ANODE

An easily corroded material deliberately installed in a pipe or tank. The intent of such an installation is to give up (sacrifice) this anode to corrosion while the water supply facilities remain relatively corrosion free.

SAFE DRINKING WATER ACT | SAFE DRINKING WATER ACT

Commonly referred to as SDWA. An Act passed by the U.S. Congress in 1974. The Act establishes a cooperative program among local, state and federal agencies to ensure safe drinking water for consumers. The Act has been amended several times, including the 1980, 1986, and 1996 Amendments.

SAFE WATER | SAFE WATER

Water that does not contain harmful bacteria, or toxic materials or chemicals. Water may have taste and odor problems, color and certain mineral problems and still be considered safe for drinking.

SAFE YIELD | SAFE YIELD

The annual quantity of water that can be taken from a source of supply over a period of years without depleting the source permanently (beyond its ability to be replenished naturally in "wet years").

SALINITY | SALINITY

(1) The relative concentration of dissolved salts, usually sodium chloride, in a given water.

(2) A measure of the concentration of dissolved mineral substances in water.

SANITARY SURVEY | SANITARY SURVEY

A detailed evaluation and/or inspection of a source of water supply and all conveyances, storage, treatment and distribution facilities to ensure protection of the water supply from all pollution sources.

SAPROPHYTES (SAP-row-FIGHTS) | SAPROPHYTES

Organisms living on dead or decaying organic matter. They help natural decomposition of organic matter in water.

SATURATION | SATURATION

The condition of a liquid (water) when it has taken into solution the maximum possible quantity of a given substance at a given temperature and pressure.

SATURATOR (SAT-you-RAY-tore) | SATURATOR

A device which produces a fluoride solution for the fluoridation process. The device is usually a cylindrical container with granular sodium fluoride on the bottom. Water flows either upward or downward through the sodium fluoride to produce the fluoride solution.

SCHEDULE, PIPE — SCHEDULE, PIPE

A sizing system of numbers that specifies the I.D. (inside diameter) and O.D. (outside diameter) for each diameter pipe. The schedule number is the ratio of internal pressure in psi divided by the allowable fiber stress multiplied by 1,000. Typical schedules of iron and steel pipe are schedules 40, 80, and 160. Other forms of piping are divided into various classes with their own schedule schemes.

SCHMUTZDECKE (sh-moots-DECK-ee) — SCHMUTZDECKE

A layer of trapped matter at the surface of a slow sand filter in which a dense population of microorganisms develops. These microorganisms within the film or mat feed on and break down incoming organic material trapped in the mat. In doing so the microorganisms both remove organic matter and add mass to the mat, further developing the mat and increasing the physical straining action of the mat.

SECCHI (SECK-key) DISC — SECCHI DISC

A flat, white disc lowered into the water by a rope until it is just barely visible. At this point, the depth of the disc from the water surface is the recorded Secchi disc transparency.

SEDIMENTATION (SED-uh-men-TAY-shun) — SEDIMENTATION

A water treatment process in which solid particles settle out of the water being treated in a large clarifier or sedimentation basin.

SEIZE UP — SEIZE UP

Seize up occurs when an engine overheats and a part expands to the point where the engine will not run. Also called “freezing.”

SENSITIVITY (PARTICLE COUNTERS) — SENSITIVITY (PARTICLE COUNTERS)

The smallest particle a particle counter will measure and count.

SENSOR — SENSOR

A device that measures (senses) a physical condition or variable of interest. Floats and thermocouples are examples of sensors. Also called a PRIMARY ELEMENT.

SEPTIC (SEP-tick) — SEPTIC

A condition produced by bacteria when all oxygen supplies are depleted. If severe, the bottom deposits produce hydrogen sulfide, the deposits and water turn black, give off foul odors, and the water has a greatly increased chlorine demand.

SEQUESTRATION (SEE-kwes-TRAY-shun) — SEQUESTRATION

A chemical complexing (forming or joining together) of metallic cations (such as iron) with certain inorganic compounds, such as phosphate. Sequestration prevents the precipitation of the metals (iron). Also see CHELATION.

SERVICE PIPE — SERVICE PIPE

The pipeline extending from the water main to the building served or to the consumer’s system.

SET POINT — SET POINT

The position at which the control or controller is set. This is the same as the desired value of the process variable.

SEWAGE — SEWAGE

The used household water and water-carried solids that flow in sewers to a wastewater treatment plant. The preferred term is WASTEWATER.

SHEAVE (SHE-v) — SHEAVE

V-belt drive pulley which is commonly made of cast iron or steel.

SHIM — SHIM

Thin metal sheets which are inserted between two surfaces to align or space the surfaces correctly. Shims can be used anywhere a spacer is needed. Usually shims are 0.001 to 0.020 inch thick.

SHOCK LOAD — SHOCK LOAD

The arrival at a water treatment plant of raw water containing unusual amounts of algae, colloidal matter, color, suspended solids, turbidity, or other pollutants.

SHORT-CIRCUITING SHORT-CIRCUITING

A condition that occurs in tanks or basins when some of the water travels faster than the rest of the flowing water. This is usually undesirable since it may result in shorter contact, reaction, or settling times in comparison with the theoretical (calculated) or presumed detention times.

SIMULATE SIMULATE

To reproduce the action of some process, usually on a smaller scale.

SINGLE-STAGE PUMP SINGLE-STAGE PUMP

A pump that has only one impeller. A multi-stage pump has more than one impeller.

SLAKE SLAKE

To mix with water so that a true chemical combination (hydration) takes place, such as in the slaking of lime.

SLAKED LIME SLAKED LIME

See HYDRATED LIME.

SLOPE SLOPE

The slope or inclination of a trench bottom or a trench side wall is the ratio of the vertical distance to the horizontal distance or "rise over run." Also see GRADE (2).

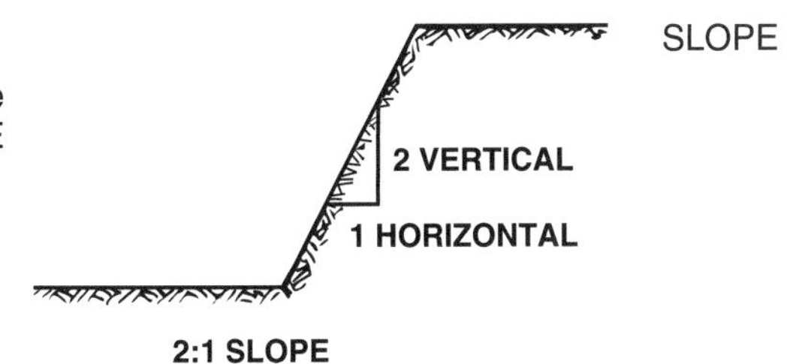

SLUDGE (sluj) SLUDGE

The settleable solids separated from water during processing.

SLURRY (SLUR-e) SLURRY

A watery mixture or suspension of insoluble (not dissolved) matter; a thin, watery mud or any substance resembling it (such as a grit slurry or a lime slurry).

SOFT WATER SOFT WATER

Water having a low concentration of calcium and magnesium ions. According to U.S. Geological Survey guidelines, soft water is water having a hardness of 60 milligrams per liter or less.

SOFTWARE PROGRAMS SOFTWARE PROGRAMS

Computer programs; the list of instructions that tell a computer how to perform a given task or tasks. Some software programs are designed and written to monitor and control water distribution systems and treatment processes.

SOLENOID (SO-luh-noid) SOLENOID

A magnetically (electric coil) operated mechanical device. Solenoids can operate small valves or electric switches.

SOLUTION SOLUTION

A liquid mixture of dissolved substances. In a solution it is impossible to see all the separate parts.

SOUNDING TUBE SOUNDING TUBE

A pipe or tube used for measuring the depths of water.

SPAN SPAN

The scale or range of values an instrument is designed to measure. Also see RANGE.

SPECIFIC CAPACITY SPECIFIC CAPACITY

A measurement of well yield per unit depth of drawdown after a specific time has passed, usually 24 hours. Typically expressed as gallons per minute per foot (GPM/ft or cu m/day/m).

SPECIFIC CAPACITY TEST SPECIFIC CAPACITY TEST

A testing method used to determine the adequacy of an aquifer or well by measuring the specific capacity.

SPECIFIC CONDUCTANCE SPECIFIC CONDUCTANCE

A rapid method of estimating the dissolved solids content of a water supply. The measurement indicates the capacity of a sample of water to carry an electric current, which is related to the concentration of ionized substances in the water. Also called CONDUCTANCE.

SPECIFIC GRAVITY SPECIFIC GRAVITY

(1) Weight of a particle, substance, or chemical solution in relation to the weight of an equal volume of water. Water has a specific gravity of 1.000 at 4°C (39°F). Particulates in raw water may have a specific gravity of 1.005 to 2.5.

(2) Weight of a particular gas in relation to the weight of an equal volume of air at the same temperature and pressure (air has a specific gravity of 1.0). Chlorine has a specific gravity of 2.5 as a gas.

SPECIFIC YIELD SPECIFIC YIELD

The quantity of water that a unit volume of saturated permeable rock or soil will yield when drained by gravity. Specific yield may be expressed as a ratio or as a percentage by volume.

SPOIL SPOIL

Excavated material such as soil from the trench of a water main.

SPORE SPORE

The reproductive body of an organism which is capable of giving rise to a new organism either directly or indirectly. A viable (able to live and grow) body regarded as the resting stage of an organism. A spore is usually more resistant to disinfectants and heat than most organisms.

SPRING LINE SPRING LINE

Theoretical center of a pipeline. Also, the guideline for laying a course of bricks.

STALE WATER STALE WATER

Water which has not flowed recently and may have picked up tastes and odors from distribution lines or storage facilities.

STANDARD STANDARD

A physical or chemical quantity whose value is known exactly, and thus is used to calibrate instruments or standardize measurements. Also called a REFERENCE.

STANDARD DEVIATION STANDARD DEVIATION

A measure of the spread or dispersion of data.

STANDARD METHODS STANDARD METHODS

STANDARD METHODS FOR THE EXAMINATION OF WATER AND WASTEWATER, 20th Edition. A joint publication of the American Public Health Association (APHA), American Water Works Association (AWWA), and the Water Environment Federation (WEF) which outlines the accepted laboratory procedures used to analyze the impurities in water and wastewater. Available from American Water Works Association, Bookstore, 6666 West Quincy Avenue, Denver, CO 80235. Order No. 10079. Price to members, $166.00; nonmembers, $211.00; price includes cost of shipping and handling.

STANDARD SOLUTION STANDARD SOLUTION

A solution in which the exact concentration of a chemical or compound is known.

STANDARDIZE STANDARDIZE

To compare with a standard.

(1) In wet chemistry, to find out the exact strength of a solution by comparing it with a standard of known strength. This information is used to adjust the strength by adding more water or more of the substance dissolved.

(2) To set up an instrument or device to read a standard. This allows you to adjust the instrument so that it reads accurately, or enables you to apply a correction factor to the readings.

STARTERS (MOTOR) STARTERS (MOTOR)

Devices used to start up large motors gradually to avoid severe mechanical shock to a driven machine and to prevent disturbance to the electrical lines (causing dimming and flickering of lights).

STATIC HEAD STATIC HEAD

When water is not moving, the vertical distance (in feet) from a specific point to the water surface is the static head. (The static pressure in psi is the static head in feet times 0.433 psi/ft.) Also see DYNAMIC PRESSURE and STATIC PRESSURE.

STATIC PRESSURE — STATIC PRESSURE

When water is not moving, the vertical distance (in feet) from a specific point to the water surface is the static head. The static pressure in psi is the static head in feet times 0.433 psi/ft. Also see DYNAMIC PRESSURE and STATIC HEAD.

STATIC WATER DEPTH — STATIC WATER DEPTH

The vertical distance in feet from the centerline of the pump discharge down to the surface level of the free pool while no water is being drawn from the pool or water table.

STATIC WATER LEVEL — STATIC WATER LEVEL

(1) The elevation or level of the water table in a well when the pump is not operating.

(2) The level or elevation to which water would rise in a tube connected to an artesian aquifer, basin, or conduit under pressure.

STATOR — STATOR

That portion of a machine which contains the stationary (non-moving) parts that surround the moving parts (rotor).

STERILIZATION (STARE-uh-luh-ZAY-shun) — STERILIZATION

The removal or destruction of all microorganisms, including pathogenic and other bacteria, vegetative forms and spores. Compare with DISINFECTION.

STETHOSCOPE — STETHOSCOPE

An instrument used to magnify sounds and convey them to the ear.

STORATIVITY (S) — STORATIVITY (S)

The volume of groundwater an aquifer releases from or takes into storage per unit surface area of the aquifer per unit change in head. Also called the storage coefficient.

STRATIFICATION (STRAT-uh-fuh-KAY-shun) — STRATIFICATION

The formation of separate layers (of temperature, plant, or animal life) in a lake or reservoir. Each layer has similar characteristics such as all water in the layer has the same temperature. Also see THERMAL STRATIFICATION.

STRAY CURRENT CORROSION — STRAY CURRENT CORROSION

A corrosion activity resulting from stray electric current originating from some source outside the plumbing system such as D.C. grounding on phone systems.

SUBMERGENCE — SUBMERGENCE

The distance between the water surface and the media surface in a filter.

SUBSIDENCE (sub-SIDE-ence) — SUBSIDENCE

The dropping or lowering of the ground surface as a result of removing excess water (overdraft or overpumping) from an aquifer. After excess water has been removed, the soil will settle, become compacted and the ground surface will drop and can cause the settling of underground utilities.

SUCTION LIFT — SUCTION LIFT

The *NEGATIVE* pressure [in feet (meters) of water or inches (centimeters) of mercury vacuum] on the suction side of the pump. The pressure can be measured from the centerline of the pump *DOWN TO* (lift) the elevation of the hydraulic grade line on the suction side of the pump.

SUPERCHLORINATION (SUE-per-KLOR-uh-NAY-shun) — SUPERCHLORINATION

Chlorination with doses that are deliberately selected to produce free or combined residuals so large as to require dechlorination.

SUPERNATANT (sue-per-NAY-tent) — SUPERNATANT

Liquid removed from settled sludge. Supernatant commonly refers to the liquid between the sludge on the bottom and the scum on the water surface of a basin or container.

SUPERSATURATED — SUPERSATURATED

An unstable condition of a solution (water) in which the solution contains a substance at a concentration greater than the saturation concentration for the substance.

SURFACE LOADING — SURFACE LOADING

One of the guidelines for the design of settling tanks and clarifiers in treatment plants. Used by operators to determine if tanks and clarifiers are hydraulically (flow) over- or underloaded. Also called OVERFLOW RATE.

$$\text{Surface Loading, GPD/sq ft} = \frac{\text{Flow, gallons/day}}{\text{Surface Area, sq ft}}$$

SURFACTANT (sir-FAC-tent) — SURFACTANT

Abbreviation for surface-active agent. The active agent in detergents that possesses a high cleaning ability.

SURGE CHAMBER — SURGE CHAMBER

A chamber or tank connected to a pipe and located at or near a valve that may quickly open or close or a pump that may suddenly start or stop. When the flow of water in a pipe starts or stops quickly, the surge chamber allows water to flow into or out of the pipe and minimize any sudden positive or negative pressure waves or surges in the pipe.

[SEE DRAWING ON PAGE 615]

SUSPENDED SOLIDS — SUSPENDED SOLIDS

(1) Solids that either float on the surface or are suspended in water, wastewater, or other liquids, and which are largely removable by laboratory filtering.

(2) The quantity of material removed from water in a laboratory test, as prescribed in *STANDARD METHODS FOR THE EXAMINATION OF WATER AND WASTEWATER*, and referred to as Total Suspended Solids Dried at 103–105°C.

T

TCE — TCE

See **T**RI**C**HLORO**E**THANE.

TDS — TDS

See **T**OTAL **D**ISSOLVED **S**OLIDS.

THM — THM

See **T**RI**H**ALO**M**ETHANES.

THM PRECURSOR — THM PRECURSOR

See PRECURSOR, THM.

TAILGATE SAFETY MEETING — TAILGATE SAFETY MEETING

Brief (10 to 20 minutes) safety meetings held every 7 to 10 working days. The term *TAILGATE* comes from the safety meetings regularly held by the construction industry around the tailgate of a truck.

TELEMETRY (tel-LEM-uh-tree) — TELEMETRY

The electrical link between the transmitter and the receiver. Telephone lines are commonly used to serve as the electrical line.

TEMPERATURE SENSOR — TEMPERATURE SENSOR

A device that opens and closes a switch in response to changes in the temperature. This device might be a metal contact, or a thermocouple that generates minute electric current proportional to the difference in heat, or a variable resistor whose value changes in response to changes in temperature. Also called a HEAT SENSOR.

THERMAL STRATIFICATION (STRAT-uh-fuh-KAY-shun) — THERMAL STRATIFICATION

The formation of layers of different temperatures in a lake or reservoir. Also see STRATIFICATION.

THERMOCLINE (THUR-moe-KLINE) — THERMOCLINE

The middle layer in a thermally stratified lake or reservoir. In this layer there is a rapid decrease in temperature with depth. Also called the METALIMNION.

THERMOCOUPLE — THERMOCOUPLE

A heat-sensing device made of two conductors of different metals joined at their ends. An electric current is produced when there is a difference in temperature between the ends.

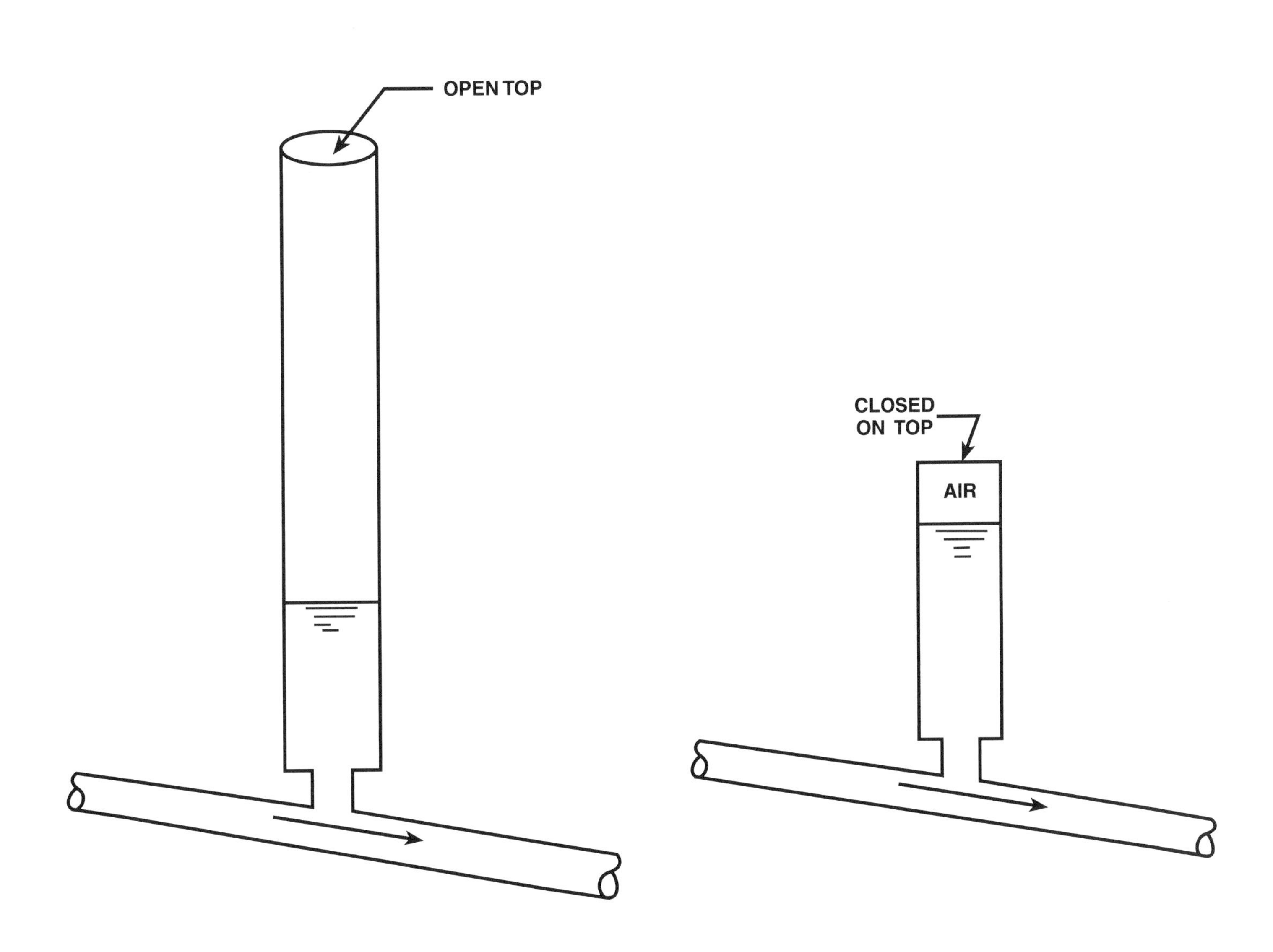

TYPES OF SURGE CHAMBERS

THICKENING — THICKENING

Treatment to remove water from the sludge mass to reduce the volume that must be handled.

THRESHOLD ODOR — THRESHOLD ODOR

The minimum odor of a water sample that can just be detected after successive dilutions with odorless water. Also called ODOR THRESHOLD.

THRESHOLD ODOR NUMBER (TON) — THRESHOLD ODOR NUMBER (TON)

The greatest dilution of a sample with odor-free water that still yields a just-detectable odor.

THRUST BLOCK — THRUST BLOCK

A mass of concrete or similar material appropriately placed around a pipe to prevent movement when the pipe is carrying water. Usually placed at bends and valve structures.

TIME LAG — TIME LAG

The time required for processes and control systems to respond to a signal or to reach a desired level.

TIMER — TIMER

A device for automatically starting or stopping a machine or other device at a given time.

TITRATE (TIE-trate) — TITRATE

To *TITRATE* a sample, a chemical solution of known strength is added drop by drop until a certain color change, precipitate, or pH change in the sample is observed (end point). Titration is the process of adding the chemical reagent in increments until completion of the reaction, as signaled by the end point.

TOPOGRAPHY (toe-PAH-gruh-fee) — TOPOGRAPHY

The arrangement of hills and valleys in a geographic area.

TOTAL CHLORINE — TOTAL CHLORINE

The total concentration of chlorine in water, including the combined chlorine (such as inorganic and organic chloramines) and the free available chlorine.

TOTAL CHLORINE RESIDUAL — TOTAL CHLORINE RESIDUAL

The total amount of chlorine residual (value for residual chlorine, including both free chlorine and chemically bound chlorine) present in a water sample after a given contact time.

TOTAL DISSOLVED SOLIDS (TDS) — TOTAL DISSOLVED SOLIDS (TDS)

All of the dissolved solids in a water. TDS is measured on a sample of water that has passed through a very fine mesh filter to remove suspended solids. The water passing through the filter is evaporated and the residue represents the dissolved solids. Also see SPECIFIC CONDUCTANCE.

TOTAL DYNAMIC HEAD (TDH) — TOTAL DYNAMIC HEAD (TDH)

When a pump is lifting or pumping water, the vertical distance (in feet) from the elevation of the energy grade line on the suction side of the pump to the elevation of the energy grade line on the discharge side of the pump.

TOTAL ORGANIC CARBON (TOC) — TOTAL ORGANIC CARBON (TOC)

TOC measures the amount of organic carbon in water.

TOTALIZER — TOTALIZER

A device or meter that continuously measures and calculates (adds) a process rate variable in cumulative fashion; for example, total flows displayed in gallons, million gallons, cubic feet, or some other unit of volume measurement. Also called an INTEGRATOR.

TOXAPHENE (TOX-uh-FEEN) — TOXAPHENE

A chemical that causes adverse health effects in domestic water supplies and also is toxic to freshwater and marine aquatic life.

TOXIC (TOX-ick) — TOXIC

A substance which is poisonous to a living organism.

TRANSDUCER (trans-DUE-sir) TRANSDUCER

A device that senses some varying condition measured by a primary sensor and converts it to an electrical or other signal for transmission to some other device (a receiver) for processing or decision making.

TRANSMISSION LINES TRANSMISSION LINES

Pipelines that transport raw water from its source to a water treatment plant. After treatment, water is usually pumped into pipelines (transmission lines) that are connected to a distribution grid system.

TRANSMISSIVITY (TRANS-miss-SIV-it-tee) TRANSMISSIVITY

A measure of the ability to transmit (as in the ability of an aquifer to transmit water).

TRANSPIRATION (TRAN-spur-RAY-shun) TRANSPIRATION

The process by which water vapor is released to the atmosphere by living plants. This process is similar to people sweating. Also see EVAPOTRANSPIRATION.

TREMIE (TREH-me) TREMIE

A device used to place concrete or grout under water.

TRICHLOROETHANE (TCE) (try-KLOR-o-ETH-hane) TRICHLOROETHANE (TCE)

An organic chemical used as a cleaning solvent that causes adverse health effects in domestic water supplies.

TRIHALOMETHANES (THMs) (tri-HAL-o-METH-hanes) TRIHALOMETHANES (THMs)

Derivatives of methane, CH_4, in which three halogen atoms (chlorine or bromine) are substituted for three of the hydrogen atoms. Often formed during chlorination by reactions with natural organic materials in the water. The resulting compounds (THMs) are suspected of causing cancer.

TRUE COLOR TRUE COLOR

Color of the water from which turbidity has been removed. The turbidity may be removed by double filtering the sample through a Whatman No. 40 filter when using the visual comparison method.

TUBE SETTLER TUBE SETTLER

A device that uses bundles of small-bore (2 to 3 inches or 50 to 75 mm) tubes installed on an incline as an aid to sedimentation. The tubes may come in a variety of shapes including circular and rectangular. As water rises within the tubes, settling solids fall to the tube surface. As the sludge (from the settled solids) in the tube gains weight, it moves down the tubes and settles to the bottom of the basin for removal by conventional sludge collection means. Tube settlers are sometimes installed in sedimentation basins and clarifiers to improve particle removal.

TUBERCLE (TOO-burr-cull) TUBERCLE

A protective crust of corrosion products (rust) which builds up over a pit caused by the loss of metal due to corrosion.

TUBERCULATION (too-BURR-cue-LAY-shun) TUBERCULATION

The development or formation of small mounds of corrosion products (rust) on the inside of iron pipe. These mounds (tubercles) increase the roughness of the inside of the pipe thus increasing resistance to water flow (decreases the C Factor).

TURBID TURBID

Having a cloudy or muddy appearance.

TURBIDIMETER TURBIDIMETER

See TURBIDITY METER.

TURBIDITY (ter-BID-it-tee) TURBIDITY

The cloudy appearance of water caused by the presence of suspended and colloidal matter. In the waterworks field, a turbidity measurement is used to indicate the clarity of water. Technically, turbidity is an optical property of the water based on the amount of light reflected by suspended particles. Turbidity cannot be directly equated to suspended solids because white particles reflect more light than dark-colored particles and many small particles will reflect more light than an equivalent large particle.

TURBIDITY METER TURBIDITY METER

An instrument for measuring and comparing the turbidity of liquids by passing light through them and determining how much light is reflected by the particles in the liquid. The normal measuring range is 0 to 100 and is expressed as Nephelometric Turbidity Units (NTUs).

TURBIDITY UNITS (TU) TURBIDITY UNITS (TU)

Turbidity units are a measure of the cloudiness of water. If measured by a nephelometric (deflected light) instrumental procedure, turbidity units are expressed in nephelometric turbidity units (NTU) or simply TU. Those turbidity units obtained by visual methods are expressed in Jackson Turbidity Units (JTU) which are a measure of the cloudiness of water; they are used to indicate the clarity of water. There is no real connection between NTUs and JTUs. The Jackson turbidimeter is a visual method and the nephelometer is an instrumental method based on deflected light.

TURN-DOWN RATIO TURN-DOWN RATIO

The ratio of the design range to the range of acceptable accuracy or precision of an instrument. Also see EFFECTIVE RANGE.

U

UNCONSOLIDATED FORMATION UNCONSOLIDATED FORMATION

A sediment that is loosely arranged or unstratified (not in layers) or whose particles are not cemented together (soft rock); occurring either at the ground surface or at a depth below the surface. Also see CONSOLIDATED FORMATION.

UNIFORMITY COEFFICIENT (U.C.) UNIFORMITY COEFFICIENT (U.C.)

The ratio of (1) the diameter of a grain (particle) of a size that is barely too large to pass through a sieve that allows 60 percent of the material (by weight) to pass through, to (2) the diameter of a grain (particle) of a size that is barely too large to pass through a sieve that allows 10 percent of the material (by weight) to pass through. The resulting ratio is a measure of the degree of uniformity in a granular material such as filter media.

$$\text{Uniformity Coefficient} = \frac{\text{Particle Diameter}_{60\%}}{\text{Particle Diameter}_{10\%}}$$

UPPER EXPLOSIVE LIMIT (UEL) UPPER EXPLOSIVE LIMIT (UEL)

The point at which the concentration of a gas in air becomes too great to allow an explosion upon ignition due to insufficient oxygen present.

V

VARIABLE COSTS VARIABLE COSTS

Costs that a utility must cover or pay that are associated with the production and delivery of water. The costs vary or fluctuate on the basis of the volume of water treated and delivered to customers (water production). Also see FIXED COSTS.

VARIABLE FREQUENCY DRIVE VARIABLE FREQUENCY DRIVE

A control system that allows the frequency of the current applied to a motor to be varied. The motor is connected to a low-frequency source while standing still; the frequency is then increased gradually until the motor and pump (or other driven machine) are operating at the desired speed.

VARIABLE, MEASURED VARIABLE, MEASURED

A factor (flow, temperature) that is sensed and quantified (reduced to a reading of some kind) by a primary element or sensor.

VARIABLE, PROCESS VARIABLE, PROCESS

A physical or chemical quantity which is usually measured and controlled in the operation of a water treatment plant or an industrial plant.

VELOCITY HEAD VELOCITY HEAD

The energy in flowing water as determined by a vertical height (in feet or meters) equal to the square of the velocity of flowing water divided by twice the acceleration due to gravity ($V^2/2g$).

VENTURI METER VENTURI METER

A flow measuring device placed in a pipe. The device consists of a tube whose diameter gradually decreases to a throat and then gradually expands to the diameter of the pipe. The flow is determined on the basis of the difference in pressure (caused by different velocity heads) between the entrance and throat of the Venturi meter.

NOTE: Most Venturi meters have pressure sensing taps rather than a manometer to measure the pressure difference. The upstream tap is the high pressure tap or side of the manometer.

VENTURI METER

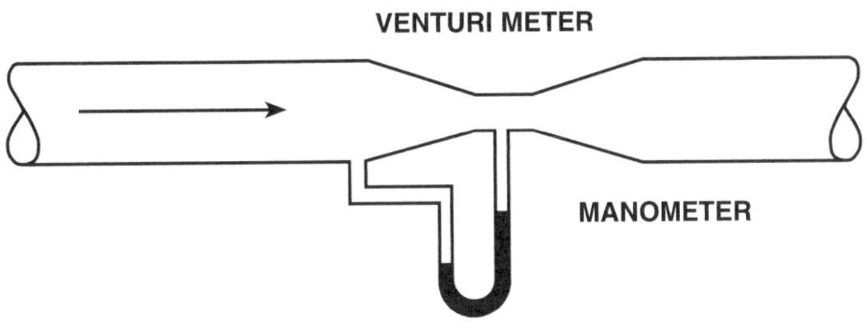

VISCOSITY (vis-KOSS-uh-tee) VISCOSITY

A property of water, or any other fluid, which resists efforts to change its shape or flow. Syrup is more viscous (has a higher viscosity) than water. The viscosity of water increases significantly as temperatures decrease. Motor oil is rated by how thick (viscous) it is; 20 weight oil is considered relatively thin while 50 weight oil is relatively thick or viscous.

VOID VOID

A pore or open space in rock, soil or other granular material, not occupied by solid matter. The pore or open space may be occupied by air, water, or other gaseous or liquid material. Also called an INTERSTICE, PORE, or void space.

VOLATILE (VOL-uh-tull) VOLATILE

(1) A volatile substance is one that is capable of being evaporated or changed to a vapor at relatively low temperatures. Volatile substances also can be partially removed by air stripping.

(2) In terms of solids analysis, volatile refers to materials lost (including most organic matter) upon ignition in a muffle furnace for 60 minutes at 550°C. Natural volatile materials are chemical substances usually of animal or plant origin. Manufactured or synthetic volatile materials such as ether, acetone, and carbon tetrachloride are highly volatile and not of plant or animal origin. Also see NONVOLATILE MATTER.

VOLATILE ACIDS VOLATILE ACIDS

Fatty acids produced during digestion which are soluble in water and can be steam-distilled at atmospheric pressure. Also called organic acids. Volatile acids are commonly reported as equivalent to acetic acid.

VOLATILE LIQUIDS VOLATILE LIQUIDS

Liquids which easily vaporize or evaporate at room temperature.

VOLATILE MATTER VOLATILE MATTER

Matter in water, wastewater, or other liquids that is lost on ignition of the dry solids at 550°C.

VOLATILE SOLIDS VOLATILE SOLIDS

Those solids in water or other liquids that are lost on ignition of the dry solids at 550°C.

VOLTAGE VOLTAGE

The electrical pressure available to cause a flow of current (amperage) when an electric circuit is closed. Also called ELECTROMOTIVE FORCE (E.M.F.).

VOLUMETRIC VOLUMETRIC

A measurement based on the volume of some factor. Volumetric titration is a means of measuring unknown concentrations of water quality indicators in a sample *BY DETERMINING THE VOLUME* of titrant or liquid reagent needed to complete particular reactions.

VOLUMETRIC FEEDER VOLUMETRIC FEEDER

A dry chemical feeder which delivers a measured volume of chemical during a specific time period.

VORTEX
VORTEX

A revolving mass of water which forms a whirlpool. This whirlpool is caused by water flowing out of a small opening in the bottom of a basin or reservoir. A funnel-shaped opening is created downward from the water surface.

W

WARNING
WARNING

The word *WARNING* is used to indicate a hazard level between *CAUTION* and *DANGER*. Also see CAUTION, DANGER, and NOTICE.

WASTEWATER
WASTEWATER

A community's used water and water-carried solids (including used water from industrial processes) that flow to a treatment plant. Storm water, surface water, and groundwater infiltration also may be included in the wastewater that enters a wastewater treatment plant. The term "sewage" usually refers to household wastes, but this word is being replaced by the term "wastewater."

WATER AUDIT
WATER AUDIT

A thorough examination of the accuracy of water agency records or accounts (volumes of water) and system control equipment. Water managers can use audits to determine their water distribution system efficiency. The overall goal is to identify and verify water and revenue losses in a water system.

WATER CYCLE
WATER CYCLE

The process of evaporation of water into the air and its return to earth by precipitation (rain or snow). This process also includes transpiration from plants, groundwater movement, and runoff into rivers, streams and the ocean. Also called the HYDROLOGIC CYCLE.

WATER HAMMER
WATER HAMMER

The sound like someone hammering on a pipe that occurs when a valve is opened or closed very rapidly. When a valve position is changed quickly, the water pressure in a pipe will increase and decrease back and forth very quickly. This rise and fall in pressures can cause serious damage to the system.

WATER PURVEYOR (purr-VAY-or)
WATER PURVEYOR

An agency or person that supplies water (usually potable water).

WATER TABLE
WATER TABLE

The upper surface of the zone of saturation of groundwater in an unconfined aquifer.

WATERSHED
WATERSHED

The region or land area that contributes to the drainage or catchment area above a specific point on a stream or river.

WATT
WATT

A unit of power equal to one joule per second. The power of a current of one ampere flowing across a potential difference of one volt.

WEIR (weer)
WEIR

(1) A wall or plate placed in an open channel and used to measure the flow of water. The depth of the flow over the weir can be used to calculate the flow rate, or a chart or conversion table may be used to convert depth to flow.

(2) A wall or obstruction used to control flow (from settling tanks and clarifiers) to ensure a uniform flow rate and avoid short-circuiting.

WEIR (weer) DIAMETER
WEIR DIAMETER

Many circular clarifiers have a circular weir within the outside edge of the clarifier. All the water leaving the clarifier flows over this weir. The diameter of the weir is the length of a line from one edge of a weir to the opposite edge and passing through the center of the circle formed by the weir.

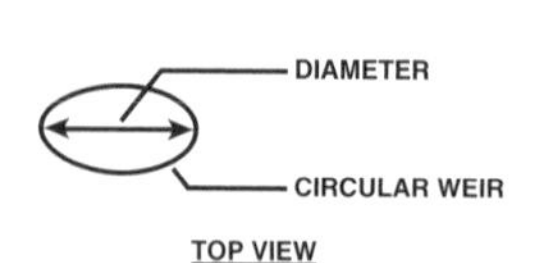

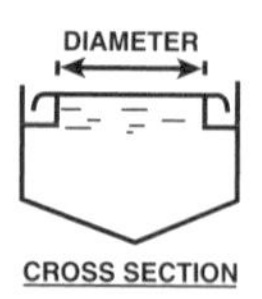

WEIR LOADING WEIR LOADING

A guideline used to determine the length of weir needed on settling tanks and clarifiers in treatment plants. Used by operators to determine if weirs are hydraulically (flow) overloaded.

$$\text{Weir Loading, GPM/ft} = \frac{\text{Flow, GPM}}{\text{Length of Weir, ft}}$$

WELL ISOLATION ZONE WELL ISOLATION ZONE

The surface or zone surrounding a water well or well field, supplying a public water system, with restricted land uses to prevent contaminants from a not permitted land use to move toward and reach such water well or well field. Also see WELLHEAD PROTECTION AREA (WHPA).

WELL LOG WELL LOG

A record of the thickness and characteristics of the soil, rock and water-bearing formations encountered during the drilling (sinking) of a well.

WELLHEAD PROTECTION AREA (WHPA) WELLHEAD PROTECTION AREA (WHPA)

The surface and subsurface area surrounding a water well or well field, supplying a public water system, through which contaminants are reasonably likely to move toward and reach such water well or well field. Also see WELL ISOLATION ZONE.

WET CHEMISTRY WET CHEMISTRY

Laboratory procedures used to analyze a sample of water using liquid chemical solutions (wet) instead of, or in addition to, laboratory instruments.

WHOLESOME WATER WHOLESOME WATER

A water that is safe and palatable for human consumption.

WIRE-TO-WATER EFFICIENCY WIRE-TO-WATER EFFICIENCY

The combined efficiency of a pump and motor together. Also called the OVERALL EFFICIENCY.

WYE STRAINER WYE STRAINER

A screen shaped like the letter Y. The water flows in at the top of the Y and the debris in the water is removed in the top part of the Y.

X

(NO LISTINGS)

Y

YIELD YIELD

The quantity of water (expressed as a rate of flow—GPM, GPH, GPD, or total quantity per year) that can be collected for a given use from surface or groundwater sources. The yield may vary with the use proposed, with the plan of development, and also with economic considerations. Also see SAFE YIELD.

Z

ZEOLITE ZEOLITE

A type of ion exchange material used to soften water. Natural zeolites are siliceous compounds (made of silica) which remove calcium and magnesium from hard water and replace them with sodium. Synthetic or organic zeolites are ion exchange materials which remove calcium or magnesium and replace them with either sodium or hydrogen. Manganese zeolites are used to remove iron and manganese from water.

ZETA POTENTIAL

ZETA POTENTIAL

In coagulation and flocculation procedures, the difference in the electrical charge between the dense layer of ions surrounding the particle and the charge of the bulk of the suspended fluid surrounding this particle. The zeta potential is usually measured in millivolts.

ZONE OF AERATION

ZONE OF AERATION

The comparatively dry soil or rock located between the ground surface and the top of the water table.

ZONE OF SATURATION

ZONE OF SATURATION

The soil or rock located below the top of the groundwater table. By definition, the zone of saturation is saturated with water. Also see WATER TABLE.

ZOOPLANKTON (ZOE-PLANK-ton)

ZOOPLANKTON

Small, usually microscopic animals (such as protozoans), found in lakes and reservoirs.

SUBJECT INDEX

A

B

C

E

F

M

X

Y

Z

NOTES

NOTES

NOTES

NOTES

NOTES

NOTES

NOTES